The Structure and Chemistry of Solid Surfaces

The Structure and Chemistry of Solid Surfaces

Edited by
GABOR A. SOMORJAI

Inorganic Materials Research Division
Lawrence Radiation Laboratory
University of California
Berkeley

Proceedings of the Fourth International Materials Symposium
"The Structure and Chemistry of Solid Surfaces," held at the
University of California, Berkeley, June 17-21, 1968

John Wiley & Sons, Inc.
New York · London · Sydney · Toronto

Library of Congress Catalogue Card Number: 71-90401

SBN 471 81320 6

Printed in the United States of America

Contributors

M. Abon
   Institut de Recherches sur la Catalyse, Faculte des Sciences, Universite de Lyon, Villeurbanne, France

Richard O. Adams
   The Dow Chemical Company, Rocky Flats Division, PO Box 888, Golden, Colorado 80402

Roland E. Allen
   Department of Physics, University of Texas, Austin, Texas 78712

Gilbert F. Amelio
   Georgia Institute of Technology, Engineering Experiment Station, 225 North Avenue, N.W., Atlanta, Georgia 30332

Dagmar Andersson
   Institute of Physics, Chalmers University of Technology, Gothenburg, Sweden S40220

S. Andersson
   Institute of Physics, Chalmers University of Technology, Gothenburg, Sweden S40220

J. F. Antonini
   Battelle Memorial Institute, Geneva Research Center, Carouge, Geneva, Switzerland

Robert A. Armstrong
   Radio and Electrical Engineering Division, National Research Council, Ottawa, Canada

John R. Arthur, Jr.
   Bell Telephone Laboratories, Murray Hill, New Jersey 07971

M. Balkanski
   Laboratoire de Physique des Solides de la Faculte des Sciences, Paris, France

R. C. BANSAL
Materials Science Department, Pennsylvania State University, University Park, Pennsylvania 16802

E. BAUER
Michelson Laboratory, China Lake, California 93555

G. E. BECKER
Bell Telephone Laboratories, Murray Hill, New Jersey 07974

ALAN J. BENNETT
General Electric Company, General Physics Laboratory, P.O. Box 8, Schenectady, New York 12301

JOAN BERKOWITZ
Arthur D. Little, Inc., 16 Acorn Park, Cambridge, Massachusetts 02140

WILLIAM V. BEST
Physical Chemical Laboratory, University of Missouri, Columbia, Missouri 65201

MICHAEL N. BISHARA
Department of Aerospace Engineering, University of Virginia, Charlottesville, Virginia 22901

HUGH E. BISHOP
United Kingdom Atomic Energy Authority, Metallurgy Division, A.E.R.E., Harwell, Berks., England

M. BOUDART
Department of Chemical Engineering, Stanford University, Stanford, California 94305

DARYL BOUDREAUX
Department of Physics, 333 Jay Street, Polytechnic Institute of Brooklyn, Brooklyn, New York

JOHN M. BLAKELY
Department of Materials Science and Engineering, Cornell University, Ithaca, New York 14850

L. E. BRADY
Eastman Kodak Company, Research Laboratory, Kodak Park, Rochester, New York 14650

JAMES J. BURTON
Department of Physics, University of Illinois, Urbana, Illinois

BRUCE D. CAMPBELL
Los Alamos Scientific Laboratory, Los Alamos, New Mexico 87554

J. W. CASTLE
Eastman Kodak Company, Research Laboratory, Kodak Park, Rochester, New York 14650

CHUAN C. CHANG
Bell Telephone Laboratories, Murray Hill, New Jersey 07974

JOHN M. CHARIG
Allen Clark Research Center, The Plessey Company, Ltd., Caswell, Towcester, Northants. England

JAR-MO CHEN
Bartol Research Foundation, Swarthmore, Pennsylvania 19801

B. C. CLARK
Research Laboratories, General Motors Corporation, Warren, Michigan 48090

R. N. COLTHARP
Department of Chemistry, University of Florida, Gainesville, Florida

C. COROTTE
Laboratoire de Spectrometrie, Physique, Faculte des Sciences de Grenoble, Grenoble, France

JOHN M. COWLEY
School of Physics, University of Melbourne, Parkville, Victoria, 3052, Australia

LYN C. CROUSER
Field Emission Corporation, Melrose Avenue, McMinnville, Oregon 97128

W. E. DANFORTH
Bartol Research Foundation, Swarthmore, Pennsylvania 19081

J. G. DASH
Department of Physics, University of Washington, Seattle, Washington 98105

LOUIS DE BERSUDER
Laboratoire de Chimie Physique, Faculte des Sciences d'Orsay, Orsay, France

FREDRICH W. DE WETTE
Department of Physics, University of Texas, Austin, Texas 78712

P. DUCROS
Laboratoire de Spectrometrie Physique, Faculte des Sciences de Grenoble, Grenoble, France

ROBERT P. EISCHENS
Texaco Research Center, Beacon, New York 12508

GERT EHRLICH
General Electric Research and Development Center, PO Box 8, Schenectady, New York 12301

PEDER J. ESTRUP
    Department of Physics and Chemistry, Brown University, Providence,
    Rhode Island 02912

H. E. FARNSWORTH
    Brown University, Providence, Rhode Island 02912

DONALD G. FEDAK
    Hammond Laboratory, Yale University, New Haven, Connecticut 06520

JOSEPH FINE
    National Bureau of Standards, Washington, D. C. 20234

SAM S. FISHER
    Department of Aerospace Engineering, University of Virginia
    Charlottesville, Virginia 22901

JOHN V. FLORIO
    Hammond Laboratory, Yale University, New Haven, Connecticut 06520

J. BARRY FRENCH
    Institute for Aerospace Studies
    University of Toronto
    Toronto 5, Ontario, Canada

JOHN W. GADZUK
    National Bureau of Standards, Atomic Physics Division, Washington, D.C.

GEOFFREY GAFNER
    Chemical Physics Group, National Physical and National Chemical
    Research Laboratories, South African Council for Scientific and Industrial
    Research, Pretoria, South Africa

ROBERT L. GERLACH
    Sandia Corporation, Albuquerque, New Mexico

LESTER H. GERMER
    Department of Applied Physics, Cornell University, Ithaca, New York 14850

A. GERVAIS
    Polytechnic Institute of Brooklyn, 333 Jay Street, Brooklyn, New York 11201

J. J. GILMAN
    Department of Metallurgy and Materials Research Laboratory, University
    of Illinois, Urbana, Illinois 61801

FRANK O. GOODMAN
    Daily Telegraph Theoretical Department, School of Physics, University
    of Sydney, Sydney, N.S.W., Australia

JOHN C. GREGORY
NASA—R-SSL-NP, Huntsville, Alabama 35812

TRICE W. HAAS
Aerospace Research Laboratories, Bldg. 450, Wright Patterson Air Force Base, Ohio 45433

HOMER D. HAGSTRUM
Bell Telephone Laboratories, Murray Hill, New Jersey 07974

J. F. HAMILTON
Eastman Kodak Company, Research Laboratory, Kodak Park, Rochester, New York 14650

DAN HANEMAN
School of Physics, University of New South Wales, PO Box 1, Kensington, Sydney, NSW, Australia 2033

C. A. HAQUE
Bell Telephone Laboratories, Murray Hill, New Jersey 07971

D. O. HAYWARD
Chemistry Department, Imperial College of Science and Technology, London, S.W. 7., England

ROGER HECKINGBOTTOM
Post Office Research Station, Dollis Hill, London, N.W. 2., England

VOLKER HEINE
Cavendish Laboratory, Free School Lane, Cambridge, England

ROBERT HERMAN
Research Laboratories, General Motors Corporation, Warren, Michigan 48090

D. L. HERON
School of Physics, University of New South Wales, PO Box 1, Kensington, Sydney, NSW, Australia 2033

HANSJURGEN HEYNE
General Electric R and D Center, PO Box 8, Schenectady, New York 12301

GILBERT HOCHSTRASSER
Battelle Memorial Institute, Geneva Research Center, Carouge, Geneva, Switzerland

FERDINAND HOFMANN
Space Science Laboratory, University of California, Berkeley, California

V. HOFFSTEIN
Department of Physics, Polytechnic Institute of Brooklyn, 333 Jay Street, Brooklyn, New York

MICHAEL P. HOOKER
Systems Research Laboratories Inc., 500 Woods Drive, Dayton, Ohio 45432

B. J. HOPKINS
    Surface Physics, The University, High Field, Southampton, England

A. M. HORGAN
    School of Chemical Sciences, University of East Anglia, Norwich, Norfolk, England

ALAN J. HOWSMON
    Institute for Aerospace Studies, University of Toronto, Toronto, Ontario, Canada

FRANKLIN C. HURLBUT
    Aeronautical Science and Research Engineering, University of California, Berkeley, California

ALLEN G. JACKSON
    Systems Research Laboratories, Inc., 500 Woods Drive, Dayton, Ohio 45432

K. JAKUS
    Aeronautical Science, University of California, Berkeley, California

PHILIP J. JENNINGS
    Bell Telephone Laboratories, Murray Hill, New Jersey 07974

R. O. JONES
    Laboratory of Atomic and Solid State Physics, Cornell University, Ithaca, New York 14850

GEORGE JURA
    Department of Chemistry and Lawrence Radiation Laboratory, University of California, Berkeley, California

DAVID A. KING
    School of Chemical Sciences, University of East Anglia, Norwich, Norfolk, England

C. F. KIRK
    General Electric R and D Center, PO Box 8, Schenectady, New York 12301

A. R. KUHLTHAU
    Department of Aerospace Engineering, University of Virginia, Charlottesville, Virginia 22901

D. LAFEUILLE
    Laboratoire de Spectrometric Physique, Faculte des Sciences de Grenoble, Grenoble, France

MAX G. LAGALLY
    Physics Department, University of Wisconsin, Madison, Wisconsin 53706

REGINALD F. LEVER
    IBM Watson Research Center, PO Box 218, Yorktown Heights, New York 10598

B. G. LINSEN
    Unilever Research Laboratory, Vlaardingen, The Netherlands

THEODORE E. MADEY
Surface Chemistry Section, National Bureau of Standards, Washington, D.C. 20234

A. A. MARADUDIN
Department of Physics, University of California, Irvine, California 92664

I. MARKLUND
Institute of Physics, Chalmers University of Technology, Gothenburg, Sweden S40220

JOHN W. MAY
Department of Applied Physics, Cornell University, Ithaca, New York 14850

BRUCE MCCARROLL
Research and Development Center, PO Box 8, General Electric Company, Schenectady, New York 12301

EION G. MCRAE
Bell Telephone Laboratories, Murray Hill, New Jersey 07974

JAMES F. MENADUE
Department of Applied Physics, Cornell University, Ithaca, New York

F. P. MERTENS
Texaco Research Center, Beacon, New York 12508

DOUGLAS L. MILLS
Department of Physics, University of California, Irvine, California 92664

KURT MOLIERE
Fritz-Haber Institut der Max-Planck-Gesellschaft, West Berlin, Germany

JOSEPH M. MORABITO, JR.
Inorganic Materials Research Division, Lawrence Radiation Laboratory, Berkeley, California

ERWIN W. MUELLER
Department of Physics, The Pennsylvania State University, University Park, Pennsylvania 16802

KLAUS MULLER
Bell Telephone Laboratories, Murray Hill, New Jersey

ROLF MULLER
Department of Chemical Engineering and Lawrence Radiation Laboratory, University of California, Berkeley, California

EARLE E. MUSCHLITZ, JR.
Department of Chemistry, University of Florida, Gainesville, Florida

DONALD R. OLANDER
Department of Nuclear Engineering and Lawrence Radiation Laboratory, University of California, Berkeley, California

DAVID F. OLLIS
Department of Chemical Engineering, Stanford University, Stanford, California 94305

THOMAS J. OSINGA
Unilever Research Laboratory, Vlaardingen, The Netherlands

PAUL W. PALMBERG
North American Rockwell Science Center, Thousand Oaks, California 91360

ROBERT L. PALMER
Gulf General Atomics, Inc., PO Box 608, San Diego, California 92112

ROBERT L. PARK
Sandia Laboratory—Div. 5123, PO Box 5800, Albuquerque, New Mexico 87115

HOWARD L. PETERSEN
Physical Chemical Laboratory, University of Missouri, Columbia, Missouri 65201

R. PINCHAUX
Laboratoire de Physique des Solids, Faculte des Sciences, Paris, France

E. W. PLUMMER
Department of Applied Physics, Ithaca, New York 14850

JOHN H. POLLARD
Bartol Research Foundation, Swarthmore, Pennsylvania 19081

F. PORTELE
Fritz-Haber Institut der Max-Planck Gesellschaft, West Berlin, Germany

JAMES O. PORTEUS
Michelson Laboratory—Code 6017, China Lake, California 93555

R. H. PRICE
Institute for Aerospace Studies, University of Toronto, Toronto, Ontario, Canada

HEINZ RAETHER
Institute fur Angewandte Physik, University of Hamburg, Hamburg, West Germany

THOR N. RHODIN
Department of Applied Physics, Ithaca, New York 14850

J. C. RIVIERE
United Kingdom Atomic Energy Authority, Metallurgy Division—AERE, Harwell, Berks., England

WILLIAM D. ROBERTSON
Hammond Laboratory, Yale University, New Haven, Conn. 06520

GERD M. ROSENBLATT
Department of Chemistry, The Pennsylvania State University, University Park, Pennsylvania 06802

HOWARD SALTSBURG
Gulf General Atomics, Inc., PO Box 608, San Diego, California 92112

MILTON D. SCHEER
　　Surface Chemistry Section, National Bureau of Standards, Gaithersburg, Maryland 20234

J. T. SCOTT
　　Department of Chemistry, University of Florida, Gainesville, Florida

J. E. SCOTT, JR.
　　Department of Aerospace Engineering, University of Virginia, Charlottesville, Virginia 22901

CLAUDE SEBENNE
　　Laboratoire de Physique des Solides, Faculte des Sciences, Paris, France

BENJAMIN M. SIEGEL
　　Department of Applied Physics, Cornell University, Ithaca, New York

D. K. SKINNER
　　Allen Clark Research Center, The Plessey Company Limited, Caswell, Towcester, Northants, England

JOE N. SMITH, JR.
　　Gulf General Atomics Inc., PO Box 608, San Diego, Calif. 92112

HAROLD P. SMITH, JR.
　　Space Science Laboratory, University of California, Berkeley, California

GABOR A. SOMORJAI
　　Department of Chemistry and Lawrence Radiation Laboratory, University of California, Berkeley, California

ROLF STEIGER
　　CIBA Photochemical Ltd., Fribourg, Switzerland

RICHARD M. STERN
　　Polytechnic Institute of Brooklyn, 333 Jay Street, Brooklyn, New York 11201

EDWARD A. STERN
　　Department of Physics, University of Washington, Seattle, Washington 98105

G. ALEC STEWART
　　Department of Physics, University of Washington, Seattle, Washington 98105

ROBERT E. STICKNEY
　　Department of Mechanical Engineering and Research Laboratory of Electronics, Massachusetts Institute of Technology, Cambridge, Massachusetts

LYNWOOD W. SWANSON
　　Field Emission Corporation, Melrole Avenue, McMinnville, Oregon 97128

B. Tardy
Institut de Recherches sur la Catalyse Faculte des Sciences, Universite de Lyon, Villeurbanne, France

Howard Taub
Polytechnic Institute of Brooklyn, 333 Jay Street, Brooklyn, New York 11201

S. J. Teichner
Institut de Recherches sur la Catalyse, Faculte des Sciences, Universite de Lyon, Villeurbanne, France

Lloyd B. Thomas
Physical Chemical Laboratory, University of Missouri, Columbia, Missouri 65201

M. Tomasek
Institute of Physical Chemistry, Czechoslovak Academy of Sciences, Prague, Czechoslovakia

S. Y. Tong
Department of Physics, University of California, Irvine, California 92664

Hua-Ching Tong
Department of Metallurgy and Materials Research Laboratory, University of Illinois, Urbana, Illinois 61801

J. Charles Tracy
Department of Materials Science and Engineering, Cornell University, Ithaca, New York 14850

Franz Trautweiler
Eastman Kodak Company, Research Laboratory—Kodak Park, Rochester, New York 14650

Charles W. Tucker, Jr.
General Electric Research and Development Center, Schenectady, New York 12301

Seijl Usami
Department of Surface Physics, The University—High Field, Southampton, England

F. J. Vastola
Materials Science Department, Pennsylvania State University, University Park, Pennsylvania 16802

Jacobus S. Vermaak
Department of Physics, University of Port Elizabeth, Port Elizabeth, Republic of South Africa

PHILIP J. WALKER
    Materials Science Department, Pennsylvania State University, University
    Park, Pennsylvania 16802

RICHARD F. WALLIS
    Naval Research Laboratory Code 6470, Washington, D.C. 20390

P. M. WARBURTON
    School of Physics, University of Melbourne, Parkville, Victoria, 3052,
    Australia

HARRY W. WEART
    Department of Metallurgical and Nuclear Engineering, University of
    Missouri, Rolla, Missouri 65401

ERIC P. WENAAS
    Bell Aerosystems, Buffalo, New York 14214

MAURICE B. WEBB
    University of Wisconsin, Department of Physics, Madison, Wisconsin 53706

JAMES R. WOLFE
    AVCO Corporation, 10700 Independence, Tulsa, Oklahoma

WYMOND J. WOSTEN
    Unilever Research Laboratory, Vlaardingen, The Netherlands

JOHN T. YATES, JR.
    Surface Chemistry Section, National Bureau of Standards, Washington, D.C.
    20234

Preface

An International Conference on the "Structure and Chemistry of Solid Surfaces" was held at the University of California, Berkeley Campus on June 19-21, 1968. This book contains the papers which were presented at this symposium.

Surface chemistry has been in a stage of rapid development in recent years. This progress is due, in part, to the application of experimental techniques which can probe the nature of surfaces on an atomic scale. The excess to high purity materials in single crystal form and ultra high vacuum technology creates clean, well-defined surfaces available for definitive investigations. The need for small devices with large surface to volume ratio in many fields of science and technology, increased research activity in aerospace science, solid state physics, biochemistry and high temperature chemistry added further impetus to the progress in surface science.

In the past, progress in our understanding of the properties of surfaces has been hampered by the lack of knowledge of the surface structure. Wide application of low energy electron diffraction, (LEED), now provides us with the needed structural data. LEED plays the same role in the study of surfaces as x-ray diffraction in the study of the bulk structure. Experimental and theoretical studies of surface phase transformations and the structures of adsorbed gases on solid surfaces shed new light on the relationship between surface structure and the nature of surface reactions. Other experimental techniques, among them field ion and field emission microscopy, electron spectroscopy, molecular beam scattering studies from surfaces and ellipsometry all provide valuable information about the physical-chemical properties of surfaces and the interaction of gases with surfaces. Each experimental method makes unique contributions which complement the experimental data obtained by the other techniques. From the application of all of the techniques to the investigation of well-defined surfaces and from theoretical studies which are carried out concurrently, a new physical picture of the atomic properties of surfaces emerges. This, I believe, is readily apparent from the reading of the papers in this volume.

Since the greatest advances in surface science in recent years came through the studies of well-defined or single crystal surfaces in a controlled ambient (ultra-high vacuum or in the presence of pure gases), this conference was organized to concentrate on such studies with the emphasis on the correlation of atomic and electronic structure and transport properties of clean solid surfaces with the chemistry of surface reactions. The papers which are concerned with the properties of clean surfaces are collected in the first part of the book. Studies of gas covered surfaces and of chemical surface reactions are discussed in papers which are grouped in the latter part of the volume.

This international conference was the fourth in the series of Materials Symposia of the Inorganic Materials Research Division (IMRD) of the Lawrence Radiation Laboratory which provide an outstanding scientific forum for discussions to further the progress of materials science. The division is engaged in interdisciplinary research in materials science under the sponsorship of the Atomic Energy Commission of the United States. The conference itself was sponsored by the College of Chemistry and the Inorganic Materials Research Division of the Lawrence Radiation Laboratory, University of California, Berkeley, California.

I would like to extend my sincere thanks to the two co-chairmen of the conference, Professor R. Gomer and Dr. R. F. Wallis for their help in organizing the meeting. We are all grateful to the Arrangements Chairman, Mr. C. V. Peterson. Without his untiring efforts the conference could not have been successful. Our sincere thanks to the administrative staff of the Inorganic Materials Research Division for their help. I am grateful to Professor Leo Brewer, Director of IMRD, and to Professor Victor F. Zackay, Assistant Director of IMRD, for suggesting to me the organization of a symposium on surfaces and for their continued support. The continued support of the Atomic Energy Commission of Materials Science through the Lawrence Radiation Laboratory is greatly appreciated.

GABOR A. SOMORJAI

Berkeley, California
July 1968

CONTENTS

The Structure and Chemistry of Solid Surfaces

ELECTRONS AT CLEAN SURFACES*

V. Heine

Cavendish Laboratory
Cambridge, England

1. Introduction

It is now possible to solve the Schrödinger equation for electrons at solid surfaces, with particular application to surface states on semiconductors and low energy electron diffraction (LEED). Although these are experimentally quite unrelated, the same theoretical formalism can be applied to both. In Cambridge (England) we have been interested in such calculations for several years, and this conference is an appropriate time to review the physical significance of the calculations and to discuss some outstanding problems.

The formalism [1,2]) is basically as follows. The potential is assumed to be equal to the periodic potential in the bulk up to a matching plane half an atomic spacing beyond the surface layer of atomic sites. Outside this plane the potential is assumed to be zero. Are the errors in this model of the potential serious? For almost all purposes, no. If an experimental value of the work function is used to fix the absolute energy scale in the bulk, then the errors cancel to the first order of approximation. Jones [3]) has shown the remaining errors are quite small in the case of the surface state calculations. In LEED the sharp discontinuity of the potential will give rise to some CO beam scattering, which will have to be corrected for because the real potential is sufficiently

---
*Supported in part by a contract with the United Kingdom Atomic Energy Research Establishment, Harwell.

smooth for this source of scattering to be virtually negligible [5]).
Similarly other deviations of the potential near the surface from the
bulk will probably be responsible for no more than a few percent of the
observed LEED, unless there is a significant "reconstruction" of the
atomic positions.

With this model of the potential, then, the problem of solving the
wave equation falls into three parts, exactly as in the transmission of
any classical wave from one medium to another. Firstly the wave function in the vacuum outside the surface may be written as simple exponentials. Secondly the wave function inside the solid can be expressed in
terms of Bloch functions obtained from the Schrödinger equation with a
periodic potential. It is important to note here that a single Bloch
function is in general inadequate [6]). We need an infinite sum of Bloch
function, all with the same energy, of which only a finite number are the
usual propagating waves and all the rest are evanescent solutions. Thirdly we have to match the solutions in the two regions at the boundary
between them.

The heart of the matter clearly lies in generating the Bloch functions in the material that are required in the matching. Here there are
two problems: the setting up of the potential and then the solution of
the Schrödinger equation. In both directions one may draw on much val-

uable experience gained with ordinary band structure calculations, but
all the methods used there need extension in various ways. The main point
is that solid state physics has advanced beyond crude, qualitative models,
and to understand the variety of behaviour of real solid surfaces, we
have to insert a realistic band structure into the calculation. Moreover if one knows the band structure, one can already say qualitatively
quite a lot about possible surface states or interpretation of LEED
intensities.

2. Surface states on semiconductors

The formalism just outlined has been used by R. O. Jones [3, 4]) to
calculate the intrinsic surface states on a clean perfect surface of
silicon. The most complete results are for a (110) face. Fig. 1a shows
the 2-dimensional Brillouin zone and Fig. 1b the energy [4]). There are
two bands, corresponding to a total of two states (counting spin degeneracy) per broken bond at the surface, as expected from Shockley's
work [7]). The two bands are degenerate along $\overline{XM}$, and in the electrically
neutral condition are exactly half filled [7]). Since the two bands are
joined together along $\overline{XM}$, it may be easier to picture a single energy
band extending throughout the double zone of Fig. 2. Here there is an
energy gap across the lines $\overline{MX'}$ but not along $\overline{MX}$. With the band half

filled, there will therefore be a free Fermi surface near $\overline{MX}$ (Fig. 2)
with an energy about one-third of the way up the usual band gap of the
bulk silicon, in good agreement with experiment [8]. Indeed this feature
of the calculations seems not to be very sensitive to the detailed approximations, which correlates with the fact that the figure of one-third
is reasonably representative for quite a number of semiconductors and
conditions (see references in Heine [9])).

    Two points call for comment. Firstly, how far may these states
be regarded as "dangling bonds" on the surface? In some respects 'yes',
and others 'no'. They almost certainly do extend out from the material
a little further than the valence band states. However they are not
confined to the surface atoms, decaying as they do over several atomic
layers into the bulk (Fig. 3a). Of course a broken bond is a favorable
site for attaching another atom, but this is because all the states of
the valence band stick out there, and the formation of a new bond involves all the occupied states of the material. The most important role
of the surface states in surface chemistry may be their screening action.
They form a conducting system with a free Fermi surface like a metal,
and they can screen out any electrostatic surface perturbations. Such
flow of charge always lowers the energy, and can help to make the surface an energetically favourable place for an ion or a polarized or

polarisable atom.

The second point concerns oxide coated surfaces. For many purposes, e.g., tunnelling, oxide behaves very much like vacuum. But
there are no surface states at a silicon-silicon oxide interface, except for a few due to impurities or imperfections. In this respect the
oxide behaves differently from vacuum. We can probe this equation
more deeply by means of Shockley's theorem [7]. This states that when
a band gap is "inverted", as it is in a group IV semiconductor, then
the solutions of the Schrödinger equation in the gap have the form of
Fig. 3a, and one can always obtain a surface state by fitting on an
expotentially decreasing tail which is the correct solution in the
vacuum. If the band gap is not "inverted", then the wave functions are
as in Fig. 3b and no surface state is possible. Now oxides and other
ionic-type compounds do not usually show the characteristic behavior
of surface states, and one might conclude that their band gap is of the
latter type. In that case however one would expect states at an oxide-
silicon interface because the two types of solution Figs. 3a and 3b
could be matched together if we turn Fig. 3b around to represent ψ in
the oxide. We can now suggest a resolution of this paradox which has
worried me for some time. Jones [10] has shown that if we go from silicon to a III-V compound, then the degeneracy along $\overline{XM}$ is lifted and a

band gap appears. If the atoms are different enough, e.g., in a II-VI
material, the free Fermi surface is completely eliminated, and with it
a lot of the behavior characteristic of surface states. The Fermi level
is no longer pinned but can ride up and down in the gap between the occupied and unoccupied bands of surface states. This appears to correspond to the experimental situation [10]), and allows us to suggest
that the oxide coating on silicon also has an "inverted" band gap
with a full and an empty band of surface states quite widely separated.
In that case the wave functions at the oxide-semiconductor interface
would appear as in Fig. 3c, with no matching and no interfacial states
being possible.

3. Atomic scattering at the energies of LEED

Experimentally the outstanding feature of LEED is the richness in
the structure observed, including the strong peaks not associated with
simple Bragg reflection. What is this due to? In some cases it may be
due to some reconstruction or other imperfection at the surface, but for
clean close-packed metal surfaces I believe it is due simply to strong
atomic scattering. Certainly strong secondary Bragg peaks require
strong scattering so that the peaks due to multiple scattering are comparable with single Bragg scattering. Conversely McRae [11]) has shown
that strong scattering does give rise to some of the features seen in

LEED. I regard his "computor experiment" as having particular importance because it is the only case where we have LEED intensities (calculated virtually without approximation) for a crystal that we know for certain has a clean perfect surface and spherical atomic potentials. The calculations also showed that when the scattering becomes weak, the secondary Bragg peaks disappear rapidly in intensity compared with the primary Bragg peaks.

The scattering amplitude from a single atom can be expressed as [12])

$$\Sigma_l \; [1 - \exp(2i\eta_l)] \; \text{(other factors)} \tag{1}$$

in terms of the phase shifts $\eta_l(E)$ for all angular momenta l.
If we write

$$\eta_l = n\pi + \delta_l \quad (0 \leqslant \delta_l < \pi), \tag{2}$$

we note that the integer multiples (n) of pi do not affect the scattering amplitude (1), and that the scattering is a maximum when $\delta_l = \frac{1}{2}\pi$. 'Strong' scattering in fact means a phase shift of around $(n + \frac{1}{2})\pi$, 'weak' scattering one around $n\pi$.

By way of a typical example, the $l = 0$ phase shift for aluminium is plotted very schematically in Fig. 4. At zero energy it starts around 2π, where the integer 'two' means there are two bound states, the 1s and 2s, at negative energies. The fact that η_0 is so near 2π for energies up to 20 eV or more, means that the scattering is small and the band gaps around the Fermi energy are small, as is well know experimentally for all non-transition metals. As the energy increases to $+\infty$, η_0 must tend to zero, and Fig. 4 shows roughly how this happens. The first major drop from 2π to π occurs when the kinetic energy of the electron becomes comparable with the kinetic energy of an electron in the 2s orbital, i.e., at an energy about as much above zero as the bound state lies below zero. This is of the order of 100 eV. Similarly there is a further drop at around 1000 eV connected with the 1s shell. In between, η_0 will be around π and the scattering weak. The three regions of weak scattering are marked along the energy axis of Fig. 4, and it is only in these that we may think about the band structure of the material in nearly-free-electron terms or the scattering of electrons in terms of a 2 or 3 beam theory. On the other hand when η_l is around $(n + \frac{1}{2})\pi$, the band gaps become comparable with the spacings between bands, and we have multiple scattering and strong secondary Bragg peaks, etc., in LEED.

The suggestion from Fig. 4 therefore is that there may be a plateau around 200-500 eV where the LEED patterns are simpler than at lower energies. It is not known how clearly separated the two regions of strong scattering around 100 and 1000 eV are, exactly where they are, nor how much summing over all l complicates the picture, but there is some evidence in the experimental data for the plateau. Fig. 5 shows the I_{00} beam for tungsten: it is only above 350 eV that a clear distinction between primary and secondary peaks becomes meaningful. The energies of the outer core levels are given in table 1. It remains to be investigated wheter we should regard 350-600 eV as lying beyond the influence of the 5p, 5s, 4f states and before the 4d, 4p, 4s, or whether beyond the influence of the latter.

The tungsten results show another related feature, pointed out by Taylor [13]. The low energy portion of the curve is shown enlarged in Fig. 6 and it is seen that there is a peak missing at around 80 eV. According to the Friedel sum rule [14], we ought to loose $2l + 1$ bands from the band structure compared with free electron bands, when η_l drops from $n\pi$ to $(n - 1)\pi$. With less bands, there will be less band gaps, and hence missing peaks are quite possible although the whole phenomenon is really spread out over the range of energy in which η_l drops significantly.

With graphite (Fig. 7) the pattern already becomes clear at low energies, and we are probably in the region of weak scattering before the influence of the 1s state at -286 eV.

How far does the scattering potential in one cell of the solid differ from that of a free atom? The charge densities of the atoms of course overlap in the solid, and it has been found empirically in band structure calculations that good answers are obtained by calculating

the total charge density, averaging it spherically in each call, and

calculating the potential from that. It is important not to use free-

atom phase shifts. In fact the phase shifts in the solid can be cal-

culated much more reliably than the free-atom ones, because the lat-

ter contain substantial contributions from the outer edges of the atom

where there are many uncertainties about charge density, exchange and

correlation. When the potential is confined as it is in the solid,

the bulk of the scattering for E = 50 to 500 eV is determined by outer

core shell region of the atom where the potential is of the order of

-100 eV. The deviations of the potential from the spherical form

in the corners of the cell is of order 1 eV, even in covalently bonded

materials like silicon, and probably negligible as far as LEED is

concerned.

An important contribution to the pseudopotential matrix elements

comes from exchange (Fig. 8). This was first calculated using Slater's

first approximation [16]) (not the more common second $\rho^{1/3}$ approximation).

I.e., the total electron density at a point was calculated, and the ex-

change potential taken for a free electron gas and an extra electron

of the energy k^2. This gives quite good results for the mean exchange

energy, i.e., q = 0 in Fig. 8. However for the off-diagonal matrix

elements q $\neq$ 0 it differs markedly from the true value calculated from

the Hartree-Fock exchange operator (Fig. 8). The Hartree part and

repulsive pseudopotential term add to about -4 eV for q < 2, tailing off
to less than -1 eV for q >5 units. Thus the difference between the
Slater approximation and the true exchange can give an error of the
order of =00% of the value of the final matrix element, even for a
low q such as g (111).

Incidentally the final matrix elements are highly non-local, i.e.,
not a function of q alone, and we do not believe that any method of
calculation based on the assumption of a local potential will ever
serve for real materials.

So far we have discussed only the elastic component of the scattering. There is of course also a strong inelastic component which
we are now starting to investigate. Preliminary calculations by McRae [11])
as well as ourselves indicate that an imaginary component to the potential reduces the heights of the peaks and can affect their shape
according to the symmetries of the wave functions at the corresponding
band edges, but appears to introduce no major new phenomena.

4. Matching calculations of LEED

The matching formalism for calculating LEED intensities has been
outlined in §1 and elaborated by Boudreaux and Heine [2]).

The first step is to test the formalism on the same potential
already calculated by McRae [11]) with his multiple scattering method.
The (00), (10), and (11) beams have been calculated by Capart [17])

and the comparison with McRae [11] for I_{00} and $\delta_o = \frac{1}{2}\pi$ is shown in
Fig. 9, the agreement for the other beams [18] being similar.

The most important contribution of the matching formalism is the
physical picture it gives of LEED. The intensity reflected or diffracted into any particular beam depends markedly on what Bloch states
are available in the band structure of the crystal for the electrons to
go into. For example if there is a complete band gap over some range
of energy, then there are no propagating solutions in the crystal for
the electrons to go into, and conservation of the number of electrons
demands that the total intensity in all reflected beams adds to 100%.
The actual form of the wave functions around the dap determines the
distribution among the different beams.

When the scattering is weak, the band structure and the matching
can be calculated algebraically [2] by perturbation theory, and Capart [17]
has obtained good agreement with the results of McRae [11] for small
phase shifts. For a large phase shift $(\delta_o = \frac{1}{2}\pi)$, he has calculated
the band structure in detail. Fig. 10 shows how changed it is from
the free electron model, and gives the total intensity in all beams.
We note that at $E = 2.7$ units, $E = 1.85$, etc., there are complete band
gaps resulting in total reflection. The truncation of the matching
equation allows the total intensity to exceed 100% slightly. When
there is a band in the crystal which is free-electron-like as at $E = 2.4$,

then the wave function also approximates a free wave and there is good
transmission into the crystal with small reflection. The reason why one
gets strong secondary Bragg peaks is because the scattering matrix
elements of the potential are large enough for the band gaps from different Bragg reflections to interfere. The corresponding Fourier components of the wave functions get mixed with coefficients or order unity,
leading to a similar intensity in the secondary Bragg peaks.

The number of free electron bands at a particular energy (for given
component of k parallel to the surface) increases linearly with E, so
that above 20 eV one is very unlikely to find an absolute energy gap
as that in an insulator. Capart [17]) has also discussed the resonance [11])
in greater detail than in reference 2. An absolute minimum in the bands,
coupled with the inability of the corresponding beam to escape from the
crystal, gives an additional contribution to the other reflected intensities (particularly the OO beam) which can be large in some cases
such as those considered by McRae [11]). It is not expected to be so important in real cases.

We estimate these matching calculations are approximately five
times faster in computing time than those of McRae [11]).

Similar calculations of LEED intensities by Pendry [19]) with a
realistic potential for nickel are not yet complete, but one can already
obtain a good picture from the band structure. From 40-80 eV the band
structure appears reasonably free-electron-like (Fig. 11). The first

peak, shoulder, and peak correspond with the band gaps at 52, 58, and 65 eV, the lowest giving the most intense reflection because it is formed from the (331) and (000) wave in the solid. (Note that these co-ordinates are expressed relative to the cubic axes and that these waves give a primary Bragg peak in the (0$\bar{1}$) beam referred to the sur-face co-ordinates on a (111) face.) In Fig. 12 the main peak is as-sociated with the gap produced by the (000) and (442) waves in the crystal. The peak is reduced in amplitude by the presence of the band running through the middle of the gap, its effect being greatest at those energies where it lies closest (measured vertically) to the band of predominantly (000) type. This produces the shoulder on the high energy side of the peak.

5. Representation of the potential

In the following, we shall assume that we are setting up a secular equation for the band structure in terms of plane waves (or rather pseudo plane waves), but since any exact calculation of LEED intensities involves the solution of the Schrödinger equation in the solid, much of it is relevant to the other approaches.

We shall adopt the basic pseudopotential philosophy [14, 21], namely, there are many potentials which all give the same scattering and hence identically the same band structure. Given a real atomic potential, we construct from it some pseudopotential with this property. We can follow either the formalism of Austin, Heine and Sham [22] (AHS) related to the orthogonalized plane wave method, or we can use the phase shifts (or logarithmic derivatives of the wave function) as was first done by Slater in the augmented plane wave (APW) method. In either direction there are whole families of possible pseudopotentials, and it is a matter of choosing the most suitable. These pseudopotentials all throw out the $n\pi$ of eq. (2), and for low energies there is relatively little difference between them. But at the energies of LEED we have two new problems. The energy spacing between consecutive bands decreases on the average as $E^{-3/2}$ so that the number of plane waves needed in the secular equation increases. Also the wavelength decreases, one is involved in much higher angular momenta, nd the Bessel functions involved in the denominator of some of the pseudopotentials pass through zero.

Pendry [23] has investigated the APW pseudopotential of Slater [14, 21], the KKRZ pseudopotential derived by Ziman from a transformation of the Korringa Kohn Rostoker method for band structure calculations [21, 24], Johnson's modification [25] of the APW form, as well as two whole families of pseudopotentials based on extensions of these using the ideas of Johnson [25], Lloyd [26], Morgan [27] and others. He has also considered

an extension of them in another direction. When using the APW method,
for instance, one sets up the potential, and makes it a constant outside
the "muffin-tin" radius R and spherically symmetric inside. Clearly
there is least error involved in the process if R is chosen as large as
possible, equal to the radius R_i of the sphere inscribed in the atomic
polyhedron. The pseudopotential is then set up using the logarithmic
derivative

$$L \ (E,R,l) \ = \ \psi_l'(R)/ \ \psi_l(R) \tag{3}$$

of the wave function at R_I. However, in this last step we have a degree
of freedom not used heretofore. Given L at R_I, we can analytically continue ψ inwards using spherical Bessel functions for zero potential to
obtain L at some smaller radius R, and then use this L(R) to set up
the pseudopotential. Fig. 13 shows the results of calculation with three
pseudopotentials for a whole range of R/R_I. The potential is that of
McRae [11]) with δ_o = ½ π and the energy being calculated that at k = 0 in
the third band. The correct energy is thought to be around 1.46 units:
at least comparison of the results in the figure with those from 81 plane
waves shows convergence of the KKRZ and the APW methods to this value
for R/R_1 in the range 0.3 to 0.5. The results from the Johnson pseudopotential are actually diverging away from the right energy, and Pendry [23])

has explained this behavior in terms of its analytic properties. We note that neither the APW nor the KKRZ pseudopotentials appear to give satisfactory convergence for around $R/R_I = 0.6$ to 1.0, so that choice of R_I is important. At $R/R_I < 0.7$ the APW form appears to give no roots to the secular equation. For $0.71 < R/R_I < 0.82$ the KKRZ form will converge to $E = \frac{1}{2}(\pi/R)^2$ instead of the correct energy, where the π comes from the first zero of $j_o(x)$. An advantage of the KKRZ form is that the summation over l involves only the difference of ψ from $j_l(\kappa r)$ and so converges more rapidly in l than the APW formula for the matrix elements: for large R/R_I this might be an important consideration at the energies of LEED, though perhaps no more serious than at ordinary band structure energies if R is chosen small. Taking all factors into consideration, Pendry [23] has suggested the KKRZ form as the best pseudopotential

$$< \underset{\sim}{k} \mid V_{ps} \mid \underset{\sim}{k'} >$$

$$= - \frac{4\pi}{\Omega\kappa} \sum_l (2l + 1) \frac{j_l (kR_l) \; j_l (k'R_l)}{j_l (\kappa R_l) \; j_l (\kappa R_l)} \; \tan \eta_l' \qquad (4)$$

with R_l chosen for each l such that κR_l is slightly inside the first zero of $n_l(x)$. Here Ω is the atomic volume, $\frac{1}{2}\kappa^2 = E$, and η_l' the modified phase shift defined by Ziman [24].

The other family of pseudopotentials investigated by Pendry [28])
is that defined by Austin et al [22]) which for the simple case of a
single core state $|c\rangle$ takes the general form

$$V_{ps} = V + A \; |c\rangle\langle c|. \tag{5}$$

where A is arbitrary. The state $|c\rangle$ is still an eigenstate of
V_{ps}, with an energy

$$E_c' = E_c + A \tag{6}$$

instead of its proper energy E_c. The orthogonalized plane wave (OPW)
form has

$$A(OPW) = E - E_c, \qquad E_c'(OPW) = E. \tag{7}$$

Pendry's choice is

$$A(Pendry) = -E_c, \qquad E_c'(Pendry) = 0. \tag{8}$$

Fig. 14 shows the energy levels in a certain one-dimensional potential
compared with the diagonal matrix elements $\langle k \mid T + V_{ps} \mid k\rangle$. The

smaller this difference, the better presumably the convergence of the secular equation and the "better" the choice of pseudopotential. In three dimensions, it can be shown that (8) gives the best convergence at high energies, i.e., energies greater than

$$<c \mid T \mid c> \approx |E_o| \qquad\qquad (9)$$

where the phase shift drops rapidly (Fig. 4). At low energies around the Fermi level, the form (8) is nearly the same as the OPW form which has long proved its usefulness in band structure calculations. Pendry has also shown that in the difficult region of energy around (9), the choice (8) gives nearly the best possible convergence. It is being employed in the calculations on nickel and other solids mentioned in §4.

In the course of this work, considerable use has been made of the formalism of Weinberg [29]) to establish mathematical criteria for the speed of convergence of the secular equation. Actually what was taken was the convergence of perturbation theory, although secular equations will converge even when perturbation theory will not. Perturbation series have the form

$$V_{ps} + V_{ps}(G_oV_{ps}) + V_{ps}(G_oV_{ps})(G_oV_{ps}) + \dotsb \qquad (10)$$

with

$$G_o(E) = \frac{1}{E-k^2} \qquad (11)$$

The rate of convergence is controlled by the eigenvalues $\eta_i(E)$ of the operator G_oV:

$$(G_oV_{ps})|i\rangle = \eta_i(E)|i\rangle \qquad (12)$$

The complex eigenvalue η_i are functions of the parameter E occurring in G_o, and their variation with E may be plotted as trajectories in the complex plane (Fig. 15). They start at $\eta = 0$ for $E = -\infty$ and loop back to the origin at $E = +\infty$. One of them is connected with the core state $|c\rangle$: in fact we have

$$\eta_c(E) = 1 \quad \text{and} \quad |i(E)\rangle = |c\rangle$$

$$\text{at } E = E_c' \qquad (13)$$

There are three criteria for good convergence:

(i) $E_c' \ll \langle c|T|c\rangle$, at which there occurs an essential singularity.

(ii) The η-trajectories should keep as close to, or inside the unit circle.

(iii) The amplitude a_i of $|i\rangle$ in an expansion of plane waves and of the energy eigenstates for $E > 0$ in terms of the $|i\rangle$, should be small.

These have been applied in arriving at the choice (8). Actually in the energy region of interest in LEED or ordinary band structure calculations, probably no pseudopotential of the family (5) gives a convergent perturbation series. What a "good" pseudopotential does is to give a perturbation series which is asymptotically convergent because there is one or more $|i\rangle$ with η_i outside the unit circle but of only quite small amplitude in the wave function. This means one may get a good estimate of the eigenvalue from quite a small secular equation, or even from the diagonal matrix element $\langle k|H_{ps}|k\rangle$, but that the convergence from there to the correct eigenvalue is slow. It has long been known [30] that this is precisely what the OPW secular equation does if one makes an error in the calculation of E_c.

Finally we may compare the best (4) and (8) of the two types of pseudo-potential. The advantage of the form (4) is that one summarises the potential once and for all in terms of a few curves $\eta_\ell(E)$, from which one can also see physically what the various components of the scattering are doing, e.g. in relation to the discussion of Fig. 4. However we have seen in §3 that a proper non-local Hartree-Fock treatment of the exchange is needed, and this is much less convenient to transcribe into phase shifts than to incorporate with the $\langle k|V_{ps}|k'\rangle$ matrix elements of (8). Also the form (8) has matrix elements independent of energy, so that one gets all eigenvalues (for given $\underset{\sim}{k}$) from one diagonalisation instead of hunting for each one separately, a not inconsiderable factor when spanning an energy of several hundred eV. For these reasons the form (8) is being used in the calculations on real materials, although the form (4) was used for discussing the model of McRae[11].

Table 1. Energies of core levels in tungsten (eV).

5p	5s	4f	4d	4p	4s
-46	-71	-54	-258	-422	-496

References

1.   V. Heine, Surface Science $\underline{2}$ (1964), p. 1.

2.   D. S. Boudreaux and V. Heine, Surface Science $\underline{8}$ (1967), p. 426.

3.   R. O. Jones, Thesis submitted to Cambridge University, and to be published.

4.   R. O. Jones, Phys. Lett. $\underline{20}$ (1968), p. 992.

5.   G. Capart, Thesis submitted to University of Louvain, and to be published.

6.   H. Bethe, Ann. Phys. $\underline{87}$ (1928), p. 55.

7.   W. Shockley, Phys. Rev. $\underline{56}$ (1939), p. 317.

8.   F. G. Allen and G. W. Gobeli, Phys. Rev. $\underline{127}$ (1962), p. 150.

9.   V. Heine, Phys. Rev. $\underline{138}$ (1965), p. A1689.

10.   R. O. Jones, later this conference.

11.   E. G. McRae, J. Chem. Phys. $\underline{45}$ (1966), p. 3258.

12.   L. I. Schiff, Quantum Mechanics (McGraw-Hill Book Co., New York, 1955), p. 105.

13.   N. J. Taylor, Surface Science $\underline{4}$ (1965), p. 175.

14.   J. M. Ziman, Principles of the Theory of Solids (Cambridge Univ. Press, 1964).

15.   J. J. Lander and J. Morrison, J. Appl. Phys. $\underline{35}$ (1964), p. 3593: K. Hirabayashi and Y. Takeishi, Surface Science $\underline{4}$ (1965), p. 150.

16.   J. C. Slater, Phys. Rev. $\underline{81}$ (1951), p. 385.

17.   G. Capart, to be published.

18.   Reference 11 and private communication.

19.   J. B. Pendry, unpublished.

20.   R. L. Park and H. E. Farnsworth, Surface Science $\underline{2}$ (1964), p. 527.

21.   V. Heine, in: The Physics of Metals: I. Electrons. Ed. J. M. Ziman (Cambridge University Press, 1968).

22. P. J. Austin, V. Heine, and L. J. Sham, Phys. Rev. <u>127</u> (1962), p. 276.

23. J. B. Pendry, to be published.

24. J. M. Ziman, Proc. Phys. Soc. <u>86</u> (1965), p. 337.

25. K. H. Johnson, Phys. Rev. <u>150</u> (1966), p. 429.

26. P. Lloyd, Proc. Phys. Soc. <u>86</u> (1965), p. 825.

27. G. J. Morgan, Proc. Phys. Soc. <u>89</u> (1966), p. 365.

28. J. B. Pendry, J. Phys. C. (to appear).

29. S. Weinberg, Phys. Rev. <u>130</u> (1963), p. 776: ibid <u>131</u> (1963), p. 440.

30. V. Heine, Proc. Roy. Soc. A <u>240</u> (1957), p. 354.

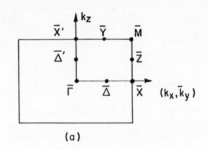

(a)

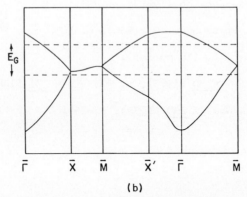

(b)

Fig. 1. (a) The two-dimensional Brillouin zone for surface states on a (110) surface of silicon. (b) The energy $E(\underset{\sim}{k})$ of the surface states.

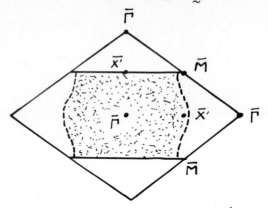

Fig. 2. The double zone, showing the filled states (stippled) and the Fermi "surface" (dashed).

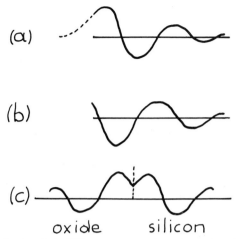

(a)

(b)

(c)

oxide   silicon

Fig. 3. Solution of the Schrödinger equation in a band gap: (a) for an
'inverted' band gap leading to surface states: (b) for a 'non-
inverted' band gap: (c) for silicon and silicon oxide where both
are assumed to have 'inverted' gaps.

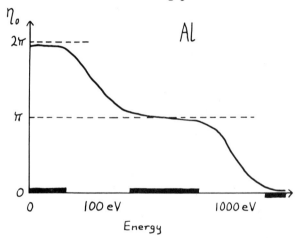

Fig. 4. The l = 0 phase shift for an aluminum atom in the solid metal (highly
schematic). The regions indicated dark on the energy axis represent
regions of small scattering where a nearly-free-electron theory might
be expected to hold.

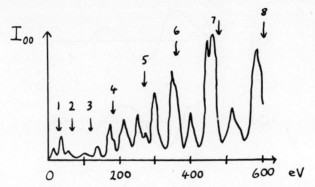

Fig. 5. The (00) LEED intensity for tungsten (after Taylor [13]). The arrows denote the expected positions of Bragg reflections.

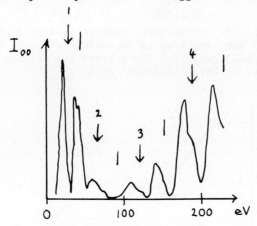

Fig. 6. The (00) LEED intensity for tungsten with expanded scale (after Taylor [13]). The lines mark half-integer reflections seen strongly in fig. 5.

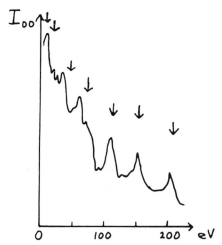

Fig. 7. The (00) LEED intensity for graphite (after reference 15). The calculated peaks come at the positions shown.

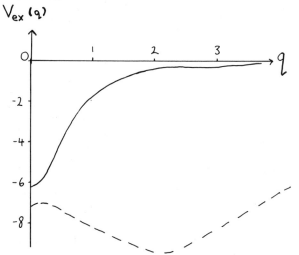

Fig. 8. The exchange contribution to the matrix element $\langle k| V_{ps} |k+a\rangle$ where $\frac{1}{2}k^2 = 100$ eV and a is directed backwards from k. The (111) reciprocal lattice vector comes at q = 1.63 atomic units. Full curve – Slater approximation: broken curve – Hartree-Fock. (After Pendry[19])).

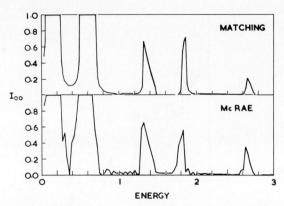

Fig. 9. The (OO) LEED intensity for the model of McRae [11]) with $\delta_o = \frac{1}{2}\pi$ cal-
culat•d with the matching formalism by Capart [17]) and compared with
McRae [11]).

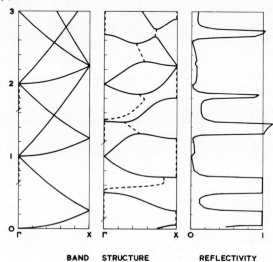

BAND STRUCTURE   REFLECTIVITY

Fig. 10. Calculations on the model of McRae [11]) with $\delta_o = \frac{1}{2}\pi$ (after Capart [17])).
(left) Free electron band structure. (center) Calculated band struc-
ture. (right) Total LEED reflected intensity from all beams. The energy
is plotted vertically in the units of McRae [11]).

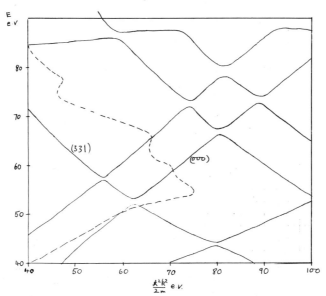

Fig. 11. Full curves: band structure E($\underline{k}$) for nickel for $\underline{k}$ perpendicular to a (111) face. Note that the horizontal axis for k has been distorted with $\hbar^2 k^2/2m$ given in eV. Broken curve: intensity of the ($0\bar{1}$) LEED beam [20]) in arbitrary units.

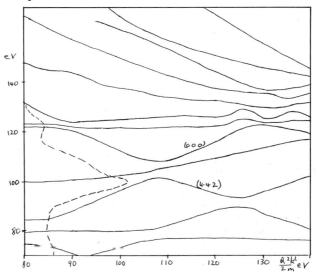

Fig. 12. The same as Fig. 11 at higher energy.

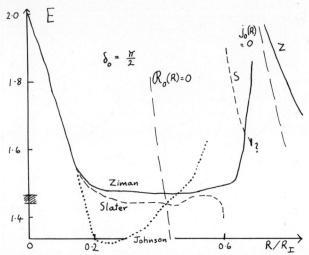

Fig. 13. Energy from 251 x 251 secular equation and Ziman (KKRZ), Slater (APW) and Johnson pseudopotentials, for various values of R/R_I. The correct eigenvalue lies around 1.46. The zeros of the radial wave function R_o and of j_o are shown.

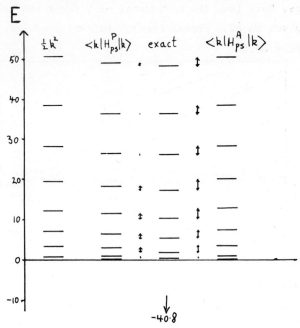

Fig. 14. Energy levels at k = 0 in a one-dimensional periodic potential of square wells such that there is one bound state at E = -40.8 units. The exact energy values in the third column are compared with the zero order starting estimates with Pendry (P) and Austin (A) forms for V_{ps}. The first column gives the free electron eigenvalues.

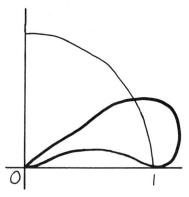

Fig. 15.   Trajectory of an eigenvalue $\eta(E)$ in the complex η plane.   A portion of
the unit circle is also shown.

CALCULATION OF BACK-REFLECTED LEED INTENSITIES
USING A PLANE-WAVE MULTIPLE SCATTERING MECHANISM*

G. Gafner

Chemical Physics Group of the National
Physical and National Chemical Research Laboratories
South African Council for Scientific
and Industrial Research, Pretoria

A cardinal need of surface science is a theory which would allow the derivation of surface structures from LEED patterns. Until such a theory becomes available the field of surface crystallography must remain in its present state of sterility in which only two-dimensional arrangements can be deduced from LEED patterns and these only ambiguously when the surface is complicated, as most of interest seem to be.

Much attention has recently been given to multiple scattering theory as the invalidity of the kinematical theory, and of minor variations thereof, now seems to be generally accepted. McRae [66,67] has been responsible for a profound attempt to build onto existing diffraction theory so as to increase its relevance to LEED. This theory has not, as yet, been used to attempt analyses of real surface structures but would appear to be heading rapidly in that direction. Another contribution to LEED interpretation is that of Gerlach and Rhodin [67] whose simple asymmetric interaction theory has successfully reproduced experimental I-V curves of (0,0) beams from simple surfaces.

The present study is a continuation of earlier work, Gafner [64] in which the kinematical theory was used to calculate the I-V dependence of beams from Ni(111). The striking disparities which ensued indicated the need for multiple diffraction calculations on the same system – these calculations are reported here.

The complexity of dynamical diffraction theory needs little emphasis – its reduction to practicality depends on how far simplifying assumptions can be made without introducing effects which are detrimental to its accuracy. Most theorists have initiated their work with as comprehensive a theory as possible and have suggested simplifications

* Sponsored by the South African Coal, Oil and Gas Corporation.

which might be acceptable. The approach adopted here has been to base exact calculations on the simplest plausible model of the multiple scattering process in the hope that these would both prove the relevance of multiple scattering to LEED and demarcate the areas of oversimplification. As will be seen, a highly encouraging measure of success has been obtained.

ASSUMPTIONS ON WHICH THE STUDY IS BASED

The following assumptions have been incorporated into this work:

(1) The electron beam is ideally monochromatic, has infinite lateral and longitudinal coherence and is not polarized by scattering.

(2) The crystal is perfect and can be represented by a stationary point lattice with the geometry established for the bulk lattice.

(3) Refraction effects, i.e. change of beam wavelength with position in the cell and phase change on scattering, can be simulated by the application of an overall inner potential.

(4) Pathlengths can be calculated from plane-wave theory.

(5) The scattering factor is a function of scattering angle alone and thus independent of wavelength for the range of voltages considered here.

(6) Experimental results which are used for comparison here represent the elastically scattered components exclusively.

DISCUSSION OF THE ASSUMPTIONS

The first assumption is unlikely to lead to serious error as the beam properties mentioned are reasonably well reproduced in practice. The ignoring of polarization might have to be studied more carefully at a later date as such effects are present, references are given by Gervais, Stern and Menes,[68] and might have significant influences. Assumption (2) is acceptable as the evidence is against large spacing changes occurring at the surface of crystals and thermal motion, if reasonably isotropic, can be compensated by suitable modification of the scattering factor against scattering angle curve.

The crucial assumption (3) is most likely to cause gross differences between the calculations done here and experiment. The striking experimental proof of the relevance of multiple scattering to LEED which has been provided by the observation of marked Renninger effects by Stern, Gervais and Menes[68] leaves no doubt as to the large scattering cross-sections which apply. The simple treatment given here tests one possible way of attempting to take some of the complicating factors into account. It is not clear how phase change on scattering should be introduced as a function of scattering angle as given by the scattering curves of Lander and Morrison[63] and, if this is done, what wavelength geometry must be used. Let the justification for this assumption and for (4) thus depend on whether these calculations check with experiment or not.

The scattering factors of Lander and Morrison[63] validate (5) to sufficient degree for the voltage range in question and (6) is unlikely to lead to complications.

METHOD   USED   IN   THE   CALCULATION

The method used in the calculation of multiple scattering is illustrated two-dimensionally in figures 1 and 2.   Scattering angles and, in the three-dimensional case, azimuth angles are calculated from $\lambda = d \sin\theta$ and reciprocal lattice theory respectively.   Relative amplitudes are read from a prescribed curve of θ against diffracted amplitude.   The amplitudes of waves which are formed at each diffraction event are adjusted to make their sum equal to the product of an absorption multiplier and the incident amplitude.   Statistical consi- derations make this multiplier equal to $\pi/2$ when absorption is zero. This ensures that energy which is lost during destructive interference is returned to beams which are not annihilated.   In-plane resonances, as proposed by McRae[66] are excluded from the calculations.

In figure 1, part A, the normally incident beam (1) splits into six components at layer 1.   Three of these are back-reflected and their amplitudes (deduced as described above) and phase angles (zero) are the initial components of the calculated diffraction pattern.   The three forward diffracted beams, 2, 3 and 4, whose geometry is related to that of the back-reflected beams by a mirror plane in the surface, each forms a diffraction pattern at layer 2.   These are geometrically similar to that formed by the incident beam on striking layer 1, but involve different phase angles resulting from the different pathlengths traversed by the beams before reaching an atom in layer 2.   The parallel components of these three patterns (B, C and D) are combined vectorially to give the resultant pattern E.   Each of these six beams are treated as incident beams in the next stage of the multiple diffraction process.

Figure 2 illustrates how the multiple scattering events progress through the crystal in part of an actual calculation. The incident beam (A) diffracts to form a back-reflected set (C) and a forward-reflected set represented by the continuation of (A) past layer 1. Oscillatory diffraction occurs between layers 1 and 2 until all beams have amplitudes of less than a prescribed value (0.1% of the incident beam amplitude suffices). The back-reflected set (B) represents such an exhausted set (each traverse involves the reconstitution of many sets into one resultant set as illustrated in figure 1). During this process sets (C), (D) and (E) are combined vectorially into the main back-reflected pattern. Sets (F), (G) and (H) are then combined and oscillatory diffraction between layers 2 and 3 is initiated at (I) and continues to exhaustion at (J). Sets (K) and (L) are then combined and oscillatory diffraction between layers 1 and 2 is started at (M) and continues to exhaustion at (N). Sets (O) and (P) are added into the main back-reflection pattern and an abortive attempt is made, in this example, to start diffraction between layers 2 and 3 with set (R) at (Q). All components of set (S) are smaller than the prescribed minimum. Sets (U), (V) and (W) are then combined to start diffraction between layers 3 and 4 at (T). Subsequent events lead to the generation of sets (X) and (Y) which are added into the main back-reflected set.

COMPUTER PROGRAM

A Fortran computer program has been developed to do the calculations for systems with surface cells containing one atom. It is thus only applicable to simple systems such as metals. Input consists of the surface cell constants and the positional co-ordinates of as many atoms as there are layers in the model crystal. Absorption is introduced by

a multiplier whose product with the amplitude of the beam under
consideration is used to scale the amplitudes of the diffracted beams
which it creates so that their sum equals this product. The amplitude
scattering function is entered in tabular form as a function of scattering
angle. A cutoff amplitude value must be specified and when all
amplitudes in one oscillatory diffraction cycle fall below this value,
this cycle is considered to be complete and calculation is transferred
to diffraction between the next higher or lower layers depending on
whether the cycle concerned the uppermost layers or not.

Calculation starts with the establishing of beam indices, azimuth-
and diffraction angles for all beams within the reflecting sphere
applicable to the longest wavelength to be studied. These are then
used to calculate the back-reflected multiple diffraction pattern as
described previously. Output consists of wavelength, voltage and a
table of indices and intensities. The latter shows the expected threefold symmetry for Ni(111). A prescribed decrement is then applied to
the wavelength and the process restarted. Calculation finally halts
when the specified minimum wavelength is reached. Approximately one
hour of I B M 360/40H computer time is needed for a satisfactory range
of wavelengths between the equivalent of 34 and 230 volts.

CALCULATIONS ON Ni(111)

The first calculations were directed at obtaining an optimum value
for the absorption multiplier (AM). The relative scattering factor was
assumed to be unity for all scattering angles other than zero where
it was given the value of 9. This ratio was estimated from the shapes
of the experimental curves of Park and Farnsworth.[64] The results for
various AM values are given in figure 3 together with the   experimental

curve for the (0,1) beam. Best fit is obtained with AM=1.3 and an unexpectedly excellent profile agreement occurred. The peak to background ratio is, however, incorrect and an even higher AM is indicated. The maximum value of the AM is $\pi/2$ and calculations with a higher value become lengthy and thus too expensive to undertake. Furthermore, the elastically scattered fraction is already unreasonably high with AM=1.3. Attention was thus turned to the AM-dependent variable Fo/F the ratio of the scattering factor for zero scattering angle to that for all other angles. Any attempt to specify the scattering curve more closely than this is considered premature at this stage.

Calculations were done with various Fo/F ratios, keeping AM=1.3. The results are given in figure 4. The (0,1) and (1,0) curves are almost identical for Fo/F=4.5, indicating that too much multiple scattering occurs. With a ratio of 18, the relative peak heights have changed beyond those of the experimental results and the ratio is thus too large (the first peak in the calculated (1,0) curve should be ignored as it is absent from the experimental curve due to refraction). The best fit is obtained with the intermediate ratio of 12. Results for both beams have acceptably close profiles for this ratio but loss of resolution occurs at higher voltages - the opposite would have been expected as the assumptions favour experimentally unattainable detail.

CONCLUSIONS

The results obtained give reason for optimism regarding the proximity of a workable method for converting LEED results to surface structures. The method used is a first approximation and is oversimplified but nevertheless shows such characteristic features as the non-Bragg peak at about 100 volts in the (1,0) beam and goes a long way towards reproducing peak heights.   The higher observed intensity of the (0,1) beam at 50 volts is not reproduced.   Preliminary calculations incorporating phase changes with scattering have shown that these result in the desired increase in the intensity of this beam. Inexplicable differences in the peak positions, similar to those found when the kinematical theory is used, remain.

Various modifications to the program are under consideration and the next step will be to do calculations with values of AM and Fo/F as indicated by experiment.

ACKNOWLEDGMENTS

The author expresses his sincerest gratitude to the South African Coal, Oil and Gas Corporation for their generous and continued support of this research, to Mrs. M.C. Pistorius who programmed the theory and in so doing introduced many novel ideas of her own, and to Professor Somorjai and the organisers of this conference for their sponsored invitation to participate.

REFERENCES

Gafner, G. (1964), Surface Science, $\underline{2}$, 534.

Gerlach, R.L. and Rhodin, T.N. (1967), Surface Science, $\underline{8}$, 1.

Gervais, A., Stern, R.M. and Menes, M. (1968, Acta Cryst., $\underline{A\ 24}$, 191.

Lander, J.J. and Morrison, J. (1962), J. Chem. Phys., $\underline{37}$, 729.

McRae, E.G. (1966), J. Chem. Phys., $\underline{45}$, 3258.

(1967), Surface Science, $\underline{8}$, 14.

Park, R.L. and Farnsworth, H.E. (1964), Surface Science, $\underline{2}$, 527.

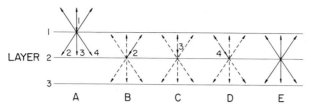

Fig. 1. Reconstruction of the set of waves formed by diffraction of a normally incident beam.

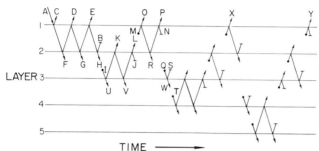

Fig. 2. Progress of a multiple scattering calculation. Arrows indicate sets of waves which are taken into account; bars (such as at B) indicate sets of waves in which all waves have amplitudes which are smaller than the prescribed minimum; dots indicate the points at which new cycles of diffraction are started.

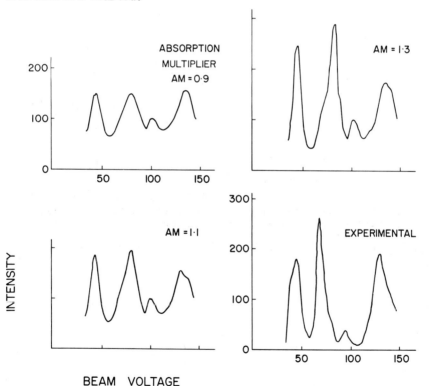

Fig. 3. Calculated I-V curves for Ni(111) (0,1) for various values of the absorption multiplier compared with the experimental result of Park and Farnsworth[64]. The F_o/F ratio used is 9.0.

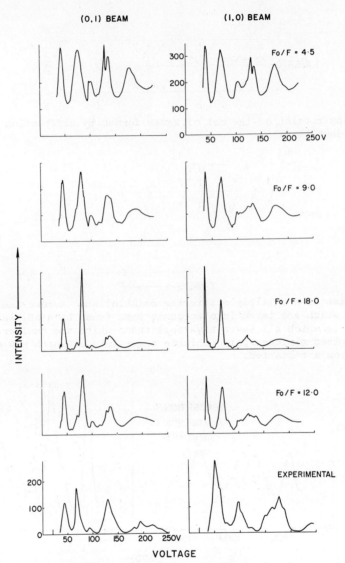

Fig. 4. Calculated I-V curves for Ni(111) (0.1) and (1.0) for various values of the Fo/F ratio compared with the experimental results of Park and Farnsworth[64]. The absorption multiplier used is 1.3.

LEED INTENSITY CALCULATIONS USING KAMBE'S THEORY[+]

J. M. Chen[*]

Department of Electrical Engineering
University of Minnesota
Minneapolis, Minnesota

L. Introduction

Recently, some surface structure studies have been made on clean and alkali-metal covered Germanium surfaces using LEED, electron reflectance, and work function measurements.[1] In these studies, the electron reflectance spectra (00 beam intensity curves) were measured electronically with a technique used by Fox et al.[2] and more recently by Zollweg[3]. This technique used by crossed B and E fields to separate the incident and the reflected beams so that reflectance spectra can be measured with the primary beam at normal incidence, and it makes use of an ac modulation to reduce the energy spread of the primary beam so that reflectance can be accurately measured for primary energies from 0.1 eV to 140 eV. Pronounced peaks were observed in the 0 to 20 eV range of the reflectance spectra and they were extremely sensitive to surface conditions. This paper presents some efforts toward identifying the origin of these peaks.

2. Deductions from the Experimental Results

Usually, the peaks in an intensity curve are correlated with the constructive interference of electron waves reflected from different layers of the crystal, so that the important parameter is the spacing between layers. However, if the energy of the electrons is low enough such that the wavelength of the electrons become comparable to or larger than the atomic

---
[+] Supported in part by the Air Force Avionics Laboratory, Research and Technology Division, Air Force System Command, United States Air Force.
[*] Present address: Bartol Research Foundation, Swarthmore, Pennsylvania 19081

spacings, the incident electrons would interact with more than one atom before leaving the surface, so that multiple scattering effects become important. At low energies, it is no longer possible to consider the atoms as
discrete independent scatters (kinematic theory) and one must consider the
crystal potential as a whole interacting with the incoming electron wave
according to quantum theory (dynamical theory). In the latter theory, atoms
in the same layer, as well as atoms in different layers, can contribute to
peaks in the intensity spectrum.

The peaks that are most pronounced and most sensitive to surface structure
transformations as measured in the present study, are all in the 0 to 20 eV
range. Palmberg and Peria[4] showed that for a Ge(111) surface, the kinematical
and pseudo-kinematical approximations were not valid for energies below 150
eV. Cordes[5] tried with little success to match the positions of the peaks
with the resonances of plane waves reflected from different layers of the
crystal. Thus it appears that in the 0 to 20 eV range, a detailed calculation
with dynamical theory will have to be used even for a qualitative understanding of the experimental results.

In the energy range of LEED, the atomic scattering cross section is of
the same order of magnitude as the atomic area; this means that low energy
electrons are strongly scattered by the top few layers of the crystal. One
commonly used method to take this into account is to assume that the crystal
is only a few layers thick. Palmberg and Peria estimated that 4 to 6 layers
contributed to their intensity peaks in the 150 to 400 eV range. Haque and
Farnsworth[6] showed that most of the diffracted intensity below 250 eV is
due to the first two, or at most three, atomic layers. Of the few layers, the
surface monolayer of atoms should play a dominant role, because it is exposed
directly to the electron beam. The present experimental data suggest that
the peaks in the 0 to 20 eV range could be due to multiple scattering effects

in the top layer atoms. This is inferred from the following observations.

a. A drastic change in the reflectance spectrum (Fig. 1, (a) and (b) was observed for the Ge(111) surface when it was covered with 0.05 monolayer of Cs. Although the surface structure model for the (111) surface of Ge has not been agreed on, all proposed models[4,7,8,9] assume some displacement or rearrangement of top layer atoms. Since the scattering of 0.05 monolayer of Cs atoms is not enough to account for such a drastic change in the reflectance spectrum it is interpreted as due to relaxation of the Ge surface induced by the deposition of Cs. This means that the positions of the top layer Ge atoms play a major role in determining the reflectance spectrum.

b. Sputtering of the surface with 150 eV argon ions caused the peaks to disappear (Fig. 1(c)). Presumably, sputtering disrupted the symmetry of the surface atomic arrangement more severely than the layer structure near the surface. Thus it appears that it is the symmetry of the surface, not the existence of layers, which generates the intensity peaks.

c. Heating of a Cs-covered Ge(111) surface to about $420°C$ also caused the peaks in the 0 to 20 eV range to disappear (Fig. 2(a)). If one argues that the sputtered crystal is highly disrupted, lacking layer structure near the surface, then the heated surface should have as much layered structure as the clean surface. So the layer structure is not responsible for the peaks in the spectrum. On the other hand, after heating to $420°C$ the Cs-covered Ge(111) surface was in a stage of transition from a Ge(111) 1 × 3-Cs structure (with a reflectance spectrum shown in Fig. 2(b)) to the Ge(111) 2 × 8 structure of the clean surface, the surface atoms were highly disordered, and this correlated nicely with the lack of peaks in the spectrum.

Even if the topmost layer of atoms were not solely responsible for the low energy peaks, it is still meaningful to study the monolayer diffraction problem, because in some cases it is much more convenient to calculate the transmission and reflection coefficients of a monolayer first, and to then take the effects of other layers into account.

Following the above considerations, it was decided to do an intensity calculation on a monolayer crystal using dynamical theories.

3.   Intensity Calculation Using McRae's Theory

The multiple scattering method of McRae[10] was first used to calculate the reflectance from a monolayer of muffin-tin potential. The atoms are assumed to lie at the centers of non-overlapping spheres such that the potential is spherically symmetric within the sphere and has a constant value between the spheres.   The crystal is assumed periodic in two dimensions and one layer thick in the z direction.   The incident electron beam is represented by a plane wave.   If one further assumes that the atoms scatter isotropically (s-wave scattering) and considers only the case of normal incidence and specular reflection, then the reflectance can be expressed as:

$$C_o = \frac{i2\pi}{AK} \; \frac{f_o}{1 - S'(0,0)f_o} \tag{1}$$

Where $|C_o|^2$ is the specular reflectance, A is the area of the surface unit mesh, K the magnitude of the wave vector of the incident beam, f_o the s-wave scattering factor, and

$$S'(0,0) = \sum_{\underline{g} \neq 0} \frac{\exp(iK|\underline{g}|)}{|\underline{g}|}, \tag{2}$$

where $\underline{g}$ is a direct lattice vector.

The summation in Eq. (2) is very slowly convergent.   McRae showed that to sum over an infinite two-dimensional lattice, it is advantageous to use the Ewald method[11] in splitting the summation into two series, one in the direct space and the other in the reciprocal space; both series converge more rapidly than the original one.   However, the actual surface is most likely a mosaic structure with the size of coherent patches determined by the density of dislocations, steps, and other defects on the surface. Therefore it seems more realistic to calculate Eq. (2) over an area of finite size. In that case, the summation can be done directly with a high speed computer. Using Eqs. (1) and (2), the electron reflectance spectrum of a monolayer of rectangular array of atoms with a 4 Å $\times$ 8 Å unit mesh was computed and is shown in Fig. 3. The atoms are represented by a single phase shift $\delta_o = 36°$ (this corresponds to an isotropic scatterer of medium mass), and the coherent patch consists of 8×10^4 unit meshes. Increasing the size of the coherent patch to 7.2×10^5 unit meshes did not cause any appreciable change in the spectrum, but decreasing it to 4×10^4 meshes caused all of the peaks to decrease in amplitude and some of them disappeared altogether. Figure 4 shows the computed reflectance spectrum from a monolayer crystal with a different lattice symmetry, the atoms are arranged in a close-packed hexagonal array with an atomic spacing of 6.928 Å.

Figures 3 and 4 show that peaks do appear on the reflectance spectrum of a monolayer crystal and that they have approximately the same average width and amplitude as those measured. The position of these peaks on the energy scale depends on the atomic potential (σ_o) and on the dimensions and symmetry of the lattice structure. All these results are consistent with the experimental result. The average intensity of the computed reflectance however decreases approximately as $1/E^2$ with increasing energy E; this differs from the measured approximately $1/E$ dependence. The most probable reasons for the rapid decrease in computed reflectance with increasing energy are:

a. As E increases, phase shifts other than δ_o become appreciable, so that s-wave scattering is no longer a good approximation.

b. As E increases, the coherent wave representing the primary electrons penetrates deeper into the crystal, so that the assumption of scattering by a single layer of atoms is no longer valid.

The next logical step then is to calculate the intensity curves with more than one phase shift. In this case, the method proposed by Kambe[12] is more appropriate.

4. Intensity Calculation Using Kambe's Theory

Kambe considers the interaction of a plane wave with a physical model that is exactly the same as that used for computation with McRae's theory. However, the mathematical formulations are considerably different. Kambe applied the Green's function method used by Kohn and Rostoker[13] for band structure calculations. This is a very appropriate approach, because both problems deal with the interaction of low energy electrons with a periodic potential, they only differ in boundary conditions. By modifying the band theory one can take advantage of the mathematical techniques developed by band theorists. Kambe showed that in the case of s-wave scattering, his method is identical to that of McRae.

In Kambe theory, the solution of the Schrodinger equation within the spherically symmetric atomic potential is assumed to be already obtained for a given value of $\underset{\sim}{K}$ by the method of partial waves. We have

$$\Psi(R) \;=\; \underset{\ell,m}{\Sigma} \;\; C_{\ell m} \, R_\ell \, (R) \, Y_{\ell m} \, (\theta,\phi) \tag{3}$$

where $Y_{\ell m}$ are spherical harmonics. The radial function $R_\ell(R)$ should be calculated for each atom. The space outside the spheres is assumed to be vacuum, the potential being zero. The problem then is to match $\Psi(\underset{\sim}{R})$ and $\dfrac{\partial \Psi}{\partial R}$ at the surface of the spheres $R = R_i$ (to determine the values $C_{\ell m}$). Because of the two-dimensional periodicity of the crystal potential, $\Psi(R)$ should be a Bloch wave:

$$\Psi(\underset{\sim}{R} + \underset{\sim}{g}) \;=\; \exp(i\underset{\sim}{k} \cdot \underset{\sim}{g}) \, \Psi(\underset{\sim}{R}). \tag{4}$$

Therefore $\Psi(R)$ need only be solved in one unit cell of the monolayer crystal.

Kambe showed that the solution in the unit cell should satisfy the integral equation:

$$\Psi(\underset{\sim}{R}) = e^{i\underset{\sim}{K} \cdot \underset{\sim}{R}} + \int_{sphere} (G \frac{\partial \Psi}{\partial R'} - \Psi \frac{\partial \Psi}{\partial R'})dS' \tag{5}$$

where the integration is over the surface of the sphere, and the Green's function $G(\underset{\sim}{R},\underset{\sim}{R}')$ contains summation over all direct lattice vectors. Following Kohn and Rostoker[13], Kambe expanded the plane wave and the Green's function into spherical harmonics and showed that Eq. (5) can be expressed as a system of algebraic equations:

$$\beta_{\ell m} = X_{\ell m} + \sum_{\ell' m'} \alpha_{\ell m \ell' m'} X_{\ell' m'} \tag{6}$$

where

$$X_{\ell m} = C_{\ell m} R_{\ell}(R_i) \tag{7}$$

$$\beta_{\ell m} = 4\pi i^{\ell} Y_{\ell m}^* (\theta_{\underset{\sim}{K}} \cdot \phi_{\underset{\sim}{K}}) j_{\ell} (KR_i) \tag{8}$$

$$\alpha_{\ell m \ell' m'} = R_i^2 [\delta_{\ell \ell'} \delta_{mm'} Kn_{\ell} (KR_i) j_{\ell}(KR_i)$$

$$+ A_{\ell m \ell' m'} j_{\ell}(KR_i) j_{\ell'}(KR_i)] \epsilon_{\ell'} \tag{9}$$

where j_{ℓ} and n_{ℓ} are the spherical Bessel functions of the first and second kind respectively.

So the diffraction problem is reduced to solving the Eq (6), where the unknown are $X_{\ell m}$. The coefficients $\alpha_{\ell m \ell' m'}$ are determined by ϵ_{ℓ} which is a property of the atomic potential, and by $\Lambda_{\ell m \ell' m'}$ which is determined by the symmetry of the atomic array. Equation (6) is an infinite system. However if the energy of the incident electrons is low enough and the atoms are not too heavy, then one would expect that ϵ_{ℓ} would approach zero with increasing

ℓ, so that only a finite number of unknowns $X_{\ell m}$ need be taken into account.

Having obtained $X_{\ell m}$, the specular reflectance with normal incidence can be expressed as

$$C_o = \frac{2\pi R_i^2}{iK\ A} \sum_{\ell} (-i)^{\ell} j_{\ell}(KR_i) \epsilon_{\ell} \sum_{m} Y_{\ell m} X_{\ell m} \tag{10}$$

Here again only those $X_{\ell m}$ which correspond to nonvanishing ϵ_{ℓ} need be taken into account.

The actual calculation consisted of two parts, the first part was the evaluation of the structure constants $A_{\ell m \ell' m'}$, and the second part the evaluation of the atomic factors ϵ_{ℓ}. The calculation of $A_{\ell m \ell' m'}$ followed closely the method used by Kohn and Rostoker[13] and Ham and Segall[14] in band structure calculations. The atomic factors ϵ_{ℓ} can be expressed as:

$$\epsilon_{\ell} = K\left(\frac{j_{\ell+1}}{j_{\ell}} - \frac{n_{\ell+1}}{n_{\ell}}\right)_{R=R_i} \times \frac{n_{\ell}\tan\delta_{\ell}}{(j_{\ell} - n_{\ell}\tan\delta_{\ell})} \tag{11}$$

Calogero[15] derived a first order nonlinear differential equation for calculating the partial phase shifts δ_{ℓ} by numerical integration. The phases are obtained by integration of the equation

$$\delta(\ell,K;R) = \frac{1}{K} \int_o^R U(R') \left[\cos\delta(\ell,K;R')\hat{j}_{\ell}(KR')\right.$$
$$\left. - \sin\delta(\ell,K;R)\hat{n}_{\ell}(KR')\right]^2 dR' \tag{12}$$

where the functions $\hat{j}_{\ell}$ and $\hat{n}_{\ell}$ are defined by

$$\hat{j}_{\ell}(KR) = KRj_{\ell}(KR) \tag{13}$$

$$\hat{n}_{\ell}(KR) = KRn_{\ell}(KR) \tag{14}$$

The ℓth order phase is given by

$$\delta_{\ell} = \lim_{R \to \infty} \delta(\ell,K;R) \tag{14}$$

To calculate ϵ_ℓ, the statistical Thomas-Fermi-Dirac potential[16] was used in Eq. (12) for evaluating the partial phase shifts δ_ℓ. Equation (11) was then used for calculating ϵ_ℓ.

The results of these calculations were then substituted into Eq. (9). In the present approximation, partial waves with angular momentum quantum number ℓ up to 3 were included. This resulted in 16 simultaneous linear algebraic equations in the form of Eq. (6). The solutions of this set of equations were then substituted in Eq. (10) and the reflectance $|C_o|^2$ calculated.

It is found that although δ_ℓ decreases rapidly with increasing ℓ, ϵ_ℓ decreases very slowly. This does not cause severe difficulty in the calculation however, because in Eq. (10) ϵ_ℓ is always accompanied by a j_ℓ which decreases by approximately a factor of 5 when ℓ is increased by 1. The slow decrease of ϵ_ℓ does require that a fairly large number of $X_{\ell m}$'s must be included for any accurate calculation.

The calculated reflectance spectrum for a monolayer of Cs atoms in a close-pack hexagonal array with an atomic spacing of 6.928 Å is shown in Fig.5. Figure 6 shows the spectrum for a monolayer of singly ionized Cs in a hexagonal array with an ionic spacing of 4 Å. The qualitative results of Kambe theory are very similar to those of McRae theory, i.e., the amplitude and width of peaks are comparable to measured values, and the position of these peaks depends on the atomic potential as well as on the symmetry and dimension of the atomic array. The main difference between these two calculated results is that in Kambe theory, by using $\ell = 0,1,2$ and 3, the reflectance decreases at a much slower rate with increasing energy. Thus it appears that the $1/E^2$ dependence of reflectance in McRae theory is due to the s-wave approximation, rather than the monolayer approximation, and that the peaks in the 0 to 20 eV range could be due to multiple scattering effect among top layer atoms.

5.   Discussion

The purpose of the present calculation is to fine a possible origin of the peaks in the low energy region. No attempt has been made to fit the measured intensity data because the present theories assume a muffin-tin model; every atom is represented by a spherically symmetric potential. This might be a good approximation for metal atoms in the bulk, but would not be appropriate for semiconductors like Ge or Si where the bonds between atoms are highly directional covalent bonds. The model would not be appropriate for alkali-covered semiconductor surfaces either, because the bonds to the alkali atoms are partially covalent thus making the bonds again directional. Even for metal surface atoms, the muffin-tin potential might still be a poor approximation because the very existence of the surface means that the atomic potential for surface atoms would be short range in the plane of the surface and also toward the bulk of the crystal but it would be long range in the direction away from the surface.

Segall[17], in using the Green's function method for energy band calculations on Ge, treated the anisotropic part of the potential as a perturbation. Presumably, this kind of perturbation method will have to be used for a serious fitting of experimental data.

ACKNOWLEDGMENT

The author is very grateful to his advisor, Professor W. T. Peria, for encouragement and many helpful discussions.

REFERENCES

1. J. M. Chen, Ph.D. Thesis, Electrical Engineering Department, University of Minnesota, 1968.

2. R. E. Fox, et al., Rev. Sci. Instr. 26 (1955) 1101.

3. R. J. Zollweg, J. Appl. Phys. 34 (1963) 2590.

4. P. W. Palmberg and W. T. Peria, Surface Science 6 (1967) 57.

5. L. F. Cordes, Ph.D. Thesis, Electrical Engineering Department, University of Minnesota, 1966.

6. C. A. Haque and H. E. Farnsworth, Surface Science 4 (1966) 195.

7. J. J. Lander and J. Morrison, J. Appl. Phys. 34 (1963) 1403.

8. D. Haeman, Phys. Rev. 121 (1961) 1093.

9. R. Seiwatz, Surface Science 2 (1964) 473.

10. E. G. McRae, J. Chem Phys. 45 (1966) 3467.

11. P. P. Ewald, Ann. Physik 49 (1916) 1.

12. K. Kambe, Z. Naturforsch. 22a (1967) 322, 422.

13. W. Kohn and N. Rostoker, Phys. Rev. 94 (1954) 1111.

14. F. S. Ham and B. Segall, Phys. Rev. 124 (1961) 1768.

15. F. Calogero, Nuovo Cimento 27 (1963) 261.

16. L. H. Thomas, J. Chem. Phys. 22 (1954) 1758.

17. B. Segall, J. Phys. Chem. Solids 8 (1959) 371.

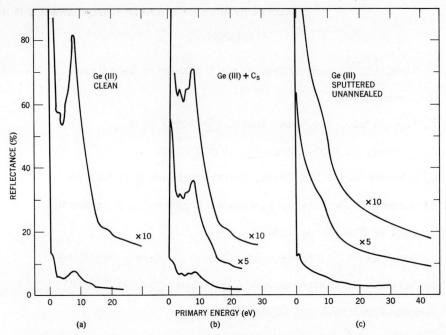

Fig. 1. Electron reflectance spectra for the Ge(111) surfaces. (a) Clean and annealed. (b) When covered with 0.05 monolayer of Cs. (c) Sputtered, unannealed.

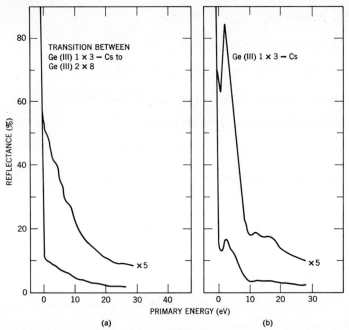

Fig. 2. Electron reflectance spectra for Cs-covered Ge(111) surfaces. (a) Transition between Ge(111)1x3-Cs to Ge(111)2x8. (b) Ge(111)1x3-Cs structure.

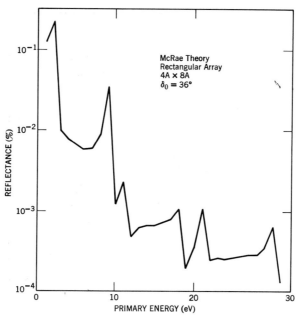

Fig. 3. Calculated electron reflectance spectrum of a rectangular array using McRae theory.

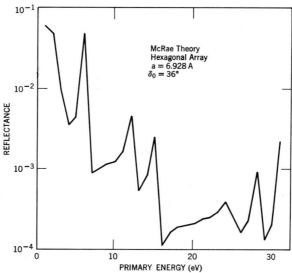

Fig. 4. Calculated electron reflectance spectrum of a hexagonal array using McRae theory.

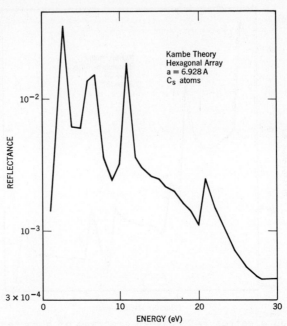

Fig. 5.   Calculated electron reflectance spectrum of a hexagonal array of Cs atoms using Kambe theory.

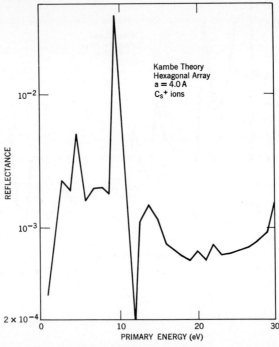

Fig. 6.   Calculated electron reflectance spectrum of a hexagonal array of Cs$^+$ ions using Kambe theory.

SOME NEW RESULTS FROM THE MATCHING-BAND STRUCTURE TREATMENT OF LEED INTENSITIES

D. S. Boudreaux and V. Hoffstein

Department of Physics
Polytechnic Institute of Brooklyn
Brooklyn, New York

1. Introduction

A new method for calculating the intensities of beams of low-energy electrons diffracted from clean crystal surfaces was recently formulated.[1] A plane wave expansion of the wave function outside the crystal (representing the sum of incident and diffracted electron beams) is matched at the crystal surface to the wave function inside (which has been determined in an electronic band structure calculation). The coefficients in the plane wave expansion then give the amplitudes of the observed beams of electrons (for the selected energy). The theory on which this method of calculation is based has been the subject of additional study, and some specific calculations were performed which have led to the developments discussed in this paper.

In the original proposal[1] the position and indexing of the secondary (or fractional order) peaks in intensity _vs_ energy curves were explained; the intensity of these peaks was expected to be large only when positioned on the shoulders of Bragg peaks. Both experimental observation and other theoretical models,[2] however, find very intense secondary maxima well separated from Bragg peaks (in the same beam). Boudreaux and Heine[1] based their discussion on a bulk allowed wave function inside the solid, i.e., a wave function which has the full three-dimensional translational periodicity of the lattice plus evanescent waves. It is now shown how diffracted waves violating one of the Laue conditions, (and hence forbidden deep in the crystal) are significant additions near the surface to the bulk allowed terms in the one-electron wave function inside the crystal. Thus the momenta of these diffracted waves differ from the momentum of the incident plane wave by vectors of the reciprocal surface net but not by an integral multiple of $2\pi/c$, where c is the layer spacing. A mathematical discussion of their origin is given and it is shown that their amplitudes grow as the surface is approached from inside, the amplitudes at the surface being approximately proportional to the matrix elements

$$< s \mid V \mid e >$$

summed over layers. V is the atomic pseudopotential. $\mid e >$ is the evanescent standing wave representative of a Bragg reflection and existing in the associated band gap (therefore it is confined to the surface region). $\mid s >$ is the secondary wave described above. These matrix elements are the amplitudes of the secondary beams and it is shown that they can be in general quite large. Secondary beams produced in this manner are also seen to be daughters of Bragg reflections in neighboring beams and hence occur at the same energy and carry the same indexing as those previously discussed.[1]

II. The Matching-Band Structure Model

Consider an electron beam normally incident on the (001) face of an orthohombic crystal. A two-dimensional section of its Brillouin Zone structure is illustrated in Figure 1. As the beam energy is raised, the wave vector of the incident plane wave increases in the K_z direction. Let us restrict attention to low-energies for simplicity. At π/c a Bragg reflection to $-\pi/c$ occurs. Associated with it, there is an energy band gap; in this energy range the total wave field is a standing wave decaying exponentially into the solid.[4] Matching conditions at the surface imply a standing wave outside the crystal also; and it is concluded that the specularly reflected (0,0) beam exhibits a unit intensity peak whose width in energy is equal to the band gap.

Increasing the energy to the neighborhood of the point $\vec{K}$, one finds similarly Bragg reflections to the points $\vec{K}_1$ and $\vec{K}_1'$. This corresponds, after
doing the matching, to intensity maxima in the (01) or (0$\bar{1}$) spots. An additional feature is the virtual mixing of other waves, e.g., at $\vec{K}_2$; they are
designated as weak beams for they are not degenerate with waves at $\vec{K}$, $\vec{K}_1$, and
$\vec{K}_1'$. Nevertheless the surface boundary conditions allow for real electron
beams outside the crystal which have the same component of momentum, parallel
to the surface, as $\vec{K}_2$ and whose intensity is just the amplitude of virtual mixing inside the crystal.[1] The indexing of such an intensity maxima in the (0,0)
beam is fractional: $2K = (1 + \frac{c^2}{a^2})\frac{2\pi}{c}$. Its intensity is small unless it occurs
on the shoulder of a Bragg peak.[1]

For an incident beam at $\vec{K}$, in Figure 1, it has not previously been realized
that it is possible to have some mixing with a decaying wave at $\vec{K}_2'$. The wave
$\vec{K}_2'$ is degenerate with $\vec{K}$, $\vec{K}_1$, and $\vec{K}_1'$, but in a three-dimensional crystal the
mixing is forbidden by a structor factor $S(\vec{K}_2' - \vec{K}) = 0$. Near the surface,
however, and in the energy gap $\frac{\hbar^2 K^2}{2m}$, $\vec{K}$, $\vec{K}_1$, and $\vec{K}_1'$, mix to form an evanescent
wave, and the structure factor for a matrix element between $\vec{K}_2'$ and this
evanescent wave is no longer zero. (Within certain limitations it is still
formally possible to define a structure factor). It is not clear how one can
apply degenerate perturbation theory to this semi-quantitative reasoning. To
make a calculation of the contribution of $\vec{K}_2'$ to the total wave field, consider
the integral equation which describes electron diffraction.

III.   Integral Equation for Electron Diffraction and Fractional Order Peaks

An exact integral equation describing electron-diffraction can be written[1]:

$$\psi(\vec{R}) = \psi_i(\vec{R}) + \int G_0(\vec{R}, \vec{R}'; E)V(\vec{R}')\,\psi(\vec{R}')d\vec{R}' . \tag{1}$$

$\psi_i(\vec{R})$ is the incident wave, $G_0(\vec{R}, \vec{R}')$ is the free particle Green's function, and $V(\vec{R})$ is the potential field responsible for the diffraction (in the case at hand, the crystal potential).

The solutions must satisfy the Bloch condition parallel to the surface, permitting Eq. (1) to be transformed to[1]

$$\psi(\vec{R}) = \psi_i(\vec{R}) - \frac{i}{4\pi A} \sum_n \frac{1}{K_n} e^{i(\vec{k}_i{}''+\vec{g}_n{}'')\cdot\vec{r}}$$

$$\cdot \sum_\nu \int e^{-i(\vec{k}_i{}''+\vec{g}_n{}'')\cdot\vec{r}'+ik_n|Z-Z'|} V(\vec{R}'-Z_\nu)\psi(\vec{R}')d\vec{R}'. \tag{2}$$

Here A is the area of the unit mesh on the surface plane; $\vec{k}_i{}''$ and $\vec{g}_n{}''$ are projections onto the surface plane of $\vec{K}_i$, the incident wave vector, and of the reciprocal lattice vector of the crystal, respectively. $\vec{r}$ is the projection of $\vec{R}$ on the surface.

$$K_n \equiv \sqrt{K_i{}^2 - (\vec{k}_i{}'' + \vec{g}_n{}'')^2} \tag{3}$$

which represents the normal component of momentum of a diffracted beam $n \cdot Z$ is the projection of $\vec{R}$ on the inward surface normal and $V(\vec{R}' - Z_\nu)$ is the screened ionic potential centered on a site in the νth layer.

Considering the normally incident beam to have wave vector $\vec{K}$ of Fig. 1, Eq. (2) can be simplified:

$$\psi(\vec{R}) = e^{iKz} - \frac{i}{4\pi A} \left[\frac{1}{K} \sum_\nu \int e^{iK|Z-Z'|} V(\vec{R}'-Z_\nu)\psi(\vec{R}')d\vec{R}' \right.$$

$$\left. + \frac{1}{K_1} e^{\pm i\vec{g}_i{}''\cdot\vec{r}} \sum_\nu \int e^{-ik_1|Z-Z'|\mp i\vec{g}_1{}''\cdot\vec{r}'} V(\vec{R}'-Z_\nu) \psi(\vec{R}')d\vec{R}' \right] \tag{4}$$

plus terms for $n > 1$ which cannot correspond to real diffracted waves at this energy. Further, we are only interested in the form of the solution at $Z = -\infty$, i.e., at the detector; therefore analysis need only be made on the following form of Eq. (4) which is valid for z on the surface or outside.

$$\psi(\vec{R}) = e^{ikZ} - \frac{i}{4\pi A} \left[\frac{1}{K_i} e^{-i(K_1 Z + \vec{g}_1'' \cdot \vec{r})} \sum_\nu \int e^{i(K_1 Z + \vec{g}_1'' \cdot \vec{r}')} \right.$$

(5)

$$\left. \cdot V(\vec{R}' - Z_\nu)\psi(\vec{R}')d\vec{R}' + \frac{1}{K} e^{-ikZ} \sum_\nu \int e^{iKZ'} V(\vec{R}' - Z_\nu) \psi(\vec{R}')d\vec{R}' \right] .$$

In the first two terms of the square brackets of Eq. (5), the pair $\left\{ \vec{g}_1'', (K + K_i) \right\}$ form a reciprocal lattice vector $\vec{G}_{101}$. However, in the third term $\left\{ 0, (K + K) \right\}$ does not. If the Born approximation is used in attempting a solution to Eq. (5), one may translate all the integrals in the sum over ν to a single unit cell within the crystal. This process leads to the introduction of a multiplicative structure factor which is unity for the first two terms but zero for the last (because 2K is not a reciprocal lattice vector). But the Born approximation is not good. A better guess for $\psi(\vec{R})$ as the first iteration would be the evanescent wave, known to exist in the band gap at $\vec{K}$:

$$\psi(R) \approx e^{-QZ} \left[e^{i(KZ-\phi)} \pm \frac{2}{\sqrt{2}} \cos (\vec{g}_1'' \cdot \vec{r}) e^{i(K_2 Z + \phi)} \right]$$

(6)

Here $Q - \phi = \tan^{-1} \left\{ \dfrac{Q}{\sqrt{2} V_{101}} \dfrac{\pi}{c} \right\}$ and $Q_{max} = \sqrt{2} V_{101}/2\pi/c$.

Now the first iteration of (5) does not give zero for the third term because the structure factor is no longer zero.

Thus the expression

$$\frac{1}{K} \sum_\nu \int e^{iKZ} V(\vec{R}' - Z_\nu)\psi(\vec{R}')d\vec{R}'$$

(7)

with ψ given by Eq. (6) is a first approximation to the amplitude of the specularly reflected beam at the energy $E \approx \dfrac{\hbar^2 K^2}{2m}$. Inside the crystal this wave must be damped out for it is not a bulk allowed solution. Also the expression (7) becomes zero when Q is zero for the same reason as in the Born approximation.

Specific calculations were done to evaluate (7). V was chosen to be a spherically symmetric model potential[5] whose value is zero for radii less than 4 Bu and $-\frac{2}{R}$ for radii greater than 4 Bu; Thomas Fermi screening was used. A tetragonal lattice was chosen with constants a = 8 Bu and c = 6 Bu. The integral was evaluated numerically in the atomic sphere approximation. The first iteration implied intensities in the (0, 0) beam about 10 times as large as those in the (0, 1) or (0, $\bar{1}$) beams. The actual numerical results for an electron beam at $\vec{K}$ are (0, 1) and (0, $\bar{1}$) intensities of 0.5 over an energy range of about 0.5 ryd.; the first iteration (0, 0) beam intensities are 0 at the top and bottom of the gap above rising to 9.01 at the center.

Further iteration is required for quantitative predictions of spot intensities, but it is clear that the fractional order peak will be of magnitude comparable to Bragg peaks. Iteration is not very easy and we are presently investigating more efficient ways of deducing the required amplitudes, so that the technique may be applied to real metals.

REFERENCES

1. Boudreaux, D. S. and Heine, V., Surface Sci. $\underline{8}$, (1967), p. 426.

2. McRae, E. G., J. Chem. Phys. $\underline{45}$, (1966), p. 3258.
    Gervais, A., Stern, R. M., and Menes, M., Proceedings of 22 MIT
    Physical Electronics Conference (1967).

3. Hofmann, F. and Smith, H. P., Jr., Phys. Rev. Let. $\underline{19}$, (1967),
    p. 1472.

4. V. Heine, Surface Sci. $\underline{2}$, (1964), p. 1.

5. V. Heine and I. Abarenkov, Phil. Mag. $\underline{9}$, (1964), p. 451.

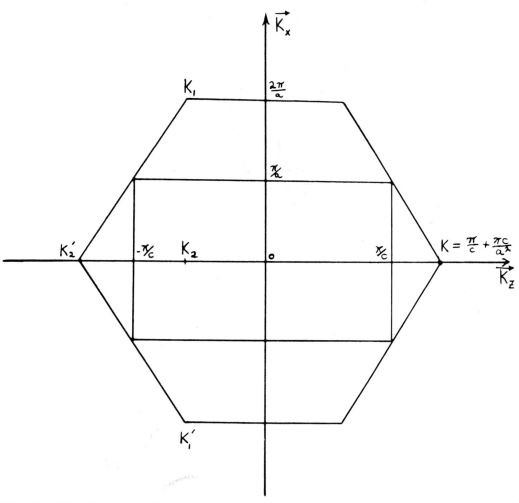

Fig. I. Two dimensional slice of the Brillouin Zone diagram for an orthohombic crystal. The crystal surface is parallel to the $\hat{k}_x$, $\hat{k}_y$ plane.

CALCULATION OF LOW ENERGY ELECTRON DIFFRACTION INTENSITIES USING DYNAMICAL THEORY

F. Hofmann and Harold P. Smith, Jr.

Space Sciences Laboratory
University of California
Berkeley, California

I. Introduction

It is generally agreed that, up to the present time, the use
of Low Energy Electron Diffraction (LEED) as a tool for surface
structure analysis has met with only partial success, due to the
lack of a sufficiently accurate, numerically tractable, theoretical
model. Thus, it has not been possible to fully exploit the rapidly
accumulating experimental information, especially beam intensity
measurements.

Recently, a number of theoretical treatments of LEED have been
suggested. They may be divided into two categories, i.e., kinematical and dynamical theories. In the former, multiple scattering
is usually neglected, whereas in the latter, this phenomenon is
treated, more or less, rigorously. Several authors have shown,[1-3]
however, that multiple scattering is a vitally important process
in LEED and that only the dynamical theories may be able to explain
in detail the dependence of the beam intensities on electron energy.

Dynamical effects in LEED have been investigated by a number
of authors,[3-7] using various models, methods, and approximations.
Our method[6] is based on the Bethe theory[8] which assumes that the
incident and the diffracted electron beams are plane waves and that
the crystal can be represented by a three-dimensional, periodic
potential, cut off abruptly at a plane surface.

In the present paper, we first present a brief review of the
theory, including the numerical methods for solving the matrix
equations, then, we discuss the results of a detailed beam-intensity
calculation for a simple fcc lattice, and finally, we attempt a
comparison with experimental results.

II. Theory

A. Wave Function in the Crystal

Let us consider a perfect crystal, represented by a complex, periodic potential,

$$V(\vec{r}) = \sum_{\ell,m,n} v_{\ell,m,n}\, e^{2\pi i(\vec{g}_{\ell,m,n}\cdot\vec{r})} \qquad , \ z < 0$$

$$V(\vec{r}) = 0 \qquad\qquad\qquad\qquad , \ z \geq 0$$

(1)

where $\vec{g}_{\ell,m,n}$ is the reciprocal lattice vector. It can then be shown [8] that the wave function,

$$\psi(\vec{r}) = \sum_{\ell,m,n} \psi_{\ell,m,n}\, e^{i(\vec{k}_{\ell,m,n}\cdot\vec{r})} \qquad , \ z < 0 \qquad (2)$$

is a solution of the Schrödinger equation, $\nabla^2\psi(\vec{r})+(K^2+V(\vec{r}))\psi(\vec{r}) = 0$, if

$$\vec{k}_{\ell,m,n} = \vec{k}_o + 2\pi\vec{g}_{\ell,m,n} \tag{3}$$

where $\vec{k}_o$ is, at this stage, an arbitrary vector, and if the expansion coefficients $\psi_{\ell,m,n}$ satisfy the equations

$$\psi_{\ell,m,n}(K^2 - |\vec{k}_{\ell,m,n}|^2) + \sum_{\ell',m',n'} v_{\ell',m',n'}\, \psi_{\ell-\ell',m-m',n-n'} = 0. \tag{4}$$

Equation (4) may be written in matrix form,

$$K^2\underset{\sim}{\psi} + A\underset{\sim}{\psi} = 0 \tag{5}$$

where $\underset{\sim}{\psi}$ is a vector whose components are the coefficients, $\psi_{\ell,m,n}$ The dimensionality of $\underset{\sim}{\psi}$, i.e., the number of $\psi_{\ell,m,n}$'s, is, in general, infinite. It becomes finite only if the sums in equations (1) and (2) are truncated. Thus, if L, M, and N are the numbers of values of ℓ, m, and n, respectively, $\underset{\sim}{\psi}$ will have L.M.N components, and A will be a L.M.N by L.M.N square matrix.

B. Wave Function in the Vacuum

According to the assumption that both the incident and the diffracted electron beams are plane waves, the wave function in the vacuum is written as

$$\Phi(\vec{r}) = \sum_{\ell,m,n} \Phi_{\ell,m,n}\, e^{i(\vec{h}_{\ell,m,n}\cdot\vec{r})} \quad , \quad z \geq 0 \tag{6}$$

where

$$\left|\vec{h}_{\ell,m,n}\right| = K. \tag{7}$$

C. Boundary Conditions

The continuity of the wave function and its derivative with respect to the surface normal (z-direction), at the surface (z=0), may be stated as

$$\sum_{\ell,m,n} \psi_{\ell,m,n}\, e^{i\vec{k}_{\ell,m,n}\cdot\vec{r}} = \sum_{\ell,m,n} \Phi_{\ell,m,n}\, e^{i\vec{h}_{\ell,m,n}\cdot\vec{r}} \tag{8}$$

$$\sum_{\ell,m,n} \psi_{\ell,m,n}(\vec{k}_{\ell,m,n}\cdot\vec{1}_z)e^{i\vec{k}_{\ell,m,n}\cdot\vec{r}} = \sum_{\ell,m,n} \Phi_{\ell,m,n}(\vec{h}_{\ell,m,n}\cdot\vec{1}_z)e^{i\vec{h}_{\ell,m,n}\cdot\vec{r}} \tag{9}$$

where $\vec{1}_z$ is a unit vector in the z-direction. Equations (8) and (9) must be satisfied for all values of x and y, and, therefore, we must require that

$$
\left.\begin{array}{rcl}
\vec{k}_{\ell,m,n} \cdot \vec{1}_x &=& \vec{h}_{\ell,m,n} \cdot \vec{1}_x \\[2em]
\vec{k}_{\ell,m,n} \cdot \vec{1}_y &=& \vec{h}_{\ell,m,n} \cdot \vec{1}_y
\end{array}\right\} \tag{10}
$$

At this point, it is convenient to consider a specific case, e.g., a cubic lattice, in order to simplify the equations. This can be done without loss of generality, since the extension of the treatment to other lattices is straight forward. For a cubic lattice, with lattice constant a, we have

$$
\vec{g}_{\ell,m,n} = \frac{1}{a}(\ell,m,n) \tag{11}
$$

Combining equations (3), (10), and (11) yields

$$
\left.\begin{array}{rcl}
\vec{h}_{\ell,m,n} \cdot \vec{1}_x &=& k_{o_x} + \dfrac{2\pi\ell}{a} \\[2em]
\vec{h}_{\ell,m,n} \cdot \vec{1}_y &=& k_{o_y} + \dfrac{2\pi m}{a}
\end{array}\right\} \tag{12}
$$

where k_{o_x} and k_{o_y} are the x and y-components of the vector $\vec{k}_o$, and equation (7) then reads

$$
\vec{h}_{\ell,m,n} \cdot \vec{1}_z = \pm\left[K^2 - (k_{o_x} + \frac{2\pi\ell}{a})^2 - (k_{o_y} + \frac{2\pi m}{a})^2\right]^{1/2} \tag{13}
$$

The index, n, in equation (13) can now be dropped and the wave function in the vacuum, equation (6), may be written as

$$\Phi(\vec{r}) = \sum_{\ell,m} \left[\Phi_{\ell,m}^{-} e^{i\vec{h}_{\ell,m}^{-} \cdot \vec{r}} + \Phi_{\ell,m}^{+} e^{i\vec{h}_{\ell,m}^{+} \cdot \vec{r}} \right] , \quad z \geq 0 \tag{14}$$

where $\vec{h}_{\ell,m}^{\pm}$ is a vector whose components are identical to those

of $\vec{h}_{\ell,m,n}$ (equations (12) and (13)), with the exception that the

superscript in $\vec{h}_{\ell,m}^{\pm}$ indicates the sign of its z-component.

Assuming that there is only one incident electron beam, whose

amplitude, $\Phi_{o,o}^{-}$, is equal to unity, we can simplify equation (14)

to

$$\Phi(\vec{r}) = e^{i\vec{h}_{o,o}^{-} \cdot \vec{r}} + \sum_{\ell,m} \Phi_{\ell,m}^{+} e^{i\vec{h}_{\ell,m}^{+} \cdot \vec{r}} , \quad z \geq 0 \tag{15}$$

where the first and second terms on the right hand side represent

the incident and the diffracted electron beams, respectively.

It follows from equations (12) and (15) that the real parts

of k_{o_x} and k_{o_y} are determined by the direction of incidence of

the primary electron beam, whereas their imaginary parts must be

zero, due to the boundary conditions at infinity. The z-component

of $\vec{k}_o$, however, cannot be obtained as easily. Methods for com-

puting k_{o_z} will be given below.

Equations (8) and (9), i.e., the boundary conditions at the

surface, are now decomposed into a set of equations relating the

coefficients of the various harmonics in x and y,

$$\sum_{n} \psi_{\ell,m,n} = \Phi_{\ell,m}^{+} \qquad\qquad , \quad \ell \neq 0 \text{ or } m \neq 0$$

$$\left. \begin{array}{l} \\ \\ \\ \sum_{n} \psi_{\ell,m,n} = \Phi_{\ell,m}^{+} + 1 \qquad\qquad , \quad \ell = m = 0 \end{array} \right\} \tag{16}$$

$$\sum_n (\vec{k}_{\ell,m,n} \cdot \vec{1}_z)\psi_{\ell,m,n} = (\vec{h}^+_{\ell,m} \cdot \vec{1}_z)\Phi^+_{\ell,m} \qquad\qquad , \ \ell\neq0 \text{ or } m\neq0$$

$$\sum_n (\vec{k}_{\ell,m,n} \cdot \vec{1}_z)\psi_{\ell,m,n} = (\vec{h}^+_{\ell,m} \cdot \vec{1}_z)(\Phi^+_{\ell,m}-1) \qquad\quad , \ \ell=m=0$$

$$(17)$$

The $\Phi^+_{\ell,m}$'s may be eliminated from equations (16) and (17), and we obtain

$$\sum_n \left[(\vec{k}_{\ell,m,n}-\vec{h}^+_{\ell,m})\cdot\vec{1}_z\right]\psi_{\ell,m,n} = 0 \quad , \quad \ell\neq0 \text{ or } m\neq0$$

$$\sum_n \left[(\vec{k}_{\ell,m,n}-\vec{h}^+_{\ell,m})\cdot\vec{1}_z\right]\psi_{\ell,m,n} \quad + 2(\vec{h}^+_{\ell,m}\cdot\vec{1}_z) = 0 \quad , \quad \ell=m=0$$

$$(18)$$

Equation (18) is finally written in matrix form,

$$B \underset{\sim}{\psi} = \underset{\sim}{\chi} \tag{19}$$

where $\underset{\sim}{\chi}$ is a L.M dimensional vector, and the matrix, B, has L.M.N columns and L.M rows.

D. Numerical Solution

Equations (5) and (19) may be combined into one single equation,

$$G \underset{\sim}{\psi} = \underset{\sim}{H} \tag{20}$$

where $\underset{\sim}{H}$ is a vector with L.M.(N+1) components, and G has L.M.N columns and L.M.(N+1) rows. Equation (20) cannot be solved exactly, since the number of equations, L.M.(N+1), is larger than the number of unknowns, L.M.N. The optimum solution is found, according to the principle of least squares, by minimizing the quantity

$$S = (G\underset{\sim}{\psi} - \underset{\sim}{H}) . D(G\underset{\sim}{\psi} - \underset{\sim}{H})^* \tag{21}$$

where D is a positive, real, diagonal matrix, containing suitable weighting coefficients, and * indicates complex conjugation. Since S is positive and quadratic in any of the unknowns, $Re(\psi_{\ell,m,n})$ and $Im(\psi_{\ell,m,n})$, the minimum is characterized by the conditions

$$\frac{\partial S}{\partial Re(\psi_{\ell,m,n})} = 0, \quad \frac{\partial S}{\partial Im(\psi_{\ell,m,n})} = 0, \quad \text{any } \ell,m,n \tag{22}$$

or, equivalently, in matrix form,

$$P\underset{\sim}{\psi} = \underset{\sim}{Q} \tag{23}$$

where $\underset{\sim}{Q}$ is a L.M.N dimensional vector, and P is a L.M.N by L.M.N square matrix. Equation (23) is solved, not by straight-forward inversion of P, but by an iterative technique, which has the advantage of not being affected by round-off errors in the computer.

It should be noted that the elements of both P and $\underset{\sim}{Q}$ contain the components of the vector $\vec{k}_o$. As we have seen above, the x and y-components of $\vec{k}_o$ are determined by the angles of incidence of the primary beam. The z-component, however, must, in general, be treated as an unknown. It may be computed by solving the equations

$$\frac{\partial S}{\partial Re(k_{o_z})} = 0, \quad \frac{\partial S}{\partial Im(k_{o_z})} = 0 \tag{24}$$

simultaneously with equation (23). If this is done, we find

that $Re(k_{o_z})$ is very nearly equal to $-\left[K^2 + Re(v_{o,o,o})\right]^{1/2}$, whereas

$Im(k_{o_z})$ approaches $-Im(v_{o,o,o})/2 \left[K^2 + Re(v_{o,o,o})\right]^{1/2}$ for large

values of $Im(v_{o,o,o})$.

Finally, once the vector $\underset{\sim}{\psi}$ is known, the intensities of the

diffracted beams, i.e., $\left|\Phi_{\ell,m}^{+}\right|^2$, are readily computed from

equations (16) and (17).

III. Results

A. The Potential

In the present paper, we have chosen aluminum as the object

of our investigation, for the following two reasons: (1) the

aluminum crystal potential is fairly well known thanks to a

number of excellent theoretical contributions, [9, 10] and (2)

LEED beam intensity measurements on aluminum single crystals are

now under way at our laboratory. [11]

In choosing a suitable potential for beam intensity

calculations of the nature outlined in the last section, a

compromise has to be sought between mathematical simplicity and

physical reality. Mathematical simplicity is necessary to keep

the computation time within tolerable limits. By physical

reality we mean a "self consistent" potential, which has been

shown to be successful in band structure calculations. [9] Our

potential has the following properties: (1) Since we are con-

sidering a fcc lattice (with lattice constant, $a = 4.04$ Å), all

expansion coefficients, $v_{\ell,m,n}$, with "mixed" indices are zero.

(2) The "inner potential", i.e., the real part of the coefficient,

$v_{o,o,o}$, is assumed (in all cases except in one of the curves of Fig. 3) to be equal to the average energy shift of the Bragg peaks, as observed experimentally (Fig. 1).   (3)   The shape of the real part of the potential is obtained by Fourier analyzing the "self consistent" aluminum potential given in reference 9. (4)   The "amplitude" of the potential, here defined as the differ- ence between the maximum and minimum values of $V(\vec{r})$, may be varied by introducing a scaling factor, by which all expansion coefficients, $v_{\ell,m,n}$, except $v_{o,o,o}$, are multiplied.   (5)   The imaginary part of the potential is assumed independent of position.

The above description of the crystal potential suggests investigation of the following parameters: (1)   The magnitude of the "inner potential", (2)   the number of terms retained in the Fourier expansion of the potential, i.e., the degree of "mathematical simplicity", (3) the "amplitude" of the potential, and (4) the magnitude of its imaginary part.

The effects of these four parameters on the beam intensity versus energy curves will be discussed below.

B.   A Set of Typical Beam Intensity vs. Energy Curves

Fig. 2 shows the result of a typical calculation.   Intensities of all electron beams emerging from the crystal are plotted as functions of the primary beam energy.   Apart from the ordinary Bragg peaks, there is evidence for at least two additional fea- tures, (1) secondary Bragg peaks (dashed arrow), and (2) a resonance peak, followed by an intensity minimum, which happens to eliminate one of the ordinary Bragg peaks, i.e., the $N = 4$ peak of the (1,1) beam. (The resonance peak is not very pronounced, in this particular case; it will become more evident in Fig. 5, below). Both of these phenomena have been predicted[3] and observed[1] by McRae. However, the resonance phenomenon has, in the past, been attributed[3,5] to the specular reflectivity, i.e., the (0,0) beam, whereas it is here seen to occur in a non-specular beam as well.

It is also seen from Fig. 2 that the diffracted beams become particularly intense when their angle of emission approaches 90°, i.e., when they are travelling almost parallel to the surface (the (1,1) beam at $\sim$ 19 volts, and the (0,2) beam at $\sim$ 38 volts). This effect has also been predicted by Boudreaux and Heine.[5]

C.   Effect of the "Inner Potential"

Fig. 3 shows that, when the "inner potential", i.e., $Re(v_{o,o,o})$, is decreased from 10 to 5 volts, the intensity curves are shifted towards higher energies by about 5 volts, their shape remaining almost unchanged. Of course, this is to be expected since the Bragg peaks are caused by resonances inside the crystal where the electron energy depends on the "inner potential".

D.   Truncation of the Potential Expansion

Fig. 4 shows a comparison between two calculations, performed under identical conditions, except that the Fourier expansion of the potential has been truncated at different points. The fact that there is little difference between the two curves indicates that the intensity of the (1,1) beam is not very sensitive to small changes in the shape of the potential. It also provides some justification for the use of these relatively simple potentials in calculations of this type.

E.   The "Amplitude" of the Real Part of the Potential

The effect of the potential amplitude is, in some sense, equivalent to the effect of the atomic scattering cross section, since these two quantities are related. When the cross section is increased, multiple scattering effects should become more pronounced. This is precisely what happens to the secondary Bragg peaks, as seen in Fig. 5, demonstrating that they are caused by multiple scattering.

Fig. 5 also shows that the resonance phenomenon, which was discussed earlier, only appears for large scattering cross sections. If these cross sections are small, both the secondary Bragg peaks and the resonance effects are absent. In this case, our results, as expected, approach those of the modified kinematical theory.

F.   The Imaginary Part of the Potential

Inelastic scattering of electrons is, in our model, simulated by introducing an imaginary potential. The effect of changing the magnitude of this potential is shown in Fig. 6. It is seen that, as the imaginary potential (i.e., the inelastic scattering cross section) decreases, the amount of "structure", i.e., the peak-to-valley ratio, increases considerably. This behaviour is to be expected. A decrease in the inelastic scattering cross section causes an increase in the "penetration depth" of the primary beam and, therefore, enhances all three-dimensional effects. Clearly, the very existence of Bragg peaks (i.e., "structure") is a three-dimensional effect par excellence.

The disappearance of the secondary Bragg peaks for large values of the imaginary potential, as seen in Fig. 6, indicates that these peaks probably originate from multiple scattering events involving atoms in different layers, rather than atoms in a single layer.

IV. Comparison with Experiment

Beam intensity measurements on the (100) surface of aluminum single crystals are being performed by Mr. S. M. Bedair at our laboratory. Fig. 7 compares his measurements[11] with a typical set of our theoretical curves (same as Fig. 2). All three experimental curves are normalized to the same primary electron beam current. Intensity ratios between different beams are, therefore, represented correctly in the Figure, even though the experiment gives no absolute values for the intensities.

The comparison between theory and experiment, as presented in Fig. 7, is somewhat unfair, for two reasons. First of all, no attempt has been made to optimize the parameters of the potential, i.e., the expansion coefficients, $v_{\ell,m,n}$. Secondly, the imaginary part of the potential has been assumed to be constant, whereas it should actually be allowed to vary with electron energy such that it could properly represent inelastic scattering cross sections.

A thorough parameter optimization, along the lines indicated above, could certainly improve the agreement between theory and experiment. However, this would represent a large investment in computer time.

V. Conclusions

(1)   We have demonstrated that the Bethe theory is well suited for the computation of LEED intensities, if the capabilities of large scale, high speed digital computers are utilized.

(2)   The calculated intensity versus energy curves exhibit, in addition to the well known integer order Bragg reflection peaks, secondary Bragg peaks and resonance effects.   The latter two phenomena disappear when the elastic scattering cross section is made very small, or when the inelastic scattering cross section (imaginary potential) is made very large.

(3)   Considering the fact that there are at least two assumptions, inherent in our model, which have to be seriously questioned,[3] i.e., the incident plane wave and the perfectly periodic crystal surface, and taking into account that we have not tried to optimize our potential, the agreement between theoretical and experimental[11] results, as exemplified in the beam intensity versus energy curves for an aluminum (100) surface, may be considered good.

References

[1]E. G. McRae and C. W. Caldwell, Surface Sci. 2, 509 (1964)

[2]G. Gafner, Surface Sci. 2, 534 (1964)

[3]E. G. McRae, J. Chem. Phys. 45, 3258 (1966)

[4]K. Hirabayashi and Y. Takeishi, Surface Sci. 4, 150 (1966)

[5]D. S. Boudreaux and V. Heine, Surface Sci. 8, 426 (1967)

[6]F. Hofmann and Harold P. Smith, Jr., Phys. Rev. Letters 19, 1472 (1967)

[7]Paul M. Marcus and Donald W. Jepsen, Bull. Amer. Phys. Soc. Series II, 13, 367 (1968)

[8]H. Bethe, Ann. Physik 87, 55 (1928)

[9]E. C. Snow, Phys. Rev. 158, 683 (1967)

[10]V. Heine, Proc. Roy. Soc. (London) A240, 361 (1957)

[11]S. M. Bedair, private communication

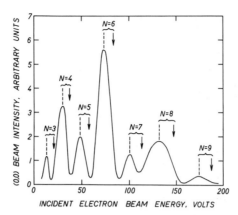

Fig. 1. Measured[11] intensity of (0,0) beam vs. energy. Conditions: aluminum (100) surface, normal incidence. The arrows indicate the locations of Bragg reflections, as they would appear if the "inner potential" were zero. N designates the order of the Bragg peaks. The energy difference between the arrows and the observed peaks, and, thus, the effective "inner potential," shows a slight increase with energy (i.e., with penetration depth) of the primary beam. The average energy shift is ∼ 10 volts.

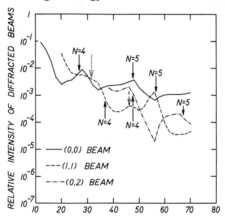

Fig. 2. Calculated beam intensities vs. energy. Conditions: primary beam intensity = 1, normal incidence, fcc lattice, a = 4.04 Å, (100) surface (potential cut off half-way between two layers of atoms), "inner potential" = 10 volts, imaginary part of potential = 2.5 volts, potential "amplitude" = 28 volts, potential expansion truncated such the $l^2 + m^2 + n^2 \leq 4$, wave function expansion truncated such that $-3 \leq l \leq 3$ $-3 \leq m \leq 3$ $-4 \leq n \leq 4$. The solid arrows indicate Bragg reflections of order N (including an "inner potential" correction of 10 volts). The dashed and dotted arrows mark the positions of a secondary Bragg peak and a resonance peak, respectively. Beams with "mixed" indices are absent due to symmetry of the fcc lattice.

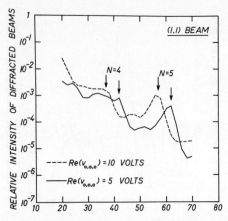

Fig. 3. Calculated intensity of (1,1) beam vs. energy, showing the dependence on the "inner potential". Same conditions as in Fig. 2, except: potential "amplitude" = 20 volts, potential expansion truncated such that $l^2 + m^2 + n^2 \leqslant 3$, "inner potential" = 5 (10) volts for the solid (dashed) curve.

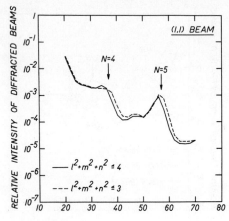

Fig. 4. Calculated intensity of (1,1) beam vs. energy, showing the effect of the number of terms retained in the potential expansion. Same conditions as in Fig. 2, except: potential "amplitude" - 20 volts, potential expansion truncated such that $l^2 + m^2 + n^2 \leqslant 3$ (4) for the dashed (solid) curve.

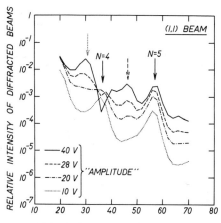

Fig. 5. Calculated intensity of (1,1) beam vs. energy, showing the dependence on the potential "amplitude" (i.e., on the atomic scattering cross section). Same conditions as in Fig. 2, except: potential expansion truncated such that $l^2 + m^2 + n^2 \leq 3$, potential "amplitudes" are 10, 20, 28, and 40 volts, as indicated. The figure shows the appearance of a secondary Bragg peak (dashed arrow) and a resonance peak (dotted arrow).

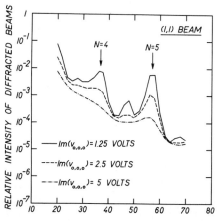

Fig. 6. Calculated intensity of (1,1) beam vs. energy, showing the dependence on the imaginary potential (i.e., on the inelastic scattering cross section). Same conditions as in Fig. 2, except: potential "amplitude" = 20 volts, potential expansion truncated such that $l^2 + m^2 + n^2 \leq 3$, imaginary potentials are 1.25, 2.5, and 5 volts, as indicated.

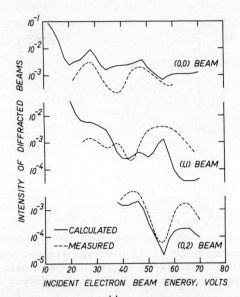

Fig. 7. Comparison between measured[11] anc calculated (same as Fig. 2) beam
intensities vs. energy. Conditions: aluminum (100) surface, normal
incidence.

INTENSITIES IN REFLECTION ELECTRON DIFFRACTION PATTERNS

J. M. Cowley and P. M. Warburton

School of Physics
University of Melbourne
Parkville, Victoria, Australia

1. Introduction

The technique of 'reflection' diffraction of high energy electrons (in the range 20 to 100 keV) played an important part in the early investigation of the structure of the surfaces of solids during the 1930's. Before the advent of electron microscopy, the method was seriously limited by a lack of information on the morphology of the surfaces studied. It was possible to decide only by inference from the diffraction patterns, or subsidiary techniques, whether the idealized conditions of small-angle incidence on an atomically flat surface was realized, whether the surface was so rough on a micron scale that the diffraction pattern was produced by transmission through the tips of projections, or whether there was some intermediate state of roughness with an indecipherable mixture of diffraction conditions.

For the interpretation of the observed intensities the only theoretical approximations available were the kinematical and the two-beam dynamical approximations. It was adequately demonstrated that neither of these could be applied other than in exceptionally favorable circumstances, and even then as rough indications only.

More recent work has shown that meaningful information on very thin surface layers can be obtained only under strictly controlled conditions of surface composition and cleanliness. The introduction of ultra-high vacuum techniques and improved methods for producing flat surfaces has recently increased the reliability and sensitivity of the technique so that it has become comparable and complementary to the LEED technique for the study of surface structures

(Sewell and Cohen 1965, Siegel and Menadue, 1967).

The similarity with LEED in the information obtainable for the ideal case of atomically flat surfaces comes because the component of the momentum of the electrons perpendicular to the surface, and hence the penetration of the electrons into the crystal, is roughly the same. The major differences in the fields of application and interest for the two techniques come from the very great sensitivity of the reflection method to departures from ideal surface flatness.

For both techniques progress is currently impeded by the lack of a theoretical basis for the rapid, accurate calculation of diffracted beam intensities. A number of approaches have been made to a theoretical treatment of LEED (see, for example, this conference) but all, to date, are seriously limited in their ability to provide accurate computations of diffraction intensities for situations of practical interest. It is possible, in principle, to adapt any of these approaches to the high energy reflection case, which should always be simpler and more accurate in practice because the electron scattering factors for individual atoms are known much more precisely for the higher voltages involved.

For example, it would be possible to set up the dispersion equations for an electron in an infinite periodic lattice, following Bethe (1928), and then apply the boundary conditions to give a matrix equation which can be solved by standard methods (compare the approach to LEED by Hofmann and Smith, 1967). However, it would be difficult to modify such an approach to include the effects of the perturbations of the crystal structure at the surface which are usually the objects of study.

Instead we have chosen to explore the application of the dynamical theory of electron diffraction which takes into account, from the beginning, the experimental situation of reflection diffraction and makes use of the approximations valid for high-energy electrons. It employs the concepts of the development of n-beam dynamical theory of Cowley and Moodie (1957) which has formed the basis for the 'slice method' of calculation of intensities for transmission electron diffraction, due to Goodman and Moodie, which has recently been applied widely and successfully in comparison with experiment (see for example, Goodman and Lehmpfuhl, 1967, Cowley, 1968, 1969). An extension of the slice method of calculation to LEED is possible if small-angle scattering approximations are removed and back-scattering is taken into account (A. F. Moodie, private communication). However, for reflection diffraction of high energy electrons we may retain the small angle approximations and consider forward scattering only.

We deal here with the idealized case of an electron beam incident at a small angle to a plane surface of a crystal. We assume periodicity in the sample in directions parallel to the surface but not necessarily in the direction of the surface normal.

2. Principle of the Method

We describe here the formulation of the computational scheme in its simplest form in order to make clear the principle of the method. A more detailed and complete account of the mathematical procedures will be given elsewhere.

For the reflection case, it is impracticable and unnecessary to follow the procedure used for transmission slice-method calculations of following the modification of the electron wave as it traverses the crystal from an entrance to an exit face. We may, in fact, consider a plane wave striking a crystal face which is effectively infinite in all dimensions, since the limitations of beam size in practice do not affect the observed diffraction pattern.

The incident beam enters the crystal and sets up wave fields inside and outside the crystal which must represent a steady state, with the same periodicities as the crystal sample. We obtain a relationship between the amplitudes of the wave fields, and so between the intensities of diffracted beams, by equating the wave functions on two plane separated by the periodicity of the crystal in the beam direction.

For example we consider an orthorhombic crystal with the c-axis parallel
to the surface and almost parallel to the incident beam. There will be no
periodicity in the x-direction perpendicular to the crystal surface although a
periodic potential variation with the unit cell dimension a will be approached
for x positive (inside the crystal). In the y direction periodicity may be
assumed or not, according to whether a one- or two- dimensional diffraction
pattern is to be calculated. Figure 1 shows a diagrammatic sketch of the x-z
plane.

We consider the wave function on x-y planes perpendicular to the crystal.
The steady-state hypothesis requires that the wave function at z=c is identical
with that at z=0. But if we consider the propagation of the wave between these
two planes we must take account first of the phase and amplitude change due to
the potential distribution, $\phi(xyz)$, which is made complex to include absorption
effects, by multiplying the wave function by the transmission function,

$$q(xy) \quad = \quad \exp \{i\sigma\phi_c (xy)\} , \tag{1}$$

where $\sigma = 2\pi me\lambda/h^2$ is the interaction constant, and

$$\phi_c(xy) \quad = \quad \int_0^c \phi(xyz) \ dz, \tag{2}$$

(see Cowley and Moodie, 1967). For this case $\phi(xyz)$ is the potential distribution in the crystal lattice for x positive and zero for x negative (i.e.
in vacuum).

Then we must take into account Fresnel diffraction between the planes by
convolution with the propagation function, given in the small-angle approximation by

$$p(xy) \quad = \quad \exp \{ik(x^2 + y^2)/2c\}. \tag{3}$$

Also we must take into account that, for each increment in z, there is an
additional contribution from the incident beam, assumed to be a plane wave, so

that we must add

$$\psi_0(xy) \;=\; 0 \text{ for } x > 0. \tag{4}$$
$$K \exp \{ikx \sin \alpha\} \qquad \text{for } x < 0,$$

where K is a constant which may be given any convenient value.

Thus our basic equality is

$$\psi(xyc) = \psi(xy0) = \psi(xy),$$

and

$$\psi(xy) = \{\psi(xy) \cdot g(xy) + K\psi_0(xy)\} * p(xy) \tag{5}$$

It is convenient to work with the Fourier transforms of these functions, which must then be related by the equation given by Fourier transforming (5);

$$\psi(uv) \;=\; [\psi(uv') * Q(uv)] \cdot P(uv) + K\psi_0(uv) \cdot P(uv) \tag{6}$$

where u, v are reciprocal space coordinates. For computational purposes it is necessary to sample the continuous functions at a finite set of points. The periodicity in the y direction gives us the usual set of integral indices, k, corresponding to the unit cell dimension, b. For the x-direction we impose a large artificial periodicity, L, extending from x = –A to +B as indicated in Fig. 1, in order to give a fine sampling of the reciprocal space coordinate u.

The possibility then arises that, since we have artificially repeated the crystal vacuum system at regular intervals in the x-direction, there will be additional interference between beams 'leaking' from one periodicity to another. This is avoided by multiplying the transmission function q(xy) for a slice by a function A(x), such as that shown in Fig. 2, which is zero at x = –A and at x = B so that electron waves arriving at these planes are automatically set to zero. With this modification of Q(u,v), the convolution integral of (b) is rewritten as a sum over indices to give

$$\psi(hk) \;=\; \Sigma_{h'} \, \Sigma_{h'} \, \psi(h'K') \cdot Q(h-h',\, k-k') \cdot P\!\left(\frac{h}{L}\,\frac{k}{b}\right) + K\psi_0(hK) \cdot P\!\left(\frac{h}{L},\frac{k}{b}\right) \tag{7}$$

This is, in fact, a matrix equation, and may be solved by one of the computer techniques recently developed for handling general complex matrices to give values for $\psi(h,k)$ and also by summing the Fourier series, for the wave function $\psi(xy)$. From this, the part corresponding to the waves in vacuum is extracted and Fourier transformed to give the observed Fraunhofer diffraction patterns.

3.   Some Preliminary Results

Initial tests of the method have been made with calculations for various angles of incidence of a 100 keV electron beam on a silicon (100) face, with the z axis in the [001] direction. The space-group forbidden reflections limit the observable Bragg reflections to h = 4n for k = 4n and H = 4n+2 for k = 4n + 2. Figures 3(a), (b) show results calculated for k = 0 and k = ±2 for incident angles α = 0.4 degree and 0.8 degrees. In order to limit these initial calculations to a few minutes on an IBM7044 computer the spacing between sampling points along the lines of reflections was too great to give any accurate representation of the intensities, but was sufficient only to indicate the general features of the intensity variations. In each case, a 'shadow edge' is visible (at h = 1.1 for 3(a) and at h = 2.2 for 3(b)). Peaks appear for the expected Bragg reflections, with subsidiary peaks near the forbidden or fractional order positions such as 200.

With no increase in computing times, greater accuracy can be achieved in one-dimensional calculations relating to the k = 0 line only, corresponding to a situation in which the crystal is rotated around the surface normal through such an angle that the periodicity of the projection ϕ (xy) in the y direction is suppressed.

4.   Conclusion

The initial results suggest that this method of computation is adequate for the calculation of at least the simpler patterns given by reflection from ideally smooth, flat crystal faces. To the extent that these ideal conditions are experimentally attainable, detailed comparison with observations should be possible.

In reflection electron patterns, as usually recorded, most of the intensity comes from electrons which have undergone a number of inelastic collisions and lost an appreciable amount of energy. Our calculations can be made to apply to elastically scattered electrons only, as measured by the use of an energy filter, by inserting the appropriate absorption function in the imaginary part of $\phi(xy)$. Or, with less precision, and with different assumptions as to the absorption functions, they could be made appropriate to measurements of the total intensity of the diffraction spots or streaks, including those electrons inelastically scattered with very little change of direction.

It is clear that with some slight modifications of the programs, it should be possible to take into account small departures from the ideal surface conditions. For example, for a surface which is smooth but wavy, some integration over the angle of incidence would be necessary. Or if the surface contains steps, the contribution of electrons entering the crystal through the top surface $(x=0)$ and leaving through a step edge $(z=\text{constant})$ could be included by retaining part of the crystal wave field in the final calculation of intensities from (7).

For the other extreme case of a very rough surface, when the diffraction pattern is given by transmission through the relatively thin extremities of projections from the surface, a treatment is being developed as an extension of the current methods for calculating the intensities of transmission diffraction patterns. It seems probable that for cases of intermediate degrees of roughness the difficulties of determining experimentally the large number of parameters required to describe the diffracting conditions sufficiently well will in general preclude the possibility of accurate calculations.

ACKNOWLEDGMENT

This work has been supported by a grant from the Australian Research Grants Committee.

REFERENCES

Bethe, H. A. (1928), Ann Phys. Lpz. 87, 55.

Cowley, J. M. (1968), Progress in Materials Science, 13 No. 6, 269.

Cowley, J. M. (1969), Acta Cryst. (In Press).

Cowley, J. M. and Moodie, A. F. (1957), Acta Cryst . 10, 609.

Goodman, P. and Lehmpfuhl, G. (1967), Acta Cryst. 22, 14.

Hofmann, F. and Smith, H. P. (1967), Phys. Rev Letters, 19, 1472.

Sewell, P. B. and Cohen, M. (1965), Appl. Phys. Letters 7, 32.

Siegel, B. M. and Menadue, J. F., (1967), Surface Sci., 8, 206.

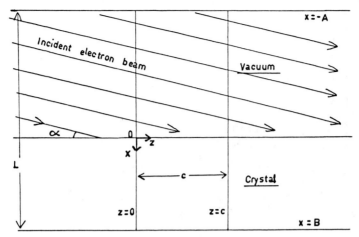

Fig. 1. The crystal-vacuum system. The portion shown is considered to be repeated periodically in the x direction.

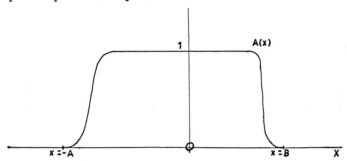

Fig. 2. The form of the function A(x).

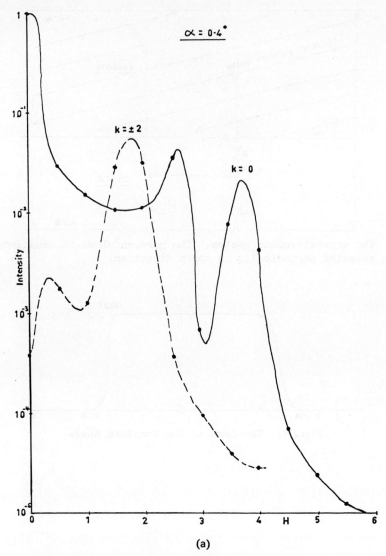

Fig. 3 (a). The intensity distributions along the lines k = 0 and k = +2, for 100 keV electrons incident at an angle of $a = 0.4°$ on $\overline{a}$ (100) face of silicon, with z-axis in the [001] direction.

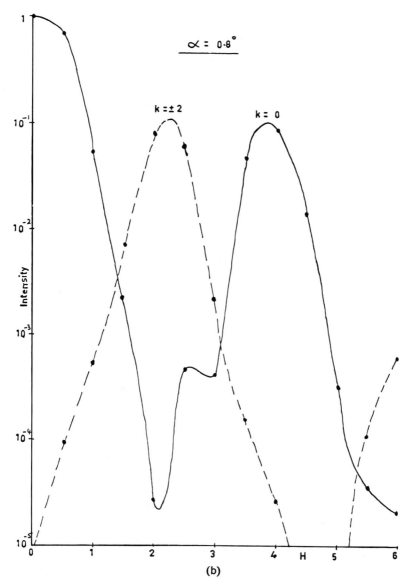

$\alpha = 0.8°$

Fig. 3 (b). As for figure 3, but with $a = 0.8°$.

MODEL COMPUTATIONS OF INELASTIC SCATTERING
OF LOW-ENERGY ELECTRONS BY CRYSTALS

E. G. McRae and P. J. Jennings

Bell Telephone Laboratories, Inc.
Murray Hill, New Jersey

1. INTRODUCTION

In low-energy electron diffraction experiments, the diffraction pattern is observed against a background that is due in part to small-loss inelastically-scattered electrons. It has been shown in several cases that the angular distribution of the intensity in the inelastic background exhibits an observable structure.[1-6] By analogy with the theory for the corresponding inelastic structure in X-ray diffraction,[7] it may be expected that a theoretical analysis of background structure in low-energy reflection experiments can give a valuable insight into the nature of the elastic electron wave field in the crystal. In this paper we discuss the background structure on the basis of a simple model, and we indicate by numerical calculation the sort of background structure that can be expected theoretically. The model and general approach are essentially the same as in a previous paper[8] dealing with the elastic reflection of low-energy electrons by crystals.

2. THEORY

2.1 Model and Assumptions

We limit our treatment to one-step inelastic processes. This means that an electron that has been scattered inelastically is supposed to escape from the crystal without further inelastic scattering. In calculating the background structure due to a one-step inelastic process, we adopt a highly simplified model, the chief provision of which is that the atoms are isotropic scatterers with respect to both elastic and inelastic scattering. This is equivalent to assuming a delta-function atomic-scattering potential.

The other essential model specifications are: the primary wave field is a plane wave; the crystal surface is perfectly periodic and of infinite extent; the effect of inelastic scattering in the boundary region of scattering potential just above the topmost atom layer may be neglected. Finally, it is assumed that there is no definite phase relationship between the inelastically-scattered waves emitted by different atoms - i.e., the atoms are assumed to be relatively incoherent emitters.

2.2 Intensity Formula

Let us imagine an idealized experiment in which a monoenergetic beam of electrons impinges on a crystal surface, and a small movable detector registers the inelastically scattered electrons that have suffered energy losses lying in a particular small range. The points of observation are all supposed to lie on a sphere with its center in the region of impingement, where the radius of the sphere is large in comparison with the dimensions of the impingement region. The point of observation may be defined by the direction of the corresponding outward radius vector of the sphere - the "direction of observation". The theoretical problem that we treat in this paper is to calculate the relative inelastic intensity, for a given value of the median electron energy loss, as a function of the direction of observation.

The main ideas of our treatment are indicated schematically in Fig. 1. We suppose that the propagation vector $\underset{\sim}{K}$ of the primary field makes an angle γ with the surface normal, and that it is required to calculate the relative inelastic intensity for a direction of observation making an angle $\bar{\gamma}$ with the surface normal and a given value ΔE of the median electron energy loss. Because of the assumption of relative incoherence, we may proceed by calculating the inelastic intensity contribution from a representative atom, and then summing the atomic contributions. A representative atom is

indicated by the filled circle in Fig. 1. For simplicity of presentation, we
treat the case in which there is just one atom per unit mesh in each atom
layer of the crystal.

The calculation consists of two parts, which correspond respectively
to the excitation of the inelastically-scattered electrons.

In a one-step inelastic process, the inelastic wave field is excited
by the effective elastic wave field. In the present model, the intensity of
the inelastic wave field emitted by an atom is proportional to the intensity
of the effective elastic wave field incident on it, at the atom center. Let
$\alpha(\underset{\sim}{K})$ denote the effective elastic wave field produced by a primary field of
unit intensity and propagation vector $\underset{\sim}{K}$. According to the above discussion,
the atom in question acts as a point source of inelastic electron waves of
source strength proportional to $|\alpha(\underset{\sim}{K})|^2$.

The escape process may be conveniently described with the aid of
the reciprocity theorem.[9] According to the reciprocity theorem, the intensity
of the inelastic wave field at a point of observation A, due to a point source
of strength $|\alpha(\underset{\sim}{K})|^2$ at the point B in the crystal, is equal to the intensity
at B due to a point source of the same strength at A.

Since by definition the point of observation is remote from the
impingement region, it follows that the wave field emitted by a source at the
point of observation behaves like a plane wave near the impingement region.
This equivalent plane wave is called the "reciprocal field". The propagation
vector $\overline{\underset{\sim}{K}}$ of the reciprocal field has the direction opposite to that of the
direction of observation, as indicated in Fig. 1, and its modulus is given
in terms of the median electron energy loss ΔE by

$$(\hbar^2/2m)(|\underset{\sim}{K}|^2 - |\overline{\underset{\sim}{K}}|^2) = \Delta E. \tag{1}$$

Let $\alpha(\overline{\underset{\sim}{K}})$ ($\overline{\alpha}$ in Fig. 1) denote the effective elastic wave field produced at the center of the representative atom by a primary field of unit

intensity and propagation vector $\underset{\sim}{K}$. Then the intensity of the inelastic field

at the point of observation, due to the atomic source of strength $\propto |\alpha(K)|^2$,

is proportional to $|\alpha(\overline{K})|^2$ as well as to $|\alpha(\underset{\sim}{K})|^2$.[10]

We choose to express the inelastic intensity as the ratio of

inelastic to primary flux across a plane parallel to the crystal surface.

This brings in the factor Θ defined by

$$\Theta = |\overline{K}|\cos \overline{\gamma}/|\underset{\sim}{K}|\cos \gamma . \tag{2}$$

The final step is to sum over all the atoms. Because of the assumption of

one atom per unit mesh-layer, the effective fields for different atoms in

the same atom layer are of the same modulus. Let N denote the total number

of atom layers included in the calculation and let α_μ denote the effective

field incident on an atom in the μth layer ($\mu = 0,1,...,N-1$). Our final

result is that the angular dependence of the inelastic intensity is described

by an <u>inelastic intensity function</u>

$$I(\underset{\sim}{K},\overline{K}) = \Theta \sum_{\mu=0}^{N-1} |\alpha_\mu(\underset{\sim}{K})\alpha_\mu(\overline{K})|^2 . \tag{3}$$

In the limit of vanishing scattering potential, the effective fields become

equal to the primary field, and the inelastic intensity reduces to $N\Theta$.

3.   RESULTS

3.1 Computational Procedure

We report computations of the inelastic intensity function

$I(\underset{\sim}{K},\overline{K})$ for a fifteen-layer simple-cubic crystal with a (100) surface. The

corresponding elastic intensity curves, in which the reciprocal field used

in the inelastic calculations plays the part of the primary field, are also

reported as an aid in interpreting the inelastic curves. In the computation of the inelastic intensity function, the effective fields were computed by a method previously described,[8] and the computed values were inserted in Eq. (3). The elastic intensities were computed, with very little extra effort, by way of the dynamical structure factor as previously described.[8]

In all computations on the inelastic intensity, the energy loss ΔE was set equal to zero. The computational results thus refer to small-loss inelastic scattering, as in scattering involving phonon excitation.

The computational results on inelastic intensities are presented as plots of $I(\underset{\sim}{K},\overline{\underset{\sim}{K}})$ versus $\overline{\gamma}$; the angle of observation, for specific azimuthal directions. For each such inelastic intensity profile, the fixed parameters are: the atomic-scattering phase shift δ; the electron energy parameter κ^2 ($\kappa = a/\lambda$, a = lattice parameter, λ = wavelength); and the azimuthal angle ϕ defined as the angle between the surface projection of $\overline{\underset{\sim}{K}}$ and the 10 direction in the surface.

All of the inelastic intensity profiles reported in this paper refer to normal incidence. The atomic-scattering phase shift δ is taken to be real throughout; this means that the effect of inelastic scattering on the elastic wave field is neglected. Though it is not strictly consistent with the rest of the treatment, this simplifying assumption cannot be a critical source of error as it is known from previous work that the form of the computed wave field is not materially affected by the introduction of an imaginary part in δ.[8]

3.2 Behavior of the Effective Field

In interpreting the computational results, it is helpful to examine the separate effective-field factors, $|\alpha_\mu(\underset{\sim}{K})|^2$ and $|\alpha_\mu(\overline{\underset{\sim}{K}})|^2$, whose product appears under summation in the expression for the inelastic intensity function (Eq. (3)).

The properties of the effective-field factors may be described with reference to diffraction conditions for the primary and reciprocal beams. Three types of diffraction conditions are of importance in the present context; they are Bragg conditions, Laue conditions and surface-wave resonance conditions. Let $\underset{\sim}{K}'$, $\underset{\sim}{K}''$ denote real propagation vectors with the same modulus and with surface projections $\underset{\sim}{k}''$, $\underset{\sim}{k}''$, let $\underset{\sim}{n}$ denote a vector in the direction of the inward surface normal of the crystal, let $\underset{\sim}{g}'$, $\underset{\sim}{g}''$ denote vectors of the reciprocal lattice and let $\underset{\sim}{v}$ denote a vector of the reciprocal net. Then in the limit of vanishing scattering potential the diffraction conditions for propagation vector $\underset{\sim}{K}'$ are defined by

$$\underset{\sim}{K}'' - \underset{\sim}{K}' = 2\pi\underset{\sim}{g}' \; , \quad \underset{\sim}{K}'' \cdot \underset{\sim}{n} < 0 \qquad \text{(Bragg)}$$

$$\underset{\sim}{K}'' - \underset{\sim}{K}' = 2\pi\underset{\sim}{g}'' \; , \quad \underset{\sim}{K}'' \cdot \underset{\sim}{n} > 0 \qquad \text{(Laue)}$$

$$\underset{\sim}{k}'' - \underset{\sim}{k}' = 2\pi\underset{\sim}{v} \; , \quad \underset{\sim}{K}'' \cdot n = 0 \qquad \text{(resonance)}. \qquad (4)$$

The above conditions are indicated in an Ewald-sphere diagram in Fig. 2.

In real cases (nonvanishing scattering potential) the diffraction features associated with the above conditions are always displaced somewhat from the nominal locations indicated by Eq. (4). For precise location of the diffraction features, we have computed the intensity curves for all possible diffracted beams in both transmission and reflection. The locations of Bragg and Laue features are given by the minima of the transmitted intensity in the 00 beam. The locations of Bragg features are given separately by peaks in the total reflected intensity (Bragg peaks). The locations of resonance features are given by special minima in the total reflected intensity.[8] For some purposes we find it necessary to make a distinction between Bragg peaks for the 00 (specular) reflected beam and Bragg peaks for nonspecular beams (the latter usually appear as "secondary peaks"[8] in the 00 reflectivity curve).

Our computations show that the value of the effective-field factor $|\alpha_\mu(\underset{\sim}{K}')|^2$ generally changes rapidly as $\underset{\sim}{K}'$ is varied in the vicinity of one of the diffraction conditions defined above. For Bragg and Laue conditions, the fluctuations of the effective-field factor are probably due to a wave-interference process of the general type described by existing dynamical theories of Kikuchi lines and Kikuchi bands.[11]

In general we find that the effective-field factor exhibits forms of dependence on the layer-depth that are characteristic of the specific diffraction conditions satisfied by the incident beam. In cases of strong elastic scattering ($\delta \gtrsim 1.0$) these properties are most pronounced and the variation with respect to changes in energy and orientation of the incident beam are greatest. There are four kinds of depth-dependence that we judge to be especially significant for interpretation of the computed inelastic intensity profiles. These are illustrated in Fig. 3 by the results of computations for $\delta = \pi/2$ (this is the value of δ for which an isotropic scatterer assumes its maximum cross section for a given energy) and may be described briefly as follows.

(a)   Bragg Peak in 00 Beam

For conditions corresponding to a Bragg peak in the reflectivity curve for the 00 beam, $|\alpha_\mu|^2$ is large for small layer-depth z_μ and decreases rapidly and monotonically as z_μ increases (Fig. 3a).

(b)   Bragg Peak in Nonspecular Beam

For conditions corresponding to a Bragg peak for a nonspecular beam, $|\alpha_\mu|^2$ is again large for small z_μ, and decreases rapidly but non-monotonically as z_μ increases (Fig. 3b).

(c)   Surface-Wave Resonance

In the vicinity of a surface-wave resonance, $|\alpha_\mu|^2$ is small for $z_\mu = 0$, and is close to unity for other values of z_μ up to the last layer

(Fig. 3c).

(d)   <u>Surface-Wave Resonance Plus Bragg Peak</u>

In the case of a near-coincidence between a surface-wave resonance and a Bragg peak, $|\alpha_{\mu}|^2$ can attain exceedingly large values over a very limited range of energy and orientation. An example of this type of depth-dependence is shown in Fig. 3d. As the energy and orientation of the primary beam are altered the shapes of the corresponding curves vary rapidly. In most cases the plots exhibit a series of irregular, high peaks until a depth-dependence similar to that in Figs. 3a, b or c is reached. The precise conditions under which these "giant" fields are generated is not known, and this question is being studied further by us.

3.3 <u>Weak-Scattering Case</u>

The inelastic intensity profile for a case of relatively weak elastic scattering ($\delta = 0.1$) is shown in Fig. 4, middle frame, together with the corresponding reflected intensity curve (top) and the 00 transmitted intensity curve (bottom).

In the absence of elastic scattering in the excitation and escape processes, the inelastic intensity profile for the present model with N = 15 would be simply

$$N\Theta = 15 \cos \bar{\gamma} \ . \tag{5}$$

The computed inelastic intensity profile shown in Fig. 4 shows structure resulting from elastic scattering. This structure is superimposed on a background of the general shape indicated by Eq. (5).

The structural features in the inelastic intensity profile in Fig. 4 can all be identified with Bragg or Laue conditions by comparison with the top and bottom curves. A feature associated with reciprocal-lattice vector g corresponds in a background display to a Kikuchi line or band generated by rotating the Ewald-sphere construction of Fig. 2 about an axis parallel to g.

The pair of peaks near $\bar{\gamma} = 0.8$ in Fig. 4 is the profile of a Kikuchi band associated with a Bragg condition for the 00 beam. The strong intensity of the peak at $\bar{\gamma} \sim 0.8$ and $\phi = 0$ in the inelastic intensity function is a result of the "accidental" interaction of Bragg peaks in the 00, 11, 1-1 and 20 beams with nearby resonances in the 01, 0-1, 2-1 beams. Multiple interactions of this type are expected only for special values of k^2 and with the reciprocal beam incident along axes of high symmetry in the reciprocal lattice. Therefore bright circles as in Fig. 4 will occur only for these special values of the incident energy, and when present they will have strongly varying intensity as a function of ϕ. This band would appear as a doubled bright circle in a background display. There is a band rather than a line in this case because as $\bar{\gamma}$ increases the Ewald sphere touches the same reciprocal-lattice point twice in an angular interval smaller than the natural width of a Bragg peak.

The inelastic profile exhibits in addition a sequence of less prominent features. Some of these correspond to Bragg and others to Laue conditions, and they may be identified as Kikuchi lines.

The shape of the Kikuchi band in Fig. 4 is similar to that of the reflected intensity peak, indicating that the chief mechanism at work here is the build-up of the effective field as indicated for an extreme case in Fig. 3a. The Kikuchi line profiles have irregular shapes which do not seem to have any simple interpretation in existing theory.

3.4   Strong-Scattering Case

Computations for cases of relatively strong elastic scattering ($\delta \gtrsim 1$) generally result in inelastic intensity profiles more complicated than those of the weak scattering case. In the strong-scattering case, the features associated with Bragg conditions are much more prominent than those associated with Laue conditions; the profiles depend more critically on the electron energy; and new features appear that do not seem to have any noticeable counterpart for weak scattering. The key to interpretation seems to lie in the depth-dependence of the effective-field factors, four types of which are described in Section 3.2. These cases can be combined in incident or reciprocal beams in different ways to give qualitatively different inelastic intensity profiles. We describe two combinations for which the inelastic intensity function can attain especially large values.

The first case is that in which the primary energy coincides with that of a Bragg peak in the 00 reflectivity curve. If a Bragg condition is satisfied for the reciprocal field, the inelastic intensity function is essentially an overlap integral between two functions of Type (a) described in Section 3.2, or between functions of Types (a) and (b). Thus a background display in this case consists of Kikuchi bright lines, and the inelastic intensity profile exhibits high peaks coinciding with peaks in the reflected intensity curve. This result is illustrated by computed curves in Fig. 5. Similar but less striking regularities are obtained in cases in which the primary energy coincides with that of a Bragg peak in a nonspecular reflectivity curve (secondary peak in the 00 curve).

The second case in which strong inelastic features can occur is one for which the effective field produced by the primary field is deeply-penetrating and close to unit modulus. Such fields are obtained in the neighborhood of surface-wave resonances in the primary beam and Fig. 6 refers to one of these situations. Since the primary effective-field factor is close to unity over many layers (Type (c) depth-dependence, Fig. 4c) there is a large overlap with reciprocal effective-field factors of Type (d) (shown in Fig. 4d). The overlap of a Type (c) function with functions of Types (a),

(b) and (c) are smaller, and the maxima corresponding to Bragg conditions for the reciprocal beam are thus relatively weak.

The conditions for the generation of "giant" fields are very special ones. They are most likely to be encountered along azimuths corresponding to high-symmetry directions in the crystal surface, and persist only over small ranges of angle. Therefore the theory indicates bright spots or short arcs in the background structure. Correspondingly, the inelastic intensity profile contains high, narrow peaks corresponding to near-coincidences between surface-wave resonances and Bragg peaks for the reciprocal field. This result is illustrated by computed curves in Fig. 6. The inelastic intensity profile shows two main peaks. The more prominent peak is associated with the secondary Bragg peak at $\overline{\gamma} = 0.65$ and the resonance minimum at $\overline{\gamma} = 0.61$. For the other peak the associated Bragg and resonance conditions are at $\overline{\gamma} = 0.22$ and $\overline{\gamma} = 0.25$, respectively.

4. DISCUSSION

In the model discussed in this paper, elastic scattering is handled by solving the effective-field equations for an array of isotropic scatterers. This procedure is known to give qualitatively correct results in application to low-energy electron diffraction intensities.[8] Inelastic scattering is treated by providing relatively incoherent isotropic emitters at the atom centers. This picture might apply fairly accurately to inelastic processes involving atomic excitations, but for phonon and plasmon excitations it gives at best only a partial description. If the present method of treatment were pursued in more detail, the inelastic mechanism would be represented by a distribution of sources with relative coherence. In the case of electron-phonon inelastic scattering, for example, this would bring in anisotropic thermal diffuse scattering effects as well as the effects discussed in this paper.

The chief properties of the model may be summarized as follows. For weak scattering, the background structure may display Kikuchi bands associated with Bragg conditions and Kikuchi lines associated with both Bragg and Laue conditions. For strong scattering the Kikuchi bright lines are very prominent, owing to the build-up of effective field near the crystal surface. In addition, special effects associated with the generation of "giant" fields are predicted. Previous work indicates that the strong scattering situation is the one most commonly encountered in the electron energy range below ~100 eV. In our opinion it is worthwhile to make careful experimental studies of background structure in this energy range, because the results could give new information about the nature of the electron wave field.

ACKNOWLEDGMENT

We are pleased to acknowledge the substantial contribution of D. E. Winkel to the computer programs used in this work.

REFERENCES

1. E. G. McRae and C. W. Caldwell, Surf. Sci. 2 (1964) 509.

2. C. W. Caldwell, Rev. Sci. Inst. 26 (1965) 1500.

3. D. C. Johnson and A. U. MacRae, J. Appl. Phys. 37 (1966) 1945.

4. J. L. Robins, R. L. Gerlach and T. N. Rhodin, Appl. Phys. Lett. 8 (1966) 12.

5. H. Taub and R. M. Stern, Appl. Phys. Lett. 9 (1966) 261.

6. R. M. Stern, A. Gervais and H. Taub, this Symposium.

7. R. W. James, "The Optical Principles of the Diffraction of X-Rays," (G. Bell and Sons, Ltd., London, 1962) p. 413.

8. E. G. McRae, J. Chem. Phys. 45 (1966) 3258.

9. The reciprocity theorem was used by von Laue in his treatment of the

analogous process in X-ray scattering. See R. W. James, _The Optical Principles of the Diffraction of X-Rays_, (G. Bell and Sons, London, 1962) p. 439 and references given there.

10. The reciprocity theorem applies to the total wave field - the effective field incident on the atom plus the field emitted by the atom - but the distinction between total and effective fields is immaterial in the present contest because the emitted field at the atom center is proportional to the effective field at the same point.

11. Y. Kainuma, Acta Cryst. <u>8</u> (1955) 247.

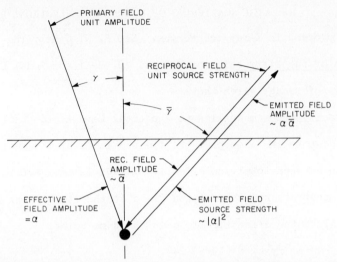

Fig. 1. Schematic indication of the method of treatment. The shaded line represents the crystal surface, the broken line the surface normal, and the filled circle denotes a representative atom.

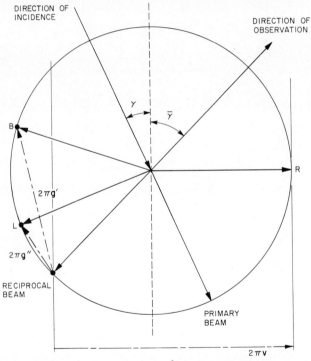

Fig. 2. Bragg (B), Laue (L) and surface-wave resonance (R) conditions for the reciprocal beam, illustrated with reference to an Ewald-sphere construction. The vertical broken line represents the surface normal, the vertical full lines represent reciprocal-net normals, and the filled circles denote reciprocal-lattice points. For a compact presentation, the three conditions are shown to be satisfied simultaneously.

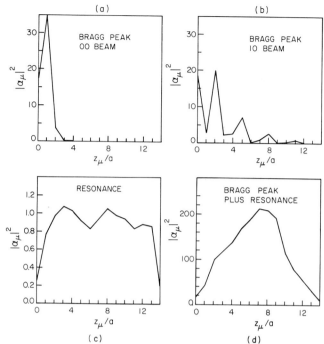

Fig. 3. Characteristic types of depth-dependence of the effective-field factor. a_μ denotes the effective field incident on an atom in the μth layer, z_μ denotes the depth of this layer and a denotes the lattice parameter. All curves are computed with $\delta = \pi/2$. (a) Normal incidence, $\kappa^2 = 1.80$; (b) Normal incidence, $\kappa^2 = 2.64$; (c) Normal incidence, $\kappa^2 = 2.01$; (d) Angle of incidence $34°$, $\kappa^2 = 2.01$, $\varphi = 45°$.

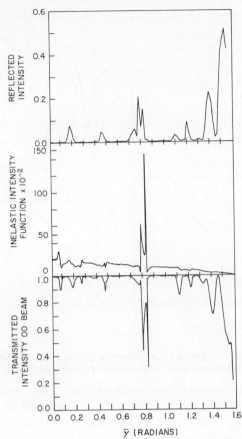

Fig. 4. Computed inelastic intensity profiles (middle frame), reflected intensity curve (top) and transmitted intensity curve for the OO beam (bottom) for a case of weak scattering. The inelastic intensity is plotted as a function of angle of observation $\bar{\gamma}$ along the azimuth designated by φ. The reflected and transmitted intensities are expressed as ratios to the intensity of the primary beam, and they are plotted as a function of the angle of incidence of the reciprocal beam $\bar{\gamma}$. The total reflected intensity is obtained by summing over the intensities of all reflected beams at each angle of incidence.

Parameters: $\delta = 0.1$, $\kappa^2 = 1.95$, $\varphi = 0$.

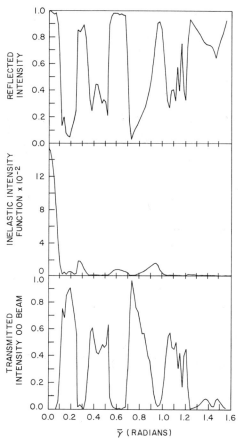

Fig. 5. Computed inelastic intensity profile (middle frame) reflected intensity curve (top) and transmitted intensity curve (bottom) for a strong-scattering case corresponding to Figs. 3a, b. For other details see Fig. 4 caption.

Parameters: $\delta = \pi/2$, $\kappa^2 = 1.80$, $\varphi = 0$.

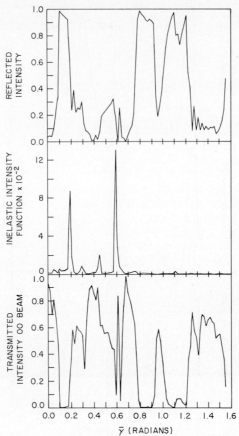

Fig. 6. Computed inelastic intensity profile (middle frame), reflected intensity curve (top) and transmitted intensity curve for the OO beam (bottom) for a strong-scattering case corresponding to Figs. 3c, d. For other details see Fig. 4 caption.

Parameters: $\delta = \pi/2$, $\kappa^2 = 2.01$, $\varphi = 45°$.

DYNAMICAL INTERPRETATION OF THREE DIMENSIONAL LOW ENERGY ELECTRON DIFFRACTION INTENSITIES[+]

R. M. Stern, H. Taub* and A. Gervais

Department of Physics
Polytechnic Institute of Brooklyn
Brooklyn, New York

INTRODUCTION

Low energy electron diffraction has until recently been considered primarily as a tool for the study of surfaces because of the assumed two dimensional character of the diffraction region. Treated as such the observed dependence of diffracted intensities with variation of the incident beam direction must be considered to be anomalous. If treated as an infinite three dimensional problem, the appearance of forbidden fractional order reflections at constant incident direction but varying electron wavelength is likewise anomalous. By introducing the theoretical approach usually applied to high energy electron diffraction, together with the techniques of x-ray structure analysis, it is possible to identify the proper three dimensional nature of the diffracting lattice. In low energy electron diffraction, very strong coupling may exist between diffracted beams. Because the strength of the coupling is extremely sensitive to geometry, it is possible that only one strong beam is excited at a time. This fact will permit the identification of individual reflections in terms of a two beam model.

The diffraction of electrons from the surface of a periodic solid can be treated as a boundary value problem where the incident wave is expanded at the surface in terms of the appropriate solutions to the wave equation in both the vacuum (Bragg reflections) and the solid (transmitted

* John Dubrovin Scholar, supported by the Aares Institute.

[+] Supported by USAFOSR GRANT 1263-67.

beams)[1]. Both the vacuum waves outside and the crystal waves inside the

solid must satisfy the Schrodinger equation

$$- \frac{\hbar^2}{2m} \, \nabla^2 + V(\underline{r}) \, \phi \, (\underline{r}) \; = \; E\phi(\underline{r})$$

In a perfect crystal the potential $V(\underline{r})$ can be represented by a Fourier series

$$- \frac{2m}{\hbar^2} \, V \, (\underline{r}) \; = \; v_0 + \underset{H}{\Sigma'} \, v_H e^{2\pi i \, \underline{H} \cdot \underline{r}}, \quad \underline{H} \; = \; h\underline{a} + k\underline{b} + \ell\underline{c}$$

In the vacuum $V(\underline{r}) = 0$. The incident beam sets up a scattered wave field

which is a superposition of reflected plane waves and evanescent waves in

the vacuum, and a superposition of Bloch waves (both travelling and evane-

scent) with the solid.[2] In a completely elastic theory as will be discussed

here the usual conditions of continuity at the crystal boundary apply. Total

energy and momentum parallel to the crystal surface (to within a reciprocal

lattice vector) are conserved. The introduction of the crystalline boundary

will relax the periodic boundary conditions normal to the surface.[3,4] This

may result in the introduction of selection rules with respect to the allowed

electron momenta normal to the crystal surface which are different than those

associated with that direction in an infinite crystal (i.e., permitting certain

forbidden fractional order reflections) but otherwise the problem can be treated

as for the general three dimensional case.

The details of the solutions for internal electrons $(E < 0)$ in an infinite

solid are well known: this is just the calculated band structure for the solid

being considered.[5,6] The diffraction problem concerns itself with the internal

solutions for a given magnitude and direction of the external electron wave

vector $\underline{K}_o$, which is determined by the energy and direction of the incident,

external electron beam. The relation between internal momentum $(\underline{k}_o)$ and $\underline{K}_o$

depends on the orientation of the crystal surface with respect to the reciprocal lattice because of the refraction due to the constant part of the crystal potential,[7-9] and the dispersion.

In low energy electron diffraction the problem is complicated by the fact that only the back reflected (Bragg) intensities can be directly observed. Thus the object of any low energy electron diffraction theory is to predict the reflection coefficients in terms of a crystalline wave field which cannot be directly measured. In high energy electron or x-ray diffraction the wave field within the crystal can be deduced from the details of the fine structure of the transmitted beam since the finite diffracting crystal has both an entrance and an exit surface.[10-12]

THEORY

Inside the solid the solutions are assumed to be expanded in terms of plane waves

$$\phi(\underline{r}) = \sum_H \chi_H e^{2i\underline{k}_H \cdot \underline{r}}$$

where the problem is the determination of the coefficients χ_H. Substitution of the expansions of ϕ and V into the Schrodinger equation leads to a set of secular equations

$$\chi_H = \sum_G \frac{\chi_{H-G} V_G}{k_H^2 - k^2}$$

which can be represented by the matrix equation

$$
\begin{pmatrix}
k^2 + v_0 - k_0^2 & v_{-G_1} & v_{-G_2} & \cdots \\
v_{G_1} & k^2 + v_0 - k_{G_1}^2 & v_{G_1-G_2} & \cdots \\
v_{G_2} & v_{G_2-G_1} & k^2 + v_0 - k_{G_2}^2 & \cdots \\
\vdots & \vdots & \ddots &
\end{pmatrix}
\begin{pmatrix}
\chi_0 \\
\chi_{G_1} \\
\chi_{G_2} \\
\vdots
\end{pmatrix}
= 0
$$

The dispersion surface[*], which defines the allowed wave vectors in the crystal, is found by setting the determinant of the coefficients equal to zero:

$$\begin{vmatrix} k^2 + v_0 - k_0^2 & v_{-G_1} & v_{-G_2} & \cdots \\ v_{G_1} & k^2 + v_0 - k_{G_1}^2 & v_{G_1-G_2} & \cdots \\ v_{G_2} & v_{G_2-G_1} & k^2 + v_0 - k_{G_2}^2 & \cdots \\ \vdots & \vdots & \vdots & \ddots \end{vmatrix} = 0$$

where the order of the determinant is equal to the number of reflections excited for a particular geometry.

The number and direction of the diffracted waves in the crystal can be determined by two geometrically identical but formally different methods: the Ewald sphere of reflection[13] and the Brillouin zone construction.   The former is more suitable for three dimensional crystals but both provide the identical geometry for the determination of which reciprocal lattice points are excited by the incident wave.[14,15] This is identical to determining which strong waves are coupled to the incident wave in the crystal through the appropriate Fourier components of the crystal potential.

The crystalline waves which are excited can be determined from the geometry shown in Fig. 1 which indicates the intersection of the Ewald sphere with the (100) plane of the reciprocal lattice for tungsten, at an electron energy of 460 eV.   For the geometry of Fig. 1a, no reciprocal lattice points fall on the sphere and hence no strong reflections are excited.   This condition results in the one beam case, only the incident wave vector $\underline{k}_o$ propagating in the crystal.   If the incident direction is now changed with respect to the crystal, for example by a rotation about an axis normal to the plane of the figure, then the sphere of reflection rotates about a

---

[*]   The dispersion surface is the surface of constant energy.

parallel axis, passing through the origin of reciprocal space. For some orientation as shown in Fig. 1b, a reciprocal lattice point (H) may fall on the sphere. For this geometry, the reflection H will be excited. Further rotation to the orientation of Fig. 1c will move the point H away from the sphere.

It is now convenient to introduce the parameter $\xi_H = |\underline{k}_H| - |\underline{k}_H|$, the excitation error, as shown in Fig. 1. For a point on the sphere of reflection $\xi = 0$. For a point inside the sphere $\xi < 0$. In Fig. 1a second reflection L is also indicated which is excited for the orientation of Fig. 1b, although the minimum excitation error is greater than zero. A given reflection is found to be excited over a range of values for the excitation error $-w \leq \xi \leq w$.[16] The magnitude of w is determined by the number and strength of the simultaneous reflections for a particular diffraction geometry.

In the most general treatment of this problem all the reciprocal lattice points must be considered, divided into three groups. Those far from the sphere of reflection can be ignored, but there will always be a large number of reflections which, although they do not lie on the sphere, lie sufficiently close to it to affect the shape of the dispersion surface and hence the strength of the diffracted beams arising from the reflections on the sphere itself (the weak beams). As will be shown below, a good approximation can be made by considering cases with only two or three strong beams, all the weak beams being ignored. For the two beam case the dispersion equation reduces to

$$(k^2 + v_0 - k_0^2)(k^2 + v_0 - k_G^2) = v_G v_{-G}$$

In the three beam case the dispersion equation is

$$\begin{vmatrix} k^2 + v_0 - k_0^2 & v_{-G} & v_{-H} \\ v_G & k^2 + v_0 - k_G^2 & v_{G-H} \\ v_H & v_{H-G} & k^2 + v_0 - k_H^2 \end{vmatrix} = 0$$

Figure 2 shows a machine calculation and resulting machine drawing for the dispersion surface for the reflection H = (002).[17] Here the two Fourier coefficients are taken as v_0 = 20 eV[9] and v_H = 10 eV. The incident energy is 20 eV, the crystal is assumed to be tungsten with a lattice spacing of 3.1624 Å. The determination of the waves excited in the crystal for a particular orientation of the crystal surface normal $\underline{n}$, is shown in Fig. 2b.[16] It can be seen that one of the results of the dynamical theory is the excitation of two waves, one from each dispersion sheet, associated with each reflection. These two waves, which will be referred to as the type one wave for the dispersion sheet nearest to the origin and the type two wave for the dispersion sheet farthest from the origin, have the same energy but different propagation directions. This is a manifestation of considering the dispersion in $\underline{k}$ at constant E; in band theory the dispersion is usually associated with E at constant $\underline{k}$. It should be pointed out that in x-ray diffraction this angular deviation is of the order of several seconds of arc, but as can be seen from Fig. 2, in the case of low energy electron diffraction the deviation is of the order of several degrees. The angular width of the dispersion is also much larger, being of the order of 10°.

The type one wave is out of phase with the type two wave, the former having nodes at the atomic planes, the latter having antinodes at the atomic planes. The origin of the absorption within the solid is expected primarily to be atomic excitations. The probability for these excitations is proportional to the wave field intensity at the atomic sites. For a rigid point lattice the type two wave should suffer an absorption larger than the average value in the crystal. Thermal vibrations and the extension of atomic wave functions introduce a finite absorption for the type one wave but it will still be considerably smaller than the average absorption. Figure 3a shows a plot of the intensity of each of the two types of waves as a function of excitation error. Figure 3b shows the absorption of each wave field vs. excitation error. The net absorption μ of the total wave field as a function of excitation error[19] ($\mu = \mu^{(1)} I^{(1)} + \mu^{(2)} I^{(2)}$) is shown in Fig. 3c. It can be seen that for some negative excitation error the net absorption in the crystal is greatly reduced due to the presence of the weakly attenuated beam. The importance of this beam is consistent with the observation of anomalous transmission (Borrmann effect) of x-rays[20,21] and high energy electrons.[22,23]

In low energy electron diffraction, the excitation of a Laue reflection can be expected to result in a reduction of the total secondary electron current, since the secondary electron current is a measure of the net absorption of the wave field in the crystal.[24] If the absorption is large, secondary electrons are produced near the surface and may escape from the crystal. If the absorption is low, the secondary electrons are produced within the volume of the crystal: their escape probability is exponentially dependent on the depth below the surface where they are produced, the emission process being characterized by some average escape length which takes any anisotropy into account. From Fig. 3b it can be seen that far from zero excitation error the crystal exhibits some average value of the absorption. As the crystal is rotated through a Laue reflection such that the reciprocal lattice point passes through the sphere of reflection from outside to inside, the net absorption of the wave field increases to above average value, then decreases below the average value before returning again to the average level. The initial increase in absorption is due to the large amplitude of the strongly attenuated wave field relative to the weakly attenuated one. The subsequent decrease in net absorption occurs when the weakly attenuated wave field has the large amplitude.[25] Since the average elastic mean free path for electrons must be less than the escape depth for secondary electrons, the increase in net absorption will not significantly affect the secondary emission while the decrease will have an appreciable effect. The secondary electron current should therefore exhibit the same behavior as the curve in Fig. 3b for each reflection excited except that the rise above the average value of the emission will be suppressed.

The excitation of a Bragg-reflection is not accompanied by the same absorptive processes as for a Laue reflection. For a single Bragg condition no travelling waves are allowed to propagate within the crystal, corresponding to total reflection.[1,2] If the incident direction is maintained while the energy is varied, total reflection occurs over a range of energies equal to the energy gap in the solid, which for a reflection H is equal to twice the Fourier component of the crystal potential for the reflection $2v_H$.

In the gap the only solutions of Eq. 1 which are allowed are for imaginary k. These exponentially damped (evanescent) waves are solutions to the Schrodinger equation at all energies, but are forbidden in the infinite periodic solid because their amplitudes become infinite in one direction or another. Their existence in low energy electron diffraction is a direct result of the existence of the crystal surface.

The excitation of simultaneous Bragg reflections will affect the intensity of all reflections since total current must be conserved. The diffracted intensities will then vary continuously as the diffraction conditions are varied in any experiment: the excitation of any new beam affecting the intensity of all others through the multiple scattering coupling.[14,26,27]

At low energies the total secondary emission current is predominantly elastic. The conservation of current requires that as the diffraction conditions are varied, the exchange between Laue and Bragg reflections results in variations of the total (elastic) secondary emission. At high energies, the excitation of a strong Laue reflection will result in the reduction of the escape probability of electrons due to the anomalous transmission of one of the wave fields. The secondary current will not be sensitive to Bragg reflections, especially at high energies.

Having demonstrated that the measurement of the secondary electron current
provides a method for the indirect determination of the internal wave fields
in the crystal, it is now possible to examine a number of experimental results
and in particular to compare measurements of the reflection coefficient and the
secondary emission. This will allow a determination of the significant diffraction processes for particular values of the diffraction parameters, and indicate which crystalline wave fields determine the diffracted intensities.

EXPERIMENTAL OBSERVATIONS

Diffraction cameras commonly used in LEED studies display at constant
energy only sufficient information to allow the determination of the periodicity
of the two dimensional lattice net parallel to the exposed crystal surface; a
photograph of the diffraction pattern at constant voltage representing the orthographic projection of the intersection of the sphere of reflection with the
reciprocal lattice. If a photograph of the diffraction pattern is made using
a polychromatic beam of electrons, the resulting diagram is a projection of the
volume of reciprocal space enclosed by the limiting spheres of reflection at
high and low energy. Figure 4 shows such a photograph for electrons between 75
and 150 eV.[28] This is the electron analog of the x-ray back Laue diagram.[29]
The important feature to note is that over this energy range the diffraction
pattern is composed of spots indicating the three dimensional nature of the
lattice. Electron Laue diagrams made at lower voltages show radial streaks
indicating a two dimensional diffracting lattice below about 50 eV.[*] Figure
4 also shows a standard x-ray Laue diagram from the same surface (tungsten
110) for comparison. The similarity of the two diagrams should be noted.

---

[*] An alternate dynamical interpretation can be made which attributes this
observation to the large magnitude of the dispersion and the corresponding
large range of excitation error over which a reflection is excited for low
energies.

Figure 5 shows a plot of the specularly reflected intensity as a function of electron energy for the (110) surface of tungsten. This intensity is proportional to the shape of the reciprocal lattice row passing through the origin containing the reciprocal lattice points of the type $(n,n,0)$. The incident beam direction, here about $4°$ from the (110) normal, has been carefully chosen so as to produce the curve shown with no dominant features except for the integral order Bragg reflections and forbidden half-order reflections typical for these measurements near to (110) incidence. These strong Bragg reflections from the crystal planes parallel to the surface indicated the strong three dimensional character of the reciprocal lattice. The fractional order beams are a result of the relaxation of the selection rules for diffracted momenta normal to the surface: their position (i.e., order) depending on the orientation of the incident beam with respect to the crystal surface.[27]

The total secondary electron current (measured from crystal to ground) is also shown in Fig. 5. The variation of the secondary emission over this range of energy at constant angle of incidence is seen to have structure of less than one percent compared with the 75 percent variation of intensity found for a typical diffracted beam over the same voltage range.[30] At low energy the increase in the secondary emission is due to the increase of the unregulated primary current with increasing voltage, together with the continual excitation of additional secondary processes over this voltage range. From the lack of structure in the secondary emission comparable to that of the Bragg reflected intensity, it may be concluded that the absorption of the total wave field in the crystal does not vary rapidly. The dynamical diffraction conditions therefore do not change rapidly in measurements where the incident beam direction is maintained constant and the electron energy is varied.

There exist two other methods of systematically exploring the reciprocal lattice typically used in x-ray analysis and also appropriate for LEED measurements.

The intensity of the specularly reflected beam can be monitored as the crystal is rocked about an axis perpendicular to the incident beam direction and lying in the plane of the crystal surface, while the incident electron energy is maintained constant. The shape of the same reciprocal lattice row normal to the surface is now studied as when the voltage is varied at constant incident angle. The intersections of the reciprocal lattice points of this row with the sphere of reflection now occur at continuously varying angles of incidence. Figure 6 shows a plot of both the total secondary emission and the intensity of the specularly reflected beam as a function of incident direction (a rocking curve). The electron energy (458 eV) is chosen so that at normal incidence the 880 reflection (see Fig. 5) is excited.

The reflected intensity shows a broad maximum which contains regular fine structure: the angular width of the envelop of the fine structure is consistant with the energy width of the same reflection. This means that the same range of excitation error, w_H, is found from both types of measurements. For a single scattering model the magnitude of w_H is determined by the natural width of the gap $(2\nu_H)$ for an infinite non-absorbing crystal. For a finite crystal, or a crystal where the attenuation of the incident beam determines the effective diffracting volume, resolution broadening will occur. For a model which considers simultaneous reflections the range of the excitation error (defined in Fig. 1) in voltage may be as large as the magnitude of the average inner potential of the crystal V_o. A very crude estimate of the volume of the diffracting crystal can be made by subtracting the effect of the estimated width of the gap from the magnitude of w and assuming as exponential decay of the wave field. For $2\nu_H = 8$ eV, the wave field is found to decay to 1/e in about 10-20 lattice planes for the (880) reflection from tungsten (110) at normal incidence.

The total secondary emission plot accompanying the rocking curve of Fig. 6 begins to show strong angular variation. This indicates that the rocking curve is accompanied by variations in the total absorption and hence in the dynamical diffraction conditions. This conclusion is reinforced by measuring the secondary emission at higher energies, as is shown in Fig. 7.[20,24,31] Here the elastic secondary current (to within 5 volts) is measured using the grid system of the diffraction camera. As the incident energy is increased and more inelastic processes are excited, the magnitude of the structure in the secondary emission increases.

It is also possible to explore the reciprocal lattice by rotating the crystal about its normal for some constant electron energy and diffraction angle (as measured between the incident beam direction and that of the specularly reflected beam).[26] The intensity of the specularly reflected beam can then be measured as a function of the azimuthal angle which lies in the surface and defines the orientation of the plane of diffraction containing the incident and specularly reflected beam, with respect to some reference direction in the surface.

Figure 8a shows such a rotation diagram,[27] the voltage being set so that the specularly reflected beam exhibits a maximum in intensity corresponding to a Bragg reflection from the planes parallel to the surface at the chosen angle of incidence (see the intensity vs. voltage curve at the right hand side of the figure.) From Fig. 1, it can be seen that the rotation of the crystal about the surface normal maintains the diffraction conditions unchanged for the specularly reflected beam. In Fig. 8a the total secondary emission is also shown.[30] An extraordinary amount of fine structure is found in the rotation diagram, the half width of the maxima and minima in some cases being less than 1° which is the magnitude of the divergence of the incident beam.

It is important to note that the structure of the total secondary emission strongly resembles that of the intensity of the specularaly reflected beam.   In particular the minima in each curve occur at the same orientation, except for the lack of secondary emission structure about the (002) orientation.   Figure 8b shows the rotation diagram for a voltage $(V_{1/2})$ where no Bragg maximum appears in the specularly reflected beam, the diffracted intensity being near to the background.   Except for the reduction in amplitude of the modulation of the intensity, the structure in the rotation diagram is similar to that of the neighboring 550 and 660 Bragg reflections.

INTERPRETATION

As the electron energy is raised from zero to the order of 500 eV, (exciting the 880 reflection at normal incidence) the sphere of reflection passes through about 800 reciprocal lattice points.   Rocking the sphere at this voltage through 180° intersects about 1,200 reciprocal lattice points. Rotating the sphere at this voltage and 45° incident angle passes the sphere through 1,600 reciprocal lattice points.   The strong difference between the three types of measurements described above are therefore not due to the number of reciprocal lattice points intersected by the sphere, but are a result of the specific reflections which are excited during the measurement.   For each type of measurement there are certain reflections which are inaccessable for a given set of boundary conditions.

In order to determine which reflections are important it is necessary to identify those reflections excited for any given set of values of diffraction parameters.   This is most easily done by maintaining at least one of the diffraction parameters constant: it is particularly convenient to plot the locus of a particular reflection at constant energy.[24,31]   Figure 9 shows the loci at 1000 eV of the incident beam directions for the excitation of reflections H =

$h^2 + k^2 + \ell^2 \leq 6$ (left hand diagram) and $6 < H \leq 12$ (right hand diagram). These

diagrams are an orthographic projection of the intersection of the extension

of the incident wave vector $\underline{K}_o$ with the sphere centered at the crystal, drawn

to an included angle of $120°$ (the aperture of the diffraction screen in the

post-accelerated camera). Examination of the diffraction geometry will show

that these diagrams also represent the directions of the diffraction cones

for the reflection of an internal source from the planes of the zone of the

reflection H (the optical reversibility or reciprocity theorem) the definition

of Kikuchi lines:[*] they are also the Brillouin zone boundaries.

It is instructive to first examine the rotation to determine which reflec-

tions are important in the diffraction process. If there are no dynamical

interactions excited during the rotation about a given Bragg reflection, the

intensity of the specularly reflected beam should remain constant. Analysis of

the rotation diagram about the 550 reflection shows that the reflected inten-

sity and the secondary emission undergo strong reduction whenever a low index

Laue beam is excited (for $H \leq 12$ all the reflections are in the forward direc-

tion). These Laue reflections represent the excitation of three simultaneous

beams, (550), (000), (h,k,ℓ). The orientation and indices of the Laue reflec-

tions are shown in Fig. 10a. If the rotation is not made about a Bragg reflec-

tion, then the Laue reflections will be accompanied only accidentally by

additional reflections, and hence the two beam case is excited (see the curve

$V_{1/2}$ in Fig. 8).

The similarity between the general two and three beam case indicates that

the excitation of a Bragg reflection is relatively unimportant, the common pro-

cess which determines the absorption in the crystal being the excitation of

the forward Laue Beams.

---

[*]   Strong Kikuchi diagrams are observed in the diffraction pattern above 600 eV
from tungsten and other crystals.[32]

The importance of the excitation of a Laue beam, Bragg reflections notwithstanding is understood by considering the wave fields within the crystal. The accidental simultaneous excitation of a Bragg reflection generally does not produce a large affect on the strength of the propagating wave field because the intensity of such reflections is small compared to that of the Laue reflections. The Bragg intensity is proportional to the product of the structure factor and the Debye-Waller factor, both of which are small for large values of H (a characteristic of Bragg reflections at these energies) and leading to an upper limit of the magnitude of the index H of reflections which must be considered.

During the rotation, the structure in the reflected intensity which is not accompanied by similar structure in the secondary emission, (for example about the (002) orientation in Fig. 8a) is found to be the result of the exchange of energy between several Bragg reflections, a process which does not affect the absorption of the wave field in the crystal. These orientations are shown in Fig. 10b.

The variation of the secondary emission with incident direction during the rocking curves can be analyzed in a similar manner. Figure 11 shows the loci of the Laue reflections at 2000 eV.[24] The regions of the diagrams scanned by the incident beam are shaded. These equatorial regions are redrawn above the diagrams on a linearized angular scale. The secondary emission is shown in the center of the figure. Where the calculated positions of the Laue excitations (as indicated by the vertical lines) are closely spaced, the total secondary emission is found to be strongly attenuated. The secondary emission is therefore sensitive to the excitation of a diffracted beam having a low index, the same conclusion reached from an analysis of the rotation diagram.

The analysis of the specularly reflected intensity of the rocking curve is complicated by the fact that the excitation error of the Bragg reflection

is continually varying. The scan is also not equatorial in order to allow
observation of the specular reflection near normal incidence. The intensity
is affected by both simultaneous Bragg and simultaneous Laue reflections, the
latter being the mixed three beam case which is now being investigated.[16,33]
Over the limited range of the excitation of the Bragg reflection ($\pm$ 10° for
the 880 reflection: Fig. 6) little sensitivity in the secondary emission is
observed. This indicates that the coupling is to other Bragg reflections. The
first Laue reflection (002) is excited at 12.5°.

CONCLUSIONS

It has been shown that the significant structure in the intensities measured in low energy electron diffraction can be interpreted in terms of a
simple geometrical model.

The inadequacy of intensity vs. voltage measurements for exploring a
three dimensional reciprocal lattice is now apparent. Since low index Laue
reflections are in general inaccessable to the sphere of reflection for constant but arbitrary incident beam direction, the measurement of the intensity
of a reflected beam as a function of voltage does not necessarily result in the
excitation of strong wave fields in the crystal, or in the strong variation
of the dynamical conditions for the diffracted waves in the crystal. Such
measurements, therefore, as far as Laue (transmitted) reflections are concerned correspond to the one beam case.

If however a low index Laue beam is excited, it remains excited over a l
large range of voltage since the position of the sphere of reflection near
to the origin is relatively insensitive to the diameter of the sphere[*]. Laue

---

[*] On the other hand rocking the crystal maintains certain Bragg reflections over
a range of angle large compared to the range of excitation of the Laue reflection. Exciting the mixed three beam case in the rocking curve involves a transition from the two beam Bragg case, while in an I vs. V diagram the transition
is from the two beam Laue case.

reflections are however extremely sensitive to small variations of the incident

beam direction as may be seen from the geometry.   Because of stray magnetic

fields it is extremely difficult to maintain the incident beam direction con-

stant over a large range of electron energies which may result in the accidental

excitation of new beams: a situation which makes interpretation of intensity

vs. voltage measurements difficult.[*]

It has been mentioned that the wave matching approach at the crystal sur-

face must take into consideration all of the solutions to the wave equation

including the evanescent waves which decay away from the surface both in the

crystal and in the vacuum.   The extraodinary sensitivity of the surface re-

lated structure in the intensity measurements (the fractional order peaks

and much of the rotational diagram structure disappear for contaminated sur-

faces[27,28,32,33]) indicates that these waves, especially the evanescent vacuum

waves, usually ignored in LEED theory,[3] must be considered in detail. The two

beam approximation outlined above is applicable only to those effects which

are the result of the three dimensional nature of the diffracting lattice: evi-

dently the major effects.   The surface must now be considered a perturbation

of the three dimensional problem and surface effects studied in detail sepa-

rately.

It may be concluded that much of the variations in the intensity of the

diffracted beams observed in LEED measurements can be interpreted in terms

of the expected bulk, three dimensional reciprocal lattice of the crystal

being studied, regardless of the nature of the crystal. This conclusion is

also justified down to relatively low energies.   Under these circumstances

the classical x-ray techniques for crystal structure analysis are seen to be

---

[*]   Such beams can of course be excited by careful choice of the diffraction
conditions, the excitation of a Laue reflection at 500 eV being sensitive
to about 1/2 degree.   The minima in the center of the (550) Bragg maximum in
Fig. 8 and in the series of Bragg reflections in Fig. 5, are due to just such
three beam cases. A detailed and systematic study of such excitations in now
underway.[33]

more appropriate than those commonly used in low energy electron diffraction. These techniques allow the determination of the strength of the dynamic interactions in the crystal, a measurement heretofore considered unavailable from LEED observations.

ACKNOWLEDGMENT

We would like to acknowledge helpful discussion with Professors D. S. Boudreaux and H. Wagenfeld, Miss J. Perry, and Mr. S. Friedman.

REFERENCES

1. H. A. Bethe, Ann. Phys. $\underline{87}$ (1928) 55.

2. L. Brillouin, Wave Propagation In Periodic Structures, McGraw-Hill Book Co. Inc., N. Y., 1946.

3. D. Boudreaux and J. Heine, Surf. Sci. $\underline{8}$ (1967) 426.

4. E. G. McRae, J. Chem. Phys., $\underline{45}$ (1966) 1258.

5. P. M. Morse , Phys. Rev. $\underline{35}$, (1930) 1310.

6. A. Somerfeld and H. A. Bethe, Handbuch der Physik, Bd. $\underline{24}$, Berline (1933).

7. G. P. Thomson and W. Cochran, Theory and Practice of Electron Diffraction, MacMillan and Co., London (1939).

8. W. E. Laschkarew, Trans. Far. Soc., $\underline{31}$ (1935) 1081.

9. A. Gervais and R. M. Stern—Surf. Sci., to be published.

10. G. Lehmpfuhl and A. Reiszland - to be published in Z. Naturforschg.

11. P. Hirsch, A Howie, and M. J. Whelan, Proc. Roy. Soc., $\underline{A252}$ (1960) 499.

12. B. Batterman and H. Cole, Rev. Mod. Phys., $\underline{36}$ (1965) 681.

13. P. P. Ewald, Z. Kristallogr., $\underline{56}$ (1921) 129.

14. H. D. Heidenrecih, Phys. Rev., $\underline{77}$ (1950) 271.

15. D. Boudreaux, R. M. Stern and J. Perry - to be published.

16. R. W. James, The Optical Principles of the Diffraction of X-rays, G. Bell and Sons, Ltd., London, England (1965).

17.   H. Taub, H. Wagenfeld and R. M. Stern - to be published.

18.   H. D. Heidenreich, Fundamental of Transmission Electron Microscopy, Inter-
      science, New York (1964).

19.   H. Taub, Ph. D. Thesis, P. I. B., 1969.

20.   G. Borrmann, Physik. X. 49, (1941) 157.

21.   E. J. Saccocio and A. Zajac, Phys. Rev. 139 (1965) 255.

22.   H. Hashimoto, A. Howie and M. J. Whelan, Proc. Roy. Soc., A269 (1962) 80.

23.   K. Kohra and H. Watanabe, J. Phys. Soc. Jap., 14 (1959) 1119.

24.   R. M. Stern and H. Taub, Phys. Rev. Lett., 20, (1968) 1340.

25.   S. Miyake, K. Hayakawa and R. Midda, Acta Cryst. A24 (1968) 182.

26.   M. Renninger, Z. Physik, 106 (1937) 141.

27.   A. Gervais, R. M. Stern and M. Menes, Acta Cryst. A24 (1968) 19.

28.   R. M. Stern, Proc. Am. Cryst. Association 4 (1968) 14.

29.   A. Guinier - Theorie et Technique de la Radiocristallographie, Dunod, Paris
      (1956).

30.   R. M. Stern, A. Gervais, M. Menes, Acta Cryst. A25 (1969) to be published.

31.   H. Taub and R. M. Stern - to be published.

32.   R. Baudoing, R. M. Stern and H. Taub - Surf. Sci. 11 (1968) 255.

33.   S. Friedman, Ph.D. Thesis, P. I. B. 1969 - to be published.

34.   R. Baudoing - unpublished thesis, University of Grenoble, 1967.

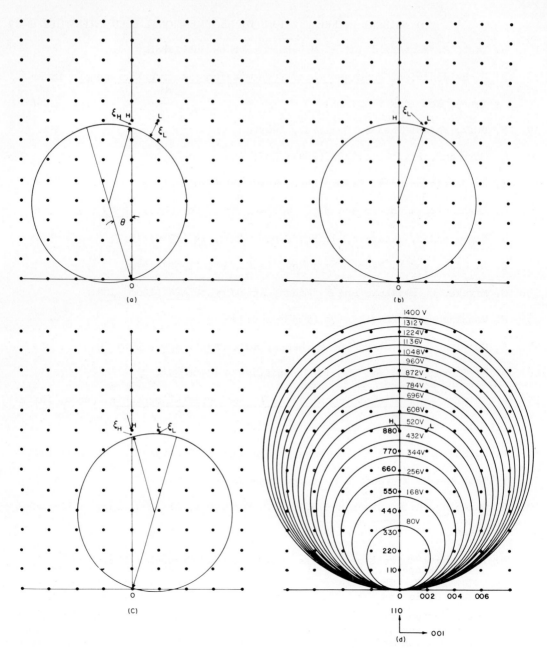

Fig. 1. Ewald construction showing excitation of the reflections corresponding
to the reciprocal lattice points H and L. a) Incident beam inclined
θ degrees from normal. The excitation error for the reflection H is
shown. b) Normal incidence, ξ_H = 0. The reflection L is also excited.
c) Incident direction has passed through normal. The reflections H and
L are both excited. d) Spheres of reflection for several energies at
normal incidence. The indexing of the reciprocal lattice points in the
plane is also shown.

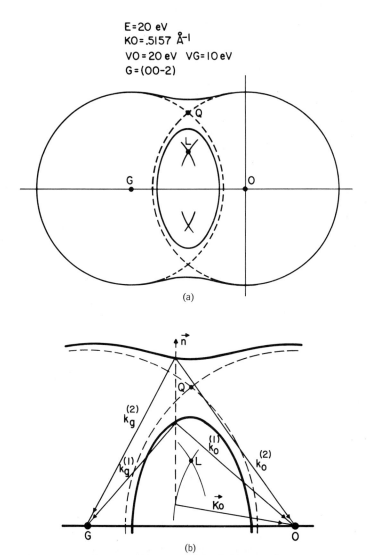

E = 20 eV
KO = .5157 Å⁻¹
VO = 20 eV   VG = 10 eV
G = (00-2)

(a)

(b)

Fig. 2.   Dispersion surface for the two beam case.   a) Machine calculated and
drawn dispersion surface.   Inner potential V_O = 20 eV. V_G = 10 eV,
E = 20 eV.   The constant energy surface for no interaction (i.e.,
V_G = 0) is the spherical surface about O and G.   b) One half of the
dispersion surface showing the effect of boundary conditions. The
crystal surface is oriented such that its normal $\underline{n}$ is parallel to the
orientation shown. The tie points are the intersection of the normal
passing through the end of $\underline{K}_O$) and each of the sheets of the dis-
persion surface.   The wave vectors in the crystal $\underline{k}_O{}^{(1)}$, $\underline{k}_G{}^{(1)}$, $\underline{k}_O{}^{(2)}$,
$\underline{k}_G{}^{(2)}$ are drawn from the tie points to each excited reciprocal lattice
point.   This construction conserves electron momentum parallel to the
surface.   The Laue point Q defines the point at which simultaneous dif-
fraction takes place according to kinematical theory and $|\underline{k}_G| = |\underline{k}_O|$.
The Lorentz point L corresponds to the condition $|\underline{K}_G| = |\underline{K}_O|$.

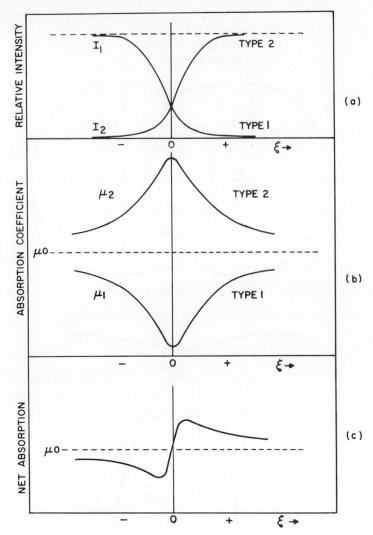

Fig. 3.   (a) Intensity of the type one and type two waves as a function of
deviation parameter X. At zero deviation parameter (the symmetrical
Laue point), the amplitude of the two waves are equal. (b) Absorption coefficient of each of the two waves as a function of deviation
parameter. (c) Net absorption (μ) of the total wave field in the two
beam case, as a function of deviation parameter. $\mu = \mu^{(1)}I^{(1)}$
+ $\mu^{(2)}I^{(2)}$. The deviation parameter is proportional to the excitation
error, as defined in reference 22.

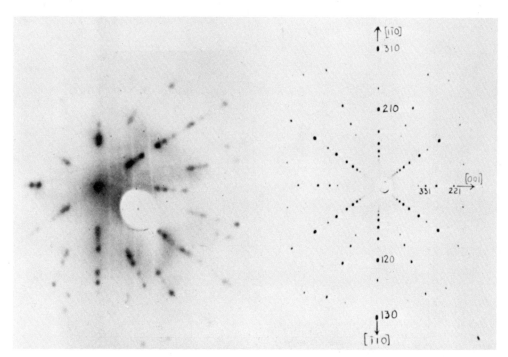

Fig. 4.   Right hand figure: X-ray back Laue diagram. Left hand figure: Electron back Laue for the range 75-150 eV.

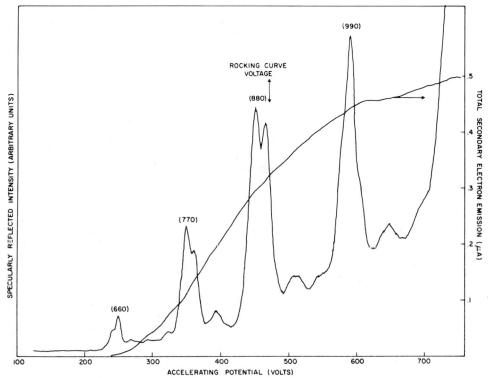

Fig. 5.   Intensity vs. voltage for the specularly reflected beam near to normal incidence. Also shown is the total secondary current as a function of voltage for the same orientation. If the reflections are due to the simple two beam case their half width should be of the order of 2-4 volts.

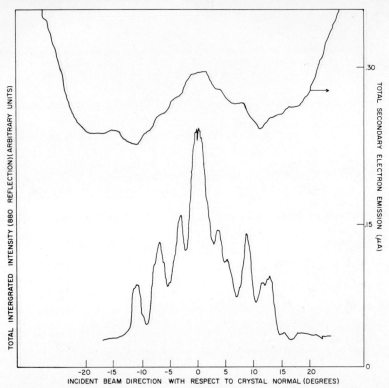

Fig. 6. Rocking curve at 467 volts. Intensity of the specularly reflected beam as a function of incident direction. The 880 reflection has zero excitation error at normal incidence. The reflection is excited over about 10°. The total secondary electron emission is also shown as a function of angle. The minimum at 12.5° corresponds to the excitation of the Laue reflection 002.

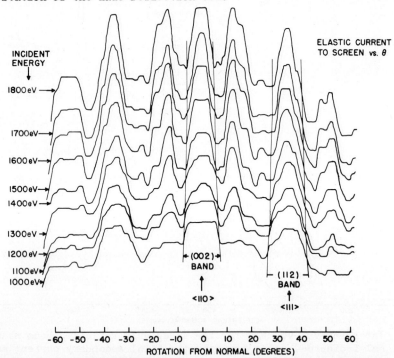

Fig. 7. Elastic secondary emission (as analyzed with the diffraction camera grid system to within 5 eV) as a function of incident direction for energies between 1000-1800 eV. The orientation corresponding to the excitation of the Laue reflections 002, 00-2, 112, and 1-1-2 are shown. These correspond to the position of the Kikuchi lines having the same index which occur as the limits of the Kikuchi bands shown.

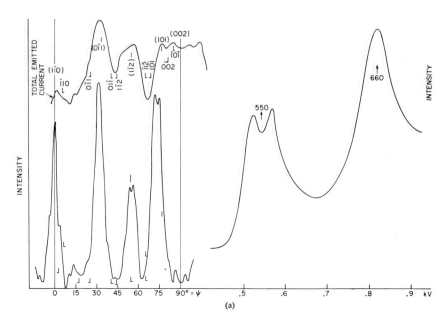

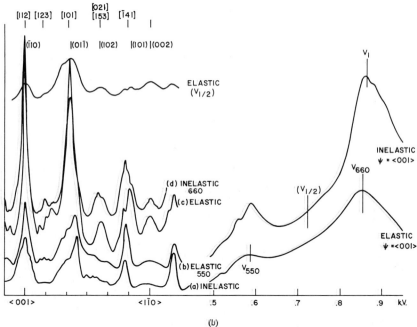

Fig. 8. a) Right hand diagram. Intensity vs. voltage showing the 550 and 660 Bragg reflections. The diffraction angle is 56.4°. Left hand diagram: Lower curve shows the specularly reflected intensity as a function of azimuthal angle. Upper curve shows the total secondary current as a function of azimuthal angle. The rotation is made about the 550 reflection. The indices show the orientations for dense planes (with brackets) and for the excitation of Laue reflections (no brackets). b) Right hand diagram. Intensity vs. voltage curves at $\theta = 57.5°$ showing the 550 and 660 reflection. Both the elastic and inelastic rotation diagrams for the 660 and 550 reflection. Also shown is a rotation diagram about a forbidden reflection $V_{1/2}$ which corresponds to the point (11/2, 11/2, 0). Except for the magnitude of the current, the structure of the curve is similar to that of the integral order Bragg reflections.

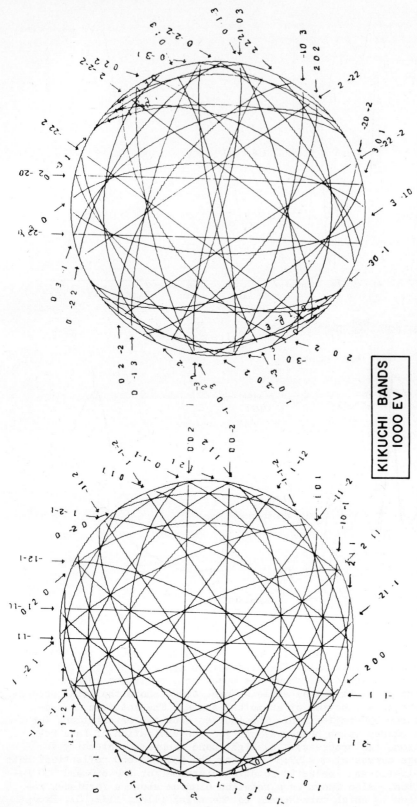

KIKUCHI BANDS
1000 EV

Fig. 9. Machine drawings of Brillouin Zone Boundaries for $H = h^2 + k^2 + l^2 \leq 6$ (left) and $6 < H \leq 12$ (right). The Brillouin Zone Boundary diagrams may be considered to be plots of the loci of the incident beam directions for the excitation of the reflection having the same index as the boundary. The intersection of several Brillouin Zone Boundaries corresponds to the excitation of several simultaneous reflections. The Brillouin Zone Boundaries are identical to the positions of the Kikuchi lines reflected from planes having the same index as the zone boundaries.

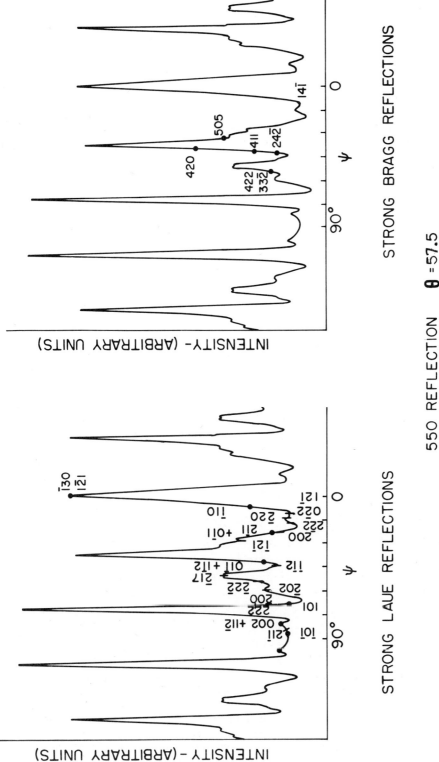

Fig. 10. Rotation diagrams ($\theta = 57.5°$) about the 550 reflection showing the orientation for the excitation of the low index Laue reflections (left) and the Bragg reflections (right) which contribute to a reduction of the specularly reflected intensity. The Bragg reflections can be distinguished because they do not reduce the secondary emission.

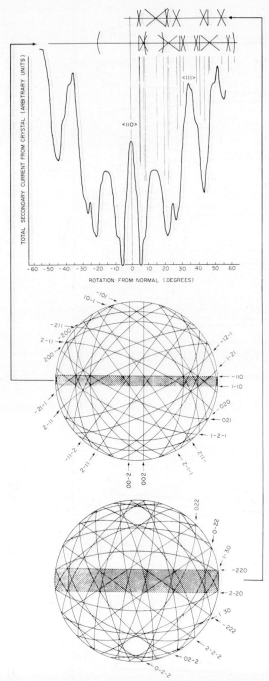

Fig. 11. Analysis of the variation of the total secondary electron emission as a function of incident direction in terms of the excitation of Laue reflections. The uppermost curves show the positions of the Laue reflections as determined from the Kikuchi diagrams shown at the bottom of the figure. The area scanned during the rotation is shaded in the diagrams. The secondary emission shows minima associated with the excitation of each Laue.

ELECTRONIC STATES OF ORDERED SURFACE MONOLAYERS

H. D. Hagstrum and G. E. Becker

Bell Telephone Laboratories, Inc.
Murray Hill, New Jersey

I.   INTRODUCTION

When foreign atoms are adsorbed on the clean surface of a metal crystal
we expect changes in the electronic states of the surface region of the crystal.
Changes in both the density of states and wave function magnitude in the vicinity
of the adsorbed atom occur. Some aspects of the matter have been discussed in
the literature for surface atoms, particularly the alkalis.[1,2] We are dealing
here with the surface analog of the virtual bound states discussed for impurities dissolved in the bulk.[3,4] In this paper we present some of the experimental work in which we have determined the energy range or contour of the virtual
bound state for ordered monolayers on metals. We shall discuss the nature of
the virtual bound state, the method of its experimental study and the results
for oxygen, sulfur, and selenium adsorbed on the (100) face of nickel. We
believe these to be the first direct measurements of the position and breadth
of a virtual bound state. This has been possible because of the unique
sensitivity of the experimental method to surface wave function magnitude.

II.   VIRTUAL BOUND STATE OR SURFACE-ADSORBATE BAND

Let us first examine what we expect to happen when a foreign atom is
adsorbed on a metal surface. In Fig. 1 is drawn a schematic electron energy-
level diagram for a metal and a foreign atom. The atom is shown at two positions relative to the metal surface: distant and adsorbed.

When the atom is distant from the surface the metal and atom form two
separate quantum-mechanical systems and the atomic states are sharp. When the
atom is adsorbed the situation is quite different. Now a large number of

electronic states of the metal-atom system pass through the atom position.
The discrete states of the free atom corresponding to the lowest energy electron configuration have now been replaced by a broad energy region in which
wave function magnitude at the atom is larger than it would be at this position
in the absence of the absorbed atom. The contour of wave function enhancement
at the absorbed atom, $\Psi^2_{A_s}$, is shown schematically in Fig. 1 centered at the
band energy ζ_A and of full width Δ_A at half maximum. This region has been
called the surface-adsorbate or A band.[2] It is a virtual bound state because
electrons in it are in quite intimate contact with the continuum of filled
electronic states of the metal. The point of maximum wave function magnitude
in the virtual bound state has been shifted by an amount S_A from the energy
of the free-atom ground state which lies below the vacuum level by the free-
space ionization energy, $E_i(\infty)$.

In this work we have used electronegative atoms of relatively large
free-space ionization energy. This puts the virtual bound state relatively
deep in the band completely below the Fermi level. This is in contrast with
the virtual bound state produced by the adsorption of an alkali atom in which
case the state straddles the Fermi level and is only partially occupied.
Ordered structures of O, S, and Se are formed on the clean Ni(100) face in a
convenient, chemically self-limiting manner. These structures have been
studied and documented by means of low-energy electron diffraction.[5]

III. METHOD OF OBSERVATION

The experimental method of observing the virtual bound state of adsorbed
atoms is the ion-neutralization spectroscopy (INS) discussed extensively in
a previous publication.[6] INS is a means of extracting band structure information from the kinetic energy distributions of electrons ejected from the
solid on the neutralization at the surface of a slowly moving noble gas ion

such as He^+. Figure 2 is an energy level diagram which illustrates the
electronic transitions accompanying ion neutralization as well as the expected
effect of the virtual bound state on the ion-neutralization process. In Fig.
2 we depict the substrate metal with a foreign atom adsorbed at its surface.
Also shown is a second atomic well, namely that of the incoming He^+ ion.

When the ion is sufficiently close to the surface for wave functions from
the solid to overlap the ground state wave function of the ion, a two-electron,
Auger-type process occurs. One electron drops by transition 1 from a band
state, say at $\zeta_1 = \zeta + \Delta$, into the atomic ground state neutralizing the ion.
A second electron receives the energy thus released and is excited from
another band state, say at $\zeta_2 = \zeta - \Delta$, to the energy E above the vacuum level.
Some of these excited electrons leave the solid and their kinetic energy
distribution is measured. Ejected electrons appear over a band of kinetic
energies E since the initial energies ζ_1 and ζ_2 may vary through the filled
band. The relative probability of a given elemental transition is governed,
among other things, by the amount of wave function overlap between the initial
bound state at ζ_1 and the atomic ground state.

We can now see why the virtual bound state of the adsorbed atom makes
its presence felt in this experiment. Because the wave function magnitude is
greater at the adsorbed atom over the energy range of the virtual bound state,
the magnitude of the wave function tail at the ion position will also be
enhanced over the same energy range. Thus we expect that ion-neutralization
processes which involve electrons in the energy range of the virtual bound
state will be more probable in the presence of the adsorbed atom than in its
absence.

We need now to discuss briefly the connection between the measured
kinetic energy distribution, $X(E)$, and a function of band energy called the
transition density, $U(\zeta)$, which will be altered by the presence of the virtual

bound state. If the matrix element variation with energy ζ can be factored
into two functions of the initial electron energies ζ_1 and ζ_2 we may write the
probability of an elemental Auger process as proportional to the product
function $U(\zeta_1)U(\zeta_2) = U(\zeta+\Delta)U(\zeta-\Delta)$. $U(\zeta)$ contains both matrix-element and
density-of-states factors. We have called $U(\zeta)$ the transition density because
it represents the state density weighted by transition probability. The final
state at E is independent of Δ for constant ζ so integration over Δ is
required to get the number of excited electrons produced at E.

   We define a fold or self-convolution function

$$F(\zeta) = \int_0^\zeta U(\zeta+\Delta)U(\zeta-\Delta)ds \qquad (1)$$

which is the total probability that all pairs of initial electrons symmetrically disposed in energy with respect to ζ will be involved in the ion-neutralization process and produce excited electrons at the final energy E. Using the
relation between ζ and E:

$$E = E_i'(s_t) - 2(\zeta+\Phi) \qquad (2)$$

$F(\zeta)$ becomes F(E) which is proportional to the density of excited electrons
inside the solid. In Eq. (2) E_i' (s_t) is the effective ionization energy of
the parent noble gas atom at the distance from the surface, s_t, at which the
Auger transitions occur. The external distribution of ejected electrons,
X(E) is related to F(E) by the probability of escape, P(E), thus:

$$X(E) = F(E) \cdot P(E) \qquad (3)$$

   The method of INS involves the determination of $U(\zeta)$ given experimental
measurements of X(E) for two small incident ion energies. The method is
discussed in detail in Ref. 6. F(E) is determined from Eq. (3) and $F(\zeta)$ using
Eq. (2). $U(\zeta)$ is the result of unfolding Eq. (1) which is done by a digital
sequential method.

IV.   EXPERIMENTAL RESULTS

In Figs. 3, 4, and 5 we illustrate functions involved in the procedure outlined above for surfaces which are clean or covered with ordered structures involving oxygen.   Figure 3 shows experimental kinetic energy distributions for 5 eV He$^+$ ions incident on clean Ni(100) and on Ni(100) with primitive (2×2)0 and centered C(2×2)0 surface structures involving oxygen.   We note the large changes in the form of the function.   In Fig. 4 the corresponding F(ζ) functions are shown.   These are derived from an extrapolation of X(E) distributions for 5 and 10 eV He$^+$ ions to essentially zero incident ion velocity.

The unfolds of the F(ζ) functions of Fig. 4 are given in Fig. 5.   Here we see that the clean surface produces a U(ζ) function showing the nickel d band lying between the Fermi level (ζ = 0) and $\zeta \sim 2$ eV as found earlier.[7] The unfold functions for the oxygen structures each shows a much attenuated d band and a large broad peak appearing deeper in the band.   We interpret this peak as giving the contour of the oxygen virtual bound state since it is in this energy region [Fig. 2] that the transition probability of the Auger process is increased due to the enhanced wave function tunneling into the He well.   The nickel d band is reduced in intensity because it lies outside the range of the oxygen virtual bound state.   The ion-neutralization process occurs on the average farther from the surface metal atoms when oxygen is present accounting for the reduced d band intensity.

In Fig. 6 we plot U(ζ) functions for clean Ni(100) and for the surface structures Ni(100) C(2×2)0, C(2×2)S, and C(2×2)Se.   The oxygen surface is produced from O_2 gas, the S and Se surfaces from the breakup of H_2S and H_2Se, respectively.   We note that the S and Se structures also produce broad virtual bound states.   Note also the small, relatively sharp peak at $\zeta \sim 7.7$ eV for Se.   This peak appears to be present for the C(2×2)Se structure but absent when the surface structure is predominantly (2×2)Se.

V. SUMMARY

We summarize and discuss briefly our principal results for O, S, and Se on Ni(100).

(1) With the foreign-atom structure present we have observed the development in the transition density function of a broad peak lying deep in the filled band. This we interpret as essentially the contour of the virtual bound state developed from the lowest electronic configuration of the free adsorbate atom.

(2) The virtual bound states are about 3 eV wide at half maximum. This work and work on other systems shows this width to be essentially concentration independent indicating strong electronic interaction of adsorbate and substrate but much weaker interaction among adsorbate atoms themselves.

(3) Measurement of work function change accompanying adsorption combined with the clean-metal work function enables us to estimate the energy shifts S_A toward the vacuum level from the free-atom ground state to the energy position of the maximum of the virtual bound state. Whereas the oxygen shift, so defined, is near 3 eV, the shifts for sulfur and selenium are each only a few tenths of an eV. Another way of stating this result is that the virtual bound states for C(2×2) O, S, Se lie much closer together than do the free-space ground levels. Thus there appears to be a definite difference in positions of the virtual bound states relative to the free-atom states for oxygen on the one hand and sulfur and selenium on the other.

(4) The virtual bound state determined in this work specifies where the most loosely bound electrons on the adsorbate atom or in the metal adsorbate complex lie. It does not tell us, however, the number of electrons on the adsorbate atom or its ionicity. From the fact that

work function increases in the range 0.3 to 0.6 eV on adsorption it is clear that the adsorbate is charged negatively. The degree of ionicity, however, depends on the nature of the reconstruction of the metal surface.

The authors gratefully acknowledge the technical assistance of Philip Petrovich.

REFERENCES

1. R. W. Gurney, Phys. Rev. 47, 479 (1935).

2. R. Gomer and L. W. Swanson, J. Chem. Phys. 38, 1613 (1963).

3. J. Friedel, Chapter XIX, Metallic Solid Solutions, Eds. J. Friedel and A. Guinier, (W. A. Benjamin, Inc., New York, 1963).

4. P. W. Anderson, Phys. Rev. 124, 41 (1961).

5. H. E. Farnsworth and H. H. Madden, Jr., J. Appl. Phys, 32, 1933 (1961); A. U. Mac Rae, Surface Science 1, 319 (1964).

6. H. D. Hagstrum, Phys. Rev. 150, 495 (1966).

7. H. D. Hagstrum and G. E. Becker, Phys. Rev. 159, 572 (1967).

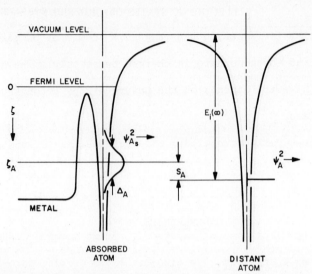

Fig. 1. Diagram showing energy levels of a solid and a foreign atom in two positions, distant from the surface and adsorbed.

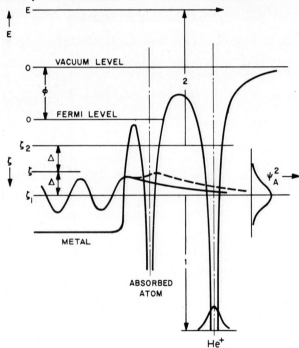

Fig. 2. Energy level diagram showing electronic transitions by which the ion is neutralized and the enhancement of wave function magnitude outside a solid in the energy range of the virtual bound state of the absorbed atom.

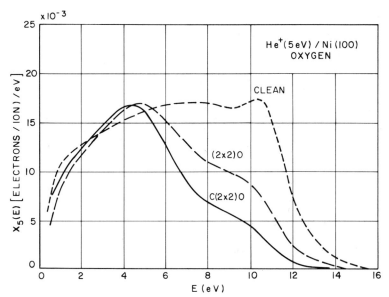

Fig. 3. Kinetic energy distributions of electrons ejected by 5 eV He$^+$ ions from the Ni(100) surface when clean and when covered with a primitive (2x2) and a centered C(2x2) structure involving oxygen.

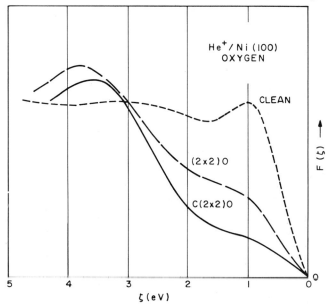

Fig. 4. Fold functions derived from electron kinetic energy distributions.

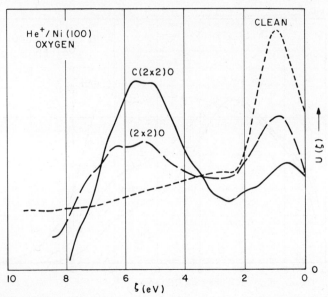

Fig. 5. Unfold or transition density functions derived from the fold functions of Fig. 4.

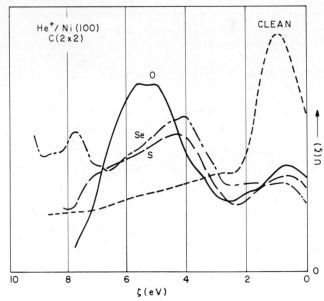

Fig. 6. Transition density functions for clean Ni(100) and for the surface having upon it C(2x2) structures involving oxygen, sulfur, and selenium.

STUDY OF SURFACES BY MEANS OF SURFACE PLASMONS
DETECTION OF THIN FILMS, SURFACE ROUGHNESS

H. Raether
University of Hamburg, Germany

1. Detection of Thin Films on Surfaces by Surface Plasmons

The surface plasmons represent the quanta of the oscillations of the electron density in the boundary surface of a solid (or a liquid). If the electrons of this solid can be regarded as quasi free with a plasma frequency ω_p, the frequency of the surface oscillations ω_s is given by $\omega_s = \omega_p(1 + \eta)^{-1/2}$, $\eta = \eta_1 + i\eta_2$ means the complex dielectric constant outside the solid[1]. The wave vector K_s of the plasma wave is connected with its frequency ω_s by a dispersion relation $\omega_s(K_s)$ due to the physical conditions. If we have not too thin a film of the solid, we can neglect the coupling dispersion due to the interaction of the two surfaces of the film. Retardation effects (finite velocity of the electromagnetic field) have the effect that ω_s approaches the light line $\omega = c \cdot K$ for small values of K_s. In Fig. (1) the dispersion relation for ω_s of a pure Al surface (a thick, evaporated in UHV) and for ω_s of an oxidized surface of Al (very thick oxide film with $\eta_1 = 3$) is reproduced[2]. If we excite the surface plasmons by electrons, the electrons with the momentum $mv = \hbar K_{e\ell}$, passing the surface of the Al film, loose the energy $\hbar\omega_s$ and transfer the momentum $\hbar K_s$ to the surface plasmon and are thus deflected by $v = K_s/K_{e\ell}$. By measuring $\hbar\omega_s$ at different scattering angles v as in Fig. (1), one obtains $\Delta E = \hbar\omega_s$ as function of K_s.

Since the electric field of these oscillations is the more extended into the space the longer its wave length Λ_s (or the smaller $K_s = 2\pi/\Lambda_s$), a thin film of thickness τ on a pure substrate influences the plasma oscillations the less the smaller τ/Λ_s or $K_s\tau$. For small values of $K_s\tau$, ω_s of a contaminated surface therefore should approach ω_s of a clean surface: $\omega_p / \sqrt{2}$ (at very

small values of $K_s \tau$ the retardation drops ω_s to zero). In Fig. (2a) the

dependence $\omega_s(K_s)$ or $\omega_s(\nu)$ with different values of τ is reproduced for Al[2]).

By measuring the $\Delta E(\nu)$ dependence, one can thus deduce the thickness τ of the

oxide film on an Al surface by comparing $\omega_s(\nu)$ with the theoretical curves.

Figure (2b) shows the result: An Al surface, evaporated at $3\text{-}6\cdot 10^{-9}$ torr,

had been contained for hours in this vacuum (without liquid Nitrogen trap);

at different times the $\omega_s(\nu)$ curve has been remeasured. The thickness τ of

the oxide film as function of time t (in hours) is plotted in Fig. (3). If the

vacuum pressure is raised to 10^{-6} torr, the growth of the oxide film increases

rapidly[3]).

The case of Al is an example for the behavior of a free electron gas

where the displacement of the value of ω_s due to the contaminating oxide film

is strong. In the case of a non-free electron plasma, e.g. Silver, the loss

displacement is in general much smaller and more difficult to measure. Here

the decrease of intensity of the surface loss due to the imaginary part η_2

of the film can be used as a sensitive indication of a contaminating film.

This is demonstrated by a Silver film coated with Carbon films of

different thicknesses τ. The value η of Carbon has been taken as constant

in the region of 2-4 eVolt $(\eta_1 = \eta_2 = 3)$[4]). The Silver loss spectrum obtained

in transmission has two peaks; one at 3,8 eV due to the volume plasma loss,

which is independent on what happends on the surface, the other at 3,6 eV

is due to the surface loss. If one covers this Silver surface on both sides[*]

with Carbon films of different thicknesses, one obtains the spectra of Fig.

(4) (full line). The dotted line shows the calculated values, indicating by

---

[*] The symmetric coating of the Silver film with Carbon makes the quantitative evaluation more convenient.

the agreement that the spectral <u>intensity</u> of a surface loss gives information

of the thickness of the contaminating film[5].

In these cases the electron beam passes the film in the direction of the

foil normal (transmission method). Using the reflexion method - the beam

leaves the surface on the same side where it enters - the sensitivity can be

increased essentially. The experiments have shown that the intensity of the

surface loss, excited by electrons, is the higer the more grazing the electrons

hit the surface[6]. Measurements of the intensity of the Ag surface loss as

function of the incidence angle α (angle between the electron beam and the

foil normal) confirmed the theoretical dependence with $1/\cos\alpha$[4]. This effect

should be important for studies of surfaces with plasmons.

II. The Roughness of a Surface Studied by the Excitation of the
Radiative Surface Plasmons

The surface modes excited by electrons on a thick film show a dispersion

as given in Fig. (1). Lying right of the line $\omega = c \cdot K$ (light line) it follows

that the phase velocity of these plasmons $v_s = \omega_s/K_s$ is smaller than the light

velocity c or $K_s > K$ (for the same frequency ω). These RITCHIE modes, in

spite of their transverse components, can therefore not couple with photons.

Now FERRELL has shown that there exist "radiative" surface modes (with

$K_s < K$, left from the light line), which frequency approaches ω_p for $K_s = 0$.

Figure (5) shows as an example the dispersion curve of these modes for (clean)

Al[7]. These modes being able to couple with the electromagnetic field, can

thus be excited by light and should decay into photons[8]. If light, contain-

ing the frequency region about ω_p and being polarized in the plane of incidence,

passes through a plane surface of a thin film, e.g. of Silver, under oblique

incidence (θ_o) against the foil normal, the normal component of the incoming

electric field of light should excite this mode, see Fig. (6). It travels

with the phase velocity $c/\sin \theta_o$ along the surface and excites the radiative surface mode with the wave vector $K_s = \omega/c \cdot \sin\theta_o$. However, the reemission

of light by decay of this mode is cancelled by interference in all directions

except that of the reflected and the transmitted light (application of FRESNEL's

equation to a plane surface). The experiments, however, show that an emission

of light with frequency ω_p is observed in nearly all directions, due to the

excitation of this radiative mode[9]. The emitted light is polarized in the

plane of incidence. (Figure (6) shows the experimental arrangement. The

spectral distribution of the emitted light and its spatial distribution as

function of the incidence angle θ and the azimuth ϕ is reproduced in Fig.

(7a,b,c).

This phenomenon is ascribed to the fact that the surface is not a plane

one but contains irregularities, e.g., a certain roughness which makes that

the phases of the light scattered into all directions (different from the

reflected K_{refl} and the transmitted beam K_{trans}, see Fig. (8), do not cancel

each other perfectly[10-12].

If the tangential component of the incoming beam is labelled $K_{\perp_o}$, that

of the scattered $K_\perp$, one obtains intensity on a plane surface only for those

directions for which $K_{\perp_o} = K_\perp$ (reflexion and transmission). However, if there

exists e.g. in a periodic distance s a deviation from a plane surface, addi-

tional intensity directions appear, in other words $K_{\perp_o} - K_\perp$ is no more zero,

but may be $n\frac{2\pi}{s}$ with $n = \pm 1, \pm 2$, i.e., the radiative mode excited by the

incoming light decays and radiates into directions different from $\theta = \theta_o$.

If a continuous spectrum of s values with vectors $n\frac{2\pi}{s}$ lying outside

the plane of incidence exists as on a rough surface with a most probable

values $\bar{s}$, a continuous distribution of the scattered light in space in respect

to θ and ϕ is expected. The evaluation of some measurements on a vaporized

Silverfilm (600 Å thickness, vaporization rate 100 Å/min on Quartz glass)

gives: $\bar{s} \sim 1500$ Å and a roughness δ (root of the mean square of the elevations) of the order of 50 Å. These figures are roughly confirmed by electron micrographs made by shadowing. If one measures $\bar{s}$ and δ as function of the thickness of the Silverfilm, one obtains nearly constant values of $\bar{s}$, but increasing values of δ (dotted line), see Fig. (9). This is again confirmed by electron microscopic studies[13].

A very smooth Silver surface, produced by epitaxial growth of an evaporated Silver film on cleaved mica or by melting and cooling afterwards a surface, does not radiate in agreement with the above conception.

These PREL experiments need the condition $|K_{\perp} + K_{f\perp}| < K = \omega/c$ for the decay of the radiative surface plasmons into photons. If however $K_{\perp} \pm K_{f\perp}| > K$, it can occur that the vector $K_{\perp} + K_{f\perp}$, and the frequency ω of the light with which we observe, fulfill the dispersion relation of the RITCHIE modes $\omega_s(K_s)[\omega = \omega_s$ and $K_s = K_{\perp} \pm K_{f\perp}]$ and excite thus these plasmons. Observing the specular reflexion of light $(\theta = \theta_o)$ on a surface with a grating of periodicity s, one should obtain by changing the reflexion angle θ or the value of $K_{\perp}$ in the dependence of the reflected intensity as function of θ, a dip at certain values of θ. This indicates absorption of light by excitation of the RITCHIE modes. This effect has been observed recently[14].

These experiments demonstrate the role of the surface roughness for the surface plasmons. There are indications that the above dispersion relation $\omega_s(K_s)$, which is derived for plane surfaces is changed by this effect. A further effect of the surface roughness may be that the angular dependence of the intensity of $\hbar\omega_s$ as function of the scattering angle ν or K_s shows at angles of $6-8\cdot10^{-4}$ a steeper decrease than the ν^{-3} dependence[15]. This angle corresponds to a $\Lambda_s \sim 100$ Å which is comparable with the roughness of ~ 50 Å. The roughness damps apparently the oscillations with higher K_s values.

Naturally one can obtain these optical results with Maxwell's equations if one replaces the boundary conditions of a plane surface and takes into account its roughness. But introducing the conception of surface plasmons as an intermediate state gives an understanding of the phenomenological optical results by electronic processes.

References

1. R. H. Ritchie, Phys. Review 106, (1957), p. 874.

2. Taken from T. Kloos, Z. Physik 208, (1968), p. 77.

3. Unpublished results of T. Kloos.

4. J. Daniels, Z. Physik 203, (1967), p. 235.

5. J. Daniels, Z. Physik 213, (1968), p. 227.

6. H. Raether, 5th International Congress of Electron Microscopy, Philadelphia, 1962, AA3; M. Creuzburg and H. Raether, Z. Physik 171, (1963), p. 436; J. Lohff, Z. Physik 171, (1963), p. 442.

7. Calculated by E. Kröger (Hamburg).

8. H. Raether, Solid State Excitations by Electrons (Springer Tracts in Modern Physics, Vol. 38, (1965), p. 153.

9. J. Brambring and H. Raether, Phys. Review Letters 15, (1965), p. 882; Z. Physik 199, (1967), p. 118.

10. E. A. Stern, Phys. Review Letters 19, (1967), p. 1321.

11. E. Kretschmann and H. Raether, Z. Naturforsch. 22a (1967), p. 1623.

12. R. E. Wilems and R. H. Ritchie, Phys. Review Letters 19, (1967), p. 1325.

13. Observed by E. Schröder (Hamburg).

14. Y. Y. Teng and E. A. Stern, Phys. Review Letters 19, (1967), p. 511.

15. C. Kunz, Z. Physik, 180, (1964) p. 127; P. Schmuser, Z. Physik, 180, (1964) p. 105.

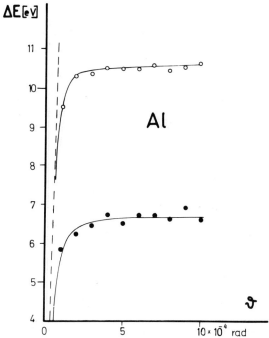

Fig. 1. Dispersion curve of the surface loss of Al without (10,5 eV) and with a thick oxyde film (6,8 eV) including retardation (circles: measured, full line: calculated).

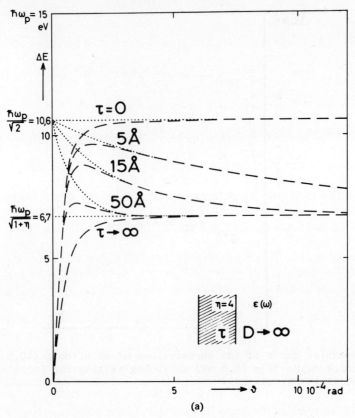

Fig. 2a. Calculated dispersion curve of the surface loss of Al with oxyde films of different thicknesses τ (retardation included).

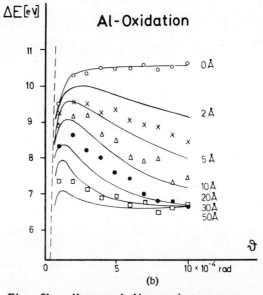

Fig. 2b. Measured dispersion curves.

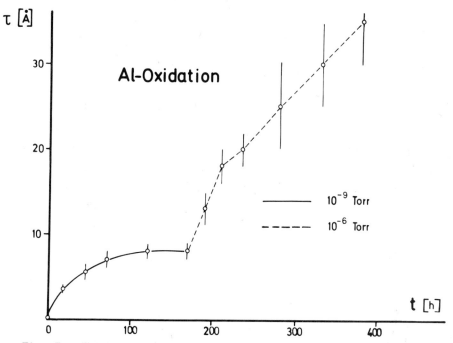

Fig. 3. Thickness of the oxyde film on Al as function of time.

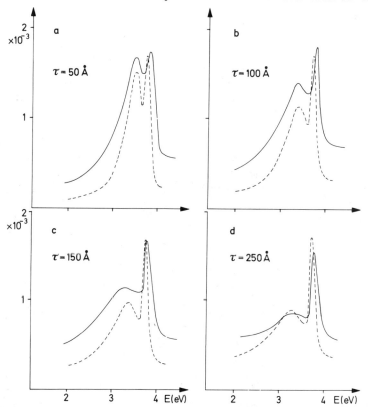

Fig. 4. Energy loss spectrum of a Silver film in transmission with the volume (3,8 eV) and the surface loss (3,6 eV) covered with Carbon film of different thicknesses τ.

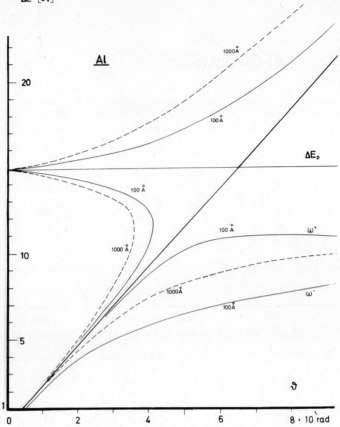

Fig. 5. Dispersion of plasmons on Al films of different thicknesses. Right of the light line ($\omega = c \cdot K$) one finds the non-radiative surface RITCHIE modes, split up in ω_+ and ω_- by the coupling of the two surfaces. Left of the light line there are the radiative modes approaching ω_p for $K_s \to \sigma$ (FERRELL mode).

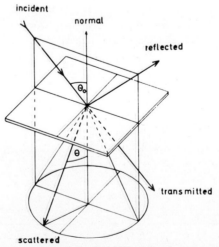

Fig. 6. Experimental arrangement to measure the plasma resonance light emission, excited by light (PREL).

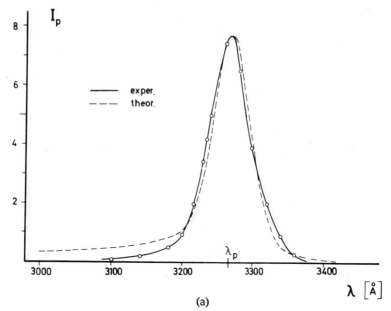

(a)

Fig. 7a. Spectral distribution of PREL (dotted line: distribution calculated with the optical constants of the thin Silver film, points: measured, by E. Kretschmann).

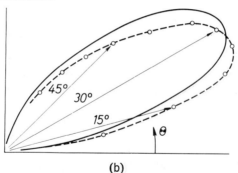

(b)

Fig. 7b. Intensity distribution of PREL as function of the angle θ between between foil normal and direction of observation ($\theta_0 = 30°$, $\phi = 90°$).

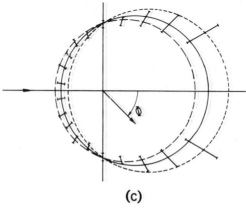

(c)

Fig. 7c. The same for the azimuthal distribution around the foil normal ($\theta_0 = 30°$).

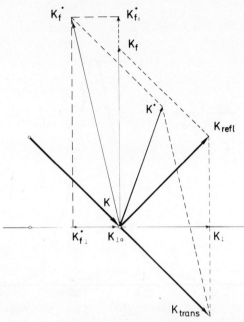

Fig. 8. Demonstrates the components of the momentum $\hbar K$ of light in the case of reflexion and transmission at a smooth surface: $K_{\perp 0} = K_{\perp}$; continuous values of the momentum transferred from the film $\hbar K_f$ to the photon are allowed. On a surface with a periodic deviation s from a smooth surface (grating) the parallel component of $\hbar K_f^*$ may be continuous again, but the tangential component $\hbar K_{f\perp}^*$ is restricted to the values $n\frac{2\pi}{s}$, $n = \pm 1, \pm 2$, etc. Thus additional intensity directions K^* are possible (these lie outside the plane incidence if the vector $K_{\perp}$ is not parallel to the vector $K_{f\perp}^*$).

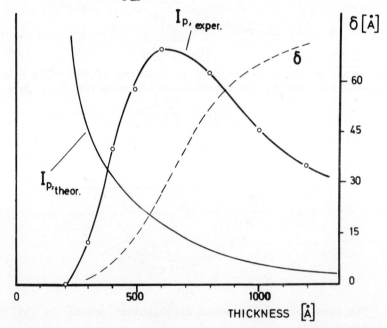

Fig. 9. Intensity of PREL as function of the thickness of the Silver film. The dotted line gives the roughness of the film calculated from the experimental values ($\mathscr{F}_{p, \, exper}$).

TRUE SECONDARY ELECTRON ENERGY DISTRIBUTIONS

G. F. Amelio and E. J. Scheibner

Physical Sciences Division
Georgia Institute of Technology
Atlanta, Georgia

Energy distribution measurements of scattered electrons have been shown to provide information about the properties of solid surfaces if the mechanisms producing the distribution are understood.[1] Such energy distributions have been historically divided into three regions -- the elastic peak, the characteristic loss region (plasmon and interband excitations), and the true secondary region. Whereas the first two regions noted have received extensive investigation in the literature, the third region has been largely overlooked primarily because of a lack in understanding of the physical phenomenon contributing to observed distributions. Qualitatively the true secondary electron energy distribution is characterized by a large, broad peak of very low energy (termed the "slow peak") on which is superimposed various subsidiary maxima. The slow peak is attributed to multiple secondary scattering of excited crystal electrons to lower energies thus resulting in a piling-up of electrons in the one to five electron volt region.[2] Some theoretical arguments to describe this process have been attempted in the case of metals with limited success.[3,4,5] The subsidiary maxima, usually ascribed to Auger processes[6] and, more recently, to single electron "umklapp" mechanisms,[7] however, have not been included in these considerations. In fact, except for some recent work[8,9,10,11], details of the Auger mechanisms have been largely unexplored.

In this paper a reasonably complete interpretation of the true secondary spectrum will be outlined. By including necessary empirical considerations, theoretical predictions are obtained which are then compared with experimental observations in copper and silicon. Applications of such a formalism are discussed.

Model

Theoretically, the phenomenon of secondary electron emission is conveniently divided into two steps: (1) the production of the secondaries and (2) the diffusion of these electrons to and across the surface. Whereas the factors influencing (1) are often reasonably well treated, the effect of (2) is frequently overlooked or ignored. The highly interacting secondary electron is likely to suffer many inelastic collisions before emission as indicated by mean free path studies.[12] One might expect then that the "true" internal distribution of secondaries will be significantly different from what is experimentally observed. It is therefore essential that both points (1) and (2) be intrinsically included in any satisfactory theory.

In attempting to write down a dynamical equation governing the production and diffusion of secondary electrons, we observe two important points: (1) in any sample which is being continually bombarded by primary electrons there is expected to be a larger number of internal secondary electrons; (2) because of their highly interacting nature, these secondaries will inelastically scatter frequently. These two considerations suggest that a statistical approach should be of value. Assuming each secondary electron in the crystal can be characterized by its position and momentum, a phase space ensemble is constructed in which all electrons with position between $\vec{r}$ and $\vec{r} + \vec{dr}$ and momentum between $\vec{p}$ and $\vec{p} + \vec{dp}$ belong to that element of the ensemble labeled $\vec{r}$, $\vec{p}$. The population density of electrons in this phase space cell at time t is designated $N(\vec{r}, \vec{p}, t)$ and, if the number of electrons in the cell is sufficiently large so that N can be treated as a continuous quantity, then N is governed by a continuity equation of the form

$$\frac{\partial N}{\partial t}(\vec{r},\vec{p},t) + \frac{\vec{p}}{m} \cdot \vec{\nabla}N(\vec{r},\vec{p},t)$$

$$= S(\vec{r},\vec{p},t) - \frac{|\vec{p}|N}{m\lambda(\vec{p})}(\vec{r},\vec{p},t) + \int \frac{|\vec{p}'|N(\vec{r},\vec{p}',t)}{m\lambda(\vec{p}')} \, F(\vec{p},\vec{p}')dp' \qquad (1)$$

where m is the electron mass. The non-homogeneous right hand side has
three contributions. $S(\vec{r},\vec{p},t)$ is the source term and represents those
secondary electrons which enter the phase space ensemble as a result of
some physical excitation. The second term represents the rate at which
electrons are lost from the phase space cell $\vec{r},\vec{p}$ as a result of inelastic scattering. Finally the third term, the collision integral, describes
the rate of increase in population as a result of scattering into the cell.
$F(\vec{p},\vec{p}')$, the scattering function, is the probability that, given an electron at $\vec{p}'$, one will be found at p after scattering.

Approximations and Simplifications

As it stands, equation (1) offers little hope of analysis. This
expression can be simplified somewhat by considering the typical experimental geometry (Figure 1). Here the primary electrons are normally incident on a plane crystal surface which is considered to be of infinite
extent. The problem is additionally simplified by ignoring the details
of the crystal field and assuming azimuthal symmetry. The validity of
this conjecture is based on the observation that the phase space cell is
large compared to atomic dimensions (as it must for equation (1) to be
applicable). Further, the angular dependence of the population density
can be most easily handled by expanding N, S, and F in spherical harmonics. Inserting the resultant expansions in the geometrically simplified

expression, using the orthogonality of the spherical harmonics, and employing the addition theorem for the associated Legendre functions, one
obtains the following expression in terms of the expansion coefficients.

$$\psi_k(Z,E) = \frac{\ell(E)}{2k+1}\left[k\,\frac{\partial\psi_{k-1}}{\partial Z} + (k+1)\,\frac{\partial\psi_{k+1}}{\partial Z}\right]$$

$$+ \int_E^\infty F_k(E,E')\psi_k(Z,E')dE' + S_k(Z,E); \quad k = 0, 1, \ldots \qquad (2)$$

$$\psi_k = \frac{|\vec{p}|N_k}{m\ell(E)}$$

Equations (2) are a set of coupled integro-differential equations which
are in general very cumbersome. Note, however, that if the Z dependence
is rather weak the equations uncouple and become much simplified. This
is a valid approximation if the screening length is sufficiently small
that the volume within that distance of the surface is considerably
smaller than the total volume of interest. For moderately energetic primaries this is indeed the case. Equations (2) are then written

$$\psi_k(E) = \int_E^\infty F_k(E,E')\,\psi_k(E')dE' + S_k(E); \quad k = 0, 1, \ldots \qquad (3)$$

Solutions to the Simplified System

Solutions to equations (3) can most usefully be found by investigating the form

$$\psi_k(E) \;=\; \int_{E_a}^{E_b} G_k(E,E'') \; S_k(E'') \; dE'' \tag{4}$$

G_k is a Green's function of order k which allows (4) to be a solution to (3) if the following relation is satisfied.

$$G_k(E,E'') \;=\; \delta(E - E'') + \int_E^{\infty} G_k(E',E'') \; F_k(E,E') \; dE' \tag{5}$$

where $\delta(E - E'')$ is the familiar Dirac distribution. Note that the Green's function depends only on the scattering function $F_k(E,E')$. Once the F_k's are available, the Green's functions can be found and solution (4) gives the internal secondary distribution as soon as the appropriate source function or functions are employed.

Source Functions

In the current work, only the two most important source functions are considered. These are single electron collisions and the Auger process.

In the case of electron excitations, the primary beam interacts in a Coulombic fashion with the crystal electrons and some are excited to available states above the fermi level, leaving a hole in the fermi sea. In semiconductors, such a process is often called pair production because the hole can be treated in many respects like a positive electron. Restricting attention to the conserved wavevector situation (as opposed to the "umklapp" processes) the following angular excitation function is found[13]

$$S(E',\theta) = \frac{e^4 k_f^3}{3\pi \epsilon^2 E_p (E'-E_f)} \cdot \frac{a+b \cos\theta - \cos^2\theta}{\frac{1}{6}(b^2+4a)^{3/2}} \tag{6}$$

where

$$a = \frac{k_f^2}{k'^2} - \frac{k'^2 - k_f^2}{p^2}$$

$$b = \frac{2(k'^2 - k_f^2)}{pk'}$$

$p \equiv$ primary momentum

$k_f \equiv$ fermi momentum

$k' \equiv$ secondary momentum

If (6) is expanded in the following form

$$S(E',\theta) = \frac{1}{4\pi} \sum_{\ell=0}^{\infty} (2\ell+1)\, S_\ell(E') P_\ell(\cos\theta)$$

the expansion coefficients are

$$S_o(E') = \frac{e^4 k_f^3}{3\pi\, \epsilon^2\, E_p\, (E' - E_f)^2} \tag{7}$$

$$S_1(E') = S_o(E') \frac{2b(a - \frac{b^2}{4})}{b^2 + 4a}$$

$$S_2(E') = S_o(E') \frac{6(\frac{a^2}{5} - \frac{a}{3} + \frac{ab^2}{2} + \frac{3b^3}{8} - \frac{3b^4}{10} - \frac{b^2}{12}}{b^2 + 4a}$$

.
.
.

Except for $S_o(E')$, these expressions are mathematically prohibitive. They can be simplified by considering two very important cases. These are (1) $p \gg k' \sim k_f$ (low energy secondaries) and (2) $p \gg k' \gg k_f$ (high energy secondaries). For case (1), $S_1(E')$ vanishes and $S_2(E')$ approaches $- 1/5 \, S_o(E')$. For case (2), $S_1(E')$ again vanishes and $S_2(E')$ approaches $- \frac{1}{2} S_o(E')$.

In the case of the Auger effect the excitation function is just the true Auger distribution (the convolved transition density[9]). It is assumed to have no angular dependence which is perhaps reasonable since the Auger mechanism is a two electron process. Moreover, such an assumption is consistent with the basic statistical approach in which the details of the crystal field are ignored.

Scattering Mechanism and Green's Functions

There are several ways by which a secondary electron can lose energy as it cascades through the crystal. However, there are two mechanisms which appear to be particularly important and discussion will be restricted to them. These are electron-electron scattering and plasmon creation.

It is clear that a sufficiently energetic secondary should be able to excite a crystal electron from its occupied state into an unoccupied

level in the conduction band.   Investigation of the screened coulombic

interaction during such "collisions" reveals information about the energy

dependence of the scattering symmetry.[3]   To a rough approximation it turns

out that for secondaries with energy less than about 100 ev the scattering

is primarily spherically symmetric in nature.   Above 100 ev the scattering

is more accurately described as Rutherford.   For sake of simplicity in the

theory, a cutoff energy of 100 ev is defined such that below this value the

secondaries strictly scatter in a spherically symmetric manner and above it

they exhibit a Rutherford behavior.   For the former case the scattering

function coefficients are given by[3]

$$F_\ell \ (E,E') \ = \ \frac{2}{E'} \ P_\ell \left(\sqrt{\frac{E}{E'}} \ \right) \tag{8}$$

where P_ℓ is the ℓth Legendre polynomial.   The factor of two appears because

the scattering function has normalization two rather than unity.   This

amounts to a cognizance of the fact that for every particle which scatters

there are two electrons in the cascade after the collision.   The associated

Green's functions are found by putting (8) into (5) and solving the result-

ant expression by use of the Mellin transform.   The result is

$$G_0(E,E'') \ = \ \begin{cases} 2 \ \dfrac{E^4}{E^2} \ + \ \delta \ (E \ - \ E'') \ ; & E'' \geqq E \\[2em] 0 & ; \ \ E'' < E \end{cases}$$

$$
G_2(E,E'') = \begin{cases} \dfrac{2}{\sqrt{EE''}} \cos\left[\dfrac{\sqrt{3}}{2}\ln\left(\dfrac{E''}{E}\right)\right] + \delta(E - E''); & E'' \geq E \\[20pt] 0 & ; \ E'' < E \end{cases}
$$

$$
G_\ell(E,E'') = 0 \quad \text{for} \quad \ell \neq 0,\, 2 \ .
$$

The high energy (> 100 ev) case for which Rutherford scattering is applicable is handled analogously. For the sake of brevity the results are not listed here. The interested reader is referred to reference 13.

The second loss mechanism is plasmon excitation by the cascading secondary. As was mentioned earlier this turns out to be an important contribution, as one might suspect if mean free path lengths are considered.[14] Moreover, energy spectrum data from the characteristic loss region indicates that the plasma loss contribution is of fundamental importance. The effective cross section for plasmon production has been calculated in the literature[15] from which an expression for the scattering function can be obtained. In this case normalization is to unity because electron multiplication does not apply. The scattering coefficients and applicable Green's functions are found as before.

Internal and External Electron Energy Distributions

With the appropriate source functions and Green's functions available equation (4) can be solved for the internal distribution coefficients $N_k(E)$ for each case. One thus obtains a set of contributions for the internal secondary electron distribution $\left\{N_k^{(i)}(E)\right\}$ with the total internal spectrum given approximately by

$$
\eta(E) = \sum_k \sum_i \lambda_{ki}(E)\, N_k^{(i)}(E)\, P_k(\cos\theta) \tag{9}
$$

where the λ_{ki} (E) are empirical parameters dependent on the material. The k expansion in (9) is found to converge rapidly and it is sufficient to consider k = 0, 1, 2 only.

Experimentally, one does not observe $\eta(E)$, the internal secondary electron energy distribution, but rather an external current which is a function of the energy $J(E')$ where E' denotes the external energy of an electron which, because of the work function, had energy $E(\neq E')$ internal to the sample. Careful consideration of the effect on the internal distribution as it crosses the surface boundary and is collected leads to the following theoretical expression for the external secondary electron energy distribution experimentally observed.[13]

$$J(E') = \delta \left\{ \frac{E'}{E' + W} \sum_i \lambda_{oi}(E) N_o^{(i)} (E' + W) \right.$$

$$\left. - \frac{5}{2} \left[\frac{3}{2} \left(1 - \frac{W^2}{(E' + W)^2} \right) - \frac{E'}{E' + W} \right] \sum \lambda_{21}(E) N_2^{(i)} (E'+W) \right\} \qquad (10)$$

where

$$\delta = 1/2 \sqrt{\frac{E' + W}{2m}}$$

and

$$W = \text{energy of the vacuum level} .$$

Comparisons with Experimental Results

In order to compare the theory with experimental results (10) must
be evaluated using the appropriate parameters for the material under consideration. For metals these include mean free path to single electron
excitation, mean free path to plasmon creation, the characteristic Auger
distribution and amplitude, the work function, the fermi energy, the primary energy and plasmon energy.

Copper

Based on Quinn's[14] results it is estimated that the mean free paths
of electron collisions and plasmon creation are each about 10Å. Actually
these mean free paths are functions of initial and final state energy and
become significantly greater than 10Å in the vicinity of zero. However,
for simplicity of calculation and because the primary energy to be considered is much greater than one electron volt a constant value of 10Å is
assumed. The characteristic Auger spectrum as given in ref. 9 is used.
The amplitude is adjusted to agree with empirical observations. The work
function is assumed to be 4.5 ev[16] and the fermi level 6 ev above the bottom of the conduction band[17] which is chosen as the zero of energy.

For these values and a primary energy of 250 ev the distribution
shown in figure 2 is predicted. The experimental results are also shown.
The greater spread in the data curve than the theory can be partially
attributed to resolution limitations in the experimental apparatus.

Silicon

When applying the theory to semiconductors it is no longer suitable
to choose the bottom of the conduction band as the zero of energy. The
choice of the zero of energy here is tantamount to selection of an effective

penetration of the fermi volume by the secondary electrons. This value
is chosen to be five volts below the bottom of the conduction band. This
choice of penetration depth is hardly more than an educated guess based
on a review of Kane's[18] work among other things. However, the results
are not overly sensitive to this choice and a nominal value of five volts
below the bottom of the conduction band seems to be a good choice for the
semiconductors investigated thus far.

For the mean free paths the values chosen are 10Å (plasmon creation)
and 270Å (electron collision) again based on the work of Quinn and Kane.
The Auger distribution is fitted empirically and work function and fermi
level are 4.85 ev and 4.45 ev respectively. The latter value was based on
a choice of the band gap mid-point as the fermi level.

The results for silicon are shown in figure 3 for a primary energy
of 950 ev. The deviation between the theory and the experimental results
are similar in nature to the deviation witnessed in copper.

Discussion

The consistent deviation between the experimental and theoretical
results are probably in some part due to experimental limitations as mentioned earlier. However, the majority of the discrepancy must be attributed to the simplicity of the model. Umklapp (phonon assisted) processes
were omitted completely. The mean free path was taken as a constant. Cut-
off and effective values were chosen where monotomic transitions were the
actual case. Nonetheless the agreement is reasonably good and it is probably safe to say that no crucial physical consideration has been excluded
from the formalism.

The value of such a theory as a means for correcting for the effects of background and inelastic scattering in true secondary processes should be evident. Using this approach it is now possible to obtain "true" characteristic Auger distributions by requiring consistency between the theoretical and experimental results. These Auger distributions, in turn, are proving to be an extremely valuable tool for study of solid surfaces (see, for example, ref. 8). By virture of the approximate linearity of the theory it should be possible to include other effects and apply the formalism in the analysis of other phenomenon occurring in the true secondary region of the electron energy distribution resulting from metal and semiconductor materials as demonstrated here in the cases of copper and silicon.

References

1. E. J. Scheibner and L. N. Tharp, Surface Science $\underline{8}$, 247 (1967).

2. A. J. Dekker, "Secondary Electron Emission," Solid State Phys., Seitz and Turnbull, ed., Vol. 6, 251 (1958).

3. P. A. Wolff, Phys. Rev. $\underline{95}$, 56 (1954).

4. H. Stolz, Annalen der Physik $\underline{3}$, 197 (1959).

5. A. I. Guba, Soviet Phys.-Solid State $\underline{4}$, 1197 (1965).

6. J. J. Lander, Phys. Rev. $\underline{91}$, 1382 (1953).

7. L. N. Tharp, Ph.D. Thesis, Georgia Institute of Technology, June 1968.

8. G. F. Amelio and E. J. Scheibner, Surface Science $\underline{11}$, 242 (1968).

9. G. F. Amelio, Journal Mathematical Phys., to be published.

10. G. F. Amelio, Surface Science, to be submitted.

11. H. D. Hagstrum, Phys. Rev. $\underline{150}$, 495 (1966).

12. H. Seiler, Z. Angew. Physik $\underline{22}$, 249 (1967).

13. G. F. Amelio, Ph.D. Thesis, Georgia Institute of Technology, 1968.

14. J. J. Quinn, Phys. Rev. $\underline{126}$, 1453 (1962).

15. D. Pains, UFN, $\underline{62}$, 399 (1957).

16. American Inst. Phys. Handbook, 2nd ed., p. 9-148.

17. B. Segall, Phys. Rev. $\underline{125}$, 109 (1962).

18. E. O. Kane, Phys. Rev. $\underline{159}$, 624 (1967).

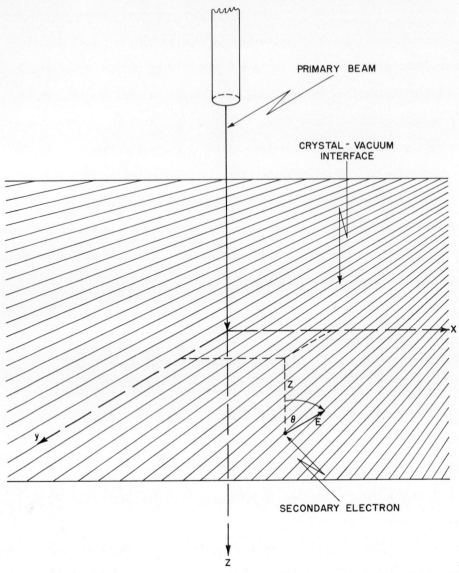

Fig. 1. Typical experimental geometry showing coordinates and energy of a
secondary electron.

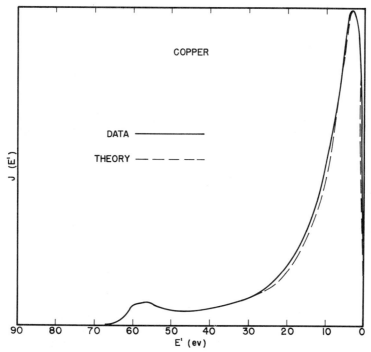

Fig. 2. True secondary electron energy distribution obtained from a clean copper surface with 250 eV primary electrons. The dashed line gives the theoretical prediction.

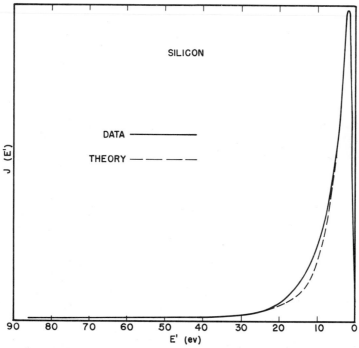

Fig. 3. True secondary electron energy distribution obtained from a clean silicon surface using 1000 eV primaries. The theoretical distribution is given by the dashed lines.

DIFFRACTION OF INELASTICALLY SCATTERED ELECTRONS
IN TUNGSTEN AT LOW ENERGIES

J. O. Porteus
Michelson Laboratory
China Lake, California

INTRODUCTION

The study of electron scattering by plasmons at low energies ordi-
narily requires diffraction or an intervening scattering event to return the
electrons from the sample for observation. In addition to information on
inelastic interactions, such multiple processes may also provide a useful
adjunct to elastic scattering in studies of the diffraction process itself.
Previous work[1,2] on the distribution of low energy electrons inelastically
scattered from single-crystal surfaces indicates that diffraction plays a
significant role. These measurements were generally limited to the immediate neighborhood of a few elastic diffraction maxima, where intense inelastic scattering with a relatively wide distribution of energy is commonly
observed. In the present work, measurements are extended to regions of the
diffraction pattern remote from elastic maxima, where the plasmon loss
maxima are found to be more distinct. In addition to a broader scope, a
clearer view of the diffraction phenomena is obtained. Tungsten (110) was
chosen as the surface for investigation because it has been extensively
studied, can be easily cleaned, and because a recent study of the total
scattering[3] identifies two pronounced loss maxima from a surface and a volume plasmon loss. A relevant optical study on this surface has also been
reported.[4]

EXPERIMENTAL METHOD

1. Apparatus

A schematic view of the diffractometer is given in Fig. 1. Electrons
thermionically emitted at the tungsten filament 1 are extracted from the
space charge region by the electrode 3, focused by the Wehnelt cylinder 2,
and decelerated at the anode 4. Two pairs of vertical deflectors, 6 and 7,
permit accurate positioning of the beam in the horizontal plane of rotation

of the collector 9, or slightly out of this plane to correct for minor sample misalignment. The vertical deflection obscures the filament from the sample and also provides some monochromatization, giving a resulting beam energy-width of about 0.5 eV. An additional pair of deflectors 5 permit horizontal aiming. After traversing a coarse limiting slit the beam strikes the sample surface 8, which ideally contains the collector rotation axis. Fine collimation is provided by a pair of circular collector aperatures 10, which select a 1.3° average cone of electrons emerging from a limited area of the sample. The effective incident beam divergence is thereby restricted to about 0.5°, depending somewhat on the collector position. A high-resolution retarding lens immediately preceding the Faraday cylinder excludes electrons with energies less than a preselected value. The collected electron signal is amplified by a Cary model 36 vibrating-reed amplifier and, after filtering, is displayed on an XY recorder as a function of either retarding, or primary beam voltage. A synchronized programming of anode and vertical deflection voltages with collector motion provides automatic recording of intensity vs. voltage data. The sample azimuth can either be set or varied continuously for rapid diffraction pattern scanning at normal incidence by rotating the sample about its normal, using the rotating drive 15 coupled through the manipulating rod 14. With the rod disengaged by means of the sliding drive 16, the sample can be rotated about the vertical axis to select the desired angle of incidence. For direct observation of the incident beam, or for sample cleaning, the sample may be withdrawn to the position 11. Electron-bombardment heating on the reverse side of the sample wafer is provided by the gun 13, while a gas stream can be directed onto the front surface through the nozzle 12. All drives are magnetically coupled through the vacuum

envelope with well concentrated fields, permitting angular settings
accurate to 0.2°. All drive magnets are either field-compensated or
can be operated intermittently to eliminate magnetic field distortions.
All parts except the electromagnets are of nonferromagnetic materials,
principally stainless steel. A freon-cooled Orb-Ion pump and baffle
provide a field-free working vacuum in the low 10^{-10} torr range after
system bakeout. With the aid of Helmholtz coils the residual magnetic
field in the diffraction region is maintained at less than 20 milligauss.

2.   Sample

The sample is a 3/8" diameter by .020" thick wafer, which has been
spark-cut from 110 oriented, single floating-zone refined, tungsten single
crystal obtained from Aremco. The surface for study was mechanically
ground to within 0.05 degrees of the (110) planes with the aid of a Siemens
x-ray diffractometer and a specially constructed jig. This was followed by
mechanical polishing and electropolishing to a mirror finish. Final cleaning was performed in ultrahigh vacuum by intense heating, followed by oxidation and desorption in the usual manner[3,5] until the clean tungsten
diffraction pattern was observed. As an additional check on cleanliness
the secondary electron spectrum was examined for Auger peaks from possible
contaminants,[3] and none were detected. However, the sensitivity was considerably limited by the small collector acceptance angle.

DISCUSSION AND RESULTS

1.   Elementary models

A model first proposed by Davisson and Germer[6] for inelastic scattering accompanied by diffraction is illustrated for the case of normal inci-
dence in Fig. 2. The incident beam $\underset{\sim}{K}$ is first diffracted into $\underset{\sim}{K}'$ and is
then scattered into a cone of inelastic loss beams such as $\underset{\sim}{K}''$ surrounding

the diffracted elastic beam. Since the primary beam energy governs the
diffraction in this model, this type of double process will be designated
as I. Note that the loss beam maxima coincide with those of the diffracted
elastic beam, i.e. when $\underset{\sim}{K}'$ terminates on a reciprocal lattice point P.

A second type of model, which was first proposed by Turnbull and
Farnsworth,[2] is shown for normal incidence in Fig. 3. Here the incident
beam $\underset{\sim}{K}$ is first inelastically scattered into a cone about the forward
direction. Certain component beams in this inelastic cone, such as $\underset{\sim 1}{K}'$
and $\underset{\sim 2}{K}'$, are then selectively diffracted into respective beams such as $\underset{\sim 1}{K}''$
and $\underset{\sim 2}{K}''$, which form a corresponding emergent cone (not shown). Since diffraction is governed by the energy of the inelastic beams (secondary
energy), this type of double process will be designated as II. Because of
the difference in length of primary and secondary wave vectors the emergent
loss beams do not generally approximate the direction of the diffracted
elastic beam unless the energy loss is small. However, they do coincide
with the position the elastic beam would have at the corresponding primary
energy and angle of incidence. If the forward scattering cone is strongly
concentrated about $\underset{\sim 1}{K}'$ one might expect diffraction by the reciprocal lattice rod, as illustrated by $\underset{\sim 1}{K}''$, in analogy with elastic diffraction. However, if the cone is broad one might expect a type of diffraction more
closely resembling Kikuchi line phenomena. The beam $\underset{\sim 2}{K}'$, for example, which
satisfies the condition for Bragg reflection, can give rise to the component
$\underset{\sim 2}{K}''$ of a Kikuchi excess line associated with the inelastic scattering.

Although both models can easily include dynamical effects in the diffraction stage of the double process, the assumed independence of the diffraction and inelastic scattering is basically a kinematic concept. In
spite of this artifical aspect, the models offer a convenient basis for
discussion of present results.

An important difference between I and II lies in the fact that in II
the diffraction is dispersive in the inelastic scattering whereas in I it
is not. Thus features of the inelastic scattering corresponding to different energy losses, which appear at essentially the same diffraction
positions in I, may appear in quite different positions in II. The dispersion associated with the inelastic scattering event must, of course, be
present in both I and II, but this may be comparatively small. Also, II
requires coherence of the inelastically scattered electrons so that a
smaller incoherent background would be expected to accompany II.

2. Observations

When the collector is near an elastic beam, retarding-field scans
of the electron energy distribution generally show, in agreement with
previous observers, a strong inelastic background scattering which increases with decreasing energy loss. Some evidence of discrete-loss
scattering was observed but these features are usually broad and largely
obscured by the background. An increase of background intensity with
increasing elastic intensity as predicted by I was also observed. However, in some situations a dispersive redistribution of background
energy with varying collector angle was seen, as predicted by II. In
addition to this apparent mixture of double processes near the elastic
beam, two experimental complications were generally associated with the
large elastic component. One of these was a slight reduction in collector resolution, probably due to space charge distortion of the retarding field by the relatively large accumulation of retarded electrons. A more serious problem of possibly similar origin was an
instability in the collected current, which made inelastic measurements
difficult within 2° of the elastic beam.

Figure 4 shows the results of a sequence of retarding-field scans made at a fixed collector position in the 01 azimuth and away from the 02 elastic beam. Each of the heavy solid curves represents an electron energy distribution, where a localized rise in collector current corresponds to a peak in the energy spectrum. The sharp rise at zero loss results from elastic and thermal scattering, while the more gradual rise which shifts (dashed line) as the primary energy is changed is associated with inelastic scattering. If one compares the relative heights of the inelastic rise with the aid of the tangent lines, a maximum is observed at a primary energy of about 54 eV and a corresponding energy loss of 10.5 eV measured from the midpoint of the elastic rise. Scans at higher primary energies with the same collector setting reveal an additional loss maximum at an energy loss of 22 eV. These values are somewhat lower than the 12.5 and 23.5 eV energy loss values reported for surface and volume loss maxima, respectively, in the total scattering.[3] The different average dispersion resulting from the larger angular distribution in the total scattering measurements may   account for these differences.[7] The optical data on tungsten[4] is questionable regarding sample contamination and, furthermore, shows a large unexplained discrepancy with electron loss measurements concerning the volume loss. For these reasons the interpretation of the losses given in the total scattering study[3] is adopted here.

The observed shift of the inelastic rise in Fig. 4 over the 8 eV range of primary energies shown amounts to 6.5 eV, as compared to an 8 eV maximum shift from diffractive dispersion predicted by model II under the following conditions: (1) infinitely narrow scattering cone, and (2) infinite angular resolution, including zero collector acceptance angle and zero thickness of the reciprocal lattice rod. Departure from these ideal conditions could easily explain the smaller observed shift. The dispersion contribution from scattering by the surface plasmon, which should contribute to the curvature of the dashed line of Fig. 4, is evidently too small to be detected here.

A search for loss maxima was made in two azimuths of the diffraction pattern at normal incidence. Either the primary energy or the collector position was held fixed and the other varied for each scan sequence. Results for the 11 azimuth are summarized in Fig. 5, where observed maxima are compared in terms of colatitude angle and secondary energy, i.e. primary energy minus energy loss. Surface and volume loss maxima largely coincide on this plot in agreement with model II and, except for some isolated maxima, form two inelastic 11 beams running parallel to the 11 elastic beam. However, the inelastic beams show a definite shift to higher energy and/or angle relative to the normally incident elastic beam, which is in disagreement with II if the scattering is symmetrical in the forward direction. Although the overall scattering contributing to all inelastic beams must show this symmetry at normal incidence, an asymmetric scattering, if compensated, can be associated with a particular beam. This, however, implies coupling between the two elements of the double process. Coupling between diffraction and inelastic scattering is regarded as the most plausible explanation of the shift. An explanation on the basis of different effective inner potentials for the diffraction of elastic vs. inelastic electrons can be ruled out, as will be shown.

Little evidence is found of beams associated with inelastic Kikuchi lines, which might be expected with a wide scattering cone. In terms of Fig. 5 these beams should lie along the paths followed by elastic beam maxima as the angle of incidence is varied. Such a path with a possibly associated beam is indicated by the dashed line in the figure. Figure 6 shows loss beam positions observed in the 01 azimuth relative to the 02

elastic beam. The results are essentially the same as for the 11 azimuth
with even less evidence for inelastic Kikuchi beams.

Measurement of the inelastic scattering cone angle in the plane
of a given azimuth is complicated by the dispersion and by diffracted
intensity variations along the reciprocal lattice rod. However, the
scattering cone angle normal to the azimuth plane should be more directly related to the corresponding cone angle of the observed loss beams.
Accordingly, retarding field scan sequences of the 11 and the 02 loss
beams were made across their respective azimuths, and the angles
subtended by their half-maximum intensity cones were determined without
correction for instrumental broadening effects. The 11 volume and
surface loss beams measured at 80 and 70 eV primary energies, respectively,
both gave a value of roughly 6°, while the corresponding 02 beams measured at 65 and 55 eV, respectively, gave a value of roughly 8°. No
significant differences were noted between the surface and volume loss
angular distributions. These results, like the shift in inelastic
beam positions, may be influenced by coupling between scattering by
plasmons and diffraction. This needs verification, however, by a
theoretical calculation of the inelastic scattering angular dependence
appropriate to the present experimental conditions. Such a calculation
is in progress,[8] but values are not yet available for comparison.

By simultaneously varying the primary energy and collector position, it
is possible to track a loss beam in a sequence of retarding-field scans. The
varying intensity of the loss beam in this sequence provides an intensity vs.
voltage curve for this beam. Figure 7 compares the observed peaks in such
curves for the 11 surface and volume loss beams with the 11 elastic curve.
When compared in terms of secondary energy the agreement of peak positions
is good--too good in fact to permit explanation of the anomalous shift in

loss beam positions as an inner-potential effect. Figure 8 shows similar results for the 02 beams, again with generally good agreement of peak positions. Although a few possibly important differences are apparent, the essential similarity of the elastic and inelastic data is in support of II, at least as a first approximation.

The similarities apparent in Figs. 7 and 8 suggest that plasmon scattering studies may have an important bearing on the interpretation of elastic intensity vs. voltage data. This would not be the case if the scattering involved in II were predominantly by image forces before penetration, as Heidenreich has proposed for fast electrons,[9] since elastic and inelastic intensity vs. voltage data would then contain basically the same information regarding diffraction. However, the present evidence for coupling between inelastic scattering and diffraction and minor differences which appear in the last two figures do not favor this interpretation. More definitive experiments with varying surface conditions are needed to explore this interesting possibility.

SUMMARY

The double process of electron scattering by plasmons and subsequent diffraction has been investigated at low energies on a tungsten (110) surface. Two well-defined inelastic beams corresponding to previously observed plasmon losses, and showing diffractive dispersion, were found to accompany an elastic beam. The positions of the loss beams indicate that an adequate theory of scattering by plasmons must include coupling of the scattering with diffraction. Intensity vs. voltage characteristics of the loss beams were studied and were found to be similar to those of the associated elastic beam when plotted as a function of secondary energy. This is expected on the basis of the double process with dynamical diffraction. Possible

application of the present results include studies of electron–plasmon interaction at low energies and additional information on LEED interpretations.

ACKNOWLEDGMENTS

Grateful acknowledgments are due to Dr. E. Bauer for many helpful
discussions and for reviewing the manuscript; also to Dennis K. Burge for
calling author's attention to the optical data. This work was supported
in part by the National Aeronautics and Space Administration under Contract
No. R-05-030-001.

REFERENCES

1. Paul P. Reichertz and H. E. Farnsworth, Phys. Rev. $\underline{75}$, 1902 (1949).

2. John C. Turnbull and H. E. Farnsworth, Phys. Rev. $\underline{54}$, 509 (1938).

3. L. N. Tharp and E. J. Scheibner, J. Appl. Phys. $\underline{38}$, 3320 (1967).

4. D. W. Juenker, L. J. LeBlanc and C. R. Martin, J. Opt. Soc. Am. $\underline{58}$,
    164 (1968).

5. R. M. Stern, Appl. Phys. Letters $\underline{5}$, 218 (1964).

6. C. Davisson and L. H. Germer, Phys. Rev. $\underline{30}$, 705 (1927).

7. H. Raether, in Springer Tracts in Modern Physics (Springer-Verlag,
    Berlin, 1965), Vol. 38, p. 140.

8. H. N. Browne and E. Bauer, private communication.

9. R. D. Heidenreich, J. Appl. Phys. $\underline{34}$, 964 (1963).

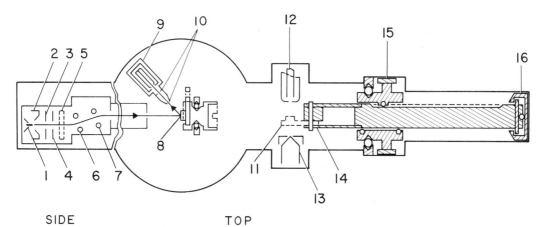

SIDE   TOP

Fig. 1.   Schematic of the diffractometer. The electron gun is shown rotated 90°
          relative to the rest of the apparatus.

TYPE I DOUBLE PROCESS

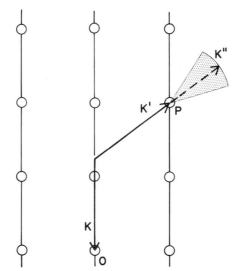

Fig. 2.   The model of Davisson and Germer for combined diffraction and inelastic
          scattering.

TYPE II DOUBLE PROCESS

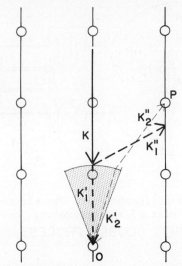

Fig. 3. The model of Turnbull and Farnsworth for combined diffraction and inelastic scattering.

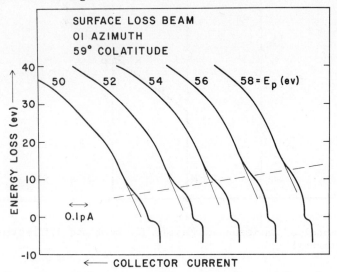

Fig. 4. Retarding-field scan sequence through a surface loss beam.

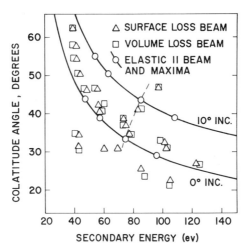

Fig. 5. Distribution in secondary energy and colatitude angle of loss beams at normal incidence in the 11 azimuth. The 11 elastic beam at normal and 10° incidence is shown for comparison.

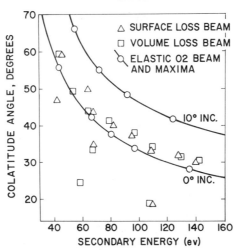

Fig. 6. Distribution in secondary energy and colatitude angle of loss beams at normal incidence in the 01 azimuth. The 11 elastic beam at normal and 10° incidence is shown for comparison.

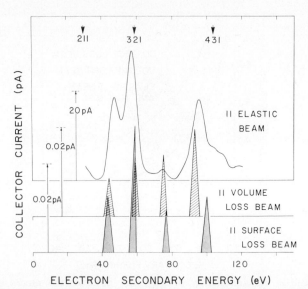

Fig. 7. Intensity vs. voltage features of 11 elastic and loss beams compared in terms of secondary energy. The heights of the triangles indicate maximum collector signal at peaks in the loss beams. Bases of the triangles indicate uncertainty in peak positions. All measurements were made at normal incidence.

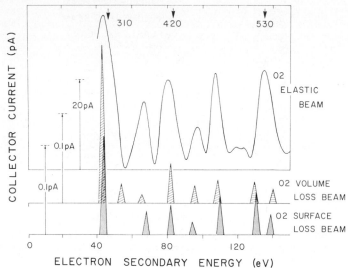

Fig. 8. Intensity vs. voltage features of 02 elastic and loss beams compared in terms of secondary energy. Triangles have the same significance as in Fig. 7. All measurements were made at normal incidence.

THEORY OF AUGER EJECTION OF ELECTRONS
FROM METALS BY IONS

Eric P. Wenaas
State University of New York at Buffalo

A. J. Howsmon
Institute for Aerospace Studies
University of Toronto, Canada

I. Introduction

A beam of low energy ions incident on a metal surface can be neutralized by various processes including Auger neutralization with subsequent secondary electron emission, and resonance neutralization followed by secondary electron ejection due to Auger de-excitation of the excited atom. Of these two processes, Auger neutralization is of greater interest because in most cases (e.g., He^+, Ar^+, Kr^+, and Xe^+ incident on tungsten) this is the only process possible. Many authors[1-7] have treated this process theoretically, but undoubtedly the most comprehensive treatment is one given by Hagstrum[5]. There are several objections to this treatment which uses perturbation theory to obtain transition matrix elements as a function of ion-surface separation including the following: it is necessary to fit to experimental data rather than to evaluate transition matrix elements from first principles; the total secondary electron yield as a function of incident ion energy cannot be predicted; the depth of origination of the Auger electrons cannot be determined; the spatial distribution of the secondary electrons cannot be found; and the predicted yields for Ar^+ and Kr^+ are almost 50% lower than experimental yields.

Apart from these functional difficulties mentioned, there is a more fundamental problem in the theory arising from the use of a transition matrix element which is a function of ion-surface separation -- a physically nonmeasurable quantity. The approach requires the subsequent use of classical ideas to arrive at the physically measurable transition probability (i.e., secondary electron energy distribution and total yield). In order to enhance agreement of the theory with experiment it is necessary to interject some quantum ideas such as energy level perturbations, effective ionization potentials, and uncertainty principle effects into the semiclassical approach.

The theory presented here will attempt to overcome these difficulties by using a quantum scattering approach instead of the perturbation approach (Section II). A formal derivation of the scattering matrix elements will not be repeated here but the Hamiltonian, the appropriate wave functions, and the forms of the matrix elements will be given (Section III). An approximate model useful for calculations will be discussed (Section IV). The expression for the secondary electron energy distribution and total yield will be presented in terms of the matrix elements of the previous section. The results of a simple calculation using constant matrix elements are shown to be in exact agreement with the theory of Hagstrum (Section V). Finally, the matrix elements found from the scattering theory will be used to calculate the secondary electron energy distribution and total yield for the case of He^+ on tungsten. The results and a discussion of these results will also be given (Section VI).

II. Scattering Approach

Consider the system of Fig. 1 comprised of an ion and two of the many conduction electrons that exist below the Fermi level in a crystal. The incident ion and the two electrons in the crystal will be designated the initial state of the system, while the reflected atom and secondary ejected electron will be designated the final state of the system. The problem is to find a method to determine the probability of a transition from the initial state to the final state.

The scattering and perturbation approach both describe transitions occuring in the system and both can be formally derived from the same time-dependent quantum theory, but they convey different information about transitions. The choice of the approach depends upon the information desired. The perturbation approach describes the transition of the initial state with an ion at a distance, s, in front of the crystal surface to the final state with an atom also at a distanct, s. This approach does not describe the total or cumulative transition probability as the ions move toward the surface and the atom scatters away from the surface. The scattering approach, on the other hand, describes the total probability that a system in an initial state with an ion far from the crystal will scatter from the crystal into a final state with a secondary electron and an atom again far from the crystal.

Because the measurable quantities--including incident ion energy, secondary electron energy distributions, and total electron yield--are measured at points far removed from the scattering center (i.e., the crystal surface), it is natural that the scattering approach, which describes this situation, should be used. The following difficulties arise if the perturbation approach is used:

1) A classical calculation is necessary to find the total transition probability from the transition probability as a function of ion-surface separation.

2) The initial and final states have not really existed for a long time at each point, s, and therefore have finite life times.

3) The initial and final ion and surface states are mutually perturbed because they are separated by a finite distance, s. This causes a shift of the energy levels in the crystal , the ion, and the

atom, so that initial and final wave functions are thus no longer known. Because of the level shifts, the concept of an effective ionization energy must be introduced.

The objections are removed by using scattering theory. The initial and final particle energies and wave functions need to be known only for large ion-surface separations where the mutual perturbation is absent, and the energies are well specified because they are measurable. The calculation is completely within the framework of quantum theory, and therefore, uncertainty principle effects need not be considered separately. Finally, and most important, the scattering matrix elements, which are not a function of ion-surface separation, describe the total probability of the transition.

III. The Scattering Matrix

Very few problems in quantum mechanics can be solved exactly, and whether perturbation theory or scattering theory is used, one generally must approximate at least part of the solution to the problem. The Born approximation is a common method used in scattering theory and consists of approximating the total solution by the first term in a series solution. If one naively attempts to apply the scattering theory for a particle in a potential using the Born approximation to the Auger neutralization problem, certain difficulties arise. First, the Born approximation will not be valid because of the strong interaction potential between the ion and the surface. Second, the emerging particles (the atom and electron) are not the same as the incident particle (ion), and therefore the Hamiltonians for the initial and final states are not the same. Situations of this type are known as rearrangement collisions. The difficulty associated with the strong ion–surface interaction can be treated using a distorted Born approximation in which the potentials are divided into strong and weak. The strong potential is then treated exactly and the weak one approximately. The formal development of the scattering theory involving rearrangement collisions has been considered by Lippmann[8] and can be applied to this problem to remove the second objection.

If the Hamiltonian for a system can be written as

$$H = H_o + V_o + V_1 = H'_o + V'_o + V'_1 \tag{1}$$

where H_o is that part of H for which the solution is known, V_o is that portion of the potential which can be solved exactly, V_1 is the portion for which approximate methods must be used, and the primes refer to a final or rearranged state, then the scattering matrix to the first distorted Born approximation for a rearrangement collision is

$$S_{f'i} = -2\pi i \delta(E_{f'} - E_i) H_{f'i}. \tag{2}$$

The matrix elements, $H_{f'i}$, are given by

$$H_{f'i} = \left\langle \chi_{f'}^{(-)} \middle| V_1 \middle| \chi_i^{(+)} \right\rangle \tag{3}$$

in the unrearranged or initial-state form, and by the equivalent relation

$$H_{f'i} = \left\langle \chi_{f'}^{(-)} \middle| V'_1 \middle| \chi_i^{(+)} \right\rangle \tag{4}$$

in the rearranged or final-state form. The wave functions $\chi_i^{(+)}$ and $\chi_{f'}^{(-)}$ are solutions to the integral equations

$$\chi_i^{(+)} = \varphi_i + \frac{1}{E - H_o + i\epsilon} V_o \chi_i^{(+)}, \tag{5}$$

$$\chi_{f'}^{(-)} = \varphi_{f'} + \frac{1}{E - H_o' - i\epsilon} V_o' \chi_{f'}^{(-)} , \tag{6}$$

where φ_i and $\varphi_{f'}$ are eigenfunctions of H_o and H_o' respectively, and E_i and $E_{f'}$ are eigenvalues of H_o and H_o'. The (+) refers to an outgoing-wave solution and the (−) refers to an ingoing-wave solution.

The Hamiltonian of interest in the Auger neutralization problem is

$$H = \frac{P_1^2}{2m_1} + \frac{P_2^2}{2m_2} + \frac{P_3^2}{2m_3} + V_{1-c} + V_{2-c} + V_{3-c} + V_{1-2} + V_{1-3} + V_{2-3} \tag{7}$$

where the subscripts 1, 2, 3, and c refer to electron 1, electron 2, ion, and crystal respectively. Notice that particles 1 through 3 interact with the rest of the crystal as a whole through potentials V_{1-c}, V_{2-c}, and V_{3-c}. The Hamiltonian in the initial-state form is divided according to Eq. (1)

$$H_o = \underbrace{\frac{P_1^2}{2m_1} + V_{1-c}}_{\varphi(1)} + \underbrace{\frac{P_2^2}{2m_2} + V_{2-c}}_{\varphi(2)} + \underbrace{\frac{P_3^2}{2m_3}}_{\varphi(3)} \tag{8}$$

$$V_o = V_{3-c} \tag{9}$$

$$V_1 = V_{1-2} + V_{1-3} + V_{2-3} \tag{10}$$

where the φ's are solutions of the wave equation for the portion of the Hamiltonian appearing in brackets above each respective φ. The wave functions $\varphi(1)$ and $\varphi(2)$ represent the two electrons moving within the potential of the crystal, and $\varphi(3)$ represents the ion incident on the crystal. The total wave function for the initial state described by H_o is the product of the individual φ's

$$\varphi_i = \varphi(1)\varphi(2)\varphi(3) \tag{11}$$

The Hamiltonian expressed in the final-state form is

$$H_o = \underbrace{\frac{P_1^2}{2m_1} + V_{1-c}}_{\varphi'(1)} + \underbrace{\frac{P_2^2}{2m_2} + \frac{P_3^2}{2m_3} + V_{2-3}}_{\varphi'(2,3)} \tag{12}$$

$$V_o' = V_{3-c} + V_{2-c} \tag{13}$$

$$V_1' = V_{1-2} + V_{1-3} \tag{14}$$

where $\varphi'(1)$ represents the wave function for a secondary electron and $\varphi'(2,3)$ is a hydrogenic wave function representing an atom reflected from the crystal. Again the product of the φ'(s) is the total wave function for the final state described by H_o'

$$\varphi_{f'} = \varphi'(1)\varphi'(2,3). \tag{15}$$

Notice that as a result of the distorted Born approximation, $X_i^{(+)}$ and $X_{f'}^{(-)}$ appear in the expressions (3) and (4) for the matrix elements rather than φ_i and $\varphi_{f'}$. The wave functions $X_i^{(+)}$ and $X_{f'}^{(-)}$ are no more than φ_i and $\varphi_{f'}$ distorted by the presence of the crystal, represented by potential V_o and V_o'. Equivalently, the X's represent solutions to the problems of the ion scattering from the crystal <u>without</u> neutralization and the atom scattering from the crystal <u>without</u> ionization.

The perturbation potential, V_1 or V_1', depending on the form of the matrix elements chosen, contains potential V_{1-3}, V_{2-3}, and V_{1-2}. Because V_{1-3} and V_{2-3} yield small or zero matrix elements with respect to V_{1-2}, it is apparent that the perturbation causing the transition is the interaction potential

between the two crystal electrons and not between the ion or atom and
crystal electrons. Also note that the initial and final energies in the energy-
conserving delta function in Eq. (2) include not only the two electron kinetic
energies and the ion potential energy, but also the ion and atom kinetic
energies which are not necessarily the same. This allows for non-adiabatic
or kinetic ejection of the electrons in the Auger process.

IV. The Model and Calculations

The Hamiltonian and general form of the wave functions for the Auger
neutralization process were presented in the last section. A model of the
crystal, ion, and atom is necessary to specify wave functions and interaction potentials so that the necessary calculations can be performed. A
free electron model of the crystal was chosen in which the exponential tails
of the wave functions outside the crystal were retained. The incident ion
is represented by a plane wave and the reflected atom is represented by a
hydrogenic wave function including a plane wave representing center of mass
motion. The two Auger electrons interact through a Coulomb potential which
is effectively unscreened as discussed by Heine[9]. The ion and atom are
reflected from the surface by an effective repulsive potential of the type
suggested by Hagstrum[5]. The wave functions were antisymmetrized to include
exchange effects. The calculations indicated by Eqs. (3) or (4) for the
matrix elements, $H_{f'i}$, are lengthy, but can be completed in closed form. The
result is used in the next section to find the electron distribution and total
yield.

V.   Secondary Electron Energy Distribution and Total Yield

The secondary electron energy distribution and total yield can be found from the scattering matrix elements using the Golden Rule of scattering theory. The energy distribution is

$$\eta_o(E_k,\vec{K}) = \int_{E_{k'}}\int_{\Omega_{k'}}\int_{E_{k''}}\int_{\Omega_{k''}}\int_{E_{k'}}\int_{\Omega_{k'}}\int_{\Omega_k} \eta(E_{k'})\ \eta(E_{k''})\left|H_{f'i}\right|^2$$

$$\eta(E_k)\ \delta(E_{f'}-E_i)\ dE_{k'}\ d\Omega_{k'}\ dE_{k''}\ d\Omega_{k''}\ dE_{K'}\ d\Omega_{K'}\ d\Omega_k \ , \qquad (16)$$

with

$$H_{f'i} = H_{f'i}(\vec{k'},\vec{k''},\vec{k},\vec{K},\vec{K'}) \qquad\qquad (17)$$

$$E_{-i} = E_k + E_{K'} \qquad\qquad (18)$$

$$E_i = E_{k'} + E_{k''} + E_K + I - 2D \ . \qquad\qquad (19)$$

The following notation has been used

$\eta_o(E_k,\vec{K})$	secondary electron distribution
$\eta(E_{k'})$, $\eta(E_{k''})$	electron density of occupied states
$\eta(E_k)$	electron density of unoccupied states
$\vec{k'}$, $\vec{k''}$	crystal electron wave vectors
$\vec{k}$	ejected electron wave vector.
$\vec{K}$	incident ion wave vector
$\vec{K'}$	reflected atom wave vector
I	ionization potential

D   surface barrier potential

μ   Fermi level.

The total yield is the integral of the distribution

$$\gamma(\vec{K}) \;=\; \int \eta_o(E_k,\vec{K}) \; dE_k \; . \tag{20}$$

The expressions for secondary electron distribution and total yield are normalized by using the experimental result that virtually all incident ions are neutralized. Of the Auger electrons produced by this neutralization process, only those electrons having sufficient momentum in the normal direction to overcome the potential barrier of the crystal can escape to the vacuum and become true secondaries.

Before considering the secondary electron distribution predicted by using the matrix elements of Eqs. (3) or (4), it is of interest to find the secondary electron distribution and total yield assuming the matrix elements are constant. This is equivalent to assuming that the transition probability from any initial state to any final state is the same for all states. Hagstrum[5] made a calculation for He^+ on tungsten using perturbation theory, and the results of these two calculations will be compared. The densities of initial and final electron states are assumed, as Hagstrum also did, to be proportional to the square root of their energies. The result of the integration of Eq. (16) assuming $H_{f'i}$ constant is

$$\eta_o(E) \;=\; \frac{1}{C} \, F(E) \, \eta_2(E) \tag{21}$$

where

$$F(E) \;=\; 0 \qquad\qquad\qquad E \le I\text{-}D \qquad (22)$$

$$F(E) = (E - I + D)^2 \frac{\pi}{4} \qquad\qquad I-D \le E \le I-D+\mu \qquad (23)$$

$$F(E) = (2\mu + I - D - E)\left[(E - I + D)\mu - \mu^2\right]$$
$$+ .5(E - I + D)^2 \text{SIN}^{-1}\left[\frac{2\mu - (E - I + D)}{E - I + D}\right] \qquad I-D+\mu \le E \le I-D+2\mu \quad (24)$$

$$F(E) = 0 \qquad\qquad I-D+2\mu \le E \qquad (25)$$

where the normalization constant, C is

$$C = 2\int_{\mu}^{I-D+2\mu} \eta_1(E)\ F(E)\ dE + 2\int_{D}^{I-D+2\mu} \eta_2(E)\ F(E)\ dE \qquad (26)$$

and

$$\eta_1(E) = \sqrt{E} \qquad E \le D$$
$$= \sqrt{D} \qquad E \ge D \qquad\qquad (27)$$

$$\eta_2(E) = 0 \qquad E \le D$$
$$= \sqrt{E} - \sqrt{D} \qquad E \ge D \qquad\qquad (28)$$

The secondary electron energy, E is referenced to the bottom of the conduction band inside the crystal, and $C = .215 \times 10^4$ for He^+ on tungsten.

The closed form solution consisting of Eqs. (21–25) is shown in Fig. 2. This is identical to the result found by Hagstrum* when he assumes no variation

---

*See Fig. 13, page 349 of Reference 5.

of energy levels or uncertainty effects, and an isotropic probability
distribution. The total yield is the integral of Eq. (21) and the result
for He^+ is .1423, which is also identical to the result of Hagstrum. Exact
agreement has been obtained for all other noble gas ions incident on tungsten.

The actual matrix elements are not constant but are complicated functions
of electron and ion momentum represented by Eqs. (2) and (3). Insertion of
these matrix elements in Eq. (16) results in an integral over six independent
variables and can be carried out without approximation only on a computer.
Each of the resulting calculations requires 90 seconds of CDC 6400 computer
time**. The six-dimensional integration was carried out using Monte Carlo
techniques and accuracy of the results is approximately within $\pm 5\%$. The
next section describes some of the results obtained.

VI.   Results

The results for the secondary electron energy distribution and total
yield predicted by scattering theory for He^+ incident on tungsten will be
presented here. Results for Ar^+, Kr^+, and Xe^+ have also been obtained and
will be available elsewhere. The theoretical and experimental[10] electron
yields as a function of ion kinetic energy are shown in Fig. 3. Initial and
final electron densities were assumed to be proportional to $E^{1/2}$. The total yield
is fairly insensitive to the form of the occupied density of states for
electrons in tungsten as will be shown later. The magnitude and shape of
the curve is, however, somewhat sensitive to the form of the repulsive atom-
surface interaction. The results of Fig. 3 were obtained using an atom-

---

**I would like to express my appreciation for computer time supplied by
the Computing Center at the State University of New York at Buffalo.

surface interaction potential very similar to the one suggested by Hagstrum[5]

The rise of the experimental curve above 500 eV has been related[11] to the

contribution to secondary emission by processes other than Auger ejection,

and therefore should not and does not appear in the theoretical curve for

Auger ejection. The resulting theoretical curve is seen to be in excellent

agreement with experiment.

The secondary electron energy distributions used to calculate the total

yield curve are shown in Fig. 4., and the corresponding experimental distri-

butions according to Hagstrum[11] are shown in Fig. 5. The two sets of

distributions are strikingly similar in almost all respects. Both distribu-

tions broaden significantly more at higher electron energies than at lower

energies, both distributions broaden through a point similarly located on the

high energy portion of the distribution, and the peaks of both distributions

decrease by approximately the same percentage as a function of incident ion

kinetic energy. The broadening in the theoretical curve is caused by kinetic

energy exchange between the ejected electron and the ion. This source alone

seems to account for all the broadening present. The broadening is a linear

function of velocity (see the lower scale in Fig. 4), a result Hagstrum[12]

has confirmed experimentally. Note that the maximum energies of the theor-

etical distributions correspond well to the maximum energies of the experi-

mental distributions for the various incident ion energies.

The main discrepancy between theory and experiment is the occurrence in

the theoretical distribution of too many high energy electrons and too few

low energy electrons. It was thought at first that this discrepancy could

be accounted for by using a more realistic density of occupied states for

the electrons in tungsten. The results for various densities of states[13,14]

are shown in Fig. 6. The form of the secondary electron distribution is

sensitive to the density of electron states, but the number of high energy

secondary electrons is still too large. It is apparent that another mechanism must exist which acts to redistribute the secondary electron energies.

Probst[6] suggested that collisions between the emerging Auger electrons and the crystal electrons might play an important role in shaping the
final electron distribution. He estimated that as many as 50% of the secondary electrons are invloved in electron-electron collisions. According
to the present theory, this percentage seems reasonable. From Fig. 7, in
which theoretical and experimental electron distributions are displayed together, it can be seen that more than 40% of the electrons must be involved
in such a process to bring theory and experiment into agreement. Probst[6],
using experimental electron scattering results of Harrower[15], calculated
an energy distribution in which he attempted to include the effect of inelastic electron collisions.

Hagstrum[16] developed a method for determining the effect of inelastic electron scattering and found the percentage of scattered electrons
to be in the range of 5 to 10%, significantly less than the amount suggested
by Probst. His method requires a knowledge of the distribution of secondary
electron energies as a function of incident ion energy, a quantity he ob-
obtains from experimental data.

Further theoretical work now in progress supports Hagstrum's conclusion that only a small number of electrons are inelastically scattered.
The discrepancy between the experimental and theoretical electron distributions has been accounted for by considering the effect of the reflected
atom on the secondary electrons, rather than the effects of inelastic
collisions. The secondary electron energy distributions of Fig. 4 were

calculated using a sinusoidal approximation for the distorted initial ion

and final atom wave functions given by Eqs. (5) and (6). This has the

effect of unduly restricting the reflected atom kinetic energy to values

approximately equal to the incident ion kinetic energy. When the correct

distorted waves are substituted for the sinusoidal approximations, the

kinetic energy of the reflected atom may be significantly higher, but not

significantly lower, than that of the incident ion.

This result can be seen from the following heuristic argument util-

izing the interaction potential curves calculated by Hagstrum[5] as shown in

Fig. 8. The ion will be attracted to the surface by the image charge attrac-

tive potential which will not act to slow down the reflected atom. Also,

even without considering the image potential, the repulsive force on an atom

at a given position is greater than the repulsive force on an ion at the

same position. Therefore, as a result of the neutralization process, the

reflected atom will be repulsed from the surface with a greater force than

the ion would have been had it not been neutralized. These two effects

can cause the atom to leave the surface with a greater kinetic energy than

that of the incident ion. Thus, the effect of the surface is to convert

some of the potential energy of the ion into kinetic energy.

The final atom will actually have a spread of energies E_K, consistant

with the delta function of Eq. (16), and the probability of occurrence of

each of these final atom energies is determined by the matrix elements of

Eq. (17).

The higher range of reflected atom energies will effect the electron

distribution through the energy conserving delta function of Eq. (16).

Because the energy must be conserved, the resulting distribution of

higher energy atoms will cause a lower distribution of secondary electron energies. This allows violation of the lower and to a lesser extent
the upper limits of secondary electron energies as claculated without considering ion and atom motion. This is also the cause of secondary electron
energy broadening, even at low incident ion energies.

One additional effect favors the occurrence of lower electron energies
and higher reflected atom energies. The probability that an electron will
make a transition from the metal to a free state in the vacuum is approximately proportional to the inverse of the change in electron energy required. Thus, transitions to lower electron kinetic energy states where
allowed are favored because the change in energy is less.

The distribution of secondary electron energies is determined in the
same manner as before except that the calculations become more difficult.
Preliminary results show that the utilization of proper distorted waves
bring the experimental and theoretical curves into good agreement.

This result has important implications for momentum transfer calculations at an ion-gas - solid interface. Because the most probable reflected atom energy is higher than the incident ion energy, the normal
momentum accommodation coefficient may actually be negative for slow ion
encounters with a metal surface. Further work on this point is in progress.

The region of most probable neutralization can easily be found. This
is simply that region of ion-atom coordinate space that contributes most
to the integral for the matrix elements. Approximately 90% of the contribution comes from a region approximately .5 to 1.5 Ångstroms wide
located in front of the classical turning point of the atom.

Typical regions in Ångstroms for He^+ incident on tungsten are shown in Table I along with the point of most probable neutralization, S_m, predicted by Hagstrum[5].

Table I. Region of Neutralization

Ion energy	Region	S_m
40 eV	1.8 - 2.8	2.20
200 eV	1.4 - 2.4	2.02
1000 eV	1.0 - 2.0	1.84

The classical turning points for atoms on surfaces are not specifically known so that these regions must be taken as approximate, but the agreement with Hagstrum is good.

The depth of origination of the Auger electron from within the crystal may also be determined by finding that region of initial electron coordinate space that contributes the most to the integral for the matrix elements. This region has not yet been mapped out in detail and depends upon ion kinetic energy and ionization energy. The main contribution comes from a region beginning at 2 to 3 Ångstroms below the surface to a region extending an Angstrom or two outside the surface. A small percentage of electrons originate from as deep as 6 to 8 Ångstroms, however.

VII. Conclusions

A theory of Auger neutralization has been presented based upon quantum scattering theory. This approach yields scattering matrix elements independent of ion-surface separation from which the secondary electron energy distribution and total yield may be calculated in a straightforward manner totally within the framework of quantum theory. The results of the theory for He^+ incident on tungsten agree well with experimental results, except

for the occurrence of too many high energy electrons in the secondary electron distribution. This discrepancy has been accounted for by using the
proper distorted ion and atom wave functions. The average depth of origination of Auger electrons and the most probable region of neutralization
have been found. The electron distribution and total yield have been found
for Ar^+, Kr^+, and Xe^+ incident on tungsten as well as for He^+ as reported
in this work. The spacial distribution of the secondary electrons can
also be calculated and work in this area is continuing. These results
along with the details of the further calculations outlined here will be
reported later.

This work has been part of a continuing theoretical effort to investigate phenomena occurring at gas-solid interfaces utilizing the two potential
formalism of quantum scattering theory.

References

1. M. L. E. Oliphant and P. B. Moon, Proc. Roy. Soc. (London) A127,388 (1930).

2. H. S. W. Massey, Proc. Cambridge Phil. Soc.,26,386 (1930).

3. S. S. Shekhter, J. Exptl. Theoret. Phys. (U.S.S.R.), 7, 750 (1937).

4. A. Cobas and W. B. Lamb, Jr., Phys. Rev., 65, 327 (1944).

5. H. D. Hagstrum, Phys. Rev., 96, 336 (1954).

6. F. M. Probst, Phys. Rev., 129, 7 (1963).

7. V. I. Rydnik and B. M. Yavorskii, Sov. Phys. Doklady, 7, 533 (1962).

8. B. A. Lippmann, Phys. Rev., 102, 264 (1956).

9. V. Heine, Phys. Rev., 151, 561 (1956).

10. H. D. Hagstrum, Phys. Rev., 104, 317 (1956).

11. H. D. Hagstrum, Phys. Rev., 96, 325 (1954).

12. H. D. Hagstrum, Phys. Rev., 139, 526 (1965).

13. M. F. Manning and M. I. Chodorow, Phys. Rev., 56, 787 (1939).

14. H. Claus and K. Ulmer, Zeit. Phys., 173. 462 (1963).

15. G. A. Harrower, Phys. Rev., 104, 52 (1956).

16. H. D. Hagstrum and Y. Takeishi, Phys. Rev., 137, A304 (1965).

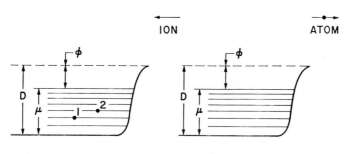

ELECTRON

ION   ATOM

a) INITIAL STATE   b) FINAL STATE

Fig. 1 Schematic diagram of Auger neutralization process showing
a) the initial state with two electrons in the metal and
an ion incident on the metal, and b) the final state with
a reflected atom and secondary electron.

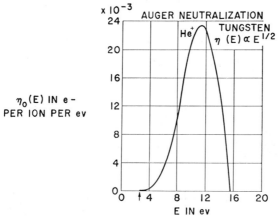

Fig. 2 Theoretical secondary electron distribution, $\eta_o(E)$, for He^+
incident on tungsten from Eq. (21) when the matrix elements
$H_{f'i}$ are assumed to be constant and the initial and final
densities of electron states are assumed to be proportional
to $E^{1/2}$.

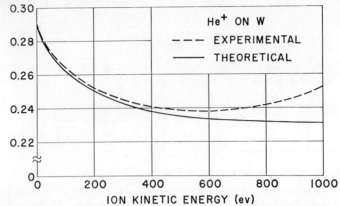

Fig. 3   Total electron yield, γ, versus incident ion kinetic energy for He$^+$ on atomically clean tungsten. The initial and final electron states are assumed to be proportional to $E^{1/2}$. The experimental curve is according to Hagstrum (ref. 11).

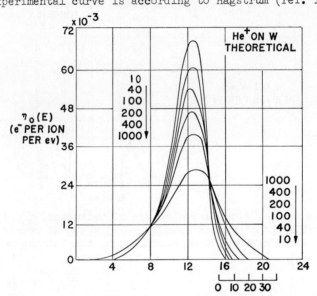

Fig. 4   Theoretical secondary electron distributions ejected by 10, 40, 100, 200, 400, and 1000 eV He$^+$ ions incident on tungsten. Initial and final electron states are assumed to be proportional to $E^{1/2}$. The lower scale is explained in the text.

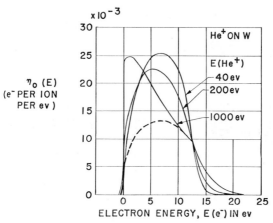

Fig. 5   Experimental secondary electron distributions ejected by 40,
200, and 1000 eV He[+] ions incident on tungsten according to
Hagstrum (ref. 11). The dashed curve is the portions of the
1000 eV distribution thought to be due to the Auger process
alone.

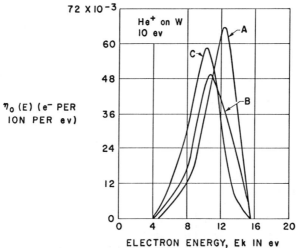

Fig. 6   Theoretical secondary electron distributions ejected by 10 eV He[+]
ions incident on tungsten. Initial electron states are assumed
to be a) proportional to $E^{1/2}$, b) according to the theory of
Manning and Chodorow (Ref. 13), c) according to the experimental
work of Claus and Ulmer (Ref. 14). All final states are assumed
to be proportional to $E^{1/2}$.

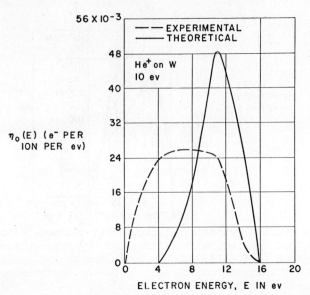

Fig. 7 Experimental and theoretical secondary electron distributions
ejected by 10 eV He$^+$ ions incident on tungsten. The experimental
curve is according to Hagstrum (Ref. 10), and the theoretical
curve is from Fig. 6b.

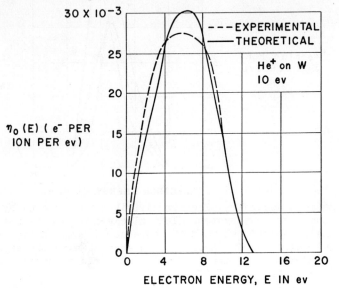

Fig. 8 Energy of interaction of helium atoms and ions with a tungsten surface
according to Hagstrum.

SURFACE STATES IN III-V SEMICONDUCTORS

R. O. Jones

Laboratory of Atomic and Solid State Physics
Cornell University
Ithaca, New York

I. Introduction

The energy bands and density of surface states are fundamental properties of semiconductor surfaces and have attracted considerable theoretical and experimental attention. Heteropolar semiconductors, in particular, are currently receiving extensive study by a variety of experimental techniques and much interesting information concerning both these properties is now available. Ambiguities in the interpretation of the data arise, however, not only from experimental uncertainties such as impurities and defects, but from difficulties in constructing realistic models for the calculation of surface properties. A number of qualitative estimates of surface state energies of III-V semiconductors have been given in the past,[1-4] but the nature of the potentials used makes direct comparison with experimental data impracticable.

Calculations presented here provide approximate energy bands for localized states at the cleavage plane of GaAs, InP and ZnSe. The model of the interface is that due to Shockley,[5] in which the bulk crystal potential is unperturbed up to the plane midway between two atomic layers, where it changes abruptly to the vacuum level determined from photoemission experiments.[6] For a given parallel component of the crystal momentum, which remains a good quantum number, localized states may be calculated by matching appropriate vacuum states to eigenfunctions of the bulk Hamiltonian with <u>complex</u> <u>k</u>-vectors normal to the surface plane. A calculation of electronic energy bands for complex crystal momentum vectors is therefore a necessary preliminary to a calculation of surface states. The empirical pseudopotential method has proved successful in describing the band structure of a number of semiconductors[7] and is the basis of the present calculation.[8] Recent work has shown that a simple form of this method accounts for the main features of covalency in diamond and zinc blende crystals.[9,10] These simplifications are adopted here and are described in the following paragraphs. The results of the matching calculation are then presented, together with a short discussion of relevant experimental information.

II. Pseudopotential Method and Complex Band Structure

For a valence electron in a crystal, the strongly attractive core potential is largely cancelled by a repulsive potential due to the requirement that the valence state be orthogonal to the states in the core. The weak effective potential which results

can be Fourier analyzed and written as the product of a structure

factor $S(g)$ and a pseudopotential form factor V_g. The potential

is conveniently separated into symmetric and antisymmetric parts;

$$V(r) = \sum_g [S^s(g)V_g^s + iS^a(g)V_g^a]e^{-ig \cdot r} \tag{1}$$

where the structure factors $S^s(g)$, $S^a(g)$ have particularly simple

forms if the origin is midway between the two atoms in the unit

cell.[7] The determination of form factors from experimental

information has been carried out by Cohen and Bergstresser,[7] and

it is possible to carry out calculations similar to theirs for

complex momenta. Such calculations have, in fact, been carried

out for specific symmetry directions in silicon[11-13] and for

GaAs.[14] However, since the calculation of surface states requires

such a calculation to be performed for all values of the component

of k parallel to the surface, it is necessary to effect some simpli-

fication and generalization.

It was noted by Heine[15] that the energy bands in diamond-

type semiconductors could be described in the neighborhood of the

point X by truncating the secular determinant to second order and

taking the influence of other waves into account by perturbation

theory. Subsequently,[10] it has been found that second order

Brillouin-Wigner perturbation theory provides a better description
by predicting, with good accuracy, the position of the conduction
band minimum in diamond-type semiconductors and the form of the
band structure in a substantial fraction of the Brillouin zone.
At points in the band structure where $\nabla_k E(\underline{k})$ vanishes, real energy
bands will occur with complex momentum values. Since these are
of interest in the present context, it has been found useful to
visualize the energy bands in the nearly free electron model using
the Jones zone.[16] The Jones zone for the diamond structure,
shown in figure 1, contains four valence electrons per atom and
is particularly convenient for the (110) face of diamond and zinc
blende crystals, since for both types, the lines Σ, Δ are points
of zero slope for $\underline{k}$ in the direction (110). In the calculations
which follow, complex band structures for all $k_{||}$ will be described
in terms of a two-band model in essentially the same way as the
Phillips theory of covalency assumes at the outset the existence
of an energy gap and two-component bonding and antibonding wave
functions of the form $\alpha|k> \pm \beta|k - 2k_F>$.

The energy band structure for complex $\underline{k}$ takes a particularly
simple form if the two-band approach is adopted.[17] If the wave
vectors $\underline{k}$, $\underline{k}-\underline{g}$ are connected by a Fourier component of the potential V_g, the band structure has the form shown in figure 2. The
energy may be followed as a monotonically increasing function if

the momentum is allowed to take the complex values shown, and the wave function for increasing energy is

$$\psi \sim e^{qx} \cos[\frac{gx}{2} + \theta]$$

$$\theta = \frac{1}{2} \sin^{-1}[\frac{-qg}{V_g}] \qquad , \qquad (2)$$

if the crystal is taken to occupy the negative half-space. If V_g is negative, it will be possible to match this wave function and its normal derivative to an exponentially damped vacuum state. It may be remarked in passing that the characteristic bonding valence states in covalent crystals arise from such a negative Fourier component.

III. Calculation of Surface States

In applying the two-band model to actual crystals, it is necessary to know the real band structure and, in particular, the energy splittings in the neighborhood of the Fermi level. Furthermore, the leading terms in the expansion of the wave function and the size of the loops in the complex band structure must also be known. For the crystals under discussion, the necessary energy levels have been taken from Cohen and Bergstresser.[7] The wave functions determined in an earlier calculation of surface states in silicon[18] have been used in the calculations for germanium

presented here.   The earlier work verified that the wave functions
in the neighborhood of the band gap are dominated by states with
wave vectors on the surface of the Jones zone, and basis plane
waves for particular values of $k_{||}$ are chosen by symmetry from
such wave vectors.

The band structures of heteropolar semiconductors may be
described by introducing an antisymmetric component into the
pseudopotential (equation (1)).   The compounds discussed here have
been chosen because the symmetric part of the potential is, in
each case, the same as that in germanium, and it is therefore
possible to assess the effect of changes in the antisymmetric part.
In the simple two-band model described here, the asymmetry causes
the phase of the wave function to shift by an amount proportional
to the antisymmetric component V_g^a.

In performing the matching calculations, a number of alternative approaches are possible.   The introduction of V_g^a causes the
site of the covalent bonding charge[9] to shift towards the cation.
Moving the matching plane to this site and using the homopolar
wave functions should lead to the heteropolar surface state
energies.   In a similar spirit, we may attribute the increase in
energy splittings near the Fermi level to the inclusion of a single
purely imaginary component in V_g, and from this determine the shift

in the phase of the wave function and the new surface state

energies.   Neither of these approaches is applicable.   The effect

of the antisymmetric potential on the valence wave function is

strongly $\underline{k}$-dependent and is weakest in the neighborhood of X, the

point of greatest interest in these calculations.   At X, the main

effect is to split the conduction band degeneracy $X_1 \rightarrow X_1 + X_3$.

The effective matrix elements which cause the band gaps at X, [10,15]

although dependent on the V_g^a, remain real and the phases of the

wave function at the extremities of the gap are unchanged, although

the gap increases.   As a result the effect of the antisymmetric

potential is less pronounced than the other approaches would suggest.

The method adopted here has been to calculate wave functions within

the framework of the Jones zone model, to take the Shockley

matching plane and the additional approximation that, if the

matching cannot be performed at all points in the surface plane,

it should be carried out at points in the plane where the surface

state wave function has maximum modulus.

The glide plane present in (110) surfaces of crystals with

the diamond structure causes a degeneracy along the edge $\bar{Z}$ of the

two-dimensional Brillouin zone[18] (figure 3).   For zinc blende

crystals, the degeneracy is split and the separation should increase

with increasing asymmetry.   This is shown to be the case by the

results of the matching calculation shown in figure 4.   The following points should be noted:

(1) Spin-orbit coupling has been neglected throughout.   The
energy splittings in the experimental data consulted by Cohen and
Bergstresser were corrected to remove this effect.   Spin-orbit
effects are important in all the crystals discussed here and the
surface state bands will be modified by its inclusion.   The dominant waves used in this calculation are related to states at X
and, since spin-orbit splittings are zero at that point in diamond,
the changes should not be great in germanium.   In the other
crystals, spin-orbit coupling should increase the separation
between the bands.

(2) A more complete calculation should determine the potential in the neighborhood of the surface in a self-consistent
fashion,[19] an effect which is likely to be particularly important in strongly heteropolar crystals.[9]   The calculated imaginary
components of $k_\perp$ indicate, however, that the states are less
localized than the region of expected disorder, particularly for
the III-V compounds and germanium, thus providing intuitive
support for the model.

(3) Following calculations of Shockley[5] and Goodwin,[17]
there should be a total of one surface state per surface atom and,

for a charge neutral surface, this state will be detached from the valence band. The lower band of the surface state spectrum should be occupied and the upper band unoccupied. Fermi level pinning should occur in the lower half of the band gap for germanium and, less markedly, at higher energies in the intermetallic compounds.

Earlier model calculations of surface states in alternant crystals include those of Aerts[1] and Phariseau,[2] which are one dimensional Kronig-Penney in character, and Amos and Davison,[3] who adopted a one dimensional tight-binding model. Koutecky and Tomasek[4] have performed tight-binding calculations for the (100) and (111) faces. The work of Levine and Mark[20] was the first to give surface state parameters for II-VI compounds and adopted a purely electrostatic model for the ionic interactions.

IV. Survey of Experiment

Experiments on semiconductor surfaces are extensive and there has been a number of recent reviews.[21] The situation is still unclear, however, and the continuing interest in heteropolar compounds, in particular, is very welcome. It appears to be possible to separate clean surfaces into two categories; those in which the Fermi level is stabilized at the surface by a high density of surface states (Si, Ge) and more ionic crystals in which the surface state density in the energy gap is much lower. Photoemission experiments on III-V semiconductors generally support this assign-ment,[22] though recent measurements[23] indicate that emission from occupied surface and/or impurity states may be observed in InP and that a high density of surface states may exist on GaAs.[24] From an experimental viewpoint, two predictions of the present calculations should be checked.

(1) The calculated intrinsic surface state spectra of GaAs and InP are similar, a similarity which should be manifest in experimental data if very pure crystals are used.

(2) Some of the states below the bulk energy gap will overlap states which have the same $k_{||}$ and the same symmetry. Resonant states will result. However, a fraction of the states outside the bulk forbidden gap will be genuinely localized states and should contribute to photoemission if, as suggested by Fischer, emission from surface states occurs.

(3) Under appropriate conditions, interband transitions could be excited optically.

The nature of the approximations made in this work mean that quantitative results should not be expected. Two trends are, nevertheless, clear. Firstly, surface states do exist on III-V compounds with a gap in the spectrum which is proportional to the antisymmetric component of the potential. In addition, with increasing asymmetry, the Fermi level at an intrinsic surface should rise towards the conduction band.

I am grateful to Dr. V. Heine, Dr. T. K. Bergstresser and Dr. T. E. Fischer for stimulating discussions.

Note:
        An error has been detected in the calculations reported here. The qualitative nature of the results and the discussion are unchanged, as are the calculations for the case of germanium. Corrected calculations of the band splittings at $\overline{Z}$, in which the self-consistency requirement is more carefully considered, will be published elsewhere.

References

Work supported by the Advanced Research Projects Agency through the Materials Science Center at Cornell University, Report #970.

1. E. Aerts, Physica 26, 1057 (1960).

2. P. Phariseau, Physica 30, 608 (1964).

3. A. T. Amos and S. G. Davison, Physica 30, 905 (1964).

4. J. Koutecký and M. Tomášek, Surface Science 3, 333 (1965);
    M. Tomášek, Czech. J. Phys. B16, 828 (1966).

5. W. Shockley, Phys. Rev. 56, 317 (1939).

6. Ge: G. W. Gobeli and F. G. Allen, Surface Science 2, 402 (1964);
    GaAs: G. W. Gobeli and F. G. Allen, Phys. Rev. 137, A245 (1965);
    InP: T. E. Fischer, Phys. Rev. 142, A519 (1966);
    ZnSe: R. K. Swank, Phys. Rev. 153, 844 (1967). A survey of the photoemission technique as applied to semiconductors is in an article by T. E. Fischer (to be published).

7. M. L. Cohen and T. K. Bergstresser, Phys. Rev. 141, 789 (1966); T. K. Bergstresser and M. L. Cohen, Phys. Rev. 164, 1069 (1967).

8. The first application of pseudopotential theory to surface state calculations is that of E. Antončik, J. Phys. Chem. Solids 21, 137 (1961).

9.   J. C. Phillips, Phys. Rev. 166, 832 (1968); 168, 905 (1968).

10.   V. Heine and R. O. Jones, to be published.

11.   R. O. Jones, Proc. Phys. Soc. (London) 89, 443 (1966).

12.   I. Bartos, Czech. J. Phys. B17, 481 (1967).

13.   C. M. Chaves, N. Majlis and M. Cardona, Solid State Comm. 4, 271 (1966).

14.   C. M. Chaves, N. Majlis and M. Cardona, Solid State Comm. 4, 631 (1966).

15.   V. Heine, in "Phase Stability in Metals and Alloys," eds. P. Rudman, J. Stringer and R. I. Jaffee, (McGraw-Hill, New York, 1967).

16.   H. Jones, Proc. Roy. Soc. (London) A144, 225 (1934).

17.   E. T. Goodwin, Proc. Camb. Phil. Soc. 35, 205 (1939).

18.   R. O. Jones, Phys. Rev. Letters 20, 992 (1968).

19.   A. J. Bennett and C. B. Duke, (this volume).

20.   J. D. Levine and P. Mark, Phys. Rev. 144, 751 (1966).

21.   See, for example, D. R. Frankl, "Electrical Properties of Semiconductor Surfaces," (Pergamon, Oxford, 1967).

22.   J. van Laar and J. J. Scheer, Surface Science 8, 342 (1967).

23.   T. E. Fischer, to be published.

24.   T. E. Fischer, to be published.

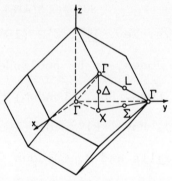

Fig. 1. The Jones zone for the diamond structure, showing the symmetry of a number of points on the surface.

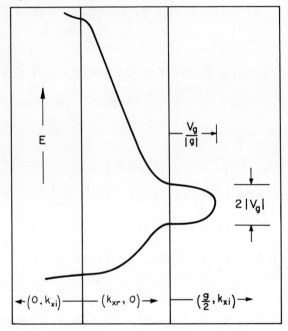

Fig. 2. Complex band structure--two band, nearly free electron model.

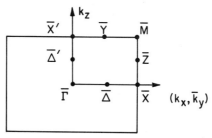

Fig. 3. Two dimensional Brillouin zone for (110) surface of diamond.

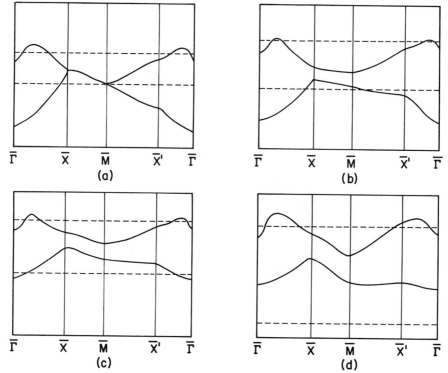

Fig. 4. Surface state energy bands--ideal (110) face of (a) Ge, (b) GaAs, (c) InP, and (d) ZnSe. The dotted lines represent the bulk energy gap, respectively 0.9 eV, 1.4 eV, 1.6 eV, and 2.9 eV.

LOCALIZED ELECTRONIC STATES ON CRYSTAL SURFACES

M. Tomasek

Institute of Physical Chemistry
Czechoslovak Academy of Sciences
Prague, Czechoslovakia

1. INTRODUCTION

In this paper, we propose to give a detailed discussion of general properties of a method which was devised for the calculation of localized electronic
states on crystal surfaces and first used for this purpose in 1964.[1] One-
electron theory of solids and the underlying approximations are utilized without
any comment. Under the name of localized electron states both surface and
chemisorption electronic states of Shockley and Tamm type are understood.

The calculation of energy bands of finite crystals is a rather difficult
task. This is because the limitation of the crystal causes a very strong
perturbation to the original infinite crystal. Non-perturbational methods
represent an indispensable working tool in this case. Out of these methods, the
one-electron Green function method seems to be extremely useful. In the rest
of this section, main features of the latter are reviewed in a concise way to
provide the basis for further discussion on symmetry properties and complex
band structure, given in section 2. Section 3 contains the application of the
theory to the problem of localized states on the (111) surface of diamond-like
semiconductors with a realistic potential.

Let us imagine an ideal system formed by two non-interacting subsystems,
the first being the infinite "cyclic" crystal with Born-v.Karman periodic
boundary conditions, the other a collection of atoms or molecules. In the
abstract space Π_o of this system, the respective Hamiltonian operator $\mathcal{H}_o$ is
defined

$$\mathcal{H}_o \, | \, \varepsilon_i > = \varepsilon_i | \varepsilon_i > \tag{1}$$

A real finite crystal with chemisorbed species is obtained from this ideal system

1) By removing a certain part of the infinite crystal, in which way at least two surface planes are formed. The removed part of the abstract space is denoted as $\tilde{H}_o$ and the corresponding Hamiltonian as $\tilde{\mathcal{H}}_o$.

2) By switching on the operator $\mathcal{H}_1$ which includes both perturbations caused by surface formation and interactions between the surface and chemisorbed species. Let us denote as H_1 that part of the abstract space where this operator is defined.

The Hamiltonian $\mathcal{H}$ of this real system fulfills the equation

$$(\varepsilon - \mathcal{H}) | \varepsilon > = (\varepsilon - \mathcal{H}_o - \mathcal{V}) | \varepsilon > = 0 \tag{2}$$

with

$$\mathcal{V} = \mathcal{H}_1 - \tilde{\mathcal{H}}_o \tag{3}$$

The space H where $\mathcal{H}$ is defined is a subspace of H_o. By expanding $| \varepsilon >$ in the complete set of $| \varepsilon_i >$

$$| \varepsilon > = \sum_i a_i | \varepsilon_i > \tag{4}$$

and using (2) one has

$$(\varepsilon - \mathcal{H}_o) | \varepsilon > \equiv \sum_i a_i (\varepsilon - \varepsilon_i) | \varepsilon_i > = \mathcal{V} | \varepsilon > \tag{5}$$

whence

$$a_i = \frac{< \varepsilon_i | \mathcal{V} | \varepsilon >}{\varepsilon - \varepsilon_i} \tag{6}$$

(4) and (5) give

$$|\mathcal{E}> = G\gamma|\mathcal{E}>$$ (7)

G is the one-electron Green function of our abstract space H_0. Because (7) is also obtained by multiplying (5) with the operator $(\mathcal{E} - \mathcal{H}_0)^{-1}$ from the left, one has

$$G = (\mathcal{E} - \mathcal{H}_0)^{-1} = \sum_i \frac{|\mathcal{E}_i \times \mathcal{E}_i|}{\mathcal{E} - \mathcal{E}_i}$$ (8)

An arbitrary basis (representation) could be used to solve (7) directly. However, an artifice described below has proved to be most useful. Though approximate, it has the virtue of making possible a very simple formulation of boundary conditions both on the pure surface of the crystal and on the surface covered by the chemisorbate. This would generally be the largest difficulty in solving (7) directly.

Let us construct an identity operator $\mathfrak{I}$, defined in the whole abstract space H_0, as a sum of three projection operators corresponding to three disjoint parts of H_0

$$\mathfrak{I} = \sum_{n \in H_0} |\phi_n><\phi_n| \equiv P_{\tilde{H}_0} + P_{H_1} + P_{H-H_1}$$ (9)

where e.g.

$$P_{\tilde{H}_0} = \sum_{n \in \tilde{H}_0} |\phi_n><\phi_n|$$

The $|\phi_n>$ represent mutually orthogonal "localized" functions. The main role of n is to number layers of either elementary cells of the crystal or atoms of the chemisorbate in the direction perpendicular to the crystal surface. Along planes parallel to the surface, $|\phi_n>$ are delocalized.

The artifice consists in supposing that[2,3]

1)
$$P_{\widetilde{H}_o} |\mathcal{E}> = 0$$

This seemingly trivial assumption has an important consequence, namely

$$P_{\widetilde{H}_o} \mathcal{H} = 0$$

and

$$P_{\widetilde{H}_o} \mathcal{V} = - P_{\widetilde{H}_o} \mathcal{H}_o \tag{10}$$

This also implies that for $n\epsilon\widetilde{H}_o$

$$<\phi_n|\mathcal{E}> = 0 \tag{10'}$$

must hold.

2)
$$\mathcal{V} P_{H-H_1} = P_{H-H_1} \mathcal{V} = 0$$

which means that $\mathcal{V}$ has non-zero matrix elements only there where it is de-fined, i.e., in the part $H_1 + \widetilde{H}_o$ of the whole abstract space H_o.

By putting (9) into (7), applying assumptions 1,2) above and using (10) one has

$$|\mathcal{E}> = G \mathcal{V} \mathcal{S}|\mathcal{E}> = G(P_{H_1} \mathcal{V} P_{H_1} + P_{\widetilde{H}_o} \mathcal{V} P_{H_1})|\mathcal{E}> =$$

$$= G(P_{H_1} \mathcal{V} P_{H_1} - P_{\widetilde{H}_o} \mathcal{H}_o P_{H_1})|\mathcal{E}> \tag{11}$$

Multiplication of (11) by $<\phi_s|$ from the left gives finally[2,3]

$$<\phi_s| \mathcal{E}> = \sum_{n\epsilon H_1}\left\{ \sum_{n\epsilon H_1} <\phi_s|G|\phi_r><\phi_r|\mathcal{V}|\phi_n> \right.$$

$$\left. - \sum_{r\epsilon \widetilde{H}_o} <\phi_s|G|\phi_r><\phi_r|\mathcal{H}_o|\phi_n| > \right\} <\phi_n| \mathcal{E}> \tag{12}$$

Equation (12) determines the expansion coefficients $\langle \phi_n | \mathcal{E} \rangle$ of the exact wave function in terms of $| \phi_n \rangle$. It is a system of equations which is non-linear in energy. From its solvability condition, the energy $\mathcal{E}$ is obtained. Note that symmetry considerations have not yet been used.

2.   SURFACE SPACE GROUPS AND COMPLEX BAND STRUCTURE

In this section, we shall take symmetry arguments into consideration and we shall show how to use them to simplify the problem of solving (12). The abstract space H_o contains two subsystems: the crystal and the chemisorbate. Their symmetries are usually different but the advantage of the present method is that they can be treated separately. To explain the method clearly, we shall limit ourselves to the crystal subsystem. Mutatis mutandis, the case of chemisorption can easily be obtained. When speaking about H_o below we mean the infinite crystal with Born-v.Kármán boundary conditions which is charac-terized by a certain symmetry group, called space group. The crystal surface formation generally reduces the number of symmetry elements of this group; the remaining symmetry operation form a subgroup which will be called the surface space group or shortly the surface group.

Contrary to the infinite crystal space group, the surface group contains only the two-dimensional translation group as a subgroup. The non-translational part of the surface group is also smaller: there are rotations about axes normal to the surface, mirror planes or glide planes perpendicular to the surface. The latter planes have glide directions parallel to the surface; there are no screw axes. The two-dimensional translation group consists of elementary translations parallel to the crystal surface only. It determines both a two-dimensional reciprocal space spanned by non-collinear wave vector components $\vec{k}_2$ and $\vec{k}_3$ lying in the plane of the surface and a two-dimensional Brillouin zone called surface Brillouin zone (SBZ).

Let us denote S a symmetry operation from the surface group, which can naturally be also an arbitrary combination of the operations mentioned above. All the S commute with the Hamiltonian $\mathcal{H}$ from (2)

$$[s,\mathcal{H}] = s\mathcal{H}-\mathcal{H}s = 0 \tag{13}$$

and it follows that irreducible representations (1Rs) of the surface group can serve as a tool for the classification of electronic states of limited crystals including localized states on their surfaces.

Any S commutes also with $\mathcal{U}$ and $\mathcal{H}_o$

$$[s,\mathcal{U}] = 0 \tag{14}$$

$$[s,\mathcal{H}_o] = 0 \tag{15}$$

From (8) and (15) one has

$$[s,G] = 0 \tag{16}$$

Actually, G is a function of $\mathcal{H}_o$ and (16) follows immediately if (15) is true. Another way to make sure that (16) holds is to realize that degenerate states $|\mathcal{E}_i>$ of a given energy level $\mathcal{E}_i$ mutually transform under the action of any symmetry operator S.

To be able to classify characteristic states $|\mathcal{E}>$ of the equation (2), we have to know how to construct IRs of surface groups. The procedure does not differ from the usual one for space groups of infinite crystals[4] except that now we work with two-dimensional space groups. The notion of SBZ is of fundamental importance here. Each point of SBZ k-space is either a "general" or a "special" point. The latter are either high symmetry points of SBZ or lie on one of the SBZ symmetry lines. For a "general" point, group theoretical arguments do not help to simplify the problem of solving (12). We shall leave them out from further considerations. An opposite situation occurs with "special" points. Here, a group $\mathfrak{S}(k_2,k_3)$ can be formed, which is called the little group of the wave vector. It does not contain the two-dimensional subgroup of the pure translations. It comprises all those symmetry operations of the surface group which either leave a particular SBZ "special" point (k_2, k_3) invariant or send it into an equivalent point when being applied to our two-dimensional k-space. If the surface group is a non-symmorphic group, the construction of $\Im(k_2, k_3)$ deserves further specification. For the lack of space, however, this has to be left to the reader.[4]

The theory shows[4] that for "special" point, IRs of a surface group are obtained from the IRs of $\Im(k_2, k_3)$ in a unique way. Therefore, the classification of electronic states of (2) is completely determined by the IRs of the little group of the wave vector $\Im(k_2, k_3)$. The latter are also sufficient to perform symmetrization of basis functions $|\phi_n\rangle$ of the special representation chosen in (9). Before doing it, however, one has to be sure that the $|\phi_n\rangle$ are Bloch functions. If they are not, it is better to symmetrize them by using IRs of surface groups directly.

Suppose that the $|\phi_n\rangle$ have been symmetrized to transform according to an IR of either $\Im(k_2, k_3)$ or the surface group. As a consequence, great simplification results in (12). Because (14), (15), and (16) hold, matrix elements of the operators $\mathcal{U}$, $\mathcal{H}_0$ and G taken with respect to $|\phi_n\rangle$ belonging to different IRs are all zero. This leads to the splitting of the matrix of (12) into separate blocks, associated with different IRs. Hence, localized states on crystal surfaces can be treated separately for individual IRs.

We have not yet mentioned the wave vector component $\vec{k}_1$ which together with $\vec{k}_2$ and $\vec{k}_3$ spans the three-dimensional k-space of $\mathcal{H}_0$, i.e., of our infinite crystal. $\vec{k}_1$ is perpendicular to the crystal surface and consequently, its scalar value k_1 does not represent a good quantum number for the operator $\mathcal{H}$ of a finite crystal. However, k_1 necessarily enters our considerations, if we want to calculate Green function matrix elements of (12). To solve (12), a "special" point (k_2, k_3) has first to be fixed. This is natural, because matrix elements taken with respect to $|\phi_n\rangle$ belonging to the same IR of the surface group are the only ones which will remain in (12). However, if k_2, k_3 are fixed, the summation over the index i in (8) includes summation over individual energy bands j (j = 1, ...t) and allowed values of k_1 only.

As $\mathcal{E}_i$ from (1) is a periodic function in k-space, it can be Fourier analyzed with respect to k_1 and therefore expressed in terms of powers of the quantity

$$z = \exp(i\vec{k}_1 \cdot \vec{a}_1) = \exp(i\,\xi) \tag{17}$$

where $\vec{a}_1$ represents the elementary translation of the direct lattice associated with $\vec{k}_1$. Hence, by changing in (8) the summation over k_1 into integration along a unit circle in the complex z-plane, one obtains for the Green function matrix element

$$<\phi_s|G|\phi_r> = \frac{1}{2\pi i} \oint F_{sr}(z,k_2,k_3,\mathcal{E})dz \tag{18}$$

where

$$F_{sr}(z,k_2,k_3,\mathcal{E}) = \sum_j \frac{b_{js}(z,k_2,k_3)\bar{b}_{jr}(z,k_2,k_3)}{[\mathcal{E}- \mathcal{E}_j(z,k_2,k_3)]\,z} \tag{19}$$

and

$$b_{jn}(z,k_2,k_3) = <\phi_n|\,\mathcal{E}_j(z,k_2,k_3)> \tag{20}$$

Following Cauchy's residual theorem, (18) is easily evaluated as a sum of residua of (19) in those points z_p inside the unit circle of the z-plane, in which (19) has its poles. The poles z_p of F_{sr} in their turn are mainly determined by the roots of the equations (j = 1, ...t)

$$\mathcal{E}- \mathcal{E}_j(z_p,k_2,k_3) = 0 \tag{21}$$

It is interesting to see that (21) determine at the same time the so-called complex band structure,[5-8] to which each of the z_p contributes a single branch. Real and imaginary parts of ξ_p from (17) are given by the expressions

$$\text{Re } \xi_p = \omega_p$$

$$\text{Im } \xi_p = - \ln |z_p| \tag{22}$$

with the notation $z_p = |z_p| \exp(i \omega_p)$ utilized. Because of time-reversal symmetry and complex conjugation, any particular ξ_p fulfills the following relations

$$\mathcal{E}_j(\xi_p) = \mathcal{E}_j(-\xi_p) = \mathcal{E}_j(\bar{\xi}_p) = \mathcal{E}_j(-\bar{\xi}_p)$$

where k_2, k_3 have been omitted and $\bar{\xi}_p$ given the meaning of the complex conjugate of ξ_p. Residua of (18) are functions of the z_p which, according to (21), depend on $\mathcal{E}$, k_2 and k_3. The same dependence appears in (12) when (18) is inserted into it. An example may be found in section 3, which illustrates all the above considerations and shows how to solve (12) in a particular case.

It is worth notice, that out of the complete set of z_p, all the z_p fulfilling the condition

$$|z_p| < 1 \tag{23}$$

are used to calculate (18). This is an advantage over direct methods of calculating surface states,[9-12] in which complete sets of decaying wave functions compatible with the complex band structure can hardly be included in such a direct way. Apart from that, the use of direct methods runs into serious difficulties in case of study of chemisorption. The Green function method is free from this disadvantage.

3. THE (111) SURFACE OF CRYSTALS WITH DIAMOND-LIKE STRUCTURE

In this section, localized electronic states on the (111) surface of diamond-like crystals with a realistic potential will be examined. The results will be illustrated numerically on diamond, silicon and germanium. The surface group of the diamond-like crystal delimited by a (111) surface is a symmorphic group and little groups $\mathfrak{S}(k_2, k_3)$ of "special" points in SBZ are therefore point groups. The "special" point $k_2,k_3 = 0$ will be chosen here because in that case largest simplifications occur. The point $k_2,k_3 = 0$ forms the center of the SBZ and the corresponding little group $\mathfrak{I}(0,0)$ is the point group C_{3v}. This group has three IRs of which only the one-dimensional IR denoted as Λ_1 and the two-dimensional IR denoted as Λ_3 intervene in physically interesting parts of the energy bands of diamond-like semiconductors. Localized states of Λ_3 character will be treated in detail in this section, whereas for Λ_1 states, only the results will be quoted.

Let us first describe the crystal model utilized as well as the model of the pure or contaminated surface. The representation of localized functions is used with each crystal atom carrying four localized functions - sp^3 hybrids (see Fig. 1). Any elementary cell contains two atoms. The function $|\phi_n>$ from (9) is supposed now to be a Bloch function formed as the linear combination of localized functions $|X_\ell(\vec{r}-\vec{r}_m)>$ of a particular type ℓ from the layer m which has one elementary cell thickness. n stands for two indices ℓ and m. Choosing Fig. 1 as an example, ℓ can take the values $\ell = 1,...4$, $4',...1'$, thus exhausting all eight possibilities in the first (m = 1) layer of elementary cells of the crystal. It is clear that $|\phi_{\ell m}>$ is again a "two-dimensional" Bloch function parallel to the plane of the surface and that it depends on k_2, k_3.

For the crystal model suggested, $|\mathcal{E}_i>$ in (1) is written as

$$| \mathcal{E}_{jk_1} > = \sum_{\ell=1}^{8} \sum_{m} |\phi_{\ell m}\rangle\langle \phi_{\ell m}| \mathcal{E}_{jk_1} > \qquad (24)$$

where (cf.(20))

$$< \phi_{\ell m}|\mathcal{E}_{jk_1} > = C_{j\ell}\, e^{i\xi \cdot m} \qquad (25)$$

$|\mathcal{E}_{jk_1}>$ as well as $C_{j\ell}$ are functions of k_2,k_3. The formulation of the secular equation of $\mathcal{H}_o$ in terms of $|X_{\ell m}>$ takes into account all interactions between

$|X_{\ell m}>$ up to those of second-nearest neighboring atoms. The resulting thir-

teen interaction parameters are listed in Table I. They were obtained[13] by

Slater and Koster procedure[14] from data published by several authors. An example

will clarify the notation (cf. Fig. 1)

$$\gamma \equiv (4,4') = <X_{41}|\mathcal{H}_o|X_{4'1}>$$

Parameters in Table I are sufficient to treat Shockley surface states on pure

crystal surfaces. For Tamm surface states or chemisorption localized states,

additional parameters are needed. We shall introduce only the most rudimentary

ones. Namely, the coulomb integral $\rho = <X_{11}|\mathcal{V}|X_{11}>$ accounting for the per-

turbation (3) in the region of the surface hybrid orbital $|X_{11}>$, the coulomb

integral δ of a chemisorbing atom A which is the measure of its electronega-

tivity of ionization potential and finally the resonance integral

$\sigma = <X_{oo}|\mathcal{V}|X_{11}>$ between $|X_{oo}>$, the localized function of the chemisorbed

atom and the surface hybrid orbital $|X_{11}>$. As is seen from below, in this

model of the surface, the treatment of Tamm and Shockley surface and chemi-

sorption states is meaningful for the Λ_1 IR, whereas for the Λ_3 IR, the model

provides no reasons to examine Tamm surface and chemisorption states at all.

Now, we shall put everywhere $k_2, k_3 = 0$ and symmetrize the $|\phi_{\ell m}>$ according

to $\mathfrak{S}$ (0,0). For each particular m, the eight functions $|\phi_{\ell m}>$ from (24) are

combined to new functions, $|\phi_{\ell m}^{\Lambda i}>$ (i = 1,3), the first four forming the bases

of IR Λ_1, the remaining four transforming according to IR Λ_3. The latter

functions (ℓ=I,II,III,IV) represent two equivalent bases (ℓ=I,III and ℓ=I,IV)

because of the two-fold degeneracy of IR Λ_3 in diamond-like semiconductors.

As an example, the first layer of elementary cells of the crystal is chosen

by putting m = 1. Using the notation of Fig. 1, one has

$$|\phi_{I,1}^{\Lambda_3} > = \sqrt{\frac{2}{3}} \left[|\phi_{31}> - \frac{1}{2} \left(|\phi_{21}> + |\phi_{41}> \right) \right]$$

$$|\phi_{II,1}^{\Lambda_3}> = \sqrt{\frac{2}{3}} \left[|\phi_{3'1}> - \frac{1}{2} \left(|\phi_{2'1}> + |\phi_{4'1}> \right) \right]$$

(26)

$$|\phi_{III,1}^{\Lambda_3}> = \sqrt{\frac{1}{2}} \left(|\phi_{41}> - |\phi_{21}> \right)$$

$$|\phi_{IV,1}^{\Lambda_3} > = \sqrt{\frac{1}{2}} \left(|\phi_{4'1}> - |\phi_{2'1}> \right)$$

(27)

Naturally, the form of (26) and (27) is true for an arbitrary m $\geq$ 1.

The functions (26) do not combine with those of (27) in matrix elements of $\mathcal{U}$, $\mathcal{H}_o$ and G. This leads to an additional splitting of the matrix (12) of localized states into separate blocks associated with (26) or (27), respectively. The same is true for volume states: By expressing (24) in terms of (26), (27), one finds the splitting of the characteristic problem of (1) in two two-dimensional problems with the same secular determinant

$$\Delta = \begin{vmatrix} A + p + q(e^{i\xi} + e^{i\xi}) & r + se^{-i\xi} \\ r + se^{i\xi} & A + p + q(e^{i\xi} + e^{-i\xi}) \end{vmatrix}$$

(28)

which again show the two-fold degeneracy of Λ_3. Equation (28) describes a typical two-band system[11]. To derive it, (25) has been used and the following notation introduced

$$A = \alpha_o - \mathcal{E}$$

$$p = -\gamma' - \kappa - \omega + 2(\nu + 2\mu - \nu - \phi)$$

$$q = \mu + 2\nu - \nu - \phi - \lambda$$

$$r = \gamma - \beta + 2(\mathcal{E} - \alpha)$$

$$s = \mathcal{E} - \beta$$

(29)

Let us examine Shockley surface states associated with (26). The results for Shockley states of (27) are the same. Equation (12) gives

$$< \phi_{\ell m}^{\Lambda_3} |\mathcal{E} > = - \left[qG_{\ell m; I, 0} + sG_{\ell m; II, 0} \right] < \phi_{I,1}^{\Lambda_3} |\mathcal{E} >$$
$$-qG_{\ell m; II, 0} < \phi_{II,1}^{\Lambda_3} |\mathcal{E} > \tag{30}$$

Because $m\tilde{H}_0$ for $m = 0$, we can use (10') as the boundary condition of our problem and put

$$< \phi_{\ell 0}^{\Lambda_3} |\mathcal{E} > = 0, \qquad \ell = I, II$$

in (30). The resulting system of equations

$$\left[qG_{Io; Io} + sG_{Io; IIo} \right] < \phi_{I,1}^{\Lambda_3} |\mathcal{E} > + qG_{Io; IIo} < \phi_{III,1}^{\Lambda_3} |\mathcal{E} > = 0$$
$$\left[qG_{IIo; Io} + sG_{IIo; IIo} \right] < \phi_{I,1}^{\Lambda_3} |\mathcal{E} > + qG_{IIo; IIo} < \phi_{II,1}^{\Lambda_3} |\mathcal{E} > = 0 \tag{31}$$

has the solvability condition

$$G_{Io; Io} \cdot G_{IIo; IIo} = G_{Io; IIo} \cdot G_{IIo; Io} \tag{32}$$

where the notation (cf. (18))

$$< \phi_{\ell m}^{\Lambda_3} | G | \phi_{\ell'm'}^{\Lambda_3} > = G_{\ell m; \ell'm'} \tag{33}$$

has been introduced.

To calculate (33), we use in (18) equation (25) together with a convenient relation[3], which is true for any point (k_1, k_2, k_3) of the k-space

$$\sum_j \frac{C_{j\ell} \bar{C}_{j\ell'}}{\mathcal{E} - \mathcal{E}_j} = (-1)^{\ell + \ell' + 1} \frac{\Delta_{\ell'\ell}}{\Delta} \tag{34}$$

$\Delta_{\ell'\ell}$ is a subdeterminant produced by omitting the ℓ'th row and the ℓth column from the determinant Δ, defined by (28). We get the following results

$$G_{Io;Io} = G_{IIo;IIo} = (A + p) \phi_o + q(\phi_1 + \phi_{-1})$$

$$G_{Io;IIo} = r\phi_o + s\phi_1 \qquad (35)$$

$$G_{IIo;Io} = r\phi_o + s\phi_{-1}$$

where ϕ_k is given by equation (22) and also (3), (10), (11) and (13) to (16) of Ref. 15. The quantities w, $\mathcal{E}_1$, $\mathcal{E}_2$ involved there are now expressed as

$$w = (A + p)^2 + 2q^2 - r^2 - s^2$$

$$\mathcal{E}_1 = rs - 2q(A + p) \qquad (36)$$

$$\mathcal{E}_2 = -q^2$$

By inserting (35) into (32), expressing the ϕ_k by means of the poles z_p ($p = 1, \ldots 4$) according to (22) of Ref. 15, carrying out some manipulations and exploiting the fact that the poles satisfy (13) of the mentioned paper, we have

$$s^2[z_1 z_4 + z_2 z_3 - 2] - \mathcal{E}_2 [z_3 + z_4 - (z_1 + z_2)]^2 = 0 \qquad (37)$$

The poles z_p given by (15) of Ref. 15 determine directly the complex band structure of our two-band system (28) when (22) is used. Substitution of (15) and (16) of Ref. 15 into (37) gives the final relation

$$(w + 2\mathcal{E}_2) \left\{ 1 + 8 \mathcal{E}_2 s^{-2} - \left[\frac{(w-2\mathcal{E}_2)^2 - 4\mathcal{E}_1^2}{(w + 2\mathcal{E}_2)^2} \right]^{1/2} \right\} \qquad (38)$$

$$+ 2(\mathcal{E}_1^2 s^{-2} + 2\mathcal{E}_2) = 0$$

Raising (38) to the second power, we get

$$-w = \frac{\varepsilon_1^2}{4\varepsilon_2 + s^2} + 2\varepsilon_2 + s^2$$

The energy of Shockley surface state is then found to be

$$\mathcal{E} = \alpha_o + p - 2qrs^{-1} \tag{39}$$

It is worth noticing that (39) is a double root. By putting it back into (38), the "existence condition" (38) of Shockley surface states is not fulfilled for any set of parameters from Table I as well as for those of Dresselhaus[16]. We find, in agreement with Ref. 1, that Λ_3 type Shockley surface states do not exist on the (111) surface of diamond, silicon and germanium for the center of SBZ. If they existed, Jahn-Teller distortion could take place[17] on this surface because of the two-fold degeneracy of IR Λ_3.

Now let us discuss the results of calculations of surface and chemisorption localized states corresponding to the Λ_1 IR. They are illustrated for diamond in Table II, where volume state parameters from the fourth column of Table I have been used in calculating several particular models of the pure and contaminated surface on the Minsk 22 electronic computer. Chemisorption localized states have been treated assuming that the layer of chemisorbed atoms had the same two-dimensional symmetry as the crystal surface.

First of all, a Shockley state exists on the pure crystal surface (δ, ρ, $\sigma = 0$). It is located slightly above the middle of the direct gap and corresponds to a double root. It has the right distribution of surface electron density. Most of the electron density is concentrated on the surface hybrid orbital 1 (see Fig. 1). Only a small fraction can be found on hybrids $\ell = 2, \ldots 4, 4'4, \ldots1'$. By perturbing the pure surface (δ, $\sigma = 0$, $\rho \neq 0$), the double root of the Shockley state splits. One surface state moves in the expected direction, the other one remains unchanged. Same situation occurs for chemisorption $(\delta, \rho, \sigma \neq 0)$. Here, bonding and antibonding states are formed by the interaction of the adsorbing atom energy level with one of the states only. The remaining one does not change. This inert state belongs to the unperturbed "complementary" surface which is always created by the finite crystal formation.

When surface parameters δ, ρ and σ are changed, the general behavior of various localized state energy levels is the same as described in Ref. 1. Two things, however, have to be mentioned in the light of the present paper. First, the existence of Shockley surface states on the (111) surface of diamond has been doubted in Ref. 1. Another parameterization corresponding to a realistic potential gives optimistic results. This shows again that LCAO method need not converge steadily when going from less to more sophisticated models. Second, the Shockley state found in Ref. 1 was a single, not a double root. This is because, contrary to Ref. 1, the "complementary" surface has same Miller indices as the one investigated in this paper.

Full description of the results for localized states of Λ_1 type together with those for the (100) surface will appear elsewhere[18].

REFERENCES

1. M. Tomásek, Surface Sci. 2, (1964), p. 8; Solid Surfaces, Ed. H. C. Gatos (North-Holland Publishing Co., Amsterdam 1964), p. 8.

2. J. Koutecký, Phys. Rev. 108, (1957), p. 13; Adv. Chem. Phys. 9, (1965) p. 85.

3. J. Koutecký, M. Tomásek, Phys. Rev. 120, (1960), p. 1212.

4. G. F. Koster, Solid State Physics, Eds. F. Seitz and D. Turnbull, 5, (1957), p. 173.

5. W. Kohn, Phys. Rev. 115, (1959), p. 809.

6. E. I. Blount, Solid State Physics, Eds. F. Seitz and D. Turnbull, 13, (1962) Appendix C.

7. V. Heine, Proc. Phys. Soc. (London) 81, (1963), p. 300; Surface Sci. 2, (1964), p. 1.

8. R. O. Jones, Proc. Phys. Soc. (London) 89, (1966), p. 443.

9. A. W. Maue, Z. Phys. 94, (1935), p. 717.

10. E. Antončik, J. Phys. Chem. Solids 21, (1961), p. 137.

11. C. M. Chaves, N. Majlis, M. Cardona, Solid State Commun. 4, (1966), p. 271.

12. R. O. Jones, Phys. Rev. Letters 20, (1968), p. 992.

13. M. Tomásek, Unpublished Results (1962).

14. J. C. Slater, G. F. Koster, Phys. Rev. 94, (1954), p. 1498.

15. M. Tomásek, J. Koutecký, Czechosl. J. Phys. B10, (1960) p. 268.

16. G. Dresselhaus, M. S. Dresselhaus, Phys. Rev. 160, (1967), p. 649.

17. P. Ducros, Surface Sci. 10, (1968) p. 295.

18. M. Tomásek, to be published.

TABLE I.

		diamond[14]	diamond[a]	silicon[b]	germanium[c]
α_o	(1,1)	-0.626[d]	-0.72531	-0.81180	-0.04387
γ'	(1,2)	-0.248	-0.17187	-0.07747	-0.03615
γ	(4,4')	-0.721975	-0.63742	-0.23518	-0.29633
α	(1,4')	0.0103	-0.04810	-0.04416	-0.01932
β	(1,2')	-0.1447	-0.09628	-0.02212	-0.02307
ε	(1,1')	0.076021	0.04122	-0.00490	0.02769
ν	(2,$\bar{1}$)	0.01025	0.02585	0.00121	0.00387
ω	(1,$\bar{4}$)	0.08325	-0.01138	0.01731	-0.04960
$\mathcal{H}$	(4,$\bar{1}$)	0.0415	0.06484	0.01731	-0.07786
φ	(1,$\bar{2}$)	0.03125	-0.01226	0.00121	0.01026
λ	(2,$\bar{3}$)	-0.07875	-0.02486	-0.01472	0.00126
μ	(1,$\bar{1}$)	0.03425	0.00720	0.02107	0.00001
ν	(2,$\bar{2}$)	-0.06275	-0.00111	-0.00290	0.00614

[a] Input data for Slater and Koster procedure[14] were taken from L. Kleinman, J. C. Phillips, Phys. Rev. 116, (1959), p. 880.

[b] Taken from F. Bassani, M. Yoshimine, Phys. Rev. 130, (1963), p. 20.

[c] Taken from F. Herman, S. Skillman, Proc. Int. Conf. Semicond. Phys., Prague 1960, p. 20.

[d] All energy values are given in Ry (rydberg) units.

TABLE II.

Values of surface parameters	$\delta, \rho, \sigma = 0$	$\delta, \sigma = 0; \rho \neq 0$				$\sigma \neq 0; \delta = -0.725; \rho = 0$	
		$\rho = 0.1$	$\rho = -0.1$	$\rho = 0.2$	$\rho = -0.2$	$\sigma = 0.1$	$\sigma = 0.2$
Bottom of the conduction band (Λ direction)	-0.35[a]	-0.35	-0.35	-0.35	-0.35	-0.35	-0.35
Antibonding chemisorption state	--	--	--	--	--	-0.519	-0.459
Tamm surface state	--	-0.488	-0.608	-0.431	-0.664	--	--
Shockley surface state	-0.548(2x)	-0.548(1x)	-0.548(1x)	-0.548(1x)	-0.548(1x)	-0.548(1x)	-0.548(1x)
Bonding chemisorption state	--	--	--	--	--	-0.751	-0.775
Top of the valence band (Λ direction)	-0.8	-0.8	-0.8	-0.8	-0.8	-0.8	-0.8

[a] All energy values are given in Ry (rydbergs) units.

15–19

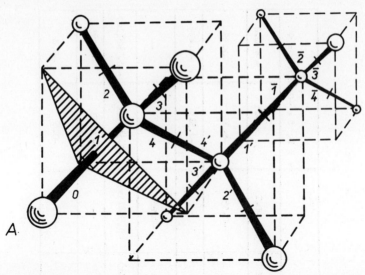

Fig. 1. Diamond-like lattice with (111) surface represented by shaded area. Localized functions $| \chi_{lm} >$ (m = 1) of the crystal described by indices $l = 1, \ldots 4, 4', \ldots 1$, that of chemisorbed atom A by index 0. Layers of elementary cells in the finite crystal denoted $m = 1, 2, \ldots$. From the surface outside $m = 0, -1, -2, \ldots$.

VIBRATIONS OF ATOMS IN CRYSTAL SURFACES*

A. A. Maradudin, D. L. Mills, and S. Y. Tong

Department of Physics
University of California
Irvine, California

I. Introduction

There is considerable interest today in the dynamical properties of atoms in finite or semi-infinite crystals on the part of theorists and experimentalists alike.

In the case of ionic crystals a good deal of this interest stems from the fact that the frequencies of the long wavelength optical vibration modes, which enter in an important way into the optical properties of such crystals in the infrared, are determined to a large extent by the Coulomb interactions between ions. Because of their long range, these interactions are sensitive to the size and shape of the crystal specimen, and accordingly so are the optical properties. The results of recent theoretical calculations by Fuchs and Kliewer[1], by Ruppin and Englman[2], and by Lucas[3], show that the spectrum of normal modes of a finite ionic crystal can contain optical surface modes, which have no counterpart in an infinitely extended crystal. Such modes, unlike Rayleigh waves[4] which are acoustic surface modes, are characterized by frequencies which tend to nonzero values in the limit of long wavelengths, and by the vibrations of the constituent sublattices against each other rather than in parallel, with amplitudes which are wavelike parallel to the free surface but decay exponentially with increasing distance

* Research partially supported by the Air Force Office of Scientific Research, Office of Aerospace Research, U. S. Air Force under AFOSR Grant Number 68-1448.

into the crystal. These optical surface modes are infrared active and should
be observable in infrared absorption experiments.

At the low frequency end of the spectrum of vibrations of finite or
semi-infinite crystal the study of properties of Rayleigh surface modes has
been stimulated by the development of techniques for generating[5] and detecting[6] such modes in the laboratory. Initially, the property of Rayleigh
waves studied experimentally was their speed of propagation. However, recently their attenuation has become the object of experimental study[7], a property
which, unlike the speed of propagation[8], has not yet been studied theoretically.

In this paper we present the results of calculations which represent the
first step toward an understanding of the damping of Rayleigh surface waves,
and of calculations of the frequencies and displacement patterns of optical
surface modes in alkali-halide crystals which are free from the approximations
made in previous studies of this problem.

II. Normal Modes of an Ionic Crystal Slab

A. Theory

The equations of motion of a semi-infinite ionic crystal slab of the
rocksalt structure, bounded by a pair of (100) surfaces normal to the z-
direction but infinite in the x- and y-directions, are

$$M_\kappa \, \ddot{u}_\alpha \, (\ell\kappa) \; = \; - \sum_{\ell'\kappa'\beta} \Phi_{\alpha\beta}(\ell\kappa;\ell'\kappa') \, u_\beta \, (\ell'\kappa'), \tag{2.1}$$

where M_κ is the mass of the κth kind of ion and $u_\alpha \, (\ell\kappa)$ is the α cartesian
component of the displacement of the ion $(\ell\kappa)$. With the substitution

$$u_\alpha(\ell\kappa) \; = \; \frac{v_\alpha(\ell\kappa)}{M_\kappa^{1/2}} \; e^{-i\omega t}$$

Equation (2.1) becomes

$$\omega^2 v_\alpha(\ell\kappa) = \sum_{\ell'\kappa'\beta} \frac{\Phi_{\alpha\beta}(\ell\kappa;\ell'\kappa')}{(M_\kappa M_{\kappa'})^{1/2}} \, v_\beta(\ell'\kappa'). \qquad (2.2)$$

If we assume a (+) ion at the origin (0,0,0) of the coordinate system, then

$$\underset{\sim}{x}(\ell+) = r_o(\ell_1,\ell_2,\ell_3) \quad \ell_1 + \ell_2 + \ell_3 = \text{even}$$

$$\underset{\sim}{x}(\ell-) = r_o(\ell_1,\ell_2,\ell_3) \quad \ell_1 + \ell_2 + \ell_3 = \text{odd}$$

where $\underset{\sim}{x}(\ell\kappa)$ is the equilibrium position vector of the κth ion in the ℓth unit cell, and r_o is the distance between nearest neighbor ions.

Due to the pair of free surfaces perpendicular to the z-direction, we can assume wave-like solutions satisfying periodic boundary conditions in the x- and y-directions only. We therefore write $v_\alpha(\ell\kappa)$ as the product of a wavelike function of ℓ_1 and ℓ_2 and an unknown function of ℓ_3,

$$v_\alpha(\ell\kappa) = v_\alpha^{(\kappa p)}(\ell_1\ell_2\ell_3) = e^{i\ell_1\Phi_1 + i\ell_2\Phi_2} \, \xi_\alpha^{(\kappa p)}(\ell_3), \qquad (2.3)$$

where the symbol κ is (+) or (-) depending on whether $\ell_1 + \ell_2 + \ell_3$ is an even or odd integer, respectively, and the symbol p is either (e) or (0) depending on whether ℓ_3 labels an even or an odd layer, respectively. Using Eq. (2.3), we can rewrite Eq. (2.2) as a set of four equations with the general form

$$\omega^2 \xi_\alpha^{(\kappa p)}(\ell_3) = \sum_{\ell_3'\beta} \sum_{\kappa'p'} D_{\alpha\beta}^{(\kappa p;\kappa'p')}(\Phi_1\Phi_2;\ell_3\ell_3') \xi_\beta^{(\kappa'p')}(\ell_3'), \qquad (2.4)$$

where

$$D_{\alpha\beta}^{(\kappa p;\kappa'p')}(\Phi_1\Phi_2;\ell_3\ell_3') = \frac{1}{(M_\kappa M_{\kappa'})^{1/2}} \sum_{\ell_1'\ell_2'} \Phi_{\alpha\beta}^{(\kappa p;\kappa'p')}(\ell_1\ell_2\ell_3;\ell_1'\ell_2'\ell_3') \times$$

$$\times e^{i(\ell_1'-\ell_1)\Phi_1+i(\ell_2'-\ell_2)\Phi_2} \qquad (2.5)$$

We assume that the potential function depends only on the distance between the ions at the sites labeled by $(\ell_1\ell_2\ell_3)$ and $(\ell_1',\ell_2',\ell_3')$. With this assumption $\Phi_{\alpha\beta}(\ell_1\ell_2\ell_3;\ell_1'\ell_2'\ell_3')$ is a function of $\ell_1,\ell_2,\ell_1',\ell_2'$ through the differences $(\ell_1-\ell_1')$ and $(\ell_2-\ell_2')$ only. We can therefore rewrite Eq. (2.5) as

$$D_{\alpha\beta}^{(\kappa p;\kappa'p')}(\Phi_1\Phi_2;\ell_3\ell_3') = \frac{1}{(M_\kappa M_{\kappa'})^{1/2}} \sum_{\bar\ell_1\bar\ell_2} \Phi_{\alpha\beta}^{(\kappa p;\kappa'p')}(\bar\ell_1\bar\ell_2;\ell_3\ell_3') \times$$

$$\times e^{-i\bar\ell_1\Phi_1-i\bar\ell_2\Phi_2} \qquad (2.6)$$

where we have put $\bar\ell_1 = \ell_1 - \ell_1', \ell_2 = \ell_2-\ell_2'$.

The matrix $D_{\alpha\beta}^{(\kappa p;\kappa'p')}(\Phi_1 \Phi_2;\ell_3\ell_3')$, called the dynamical matrix, is a $6N \times 6N$ matrix, where N is the number of layers in the slab, whose eigenvalues are the squares of the normal mode frequencies. We can divide $D_{\alpha\beta}^{(\kappa p;\kappa'p')}(\Phi_1\Phi_2;\ell_3\ell_3')$ into two parts, a part corresponding to the Coulomb interactions between the ions and a part corresponding to the short-range interactions.

For nearest neighbor short-range interactions, the dynamical matrices $D_{\alpha\beta}^{s(\kappa p;\kappa'p')}(\Phi_1\Phi_2;\ell_3\ell_3')$ are all diagonal in form. We define two constants A and B[9] by

$$A = \frac{4r_o^3}{e^2} \left. \frac{d^2v(r)}{dr^2} \right|_{r=r_o} \qquad (2.7a)$$

$$B = \frac{4r_o^2}{e^2} \left. \frac{dv(r)}{dr} \right|_{r=r_o} , \qquad (2.7b)$$

where $v(r)$ is the short-range interaction potential and e is the magnitude
of the electronic charge. Due to the smaller number of neighbors an ion at
the surfaces has, compared with an ion in the bulk of the slab, some of the
matrices $D_{\alpha\beta}^{s\ (\kappa p;\kappa' p')}(\Phi_1,\Phi_2;\ell_3\ell_3')$ have different elements for the surface
layers and for the bulk layers. Thus, for example, for the case $\kappa = \kappa'$;
$p = p'$,

$$D_{\alpha\beta}^{s(\kappa p;\kappa p)}(\Phi_1\Phi_2;\ell_3\ell_3') = \frac{e^2}{M_\kappa r_o^3}\ \delta_{\ell_3,\ell_3'}\ \delta_{\alpha\beta}\ (\tfrac{A}{2} + B) \qquad (2.8)$$

for a bulk layer, while for a surface layer,

$$D_{\alpha\beta}^{s(\kappa p;\kappa p)}(\Phi_1\Phi_2;\ell_3\ell_3') = D_{yy}^{(\kappa p;\kappa p)}(\Phi_1\Phi_2;\ell_3\ell_3') = \frac{e^2}{M_\kappa r_o^3}\ \delta_{\ell_3\ell_3'}\ (\tfrac{A}{2} + \tfrac{3B}{4})$$

$$\qquad (2.9)$$

$$D_{zz}^{s\ (\kappa p;\kappa p)}(\Phi_1\Phi_2;\ell_3\ell_3') = \frac{e^2}{M_\kappa r_o^3}\ \delta_{\ell_3\ell_3'}\ (\tfrac{A}{4} + B) \qquad (2.10)$$

The coulomb contribution to $D_{\alpha\beta}^{(\kappa p;\kappa' p')}(\Phi_1\Phi_2;\ell_3\ell_3')$ is

$$D^{c(\kappa p;\kappa'\mathbf{p}')}(\Phi_1\Phi_2;\ell_3\ell_3') = \frac{-e_\kappa e_{\kappa'}}{(M_\kappa M_{\kappa'})^{1/2} r_o^3} \sum_{\bar{\ell}_1\bar{\ell}_2} \frac{e^{-\bar{\ell}_1\Phi_1 - \bar{\ell}_2\Phi_2}}{(\bar{\ell}_1^2+\bar{\ell}_2^2+\bar{\ell}_3^2)^{5/2}} \times$$

$$\times \left[3\ell_\alpha \ell_\beta - \delta_{\alpha\beta}(\bar{\ell}_1^2+\bar{\ell}_2^2+\bar{\ell}_3^2) \right] , \qquad (2.11)$$

where e_κ is the charge of the κth ion, and $\bar{\ell}_3 = \ell_3 - \ell_3'$. As it stands,

the lattice sum in Eq. (2.11) is slowly convergent, and must be transformed
into a much more rapidly convergent sum before its numerical evaluation becomes
feasible. We have carried out such a transformation, using a method due
originally to Mackenzie[10]. Thus, for example, when $\kappa = \kappa'$; $p = p'$, we can
re-express $D_{xx}^{c\ (\kappa p;\kappa p)}(\Phi_1\Phi_2;\ell_3\ell_3')$ in terms of sums of exponential functions
as

$$
D_{xx}^{c\ (\kappa p;\kappa p)}(\Phi_1\Phi_2;\ell_3\ell_3') = \frac{e^2}{M_\kappa r_o^3}\ \frac{\pi}{2}\ \sum_{m=-\infty}^{\infty}\ \sum_{n=\infty}^{\infty}(1+(-1)^{m+n})\ \frac{(\Phi_1 + \pi m)^2}{\{(\Phi_1+\pi m)^2+(\Phi_2+\pi n)^2\}^{\frac{1}{2}}}\ \times
$$

$$
\times\ e^{-|\bar{\ell}_3|\{(\Phi_1+\pi m)^2 + (\Phi_2+\pi m)^2\}}\ . \qquad (2.12)
$$

For non-zero $|\bar{\ell}_3|$, the sums in Eq. (2.12) converge very rapidly for general
Φ_1 and Φ_2.

When $|\bar{\ell}_3| = 0$, the sums in Eq. (2.12) are no longer useful since they
become slowly convergent. By a slightly different transformation[11], we
can express $D_{\alpha\beta}^{c(\kappa p;\kappa'p')}(\Phi_1\Phi_2;\ell_3\ell_3)$ in terms of rapidly convergent sums of
modified Bessel functions $K_n(x)$ with n = integer. Thus, for example, for the
case $\kappa \neq \kappa'$; $p = p'$, we obtain

$$
D_{xx}^{c\ (\kappa p;\kappa'p)}(\Phi_1\Phi_2;\ell_3\ell_3) = \frac{e^2}{(M_\kappa M_{\kappa'})^{1/2}r_o^3}\left\{\sum_{n=1}^{\infty}\ \sum_{m=-\infty}^{\infty} 2\cos\left((2n-1)\ \Phi_1\right)\ \times\right.
$$

$$
\left[(\pi m + \Phi_2)^2\ K_o\left((2n-1)|\pi m + \Phi_2|\right) + \frac{|\pi +\Phi_2|}{(2n-1)}\ K_1\left((2n-1)|\pi m + \Phi_2|\right)\right]\ -
$$

$$
- \sum_{n=1}^{\infty}\ \sum_{m=-\infty}^{\infty} 2\cos\left((2n-1)\Phi_2\right)\left[(\pi m + \Phi_2)^2\ K_o\left((2n-1)|\pi m + \Phi_2|\right)\right]\right\}\ . \qquad (2.13)
$$

B.   Normal Mode Frequencies for a 15 Layer Slab

The eigenvalue equation (2.4) was solved numerically for a 15 layer slab of an ionic crystal chosen to represent NaCl.   The physical constants on which these calculations were based are:

$$M_+(Na) = 38.16 \times 10^{-24} gm$$
$$M_-(Cl) = 58.851 \times 10^{-24} gm$$
$$r_o = 2.814 \times 10^{-8} cm,$$
$$e = 4.8 \times 10^{-10} e.s.u.$$
$$A = 9.288$$
$$B = -1.165 .$$

The normal mode frequencies $\{\omega_j(\Phi_1,\Phi_2)\}$ and the corresponding eigenvectors $\{\xi_\alpha^{(j)(\kappa p)}(\Phi_1,\Phi_2;\ell_3)\}$ $(j = 1,2,3,...,6N = 90)$ were obtained for values of (Φ_1,Φ_2) distributed uniformly throughout the irreducible element of the two-dimensional first Brillouin zone for the slab.

Localized surface modes of both optical and acoustic types are found. The limiting frequencies at infinite wavelength of the highest frequency longitudinal optical (LO) and lowest frequency transverse optical (TO) bulk modes of the slab are found to be slightly shifted from the frequencies of the limiting LO and TO modes of an infinite crystal.   Two Rayleigh surface waves with slightly different frequencies are found lying below the lowest acoustic bulk mode (Fig. 1).   For the optical modes, we find at $\Phi_1 = \Phi_2 = 0$ two transverse optical surface modes whose eigenvectors have opposite parities, and whose frequencies are nearly degenerate with each other and lie below the limiting TO frequency of the infinitely extended crystal (Fig. 2).   Each of these two surface optical modes is doubly degenerate and has ionic displacement amplitudes that attenuate exponentially at $\Phi_1 = \Phi_2 = 0$.   When we go away from the point $\Phi_1 = \Phi_2 = 0$, we find two nearly degenerate higher frequency surface modes whose limiting frequencies at infinite wavelength differ from the limiting LO frequency of the infinitely extended crystal. The displacement amplitudes of these two upper surface modes have very little attenuation at the point $\Phi_1 = \Phi_2 = 0$.

The fact that surface modes comprise two branches, which are nearly degenerate in frequency, is due to the presence of two free surfaces and a plane of reflection symmetry midway between them. The surface modes in the presence of two free surfaces are essentially linear combinations of the surface modes associated with each of the surfaces separately of even and odd parity with respect to the midplane of the slab, and consequently, they have slightly different frequencies. As the thickness of the slab is increased, the frequency of each of these surface modes approaches that of a surface mode in a semi-infinite crystal.

C.   The Imaginary Part of the Crystal Dielectric Tensor

Optical surface modes have a dipole moment associated with them, and should absorb electromagnetic radiation at the frequencies of these modes. The infrared absorption coefficient of a crystal is directly related to the imaginary part of the crystal dielectric response tensor, which is given by[12]

$$\epsilon_{\mu\nu}^{(2)}(\omega) = 2\pi \left(\frac{e^{\beta\hbar\omega}-1}{\hbar V} \right) \int_{-\infty}^{\infty} dt \, e^{-i\omega t} \, \langle M_{\mu}(t) M_{\mu}(0) \rangle \qquad (2.14)$$

where $M_{\mu}(t)$ is the Heisenberg representation operator for the μ component of the crystal dipole moment. For an ionic diatomic slab, we can express Eq. (2.14) as[11]

$$\epsilon_{\mu\nu}^{(2)}(\omega) = \frac{e^2 \pi^2}{N(2r_o^3)} \left\{ \sum_j \frac{\delta\left(\omega - \omega_j(\underset{\sim}{0})\right)}{\omega_j(\underset{\sim}{0})} \left[\left(\sum_{\ell_3} \frac{\xi_\mu^{(j)(+e)}(\underset{\sim}{0}; \ell_3)}{M_+^{1/2}} \right. \right. -$$

$$-\sum_{\ell_3}\frac{\xi_\mu^{(j)(-e)}(\underset{\sim}{0};\ell_3)}{M_-^{1/2}}+\sum_{\ell_3}\frac{\xi_\mu^{(j)(+0)}(\underset{\sim}{0};\ell_3)}{M_+^{1/2}}-\sum_{\ell_3}\frac{\xi_\mu^{(j)(-0)}(\underset{\sim}{0};\ell_3)}{M_-^{1/2}}\Big)$$

$$\Big(\sum_{\ell_3'}\frac{\xi_\mu^{*(j)(+e)}(\underset{\sim}{0};\ell_3')}{M_+^{1/2}}-\sum_{\ell_3'}\frac{\xi_\mu^{*(j)(-e)}(\underset{\sim}{0};\ell_3')}{M_-^{1/2}}+\sum_{\ell_3'}\frac{\xi_\mu^{*(j)(+0)}(\underset{\sim}{0};\ell_3')}{M_+^{1/2}}-$$

$$-\sum_{\ell_3'}\frac{\xi_\mu^{*(j)(-0)}(\underset{\sim}{0};\ell_3')}{M_-^{1/2}}\Big)\Big]\Big\} \tag{2.15}$$

where the index j runs from 1 to 6N, the total number of normal modes of the slab. For the crystal slab the tensor $\epsilon_{\mu\nu}^{(2)}(\omega)$ is diagonal in μ and ν, with $\epsilon_{xx}(\omega)=\epsilon_{yy}(\omega)$. In Fig. (3a) and Fig. (3b), we show the values of $\epsilon_{zz}(\omega)$ and $\epsilon_{xx}(\omega)$, respectively. Since the surface optical modes at $\underset{\sim}{\Phi}=0$ are transverse in nature, they contribute only to $\epsilon_{xx}(\omega)$. The line (A) in Fig. (3b) is due to the transverse surface optical modes while the other lines (B), (C), and (D) are due to the transverse optical bulk modes. We see from this figure that, at least in the case of a slab of 15 layers, the absorption by the optical surface modes is comparable with that by bulk transverse optical modes, and this result suggests that surface optical modes may be experimentally observable.

D.   Relaxation Effects in a Slab

In the preceding sections, we have not included the effects of relaxation of the spacings between ionic layers due to the existence of the pair of free surfaces of the slab. As a consequence, the layers (especially those near the surfaces) may not be at their true equilibrium separations. In this section, we vary the spacings between the layers, and allow the inter-ionic distance in each layer to change. We then minimize the total potential energy of the slab as a function of the inter-ionic distance and the separations between the layers, and estimate the magnitudes of the effects due to relaxation on the surface modes.

Let r be the separation distance between two adjacent ions in the same layer of the slab and let the $(i+1)^{th}$ layer be separated from the ith layer by a distance $(1+\epsilon_i)r$, where ϵ_i is a parameter to be determined. The total potential energy of the slab can be separated into two parts, a part due to Coulomb interactions and a part due to short range interaction. Thus, for a slab of N layers, we write

$$\Phi^T(r;\epsilon_1,\ldots,\epsilon_{N-1}) = \Phi^C(r;\epsilon_1,\ldots,\epsilon_{N-1}) + \Phi^S(r;\epsilon_1,\ldots,\epsilon_{N-1}) \;.$$

For a given N, the quantity $\Phi^S(r;\epsilon_1,\ldots,\epsilon_{N-1})$ has finite number of terms and can be easily handled. The quantity $\Phi^C(r;\epsilon_1,\ldots,\epsilon_{N-1})$ involves two-dimensional infinite lattice sums that can be transformed into rapidly convergent sums of modified Bessel functions by using the same technique as described in Section 1. We then minimize $\Phi^T(r;\epsilon_1,\ldots,\epsilon_{N-1})$ with respect to ϵ_i and r by solving the set of N-1 simultaneous equation $\partial \Phi^T/\partial \epsilon_i \; (r;\epsilon_1,\ldots,\epsilon_{N-1}) = 0$, $i = 1,\ldots N-1$; for different values of r. The value of r min. at which $\Phi^T(r;\epsilon_1,\ldots,\epsilon_{N-1})$ is a minimum is then determined. For a NaCl crystal slab of 15 layers, and using the Born-Mayer potential function for the short-range interaction potential, we find the minimum value of $\Phi^T(r;\epsilon_1,\ldots,\epsilon_{N-1})$ occurs at $r = 2.798 \times 10^{-8}$ cm., giving a decrease of 0.57% from the value of $r_0 = 2.814 \times 10^{-8}$ cm. of the infinitely extended crystal. The shifts in the interplanar separation distances at minimum potential and at other values of r are listed in Table 1. Using the results for r and ϵ_i (i = 1,...N-1) at minimum potential, the normal mode frequencies of a relaxed slab of 15 layers are calculated. The number of Rayleigh and optical surface modes remains the same, but their frequencies and those of the bulk modes are shifted from the corresponding frequencies of an unrelaxed slab of 15 layers.

III.   The Damping of Surface Acoustical Phonons in Anharmonic Crystals

Recently it has proved possible to generate high frequency (300Mc) surface acoustical waves by direct methods[5], and to observe Brillouin scattering of laser light[6] from these modes.   Since either of these methods in principle may be employed to measure the mean free path of the surface waves, we have undertaken a theoretical investigation of the contribution of the mean free path of Rayleigh waves in an anharmonic model crystal.   Our calculation applies in the high frequency, low temperature regime where the thermal phonon mean free path is long compared to the wavelength of the surface mode in question.. The computation for surface phonons is thus the analogue of the earlier work of Landau and Rumer[13] on the attenuation rate of bulk ultrasonic waves.

We begin with a monatomic, simple cubic crystal.   In the harmonic approximation, the atoms are assumed to interact by means of short range, nearest and next neighbor central force interactions.   The ratio of the nearest and next nearest neighbor force constants is adjusted so the crystal is elastically isotropic in the long wavelength limit.   Two free surfaces are then created by setting to zero all the interactions associated with bonds that cross a fictitious plane that passes between two (100) atomic planes.   In the harmonic approximation, the effect of the free surfaces on the lattice vibrations of this model has been discussed previously by Maradudin and Wallis in their study of the surface contribution to the low temperature specific heat[14].

Cubic anharmonic terms are introduced into the Hamiltonian by assuming that only anharmonicity in the coupling between nearest neighbors is important. This leads to a contribution of the form

$$V_3 = \frac{1}{6} \sum_{\ell \ell' \ell''} \sum_{\alpha\beta\gamma} \Phi_{\alpha\beta\gamma}(\ell\ell'\ell'') \, u_\alpha(\ell) u_\beta(\ell') u_\gamma(\ell'') \,,$$

where we include only the contribution from the nearest neighbor interactions. In this expression, we ignore the fact that the contribution to V_3 from the atoms in the surface layer differs from that of atoms in the bulk, as a consequence of the broken bonds. An argument that suggests this is not a serious approximation for the present purposes will be sketched below.

We compute the attenuation rate of a surface acoustical mode (Rayleigh wave) by calculating the imaginary part of the proper self energy to second order in V_3. In lowest order of perturbation theory, the Fourier coefficient of the diagonal element of the proper self energy matrix for the sth normal mode of a Bravais crystal may be written[15] (with $\hbar = 1$)

$$P_s(i\omega_\ell) \equiv P_{ss}(i\omega_\ell) \equiv 18\beta^2 \sum_{s_1 s_2} \sum_{n=-\infty}^{\infty} V^2_{ss_1s_2} D^{(0)}_{s_1}(i\omega_n) D^{(0)}_{s_2}(i\omega_\ell - i\omega_n). \quad (3.1)$$

In this expression, $\omega_\ell = 2\pi\ell/\beta$, where ℓ is an integer and $\beta = (k_B T)^{-1}$. The indices s_1 and s_2 run over all $3N$ normal modes of the crystal. The function $D^{(0)}_s(i\omega_\ell)$ is the free phonon propagator

$$D^{(0)}_s(i\omega_\ell) = \frac{2\omega_s}{\beta} \frac{1}{\omega_s^2 + \omega_\ell^2},$$

where ω_s is the frequency of the sth normal mode of the crystal. The coefficient $V_{s_1 s_2 s_3}$ is given by[15]

$$V_{s_1 s_2 s_3} = \frac{1}{6} \frac{1}{(2M)^{3/2}} \frac{1}{(\omega_{s_1} \omega_{s_2} \omega_{s_3})^{1/2}} \sum_{\ell_1 \alpha_1} \sum_{\ell_2 \alpha_2} \sum_{\ell_3 \alpha_3} \Phi_{\alpha_1 \alpha_2 \alpha_3}(\ell_1 \ell_2 \ell_3)$$

$$\times B^{(s_1)}_{\alpha_1}(\ell_1) B^{(s_2)}_{\alpha_2}(\ell_2) B^{(s_3)}_{\alpha_3}(\ell_3).$$

The quantity $B_\alpha^{(s)}$ (ℓ) is the α component of the displacement of atom ℓ, when the sth normal mode is excited.

If one analytically continues the function $P_s(i\omega_\ell)$ off the imaginary axis, one obtains

$$- \frac{1}{\beta} P_s(\omega \pm io) = \triangle_s(\omega) \mp i\Gamma_s(\omega).$$

The damping constant of the sth normal mode is then given by $\Gamma_s(\omega_s)$, to lowest order in perturbation theory.

In order to carry out the computation of the damping constant, we have found it convenient to introduce some approximations. First of all, in the sum over s_1 and s_2 of Eq. (3.1), we include only the contributions in which both s_1 and s_2 refer to bulk phonons. Thus, the contributions to the damping rate that comes from the interaction of the surface phonon with thermally excited Rayleigh waves is ignored. We justify this approximation with the following heuristic argument. The wavelength of the thermally excited surface waves is short compared to that of the ultrasonic surface wave. The ratio of the thermal phonon wavelength to that of the low frequency ultrasonic wave of frequency Ω is $(\hbar\Omega/k_B T) \ll 1$. It is well known that the displacement field associated with a Rayleigh wave penetrates into the crystal a distance of the order of its wavelength parallel to the surface. Now a bulk phonon is able to interact with the ultrasonic surface wave throughout the volume within which the surface phonon displacement field is large. However, the remarks above lead to the conclusion that the volume within which the ultrasonic surface wave - thermal surface wave interaction occurs is smaller than the interaction associated with the ultrasonic surface wave - bulk thermal phonon interaction by the factor $(\hbar\Omega/k_B T)$. This consideration suggests one may to first approximation ignore the contribution to the damping rate from the thermally excited surface phonons.

In this spirit, we also ignore the surface corrections to the eigenvectors and frequency distribution of the bulk modes. Hence, we assume the modes s_1 and s_2 of Eq. (3.1) to be the plane wave modes of the perfect, infinitely extended crystal in the absence of free surfaces. We employ an argument similar to the above to justify neglect of the difference in the contribution to V_3 from the atoms in the surface layer, compared to the contribution from atoms in the bulk.

Within the framework of the above mentioned approximations, it is possible to carry out the computation of the damping constant of the surface wave once the amplitude $B_\alpha^{(s)}(\ell)$ associated with the ultrasonic surface wave is known. We have obtained this amplitude from the free phonon propagator derived previously in Ref. (14). In the harmonic approximation, the one phonon propagator has the form

$$U_{\alpha\beta}(\ell\ell',\omega^2) = \frac{1}{M} \sum_s \frac{B_\alpha^{(s)}(\ell)B_\beta^{(s)}(\ell')}{\omega^2-\omega_s^2} \quad .$$

Then

$$B_\alpha^{(s)}(\ell)B_\beta^{(s)}(\ell') = 2\omega_s \operatorname*{Res}_{\omega=\omega_s} U_{\alpha\beta}(\ell\ell',\omega^2) \quad .$$

We have isolated the surface mode pole in the propagator derived by Wallis and Maradudin[14], and obtained the surface phonon amplitudes by extracting the residue at the pole. We then find that to leading order in $(k_B T/\hbar\Omega)$, the damping rate of a surface acoustical wave propagating parallel to the x axis is given by

$$\Gamma_s = \frac{0.397}{(2\pi)^3} \xi(4) \left[I_{tt} + \left(\frac{c_t}{c_\ell}\right)^2 I_{\ell\ell} \right] \left[\phi'''(a_0) \right]^2 \frac{\omega_s (k_B T)^4}{\rho^3 c_s^2 c_t^2 \hbar^4} \qquad (3.2)$$

In this expression I_{tt} and $I_{\ell\ell}$ are dimensionless numbers. For the present model a numerical computation gives $I_{\ell\ell} = 0.337$, and $I_{tt} = 0.0521$. The density of the crystal is ρ, c_t and c_s are the sound velocity of bulk transverse and surface waves respectively, and $\phi'''(a_o)$ is the third derivative of the two body potential, evaluated at the nearest neighbor separation a_o.

The result of Eq. (3.2) exhibits the ωT^4 dependence characteristic of the frequency and temperature dependence of bulk transverse phonons in the Landau-Rumer regime. We have compared the mean free path of the Rayleigh wave to that of a bulk transverse ultrasonic phonon propagating parallel to the x axis of the model crystal. We find the Rayleigh wave mean free path much shorter than that of the bulk phonon. However, for reasons discussed in detail elsewhere[16], while the value of this ratio may be rather sensitive to the details of the interatomic forces, we expect in general that the Rayleigh wave mean free path should be comparable to, or shorter than that of a bulk transverse phonon of the same frequency.

REFERENCES

1. R. Fuchs and K. L. Kliewer, Phys. Rev. $\underline{140}$, A2076 (1965).

2. R. Englman and R. Ruppin, Phys. Rev. Letters $\underline{16}$, 898 (1966).

3. A. A. Lucas (preprint).

4. Lord Rayleigh, Proc. Lond. Math. Soc. $\underline{17}$, 4 (1887).

5. R. Arzt, E. Salzmann, and K. Dransfeld, Appl. Phys. Letters $\underline{10}$, 165 (1967).

6. J. Krokstad and L. O. Svaasand, Appl. Phys. Letters $\underline{11}$, 155 (1967).

7. K. Dransfeld (private communication).

8. R. Stonely, Proc. Roy. Soc. A232, 447 (1955); D. C. Gazis, R. Herman, and
   R. F. Wallis, Phys. Rev. $\underline{119}$, 533 (1960).

9. E. W. Kellermann, Phil. Trans. Roy. Soc. London, $\underline{A238}$, 513 (1940).

10. J. K. Mackenzie, Ph.D. thesis, University of Bristol (1949) (unpublished).
    See also B.M.E. van der Hoff and G. C. Benson, Canad. J. Phys. $\underline{31}$, 1087
    (1953).

11. A. A. Maradudin and S. Y. Tong (to be published).

12. This result follows from Eq.(8.14) of the article by A. A. Maradudin in
    Astrophysics and the Many-Body Problem (Benjamin, New York, 1963), p. 107,
    if the dielectric tensor is related to the dielectric susceptibility tensor
    by $\epsilon_{\mu\nu}(\omega) = \delta_{\mu\nu} + 4\pi\chi_{\mu\nu}(\omega)$.

13. L. Landau and G. Rumer, Physik Z. Sowyetunion $\underline{11}$, 18 (1937).

14. R. F. Wallis and A. A. Maradudin, Phys. Rev. $\underline{148}$, 945 (1966).

15. A. A. Maradudin, Annals of Physics (N.Y.) $\underline{30}$, 371 (1964).

16. A. A. Maradudin and D. L. Mills, Phys. Rev. $\underline{173}$, 881 (1968).

r (10^{-8} cm)	A	B	$\epsilon_1=\epsilon_{14}$	$\epsilon_2=\epsilon_{13}$	$\epsilon_3=\epsilon_{12}$	$\epsilon_4=\epsilon_{11}$	$\epsilon_5=\epsilon_{10}$	$\epsilon_6=\epsilon_9$	$\epsilon_7=\epsilon_8$	$\Phi^T(r; \epsilon_1,\cdots,\epsilon_{14})$
2.818	9.223	-1.155	-0.0059	-0.0023	-0.0024	-0.0024	-0.0024	-0.0024	-0.0024	-4.06317
2.814	9.288	-1.165	-0.0034	0.0001	0	0	0	0	0	-4.06329
2.813	9.304	-1.167	-0.0028	0.0006	0.0005	0.0005	0.0005	0.0005	0.0005	-4.06332
2.798	9.547	-1.204	0.0059	0.0089	0.0088	0.0088	0.0088	0.0088	0.0088	-4.06351
2.740	10.574	-1.362	0.0343	0.0361	0.0360	0.0360	0.0360	0.0360	0.0360	-4.05968
2.700	11.332	-1.481	0.0496	0.0508	0.0508	0.0508	0.0508	0.0508	0.0508	-4.05153
2.600	13.433	-1.823	0.0786	0.0791	0.0791	0.0791	0.0791	0.0791	0.0791	-4.00351

$r = r_0$ (INFINITE CRYSTAL) → (points to $r = 2.814$)

$r = r_{MIN}$ → (points to $r = 2.798$)

Table 1. Values of A, B, $\bar{\epsilon}_1$ ($i=1,\ldots,N-1$), and $\Phi^T(r;\epsilon_1,\ldots,\epsilon_{N-1})$ for different values of r for a 15 layer NaCl crystal slab with relaxation.

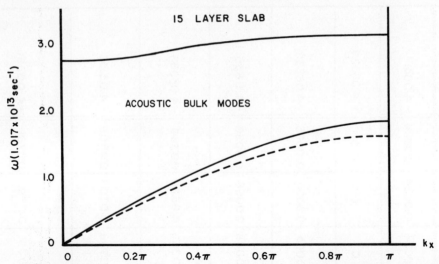

Fig. 1. The acoustic bulk modes and the Rayleigh waves propagating in the x-direction ($\kappa y = 0$) for a 15 layer NaCl crystal slab. The dashes curve represents two nearly degenerate Rayleigh waves.

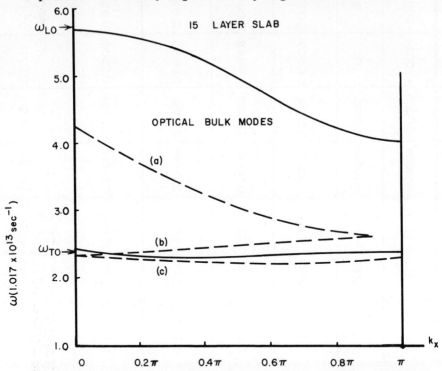

Fig. 2. The optical bulk modes and the optical surface modes propagating in the x-direction ($\kappa y = 0$) for a 15 layer NaCl crystal slab. Each of the three dashed curves marked (a), (b) and (c) represents two nearly degenerate optical surface modes.

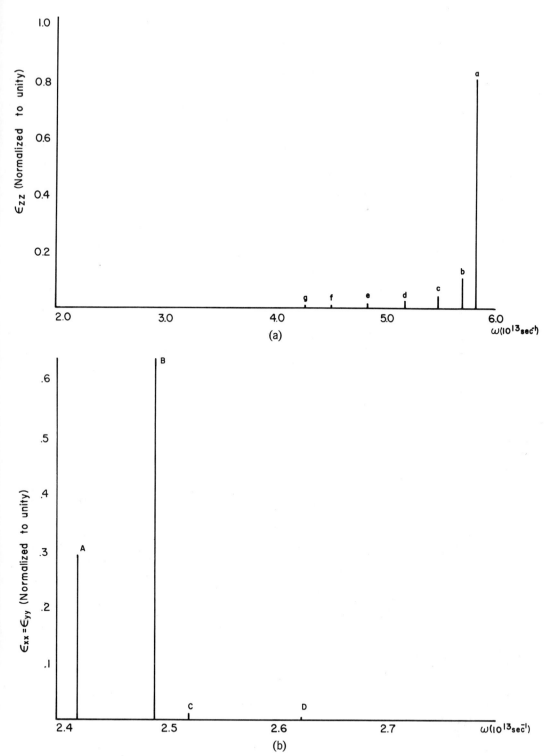

Fig. 3. The imaginary part of the crystal dielectric tensor for a 15 layer NaCl crystal slab. $\epsilon_{zz}(\omega)$ in a Fig. 3a and $\epsilon_{xx}(\omega)$ in Fig. 3b are normalized such that $\int_0^\infty \epsilon_{\mu\mu}(\omega)d\omega = 1$.

THEORY OF THE DEBYE-WALLER FACTOR FOR SURFACE ATOMS

R. F. Wallis

Naval Research Laboratory, Washington, D. C.

B. C. Clark and Robert Herman

Research Laboratories, General Motors Corporation
Warren, Michigan

1. Introduction

The mean square displacements of the atoms of a crystal determine to a large extent the temperature dependence of the intensities of electrons or x-rays scattered from the crystal. For very low energy electrons which are scattered primarily from the surface layer of atoms, the temperature dependence of the peak intensities is indicative of the mean square displacements of the surface atoms.

The interatomic forces acting upon an atom determine its mean square displacement. A surface atom is acted upon by fewer neighbors than an interior atom, and this in itself will generally cause the mean square displacement to be larger for the surface atom. The environment of a surface atom has different symmetry than that of an interior atom. This may lead to an anisotropy in the mean square displacements at the surface even in cubic crystals. Another consequence of the different symmetry at the surface is that the interatomic coupling constants for surface atoms may be different from those for interior atoms, and this may lead to further anisotropy of the surface Debye-Waller factor. The anisotropy will also be affected by the relative contributions of central forces and of angular stiffness forces.

The linear dependence of mean square displacement on absolute temperature observed experimentally generally extends down to temperatures below the Debye temperature. The larger mean square displacements of surface atoms implies a smaller Debye temperature, and one may expect the linear temperature region for surface mean square displacements to extend to lower temperatures than for the bulk case. The anisotropy of the mean square displacements may also vary with temperature.

Specific calculations to illustrate these ideas have been carried out for a nearest-neighbor central-force model of a face-centered cubic lattice. Detailed results are presented for the (110) surface of nickel. An improved model consisting of nearest and next-nearest neighbor central forces plus angle-bending forces involving triples of nearest neighbors has been employed to calculate mean square displacements at (100) surfaces of a number of face-centered cubic metals. The results are compared with available experimental data.

2. Theoretical Formulation

The scattering of X-rays or electrons by a crystal is determined by a modified scattering factor f_k given by

$$f_k = f_k^{\,o} \exp(-M_k) \tag{1}$$

where $f_k^{\,o}$ is the atomic scattering factor for atom k and $\exp(-M_k)$ is the square root of the Debye-Waller factor. The quantity M_k is related to the atomic displacement from equilibrium u_k by

$$M_k = (8\pi^2/\lambda^2)\,\langle (u_{k\Delta s})^2\rangle \cos^2 \psi \tag{2}$$

where λ is the wavelength of the radiation, $\Delta s = s' - s$ with s' and s unit vectors in the directions of the scattered and incident radiation, respectively, ψ is the angle between s' and Δs, and the angular brackets denote a thermal average over a canonical ensemble. By varying Δs one can obtain mean square displacement components which make various angles with the surface. In the high temperature limit a Debye temperature can be defined in terms of the mean square displacement by means of the equation

$$\Theta_{k\Delta s}^2 = m_k k_B \Big/ \Big[3\hbar^2 T\langle (u_{k\Delta s})^2\rangle\Big] \tag{3}$$

where m_k is the mass of atom k and k_B is Boltzmann's constant.

To calculate the mean square displacements we assume that the displacement components satisfy harmonic equations of motion

$$m_k \ddot{u}_{ki} = -\sum_{k'j} \alpha_{ki,k'j} u_{k'j} \tag{4}$$

where i denotes a Cartesian component and the $\alpha_{ki,k'j}$ are the harmonic force constants. After making the transformation

$$v_{ki} = m_k^{\frac{1}{2}} u_{ki} \tag{5}$$

one can write the equations of motion in the form

$$\ddot{v}_{ki} = -\sum_{k'j} D_{ki,k'j} v_{k'j} \tag{6}$$

where the $D_{ki,k'j}$ are the elements of the dynamical matrix defined by $D_{ki,k'j} = \alpha_{ki,k'j}/(m_k m_{k'})^{\frac{1}{2}}$. The presence of the free surface is reflected in the values of appropriate elements of either the force-constant matrix or the dynamical matrix.

The mean square displacement components can be expressed[1] in terms of the eigenvalues ω_p^2 and eigenvectors $e_{ki}(p)$ of the dynamical matrix

$$\langle u_{ki}^2 \rangle = (1/m_k) \sum_p |e_{ki}(p)|^2 \, \bar{\epsilon}(p)/\omega_p^2 \tag{7}$$

where $\bar{\epsilon}(p)$ is the mean energy of the p th normal mode of vibration given by

$$\bar{\epsilon}(p) = \hbar\omega_p (n_p + \tfrac{1}{2}), \tag{8}$$

$$n_p = \left[e^{\hbar\omega_p/k_BT} -1 \right]^{-1} . \tag{9}$$

By using a well-known theorem of matrices[1] Eq. (7) can be expressed directly in terms of the dynamical matrix

$$\langle u_{ki}^2 \rangle = (\hbar/2m_k) \left[D^{-\frac{1}{2}} \coth (\hbar D^{\frac{1}{2}}/2k_BT) \right]_{ki,ki} . \tag{10}$$

For computational purposes it is convenient to consider the high temperature expansion given by

$$\langle u_{ki}^{2}\rangle = \left[(k_{B}T/m_{k})D^{-1} + \hbar^{2}I/(12m_{k}k_{B}T) - \hbar^{4}D/(720m_{k}(k_{B}T)^{3}) + \dots \right]_{ki,ki} \qquad (11)$$

and the 0°K result given by

$$\langle u_{ki}^{2}\rangle = (\hbar/2m_{k})\left[D^{-\frac{1}{2}}\right]_{ki,ki} . \qquad (12)$$

In this paper we consider crystals in the form of thin flat plates. By using cyclic boundary conditions in the two directions of large dimension, one can work with reduced dynamical matrices of significantly smaller size than the original matrices. One must, however, then sum over the two components of wave vector thus introduced.

3. Anisotropy of Surface Mean Square Displacements

The first theoretical discussion of the anisotropy of surface mean square displacements was given by Kalashnikov[2] who used the isotropic continuum model of Debye. When applied to silver, Kalashnikov's theory predicts that the normal mean square displacement component at the surface is smaller than the tangential component, although both are larger than the bulk value. This result is in qualitative but not quantitative agreement with the experiments of Zamsha and Kalashnikov[3] on the (001) surface of silver.

The recent lattice dynamical calculation of Rich[4] based on a nearest-neighbor model of a simple cubic lattice does not exhibit anisotropy because of the special character of the model. Maradudin and Melngailis[5] using a nearest and next-nearest neighbor central force model of a simple cubic lattice found that the normal component at a (001) surface is larger than the tangential component for the case of isotropy.

Calculations based on the high temperature limit of Eqs. (7) and (11) have been carried out by Clark, Herman, and Wallis[6] for a nearest-neighbor central-force model of a face-centered cubic lattice. The results for the mean square displacement components at (100), (110), and (111) surfaces are given in Table I for a crystal twenty layers thick. The units are kT/α where α is the force constant.

From Table I one sees that the normal components are very close to twice the bulk value for each surface. For the (100) and (111) surfaces the tangential components are smaller than the normal components, but still larger than the bulk. The (110) surface exhibits a new type of anisotropy due to the non-equivalence of the two tangential directions. In one of the tangential directions the mean square displacement is larger than in the normal direction. The results for the (110) surface can be compared with MacRae's[7] experimental low energy electron diffraction (LEED) data as shown in Table II.

Table I. Mean square displacement components of atoms at (100), (110), and (111) surfaces of f.c.c. crystals twenty layers thick. The bulk values are those for atoms in the tenth layer from the surface.

		normal	tangential
(100)	surface	0.82	0.60
	bulk	0.40	0.40
(110)	surface	0.80	0.64 [$1\bar{1}0$]
			0.86 [001]
	bulk	0.40	0.39 [$1\bar{1}0$]
			0.39 [001]
(111)	surface	0.82	0.53
	bulk	0.40	0.41

Table II. Theoretical and experimental mean square displacement components in units of kT/α for atoms at a (110) surface and in the bulk of a face-centered cubic crystal.

	[110]	[1$\bar{1}$0]	[001]	bulk
theoretical	0.80	0.64	0.86	0.40
experimental	1.41	0.63	1.41	0.45

The nearest-neighbor force constant was chosen so that the maximum vibrational frequency would agree with that obtained by Birgeneau et al.[8] from neutron scattering. One sees from Table II that the theoretical values agree qualitatively, but not quantitatively, with experiment.

4. Changes in Surface Force Constants

The surface anisotropies discussed in the preceding section are not a consequence of changes in the force constants coupling surface atoms to their neighbors. There is no reason, however, why the surface force constants should be the same as the bulk values. This problem has been discussed in general terms by Feuchtwang.[9] In a recent publication, Clark, Herman, Gazis, and Wallis[10] calculated the change in surface force constants for a body-centered cubic lattice with nearest and next-nearest Lennard-Jones interactions. They found that the force constants at the surface were smaller than the corresponding bulk values by 25-35%. These changes are an anharmonic effect and are surprisingly large when compared to the 3% change found for the equilibrium interlayer spacing at the surface.

Another recent investigation of changes in surface force constants is that of Vail[11] who considered a simple-cubic lattice with Morse-type interactions out to seventh neighbors. Vail found decreases in the nearest-neighbor surface force constants affecting atomic motion normal to the surface of about 40%, but increases in the surface force constants coupling two atoms in the surface layer of about 7%.

The effect of surface force constant changes on the mean square displacements of surface atoms has been investigated by Wallis, Clark, and Herman[12] for the case of a (110) surface of a face-centered cubic crystal with nearest-neighbor central forces. An atom in such a surface is coupled to atoms in the adjacent interior layer by a force constant α_1, to atoms two layers away by a force constant α_2, and to atoms within the surface layer by a force constant α_3. The force constants α_1, α_2, and α_3 need not be equal to one another or to the bulk force constant α.

By varying α_1, α_2, and α_3 one can improve the agreement between the theoretical and experimental results for nickel. Table III shows two cases for a crystal thirty layers thick together with MacRae's experimental data. The indicated decreases in the surface force constants α_1 and α_2 do not seem unreasonable, but the magnitude of the increase in α_3 may not be realistic.

Table III. Theoretical mean square displacement components of atoms at a (110) surface for various values of the surface force constants. MacRae's experimental results are also shown.

α_1/α	α_2/α	α_3/α	[110]	[1$\bar{1}$0]	[001]
0.5	0.5	1.0	1.33	0.84	1.41
0.5	0.5	2.2	1.33	0.61	1.41
experimental (Ni)			1.41	0.63	1.41

5. Temperature Dependence of Surface Mean Square Displacements.

Low energy electron diffraction experiments are frequently done at temperatures not much larger than the bulk Debye temperature. One may ask to what extent it is reasonable to use the high temperature limit of Eq. (11) at temperatures near or below the Debye temperature. We have investigated this point by making calculations of the temperature dependence of the mean square displacements for a (110) surface of a face-centered cubic crystal. The nearest-neighbor central force model was used in the harmonic approximation

for a crystal twenty layers thick. Up to twenty terms in the expansion of
Eq. (11) were used. The convergence was good down to about one-quarter the
bulk Debye temperature. The mean square displacements at 0°K were calculated
using Eq. (12).

The results for the temperature dependence are shown in Fig. 1. The
points are the computed values, and straight lines have been drawn through
the high temperature portions. The force constants in the bulk and at the
surface are those for nickel given by the first line in Table III. One sees
from Fig. 1 that the high temperature linear behavior persists down below
200°K or one-half the Debye temperature thus justifying the use of the high
temperature limit in analyzing experimental data near the Debye temperature.
Although it is difficult to discern in Fig. 1, the deviation from linear
behavior at the lower temperatures is percentage-wise about twice as great
for the bulk displacements as for the surface displacements.

The series expansion of Eq. (11) seems to be good for temperatures as
low as 100°K for nickel. When twenty terms in the series are used at 100°K,
there is a factor of 10^8 difference in the leading and tenth term of the
series for the surface layer displacements and another factor of 10^6 between
the tenth and twentieth terms. For temperatures above 200°K, ten terms in
the series are clearly adequate. At 200°K, the ratio of the tenth term to
the leading term is approximately 10^{-12}.

In order to exhibit the behavior of the anisotropy in a little clearer
fashion, we plot the ratios of the parallel mean square displacement components
to the perpendicular component as a function of temperature in Fig. 2. The
anisotropy is essentially constant down to temperatures around 200°K and then
decreases at lower temperatures. The maximum change in anisotropy in going
from high temperatures to 0° is about 10%.

The present treatment neglects anharmonicity. Inclusion of anharmonicity may be expected to produce deviations from the linear dependence of mean square displacement on temperature at high temperatures.

6.   Calculations for Various Face-centered Cubic Metals.

The nearest-neighbor central force model gives a fairly good representation of nickel, but one cannot expect such a model to be universally applicable. In this section we present results for face-centered cubic crystals based on a model with nearest and next-nearest neighbor central interactions together with angle-bending interactions involving triplets of nearest-neighbors. The force constants for these interactions are designated α, β, and γ, respectively, and are chosen to fit the three elastic constants (room temperature) of the metals considered according to the equations

$$\alpha = (a/6)\ (c_{44} + 2c_{12}) \tag{13}$$

$$\beta = (a/12)\ (3c_{11} - 2c_{12} - 4c_{44}) \tag{14}$$

$$\gamma = (a/18)\ (c_{44} - c_{12}) \tag{15}$$

where a is the edge of the fundamental cube.

The calculations have been carried out for the (100) surface of crystals twenty layers thick. The results are given in Table IV in terms of the ratios of the normal and tangential mean square displacement components to the bulk value.

One sees immediately the rather striking result that the normal-to-bulk ratio is very close to 2.0 for all cases. The tangential-to-bulk ratio shows greater variation from metal to metal. The values for nickel are not much different from those derived from the nearest-neighbor model.[6]   The normal value for silver agrees well with Kalashnikov's isotropic continuum calculation,[2] but the tangential value is less than Kalashnikov's.

Table IV. Ratios of normal and tangential mean square displacement
components to the bulk value for the (100) surface of several f.c.c. metals.

metal	normal-to-bulk ratio	tangential-to-bulk ratio
copper	2.02	1.50
silver	2.02	1.40
gold	2.02	1.21
aluminum	2.01	1.32
lead	2.01	1.31
nickel	2.03	1.51
palladium	2.02	1.34
platinum	2.01	1.22

Experimental data for the (100) surface of f.c.c. metals is still sparse.
Lyon and Somorjai[13] have studied the (100) surface of platinum and find a
normal-to-bulk ratio of about 4.0, much larger than our calculated value. A
very similar situation obtains for the (100) surface of palladium studied by
Goodman, Farrell, and Somorjai.[14] It would appear from these results that
the surface force constants for these two metals are significantly smaller
than the bulk values. The indicated weak binding of the surface atoms may be
related to the ease with which surface transformations occur in platinum[13]
and palladium.[14]

Acknowledgment: It is a pleasure to acknowledge helpful discussions
with Mr. Michael Marcotty on certain aspects of the numerical calculations.

References

1. M. Born, Rept. Prog. Phys. $\underline{9}$, 294 (1942).

2. S. G. Kalashnikov, Zh. Eksperim. i Teor. Fiz. $\underline{13}$, 295 (1943).

3. O. I. Zamsha and S. G. Kalashnikov, Zh. Eksperim. i Teor. Fiz. $\underline{9}$, 1408 (1939).

4. M. Rich, Phys. Letters $\underline{4}$, 153 (1963).

5. A. A. Maradudin and J. Melngailis, Phys. Rev. $\underline{133}$, A1188 (1964).

6. B. C. Clark, R. Herman, and R. F. Wallis, Phys. Rev. $\underline{139}$, A860 (1965).

7. A. U. MacRae, Surface Sci. $\underline{2}$, 522 (1964).

8. B. J. Bergeneau, J. Cordes, G. Dolling, and A. B. D. Woods, Phys. Rev. $\underline{136}$, A1359 (1964).

9. T. E. Feuchtwang, Phys. Rev. $\underline{155}$, 715, 731 (1967).

10. B. C. Clark, R. Herman, D. C. Gazis, and R. F. Wallis, in Ferroelectricity, edited by E. F. Weller, Elsevier Publishing Corporation, Amsterdam, 1967, p. 101.

11. J. M. Vail, Can. J. Phys. $\underline{45}$, 2661 (1967).

12. R. F. Wallis, B. C. Clark, and R. Herman, Phys. Rev. $\underline{167}$, 652 (1968).

13. H. B. Lyon and G. J. Somorjai, J. Chem. Phys. $\underline{44}$, 3707 (1966).

14. R. H. Goodman, H. H. Farrell, and G. A. Somorjai, to be published.

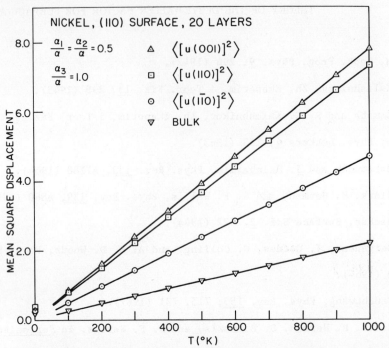

Fig. 1. Temperature dependence of the mean square displacement components at a (110) surface of nickel. The units of the ordinate are 10^{-18} cm^2.

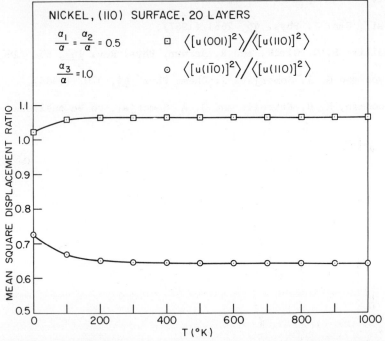

Fig. 2. Temperature dependence of mean square displacement ratios at a (110) surface of nickel.

CALCULATION OF MEAN-SQUARE AMPLITUDES
AND THERMAL DIFFUSE SCATTERING FOR
SURFACES OF NOBLE GAS CRYSTALS*

F. W. de Wette and R. E. Allen
Department of Physics, University of Texas
Austin, Texas

I. INTRODUCTION

Beginning with the work of MacRae and Germer[3] and MacRae[4] on surfaces

of nickel, there have been a number of experimental studies of the dynamics of

crystal surfaces in recent years. Lyon and Somorjai[5] have measured the mean-

square amplitudes for platinum surfaces, Aldag and Stern[6] have observed the

thermal diffuse scattering from tungsten, and Jones, McKinney, and Webb have

carried out careful measurements of both the mean-square amplitudes[7] and the

thermal diffuse scattering intensity[8] for the (111) surface of silver. Since these

experiments have all been carried out on metal surfaces, it would be very difficult

to perform first-principle calculations of the measured quantities. Theoretical

studies have been made of simple models, however, and these show qualitative

agreement with the experimental results. In particular, Maradudin and Melngailis[9]

calculated the mean-square amplitudes for a model with nearest- and next-nearest-

neighbor interactions using a Green's function approach, and Clark, Herman, and

Wallis[10] performed a numerical calculation of these quantities for nickel assuming

a nearest-neighbor interaction. Theoretical studies of the thermal diffuse scattering

have been made by Wallis and Maradudin[11] and Huber[12]. All of the theoretical

results have been obtained for the high-temperature limit and have involved other

approximations, such as neglecting the change in force constants near the surface.

The theoretical calculations of the thermal diffuse scattering have also been based

on the continuum approximation and the assumption that scattering takes place at

the surface plane only.

*   Work supported by the U.S. Air Force Office of Scientific Research under
        Grant No. AF - AFOSR 1257-67.

The present work was motivated by a desire to avoid such approximations, so that a detailed comparison between theory and experiment might be possible. We have previously reported results for various surface quantities for the (100), (111), and (110) surfaces of noble gas crystals which were obtained by means of the quasiharmonic approximation[1] and also by means of an independent method, molecular dynamics[2], which takes anharmonic effects into account completely. In the present paper we are concerned primarily with the thermal diffuse scattering. In Section II the model and the method of calculation are described, and in Section III some results are given for the surfaces of noble gas crystals.

II. PROCEDURE FOR CALCULATIONS

The general model used in these and the earlier calculations[1,2] consists of a crystal of finite thickness, with two free surfaces, in which the atoms interact through a Lennard-Jones (6, 12) potential. The z axis is chosen to be perpendicular to the surfaces, and cyclic boundary conditions are taken in the x and y directions. The position of a plane of atoms parallel to the surface is given by ℓ_3 (not necessarily an integer, because the atoms are displaced from their bulk positions). The position of an atom within a plane is specified by ℓ_1 and ℓ_2.

The approximations involved in our model, such as the finite thickness, are justified in Ref. 1. It is believed that, for argon and the heavier noble gases at low temperatures, the only important assumption involved in calculating the dynamical properties of the crystal is the validity of the quasi-harmonic approximation, and this approximation gives realistic results at low temperatures for these substances. In calculating scattering intensities we also need to use the kinematical approximation--that is, to assume that the radiation field seen by an atom is the original field incident upon the crystal, rather than the original field plus a scattered field. The validity of this approximation for describing low-energy electron diffraction is not certain, but McKinney, Jones, and Webb[7,8] found that their data agree rather well with the results of an analysis based on the kinematical approximation.

Under these assumptions (quasiharmonic and kinematical approximations), the intensity of low-energy electrons scattered elastically from the crystal is given by[11]

$$I(\underset{\sim}{Q}) = I_{inc}\, \sigma_0\, f_0^{\;2} \sum_{\ell_3 \ell_3'} \sum_{\ell\ell'} \alpha\,(\ell_3)\,\alpha\,(\ell_3')\, e^{i\underset{\sim}{Q}\cdot(\underset{\sim}{\ell}-\underset{\sim}{\ell}')}$$

$$\times\; e^{-\tfrac{1}{2}\,\langle\,(\underset{\sim}{Q}\cdot[\underset{\sim}{u}\,(\ell)-\underset{\sim}{u}\,(\ell')\,])^2\rangle} \tag{1}$$

where I_{inc} is the incident intensity, σ_0 is the Thomson factor[13], f_0 is the

atomic scattering factor, $\alpha\,(\ell_3)$ is the transmission factor for the plane labeled

by ℓ_3, $\underset{\sim}{u}\,(\ell)$ is the displacement of the atom labeled by $\ell = (\ell_1,\, \ell_2,\, \ell_3)$ from

its equilibrium position, $\underset{\sim}{Q}$ is the difference between the wavevectors of the

scattered and incident electrons, and the brackets indicate a thermal average.

(For a derivation of Eq. (1), cf. Ref. 13 or 14.) We write two-dimensional

vectors without boldface, so that $\ell = (\ell_1, \ell_2)$ and $Q = (Q_1, Q_2)$. The

exponential in Eq. (1) can be expanded (the so-called phonon expansion) to

give a series of terms in the expression for I,

$$I = I_1 + I_2 + I_3 + \; . \; . \; . \; . \; . \; . \tag{2}$$

I_1 and I_2 are respectively the intensities of the Laue scattering and the first

order thermal diffuse scattering. I_2 may be expressed as follows[11]:

$$I_2\,(\underset{\sim}{Q}) = I_{inc}\,\sigma_0\,f_0{}^2\,\frac{\hbar N}{2M} \sum_{\ell_3 \ell_3'} \alpha\,(\ell_3)\,\alpha(\ell_3')\,e^{-M(\ell_3)}\,e^{-M(\ell_3')}$$

$$\times\,e^{iQ_3\,(\ell_3 - \ell_3')} \sum_{p} [\,\underset{\sim}{Q}\cdot\underset{\sim}{\xi}^*\,(\ell_3;\, Q,\, p)\,]\;[\,\underset{\sim}{Q}\cdot\underset{\sim}{\xi}\,(\ell_3';\, Q,\, p)\,]$$

$$\times\,\frac{1}{\omega_p(Q)}\;\frac{\exp\,(\hbar\omega_p(Q)/k_B T) + 1}{\exp\,(\hbar\omega_p(Q)/k_B T) - 1}\;, \tag{3}$$

where N is the number of atoms in one plane, M is the mass of an atom, $e^{-2M(\ell_3)}$
is the Debye-Waller factor for the ℓ_3 plane, k_B is the Boltzmann constant, T is

the temperature, and $\underset{\sim}{\xi}(\ell_3;\, Q,\, p)$ is the pth eigenvector of the dynamic matrix;

p labels the branches of the function $\omega(Q)$. (All the calculations reported here are

for a crystal 11 layers thick, so that the dynamic matrix is of dimensions 33 by 33

and p = 1, 2, ... , 33.)

For a Lennard-Jones potential, with potential parameters σ and ϵ, one can define the dimensionless temperature

$$T^* = \frac{k_B}{\hbar} \sqrt{\frac{M\sigma^2}{\epsilon}}\; T \qquad , \qquad\qquad (4)$$

and the dimensionless frequency

$$\Omega = \sqrt{\frac{M\sigma^2}{\epsilon}}\; \omega \qquad . \qquad\qquad (5)$$

If in addition we take α_0 to be the transmission factor for the surface plane and a to be the lattice spacing ($\sqrt{2}\, a = $ nearest-neighbor distance), then we can write

$$I_2 = c\, I_2^* \qquad , \qquad\qquad (6)$$

where

$$c = I_{inc}\, \sigma_0\, f_0^{\;2}\, \frac{\hbar N}{2M}\, \sqrt{\frac{M\sigma^2}{\epsilon}}\; \alpha_0^{\;2}/a^2 \qquad , \qquad\qquad (7)$$

and where the dimensionless quantity I_2^* is defined by

$$I_2^*\,(\underset{\sim}{Q}) = \frac{a^2}{\alpha_0^{\;2}} \sum_{\ell_3 \ell_3'} \alpha\,(\ell_3)\,\alpha\,(\ell_3')\, e^{-M(\ell_3)}\, e^{-M(\ell_3')}\, e^{iQ_3(\ell_3-\ell_3')}$$

$$\times \sum_p [\,\underset{\sim}{Q}\cdot\underset{\sim}{\xi}^*\,(\ell_3;\, Q,\, p)\,]\, [\,\underset{\sim}{Q}\cdot\underset{\sim}{\xi}\,(\ell_3;\, Q,\, p)\,]\; \frac{1}{\Omega_p(Q)}$$

$$\times\; \frac{\exp(\,\Omega_p\,(Q)/T^*\,) + 1}{\exp(\,\Omega_p\,(Q)/T^*\,) - 1} \qquad . \qquad\qquad (8)$$

The Debye-Waller factors are related to the mean-square amplitudes by the equation

$$M\,(\ell_3) = \frac{1}{2} \sum_\alpha Q_\alpha^{\;2}\langle u_\alpha^{\;2}\,(\ell_3)\rangle \quad , \quad (\alpha = x,\, y,\, z) \quad (9)$$

and the mean-square amplitudes are given by the equation[14]

$$\langle u_\alpha^{\;2}\,(\ell_3)\rangle = \frac{\hbar}{2NM} \sum_{q,p} \frac{|\xi_\alpha\,(\ell_3;\, q,p)|^2}{\omega_p\,(q)}\; \coth\left(\frac{\hbar\omega_p(q)}{2k_B T}\right) \qquad . \qquad\qquad (10)$$

Once the eigensystem of the dynamic matrix has been determined, therefore,
it is a rather straightforward matter to calculate I_2^* given Q, the temperature,
and the transmission factors $\alpha(\ell_3)$.

In an actual scattering experiment it is convenient to hold $|Q|$ constant
while varying Q_x and Q_y, since this procedure corresponds to tilting and
rotating the crystal while keeping the rest of the apparatus fixed (cf. Ref. 7).
Our calculations, therefore, have been performed for fixed values of $|Q|$,
with Q_x and Q_y ranging over an area which includes the first, or first and
second, two-dimensional Brillouin zones associated with a typical surface
plane of atoms. The regions in which the values of Q were taken to lie are
shown in Fig. 1. The inner region in each case is the first Brillouin zone.
Q ranges over the first and second zones for the (100) surface, the first zone
only for the (110) surface, and a rectangular area including the first zone for
the (111) surface. In each case the values of Q lay on a 40×40 grid, so
that there were 1600 mesh points uniformly distributed throughout the region
shown.

In the calculations for multi-layer scattering, $\alpha(\ell_3)$ was taken to
decrease geometrically with increasing distance from the surface, and the
ratio $\alpha(\ell_3 - 1)/\alpha(\ell_3)$ was taken to be 0.5.[15] In all the calculations the
values used for M, σ, and ϵ were those for argon (taken from Horton and
Leech[16]). The density was taken to be given by $\sigma/a = 1.28$, which is
appropriate for argon near absolute zero.

It should be pointed out that our model does not give physically
realistic results for the scattering intensity for very small values of $|Q| =
(Q_x^2 + Q_y^2)^{\frac{1}{2}}$, because of the finite thickness of the crystal. In a semi-

infinite crystal, there are an infinite number of modes in the summation over

p in Eq. (3). In our model crystal with N_3 planes, however, there are only

$3N_3$ modes. For all values of Q except those very near the reciprocal lattice

points, the frequencies of these modes are rather evenly spaced and so our

procedure gives a rather accurate sampling of the modes in a semi-infinite

crystal. Near a reciprocal lattice point, however, the three acoustic modes

dominate the summation of Eq. (3); since there is no longer an evenly spaced

set of sample modes, the results for the semi-infinite crystal and the crystal

of finite thickness will be considerably different. For example, by using

Eq. (3) and the fact that for the three acoustic modes $w(Q)$ is proportional

to $|Q|$ when $|Q|$ is small, one can easily show that, when Q is near a

reciprocal lattice rod, $I_2(Q)$ for our model is inversely proportional to the

distance to the rod, $|Q - G|$, at absolute zero, and is inversely proportional

to $|Q - G|^2$ in the high-temperature limit. Wallis and Maradudin[11] and

Huber[12] have shown that for a semi-infinite crystal, in the continuum approxi-

mation and the high-temperature limit, $I_2(Q)$ is inversely proportional to

$|Q - G|$, and this behavior has been observed experimentally by McKinney,

Jones, and Webb[8]. Our slab-shaped model and the semi-infinite crystal

thus differ in the values of $I_2(Q)$ when Q is very near a reciprocal lattice

rod, and our method is consequently invalid for calculating the scattering

intensities for a semi-infinite crystal in these regions of reciprocal space.

A related peculiarity of slab-shaped models is the fact that if points

very close to the origin of the first Brillouin zone are included in the sum-

mation over q of Eq. (10), the expression for the mean-square amplitude

diverges logarithmically at finite temperatures[17], since the contribution of

a point very close to the origin is proportional to $1/|q|^2$ and the area of a ring about the origin is proportional to $|q|$. In order for this feature to have an important effect on the results, however, it is necessary to take sample points in the summation of Eq. (10) extremely close to the origin[18] (which corresponds to taking the periodicity length of the crystal very large). In Ref. 1 it is shown that the effect of the finite thickness on the calculated values of the mean-square amplitudes is only a fraction of one per cent at absolute zero and rises to a maximum of a few per cent at high temperatures. The effect increases with temperature because the contribution of the lower frequencies in Eq. (10) increases: at absolute zero the contribution of a term is proportional to $1/\omega$, whereas in the high-temperature limit it is proportional to $1/\omega^2$.

III. RESULTS AND DISCUSSION.

In this section we present some of the results which have been

obtained for the thermal diffuse scattering. In each case the results

are for fixed magnitude of the momentum transfer vector, i.e. for fixed

$| \underset{\sim}{Q} | = (Q_x^2 + Q_y^2 + Q_z^2)^{\frac{1}{2}}$. In Fig. 2 we show for the (100) surface

a doubly logarithmic graph[19] of the calculated thermal diffuse scattering

intensities $I_2 (\underset{\sim}{Q})$ versus $| Q | = (Q_x^2 + Q_y^2)^{\frac{1}{2}}$. The intensities plotted

are for 1600 mesh points uniformly distributed over the first and second

Brillouin zones; i.e., the points lie in all directions about the (00)

reciprocal lattice rod. The intensities shown are for $T = 0^{\circ}K$ and $| \underset{\sim}{Q} | =$

$3 \pi / a$; this value of $| \underset{\sim}{Q} |$ corresponds to the second minimum along the

(00) rod.

The plus marks in the figure represent the intensities for surface

scattering only, and the squares represent those for multi-layer scattering

with $\alpha (\ell_3 - 1)/\alpha (\ell_3) = 0.5$. It is evident that there are important

qualitative differences between the two cases: The intensity for multi-

layer scattering is smaller near the (00) rod than that for single-layer

scattering, but decreases more slowly as $| Q |$ increases and reaches a

weak maximum near the zone boundary. This behavior can be understood

by considering the variation of Q_z as a function of $| Q |$ for fixed $| \underset{\sim}{Q} |$:

Near the (00) rod (small $| Q |$) the interference of radiation scattered by

successive layers is more destructive than it is for larger values of $| Q |$,

while there are, of course, no interplanar interference effects for scattering

from the surface only. (There is a sharp increase in I_2 for the largest

values of $| Q |$ shown, because these values correspond to points close to the nearest reciprocal lattice rods, as can be seen from Fig. 1.)

In Fig. 3 a similar graph for the (110) surface is shown, for $T = 0^{\circ}K$ and $|\underset{\sim}{Q}| = \sqrt{2} \pi / a$ (first minimum along (00) rod). In the (110) case there is a stronger anisotropy (represented by a spread in the intensities at a given $|Q|$) for the larger values of $|Q|$, because the first Brillouin zone is rectangular rather than square.

In order to see how the scattering intensity depends on Q_x and Q_y, rather than just on the magnitude of Q, one can determine the curves of isodiffusion – i.e., the curves mapped in the two-dimensional reciprocal space along which I_2 is constant. We have determined such curves numerically by interpolating between the points (in two-dimensional reciprocal space) for which I_2 was actually calculated.[20] Isodiffusion curves for multi-layer scattering from the (100) and (110) surfaces, under the conditions stated above, (i.e. $T = 0^{\circ}K$, and $|Q| = 3 \pi / a$ and $\sqrt{2} \pi / a$ respectively) are shown in Figs. 4 and 5. The contour lines are spaced logarithmically, with the larger numbers corresponding to the higher intensities. (Numbers larger than 9 are represented by letters, with $A = 10$, $B = 11$, etc.)

In Figs. 6, 7 and 8 results are shown for the (110) surface at a finite temperature, namely $T^* = 10^{22}$, with $|\underset{\sim}{Q}| = 2 \pi\sqrt{2}/a$. Since this value of $|\underset{\sim}{Q}|$ corresponds to the first maximum along the (00) rod, the interference is more constructive for $|Q| = 0$ than for larger values of $|Q|$. The curve for multi-layer scattering consequently drops more rapidly as $|Q|$ increases than that for surface scattering only. Both curves drop more rapidly than those of Fig. 4 because the frequencies, which increase with $|Q|$,

have a larger influence at higher temperatures. (See Eq. (3)). In

Figs. 7 and 8, in which the results for multi-layer and for surface

scattering are respectively shown, it can be seen that there is

again considerable difference in the behavior of $I_2(Q)$ for the two

cases. In addition, comparison of Figs. 5 and 7 shows that there is

considerable difference between the patterns of the curves of iso-

diffusion for different values of $|Q|$.

Recently, McKinney, Jones, and Webb[8] have measured $I_2(Q)$

for values of $|Q|$ corresponding to intensity maxima along the (00)

rod and Q lying within the first Brillouin zone. Their measurements,

which were carried out for the (111) surface of silver, indicate

that $I_2 \propto |Q|^{-1}$ in accordance with the theoretical result obtained

for the continuum approximation in the high temperature limit.[8,11,12]

In comparing these results with the present calculated results we have

to keep in mind that near the zone boundary the continuum approxi-

mation is no longer valid, and that in this region there are also

large experimental uncertainties in the experimental first order

thermal diffuse scattering due to the existence of a temperature-

independent background (resulting from multiphonon scattering[23]).

On the other hand, as pointed out in Sec. II, our results are not

valid for the case of a semi-infinite crystal in the immediate

neighborhood of the reciprocal lattice rods. For intermediate

values of $|Q|$, however, our model is valid and one expects the

continuum approximation also to be valid.

For comparison with the results of Jones, McKinney, and

Webb, it is of interest to calculate I_2 for the (111) surface at

high temperatures. Results for $T^* = 70$, which is well into the

high temperature regime since $\Omega_{max} \simeq 25$, are shown in Figs. 9-11.
It can be seen from Fig. 9 that the curves of isodiffusion are
roughly parallel to the edges of the Brillouin zone. These results
are for $|\underset{\sim}{Q}| = 6\pi\sqrt{3}/2a$ (third maximum along (00) rod), but the
same behavior is observed for other values of $|\underset{\sim}{Q}|$ and other tempera-
tures.   The small departures from hexagonal symmetry in Fig. 9
are caused by the use of a rectangular mesh.

Figures 10 and 11 are respectively for the second maximum
and fourth minimum along the (00) rod.   ($|Q| = 4\pi\sqrt{3}/2a$ and
$7\pi\sqrt{3}/2a$).   It can be seen from these figures that the calculated
intensities for surface scattering indeed lie roughly along a line
with slope -1 for values of Q which are not close to either the
center of the two-dimensional Brillouin zone or the zone boundary.
Near the zone boundary, however, the slope of the curve increases
(because of the presence of other reciprocal lattice rods beyond
the boundary).   In addition, in Figs. 10 and 11 it can be seen that
the intensities are much larger for multi-layer scattering than for
surface scattering, because the mean-square amplitudes are larger
for the first plane of atoms than for the deeper planes, and the
Debye-Waller factor of the first plane is consequently smaller than
those of the deeper planes.

The results given in this paper were obtained for argon,
but one expects the qualitative features, at least, to be characteristic of the crystal structure and not to depend very strongly
on the details of the interaction between the atoms.   It would be
of interest, therefore, if more extensive experimental investigations
could be carried out for the various surfaces of silver, nickel,
platinum, or other materials with an fcc structure.   From the point

of view of the present work, it would, of course, be of extreme interest if measurements of surface effects could be performed for argon or the heavier noble gases, since a direct and detailed comparison with the results of this paper would then be possible.

REFERENCES

1. R. E. Allen and F. W. de Wette, Phys. Rev. 179, (1969).

2. R. E. Allen, F. W. de Wette, and A. Rahman, Phys. Rev. 179 (1969).

3. A. U. MacRae and L. H. Germer, Phys. Rev. Letters 8, (1962), p. 489.

4. A. U. MacRae, Surface Science 2, (1964), p. 522.

5. H. B. Lyon and G. A. Somorjai, J. Chem. Phys. 44, (1966), p. 3707.

6. J. Aldag and R. M. Stern, Phys. Rev. Letters 14, (1965), p. 857.

7. E. R. Jones, J. T. McKinney, and M. B. Webb, Phys. Rev. 151, (1966), p. 476.

8. J. T. McKinney, E. R. Jones, and M. B. Webb, Phys. Rev. 160, (1967), p. 523.

9. A. A. Maradudin and J. Melngailis, Phys. Rev. 133, (1964), p. A1188.

10. B. C. Clark, R. Herman, and R. F. Wallis, Phys. Rev. 139, (1965), p. A860.

11. R. F. Wallis and A. A. Maradudin, Phys. Rev. 148, (1966), p. 962.

12. D. L. Huber, Phys. Rev. 153, (1967), p. 772.

13. M. Born, Rept. Progr. Phys. 9 (1942), p. 294.

14. A. A. Maradudin, E. W. Montroll, and G. H. Weiss, Theory of Lattice Dynamics in the Harmonic Approximation (Academic Press, New York, 1963), p. 237.

15. Jones, McKinney, and Webb found that for an angle of incidence of 68°, the ratio was 0.38 for 50 eV electrons and 0.63 for

271 eV in their experiments.

16. G. K. Horton and J. W. Leech, Proc. Phys. Soc. 82, (1963), p. 81

17. We wish to thank Dr. B. J. Alder for drawing our attention to this fact. The corresponding behavior for two-dimensional crystals is well-known. See, for example, Ref. 14, p. 241-242.

18. In Ref. 1, the effects of changing the thickness and the number of sample points were tested separately, and an independent calculation was performed for the bulk. It was found that the results of surface calculations with 9, 30, and 64 sample points in the irreducible element and the result of the bulk calculation all agreed to within a few percent even at very high temperatures.

19. All the graphs of this paper were plotted by the CDC 6600 computer at the University of Texas.

20. The interpolation was performed by the program CONTOUR, written by J. A. Downing and E. M Greenawalt of the Computation Center of the University of Texas.

21. The regions shown in these figures are slightly smaller than those of Fig. 1 because the outermost mesh points lie on the boundaries. (The linear dimensions are reduced in all cases by 1/40).

22. $T/T^* = 3.5°K$ for argon, so $T^* = 10$ corresponds to 35°K. Changes in density with temperature were not taken into account.

23. R. F. Barnes, M. G. Lagally, and M. B. Webb, Phys. Rev. 171, (1968), p. 627.

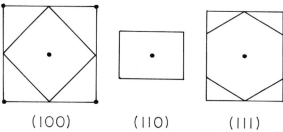

(100)   (110)   (111)

Fig. 1. Regions in two-dimensional reciprocal space for which intensities
were calculated.

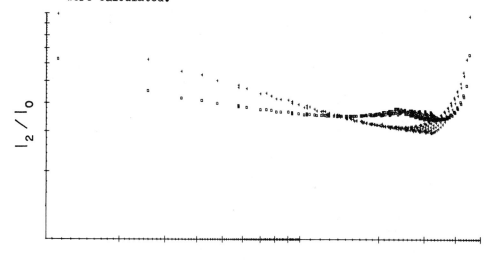

Fig. 2. I_2 versus $|Q| = (Q_x^2 + Q_y^2)^{1/2}$ for (100) surface at $T = 0°K$ with
$|Q| = 3\pi/a$ (second minimum along (00) rod). $I_0 = 21.6$ c, where c is
defined by Eq. (7).

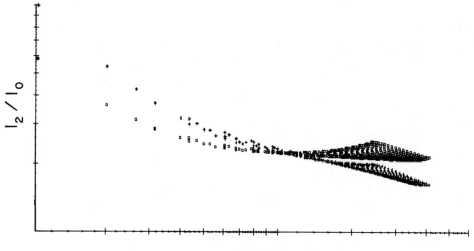

Fig. 3. I_2 versus $|Q|$ for (110) surface at $T = 0°$ K with $|Q| = \sqrt{2}\,\pi/a$ (first
minimum along (00) rod). $I_0 = 10.3$ c.

Fig. 4. Curves of equal intensity I_2 for (100) surface at $T = 0^{\circ}$ K with $|\underset{\sim}{Q}| = 3\pi/a$ (multi-layer scattering).

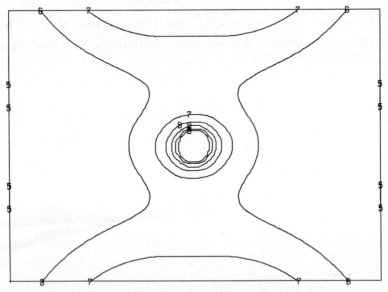

Fig. 5. Curves of equal intensity for (110) surface at $T = 0^{\circ}$ K with $|\underset{\sim}{Q}| = \sqrt{2}\,\pi/a$ (multi-layer scattering).

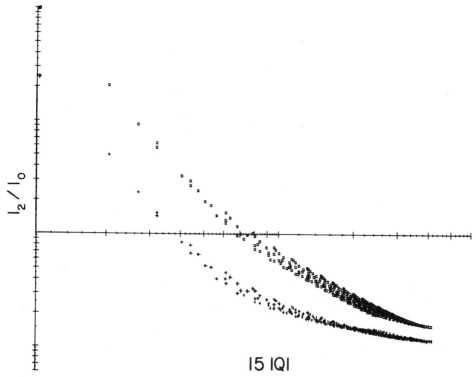

Fig. 6. I_2 versus $|Q|$ for (110) surface at T = 35° K with $|Q| = 2\sqrt{2}\pi/a$ (first maximum along (00) rod). $I_0 = 8.16 \times 10^3$ c.

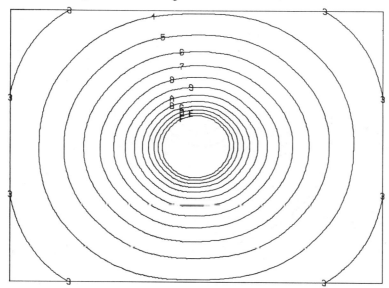

Fig. 7. Curves of equal intensity for (110) surface at T = 35° K with $|Q| = 2\sqrt{2}\pi/a$ (multi-layer scattering).

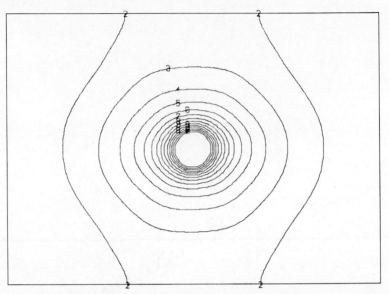

Fig. 8. Curves of equal intensity for (110) surface at T = 35° K with
$|\underset{\sim}{Q}| = 2\sqrt{2}\,\pi/a$ (surface scattering).

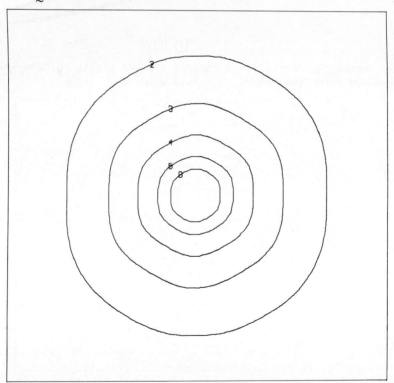

Fig. 9. Curves of equal intensity for (111) surface at T* = 70 with
$|\underset{\sim}{Q}| = 6\pi\sqrt{3}/2a$ (third maximum along (00) rod, multilayer scattering).

18—18

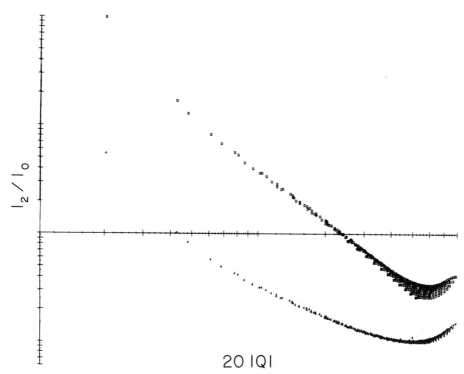

Fig. 10. I_2 versus $|Q|$ for (111) surface at $T^* = 70$ with $|\underset{\sim}{Q}| = 4\pi\sqrt{3}/2a$ (second maximum along (00) rod). $I_o = 7.23 \times 10^2$ c.

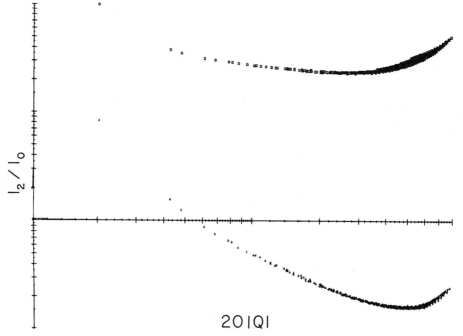

Fig. 11. I_2 versus $|Q|$ for (111) surface at $T^* = 70$ with $|\underset{\sim}{Q}| = 7\pi\sqrt{3}/2a$ (fourth minimum along (00) rod). $I_o = 1.75 \times 10^{-2}$ c.

THE EFFECT OF TEMPERATURE ON LEED INTENSITIES FROM ADSORBED STRUCTURES

Peder J. Estrup

Departments of Physics and Chemistry
Brown University
Providence, Rhode Island

1. Introduction

Thermal treatment is used routinely in low energy electron diffraction (LEED) studies of surfaces to clean and anneal samples, to induce the formation of different surface structures, to accelerate any surface reactions and to control rates of adsorption and desorption. In addition it is frequently possible to use temperature variations to obtain quantitative information concerning the atomic processes at the surface, the main limitations being the initial order of the given surface structure and the extent to which the diffraction data can be interpreted unambiguously.

An important example of this type of study is the measurement of the intensity, I, of a diffracted beam as a function of temperature, which for clean surfaces (in cases where no structural change occurs) can provide data on surface lattice dynamics[1]. Results are now available for surfaces of Ni[2], Pt[3], Ag[4], Pd and Pb[5] and in all cases it has been found that the intensity decrease with temperature to a good approximation follows the Debye-Waller relation

$$I \propto \exp(- 2M) \tag{1}$$

with $M \propto \dfrac{T}{m\theta^2}$ at high temperatures. T is the temperature, m the mass of the scatterer and θ the effective Debye temperature. In the Debye model of lattice vibrations θ is inversely proportional to the root mean square displacement of the atoms; this displacement is expected to be relatively large for surface atoms and the fact that $\theta_{(surface)}$ consistently is found to be smaller than $\theta_{(bulk)}$ (as determined by e.g., X-ray diffraction) is thus readily interpreted. There are still difficulties in the quantitative prediction of θ but the main features of I(T) for clean surfaces can be accounted for.

For adsorbed structures the situation is more complex: The diffraction from such layers is incompletely understood and conflicting models for the formation of these structures have been proposed[6]; both the substrate and the adsorbate may contribute to the diffracted intensity and processes other than surface atom vibration may become highly probable as the temperature is raised. Some of these problems have been discussed by Lander[7] and have been illustrated by the behaviour of the Si(111) + Aℓ system[8]. In this paper the temperature effect on LEED intensities from some adsorbed layers on the tungsten (100) surface is examined and the application to chemisorption problems is discussed.

2. Adsorbed Structures

LEED studies of the W(100) surface and its interaction with a number of adsorbates have been described previously[6,11,13-16].

Figure 1 shows the dependence on temperature of the 10 beam intensity for the clean surface. As expected from the Debye-Waller equation (1) a linear relationship is found between the logarithm of the intensity and the temperature, the slope giving a surface Debye temperature of $183°K$[9]. This result may be compared with the value $\theta = 280°K$ for bulk tungsten[10].

Adsorption on this surface usually leads to the appearance of "extra" spots in the LEED pattern. For the simpler surface structures the location of these spots suffices to determine the new (lateral) periodicity but it is difficult to establish the nature of the scattering centers induced by the adsorption and different interpretations have therefore been given of the LEED results. In one model it is proposed that the diffraction is due to relocated substrate atoms with negligible scattering by the adsorbate; another model assumes that the adsorbed species themselves produce a sufficient change in the scattering potential at the surface[6,11]. Multiple scattering[12] may be important in either case but it is believed that the distribution of diffracted intensity over a reciprocal unit mesh (at a given wavelength) can be treated by the kinematical theory.

The effect of temperature on the intensity of the extra beams will depend on the nature of the scatterers in the adsorbed layer. If substrate atoms are res- ponsible for the effect, the Debye-Waller factor may be expected to have the same order of magnitude as for beams from the clean surface. On the other hand, if the adsorbed species are directly involved in the diffraction process, the tem- perature effect should show large variations from one adsorbate system to another.

A qualitative test is provided by the W(100) + O_2 system[13]. The structures produced by oxygen interaction include the (4 X 1) structure, formed at room tem- perature, and the (2 X 1) structure which results after heat treatment. Studies of the associated work function changes and the electron desorption cross-sections indicate that in the (4 X 1) structure the oxygen is located on top of the tungsten but that formation of the (2 X 1) structure involves penetration of oxygen into the substrate. The temperature effect on the LEED patterns leads to the same conclusion if it is assumed that the room temperature pattern is due to scat- tering by oxygen. This may be seen from Figure 2 where the $\frac{1}{2}$ O beam intensities from both structures are shown. The (4 X 1) intensity exhibits a rapid decay with temperature whereas the (2 X 1) intensity decreases much more slowly, at a rate comparable to that for clean W(100).

This result suggests that the temperature effect may be generally useful in distinguishing between "reconstruction" and simple adsorption. By this criterion most of the   adsorbates which we have studied interact with the W(100) surface without reconstruction. As examples, Figure 2 shows the behaviour of the inten- sity, $I(\frac{1}{2}\frac{1}{2})$ from the C(2 X 2) - Th[14] and the C(2 X 2) - H[11] structures and the curves for the C(2 X 2) - N[15] and the C(2 X 2) - CO[16] structures (not shown) fall between those for Th and (4 X 1) - O. It is clear that different adsorbates lead to quite different intensity dependence but it is also seen that the Debye-Waller equation is not obeyed. The decay is not exponential and the mass dependence predicted by (1) is not obtained. Additional evidence is provided by a comparison

of LEED intensities from adsorbed hydrogen and deuterium structures. The latter

adsorbate[17] produces the same LEED patterns as hydrogen[15] (including the beam

splitting at coverages above half a monolayer) and within the experimental uncer-

tainty no difference is observed in the temperature effect on $I(\frac{1}{2}\frac{1}{2})$ from the

two C(2 X 2) structures. It appears, therefore, that in these systems the vibra-

tion of surface atoms is a minor factor in the intensity decrease.

3. Order-Disorder Model

In order to describe the temperature effect we consider the mechanism used

to discuss order-disorder phenomena in alloys[18,19,20]. The application to the

present problem is illustrated schematically in Figure 3 A and B. A represents

the perfect C(2 X 2) arrangement with all the hydrogen atoms on "right" sites.

At an elevated temperature a fraction w = 1 - r of the atoms have jumped to "wrong"

sites and a disordered structure, B, is the result. In the Bragg-Williams (B. - W.

approximation the structure is characterized by a single long range order parameter

S, given by

$$S = 2r - 1 \qquad\qquad (2)$$

which is unity for the perfectly ordered arrangement and zero for the random ar-

rangement. We assume that the diffracted intensity can be calculated by the ex-

pression[6]

$$I \propto \sum_{m}^{N_o} (N_o - |m|) \, P_m \, \exp(i \, m \, \overline{k}\cdot\overline{a}) \qquad\qquad (3)$$

where N_o is the number of sites ($\frac{N_o}{2}$ the number of atoms), $\overline{a}$ is the lattice vector

and P_m the probability that a given site and a site at a distance $m\overline{a}$ are both oc-

cupied. (For simplicity the one-dimensional expression is given; generalization

to two dimensions is straightforward). When m is even it is seen that

$P_m = \frac{1}{2} r^2 + \frac{1}{2} w^2$ which, from (2), is equal to $\frac{1}{4} (1 + S^2)$. Similarly, for m odd,
we obtain $P_m = \frac{1}{2} rw + \frac{1}{2} wr = \frac{1}{4} (1 - S^2)$. Insertion into (3) and summation yields

$$I \propto \frac{1}{4} \frac{\sin^2(\frac{1}{2} N_o \, \bar{k} \cdot \bar{a})}{\sin^2(\frac{1}{2} \bar{k} \cdot \bar{a})} + \frac{N_o}{4} (1 - S^2) + \frac{1}{4} S^2 \frac{\sin^2(\frac{1}{2} N_o \, \bar{k} \cdot \bar{a})}{\cos^2(\frac{1}{2} \bar{k} \cdot \bar{a})} \qquad (4)$$

The first term in (4) gives the "normal" beams, the second term a uniformly distributed background and the last term the "extra" beams. Hence, for the disordered C(2 X 2) structure we have

$$I(\tfrac{1}{2} \tfrac{1}{2}) \propto S^2$$

and the temperature effect can be predicted if S(T) can be determined. The B. - W.
treatment leads to the well-known result[18,19]

$$S = \tanh (S \frac{T_c}{T}) \qquad (5)$$

for the equilibrium condition, where T_c is the critical temperature. T_c depends on
the interaction energy between neighboring species and this model therefore predicts the same temperature effect for hydrogen and deuterium layers.

In Figure 4 normalized results for $I(\tfrac{1}{2} \tfrac{1}{2})$ are plotted against temperature and
compared with the values calculated from (5) using T_c = 625°K (solid curve). As
seen, a reasonable fit is obtained and the model appears to contain the essential
features of the temperature effect. However; even though the experimental accuracy is rather poor, the data show a definite deviation from prediction as T
approaches T_c. This may be due in part to a substrate effect: it is likely that
the diffraction involves multiple scattering[11,12] and thermal agitation of the
tungsten atoms will tend to reduce the intensity of the "extra" spots as well. In
the range 500-600°K a reduction of 5-10% may be expected.

Probably the main source of the discrepancy is the approximations made in the
B. - W. method and more refined treatments predict a much steeper approach to the

critical point[20]. Of particular interest here is the Ising model for which the exact solution in two dimensions is known[20,21] and which predicts the temperature dependence $I \propto (1 - \frac{T}{T_c})^{1/4}$ (as contrasted to the B. - W. result $I \propto (1 - \frac{T}{T_c})$, near T_c.) The broken curve in Figure 4 shows this dependence for $T_c = 550°K$.

It should be possible to improve the experimental technique and obtain data that permit a more detailed comparison with theory. It may be noted, however, that most statistical-mechanical calculations consider nearest neighbor inter-actions only, whereas the adsorption experiments indicate that long range inter-actions are important[11] ("island" formation, superstructures with large unit mesh). The nature of these interactions has been discussed recently[22].

4. Other Adsorption Systems

The W(100) + H (or D) surface is one of the most convenient systems for a quantitative study of the order-disorder transition since the low critical tem-perature of the adsorbate structure and the relatively high surface Debye tem-perature make for an easy separation of adsorbate and substrate effects[23]. For other adsorbates, with higher T_c, it will be more difficult to substract the pos-sible contributions from both substrate and adatom vibrations but at least a qualitative ranking of the adsorbates (e.g., $T_c(N) > T_c(H)$) is feasible until a better understanding of the diffraction process is reached.

Future work may also involve studies of adsorbates which produce several distinct structures so that the influence of the number of nearest neighbors can be investigated. In addition it is anticipated that I(T) for some adsorbates will show hysteresis effects and that the kinetics of approach to equilibrium can be studied.

Acknowledgments

The experimental data were obtained in collaboration with Mr. J. Anderson. This work was supported in part by the Advanced Research Project Agency.

References

1. R. F. Wallis, Surface Sci. 2, 146 (1964). See also papers by R. F. Wallis, B. C. Clark, R. Herman, This Symposium; A. A. Maradudin, D. L. Mills, S. Y. Tong, ibid; M. G. Lagally, M. B. Webb, ibid.

2. A. U. MacRae, Surface Sci. 2, 522 (1964).

3. H. B. Lyon, G. A. Somorjai, J. Chem. Phys. 44, 3707 (1966).

4. J. T. McKinney, E. R. Jones, M. B. Webb, Phys. Rev. 151, 476 (1966) 160, 523 (1967).

5. R. M. Goodman, H. H. Farrell, G. A. Somorjai, J. Chem. Phys. 48, 1046 (1968).

6. P. J. Estrup, J. Anderson; Surface Sci. 8, 101 (1967).

7. J. J. Lander, Progress in Solid State Chemistry 2, 26 (1965).

8. J. J. Lander, J. Morrison, Surface Sci. 2, 553 (1964).

9. The data of Figure 1 were obtained at a beam voltage of 150 V. Measurements at lower voltages have not been completed in time for inclusion in this report.

10. A. Guinier, X-Ray Diffraction. W. H. Freeman and Co., San Francisco (1963) p. 192.

11. P. J. Estrup, J. Anderson, J. Chem. Phys. 45, 2254 (1966).

12. E. G. McRae, Surface Sci. 11, 492 (1968).

13. P. J. Estrup, J. Anderson, Rept. 27th Annual Physical Electronics Conference M.I.T. (Cambridge) 47 (1967).

14. P. J. Estrup, J. Anderson, W. E. Danforth, Surface Sci. 4, 286 (1966).

15. P. J. Estrup, J. Anderson, J. Chem. Phys. 46, 567 (1967).

16. J. Anderson, P. J. Estrup, J. Chem. Phys. 46, 563 (1967).

17. P. J. Estrup, J. Anderson, Unpublished results.

18. W. L. Bragg, E. J. Williams, Proc. Roy. Soc. A145, 699 (1934).

19. C. Domb, Advances in Physics 9, 149, 245 (1960).

20. M. E. Fisher, Rep. Progr. Phys. 30, 615 (1967).

21. L. Onsager, Nuovo Cim. (Suppl) 6, 261 (1949).

22. T. B. Grimley, Proc. Phys. Soc. 90, 751 (1967).

23. For purposes of comparison with theory a "physisorbed" layer, e.g., a noble gas on tungsten, might provide a more favorable system, but the required cooling of the sample presents additional experimental difficulties.

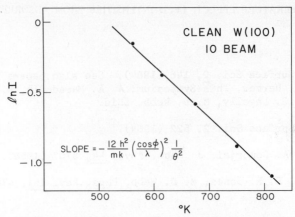

Fig. 1. Plot of $\ln$ I vs. temperature for the 10 beam from clean W(100). Normal incidence at beam voltage 150 V. The slope yields θ = 183°K.

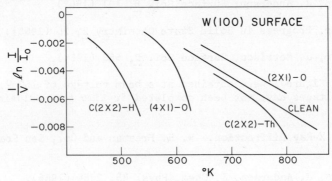

Fig. 2. Effect of temperature on various adsorbed structures on W(100). For the hydrogen and thorium structures the curves give the ½ ½ spot intensity; for the oxygen structures the ½ 0 spot intensity is shown. Beam voltages in the range 125-150 V.

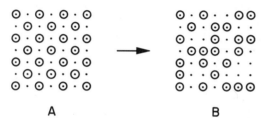

A   B

Fig. 3. Schematic representation of the disordering of the C(2 X 2) - H
structure. A: perfect (low temperature) arrangement. B: disordered
(high temperature) arrangement.

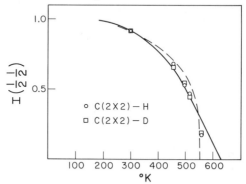

Fig. 4. Plot of $I(\frac{1}{2} \frac{1}{2})$ vs. temperature for the C(2 X 2) - H and C(2 X 2) - D
structures. Data taken at 45 Volts. The solid curve gives the B. - W.
values (S2) for T_c = 625°K, the broken curve those for the Ising
model with T_c = 550°K.

EFFECTS OF PHONON SCATTERING AND THE ATOMIC SCATTERING FACTOR IN LOW-ENERGY ELECTRON DIFFRACTION*

Max G. Lagally and Maurice B. Webb

University of Wisconsin
Madison, Wisconsin

I. INTRODUCTION

This paper presents the results of low-energy electron diffraction experiments from the (111) surface of Ni at high temperatures and is an outgrowth of work on multiphonon scattering. In an earlier paper[1] it was shown that the observed thermal diffuse scattering could be separated into one and multiphonon components. The authors used a pseudokinematic description of the diffraction and considered the scattering from a layer of atoms whose thermal motion is described by a Debye spectrum in the high-temperature limit, but with a Debye temperature appropriate to the larger surface vibrational amplitudes. They showed that

$$R = \frac{\int_{zone} I_{multiphonon} (\underline{S}) \, d^2S}{\int_{zone} I_{Bragg} (\underline{S}) \, d^2S} = (e^{2M} - 1 - 2M), \tag{1}$$

where R is the ratio of the integrated intensity in the multiphonon scattering to the integrated intensity in the associated Bragg peak and $2M$ is the exponent in the Debye-Waller factor. For fixed scattering angles, the multiphonon scattering was assumed to be uniform in the Brillouin zone. The energy and temperature dependences of the observed intensities were compared with Eq. (1) and found to agree. This result lent support to the assumption of uniform distribution of the multiphonon scattering and made it possible to account for all of the observed elastic[2] scattering.

Even at room temperature the phonon scattering contributes appreciably to the observed intensities; at elevated temperatures the multiphonon component dominates the elastic scattering. It is the purpose of this paper to describe some of the effects of the phonon scattering on other low-energy electron diffraction experiments and to use its properties to determine separately the factors which appear in a kinematic description of the diffraction, i.e., the effective atomic scattering factor and the interference function. In Sec. II we briefly describe the apparatus

* This work supported by the U.S. Air Force Office of Scientific Research Grant No. AF-AFOSR-51-66.

and pertinent experimental parameters. In Sec. III we outline some of the properties
of the phonon scattering, present results on the integrated elastically scattered
intensity and its temperature and energy dependence, and establish experimentally
the uniform distribution of the multiphonon scattering in the Brillouin zone. In
Sec. IV, we use these data to extract the effective atomic scattering factor as a
function of scattering angle and energy. Finally, in Sec. V, we show the effect
of temperature on the intensity distribution along the (00) reciprocal-lattice rod,
separate the observed intensity into phonon and Bragg components, and comment on
the rigid-lattice interference function.

II. EXPERIMENTAL

Much of the apparatus used in these experiments has been described previously. [1]
Briefly, it consists of a two-circle goniometer, moveable Faraday collector for
direct measurement of the scattered intensity, and nearly hemispherical Ehrenberg
grids[4] and a fluorescent screen for visual observation of the diffraction and
integrated-intensity measurements. The Faraday collector aperture is a .040 in.
diam. hole and subtends 23 mrad at the crystal. Observed diffraction maxima obtained
by changing the tilt of the crystal with respect to the incident beam have a full
width at half maximum of 12 mrad, which is almost entirely due to instrumental width.
The collector is biased 2.7V positive with respect to the electron gun filament to
accept elastically scattered electrons only.

For integrated-intensity measurements, the fluorescent screen is used as the
collector. The inner of two concentric grids is held at crystal potential, while
the outer grid is biased 2.7V positive with respect to the electron gun filament.
The fluorescent screen is held 6V positive with respect to the outer grid. The
collector actually subtends 140° and has a small hole to admit the incident beam.
The energy resolution of the system is 1.5 eV. Integrated elastically scattered
intensities vary between 1 and 2% of the incident beam over the energy range 25 to
450 eV.

The experiments we will report were done on the (111) surface of a Ni single crystal. Preparation of the surface was discussed previously.[1] The crystal was cleaned by heating it to 1000°C in an H_2 atmosphere of 10^{-7} Torr for 2 min., after which the system was pumped down to 5 x 10^{-10} Torr. This procedure was repeated 6 to 10 times until no fractional order maxima were observed. Peaks with intensities .05% of the clean-Ni diffraction peaks would have been detected. If the crystal was left at room temperature for several hours, especially after the H_2 treatment, an apparent amorphous impurity layer formed on the surface, which at low energies enhanced the elastic scattering and at higher energies diminished it. This layer could be removed by heating to 200°C, and its desorption observed by monitoring the elastic intensity as the temperature was increased. To assure cleanliness, all the data presented here were taken at temperatures above 200°C, or immediately after removal of this layer.

III. MULTIPHONON SCATTERING

It is clear that a kinematic theory of low-energy electron diffraction is inadequate in many respects;[5] however, this simple description does provide a very convenient, and at present the only, framework within which to discuss the experiments on phonon scattering. It should be stated in the beginning that most of the analysis of the data and the conclusions in this paper are meaningful quite independent of this theoretical description, a point that will become explicit in later sections.

Summarizing the kinematic description, we write the time-averaged intensity scattered per unit solid angle per unit incident intensity as

$$I(\underline{S}) = |f(\theta,E)|^2 \sum_{i,j} \langle e^{i\,\underline{S}\cdot(\underline{r}_i-\underline{r}_j)}\rangle = |f(\theta,E)|^2 \mathscr{L}(\underline{S}), \tag{2}$$

where $f(\theta,E)$ is the effective atomic scattering factor, $\underline{S} = \underline{k} - \underline{k}_o$ is the diffraction vector, and $\mathscr{L}(\underline{S})$ is the interference function. If the $\underline{r}$'s are the equilibrium atomic positions this becomes the interference function for the rigid lattice, $\mathscr{L}_o(\underline{S})$. Now letting $\underline{r}_i$ be the equilibrium position of the i'th atom and $\underline{u}_i$ its instantaneous displacement,

$$I(\underline{S}) = |f(\theta,E)|^2 \sum_{i,j} e^{i\underline{S}\cdot(\underline{r}_i-\underline{r}_j)} \langle e^{i\underline{S}\cdot(\underline{u}_i-\underline{u}_j)} \rangle \tag{3}$$

$$= |f(\theta,E)|^2 e^{-2M}\left\{ \mathscr{L}_o(\underline{S}) + \sum_{i,j} e^{i\underline{S}\cdot(\underline{r}_i-\underline{r}_j)}\langle(\underline{S}\cdot\underline{u}_i)(\underline{S}\cdot\underline{u}_j)\rangle \right.$$

$$\left. + \sum_{i,j} e^{i\underline{S}\cdot(\underline{r}_i-\underline{r}_j)}\left[\langle e^{(\underline{S}\cdot\underline{u}_i)(\underline{S}\cdot\underline{u}_j)}\rangle - 1 - \langle(\underline{S}\cdot\underline{u}_i)(\underline{S}\cdot\underline{u}_j)\rangle\right]\right\} \tag{4}$$

$$= |f(\theta,E)|^2\left\{\mathscr{L}_{Bragg}(\underline{S}) + \mathscr{L}_{one-phonon}(\underline{S}) + \mathscr{L}_{multiphonon}(\underline{S})\right\}, \tag{5}$$

where $2M = \langle(\underline{S}\cdot\underline{u})^2\rangle$ is the exponent in the Debye-Waller factor.

To evaluate $I(\underline{S})$ as a function of $\underline{S}$ directly would be prohibitively difficult. It has been shown,[1] however, that the interference function integrated over a Brillouin zone is given by

$$\int_{zone}\mathscr{L}(\underline{S})\, d\underline{S} = e^{-2M}\left[1 + 2M + (e^{2M}-1-2M)\right]\int_{zone}\mathscr{L}_o(\underline{S})d\underline{S} = constant, \tag{6}$$

where the integration is over the area of a Brillouin zone for nonpenetrating radiation, and over the volume of a zone for penetrating radiation.[6]

The three terms are the integrated zero , one , and multiphonon interference functions. Their relative values are plotted as a function of 2M in Fig. 1. For Ni at room temperature and an incident-electron energy of 100 eV, $2M \approx 1.2$, and the three components make nearly equal contributions to the integrated intensity. At higher energies and temperatures, the first two components decrease rapidly and the multiphonon contribution approaches one asymptotically. The sum of the three components is a constant independent of 2M and thus of the position of the atoms.

The integrated intensity is then given by the sum of the three terms in Eq. (6) each multiplied by an appropriate average of $|f(\theta,E)|^2$. Since these averages will be nearly the same, the integrated intensity should be constant. To check this experimentally, we measure the elastically backscattered current to the nearly hemispherical collector. (7)   In Fig. 2 the integrated current is plotted as a function of temperature for various energies and in Fig. 3 it is plotted as a function of energy at both a high and a low temperature. The main feature in Fig. 2 is that the elastic current is nearly temperature-independent, even though in this temperature range the Bragg intensities change by as much as a factor of thirty. Fig. 3 shows that there is considerable structure in the energy dependence of the elastic current at 357°K; this has become much less pronounced at 880°K. It is evident also that the two curves oscillate roughly about the same smooth curve, which reflects the energy dependence of the effective atomic scattering factor averaged over back angles.

Variations of the integrated intensity with temperature and the structure in its energy dependence arise because the experimental integration is not bounded by zone boundaries, nor is it over the whole volume of the 3-dimensional zones. Thus at an energy where several third-Laue-condition maxima lie within the shell of integration, the intensity which is removed from these maxima with increasing temperature is spread throughout the associated zones and some is no longer collected. Alternatively, maxima just off the Ewald sphere will add diffuse intensity to the measured current as the temperature is raised. These effects are small at high energies or temperatures, where much of the intensity is already in the multiphonon component. The reflections primarily responsible for the structure in Fig. 3 are those on the (00) reciprocal-lattice rod. This can be seen by comparing the maxima and minima with those in Fig. 6.

Incidentally, these variations in the integrated elastic current may lead to an apparent temperature dependence in the integrated inelastic scattering which is not related to the inelastic cross sections. It has been shown(8) that, at least for $\Delta E < 15$ eV, the inelastically scattered electrons arise from a two-step process: an inelastic scattering followed by a Bragg reflection or vice versa. The temperature-dependent structure in the integrated Bragg reflection can lead to a spurious temperature-dependent structure in the energy-loss spectrum. This problem can be eliminated by operating at very high temperatures or else by dividing the inelastic intensity by the appropriate elastic intensities.

We have seen then that experimentally the integrated elastically scattered intensity is nearly independent of temperature, in accord with Eq. (6). This feature is actually a general result of the kinematic theory independent of assumptions about the thermal motion of the atoms.[9] If in Eq. (2) the terms with $i = j$ are separated and the result is integrated over the Brillouin zone,

$$\int_{zone} I(\underline{S}) d\underline{S} \propto N |f(\theta,E)|^2 + \int_{zone} |f(\theta,E)|^2 \sum_{i \neq j} \langle e^{i \underline{S} \cdot (\underline{r}_i - \underline{r}_j)} \rangle d\underline{S}$$

$$\propto N |f(\theta,E)|^2, \tag{7}$$

since the oscillating terms in the sum each integrate to zero. To our knowledge, the integrated intensity as the crystal is disordered has not been examined in the more complete theoretical descriptions of low-energy electron diffraction.

We have assumed throughout that the multiphonon part of the interference function is uniform in the Brillouin zone. This can be checked simply at energies and temperatures where 2M is large and the scattering is essentially all multiphonon. If the scattered intensity is measured as a function of the scattering angle 2θ at each of several angles of incidence, any structure in the interference function will move with the orientation of the crystal. Fig. 4 shows the result for three angles of incidence at the energy corresponding to the (666) reflection and a temperature of 850°K. The residual (00) and (10) Bragg reflections, indicating the centers of the Brillouin zones, move with changes in the angle of incidence as expected. There is no other structure that moves with the orientation of the lattice; hence the multiphonon component of the interference function must be uniform to the extent to which the three curves coincide away from the Bragg peaks.

IV.   THE EFFECTIVE ATOMIC SCATTERING FACTOR

We now determine the effective atomic scattering factor for low-energy electrons scattered from Ni.   In Fig. 4, we subtract the zero and one phonon components by drawing the heavy line, leaving only the multiphonon scattering versus scattering angle 2θ.   An alternative statement, not in the language of the kinematic description, is that this procedure separates the scattering into a part which depends on the crystal and its orientation and a part which does not.

Curves like Fig. 4 have been obtained for energies between 65eV and 400eV at 850°K.   They are symmetric about $2\theta = 180°$ as expected for the atomic scattering factor.   One might expect the curves to be skewed because of the attenuation along the different path lengths on either side of the incident beam, but this is not apparent in the Ni data.

Using Eq. (6) and the fact that the multiphonon interference function is uniform gives

$$I_{multiphonon} \ (\underline{S}) \propto |f(\theta,E)|^2 \ e^{-2M}(e^{2M} - 1 - 2M). \tag{8}$$

Dividing by $e^{-2M}(e^{2M} -1- 2M)$ leaves the angular dependence of $|f(\theta,E)|^2$, which is plotted for several energies in Fig. 5. Division by $e^{-2M}(e^{2M} -1- 2M)$ effectively smears out the intensity in the residual Bragg peaks and gives the angular dependence that would obtain at infinite temperature.   At 850°K and throughout the angular range covered $e^{-2M}(e^{2M}-1- 2M)\approx 1$ for all but the lowest energies. Thus $|f(\theta,E)|^2 \approx I_{multiphonon}(\underline{S})$ and Fig. 5 is nearly independent of any analysis of the phonon scattering.

The square of the atomic scattering factor $|f(\theta,E)|^2$ is of central importance in the interpretation of diffraction experiments, but it is experimentally available for only a few atoms and then generally from gas data. (10)   The present procedure is both simple and applicable to most all materials investiaged by low-energy electron diffraction.   However, several effects remain to be considered.   These include multiple scattering, variations of the absorption coefficient with energy, and the influence of the inner potential.   We are investigating these effects for Ni, Ag, and liquid Hg.   Our estimate is that they are small in Ni.

V.   THE (00) RECIPROCAL-LATTICE ROD

We now examine the intensity distribution along the (00) reciprocal-lattice rod and its temperature dependence. Data for the intensity versus energy for a low and a high temperature are plotted in Figs. 6 and 7. As the temperature increases the peaks decrease because of the Debye-Waller factor, while the contribution of the phonon scattering to the observed intensity increases.

In Fig. 8 we replot the observed intensity at 850°K on an expanded scale and also show the multiphonon scattering determined from curves such as those in Fig. 4. Clearly the multiphonon scattering makes a large contribution along the rod; in fact between the peaks the observed intensity is nearly all multiphonon instead of Bragg scattering. This is true also for the low-temperature data at higher energies as is shown in Fig. 9. At lower energies, however, the valleys do not go to zero. Here the large one-phonon contribution along the rod makes it difficult to separate $I_{Bragg}(S)$, but any one-phonon subtraction will make the valleys deeper than in the measured data. The feature of essentially zero Bragg scattering between peaks on the reciprocal-lattice rod is inconsistent with the kinematic description using any reasonable absorption coefficient; however, it is a result of dynamical theories

The multiphonon component along the rod also affects the Debye-Waller measurements, causing the intensity to fall more slowly than e^{-2M}. This is shown fo the (666) reflection in Fig. 10. Subtracting the multiphonon component gives the lower line in Fig. 10. The accuracy of the multiphonon subtraction is insufficient to make any statement about the remaining curvature. (12) At low temperatures the slope is not sensitive to small errors in the subtraction; all values of 2M used in this paper have been determined from the limiting slopes at low temperature of plots like Fig. 10. Values of 2M have been determined for all main and subsidiary peaks and some valleys. These data and their analysis will be presented elsewhere.

Finally, we would like to comment on $\mathcal{L}_o(S)$, the interference function for the

rigid lattice. This is the function calculated in many theoretical investigations.
To determine it experimentally from the intensity distribution along the (00)
reciprocal-lattice rod, it is necessary to consider the phonon scattering, the
Debye-Waller factor, the effective atomic scattering factor, and a factor which
relates the elements of solid angle in the experiment to a volume element in
reciprocal space, analogous to the Lorentz factor in x-ray scattering. To account
for these is difficult: at energies where the intensity is small $\mathscr{A}_0(\underline{S})$ is sensitive
to errors in subtracting the multiphonon component; the Debye-Waller correction
amounts to a long extrapolation and is sensitive to small errors in 2M; and the
electron optics are not well enough known to make the Lorentz correction reliably.
Because of these uncertainties $\mathscr{A}_0(\underline{S})$ has not been determined all along the reciprocal-
lattice rod, but to illustrate the effect of these corrections, we list in Table I
the resulting magnitudes and widths of some third-Laue-condition peaks. Even for
these the uncertainties are large, and the table is presented only to indicate that
$\mathscr{A}_0(\underline{S})$ much more nearly repeats from zone to zone than is apparent in the observed
intensity.

Peak	Energy (eV)	Measured Relative Intensity (% of incident current)	$\lvert\underline{S}\rvert$ $(\mathring{A}^{-1})$	$\mathscr{A}_0(\underline{S})$ (Arbitrary units)	FWHM $\delta(\lvert\underline{S}\rvert)$ $(\mathring{A}^{-1})$
(555)	215	.113	14.7	1.0	0.6
(666)	320	.015	17.9	0.93	0.5
(777)	435	.003	20.9	0.84	0.4

VI. CONCLUSIONS

In this paper we have presented experimental results on the effects of temperature on low-energy electron diffraction. Several conclusions have been reached:

1. The multiphonon scattering is a large fraction of the observed intensity even for low energies at room temperature.

2. The elastically scattered intensity integrated over a Brillouin zone is independent of thermal disorder.

3. The multiphonon scattering is uniformly distributed in a Brillouin zone.

4. There is a diffuse component of the observed intensity which is independent of the crystal and its orientation and which gives the energy and scattering angle dependence of the square of the effective atomic scattering factor.

5. The intensity along the (00) reciprocal-lattice rod is essentially zero between the diffraction peaks after the diffuse multiphonon intensity has been subtracted.

6. Estimates of the rigid-lattice interference function along the (00) reciprocal-lattice rod show that it much more nearly repeats from zone to zone than is apparent in the observed scattering.

ACKNOWLEDGMENTS

The authors would like to thank W. N. Unertl and C. D. Gelatt, Jr. for their assistance and W. H. Weber and J. S. Schilling for helpful discussions.

REFERENCES

1. R. F. Barnes, M. G. Lagally, and M. B. Webb, to appear in Phys. Rev. July 15, 1968.

2. We include in "elastically scattered electrons" those whose energy has changed by the order of phonon energies. This corresponds to our experimental situation.

where the energy resolution is limited by the thermal spread from the electron source.

3. E. R. Jones, J. T. McKinney, and M. B. Webb, Phys. Rev. 151, 476 (1966).

4. E. J. Scheibner, L. H. Germer, and C. D. Hartman, Rev. Sci. Instr. 31, 112 (1960); 31, 675 (1960).

5. See for example: J. J. Lander and J. Morrison, J. Appl. Phys. 34, 3517 (1963),

    G. Gafner, Surface Sci. 2, 534 (1964),

    E. G. McRae, J. Chem. Phys. 45, 3258 (1966),

    D. S. Boudreaux and V. Heine, Surface Sci. 8, 426 (1967),

    K. Kambe, Z. Naturforsch. 22a, 322 (1967).

6. S. V. Semenovskaya and Ya S. Umanskil, Fiz. Tverd. Tela 6, 2963 (1964)[English transl.: Soviet Phys. - Solid State 6, 2362 (1965)].

7. The collector actually subtends 140°. The crystal was tilted 12° from normal so that the (00) beam was collected instead of being reflected back along the incident beam.

8. W. H. Weber and M. B. Webb, Bull. Am. Phys. Soc. 13, 89 (1968).
    See also L. N. Tharp and E. J. Scheibner, J. Appl. Phys. 38, 3320 (1967).

9. A. Guinier, X-Ray Diffraction (W. H. Freeman and Company, San Francisco and London, 1963) page 29.

10. A discussion of atomic scattering factors in low-energy electron diffraction experiments is given by J. J. Lander and J. Morrison, J. Appl. Phys. 34, 3517 (1963). Equivalent data for polycrystalline samples at high temperature have been obtained in experiments on spin polarization by R. Loth, Z. Physik 203, 59 (1967).

11. See for example: P.M. Marcus and D. W. Jepsen, Phys. Rev. Letters 20, 925 (1968).

12. Eq. (6) of Ref. 3.

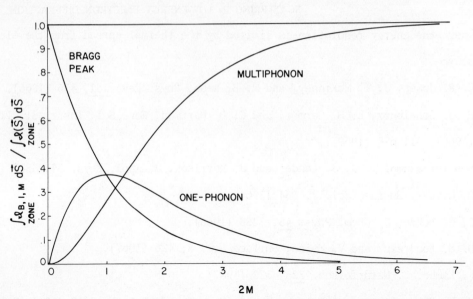

Fig. 1. Variation of the integrated interference function in the zero-, one-, and multiphonon components with 2M. At each point, their sum is equal to one.

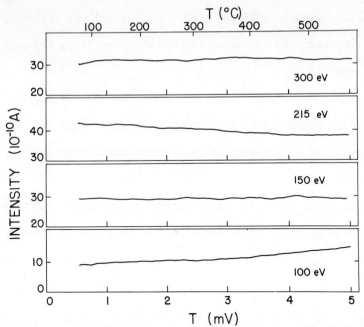

Fig. 2. Integrated elastically scattered current versus temperature. Reproduction of recorder traces of the elastically scattered current collected over back angles versus the thermocouple emf at several energies. 100 eV and 215 eV were chosen to coincide in Fig. 3 with a minimum and a maximum.

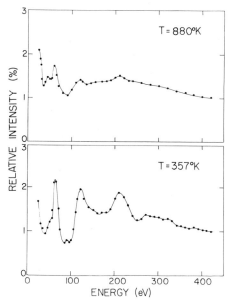

Fig. 3. Integrated elastically scattered intensity versus energy at a high and a low temperature. Relative Intensity is in per cent of incident current.

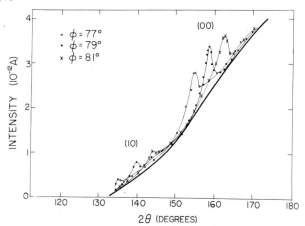

Fig. 4. Uniformity of the multiphonon scattering. Measured elastic current versus scattering angle at 850°K for three angles of incidence. E = 320 eV, corresponding to the (666) reflection for ϕ = 79°. The heavy line represents the multiphonon scattering.

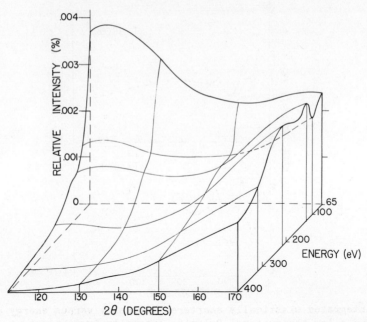

Fig. 5. $|f(\theta,E)|^2$ as a function of scattering angle and energy.

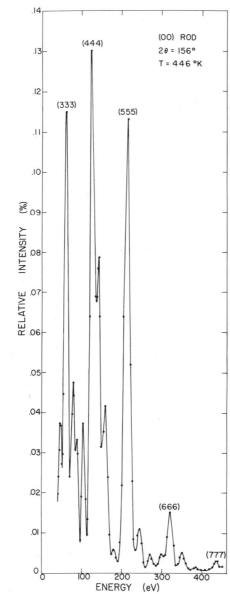

Fig. 6.   Elastic intensity along the (00) reciprocal-lattice rod for (111) Ni
at a low temperature. Relative Intensity, in % of incident current,
versus energy.

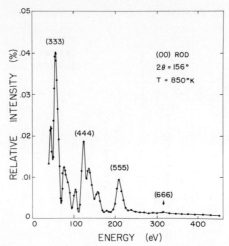

Fig. 7. Elastic intensity along the (00) reciprocal-lattice rod at a high temperature.

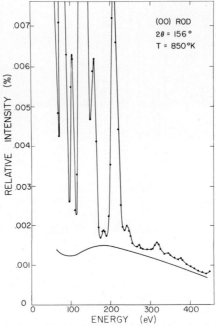

Fig. 8. Fig. 7 redrawn on an expanded scale. The solid line is the multiphonon scattering, obtained by plotting the value of the relative intensity at $2\theta = 156°$ of curves like the heavy line in Fig. 4 for various energies.

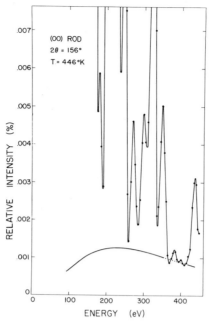

Fig. 9. Fig. 6 redrawn on an expanded scale. The solid line is the multi-phonon scattering.

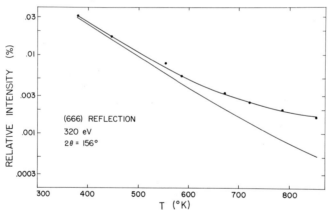

Fig. 10. Debye-Waller plot for the (666) reflection at 320 eV. The data points give the measured value of the relative intensity. The lower curve is the result after subtraction of the multiphonon scattering.

SURFACE PHASE TRANSFORMATIONS: AN INTERPRETATION OF LEED RESULTS

J. J. Burton* and G. Jura

Inorganic Materials Research Division, Lawrence Radiation Laboratory
Department of Chemistry
University of California
Berkeley, California

I. INTRODUCTION

A number of theories have been proposed to explain the Low Energy Electron Diffraction (LEED) patterns observed from the surfaces of metals. There are two basic areas of concern in understanding LEED data. Theories have been advanced to account for the variation of the spot intensities as a function of the voltage of the electron beam. These theories have been based primarily on either the kinematic model used in x-ray diffraction work or on multiple scattering; McRae[1,2] has had considerable success in explaining and predicting experimental intensity results using multiple scattering. The second area of concern has been the explanation of LEED patterns which have periodicities not corresponding to the periodicity of the substrate. Explanations of this phenomenon have been based on large concentrations of surface vacancies,[3] on surface impurities,[4,5] and on rumpled surface layers.[6] This paper will attempt to explain the origin of certain types of LEED patterns having unexpected periodicities.

In this paper we review briefly some LEED data (Sec. II), summarize a theory of the authors' on phase transformations on solid surfaces (Sec. III) and apply this theory to the interpretation of LEED data (Sec. IV). Finally, we examine some predicted properties of the transformed surfaces and indicate some possible experiments to confirm the authors' theory. We show that the transformed surface structures should have catalytic properties, ad-atom surface diffusion coefficients, and surface vibrational frequencies which differ greatly from those of the normal surface.

---

* Present address: Dept. of Physics, University of Illinois, Urbana, Ill.

II. SUMMARY OF LEED DATA

An excellent review article[7] has been written on LEED and the reader
is referred to it for general information on the technique. In this
section, we summarize those LEED results which are pertinent to this
paper.

If a back diffraction experiment is done from a two-dimensional
lattice, the observed diffraction pattern is expected to exhibit the
symmetry of the space reciprocal to the lattice. Low energy electrons
are believed to not penetrate many layers into a crystal surface. The
diffraction patterns are hence expected to exhibit a symmetry reciprocal
to that of the cut plane of the crystal. The diffraction pattern having
the expected symmetry is known as a (1×1) pattern, using the notation of
Wood;[8] this means that the repeat distance of the diffraction pattern is
equal to that expected from the surface net in both principle axis
directions. A diffraction pattern with a repeat distance of 1/n in one
direction and 1/m in the other is known as (n×m); this pattern has extra
spots not expected from the bulk structure.

LEED experiments have been carried out on a number of FCC metals.
Observed LEED patterns for various metallic crystal faces are tabulated
in Table I along with the temperature regions in which these patterns
form. The data presented in Table I are for nominally clean surfaces.
Additional structures have been observed in the presence of high impurity
concentrations. LEED patterns classifiable as (n×1) have not been observed on the (111) surface of FCC metals except, of course, the expected
(1×1) pattern. We have not included the so-called ring structures in

Table I; they are not related to the theory presented in this paper and have been extensively discussed by Somorjai.[9]

There are several interesting properties of the surface structures observed by LEED which are relevant to our future discussion.

1. The (nX1) and (nXm) structures often require careful surface preparation, ion bombardment, and annealing before they can be observed.

2. Two (nX1) patterns can coexist, perpendicular to each other on a (100) surface.

3. Once an (nX1) structure is formed the solid may be cooled below its temperature region of stability and the pattern disappears.[9] Subsequent reheating causes the pattern to reappear.

4. Two or more different LEED patterns may be formed on the same face of a particular metal in the same temperature range.

5. The (nX1) structures disappear at temperatures well below the melting point of the solid.[9]

6. The formation of surface structures on copper,[11] nickel,[7] and platinum,[10] have been found to be sensitive to the presence of gases in the system.

7. (5x1) LEED patterns are found on the (100) surface of epitaxially grown gold crystals, indicating that surface structures can be formed on the pure metal.[12]

8. Deposition of three mono-layers of gold on a (1x1) structure of a silver (100) surface cause formation of a (5x1) gold pattern.[12]

9. Deposition of a mono-layer of silver on a (5x1) structure of a gold (100) surface caused the LEED pattern to revert to (1x1).[12]

III.   THEORY

A satisfactory theory of surface structure must explain the experimental observations cited above. The theory presented in this section will be shown in Sec. IV to explain most of these data.

The authors have previously examined the possibility of a rearrangement of the argon (100) surface without a change in the structure of the bulk crystal.[13] The argon lattice was represented by a set of point atoms interacting by a pair-wise additive Lennard-Jones 6-12 potentials. Only the potential energy of the lattice was considered. The Einstein approximation was used for calculating entropies and zero-point energies. In making calculations on the properties of the argon surface, the relaxations of the ideally flat (100) argon surface were considered.[14]

This model has certain obvious disadvantages. Kinetic effects and many-body forces are neglected. The Lennard-Jones 6-12 potential predicts that the hexagonal-close-packed structure should be the stable structure of argon, whereas experimentally, face-centered-cubic is observed. These problems are discussed in the earlier publication.[13]

Though this model is not an exact representation of solid argon, it is believed to be useful for gaining insight into the real world. As will be seen, this study of argon can also yield some valuable ideas about metals, which are distinctly different from solid argon.

Using the model described above, the authors showed that it is thermodynamically possible for the normal (1×1) structure, Fig. 1, of the argon (100) surface to undergo a phase transition to a C(2×1) structure, Fig. 2a, without any corresponding transformation of the bulk.

We call this structure C(2×1) as it has a unit cell of normal length in one direction and twice as long as normal in the other direction; the "C" refers to the centered surface atom in the unit cell. In this transformation, every other row of the (1×1) structure is translated one-half an atomic distance parallel to the surface, Fig. 1 and 2a. The translating atoms relax perpendicularly outwards from the surface plane so that the C(2×1) structure has a saw tooth appearance, Fig. 2b. The C(2×1) structure is thermodynamically stable with respect to the (1×1) structure at $81.5°K$, which temperature is below the melting point of argon, $84°K$. The transformation can occur because the potential energy of the C(2×1) is not too much higher than that of the (1×1) and the C(2×1) structure's zero point energy is lower than that of the (1×1) structure while its entropy is higher.

The authors examined the effects of surface impurities on the transition temperature. Neon and krypton impurities were considered. Argon-impurity interactions were represented by Lennard–Jones 6-12 potentials based on impurity-impurity and argon-argon potentials. It was found that the effect of the impurity depended greatly on whether the impurity went into a shifted or a normal position and on the nature of the impurity. For instance, a 5% concentration of neon impurity atoms in the shifted rows lowers the transition temperature to $72°K$; but krypton impurities in the unshifted rows raise the transition temperature. Surface impurities can greatly influence the temperature at which the C(2×1) structure is formed from the (1×1).

The effect of a vacancy in the surface layer was considered. If a vacancy is introduced into a shifted row, even if the vacancy is adjacent to an impurity atom, the entire row of atoms collapses back into the unshifted configuration as the atoms are no longer located at potential minima. At high enough temperatures, the normal formation of surface vacancies could cause the $C(2\times1)$ structure to transform back to a (1×1) structure.

Many possible surface structures can be obtained by shifting some rows of surface atoms. We have carried out calculations for a $C(5\times1)$ structure in which two of every five rows are shifted, Fig. 3. The temperature at which this structure becomes stable relative to the (1×1) is essentially the same as that required for formation of the $C(2\times1)$ from the (1×1) structure. Another $C(5\times1)$ structure can be formed by shifting one row of every five; a $C(3\times1)$ could be formed by shifting one row in three. All of these structures would be expected to form at roughly the same temperature as their entropies and energies of formation should be proportional to the number of atoms shifted. Many different surface structures can be formed by shifting surface rows. All such structures must be classifiable as $C(n\times1)$ as parallel rows of atoms are shifted. When many similar structures can be formed, it is possible that the one which is formed is determined by the distribution of impurities on the surface.

IV. CORRELATION OF THEORY WITH LEED DATA

We have found that it is possible to form a variety of $C(n \times 1)$ surface structures on the (100) face of a face-centered-cubic crystal. Because of the symmetry of the (100) surface (square), these structures can all have two different orientations at 90° to each other. This is observed for all $(n \times 1)$ LEED patterns on (100) surfaces.

If we examine the appearance of (110) face of a FCC crystal, Fig. 4a, we see that it is possible to form distinct $(n \times 1)$ structures in two different fashions. These arise from shifting the rows along different axes, Figs. 4b and 4c. A shift along the long axis, Fig. 4b, brings the shifted atoms appreciably closer to the substrate atoms. A shift along the short axis, Fig. 4c, causes less crowding. Thus it should be easier to have a phase transformation involving shifts along the shorter axis than along the longer axis. Thus, we would expect it to be possible to form $(n \times 1)$ structures on (110) surfaces having only one of two conceivable orientations. Lyon[10] has, in fact, observed (2×1) and (3×1) structures on the (110) surface of platinum where only one orientation of the structure existed at a time. The patterns corresponding to shifts along our long axis were easily removed. The structures corresponding to shifts along the short axis were found to be quite stable. This is in accord with our expectation that the one structure should be more stable than the other.

We now turn to the (111) face, Fig. 5. Any attempt to shift a row of surface atoms an appreciable distance in any direction will bring the atoms into close proximity with both substrate atoms and other surface atoms. This occurs because the (111) surface layer is close packed. Furthermore, if we shift a row of surface atoms, they will be moved to positions of low symmetry which would lead us to expect the absence of a potential minimum for the shifted row. Thus we would expect that we cannot form (n×1) structures on (111) surfaces by shifting some rows of atoms. LEED patterns corresponding to (n×1) structures have not been observed on (111) faces of FCC crystals.

The theory developed in the previous section indicates that surface defects can be very important in surface phase transformations. This may explain why very careful treatment of the surface is required in order to produce LEED patterns other than (1×1). The expected sensitivity of structure to impurities may determine which of several possible surface structures is observed. On the nickel (110) surface,[7] only a (1×1) pattern is observed except in the presence of oxygen. (2×1) structures are observed on the (110) and (100) surfaces of copper[11] after exposure to oxygen.

We expect the formation of shifted row surface structures to be a normal first order phase transformations. There is no reason to expect that it should not be reversible. LEED experiments have shown that once an (n×1) pattern is formed, it disappears on cooling and reappears on heating, just as though an ordinary reversible phase change occurs.[9]

We have found that formation of surface vacancies will cause our shifted row structures to revert to (1×1). Experimentally, (n×1) struc- tures disappear as the temperature is raised.[9]

According to the theory developed above, a number of possible surface structures are quite similar in free energy. Impurities may determine which structure is formed. Lyon and Somorjai[9] have found that either a

(5×1) or a (2×1) LEED pattern can be observed from a (100) platinum surface

in the same temperature region. They found that prolonged heating of the

(2×1) pattern surface caused the (2×1) pattern to disappear and a (5×1) to

appear. After the (5×1) structure was formed, the (2×1) could not be

regenerated.

The shifted row surface structures arise from movements in the surface

layer of atoms only. No rearrangement of the bulk occurs. This is in

accord with the findings of Palmberg[12] on epitaxially grown single crystals;

deposition of a mono-layer of silver on gold destroyed the gold surface

structure and deposition of three mono-layers of gold on silver caused the

appearance of a gold structure.

The theory presented in Sec. III has been shown above to account for

many of the properties of the structures observed by LEED. There is one

piece of experimental data for (n×1) structures for which it does not account.

The extra spots of the (5×1) structure on the gold[5] and platinum[9] (100) sur-

faces (that is, spots on the (5×1) LEED pattern which are not present in the

(1×1) pattern) are not single spots. They are slightly split into pairs.

Our model does not account for this splitting. It is possible that a

careful calculation of LEED patterns based on our (5×1) structure model and

considering multiple scattering effects would give rise to these pairs.

Our model also cannot account for the (n×m) structures observed on a

number of metals and nonmetals. It is possible that these structures

arise by some surface atoms relaxing outwards perpendicularly to the

surface as predicted by Feuchtwang[15] and Haneman.[6]

V. SOME POSSIBLE EXPERIMENTAL TESTS

We have calculated the potential diagrams for adsorption of an argon atom on the (1×1), C(2×1), and C(5×1) structures of the argon (100) surface. These potential plots are shown in Figs. 6, 7, and 8. As can be seen from these figures, the adsorption energy of argon on argon is reduced from ~1380 cal/mole on the (1×1) structure to ~1200 cal/mole on the C(2×1) and 1100-1200 cal/mole on the C(5×1). All adsorption sites are equivalent on the (1×1) structure while there are two energetically different sites on the C(2×1) structure and five on the C(5×1) structure. This suggests that very careful studies of heats of adsorption at very low coverages could distinguish between these structures.

Figures 6, 7, and 8 show that the symmetry of the adsorption sites varies from one structure to another. Thus, the three structures may have very different catalytic properties.

Figures 6, 7, and 8 also show that the energy barriers to ad-atom surface diffusion vary from structure to structure. The (1×1) structure is isotropic and the diffusion barrier is ~380 cal/mole. The C(2×1) structure barriers vary from ~100 to ~270 cal/mole. The C(5×1) barriers vary from ~70 to ~260 cal/mole. The shifted rows of atoms in the C(2×1) and C(5×1) structures create pipes which allow very low energy ad-atom diffusion in the direction of the shift. Thus, it may be possible to observe very fast ad-atom diffusion on the shifted row structures. Such fast diffusion may be more readily observed on (110) surfaces than on (100) surfaces as it is possible to form shifted row structures on (110) surfaces in which all the pipes have the same crystal orientation.

We have also calculated Einstein vibrational frequencies for the surface atoms in the (1×1) and C(2×1) structures of the argon (100) surface, Table II. The vibrational frequencies parallel to the surface plane are lowered significantly in the shifted row structure--by about 30%. Such a change in surface frequency may be detectable by LEED Debye-Waller factor measurements.[16,17]

VI. CONCLUSIONS

It is possible for phase transitions to occur on crystal surfaces without any change in the structure of the bulk crystal. In these transitions, rows of surface atoms are shifted parallel to the crystal surface, forming surface structures which do not have the same symmetry as the bulk crystal. Such phase transitions can occur on (100) and (110) surfaces of FCC crystals, but not on (111) surfaces. The phase transitions are sensitive to surface impurities. At sufficiently high temperatures, the generation of surface vacancies can cause the transformed structure to disappear. These characteristics of the predicted phase transformations correlate very well with LEED data.

The transformed surface structures should behave differently than the normal structures in catalysis experiments, in adsorption experiments, and in ad-atom surface diffusion experiments. The vibrational frequencies in the transformed structures are drastically altered from the frequencies in the normal structures; this change in frequency may be detectable by LEED studies.

ACKNOWLEDGMENT

This work was performed under the auspices of the United States Atomic Energy Commission.

REFERENCES

1.   E. G. McRae, J. Chem. Phys. <u>45</u>, 3258 (1966).

2.   E. G. McRae, <u>Fundamentals of Gas Surface Interactions</u>, H. Saltsburg, J. N. Smith, and M. Royers, Eds. (Academic Press, New York, 1967) p. 116.

3.   A. M. Mattera, R. M. Goodman, and G. A. Somorjai, Surf. Sci. <u>7</u>, 26 (1967).

4.   D. G. Fedak and N. A. Gjostein, Acta Met. <u>15</u>, 827 (1967).

5.   D. G. Fedak and N. A. Gjostein, Surf. Sci. <u>8</u>, 77 (1967).

6.   D. Haneman, Phys. Rev. <u>121</u>, 1093 (1967).

7.   A. U. MacRae, Science <u>139</u>, 379 (1963).

8.   E. A. Wood, J. Appl. Phys. <u>35</u>, 1306 (1964).

9.   H. B. Lyon and G. A. Somorjai, J. Chem. Phys. <u>46</u>, 2539 (1967).

10. H. B. Lyon, Low Energy Electron Diffraction Study of Low Index Platinum Single Crystal Surfaces, (Ph. D. Thesis), University of California at Berkeley, 1967; LRL Report No. UCRL-17549.

11. D. F. Mitchell, G. W. Simmons, and K. R. Lawless, Appl. Phys. Let. <u>7</u>, 173 (1965).

12. P. W. Palmberg and T. N. Rhodin, Phys. Rev. <u>161</u>, 586 (1967).

13. J. J. Burton and G. Jura, to be published.

14. J. J. Burton and G. Jura, J. Phys. Chem. <u>71</u>, 1937 (1967).

15. T. E. Feuchtwang, Phys. Rev. <u>155</u>, 715 (1967).

16. H. B. Lyon and G. A. Somorjai, J. Chem. Phys. <u>44</u>, 3707 (1966).

17. E. R. Jones, J. T. McKinney, and M. B. Webb, Phys. Rev. <u>151</u>, 476 (1966).

Table I. Observed LEED surface structures for the face-centered-cubic metals. (The temperature region of observation of each structure is given when available. Only data for clean surfaces are included.)

Metal	Surface	Structure	Temperature	Reference
Pd	(100)	(1×1)		
		(2×1)	200°-300°C	3
		(2×2)	250°-550°C	3
Pt	(100)	(1×1)		
		(5×1)	350°-500°C	9
		(2×1)	300°-500°C	9
	(110)	(1×1)		
		(2×1)		10
		(3×1)		10
		(4×1)		10
	(111)	(1×1)		
		(2×2)	800°-1000°C	9
		(3×3)	800°-1000°C	9
Ag	(100)	(1×1)		
		(2×2)	600°-750°C	3
Au	(100)	(1×1)		
		(5×1)	150°-400°C	3
		(6×6)	350°-700°C	3

Table II. The Einstein vibrational frequencies of atoms in the (1×1) and C(1×1) and C(2×1) structures of the argon (100) surface. Frequencies are given for vibrations parallel and perpendicular to the surface plane.

Structure	Atom	Vibration	Frequency
(1×1)		Parallel	1.2×10^{12} cycles/sec
(1×1)		Perpendicular	$.9 \times 10^{12}$ cycles/sec
C(2×1)	unshifted	Parallel	.9 cycles/sec
C(2×1)	unshifted	Perpendicular	.9 cycles/sec
C(2×1)	shifted	Parallel	.8 cycles/sec
C(2×1)	shifted	Perpendicular	.9 cycles/sec

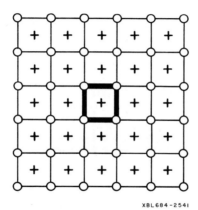

XBL684-2541

Fig. 1.   The (1x1) structure of the (100) surface of a FCC crystal. The intersections of the lines are the normal surface sites and the atoms are
circles. The unit cell is indicated with heavy lines and the atoms in
the second layer with pluses.

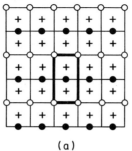

(a)

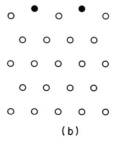

(b)

XBL684-2542

Fig. 2.   The C(2x1) structure of the (100) surface of a FCC crystal. The solid
circles represent shifted atoms. (a) Top view: the intersections of
the lines are the normal surface sites. The unit cell is shown by
heavy lines. Unshifted surface atoms are open circles and second layer
atoms are pluses. (b) Cross section: atoms in unshifted positions are
open circles.

21-15

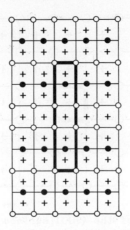

XBL 684-2543

Fig. 3. The C(5x1) structure of the (100) surface of a FCC crystal. The intersections of the lines are the normal surface sites. The unshifted surface atoms are open circles and the shifted atoms shaded circles. The
second layer atoms are pluses. The unit cell is shown by heavy lines.

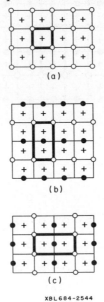

(a)

(b)

(c)

XBL 684-2544

Fig. 4. The (110) surface of a FCC crystal. The intersections of the lines are
the normal surface sites. The unshifted surface atoms are open circles. Second layer atoms are pluses. The unit cell is shown with heavy
lines. (a) The (1x1) structure. (b) A C(2x1) structure with shaded
atoms shifted along the long axis. (c) A C(2x1) structure with shaded
atoms shifted along the short axis.

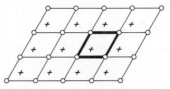

XBL 684-2545

Fig. 5. The normal (111) surface of a FCC crystal. The intersections of the
lines are the normal surface sites. The surface atoms are open circles
and the pluses second layer atoms. The unit cell is indicated by heavy
lines.

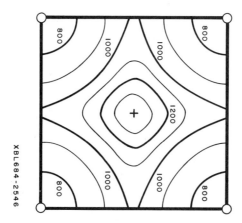

Fig. 6. The heat of adsorption of an argon atom on the (1x1) structure of the argon (100) surface. The heaviest lines are the boundaries of the unit cell. The circles are the surface atoms and the plus a second layer atom. Energy contours are given in calories per mole.

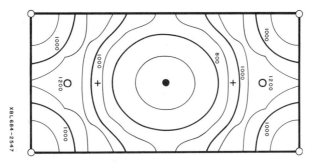

Fig. 7. The heat of adsorption of an argon atom on the C(2x1) structure of the argon (100) surface. The heaviest lines are the boundaries of the unit cell. The open circles are the unshifted surface atoms and the shaded circle is a shifted surface atom. The pluses are second layer atoms. Energy contours are given in calories per mole.

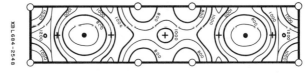

Fig. 8. The heat of adsorption of an argon atom on the C(5x1) structure of the argon (100) surface. The heaviest lines are boundaries of the unit cell. The open circles are unshifted surface atoms and the shaded circles are shifted surface atoms. The pluses are second layer atoms. Energy contours are given in calories per mole.

21–17

SOME ATOMISTIC CONSIDERATIONS OF SURFACE BINDING ON METALS[*]

T. N. Rhodin, P. W. Palmberg,[**] and E. W. Plummer[***]

Department of Applied Physics, and
Laboratory of Atomic and Solid State Physics
Cornell University, Ithaca, New York

A. INTRODUCTION

1. Atomistic Concept of a Metal Surface

Metal crystal surfaces are conveniently visualized to consist of a two-dimensional array of hard spheres in a geometric packing characteristic of some crystal plane in the bulk lattice. The spheres are generally thought to be connected together by non-directional metallic bonds of equal strength. In terms of this conventional pairwise interaction model[†] the total bonding per atom as well as the energy barrier to surface diffusion is usually considered to be directly related to the number of nearest neighbors. In addition, according to this approach the most stable (100) surface of any fcc metal, for example, would be nonreconstructed and atomically flat.[1] Many of our atomistic concepts of solid metal surfaces originate from this viewpoint. They tend to overlook the important contributions of the surface electron gas and the not unlikely possibility that the surface atoms may assume unique configurations and displacements. It is therefore becoming increasingly apparent that the atomic configuration and valency behavior specific to atoms on the metal surface must be more explicitly considered in terms of atomistic descriptions. Quantum mechanical treatment of the atomistic properties of the atom-surface interaction have been formulated by Schmidt and Gomer[2] and extended in some detail by Gadzuk.[3] Although continuing effect has been devoted to various quantum mechanical aspects of surface binding by others[4-7] as well, progress on the formulation of chemisorption from first principles has been slow. In this paper are suggested some atomistic concepts of surface bonding and surface structure from a phenomenalogical approach which may prove helpful in the development of these theoretical treatments.

---

[*]Supported by Air Force Office of Scientific Research Grant AFOSR- 586-68
[**] Present address: North American Rockwell Science Center, Thousand Oaks, California
[***] Present address: National Bureau of Standards, Gaithersburg, Maryland
[†]Pairwise potentials can also be derived for systems other than simple nearest neighbors.

2. Surface Mobility, Bonding, and Rearrangement

From the viewpoint of the experimentalist the conventional concept of simple pairwise interactions is proving to be seriously inadequate when related to atomistic mechanisms for the mobility and the bonding of atoms on metal surfaces. Of special significance in supporting this viewpoint are the measurements of Ehrlich and Hudda[8] on the atomistics of surface self-diffusion of tungsten on tungsten. They observed that pair interactions gave the wrong qualitative order of mobilities of tungsten on the (110), (321) and (211) planes.[9] Subsequent studies both by Ehrlich and Kirk[11] on the desorption of single adatoms of tungsten and by Plummer and Rhodin[10, 12] on the desorption of single adatoms of tungsten and other transition metals from atomically perfect low index planes have amplified and generalized this important conclusion. It is also particularly significant that recent results by various investigators[13-18] on the rearrangement of clean (100) surfaces of certain fee metals can also be interpreted in terms of a concept of surface valency consistent with that proposed by Plummer and Rhodin[12] for the interpretation of results on the energetics of atomic desorption. It is the objective of this paper therefore to critically review the implications of certain recent results on the nature of surface atomistic processes relating specifically to mobility, bonding and rearrange and to interpret them in terms of some new concepts of the metal surface.

Since the atom surface self-diffusion studies have been discussed in some detail in the literature,[8, 9] only the most significant conclusions will be summarized here. The implications of work on surface binding[10, 11, 12] and on the mechanism of superstructure formation[14, 17, 18] on certain clean metal surfaces is of more recent occurrence and will be presented in somewhat more detail. In all three cases the reader is referred to the literature for complete and detailed descriptions of both the experimental methods used and the arguments employed. The objective of this paper is to evaluate and to compare the common significance of the results related to these three fundamental surface phenomena on the assumption that the essential features have to some extent a common basis for interpretation.

B.   SELF-DIFFUSION AND SELF-BONDING ON TUNGSTEN SURFACES

1. Variation in Surface Bonding with Crystallography

The experimental diffuculties typical of quantitative studies of chemisorption on an atomistic scale have long been recognized. The unique suitability of the field ion microscope and associated field ionization phenomena to position, to image and to remove single atoms from perfect single crystal planes of a metal was first established by Müller[20] and used by Müller[21] for field desorption studies of thorium and barium and by Utsugi and Gomer[27] for cesium. The uniqueness of the field desorption method is illustrated by the field ion micrograph in Fig. 1 where 1 tungsten atom, placed on the center of an atomically perfect (110) tungsten plane, is about to be selectively removed by pulsed field desorption.[10] At the present time, the best justification for the validity of the field desorption model used to calculate atomic binding energies is a comparison of experimentally measured adsorption energies calculated from field desorption theory[19, 22, 23] to values obtained by thermal desorption[24-29] as indicated in Table 1. It is evident that there is good agreement between the field desorption values and the ones obtained from thermal desorption from polycrystalline surfaces. More data are required of this kind for desorption of a given adatom from specific planes of a given substrate to make a more critical evaluation of the correct field desorption model to be used.

As previously stated, quantitative measurements of the movement of individual tungsten adatoms at different temperatures was first achieved by Ehrlich and Hudda[8] on the (110), (211) and (321) planes of the tungsten lattice. Ehrlich[9] also demonstrated that the technique of depositing and subsequently field desorbing a known atom from an atomically perfect field evaporated tungsten tip could yield quantitative information on binding energies. The binding energy of individual tungsten atoms using field desorption was studied on the (110), (211), (310), (111), (321), and (411) planes of tungsten by Ehrlich and Kirk[11] and on the (110), (100), (111) and (112) planes by Plummer and Rhodin.[10, 12] These results are related directly to the variation of the surface potential on an atomistic scale since the atom's motion over the surface reflects directly the variation of surface binding with the atom-surface interaction at each specific position. Assuming a potential based on a simple pairwise interaction Drechsler[30] has calculated the energy barriers to self-surface diffusion and to self-surface adsorption. These calculated values for tungsten on tungsten are listed in Table 2 together with the diffusion data of Ehrlich and Hudda[8] and the binding energy data of Ehrlich and Kirk[11] and that of Plummer and Rhodin.[10, 12] When the experimental data is compared to the calculated values it is clearly apparent that besides the difference in magnitudes between the measured and calculated values themselves, that the order of the calculated values clearly disagrees with experiment.[9, 12] Tungsten is not the only metal which exhibits these surface properties. Maiya and Blakely[31] concluded from their measurements on the surface self-diffusion of nickel, that if they insisted on using a nearest neighbor model for the binding energy, then the strength of this bond must increase as the number of nearest neighbors decreases. Ehrlich and Hudda[8] pointed out that the quite different experimental values obtained for surface diffusion on the almost identical (211) and (321) planes is in qualitative accord with the associated greater "smoothening" behavior of conduction electrons on rougher surfaces as postulated by Smoluchowski.[32] Contrary to predictions of pairwise bonding, the influence of electron redistribution appeared to be very important in explaining the low activation energy barrier characteristic of diffusion on the rougher (211) face. The significant conclusion[9, 12] is that simple pair interactions fail to describe surface diffusion or surface bonding observations even qualitatively. It should be pointed out that the mechanisms contributing to the atomic surface diffusion are not simple and often involve additional more subtle considerations.[8] It should also be noted that the interpretation of field desorption studies in terms of surface binding energy variations with crystal face also involved critical assumptions some of which are as yet unresolved.[9, 11, 12] In the latter case two specific uncertainties arise as to the value of the effective work function of the (110) plane of tungsten and the value of the charge[*] of the tungsten ion evolved in field desorption from a given plane. However, the results of both diffusion and bonding studies of tungsten on tungsten clearly emphasize the importance of considering in more detail how the electronic overlap between the adatom and a nearest neighbor on the surface is affected by the crystallography of the substrate plane.

2. Redistribution of Surface Charge

One interpretation is to maintain the concept of a pairwise potential, but to take into account the crystallographic dependence of this potential caused by an electronic charge redistribution occurring at the surface.[32] This means that the nearest neighbor potential energy would be different on different faces. On the higher index planes charge smoothing would leave the atoms in the top layer of the surface partially denuded of their electronic charge cloud,[9,12] so that their interaction with the adatom would be less than a surface atom on a densely packed plane where very little charge smoothing would occur. It appears nearly impossible at the present time to calculate rigorously from first principles the electronic redistribution at the surface. Alternatively, it is helpful to consider a relatively simple modification of the nearest neighbor model in an

---

[*]There is a good possibility that Müller's atom-probe microscope will provide a unique method for resolving this critical question. (See E. W. Müller, et al, Rev. Sci. Instr. 39, (1968) p. 83.

attempt to account for the variation of the atom-surface interaction with surface geometry.

Basically, we must determine how the electronic overlap between the adatom and a nearest neighbor on the surface is affected by the atomic structure of the substrate plane. If the adatom is on a high index plane, some of its nearest neighbors will be protruding from the surface and consequently will be partially depleted of charge so that the overlap with the adatom will be less than normal whereas an adatom on a low index (close packed) plane will have nearest neighbors which are almost completely buried in the surface charge cloud and therefore the overlap will be unaffected by charge smoothing. One useful approach[12] to this effect is to use as an index of the magnitude of this charge depletion effect, the angle θ that the line of centers between the adatom and its nearest neighbor makes with the normal to the plane. The larger this angle the more the nearest neighbor atom in the plane protrudes from the surface (see Fig. 2). Therefore, let us assume that the overlap is maximum when this angle is zero and decreases as the angle increases. Let the nearest neighbor bond strength be a function $f(\theta)$ of the angle. Then the binding energy of an adatom on any plane is then given by:

$$H = \sum_{i=1}^{N} f(\theta_i) \qquad (1)$$

The sum is over the N nearest neighbors of the adatom.

In Table 3, we tabulate the number of nearest neighbors N and their corresponding angle, θ from the surface normal for the four low index planes.

Since $f(\theta)$ should be symmetrical about $\theta = 0$, let's expand f in terms of a Fourier cosine series.

$$f(\theta) = \frac{a_o}{2} + \sum_{n=1}^{\infty} a_n \cos n\theta \qquad (2)$$

If we carry this series out to n = 3, we have four constants which can be evaluated by solving Eq. 1 for the binding energy of the four planes (110), (100), (112) and (111), using the data in Table 3. These four simultaneous equations can then be solved to evaluate the a_i's. The values obtained are listed below:

$$a_0 = 0.594 \text{ eV}$$

$$a_1 = 5.00 \text{ eV}$$

$$a_2 = 0.701 \text{ eV}$$

$$a_3 = 0.561 \text{ eV}$$

Since a_1 is nearly an order of magnitude larger than any of the other a_i's, it is a good approximation to consider the angular dependency of the bond strength as a simple cosine function. Therefore, we can conclude that for the low index planes the binding of tungsten on tungsten can be explained satisfactorily by incorporating into the expression for the pairwise bond strength a surface structure factor, which is consistent with Smoluchowski's[32] model of the charge redistribution at the surface of a metal.

This simple model will likely be inadequate in explaining the binding energies of tungsten adatoms on high index tungsten planes. On these planes the assumption that the angle θ reflects uniquely the charge depletion around the surface atom is not valid. For example an adatom on the (123) plane will form a large angle θ with one of the surface atoms, but the charge depletion surrounding this atom may not be as large as the angle would indicate because this surface atom is really an edge atom on a (110) plane.

This admittedly oversimplified discussion of the effect of crystallographic structure of the substrate plane on the surface binding of metal atoms leads directly to an other interesting question.

What is the effect of changing the adatom on the crystallographic dependence of its binding energy and how does the binding energy for a given substrate plane

change as the electronic configuration of the adatom is varied in terms of, for
example, the transition elements of period 6 of the periodic table?
In figure 3 are plotted the binding energies for the period 6 transition metal
elements on the four low index planes of tungsten calculated from the desorption
data of Plummer and Rhodin.[12] Disregarding the significance of the absolute
values of the binding energy, the over-all relative variation of binding with
atomic density of the substrate is clearly evident. The effect of the substrate
structure is basically the same for all adatoms; that is, the order of the binding for any adatom on the four crystal faces of tungsten depends on the atomic
density of the substrate in decreasing order. The second important general conclusion emerges that, basically the electronic configuration of the adatom
determines the relative magnitude of the binding energy for a given adatom. The
consistency of the observed dependence of binding of all these different adatoms
on the atomic density of the substrate plane with that previously observed for
tungsten supports further the earlier conclusion as to the inadequacy of a
simple pair-wise model to explain surface bonding.

C.   VARIATION IN SURFACE BONDING WITH THE ELECTRONIC CONFIGURATION OF THE ADATOM

It is very interesting to consider a possible interpretation of the maximum
observed in the binding energy with atomic number apparent in Figure 3 in terms
of the possible contribution of the d-electrons to what might be approximately
referred to as an "effective surface valency"[12] for the adatom.

1. Contribution of d-Electrons to Surface Binding

The shape of the curves in Fig. 3 for the four different planes immediately
answers the first part of the question. The effect of the substrate structure is
basically the same for all the adatoms; that is, the order of the binding for
any adatom on the four different crystal faces of tungsten is (110), (100),
(112) and (111) in decreasing order. There are two exceptions to this statement,

the binding of iridium on the (100) plane is larger than on the (110) plane
and platinum is bound more tightly on the (111) plane than on the (112) plane.
Both of these cases are uncertain because of experimental difficulties in making
the measurements for platinum and in interpreting the data on the (110) plane.
Since the experimental dependability of these specific observations is not firmly
established, they will not be discussed further. The most important conclusion
to be drawn from the data in Fig. 3 is that basically the electronic configuration of the adatom determines the magnitude of the binding energy but that the
atomic structure of the surface dictates the variation with surface crystallography for a given adatom. This appears to be an important conclusion of considerable generality. Its implications are sometimes neglected in the considerations of the atomistic properties of metal surfaces.

Before we discuss the effects on binding of the number of d-electrons in
the adatom, it is necessary to consider briefly what is known about the interaction of d-electrons in the bulk and how this may be pertinent to the analysis
of the surface binding of d-electron elements. Until recently the d-band in
solids was considered to be very narrow with the d-electrons localized about
the atom.[33] This model attributed the increase in cohesive energy of the transition metals to a hybridization of the s-electrons near the ion core by the
localized d-electrons. However, more recently Brooks[34] showed that even for the
monovalent noble metals, there exists a large overlap of the d-electrons, so
that any theory which ignored this effect would yield low values of cohesive
energy. Experimental data substantiates Brooks' conclusion; in that it indicates
that all the outer electrons contribute equally to the cohesive energy for transition metals with increasing atomic number up to the Group VIB.[35] Beyond this
group the cohesive energy decreases as the number of d-electrons increases.

In 1938 Pauling[36] proposed a semiempirical model to describe the transition
metals. His description of the transition metals was based on the assumption that

there were two d-bands--one described by diffuse (bonding) wave functions and the other by localized (antibonding) wave functions. This hypothesis has been corroborated by theoretical calculations by Wood.[37] He concluded that it appeared that in transition metals there might be two "different kinds" of d-electrons present, and that those models of transitions metals which considered the d-wave functions solely as localized in nature require modification. There is one important conclusion to be drawn from all of these calculations and that is that there are two high density of states regions in the d-band, a low energy one with very diffuse bonding wave functions, and a high energy one with contracted antibonding wave functions, i.e., there are basically two mutually exclusive d-bands. The above discussions demonstrate that the relevant parameter with respect to the cohesive energy of the transition metals is the number of electrons in the anti-bonding d-band with respect to the number in the bonding d-band. Each electron in the lower energy bonding band contributes equally to increase the cohesive energy, while each electron in the upper antibonding band causes a decrease in the cohesive energy.

These concepts of d-band electrons associated with the bulk properties of the transition metals can be related to the somewhat analogous considerations of the interaction of an atom of a transition element with a crystal surface of a transition metal such as tungsten. When a single atom of a transition element is brought up to the surface of a metal, the monoenergetic atomic levels will broaden into bands.[38] The amount of interaction with the surface will determine how broad the bands are. Since at a surface the adatom has less than half as many neighboring atoms as it would have in the bulk, the d-band is probably not as wide and consequently the subbands will be very narrow. Nevertheless, the degenerate d-levels in the atom will be split into bonding and antibonding bands. The basis for this statement (summarized in Fig. 3) is the observed variation in surface binding with the d-electron concentration.

2. Concept of Effective Surface Valency

On the assumption that our simplified understanding of the variation in binding with crystal face has some validity let's normalize the binding energies of Fig. 3 with respect to tungsten for each face according to the θ function indicated in Fig. 2 and discussed in the preceding section. These values are plotted in Fig. 4 to emphasize the relative changes in bond strength with respect to the electronic configuration of the adatom. The left ordinate is the relative binding energy (with respect to tungsten) and the right ordinate is the effective number of electrons contributing to the bonding, based on the assumption that the surface binding energy of tungsten results from six equally participating electrons. The latter parameter may usefully be considered as an "effective surface valency," defined as the number of electrons contributing to the binding. The significant conclusion from all of the binding energies measured is the nearly perfect adherence to integral values of the effective number of bonding electrons! All these values can be explained in terms of the number of bonding and antibonding d-electrons with the additional assumption that in some cases the surface-adatom interaction is sufficiently large to change the ground state configuration of the adatom. For example, tungsten has either 4 bonding d-electrons and 2 s-electrons or 5 bonding d-electrons and one s-electron (total six), while rhenium has 5 bonding d-electrons and 2 s-electrons (total seven). Applying this scheme to osmium with an atomic configuration of $5d^6 6s^2$ gives 5 bonding d-electrons, 1 antibonding d-electron and 2 s-electrons for a total of 6 bonding electrons, while iridium with a $5d^7 6s^2$ configuration gives 5 electrons contributing to the binding. But with platinum and gold, the interaction of the surface with the adatom must change the configuration since the ground state of a platinum atom is $5d^{10} 6s^0$, which would mean zero bonding electrons and gold with a $5d^{10} 6s^1$ configuration should have only one bonding electron. We know that gold forms a stable trivalent compound in which electrons are removed from the 5d sub-group. In the atom the energy difference between the $5d^{10} 6s^1$ and $5d^9 6s^2$ configurations is only 1.86 eV; therefore, it is very likely that the configuration of platinum on the surface is $5d^8 6s^2$ and gold is $5d^9 6s^2$ which would give a "surface valence" of 4 and 3, respectively. It is of some interest that this change in the ground state configurations of gold and of platinum adatoms may partially explain the anomalous surface structures observed on the (100) planes of gold[17, 18] and on platinum.[15]

For the cases of iridium and platinum we can see from Fig. 4 that they have two different "surface valency numbers" depending upon the substrate plane. In all cases, the "surface valency number" is largest for the planes with the greatest binding energy. For example, the "surface valency number" of platinum is 4 on the (110) and (100) planes, but 3 on the (112) and (111) planes. This is in general agreement with our hypothesis that the electronic configuration of the adatom is influenced by interaction with the surface. It is interesting to compare the "surface valency numbers" obtained in this study with the oxidation states of the adatom. In Table 4, we list the suggested "surface valency number" of the adatom and the obser oxidation states of the atom from chemical compound formation. There are two cases where "surface valency numbers" different from the oxidation numbers are observed; these are a valence of 4 for iridium, and of 3 for platinum. In our model, it is difficult to explain a "surface valency number" of 4 for iridium and of 3 for platinum since the atomic configuration of iridium should be $5d^7 6s^2$ (valence 5) or $5d^8 6s^1$ (valence 3) and likewise platinum should be $5d^9 6s^1$ (valence 2) or $5d^8 6s^2$ (valence 4). The only possible explanation seems to be a hybridization of states or in terms of Pauling's resonance model, the ground state is in resonance between the two states, giving an effective valence somewhere between the two extremes.

In conclusion we can definitely say that the data show remarkable agreement with the "two d-band" model suggested by Pauling[36] and substantiated by Wood.[37] The binding energies are nearly perfectly quantized into integral values of the number of electrons contributing to the bonding. Although these simple observations should not be taken too literally this suggests that the adatoms on the surface still retain much of their atomic characteristics. It is well known that if we assign a valence to an atom in the bulk it will probably be a non-integer[35] because of the partially filled nature of the s-bands as they overlap.

D. SURFACE REARRANGEMENT OF CERTAIN CLEAN fcc METAL SURFACES

The results on the activation barrier to surface diffusion and on the binding energy of adsorbed metal atoms show that careful consideration of the surface electronic structure of metals is needed to adequately explain many of their surface properties on an atomistic basis. In particular, it does not seem unreasonable that deviations in the surface electronic configuration from that of the bulk could also have considerable influence on the occurrence of surface atomic rearrangement.

Low-energy electron diffraction[17, 18] (LEED) coupled with surface analysis by Auger (electron) spectroscopy[39] (SAAS) studies suggest strongly that the (100) surface of Au is reconstructed in the absence of stabilizing impurities. Since LEED patterns similar to those reported for the Au(100) surface have also been observed for the Pt(100)[15] surface, it is probable that surfaces of both of these two metals are rearranged into a similar atomic arrangement. In view of the reproducibility with which the (1x5) superstructure forms on the Au(100) surface in the absence of detectable impurities,[39] it seems reasonable to accept this structure as clean and to give attention to the nature of the rearrangement mechanism. In particular, it is of interest to consider why clean Au(100) and Pt(100) surfaces undergo re-arrangement while the (100) surfaces of very similar metals such as Ag[17, 39] and Pd[39] are characterized by a bulk atomic arrangement.

Before turning our attention to consideration of the rearrangement mech-anism, it may be helpful to review the experimental evidence for rearrangement of the Au(100) surface. In summary, the experimental observations are the following.

(1) Low-energy electron diffraction analysis of the Au(100) surface strongly indicates the existence of a surface layer with a hexagonal-type* geometry with an inter-atomic spacing about 5% less than that of bulk Au.[14, 17, 18]

(2) Analysis of "hexagonal" Au and Ag overlayer structures on the Cu(100) surface has shown that the inter-atomic spacing of Au is about 5% less than its bulk value when positioned on the Cu surface and, more significantly, is contracted by a considerably larger factor (3.3%) than Ag when placed in an identical environment.[18]

(3) Auger electron spectroscopy studies[39] on Au(100) films have failed to reveal a foreign element even though the method is sensitive to most impurities in concentration of one-tenth of a monolayer.[39, 40]

(4) Experimental conditions under which the Au(100)1x5 structure has been observed (e.g. on extremely thin Au crystals grown on Ag(100)[17, 18] indicate that the "hexagonal" overlayer structure exists when the surface impurity concentration is, at most, a few parts in 10^3.

(5) Deposition of a monolayer of Ag, Cu or Pd on the Au(100) surface stabi-lizes a (1x1) surface structure.[18]

---

*That the geometry is exactly hexagonal cannot be concluded on the basis of available measurements. Hence "hexagonal" is used with quotation marks.

(6) Results on clean single crystal films[17, 18] show that the (1x5) structure is thermally stable up to 500°C, the maximum attainable temperature. Fedak and Gjostein[13] have reported that the (1x5) structure on bulk Au crystals is stable through temperatures up to 800°C.

From these observations it can be concluded that a "hexagonal" layer of Au exists on its own (100) surface, and that the rearrangement mechanism is associated with a 5% contraction of surface Au atoms relative to their size in the bulk. Fedak and Gjostein[14] interpreted the LEED pattern for the rearranged (100) Au surface illustrated in Figure 5 in terms of the (5x20) superstructure unit mesh indicated in Figure 6. From multiple diffraction theory it can be shown that multiple diffraction between a single "hexagonal" atomic layer and the Au(100) substrate can account for the high intensity of fractional order beams in the observed pattern.[18] Although thermally stable, the "hexagonal" superstructure may be converted to a simple (1x1) structure through adsorption of foreign atoms.

Let us now consider some unique features in the electronic structure of Au which may be related to surface rearrangement.[18] In comparison with other IB metals (Ag and Cu), whose surfaces apparently do not rearrange, the most distinguishing characteristic in the valence structure of Au is the very "soft" nature of the 5d-shell. This property arises from the fact that the 5d electrons are loosely bound with an energy of only 7.07 eV according to Hartree-Foch calculations. From considerations based on pseudo-potential theory, Austin and Heine[41] argue that the loosely bound 5d-shell results in a tightly bound 6s electron. In other words, if the 5d shell is spatially extended then the 6s-electron tends to be contracted. In Au this effect is manifested by an abnormally small difference between the atomic and ionic radii (1.44 and 1.37 Å). In comparison, the difference is much greater for Ag (1.44 vs 1.13) and Cu (1.28 vs 0.96) because the outer d-shells are not so soft.

Since the 5d-electrons in atomic Au are bound with energy comparable to that of the 6s-electron, the electronic configuration is easily promoted to a higher valency state through interaction with neighboring atoms. The monovalent $5d^{10}6s^1$ ground state and the excited $5d^9 6s^2$ configuration for atomic Au are separated by only 1.1 eV, whereas the corresponding values for Ag and Cu are 2.7 and 1.4 eV, respectively. These variations are reflected in the band structure through the energy difference between the top of the d-band and the Fermi level. For Au, Cu and Ag this difference is about 2.0,[42] 2.0,[43] and 4.0 eV,[43] respectively. The lower values are correlated with greater hybridization of d-orbitals with higher lying s- and p-orbitals. That higher valency states of Au are readily promoted is also evident from the existence of stable trivalent and pentavalent Au compounds and is consistent with interpretation of the effective surface valency for gold adsorbed on tungsten.[11, 12]

Variations in the energy required for promoting higher valency electron configurations are also correlated with the dissociation energy and internuclear separation of diatomic molecules of IB metals, as shown in Table 5. The correlation is most obvious for the dissociation energy D_o. Even when normalized in terms of the sublimation energy H, the dissociation energy of Au_2 is appreciably larger than that of Cu_2 or Ag_2. The internuclear separation r_o, when normalized in terms of the lattice interatomic spacing r_1, appears to exhibit the same effect. Since an accurate value for Ag_2 is not available, the correlation is not definitely established. As Ames and Barrow[46] point out, the high dissociation energy of Au_2 is not consistent with an atomic $5d^{10}6s^1$ configuration. Hence the molecule must be bound with considerably greater valency than unity through s-d mixing.

In regard to the atomic structure of metal surfaces, the most important deduction from these observations is that through atomic contraction, the bond energies may be greatly increased as the number of nearest neighbors decreases.

For the diatomic molecule where the coordination number is drastically reduced relative to that of atoms within a bulk crystal, the internuclear spacing for Au_2 is reduced by about 14%. In view of this fact the observed 5% atomic contraction of the "hexagonal" layer on the (100) surface is not surprising.

For the (100) surface of fcc metals, the coordination number of surface atoms is eight compared to twelve for bulk atoms. According to arguments of Oriani,[47] this reduction in the number of nearest neighbor bonds should be accompanied by an increase in the strength of remaining bonds. There are two factors which must be carefully considered, however, before such a deduction is possible.

(1) Strengthening of bonds is likely associated with reduction in interatomic spacing and hence may be inhibited if the surface has a strong tendency to retain a bulk atomic arrangement. Reduction in the spacing between the atoms in the first and second layers may increase the energy of associated bonds, but since the atoms of the second layer have a bulk coordination number, this effect may be small.

(2) In cases where the effective valency of metal atoms is strongly dependent on promotion of the atomic configuration through interaction with neighboring atoms, the surface valency may be considerably reduced relative to the bulk. For IB metals the surface valency should lie somewhere between the isolated valency of unity and the higher effective valency characterizing bulk atoms. It is possible that this effect could reduce the energy associated with bonds among surface atoms, relative to those of the bulk.

If both of these effects are significant, the binding energy within the surface layer may be increased appreciably through contraction. Reduction in the interatomic spacing of surface atoms would allow the surface bonds to assume their "natural" length and also perhaps enhance the effective valency through greater interaction. If contraction does occur it is likely that loss of registry with the substrate would result in a hexagonal-type configuration of surface atoms for which mutual interaction is optimum.

Among the three IB metals, the above considerations suggest that surfaces of Au are most likely to undergo rearrangement. Present observations suggest that rearrangement does indeed occur on the Au(100) surface, but not on the (100) surface of Ag or Cu. Low energy electron diffraction analysis of the Au(100) structure shows that the surface is rearranged into a "hexagonal" configuration having an internuclear separation about 5% less than that of bulk Au. The rearranged surface structure can only be favored if the interfacial energy associated with the "hexagonal" overlayer and nonreconstructed substrate is less than the additional surface layer bonding energy gained through contraction and rearrangement. Experimental observation indicates that unique properties of Au cause the latter effect to dominate.

The absence of rearrangement on the clean Cu(100) surface is somewhat surprising in view of the close similarity in the electronic structure of Cu and Au. The experimental observation of a bulk arrangement for the Cu(100) surface suggests that the hexagonal arrangement of Au is only slightly favored over a bulk arrangement and that only those metals whose valencies are extremely sensitive to environmental interaction are candidates for rearrangement.

Of all metals whose surface structures have been studied, rearrangement has consistently been observed only on Pt[15*] and Au.[13,14,17] The (1x5) diffraction pattern reported[15**] for Pt(100) surfaces is identical to that observed from the Au(100) surface, including splitting of 1/5 order beams[11] (see Figure 5). This result strongly suggests that a contracted hexagonal layer of Pt is also present on Pt(100) surfaces. According to the present interpretation in terms of enhanced surface valency, the occurrence of

---

*On the basis of these considerations it is predicted that the clean (100) face of iridium is also likely to undergo rearrangment.

**New evidence from combined LEED and SAAS studies indicate that it may only be the (1x5)Pt(100) structure which is typical of a clean Pt surface.[48]

rearrangement on Pt surfaces is easily explained. The atomic configuration of Pt is $5d^{10}6s^{0}$ which leaves zero electrons available for bonding. Promotion of this state to high valency configurations is easily accomplished, as evidenced through the cohesive properties of bulk Pt. As in the case of Au, the valency is strongly dependent on environment. It has been found,[11, 12] in fact, from binding studies of Pt on W, that the effective valency of Pt is four when present as an adatom on the W(111) and W(100) planes, but only three when present as an adatom on the more open W(112) and W(111) planes. This result is consistent with the hypothesis that the effective valency is strongly dependent on the atomistic nature of the surface environment.

The proposed model for rearrangement of metal surfaces in terms of enhanced surface valency is consistent with LEED observations which indicate that the close-packed (111) surface of Au is not reconstructed[13] while the more open (100) and 110) surfaces are rearranged.[13, 14, 17, 18] The coordination number for surface atoms of the (111), (100), and (110) surface is respectively 9, 8, and 7. Since the surface valency is expected to decrease with decreasing co-ordination number, rearrangement is most probable for the (110) surface and least probable for the (111) surface.

Interpretation of the driving force for rearrangement in terms of a modified surface valency is consistent with the fact that the (Au(100)1x5 structure can be converted to a (1x1) structure through adsorption of a mono-layer of Ag, Pd or Cu. In the presence of the overlayer, the environment of the top layer of Au atoms is more bulk-like and hence their electronic con-figuration and atomic size approximate that of bulk atoms.

E.   SUMMARY

1. It is most significant that an almost inverse relationship exists between the binding energies of the transition element adatoms and the number

of nearest neighbors on the tungsten surface. Results from both surface diffusion binding studies support strongly the hypothesis that this results from
a charge redistribution at the surface which is crystallographically dependent.
Taking this into consideration, binding of most of the period 6 transition
metals on the low index planes of tungsten can be explained satisfactorily in
terms of a modified pairwise interaction.

2. The change in binding energy with the number of d-electrons in the
adatoms of the period-6 transition elements is interpreted in terms of a
"two d-band" model. This model leads to a concept of a surface valency number
which is a direct consequence of the influence of the surface environment on
the number of electrons in the adatom available for binding.

3. Recent experimental results on the atomic arrangement of Au(100),
Pt(100) and related metal overlayer surface structures indicates that the
bond energies at the surface may be significantly increased through interatomic contraction as the number of nearest neighbors decrease. In cases
where electronic transitions are likely to occur between the d-electrons in
one level and the next s-level, the "effective surface valency" is more likely
to be enhanced by such a transition in the surface. It is significant that it
is those metals, gold and platinum for which this transition is most probable,
that rearrangement of the (100) face has been observed.[*]

ACKNOWLEDGMENT

Support by the Air Force Office of Scientific Research and the Advanced
Research Projects Agency through the Cornell Materials Science Center is gratefully acknowledged. Opportunity to participate in the Fourth International
Materials Symposium on the Structure and Chemistry of Solid Surfaces is also
appreciated.

---

[*]It would be very interesting to investigate the rearrangement of the (100) Ir
surface which has not yet been studied from this viewpoint to the writers'
knowledge.

REFERENCES

1. J. F. Nicholas, J. Phys. Chem. Solids 24, (1963) p. 1279.

2. R. Gomer and L. W. Swanson, J. Chem. Phys. 38, (1963) p. 1613;
    L. Schmidt and R. Gomer, J. Chem. Phys. 42, (1965) p. 3573.

3. J. W. Gadzuk, Surface Science 6, (1967) p. 133; 6, (1967) p. 159;
    Phys. Rev. 154, p. 662; Solid State Comm. 5, (1967) p. 743.

4. R. W. Gurney, Phys. Rev. 47, (1935) p. 479.

5a. L. V. Dobretsov, Electron and Ion Emission. NASA Technical Translation,
    number F-73 (1952).

 b. T. Toya, J. Res. Inst. Catalogs, (Hokkaido Univ.) VI, (1958) p. 308;
    VIII, (1961) p. 209.

6a. T. B. Grimley, Proc. Phys. Soc., 90, (1967) p. 751; 92, (1967) p. 776
    and Molecular Processes on Solid Surfaces, edited by E. Drauglis and
    R. Gretz, McGraw-Hill, N. Y. (1969).

 b. T. Jansen, Molecular Processes in Solid Surfaces, edited by E. Drauglis
    and R. Gretz, McGraw-Hill, N. Y. (1969).

7. A. J. Bennett and L. M. Falicov, Phys. Rev. 151, (1966) p. 512.

8. G. Ehrlich and F. Hudda, J. Chem. Phys. 44, (1966) p. 1039.

9. G. Ehrlich, Disc. Faraday Soc. 41, (1967) p. 7.

10. E. W. Plummer and T. N. Rhodin, Abstracts of the 14th Field Emission
    Symposium, June 26, 1967, Gaithersburg, Maryland; Applied Phys. Letters
    11, (1967) p. 194.

11. G. Ehrlich and C. F. Kirk, J. Chem. Phys. 48, (1968) p. 1465.

12. E. W. Plummer and T. N. Rhodin, J. Chem. Phys. 49, (1968) 3479.

13. D. G. Fedak and N. A. Gjostein, Phys. Rev. Letters 16, (1966) p. 171;
    Acta Met. 15, (1967) p. 827.

14. D. G. Fedak and N. A. Gjostein, Surface Science 8, (1967) p. 77.

15. S. Hagstrom, H. B. Lyon and G. A. Somorjai, Phys. Rev. Letters 15,
    (1965) p. 491; L. B. Lyon and G. A. Somorjai, J. Chem. Phys. 46,
    (1967) p. 2539.

16. A. M. Mattera, R. M. Goodman and G. A. Somorjai, Surface Science 7,
    (1967) p. 26.

17. P. W. Palmberg and T. N. Rhodin, Phys. Rev. 161, (1967) p. 586.

18. P. W. Palmberg and T. N. Rhodin, J. Chem. Phys. 49, (1968) 134; IBID, 49, (1968)
    147.

19.   E. W. Müller, Adv. In Electronics and Electron Physics 8, (1960) p. 83.

20.   E. W. Müller, Zeit. Electrochemie 61, (1957) p. 43.

21.   E. W. Müller, Phys. Rev. 102, (1956) p. 618.

22.   R. Gomer and L. W. Swanson, J. Chem. Phys. 38, (1963) p. 1613.

23.   D. G. Brandom, Surface Science 3, (1964) p. 1.

24.   C. E. Moore and H. W. Allison, J. Chem. Phys. 23, (1955) p. 1609.

25.   V. M. Govrilyuk and V. K. Medveder, Soviet Physics-Solid State 7, (1963) p. 1591.

26.   I. Langmuir, J. Franklin Inst. 217, (1934) p. 543.

27.   H. Utsugi and R. Gomer, J. Chem. Phys. 37, (1962) p. 1720;
      H. Utsugi and R. Gomer, J. Chem. Phys. 37, (1962) p. 1706.

28.   I. Langmuir, Phys. Rev. 44, (1933) p. 423.

29.   R. P. Godwin and E. Luescher, Surface Science 3, (1965) p. 42.

30.   M. Drechsler, Zeit. fur Elektrochemie 58, (1954) p. 327.

31.   S. Maiya and J. Blakely, J. Appl. Phys. 38, (1967) p. 698.

32.   R. Smoluchowski, Phys. Rev. 60, (1941) p. 661.

33.   N. F. Mott and J. H. Jones, "The Theory of the Properties of Metals and Alloys," (1936).

34.   H. Brooks, Nuovo Cimento, Supplemento serio 7, (1958) p. 165.

35.   W. Hume-Rothery and B. R. Coles, Adv. in Phys. 3, (1954) p. 149.

36.   L. Pauling, Phys. Rev. 54, (1938) p. 899.

37.   J. H. Wood, Phys. Rev. 117, (1960) p. 714.

38.   R. W. Gurney, Phys. Rev. 47, (1935) p. 479.

39.   P. W. Palmberg and T. N. Rhodin, J. Appl. Phys. 39, (1968) p. 2425.

40.   R. E. Weber and W. T. Peria, J. Appl. Phys. 38, (1967) p. 4355.

41.   B. J. Austin and V. Heine, J. Chem. Phys. 45, (1966) p. 928.

42.   W. F. Krolikowski and W. E. Spicer, private communication.

43.   C. N. Berglund and W. E. Spicer, Phys. Rev. 136, (1964) p. A1030.

44.   J. Drowart and P. Goldfinger, Angewandte Chemie 6, (1967) p. 571.

45.   D. N. Travis and R. F. Barrow, Proc. Chem. Soc. 64, (1962) February.

46.   L. L. Ames and R. F. Barrow, Trans. Faraday Soc. 63, (1967) p. 39.

47.   R. A. Oriani, J. Chem. Phys. 18, (1950) p. 575.

Table 1

parison of Thermal and Field Desorption Data for Metals on Tungsten.

atom	n_o	Desorption Field T = 0	H_a	H_a(thermal)	Ref.
Ba	2	0.9 V/A	3.8-4.6 eV	3.7-4.7 eV	21, 24, 25
Th	1	2.9	7	7.7	21, 22, 26
Cs	1	0.4	2.8-3.1	2.83-2.95	27, 28
W	2	4.8	7-8*	8.6	10, 12
Au	1	2.6-2.9	3.3-3.8	3.5	12, 29

*measured for different crystal faces.

Table 2

alculated and Experimental Values of Binding Energy and Activation nergy for Surface Diffusion for W Adatoms on W.

	1	2	3	4	5
ane	$H(exp)$[12]	$H(exp)$[11]	$Q_a(exp)$[8]	$H(cal)$[30]	$Q_a(cal)$[30]
10)	8.2 eV	5.3 eV	0.96 eV	4.58 eV	1.35 eV
00)	8.0	--	--	7.82	--
12)	6.9	7.0	0.57	7.64	1.78
11)	6.7	6.0	--	9.05	--
23)	7.4	6.6	0.87	7.72	1.87

Table 3

Number and Angle of Line of Centers from the Surface
Normal for an Adatom.

Plane	Bonds	
(110)	2	$30°$ 55'
	1	$14°$ 29'
(100)	4	$54°$ 44'
(112)	2	$61°$ 50'
	1	$19°$ 27'
(111)	3	$70°$ 32'
	1	$0°$

Table 5

Experimental Results for the Ground State
of Cu_2, Ag_2 and Au_2 Molecules

Metal	D_o[44] (Kcal/mole)	D_o/H	$r_o(Å)$[45,46]	r_o/r_1
Cu_2	45.5 ± 2	0.55	2.219	0.868
Ag_2	$37.6 \pm$	0.54	--	--
Au_2	51.5 ± 2	0.63	2.4719	0.858

Table 4

Surface Electronic Configuration of Adatoms and "Surface Valency Numbers".

Adatom	Ground State Configuration	Configuration at Surface	Surface* Valence	Oxidation* States
Hf	$5d^2\ 6s^2$	$5d^2\ 6s^2$	4	4
Ta	$5d^3\ 6s^2$	$5d^3\ 6s^2$	5	5
W	$5d^4\ 6s^2$	$5d^4\ 6s^2$	6	$\underline{6}$,5,4,3,2
Re	$5d^5\ 6s^2$	$5d^5\ 6s^2$	7	$\underline{7}$,6,4,2
Os	$5d^6\ 6s^2$	$5d^6\ 6s^2$	6	8,6,$\underline{4}$,3,2
Ir	$5d^7\ 6s^2$	$5d^8\ 6s^1$--$5d^7\ 6s^2$	$\underline{5}$,4	6,$\underline{4}$,3,2
Pt	$5d^{10}\ 6s^0$	$5d^9\ 6s^1$--$5d^8\ 6s^2$	$\underline{4}$,3	$\underline{4}$,2
Au	$5d^{10}\ 6s^1$	$5d^9\ 6s^2$	3	$\underline{3}$,1

*The underlined values are the most stable.

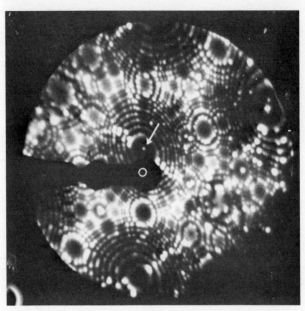

Fig. 1. Tungsten tip field evaporated at 5726 volts before a small deposition of W atoms. There is one single W atom on the central (110) plane. Imaging voltage was 5185 V. The white circle indicates the aperture of a Faraday cage probe used for supplementary **measurement of field** emitted electrons. (Plummer and Rhodin, ref. 11)

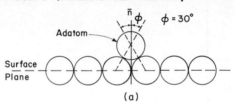

(a)

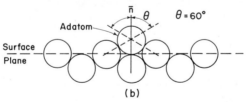

(b)

Fig. 2. A two dimensional schematic drawing of an adatom on a low index (close packed) plane (a), and a high index plane (b). In case (a) the adatom has two nearest neighbors and the line of centers between the adatom and these neighbors is 30° from the normal to the surface, $\underline{n}$. In case (b) the adatom has three nearest neighbors, two of which protrude far enough from the surface so that their lines of centers are 60° from the surface normal. The third nearest neighbor forms an angle of 0°. (Plummer and Rhodin, ref. 12)

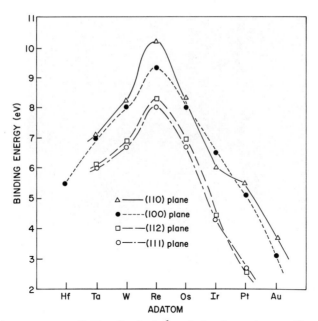

Fig. 3. Binding energy of the Period 6 metal elements on the four low index planes of tungsten. (Plummer and Rhodin, ref. 12)

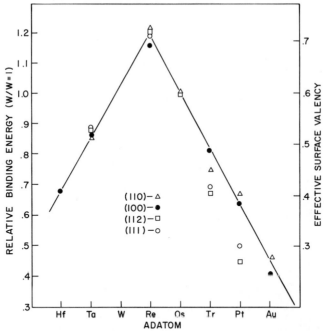

Fig. 4. Relative binding energies of the 5d-transition elements with respect to tungsten on the four low index planes of tungsten. The units to the right are the effective number of electrons contributing to the binding. (Plummer and Rhodin, ref. 12)

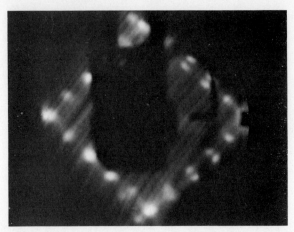

Fig. 5.   LEED pattern from Au(100) surface structure of epitaxial film il-
          lustrating splitting of 1/5 order beams (30 eV). Specular reflected
          (00) beam obscured by Au source at far right. (Palmberg and Rhodin,
          ref. 17)

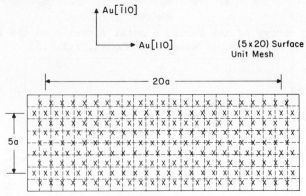

Fig. 6.   Schematic representation of Au(100) structure showing "hexagonal" net
          superposed on (1x1) orthogonal substrate. (Fedak and Gjostein, ref.
          14)

ON THE INTERPRETATION OF COMPLEX LEED PATTERNS

E. Bauer

Michelson Laboratory
China Lake, California

I.   INTRODUCTION

During the past ten years numerous complex LEED patterns have been reported, both for "clean" surfaces and surfaces covered with "adsorption" layers. Most of these patterns--when interpreted at all--were attributed to essentially two-dimensional real superstructures. The interpretation was based on the assumption that the diffraction process is essentially two-dimensional due to the low penetration depth of slow electrons and that the spot intensity may be analyzed in terms of the elementary theory of LEED. However it was pointed out some time ago [1] and is now widely accepted that the diffraction of slow electrons--as far as the intensities are con-cerned--must be considered as a three-dimensional process in which the wave field effective in the diffraction process varies strongly normal to the surface (dynamical theory of LEED). It is the purpose of this paper (1) to examine the interpretation of complex LEED patterns from the three-dimensional point of view, and (2) to demonstrate that--in spite of the difficulties in the interpretation of these patterns--surface structures can be determined with a reasonable degree of confidence, if LEED is combined with other tech-niques, especially Auger electron spectroscopy [2].

II.   GENERAL CONSIDERATIONS

The two-dimensional picture of the diffraction process is based on the assumption that the incident wave is attenuated mainly by true absorption,

i.e. inelastic scattering. This absorption is assumed to be very strong, so
that essentially only those atoms which are directly exposed to the incident
wave contribute to the (elastic) diffraction pattern [3]. Conversely, the
three-dimensional picture of the diffraction process is based on the assumption that the incident wave—like all diffracted waves—is attenuated mainly
by elastic and quasi-elastic scattering. Although the incident wave may be
considerably attenuated by scattering in the first atomic layer, the amplitude of the (elastic) wave field incident on the second and following layers
still is large because absorption is assumed to be small. Consequently many
layers can contribute to the diffraction process. That this picture is at
least as reasonable as the previous one follows from calculations [4] of
the absorption and (zero order) scattering coefficients κ and κ'_o respectively, as illustrated in Fig. 1. κ was obtained by treating the crystal
as a free electron gas with Fermi momentum $k_F = 1.1$ and with the conduction
band bottom 20 eV below the vacuum level in the example shown. κ is determined by the imaginary part $E_I(k)$ of the interaction energy of the incident
electron with the electron gas (see e.g. ref. 5):

$$\kappa = \frac{2}{k} E_I(k) = \frac{2}{k} \frac{1}{2\pi^2} \int_{k_F \leq |\underset{\sim}{k}-\underset{\sim}{p}| \leq k} \mathrm{Im} \frac{d\underset{\sim}{p}}{p^2 \epsilon(p, E(k) - E(\underset{\sim}{k}-\underset{\sim}{p}) + i\delta)} \tag{1}$$

$\kappa'_o = \rho Q$ (ρ = number of atoms per unit volume, Q = total elastic atomic scattering cross-section) was calculated by solving the Schroedinger equation
for an isolated atom using "trimmed" Thomas-Fermi-Dirac potentials. The
"trimming" was done so as to simulate the cut-off of the potential of one
atom by that of its neighbor in the solid. Figure 1 clearly shows that
attenuation by scattering is about three times as large as by absorption in
the energy range of interest, justifying the three-dimensional picture. It
has to be kept in mind, however, that in the calculations on which Fig. 1

is based rather crude simplifications were made: (1) the free electron gas approximation is frequently poor and the influence of the surface on $E_I(k)$ [6] has been neglected, and (2) the expression $\kappa'_o = \rho Q$ assumes that no phase relations exist between the waves scattered by the different atoms; in a crystal such phase relations exist and lead to a dependence of κ' on energy and direction of the wave: $\kappa'_o \to \kappa' = \kappa'(\underline{k})$ (primary extinction).

In the dynamical theory of the diffraction of a plane wave by a plane crystal surface the intensity of a diffracted beam is given by

$$I(\underset{\sim}{k},\underset{\sim}{k}_o) = |G(\underset{\sim}{k}^t - \underset{\sim}{k}^t_o)|^2 \; |F(\underset{\sim}{k},\underset{\sim}{k}_o)|^2 \quad . \tag{2}$$

Here $\underset{\sim}{k},\underset{\sim}{k}_o$, $\underset{\sim}{k}^t,\underset{\sim}{k}^t_o$ are the wave vectors of the incident and diffracted waves and their tangential components respectively. The lattice amplitude G determines the geometry of the diffraction pattern and is given by

$$G(\underset{\sim}{k}^t - \underset{\sim}{k}^t_o) = \sum_{m_1 m_2} e^{-i(\underset{\sim}{k}-\underset{\sim}{k}_o) \cdot (m_1 \underset{\sim}{c}_1 + m_2 \underset{\sim}{c}_2)} \tag{3}$$

where $\underset{\sim}{c}_1$, $\underset{\sim}{c}_2$ represent the lateral periodicity of the true or apparent superstructure of the surface [7]. The dynamical structure amplitude F determines the intensity distribution in the diffraction pattern and is given by

$$F(\underset{\sim}{k},\underset{\sim}{k}_o) = -\frac{1}{4\pi} \int_{\Omega_o} e^{-i\underset{\sim}{k}\cdot\underset{\sim}{r}} U(\underset{\sim}{r}) \; \psi(\underset{\sim}{k},\underset{\sim}{k}_o,\underset{\sim}{r}) \; d\underset{\sim}{r} \quad . \tag{4}$$

The integration extends over the whole dynamical unit cell $\Omega_o = (\underset{\sim}{c}_1 \underset{\sim}{c}_2 \underset{\sim}{c}_3)$ where $\underset{\sim}{c}_3$ ($\perp \underset{\sim}{c}_1,\underset{\sim}{c}_2$) is determined by the penetration depth of the (elastic) electron wave field. U is the effective scattering potential--which generally is complex (absorption!) and energy dependent (exchange, polarization, absorption)--and ψ the wave amplitude. If Ω_o is divided into atomic cells Ω_A (Wigner-Seitz cells) centered at the equilibrium positions $\underset{\sim}{r}_\nu$ of the atoms, then F can be written as

$$F = \sum_{\nu} e^{-i\underset{\approx}{k}\cdot\underset{\approx}{r}_{\nu}} \cdot \left(- \frac{1}{4\pi} \int_{\Omega_A} e^{-i\underset{\approx}{k}\cdot\underset{\approx}{r}'} U(\underset{\approx}{r}_{\nu}+\underset{\approx}{r}') \ \psi(\underset{\approx}{k},\underset{\approx}{k}_o,\underset{\approx}{r}_{\nu}+\underset{\approx}{r}') \ d\underset{\approx}{r}' \right) \cdot \quad (4a)$$

Expression (4a) is similar to the usual expression for the structure ampli-

tude,

$$F = \sum_{\nu} f_{\nu} e^{-i(\underset{\approx}{k}-\underset{\approx}{k}_o)\cdot\underset{\approx}{r}_{\nu}} \quad (5)$$

which is obtained when $\psi(\underset{\approx}{r})$ is written in the form of a Bloch wave:

$\psi(\underset{\approx}{k},\underset{\approx}{k}_o,\underset{\approx}{r}) = \chi(\underset{\approx}{k},\underset{\approx}{k}_o,\underset{\approx}{r}) \ e^{i\underset{\approx}{k}_o\cdot\underset{\approx}{r}}$, where $\chi(\underset{\approx}{r})$ has the lateral periodicity of the

surface. The dynamical scattering amplitude of the atomic cell ν,

$$f_{\nu}(\underset{\approx}{k},\underset{\approx}{k}_o) = - \frac{1}{4\pi} \int_{\Omega_A} e^{-i(\underset{\approx}{k}-\underset{\approx}{k}_o)\cdot\underset{\approx}{r}'} U(\underset{\approx}{r}_{\nu}+\underset{\approx}{r}') \ \chi(\underset{\approx}{k},\underset{\approx}{k}_o,\underset{\approx}{r}_{\nu}+\underset{\approx}{r}') \ d\underset{\approx}{r}' \quad (6)$$

differs considerably from the scattering amplitude of a free atom--both in

the Born and partial wave approximation--because ψ (or χ) depends upon $\underset{\approx}{k},\underset{\approx}{k}_o$,

the environment and position of the atom in the crystal. To obtain mean-

ingful values for f_{ν}, the integral equation

$$\psi(\underset{\approx}{k},\underset{\approx}{k}_o,\underset{\approx}{r}) = e^{i\underset{\approx}{k}_o\cdot\underset{\approx}{r}} - \frac{1}{4\pi} \int_{crystal} \frac{e^{i\underset{\approx}{k}|\underset{\approx}{r}-\underset{\approx}{r}'|}}{|\underset{\approx}{r}-\underset{\approx}{r}'|} U(\underset{\approx}{r}') \ \psi(\underset{\approx}{k},\underset{\approx}{k}_o,\underset{\approx}{r}') \ d\underset{\approx}{r}' \quad (7)$$

for ψ or the corresponding Schroedinger equation has to be solved. At pre-

sent this cannot be done for complex surfaces. To determine the nature of

complex surfaces now we must therefore simplify the theory of LEED intensi-

ties drastically and/or obtain additional information from other observa-

tions such as Auger electron spectroscopy [2].

The simplifications of the theory, which have been made in the past

were based on the two-dimensional picture of the diffraction process. The

trial-and-error method was used in the intensity analysis of patterns from

"clean" elemental semiconductors [8-10], and the Patterson function method

in the intensity analysis of patterns of CO and O adsorption layers on Pt and Rh [11]. In the first method it is assumed that all atoms ν which con-

tribute to the diffraction process have the same f_ν. This requires that

all atoms have the same χ in Eqn. (6). The second method requires the

additional assumption that χ is constant in all atoms contributing to the

diffraction process; otherwise the Fourier transform relationship between

structure amplitude and electron density, which is the basis of the Patterson

function method, does not exist. It is obvious that both methods represent

rather drastic simplifications of the diffraction process and that the sur-

face structures derived with them have to be taken with considerable caution

as illustrated by the various structures of elemental semiconductor surfaces

deduced with the trial-and-error method [8-10].

In this paper, which is based on the three-dimensional picture of the

diffraction process, we make a simplfication which is suggested by experi-

ments on surfaces producing LEED patterns expected from the bulk periodicity

("ideal surfaces"). The intensity versus voltage curves of many of these

surfaces show main maxima with average spacings corresponding to the period-

icity normal to the surface. This indicates that the contribution of the

imaginary part of f_ν--resulting from U and χ in Eq. (6)--to the phase term

in Eq. (5) does not change it so drastically as to destroy the third Laue

condition completely. Consequently, F and $|F|^2 \sim I$ show some periodicity

normal to the surface. The gross features of the intensity _versus_ voltage

curves--mainly the existence and average spacing of the main maxima--will

therefore be used to obtain information on the periodicity of a surface

structure normal to the surface, also in the case of complex diffraction

patterns. Another piece of information can be obtained from _relative_ inten-

sities. In pseudo-superstructure patterns, i.e. patterns produced by sur-

faces covered with a surface layer of different lateral periodicity and/or

azimuthal orientation [7], the beams produced by single scattering are frequently stronger than the beams due to multiple scattering. This is expected to happen when the third Laue condition for the surface layer is approximately fulfilled. Therefore fractional order beams which are very strong at voltages V spaced at intervals ΔV different from those of the integral order beams are usually singly scattered beams from the surface layer and can give information on the lateral periodicity of the surface layer. Thus LEED can provide at present information on the lateral and normal geometrical unit cell dimensions of the surface structure, but not on the number, nature, and distribution of the atoms in the unit cell.

In the past, the nature of the atoms present in a given surface structure had to be deduced from the experimental conditions which led to this structure. Auger electron spectroscopy, originally proposed by Lander [12] and perfected by Harris [13], when combined with LEED [2] permits determination of the nature of the atoms present on the surface. When the Auger electron signal is calibrated the number of atoms can be determined too [2]. However, it has to be kept in mind that the amplitude and width of many Auger transitions depend strongly upon the environment of the atom. Therefore the number of atoms of a given kind per unit cell of a suspected structure can in general be determined only after calibration of the Auger signal on a crystal which is known to have this structure. We will now illustrate these general considerations using two of the best known examples of complex patterns.

III.   THE "CLEAN" Si(111) SURFACE

Farnsworth et al. [14] discovered ten years ago two complex diffraction patterns on the Si(111) surface, the Si(111)-7x7 (in short, 7) and the Si(111) - $\sqrt{19}$ x $\sqrt{19}$ R(23.5) (in short, $\sqrt{19}$) patterns shown in Fig. 2. These two patterns have since been reproduced in many laboratories under a wide

variety of conditions and therefore have been generally attributed to clean
surfaces. On the other hand, it has been suggested that the patterns may
be due to double scattering between the Si substrate and a surface reaction
layer with different periodicity and/or orientation [15]. Recently it was
proposed [16] that the $\sqrt{19}$ pattern is due to a reconstructed surface layer
[8-10] stabilized by an extremely small amount of Ni on the otherwise clean
Si surface.

The electron energy spectrum--or more precisely its derivative $\frac{dN(E)}{dE}$ --
from a "clean" Si(111) surface, when measured at low resolution and sensitivity (Fig. 3a) does not show any particular features characteristic of
impurities; however at high resolution and sensitivity two peaks can be
found in the 40 to 60 eV range; one at 45 eV, one at 57 eV (Fig. 3b). In
our apparatus these are the positions of the main peaks of Fe and Ni respectively (as determined by depositing Fe and Ni onto the surface). If the
crystal is treated so as to produce a 7 pattern, the Fe peak grows with
increasing intensity of the 7 pattern and the Ni peak becomes weaker but
never disappears completely. When the crystal is treated to develop the
$\sqrt{19}$ pattern, the Fe peak disappears almost completely and the Ni peak grows
considerably. This is illustrated in Fig. 4 which shows the height of the
Ni Auger signal and of the intensity of one of the fractional order spots
(0' in Fig. 6) as a function of quenching (annealing) temperature. The
curves do not represent equilibrium conditions because the heating period
was only 30 sec which is insufficient to establish equilibrium over the
rising part of the curve. The relation between Auger signal and spot intensity, however, is clear. If the crystal is heated to 1000°C, where it

produces a 1x1 pattern, both Fe and Ni peaks have nearly disappeared.
Thus, the 7 pattern is definitely connected with the presence of Fe, the
$\sqrt{19}$ pattern with the presence of Ni in the surface layer. In what way are
now Fe and Ni connected with these two structures? We have to consider
the following possibilities:

(1)   The Fe and Ni atoms simply sit on top of the unreconstructed Si
surface forming a two-dimensional true or apparent superstructure.

(2)   They represent a trace impurity stabilizing a true or apparent
superstructure made up of Si atoms [16].

(3)   They are an essential component of a surface layer which consists
of both Si and Fe or Ni respectively and form a true or apparent superstructure.

We will examine now these possibilities by analyzing the geometry and
the gross features of the intensities of the diffraction patterns.

The 7 pattern is characterized by the high intensity of those fractional order spots which have the indices $\{40\}$, $\{44\}$ with respect to the
various integral order spots characteristic of the unreconstructed surface.
This suggests that the pattern is not due to a real superstructure but is
due to an apparent one produced by multiple scattering between the Si substrate and a surface layer with $\frac{7}{4}$ the lateral periodicity of the Si(111)
surface. The I(V) curves of the $\{40\}$ spots--which are the $\{10\}$ spots of
the surface layer--show at normal incidence intensity maxima at the following beam energies: 39.5, 56, 79, 97, 126, 151, 191, 222 eV. Assuming zero
inner potential these energies correspond to $\frac{1}{\lambda}$ values of .513, .611, .726,
.805, .917, 1.003, 1.128, and 1.217 $\overset{\circ}{A}{}^{-1}$. If these maxima of the $I(\frac{1}{\lambda})$ curve

are considered to be the main maxima of a structure periodic normal to the

surface with a_3, then a_3 can be obtained from

$$\frac{1}{a_3} = a_3^* = \frac{1}{\lambda_{n+1}} + \sqrt{\frac{1}{\lambda_{n+1}^2} - a_1^{*2}} - \frac{1}{\lambda_n} \cdot \sqrt{\frac{1}{\lambda_n^2} - a_1^{*2}} \quad \text{where } a_1^* \text{ is the lateral}$$

reciprocal periodicity: $a_1^* = \frac{4}{7} a_1^*{}_{Si} = \frac{4}{7} \cdot \frac{\sqrt{24}}{3a_{Si}} = .172 \text{ Å}^{-1}$. With the $\frac{1}{\lambda}$

values given above we obtain a mean value of $a_3^* = .204 \text{ Å}^{-1}$ with a root

mean square deviation of .012 Å^{-1} and from these values $a_3 = 4.9 \pm .3$ Å

which is definitely incompatible with the periodicity of Si normal to the

surface. The diffraction pattern always shows perfect sixfold symmetry,

even in the rounded regions of the crystal surface. This is usually a

quite reliable indication that the sixfold symmetry is not due to three

(two) equivalent structures with twofold (threefold) symmetry. The diffrac-

tion pattern has therefore to be due to a hexagonal structure with the basal

plane parallel to the Si(111) surface or due to a cubic structure parallel

to the Si substrate. In such a cubic structure the three-dimensional recip-

rocal lattice vectors corresponding to $\underset{\sim}{a_1^*}$ and $\underset{\sim}{a_3^*}$ are $\langle 1\bar{1}0 \rangle$ and $[111], \langle 110 \rangle$

and $[222]$ or $\frac{1}{3} \langle 22\bar{4} \rangle$ and $[111]$ respectively for the various possible lattices.

This corresponds to $\frac{a_3^*}{a_1^*}$ ratios of 1.225, 2.45, and .95 respectively. Clearly,

only the first ratio which gives an a-value of a = 8.3 Å is compatible with

the ratio $\frac{a_3^*}{a_1^*} = \frac{.204 \pm .012}{.172} = 1.185 \pm .070$ deduced from experiment. For a

hexagonal surface we obtain a $= \frac{2}{\sqrt{3}a_1^*} = 6.72$ Å. We now have two possible unit

cell dimensions for our surface structure which according to the Auger spec-

trum must contain both Fe and Si: a primitive cubic cell with a = 8.3 Å and

a hexagonal cell with a = 6.72 Å and c = 4.9 $\pm$ .3 Å. The choice between

them is based on the assumption that the surface layer is thick enough
that it may be considered as a bulk phase. The system Fe-Si contains no
cubic phase with a = 8.3 Å, but does have a hexagonal phase, the $Fe_5Si_3(\eta)$
phase, with unit cell dimensions which agree very well with our values:
a = 6.727, c = 4.705 or 6.742, c = 4.708 Å [17]. We conclude therefore
that a Si(111) surface which produces a 7 pattern is not clean but covered
with a surface layer with the unit cell dimensions of Fe_5Si_3 (in short
"Fe_5Si_3-structure"). The way the 7 pattern is formed by this surface layer
is indicated in Fig. 5. The thickness of this layer and the amount of Fe
in it cannot be determined at present. However, the high spot-to-background
intensity ratio of a well developed 7 pattern suggests that Fe vacancies or
other impurities when present are periodically distributed. It is hoped
that future Auger work will allow the determination of layer thickness and
Fe content.

The $\sqrt{19}$ structure is more difficult to analyze. None of the "fractional"
order beams is distinguished by a high intensity at well defined, approximately equally spaced $\frac{1}{\lambda}$ values. Thus no periodicity normal to the surface
can be derived from the diffraction pattern. The intensity of several intense fractional order beams is related to the intensity of the integral
order beams. This is a strong indication that the diffraction pattern is
due to double scattering between the unreconstructed Si substrate and a
surface layer with different periodicity and azimuthal orientation. The
diffraction pattern (Fig. 2, 6a) consists of two hexagonal patterns (Fig. 6b)
which are rotated ± 23.5° against the basic Si pattern. Any pair of vectors
$\underset{\sim}{a}_1^*, \underset{\sim}{a}_2^*$ except those of the Si 1x1 pattern can generate this network by combination with the vectors of the basic Si pattern. The observation that the

relative spot intensities in the rounded regions of the surface are the same as those in the flat regions suggests true hexagonal symmetry of the two-dimensional reciprocal net of the surface layer, thus $|\underset{\sim}{a}_1^*| = |\underset{\sim}{a}_2^*| = a_1^*$ and $\sphericalangle(\underset{\sim}{a}_1^*, \underset{\sim}{a}_2^*) = 60°$. The general aspects of the intensity distribution lead to the choice of $\underset{\sim}{a}^*$ vectors $\underset{\sim}{h}_{hk}$ leading to one of the six points (hk) shown in Fig. 6b. In a previous short communication [18], we suggested that $\underset{\sim}{a}_1^* = \underset{\sim}{h}_{23}$ ($|\underset{\sim}{h}_{23}| = |\underset{\sim}{h}_{32}| = |\underset{\sim}{a}_1^*{}_{Si}|$). This suggestion was based on the hypothesis that the surface layer has to have a structure known to occur in the system Ni-Si, a condition which is fulfilled only when $|\underset{\sim}{a}_1^*| = |\underset{\sim}{h}_{23}| = |\underset{\sim}{a}_1^*{}_{Si}|$. Two structures, θ-Ni$_2$Si and NiSi$_2$, when slightly distorted, are then compatible with the geometry of the diffraction patterns. The high temperature (θ) modification of Ni$_2$Si is hexagonal with a = 3.805 Å and c = 4.890 Å [19]; when its (001) plane is parallel to the substrate, then $|\underset{\sim}{a}_1^*| = \dfrac{2}{\sqrt{3}a} = .3035$ Å^{-1} which differs 1% from $|\underset{\sim}{a}_1^*{}_{Si}| = \dfrac{\sqrt{24}}{3a_{Si}} = .3007$ Å^{-1}.

NiSi$_2$ is cubic with a = 5.395 Å [20]; when present in parallel orientation on the Si(111) surface $|\underset{\sim}{a}_1^*| = .3027$ Å^{-1} which differs only .7% from $|\underset{\sim}{a}_1^*{}_{Si}|$. A distinction between these two structures is possible on the basis of the following observations made on Si(111) surfaces producing a 1x1 pattern:

(1) The 1x1 pattern can be obtained in various ways, e.g.: (a) by deposition of Ni to a thickness of several to many atomic layers followed by a short anneal of the crystal at 700°C or lower (surface A), and (b) by annealing of a "clean" crystal for several minutes at 700°C after it had been heated sufficiently long at 1200 to 1300°C to desorb most of the Fe from crystal and leads (surface B).

(2) The surface A is characterized by a strong Ni Auger peak (Fig. 7d), the Ni Auger peak of the surface B is barely detectable.

(3)   The I_{oo}(V) curves of both surfaces generally have maxima at nearly the same voltages (or $\frac{1}{\lambda}$ values).   The $\frac{1}{\lambda}$ intervals between the main maxima agree in both cases qualitatively with the periodicity of Si(NiSi$_2$) normal to the surface.   However there are significant differences in the relative height of the maxima.   In particular, the peak which occurs on surface A at 177 eV is very strong, while the corresponding peak at 181 eV of surface B is very weak.

(4)   The spot-to-background intensity ratio of surface A is lower than that of surface B, the Kikuchi pattern of surface A is sharper than that of surface B.

From (2) we conclude that the surface A has a high Ni content, from (3) that its periodicity normal to the surface is essentially the same as that of Si in spite of the high Ni content, and from (4) that the Ni atoms are distributed not at random but in a well ordered manner.   An epitaxial surface layer of NiSi$_2$ is the simplest explanation compatible with the conclusions.   Unless we invoke the assumption that NiSi$_2$ can grow in two different orientations on the Si(111) surface, e.g. depending upon cooling conditions or impurity effects, we are left with θ–Ni$_2$Si as the cause of the $\sqrt{19}$ pattern.

It must be kept in mind, however, that θ–Ni$_2$Si and NiSi$_2$ were selected as possible causes of the $\sqrt{19}$ pattern on the assumption that the surface layer has to have a structure found in bulk.   If we drop this assumption the other points (hk) shown in Fig. 6b have to be considered too.   We can reduce the number of possible $\underset{\sim}{a}_1^*$ vectors by extracting more information from the LEED pattern.   This can be done when the strongly simplifying assumption is made, that the multiple scattering process may be separated into subsequent scattering acts in surface layer and substrate.   Then the observed intensity of a doubly scattered beam should be proportional to the intensity of the substrate beam which produces the doubly scattered beam on its way out through the surface layer. One characteristic feature of the $\sqrt{19}$ pattern is that the intensities of the spots marked A'(B') in Fig. 6a increase and decrease with that of the basic Si spots marked A(B). According to the simplified picture of the double scattering process this can happen only if $\underset{\sim}{a}_1^* = \underset{\sim}{h}_{22}$, $\underset{\sim}{h}_{41}$ or $\underset{\sim}{h}_{33}$, which can be seen by superimposing the corresponding hexagonal networks--such as the one shown in Fig. 6c--with that of the unreconstructed Si surface, using points A and B as origins. Another characteristic feature of the $\sqrt{19}$ pattern is that the spots marked 0' in Fig. 6a are very strong over wide voltage ranges, while the spots 0", 0''' never are very strong. For the chosen $\underset{\sim}{a}_1^*$'s ($\underset{\sim}{h}_{22}$, $\underset{\sim}{h}_{41}$ and $\underset{\sim}{h}_{33}$) these 0 spots (0', 0", and 0''' respectively) are produced by the incident beam and by the 00 beam on its way out through the surface layer and are therefore expected to be strong, leading to the choice of $\underset{\sim}{a}_1^* = \underset{\sim}{h}_{22}$, $|\underset{\sim}{a}_1^*| = \sqrt{\frac{12}{19}} \, |\underset{\sim}{a}_1^* \,_{Si}| = .239 \, Å^{-1}$. Nothing can be deduced about $|\underset{\sim}{a}_3^*|$ because of the lack of regularity in the I(V) curves due to the strong dynamical coupling between the various beams. Thus we can say only that the surface layer has either a cubic structure with

$$a = \frac{\sqrt{2}}{|\underset{\sim}{a}_1^*|} = 5.92 \, Å \text{ or } a = \frac{\sqrt[3]{24}}{|\underset{\sim}{a}_1^*|} = 6.83 \, Å \text{ or a hexagonal structure with } a = \frac{2}{\sqrt{3} \, |\underset{\sim}{a}_1^*|}$$

$= 4.83 \, Å$. None of those structures is known to exist in the system Ni-Si. We can therefore only speculate about its nature: it could be an expanded surface layer of Ni_3Si which does not have the structure found in the bulk, but Fe_3Si structure ($a_{Fe_3Si} = 5.64 \, Å$); or more likely it could contain besides Ni other impurities which determine the structure. This second possibility can by no means be excluded at present, even in the case of the 7 pattern, for the following reason: most of the Auger electron spectra were obtained with primary energies up to 600 eV. In order to excite an Auger transition efficiently the energy of the incident electrons must be about three times the energy of the lower x-ray level involved in the transition. Consequently, one can detect efficiently only those atoms which have major Auger transitions below 200 to 300 eV. For example the Auger transitions of Ta in this energy region are much weaker than the 45 and 57 eV peaks of Fe and Ni respectively. If present in the surface layer in a state similar to that of Ni, Ta could not be detected at present. A surface layer of composition Ta_2Ni_3Si is therefore not incompatible with the Auger electron spectrum. This composition--like similar ones involving Nb, Mo or W in place of Ta-- is hexagonal with a $\approx$ 4.8 Å [21,22] which agrees well with the value a = 4.83 Å deduced from the LEED pattern. The observation that the $\sqrt{19}$ pattern forms much faster near the Ta leads than in the center of the crystal is also in favor of such an interpretation.

The results derived above, i.e. (1) a Fe-containing surface layer with Fe_5Si_3 structure as cause of the 7-pattern, and (2) a Ni-containing surface layer with $\theta-Ni_2Si$ or Ta_2Ni_3Si structure or with an unknown structure as cause of the $\sqrt{19}$ pattern, lead immediately to a number of questions:

(1) Why do the surface layers have the observed structures and not other structures, e.g., why does Ni not form the $NiSi_2$ (1x1) structure with its low mismatch?

(2) Where is the Fe and Ni coming from?

(3) Why does the Fe-containing layer form at the low temperature, the Ni-containing layer at the higher temperatures?

We have at present no answer to question (1) except the suggestion that the observed structures may have electron configurations leading to a lower surface energy. In the case of the Ni_2Si layer model, the tendency to form Ni_2Si might be attributed to the fact that Ni_2Si has the highest heat of formation in the system Ni-Si [23]. The source for the Fe and Ni may be (a) the interior of the crystal, (b) the crystal mounting, (c) the surface preparation process, (d) the handling of the crystal, and (e) the vacuum system itself. We can exclude (a) as a general source because of the wide variety of high purity crystals which have been investigated. The crystal mounting, usually made of refractory metals like Ta or Mo, may contain a considerable amount of Ni and Fe, e.g., Fansteel metallurgical grade Ta sheet contains up to .02% Fe, and is therefore a likely source in many cases. The acids used in surface preparation contain up to 10^{-3}% Fe which can lead to $2.4 \cdot 10^{15}$ adsorbed Fe atoms on the surface [24], enough to form a Fe_5Si_3 layer one unit cell thick. Variations in surface preparation procedures make this also an unlikely general source. Introduction of Ni into the crystal by improper handling with metal tweezers has already been pointed out [16]. The vacuum system, usually made of stainless steel, or at least containing Fe and Ni parts (e.g., Kovar), can be a continuous source of Fe and Ni via the following process: CO, which is one of the main residual gas components, interacts with Fe(Ni) surfaces to form a Fe(Ni) carbonyl layer. The adsorbed carbonyl molecules can be desorbed either thermally or by collisions of atoms, ions or electrons with the surface. When they hit the surface of the Si crystal or of the crystal mounting they can dissociate depending upon surface condition and temperature leaving

Fe(Ni) on the surface. We believe this process to be the major source in many of our experiments; without it the regeneration of the Fe_5Si_3 structure (with the accompanying Fe Auger Peak) after excessive heating to desorb all Fe would be difficult to understand.

The occurrence of the Fe(Ni)-containing layer at the lower (higher) temperature can be attributed to the relative solid solubilities of Fe and Ni. The solid solubilities of Fe and Ni in Si at 900°C are $6 \cdot 10^{12}$ and $3 \cdot 10^{16}$ atoms/cm^3, respectively; at 1300°C $5 \cdot 10^{16}$ and $8 \cdot 10^{17}$ atoms/cm^3, respectively [25,26]. Consequently, when Fe and Ni are supplied simultaneously to the crystal, either by diffusion from the crystal mounting or by carbonyl decomposition, most of the Fe has to stay at the surface in the form of Fe or an Fe–silicide while a considerable amount of Ni can go into solid solution. In particular, at the low annealing temperatures used to produce the 7-pattern essentially all Fe remains at the surface, thus forming the 7 pattern. (A Fe_5Si_3 layer 1 unit cell thick contains $2.5 \cdot 10^{15}$ Fe atoms/cm^2.) In order to produce the $\sqrt{19}$ pattern the Fe has to be desorbed or dissolved in the bulk of the crystal or in the crystal mounting and replaced by a sufficient amount of Ni. This requires first that the crystal is heated high enough, e.g., to 1000°C, where the vapor pressure of Fe is $5 \cdot 10^{-7}$ torr so that desorption becomes possible, or that Fe can diffuse into the crystal mounting. Significant dissolution in the bulk of the Si crystal is only possible at higher temperatures; for example, dissolution of the Fe contained in a Fe_5Si_3 layer one unit cell thick can occur only at temperatures above 1180°C. However, if the crystal contains precipitation centers at which Fe can precipitate out, the crystal itself can also act as a sink for Fe. The second prerequisite for the production of the $\sqrt{19}$ pattern is proper pretreatment of the crystal (and/or crystal mount) such that it contains a sufficient amount of dissolved or precipitated Ni, which can dissolve at the annealing temperature, e.g. at 1000°C where the saturation concentration is $1.5 \cdot 10^{17}$ Ni atoms/cm^3 [26]. Upon rapid cooling some of the Ni can then precipitate out at the surface to form the $\sqrt{19}$ structure. These considerations show that the formation of the observed surface layers depends upon many parameters. This is probably to a large extent responsible for the differences between the required annealing conditions as found in various laboratories [8,13,16,27].

SUMMARY

In this paper we have proposed a view of the diffraction process which is diametrically opposite to that which has been the basis of the methods used in the past to interpret complex LEED pattern. Then absorption was assumed to be so strong that an essentially two-dimensional structure analysis could be made. Here absorption is considered to be so weak that the elastic wave field can penetrate considerably into the crystal and a three-dimensional approach appears more appropriate. We have applied this approach to the problem of the annealed Si(111) plane, not so much to come up with unequivocal structures; rather, we wanted to illustrate the procedures which can be followed to arrive at surface models which have to be considered in a detailed dynamical structure analysis with numerical methods. More detailed studies should allow reduction of the number of likely surface models still more. For example, Auger electron spectroscopy studies as a function of primary energy at fixed angle of incidence and vice versa which are in progress are expected to give information on the distribution of Fe and Ni normal to the surface. Increased sensitivity should allow detection of atoms with weaker Auger peaks, e.g. of Ta. A better understanding of the Auger transition probabilities as a function of environment should allow one to obtain information on the area concentration of impurities and their environment. Quantitative studies of the formation kinetics of the various LEED patterns on crystals with controlled impurity and imperfection content also promises to be a useful tool. In conclusion, much experimental work needs to be done before a mathematical structure analysis should be attempted.

ACKNOWLEDGMENT

The author wishes to thank Dr. J. O. Porteus for critically reading the manuscript and for valuable comments. Financial support of the National Aeronautics and Space Administration under Contract No. R-05-030-001 is appreciated.

REFERENCES

[1] E. Bauer, Phys. Rev. 123, 1206 (1961).

[2] R. E. Weber and W. T. Peria, J. Appl. Phys. 38, 4355 (1967).

[3] J. J. Lander, Progr. Solid State Chem. 2, 26 (1965).

[4] H. N. Browne and E. Bauer, unpublished.

[5] J. J. Quinn, Phys. Rev. 126, 1453 (1962).

[6] P. A. Fedders, Phys. Rev. 153, 438 (1967).

[7] E. Bauer, Surface Sci. 7, 351 (1967).

[8] J. J. Lander and J. Morrison, J. Chem. Phys. 37, 729 (1962);
     J. Appl. Phys. 34, 1403 (1963).

[9]   R. Seiwatz, Surface Sci. 2, 473 (1964).

[10]   N. R. Hansen and D. Haneman, Surface Sci. 2, 566 (1964).

[11]   C. W. Tucker, Jr., Surface Sci. 2, 516 (1964); J. Appl. Phys. 37, 3013, 4147 (1966).

[12]   J. J. Lander, Phys. Rev. 91, 1382 (1953).

[13]   L. A. Harris, G. E. Res. Develop. Rept. #67C201 (1967); J. Appl. Phys. 39, 1419 (1968).

[14]   R. E. Schlier and H. E. Farnsworth, J. Chem. Phys. 30, 917 (1959).

[15]   E. Bauer, Colloque Internat. CNRS 1965, No. 152, p. 19.

[16]   A. J. van Bommel and F. Meyer, Surface Sci. 8, 467 (1967).

[17]   A. R. Weill, Nature 152, 413 (1943).

[18]   E. Bauer, Phys. Letters 26, 530 (1968).

[19]   K. Toman, Acta Cryst. 5, 329 (1952).

[20]   K. Schubert and H. Pfisterer, Z. Metallk. 41, 433 (1950).

[21]   D. I. Bardos, K. P. Gupta and P. A. Beck, Trans. Met. Soc. AIME 221, 1087 (1961).

[22]   M. Yu. Teslyuk, V. Ya. Markiv, and E. I. Gladyshevskii, J. Struct. Chem. 5, 364 (1964).

[23]   U. Dehlinger, Z. Metallk. 43, 109 (1952).

[24]   V. S. Sotnikov and A. S. Belanovskii, Russ. J. Phys. Chem. 34, 1001 (1960).

[25]   J. D. Struthers, J. Appl. Phys. 27, 1560 (1956).

[26]   J. H. Aalberts and M. L. Verheijke, Appl. Phys. Letters 1, 19 (1962).

[27]   F. Jona, Appl. Phys. Letters 6, 205 (1965); IBM J. Res. Develop. 9, 375 (1965).

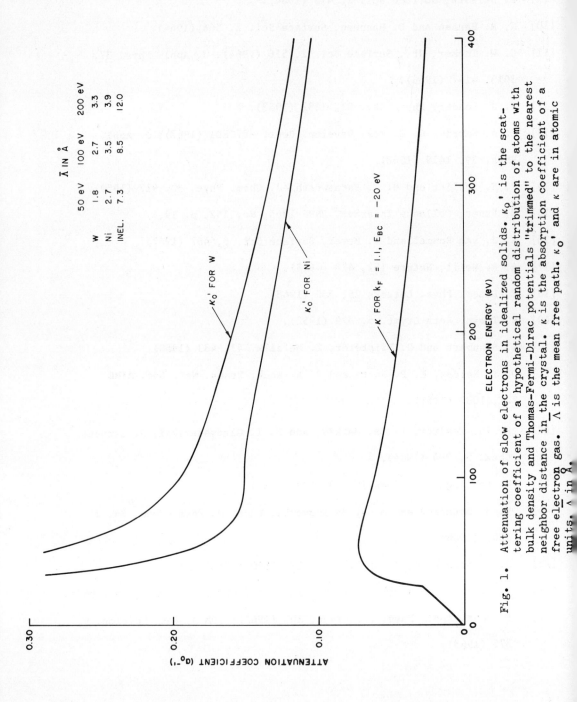

Fig. 1. Attenuation of slow electrons in idealized solids. κ_0' is the scattering coefficient of a hypothetical random distribution of atoms with bulk density and Thomas-Fermi-Dirac potentials "trimmed" to the nearest neighbor distance in the crystal. κ is the absorption coefficient of a free electron gas. $\bar{\Lambda}$ is the mean free path. κ_0' and κ are in atomic units. $\bar{\Lambda}$ in Å.

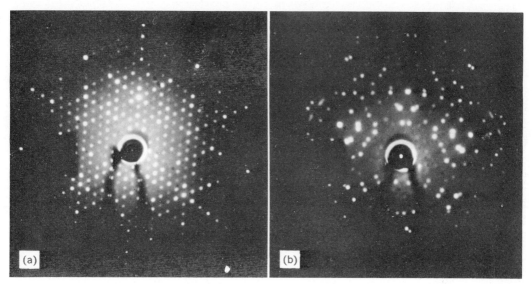

Fig. 2. LEED patterns from annealed Si(111) planes; (a) 7x7 pattern (80 eV),
(b) $\sqrt{19}$ x $\sqrt{19}$ R(23.5°) pattern (55 eV).

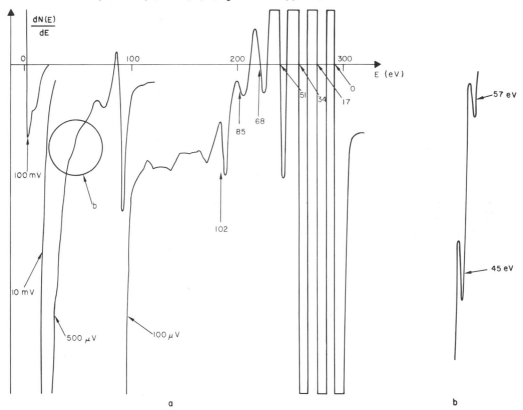

Fig. 3. Derivative $\dfrac{dN(E)}{dE}$ of the total energy distribution of electrons from a
Si(111) surface bombarded with slow electrons; (a) total spectrum at
low resolution (300 eV primary electrons), numbers on left denote
detector sensitivity, numbers on right energy loss values, (b) section
of spectrum at high resolution (600 eV primary electrons).

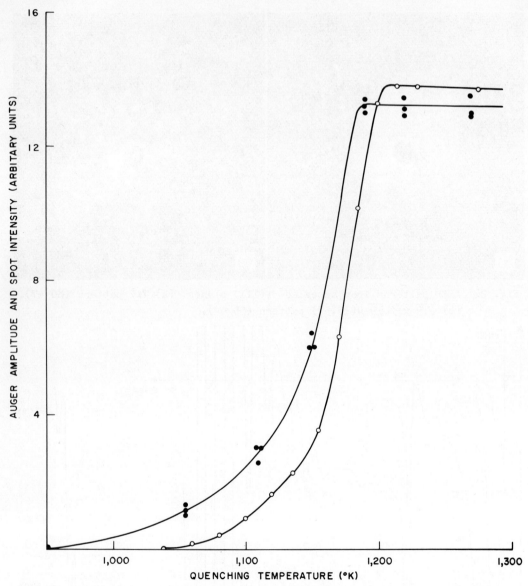

Fig. 4. Amplitude of Auger signal at 56 eV (full circles) and of fractional order spot intensity (open circles) in $\sqrt{19}$ pattern as a function of quenching temperature for 30 sec annealing periods.

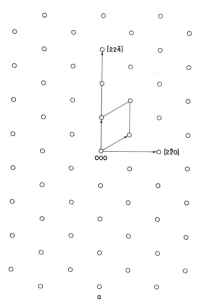

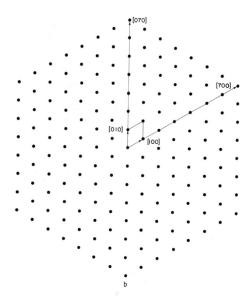

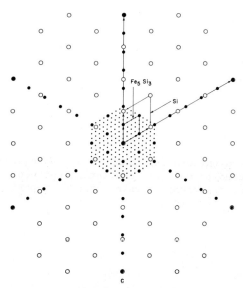

Fig. 5. Explanation of 7 pattern in terms of multiple scattering between a surface layer with Fe$_5$Si$_3$ unit cell dimensions and the unreconstructed Si substrate; (a) basic Si pattern, (b) Fe$_5$Si$_3$ pattern, (c) multiple scattering pattern from superimposed Fe$_5$Si$_3$ and Si.

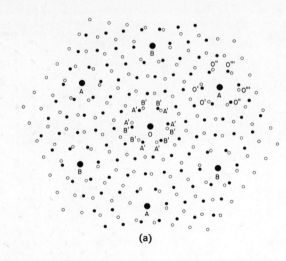

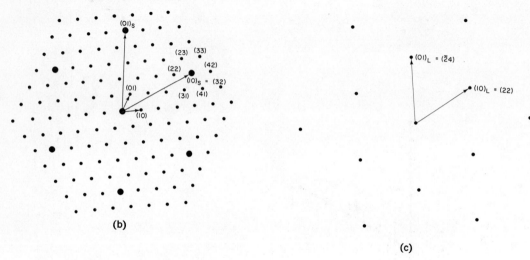

Fig. 6. Analysis of $\sqrt{19}$ pattern. The observed pattern (a) consists of two equivalent patterns (b), which are produced by multiple scattering between the substrate and a surface layer of different orientation and/or periodicity, e.g., like that producing pattern (c).

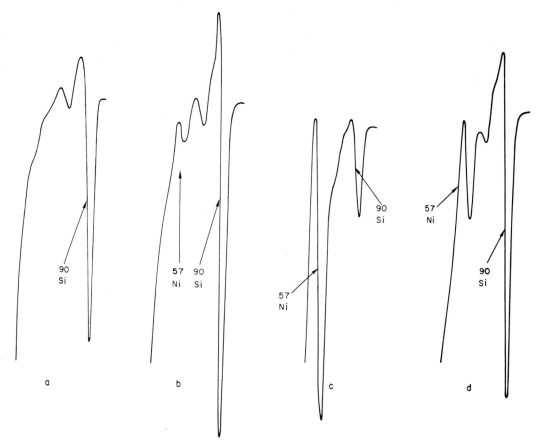

Fig. 7. Partial energy spectrum of Si(111) surfaces; (a) at 1000°C (1x1 pattern), (b) after quenching to room temperature ($\sqrt{19}$ pattern), (c) after deposition of Ni (LEED only background), (d) after short anneal at 700°C (1x1 pattern).

NATURE OF CLEAN CLEAVED SILICON SURFACES
WITH WAVE FUNCTION OVERLAP CALCULATIONS

D. Haneman and D. L. Heron

School of Physics
University of New South Wales
Kensington, Australia

1.   Introduction.

A combination of various measurements of the properties of Si surfaces produced by cleavage in ultra high vacuum, led to a proposed model for the surface based on a furrowed structure involving two kinds of atom site. [1]) In this work the surface wave functions necessary to account for the experimental results are calculated. By computing the overlap between the wave functions, the spin-spin interactions are estimated. The wave functions used are precisely specified and may be used as a basis for calculation of surface properties.

The salient features of the experimental measurements are listed below.

(a) LEED.

1)   The cleaved (111) surface has a 2 x 1 unit mesh ($\frac{1}{2}$ orders in one preferred (112) azimuth). [2,3])

2)   The preferred (112) azimuth is the one nearest to the direction of the crack that caused the cleavage. [3])

3)   Exposure to oxygen at 10^{-7} torr min removes the $\frac{1}{2}$ order beams but the integral order beams remain (with weakening intensity) clearly visible till approximately 10^{-1} torr min when the LEED patterns disappear. [3])

4)   Annealing at approximately 400°C removes the half order beams, [3]) and at approximately 800°C the pattern converts to the 1/7 order one characteristic of ion bombarded and annealed surfaces. [2,3])

(b) Work function and surface conductivity

1)   Exposure to oxygen in the region of 10^{-6} to 10^{-5} torr min increases the

work function [4]) and markedly affects the surface conductivity. [5,6])

(c) E.P.R.

1) The clean cleaved surface has a density of spins of approximately 1 per 10 surface atoms (single line at g = 2.0055, line width 6 gauss, partly inhomogeneously broadened). [1])

2) Exposure to oxygen in the region of 10^{-5} torr min has no detectable effect on the e.p.r. signal (although work function and surface conductivity are affected). However exposure in the region of 10^{-1} torr min causes a marked increase in the signal (and in number of spins). [1])

The above facts can all be correlated on the basis of the model shown in Fig. 1. The atoms lie in two kinds of row as shown. This gives a 2 x 1 unit cell, satisfying the LEED symmetry. (Simple kinematic intensity calculations are compatible with the observed intensities). The lowered atoms ("p dangling bonds") have unpaired electrons overlapping sufficiently so that no contribution to the e.p.r. signal is observed. However oxygen adsorbs onto these sites in the 10^{-6} torr min range, accounting for the effects on work function, surface conductivity, and LEED intensities in this range, whereas no change in e.p.r. is observed. The raised atoms ("s dangling bonds") have unpaired electrons which do not overlap sufficiently to "pair off", hence they give an e.p.r. signal. The particular nature of the sites is such that the oxygen sticking coefficient is low, and hence no effects are observed until exposures in the 10^{-1} torr min range are applied. This corresponds to complete coverage of all surface sites with disordering so that the LEED pattern is lost at this stage.

The above picture accounts qualitatively for the various kinds of experimental observation. Reasons for the occurrence of the raised and lowered atom rows have been described previously [1]) (based on a similarity to the annealed structure containing raised and lowered atoms, which has been pulled into longitudinal regions by the stress wave associated with the advancing

crack that caused the cleavage). It is however necessary to check that wave

function overlap of electrons on the surface atoms can be sufficient to cause

spin ordering. This result is far from obvious since the distance between

surface atoms on the rows is 3.84 Å, compared with the bulk atom spacing of

2.35 Å.

2.   Calculation of Surface Electron

Wave Function Overlaps.

Since the spacing between surface atoms is over 50% greater than the

bulk spacing, an approach based on the use of atomic orbitals is a good

starting point. We set up wave functions as linear combinations of atomic

orbitals, and calculate the overlaps between adjacent neighbours, labelled

atoms 1 and 2. The wave function overlap is

$$\alpha = \int f_1(r) \; f_2(r) \; dr \tag{1}$$

For a normal bulk type sp^3 wave function

$$f = \tfrac{1}{2} s + \frac{3^{\frac{1}{2}}}{2} p_z \tag{2}$$

(z is normal to the surface). However the surface electrons are in a condition

different from the bulk due to the absence of a neighbouring atom. As explained

earlier [7]) the total strain in the surface can be lowered if some surface

dangling bonds become more s type and others become more p type, leading to a

buckling of the surface due to the consequent changes in bonds to the second

layer. This buckling results in large surface unit meshes, as observed. An

energy calculation [8]) (for annealed surfaces) indicates that the surface

energy is lowered by this effect.

In the present case of cleaved surfaces let us consider the surface wave

functions. If the dangling bonds were pure $3p_z$ the bonds to the second layer

would be sp^2 which are planar and this is clearly an extreme case. However

out of interest, the overlap has been calculated for this pure $3p_z$ case, and for the cases of a pure 3s dangling bond, a sp^3 bond, and a $3p_y$ bond at the

surface spacing of 3.84 Å. The atomic functions used were those of Watson

and Freeman. [9)]

The results are shown in Table 1.

Table 1.

Overlap integral 2α for a dangling bond with a like neighbour on 2 opposite

sides in y direction.

Wave function	2α for surface spacing of 3.84 Å	2α for bulk spacing of 2.35 Å	Symbol used in text
$3p_z$	0.12		$2(p_z \mid p_z)$
$3p_y$	0.36	0.64	$2(p_y \mid p_y)$
3s	0.09	0.54	$2(s \mid s)$

As a basis for comparison of magnitudes, the overlap for the bulk spacing

was also calculated, and is shown in Table 1. For sp^3 functions, the bulk

spacing overlap $\alpha = 0.67$. These bulk spacing figures are not closely accurate

estimates for the true overlap at the bulk spacing since the LCAO method is

too simple at this spacing. However they are approximate and the figures do

demonstrate that the overlap of s functions at the surface spacing is in the

region of 1/6 of the value at the bulk spacing. Although this fraction is

small, it is nevertheless still quite significant.

The overlap is largest for p_y functions because these are directed along

the rows ie. the line joining neighbours. However the adoption of pure p_y

dangling bond wave functions imposes severe conditions on the bond types to

the second layer atoms. It is therefore more reasonable to consider the

effect of adding a small proportion of p_y to the original sp^3 function. This

will clearly cause a tilting of the wave function, see Fig. 2, along the row

direction and increase the overlap, as required by the e.p.r. results.
(Pertinently, that (1) there are of order 1 spin per 10 atoms, (2) some surface sites which adsorb oxygen give no spin resonance and hence have sufficient
overlap for spin pairing).

We calculate the overlap in this case by considering three adjacent surface atoms 1, 2, and 3, and compute the overlap between the centre orbital 2
and its two neighbours. For the orbitals we assume LCAO's:

$$f_1 = a_s s - a_y p_y + a_z p_z$$
$$f_2 = a_s s + a_y p_y + a_z p_z \tag{3}$$
$$f_3 = a_s s - a_y p_y + a_z p_z$$

which contain an admixture of p_y (the y-axis is along the row) such that p_y
lobes of like sign point towards each other. Writing the overlap of functions
u_1 and u_2 centred on adjacent atoms, as

$$(u_1 \mid u_2) = \int u_1 u_2 \, dr \tag{4}$$

the total overlap of an electron in orbital f_2 is then

$$\alpha_t = (f_1 \mid f_2) + (f_2 \mid f_3) \tag{5}$$

This may be expressed in terms of the atomic orbital overlaps as

$$\alpha_t = 2(a_s^2 (s|s) + a_y^2 (p_y \mid p_y) + a_z^2 (p_z \mid p_z) \;) \tag{6}$$

since

$$(s \mid p_z) = (p_y \mid p_z) = 0$$

and the contribution of $(s \mid p_y)$ to $(f_1 \mid f_2)$ cancels its
contribution in $(f_2 \mid f_3)$, that is,

$$(a_2 \mid p_{y3}) = - (p_{y1} \mid s_2)$$

It is useful to calculate the overlap α_t as a function of the fraction
of p_y orbital that is admixed. We call this fraction $\beta/(1+\beta^2)^{\frac{1}{2}}$.
Normalization requires

$$f = \frac{1}{(1 + \beta^2)^{\frac{1}{2}}} \; (s/2 + 3^{\frac{1}{2}}/2 \; p_z \pm \beta p_y) \tag{7}$$

whence

$$a_y = \beta/(1 + \beta^2)^{\frac{1}{2}}, \quad a_z = 3^{\frac{1}{2}}/(2(1 + \beta^2)^{\frac{1}{2}}), \quad a_s = 1/(2(1 + \beta^2)^{\frac{1}{2}}) \tag{8}$$

The values of α_t were calculated for various values of β, the results being shown in Fig. 3.

We note that the total overlap increases up to 0.24 for $\beta = 1$, corresponding to 50% of the p_y charge density being admixed into the dangling surface orbital. (The proportion is $a_y^2 = \beta^2/(1 + \beta^2)$). At $\beta = 0.5$, corresponding to only 0.2 of the p_y charge density being added, the overlap integral is 0.16. This may be contrasted with the bulk sp^3 overlap integral which is calculated by the above methods to be 0.67 for a pair of atoms at the bulk spacing. The overlap is therefore appreciable.

We now estimate the degree of spin ordering that can result from interactions of these magnitudes. Discussions have been given for many systems of neighboring paramagnetic entities and the requirement for spin ordering is that the exchange integral be appreciably larger than kT. This means that the spins are paired at the temperature T, thermal activation being insufficient to excite many electrons into upper, unpaired states. We thus estimate the exchange interactions along the rows with p_y - admixed orbitals ("p row").

The exchange integral between electrons i and j, in terms of the wave function f_1 on atom 1 and the wave function f_2 on atom 2, is

$$J_{ij} = \iint f_1(r_i) \, f_2(r_j) \, \frac{e^2}{r_{ij}} \, f_1(r_j) \, f_2(r_i) \, dr_i \, dr_j \tag{9}$$

where $r_{ij} = |r_i - r_j|$. This double integral is difficult to compute. It may however be estimated by the formula

$$J_{ij} \sim \frac{\alpha^2 e^2}{r} \tag{10}$$

where r is the distance between the atoms, 3.84 Å. This approximation, which is more exact the less the overlap, was tested by applying it to the hydrogen molecule for which the exchange integral has been accurately calculated as 0.65 rydbergs. [10]) Applying the above formula (10), the exchange integral was calculated as 0.81 rydberg, which is reasonably close to 0.65 ry, considering that the overlap in the hydrogen molecule is large, 0.75. For the case considered here, with much smaller overlap, the approximation is expected to be at least as good, and probably better than for the hydrogen molecule. We now calculate the exchange integral J for the case of $\beta = 0.5$ mentioned above, corresponding to $\alpha_t = 0.16$ (Fig. 3). For this case the overlap between a surface electron and the neighbour whose wave function is tilted towards it is computed to be 0.13. Then one obtains

$$J = 0.064 \text{ ev} \tag{11}$$

It is correspondingly larger for larger β and α.

The above value of J is about $2\frac{1}{2}$ times kT at room temperature. Therefore as discussed above one expects the dangling electron spins to be paired, and no e.p.r. signal to be observed from rows of these atoms. This is in accord with the model of Fig. 1. It is possible that the spin pairing resonates between different pairs, so that a spatially fixed set of pairs cannot be identified.

It is necessary to check whether the wave functions used are compatible with the bond requirements to the nearest neighbour second layer atoms. The appearance of the surface wave functions is sketched approximately to scale (lines of equal charge density) in Fig. 2, for the particular case $\beta = 0.5$, i.e. 20% of the p_y charge density is added to the surface orbital (with renormalization as in Eq. (7)). If the bonds to the second layer atoms had no other constraints and were directed at angles required only by their new wave functions, the point of intersection would be at a distance of 0.13 Å further from the p row than the actual second layer atom distance of 1.24 Å from this row. This difference is small and indicates that only small forces would be needed from the bonds that actually tie the second layer atom to the rest of the lattice, to retain this atom at or near its normal position.

Rather larger shifts of second layer atoms in the z direction (-0.74 and

0.93 Å) would occur if constraints from the rest of the lattice were ignored.

The true z shifts, taking into account the lattice ties, would be much

smaller. Such z shifts would not affect the symmetry of LEED patterns for

normal incidence measurements. Their presence would affect LEED intensities

and also symmetries in off-normal incidence experiments, but would be dif-

ficult to establish under present knowledge of LEED analysis.

It remains now to consider the "s row" wave functions. The atoms in

these rows were assumed raised, with wave functions having more s content

than in the "ideal" sp^3 case. This both resulted in an energy reduction [8]

(in the annealed surface model) and also helped explain the small anisotropy

of the e.p.r. signal from the aligned cleaved Si samples.

Referring to Table 1, the overlap along a row between a p_z orbital and

two surface neighbours is 0.12, compared to 0.09 for pure 3 s orbitals. The

difference is small and no preference on grounds of overlap is indicated for

either function, or, therefore, for any hybrid function of them. Estimating

the exchange integral by Eq. (10), one obtains J = .008 ev for any pair of s

orbitals. This is less than kT at room temperature and indicates that spin

ordering will be far from complete, leading to the possibility of spin

resonance. The observed spin density of 1 per 5 atoms (only the s rows con-

tributing), appears therefore quite reasonable on this picture. The present

calculations show that the overlap is about 0.1, which is not inconsistent

with the electrons forming a kind of one dimensional band, showing conduction

electron resonance, as discussed earlier. [1]

The wave functions calculated here provide a specific basis for account-

ing for the properties of cleaved Si surfaces. The model of Fig. 1 is con-

sistent with all current experimental data including e.p.r. data, if it is

assumed that the orbitals along the p rows are mixed sp_z^3 - p_y functions with at least 20% p_y content. The s rows have dangling orbitals with no appreciable p_y content, hence forming a low-overlap band. The two kinds of rows are possible sources of the two surface state bands on cleaved Si surfaces. [4])

Acknowledgments

Stimulating discussions have been had with members of the laboratory, particularly Mr. D. J. Miller, who made many useful suggestions.

Added note: A paper at this conference by Hochstrasser and Antonini reports data for surface e.p.r. centers on crushed silica. As discussed with the authors, the differences between the silica signal, ascribed to a dangling Si bond, and the dangling Si bonds on the cleaved Si surfaces, are probably due to two factors. 1) The Si atom with the dangling bond has three oxygen nearest neighbours in the case of silica, instead of three Si neighbours, hence the dangling wave function is quite different, 2) the distance between dangling Si bonds in silica surfaces is much larger than on silicon, hence there is negligible overlap and a much sharper, structured line results.

References

1. D. Haneman, Phys. Rev. 170, 705 (1968).

2. J. J. Lander, G. W. Gobeli and J. Morrison, J. Appl. Phys. 34, 2298 (1963).

3. J. W. T. Ridgway and D. Haneman, Appl. Phys. Letters, in press.

4. F. G. Allen and G. W. Gobeli, Phys. Rev. 127, 150 (1962).

5. M. Henzler, phys. stat. sol. 19, 833 (1967).

6. D. E. Aspnes and P. Handler, Surface Sci. 4, 353 (1966).

7. D. Haneman, Phys. Rev. 121, 1093 (1961).

8. A. Taloni and D. Haneman, Surface Sci. 10, 215 (1968).

9. R. E. Watson and A. J. Freeman, Phys. Rev. 123, 521 (1961).

10. J. C. Slater, Quantum Theory of Molecules and Solids, Vol. 1, (McGraw Hill, 1963, New York, N. Y.).

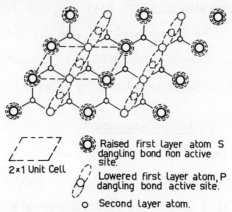

| Raised first layer atom S dangling bond non active site.
| Lowered first layer atom, P dangling bond active site.
o Second layer atom.

2×1 Unit Cell

Fig. 1. Model of surface structure of cleaved Si as deduced previously.[1] Alternate rows are vertically displaced. The p and s terminology refers to admixed p - sp³ orbitals and partial s orbitals, respectively. Only the s rows have a net paramagnetism.

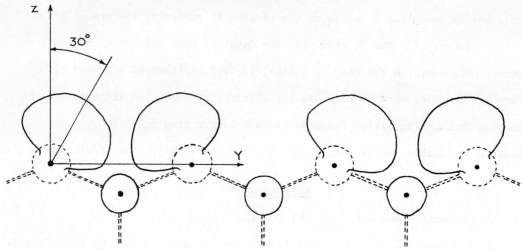

Fig. 2. Surface model, with "p row" viewed in elevation. The orbitals are drawn as lines of constant probability density and are the correct shape for $\beta = 0.5$, corresponding to 20% of the p_y charge density being admixed to the sp^3 function, the whole function being renormalized (Eq.(7) in text). Note that the functions are tilted, increasing the overlap. They may resonate to either neighbour.

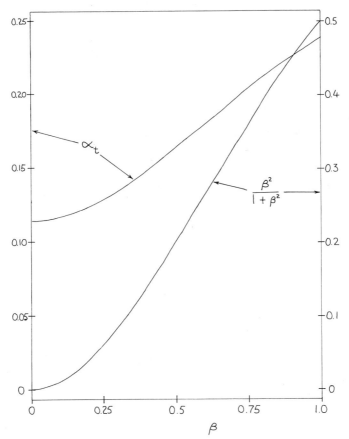

Fig. 3. Overlap integral a_t plotted as function of fraction β of p_y orbital admixed into sp³ dangling orbital, for atom with two opposite neighbours spaced 3.84 Å. The function $\beta^2/(1 + \beta^2)$, which is the fraction p_y charge density used, is also plotted as function of β.

SELF-CONSISTENT CHARGE DENSITIES, POTENTIALS, AND WORK FUNCTIONS AT METALLIC INTERFACES

Alan J. Bennett and C. B. Duke

General Physics Laboratory
General Electric Research and Development Center
Schenectady, New York

Using a jellium model,[1,2] we calculate self-consistently the dipole, exchange, and correlation potentials and the charge distribution at a metal-vacuum interface, thereby obtaining the work function of the system. A comparison of our present results with those for bimetallic junctions[1,2] demonstrates that the barrier height is not simply related to the work functions of the constituent materials due to the redistribution of the charge density when the components are in intimate contact. The metal-vacuum calculation is performed following a procedure similar to that used for the bimetallic interface,[1,2] which enables us in some ways to improve and to extend the previous calculation[3] of the dipole layer at a metal-vacuum interface.

The self-consistent procedure consists of:

1. Assuming a model distribution of electronic charge near a surface specified by a step discontinuity in a uniform positive background charge density;

2. Using the Wigner[4] interpolation formula for the correlation potential and evaluating the exchange and correlation potentials in the local-density approximation;[5]

3. Calculating the Coulomb Hartree potential from Poisson's equation;

4. Using the sum of the Coulomb, exchange, and correlation potentials in the Schrödinger equation to find the one-electron wave functions;

5. Calculating the electron density by assuming that all of the one-electron eigenstates whose eigenvalues lie below the Fermi energy are filled;

6. Approximating the calculated (total) charge density with a parameterized analytical form subject to the requirement of charge neutrality; and

7. Using this model charge density in steps 2 through 6 and repeating those operations until the charge density in two successive iterations differs by less than a prescribed amount.

Our calculation includes all contributions to the one-electron potential in the self-consistent procedure. Changes in the charge distribution cause changes in both the exchange-correlation and Hartree potentials. Previous work,[3] while attempting to include some momentum dependence of the one -electron effective potential, neglected the changes in the exchange and correlation contributions during successive stages of the iteration procedure.

Figure 1 shows the calculated electron density and potential (obtained from the model charge density of the previous iteration) used to obtain that density at various stages of the calculation for a 10^{22} cm^{-3} metal-vacuum interface. The iterations converge rapidly. The converged results shown are those obtained in the third and fourth iterations. They are, however, only semiquantitative because of the simplicity of the analytical form assumed for the model charge density and the nature of the coupling between the Schrödinger and Poisson equations.

The model charge density used in step 6 is given by

$$n(x) = n_0 (1 + h_L/h_R)^{-1} \exp(-h_R x) ; \qquad x > 0 \qquad (1)$$

$$n(x) = n_0[1 - (1 + h_R/h_L)^{-1} \exp(h_L x)]; \qquad x < 0 \qquad (2)$$

in which n_0 is the uniform positive background charge of the bulk metal. The decay length h_R is fitted to the calculated charge in the asymptotic region $x \gg 0$. The decay length h_L is determined in the region $x < 0$. Charge neutrality is guaranteed by moving the boundary of the uniform positive charge.[3] An error, arising from our use of a model charge density, is caused by the uncertainty in the value of h_L by about ±10%.

The use of the model charge distribution to represent the charge distribution at various stages of the calculation is an intrinsic limitation of our procedure. By distributing the positive background charge to fulfill the condition of over-all charge neutrality, however, the final value for the Hartree potential at the barrier can be obtained numerically from our (converged) calculated charge density without recourse to the model. Except for very-low-density electron fluids, our model represents the calculated charge distribution reasonably well, as illustrated in Figs. 2a, b, and 3a. Even a graphical procedure, which avoids the use of a model density, can lead to large errors in the Hartree potential. Small changes in the density strongly affect the Coulomb Hartree potential via Poisson's equation but, unfortunately, the density itself is not a very sensitive functional of the Hartree potential in the Schrödinger equation.

In Fig. 2a, (b) the converged electron density, model charge density, total potential, and exchange-correlation potential are shown for the $n = 10^{22}$ cm^{-3} [$n = 10^{21}$ cm^{-3}] metal-vacuum interface. Figure 3a shows the same quantities for the $n = 5 \cdot 10^{22}$ cm^{-3} system. Previous work,[6,7] which did not include self-consistency, indicated that the exchange-correlation potential oscillates inside the metal. Our self-consistent calculation shows that (within the local density approximation) these oscillations are greatly diminished in the converged solution because of the smoothing of the density near the interface. Bardeen (Ref. 3) used the fact that the exchange-correlation potential becomes equal to the usual image expression far from the metal interface as a boundary condition in determining that potential. In our work, the potential far from the metal is a consequence of the self-consistent procedure. In the local density approximation

$$V_{ex-c} \xrightarrow{\;x \to \infty\;} \frac{1.20}{r_s} \;\text{(Hartrees)} \qquad\qquad (2a)$$

and hence the image potential is obtained if

$$n(x) \xrightarrow{\;x \to \infty\;} \frac{2.18\times10^{-3}}{x^3} \;. \qquad\qquad (2b)$$

Our simple model charge density drops off more rapidly than x^{-3}. A more complicated model is required to obtain analytically the image force far from the interface. However, the construction of such a model is unwarranted because far from the interface the image potential is caused primarily by the non-local polarization by the external charge of the electronic charge near the interface rather than by the

interaction of the external charge with the evanescent contribution
to the electron density.

The Coulomb Hartree potentials obtained are physically reasonable
(with the exception of that of the $5 \cdot 10^{22}$ case which seems large).
They reflect both the magnitude of the total potential and the number
of electrons available to form the dipole layer.

Figures 2c and 3c show the charge densities and potentials of
10^{22} cm^{-3} – 10^{21} cm^{-3} and $5 \cdot 10^{22}$ cm^{-3} – 10^{22} cm^{-3} bimetallic junctions,[1,2] respectively. These figures should be compared with Figs.
2a, 2b and 3a, 3b which give the corresponding quantities for the constituent materials of the junctions at vacuum interfaces. We note that
the size of the dipole potential across the interface is a result of the
metal–vacuum calculation but is an essential input to the bimetallic
junction calculation which ensures that the Fermi levels of the two
materials are equal. The maximum value of the potential in the bimetallic
contact is not simply equal to the difference in the work functions of
its constituent materials as suggested by Mott and Gurney,[8] despite the
fact that the model does not predict charge accumulation in surface
states near the interface. In order to equalize the Fermi energies
of the two materials, the Hartree potential must cause a net change in
potential energy across the junction equal to the difference in their
separation energies (the separation energy of a material is defined as
the difference between its Fermi energy and its bulk exchange-
correlation potential). However, the shape of the Hartree potential and
its maximum height are determined by the self-consistent charge distribution in the junction region as discussed in Refs. 1 and 2.

REFERENCES

1. A. J. Bennett and C. B. Duke, Phys. Rev. <u>160</u>, 540 (1967).

2. A. J. Bennett and C. B. Duke, Phys. Rev. <u>162</u>, 578 (1967).

3. J. Bardeen, Phys. Rev. <u>49</u>, 653 (1936).

4. E. P. Wigner, Phys. Rev. <u>46</u>, 1002 (1934).

5. W. Kohn and L. J. Sham, Phys. Rev. <u>140</u>, A1133 (1965).

6. H. J. Juretschke, Phys. Rev. <u>92</u>, 1140 (1953).

7. T. L. Loucks and P. H. Cutler, J. Phys. Chem. Solids, <u>25</u>, 105 (1964).

8. N. F. Mott and R. W. Gurney, <u>Electronic Processes in Ionic Crystals</u>,
    Clarendon Press, Oxford, England (1948).

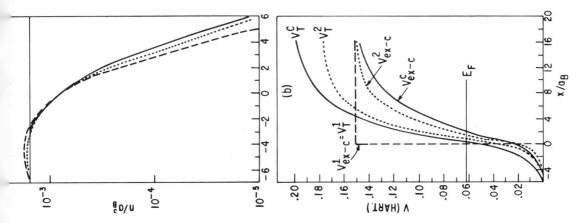

Fig. 1. The electron density (a) and the potential (as calculated from the model density) used to obtain that density (b) in the vicinity of a 10^{22} cm^{-3} metal-vacuum interface. The horizontal solid line in (a) denotes the uniform positive background inside the metal. The electron densities are normalized so that their half dropoff values occur at the same value of the abscissa. All distances are measured in units of the Bohr radius. The dashed line in part (a) indicates the electron density calculated in the first iteration. The potential V_T^1 used to obtain that density is shown by the dashed line in (b). The dotted line in part (a) indicates the electron density calculated in the second iteration. The total potential V_T^2 and exchange-correlation potential V_{ex-c}^2 used to obtain that density are shown by the dotted lines in (b). The solid line in part (a) indicates the converged electron density calculated in the fourth iteration. The converged total potential V_T^c and exchange-correlation potential V_{ex-c}^c are shown by the solid lines in (b).

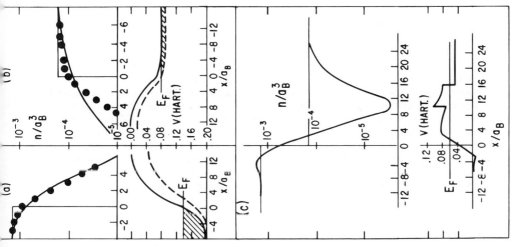

Fig. 2. (a) The (converged) calculated electron density and model electron density (solid circles) at a 10^{22} cm^{-3} metal-vacuum interface. The total potential (solid line) and exchange-correlation potential (dashed line) used to obtain this density are given in the lower part of the figure. (b) The corresponding results for a 10^{21} cm^{-3} metal-vacuum interface. (c) The (converged) calculation electron density in the neighborhood of a bimetallic junction (10^{22} cm^{-3} - 10^{21} cm^{-3}) as given in Fig. 4 of Ref. 2. The total potential used to obtain that density is shown in the lower figure (from Fig. 5 of Ref. 2). All distances are measured in units of the Bohr radius.

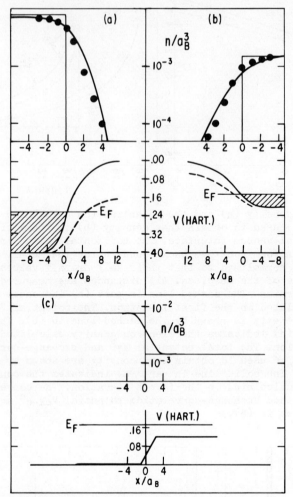

Fig. 3. (a) The (converged) calculated electron density and model electron density (solid circles) at a $5 \cdot 10^{22}$ cm^{-3} metal-vacuum interface. The total potential (solid line) and exchange-correlation potential (dashed line) used to obtain this density are given in the lower part of the figure. (b) The corresponding results for a 10^{22} cm^{-3} metal-vacuum interface. (c) The (converged) calculated electron density in the neighborhood of a bimetallic junction ($5 \cdot 10^{22}$ cm^{-3} - 10^{22} cm^{-3}) as given in Fig. 9 of Ref. 2. The total potential used to obtain that density is shown in the lower figure (from Fig. 10 of Ref. 2). All distances are measured in units of the Bohr radius.

CLEAN SURFACE EFFECTS
ON THE OPTICAL PROPERTIES OF METALS

Edward A. Stern

University of Washington
Seattle, Washington

I.   INTRODUCTION.

The standard theory of the optical properties of metals
is based on the assumption that a clean surface does not
give a significant contribution. The optical constants are
therefore calculated from the bulk properties. To this
standard theory an absorption contribution due to diffuse
scattering of electrons from the surface was added by Holstein[1]
using a classical calculation. There have been recent quantum
mechanical calculations treating the boundary as flat and
specular reflecting the enclosed free electron gas[2]. More
recently it has been realized that surface roughness can
contribute to the absorption of light by coupling to the
surface plasmons[3-5].

For the most part, the various calculations confirm
that the boundary effect is small. Yet, calculations[6] of
the optical absorptions using bulk properties of the metals
most completely understood, namely the alkali metals, give
very poor agreement with experiment[7]. Not only is the
quantitative agreement poor, but even qualitatively the

agreement is poor. The theory cannot explain anomalous absorptions occurring in some of the most careful data[7] and not appearing in other apparently as reliable data[8,9]. These anomalous absorptions occur in the energy region above the main part of the Drude absorption and below those of the predicted interband absorptions.

In this paper we consider a surface effect which does not occur in a free electron gas. We consider a nearly free electron model of a metal and calculate the effects on the optical absorption of the change in wave function produced by the presence of an energy gap and a surface. We find that for a flat surface oriented parallel to one of the principal crystal planes, the surface states that then exist cause optical absorptions of the same magnitude as interband ones for Na. These absorptions occur at energies, and have a frequency variation, which agree as poorly with the measured anomalous absorptions in Na as is the case for the interband absorptions.

However, just as the known case for interband absorptions, the values of the calculated anomalous absorptions are also about one order of magnitude too small. For the other alkali metals, the location of the anomalous absorptions is reasonably described by the theory but the quantitative agreement is even poorer than for Na.

For a flat surface oriented at an angle to the principle crystal planes, no surface states exist but an evanescent

wave is added to some of the electron states of energy
where the constant energy surface touches the brillouin
zone boundary. This also introduces absorptions in the
anomalous region but their magnitude is down by at least
one order of magnitude compared to the previous case and
are small compared to interband absorptions. This suggests
that the Mayer data[7] showed the anomalous absorptions
because his method of preparing surfaces by solidifying
bulk samples produced an unstrained surface parallel to a
principal crystal plane. The other data did not show the
anomalous absorptions because the surfaces were prepared
under strain and were polycrystalline without matching
any single principal crystal plane.

The measured anomalous absorptions in Na show a large
temperature variation, increasing in strength with temperature while those of the other alkali metals show a much
smaller variation and in the opposite direction. The large
temperature variation for Na can be explained in this
picture by assuming that the thermal stresses inevitably
involved in cooling distort the surface and destroy its
parallelness to a principal crystal plane. In this same
picture it must be assumed that for the other alkalis
cooling does not distort the surface. The small temperature
variation in the other alkali metals can be explained as
a Debye-Waller factor decrease of the energy gap with
increasing temperature.

In the next section (Sect. II) the properties of the surface states are reviewed and the absorptions caused by them are calculated. Sect. III discusses the situation where the surface is not parallel to a principal crystal plane and no surface states exist. Section IV concludes with a discussion.

II. SURFACE STATES.

In this section we review the properties of surface states. Consider the first brillouin zone of the alkali metals shown in Fig. 1. We assume that the sample surface is in the x-y plane perpendicular to the reciprocal lattice vector $(0, 0, 2\pi/a)$. Note that our coordinate system does not match with the cubic coordinate system of the real space lattice. Using reasoning similar to that in references 10 and 11, we find if $V_0 < 0$ that surface states of energy $2K\left(\frac{\pi}{a}\right)^2$ and crystal momentum centered at the points N in Fig. 1 can exist of the forms

$$\chi_1 = \cos\left[\frac{\pi}{a}\left(x + \frac{y}{\sqrt{2}} + \frac{z}{\sqrt{2}}\right) + \delta\right]\exp\left(\frac{V_0 az}{2^{3/2}\pi K}\right)$$

$$\chi_2 = \cos\left[\frac{\pi}{a}\left(x - \frac{y}{\sqrt{2}} + \frac{z}{\sqrt{2}}\right) + \delta\right]\exp\left(\frac{V_0 az}{2^{3/2}\pi K}\right)$$

$$\chi_3 = \cos\left[\frac{\pi}{a}\left(x - \frac{y}{\sqrt{2}} - \frac{z}{\sqrt{2}}\right) + \delta\right] \exp\left(\frac{V_0 az}{2^{3/2}\pi K}\right)$$

$$\chi_4 = \cos\left[\frac{\pi}{a}\left(x + \frac{y}{\sqrt{2}} - \frac{z}{\sqrt{2}}\right) + \delta\right] \exp\left(\frac{V_0 az}{2^{3/2}\pi K}\right)$$

$$\chi_5 = \cos\left[\frac{\pi}{a}\sqrt{2}\,z + \delta\right] \exp\left(\frac{V_0 az}{2^{3/2}\pi K}\right) \tag{1}$$

where δ is a phase factor which, to match typical boundary conditions, must be approximately equal to $-\frac{\pi}{4}$, $2V_0$ is the energy gap across the brillouin zone faces at point N in Fig. 1, a is the lattice constant, $K = \frac{\hbar^2}{2m}$, and the metal is in the half space $z > 0$. In addition to the surface state of (1), surfaces states exist for all energies $E > 2K\left(\frac{\pi}{a}\right)^2$. The form of the surface states of $E = K\left[2\left(\frac{\pi}{a}\right)^2 + k_\perp^2\right]$ is given by

$$\chi_i\, e^{i k_{\perp j}\cdot r} \tag{2}$$

for $k_\perp^2 \ll 2\left(\frac{\pi}{a}\right)^2$, where j has the values 1 to 5, $k_{\perp j}^2 = k_\perp^2$ and $k_{\perp j}$ is in a direction parallel to the brillouin zone face corresponding to the state χ_j. Geometrically $k_\perp$ is the radii of the 12 circles formed by the intersection of the

sphere of radius

$$\left[2\left(\frac{\pi}{a}\right)^2 + k_\perp^2\right]^{\frac{1}{2}}$$

and the 12 brillouin zone faces of Fig. 1.

It is a straightforward but **tedious** exercise to use

the golden rule to calculate the power absorption produced

by occupied electrons within the Fermi sphere of radius

$k_F < \frac{\sqrt{2}\pi}{a}$ making transitions to the surface states of (2)

induced by the assumed normally incident light with an

electric field of the form

$$\vec{\varepsilon} = \hat{x}\varepsilon_o\, e^{-\lambda z}\, \sin \omega t \tag{3}$$

where, specializing for the cases of the alkali metals,

$$\lambda = \frac{\omega}{c}\sqrt{-\varepsilon_1}\ ,$$

$\hat{x}$ is a unit vector in the x-direction, ω is the angular

frequency of the light, c is the light velocity, and ε_1 is

the real part of the dielectric constant.

This power absorption can be, in the usual manner,

expressed as a contribution to the conductivity σ, although

it must be remembered that we really have a surface effect

and we express it as a bulk conductivity only because the

experimental data is analyzed in this way. We find for

this contribution to σ

$$\sigma \approx \begin{cases} \dfrac{\lambda e^2 V_0 E_F}{2\hbar (\hbar \omega_c)^2} \dfrac{x-1}{x^3} \ , & x > 1 \\[20pt] 0 \ , & x < 1, \end{cases} \qquad (4)$$

where $\quad x = \dfrac{\omega}{\omega_c} \ ,$

$$\hbar \omega_c = 2K \left(\frac{\pi}{a} \right)^2 - E_F,$$

and E_F is the Fermi energy. Eq. (4) is valid for $\dfrac{\hbar(\omega - \omega_c)}{E_F} \ll 1.$

Fig. 2 shows a plot of σ for Na where $V_0 = 0.12$eV. The value of V_0 was chosen to agree with the calculations of Ham[12]. In the same figure the interband absorption calculated using the formula of Butcher[13] for the same values of V_0 are also plotted. Note that the calculated values are multiplied by a factor of 10. The experimental values at $20°C$ are also plotted in Fig. 2. The experimental curves at lower temperatures actually give better agreement with the calculated shapes, but since in this model the temperature dependence argument requires that the higher temperature results be those of the good surface, comparison is made against the higher temperature data.

III.   SURFACE NOT PARALLEL TO MAIN CRYSTALLINE PLANE.

In the case that the surface is not parallel to a main crystalline plane, surface states do not exist[10]. In this case, the combination of the surface and energy gaps add an evanescent part of the wave function near the surface to those states which have an energy greater than the critical energy corresponding to the constant energy surface just touching the Brillouin zone boundaries[11]. The evanescent part of the wave function can cause absorption of light by transitions of the occupied electrons to these states with the evanescent parts. The evanescent parts produce non-zero matrix elements because of the spread in momentum associated with them. These transitions occur in the anomalous region and they correspond to a σ given approximately by

$$
\sigma \approx \begin{cases} \dfrac{2}{3} \dfrac{e^2}{\pi^3} \dfrac{\lambda V_o E_F^{1/2}}{\hbar(\hbar\omega_c)^{3/2}} \dfrac{(x-1)^{3/2}}{x^3} , & x > 1 \\[2em] 0 & x < 1 \end{cases}
\tag{5}
$$

This expression is valid when $\hbar(\omega - \omega_c) \ll E_F$.

For both Na and K the σ in (5) is at least one order of

magnitude smaller than that in (4). This implies that the
anomalous absorptions should be experimentally observed
only for those surfaces where surface states can exist,
namely, specular reflecting surfaces parallel to a main
crystalline plane.

IV. DISCUSSION.

Although surface states contribute an anomalous absorption mechanism of the same order of magnitude as the interband ones for Na, it is clear from Fig. 2 that the calculated
absorptions are much smaller than the experimental values.
In addition, the ratio of the calculated anomalous absorptions
to the calculated interband ones in the other alkali metals
is much too small compared to the experimental values. We
have thus not contributed to understanding the quantitative
discrepancy between theory and experiment. However, we
have a possible qualitative explanation for the anomalous
absorptions. The energy of initiation of the absorptions
and the energy of the peak value is as satisfactorily
described by the theory presented here, as are the interband
ones by the standard theory. In addition the reason why
these anomalous absorptions have been observed in only one
type of sample also has a natural explanation in this theory.
Certainly the quantitative disagreement is most serious

and still awaits solution, yet the anomalous absorptions have been put on the same footing as the interband absorptions. There is no more reason to consider the "anomalous" absorptions any more anomalous than interband absorptions, at least for Na.

The theory presented here suggests an obvious experimental check. Electron diffraction studies on various surfaces could verify the predicted correlation between surface orientation and the appearance of the anomalous absorptions.

It is worthy of note that the experiments which observed the anomalous absorptions used light incident at a 75° angle from the normal while the calculation employed here assumed normally incident light. It is unlikely that a calculation for light incident at the experimental angle would change the magnitude of the anomalous absorption by an order, though it is not completely ruled out in the sense that the surface absorption, not having the bulk symmetry, would give varying values of σ with angle of incidence. Such an experiment, measuring the absorption as a function of the angle of incidence on a sample that shows the anomalous absorption, would be a simple and definitive way to determine if the anomalous absorptions are a surface effect as proposed here. If the measured σ depends on angle of incidence for cubic metals, it is a surface effect since if the σ depended on bulk properties only, it would be a scalar and would be independent of the angle of incidence.

REFERENCES

Research sponsored by the Air Force Office of Scientific Research, Office of Aerospace Research, U. S. Air Force, under AFOSR Grant No. AF-AFOSR-1270-67.

1. T. Holstein, Phys. Rev. **88**, 1425 (1952).

2. P. A. Fedders, Phys. Rev. **153**, 438 (1967).

3. E. A. Stern, remark in Optical Properties and Electronic Structure of Metals and Alloys (North-Holland Publishing Co., Amsterdam, 1966) edited by F. Abeles, p. 396-397; H. Boersch and G. Sauerbrey, ibid. p. 386.

4. P. A. Fedders, Phys. Rev. **165**, 580 (1968); S. N. Jasperson and S. E. Schnatterly, Bull. Am. Phys. Soc. **12**, 399 (1967); and (private communication).

5. Y. Teng and E. A. Stern, Phys. Rev. Letters **19**, 511 (1967).

6. J. Applebaum, Phys. Rev. **144**, 435 (1966).

7. H. Mayer and B. Hietel, in Optical Properties and Electronic Structure of Metals and Alloys (North-Holland Publishing Co., Amsterdam, 1966) edited by F. Abeles, p. 47.

8. M. Dresselhaus, remark in ibid p. 59.

9. N. V. Smith, Bull. Am. Phys. Soc. **13**, 387 (1968).

10. E. T. Goodwin, Proc. Cambridge Phil. Soc. **35**, 205 (1939); V. Heine, Proc. Phys. Soc. (London) **81**, 300 (1963).

11. E. A. Stern, Phys. Rev. **162**, 565 (1967).

12. F. S. Ham, Phys. Rev. **128**, 82 (1962).

13. P. N. Butcher, Proc. Phys. Soc. (London) **A64**, 50 (1951).

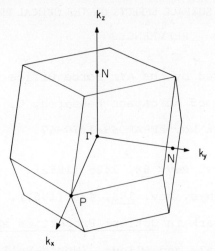

Fig. 1.   First brillouin zone of the alkali metals. Note that the coordinate
system used does not match with the cubic coordinate system of the
real space lattice.

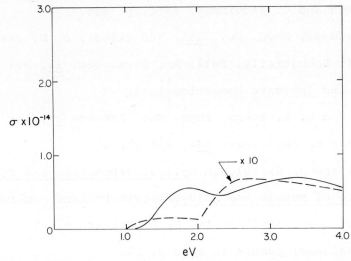

Fig. 2.   "Experimental" (solid curve) and calculated (dashed curve) conductivity
of Na, as a function of the energy of the light. The "experimental"
curve is obtained by subtracting the extrapolated Drude tail from the
values of Mayer[7]. Note that the calculated values are multiplied by
a factor of 10.

FIELD IONIZATION AND ELECTRONIC STRUCTURE
OF METAL SURFACES

Erwin W. Müller

Department of Physics
The Pennsylvania State University
University Park, Pennsylvania

Field ion microscopy[1] provides the only known means of viewing a surface in atomic detail. The interpretation of images requires an understanding of the mechanism of image formation, which, beyond the simple geometrical projection principle depends upon the interaction of the image gas atom with the electronic structure of the surface. The difficulty lies in the fact that for a typical helium ion image the field strength at the surface is of the order of 4.5 V/$\overset{\circ}{A}$. Thus the potential changes over a distance of one atomic diameter by about twice the surface atom's ionization potential. Description of the surface-gas atom interaction from first principles remains complicated. The purpose of this paper is to point to some experimental facts that need to be considered for an advanced image interpretation.

Field ionization[1,2] occurs when the potential well around the image gas atom is so much reduced that an electron may tunnel out. Near a metal surface this occurs preferentially where the field is locally enhanced above a protruding surface atom. The need for accommodating the electron above the Fermi level suggests a minimum distance of possible ionization $x_{min} = (I-\phi)/eF$, which at F = 4.5 V/$\overset{\circ}{A}$, I = 24.5 eV and ϕ = 4.5 eV places x_{min} at 4.5 $\overset{\circ}{A}$. Measurements of the energy distributions of emitted ions,[3] performed by the retarding potential method, can be interpreted to affirm this condition. Somewhat surprising, however, was that the half-width of the distribution is typically only 0.8 eV, corresponding to a thickness of the ionization zone of 0.2 $\overset{\circ}{A}$, one tenth of the diameter of the atom. Calculating the barrier penetration probability by applying the WKB method to a Coulomb potential funnel deformed by the superposition of a homogeneous field and an image force potential does not yield the observed narrow distribution.

The field above a protruding atom is locally enhanced, and this can account for the reduction of the depth of the ionization zone. As not all surface atoms are imaged, the degree of protrusion required for imaging must somehow be defined, and this is done by Moore's computer model.[4] A spherical crystal is defined by removing all atoms whose centers lie outside a reference sphere through the crystal. The locations of all atoms whose centers lie within a thin shell below this reference surface are computed and plotted in a suitable, usually stereographic, projection. The shell thickness is adjusted until the plot closely resembles a real field ion micrograph. It is found that the shell thickness varies between about 0.1 lattice parameters for a specimen radius of 200 Å to about 0.05 lattice parameters for a specimen radius of 600 Å.

Based on this model, the FIM patterns of spherical tips of various metals having the same crystal structure should look alike. This is certainly not the case. Figures 1 to 4 show tips of bcc W, Ta, Mo and Nb displaying considerably different regional brightness patterns. The difference between the various representatives of the fcc lattice, shown here in the examples of Ir and Pt (Figs. 5 and 6), are somewhat less pronounced but are still quite distinct. The same is true for hcp Re and Ru. These specific regional brightness patterns cannot be accounted for by the slight deviation of the tip shapes from a sphere which is due to the dependence of field evaporation energy upon crystallographic orientation. A concentric set of circular net plane edges around a pole, such as (113) on Ir and Pt, indicates local sphericity. Yet an abrupt change of image intensity goes right through one side of this set of planes. Also, a narrow slice of a twin crystal inserted in the tip matrix

(Fig. 7) indicates the wide variation of regional brightness while the over

all curvature and thus the externally applied field are fairly uniform.

Another type of peculiar local brightness feature is displayed around the

0001 pole of hcp Re[5], and the same effect can be seen on Ru[6] and Co[7]. At

these surfaces, 60° sections of edges of basal planes are alternatingly

visible and invisible. Metastable atom sites[8] making up the zone decoration

(Fig. 8) also mark the presence of invisible lattice steps.

The study of imaging properties of ordered binary alloys was undertaken

in the hope of being able to distinguish the two atom species.[9] Surprisingly,

it was found that a consistent image interpretation of patterns of equatomic

Pt-Co and of Pt_3Co is only possible with the assumption that the Co atom is

entirely or almost invisible, although this atom species occupies sites geo-

metrically fully equivalent to the ones of the neighboring Pt atoms. In the

case of ordered Ni_4Mo[10] the concept of the invisibility of one species, the

Ni atom, was also found to be necessary for the explanation of the FIM patterns.

The most striking evidence of how much the ionization probability can be

modified by the local electronic surface structure is given by the observation

of a well defined helium ion image of a tungsten surface at only 3 V/Å applied

field strength when a trace of hydrogen is added to the image gas.[11] Subsequent

work showed that hydrogen promotion of field ionization also occurs with other

metals[12] such as Mo, Ta, Ir, Pt, Fe,Ni and Co, although for the three latter

metals no accurate measurements of the field reduction can be made because of

the marginal stability of the surfaces at the image field.

Another, still poorly understood problem of image formation is the ap-

pearance of bright image spots upon the adsorption of gases such as oxygen[13],

nitrogen and carbon monoxide on various metals. Here it is not yet sure

whether these spots represent the adsorbed atomic or molecular species as suggested by some observers[14,15], or whether they indicate metal atoms displaced by corrosion to a more protruding position[13,16]. A possibility which is still open is the existence of surface entities as molecular units. There is some evidence for this obtained in the mass spectrometric analysis of field evapora-tion products[17], and this problem will probably be solved experimentally by the forthcoming application of the atom probe FIM[18]. Closely related to the case of bright image spots due to adsorption may be the situation encountered with impurity and self interstitials both of which appear as bright spots[1].

All the observations described indicate the inadequacy of considering field ionization at the surface simply as topographical problem. For a proper treatment one would have to know the electron wave functions for each crystal plane of the tip metal as well as at individual atomic surface sites, and then consider the overlap of these wave functions with that of the image gas atom. The field ionization rate should be proportional to the electron transition probability

$$P = \frac{4\pi^2}{h} N(E_s) \mid < \psi_s \mid V \mid \psi_a > \mid^2,$$

where h is Planck's constant, $N(E_s)$ the density of surface states, ψ_s and ψ_a are the electronic wave functions of the surface and the gas atom, respectively, and V is the interaction potential[19]. Since we are unable to carry out the cal-culation we must confine our efforts to a piece by piece evaluation of more ac-cessible terms. It appears that the two concepts of charge redistribution[20] in the surface atoms and of directional orbitals[21] extending from the surface atom may lead to more general explanations of the observed effects.

The obvious concentration of the electric field in the protruding surface atoms means that the depression of electronic surface charge $\sigma(x,y)$ by the field according to Gauss' law,

$$F(x,y) = 4\pi\sigma(x,y) = 4\pi\int\Delta\rho(x,y,z) \, dz,$$

($\Delta\rho$ is the excess charge density in the volume) should be much more pronounced
in the single protruding atom than what is calculated for a homogeneous surface. By assuming that electrons are depressed from the protruding surface
atoms only, the degree of depletion per surface atom depends upon the density
of atoms only, the degree of depletion per surface atom depends upon the density of atoms on each crystal plane. In fractions of an electronic charge the
depletion amounts to 0.176 at the closed packed (011) plane of tungsten, to
0.430 at the (111) plane, and to 0.727 at the more loosely packed (334) plane[8].
Thus even more exposed atoms such as the ones on the metastable sites causing
the [100] zone decorations, must be fully ionized[20].

There seems to be no well defined limit as to how far the ionization may
extend when the surface atom is further exposed to the applied field during
the act of field evaporation. Mass spectrometric measurements with the atom-
probe FIM[18] surprisingly show W and Ta atoms to field evaporate as three or
four fold charged ions, while Ir and Pt carry either two or three positive
charges[22].

The reduced ionization probability above one species of atoms in solute
alloys such as Pt-Co, Ni_4Mo and Pt 2% Au[23] must result from the distinct surface states and interaction potentials above the solute. The effect is expected to be larger when the atoms are further apart in the periodic system.
For alloy consitituents in the same period such as W, Re, Ir, Pt and Au alloys,
Tsong[24] suggests that the species with more s electrons, or if this number is
equal, with more unpaired d electrons, is more visible because the charge
depression by the applied field will be larger. Of course, it is well possible
that the configuration of the outer electron shells of the metals involved is
changed in the surface atoms. For instance, the close resemblance of the
regional brightness patterns of Ir, Pt and Au could indicate that the surface
atoms of these metals all may have two 6s electrons and a correspondingly

adjusted number of d electrons.

The most obvious example of charge redistribution due to adsorption and possibly formation of surface molecules is the case of hydrogen promotion of field ionization[8,25]. As suggested in Fig. 9a the conduction electrons at a bare metal surface are sufficiently depressed by the applied over-all field of 4.5 V/$\overset{\circ}{A}$ that the local field enhancement permits ionization of the helium atom beyond the critical distance x_c. At 3V/$\overset{\circ}{A}$ the charge depression is so small that little ionization takes place, and the spatial separation of the origin of ions is poor due to the increased critical distance. When hydrogen is adsorbed in the interspaces between the loosely packed atoms, Fig. 9b, the charge redistribution due to the formation of hydride like ionic molecules results in a more sharply peaked, high ionization probability at the very low external field of 3 V/$\overset{\circ}{A}$. This promotional effect only occurs at net planes which can spatially accommodate the negative hydrogen ion of 1.3 $\overset{\circ}{A}$ radius. The unexpected formation of unusual hydride molecules is also evidenced by the mass spectrometric observation[17] of field desorbed ions such as BeH^+, BeH^{++}, CuH_2^+, FeH_2^{++} and NiH^{++}. In this investigation, the heavier metals were, unfortunately, inaccessible due to instrumental limitations.

Because of the electronegativity of oxygen, nitrogen and possibly carbon monoxide, it appears highly unlikely that the bright spots seen in the field ion micrographs of surfaces exposed to these gases represent the absorbate. The obvious rearrangement of the metal surface following such adsorption rather seems to indicate the weakening of metal metal bonds and the formation of molecular entities, in which the metal atom in the presence of the field is so highly polarized as to have ionic character and thereby ionizing the image gas above its site. It is also proposed that interstitial oxygen, known to appear as bright spots[1,13,26,27], actually depletes the surface atom above it of a good fraction of its electronic charge, thereby making it highly visible.

The concept of increased field ionization due to charge redistribution can be refined by considering the possibility of directional dangling bonds or atomic orbitals affecting the probability of ionizing the image gas[21]. It has been suggested that effects of chemisorption and catalysis be described as being due to the overlap of orbitals of the adsorbate with those of the surface atoms, forming molecular orbitals. It is assumed that the direction and the type of the orbitals emerging from the surface atoms are closely related to the orbitals causing the metal metal bonds in the bulk of the crystal, that is, the surface orbitals have the direction of nearest and next nearest neighbor bonds. The assumption to be made in our case is to explain field ionization as occurring preferentially where exposed and unoccupied orbitals of the surface atom extend into space where they can overlap with the occupied orbitals of the image gas atom.

Field ionization into an unoccupied nearest neighbor orbital should be particularly effective when it extends into the direction of the applied field. This is the case for alternating layers in $60°$ sections around the basal plane of the hcp lattice of Re, Ru and Co, or in the successive layers in $120°$ alternating [011] zones around the (111) poles of fcc platinum. The latter crystallographic region of other fcc metals such as Ir does not show this simple behavior, indicating that the integral interaction of several extending orbitals to near neighbors ($t2_g$) as well as to next nearest neighbors (e_g) must be considered. Chemical specificity is introduced by the variation of occupancy and hybridization of the extended orbitals. The unraveling of the contributions of the various orbitals at different crystal planes is still a task for the future, but with the considerable selection of field-ion regional brightness patterns available it seems to be mostly a problem of diligent analysis of the orbital combinations conceivable at the various crystallographic regions. This possibility raises the hope that field ion microscopy may contribute to the elucidation of the relation between the electronic structure of the surface and the electron shell configuration of the free metal atoms.

REFERENCES

1. E. W. Müller, Adv. in Electronics and Electron Physics 13, Academic Press, New York (1960) pp. 83–179.

2. R. Gomer, "Field Emission and Field Ionization," Harvard Press (1961).

3. T. T. Tsong and E. W. Müller, J. Chem. Phys. 41, (1964) p. 3279.

4. A. J. W. Moore, J. Phys. Chem. Solids 23, (1962) p. 907.

5. E. W. Müller, Proc. III European Regional Conf. Electron Microscopy, Prague (1964) Vol. 1, p. 161.

6. A. J. Melmed, Surface Sci. 8, (1967) p. 191.

7. O. Nishikawa and E. W. Müller, J. Appl. Phys. 38, (1967) p. 3159.

8. E. W. Müller, Surface Sci. 2, (1964) p. 484.

9. T. T. Tsong and E. W. Müller, Appl. Phys. Lett. 9, (1966) p. 7: J. Appl. Phys. 38, (1967) p. 545: J. Appl. Phys 38, (1967) p. 3531.

10. R. W. Newman and J. J. Hren, Phil Mag. 16, (1967) p. 211.

11. E. W. Müller, S. Nakamura, O. Nishikawa and S. B. McLane, J. Appl. Phys. 36, (1965) p. 2496.

12. O. Nishikawa and E. W. Müller, Surface Sci. 12 (1968) p. 247.

13. E. W. Müller, "Structure and Properties of Thin Films", eds. Neugebauer, Newkirk and Vermilyea, John Wiley and Sons, New York (1959) p. 476.

14. G. Ehrlich, Disc. Faraday Soc. 41, (1966) p. 7.

15. J. F. Mulson and E. W. Müller, J. Chem. Phys 38, (1963) p. 2615.

16. A. A. Holscher and W. M. H. Sachtler, Disc. Faraday Soc. 41, (1966) p. 29.

17. D. Barofsky and E. W. Müller, Surface Sci. 10, (1968) p. 177.

18. E. W. Müller, J. F. Panitz and S. B. McLane, Rev. Sci. Instr. 39, (1968) p. 83.

19. D. S. Boudreaux and P. Cutler, Surface Sci. 5, (1966) p. 230.

20. E. W. Müller, Z. Physik. Chem. N. F. 53, (1967) 204.

21. Z. Knor and E. W. Müller, Surface Sci. <u>10</u>, (1968) p. 21.

22. E. W. Müller, Battelle Conference on Molecular Processes at Solid Surfaces, Kronberg, Germany, (1968) May.

23. E. S. Machlin and W. DuBroff, 12th Field Emission Symposium, Pennsylvania State University (1965).

24. T. T. Tsong, Surface Sci. <u>10</u>, (1968) p. 303.

25. E. W. Müller, Surface Sci. <u>8</u>, (1967) p. 462.

26. S. Nakamura and E. W. Müller, J. Appl. Phys <u>36</u>, (1965) p. 3634.

27. M. A. Forbes and B. Ralph, Acta Met. <u>15</u>, (1967) p. 707.

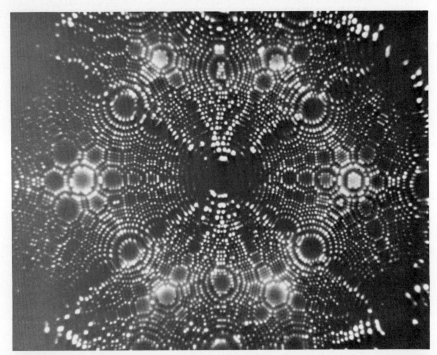

Fig. 1. Regional brightness of a [011]-oriented tungsten tip imaged with helium.

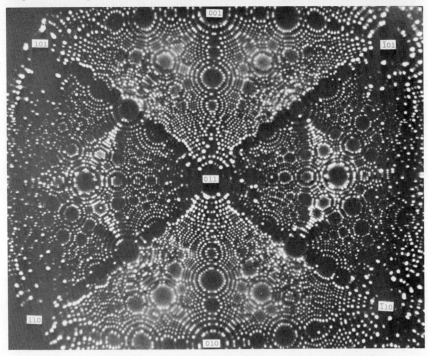

Fig. 2. Endform of a tantalum tip, [011]-orientation.

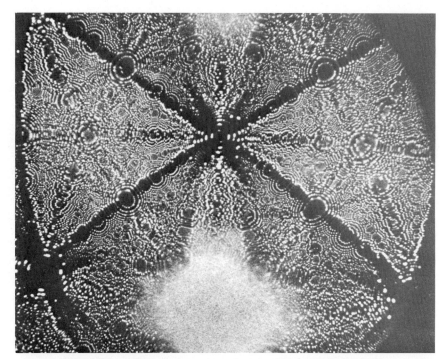

Fig. 3. Molybdenum crystal in [011]-orientation.

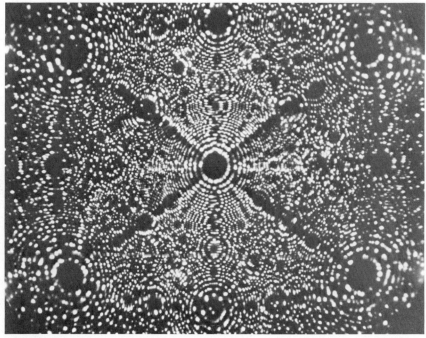

Fig. 4. Niobium crystal with [011]-axis, imaged with helium, showing some field stress induced lattice imperfections.

Fig. 5. Endform of a platinum crystal, [001]-axis.

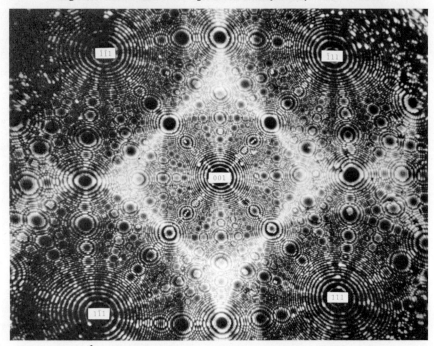

Fig. 6. Endform of an iridium crystal, [001]-axis.

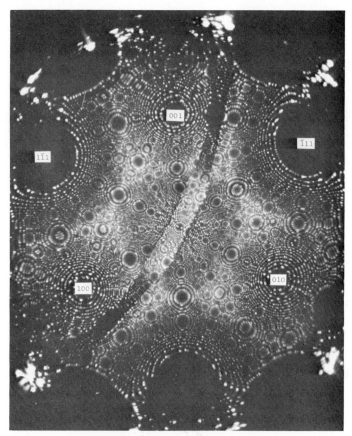

Fig. 7.   Indium tip in [111]-orientation with a slice of a twin inserted.   The
{113}-planes are common to both crystals.

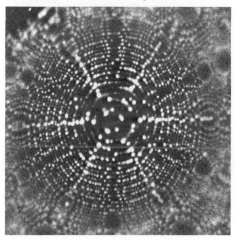

Fig. 8.   (0001) – pole of rhenium with zone decorations and alternating visible
(0001) – ledges.

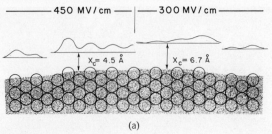

Fig. 9a. Field induced surface charge and ionization probability in the ionization zone above the surface.

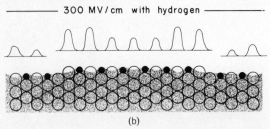

Fig. 9b. Surface charge rearranged by the interspatial adsorption of hydrogen.

LEED STUDIES OF SURFACE IMPERFECTIONS*

Robert L. Park

Sandia Laboratory
Albuquerque, New Mexico

Introduction

It is possible to use the width of a diffracted beam of low energy electrons to set a lower limit on the effective diameter of coherent scattering
domains on a single crystal surface. With present day LEED systems, however,
diffraction broadening from well annealed crystals is usually the result of
the coherent diameter of the incident electrons. A fairly frequent exception
to this occurs in the case of adsorbed impurity structures which are simply
related to the substrate lattice. In such cases, the surface structure may
be expected to form subdomains on a single coherent domain of the substrate.
Since these subdomains will have a definite antiphase relationship with one
another, beam splitting should be observed. It would be more accurate to say
that the effects of splitting should be observed; the effect in most instances
being to broaden certain diffraction beams while others are unaffected by the
subdomain structure.[1]

It will be shown that the pattern of broadening due to the antiphase domains, in some cases, permits the symmetry of the adsorption sites relative to
the substrate to be specified. That is, for example, in a C2x2 structure on a
square substrate, one can distinguish between bonding in the nearest neighbor
bridged sites which have twofold symmetry, and bonding in sites with fourfold
symmetry such as the positions of maximum coordination.

The existence of antiphase domains will, of course, affect the intensities
as well as the beam profiles. Recent dynamic theories of LEED provide an

*This work was supported by the U.S. Atomic Energy Commission.

explanation for some of the more puzzling features of observed diffraction
intensities, particularly the secondary Bragg maxima. In their present form,
however, these theories consider the interaction of perfectly coherent radiation with a defect free surface. It is important therefore to evaluate the
extent to which departures from the ideal may be expected to limit the agreement between experiment and theory.

Antiphase Domains

Impurity structures in which some beams are sharp and others broad are a
fairly common occurrence in LEED. To understand this, it is instructive to consider the problem of interference between two identical one-dimensional gratings
of N point scatterers. In analogy with the formation of double spaced surface
structures, we can construct each grating by making the separation between
points twice as large as some reference spacing a. The spacing between gratings
is constrained to be in units of a. For the case of normal incidence, the phase
difference between waves scattered from adjacent points in each grating is 2δ
where

$$\delta = 2\pi \frac{a}{\lambda} \sin \theta. \tag{1}$$

The phase difference between the two gratings is $m\delta$ where m is an integer. The
complex amplitude of the diffracted wave is of the form

$$Ae^{i\varphi} = \left\{ 1 + e^{i\,[2(N-1)+m]\,\delta} \right\} \sum_{n=1}^{N} e^{i\,2(n-1)\delta}. \tag{2}$$

Squaring to obtain the intensity pattern,

$$I \sim 4\, \frac{\sin^2 N\delta}{\sin^2 \delta} \cos^2(N-1 + \frac{m}{2})\delta. \tag{3}$$

If m is even, the two gratings will scatter in phase. In the simple case
where m = 2, it becomes a single grating of 2N points. This case is shown in
the upper plot of Fig. 1 for N = 6. For larger even values of m, the conditions

for principal maxima remain unchanged, but the widths of the principal maxima will decrease as long as the separation between gratings does not exceed
the coherence length of the incident radiation.

If the gratings are separated by an odd number of substrate spacings,
however, diffraction from the two gratings will just cancel for the half
order beams. The lower plot in Fig. 1 shows the case for m = 3. The result
is that the fractional order beams are split into two components. The amount
of splitting will, of course, decrease as N is increased. One might expect the
same effect if several small gratings are used since alternate gratings will
scatter cooperatively. The effect of this cooperation, however, as shown in
Fig. 2 is to sharpen the split components but not to decrease their separation.
The integral order beams are also much sharper since all four gratings cooperate
for these beams.

Although we would expect a subdomain structure of antiphase domains to
be a common occurrence in impurity adsorption because of multiple nucleation,
actual beam splitting is rarely observed. The reasons for this have been very
graphically demonstrated by Ellis and Campbell[2] using laser simulation. If
the surface is composed of many subdomains with irregular boundaries, the
split beams will consist of many unresolved components and would appear as
broadened peaks. If the antiphase domain boundaries run in preferred directions,
but the domains are irregular in size, streaking results. Although splitting
is not resolved in such cases, its effects should be clearly evidenced by
a broadening of certain beams while others remain sharp. The domains thus
appear large for reflections in which the subdomains cooperate, but small
for reflections in which they do not. Which beams will be broadened by the
antiphase domains will depend on the way in which the subdomains are shifted
relative to one another.

To see how this can be treated systematically, let us return for a moment to the one-dimensional case. For one domain of a double spaced structure the principal maxima occur when sin $\delta = 0$ or

$$n_1 \lambda = 2a \sin \theta \qquad (4)$$

where n_1 is an integer. This can be represented by an Ewald construction where the reciprocal lattice is an array of parallel planes 1/2a apart. For the case of antiphase gratings (m odd), splitting will occur when the zeros of the modulating function cos $(N - 1 + \frac{m}{2})\delta$ coincide with the zeros of sin δ. This occurs when cos $\delta/2 = 0$ or

$$(2n_2 + 1)\lambda = a \sin \theta \qquad (5)$$

For those with an emotional attachment to Ewald constructions, Eq. (5) can be represented by a similar construction as shown in Fig. 3. The "split" lattice must be shifted relative to the ordinary reciprocal lattice to exclude the specular beam which carries no phase information. The unit cell of this lattice in direct space is just the least spacing between domains.

A simple extension of this treatment allows us to predict which beams will be affected by antiphase domains in two-dimensional gratings. Let us first consider the C2x2 structure on a square substrate with the adsorbed specie in the positions of maximum coordination as shown in Fig. 4. The structure can exist in two phase domains represented by unit meshes B and B'. Any C2x2 unit mesh elsewhere on the surface will be in phase with either B or B'.

We can represent the surface net by the matrix

$$B = \begin{pmatrix} 1 & 1 \\ \bar{1} & 1 \end{pmatrix} \qquad (6)$$

where the rows of B are the components of the unit mesh vectors relative to the substrate unit mesh. The reciprocal unit mesh is given by

$$B^* = 1/2 \begin{pmatrix} 1 & \bar{1} \\ 1 & 1 \end{pmatrix} \qquad (7)$$

where the columns of B* are the components of the reciprocal unit mesh vectors. The reciprocal net is generated by the vectors B*n where n is the set of column vectors whose components are integers.

To determine which beams will be affected by the antiphase domains, we take a mesh G which joins B to B'. As shown in Fig. 4, G is represented by the matrix

$$G = \begin{pmatrix} 1 & 0 \\ \bar{1} & 1 \end{pmatrix}. \tag{8}$$

The mesh reciprocal to G is

$$G^* = \begin{pmatrix} 1 & 0 \\ 1 & 1 \end{pmatrix} \tag{9}$$

The beams that will see the antiphase domains are given by

$$\begin{pmatrix} h \\ k \end{pmatrix} = G^*n + \begin{pmatrix} 1/2 \\ 1/2 \end{pmatrix}. \tag{10}$$

The column vector

$$\begin{pmatrix} 1/2 \\ 1/2 \end{pmatrix}$$

excludes the (00) beam. Eq. (10) is the set of all beams containing half-integral indices just as in the one-dimensional grating.

This is no longer the case, however, if the adsorbed specie occupy sites of lower symmetry. It has been proposed in some cases, such as CO adsorption on $(100)Ni^3$ and hydrogen adsorption on $(100)W^4$, that the adsorbed specie occupy bridged bond sites between nearest neighbor substrate atoms to form a C2x2 structure. This type of site has only twofold symmetry. One would not expect to see this in the diffraction intensities, however, because it will exist on the surface in two equivalent orientations as shown in Fig. 5. Thus the apparent symmetry of the diffraction pattern is always that of the substrate. In addition to the two orientations, each may exist in two phases. The phase relationship between domains of one orientation is just that of the fourfold symmetric sites which produces splitting or broadening of the half order beams. The interference between domains of different orientation such as those shown in Fig. 5 produces a different effect. The matrix of the mesh connecting the vertically bonded domain to the horizontally bonded is given by

$$G = \begin{pmatrix} 1/2 & 1/2 \\ \bar{1} & 1 \end{pmatrix}. \tag{11}$$

In the reciprocal representation

$$G^* = \begin{pmatrix} 1 & \overline{1/2} \\ 1 & 1/2 \end{pmatrix}. \tag{12}$$

The beams which see the antiphase relationship between V and H are given by Eq. (10). The pattern of broadening is shown in Fig. 5. As in the case shown in Fig. 4, alternate rows are affected, but now the rows are diagonal and include integral order beams. When the antiphases of V and H are considered, the result is that only those beams for which the sum of the indices is an even integer are unaffected by the antiphase domains.

Therefore the pattern of broadening allows us to distinguish between adsorption in sites with fourfold symmetry and adsorption in the nearest neighbor bridged sites. This is shown very graphically by the laser simulation method if care is taken with the optics. A single large C2x2 domain produces a pattern in which all the spots are sharp as shown in Fig. 6a. A grating of the same overall dimensions but composed of many irregular shaped C2x2 subdomains of the fourfold symmetry type gives a pattern of sharp integral order spots and diffuse half orders, as shown in Fig. 6b. A grating composed of many irregular subdomains of the nearest neighbor bridged type results in a pattern in which all the beams except those for which the sum of the indices is an even integer are diffuse.

Actually since our simulation employed a single substrate domain, the "broadened" beams have a very definite structure. J. E. Houston[5] has demonstrated that this structure is an out of focus image of the antiphase domain structure. If one of these beams is selected with an aperture and properly focused, it gives a very sharp image of the subdomain structure in which the antiphase boundaries appear as dark lines. R. L. Schwoebel[6] has called attention to the close analogy to dark field electron microscopy.

Impurity Structure on (100) Pd

An example of the type of selective broadening discussed above was observed during the cleaning of a (100) Pd surface. In the early stages of cleaning the (100) face of a "high purity" palladium single crystal, the diffraction pattern from a surface prepared by argon ion bombardment followed by an anneal just sufficient to remove the sputtering damage was found to be that expected for the clean, unmodified (100) surface.[7] A typical diffraction pattern from this surface together with an intensity plot of the (11) beam is shown in Fig. 7a. The scanning LEED system used in these measurements has been described elsewhere.[7] In view of the sputter cleaning used and the absence of any unusual diffraction features, it seems reasonable to conclude that whatever portion of the surface contributed to the pattern was atomically clean.

Annealing for longer periods or at higher temperatures caused the 1x1 structure to be replaced by a P2x2 structure. Continued heating produced a change to a very stable C2x2 structure which could not be removed by heating at temperatures up to 1000°C. A typical pattern from the C2x2 structure is shown in Fig. 7b with the intensity plot of the (11) beam. The intensity plot shows very distinctive changes from that of the clean surface.

The C2x2 structure was easily removed by argon ion bombardment but returned under approximately the same heating conditions. Because of this

persistence, Mattera, Goodman, and Somorjai[8] proposed that the C2x2 is a stable
phase of the clean (100) Pd surface. After a repeated cycling of ion bombardment and high temperature heating, however, the half order beams were found
to be broadened as shown in Fig. 7c. The integral order beams on the contrary
remained sharp. Plots of their intensity moreover were identical in shape
to those from the sharp C2x2 structure, although some variation in scale was
observed depending on the amount of heating. The distortion of the integral
order beams in Fig. 7c is due to the fact that the pattern was taken at a
slightly lower voltage and the (11) beams in particular were diffracted at
almost grazing angles. In addition, the gain was set very high to bring out
the relatively weak half order beams.

  With continued cycling the fractional orders became still more diffuse as
shown in Fig. 8. After about 20 such cycles of ion bombardment and heating,
the half order features could not be regenerated. Changes in the intensity
plots of the integral orders such as those represented by the (11) plots in
Fig. 7 could still be produced by sufficient heating, however. Thus the
intensity plots seemed to indicate a change in surface structure to the C2x2,
but the half order beams were either very diffuse or absent. The absence of
"extra" reflections is clearly a very imperfect criteria for cleanliness. The
intensity plots are far more sensitive to contamination. Depending on the
nature of the impurity adsorption sites, the sharpness of the normal reflections may also be insensitive to contamination.

  From the analysis of the previous section it seems that as a result of
the continued cycling of ion bombardment and heating, the C2x2 structure
formed in smaller and smaller domains. It is also clear from the fact that only
the half order beams became diffuse, that the surface specie occupied sites
with fourfold symmetry. Because of the changes observed, Park and Madden[7]

concluded that the C2x2 was an impurity structure. This seems to have been confirmed recently by Palmberg[9] using Auger electron spectroscopy.

Coherence

The broadening of the (1/2 1/2) beams in Fig. 7c corresponds to sub-domains which are of the order of 25 Å in diameter. It is pertinent therefore to ask how small the domains must be for the broadening to be detected with present day LEED apparatus.

Suppose that a broad parallel beam of monochromatic radiation from a point source is normally incident on a grating shown in Fig. 9. Because of the finite size of the grating, the diffracted beam will not be parallel, the half-angular broadening being given by

$$\Delta\theta = \frac{\tan\,\theta}{n\,N}$$

(13)

where n is the order of diffraction and N is the effective number of scatterers. At small diffraction angles, the broadening will be difficult to detect since the effect is small and superimposed on the width of primary beam multiplied by the cosine of the diffraction angle. As one goes to larger diffraction angles, however, the contribution of the primary beam width diminishes and the broadening due to finite grating size increases as $\tan\,\theta$. Therefore to detect diffraction broadening where the effect is small, we must go to large diffraction angles.

Even at comparatively large diffraction angles, however, it is not likely with most present day LEED systems that the coherence of the surface grating will measurably affect the beam profiles from well prepared clean surfaces. We are much more likely to be limited by the coherence of the incident electrons, except in such cases as the antiphase broadening described above, or where the degree of coherence is reduced due to occluded inert gas atoms.[10]

At low energies ($\ll 50$ eV) we are limited primarily by the time coherence of the electrons, that is, by the energy spread in the primary electron beam. Thus the width in energy of a diffracted beam of electrons at a given angle can never be less than the width of the primary electron energy distribution.

At higher energies, we are restricted primarily by the spatial coherence of the electrons. Chutjian[11] has pointed out that the coherence diameter of the incident electrons is limited by the source dimensions. The effect of an extended source on the coherence width of the electrons can be estimated from elementary considerations.

Consider electrons arriving at a point P on a plane surface from a source of monoenergetic electrons which subtends an angle a when viewed from P. If a is small, there will be a spread in the transverse component of momentum p_y given by

$$\Delta p_y = a \frac{h}{\lambda} \tag{14}$$

where λ is the wavelength of the incident electrons. From the uncertainty relation, the coherence diameter Δy is just

$$\Delta y \sim \frac{\lambda}{a} . \tag{15}$$

This is the same relation one obtains from the Van Cittert–Zernike theorem of physical optics.

In general, we do not know a and it can be expected to show some variation with wavelength. However, we can expect the coherence diameter to increase with wavelength. We can, of course, determine the coherence diameter experimentally. As we have pointed out, such a measurement must be made at large diffraction angles. To obtain data over a range of wavelengths therefore requires that we make observations on a number of diffraction beams.

The clean surface of (110) Pd is well suited to this measurement since grazing diffraction beams occur at 10, 20, 30, 40, 60, 80, and 90 volts. At

higher energies the beams are generally too weak to permit accurate measurements at very large diffraction angles. The coherence diameter at these energies
calculated from the measured broadening at $85°$ are plotted in Fig. 10. Despite
the scatter, the dependence of the coherence diameter on wavelength is apparent.
The decreasing coherence diameter as the electron energy increases should be
taken into account in comparing theoretical and experimental LEED results. From
the broadening of the (01) beam, we can place a lower bound of about 300 Å on
the width of coherent scattering domains on this surface.

Conclusions

Studies of surface crystallography by LEED have benefited from the much
more extensive literature of high energy electron and x-ray diffraction. There
are surface characteristics such as steps, however, which have no obvious three-
dimensional analogue but which may have a pronounced effect on diffraction
features.[12] Other characteristics such as vacancies may be quantitatively very
different than in the bulk.[13] Although relatively uncommon in three-dimensions,
being observed only in a few ordered alloys, one would expect antiphase domains to be a very common occurrence in the formation of surface impurity
structures.

This may seriously complicate the problem of theoretical interpretation
of LEED intensities from adsorbed structures. One of the principal motivations
for such an interpretation has been to determine the positions of adsorbed
specie relative to the substrate. It seems possible in many cases, however,
to solve the two-dimensional part of the problem from a study of the diffraction beam profiles. To make full use of this approach, it is clear that more
attention will have to be given to electron coherence in the design of LEED
systems.

REFERENCES

1. P. J. Estrup, J. Anderson, and W. E. Danforth, Surface Science 4, (1966) p. 286.

2. W. P. Ellis and B. D. Campbell, Trans. Amer. Cryst. Assn., Proc. of the Symp. on "Low Energy Electron Diffraction" at Tucson, Feb. 4-7, 1968 (in press).

3. R. L. Park and H. E. Farnsworth, J. Chem. Phys. 43, (1965) p. 2351.

4. P. J. Estrup and J. Anderson, J. Chem. Phys. 45, (1966) p. 2254.

5. J. E. Houston (private communication).

6. R. L. Schwoebel (private communication).

7. R. L. Park and H. H. Madden, Jr. Surface Science 11, (1968) p. 188.

8. A. M. Mattera, R. M. Goodman, and G. A. Somorjai, Surface Science 7, (1967) p. 26.

9. P. W. Palmberg (private communication).

10. R. L. Park, J. Appl. Phys. 37, (1966) p. 295.

11. A. Chutjian, Phys. Letters 24A, (1967) p. 615.

12. W. P. Ellis and R. L. Schwoebel, Surface Science 11, (1968) p. 82.

13. R. L. Schwoebel, J. Appl. Phys. 38, (1967) p. 3154.

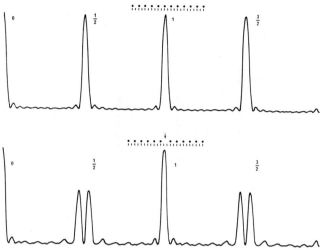

Fig. 1. Intensity pattern of two one-dimensional diffraction gratings of six points. In phase in the upper plot and out of phase in the lower.

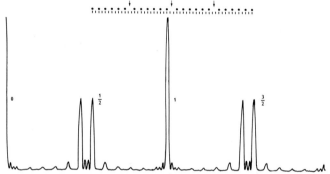

Fig. 2. Intensity pattern of four diffraction gratings. The first and third are in phase with one another and out of phase with the second and fourth.

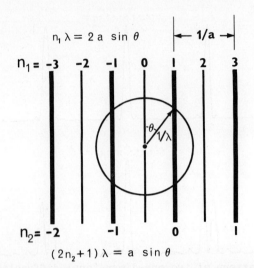

$$n_1\lambda = 2a\sin\theta$$

$$(2n_2+1)\lambda = a\sin\theta$$

Fig. 3. Ewald construction for a double spaced one-dimensional grating of point scatterers. The broad lines indexed by n_2 represent the reciprocal lattice planes of those beams which see the antiphase domains.

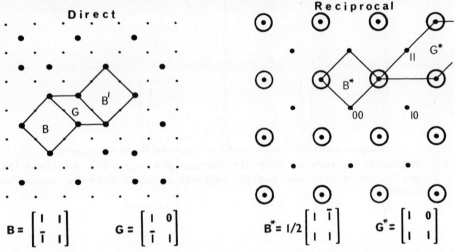

Fig. 4. Direct and reciprocal nets for a C2x2 structure with the adsorbed specie (large dots) in the positions of maximum coordination. G is a mesh joining antiphase unit meshes of the surface structure. The circled dots in the reciprocal net form unit meshes reciprocal to G and represent those beams which will appear split or broadened, due to interferences between B and B'.

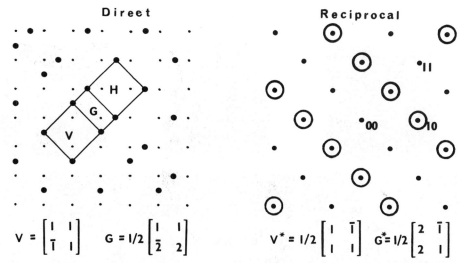

Direct

Reciprocal

$$V = \begin{bmatrix} 1 & 1 \\ \bar{1} & 1 \end{bmatrix} \qquad G = 1/2 \begin{bmatrix} 1 & 1 \\ \bar{2} & 2 \end{bmatrix} \qquad V^* = 1/2 \begin{bmatrix} 1 & \bar{1} \\ 1 & 1 \end{bmatrix} \qquad G^* = 1/2 \begin{bmatrix} 2 & \bar{1} \\ 2 & 1 \end{bmatrix}$$

Fig. 5. Direct and reciprocal nets for a C2x2 structure with the adsorbed specie (large dots) in nearest neighbor bridged sites. G is a mesh connecting a vertically bonded unit mesh to a horizontally bonded unit mesh. The circled dots in the reciprocal net form unit meshes reciprocal to G and represent those beams that will be split or broadened by interference between V and H.

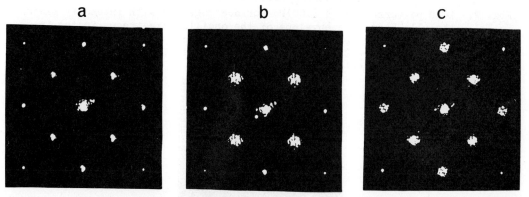

Fig. 6. Optical diffraction patterns from C2x2 gratings. (a) A single large domain. (b) Many irregular four-fold symmetric subdomains. (c) Many irregular twofold symmetric subdomains.

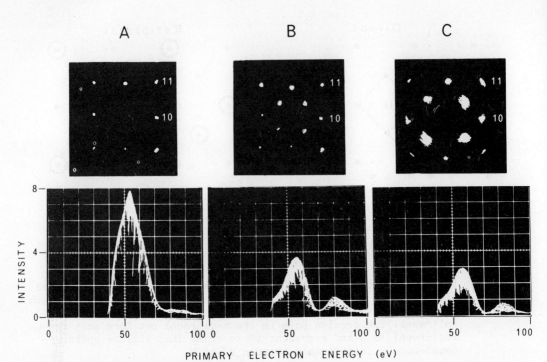

Fig. 7. LEED patterns from a (100)Pd surface together with intensity plots of the (11) beam. The envelope of the oscilloscope traces is the intensity of the beam as a function of energy. (a) Clean surface at 46 V. (b) Sharp C2x2 impurity structure at 47 V. (c) Diffuse half-order C2x2 impurity structure at 44 V.

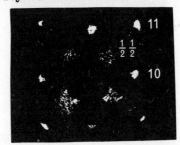

Fig. 8. Pattern from C2x2 impurity structure with just detectable half order features.

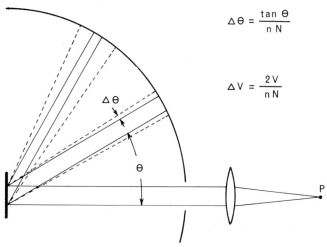

$$\Delta \Theta = \frac{\tan \Theta}{n N}$$

$$\Delta V = \frac{2V}{n N}$$

Fig. 9. Effect of finite grating size on beam width at different diffraction angles.

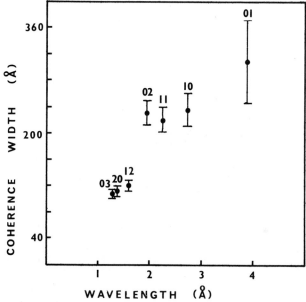

Fig. 10. Measured coherence diameter as a function of wavelength showing the increase of coherence with wavelength.

AUGER ELECTRON SPECTROSCOPY IN LEED SYSTEMS

P. W. Palmberg

North American Rockwell Science Center
Thousand Oaks, California

1. Introduction

In 1953 Lander[1] observed small peaks in the secondary-electron energy distribution function from a variety of materials and was able to relate these peaks directly to Auger transitions. Although he pointed out that detection of Auger electrons provided a means for chemical analysis of surfaces, general interest did not develop because of low sensitivity. Recently Harris[2] has demonstrated that the sensitivity is greatly enhanced by electronic differentiation of the energy distribution function. With this improved sensitivity, he has effectively applied Auger electron spectroscopy to a wide variety of surface segregation problems[2,3].

Another significant advance was made by Weber and Peria[4] who demonstrated that the display type LEED system is well suited for obtaining differentiated Auger spectra. They also made the first quantitative evaluation of the sensitivity by depositing known amounts of alkali metals on Ge and Si substrates. For K and Cs, they have shown[4,5] that less than one-tenth of a monolayer can be detected.

In addition to sensitivity, another important parameter in Auger electron spectroscopy is the Auger electron mean escape depth, which is directly related to the electron-electron mean free path. Since the resolution of the energy analyzer does not discriminate against phonon losses, the mean escape depth is determined by electron-electron or plasmon losses. In the energy range of interest in Auger electron spectroscopy, inelastic cross section data are not available, but can be evaluated by measuring the attenuation rate of

Auger peaks from a substrate during uniform deposition of an overlayer. Such an experiment has been carried out by Palmberg and Rhodin[6] who found that the escape depth for 72 eV and 362 eV Auger electrons in Ag is only about 4A and 8A, respectively. It is this characteristic which makes the technique so suitable for chemical analysis of surfaces.

Although the sensitivity achieved in conventional LEED apparatus is very high for certain elements, including Cs, K, Ag, and Pd, other elements such as Mg and Na have been extremely difficult to detect.[5,6] In general the sensitivity is lowest for elements which exhibit major Auger peaks in the high energy range because the cross section for ionization of the inner orbital giving rise to the Auger electron decreases rapidly with increasing energy. The primary experimental factors which limit the sensitivity of Auger electron spectrosocpy in standard LEED apparatus are low available primary current and restriction to bombardment at normal incidence. Because the escape depth of Auger electrons is very short compared with the penetration depth of the primary beam, grazing incidence is expected to increase greatly the number of Auger electrons ejected into the vacuum. Another factor which contributes to reduced sensitivity of the three-grid LEED system in the high energy range is rapid degradation of the energy resolution with increasing energy. It has been proposed[6] that this effect, which results from field penetration of the cut-off grid, could be reduced by adding a fourth grid.

The purposes of the present paper are: (1) to demonstrate that incorporating a high current grazing incidence, gun and a fourth grid greatly improve the sensitivity and resolution of Auger electron spectroscopy in LEED system; (2) to illustrate the power of combined Auger electron spectroscopy and LEED for determining the structure and chemistry of "anomalous" phases on the Pt(100) surfaces.

2.   Experimental Technique

The experimental scheme for obtaining Auger spectra in a LEED system is illustrated in Fig. 1. With the exception of a fourth grid and an auxiliary electron gun, the approach is identical to that described earlier.[6] In the four-grid system, the two intermediate grids are electrically connected and used for energy discrimination. The inner grid is grounded to maintain a field-free region surrounding the specimen. The outer grid is also grounded to minimize capacitive coupling between the cut-off grids, to which the perturbing signal is applied, and the fluorescent screen which serves as collector. With the target normal directed at the four-grid analyzer, the electron beam from the auxiliary gun strikes the surface at $15°$ with respect to grazing incidence. Rotation of the sample permits variation in the incident beam angle from $0-30°$ without appreciably altering the collection efficiency. The Superior Electronics model SE-3K/5U electron gun, modified with a tungsten cathode to permit repeated exposure to atmosphere, produces a well-focused, 3 keV, 200μA beam. The deflection plates are used to position the source of secondary electrons at the center of the spherically-shaped, four-grid analyzer. The shadow cast on the fluorescent screen by the LEED gun drift tube is used for locating this position precisely.

To show how the secondary-electron energy distribution function, $N(E)$ or its derivative, $dN(E)/dE$, is obtained in a LEED system, the collector current, $I(E)$ is expanded in a Taylor series about the dc retarding potential E_o:

$$I(E) = I(E_o) + \frac{dI\,(E)}{dE}\bigg|_{E=E_o}(E-E_o) + \frac{1}{2!}\frac{d^2I(E)}{dE^2}\bigg|_{E=E_o}(E-E_o)^2$$

$$+ \frac{1}{3!}\frac{d^3I(E)}{dE^3}\bigg|_{E=E_o}(E-E_o)^3 + \;.\;.\;.\;.\;.\;.\;.\;.\; \tag{1}$$

The voltage, $E - E_o$, is made to vary sinusoidally by superimposing a small signal, $k\sin \omega t$, on the retarding voltage E_o. Since $dI(E)/dE$ is equivalent to $N(E)$, Eq. (1) may be expressed as follows:

$$I(E) = I(E_o) + kN(E) \sin \omega t + \left. \frac{k^2}{4} \frac{dN(E)}{dE} \right|_{E=E_o} (1 - \cos 2\omega t)$$

$$+ \left. \frac{k^3}{24} \frac{d^2 N(E)}{dE^2} \right|_{E=E_o} (3 \sin \omega t - \sin 3 \omega t) + \ldots \quad (2)$$

From this expression it is apparent that $N(E)$ is obtained in the scheme of Fig. 1 by tuning the reference channel of the lock-in amplifier to the frequency of the perturbing voltage. The derivative results when the reference channel is tuned to double the frequency of the perturbing signal. Although the second harmonic component of the collector current is generally smaller than the fundamental signal, it is evident from Eqs. (1) and (2) that a larger perturbing signal may be used when operating in the second harmonic mode.

That the sensitivity of Auger spectroscopy is greatly increased by double differentiation of the retarding field characteristic, $I(E)$, is apparent from Fig. 2. The secondary-electron energy distribution function and its derivative are compared for a Pd(100) target under bombardment by a 1 keV, 100 μA beam. In the energy distribution function, $N(E)$, the three major Auger peaks are superimposed on a large continuous background. Differentiation of the energy distribution function enhances the sensitivity of Auger electron spectroscopy by reducing the background and converting inflection points into peaks

3. Optimization of Resolution and Sensitivity

Incorporation of a fourth grid into a standard three-grid LEED system greatly improves the resolution and extends the accessible energy range for Auger spectroscopy. Three-grid and four-grid operation are compared in Fig. 3 for a peak from 270 eV elastically scattered electrons and for a 270 eV carbon peak in an Auger spectrum. In addition to degrading the energy resolution, field penetration of the single cut-off grid in the three-grid system causes severe distortion and a 4 eV energy shift of the peaks. It is apparent

from Fig. 3 that the energy resolution is sufficiently improved in the four-

grid system to produce an accurate representation of the carbon Auger peak.

While the effective resolution of the three-grid system deteriorates to 20

eV at 1000 eV and the energy shift increases to 17 eV, the resolution of the

four-grid system remains below 2 eV and the energy shift is negligible. Since

the natural width of Auger peaks is typically 5-10 eV, the four-grid system

is capable of producing accurate Auger spectra over an energy range from zero

to at least 2000 eV.

It is evident from Fig. 4 that grazing incidence substantially increases

the sensitivity of Auger electron spectroscopy to surface impurities. The

LEED gun served as the source of primary electrons for curve (a), while curves

(b) and (c) were obtained with the grazing incidence gun. Both guns were

operated at the same voltage and current so that the only variable was the

angle of incidence. The Auger spectra reveal carbon, sulphur and nitrogen on

the (100) surface of a Pd sample. In addition to an overall amplification of

the Auger spectrum, grazing incidence preferentially increases the magnitude

of the sulphur and carbon peaks. If the incident electrons retained their

initial direction during passage through the first 10Å of the surface, the

magnitude of all Auger peaks should vary as $\sin^{-1}\theta$ with incident beam angle.

Comparison of Figs. 4a and 4b shows that the sulphur and carbon peaks follow

this predicted variation, but that a considerably weaker dependence on inci-

dent beam angle occurs for the Pd peaks. If the carbon and sulphur reside in

the immediate surface layer, as expected, it must be concluded that appreciable

scattering of 1400 eV electrons incident at 15° occurs within the first one or

two atomic layers. Decreasing the incident beam angle to 7° further increases

the relative magnitude of the sulphur and carbon Auger peaks but attenuates

the over-all spectrum. This attenuation may be caused largely by surface roughness which decreases the probability for electrons near grazing incidence to strike atomically flat regions of the surface. From these observations, it is apparent that variation of the angle of incidence may be used to obtain an indication of the chemical profile in the direction normal to the surface. Combined with LEED observations, this capability is extremely useful in deriving the atomic arrangement of complex surface structures.

Although incorporation of a high current, grazing incidence gun into a standard LEED system improves the signal-to-noise ratio considerably, a corresponding increase in the sensitivity is not always possible. As Weber and Peria[4] have pointed out, it is high background rather than noise that limits sensitivity at low energies. The energy range over which background is a limiting factor increases with increasing signal-to-noise ratio, and approaches 500 eV for the presently described system. Above 500 eV the improved sensitivity may be fully utilized. This increase in sensitivity coupled with improved resolution has greatly increased the capability of the LEED system for detection of elements exhibiting Auger peaks in the range from 500-2000 eV.

The Auger spectrum from a Pd(100) surface (Fig. 5) illustrates the improved sensitivity of the presently described system over that achieved in conventional LEED systems. The portion of the spectrum labeled (x1) is representative of the maximum sensitivity attainable in a LEED system having available a beam current of only 3 μA. As indicated, amplification of the Auger spectrum by a factor of 100 is possible with presently described modifications. One notes that Mn is detected on the "clean" surface. Absolute calibration of the technique has not been carried out, but the quantity indicated in the Auger spectrum is believed to represent, at most, a few percent of a monolayer.

4.   Application to "Anomalous" Surface Structures

Frequently, low-energy electron diffraction studies reveal complicated,

"anomalous" surface phases which defy analysis because of their unknown
chemical composition. Superstructures may be stabilized by impurities from
the bulk or from the residual gas environment. Less commonly, they represent
phases of the clean surface. In the following sections chemical analysis
of a few such structures by Auger electron spectroscopy is discussed.

4.1 Identification and Control of Carbon on Pt(100)

The occurrence of "ring" structures has been reported for various faces
of Au, [7,8] Pt, [8,9] and Ag [9] single crystals. The structure forms [7-10] irreversibly
when these metals are heated in the temperature range from 600-800°C. Extensive ion bombardment removes the structure, but heating in the 600-800°C
temperature range results in its repeated formation. Auger electron spectroscopic analysis has now shown unequivocally that the ring structure formed on
the Pt(100) surface consists of graphite crystals having their basal plane
parallel to the substrate.

The Auger spectrum of Fig. 6(a), obtained from a Pt(100) surface immediately after insertion into vacuum, reveals the presence of copious amounts of
carbon on the surface. Almost no platinum is observed, indicating that the
carbon layer was sufficiently thick to prevent escape of Auger electrons from
the platinum substrate. Observation of a very weak 1x1 LEED pattern shows
that the surface layer was highly disordered. After heating the sample at
800°C, the Auger spectrum of Fig. 6(b) and the LEED pattern of Fig. 7 resulted.
Enhancement of the substrate Auger peaks coupled with the ring diffraction
pattern shows that heating causes the carbon to aggregate into graphite islands.
Observation of the LEED pattern from the substrate, superimposed on the ring
pattern, provides further evidence for this conclusion.

Close agreement between the lattice parameter derived from the ring
pattern and that of graphite is convincing evidence that the ring pattern
results from graphite. From superimposed substrate and ring patterns, the

lattice spacing giving rise to the rings was found to be 2.13 Å, with approximately 2% accuracy. The lowest order lattice spacing in the basal plane of graphite is 2.135 Å. Distribution of the graphite pattern into the segmented ring of Fig. 7 indicates a nearly random azimuthal orientation for the graphite crystals.

Complete removal of carbon from the surface was achieved by heating the Pt crystal at 1000°C in 10^{-5} Torr of oxygen for several minutes. As illustrated in Fig. 8, the Auger spectrum obtained subsequent to this treatment does not include a carbon peak. Once the carbon was removed, heating at 1500°C for extended periods did not result in further accumulation of carbon at the surface. With the exception of energy shifts of a few electron volts, salient features of the Auger spectrum from Pt correspond well to those reported for Au by Palmberg and Rhodin[6]. This close correspondnece is expected because of their nearly equivalent electronic structure, and supports an earlier[6] conclusion that the peaks observed from Au originate from Auger transitions within the metal itself, rather than from surface impurities. Because of greatly improved sensitivity, additional minor peaks were observed from Pt. To test whether these peaks originated from Pt or surface impurities, Auger spectra were taken during heating at temperatures up to 1500°C in 10^{-5} Torr in oxygen. Failure of this treatment to alter the magnitude of any of the observed peaks, major or minor, suggests very strongly that the Auger spectrum of Fig. 8 is representative of pure platinum. Detailed analysis of the spectrum is the subject of continued investigation.

The clean Pt(100) surface produces LEED patterns, given in Fig. 9, that reveal details of the surface structure which have not been reported previously. The pattern of Fig. 9(a) is identical to that reported previously for Pt(100) by Lyon and Somorjai[8] and also bears striking similarity to that from the Au (100) surface reported by Fedak and Gjostein[7,10] as well as by Palmberg and

Rhodin.[11] Previously, analysis of the Pt(100) and Au(100) structures has been carried out under the assumption that the surface structure possessed mirror symmetry as well as two-fold rotational symmetry. Present results show that while the surface structure is characterized by two-fold rotational symmetry, it does not possess mirror symmetry. Thus, there are four, rather than two, equivalent orientations for the surface structure on the cubic substrate. Because of a thermal or stress gradient, heating the sample above 1000°C caused preferential growth of a single orientation, in which case the unidrectional pattern of Fig. 9(b) was obtained. Destruction of the superstructure by adsorption of CO or by sputtering, followed by heating to 500°C, gave the pattern of Fig. 9(a). This pattern may be generated by superimposing the four possible orientations of Fig. 9(b).

Because the most intense beams in the Pt(100) and Au(100) diffraction patterns lie near one-fifth order positions, the structures have frequently been termed (1×5). The actual translational symmetry is considerably more complicated, however, as illustrated in Fig. 10. It is not clear at present whether the previously proposed[10,11] hexagonal layer model can be appropriately modified to produce a surface unit mesh having dimensions illustrated in Fig. 10.

Although carbon can be removed from the Pt crystal by high temperature treatment in oxygen, accumulation of carbon was detected during LEED or Auger measurements. As demonstrated in Fig. 11, electron beam cracking of adsorbed CO was responsible for this effect. These spectra were taken during exposure of the crystal to 10^{-5} Torr of CO. Curve (a) was taken immediately after initiating electron impact with the target surface while curves (b) and (c) were taken after irradiation for 4 and 7 minutes, respectively. Although conditions during LEED observations are generally much less extreme, significant quantities of carbon accumulated over periods of several hours. An interesting manifestation of this effect was stabilization of either (1×1) or (2×2) surface structure. A detailed discussion of these and other impurity-stabilized surface structures will appear elsewhere.

4.2 Silicon on Pt(100)

The case of silicon on Pt(100) provides an interesting example of reversible impurity segregation at surfaces. The Auger spectrum of Fig. 12 was taken while heating the Pt(100) crystal at 1500°C. The small peak at 91.5 eV, identified as originating from silicon, disappeared below 1000°C and then reappeared upon heating to higher temperatures. Although chemical analysis of the specimen used here revealed only 4 ppm silicon,[12] the solubility is apparently sufficiently reduced above 1000°C to cause segregation at the surface. The silicon was removed irreversibly by heating in 10^{-5} Torr oxygen for 30 minutes.

5. Acknowledgment

The author is indebted to G. A. Somorjai who generously provided the Pt and Pd samples employed for these studies.

REFERENCES

1. J. J. Lander, Phys. Rev. 91, 1382 (1953).

2. L. A Harris, J. Appl. Phys. 39, 1419 (1968).

3. L. A. Harris, J. Appl. Phys. 39, 1428 (1968).

4. R. E. Weber and W. T. Peria, J. Appl. Phys. 38, 2425 (1968).

5. R. E. Weber, private communication.

6. P. W. Palmberg and T. N. Rhodin, J. Appl. Phys 39, 2425 (1968).

7. D. G. Fedak and N. A. Gjostein, Acta Met. 15, 827 (1967).

8. H. B.Lyon and G. A. Somorjai, J. Chem. Phys. 46, 2539 (1967).

9. A. B. Mattera, R. M. Goodman, and G. A. Somorjai, Surface Science 7, 26 (1967).

10. D. G. Fedak and N. A. Gjostein, Surface Science 8, 77 (1967).

11. P. W. Palmberg and T. N. Rhodin, Phys. Rev. 161, 586 (1967); J. Chem. Phys. (in press).

12. Chemical analysis made available by G. A. Somorjai who also provided the crystal.

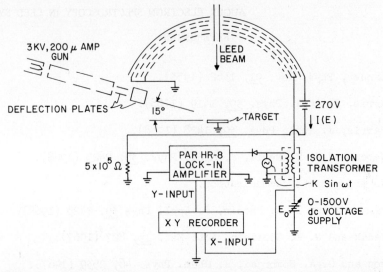

Fig. 1. Scheme for obtaining Auger spectra in LEED system.

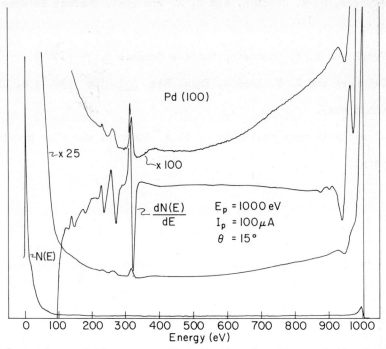

Fig. 2. Comparison of the energy distribution function and its derivative for Pd(100) surfaces.

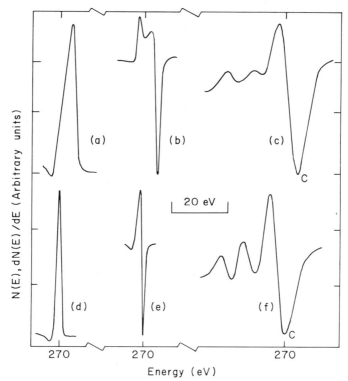

Fig. 3. Comparison of elastic peak, differentiated elastic peak and carbon peak obtained with three-grid (a, b, and c) and four-grid (d, e, and f) analyzer.

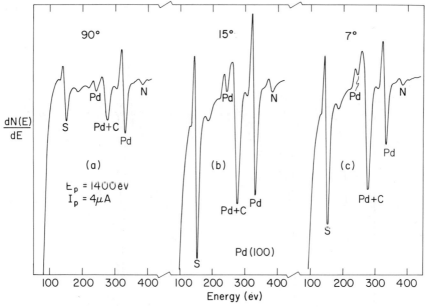

Fig. 4. Auger spectra from Pd(100) surface for incident beam angles of (a) 90°, (b) 15°, and (c) 7°.

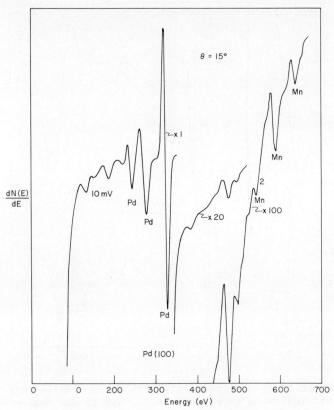

Fig. 5. Auger spectrum from Pd(100) surface.

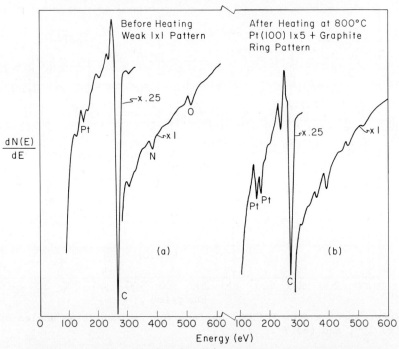

Fig. 6. Auger spectra from Pt(100): (a) immediately after insertion into vacuum; (b) after heating to 800°C.

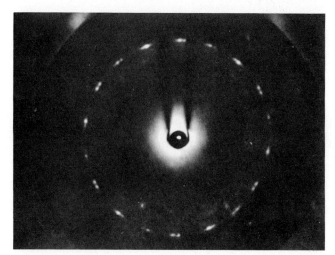

Fig. 7. Ring pattern from graphite crystals on Pt(100). 75 eV.

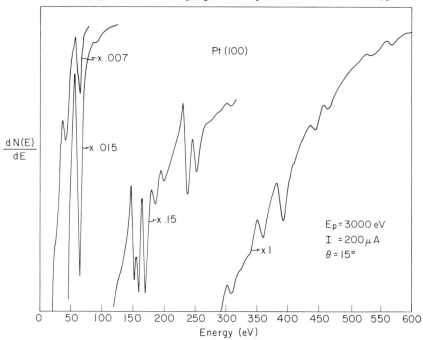

Fig. 8. Auger spectrum from clean Pt(100) surface.

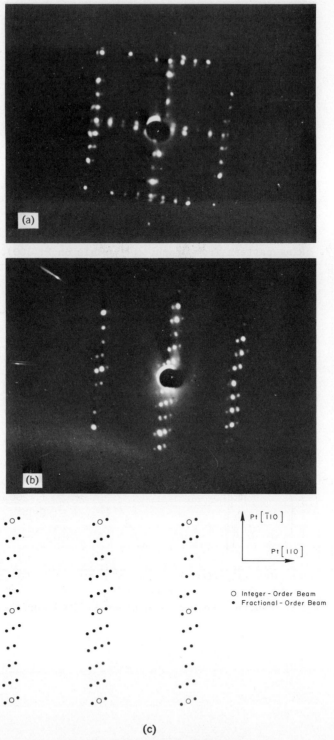

(c)

Fig. 9. LEED patterns from clean Pt(100): (a) multi-oriented domain pattern.
92 eV; (b) singly-oriented domain pattern. 76 eV; (c) Schematic
representation of Pt(100) LEED pattern.

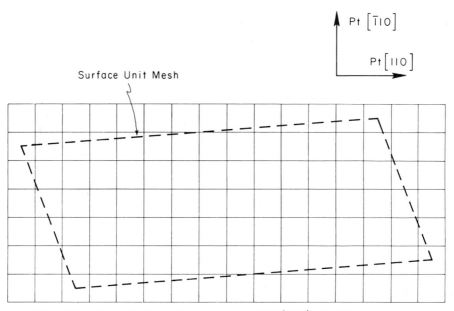

Fig. 10. Translational symmetry of Pt(100) "1x5" structure.

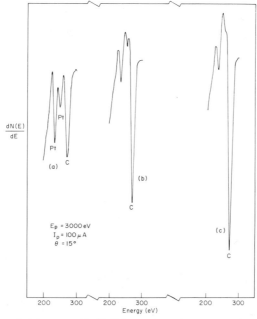

Fig. 11. Carbon accumulation on Pt(100) during 3 keV, 100 μA beam irradiation in 10-5 Torr of CO. Auger spectra (a), (b), and (c) were obtained after irradiation periods of ~ 0 mins, 4 mins and 7 mins, respectively.

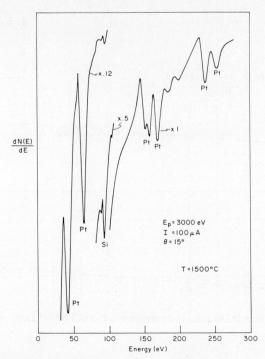

Fig. 12. Auger spectrum for Pt(100) at 1500°C. Note Si peak at 91.5 eV.

LOW-ENERGY ELECTRON DIFFRACTION STUDIES OF THE (001) FACE
OF ALUMINUM AND THE (111) FACE OF SILVER

L. De Bersuder,* C. Corotte, P. Ducros, and D. Lafeuille

Laboratoire de Spectrometrie Physique
Faculte des Sciences de Grenoble, France

APPARATUS

For the study of the (111) face of silver the apparatus used was a Varian low-energy electron diffraction system. The experiments on the (001) face of aluminium were made using a diffractòmeter designed and constructed by one of us (1). With the latter apparatus, it is possible to vary the angle of incidence of the electron beam, to rotate the crystal about an axis normal to its surface, and to observe a diffraction pattern for all angles of incidence and observation from grazing to normal incidence (Fig. 1).

The quantitative study of the intensity variation, as a function of the orientation of the incident beam, of a Bragg peak or of the background scattering was done by photometry of the fluorescent screen.

SURFACE CLEANLINESS

The methods for decontaminating the surfaces of Ag and Al in order to obtain the clean surfaces consisted essentially of the classic method of an ion bombardment followed by an annealing, and will be described in detail elsewhere (2.3).

INFLUENCE OF THE INCIDENT BEAM DIRECTION OF THE KIKUCHI SCATTERING AND THE SECONDARY EMISSION

The intensity and contrast of both the Kikuchi pattern and the secondary emission were observed to vary strongly with the direction of the incident electron beam. The variations in the intensity and contrast always affected

---

* On leave from the Laboratoire de Chimie Physique, Faculte des Sciences d'Orsay where he did the work on the (001) face of aluminium.

the whole Kikuchi pattern and for a given maximum of intensity and contrast there is a corresponding maximum in the secondary emission. Only the elastically scattered electrons were used to form the diffraction patterns observed in these experiments.

The Kikuchi patterns obtained from the Ag(111) surface have low contrast with little apparent detail at room temperature (4). The total luminous intensity of the fluorescent screen was recorded as a function of the angle of incidence θ, this angle being measured with respect to the normal to the crystal surface, the plane of incidence being parallel to the $(1\bar{1}0)$ crystal planes. The electron energy was 1200 eV, and the suppressor grid potential was such that only elastically scattered electrons were able to reach the screen (see Fig. 2). Under these conditions no Bragg spots were observable. The luminosity of the screen was seen to pass through successive maxima as θ was varied. Because of the geometrical arrangement of the elements in the Varian apparatus, only a certain fraction of the scattered electrons are able to reach the screen, which causes the measured intensity to vary slowly with θ, having a maximum at $\theta=0°$. Superimposed on this slowly varying function are the successive maxima of the fine structure. These maxima in the elastically backscattered intensity were found to occur when the incident beam direction was parallel to one of the close packed rows of the crystal. It was noted that the directions of these close-packed planes are also those which produce a maximum in the Kikuchi scattering.

The Kikuchi patterns from Al contained much more detail, which permitted a more complete study to be effected. In this case, we measured the current flowing between the crystal and the ground. This current serves to measure directly the total secondary emission including both elastically and inelastically scattered electrons.

By varying θ, curves similar to those obtained from Ag were obtained, but a monotonic increase of the secondary emission with θ is superposed on the curves in this case. This increase corresponds to a variation of the form (1-cos θ) for the secondary emission (5); we have attributed it to the inelastic electrons and subtracted it from the experimental curves. We have compared the curves obtained in this way with microphotometer traces taken from photographic negatives of the Kikuchi diagrams obtained under the same conditions. The directions in which these traces were taken correspond respectively to the planes in which the incident beam was moved during the secondary emission measurement. Figures 3 and 4 show the curves obtained at electron energies of 2000 and 500 eV, taken parallel to the (110) and (100) planes respectively. Within the limits of the experimental error, the curves for δ and the microphotometer curves are well correlated.

The correspondence of the maxima in the secondary emission with the directions of the dense-packed rows in the crystal was observed to be only approximate in the cases where the Kikuchi patterns were reasonably detailed.

These results may be interpreted in terms of the reciprocity theorem of Von Laue (7).

If $\delta(\theta,\alpha)$ is the spatial distribution function for the scattered electrons originating from a source of intensity K, situated at a point A inside the crystal (α being the asimuth angle), this source will produce at a point B outside the crystal a scattered intensity $K\,\delta(\theta,\alpha)$. The reciprocity theorem states that a source identical to the previous one, but situated at B, will produce at A an electron density equal to $K\,\delta(\theta,\alpha)$. This will give rise to an internal source with an amplitude proportional to $\delta\,(\theta,\alpha)$ and consequently a Kikuchi scattering or elastic secondary emission also proportional to $\delta\,(\theta,\alpha)$.

KIKUCHI PATTERNS OBTAINED AT NEARLY GRAZING EMERGENCE

Kikuchi patterns usually observed by LEED are of the type K III, that is, they correspond to the diffraction by a 3-dimensional lattice of a wave emitted by a source inside the crystal. Their appearance and geometry are well known. Germer and Chang have, however, reported the existence of Kikuchi lines, K I, geometrically linked to diffraction (of waves from an internal source) by one-dimensional rows of atoms.

From the (001) face of aluminum, for incident and scattered beam directions away from the normal we have observed Kikuchi lines, K II of a new type. Their geometry is linked to diffraction by a two-dimensional array, and we have observed them at primary beam energies between 200 and 2000 eV. These lines may be both light and dark as for the K I. Their geometry corresponds to scattering directions such that the scattered wave may give rise, after diffraction by the two dimensional array, to a wave diffracted parallel to the surface of the crystal (9). The lines occur in positions intermediate between those calculated without taking account of an inner potential, and those calculated using a potential of $18,5$ V . Figure 5 shows a stereographic projection of the calculated K I and K II lines for 500 eV. Figure 6 shows the observed K I and K II lines at 500 eV along with the proposed indices. $\theta=60°$ and $\alpha=0$ (for case A) and 20° (for case B). ($\alpha=0$ in the (100 plane)).

RENNINGER DIAGRAMS AT OBLIQUE INCIDENCE FROM Al (001)

With the beam incident at a given angle to the surface normal, the crystal was rotated about an axis normal to the surface. It was found that the intensity of all of the Bragg spots varied strongly with the asimuth angle, α. The variations in intensity of a spot always coincided with the passage of a Kikuchi line, bright or dark, across the position of the spot. The passage of a bright line corresponded to an increase in intensity, and that of a dark line to a decrease in intensity. This correspondance of the variations of the Bragg spot intensity with the passage of a Kikuchi line across the spot was the same for all incident beam energies and for all three types of Kikuchi lines (K I, K II and K III).

The curve of Fig. 7 shows the variation of the specularly reflected beam intensity as a function of the asimuth α, with $\theta=65.5°$ and a beam energy of 2000 eV. Figure 8 shows similar curves obtained for different values of θ at an energy of 200 eV.

The variations in intensity can be indexed in the same way as Kikuchi lines. In Fig. 7 only K III lines are involved and they have been indexed by taking into account the fact that the calculated position of a K III line with respect to the line profile is as shown in Fig. 9. This latter has been confirmed by the micro-photometer traces of Figs. 3 and 4. In Fig. 8 only the K I and K II lines are involved. Their positions are indicated by the strokes, the small circle at the extremity of the stroke indicating the line position calculated using an inner potential, the other end of the stroke indicates the line position calculated using a zero inner potential. In the case of a line of type K I the index is preceded by the letter R.

There is good agreement between the positions at which the intensity maxima and minima occur, and the calculated position of the Kikuchi lines presumed to be responsible for these fluctuations. This agreement was particularly good in the case of low index Kikuchi lines and in the absence of intersecting line.

The Kikuchi lines are indicative of particular geometric conditions. In effect, when a diffracted spot is on a Kikuchi line it is at the same time in a position where the diffracted beam can give rise to the appearance of a new beam. This is the case whatever type of Kikuchi line is involved. For the K III lines, the observed phenomena is known as the Renninger effect, and for the K I and K II we have an effect similar to a McRae resonance (10).

CONCLUSION

The Kikuchi patterns at a given angle of incidence provide an indication
of the quasi-elastic excitations for the crystal at various crystal directions.
When a Bragg spot coincides with a Kikuchi line the Bragg intensity is seen to
be modified by the Kikuchi scattering at the point. Since the coincidence
of a Bragg spot with a Kikuchi line occurs under the same geometric conditions
as those necessary for strong dynamic coupling between that spot and another
Bragg spot, it is at present difficult to know if this last named is the dominant one, or if it is the effect of interaction between elastically and inelastically scattered waves.

For very large angles of incidence the principal modes excited in the
Kikuchi scattering seem to be surface modes. In the same conditions the
Renninger diagrams seem to be largely influenced by dynamic resonance effects,
that is, once more by surface modes.

REFERENCES

1. L. De Bersuder, C. R. Acad. Sci. 262 (1966) p. 1055.

2. D. Lafeuille, Thèse de Docteur-Ingénieur, Grenoble, 1968.

3. L. De Bersuder, Thèse de Docteur-Ingénieur, Paris, 1968.

4. C. Corotte, P. Ducros, D. Lafeuille, C. R. Acad. Sci. 265, (1967)
    p. 1040.

5. H. Bruining, Physica (1938) 5, p. 901.

6. L. De Bersuder, C. R. Acad. Sci. 265 (1967) p. 885.

7. M. Von Laue, Ann. Physik 23, (1935) p. 705.

8. L. H. Germer and C. C. Chang. Surf. Sci. 4, (1966) p. 498.

9. L. De Bersuder, C. R. Acad. Sc. 266, (1968), p. 1489.

10. E. G. McRae, J. Chem. Phys. 45, (1966) p. 3258.

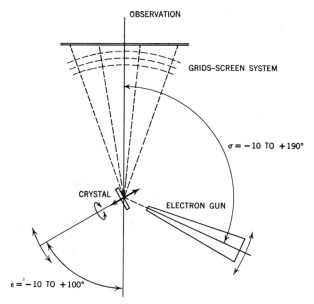

OBSERVATION

GRIDS-SCREEN SYSTEM

$\sigma = -10$ TO $+190°$

CRYSTAL

ELECTRON GUN

$\epsilon = -10$ TO $+100°$

Fig. 1.   Schematic diagram of the LEED apparatus.

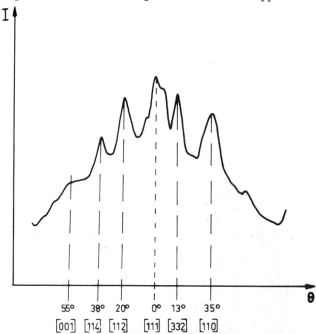

55°	38°	20°	0°	13°	35°
[001]	[114]	[112]	[111]	[332]	[110]

Fig. 2.   Ag (111). Total luminous intensity of the screen as a function of the
incident beam direction at an energy of 1200 eV. The suppressor grid
is at cathode potential. For the principal maxima the corresponding
angles of incidence are marked, along with the indices [u,v,w] of the
crystal directions which are closely parallel to the incident beam.

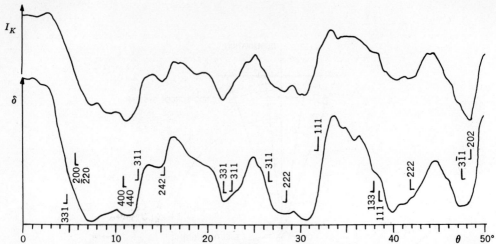

Fig. 3. Al (001). Energy 2000 eV. Intensity distribution, I_K, of a Kikuchi
pattern in the (110) plane, and variation of the secondary emission
δ with the angle of incidence, θ, the incidence beam being parallel
to the (110) planes.

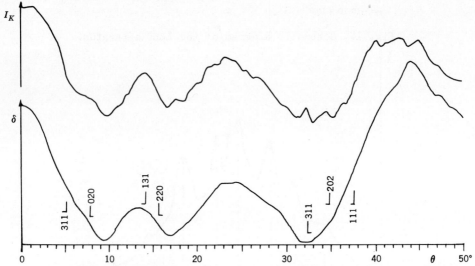

Fig. 4. Al (001). Energy 500 eV. Intensity distribution, I_K, of a Kikuchi
pattern in the (100) plane, and variation of the secondary emission,
δ, with the angle of incidence, θ, the incident beam being parallel
to the (100) planes.

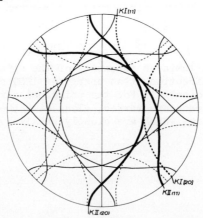

Fig. 5. Al (001). Kikuchi lines of type K I and K II. Calculated positions
for an energy of 500 eV. Stereographic projection.

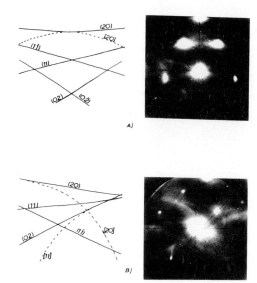

Fig. 6. Al (001). Observed K I and K II Kikuchi lines with the proposed indices. Energy 500 eV. $\theta = 60^\circ$, $a = 0^\circ$ (A) $a = 20^\circ$ (B).

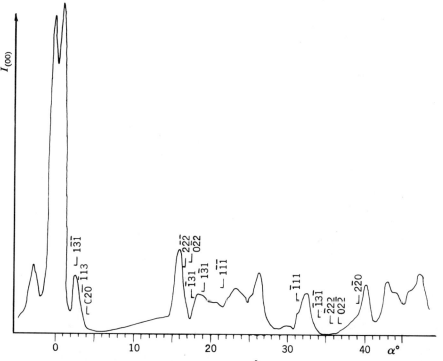

Fig. 7. Al (001). Renninger plot, $I_{oo} = \int (a)$ $\theta = 65.5^\circ$ Energy 2000 eV.

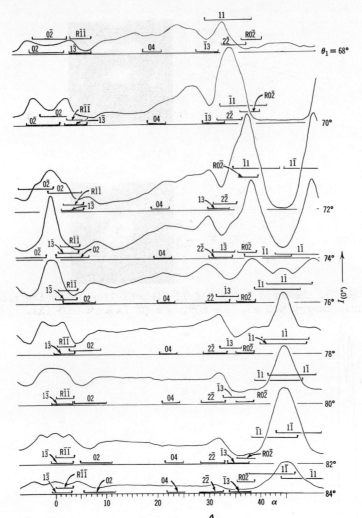

Fig. 8. Al (001) Renninger plots, $I_{00} = \int (a)$ $\theta = 68$ to $84°$ Energy 200 eV.

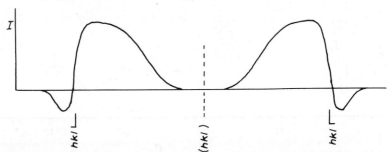

Fig. 9. Kikuchi line profile (K III type line) showing the calculated line positions.

LOW ENERGY ELECTRON DIFFRACTION STUDY
OF THE TANTALUM (112) SURFACE

T. W. Haas

Aerospace Research Laboratories
Wright-Patterson Air Force Base, Ohio

1. Introduction

The body centered cubic refractory metals, viz., V, Nb, Ta, Cr, Mo, W, form an interesting group of materials for low energy electron diffraction studies. To this point the studies have been primarily on the (110) surfaces[1-8]. This surface is the most densely packed in the bcc system and in studies thus far shows little tendency to facet due to chemisorption. The one exception appears to be the Cr(110) surface which facets in the presence of oxygen into well defined (100) planes[8]. One interesting observation has been that in many cases structures and reactivities seem to be similar for materials in the same group in the periodic table. Although some surfaces other than (110) planes have been studied[9-15], enough work has not as yet been published to establish this trend with any certainty for other surfaces. Sproull[16] and Germer and Chang[17] have studied the W(112) surface and report a number of interesting results including the faceting of this surface into (110) planes in the latter stages of oxidation. It is interesting to contrast these results with the Ta(112) work reported here. Ta and W are side by side in the periodic table, Ta being in group Vb while W is in VIb. Size effects are not great, the lattice constants being 3.30Å for Ta and 3.16Å for W. Chemical effects may therefore be the determining factor in the observed differences.

Although results for the Nb(112) surface will not be reported here it is worth mentioning that preliminary studies show great similarities between the Ta(112) and Nb(112) surfaces just as observed for the (110) surfaces reported earlier[1].

The purpose of this study, then, is to determine properties of the clean Ta(112) surface and then to determine the reactivity of this surface with H_2, N_2, O_2, and CO. It is found that all these gases chemisorb quite readily at room temperature. The interesting result is found that nitrogen chemisorption can cause faceting of this surface into (113) planes while the chemisorption of the other gases does not cause faceting.

2. Experimental

The LEED apparatus used was the post diffraction acceleration three concentric grid system available from Varian associates. No particular new or novel techniques were used, hence a detailed discussion of experimental techniques will not be given. The typical operating pressure for clean surface studies was about 2×10^{-11} torr, temperatures of the sample were measured using optical and IR pyrometers, and residual gas analysis made with an Ultek quadrapole type analyzer.

3. Clean Surface Studies

The clean surface of Ta(112) is produced simply by heating the sample in vacuum to a temperature in excess of 2000°C. This is similar to earlier results on the Ta(110) surface. In both cases it is imperative that the vacuum be very good as this surface, even at elevated temperature, is quite reactive with CO and H_2. If the vacuum is poor then the recontamination rate will exceed desorbtion rates and no real cleaning results. One can pretty well be certain that a clean surface is obtained when one can heat the sample to

2100°C and maintain a pressure in the low 10^{-10} torr range for a period of a minute or so.

The clean surface produced by this treatment gives the expected diffraction pattern as shown in fig. (1). With the photograph is a ball model drawing of the (112) surface in the same orientation as the photograph. One can see that the (h0) direction is not symmetrically related to the ($\bar{h}$0) direction so that one expects in LEED to see a different intensity curve for the (10) beam than for the ($\bar{1}$0) beam. This is so since Friedels law does not appear to hold in LEED and indeed this is verified experimentally in this case.

The (00) intensity curve from this surface is interesting. It does not show the usual multitude of major and minor peaks found with the Ta(110) surface. In fact, there is only one major peak, together with a number of minor peaks as can be seen in fig. 2. The interpretation of these intensity curves is an active area of interest at present, as many of the papers presented in this volume attest. It would seem that one should be able to classify the peaks obtained in these curves according to the way they change in voltage and amplitude as the angle between the surface normal and incident beam is varied. To be complete one should obtain this information for rotations around all the important azimuths in the surface (for example all the densely packed, close spaced rows). In a simple kinematic treatment where one looks upon the (00) curve as reflections from a stack of mirrors, more or less, one expects only simple effects. This does not obtain experimentally. The peaks in these intensity curves do seem to fall into four categories. In the first case we

have peaks which simply move to lower voltages as the incident angle
is increased. These are usually found at low voltages and probably
are the resonance peaks described by E. G. McRae[18] in his multiple
scattering theory. They are associated with the first emergance of
a non-specular diffracted beam and, as simple analysis shows, will
appear at lower voltages as the incident angle increases.

A second type of peak would be due to the simple kinematic,
single scattering event. This type of peak moves to higher voltages
as the incident angle is increased. Very few peaks observed in an
intensity curve fall into this category. One such peak in the Ta(112)
curve appears to be at about 480 volts. This peak moves up in voltage
in just the manner expected and may be due to simple kinematic
scattering.

A third set of peaks, by far the most prevalent, exhibit
complex behavior. These peaks may go up in voltage at first and
then go back down again; they may split into two peaks, they may grow
in amplitude, shrink, then grow again. This type of peak is undoubtedly due to multiple scattering. An analysis of these peaks is quite
complex and will involve a fairly complete dynamical treatment.
Calculations in our laboratory made by A. G. Jackson[19] have shown
that simple multiple scattering, using only s-waves atomic scattering,
does not reproduce the observed data too well.

The fourth type of peak does not change voltage as the incident
angle is varied, but does decrease in amplitude as this angle is
increased. In the Ta(112) curve the lone major peak at near normal
incidence has this observed behavior. The intensity curves at two
angles of incidence, 4° and 8°, show that this peak does not change

shape or change in voltage, but does decrease in amplitude. It is tentatively suggested that this peak corresponds to a maximum in the atomic scattering factor for scattering through angles of approximately 180°. Such an event may be associated with a p- or higher-wave resonance[20]. Khan, Hobson and Armstrong[21] have given some rough polar diagrams for the cross sections expected for low energy electrons. What we are suggesting is that the cross section for scattering through 180° becomes quite large for a small range of voltage. Hence a very large peak should appear in the (00) curve whenever the incident beam voltage is in the correct range. One can see that the magnitude of this peak should drop off as the incident angle is increased. This assumes, of course, that one can treat the scattering as a central potential, spherically symmetric problem.

The question of actually calculating locations of these peaks runs into the difficulty of the potential problem. Recent work on the scattering of low energy electrons by heavy metals has shown that the results depend very strongly on the type of potential used to represent the scattering center[20].

There is a great deal of interest at present in studies of the inelastic scattering of electron from surfaces[22-25]. As is by now well known, the LEED instrument is capable of providing information on the energy distribution of inelastically scattered electrons by using electronic differentiation techniques. In fact derivatives of the energy distribution curve have been found to provide even more information. The energy distribution curve from the clean Ta(112) surface is shown in fig. (3a). There is a prominent surface and

bulk plasmon peak, the surface peak being about a 14 volt loss, and
the bulk peak being about a 20 volt loss. A number of side peaks
are also observed on the secondary peak. Taking the derivative of
the energy distribution gives the curve seen in fig. (3b). Here we
see several peaks in the secondary region corresponding to Auger
transitions of 28 and 50 volts. If the gain is upped by a factor of
500 or so then another set of Auger peaks are observed. These appear
as a doublet at 168 and 180 volts.

The Auger technique is very useful in adding evidence that a
clean surface has been obtained. A principle impurity expected in
tantalum is carbon. Carbon shows up well in Auger spectroscopy and
would give a peak at around 270 volts. No peak, for the cleaned
surface, is observed in this region lending support to the assertion
that simple heating can produce a relatively clean surface.

4. Chemisorption Studies

Having determined the techniques for cleaning the Ta(112) surface,
and with a knowledge of some clean surface properties, it is then of
interest to measure the reactivity of this surface toward reactive
gases. The gases studied were N_2, H_2, O_2 and CO. It is interesting
to contract these results with those obtained from the Ta(110) surface
and where possible, with the W(112) results.

Hydrogen chemisorbs readily at room temperature on both the
Ta(110) and Ta(112) surfaces. The adsorption of hydrogen produces
no changes in the diffraction pattern hence the surface has not
rearranged or faceted. The hydrogen may be in one of two states on
the surface, either in an amorphous layer or in a lattice the same

as the underlying substrate. The intensity distribution curve for

the former case would show a simple diminishing in amplitude for all

peaks but such is not observed experimentally as can be seen in

fig. (4). Since many peaks actually increase in size one would

conclude that the hydrogen is in a crystalline layer. The suggestion

that this lattice is the same as the underlying tantalum substrate

is based on the assumption that the hydrogen layer would scatter

sifficiently to give rise to new diffracted beams if it had some

different structure than the substrate. There seems to be some

evidence for this assumption from the work of Estrup and Anderson on

H_2 chemisorption on W(100)[13].

Nitrogen does not chemisorb on the Ta(110) surface at room or

elevated temperatures. It does chemisorb onto the Ta(112) surface

quite readily at room temperature. In this case the adsorption is

amorphous in that no new diffracted beams appear and the intensities

of all observed beams diminish more or less uniformly. If the surface

with a chemisorbed layer of nitrogen is heated to about 500°C and

then cooled then a new diffraction pattern is observed. This new

pattern is shown in fig. (5). The (00) spot in this pattern is

shown below the center in the (h0) azimuth but the macroscopic surface

is still normal to the incident beam, indicating that faceting has

taken place. The new (00) beam is 22° below center in a (10) direction

indicating that the tantalum nitride layer is growing epitaxially on

the (113) tantalum planes. Only one (00) beam is observed so that one

can be sure that pyramids of a nitride are not forming on the surface.

The faceting can be made complete so that no traces of the (112) pattern remains. The tantalum nitride layer which is formed gives rise to a square pattern. It is very likely that this layer is a cubic tantalum nitride with the rocksalt structure. Such a nitride is observed for niobium[26] but data for the corresponding tantalum nitride could not be found.

Exposure of the clean Ta(112) surface to O_2 produces a series of very complex patterns. The first pattern corresponding to the lowest coverage observed is shown in fig. (6a). This pattern is a (1 x 3) oxide. Further exposure to oxygen produces a series of very complex patterns as shown in fig. (6b) and fig. (6c). These patterns undoubtedly represent the epitaxial growth of one of the well known oxides of tantalum and the complexity of the pattern is probably due to multiple scattering. A successful fit of known oxides into this pattern has not been made as yet. However the chemistry of the oxidation of tantalum is very complex and several oxides are known to grow epitaxially on the metal[27]. From the temperature of its formation and stability and from the epitaxity with the metal it would appear that the oxide is either TaOy or TaOz. TaOy is a cubic oxide with sides 4 times that of the tantalum and approximate composition of $Ta_{3.5}O$ to Ta_4O. TaOz is also cubic with a lattice constant two times that of tantalum, and an approximate composition between Ta_2O and TaO. The TaOz seems the most likely as its region of formation is from 300°C to 1200°C. It is also likely that the same oxide is growing on both the Ta(110) and Ta(112) as the conditions for the formation of the oxides observed are very similar.

Exposure of the surface at room temperature to CO produces an amorphous adsorption in the same way as did nitrogen. Heating of the surface after or during exposure to CO produces the same set of patterns as were observed during the oxidation of tantalum. This is similar to the result obtained with the (110) surface and was interpreted in this case to be due to decomposition of the CO into a tantalum oxide plus carbon. The question quite naturally comes up as to the fate of the carbon produced. The Auger technique is an ideal way to explore this problem. Several possibilities exist for the residual carbon. Evaporation of the carbon is unlikely as the transition takes place at 400°C. Tantalum carbides are also non-volatile at these temperatures. Another possibility is that the carbon forms clumps on the surface which are not evident in the diffraction data. In this case we should expect to be able to find a carbon Auger peak both before and after heating the sample. A peak at around 270 volts is found before heating but not afterward. This suggests that the carbon dissolves in the bulk and due to the limited penetration of the electron beam is no longer detectable. It must be admitted that the present measurements were made at gains which were at the limit imposed by present noise levels. Hopefully another factor of improvement in noise by two or threefold will allow more definite experiments to be carried out.

5. Conclusions

From the results of this study one would conclude that a very important factor in the interaction of gases with clean surfaces can be described as due to chemical effects. The results show a much greater correlation between the Ta(110) and Ta(112) surface than between the Ta(112) and W(112) surface. Geometry, however, does play a role, as the results with nitrogen adsorption show.

References

1.   T. W. Haas, A. G. Jackson, and M. P. Hooker, J. Chem. Phys.
      46, 3025 (1967).

2.   T. W. Haas, Surf. Sci. 5, 345 (1966).

3.   T. W. Haas and A. G. Jackson, J. Chem. Phys. 44, 2921 (1966).

4.   J. W. May and L. H. Germer, J. Chem. Phys. 44, 2895 (1966).

5.   L. H. Germer and J. W. May, Surf. Sci. 4, 452 (1966).

6.   J. E. Boggio and H. E. Farnsworth, Surf. Sci. 1, 399 (1964);
      Surf. Sci. 3, 62 (1964).

7.   J. W. May, L. H. Germer, and C. C. Chang, J. Chem. Phys. 45,
      2383 (1966).

8.   T. W. Haas, to be published.

9.   H. E. Farnsworth and K. Hayek, Supp. Nuovo Cimento V, 451 (1967).

10.   K. Hayek, H. E. Farnsworth, and R. L. Park, Surf. Sci. 10,
      429 (1968).

11.   N. Taylor, Surf. Sci. 2, 544 (1964).

12.   P. J. Estrup and J. Anderson, Surf. Sci. 8, 101 (1967).

13.   P. J. Estrup and J. Anderson, J. Chem. Phys. 45, 2254 (1966).

14.   P. J. Estrup and J. Anderson, J. Chem. Phys. 46, 567 (1967).

15.   J. Anderson and W. E. Danforth, J. Franklin Inst. 279, 160 (1965).

16.   W. T. Sproull, Phys. Rev. 43, 516 (1933).

17.   C. C. Chang and L. H. Germer, Surf. Sci. 8, 115 (1967).

18.   E. G. McRae, J. Chem. Phys. 45, 3258 (1966).

19.   A. G. Jackson, to be published.

20.   E. Merzbacher, Quantum Mechanics, (John Wiley and Sons, Inc.,
      New York, 1961),

21. I. H. Khan, J. P. Hobson, and R. A. Armstrong, Phys. Rev. $\underline{129}$, 1513 (1963).

22. E. J. Scheibner and L. N. Tharp, Surf. Sci. $\underline{8}$, 247 (1967).

23. R. E. Weber and W. T. Peria, J. App. Phys. $\underline{38}$, 4355 (1967).

24. P. W. Palmberg and T. N. Rhodin, J. App. Phys. $\underline{39}$, 2425 (1968).

25. L. A. Harris, J. App. Phys. $\underline{39}$, 1419 (1968).

26. C. J. Smithells, Metals Reference Book (Butterworth, Inc., Washington, D.C., 1962) Vol. I.

27. J. Niebuhr, J. Less. Comm. Met. $\underline{10}$, 312 (1962).

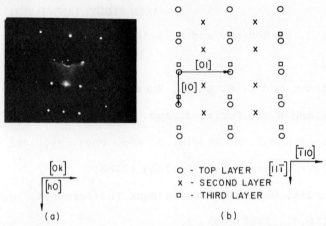

O - TOP LAYER
x - SECOND LAYER
□ - THIRD LAYER

(a)   (b)

Fig. 1. (a) Diffraction pattern taken at 165v near normal incidence from the
clean Ta(112) surface. The (0k) and (h0) directions of this reciprocal
lattice are indicated below the pattern. All photographs are oriented
in this same way. (b) Outline drawing of arrangement of tantalum atoms
in the direct lattice. Crystal directions in the bulk are indicated
below the drawing, and the unit cell used to index beams is outlined.

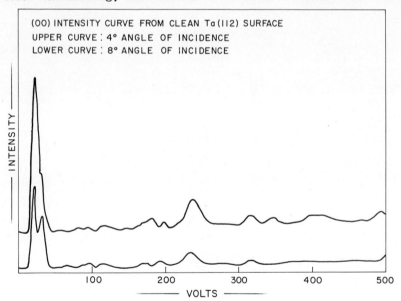

Fig. 2. Intensity distribution curves from the clean Ta(112) surface for the
(00) beam at two angles of incidence (angle between the incident beam
and the inward surface normal).

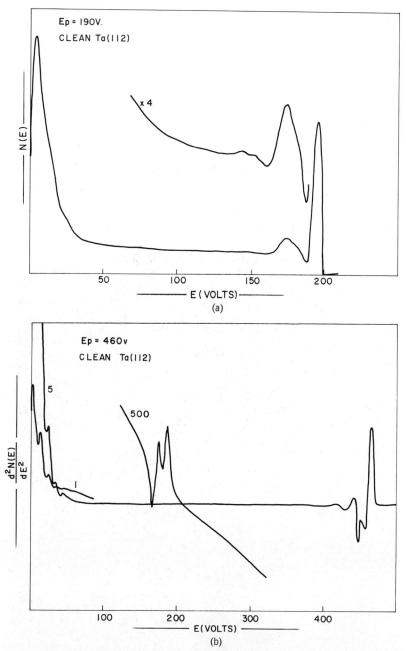

Fig. 3. (a) Energy distribution curve of inelastically scattered electrons
from clean Ta(112) surface. Incident beam at 190v. Top curve is fourfold amplification of lower to show plasmon peaks in more detail. (b)
Derivative of the energy distribution curve for 460v primary. Numbers
in figure represent gains with the highest gain being 500:1.

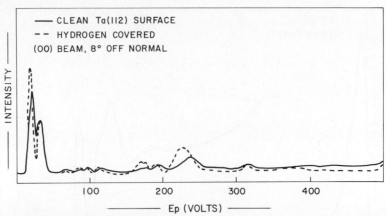

Fig. 4. Intensity distribution curve of (00) beam from clean and hydrogen covered Ta(112) surface.

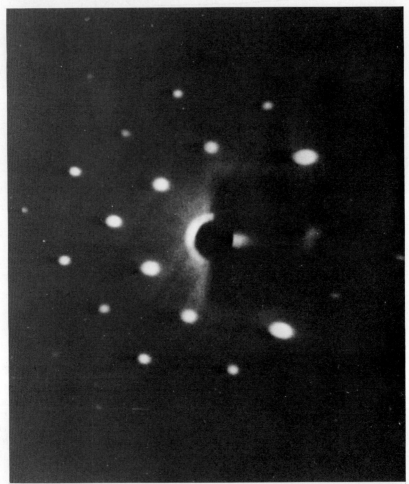

Fig. 5. Diffraction pattern resulting from reaction of clean Ta(112) surface with nitrogen at 500°C. The specularly reflected (00) beam is the first spot directly below (22°) the gun barrel in the (h0) azimuth. Photograph taken at 30v.

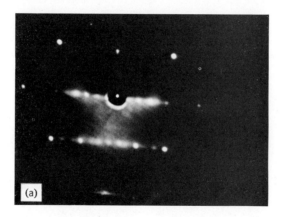

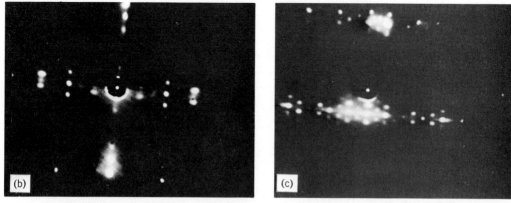

Fig. 6. Diffraction patterns resulting from progressive stages in the oxidation
of the clean Ta(112) surface. a) Initial pattern observed, a (1 x 3)
oxide 156v. b) Intermediate stage in oxidation, apparently an (8 x 4)
pattern. 51v. c) Highest coverage oxide pattern 54v. None of these
patterns show any sign of faceting due to oxidation.

SURFACE SELF-DIFFUSION ON NICKEL (111) BY RADIOACTIVE TRACERS*

J. R. Wolfe** and H. W. Weart***

INTRODUCTION

The thermal motion of individual atoms at the surface of a solid is probably the simplest atomic kinetic process that occurs there. The mechanisms and rates of the surface diffusion process remain indefinite, because of the experimental difficulties involved in producing and maintaining surfaces of the necessary degree of cleanliness on the one hand, and because of inadequate knowledge of surface structure on the other. The experiments described here were done to bring the recent substantial progress in clean surface technology and the accompanying improvement in surface structural knowledge to bear on the problem of the mechanism and kinetics of surface diffusion. More specifically, it was hoped that new data on atomic movement on well-characterized surfaces would aid in identifying those structural features that determine diffusion rates in the same way that lattice diffusion measurements aided in identifying the lattice defects responsible for mass transport within the solid.

This objective dictated careful consideration of three major

* Contribution No.43 of the Graduate Center for Materials Research, University of Missouri, Rolla, Missouri.
** AVCO Corporation, Electronics Division, Tulsa, Oklahoma.
*** Dept. of Metallurgical & Nuclear Eng., University of Missouri, Rolla, Missouri.

questions, namely, the technique to be used for measuring surface diffusivity, the material to be used and the method of characterizing the surface structure.

A number of techniques have been devised to measure the surface diffusivity D_S of crystalline materials. These include mass transport studies, field ion and field electron microscopy, and radioactive tracer methods.

Mass transport studies involve measuring a change in surface morphology due to surface energy gradients. By measuring the rate of decay of a perturbed surface, D_S can be calculated using the theory of Mullins[1]. An inherent difficulty with this technique is that curved surfaces are necessary. This makes it difficult to correlate the kinetic data with the surface structure. Furthermore, mass transport studies can only be used to study self-diffusion at the present time. Recent advances by two groups of investigators [2,3] have resulted in techniques that hopefully overcome the surface cleanliness problem.

Several methods for measuring surface diffusion coefficients utilizing the field ion or field electron microscope are available. Unlike other methods, surface cleanliness is no major problem as experiments are typically conducted in a 10^{-12} torr vacuum. The nature of the surface is difficult to specify, however, as the diffusion takes place on an emitter tip having a radius of approximately 2500 angstroms. It is possible to measure impurity surface diffusion coefficients, but the choice of surfaces is limited.

The radioactive tracer method of determining surface diffusivities, which was chosen for this study, is unique in several respects. First, the surfaces employed are as flat and smooth as it
is experimentally possible to make them. Secondly, the method can
be used for either impurity or self-diffusion determinations, within the obvious limitation of tracer availability. Until recently,
tracer studies were subject to the surface cleanliness and characterization problem. However, the availability of ultra-high vacuum
technology and low energy electron diffraction (LEED) makes it possible to work on characterized surfaces.

Once the tracer method had been chosen, the two remaining
questions concerning material and surface characterization, which
are interrelated, were readily settled. Nickel, besides having an
experimentally convenient isotope, has been extensively studied by
LEED, and its low-index planes are known to retain the atomic arrangement characteristic of the bulk all the way to the melting
point[4].   LEED surface characterization was considered adequate
for the present study, because it will detect gross contamination
and most changes in surface topography, including faceting.

The main precaution necessary in using the tracer method is
properly accounting for the various losses of tracer from the surface as illustrated in Figure 1. Shewmon[5] has emphasized the
importance of correcting for loss by diffusion into the volume and
has worked out the necessary mathematical apparatus. Loss by pipe
diffusion can be treated similarly, but the magnitude of this cor-

rection is negligible in most cases, including the present work [6].

The importance of correcting for evaporation loss appears to have gone unrecognized heretofore, but was very evident in this study. A complete analysis of the loss and a method for estimating it is reported elsewhere [7]. It will suffice to note here that evaporation loss can be treated in a mathematically similar way to loss into the bulk.

Most importantly, the estimate [7] shows that evaporation loss can become significant at temperatures as low as 0.5 T_M, depending on the material and the experimental conditions. Direct evidence of tracer evaporation was observed early in the work when difficulty was encountered in trying to establish measurable tracer spreading, although estimates indicated that the tracer should have spread over a considerable area. The time and temperature of the diffusion anneal were increased with little success until, at 1000°C, a significant amount of nickel was deposited on the chamber interior during the anneal. Measurable profiles were finally obtained by annealing below 850°C. The effect of evaporation on tracer spreading is shown in Figure 2. Two experimentally determined activity profiles, for different temperatures but equal diffusion times, are superimposed to illustrate the effective compression of tracer spreading at the higher temperature due to evaporation.

The aforementioned method of estimating evaporation loss, which is compatible with the surface diffusivity mechanism supported by the present study, was used to correct all data reported later.

Previous Studies

The work of Nickerson and Parker[8] on silver in 1951 can be credited with beginning the modern era of surface diffusion studies. This work, which employed radioactive tracers, reported a relatively small surface diffusion activation energy of approximately 10 kcal. or $1/7\Delta H_s$, the heat of sublimation.

In 1957, Mullins[1] published the first of a series of theoretical papers that make possible an alternative method of measuring surface diffusion coefficients, by mass transport   The majority of the data reported since then has been obtained by this technique, although tracer work has continued (see for example refs. 9, 10, 11, 12).

As data accumulated, a rather broad spread of activation energies Q_S (0.1 to 0.7 of the heat of sublimation) and Arrhenius pre-exponential factors D_O (10^{-7} to 10^{+6} cm^2/sec) became apparent and suspicion grew that more than one mechanism of surface diffusion was likely.   In analyzing some of the mass transport data, for example, Gjostein[13] found a break in the curve of log D_S against T/T_M (T_M is the absolute melting temperature) for selected systems which he interpreted to indicate two surface diffusion mechanisms, characterized by the following parameters:

high temperature
mechanism

$Q_S = 30T_M$ kcal/mol

$D_O = 740$ cm^2/sec } $T/T_M \gtrsim .75$

low temperature
mechanism

$Q_S = 13$ T_M

$D_O = .014$ cm^2/sec } $T/T_M \lesssim .75$

Using the terrace-ledge-kink (TLK) surface model Gjostein[13] proposed an adatom mechanism at higher temperatures and a vacancy
mechanism at lower temperatures.

Birchenall and Williams[14] have noted that while a series of
mechanisms operating in different temperature ranges could account
for some of the existing anomalies in the literature, impurity effects and methods of data analysis could also explain the non-
linearity of Arrhenius plots and widely varying activation energies.
The surfaces on which most of the experiments to date have been
done were poorly cleaned and completely uncharacterized, and can
in fact, usually be shown to be contaminated.

The most serious error in data analysis has been the failure
to account for tracer losses, as noted earlier. Choi and Shewmon[9]
were the first to carry out experiments in which loss into the bulk
was taken into account. Williams[15] recently presented an analysis
showing that tracer spreading from a depletable source will be observable only if $D_V/D_S < 10^{-9}$. He claimed that such trapping could
also account for his failure to observe spreading on alpha iron
from a constant activity source, but he took no note of the possibility that evaporation losses could cause the same effect.

Surface diffusion on nickel has been studied by tracer, mass
transport and field emission microscopy. The results of these
studies are summarized in Table I, where footnotes give some pertinent additional facts. Maiya and Blakely[2] theorize that the
lower Q_S values reported by other investigators are due to impurities. Bonzel and Gjostein[3] observed a discontinuity in the

Arrhenius plot at 800°C, which they believed to be due to faceting. They also suspected carbon surface contamination, based on LEED results.

EXPERIMENTAL PROCEDURES

The specimens used in this study were spark cut from a nickel single crystal, having a nominal purity of 99.999% and oriented within 1 degree of the (111) plane.   Platelets approximately 10 millimeters by 5 millimeters by 0.5 millimeters thick were prepared by grinding, followed by metallographic polishing.   The worked surface was removed by a final chemical polish in hot acid solution ($50\%HAC-30\%HNO_3-10\%H_2SO_4-10\%H_3PO_4$) resulting in a bright, smooth surface and a final thickness of 0.3 millimeters.   Approximately 40 0.010" diameter wires of 99.999% Ni were spot welded to each end of the specimen for current leads.   Temperature control and monitoring was accomplished by two Pt-Pt, 10% Rh thermocouples spot welded to the back of the sample.   A typical specimen is shown in Figure 3.

These experiments utilized a point source of constant radioactive strength.   The source was made by etching a 99.999% Ni needle to an 0.1 mm tip, electroplating the tip with carrier free Ni^{63}, to an activity of approximately 100,000 cpm, and finally annealing the needle in ultra-high vacuum to homogenize the activity in the tip.   It was found that each experiment only depleted the source approximately 1000 cpm and therefore the same point source was used in all experiments. The tip was cleaned by ion bombardment before each experiment.

Figure 4 shows the experimental arrangement. The specimen was mounted on a universal motion feedthrough and could be rotated to various positions for cleaning, characterization, and diffusion annealing. The samples were heated in-situ by direct resistance heating using alternating current. Temperature control was ±5°C.

LEED was used at various stages of a surface diffusion experiment to check the character of the surface. These experiments confirm Germer and co-workers[19] observation that high temperature outgassing in ultra-high vacuum alone will not usually produce a clean nickel surface. A clean nickel (111) surface could be obtained, however, by using argon ion bombardment (350 volt, .3μ amp/cm^2, 60 min.) followed by heating to 1000°C in ultra-high vacuum.

In the early experiments an impurity, believed to be carbon, was occasionally observed (by LEED) on the surface after the diffusion anneal. It was found that this contamination could be prevented by following the ion bombardment with high temperature anneals in oxygen, hydrogen and ultra-high vacuum in that order. Table II summarizes the surface cleaning procedure and Figure 5 shows LEED patterns taken at various stages of the surface cleaning.

In a typical experiment the surface of the sample was first cleaned, annealed, characterized (by LEED), and heated to the diffusion temperature. The radioactive tip was then brought into contact with the hot surface. After the diffusion anneal the radioactive needle was removed, the surface again characterized by LEED, and the sample finally removed from the vacuum chamber for analysis.

The diffusive spreading was measured by autoradiography using Kodak Type A autoradiographic plates. Activity versus distance profiles were obtained by scanning the autoradiographs with an automatic recording microdensitometer having a slit width of 10 microns. The measurements were conducted under conditions that guaranteed that film darkening at any point is proportional to the total surface and bulk activity at the corresponding specimen position[20].

Figures 6, 7, and 8 show, respectively, a typical indentation made by a point source during a surface diffusion experiment, the corresponding autoradiograph of tracer spreading from the source, and the activity profile taken from the autoradiograph.

RESULTS AND DISCUSSION

A. Data Analysis

The differential equation describing the surface tracer concentration, C_s^* (ρ, t), as a function of time (t) and distance (ρ) in cylindrical coordinates for a point source is:

$$\frac{\partial C_s^*}{\partial t} = D_s \left(\frac{\partial^2 C_s^*}{\partial \rho^2} + \frac{1}{\rho} \frac{\partial C_s^*}{\partial \rho} \right) - \frac{J_{D_v}^*}{\delta} - \frac{J_E^*}{\delta} \tag{1}$$

where $J_{D_v}^*$ and J_E^* are tracer volume and evaporation loss fluxes, respectively, D_s the surface diffusion coefficient and δ the surface layer thickness.

Assuming for the moment that $J_{D_v}^*$ and J_E^* are calculable, it is possible in principle to solve Equation 1 exactly, although extraction of D_s from the solution is difficult. On the other hand, several approximate solutions to the combined surface diffusion-parasitic loss problem are available that greatly simplify the data analysis. For the present study Shewmon's[5] approach based on Fisher's[2] classical solution to grain boundary diffusion is appropriate. According to LeClaire[22], and substantiated by the calculations of Suzuoka[23], the Shewmon (or Fisher) approach can be used with confidence providing B > 10, where B is given by:

$$B = \frac{1}{2} \frac{D_s \delta}{D_v (D_v t)^{\frac{1}{2}}} \tag{2}$$

where D_V is the volume diffusion coefficient and t the anneal time. In the present work B is typically 10^5 and the Shewmon method should be valid.

Shewmon's method leads to the following relationships:

$$J^*_{D_V} = \frac{2C^*_s(\rho,\tau)D_V}{(\pi D_V t)^{\frac{1}{2}}} \quad atoms/cm^2/sec \tag{3}$$

$$\frac{\partial C^*_s(\rho,\tau)}{\partial t} = 0 \tag{4}$$

where $C^*_s(\rho,\tau)$ is the surface tracer distribution after some anneal-ing time $t = \tau$.

Wolfe and Weart[7] have calculated J^*_E for the (111) plane of an FCC metal, assuming a TLK surface structure and using nearest neighbor bond approximations, (i.e. $6\theta = \Delta H_s$, where θ is the bond energy and ΔH_s the heat of sublimation). For J^*_E they derive the following:

$$J^*_E = \nu\delta C^*_s \lambda_1 e^{-(\frac{5}{6}\Delta H_s)/RT} \quad atoms/cm^2/sec \tag{5}$$

where λ_1 is the average number of interatomic distances between ledges, and ν the vibrational frequency.

Substituting equations 3, 4, and 5 into equation 1, the fol-lowing is obtained:

$$\frac{\partial^2 C^*_s}{\partial\rho^2} + \frac{1}{\rho}\frac{\partial C^*_s}{\partial\rho} - \alpha^2 C^*_s = 0 \tag{6}$$

where α^2 is given by:

$$\alpha^2 = \frac{2}{D_s \delta} \left[\frac{D_v}{(\pi D_v \tau)^{\frac{1}{2}}} + \frac{\lambda_1 \vartheta \delta}{2} e^{-\frac{5}{6}\Delta H_s/RT} \right] \tag{7}$$

Equation 6 is a modified Bessel equation of order zero, which for a point source of radius b and constant concentration C_o has the solution:

$$C_s^* (\rho, \tau) = \frac{C_o K_o (\alpha \rho)}{K_o (\alpha b)} \tag{8}$$

Noting that $K_o(\alpha b)$ is a constant and using the asymptotic expansion of $K_o(\alpha \rho)$, which for values of $\alpha \rho \geq 1$ can be approximated by:

$$K_o (\alpha \rho) = \left(\frac{\pi}{2\alpha \rho}\right)^{\frac{1}{2}} e^{-\alpha \rho} \tag{9}$$

Equation 8 can finally be written:

$$C_s^* (\rho, \tau) = \frac{ke^{-\alpha \rho}}{\rho^{\frac{1}{2}}} \tag{10}$$

with k given by:

$$k = \frac{C_o \left(\frac{\pi}{2\alpha}\right)^{\frac{1}{2}}}{K_o (\alpha b)} \tag{11}$$

Autoradiographic methods generally measure the total (surface plus volume) activity at any given point on the surface. The total activity A_T would be:

$$A_T (\rho, \tau) = C_s^* \delta + \int_o^\infty C_v^* (\rho, z, \tau) \, dz \tag{12}$$

where z is the distance coordinate into the volume and $C_V^*(\rho,z,\tau)$ the volume tracer distribution. The last term in Equation 12 according to Shewmon[5] is given by:

$$\int_0^\infty C_V^* (\rho,z,\tau) \, dz = C_S^* (\rho,\tau) \int_0^\infty \text{erfc} \left[\frac{z}{2(D_V\tau)^{\frac{1}{2}}}\right] \, dz \tag{13}$$

Thus, the expression for the total activity A_T becomes:

$$A_T(\rho,\tau) = \delta C_S^* + 2(D_V\tau)^{\frac{1}{2}} C_S^* \tag{14}$$

Since the $(D_V\tau)^{\frac{1}{2}}$ is typically 1×10^{-5} cm and δ is of the order of 2×10^{-8} cm, the first term can be ignored, resulting in:

$$A_T = 2 (D_V\tau)^{\frac{1}{2}} C_S^* = 2 (D_V\tau)^{\frac{1}{2}} ke^{-\alpha\rho}/\rho^{\frac{1}{2}} \tag{15}$$

Consequently, a plot of logarithm $A_T\rho^{\frac{1}{2}}$ versus ρ should be linear with a slope equal to α. The surface diffusivity D_S is found through use of the slope α, and Equation 7.

B. Surface Diffusivity Data

Surface diffusion experiments were made over the temperature range of 614°C to 840°C. Table III lists pertinent information concerning each experiment. In most cases it was possible to characterize the surface by LEED immediately before and after the diffusion anneal. Figure 9 shows before and after LEED photographs of several experiments illustrating both clean and contaminated surfaces. Below 700°C and in ultra-high vacuum it was generally found that the nickel surface could be kept clean of observable (according to

LEED) impurities for hundreds of hours. Figure 9-d, a LEED pattern of a surface held at 614° for 197 hours, shows very little
degradation compared with the original clean surface. However,
at higher temperatures extra diffraction spots, attributed to an
impurity, were occasionally observed in the final LEED patterns
after diffusion. Figure 9-b shows a typical impurity pattern observed after a 795°, 41-hour diffusion anneal. The impurity structure could usually be removed by repeating the entire surface cleaning sequence several times. The behavior of the contamination and
the experimental conditions indicate the contaminant to be carbon
diffusing from the bulk to the surface.

Activity profiles plotted as logarithm $A_T \rho^{\frac{1}{2}}$ versus ρ are
shown in Figure 10. They are linear as predicted by equation 15.
The surface diffusivities were calculated by measuring the slope
of the profiles and equating the slope with the parameter α given
in Equation 7. The following values of the parameters appearing
in Equation 7 were used to calculate the surface diffusion coefficients appearing in Table IV.

$$D_V = 1.27 \ e^{-66,900/RT} \text{cm}^2/\text{sec} \qquad (24)$$

$$\Delta H_S = 100.0 \text{ kcal/mol}$$

$$\delta = 2.5 \times 10^{-8} \text{ cm}$$

$$\nu = 1 \times 10^{13} \text{ sec}^{-1}$$

The surface diffusion coefficients shown in Table IV were calculated both with and without the evaporation correction $(-J_E/\delta)$ included. A plot of D_S versus $1/T$ for both cases is shown in Figure 11.

An error limit of ±16% was established for each D_s with the exception of the data point at 840°C, which had a considerably greater error limit due to excessive tracer evaporation. The error limit was determined by allowing a ±3% error in measuring the slope of the activity profile, a ±5°C uncertainty in the temperature measurement and a ±5% miscellaneous error. A least squares analysis of the data, ignoring the 840°C point, shows that the data can be represented by the following Arrhenius type relationship:

$$D_s = 300 \ exp[-(38\pm4 \ kcal/mol)/RT] \ cm^2/sec. \tag{16}$$

This relationship is shown as a straight line in Figure 11. Weighting the 840°C D_s value one-half as much as the other values in the least squares analysis increases the activation energy in equation 16 to $(39 \pm 4) \times 10^3$ kcal/mol and the pre-exponential factor to 700.

The tracer spreading on the nickel (111) plane was not isotropic. The elliptical nature of the spreading can be seen in Figure 7 and in the over-exposed autoradiograph shown in Figure 12. The diffusion coefficients tabulated in Table IV and plotted in Figure 11 were calculated from profiles taken along the long axis of the ellipse. Diffusion coefficients calculated from the short axis data were approximately 25% smaller than those reported. Figure 14 shows actual activity profiles for sample number 3 taken along the long and short axis.

Limited evidence, based on one experiment, suggests that surface contamination increases the measured surface diffusivity by as much as 30% and tends to eliminate the bisymmetric nature of the tracer spreading.

Some evidence was produced showing diffusion on non-close-packed planes to be considerably slower than diffusion on the close-packed (111) plane. Experiment number 6 was inadvertently conducted in a recrystallized area that according to LEED was not a low indices plane. The tracer spreading was much less than that under similar circumstances on the (111) plane and was symmetrical, as evidenced by the autoradiograph shown in Figure 13. The calculated diffusivity was 50% lower than the value obtained on the (111) plane. Additional evidence was seen in experiment number 1. In this experiment the point source was located quite close to a grain boundary on a partially recrystallized surface, as shown in Figure 6. The corresponding autoradiograph (Figure 7) shows that diffusion on the (111) side of the boundary was normal, but diffusion on the recrystallized side was greatly restricted.

C. Mechanism of Surface Diffusion on Nickel (111)

The most probable mechanisms to consider for self-diffusion on the close-packed planes of FCC metals are the adatom and vacancy mechanisms. It is proposed here that the adatom mechanism is primarily responsible for surface self-diffusion on nickel (111) over the temperature range of the present investigation, 614°C to 840°C. The single activation energy is indicative of a single transport mechanism, however, the limited temperature range precludes a more positive statement concerning the possibility of competing mechanisms as suggested by Gjostein[13]. It is suggested that the surface diffusion activation energy of 38 kcal/mol is associated with the transport mechanisms and potential energy diagram shown in Figure 15. ΔG_{36} and ΔG_{35} are the energy differences between terrace sites and kink and ledge sites, respectively; and ΔG_{3M3} is the activation energy for adatom diffusion on the terrace. Tracer atoms at kink (n=6) or ledge (n=5) sites jump to terrace sites (n=3), cross the terrace by terrace jumps, and are trapped at the next ledge.

The vacancy mechanism is discounted for several reasons. It is unlikely that tracer evaporation loss would have been observed if a vacancy mechanism were dominant. Tracer evaporation from the surface layer requires a net energy change of 9 bonds, compared with 5 for adatom evaporation. Recent calculations by Wynblatt and Gjostein[25] predict that the activation energy for vacancy self-diffusion on the Cu (111) surface is 50% greater than the adatom activation energy. They further suggest that since D_o for the adatom mechanism should always be greater than D_o for the vacancy mechanism, the single adatom mechanism should dominate.

The activation energy of 38 kcal/mol is in agreement with the predictions of a modified nearest neighbor bond energy relationship[32] as applied to the suggested adatom mechanism. Assuming the activation energy to be the energy to move the tracer

*"n" is the number of nearest neighbor atoms.

atom out of the kink or ledge trap and across the terrace (Figure 15), Q_s is given by:

$$Q_s = Q_{3M3} + \begin{cases} Q_{36} \text{ for kink site traps} \\ Q_{35} \text{ for ledge site traps} \end{cases}$$

Assuming the heat of sublimation ΔHs is $6\theta_6$ where θ_6 is the nearest neighbor bond energy for 6 neighbors[32], $Q_{35} = 31$ kcal and $Q_{36} = 42$ kcal for Ni (111). Q_{3M3} has been estimated at $\theta_6/3$ [26-28], but a recent calculation for Cu[25] gives $\theta_6/30$. Thus, Q_{3M3} for Ni (111) may vary from 0.6 to 6, which allows Q_s to range from 43 to 48 kcal/mol for kink site trapping, and from 32 to 37 kcal/mol for ledge site trapping. The 38 kcal/mol measured Q_s is in the general range for both trap sites, but its accuracy (±4 kcal/mol) will not permit any firm statement about which is preferred or on the relative contribution each site may make to trapping.

It is recognized that the values of the evaporation flux loss used to correct the higher temperature diffusivities are estimates owing to the assumptions involved in the theory. The inaccuracy of the corrections is not great enough to cause major changes in the reported activation energy. Except for the 840°C run, which was over diffused, the evaporation loss was minimized and the activation energy calculated from the uncorrected diffusivities is within a few kcal of the reported value. On the other hand, if the evaporation flux was significantly underestimated it would not have been possible to observe tracer spreading on the 840° run.

The large value of D_O found in this study is in fair agreement with the D_O proposed for adatom diffusion by Gjostein[13]. The arguments he puts forth to justify his prediction are pertinent to the present study, but will not be repeated here.

The bisymmetric nature of the tracer spreading is not understood at this time, but is very probably associated with the surface structure and the surface diffusion mechanism. It is possible that the surface is vicinal and that either the long or short axis of the ellipse is associated with the mean ledge direction. Electron microscopy revealed microscopic ledges in several areas on the surface, but monatomic ledges are, of course, beyond the resolution capabilities of the electron microscope.

The restricted mobility and the symmetrical nature of the tracer spreading on non-close-packed surfaces are further evidence of the importance of surface characterization in surface diffusion studies. These observations also tend to support the proposed adatom diffusion mechanism on a surface containing traps. It is likely that an adatom moving on a rough, open, complex surface would have its mobility sharply reduced, compared to its mobility on the relatively smooth close-packed TLK surface. The change from the bisymmetric to symmetric spreading on going from the close-packed to the complex surface would be expected for an adatom mechanism.

Direct correlation of the results of this study with previous investigations is not possible due to gross dissimilarities in the experiments. It is observed, however, that the results reported in the present study do not readily agree with Gjostein's[13] general correlation and dual mechanism hypothesis discussed in the Introduction. The activation energy determined in this study equal to $22T_M$ kcal/mol and the D_o equal to 300 seem to fall between Gjostein's predictions.

D. Low Energy Electron Diffraction Results

The primary use of low energy electron diffraction in this work was to characterize the surface on which the diffusion took place. However, a number of interesting, and as yet unreported, features of the nickel (111) surface were observed by LEED and are discussed below.

The majority of the surface diffusion data reported in the literature was obtained from experiments conducted in a hydrogen atmosphere. Figure 16-a,b shows the effect of a high temperature hydrogen anneal on an initially clean, well ordered nickel (111) surface. The present state of the art concerning LEED pattern interpretation precludes defining the nature of the surface assoc-iated with this complex pattern. It can be concluded, however, that the original well ordered (111) surface has undergone a com-plex reconstruction. Once established, the reconstructed surface was stable from ambient temperatures to 1000°C, the formation tem-perature, but could be removed by an ultrahigh vacuum anneal above 1000°C.

A number of experiments were made involving the diffusion of gold on nickel. The difficulties were such that no meaningful diffusion data were obtained; however, several experiments re- sulted in appreciable gold spreading according to LEED. Figure 16 c,d shows the nickel (111) surface before and after a gold on nickel diffusion anneal at 900°C for 100 hours. The resulting LEED pattern is similar to LEED patterns obtained by Feinstein[29] for silver monolayers deposited on nickel (111). Calculation by Feinstein[30] of the gold nickel surface lattice parameter ratio from Figure 16-d showed the measured value to be within 3% of theoretical. Feinstein [30] also claims that all of "extra" diffraction spots in Figure 16 -d can be accounted for by multiple scattering between the gold layer and nickel substrate.

An apparent faceting of the nickel (111) surface was observed after an ultra-high vacuum anneal at 1000°C for 35 minutes. The effect is shown in Figure 16-e, f. A similar phenomenon was ob- served on UO_2 by Ellis[31] who has shown the resulting "shadow" spots can be accounted for if the surface has a stepped structure with an approximately constant interledge spacing. It is also possible, however, that the shadow spots could be due to an im- purity or some unknown structural feature. The shadow spots were seen on other samples after high temperature anneals, but were not usually as distinct as those in Figure 16-f.

By proper grounding of the crystal heater power supply, it was possible to observe LEED patterns while heating and cooling the

sample. During the heating cycle a strange rotation of the diffraction spots was observed near 360°C. The reverse rotation
(back to the original positions) was observed at the same temperature upon cooling. This rotation phenomenon was recorded by photo-
graphing the LEED pattern during the rotation. Figure 17 shows the
resulting LEED pattern. The initial pattern is the stepped structure pattern shown previously in Figure 16-f. It can be seen that
each pair of spots has moved a different amount. A vector representation of the rotation is shown in Figure 18. When viewed
with the eye the rotation appears to occur quite suddenly, as
though snapping into place. The temperature at which the rotation
occurs agrees closely with the Curie point of nickel (358°C).

It is concluded that the rotation is associated with magnetic
properties of the surface.

The use of LEED in the present investigation confirms the previously suspected fact that surface characterization should be a
very important factor in any surface diffusion study. The apparent structural difference between a clean nickel (111) surface
and the same surface annealed in hydrogen as shown in Figure 16-a, b
is strong evidence in support of more and better surface characterization. The fact that even an otherwise "clean surface" can
become contaminated by impurities from within the specimen as
in experiment number 8 is additional proof of the necessity for
characterizing the surface before and after a diffusion anneal.

Although the use of LEED to characterize the surfaces of interest is a step forward in surface diffusion studies, it is not enough, because LEED is unable to supply information concerning such structural aspects of the surface as the ledge density, the average interkink distance, the vacancy concentration in the surface, and the presence of small monolayer fractions of impurities.

CONCLUSIONS

From the experiments reported here, the following conclusions may be drawn regarding Ni(111) surfaces over the temperature range 614–840°C.

1. Surfaces that appear clean and well-ordered to LEED inspection can be produced by appropriate combinations of ion bombardment and annealing in selected atmospheres including ultra-high vacuum, and can be maintained clean to LEED inspection for up to 200 hours at temperature.

2. Measured tracer diffusivities must be corrected for loss by evaporation, as well as for loss by diffusion into the bulk. A compromise must be sought between the reduced accuracy of D values if diffusion times are shortened to reduce evaporation on the one hand, and the added uncertainty of a large evaporation correction if diffusion times are lengthened in

an effort to increase tracer spreading on the
other.

3. Ni^{63} diffuses across the surface at a rate determined by:

$$D_s = 300 \exp[-(38\pm4 \text{ kcal/mol})/RT] \text{ cm}^2/\text{sec}.$$

4. The measured Q_s and D_o are most consistent with a single surface diffusion mechanism in which adatoms move with relative ease on the terraces of TLK surface, but are trapped at ledge sites.

ACKNOWLEDGMENTS

The authors gratefully acknowledge the support of AVCO Corporation, Electronics Division, who provided an assistantship held by one of us (JRW) and equipment; and the Graduate Center for Materials Research, who provided facilities and experimental equipment.

REFERENCES

1. Mullins, W. W.: J. Appl. Phys., **28** (1957) 333.

2. Maiya, P. S. and Blakely, J. M.: J. Appl. Phys., **38** (1967) 698.

3. Bonzel, H. P. and Gjostein, N. A.: "Diffraction Theory of Sinusoidal Gratings and Application to In-Situ Surface Self-Diffusion Measurements", Publication Preprint Ford Scientific Laboratory, Dec. 1967.

4. Morabito, J. M. and Somorjai, G. A.: Journal of Metals, **20** (1968) 17.

5. Shewmon, P. G.: J. Appl. Phys., **34** (1963) 755.

6. Blakely, J. M.: "Surface Diffusion" from Progress in Materials Science, Vol. 10 (1963) 396.

7. Wolfe, J.R. and Weart, H.W.: to be published.

8. Nickerson, R. A. and Parker, E. R.: Trans. ASM, **42** (1950) 376.

9. Choi, J. Y. and Shewmon, P. G.: Trans. AIME, **230** (1964) 123.

10. Shewmon, P. G. and Choi, J. Y.: Trans. AIME, **230** (1964) 449.

11. Geguzin, Y. E. and Kovalev, G. N.: Soviet Physics Solid State (English Trnsl.), **5** (1963) 1227.

12. Geguzin, Y. E., Kovalev, G. N. and Ratner, A. M.: Physics of Metals and Metallography (USSR)(English Trnsl.), **10** (1960) 45.

13. Gjostein, N. A.: "Surface Self-Diffusion in FCC and BCC Metals: A Comparison of Theory and Experiment", Surfaces and Interfaces I, Edited by J. J. Burke, N. L. Reed, V. Weiss, Syracuse University Press, Syracuse, New York, 1967.

14. Birchenall, C. E. and Williams, J. M.: "Surface Diffusion on Nearly Pure Metallic Surfaces", from Fundamental Phenomena in the Materials Sciences, Vol. 3, Edited by L. J. Bonis, P. L. deBruyn, J. R. Duga, Plenum Press, New York, 1966.

15. Williams, J. M.: Ph.D. Thesis, University of Delaware, Dept. of Chemistry, 1966.

16. Blakely, J. M. and Mykura, H.: ACTA Met, **9** (1961) 23.

17. Pye, J. J. and Drew, J. B.: Trans. Met. Soc. AIME, 230 (1964) 1500.

18. Melmed, A. J.: J. Appl. Phys., 38 (1967) 1885.

19. Germer, L. H., MacRae, A. V., and Hartman: J. Appl. Phys., 32 (1961) 2432.

20. Wolfe, J. R.: Ph.D. Thesis, University of Missouri-Rolla, Dept. of Metallurgical and Nuclear Engineering, 1968.

21. Fisher, J. C.: J. Appl. Phys., 22 (1951) 74.

22. LeClaire, A. D.: British Journal of Applied Physics, 14 (1963) 351.

23. Suzuoka, T.: J. Phys. Soc. of Japan, 20 (1965) 1259.

24. Hoffman, R.E., Pikus, F. W., and Ward, R. A.: Trans. AIME, 206 (1956) 483.

25. Wynblatt, P., and Gjostein, N. A.: "A Calculation of Relaxation, Migration and Formation Energies for Surface Defects in Copper", Publication Preprint, Scientific Laboratory, Ford Motor Company, 1968.

26. Hirth, J. P. and Pound, G.M.: J. Chem. Phys., 26 (1957) 1216.

27. Burton, W. K.; Cabrera, N., and Frank, F. C.: Phil. Trans. Royal Soc., 243A(1950) 299.

28. MacKenzie, Thesis, University of Bristol, United Kingdom, 1956.

29. Feinstein, L. G.: "LEED Studies on Rhenium", Low Energy Electron Diffraction Symposium, American Crystallographic Association Winter Meeting, Tucson, Arizona, Feb. 4-7, 1968.

30. Feinstein, L. G.: Private Communication.

31. Ellis, W. P.: "LASER Techniques in the Interpretation of LEED Patterns", Low Energy Electron Diffraction Symposium, American Crystallographic Association Winter Meeting, Tucson, Arizona, Feb. 4-7, 1968.

32. Schwobel, R. L. : J. Appl. Phys. 38 (1967) 3154

Table I.   Selected Surface Diffusion Data

Plane	Direction	D_o (cm^2/sec)	Q_s (kcal/mol)	Temp. Range °C	Method	Environment	Reference
ycrystalline		5×10^{-4}	14.3	800-1200	Scratch Smoothing	10^{-5} torr*	16
1)(110)(100)		10^{-4} **	13.8	400-1000	Tracer	H_2 ***	17
Random			21.4	237-477	Field Emission	10^{-11} torr	18
(100)	[110]	2.6	35.5	923-1452	Sinusoidal Decay	10^{-8} torr*	2
(110)	[001]	12.8	40.1	923-1452	Sinusoidal Decay	10^{-8} torr*	2
(110)	[1$\bar{1}$0]	23.9	42.7	923-1452	Sinusoidal Decay	10^{-8} torr*	2
(110)		.01	19.0	800-1080	Sinusoidal Decay	U.H.V.	3

nnealled in a nickel enclosure
stimated from reported data
racer loss into the bulk not accounted for

Table II. Typical Surface Cleaning Procedure for
Obtaining a Clean (111) Nickel Surface

Step	Treatment	Temperature °C	Atmosphere Torr	Pattern Fig. No.	Comment
1	outgassing	25	5×10^{-10}	5-a	no structure
2	outgassing-anneal	1000	1×10^{-9}	5-b	contaminated (111) surface
3	argon ion bombarded	25	4×10^{-4}	5-c	degraded surface
4	oxygen anneal	800	4×10^{-4}	5-d	oxygen superlattice
5	hydrogen anneal	1000	4×10^{-4}	5-e	complex surface
6	UHV anneal	1000	5×10^{-10}	5-f	clean nickel (111) surface

Table III. Experimental Information Concerning Surface Diffusion Experiments

Experiment Number	Temperature $^\circ C$	Time Sec.	$(D_v t)^{1/2}$ cm.	Pressure Torr	Surface (LEED)
9	614	7.1×10^5	5.3×10^{-6}	2×10^{-10}	Clean
3	655	3.7×10^5	8.6×10^{-6}	2×10^{-9}	Clean
5	695	4.1×10^5	2.0×10^{-5}	9×10^{-10}	Clean
1	755	3.5×10^5	4.9×10^{-5}	5×10^{-10}	Undetermin
8	795	1.5×10^5	6.1×10^{-5}	7×10^{-10}	Contaminat
10	795	8.3×10^4	4.5×10^{-5}	7×10^{-10}	Clean
2	840	3.5×10^5	1.9×10^{-4}	2×10^{-9}	Clean

Table IV. Tabulated Surface Self-Diffusion Coefficients of Nickel (111)

Experiment Number	Temperature $^\circ C$	D_s (cm^2/sec) uncorrected	D_s (cm^2/sec) corrected for evaporation	Surface (LEED)
9	614	1.39×10^{-7}	1.39×10^{-7}	Clean
3	655	3.62×10^{-7}	3.63×10^{-7}	Clean
5	695	8.08×10^{-7}	8.64×10^{-7}	Clean
1	755	2.17×10^{-6}	2.79×10^{-6}	Undetermine
8	795	5.38×10^{-6}	7.84×10^{-6}	Contaminate
10	795	4.04×10^{-6}	5.43×10^{-6}	Clean
2	840	5.53×10^{-6}	1.47×10^{-5}	Clean

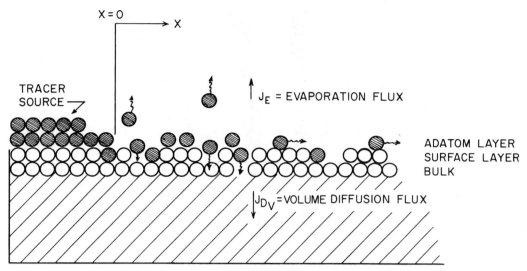

Fig. 1. Cross section of a crystalline solid having a singular surface illustrating tracer evaporation and volume diffusion losses during surface
diffusion.

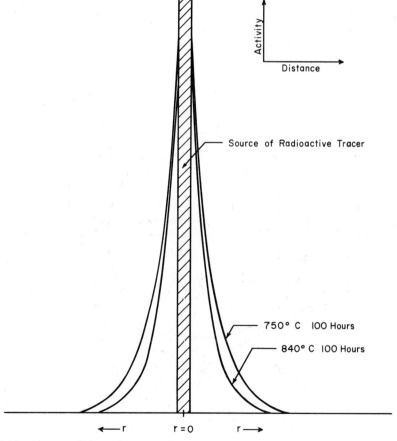

Fig. 2. Diffusion profiles for equal times but different temperatures showing the
effect of evaporation.

Fig. 3. Typical nickel sample for surface diffusion studies showing nickel lead wires, thermocouple wires and universal holder mount.

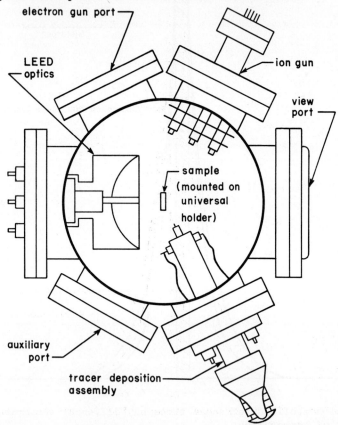

Fig. 4. Top view of ultra-high vacuum chamber showing the experimental arrangement and instrumentation.

a. Initial structureless surface.

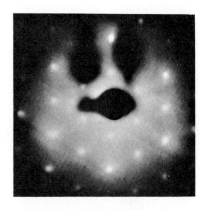

d. Annealed in oxygen, 2 x 2
2 x 2 structure.

b. After first anneal, 1000°C,
5 x 10^{-10} torr.

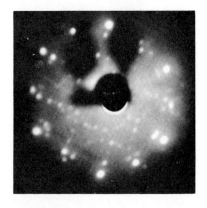

e. Annealed in hydrogen, complex
structure.

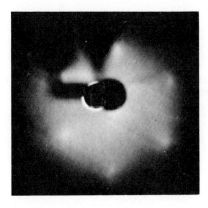

c. After ion bombardment –
surface degraded.

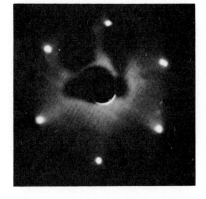

f. Final anneal 1000°C, 5 x 10^{-10}
torr, clean surface.

Fig. 5. 125 volt LEED patterns of nickel (111) at various stages of surface
preparation.

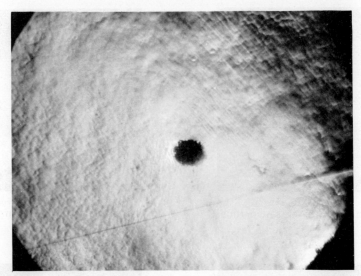

Fig. 6. Photomicrograph taken at 40X of the indentation made by the tracer point source on a nickel (111) surface.

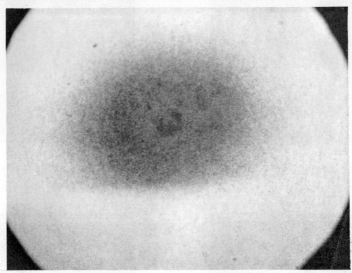

Fig. 7. Autoradiograph at a magnification of 40X of the tracer distribution on sample shown in Fig. 6 after a 96 hour diffusion anneal at 755°C.

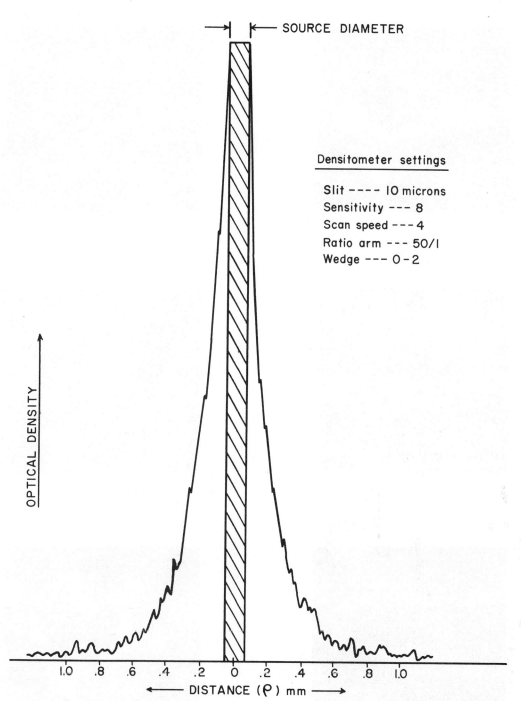

Fig. 8. Microdensitometer scan of the autoradiograph shown in Fig. 7.

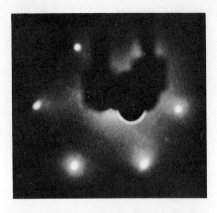

a. Experiment no. 8, clean surface before diffusion.

b. Experiment no. 8, contaminated surface after diffusion at 795° C for 41 hours.

c. Experiment no. 9, clean surface before diffusion.

d. Experiment no. 9, clean surface after diffusion at 614°C for 197 hours.

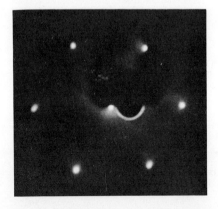

e. Experiment no. 10, clean surface before diffusion.

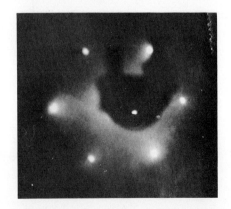

f. Experiment no. 10, clean surface after diffusion at 795°C for 23 hours.

Fig. 9. 125 volt LEED patterns of the nickel (111) surface before and after diffusion for several typical experiments.

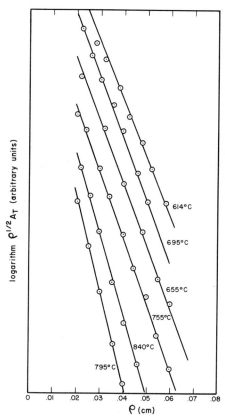

Fig. 10. Activity profiles for nickel (111) surface self-diffusion at various temperatures.

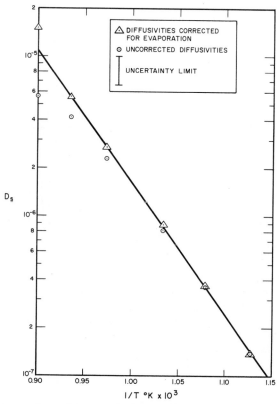

Fig. 11. Temperature dependence of the surface diffusivity for nickel (111).

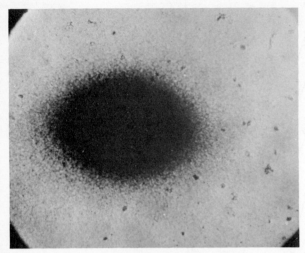

Fig. 12.

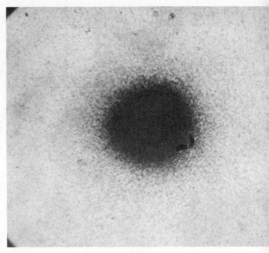

Fig. 13.

Fig. 12.   Overexposed autoradiograph of tracer spreading on nickel (III) showing
          bisymmetric diffusion.

Fig. 13.   Overexposed autoradiograph of tracer spreading on a high indices
          nickel single crystal surface.

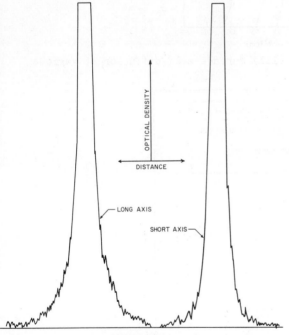

Fig. 14.

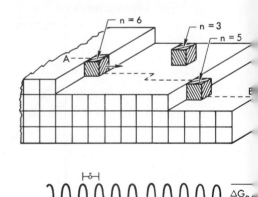

Fig. 15.

Fig. 14.   Microdensitometer tracings taken at 90° of each other along the long
          and short axes of a bisymmetrical autoradiograph from experiment No. 3.

Fig. 15.   Schematic model of a TLK surface and potential energy diagram for a
          tracer atom diffusing along the line A-B, deep energy wells correspond-
          ing to kink sites (n=6) and ledge sites (n=5), and shallow wells to
          terrace sites (n=3).

32–36

a. Clean nickel (III) surface before hydrogen anneal.

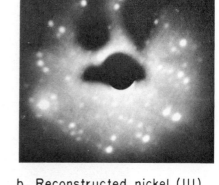

b. Reconstructed nickel (III) surface after 1000°C anneal in hydrogen.

c. Clean nickel (III) surface before gold on nickel surface diffusion

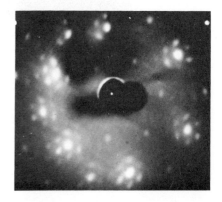

d. Gold monolayer on nickel (III) after diffusion anneal at 900°C for 100 hours.

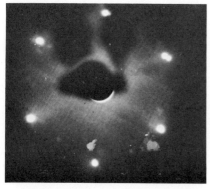

e. Clean nickel (III) surface before high temperature anneal.

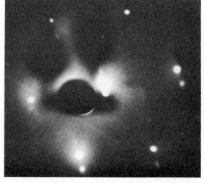

f. Possible stepped structure after 1000°C, 35 min. anneal in ultra-high vacuum.

Fig. 16.   LEED photographs of changes in the nature of the nickel (III) surface due to various treatments.

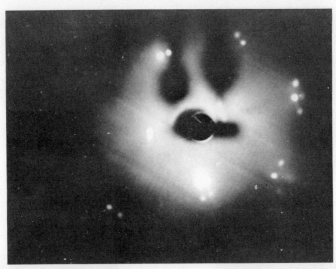

Fig. 17. Rotation of LEED diffraction spots on heating a nickel (III) surface through the Curie temperature.

Fig. 18. Vector representation of the diffraction spot rotation associated with the magnetic effect shown in Fig. 17.

LEED STUDIES OF THE POLAR {0001} SURFACES OF THE II-VI COMPOUNDS CdS, CdSe, ZnO, and ZnS*

B. D. Campbell,** C. A. Haque,*** and H. E. Farnsworth

Brown University, Providence, Rhode Island

I.   INTRODUCTION

Numerous investigations have shown that certain adsorbed gases affect the electrical properties of semiconductor surfaces.[1] Most studies of the surface properties of the II-VI compounds have been of contaminated or "real" surfaces. The purpose of the present LEED investigations was to determine (1) the techniques necessary for the production of clean surfaces of these semiconducting compounds, (2) the structure and morphology of the clean surfaces, and (3) the effects of gas adsorption on the clean surfaces.

The four materials under consideration, CdS, CdSe, ZnO and ZnS are wide band-gap, n-type semiconductors which owe their semi-conducting properties to an excess of the group II component. The four compounds have the wurtzite structure which has hexagonal symmetry. As can be seen in Fig. 1, this structure is asymmetric along the c axis (the direction perpendicular to the {0001} planes). Due to this polarity and the fact that the bonding within double layers such as A,B,C is stronger than between adjacent double layers, there are two different (0001) surfaces; the cation surface composed entirely of group II atoms and the anion surface composed of group VI atoms. The polarity has been observed in the form of different chemical and morphological properties for the two surfaces. Because of these differences, it is of interest to compare the properties of the clean anion and cation {0001} surfaces.

*   This work was supported by the U.S. Army Electronics and Development Laboratory and the Advanced Research Projects Agency. The portion of work done at Los Alamos was supported by the Atomic Energy Commission.
** Present address: Los Alamos Scientific Laboratory, Los Alamos, New Mexico 87544.
*** Present address: Bell Telephone Laboratories, Murray Hill, New Jersey 07971.

II. EXPERIMENTAL

The LEED experiments were carried out in both a rapid-scan Faraday collector system[2] at Brown University and a post-acceleration display system[3] at Los Alamos Scientific Laboratory. Most of the work was done with the Faraday collector system with which it was possible to take retarding potential curves.[2]

The surfaces were prepared by mechanical polishing followed by a chemical etch, or by polishing alone. The polarities of the crystals were determined by the chemical etch characteristics of the {0001} surfaces.[4] The crystals were heated by thermal conduction from the molybdenum mount to which they were attached. The mount was heated by either electron bombardment or a resistance coil. Crystal temperatures were determined in an auxiliary vacuum system wherein the power fed into the molybdenum mount from the electron gun assembly was calibrated against the crystal temperature, read by a Pt - Pt 13% Rh thermocouple, the junction of which was wedged into a slot in the side of the crystal. Attempts to clean the surfaces by heating in vacuum were only partially successful. Usually a diffraction pattern could be obtained by heating alone but the pattern did not have the highest intensity that was attainable. Argon ion bombardment (300–500 eV, 10–50 $\mu A/cm^2$) followed by annealing (CdS, $500^\circ C$; CdSe, $500^\circ C$; ZnO, $700^\circ C$; ZnS, $700^\circ C$) was very effective in bringing the diffraction pattern to a reproducible maximum intensity. Also, the magnitude of the maxima in the retarding potential curves attained maximum values. This condition was the cleanest state obtainable and is called the clean surface.

III. CLEAN SURFACES

The diffraction patterns from the cation or group II surfaces were intense and well resolved, but generally uncomplicated. The (0001)Zn surfaces of ZnO and ZnS produced only integral order beams and no beams due to facets or reproducible fractional orders. The

(0001)Cd surface of CdS exhibited intense half-order beams which were

similar to those observed on the (111) surface of the III-V compounds[5]

and appeared to be due to a rearrangement of the CdS surface

structure[6]. The evidence for this conclusion was the reproducibility

and stability of the pattern under extensive ion bombardment, heating

over a wide range of temperatures, and adsorption and thermal

desorption of gases. Very weak half and quarter order beams were also

observed on the (0001)Cd surface of CdSe. These beams appeared to be

due to surface rearrangement but the fact that they were very weak

left room for doubt.

The retarding potential curves that were obtained from the clean

cation surfaces are shown in Fig. 2. The voltage at which the crystal

current drops to zero is a measure of the contact potential difference

between the crystal and the hot tungsten cathode of the electron gun.

Movement of this point along the voltage axis indicates a change in

work function of the crystal. Interesting features of these curves

are the maxima and minima which are characteristic of the material.

These curves are equivalent to the primary electron current minus the

total secondary current as a function of incident electron energy.

Thus, they furnish information about the electron reflection

coefficient. The fact that the curves resemble ultraviolet

reflectivity curves suggests that they may contain information about

band structure. However, the curves also bear some resemblance to

electron reflectivity curves calculated by Gerstner and Cutler.[7]

For this case, the curve shape should depend on the form of the

potential at the surface, the atomic species involved, and lattice constants. The possibility that the structure in the curves is related to surface structure is supported by the observation that factors which disturb the surface order, such as amorphous adsorption or ion bombardment, eliminate or decrease the maxima.

The diffraction patterns from the anion or group VI surfaces were weak and not well defined, or absent. The pattern from the $(000\bar{1})O$ surface of ZnO contained only integral order beams. No diffraction pattern was obtained from the $(000\bar{1})Se$ surface of CdSe after ion bombardment and annealing. Some very weak peaks were produced by heating in H_2 but they gave no information about the surface.

Extensive faceting occurred on the $(000\bar{1})S$ surfaces of CdS and ZnS. These facets were produced by heating in vacuum but were more fully developed by ion bombardment and annealing. As reported previously,[6] $\{10\bar{1}4\}$ planes were formed on CdS by heating near $500^{\circ}C$ and $\{10\bar{1}3\}$ planes were formed near $800^{\circ}C$. Fig. 3 shows the very complicated diffraction pattern obtained from ZnS faceted by ion bombardment and annealing at $700^{\circ}C$. Because the diffraction beams converged to several different points as the voltage was increased, there appeared to be more than one type of facet. However, because of charging effects, weak intensities, and difficulties in making precise measurements, the facets were not identified. Once formed on either CdS or ZnS, the facets could be removed only by repolishing the surface.

Figures 4 and 5 show photomigrographs of the cation and anion surfaces after the crystals were studied in the LEED system. Before this study, these surfaces were polished to a smooth finish. Thus, the morphology of the surface was formed by ion bombardment and

heating in vacuum. Since the etched structure shown in the photographs did not affect the diffraction pattern, it is concluded that
no appreciable faceting parallel to other low index planes had
occurred. Hence the boundaries between (0001) planes appear to have
consisted of planes parallel to the C-axis.

In the case of the grainy (000$\bar{1}$)S surface of CdS, high
resolution electron micrographs[6] showed no morphological structure
that could readily be correlated with the facets which were detected
by LEED. Also, the light figure technique[8] showed no evidence of
facets, thus indicating that they were very small.

IV.   OXYGEN ADSORPTION PROPERTIES

In the adsorption studies, oxygen was the gas of principal
interest since it affects the electrical properties of surfaces of
these materials. Most of the work was concentrated on the cation
surfaces since the anion surfaces were poorly defined. The effects
of adsorption which were observed for the anion sides were similar to
those of the corresponding cation surfaces.

Oxygen exposures at room temperature were made with the crystal
(1) in darkness, (2) illuminated by band-gap light, (3) a few cms
from a hot tungsten filament. Before and after each exposure the
diffraction pattern and retarding potential curves were recorded.
By means of the retarding potential curves, changes in work function,
surface photovoltage, and, in some cases, surface resistance were
observed.

Oxygen exposures in darkness produced no appreciable changes in
the diffraction patterns for any of the cation surfaces. Even after
exposures of thousands of Torr-min, there was no more than a 10%

change in the intensity of the diffraction pattern. Also, no fractional order beams were found. Although the effects on the
diffraction pattern were insignificant, changes in surface electrical
properties were observed. A change in the shape of the retarding
potential curves, such as will be discussed in more detail later,
indicated an increase in surface resistivity. High conductivity CdS
crystals showed no resistance effect but did exhibit an increase in
the surface photovoltage.[6] Both the surface resistance and photovoltage could be changed to their clean-surface values by heating at
400–600°C. Thus, the observation that the oxygen exposures did not
produce a detectable effect on the diffraction pattern but did change
the electrical properties of the surface indicates that only a very
small amount of adsorption occurred even with large exposures.

Although oxygen exposures of several thousand Torr-min in darkness produced very little adsorption, two methods of inducing high-
coverage adsorption of oxygen on the (0001)Cd surfaces of CdS and
CdSe were found. In one case, an amorphous layer of oxygen, which
extinguished the diffraction pattern, was formed when exposures of a
few thousand Torr-min were made while the crystals were illuminated
by visible light. In the second case, adsorption, with a similar
effect on the diffraction pattern, was induced by placing the crystal
within a few cms of a hot tungsten filament during an oxygen exposure
at 10^{-5} Torr. Under this condition, an exposure of approximately
10^{-4} Torr-min was usually sufficient to extinguish the diffraction
pattern. Because any one of several reactions could have taken place
when tungsten was heated in the presence of oxygen, many experiments
were performed to determine whether oxygen was indeed the adsorbed

species.[6]   These included tests which eliminated effects due to CO,
CO_2, and WO_3, and those due to heat and light, as well as the
observation that the adsorption activated by the filament had the
same effect on the surface properties as photoadsorbed oxygen.   All
of them indicated that oxygen was the adsorbent.   For ZnO, large
oxygen exposures to the clean (0001)Zn surface illuminated by ultraviolet light produced no change in the adsorption results from those
which were found for exposures in darkness.   Exposures of 10^{-3} Torr-
min near a hot filament under the same conditions as with the (0001)Cd
surfaces caused some decreases in diffraction intensities but the
effect was not nearly as pronounced as with CdS and CdSe.

   For the cases of the (0001)Cd surfaces of CdS and CdSe, the
adsorption, which was activated by either light or the hot filament,
had pronounced effects on the retarding potential curves, in addition
to extinction of the diffraction pattern.   The work function increased,
as shown in Fig. 6.   In these curves, the oxygen coverage, θ, was
estimated from the diffraction beam intensities, I, for the covered,
and I_o, for the clean surfaces.

   Figure 7 illustrates a second effect which the high coverage
adsorption had on the retarding potential curves of CdS and CdSe.
Whereas the curve from the clean surface exhibited considerable
structure, the increasing oxygen coverage eliminated this structure.
The absence of structure in the retarding potential curves was also
characteristic of the surface after exposure to the atmosphere.

   For high resistivity CdS crystals and all CdSe crystals, a third
effect was produced by the adsorption.   Fig. 8A shows the retarding
potential curves from a CdSe crystal exposed to oxygen near a hot

filament.   Curve 1 was taken in darkness and curve 2 during intense
illumination by visible light.   Curve 1 shows a very low crystal
current which gradually increased with increasing incident electron
energy.   The gradual slope is attributed to a high resistance at the
surface which produced a voltage drop.   Curve 2 has a steep slope
because photoconduction had decreased the surface resistance.
The transition between curves 1 and 2 is shown in Fig. 8B in
which each curve was taken with a different light intensity.
It was concluded that the high resistance occurred only at the
surface because a very light ion bombardment (100 eV, 1 $\mu A/cm^2$)
for   a   few   minutes   eliminated   the   resistance   effect.

The change in shape of the retarding potential curves caused by high
surface resistance also occurred after O_2 exposures to the (0001)Zn
surfaces of ZnO and ZnS even though no changes in diffraction pattern
were evident.

All of the effects of oxygen exposures on the surfaces could be
eliminated by heating the crystals in vacuum close to the annealing
temperature.   An analysis of the desorbed gases with a quadrupole
mass analyzer indicated that CO and CO_2 were evolved after an oxygen
exposure.   Apparently, the adsorbed oxygen reacted with carbon before
desorption.   It is not likely that the CO and CO_2 were adsorbed
during the oxygen exposure since large exposures of these gases had
very little effect on the surface.

An additional effect, which was noted with oxygen adsorption on
(0001)Cd surfaces of CdS and CdSe, was the strong dependence of the
adsorption rate on previous surface treatment.   After ion bombardment
and annealing, an exposure of $2x10^{-4}$ Torr-min near a hot filament

was required to produce full coverage. When the adsorbed gas was removed by heating in vacuum, an exposure of only 6×10^{-5} Torr-min was required to cover the surface a second time. If the oxygen was removed a second time by heating and then a third series of exposures was made, a still lower exposure of 3×10^{-5} Torr-min was sufficient to extinguish the diffraction pattern. However, subsequent adsorption and thermal desorption of oxygen did not increase the adsorption rate further. If the crystals were ion bombarded and annealed, the adsorption rate returned to the original low rate. This effect was very reproducible and occurred with photoadsorbed oxygen. It also accounts for the wide range of adsorption rates that were observed. This effect has alternate explanations. Either the thermal desorption of oxygen created defects which increased the adsorption rate above the clean surface value, or the ion bombardment and annealing treatment deposited an adsorption-inhibiting contaminant on the surface, by diffusion from the bulk, which was subsequently removed when the oxygen was thermally desorbed. There was no difference between the diffraction patterns after ion bombardment and annealing and after thermal desorption. If the oxygen were removed as SO_2, as suggested by the experiments of Bootsma,[9] then sulfur vacancies might account for the increased rate.

Exposures in darkness of CdS, CdSe and ZnO to other gases, such as CO, CO_2 and H_2, produced results somewhat similar to those produced by O_2. No appreciable coverage was attained but there were often changes in surface resistance. Illumination during the exposures increased the magnitude of the resistance effect on CdSe in the case of CO and CO_2 exposures and on ZnO for O_2 exposures.

Otherwise, illumination had little effect on the results due to adsorption. Exposures in the presence of a hot filament also had little effect on the adsorption results with these gases.

An interpretation of the low gas coverages that were observed in these experiments can be given with the aid of the electronic theory of adsorption, as summarized by Many, et al.[1] Fig. 9 shows an energy band diagram for the surface of an n-type semiconductor. Energy is plotted along the vertical axis and distance from the surface into the bulk along the horizontal axis. E_c, E_v and E_f are the conduction band edge, the valence band edge, and the Fermi level, respectively. Assume that an electronegative gas is chemisorbed as a negatively charged ion with electron affinity A. For this chemisorption to occur the adsorbed particle must remove an electron from the bulk of the semiconductor. If A is greater than the work function, W_ϕ, there is a positive energy of adsorption. However, as the density of charged surface states increases, a potential barrier is formed at the surface and W_ϕ increases. When $A = W_\phi$ the adsorption energy is zero and chemisorption stops. For the wide band gap semiconductors, a density of charged surface states of $10^{12}/cm^2$ should be sufficient to stop the adsorption. If every adsorbed atom is ionized, this density of states amounts to a coverage of 0.1% of a monolayer.[10] Thus, if this theory is appropriate for these II-VI compounds, one would expect the maximum coverage to be too small to be detectable by LEED, yet large enough to affect the electrical properties.

With respect to the high-coverage oxygen adsorption, which was stimulated by illumination or a hot filament, on CdS and CdSe, the agreement with this theory is not complete. For ionic chemisorption, the formation of the potential barrier is such that the work function increases as the square of the coverage. However, as shown in Fig. 6, the work functions of CdS and CdSe increased as approximately the square root of the coverage as computed from the LEED results. This type of work function change is more characteristic of adsorption in a neutral form with a dipole moment. In this case, the work function increase arises from a change in the electron affinity (X) of the semiconductor rather than bending of the bands.

V.   SUMMARY

The surface structures of the II-VI compounds that were studied were relatively simple and no impurity or adsorption structures, such as have been found with many other materials, were observed. Also, although this has been known for some time, it was found that the morphology of the surface can often have little relation to the diffraction results. Gas exposures in darkness indicated that only very low coverage adsorption occurred even with high exposures. The very low gas coverages agree with the predictions of the electronic theory of adsorption and seem to indicate that most of the adsorption is ionic. However, in the case of another type of oxygen adsorption that was observed with CdS and CdSe, and was activated by light or a hot filament, the agreement with the theory is not good.

REFERENCES

1. A. Many, Y. Goldstein, and N. B. Grover, Semiconductor Surfaces (North-Holland Publ. Co., Amsterdam, 1965) p. 371.

2. R. L. Park and H. E. Farnsworth, Rev. Sci. Instr. $\underline{35}$ (1964) 1592

3. Varian Associates, Palo Alto, Calif.

4. E. P. Warekois, M. C. Levine, A. N. Mariano, and H. C. Gatos, J. Appl. Phys. $\underline{33}$ (1962) 690.
    Erratum – J. Appl. Phys. $\underline{37}$ (1966) 2203.

5. D. Haneman, J. Phys. Chem. Solids $\underline{14}$ (1960) 1962; Phys. Rev. $\underline{121}$ (1961) 1093; A. U. MacRae, Surface Sci. $\underline{4}$ (1966) 247.

6. B. D. Campbell and H. E. Farnsworth, Surface Sci. $\underline{10}$ (1968) 197.

7. J. Gerstner and P. H. Cutler, Surface Sci. $\underline{9}$ (1968) 198.

8. G. A. Wolff, J. J. Frawley and J. R. Heitanen, J. Electrochem. Soc. 111 (1964) 22.

9. G. A. Bootsma, Surface Sci. $\underline{9}$ (1968) 396.

10. R. Williams, J. Phys. Chem. Solids $\underline{23}$ (1962) 1057.

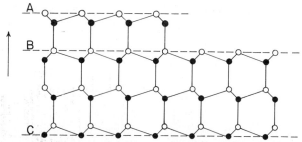

Fig. 1. Schematic drawing of wurtzite structure. Open and solid circles represent rows of group II and group VI atoms, respectively, that are perpendicular to the page. A, B, C, represent {0001} planes which are also perpendicular to the page.

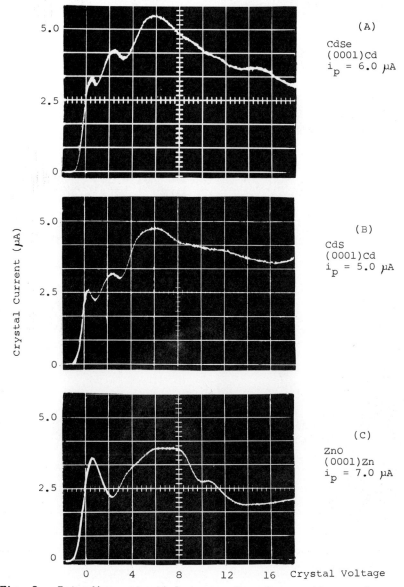

(A)

CdSe
(0001)Cd
$i_p = 6.0 \; \mu A$

(B)

CdS
(0001)Cd
$i_p = 5.0 \; \mu A$

(C)

ZnO
(0001)Zn
$i_p = 7.0 \; \mu A$

Fig. 2. Retarding potential curves for clean cation surfaces.

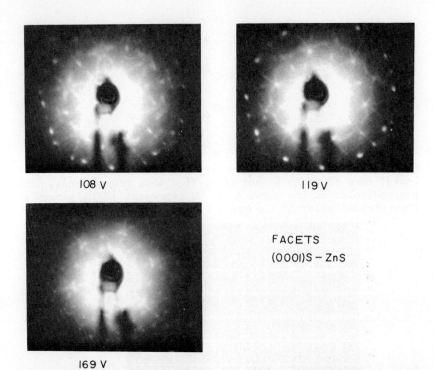

108 V

119 V

FACETS
(0001)S – ZnS

169 V

Fig. 3. Diffraction patterns from facets on (000$\bar{1}$)S surface of ZnS.

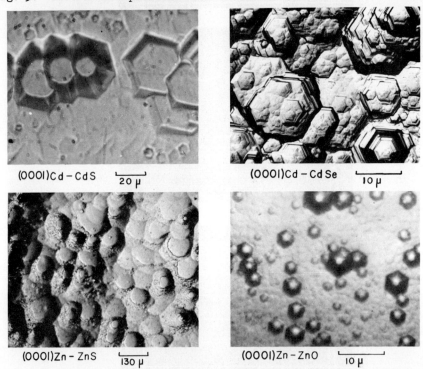

(0001)Cd – CdS   20 μ

(0001)Cd – CdSe   10 μ

(0001)Zn – ZnS   130 μ

(0001)Zn – ZnO   10 μ

Fig. 4. Photomicrographs of group II, (0001) surfaces after ion bombardment and annealing.

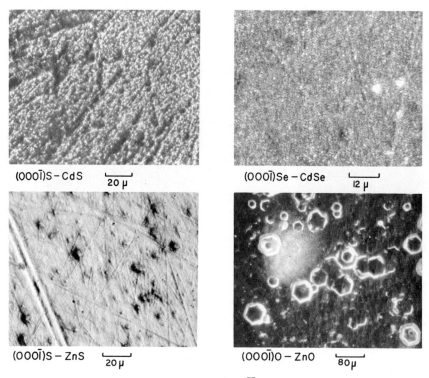

(000ī)S–CdS   ⌞ 20 μ ⌟   (000ī)Se–CdSe   ⌞ 12 μ ⌟

(000ī)S–ZnS   ⌞ 20 μ ⌟   (000ī)O–ZnO   ⌞ 80 μ ⌟

Fig. 5.   Photomicrographs of group VI, (000$\bar{1}$) surfaces after ion bombardment and annealing.

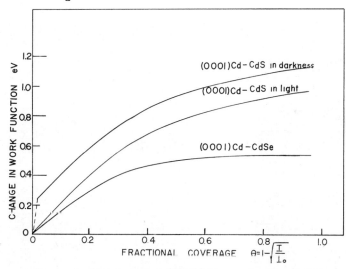

Fig. 6.   Change in work function of (0001)Cd surfaces of CdS and CdSe as a function of oxygen coverage. Coverage computed from change in diffraction intensities.

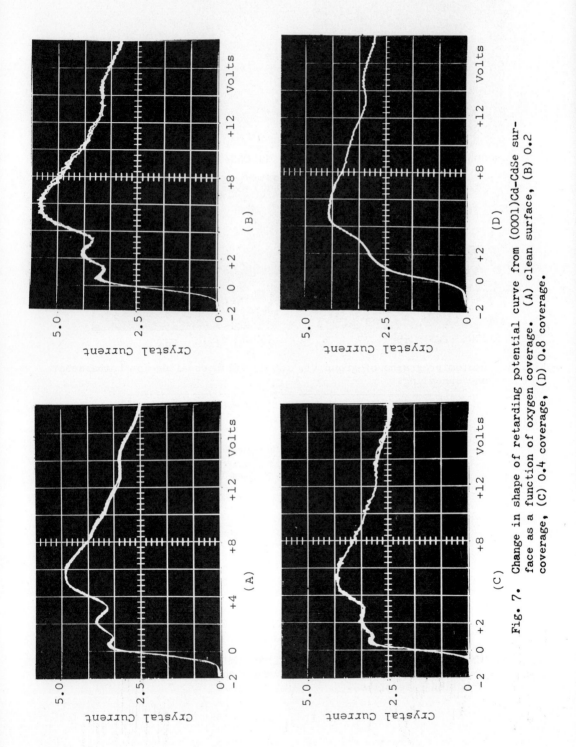

Fig. 7. Change in shape of retarding potential curve from (0001)Cd-CdSe surface as a function of oxygen coverage. (A) clean surface, (B) 0.2 coverage, (C) 0.4 coverage, (D) 0.8 coverage.

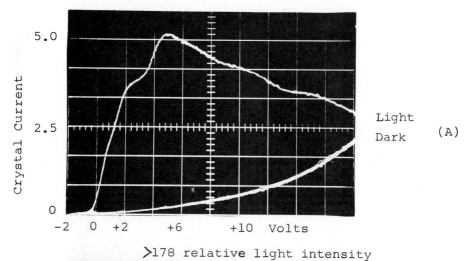

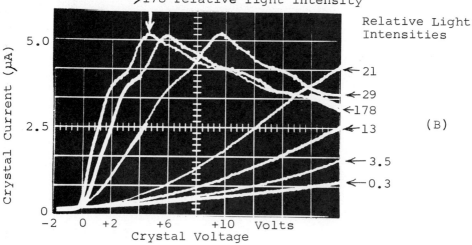

Fig. 8. Effect on retarding potential curves of (A) high surface resistance
and (B) relative light intensity for CdSe.

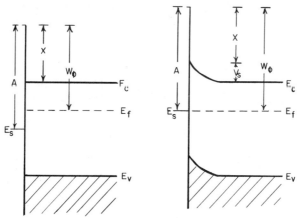

Fig. 9. Energy band diagram for surface of n-type semiconductor.

THE (0001) SURFACE OF a-ALUMINA — LEED OBSERVATIONS

J. M. Charig and D. K. Skinner

Allen Clark Research Centre
The Plessey Company Ltd.
Caswell, Towcester, Northants, England

1. INTRODUCTION

The application of LEED to the early stages of silicon epitaxy on silicon by evaporation has been reported in recent literature. [1][2] Comparatively little attention has so far been paid to other substrates. One of the more popular for silicon epitaxy is α-alumina. [3][4][5][6][7] Orientation relationships between the silicon layers and the alumina differ on different alumina faces and in certain cases multiply oriented layers occur. Only tentative explanations of these relationships and the bonding mechanisms involved have so far been possible. During an attempt to throw more light on this subject by LEED studies of the faces of α-alumina crystals prior to and during the early stages of silicon epitaxy, a structural modification of the (0001) α-alumina surface [8] has been observed to occur as the result of heat treatment above $900^{\circ}C$. This is close to the temperature most frequently used for epitaxial growth. The observation of etching of the alumina by silicon and by electron bombardment has been facilitated by the observation of this structural transformation.

2. EXPERIMENTAL

A conventional LEED system was designed and built incorporating the usual gas leak facilities and a mass spectrometer gas analyser. Targets were heated by radiation or electron bombardment (para. 2.2) and a silicon evaporation source by electron bombardment. A heat and evaporation shield was included to protect the electron optics. The ion pumped system consistently achieved an indicated base pressure of about 3.10^{-10} torr after thorough outgassing, measured by a semi-nude ionisation gauge. Mass spectrometry indicated the residual gas to be predominantly hydrogen at a pressure within a factor of two of that given by the ionisation gauge.

2.1   MATERIALS

The targets used for the LEED studies were cut from oriented Verneuil boules obtained either from the Swiss Jewel Company or from the Salford Electric Company Limited. Typical analyses for crystals from both sources gave not less than 99.95% Al_2O_3 with iron as major impurity.

Surface preparation was varied, LEED patterns being most readily obtained after chemical polishing for 2–3 minutes with a 1:1 by volume mixture of potassium persulphate and boric acid at $c.1000^{\circ}C$ with or without prior mechanical polishing. Substrates treated as for current epitaxy technique,[9] involving diamond polishing down to $\frac{1}{4}\mu$ particle size followed by firing in hydrogen at $1260^{\circ}C$ for 30 mins. were also examined with and without the hydrogen firing. Weak diffraction was often obtained from the chemically polished surfaces without further treatment in vacuo. Mechanically prepared targets only responded after lengthy outgassing.

Due to the effects observed the (0001) orientation has received most attention up to the present time. Preliminary examination of the ($\bar{1}012$) surface did not reveal a parallel effect though some extra reflections were observed.[8]

2.2   HEATING TECHNIQUE AND TEMPERATURE MEASUREMENT

Two alternative heating techniques have been employed. Electron bombardment was used for initial experiments[8] but evidence accumulated that the electron beam was decomposing and etching the alumina so an alternative radiation heater was employed. The latter was based on a similar unit used by the Post Office Research Station at Dollis Hill, London from whence the essential components of the heater were obtained. Temperature measurement is discussed in an appendix.

3.   RESULTS AND OBSERVATIONS

3.1   RADIATION HEATING

When α-Al_2O_3 is heated in vacuum by radiation, the (0001) surface desorbs gas and in the cooled condition diffraction beams are produced characteristic of the bulk alumina structure (Fig.1). If the temperature is not allowed to exceed 900°C these beams persist from the cooled target but usually remain comparatively weak and diffuse with a high background even after prolonged heating. Charging of the target occurs at low voltages under normal conditions of diffraction. Initially this may occur as high as 200 eV but generally drops to around 90 eV as cleaning proceeds. Lower voltages can sometimes be used if the target is examined while still hot.

If the temperature is raised over 900°C for even a few minutes the (0001) face of the cooled target now begins to give rise to a complex diffraction pattern (Fig.2). More prolonged heating at about 1000°C will develop the corresponding structure fully and diffraction is readily obtained from the cooled target at all voltages from about 60 eV to 1 KeV. The pattern can also be obtained up to about 800°C above which excessive light interferes with observation and under certain conditions down as low as 40 eV. The latter result was difficult to achieve and behaviour in this respect was not consistent. If heating is by radiation only, the structure from which the complex pattern derives is stable and no further diffraction changes are observed regardless of any heat treatment. Outgassing approaches completion enabling pressures close to base to be maintained during heating.

3.2   ELECTRON BEAM BOMBARDMENT

If heat treatment is carried out by means of a target accelerated beam of electrons (typically, at about 1 KeV a current of 10 mA will raise the alumina to between $900^{\circ}C$ and $1000^{\circ}C$) there is a very marked difference in diffraction behaviour. The initial outgassing followed by development of the complex pattern after heating in excess of $900^{\circ}C$ is unchanged, except that now the gas pressure in the system always remains a function of the target temperature, rising to the high 10^{-8} torr region at about $1000^{\circ}C$. This comparatively high pressure does not decrease with time at temperature. Mass spectrometry shows the predominant gases to be characteristic of a hot ion pumped system but significantly, reveals mass 32 attributed to O_2. This mass peak always appears during electron bombardment heating of alumina.

Even more marked, however, is the effect of further electron bombardment heating in the region $700–900^{\circ}C$ after first heating above $900^{\circ}C$ to produce the complex pattern. With 1 KeV electrons this corresponds to about 7–8 mA and under these conditions only 3–5 minutes heating now restores the bulk diffraction pattern (Fig.1) but with much enhanced intensity and resolution and diminished background. Diffraction is again possible from about 80 eV to 1 KeV, Kikuchi bands being evident at the high voltages.

The complex pattern can again be recovered by heating at $1000^{\circ}C$ for a few minutes.

3.3   ELECTRON BOMBARDMENT DURING RADIATION HEATING

If the alumina target is heated to a temperature in excess of $900^{\circ}C$ by radiation the complex diffraction pattern (Fig.2) develops. On reheating at a temperature in the range $700–800^{\circ}C$ the structure giving rise to this pattern is unchanged. Irradiation of the surface with electrons at this temperature will result in a return to the structure characteristic of bulk alumina (Fig.1) as when using electron bombardment heating (para 3.2). With the radiation heater the recovery of the bulk pattern can be achieved with much lower values of electron current and voltage and without the pronounced emission of gas recorded above. The range of values explored is shown in Table I.

TABLE I

Volts	(mA) Current	Time to recover bulk pattern, minutes.
1000	1	10
1000	0.5	10
1000	0.1	120
500	1.0	20
300	1.0	40
100	1.0	240

Examination of the alumina surfaces microscopically after electron bombardment showed them to be eroded. Extended heating by electron bombardment using 1 KeV electrons and currents up to 15 mA caused severe erosion and blackening of the originally transparent crystal.

3.4 DEPOSITION OF EVAPORATED SILICON

Reference to deposition of silicon is included here as it provided confirmation of the electron bombardment results. Silicon is known to etch alumina at low deposition rates without forming a deposit.[10][11] Experiments were conducted in which silicon was evaporated at various rates onto the (0001) surface at $800^{\circ}C$, having first obtained the structure yielding the complex pattern. The results are shown in Fig. 3. a, b, and c represent stages in the etching, and d shows the onset of silicon deposition when the deposition rate exceeded the rate of etching. Silicon vapour etching above $900^{\circ}C$ resulted in some enhancement of resolution and intensity of the complex pattern particularly at the lower voltages. (Fig.4).

3.5   AMBIENT EFFECTS

The effect described above takes place equally readily at 10^{-6} torr and at 5.10^{-10} torr. Both structures are essentially insensitive to ambients of oxygen, nitrogen and water vapour at room temperature. The bulk–complex transformation occurred as usual at $1000^{\circ}C$ in oxygen at pressures up to 3.10^{-4} torr.

Deterioration of the complex pattern did result from exposure to air at atmospheric pressure and to water vapour at 1 torr but the bulk structure was only slightly weakened by similar treatment. A target giving a strong, well resolved bulk pattern was heated in air at $1250^{\circ}C$ for 10 minutes and when reexamined under vacuum gave an even more intense bulk pattern. A similar experiment with the surface giving rise to the complex structure resulted in restoration of the bulk structure pattern. A good bulk pattern was obtained but with some background intensity. The transformation is thus clearly dependent on the vacuum conditions.

These observations are summarised in Table II.

TABLE II

Ambient Effects on (0001) α-Alumina

LEED pattern	Temp.	Time	Pressure	Ambient	Visual observation of Effect on pattern
Bulk	25°C	5m.	760mm	Air	Very slight deterioration
√31	25°C	5m.	760mm	Air	Marked loss of intensity and resolution
Bulk	1250°C	10m.	760mm	Air	Stronger bulk pattern
√31	1250°C	70m.	760mm	Air	Bulk pattern with background
√31	25°C	30m.	10^{-3} torr	Oxygen	No change
√31	25°C	30m.	1.5×10^{-1} torr	Oxygen	No change
Bulk	1000°C		3.10^{-4} torr	Oxygen	√31 formed
√31	25°C	30m.	10^{-4} torr	Water vapour	No change
√31	25°C	5m.	1 torr	" "	Loss of intensity. Background increased.

4. DISCUSSION AND CONCLUSIONS

The extra reflections observed in the complex pattern can all be derived from a reciprocal lattice net based on a 60° rhombic surface cell of side √31 times the ($11\bar{2}0$) inter-planar spacing rotated 8.9° from the hexagonal axes (Fig. 5a). Two domains are possible (Fig. 5b) and both are necessary to obtain all the reflections. The actual patterns of Figs. 6 and 7 may be compared with the corresponding arrays constructed from the cells of Fig.5b. The distorted spots and apparent doubles are accounted for by the overlap of reflections from the two domains, which are just resolvable in Fig. 6c, obtained after the addition of a third grid to the LEED optics and additional smoothing to power supplies. No attempt has been made to reinforce these observations by intensity measurements. An identical interpretation of similar results has recently been reported by Chang.[15] The fit of all observed reflections could not be obtained from any alternative construction attempted. Operation of the technique advocated by Bauer[12] treating the surface mesh as a rotated misfit layer and allowing for double scattering does not produce the required array.

The arrangement of the new surface can only be guessed. Since the oxygen layers are normally close-packed the suggestion due to Brennan and Pask[13] that the structural change is associated with oxygen deficiency has been examined.

Relocation of the Al atoms in tetrahedial sites as in γ-alumina[14] instead of octahedral sites would be consistent with the formation of a surface layer of AlO and several possible models can be constructed having the requisite surface mesh postulated for the observed structure. The fact that the change only takes place under vacuum and not on heating in the atmosphere seems to lend support to this idea. It must be admitted, however, that even the $\sqrt{31}$ structure is extremely insensitive to ambient conditions, oxygen included, and that it shows greatest affinity for water vapour.

The alternative origin of an impurity seems improbable on the available evidence. The $\sqrt{31}$ structure forms from the bulk in a very short time at pressures too low for external contamination to be significant. Bulk impurities of .05% are equally unlikely to diffuse in sufficient concentration in the time and this mechanism is not consistent with the rather critical temperature of the transformation.

The recovery of the bulk pattern after heating below $900°C$ under electron bombardment is attributed to etching of the surface. Without a more detailed knowledge of the erosion process it is not possible to assess the number of atom layers involved in the transformation. It is evident, however, from the fact that the recovery of the bulk pattern becomes more difficult after prolonged heating at temperatures in excess of $1000^{\circ}C$ that the $\sqrt{31}$ structure should be ascribed to a definite surface phase rather than to a simple surface layer reconstruction. The evolution of O_2 detected under rapid bombardment is associated with dissociation of the Al_2O_3. No AlO or Al_2O could be detected so it must be assumed either that excess Al is left to diffuse into the crystal, which may account for the discolouration after prolonged bombardment at the higher rates, or that the Al–containing species condenses without reaching the mass spectrometer.

The recovery of the bulk pattern during Si evaporation onto the alumina at temperatures below $900^{\circ}C$ is likewise attributed to etching away of the surface layer by a reaction of the type,

$$Al_2O_3 \;+\; 2Si \;\rightarrow\; 2SiO \;+\; 2Al_2O^{(10)(11)}$$

Again, no AlO or Al_2O could be detected mass spectrometrically and SiO, mass 44 is not readily distinguishable from CO_2. Deposition of the Si beyond the stage indicated by Fig. 3d could not be followed due to charging of the specimen under observation repelling the incident beam. The observation of alumina down to 50 eV normally is associated with the crossover potential at which the large secondary emission current exceeds the incident current. The discontinuous Si deposit interferes with the secondary emission without providing an alternative conducting path and hence charging occurs and the beam is repelled. The voltage at which this occurs increases rapidly as the surface becomes contaminated.

SUMMARY

The (0001) surface of α-alumina undergoes structural modification when heated above 900°C. The new structure can be identified with a surface mesh defined as a rhombus of side $\sqrt{31}$ x the $(11\bar{2}0)$ inter-planar spacing rotated 8.9° to the hexagonal axes. The diffraction pattern corresponding to bulk α-alumina can be recovered by etching the surface. This has been accomplished in situ by electron bombardment and by silicon evaporation.

Both the alumina surface structures are very insensitive to common gas ambients even up to atmospheric pressure.

ACKNOWLEDGMENTS

The authors are indebted to the various colleagues with whom they have had helpful discussions whilst carrying out this work. In particular they wish to acknowledge the interest shown by Professor J.A. Pask of the University of California, and the assistance with diffraction interpretation by Dr. T. Edmunds of the B.P. Research Centre, Sunbury, England.

Thanks are also due to The Plessey Company Limited for permission to submit this paper. The work was done in part under a C.V.D. contract and is published by permission of the Ministry of Defence (Navy Department).

REFERENCES

1. F. Jona, Appl. Phys. Lett., <u>9</u>, 235, (1966).

2. R.N. Thomas & M.H. Francombe, Appl. Phys. Lett., <u>11</u>(3), 108, (1967)

3. C.W. Mueller and P.H. Robinson, Proc. Inst.Elect.Electron.Engrs., <u>52</u>, 1487, (1964).

4. J.C. Porter and R.G. Wolfson, J.Appl.Phys., <u>36</u>, 2746, (1965).

5. R.L. Nolder, D. Klein and D.H. Forbes, J.App.Phys., <u>36</u>, 3444, (1965).

6. R.W. Bicknell, B.A. Joyce, J.H. Neave, G.V. Smith, Phil.Mag., <u>14</u>, 31, (1966).

7. F.H. Reynolds and A.B.M. Elliott, Solid State Electronics, <u>10</u>, 1093, (1967).

8. J.M. Charig, Appl. Phys. Lett., <u>10</u>(5), 139, (1967).

9. P.B. Hart, P.J. Etter, B.W. Jervis & J.M. Flanders, Brit. J.Appl Phys., <u>18</u>, 1389, (1967).

10. J.D. Filby, J. Electrochem. Soc., <u>113</u>, 1085, (1966).

11. F.H. Reynolds and A.B.M. Elliott, Phil. Mag., <u>13</u>, 1073, (1966).

12. E. Bauer, Surface Science, <u>7</u>, 351, (1967).

13. J.J. Brennan & J.A. Pask, to be published J.Am. Cer. Soc.

14. R.G. Frieser, J. Electrochem. Soc., <u>113</u>(4), 357, (1966).

15. C.C. Chang. Private Communication. J.App. Phys. to be published.

APPENDIX
TEMPERATURE MEASUREMENT

Temperature measurement was conducted in two ways whilst using electron bombardment heating. With the radiation heater only approximate checks of temperature were made by direct optical pyrometer reading. The more thorough temperature observations were made using a Land black body calibrated total radiation pyrometer with a manufacturer's emissivitycorrection for silicon. The instrument focusses on about a 2 mm diameter. Two separate experiments were performed using the layouts of Fig. 8.

For method (a) the pyrometer output was plotted as a function of voltage applied to the tungsten filament reading the scattered light only from the alumina. A second curve was then plotted for similar observations but with 1 KV applied throughout to the alumina. A third curve was plotted by subtracting the scattered light readings of curve one from the total radiation of curve two to give the alumina temperature. To verify the accuracy of this experiment the procedure was changed to that of method (b) (Fig. 8). In this case a layer of silicon was deposited epitaxially on one face of the alumina and this face was examined by the pyrometer. Again a curve was plotted of pyrometer output versus input to the tungsten filament and then a second plot made with 1 KV applied to the target. In the latter case the reading from the pyrometer followed the W filament initially and then dropped as the thin silicon layer became opaque as its temperature increased. With further increase in temperature the output rose again.

The results of these plots are compiled in Fig. 9 using the emissivity correction for polished silicon (0.72%). (The silicon layer was polished prior to the measurements) and with a 6% addition for adsorption loss through the pyrex window of the vacuum chamber. The agreement between the two techniques gives fair confidence in the meaningfulness of the results above about 800°C at least.

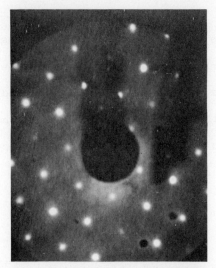

Fig. 1. a-Al_2O_3 190 eV bulk structure.

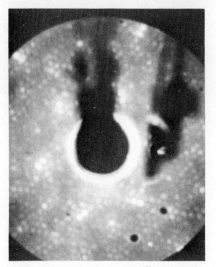

Fig. 2. a-Al_2O_3 190 eV $\sqrt{31}$ structure.

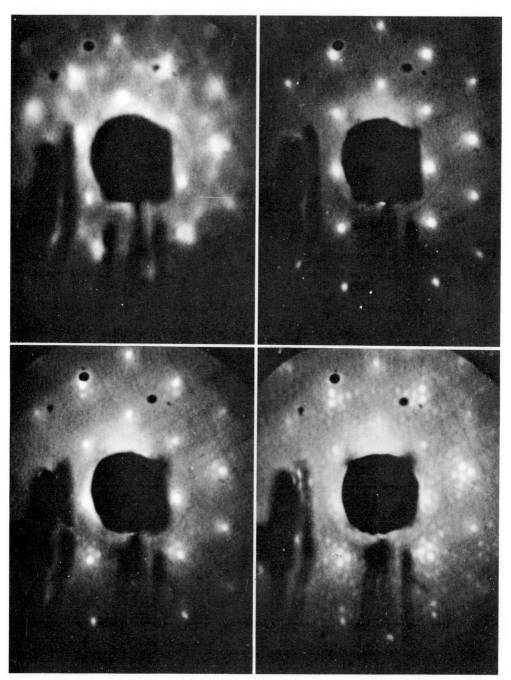

Fig. 3.   Silicon vapour etching of a-Al$_2$O$_3$ below 900°C.

Fig. 3a, b, c. $\sqrt{31}$ surface structure etched back to bulk.

Fig. 3d.   Si deposition commencing.

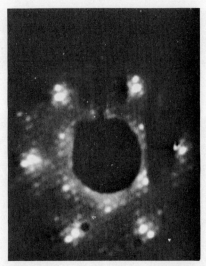

Fig. 4. a-Al$_2$O$_3$ $\sqrt{31}$ structure 60 eV. Si-vapour-etched > 900°C.

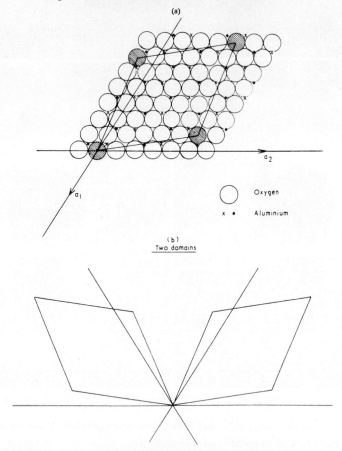

(a)

a_2

a_1

◯ Oxygen

x • Aluminium

(b)
Two domains

Fig. 5 (a) Arrangement of a-Al$_2$O$_3$ surface atoms with position of $\sqrt{31}$ structure. (b) Relationship of the 2 domains and bulk axes.

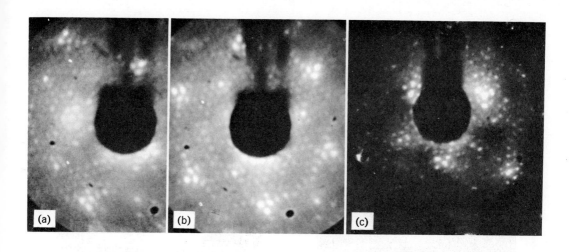

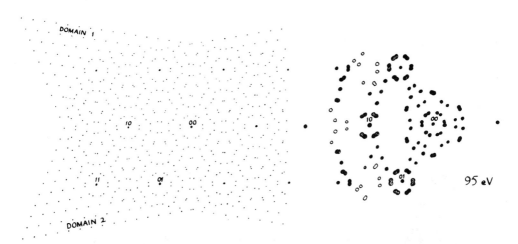

DOMAIN 1

10   00

11   01

DOMAIN 2

10   00

01

95 eV

Fig. 6a, b, c. a-Al$_2$O$_3$ $\sqrt{31}$ structure, 95 eV. & Reciprocal lattice net.

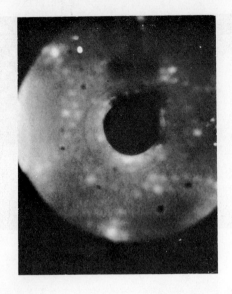
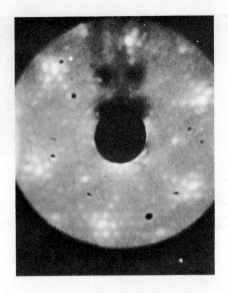

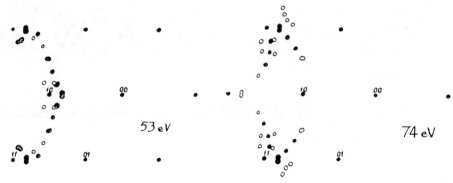

Fig. 7. a-Al$_2$O$_3$ $\sqrt{31}$ structure.   53 eV.   74 eV.

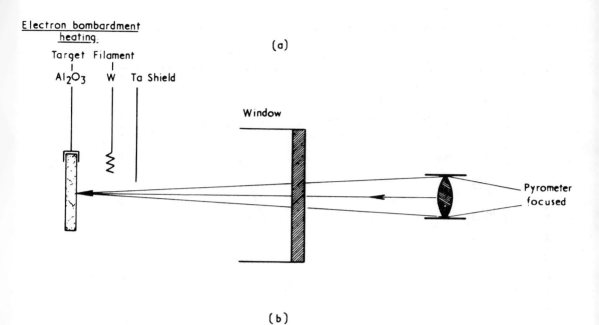

Electron bombardment heating.

(a)

Target Filament
Al₂O₃   W   Ta Shield

Window

Pyrometer focused

(b)

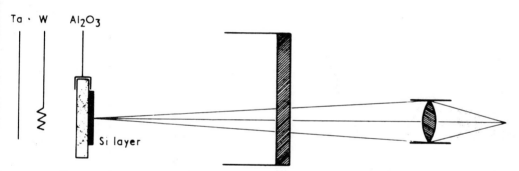

Ta · W   Al₂O₃

Si layer

Fig. 8a, b.   Schematic diagrams showing arrangement for temperature measurement of a-Al_2O_3.

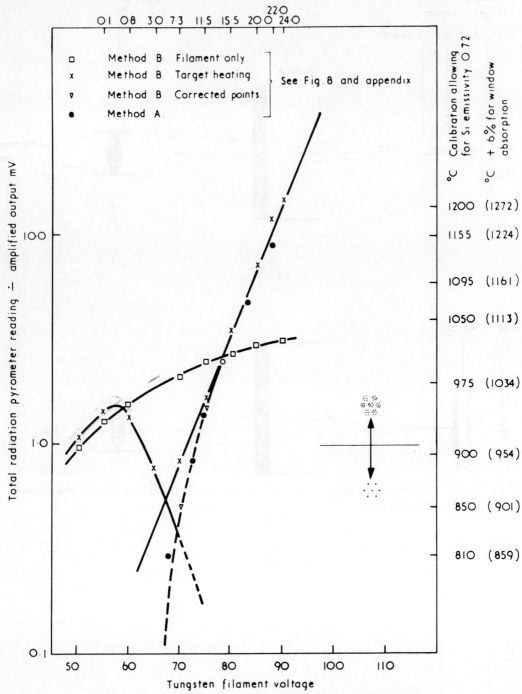

Fig. 9. Temperature measurement on a-Al_2O_3 results.

LEED OBSERVATIONS OF ELECTRIC DIPOLE FIELDS ON MICA AND ZnO

Klaus Müller

Bell Telephone Laboratories, Inc.
Murray Hill, New Jersey

INTRODUCTION

Most LEED investigations on surfaces are focussed
on 'extra' spots in the diffraction pattern or 'extra' peaks
in the intensity curves I(E) vs. energy E of diffracted
beams. In the present paper LEED experiments on samples
with surface electric dipole fields are reported in which
various unusual spot shapes are caused by those fields. Two
examples, the cleavage face of muscovite and phlogopite mica,
and the cleaved (0001) face of ZnO are discussed. The
surface fields can be either cancelled or strengthened by
adsorption of different materials at the surface. Thus LEED
provides quantitative means for the measurement of macroscopic
dipole fields at surfaces. Furthermore spot diffusivity in
patterns from insulators and semiconductors may often contain
a dipole field component.

Remarks about the Experimental Procedure

The experiments were performed in an UHV bell-jar
system with display-type LEED optics. In order to obtain clean
surfaces, all samples were cleaved in vacuum. In the case of
mica samples of about $(10 \times 10 \times 1)$ mm^3 in size, cleavage was done

by a movable knife operated from outside the bell jar. The

pencil-shaped ZnO crystals (2 ... 3 mm in diameter) were

'broken' like a glass rod, after which both surfaces could

be viewed with LEED. The primary beam diameter was about

1 mm.

The cleavage of mica in vacuum normally leads to

the formation of surface electron traps which cause the

sample to charge up to above 1000 V if the sample is

irradiated with a high-voltage beam. But this can, under

several circumstances, be avoided (for example by cleaving

the crystal at elevated temperatures) so that the crystal

shows a LEED pattern immediately after cleavage. For more

detail see Ref. 1).

Observations on mica

According to the hexagonal symmetry of the mica

surface one obtains a regular diffraction pattern with

hexagonal unit mesh and possibly with round spots, as has

been published by several authors.[2,3] If the spots are

round, scanning the beam over the entire surface or trying

another sample from the same piece of mica will generally

result in a pattern with apparently non-round spot shapes.

Most common are three-winged stars (Fig. 1a) and triangles

(Fig. 1b) and at 50 eV they are typically 5 mm on a side.

Within one pattern all spots are identical in shape and

parallel to each other, but there are more spot shapes which

may occur. Instead of showing many diffraction patterns[1]

they are just indicated in Fig. 1c. The wings of the stars
are not equally long; one may be missing. There are also
streaks which may be resolved into doublets. The triangles
may resolve into triplets. Some of them can occur with the
same sample while scanning the beam over the surface, others
appear with different samples, but all of them have been
repeatedly observed.

Electric Dipole Fields on Mica Surfaces

The variety of observations described seems to be
confusing but it actually has one cause - the presence of
electric dipole fields on the mica surface. Consider a flat
and clean surface partly covered with electric dipoles
which are parallel to the surface and ordered in domains
(Fig. 2). The primary electron beam approaching the sample
at normal incidence suffers a deflection while passing the
field lines of the domain dipole and will again be deflected
into the same direction after the scattering process at the
surface. Electron beams near domain boundaries experience
all degrees of bending between minimum and maximum deflection.
This effect is independent of the diffraction process and
produces a spot streaked in the dipole direction. Three types
of domains rotated against each other by 120° about the
surface normal causes star-like spots to arise (Fig. 1a),
provided the primary beam covers all three different types of

domains. Spots with missing wings appear when the beam only

covers two or even one type of domain. Small and somewhat

uniformly distributed domains give rise to triangular spots

where fields at boundaries between rotational domains are

relatively important and consequently deflect electrons in

all directions, but in the domain directions with maximum

deflection. However, in the case of relatively large domains

the boundary fields are of minor importance, and triple (or

double) spots appear. Even the single round spot pattern is

included in this case: the primary beam covers only one

large domain which causes the simultaneous deflection of

all spots. Taking a long photographic exposure while the

beam is scanned over a large area of the crystal generally

results in triangles or triple spots even though other

shapes appear at specific locations. This indicates that

dipole fields on mica are a general feature and that three types

of electric dipole domains are present no matter what a

particular spot shape is. The spot shape is determined by

the size of the single domains and the distribution of each

type. There is another good proof for the existence of those

field domains: as an insulator mica can be charged up so that

the primary beam is repelled and does not touch the crystal.

The mirror image of the primary beam is then a three winged

star, bearing the threefold symmetry of the dipole fields

which extend through the repeller field.

The Origin of Surface Dipoles

Mica is a sheet silicate mainly consisting of silicon-oxygen tetrahedra with strong bonds. The sheets are linked by single layers of potassium ions along which cleavage takes place (Fig. 3). There is good evidence for assuming that half a monolayer of potassium is left on each side after cleavage,[1] but the potassium ions are not ordered. In the first Si layer below the surface there are as many Al atoms replacing Si on their sites as there are K atoms on the surface, and it is the arrangement of Al[4] which determines the distribution of the surface potassium. In the Si layer each Al is the center of a charge imbalance towards which the nearest potassium ion is attracted. Thus each potassium ion moves out of its old equilibrium site by a small amount creating a dipole parallel to the surface (and rotated by about 15° from the principal crystallographic directions).

K on mica

This situation suggests an experiment in which the incomplete surface is completed by K-evaporation onto mica. If all empty K sites at the surface are filled with K atoms, induced dipoles should minimize the existing dipoles. The result is shown in Fig. 4. A mica sample is rotated out of the normal position so that during evaporation the LEED

pattern can be viewed. Starting with a triangular pattern
(slightly distorted because of rotation) the triangles
shrink in size until they become round spots when the potassium surface layer is completed. The dipole fields have
almost been cancelled and don't show in irregular spot shape
anymore.

Adsorption of Polar Molecules

Permanent dipoles have a different effect on the
diffraction pattern when adsorbed at mica surfaces.
$C_2H_4Cl_2$, $C_6H_5NO_2$ and LiCl were used. Qualitatively the
results are similar. Upon dipole adsorption, triangular
spots are resolved into triplets, streaks into doublets.
Figure 2, top, shows the domain distribution for the
appearance of streaked spots. The added dipoles apparently
adsorb and align themselves with the domain dipole at the
domain edge where the fields are strongest thus enlarging
the domains (Fig. 2, below). But as discussed earlier,
larger domains cause streaks to resolve into doublets and
for the same reason triangles into triplets.

Dipole Fields on ZnO

Another example of a crystal on which surface dipole
fields have been observed is ZnO. Pencil-shaped single
crystals, grown along the c-axis, were cleaved in UHV to expose
the (0001) surface perpendicular to the c-axis. Immediately
after cleavage a hexagonal diffraction pattern was seen (Fig. 5).

Over a wide range of energies an alternating sequence of
sharp and diffuse patterns appeared. Moreover, the diffuse
ones were always displaced in the direction in which the
cleavage proceeded. As a result all spots moved with
increasing primary beam voltage along zigzag lines towards
the center as indicated in Fig. 6. The (00) spot (here
obscured) alternates in position. In order to explain this
strange but very consistent behavior of the pattern one
may speculate that electric dipole fields are causing the
effect. Because of the structure of ZnO with alternating
layers of zinc and oxygen perpendicular to the c-axis and
the character of the bonds, partly ionic, there are expected
to be electric dipole fields normal to the (0001) surface, on
the one cleavage face directed **outward,** at the other
inward (Fig. 7). But due to irregulatities in the surface
(such as steps!) there may be domains with fields that have
a surface parallel component. Spots arising from the first
kind of domain with normal fields only will not be affected
in their position, those from the second kind domain however
will be shifted and may be broadened according to the.
azimuthal distribution of the horizontal field component.
Moreover, the primary electrons will arrive at the two
types of domains with different energies, and consequently,
the intensity curves I(E) for equivalent spots arising
from different domains are shifted by ΔV which is about 7V
according to our experiments.

This model explains the alternating sharp and diffuse spots, as well as the zigzag path of the spots and the occasional appearance of doublets, that is when the shifted I(E) curves show equal intensity. Careful intensity measurements on both surfaces, which are difficult because of the irregular movement and splitting of the spots, should lead to the surface potential difference $2\Delta V_1$ of both cleavage faces and possibly the field strength, but these measurements have yet to be made.

CONCLUSION

Electric fields have been observed with LEED on semiconductor and insulator surfaces. Both the normal and the horizontal components can be measured and with improved instrumentation this can be done with precision. Patterns seen on mica and on ZnO indicated that dipoles related to crystallographic features combine into domains to produce the long-range electric effects observed. Patterns characteristic of the clean surface can be altered by adsorption of permanent dipolar or polarizable molecules. Effects of the dipole sources on nucleation, epitaxy and other surface phenomena were not studied but are doubtlessly observable. Furthermore the relation of spot diffusivity to "domain disorder" must be extended where dipole fields are a factor contributing to the diffusivity.

ACKNOWLEDGMENT

The author should like to express his indebtedness to J. J. Lander with whom he has often had the pleasure of discussing the problems considered in this paper.   The cooperation with C. C. Chang on part of the mica work is greatly acknowledged.   Thanks are also due to E. G. McRae, M. Henzler and C. W. Caldwell for many valuable suggestions.

REFERENCES

1. K. Müller and C. C. Chang, to be published in Surface Science.

2. S. Goldsztaub and B. Lang, C. R. Ac. Sciences 258, 117 (1964):
    S. Goldsztaub, G. David, J.-P. Deville and B. Lang, C. R. Ac.
    Sciences 262, 1718-1720 (1966).

3. K. Müller, Z. Phys. 195, 105-124 (1966).

4. K. Müller and C. C. Chang, Surface Science 8, 455-58 (1968).

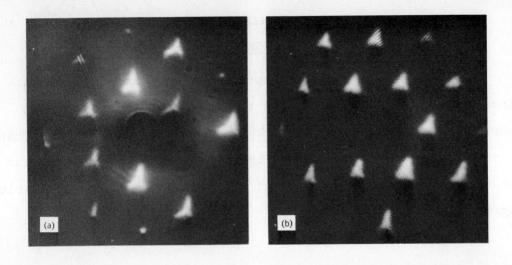

Fig. 1a.   Star-like spot shape, 50 V, mica.   b.   Triangular spots, 68 V, mica.
c.   Various spot shapes.

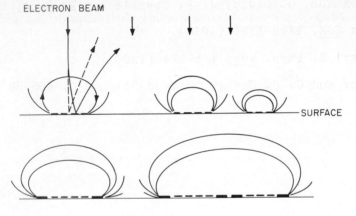

— SURFACE DIPOLES

— ADSORBED DIPOLES

Fig. 2.   Domains of surface dipoles on mica.

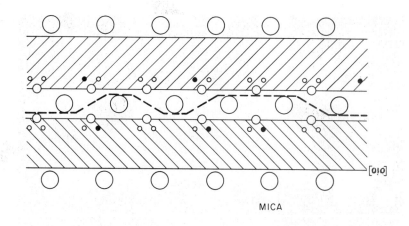

K   OXYGEN   Si   Aℓ

Fig. 3.   Cross-section through mica along [010].

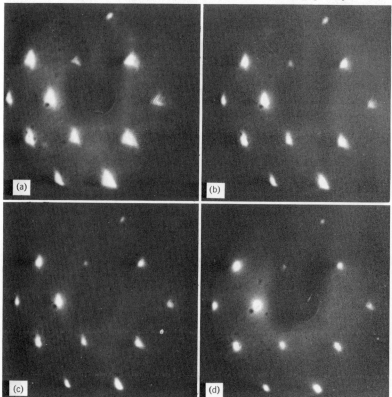

Fig. 4a.   Pattern with triangular spots before K-evaporation, 90 V, mica.
b,c.   Pattern during evaporation, 90 V.   d.   Round spot pattern after
completing the surface layer of potassium, 90 V.

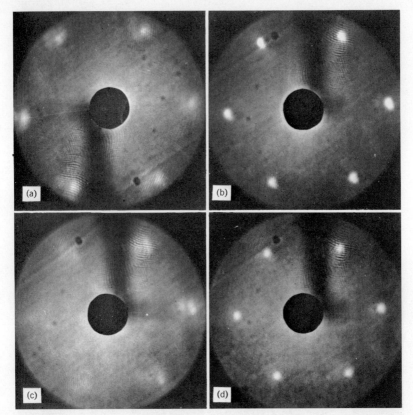

Fig. 5. LEED pattern of ZnO (0001). a. 70 V, b. 80 V, c. 90 V, d. 100 V.

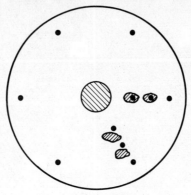

Fig. 6. Spot positions in ZnO–LEED pattern at steps of 10 V.

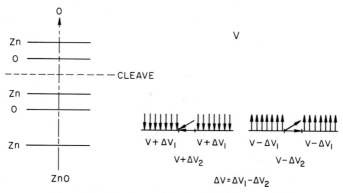

Fig. 7. Surface dipole fields on ZnO.

MS AND ESR STUDIES OF DANGLING BONDS AND ADSORBED IONS ON THE PRISTINE SURFACE OF SILICA

G. Hochstrasser and J. F. Antonini

Battelle Memorial Institute, Geneva Research Center
Geneva, Switzerland

I. Peychès

Compagnie de St.-Gobain, Paris, France

INTRODUCTION

This paper is intended to give a general idea of mass spectrometric (MS) and electron spin resonance (ESR) studies, conducted in ultrahigh vacuum (UHV), on the structure and chemistry of the pristine surface of silica.

In the first part, a general but short review of the results obtained is presented. As the description of the techniques have already been published (1) (2), this report gives only the conclusions that can be drawn from measurements.

From these results, some of the most salient ones have been chosen for detailed discussion in the second part.

1. General Review

Pristine silica surfaces, produced in UHV by fracture or grinding of quartz or silica, display a certain amount of dangling bonds. These dangling bonds, whose limited number is probably due to reconstruction processes taking place mainly during the first microsecond after the fracture, are partially paramagnetic and neutral, and partially diamagnetic but charged.

As already described (1), the so-called E_s' surface paramagnetic centers are constituted by electrons trapped in oxygen vacancies or, otherwise considered, non-bonding orbitals of superficial silicon atoms. These non-bonding orbitals are of nearly sp^3 character. They are stable in UHV for months at ambient temperature, but irreversibly and gradually damaged with increasing temperature as diffusion, from the bulk to the surface, of OH groups or alkali ions at temperatures of the order of 400°C takes place. They disappear immediately by chemisorption of gases such as CO_2 or O_2.

A statistical distribution of the ESR parameters around mean values is clearly visible, and is attributed to the creation of slightly different paramagnetic centers by the fracture process. An homogenisation of the dangling bonds ESR parameters is subsequently obtained by annealing to higher temperatures, which aids relaxation of the local stresses owing to their manner of creation.

The number of paramagnetic centers is nearly the same for silica or quartz of different origins, around 1% of the total number of Si dangling bonds theoretically obtainable by cutting the solid (2.5×10^{14} cm^{-2}). No paramagnetic centers due to SiO$^{\cdot}$ broken bonds (theoretical number 2.5×10^{14} cm^{-2}) were observed at all, even at sufficiently low temperatures to allow degenerate ground states to be observed. This may be due to another reconstruction process.

Separate UHV mass spectrometric studies, however, provide a value of 4% (that is, 8% of the theoretically available Si broken bonds) for active sites in chemisorption of CO, CO_2, and O_2, for instance. Moreover, for CO_2, there is a single thermal desorption peak, which means that there is a single type of chemisorption sites.

The apparent discrepancy between these two results has to be attributed to the non-observability by ESR of the diamagnetic but charged non-bonding orbitals. This statement is experimentally proved for oxygen by comparison between the number of molecules introduced and the number of ions chemisorbed; the number of those which are paramagnetic increase with the pressure finally to reach the total number of the molecules introduced.

The sorption of CO_2 and O_2 is due to, or followed by, a partial charge transfer from the adsorbant to the adsorbate. In the case of carbon dioxide, CO_2^- radical ions are then formed, as has been observed by means of the ESR spectrum. A same transfer occurs during O_2 adsorption, which yields O_2^- radical ions. Thus the superficial complexes may be described as being Si$^+$ CO_2^- and Si$^+$ O_2^-, respectively. These complex bonds probably are both covalent in part. Moreover, the ESR measurements of spin densities allow us to depict the configuration of the adsorbed Si$^+$ CO_2^- complex.

Thermal desorption of CO_2, completed at temperatures up to 250°C, is accompanied by an inverse charge transfer, which can be followed by the decrease of the CO_2^- signal and the increase of the E_s' signal. At temperatures higher than 250°C, the initial number of E_s' centers is completely recovered; therefore the process is quite reversible.

Such results on thermal desorption of CO_2 are confirmed by mass spectrometric measurements, giving moreover the desorption energy, which is found to be 11 kcal.mole^{-1} (or 31 kcal.mole^{-1} (*)).

In the range of experimentally accessible temperature (T_{max} = 850°C), the thermal desorption of O_2 is possible for only 10% of the sites. The desorption

---

(*) If the τ_0 value that appears in the Frenkel's equation is assumed to be equal to 10^{-13} sec.

curve has a maximum near $540^{\circ}C$, which corresponds to a desorption energy of 55 kcal.mole$^{-1(*)}$.   No other desorption takes place below $850^{\circ}C$; this means that most of oxygen bonds have a desorption energy higher than 75 kcal.mole$^{-1(*)}$.

The difference between the desorption energies is well correlated with the type of surface complexes formed, and is higher for the adsorption of oxygen, which is an oxidation process of the metal site.

It may finally be noted that the surface mobility of O_2^- ions is zero, and becomes only visible for CO_2^- at high temperatures.

2.   Some Striking Points

Of the results obtained, only five are presented here, in order to give an idea of the manner in which the main conclusions were drawn.

2.1.   The location and features of the dangling bonds

The powder obtained by crushing a piece of quartz or silica in UHV gives an ESR signal (Fig. 1), proportional to the area of the pristine surface. This observation shows that the existence of the so-called E_s' center is correlated with the surface existence; it does not prove, however, that all the E_s' centers are located just at the surface.

In fact, the intensity of the E_s' signal decreases during the adsorption of an active - even diamagnetic - gas, such as CO_2, to nearly 20% of its initial value. This percentage can be attained very rapidly, which indicates that the interaction of the E_s' sites with the gas can take place immediately, for the greatest part of these sites, without any delay due to diffusion, for instance. The reduction of the residual 20% signal to some percents, during the first few days after the gas introduction, shows that part of the E_s' centers is only accessible by some migration process, even leaving some subjacent centers permanently unaltered. It can therefore be concluded that at least 80% of the E_s' centers are true superficial centers, i.e., broken bonds.

The features of these broken bonds are given by two experimental measurements. First, the thermal desorption of preadsorbed carbon dioxide observed by MS shows that there is only one desorption peak, corresponding to a desorption energy of 11 kcal.mole^{-1}. It is concluded that there is only one type of adsorption sites at the pristine surface of silica.

Second, the nature of this type of sites is given by the ESR measurement of the spectrum hyperfine structure due to ^{29}Si. As is well known (3), if the atomic or molecular orbital, occupied by the unpaired electron responsible for

---

(*) If the τ_o value that appears in the Frenkel's equation is assumed to be equal to 10^{-13} sec.

the ESR signal, belongs partially to an atom, whose nucleus has a non-zero nuclear spin, the spectrum line is split; moreover, this splitting, called hyperfine splitting, is proportional to the spin density at the nucleus.

In the case of the E'_s centers of silica, the ^{29}Si nucleus (natural isotopic concentration : 4.7%) produces an isotropic hyperfine splitting of $A_{iso} \cong 465$ Gauss (Fig. 2).

To evaluate the spin density, this value has to be compared with the estimated splitting $A_{iso} \cong 1740$ Gauss, computed for a pure 3s orbital of silicon (4). The 3s character of the paramagnetic center thus deduced is nearly 27%, i.e., not far from the 25% s character expected for a sp^3 hybrid orbital. The E'_s centers then are dangling bonds of superficial silicon atoms, of the sp^3 type, occupied by an unpaired electron.

As the adsorption causes the E'_s centers to disappear, it can be inferred that the active centers for adsorption on the pristine surface of silica are all dangling orbitals of superficial silicon atoms, part of which are the E'_s centers, i.e., are occupied by one electron, responsible for the observed paramagnetism.

The direct participation of the superficial silicon atoms to the adsorption is definitely proved by the existence of a weak ^{29}Si hyperfine structure in the spectrum taken after adsorption of CO_2. This means that the CO_2^- orbital still involves some little part of the Si orbital precedingly dangling.

2.2. <u>The quasi-independence of shape and constants of the E'_s centers despite the various allotropic forms of SiO_2</u>

Three observations must be emphasized:

- in spite of the crude method used to create pristine surface, ESR spectra are detectable;
- surface sites, certainly present in such different orientations and environments, produce relatively sharp lines;
- the various allotropic forms of SiO_2 give rise to nearly the same ESR spectrum.

The influence of the orientation of the paramagnetic sites E'_s - the dangling bonds - is in fact detectable in the spectrum shown in Fig. 1, which is an integrated spectrum, over all the possible orientations in space, of the spectra given by individual paramagnetic centers. It is thus possible to see - by the shape of the spectrum (4 bis) -, that the paramagnetic center is of axial symmetry, what can be expected for a dangling bond.

The influence on the ESR spectrum of differences in local environments between sites on the same "crushed" surface or differences between sites of surfaces belonging to different allotropic forms of SiO_2 can indeed be explained by the result of the comparison between the E'_s surface center constants with those of the E'_1 and E'_2 bulk centers (radiation damage defects in crystalline quartz) (5) (6). As an example, the components of the g tensor for these centers are given in Fig. 3.

It can be seen that these components are nearly identical for the three types of center. It must therefore be concluded that the local order around the silicon atom possessing the dangling bond not only is the same, but moreover is nearly constant over the different kinds of centers whatever they are (bulk vacancies or surface states).

Such a conclusion is supported by the fact that a crystalline unit, the $Si(\frac{1}{2} O)_4$ tetrahedron, is the building unit of nearly all of the allotropic varieties of SiO_2 and thus appears as having a very rigid structure; the E'_s center can then be seen as a dangling orbital of the Si atom of an incomplete tetrahedron, whose dangling orbital is not too greatly influenced by the structure of the bulk, via the remaining bonding orbitals with neighboring atoms. In this manner, the pristine surface center must be nearly the same for all varieties of crystalline or amorphous SiO_2, what is actually found.

However, concerning the influence of the rough method of preparing the pristine surface, it can be experimentally seen that the ESR lines - for a given orientation, the width being extracted from the overall spectrum by computation - are at least ten times wider than expected. This broadening - usually called "inhomogenious broadening" (7) - is due to minor variations in the ESR constant from site to site. These variations may be created by non identical positions - from site to site - of the same neighboring atoms (it can, for instance, be evaluated from the spectra that the variations in the SiO distance do not exceed 2%). Moreover, these variations can be reduced by annealing, which was verified by heating silica powders to $400°C$ for a few minutes; such a heating, which decreases the local stresses, reduces also the line width, thus proving the origin of the spectrum broadening.

2.3. The surface complex with oxygen and its formation during the adsorption

By slowly introducing a selected gas in the UHV vessel, and observing the ESR signal, it is possible to follow the progress of the chemisorption. Two changes are indeed visible in the spectra.

First, the E_s' (dangling bond) signal decreases linearly with the amount of the gas introduced. Second, another signal appears which simultaneously grows up (Fig. 4). This signal differs from gas to gas and can be used to characterize the induced surface complex.

In the case of oxygen, the ESR signal is visible only at low temperatures (under $- 100^\circ C$). However, MS measurements of thermal desorption show that no desorption occurs below $500^\circ C$. Therefore the ESR lack of observation at room temperature is not due to any desorption. There is another reason, connected to the observability of some paramagnetic species: the electronic ground state of the surface complex is degenerate or nearly degenerate (8), that is, formed of more than one level of same energy. This is a first important result, because the O_2^- ion has a degenerate or nearly degenerate $2p\pi^*$ ground state (depending on the environment) and therefore is never visible at room temperature.

Moreover, the components of the g tensor are

$$g_\perp = 2.004_2 \qquad g_\| = 2.007_6, \dagger$$

then characterized by $g_\perp < g_\|$.

Amongst the different oxygen ions (9), only the superoxide O_2^- ion has g tensor components satisfying the same unequality.

The process of adsorption then involves a partial charge transfer from adsorbate to adsorbant, following the approximative equation

$$Si^\cdot + O_2 = Si^+ O_2^-$$

The chemisorption is nothing else than an oxidation. During this, the E_s^- center loses one electron (and its ESR signal disappears), and the adsorbing molecule gains it (and becomes observable as O_2^- ion).

The MS measurements, however, show that the superficial density in adsorbed particles is greater (5 to 10 times) than the number of paramagnetic E_s' centers. This difference can be understood if it is remembered that the ESR experiments are restricted to the observation of paramagnetic species; the adsorption of gas molecules on diamagnetic dangling bonds of silica is not visible by ESR, but can be followed by MS.

In the case of oxygen, exceptionally, the amplitude of the O_2^- ESR signal is pressure dependent and varies reversibly between two extremes (see Figs. 5 and 4e): the maximum amplitude is attained at relatively high pressures,

---

† As determined at the zero line cross over point, and the apparent bump maximum, respectively.

~ 10^{-3} Torr, but the minimum amplitude is constant below 5.10^{-6} Torr. Moreover, the maximum amplitude corresponds fairly well with the surface coverage as measured by MS.

A tentative explanation of this pressure dependence is the following: the O_2 molecules that adsorb on the E'_s centers immediately give radical complexes, but O_2 molecules that adsorb at the same time on similar but diamagnetic dangling bonds, and are in fact invisible by ESR, can, under external influences, become observable (by such a mechanism as that proposed by Voevodskii (10)).

2.4. <u>The kinetics of the carbon dioxide sorption-desorption process</u>

The adsorption of CO_2 takes place with the same partial charge transfer as for the O_2 adsorption. It is observed by MS, however, that the adsorbed CO_2 is completely desorbed near $250^\circ C$ and that the sorption-desorption cycle can be repeated indefinitely.

The simultaneous observation of the adsorbed ions (CO_2^- ions for carbon dioxide adsorption) and of the pristine surface E'_s centers, by ESR, confirms this observation and even reveals the mechanism. It is indeed observed that the superficial density in CO_2^- ions diminishes during the heating of the sample; simultaneously, the superficial density in E'_s centers rises up to nearly exactly the density for UHV (Fig. 6).

It can thus be inferred that the equilibrium of the adsorption reaction

$$Si^\cdot \ + \ CO_2 \ \leftrightarrows \ Si^+ \ CO_2^-$$

can be displaced from the right to the left by heating (desorption) and from the left to the right by cooling (adsorption). In other words, the desorption process is accompanied by an inverse charge transfer that renders the surface its whole sorption capacity.

Concerning the order of the desorption reaction, MS studies of thermal desorption reveal that the desorption is a first order process. This observation is confirmed by ESR, which shows that the CO_2 molecule of the gas phase is retained, when the adsorption takes place, as a single particle.

2.5. <u>The surface mobility of the adsorbing molecules</u>

In the case of no mobility of the adsorbing molecules, a simple model (11) predicts, for first order process, that the sticking probability s decreases linearly with increasing coverage θ , during the adsorption. On the contrary, in the case of high surface mobility, it indicates that this sticking probability remains constant during the adsorption, until the maximum coverage is attained, and then drops rapidly.

From the recorded p(t) curves, describing the variation of the dynamic equilibrium pressure versus time just after the sudden creation of a new surface, s(θ) curves can be deduced, from which some information concerning the mobility can thus be extracted.

The experimental results for O_2, given in Fig. 7, fit the predicted curve well in a large range of temperature and show that the mobility during adsorbtion is effectively nil. Those for CO_2 indicate that the mobility exists at temperatures near the temperature at which there is no more adsorption. At lower temperatures, such as $180^{\circ}C$, it seems that the mobility disappears, as also assumed before. However, s(θ) curves for still lower temperatures are puzzling and not yet well explained.

ACKNOWLEDGMENTS

The authors wish to thank Dr. L. A. Pétermann for many fruitful discussions, and Miss E. Pobitschka and Messrs F. Fischer and H. Guédu for their invaluable technical assistance.

REFERENCES

(1) a)   G. Hochstrasser et J. C. Courvoisier, Helvetica Physica Acta 39, 189 (1966)

   b)   G. Hochstrasser, compte rendu du Symposium de l'U.S.C.V. sur la surface du verre et ses traitements modernes (Luxembourg 1967) p. 79

(2)   J. F. Antonini, compte rendu du Symposium de l'U.S.C.V. sur la surface du verre et ses traitements modernes (Luxembourg 1967) p. 69

(3)   e. g. P. W. Atkins & M. C. R. Symons The structure of inorganic radicals (Elsevier) 1967 p. 14

(4)   W. Kohn & J. M. Luttinger, Phys. Rev. 97, 883 (1955)

(4bis)   P. W. Atkins & al., ibidem p. 269

(5)   R. H. Silsbee, J. Appl. Phys. 32, 1459 (1961)

(6)   R. A. Weeks, Phys. Rev. 130, 570 (1963)

(7)   e. g. M. Bersohn & J. C. Baird, An introduction to E.P.R. (Benjamin) 1966, p. 58

(8)   e. g. A. Carrington & A. D. MacLachlan, Introduction to magnetic resonance, (Harper) 1967 p. 200

(9)   P. W. Atkins et al., ibidem chap. 5, 6 and 7

(10)   V. V. Voevodskii, Proc. 3^d int. congr. catalysis (Amsterdam 1964) p. 88

(11)   J. F. Antonini, Proc. 4^d int. vacuum congress (Manchester 1968) (to be published).

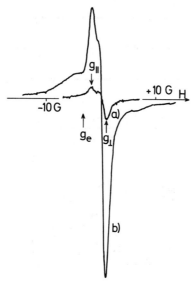

Fig. 1.   ESR signal of the E_S' centers of crushed quartz: a) after 150 pestle knocks; b) after 2000 pestle knocks.

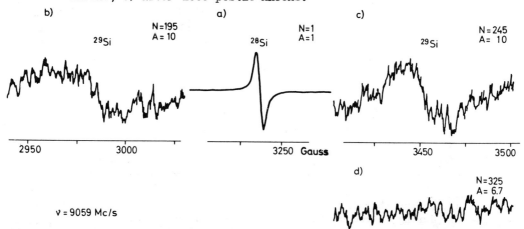

Fig. 2.   Hyperfine (hfs) structure of E_S' superficial centers: a) Same signal as in Fig. 1, but overmodulated.   b) & c) hfs due to ^{29}Si, extracted from the noise by means of a computer of average transients.   d) Same record as in c) but in air and no more in UHV,

N : number of cumulated records;
A : relative amplification.

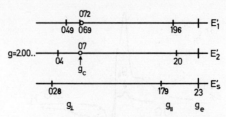

Fig. 3. Components of the g factor for the different types of E' centers

(absolute accuracy $\pm$ 0.0002; relative accuracy $\pm$ 0.00003).

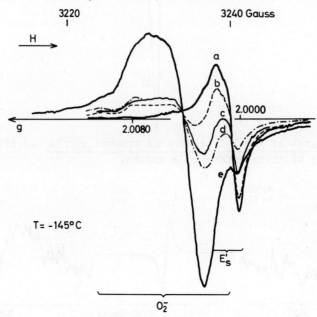

Fig. 4. Variations of the E_s' and O_2^- signals, during the introduction of

oxygen:

a) before introduction   UHV conditions

b) during introduction, 12 min. after valve opening
c) "   "   , 24 min. "   "   " $P_{source} \sim 5 \cdot 10^{-5} T$
d) "   "   , 80 min. "   "   "
e) at the end of introduction, after some hours $P_{source} \sim 10^{-2} T$

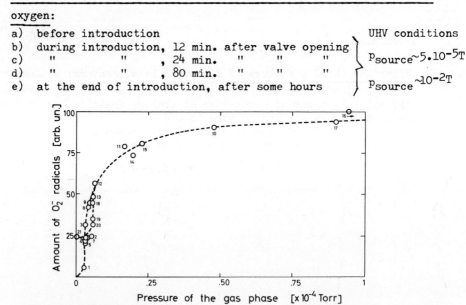

Fig. 5. Pressure dependence of the O_2^- signal: The measurements were made in
the order indicated.

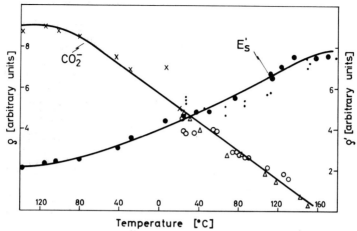

Fig. 6. Superficial density ρ of the CO_2^- adsorbed ions and ρ' of the E_s'

centers, versus the surface temperature T.

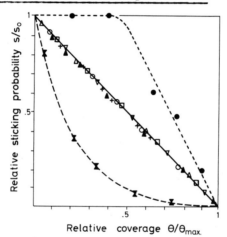

Fig. 7. Relative sticking probability s/s_0 versus relative coverage θ/θ max.

— theoretical curve for the case of no mobility.

□	experimental points for O_2				$T = -196°C$
▽	"	"	"	"	$T = 30°C$
+	"	"	"	"	$T = 212°C$
△	"	"	"	"	$T = 312°C$
○	"	"	"	"	$T = 490°C$
✗	experimental points for CO_2				$T = -55°C$
▲	"	"	"	"	$T = 180°C$
•	"	"	"	"	$T = 210°C$

LEED EXAMINATION OF THE LEAD MONOXIDE SURFACE

H. E. Bishop and J. C. Rivière

United Kingdom Atomic Energy Authority
Solid State Division, A.E.R.E.
Harwell, Berks, England

1. Introduction

Lead monoxide is used as the active semiconducting element in the vidicon tubes known as Plumbicons[1], in the form of thin evaporated layers consisting of a large number of small platelets. It has been shown[2] that the electrical properties of such layers may be dominated by surface states rather than levels due to bulk doping, which implies, in the case of a non-stoichiometric semi-conductor such as PbO, a dependence of semiconducting properties on oxygen concentration in the surface. The activity of the surface in use in the Plumbicon does in fact deteriorate with time due to loss of oxygen[1], posing a technological problem. The surface of PbO is thus of great interest; this report describes a preliminary investigation into the feasibility of using LEED as a technique in the study of this surface.

PbO exists in two allotropic forms; a tetragonal (red) low temperature form and an orthorhombic (yellow) high temperature form which evolves from the former by a displacive transformation at 488°C. The high temperature form may be stabilized at room temperature by the addition of 1% bismuth. Both forms have a layer structure parallel to the (001) plane and cleave readily along these planes.

The two structures are shown in Fig. 1. The transformation from the red to the yellow form involves a puckering of the oxygen layers and a small dis-placement of the lead atoms. The unit cell for the yellow form contains twice as many atoms as that of the red form after the transformation and the a and b axes of the yellow form are at 45° to the axis of the original red form. Physically the red form is very crumbly and it is difficult to prepare a suitable specimen for LEED. On the other hand the stabilized yellow form is much more easily handled and it is relatively easy to cleave a specimen for LEED examination.

2. Experimental

The PbO crystals were grown from the melt by slowly cooling liquid PbO in a platinum crucible*. A suitable flake of PbO was cleaved from the boule and crimped to a thin platinum backing mounted on two molybdenum supports. The specimen was heated by passing a current through the platinum backing. The temperatures quoted are those of the platinum measured by a thermocouple, and are therefore rather higher than the temperatures of the specimen itself. During heating the gases evolved from the specimen were observed with an EAI.250 Quadrupole Mass Spectrometer.

3. Results

a) Red Tetragonal form

Initially no pattern was obtained from the specimen, but it was observed to charge up at beam energies less than 64 eV, showing that there was no ohmic contact to the support. The specimen was then heated to 250°C (as measured on the support). The main gases evolved were CO_2, CO and H_2O with the CO_2 peak dominating. The CO_2 pressure with the sample at 250°C was in the region of 5×10^{-9} torr. Again no pattern was visible when the specimen was cooled.

The first diffraction patterns appeared only after the specimen had been heated to 400°C and were diffuse with a high background. They were improved a little by subsequent heating to a temperature in the region of 750°C but never became sharp. In preparing the specimen it was difficult to obtain flat areas of reasonable size and as a result there were several regions of the crystal giving patterns, some better than others. The poor quality of the patterns may well have been a result of the surface macrostructure rather than an indication of disorder on an atomic scale.

---

*The authors are grateful to Mrs. B. Wanklyn of Oxford University for growing these crystals.

At room temperature the specimen charged up for beam voltages less than about 60 volts but if the crystal were warmed (to approximately 100°C) patterns could be obtained down to low voltages. The patterns observed at 74 and 100 volts are shown in Fig. 2. They show the fourfold symmetry expected from the (001) face of tetragonal PbO, and a surface lattice spacing deduced from the patterns consistant with the bulk spacing, to the accuracy of the measurements (10%). The patterns were all rather diffuse with a relatively high background and no spot patterns were observed above 140 eV beam energy although a very diffuse Kikuchi pattern could be discerned in the region of 1000 eV.

b) Yellow Orthorhombic form

The crystals of yellow lead oxide, stabilized by bismuth, were much more easily cleaved into suitable specimens for LEED examination. Again no pattern was obtained until the specimen had been heated. The first pattern showed four fold symmetry and seemed identical to those obtained from the tetragonal sample but unfortunately in neither case were the patterns of sufficient quality to be able to check this by making intensity-voltage measurements. On further heating however (up to 650°C) a second type of pattern began to appear in some areas. During this heating a little oxygen was evolved but the amount was much less than the amounts due to the evolution of CO_2, CO and H_2O. At first both patterns appeared simultaneously (Fig. 3a) but on further heating many areas gave only the new pattern although there were always some parts of the specimen from which the original pattern could be observed. The new pattern, shown in Fig. 3, was much sharper than the initial pattern and persisted to much higher beam energies, giving a reasonably sharp Kikuchi pattern at 1000 eV. The symmetry of the pattern was rectangular as was to be expected from the (001) face of an orthorhombic crystal, while the ratio of the surface parameters was found to be 1.14 ± 0.01, agreeing well with the a/b ratio of bulk ortho-rhombic PbO which is 1.15. An indication of the c spacing near the surface

was obtained from the intensity curve measured for the 00 beams (Fig. 4).
Table 1 gives the various c spacings for the bulk forms together with spacings
deduced from the intensity voltage measurements.

Table 1.

Comparison of bulk and calculated spacings in the 'z'

direction, $\overset{\circ}{A}$

Bulk Spacing	Tet.	Ortho.	Calc. for $V_o = 17.9 \pm 0.8$
c	5.023	5.891	-
intralayer Pb-0-Pb	2.396	2.720	2.72
Interlayer Pb-Pb	2.627	3.171	3.18
			2.96

The 2.72$\overset{\circ}{A}$ and 3.18$\overset{\circ}{A}$ spacings are identical to the bulk values; the reason
for the appearance of the third apparent spacing in the "z" direction, 2.96$\overset{\circ}{A}$,
is not known at present.

From Fig. 3a and similar photographs showing both structures it was
possible to deduce the ratio between their surface parameters. The observed
ratio for $\sqrt{2} \, a_{square} / b_{rectangular}$ is 1.18 which is the same as the ratio
between the equivalent bulk spacings of the tetragonal and orthorhombic forms.

Although it was not possible to perform controlled gas adsorption experiments at the time this work was being done, the system had to be opened to
the atmosphere after the rectangular pattern had been established. After re-
evacuation and baking a pattern with fourfold symmetry similar to the initial
pattern was once again observed; however, the rectangular pattern was regenerated when the specimen was heated to 400°C.

4. <u>Discussion</u>

The observations on the orthorhombic specimen show that by heating it is possible to produce what appears to be a clean (001) surface with no surface reconstruction. They also show, in a single uncontrolled experiment, that adsorption of some contaminant from the atmosphere causes the surface to relax to the tetragonal symmetry. The relatively poor results obtained from the red form could have been due to failure to clean the surface by heating, but, as one might expect the same cleaning procedure to work for both forms, the low quality patterns more probably reflect the uneven nature of the surface. It should therefore be possible to make a more detailed study of the (001) surface of the orthorhombic form of PbO using crystals grown from a melt but suitable samples of the red form cannot be produced in this way probably as a result of the phase change from the yellow to the red form during cooling.

REFERENCES

1. Haan, E. F., van der Drift, A., and Schampers, P. P. M., Phillips Technical Review, 25, (1963-1964) pp. 133-180.

2. Van den Broek, J., Phillips Res. Repts. 22, (1967) pp. 367-374.

3. Wyckoff, R. W. G., Crystal Structures 1, 1963. (Interscience: New York and London).

4. Kay, M. I., Acta Cryst., 14, (1961) p. 80.

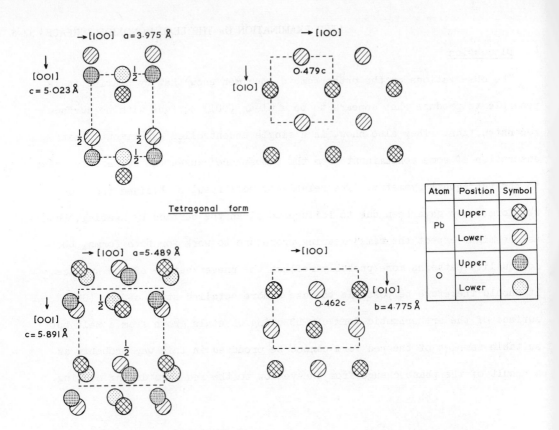

Tetragonal form

Orthorhombic form

Fig. 1. Structures (3, 4) of the two allotropes of PbO.

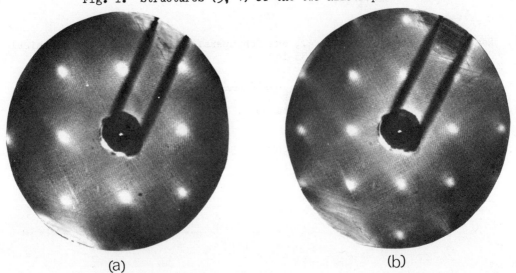

(a)   (b)

Fig. 2. LEED patterns from the (001) face of tetragonal PbO: (a) 74 volts. (b) 100 volts.

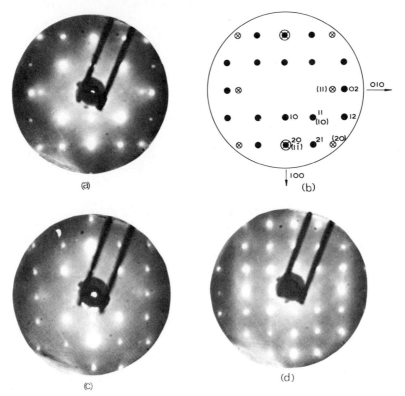

Fig. 3.   LEED patterns from the (001) face of orthorhombic PbO: (a) showing
both structure, 69 volts. (b) indexing diagram for (a), spots for the
rectangular pattern denoted by • and for the square pattern by ⊗. The
indices for the square pattern are enclosed in brackets.   (c) and (d)
rectangular pattern at 84 and 111 volts.

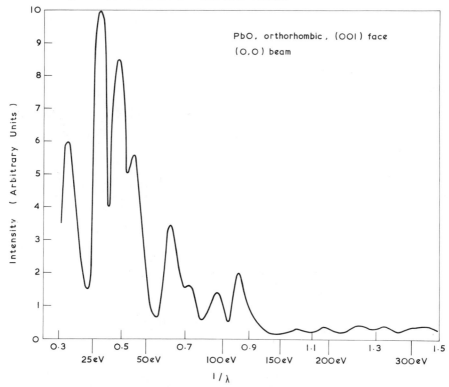

Fig. 4.   Intensity voltage curve for the (0,0) spot of orthorhombic PbO.

KINETICS AND MECHANISMS OF HIGH-TEMPERATURE VAPORIZATION PROTOTYPE STUDIES OF ARSENIC AND ANTIMONY SINGLE CRYSTALS*

Gerd M. Rosenblatt, Michael B. Dowell, Pang-Kai Lee, and H. R. O'Neal

Department of Chemistry
The Pennsylvania State University
University Park, Pennsylvania

High-Temperature Chemistry and Vaporization Kinetics

High-temperature chemistry is chemistry at temperatures far enough above room-temperature so that extrapolations of chemical behavior from room temperature data are neither reliable nor valid. At high temperatures the entropy term, $T\Delta S$, in the equation determining the position of chemical equilibrium, $\Delta G = \Delta H - T\Delta S$, can become more important than the enthalpy term, ΔH. However, most correlations of chemical behavior, such as the listing of normal valences and normal oxidation states of the elements, are based primarily upon enthalpy effects as these dominate at lower temperatures. Because entropy considerations become important at high temperatures new chemical reactions occur and new species are found. Neither the existence nor the stability of these chemical species can be predicted from room-temperature data. Consequently, a great deal of effort in high-temperature chemistry has been devoted to finding out which chemical compounds exist at high temperatures and what their relative stabilities are. Stabilities are determined by measuring equilibrium constants, $\Delta G° = -RT \ln K$. Chemical equilibria of particular interest at high temperature are those between gaseous molecules, and between gases and condensed phases, as most substances have appreciable vapor pressures at high temperature, and, at these temperatures, reactions readily occur between gases and the condensed phases with which they are in contact. Condensed phases are usually present at high temperature because experiments are carried out in some type of container. The presence of condensed phases often results in high-temperature gaseous systems

*Presented at the symposium dedicating the Graduate Center for Materials Research of the Space Sciences Research Center, University of Missouri, Rolla, October 31, 1967.

becoming more complex as the temperature increases, since the tendency of complex gaseous species to form upon an increase in vapor pressure often predominates over the tendency of gaseous molecules to dissociate with increasing temperature.

In many applications of high-temperature materials, ranging from outer space to the home kitchen, one is not interested so much in the equilibrium thermodynamic properties of the materials involved as in reaction kinetics under specified conditions. The designer of a rocket or a kitchen stove needs to know how rapidly chemical reactions occur; knowledge of equilibrium concentrations of gaseous species is insufficient. In general, intelligent use of high-temperature materials requires understanding the kinetics of those reactions which can occur at high temperature--vaporization reactions, gas-solid and gas-liquid reactions, reactions between gaseous molecules. In this paper we will consider some aspects of vaporization kinetics which have implications for high-temperature materials. In addition to their intrinsic interest, vaporization kinetics are of interest because vaporization is a particularly simple gas-solid reaction: detailed understanding of vaporization is a prerequisite for understanding many more complex gas-solid reactions. The user of high-temperature materials needs answers to practical questions about vaporization behavior, such as: How fast will a material vaporize at high temperatures? How is the rate of vaporization related to the vapor pressure and other equilibrium thermodynamic properties of the material? How can the rate of vaporization be controlled or modified?

It is well known that rates of vaporization can be related to equilibrium vapor pressures.[1] Consider a condensed sample in equilibrium with its vapor in a closed container at constant temperature. When the system is at equilibrium the gross rate of vaporization will be exactly equal to the gross rate of condensation and the net rate of vaporization will be zero.

Using the kinetic theory of gases, the number of gaseous molecules which strike unit area of surface per unit time can be calculated from the equilibrium vapor pressure. If every molecule which hits the sample surface sticks and is incorporated into the condensed phase, this number is the gross rate of condensation and, therefore, the gross rate of evaporation under equilibrium conditions. If the gross rate of vaporization is independent of the saturation of the vapor this is also the rate of vaporization into vacuum. Any deviation between an observed vaporization rate and that calculated from kinetic theory and the equilibrium vapor pressure in the above manner is described by a vaporization coefficient (α_v). Thus the problem of predicting rates of vaporization can be described as one of knowing equilibrium vapor pressures and vaporization coefficients. Vaporization coefficients are closely related to condensation coefficients (α_c) which give the fraction of gaseous molecules striking a surface which condense.

For most metals, and other simple substances, vaporization coefficients are close to unity, and rates of vaporization can be reliably predicted from equilibrium vapor pressures.[2,3] However, in molecular and ionic systems, vaporization coefficients can be very much less than unity, and observed rates of vaporization can be very much less than those predicted from equilibrium vapor pressures. In such systems the gaseous species may not exist as such in the solid lattice and chemical reactions or considerable bond rearrangement have to occur to form gaseous molecules from atoms in the solid lattice.[4] This type of behavior might be expected to be particularly important in the complex chemical rearrangements which often occur when new high-temperature molecules are formed from condensed phases.

There is another reason why vaporization kinetics are of interest in high-temperature chemistry. Because gaseous molecules and gas-solid reactions are important at high temperatures, the problem of determining high-temperature equilibrium constants often involves the measurement of vapor pressures and partial vapor pressures. At high temperature these measurements are not usually carried out by completely static means. Rather, high-temperature vapor-pressure measurements are normally carried out by either the Knudsen effusion or the Langmuir free-vaporization method.[5] In Knudsen effusion one measures the rate of escape of vapor molecules through a small orifice in a cell containing a condensed sample. This is often coupled with mass-spectro-metric analysis of the effusing vapor. In the Langmuir method the rate of vaporization of a sample into vacuum is measured. Clearly, thermodynamic data resulting from Langmuir vapor-pressure measurements will be in error unless vaporization coefficients are known and appropriate corrections are made. Errors can also arise, to a lesser but to a more complex extent, in Knudsen measurements on substances with very low vaporization coefficients. When the vaporization rate of a sample in a Knudsen cell is very slow, effusion through the orifice depletes the vapor in the cell more rapidly than vaporiza-tion can maintain the equilibrium vapor pressure. The vapor pressure calculated from the effusion data then falls somewhere between the true equilibrium vapor pressure (P_e) and that obtained from the free (Langmuir) vaporization rate ($\alpha_v P_e$). Exactly how much below P_e depends upon the vacuum vaporization coefficient of the sample and the geometry of the cell.[6,7,8] It can also depend upon the way in which α_v varies with the extent of saturation of the vapor, and with the physical state, degree of subdivision, and past history of the sample.

In order to understand those aspects of vaporization behavior parti-cularly important to high-temperature chemistry--the prediction of vaporization coefficients, the ways in which vaporization rates can be varied and controlled, the interpretation of Knudsen effusion and other dynamic and semi-dynamic vapor-pressure measurements--one has to understand the detailed, microscopic mechanism by which atoms in a solid lattice rearrange and escape to form free gaseous molecules. Of particular interest are those steps in the mechanism which are different for materials with low vaporization coefficients than for substances with vaporization coefficients near unity. There are additional reasons for being interested in the vaporization kinetics of materials with low vaporization coefficients. Vaporization is a first-order kinetic process for which enthalpies and entropies of activation can be related to standard equilibrium enthalpies and entropies of vaporization; additional insight into vaporization mechanisms may be gained by comparison of these two types of data. Detailed studies on materials with low vaporization coefficients can yield information about vaporization mechanism in general. Rate studies on systems having vaporization coefficients of unity only give back thermodynamic parameters, and no mechanistic insight can be gained by analysis of thermodynamic quantities. On the other hand, chemical systems having low vaporization coefficients give activation enthalpies or entropies which differ from equilibrium quantities and which have mechanistic implications. In systems having low vaporization coefficients there is usually one slow step in the vaporization process, and it is possible, by careful choice of experimental conditions, to vary parameters (such as concentrations) on either side of the slow step. Modification of the vaporization process in this way can further elucidate the mechanism of vaporization.

In the following sections of this paper we first outline some results from recent vaporization studies in this laboratory and then discuss their implications covering the mechanism of vaporization, the systematization of vaporization behavior, and the relationship between vaporization kinetics and high-temperature vapor-pressure measurements. The discussion

will be confined almost entirely to data obtained in our laboratory as the

present status of vaporization research, in general, is thoroughly covered

in recent reviews.[2,3,9-11] High-temperature materials have not been used

in the experiments to be described although the research is on problems of

concern to the high-temperature materials scientist. This is because precise

measurements by a variety of complementary techniques are necessary to obtain

the detailed information required to clarify vaporization processes. Many of

these measurements would be a great deal more difficult at very high tempera-

tures. In addition, interpretation of experimental results is greatly facili-

tated if the measurements are carried out on single crystals of substances

(elements or simple one and two-component systems) which have well characterized

solid and vapor phases. It is difficult to find high-temperature systems

which satisfy these requirements.

Review of Some Experimental Results.

In this work direct investigations of vaporization rates from single

crystal faces have been complemented by study of the changes in surface

morphology which occur upon vaporization, the effect of various gases,

particularly oxygen, upon the vaporization rate, and the sublimation kinetics

of minor gaseous species as well as the predominant species. A variety of

techniques have been used including vacuum microbalance measurements,

residual gas analysis, optical microscopy of vaporizing crystals, inter-

ferometry and various microscopic interferometric methods, replica electron

microscopy, mass spectrometry, and Knudsen measurements. Some of the results

obtained during investigations of the vaporization of As_4 and Sb_4 from (111)

cleavage faces of arsenic and antimony single crystals are summarized in

Table I and discussed below. Detailed experimental results will be published

elsewhere.

Arsenic and antimony were studied for a number of reasons. Vaporization studies on powdered arsenic[12] indicated that it has a very low vaporization coefficient whereas no signs of abnormal vaporization behavior were seen in equilibrium vapor-pressure measurements on antimony.[13] Yet arsenic and antimony both vaporize to form, primarily, tetrahedral gaseous molecules[13,15] and have similar, rhombohedral crystal structures.[16] Thus, they are interesting systems upon which to compare kinetic results. Bond angles and bond distances in the crystals differ appreciably from those in the gaseous molecules. The equilibrium vapor consists primarily of one gaseous species at the temperatures of interest[13,14] which greatly facilitates analysis of kinetic results. Methods for growing single crystals of both of these elements are known.[17] Single crystals are almost a necessity for detailed vaporization studies because they allow one to define the surface area precisely, to know the atomistic structure of the surface, and to minimize contamination of the surface. Another reason that arsenic and antimony were picked for vaporization studies is that these are elemental systems, which minimizes the possibility of a change in surface composition as vaporization proceeds. In a system containing more than one component, even if it is vaporizing congruently so that the solid and the vapor have the same composition, it is possible for the surface to have a quite different chemical composition--which can complicate the vaporization mechanism.

"Langmuir" vapor pressures, P_L, vaporization coefficients, α_v, and enthalpies, ΔH^*, and entropies, ΔS^*, of activation for comparison with standard equilibrium thermodynamic properties (P_E, ΔH°, ΔS°) have been derived from the measured vaporization rates, $\underline{r}$ (units of g cm^{-2} sec^{-1}), using the following equations:

$$P_L = r\ (2\pi RT/M)^{1/2}$$

$$\alpha_v = P_L/P_E$$

$$R \ln P_E \text{ (atm)} = -\Delta H^\circ/T + \Delta S^\circ$$

$$R \ln P_L \text{ (atm)} = -\Delta H^*/T + \Delta S^*$$

Rates of vaporization are measured: vaporization coefficients are derived by comparison of experimental results with literature thermodynamic data.[13,14] Therefore, uncertainties in the literature data are reflected in the vaporization coefficients quoted.

The vaporization coefficient of As_4 from arsenic single-crystal wafers cut to expose primarily (111) faces is 4.6×10^{-5} at 550°K and increases with temperature.[18] This value, which depends upon the somewhat uncertain equilibrium vapor pressure of arsenic,[14] is apparently the lowest ever demonstrated for a well defined surface. The activation enthalpy of vaporization at 550°K is 43.9 kcal/mole, about 10.8 kcal/mole higher than the equilibrium enthalpy of vaporization.[14] The activation and equilibrium entropies of vaporization agree within experimental error. The vapor pressure and equilibrium thermodynamic parameters for arsenic are somewhat uncertain. Statistical thermodynamic estimations[19] of the entropy of $As_4(g)$ yield an entropy about 2.6 cal deg^{-1} $mole^{-1}$ higher than the literature value.[14] Errors in both the entropy of $As_4(g)$ and the vapor pressure itself affect the enthalpy of vaporization. None of these errors, however, is of a magnitude to affect the conclusions below.

The vaporization coefficient of Sb_4 from (111) wafers is 0.18 at 650°K and increases with temperature. The activation enthalpy of 49.5 is about 2.1 kcal/mole higher than the equilibrium enthalpy of vaporization, and, as in the case of arsenic, the activation and equilibrium entropies of vaporization agree within experimental error. The data for antimony were corrected for the small error, about 13%, caused by molecules hitting the walls of the vaporization tube and recondensing upon the sample before they could escape from the heated area of the vacuum system.[20] The vaporization coefficient of antimony is definitely less than unity. This was demonstrated by experiments in the same apparatus on the vaporization of powdered antimony from a crucible, and Knudsen-effusion measurements on antimony. With both arsenic and antimony the rates of vaporization are constant and reproducible, although fresh-cleaved crystals may show an initial slow rate which steadily increases to the steady-state rate.

Microscopic observation of vaporizing crystals shows that vaporization of a plane, cleaved surface is accompanied by formation and subsequent growth of shallow triangular pits at the point of emergence of (spiral) dislocations. Individual pits grow at a constant rate, as expected if the rate of vaporization is proportional to the area of the pits. The rate of vaporization increases as the pits grow until the pits intersect, at which time the vaporization rate becomes constant. The surface then looks like a jagged mass of intersecting planes. Plate I illustrates these changes on a (111) arsenic surface. The time required for pit intersection depends upon the density of dislocations and the steepness of pits.

Pits are associated only with mass transport and are not formed or annealed out when the crystal is heated in equilibrium with its vapor. The pit densities are independent of time on a given surface. The angles formed by pit sides and the (111) surface were measured by polarization interferometry and are quite shallow, very much less than the 49° which would correspond to rearrangement of the (111) surface to form (110) facets. Vaporization pits on freely vaporized arsenic single crystals average about 11°, ranging between 9 and 13°, while those on freely vaporized antimony single crystals average about 5°, ranging between 2 and 9°. The vaporization

pits were identified with the point of emergence of spiral dislocations by

studies using CP-4 etchant which is known to form pits at the point of

emergence of screw dislocations on antimony,[21] and by mirror-cleavage vapori-

zation studies. Antimony crystals which were first etched with CP-4 were

observed to vaporize in a way so that the original etch pits grew and no new

pits formed. An arsenic crystal was cleaved and then both faces produced were

vaporized. The pattern of the vaporization pits on the two originally mated

faces forms an approximate mirror image relationship.

Mechanism

We now turn to some implications of these results for the microscopic

mechanism of vaporization. For clarity, we will first discuss the mechanism

in terms of the picture of a surface developed during the last forty years

and associated with the names of Volmer,[22] Cabrera,[23,24] Burton,[23,24]

Frank,[24] Kossel,[25] Stranski,[10,26] Hirth,[2,27] and Pound,[2,27] among others.

The work of these and other investigators shows that an atomic surface (shown

schematically in Figure 1) contains monatomic steps or ledges with an equili-

brium number of kinks, as well as atoms adsorbed at ledges and on the surface.

Dislocations and crystal edges can act as ledge sources. We will examine

how the experiments results enumerated above fit into a model based upon

this picture and then discuss some further experiments bearing on the mechanism

of vaporization.

All the experimental results appear to be quantitatively consistent with

a mechanism in which the unusually slow step in the vaporization process is

associated with the activation energy required to form a tetrahedral gaseous

molecule of As_4 or Sb_4 at a kink on a ledge winding out of a dislocation

etch pit. The tetrahedral molecules, perhaps, then diffuse on the surface and are subsequently desorbed in thermal equilibrium with the surface. A kink site corresponds to Volmer's so-called "half-crystal position" and, for a simple atomic lattice, the energy required to remove an atom from a kink site is just the heat of vaporization.

The kinetic data show the slow step to be an enthalpy effect and rule out mechanisms which would depend upon temperature-independent "entropy" effects, such as a required orientation for the vapor molecule before it can leave the surface,[28,29] or the molecule leaving the surface in an excited rotational-vibrational state which would correspond to partial retention of the configuration in the metal lattice.[30] The kinetic data appear inconsistent with any mechanism which requires dissociation into atoms or diatomic molecules. As Stranski[10,31] has pointed out, when activation energies (that is, activation enthalpies greater than the equilibrium enthalpy of vaporization) determine vaporization kinetics, these are expected to play their role at the kink-surface step in the stepwise vaporization process.

The ledges which are presumed to wind around the sides of the vaporization pits are too small to be seen under normal experimental conditions. Interferometric and phase contrast techniques have shown up step and ledge structure as small as 15 angstroms high.[32] Presumably, the ledges are smaller than this, and, thus, only one or two atom layers in height. Some ledge structure on antimony crystals has been seen in the electron microscope and with interference contrast but it could not be ascertained whether the structure was spiral or not. Also, some large nonspiral ledges have been built up on antimony pits under certain conditions.

It was mentioned above that fresh-cleaved crystals may show an initial slow rate which steadily increases to the steady-state rate. We now turn to a number of kinetic and microscopic experiments on the initial stages of vaporization of fresh-cleaved crystals and discuss their relation to the

postulated mechanism. From geometric considerations, pit angles, density of

pits, and size of pits on a surface at any given time can be related to the

total weight loss which has occurred on the surface at that time, assuming that

all the weight loss occurs from pits. The formulas derived (assuming uniform,

isolated pits) are:

$$w \ = \ \tan \theta \ \ell^3 \ \rho \ \eta \ /24$$

$$w \ = \ 0.146 \ \tan \theta \ \rho \ \eta^{-1/2} \ \zeta^{3/2}$$

where w is the total weight loss per unit area from the crystal surface, θ

is the average angle between the side of pits and the (111) surfaces, ℓ

is the average length of the side of a pit, ρ $(g \ cm^{-3})$ is the density of

the solid crystal, η (cm^{-2}) is the density of pits on that crystal face, and

ζ is the fraction of the surface covered by pits at a given time. Examination

of these equations shows that the weight loss which corresponds to a given

surface coverage by pits--for example, complete coverage (at pit intersection

$\zeta = 1$)--increases as the pits become steeper and decreases as the density of

dislocations increases. When cleaved at room temperature, arsenic crystals

normally show a dislocation density on the order of $5 \times 10^6/cm^2$ while the

dislocation density on antimony is usually appreciably less, on the order

of $10^5/cm^2$.

The proposed mechanism implies that the rate of vaporization is

proportional to the number of kinks and, therefore, to the length of ledge,

if there is a steady-state distribution of kinks per length of ledge. It is

expected that the rate of vaporization will continually increase if the number of kinks increases until a steady state of concentration of kinks is reached. The length of ledge winding out of a spiral dislocation pit is proportional to the projected area of the pit, that is, the area on the plane (111) surface covered by the pit. Interferometry shows the sides of pits have constant slope, and this slope is related to the distance between ledges compared to the height of the ledges. Thus, the rate of vaporization from an individual pit will be directly proportional to the area of the pit. Similarly, assuming the ledge density in the dislocation pit to be much greater than on the unpitted surface, the rate of vaporization of a crystal will be directly proportional to the fraction of the surface area which is covered by vaporization pits. It follows that the rate of vaporization is expected to increase continuously until the surface is completely covered by pits, at which time the rate of vaporization will become constant.

Examination of the data in Table I shows why an initially slow and increasing rate was always seen with antimony crystals but not usually seen with arsenic crystals. Using the above formulae, the weight loss from an arsenic crystal upon intersection of the pits, is calculated to be 0.07mg/cm^2 while that from an antimony crystal will be 0.27mg/cm^2. An "induction period" is not usually seen with arsenic because a weight loss of 0.07 mg/cm^2 is too small to show up under the usual experimental conditions. For pit intersection with antimony, on the other hand, the calculated weight loss is four times as large, and an induction period is always observed with antimony crystals. Comparison of the calculated, total weight losses before pit intersection with the measured loss in weight of antimony crystals before a steady-state rate was obtained show remarkably good agreement considering the approximations in the calculation and the experimental difficulties in specifying the end of the induction period. The measured weight loss before steady state was reached varies between 0.1 and 0.5 mg/cm^2 with an average around 0.3. The agreement between calculation and experiment indicates that during the induction period essentially all vaporization occurs from the pits, as expected when the ledge density on the unpitted surface is much less than in the pits. Recently, careful gravimetric determinations of vaporization rate from fresh-cleaved arsenic single crystals have shown an initial, slow rate in the vaporization of arsenic. Also, dislocation densities on arsenic surfaces have been reduced by cleaving at liquid-nitrogen temperatures. An induction period is then readily apparent. Other results in agreement with these conclusions include kinetic experiments on antimony crystals which have been previously treated with CP-4 etchant[21] to start pits. These crystals show a markedly reduced induction period.

The data just discussed show: that the vaporization rate continuously increases as vaporization pits grow until the whole surface is covered by pits; that the total weight loss during the period before complete pit intersection depends upon the density of dislocation pits in the predicted manner; and that, within experimental error, all weight loss which occurs during this period comes from material leaving pits. The weight-loss data are not sufficient to show, however, that the rate of vaporization is proportional to the number of kinks during the initial stages of vaporization. This is strongly indicated by analysis of time-lapse motion pictures taken of the vaporization of fresh-cleaved arsenic crystals. The above discussion shows that if the vaporization rate from a pit is proportional to the number of kinks it will be proportional to the area of the pit. Thus, during the induction period before pits intersect, the rate of vaporization will be proportional to the square of the length of a pit side. If this is so, the length of a pit side will increase at a constant rate during the induction period and a plot of length of pit side (ℓ) against time will be a straight line. On the other hand, if the rate of vaporization from a pit were constant during the period before pits intersected, a plot of ℓ^3 against time would be a straight line. Analysis of the time-lapse motion pictures clearly shows that the length of the side of a pit increases in direct proportion to time during the period before the pits intersect.

Results of field-emission observations on gallium arsenide[33] are in accord with the above mechanism. When arsenic vapor is deposited on a GaAs emitter the arsenic spreads spontaneously over the surface even when the crystal is cooled to 77°K; however, when gallium is deposited it forms a shadow facing the source. This indicates that As_4 molecules diffuse on the surface for relatively long times before being incorporated into the lattice, as expected when the incorporation step involves an activation energy. Gallium atoms, on the other hand, are immediately accommodated by the lattice and do not diffuse over the surface.

As has been noted, pit angles reflect the average spacing between ledges. A model for vaporization from spiral dislocations introduced by Cabrera and Levine[34] and considered further by Hirth and Pound[2,27,35] relates the average spacing between adjacent turns of a spiral ledge to the concentration under equilibrium conditions. From this model, it is expected that the ledge spacing will increase as the concentration of surface-adsorbed molecules increases, and decrease as the concentration of surface-adsorbed molecules decreases. The concentration of surface-adsorbed molecules depends upon the rate molecules leave kinks compared to the rate of desorption from the surface. From these considerations, one expects that the slower the kink→surface step compared to the desorption step, the steeper will be the pits. In other words, pit angles will correlate with vaporization coefficients when the slow step occurs before desorption.   This correlation is seen in the relative pit angles on arsenic and antimony surfaces.

The concentration of surface-adsorbed molecules on a given surface can be varied if the material has a low vaporization coefficient associated with a step prior to desorption. With such a material the concentration of surface-adsorbed molecules can be increased by vaporizing not into vacuum, but into conditions intermediate between equilibrium and vacuum, where large numbers of molecules in the vapor hit the surface, presumably stick to it, diffuse around on it, and then desorb from the surface because they have difficulty in being incorporated at kink sites. By this reasoning, pit angles are expected to decrease with increasing saturation of the vapor. This has been observed in Knudsen experiments on arsenic and in observations on foil-wrapped arsenic crystals. The pits on arsenic single crystals vaporized in a Knudsen cell average about 7° and have a range between less than 1° and 11°; there are many pits shallower than those seen on arsenic surfaces vaporized into vacuum. Arsenic single crystals vaporized with one side covered by foil show pits only on the exposed surface. There is still some question whether dislocation vaporization pits will be seen when the vaporization coefficient is very close to unity, although the above inter-pretation leads one to expect that pits on such surfaces will be very shallow, so shallow that they will be hard to find.

Vaporization studies on ionic crystals, KCl[31,36], KI[31,36,37], and NaCl[37,38], have shown that, in these systems also, the amount of surface roughening upon vaporization and the steepness of vaporization pits decreases as the saturation of the vapor increases. Recent measurements by Somorjai and Lester[38] of vaporization rates from sodium chloride single crystals show the vaporization coefficient to increase from $\sim$ 0.5 to 1 as the dislocation density is increased from 8×10^5 to 15×10^6 /cm^2. We interpret

these results, which differ markedly from the behavior observed with arsenic

and antimony, to suggest that, in ionic systems, surface diffusion may be the

rate-limiting step. Hirth and Pound[27] have shown that, when surface diffusion

is slow compared to the desorption rate, a concentration gradient of surface-

diffusing (adsorbed) molecules between ledges is expected, with the highest

concentration at the ledges and the lowest midway between the ledges. If the

spacing between ledges increases, the surface-diffusion gradients increase,

more molecules leave ledges, and the ledges accelerate. A limiting ledge

velocity and a terminal ledge spacing are approached asymptotically as the ledge

spacing becomes so large that it no longer affects the surface-concentration

gradient. It is expected that the spacing between ledges near the center of

a spiral dislocation pit will be less than the steady-state (terminal) spacing

on the unperturbed surface. The spacing will increase to the terminal value as

ledges recede from the center of the pit, and macroscopic pit angles are

expected to decrease accordingly as one proceeds from the center to the edge of

the pit. That is, the pit sides will show curvature. This model predicts

that the vaporization coefficient will increase with increasing dislocation

density since the number of kinks and ledges are higher near the dislocation

centers. On the other hand, one would not expect pits on surfaces where the

kink→surface step is rate determining to show curvature, because in that case,

the ledge velocity is independent of the distance between ledges, and the

ledges do not accelerate upon a change in the surface-concentration gradient.

It seems reasonable that surface-diffusion is very slow on an ionic lattice

with its large forces.

Relation to High-Temperature Vapor-Pressure Measurements.

In this section we will look at a few results which bear upon the problem

of interpreting vapor-pressure measurements on typical high-temperature samples

The results to be reviewed concern: the relationship between rates of powdered samples and rate studies on single crystals, corrections to

Knudsen-effusion experiments, and the correction of Langmuir vapor-pressure

measurements for the fact that these measurements are usually not made into a

perfect condenser.

Extensive measurements of vaporization rates from arsenic samples which

are not (111) wafers included studies of: vaporization of arsenic powder in an

open crucible; vaporization of polycrystalline arsenic from (111) faces and

other faces, with the faces not being studied wrapped under aluminum or

platinum foil; vaporization from a (110) face of arsenic with the other faces

wrapped under platinum foil. The vaporization coefficients measured under

these various conditions are shown in Table II.[18] Although the experimental

precision of a single run on a crystal covered with foil wrapping was about the

same as on (111) crystal wafers, the variation from run to run was much larger

with wrapped crystals.

The results shown in Table II can be correlated in terms of effective

vaporizing areas. The effective vaporizing area, $\underline{A}'$, of a porous sample is

defined as that area of plane surface which would vaporize at the same rate as

the sample.[8] By definition, then, the effective vaporizing area of a (111)

wafer is exactly equal to the total vaporizing area of that wafer. Also, for

any substance with a vaporization coefficient of unity, the effective vapor-

izing area will be equal to the projected plane surface area of the sample

no matter how porous or rough the sample is. However, with low vaporization

coefficients, the effective vaporizing area can be much larger than the total

projected area of the surface if it is possible for vaporization to take

place underneath foil wrapping or in cracks and crevices of a powdered or

porous sample.

Table II includes the exposed area in individual experiments and the

effective vaporizing area calculated assuming vaporization always takes place from (111) faces. The data show that, although wrapping always decreases the total rate of vaporization of a sample over that which that sample would have if it were unwrapped (that is, the effective vaporizing area is always less than total surface area), the rate of vaporization from the exposed portion of the surface appears to be higher than that calculated for an unwrapped wafer. This is caused by some molecules vaporizing in the crevices of the powder or underneath the foil. Since arsenic has a low condensation coefficient these molecules bounce around without recondensing. They eventually escape and contribute to the mass loss. This is shown very clearly by some experiments on aluminum-foil wrapped crystals in which the size of the opening in the aluminum foil was continuously increased to expose more and more of the (111) surface directly. As the hole in the foil becomes larger the total rate of vaporization of the sample increases. However, the rate of vaporization calculated per unit area of exposed surface decreases because the relative contribution from the surface underneath the foil, which is substantial when a large fraction of the sample is covered by foil, becomes smaller as the exposed area becomes larger.

The effective vaporizing areas and apparent rates of vaporization are higher on crystals wrapped in platinum foil than those wrapped in aluminum foil because aluminum foil can be wrapped more tightly. When a "(110) face" is exposed, the effective vaporizing area is greater than the total surface area because the (110) faces prepared actually consist of jagged facets of (111) faces, so that the surfaces are much rougher than (111) cleavage faces. The apparent vaporization coefficients of polycrystalline samples are higher than those of wrapped single crystals because polycrystalline samples have a much larger fraction of their total area under the foil, and also, because the surfaces of polycrystalline samples are rougher than single-crystal surfaces.

Vidale[39] has presented an interesting approximation to the effective area of a uniform, porous, powder sample by considering the problem to be one of diffusion through the powder. This leads to the equation $A' = 1.55 B (\epsilon/\alpha)^{1/2}$ where B is the cross-sectional area of the container and ϵ is the ratio of pore volume to total volume of the powder.[8] Using this equation, the effective area of the powdered sample in Table II is estimated to be about 95 cm^2. This is within an order of magnitude of the measured value.

Knudsen measurements were made on the same powdered sample. An attempt was made to obtain the equilibrium vapor pressure of arsenic from the Knudsen data using a correction of the Motzfeldt[7]-Whitman[6]-Rosenblatt[8] type, $P_E'/P_K =$ $1 + (W_a a/\alpha A')$, by substituting in the effective vaporizing area from the powder crucible studies and the measured vaporization coefficient of arsenic. The equilibrium pressure obtained from this extrapolation is about two-thirds of the literature value. The discrepancy may be caused by errors in the literature vapor pressure[14] or by systematic errors in the Knudsen measurements, or it may indicate that the correction procedure is not a wholly reliable way to extrapolate to equilibrium pressures with materials having very low vaporization coefficients. In addition to low vaporization coefficients other factors are known to contribute to difficulties in Knudsen-effusion experiments. These include temperature gradients in the cell, nonideal orifices, and surface diffusion out of the orifice.

On the other hand, if the equilibrium vapor pressure from the literature[14] and the measured Knudsen vapor-pressure are used to calculate a vaporization coefficient by a Motzfeldt-type approach, the calculated vaporization coefficient is about one-third that measured directly on the powdered sample in the crucible. Brewer and Kane[8,12] also found that the vaporization coefficient of powdered arsenic calculated from a Motzfeldt-type equation decreased as the orifice area decreased. Similar behavior has been observed with potassium iodide and other materials by Knacke, Schmolke, and Stranski,[36] and by Jaeckel and Peperle,[37] and was interpreted by these authors to mean that the vaporization coefficients (defined by $\alpha_v \, (P_E - P) = r \, (2\pi RT/M)^{1/2}$ decrease with increasing saturation of the vapor. This is a reasonable interpretation as in all these systems the steepness of vaporization pits (which reflects the distance between ledges and, presumably, the number of kinks) apparently decreases with increasing saturation. One might speculate that vaporization coefficients will normally decrease with increasing vapor saturation in materials where the kink→surface step or surface diffusion is rate-determining and spiral dislocations are the primary ledge sources-- since, under these conditions, as discussed in the preceeding section, the ledge spacing may be expected to increase as the saturation of the vapor and the number of surface-adsorbed molecules increases. This type of behavior would be directly opposite to the prediction of Hirth and Pound[27] that α_v for simple atomic systems increases smoothly from   1/3 to 1 as the ambient pressure increases from vacuum to equilibrium.

Langmuir vapor-pressure measurements also suffer from difficulties because of experimental effects and low vaporization coefficients. The relation between Langmuir vapor-pressure measurements and vaporization coefficients is apparent but less attention has been paid to the effect of tube restrictions upon molecular flow in Langmuir vapor-pressure measurements. Steady-state calculations[20] show that, in typical experimental arrangements, restrictions to molecular flow can cause appreciable errors in Langmuir vapor-pressure measurements--factors of two to five if vaporization coefficients are on the order of unity.

<u>Correlation of Vaporization Behavior.</u>

It is interesting to conjecture why the vaporization coefficients of arsenic and antimony differ and what insight can be gained from these studies in predicting the vaporization behavior of other materials. Arsenic and antimony are similar chemical systems. They both vaporize to form tetrameric gaseous molecules and they have similar crystal structures. We suspect that the difference in vaporization behavior of these two systems is associated with the unusually high stability of As_4 gas, which is indicated by the fact that arsenic sublimes at temperatures well below its melting point. Low vaporization coefficients were associated some years ago with the existence of molecular gaseous species which do not exist as such in the solid lattice, and with the rearrangement of chemical bonds during the vaporization process.[4,12] Useful as this association is, it does not appear to be either a necessary or a sufficient condition for a low vaporization coefficient. We expect very low vaporization coefficients when there is an unusually stable gaseous molecule which does not exist as such in a stable crystal lattice. That is, very low vaporization coefficients are associated with low equilibrium enthalpies of vaporization, with materials that sublime, and perhaps, with materials that have unusually high vapor pressures at their melting points.

Consider a reaction-coordinate activation-enthalpy or standard-free-energy diagram, such as Figure 2. A certain amount of energy, the activation energy for vaporization, is required to tear away the atoms which will eventually form the vaporizing molecule from the kink site. If the bonds in the gaseous molecule formed are not very stable there will not be much energy gained upon any rearrangement of the atoms as they leave the kink site, the height of the barrier (from the condensation side) will not be very high, and there will not be a very low vaporization coefficient. On the other hand, if the atoms can rearrange to form a very stable gaseous molecule, that molecule will have an energy much lower than corresponds to the top of the barrier, and the vaporization coefficient will be very small.

Such considerations suggest that very low vaporization coefficients might be seen primarily with molecules composed of the lighter atoms where the bond distances are shorter and the bonds are normally stronger. They also suggest correlation of vaporization coefficients and activation enthalpies of vaporization with the atomization energies of the gaseous molecule formed. Thus, it might be expected that vaporization coefficients of similar molecules will increase as one goes down a family in the periodic table. One might also predict that, in high temperature systems containing many vapor species, the polymeric gaseous molecule which predominates at lower temperatures will have the smallest vaporization coefficient. For example, in carbon vapor, these considerations suggest that $\alpha(C_3) < \alpha(C_2)$, in agreement with experiment.[40,41] Similarly, it might be expected that $\alpha(As_4) < \alpha(As_2) < \alpha(As)$.

Vaporization coefficients should also correlate with other physicochemical properties. When the kink $\rightarrow$ surface step is rate determining it appears reasonable to expect that the magnitude of α_v will be related to bulk self-diffusion rates and to dislocation mobilities. Such expectations are in qualitative accord with available data for arsenic and antimony. It is interesting to observe that movement of dislocations was frequently seen during vaporization of antimony; but never with arsenic.

Ackowledgments

The authors are indebted to W. R. Bitler and W. A. Steele for stimulating and helpful discussions. This research was supported by the U. S. Army Research Office - Durham.

References

1.   I. Langmuir, Phys. Rev. $\underline{2}$, 329 (1913).

2.   J. P. Hirth and G. M. Pound, Progr. Materials Sci. $\underline{11}$, (1963).

3.   B. Paul, ARS Journal $\underline{32}$, 1321 (1962).

4.   I.N. Stranski and G. Wolff, Research $\underline{4}$, 15 (1951).

5.   J. O'M. Bockris, J. L. White , and J.D. MacKenzie, Eds., Physico-Chemical Measurements at High Temperatures (Butterworths Scientific Publ., London, 1959).

6.   C. I. Whitman, H. Chem. Phys. $\underline{20}$, 161 (1952).

7.   K. Motzfeldt, J. Phys. Chem. $\underline{59}$, 139 (1955).

8.   G. M. Rosenblatt, J. Electrochem. Soc. $\underline{110}$, 563 (1963).

9.   E. Rutner, P. Goldfinger, and J. Hirth, Eds., Proc. Intern. Symp. Condensation and Evaporation of Solids, Dayton, Ohio, 1962 (1964).

10.   O. Knacke and I. N. Stranski, Progr. Metal Phys. $\underline{6}$, 181 (1956).

11.   G. A. Somorjai, Progr. Solid. State Chem. $\underline{4}$, (1967).

12.   L. Brewer and J. S. Kane, J. Phys. Chem. $\underline{59}$, 105 (1955).

13.   G. M. Rosenblatt and C. E. Birchenall, J. Chem. Phys. $\underline{35}$, 788 (1961).

14.   D. R. Stull and D. C. Sinke, Thermodynamic Properties of the Elements (Amer. Chem. Soc., Washington, D. C., 1956).

15.   L. R. Maxwell, S. B. Hendricks, and V. M. Mosley, J. Chem. Phys. $\underline{3}$, 699 (1935).   Y. Morino, T. Ukaji, and T. Ito, Bull. Chem. Soc. Japan $\underline{39}$, 64 (1966).

16.   A. J. Bradley, Phil. Mag. $\underline{47}$, 657 (1924).   R. W. G. Wyckoff, Crystal Structure (Interscience, New York, 1951).

17.   L. R. Weisberg and P. R. Celmer, J. Electrochem. Soc. $\underline{110}$, 56 (1963).

18.   G. M. Rosenblatt, P. K. Lee, and M. B. Dowell, J. Chem. Phys. $\underline{45}$, 3454 (1966).   G. M. Rosenblatt and P. K. Lee, J. Chem. Phys., $\underline{49}$, 2995 (1968).

19. R. J. Capwell, Jr. and G. M. Rosenblatt, J. Phys. Chem. 71, 1327 (1968).

20. G. M. Rosenblatt, J. Phys. Chem. 71, 1327 (1967).

21. V. M. Kosevich, Kristallografiya 5, 749 (1960); Soviet Phys.-Cryst. 5 (5), 715 (1961).

22. M. Volmer, Kinetic der Phasenbildung (Dresden and Leipzig, 1939).

23. W. K. Burton and N. Cabrera, Disc. Faraday Soc. 5, 33 (1949).

24. W. K. Burton, N. Cabrera, and F. C. Frank, Phil. Trans. Roy. Soc. A 243, 299 (1951).

25. W. Kossel, Nach. Ges. Wiss. Gottingen, 135 (1927); Ann. Phys. 33, 651 (1938).

26. I. N. Stranski, Z. Phys. Chem. 136, 259 (1928); (B) 11, 421 (1931).

27. J. P. Hirth and G. M. Pound, Acta Met. 5, 649 (1957); J. Chem. Phys. 26, 1216 (1957); J. Phys. Chem. 64, 619 (1960).

28. K. Neumann, Z. Phys. Chem. 196, 16 (1950).

29. E. M. Mortensen and H. Eyring, J. Phys. Chem. 64, 846 (1960). See also article by Eyring, Wanlass, and Eyring in ref. 9.

30. R. L. Brown, R. G. Brewer, and W. Klemperer, J. Chem. Phys. 36, 1827 (1962).

31. See article by Hirschwald and Stranski, in ref. 9.

32. O. W. Johnson, J. Appl. Phys. 37, 2521 (1966).

33. J. R. Arthur, J. Appl. Phys. 37, 3057 (1966).

34. N. Cabrera and M. M. Levine, Phil Mag. [8] 1, 450, (1956).

35. J. P. Hirth, Chap. 6 of Metal Surfaces: Structure, Energetics and Kinetics (Amer. Soc. Metals, Cleveland, 1963).

36. O. Knacke, R. Schmolke, and I. N. Stranski, Z. Kristall. 109, 184 (1957).

37. R. Jaeckel and W. Peperle, Z. Physik. Chem. 217, 321 (1961).

38. J. E. Lester and G. A. Somorjai, J. Chem. Phys. 49, 2940 (1968). J. E. Lester, Doctoral Dissertation, University of California, UCRL-17794, (1967).

38–26   G. M. ROSENBLATT, M. B. DOWELL, PANG-KAI LEE, AND H. R. O'NEAL

39.   G. L. Vidale, General Electric Missile and Space Vehicle Department
      Technical Information Series, Report No. R60 SD 468 (October 1960).

40.   R. J. Thorn and G. H. Winslow, J. Chem. Phys. 26, 186 (1957).

41.   R. P. Burns, A. J. Jason, and M. G. Inghram, J. Chem. Phys. 40, 1161 (1964).

42.   C. C. Herrick and R. C. Feber, J. Phys. Chem. 72, 1102 (1968).

Table I. Summary of Experimental Results

from Vaporization of Arsenic and Antimony (111) Faces

		Arsenic		Antimony		
Temperature	T	550°K		650°K		
Langmuir and equilibrium vapor pressure	P_L P_E	6.6×10^{-10}	1.4×10^{-5a}	1.2×10^{-9}	7.0×10^{-9b}	atm.
Vaporization coefficient	α_v	4.6×10^{-5}		0.18		
Activation and equilibrium enthalpy of vaporization	ΔH^* ΔH°	43.9	33.1^a	49.5	47.4^b	kcal/mole
Activation and equilibrium entropy of vaporization	ΔS^* ΔS°	37.9	38.0^a	35.4	35.7^a	cal/deg-mole
Average pit angle	θ	11° (9–13°)		5° (2–9°)		
Dislocation density	n	5×10^6		10^5		cm^{-2}
Density of crystal	ρ	5.7^c		6.7^c		$g\ cm^{-3}$
Weight loss during induction period	w (calc. for ζ = 1)	0.07		0.27		$mg\ cm^{-2}$
	w (obs'd)	(~ 0 – 0.18)		0.30 (0.09–0.46)		$mg\ cm^{-2}$

a. Reference 14; use of recent $As_4(g)$ entropies in ref. 19 and $As(c)$ entropies in ref. 42 leads to the following arsenic values at 550°K: $P_E = 9.3 \times 10^{-6}$ atm, $\alpha_v = 7.1 \times 10^{-5}$, $\Delta H^\circ = 34.8$ kcal/mole, $\Delta S^\circ = 40.2$ cal/deg-mole.

b. Reference 13

c. Handbook of Chemistry and Physics

38–27

Table II. Vaporization of Arsenic from Different Samples.

Sample	Run No.	Measured α 550 (10^{-5})	Exposed Area, A (cm^2)	Effective Vaporizing Area, A' (cm^2)	Total Surface Area (cm^2)
Powder, in open crucible	18	296	0.635	38.6	(~50)[a]
Polycrystalline, (111) face, other faces under aluminum foil	2	25.6	0.314	1.65	(1.57)[b]
Polycrystalline, growth face, other faces under aluminum foil	14	83.7	0.0686	1.18	(0.55)[b]
Single crystal, (111) face, other faces under aluminum foil	34	9.61	0.355	0.70	1.643
Single crystal, (111) face, other faces under platinum foil	38	10.4	0.588	1.24	1.643
Single crystal, (110) face, other faces under platinum foil	25	39.0	0.303	2.43	1.764
Single crystal wafers, (111) face, free vaporization	37	4.87	2.094	2.094	2.094

a. Very rough value based upon estimated particle size of 0.1 min.

b. Estimated projected area.

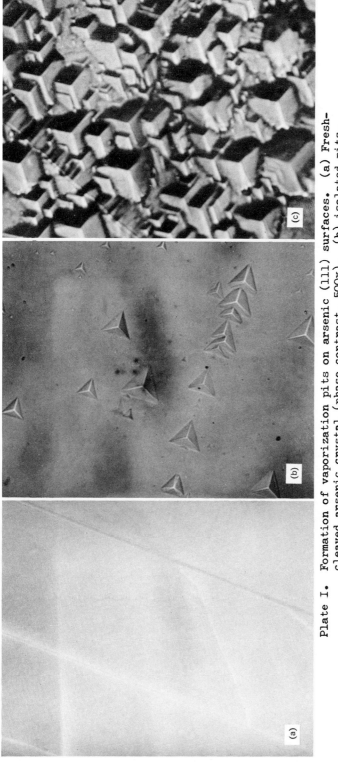

Plate I. Formation of vaporization pits on arsenic (111) surfaces. (a) Fresh-cleaved arsenic crystal (phase contrast, 500x). (b) isolated pits, after vaporization at 300°C for one hour (brightfield, 850x). (c) After extensive vaporization, at various temperatures between 240 and 310°C for about thirty hours (brightfield, 1020x). The photographs are of different crystals.

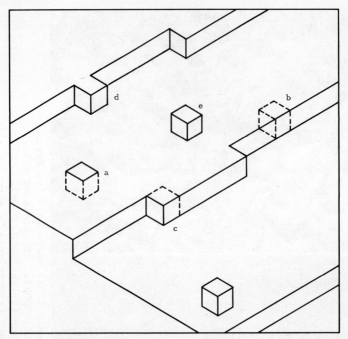

Fig. 1. Schematic drawing (from reference 2) of solid surface showing ledges, with atoms in the following positions: (a) in surface. (b) in ledge. (c) kink. (d) adsorbed at ledge. (e) adsorbed on surface.

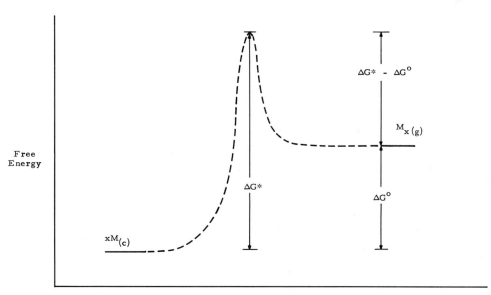

Fig. 2. Schematic standard-free-energy diagram for the retarded vaporization process $xM(c) \rightarrow M_x(g)$. The free energy changes on the diagram are related to quantities defined in the text by the equations:
$\Delta G^O = -RT \ln P_E$, $\Delta G^* = -RT \ln P_L$, $\Delta G^* - \Delta G^O = -RT \ln a$.

THE POSITIVE AND NEGATIVE SELF-SURFACE IONIZATION OF TANTALUM

Milton D. Scheer and Joseph Fine

National Bureau of Standards
Washington, D. C.

INTRODUCTION

An atomically clean metal surface can be characterized by the energies associated with a number of sublimation processes. These are:

$$(1) \quad e_s \longrightarrow e_g \quad ; \quad \phi, \text{ the electron work function}$$

$$(2) \quad M_s \longrightarrow M_g \quad ; \quad \ell_o, \text{ the atom sublimation energy}$$

$$(3) \quad M_s \longrightarrow M_g^+ + e_s; \quad \ell_+, \text{ the positive ion sublimation energy}$$

$$(4) \quad M_s + e_s \longrightarrow M_g^- \quad ; \quad \ell_-, \text{ the negative ion sublimation energy}$$

Process (4) will be observed only when M_g^- is stable or when A, the electron affinity of M, is positive. At low temperatures (less than about 2/3 of the melting point), these energies may be strongly influenced by the geometric arrangement of the surface atoms. The different work functions observed for different crystalline planes with the field emission microscope are examples of such anisotropic behavior. At high temperatures the geometric structure and symmetry characteristics of the surface become less important as a result of surface migration and rapid sublimation rates. In fact under these conditions the surface atoms can probably be thought of as a two dimensional liquid whose properties tend to be uniform even over a nominally polycrystalline surface which at low temperatures exhibits considerable anisotropy.

Process (3) has been observed for a number of refractory metals[1-8]. Early attempts at reliable measurement of ℓ_+ were limited by primitive vacuum conditions and the presence of easily ionized alkali atom impurities. Process (4) has been reported for graphite, tungsten, rhenium, and molybdenum[7,8,9]. A generalized Saha-Langmuir equation was shown to be applicable to these self-surface ionization processes[7] and reliable estimates have been made for

ℓ_+, ℓ_-, and A. Conservation of energy requires that these quantities are related to ϕ, ℓ_o and I (the 1st ionization potential) by the following equations:

$$\phi - I = \ell_o - \ell_+ \tag{1a}$$

and

$$A - \phi = \ell_o - \ell_- \tag{1b}$$

If j_+, j_- and j_m are the fluxes of positive, negative, and neutral particles subliming from the metal surface, the generalized Saha-Langmuir equation relating these quantities is given by:

$$\frac{j_\pm}{j_m} = \frac{\omega_\pm}{\omega_m} \exp\left(\frac{\pm\phi\pm q}{kT}\right) \quad \left\{ \begin{array}{l} \text{upper sign for + ionization} \\ \text{lower sign for - ionization} \end{array} \right\} \tag{2}$$

where q = -I for positive surface ionization, and q = -A for the negative ionization process. The ω's are the electronic partition functions for these particles and are given by equations of the form:

$$\omega = (2J_o + 1) + \sum_{n=1}^{\infty} (2J_n + 1) \exp(-\epsilon n/kT) \tag{3}$$

where J_o and J_n are the inner quantum numbers for the ground and n^{th} excited state respectively, while ϵ_n is the energy of the nth state relative to the ground level of the atom or ion.

In the case of tantalum, considerable difficulty was encountered in attempting to verify equation (1a). Large departures from thermal equilibrium at the sublimating surface were found and it is the purpose of this paper to report these results and to give an account of how they were resolved.

Apparatus

The apparatus used in this study consisted of a direction focusing mass spectrometer. The details of its construction have been described previously[5,7]. Positive ion (^{181}Ta$^+$) fluxes were detected with a 16 stage Cu-Be electron multiplier in the range 0.1 to 2×10^5 cm^{-2}s^{-1}. Meaningful measurements of negative ion (^{181}Ta$^-$) fluxes, on the other hand were limited to values below about one cm^{-2}s^{-1} because of the significant accumulation of a negative space charge in the region of the tantalum surface. This difficulty arose because of spatial limitations in the ion source where only about 2 mA of the thermionic electron emission could be magnetically deflected, and collected efficiently. This value for the saturation electron current was reached when the tantalum surface temperature exceeded 2100 K.

Neutral tantalum atoms were detected with a modification of the ion source that incorporated a 90 eV, 1.0 μA electron beam which was confined by a parallel 200 gauss magnetic field 1.5 cm from the tantalum surface. This electron beam produced ^{181}Ta$^+$ ions from the subliming atoms at rates between 0.1 and 10^4 per second, corresponding to surface temperatures between 2000 and 2500 K.

The output pulses of the electron multiplier were about 20 nsec wide and were amplified with a voltage gain of 5000 (multiplier and amplifier noise being discriminated away). This signal was then processed to give uniform one volt, 30 nsec pulses and used to actuate a counter with a 2.5 MHz band pass. The background count for this system was usually less than 10 pulses/minute when the magnetic analyzer was detuned or when the tantalum surface temperature was below 2000 K.

The tantalum ribbon used in this study had dimensions of 20 × 1.2 × 0.05 mm. It was zone refined and contained about 10 ppm. of C, Fe, and Mo and 30 ppm of Nb. The background pressure in the mass spectrometer was typically about 1.2× 10^{-8} torr* with carbon monoxide the major residual gas constitutent. The pressure remained below 2 × 10^{-8} torr when the tantalum ribbon was heated to 2500 K.

---

* 1 torr = 133.3 newtons/square meter.

Experimental Results: Atom and Positive Ion Sublimation

The flux of atoms, subliming under equilibrium conditions, from a heated metal surface is given by

$$j_m = \frac{\exp\left(\dfrac{\Delta S^\circ}{R} - \dfrac{\ell_o}{kT}\right)}{(2\pi M k T)^{1/2}} \tag{4}$$

where ΔS° is the entropy of sublimation, M is the mass of the tantalum atom, and T is the temperature of the surface. The rate of tantalum ionization R (Ta^+) in the mass spectrometer is proportional to the tantalum atom number density in the electron beam namely

$$R(Ta^+) = K \frac{j_m}{\bar{v}} \tag{5}$$

where $\bar{v} = \left(\dfrac{8kT}{\pi M}\right)^{1/2}$ is the mean velocity of the tantalum atoms at the temperature T. K is a proportionality constant whose value for an ion source of the type used here is between 10^{-8} and 10^{-10} $cm^3 s^{-1}$ depending on the ionization cross section, the dimensions of the electron beam, and its distance from the tantalum surface. Combining equations (4) and (5) one obtains:

$$\ln\left\{R(Ta^+)\ (T/\bar{T})\right\} = \ln\left\{\frac{K}{4kT} \exp\left(\frac{\Delta S^\circ}{R}\right)\right\} - \ell_o/kT \tag{6}$$

where $\bar{T}$ is the mean temperature of the experimental temperature range. Figure I shows the application of equation (6) to atom sublimation rates obtained with the tantalum ribbon whose dimensions and purity are given above. The ordinate j_o (corrected) $= \left\{R\ (Ta^+)\left(\dfrac{T}{\bar{T}}\right)\right\}$. There are two well-defined temperature ranges yielding widely different ℓ_o's namely, 9.5 eV for $\bar{T}$ = 2095 K and 7.6 eV for $\bar{T}$ = 2435 K. Previously reported determinations of this quantity were obtained by: (1) method of weight loss of a freely evaporating sample, and (2) free evaporation into a mass spectrometric detector as in the present study. The weight loss technique yielded values[10,11] of 7.8 and 8.1 eV at $\bar{T}$ = 2600 and 2800 K, respectively, while the earlier mass spectrometer determinations[6,12] gave the considerably different values of 7.1 and 7.3 eV at mean experimental temperatures of 2500 and 2800°K. The pulse counting electronics incorporated in the present apparatus made it possible to detect very small rates of sublimation. Consequently a change in the 2nd law slope for tantalum sublimation was noted here for the first time. It is evident that this process is inhibited below about $2200°K$; temperatures which had not been explored previously for tantalum sublimation.

Substituting equations (1) and (4) into the generalized Saha-Langmuir equation (3), one obtains an expression for positive and negative ion sublimation which is analogous to equation (6); namely,

$$ l_n \left\{ j_{\pm} \ (T/\bar{T})^{1/2} \left(\frac{\omega_m}{\omega_{\pm}}\right) \left(\exp \ \frac{-\Delta S°}{R}\right)\right\} = l_n \ (2\pi Mk)^{1/2} - l_{\pm}/kT \quad (7) $$

Figure II gives a plot of this equation for the sublimation of $^{181}Ta^+$ ions from the same tanalum ribbon used above. The quantity

$$ j_+ \ (\text{corrected}) = \left\{ j_+ \left(\frac{\omega_m}{\omega_{\pm}}\right) \ (T/\bar{T})^{1/2} \quad \exp\left(\frac{-\Delta S°}{R}\right)\right\} $$

The ω's and $\Delta S°$ were calculated using Moore's[14] "Atomic Energy Levels" and Stull and Sinke's compilation[15]. An effect analogous to that obtained for atom sublimation was observed for positive ion sublimation, with the change in slope occurring at about the same temperature (2200K). The two values for l_+ were 9.8 and 11.4 eV. Using 4.12 eV[13] for ϕ and 7.88 eV[14] for I , l_+ was calculated from equation (1a) and the experimentally determined l_o's. The calculated positive ion sublimation energies were larger than the experimental l_+'s by 1.3 and 1.6 eV in the low and high temperature ranges respectively. These discrepancies are well outside any reasonable experimental uncertainties so that it must be concluded that the sublimation rates did not achieve their equilibrium values with this tantalum sample.

Previous field emission[16] and flash filament[17] studies have shown that a heated tantalum surface dissociates carbon monoxide. Above about 2000K, the oxygen evaporates[18] as TaO while the carbon atoms diffuse into the tantalum lattice[16]. Since carbon monoxide is the predominant residual gas, extensive heating of the tantalum ribbon would result in the further accumulation of carbon in addition to that initially present (10 ppm). Consequently, it was thought that contamination of the surface with carbon was responsible for the inhibition of the atom and ion sublimation processes, particularly below 2300 K. The pronounced low temperature effect would then be due to accumulation of carbon on the surface when its diffusion rate is slow relative to this rate of formation from the reduction of CO. It was decided therefore to remove most of the carbon in the tantalum ribbon and hence provide a larger concentration gradient for more rapid diffusion into the tantalum lattice of the carbon formed on the hot surface. This then would result in a less contaminated surface for the sublimation measurements.

In order to remove the accumulated carbon, the tantalum ribbon was heated to 2500 K in an oxygen atmosphere at 10^{-6} torr for a period of about twenty hours. During this treatment the rate of CO production was monitored. Initially, the CO partial pressure was about a hundredfold larger than the original background level. After twenty hours it decreased to its original value at which time the oxygen flow was interrupted and the system evacuated to a residual gas level of about 10^{-8} torr.

The rates of atom and ion sublimation were determined once more in the 2000 to 2600 K range. The results are shown in Figures III and IV. This time no changes in slope were observed and 7.94 ± 0.09 and 11.18 ± 0.11 eV were obtained for ℓ_o and ℓ_+ respectively. The stated uncertainties are the standard deviations obtained from a least squares treatment of the data. Once more using equation (1a) to calculate ℓ_+, the following result is obtained:

$$\ell_+ = \ell_o + I - \phi = 7.94 + 7.88 - 4.12 = 11.70 \text{ eV}$$

This is only 0.5 eV larger than the experimental value and is within the limits of uncertainty of the quantities involved. Table I gives a summary of the results obtained in this laboratory for the positive ion sublimation of a number of transition metals[5,7,8]. The agreement between ℓ_+ calculated from (1a) and the experimental quantities is seen to be quite good. The conclusion that thermal equilibrium on the subliming surface is applicable to these processes is valid

provided unusual kinetic effects like carbon contamination of the tantalum

surface are absent. It is likely that the discrepancies in the values of l_o

reported in the literature[6,10-12] are due to such a surface contamination.

Negative Ion Sublimation

With the mass spectrometer biased[7] for the collection of negative ions,

^{181}Ta$^-$ was observed to sublime above 2000 K using the carbon-free tantalum rib-

bon. From equations (1) and (7), one can show that in terms of the positive

to negative ion ratio at a temperature T the electron affinity is given by

$$A = \phi + l_o - l_+ - kT \ln[\omega_- j_+/\omega_+ j_-] \qquad (8)$$

Table II shows the values obtained for $kT\ln[\omega_- j_+/\omega_+ j_-]$ in the 2000 to 2077 K

temperature range. The average of three determinations for this quantity was

found to be -0.03 ± 0.07 eV.

As mentioned previously, reliable measurements of negative ion yields at

temperatures in excess of 2100° were not possible because of space charge

limitations. The quantity ω_- was estimated by assuming that Ta$^-$ has the same

electron configuration as the ground state of W, the next higher element in

the periodic table. Using equation (8), one then obtains for the electron

affinity, of tantalum: A = 4.12 + 7.94 - 11.18 + 0.03 = 0.91 eV with an esti-

mated uncertainty of 0.3 eV. With this value for the electron affinity, equa-

tion (1a) can be used to calculate the negative ion sublimation energy; namely,

l_- = l_o + ϕ - A = 7.94 + 4.12 - 0.91 = 11.2 eV. Figure IV shows that such an

11 eV slope is consistent with the three data points obtained below 2100 K.

Conclusions

It has been shown that at temperatures in excess of 2000 K, the sublima-

tion of atoms and ions from a carbon-free tantalum surface fulfills the require-

ments of thermal equilibration at that surface. In accordance with equation

(1) the measured quantities l_o, l_+, l_-, and A are consistent with the literature

values for ϕ and I. Under the high temperature conditions of these measurements, no evidence for crystalline anisotropy was observed.

A number of years ago, Gordy and Thomas[19] noted that an empirically good linear relationship existed between electronegativity and the work function of metals. This fact can be readily understood in terms of equations (1a) and (1b) since elimination of ℓ_o yields

$$\phi = \frac{I+A}{2} + \frac{\ell_- - \ell_+}{2} = X_m + \frac{\ell_- - \ell_+}{2} \qquad (9)$$

The quantity $(I+A)/2$ is the absolute electronegativity (X_m) given by Mulliken[20]. Gordy and Thomas obtained, for most of the metals in the periodic table, the empirical relationship

$$\phi = 2.3X_p + 0.34 \qquad (10)$$

where X_p is the electronegativity given by Pauling[21]. Skinner and Pritchard[22] have shown that $X_p = X_m/3.1$, so that (10) becomes:

$$\phi = 0.74\,X_m + 0.34 \qquad (11)$$

Comparing this empirical relation with equation (9) shows that while the linear dependence of ϕ on X_m is predicted, the empirical slope is not quite unity as required by (9). However, detailed examination of the Gordy-Thomas correlation shows that a 25% increase in the coefficient of X_m would not do much of an injustice to the data they used. Further this result implies that the quantity $(\ell_- - \ell_+)/2$ is a constant, namely that

$$\ell_- = \ell_+ + 0.7 \qquad (12)$$

independent of which element is being considered. The four metals: Mo, Ta, W, and Re, for which values of ℓ_+ and ℓ_- have been determined, ℓ_- was found to be equal to or greater than ℓ_+ in each instance. The uncertainties however, were too large to demonstrate that equation (12) is obeyed. Nevertheless, it appears likely that equations (1a) and (1b) can be applied to elements other than those given in Table I.

Clementi[23] has made a set of Hartree-Fock calculations for the electron affinities of the 3rd row transition elements. Table III shows a comparison between these theoretical estimates and the self-surface ionization values determined in this laboratory for the 4th and 5th row transition metals. The value for Nb was estimated on the basis of approximate negative ion yields. Because of its low work function (~3.9 eV) extensive electron emission made it impossible to obtain accurate positive to negative ion ratios with the present apparatus. Molybdenum and chromium as well as tantalum and vanadium have ana- logous valence electron configurations and there is remarkably good agreement between the calculated and measured electron affinities. In the case of the manganese-rhenium pair however, where there is a full s and half-filled d shell in their ground state configurations, there is serious disagreement. Clementi predicts that such an electron structure should not yield a stable negative ion. In fact, his calculation gives an A value of -1.1 eV. The negative sur- face ionization of rhenium[7] has shown however, that its $5d^5 6s^2$ structure is indeed capable of producing a stable negative ion with a binding energy of about 0.1 eV.

REFERENCES

1. L. P. Smith, Phys. Rev. $\underline{35}$, 381 (1929).

2. L. L. Barnes, Phys. Rev. $\underline{42}$, 492 (1932).

3. H. B. Wahlin and L. O. Sordahl, Phys. Rev. $\underline{45}$, 886 (1934).

4. M. D. Fiske, Phys. Rev. $\underline{61}$, 513 (1942).

5. M. D. Scheer and J. Fine , J. Chem. Phys. $\underline{42}$, 3645 (1965).

6. E. Y. Zandberg, N. I. Ionov, and A. Ya. Tontegode, Zh. Tekn. Fiz. $\underline{35}$, 1504 (1965).

7. M. D. Scheer and J. Fine, J. Chem. Phys. $\underline{46}$, 3998 (1967).

8. M. D. Scheer and J. Fine, J. Chem. Phys. $\underline{47}$, 4267 (1967).

9. R. E. Honig, J. Chem. Phys. $\underline{22}$, 128 (1954).

10. D. B. Langmuir and L. Malter, Phys. Rev. $\underline{55}$, 748 (1939).

11. J. W. Edwards, H. L. Johnston, and P. E. Blackburn, J. Phys. Chem. $\underline{73}$, 172 (1951).

12. T. Babeliowsky, Physica $\underline{28}$, 1160 (1962).

13. V. S. Fomenko, "Handbook of Thermionic Properties" (Plenum Press, Inc., New York, 1966).

14. C. E. Moore, "Atomic Energy Levels" Natl. Bur. Stds. (U.S.) Circ. No. 467, Vol. 3 (1968).

15. D. R. Stull and G. C. Sinke, Advan. Chem. Ser. $\underline{18}$ (1956).

16. R. Klein and L. B. Leder, J. Chem. Phys. $\underline{38}$ 1866 (1963).

17. R. P. H. Gasser and R. Thwaites, Trans. Farday Soc $\underline{61}$, 2036 (1965).

18. M. D. Scheer and J. Fine "The Reaction of CO with a Tantalum Surface Heated above $2000°K$", Surface Science $\underline{12}$, 102 (1968).

19. W. Gordy and W. J. O. Thomas, J. Chem. Phys. $\underline{24}$, 439 (1956).

20. R. S. Mulliken, J. Chem. Phys. $\underline{2}$, 782 (1934).

21. L. Pauling, J. Am. Chem. Soc. $\underline{54}$, 3570 (1932).

22. H. A. Skinner and H. O. Pritchard, Trans. Faraday Soc. $\underline{49}$, 1254 (1953).

23. E. Clementi, Phys. Rev. $\underline{135}$, A 980 (1964).

TABLE I

POSITIVE ION SUBLIMATION ENERGIES

ION	T(K)	$*\ell_+$ (obs.)	$**(\ell_o + I - \phi)$
$^{93}Nb^+$	2100	10.4 eV	$7.5 + 6.8 - 3.9 = 10.5$ eV
$^{98}Mo^+$	1925	9.5	$6.7 + 7.1 - 4.3 = 9.5$
$^{181}Ta^+$	2280	11.2	$7.9 + 7.9 - 4.1 = 11.7$
$^{186}W^+$	2450	12.1	$8.9 + 8.0 - 4.5 = 12.4$
$^{187}Re^+$	2130	10.7	$7.8 + 7.9 - 5.0 = 10.7$

*Typical standard errors in the least square determination of ℓ_+ are between 0.1 and 0.2 eV

**Uncertainties in ℓ_o, I, and ϕ are each about 0.1 eV

TABLE II

Positive to Negative Ion Ratios for Tantalum Sublimation in the 2000 to 2100°K Temperature Range

T(°K)	j_+(ions/sec)*	j_-(ions/sec)*	(ω_-/ω_+)**	$kT\ln[\omega_- j_+/\omega_+ j_-\ ]$ (eV)
2000	0.18 ± 0.04	0.030 ± 0.015	0.129	-0.04 ± 0.08
2047	0.58 ± 0.10	0.10 ± 0.03	0.126	-0.06 ± 0.06
2077	1.67 ± 0.13	0.18 ± 0.05	0.124	+0.02 ± 0.06
				-0.03 ± 0.07

* The errors given for j_+ and j_- are average deviations from an arithmetic mean of at least six successive ion counts.

** The negative ion electronic partition function ω_- was calculated assuming that there was only one bound state with an inner quantum number equal to that of the ground state of the atom with the next highest atomic number.

Table III

ELECTRON AFFINITIES

(Comparison with theoretical estimates)

E. Clementi:
(Hartree-Fock
Calculations)
Phys. Rev. 135,
A980 (1964)

3rd row

	V	Cr	Mn	Fe
	$3d^3 4s^2$	$3d^5 4s$	$3d^5 4s^2$	$3d^6 4s^2$
A(eV)	0.94 ± 0.25	0.98 ± 0.35	-1.07 ± 0.20	0.58 ± 0.20

This work:
(self surface
ionization)

4th row

	Nb	Mo
	$4d^4 5s^1$	$4d^5 5s$
A(eV)	~ 1.0 (est.)	1.0 ± 0.2

5th row

	Ta	W	Re
	$5d^3 6s^2$	$5d^4 6s^2$	$5d^5 6s^2$
A(eV)	0.9 ± 0.3	0.5 ± 0.3	0.1 ± 0.3

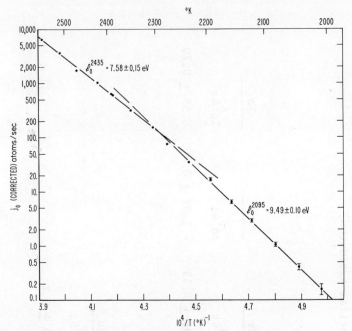

Fig. 1.   A second law plot of tantalum atom sublimation from a carbon contaminated surface. The values for l_0 were obtained by the method of
least squares in the temperature range indicated.

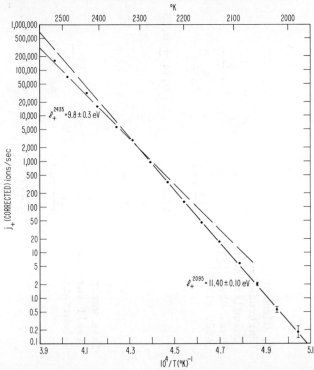

Fig. 2.   The determination of l_+ for a carbon contaminated tantalum sample. The
values for l_+ were obtained by the method of least squares in the temperature ranges indicated.

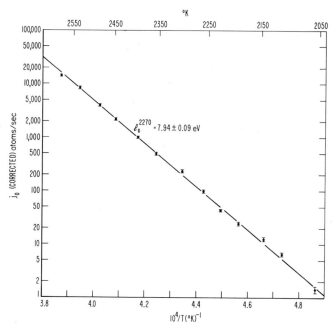

Fig. 3. Atom sublimation from a carbon free tantalum surface.

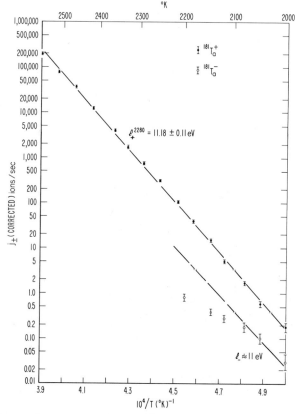

Fig. 4. Positive and negative ion sublimation from a carbon-free tantalum surface.

EMPIRICAL REPRESENTATION OF THE VELOCITY DISTRIBUTION DENSITY FUNCTION OF GAS MOLECULES SCATTERED FROM A SOLID SURFACE

Frank O. Goodman

Daily Telegraph Theoretical Department, School of Physics
University of Sydney, N.S.W., Australia
Department of Aerospace Engineering and Engineering Physics
University of Virginia, Charlottesville, Virginia *

1. INTRODUCTION

The very great importance of the specification of the velocity distribution function of gas molecules scattered from a solid surface is so well-known that it seems unnecessary to give detailed justification of its study. Suffice it to say that problems such as (i) the solution of the Boltzmann equation for a bounded gas, (ii) the calculation of accommodation coefficients and (iii) the calculation of the drag and lift on a solid body passing through a gas under free-molecule-flow conditions cannot be solved without at least partial specification of this function.

The empirical results presented here are proposed in the framework of three-dimensional classical lattice models of gas-surface scattering; the term "lattice model" is intended to be general and to include any classical lattice model of the solid. The results may be considered as unsatisfactory at present because it is not yet known how to relate every parameter specifying the scattered distribution to the parameters specifying the incident distribution. However, it is possible to make assumptions about the scattered parameters where such relations are not yet known. That this is a useful exercise may be illustrated by just two examples.

(A) Theories[1-5] of thermal accommodation coefficients which assume a one-dimensional incident gas beam normal to the surface of a three-dimensional solid and restrict discussion to "head-on" encounters between gas molecules and surface atoms may be put on a sounder basis, as the results for these very restricted encounters may be related, via the empirical results presented here, to those for more general types of encounter.

* Present address

(B) Trapping and adsorption of gas molecules are of great importance in gas-surface scattering theories and experiments. Recent theories[6] of these phenomena do not seem to be in agreement with experimental findings, the theories greatly overestimating the probability of trapping. This is a frustrating situation, particularly if only a small number of theoretical gas molecule trajectories is available[6]; the empirical results presented here will help to resolve this problem.

The results rest heavily on those of the author's hard-spheres model[7-9]; this model forms a convenient basis for such an empirical investigation because there exists a rather comprehensive set[8] of results. There are certain regimes of interest which may be quite well represented by the hard-spheres model, almost as it stands. One of these is hyperthermal free-molecule flow, and it is expected that the representation presented here should be a significant improvement on previous work in this regime. However, the results must be generalised and extended to other regimes, the appropriate generalisations being based on other theories, on experimental data, or simply on guesswork. Discussion of previous relevant work is delayed until after some definitions have been made.

A detailed description of a hypothetical theoretical model follows in Section 2, and it is as well to point out now that the results of this model have <u>not</u> been calculated, except in very special cases. The complete model is described in order that the important variables may be defined.

After spending some time calculating with the model, it is the author's opinion that such calculations are not worthwhile at this stage for many

reasons, the main one being the difficulty of adequately representing the
scattered distribution function unless very large numbers[10] of gas atom
trajectories are calculated.   It is impossible to calculate even a small
fraction of these numbers at present.   It seems, in fact, that only in
the hard-spheres limit can useful calculations of the scattered distribution
be performed with the model.   When this restriction is relaxed, even
slightly, great difficulty may be encountered with representing the
scattering[6,11-15].   Away from the hard-spheres limit, the most complex
realistic model for which fairly adequate representation of the scattering
is obtained seems to be the present form of the (basically) flat-surface
model of Logan, Stickney and Keck[16-18], called the soft-cubes model[18].   It
is for these reasons that this investigation is empirical and that it leans
so heavily on the hard-spheres model.

2.   DESCRIPTION OF A THEORETICAL LATTICE MODEL

A typical interaction is shown schematically in Fig.1.   The solid is
represented by a classical monatomic three-dimensional lattice model at
temperature T;   the type of lattice (for example, sc, bcc, etc.) is denoted
by S.   A single fundamental length, ℓ, is associated with the lattice model
(for example, ℓ may be the nearest-neighbour distance, the lattice spacing,
etc.).   Also associated with the lattice is a fundamental frequency, ω,
which may be regarded as $k\theta/h$ where θ is a characteristic lattice temperature.
The atomic mass of the lattice is m.

The gas molecule is represented by a point particle of mass M; thus the gas is assumed to be effectively monatomic and "gas molecule" may be called "gas atom" when discussing this model. The aiming point (x,y) of the gas atom trajectory (see Fig.1) is specified relative to coordinate axes fixed arbitrarily in the surface. The gas atom velocity is described in spherical polar coordinates:

$$V = (V, \theta, \phi) , \qquad\qquad (2.1)$$

where V is the gas atom speed, θ is the polar angle defined so that $\theta = 0$ is the normal outward from the surface, and ϕ is the azimuthal angle defined so that $\phi = 0$ is the positive x-direction. Incident, or initial, conditions are denoted by a subscript 0. Discussion is restricted to the case of a "monoenergetic incident beam"; this term is defined to mean an ensemble of gas atoms whose incident velocities, V_0, are identical and whose aiming points are distributed uniformly over the surface.

The gas-solid interaction energy is represented by a conventional two-parameter gas atom-solid atom pairwise potential function, characterised in the usual way by a length λ and an energy ε (for example, a Lennard-Jones 6-12 potential[19]). This results[20] in a "potential well" between the gas atom and the solid surface, the depth of which depends on S, λ/ℓ, ε and position on the surface. A suitable average well-depth is denoted by W, and it is expected[20] that $W/\varepsilon \simeq$ 5-8.

Conditions at the bottom of the well before the repulsive part of the gas-surface interaction occurs are denoted by a subscript 1, and conditions afterwards by a subscript 2; the scattered, or final, conditions are denoted

by a subscript 3 (see Fig.1).   The scattering angle at the bottom of the

well is the angle between the velocities V_2 and V_1, and is denoted by β:

$$\cos\beta = \sin\theta_1\sin\theta_2\cos(\phi_2-\phi_1) + \cos\theta_1\cos\theta_2 \quad . \qquad (2.2)$$

Limited information about the interaction is afforded by the "accommoda-

tion coefficients" (abbreviated hereafter to "acs"); these are denoted by

α_q and are defined for scalar gas atom properties, Q, as follows:

$$(Q_{\bar{o}} - Q_s)\,\alpha_q = (Q_o - \overline{Q}_3) \quad , \qquad (2.3)$$

where the bar denotes an average over the scattered distribution and Q_s is

the value $\overline{Q}_3$ would have if the scattered beam were in thermal equilibrium

with the solid.   (There is no bar over Q_o because, for all Q of physical

interest, all gas atoms in a monoenergetic beam have the same value of Q).

The acs α_q refer to the entire gas beam.   As single gas atoms are con-

sidered in the first instance, it is convenient to define an "effective ac"

of a single gas atom, denoted by γ_q, by analogy with (2.3):

$$(Q_o - Q_s)\,\gamma_q = (Q_o - Q_3) \quad ; \qquad (2.4)$$

for example, the effective speed ac, γ_v, is defined by

$$[V_o - (9\pi kT/8M)^{\frac{1}{2}}]\,\gamma_v = (V_o - V_3) \quad , \qquad (2.5)$$

where k is the Boltzmann constant.   The three acs most commonly discussed[7-9],

the energy, normal momentum and tangential momentum acs, are denoted,

respectively, by α_e, α_n and α_t.

Attention is confined to the behaviour of the gas during interaction
(for example, no attempt is made to consider which modes of the solid are
excited by the interaction); then all possible results of the interaction
may be obtained from knowledge of $\underline{M}$ and of the normalised velocity distribution density function $G_3(V,\theta,\phi)$.   The author prefers to define a new
variable, t, as follows:

$$t = MV^2/4k \quad ; \tag{2.6}$$

the parameter $\underline{t}$ may be called the "effective temperature" of a gas atom.
If M is known, information equivalent to that in $G(V,\theta,\phi)$ is contained in
$g(t,\theta,\phi)$:

$$g(t,\theta,\phi) = (2k/MV)G(V,\theta,\phi) \quad . \tag{2.7}$$

By analogy with the definition (2.6) of $\underline{t}$, an "effective well-depth
temperature", denoted by t_w, is defined as follows:

$$t_w = W/2k \quad . \tag{2.8}$$

The normalised flux distribution density function, $f(\theta,\phi)$, is obtained
from $g(t,\theta,\phi)$, and f and g are normalised, as follows

$$\int_0^{\pi/2} d\theta \int_0^{2\pi} d\phi \, f(\theta,\phi) = \int_0^\infty dt \int_0^{\pi/2} d\theta \int_0^{2\pi} d\phi \, g(t,\theta,\phi) = 1 \quad . \tag{2.9}$$

We note that $f(\theta,\phi)$ is not the ordinary flux distribution density function
in solid angle (Ω) space, which may be denoted by $F(\Omega)$; the relation

between them is

$$F(\Omega) \;=\; f(\theta,\phi)\, \mathrm{cosec}\,\theta \quad . \tag{2.10}$$

It is well-known that adequate representation of the important functions f_3 and g_3 is a very difficult problem; it is with this problem that the present work is concerned. At this stage, it may be worthwhile to give a brief survey of some of the previous work concerned with representation of these functions.

3. SOME PREVIOUS WORK CONCERNED WITH REPRESENTATION OF THE SCATTERED VELOCITY DISTRIBUTION FUNCTION

The first attempt to specify g_3 of which the author is aware is that of Maxwell[21], who assumed that a fraction α of the incident gas molecules is completely accommodated, the remaining fraction $(1-\alpha)$ being "specularly reflected". This means that:

Maxwell:
$$g_3(t,\theta,\phi) \;=\; \alpha g_T(t,\theta,\phi) \;+$$

$$(1-\alpha)\delta(t-t_o)\delta(\theta+\theta_o-\pi)\delta(\phi-\phi_o) \quad , \tag{3.1}$$

where g_T is the Maxwellian flux distribution appropriate to temperature T;

$$g_T(t,\theta,\phi) \;=\; (2/\pi T^2)\sin(2\theta)t\,\exp(-2t/T), \tag{3.2}$$

and that all the conventional acs are equal to α. The importance of this model lies, to a large extent, in its extreme simplicity; its usefulness is illustrated by the fact that the underlying ideas therein are still in use at the present time (see, for example, Ref.22).

The next attempt seems to be the hard-spheres calculation of Baule[23],
whose model was similar in many respects to the author's hard-spheres
model[7-9]; the gas atom-solid atom encounters were dealt with very
approximately, and a result of Baule's model was his now well-known
formula for the energy ac:

Baule : $$\alpha_e = 2\mu/(1+\mu)^2 , \qquad\qquad (3.3)$$

where μ is the mass ratio:

$$\mu = M/m . \qquad\qquad (3.4)$$

Some of the deficiencies in Baule's model were recognised by
Yerofeyev[24], who gave a more accurate account of the problems of
multiple encounters of a gas atom with the surface atoms, penetration of
a gas atom below the surface layer and the "shielding effect" on surface
atoms by other surface atoms.

The flat-surface theory of Logan, Stickney and Keck[16-18] begins
with the hard-cubes model[16-17] which is later modified to the soft-cubes
model[18]. These models represent the surface atoms by cubes rather than
spheres, a face of each cube being parallel to the surface; the models
are, then, essentially one-dimensional[24], although surface roughness is
considered[17,18]. Closed-form expressions for the scattering distributions
are obtained; the model has had considerable success in correlating much
of the recent low-energy scattering data.

Nocilla[27] modelled g_3 by means of a scattered gas beam in thermal
equilibrium except for a constant translational velocity superimposed on

each gas molecule;   the model was applied very successfully to some of

Hurlbut's low-energy scattering data.   However, as pointed out by

Hurlbut[28], it remains to properly relate the incident and scattered

velocity distributions in the model[28a].

Hurlbut[28] has examined the problem using an analogue computer.

His lattices were very simple, consisting of three atoms joined to each

other and to rigid walls by non-linear "springs".   A useful table of

the important qualitative trends observed in Hurlbut's model appears on

p.19 of Ref.28;   these trends are compared with those of the hard-spheres

model in Section 5 below.

Epstein[29] has constructed a model for a conditional probability

function, $Z(V_o, V_3)$, defined so that $Z(V_o, V_3) \, dV_3$ gives the probability

that a gas molecule of incident velocity V_o is scattered with a velocity

between V_3 and $V_3 + dV_3$;   the basis of this construction is the qualita-

tive behaviour of $\alpha_e(t_o)$ in the author's theory[3] of energy acs.   There

are three adjustable parameters, denoted by Epstein by θ_1, θ_2 and B, and

suitable choice of these parameters for each gas-solid system gives a

good fit of the model with the experimental data of Thomas and his

coworkers used by the author[3] and Wachman[30].   The physical significance

of the parameters θ_1 and θ_2 and the best-fit values of θ_1, θ_2 and B for

the gas-solid systems concerned seem to the author to be of the utmost

importance, and it is a pity that Epstein did not discuss this in greater

detail.   Assuming Equation (16) of Epstein's paper is (correcting the

obvious printer's error) correct, then putting $T_s = 0$ gives the following high-T_g and low-T_g expansions of $\alpha_e(T_g)$, where T_g is the incident gas temperature and where B is replaced[29] by $\alpha_e(\infty)$:

Epstein ; $T_s = 0$;   high-T_g :

$$\alpha_e(T_g) = \alpha_e(\infty) - constant/T_g^{3} + \ldots\ , \qquad (3.5a)$$

Epstein ; $T_s = 0$;   low-T_g :

$$\alpha_e(T_g) = 1 - constant \times T_g + \ldots\ . \qquad (3.5b)$$

Based on a comparison of these expansions with the corresponding ones of $\gamma_e(t_o)$ derived in Ref.4, the author's opinion is that both of (3.5) are invalid and that Epstein's model needs modification.

Considerable effort is being devoted to this problem by Oman and his group[6,11-15], whose numerical calculations on their "independent-oscillator lattice model" are carefully chosen according to a balanced statistical design; the scattering of diatomic gas molecules is also being considered by this group[13-15]. They are experiencing considerable difficulty, however, as numerical calculations such as theirs require extremely large numbers of gas atom trajectories[8,9] to obtain "smooth" distributions g_3, each trajectory requiring about 10 seconds of IBM 360-75 time[6]. Although it is at present not clear how this situation is to be remedied, the suggestion of Oman, Calia and Weiser[15] that each of their gas atom trajectories be replaced by a density distribution in angle space (this could be done in velocity space also) is a good one and gives considerable scope for development of the model[6].

Simplified calculations of the Oman type have been reported by

Raff, Lorenzen and McCoy[31], who employ a lattice model consisting of

three atoms joined to each other and to rigid walls by linear springs.

Both this model and Hurlbut's[28] restrict the problem to two dimensions

for simplicity. Care is needed when considering the results in Ref.31

on energy _acs_ as the energy ac defined therein is not the conventional

one used by other authors[32]. Apart from some of the conclusions regard-

ing the behaviour of the energy ac, which must be modified, the list in

Section IV of Ref.31 gives a useful guide to the qualitative trends to be

expected from this two-dimensional model.

A rather different, and interesting, point of view is taken by

Healy[33], who showed that scattering patterns similar to some of the

experimental ones may be obtained from the assumption of, effectively,

specular reflection of the gas molecules from a "randomly rough" solid

surface. Healy's term "rough" is a very general one and the model is

not as simple as it appears at first sight. Also, the approach is not

basically as different from other approaches as it appears at first[34].

Simple closed-form expressions for scattering distributions are obtained,

although these must be considered as very preliminary.

Although Schamberg's work[35] appeared before any of that cited above

except Maxwell's[21] and Baule's[23], it is discussed last as it seems, in

some ways, to be the most closely related to that of this paper. His

model was developed mainly for studies of hyperthermal free-molecule flow,

the regime relevant to calculations of the drag on artificial earth-satellites.   Keeping for the moment to Schamberg's notation[35] for ease

of recognition, the scattered velocity distribution is represented (in

his "conical-beam" model) by a conical beam of angular half-width ϕ_o,

the axis of the scattering cone making an angle θ_r with the surface;

the scattered speed, V_r, of the gas molecules is a constant.   The

incident beam makes an angle θ_i with the surface, and θ_r is related to

θ_i by

$$\cos\theta_r = \cos^\nu\theta_i \quad ; \quad \nu > 1 \quad , \tag{3.6}$$

and the distribution about the direction θ_r obeys a cosine law[35]. V_r

follows from the incident speed, V_i, and the energy ac, α:

$$(1-\alpha) \, V_i^{\,2} = V_r^{\,2} \quad . \tag{3.7}$$

Thus Schamberg models the scattered velocity distribution with three

(ν,ϕ_o,α) parameters, in keeping with the former[33] representation of the

interaction by means of constant energy, normal momentum and tangential

momentum acs $(\alpha_e,\alpha_n,\alpha_t)$;   this former representation is less satisfactory

because it is well-known that α_e,α_n and α_t are by no means constant.   In

the light of present-day knowledge, however, Schamberg's model, although

a considerable improvement over the former one, can no longer be called

satisfactory and should be modified.   For example, his model was developed

mainly for cases in which, in the notation of the present paper, $t_o \gg T$.

For this case it is now known that, at least for not too grazing an angle

of incidence, the maximum in the scattered distribution lies in general

below[37] the specular direction;   Schamberg's assumption is that this maximum lies above[37] the specular direction in all cases.   In other

words, it may be that $\nu < 1$ would have been better than $\nu > 1$ in (3.6)

for this model.   The present author's view is that this does not conflict

with Schamberg's observation[35] that "there is a general tendency of the

surface interaction towards obliterating the effect of the past history of

the molecules, i.e., a trend towards diffuse reflection";   the point of

view taken in the present paper is that any upward deviation of the gas

beam from its incident direction is a trend towards diffuse scattering.

4.   SIMPLIFICATION TO THE PROBLEM OF ZERO POTENTIAL WELL

It is known[2-4] that the attractive parts of the gas-solid interaction

are conservative.   The effect of the potential well, then[4], is as

follows: before the gas atom enters the significant part of the repulsive

potential, the energy associated with the normal component of its momentum

is increased by W;   after the gas atom has left this repulsive region, this

energy is decreased by W, a negative result meaning that the gas atom is

"reflected" back to the surface for a second "collision".   (The term

"collision" is used to describe the event of the gas atom entering and

leaving the repulsive region).   In this manner, the gas atom may undergo

several collisions with the solid surface, performing a motion which may

be called "hopping", until either (i) its total translational energy at

the bottom of the well is reduced below W, in which case it seems reasonable

to say that the gas atom is "trapped", or (ii) it escapes.

These very important facts enable a major simplification to be made:
that is, attention may be confined to events at the bottom of the well
during the repulsive part of the interaction.   In other words, the
problem is reduced to that of representing the distribution $g_2(t,\theta,\phi)$ in
terms of the modified incident conditions (t_1,θ_1,ϕ_1) where there is no
longer a potential well to be considered.   When this simpler problem is
solved, the actual well, W, may be used to obtain (t_1,θ_1,ϕ_1) from (t_o,θ_o,ϕ_o)
and $g_3(t,\theta,\phi)$ from $g_2(t,\theta,\phi)$;   in this way g_3 is related to (t_o,θ_o,ϕ_o).
Trapped gas atoms are assumed to be scattered according to the Maxwellian
flux distribution (3.2) appropriate to the surface temperature.   Some of
these points are discussed further in Section 8.

5.   QUALITATIVE RESULTS OF THE HARD–SPHERES THEORY

The representation of the scattered distribution to be presented rests
firmly on qualitative results of the hard-spheres model where it is thought
reasonable that they may be more general.   For this reason, a brief summary
of these results is given.   (R is the radius of a surface sphere divided by
a fundamental surface array length[7-9], where the gas atom is considered as a
point particle).

(A) t_2, and therefore, γ_e, depends strongly on the scattering angle β

(see Fig.1); t_2 decreases, and therefore γ_e increases, monotoni-

cally as β increases from 0 to π.

(B) Both t_2 and f_2 depend strongly on μ; apart from the expected

dependence of t_2 on μ, more massive gas atoms tend to be

scattered at smaller β, and less massive at larger β.

(C) The dispersion of the scattered speed,

$$\sigma^2 = \overline{v_2^2} - \overline{v_2}^2 \;,$$ (5.1)

is quite small, particularly for small μ, at any given scattering direction (θ,ϕ).

(D) f_2 is, in general, approximately an even function of $(\phi-\phi_1)$; for normal incidence (that is, $\theta_1=\pi$), f_2 is independent of ϕ_1.

(E) As tangential incidence is approached (that is, as $\theta_1 \rightarrow \pi/2$), f_2 is strongly peaked at $\phi = \phi_1$; as $\theta_1 \rightarrow \pi$, however, f_2 does not depend strongly on ϕ, at least for smaller μ.

(F) For grazing incidence (that is, θ_1 not close to π), $F_2 (= f_2 \mathrm{cosec}\theta)$ is peaked at some value, denoted by θ_m, of θ around the specular value, $\theta = \pi - \theta_1$; this peak is not very strong, except for larger μ (that is, $\mu \approx 1$). As $\theta_1 \rightarrow \pi$, there is a fairly weak peak at some intermediate value of θ, this being very weak for smaller μ. As μ becomes larger the scattering becomes more tangential (θ_m increases) for all θ_1.

(G) For restricted ranges of μ, $\underline{R}$ and $\underline{S}$, quite good correlation of α_e is obtained by the formula

$$\alpha_e(\mu,R,S,\theta_1,\phi_1) \approx - 3.6\mu\cos\theta_1/(1+\mu)^2 \;.$$ (5.2)

An example is illustrated in Fig.2. The result (5.2) is confirmed for near-tangential incidence by (A15) in the Appendix.

(H)   For similar ranges of μ, R and S, approximate qualitative correlation of α_t is obtained by

$$\alpha_t(\mu,R,S,\theta_1,\phi_1) \simeq - A(\mu,R) \cos\theta_1 \quad , \tag{5.3}$$

at least for grazing incidence;   this result is confirmed by (A17) in the Appendix.   As normal incidence is approached, however, for some systems $\alpha_t(\theta_1)$ seems to have a rather peculiar maximum at some value of $\theta_1 < \pi$;   for other systems this maximum is absent (see Fig.2).   The presence of this maximum is discussed further in Section 8.

(I)   It may be shown[38] that the qualitative dependence of $\alpha_n(\theta_1)$ is different from that stated in Refs.7-9.   For near-normal incidence, it seems that this result does hold, that is

$$\theta_1 \simeq \pi : \alpha_n(\mu,R,S,\theta_1,\phi_1) \simeq B(\mu,R) + C(\mu,R)\sec\theta_1 \quad . \tag{5.4}$$

For near-tangential incidence, however, it seems[38] that (5.4) must be replaced by

$$\theta_1 \simeq \pi/2 : \alpha_n(\mu,R,S,\theta_1,\phi_1) \simeq - D(\mu,R)(-\sec\theta_1)^{\frac{1}{2}} \quad . \tag{5.5}$$

It is difficult to distinguish between (5.4) and (5.5) from the numerical results in Refs.7-9 because (5.5) holds when $(-\sec\theta_1)^{\frac{1}{2}}$ is quite large, where calculations were sparse.

It is useful to compare the qualitative trends observed by Hurlbut (p.19 of Ref.28) for his simplified analogue model with those above for the hard-spheres model.   Hurlbut's trend (1) and the first parts of (2) and (3) are in agreement with (B) and the last part of (F);   the second

parts of (2) and (3) are intimately connected with the attractive gas-
solid potential and cannot be compared with the hard-spheres results.

Hurlbut's (4) is in partial disagreement with (A): the gas-solid

energy transfer observed in the hard-spheres results is in no way related

to the specular direction (except, of course, for very grazing incidence);

however, the _maximum_ energy transfer _does_ occur for scattering above the

specular direction, and particularly for "back-scattering". A relation

of energy transfer to the specular direction, analogous to Hurlbut's,

has also been noticed by other workers, both theoretical[11,12] and

experimental[39,40]. Hurlbut's (5) may be due to the attractive potential,

although many of the hard-spheres scattering patterns[8] could also be

called "fan-shaped".

6. EMPIRICAL REPRESENTATION OF THE VELOCITY DISTRIBUTION
FUNCTION OF THE SCATTERED GAS MOLECULES

We now consider the more general lattice model discussed in Section 2.
It is necessary to make some judicious generalisations of the hard-spheres
results in the hope that these will apply to the more general model.   It
is recalled that the problem has been reduced by the arguments in Section
4 to that of representing $g_2(\underline{t},\theta,\phi)$ in terms of (t_1,θ_1,ϕ_1), where there
is no potential well present.

On the basis of analysis of the hard-spheres results on which are
based results (A)-(C) in Section 5, we assume that, if the direction of a
gas atom were unchanged during its encounter with the surface (that is, if
$\beta = 0$), it would be completely unaccommodated; further, we assume that

maximum accommodation occurs for gas atoms whose directions are reversed during collision (that is, for $\beta = \pi$). Consistently with these assumptions, γ_v is assumed to be proportional to $(1-\cos\beta)$, where β is the scattering angle (2.2); that is, we write

$$\gamma_{v\beta} = \tfrac{1}{2}\gamma_{v\pi}(1-\cos\beta) = \gamma_{v\pi}\sin^2(\tfrac{1}{2}\beta) , \qquad (6.1)$$

where $\gamma_{v\beta}$ denotes the value of γ_v for a scattering angle of β. Thus, $\gamma_{v\pi}$ is the effective speed ac for $\beta = \pi$, which occurs for a head-on encounter normal to the solid surface; in this way, results of "head-on" theoretical calculations (for example, the lattice theories of Refs.1-5) may be related to more general cases.

Now, $\gamma_{v\pi}$ depends certainly on t_1 and perhaps on T also; for the present, however, $\gamma_{v\pi}$ is assumed to be independent of T. It may be shown from the work of Ref.4 that, under certain conditions, for T = 0 $\gamma_{v\pi}$ may be written

$$t_1 \to 0 : \gamma_{v\pi} \to \text{constant} \times (t_1/t_i)^{3/2} \qquad (6.2a)$$

$$t_1 \to \infty : \gamma_{v\pi} \to \gamma_{v\pi\infty} (1-3t_i/2t_1) \qquad (6.2b)$$

where $\gamma_{v\pi\infty}$ is the "high-speed limit" of $\gamma_{v\pi}$, given for $\mu < \sim 0.8$ by[1]

$$\gamma_{v\pi\infty} = 2\mu/(1+\mu) , \qquad (6.3)$$

and where t_i is a temperature which depends on the gas-solid interaction parameters. On the basis of these results, the following formula[41] is chosen for $\gamma_{v\pi}$:

$$\gamma_{v\pi} = \gamma_{v\pi\infty} (1+t_i/t_1)^{-3/2} \qquad (6.4)$$

If the gas or the surface in a gas-surface scattering situation is unknown, then both $\gamma_{v\pi\infty}$ and t_i are adjustable parameters; if both are known, then $\gamma_{v\pi\infty}$ may be estimated from (6.3) and a rough estimate of t_i may be obtainable from the work of Ref.4. The final representation of γ_v is, then,

$$\gamma_v = \frac{\gamma_{v\pi\infty}}{2} (1-\sin\theta_1 \sin\theta_2 \cos(\phi_2-\phi_1)-\cos\theta_1 \cos\theta_2) \left(\frac{t_1}{t_1+t_i}\right)^{3/2} . \quad (6.5)$$

The above discussion would remain substantially unaltered if γ_e were to replace γ_v in (6.1),(6.2),(6.4) and (6.5); analysis of the hard-spheres results reveals that, for small μ, either γ_v or γ_e may be used, whereas, for larger μ, γ_v gives a better, but not completely adequate, representation.

For T = 0, (6.5) determines t_2 as a function of θ_2 and ϕ_2:

$$T = 0 \; : \; t_2 = t_1(1-\gamma_v)^2 . \quad (6.6)$$

For $T \neq 0$, however, a further assumption must be made about the distribution of t_2; denoting this distribution by $Y_2(t_2)$, a simple assumption consistent with all the definitions is

$$Y_2(t) = N_t (t_T^3/t)^{\frac{1}{2}} \exp(-2t_T/T) \quad (6.7a)$$

where

$$t_T^{\frac{1}{2}} = t_1^{\frac{1}{2}} - (t_1^{\frac{1}{2}}-t^{\frac{1}{2}})/\gamma_v , \quad (6.7b)$$

and where N_t is the normalising factor:

$$N_t = 4/\gamma_v T^2 . \quad (6.8)$$

It is noted that (6.7) restricts t_2 to the range

$$t_1(1-\gamma_v)^2 < t_2 < \infty . \tag{6.9}$$

The distribution (6.7) of t expresses the result of the following procedure. The accommodation process is regarded from the point of view of accommodation of speed, V, for the reason given above; this is equivalent to accommodation of $t^{\frac{1}{2}}$ on account of (2.6). A value, $t_T^{\frac{1}{2}}$, of $t^{\frac{1}{2}}$ is chosen at random from the Maxwellian flux distribution of $t^{\frac{1}{2}}$ appropriate to the temperature T; this distribution is easily obtained from (3.2). Then, t is chosen according to (6.7b).

$Y_2(t)$ determines the distribution of t_2 in terms of $f_2(\theta,\phi)$; this latter distribution is, hopefully, written as a product of the two separate distributions $\Theta_2(\theta)$ and $\Phi_2(\phi)$, when we have

$$g_2(t,\theta,\phi) = Y_2(t) \, \Theta_2(\theta) \, \Phi_2(\phi) . \tag{6.10}$$

On the basis of result (F) of Section 5, it is assumed that $\Theta_2(\theta)$ is a cosine distribution similar to Schamberg's[35]:

$$\Theta_2(\theta) = N_\theta \cos\left(\frac{\theta-\theta_m}{2\Delta/\pi}\right) \sin\theta , \tag{6.11}$$

where θ_m is the angle at which the maximum of $\Theta_2(\theta) \operatorname{cosec}\theta$ [that is, of $f_2(\theta,\phi) \operatorname{cosec}\theta$ or of $F_2(\Omega)$; see (2.10)] occurs, Δ is the "effective angular half-width" of the distribution and N_θ is the normalising factor. Δ is not necessarily the true angular half-width of the distribution because Δ is unrestricted in value whereas θ is restricted by

$$\theta_{min} < \theta < \theta_{max} , \tag{6.12}$$

where the limits, θ_{min} and θ_{max}, are as follows:

$$\Delta \; > \; \theta_m \qquad : \; \theta_{min} \; = \; 0 \; . \qquad\qquad\qquad (6.13a)$$

$$\Delta \; < \; \theta_m \qquad : \; \theta_{min} \; = \; \theta_m - \Delta \; . \qquad\qquad (6.13b)$$

$$\Delta \; > \; \pi/2 - \theta_m \; : \; \theta_{max} \; = \; \pi/2 \; . \qquad\qquad (6.13c)$$

$$\Delta \; < \; \pi/2 - \theta_m \; : \; \theta_{max} \; = \; \theta_m + \Delta \; . \qquad\qquad (6.13d)$$

The normalising factor N_θ is defined as usual:

$$N_\theta^{-1} \; = \; \int_{\theta_{min}}^{\theta_{max}} \cos\left(\frac{\theta - \theta_m}{2\Delta/\pi}\right) \sin\theta \; d\theta \; . \qquad\qquad (6.14)$$

Thus N_θ is obtained as a series of simple integrals; the result is lengthy, however, and is not written out in detail here.   It is noted that Δ is the true angular half-width only if both (6.13b) and (6.13d) hold.

θ_m and Δ are both functions of the incident parameters; in particular, they depend very strongly on θ_1.   For the hard-spheres model it is possible to make reasonable guesses as to how θ_m and Δ depend on θ_1 (see Section 7), and similar guesses are no doubt possible for other situations; in general, however, θ_m and Δ are adjustable parameters until such relations are established.

Consideration of results (D) and (E) in Section 5 leads us to assume that (i) $\Phi_2(\phi)$ is an even function of $(\phi-\phi_1)$, (ii) at $\theta_1 - \pi$, $\Phi_2(\phi)$ is independent of ϕ, and (iii) at $\theta_1 = \pi/2$, $\Phi_2(\phi) = \delta(\phi-\phi_1)$.   Remembering that $\tan\theta_1 < 0$, an expression for $\Phi_2(\phi)$ which satisfies these conditions and gives a reasonable picture of the hard-spheres scattering results is

$$\Phi_2(\phi) = N_\phi \exp(-C \tan\theta_1 \cos(\phi-\phi_1)) \quad , \qquad (6.15)$$

where C is an adjustable parameter, expected to be of order unity in real cases, and N_ϕ is the normalisation constant:

$$N_\phi^{-1} = \int_0^{2\pi} \exp(-C \tan\theta_1 \cos(\phi-\phi_1)) \, d\phi$$

$$= 2\pi \, I_0 (C \tan\theta_1) \quad , \qquad (6.16)$$

where $I_0(x)$ is a modified Bessel function[42] of x. We note that small values of $(-C \tan\theta_1)$ imply wide scattering distributions in ϕ-space, whereas large values imply narrow distributions.

The general representation of $g_2(t,\theta,\phi)$ is obtained from (6.5) and (6.7)-(6.16); we note that in general there are five adjustable parameters, $\gamma_{v\pi\infty}$, t_i, θ_m, Δ and C. Examples of $Y_2(t)$, $\theta_2(\theta)$ and $\Phi_2(\phi)$ are shown respectively, in Figs.3-5. Fig.3 contains $TY_2(t)$ for the three cases $t_1 = T$, $\gamma_v = 0.5$ and 1; $t_1 = 2T$, $\gamma_v = 0.5$: we note that $\gamma_v = 1$ means that t forms a Maxwellian flux distribution at temperature T. $\theta_2(\theta)\mathrm{cosec}\theta$ is plotted for the three cases $\theta_m = 30°$, $\Delta = 20°$, $50°$ and $80°$ in Fig.4. In Fig.5 is shown $\Phi_2(\phi)$ for the three cases C = 2 and 4, $\theta_1 = 120°$; C = 2, $\theta_1 = 150°$: in these cases $\phi_1 = 0$ with no loss of generality.

7. THE SPECIAL CASE OF THE HARD-SPHERES MODEL

The general assumptions of Section 6 are now applied to the special case of the hard-spheres model.   Consideration of the results of this model[8] quickly leads to the relations $\Delta \rightarrow 0$ and $\theta_m \rightarrow \pi/2$ as $\theta_1 \rightarrow \pi/2$;   in other words, for tangential incidence, the scattering is tangential and the scattered beam-width is zero.   It is shown by (A18) in the Appendix that, as tangential incidence is approached,

$$\theta_1 \rightarrow \pi/2 \quad : \quad \sin\theta_m \rightarrow 1 + \text{constant} \times \cos\theta_1 \quad . \tag{7.1}$$

We assume that θ_m may be expressed in this form for all θ_1, when the constant becomes $(1-\sin\theta_{m\pi})$, where $\theta_{m\pi}$ is the value of θ_m for normal incidence $(\theta_1 = \pi)$:

$$\text{all } \theta_1 \quad : \quad \sin\theta_m = 1 + (1-s_{m\pi}) \cos\theta_1 \quad , \tag{7.2}$$

where

$$s_{m\pi} = \sin\theta_{m\pi} \quad . \tag{7.3}$$

The constant, $\theta_{m\pi}$, assumed independent of θ_1, now takes the place of θ_m as an adjustable parameter.   The dependence of Δ on θ_1 is now fixed by assuming that Δ is proportional to $\pi/2-\theta_m$:

$$\Delta = \kappa(\pi/2-\theta_m) \quad , \tag{7.4}$$

where κ, expected to be of order unity in real cases and assumed independent of θ_1, now takes the place of Δ as an adjustable parameter.

The assumed behaviour of θ_m as a function of θ_1 is of some interest, and this is illustrated in Fig.6 for the four cases $0_{m\pi} = 0$, $\pi/6$, $\pi/3$ and $\pi/2$:

for $\theta_{m\pi}$ = 0, the scattering is always above the specular direction;

for $\theta_{m\pi}$ = $\pi/2$, the scattering is always below this direction; for

intermediate values of $\theta_{m\pi}$, the scattering is below the specular direction

for near-normal incidence and above it for near-tangential incidence.

This last result seems very reasonable physically.

The general adjustable parameters $\gamma_{v\pi\infty}$ and t_i of Section 6 are not,

of course, adjustable in the hard-spheres case; $\gamma_{v\pi\infty}$ is fixed by (6.3)

and t_i by

hard-spheres: $\qquad\qquad t_i/t_1 = 0$. $\qquad\qquad\qquad$ (7.5)

Thus it is hoped that each gas-surface pair in the hard-spheres

model, defined there by the set of parameters (μ,R,S), may be associated

with a set of parameters $(\mu,\theta_{m\pi},\kappa,C)$ relevant to our empirical representa-

tion.   In terms of these latter parameters, the representation of the

distribution $g_2(t,\theta,\phi)$ is as follows: t_2 is given by (6.6) where γ_v is

now

$$\gamma_v = \left(\frac{\mu}{1+\mu}\right)(1-\sin\theta_1\sin\theta_2\cos(\phi_2-\phi_1)-\cos\theta_1\cos\theta_2) \ ; \qquad (7.6)$$

$\Theta_2(\theta)$ is given by

$$\Theta_2(\theta) = N_\theta \cos\left(\frac{\theta-\theta_m}{\kappa(1-2\theta_m/\pi)}\right)\sin\theta \ , \qquad (7.7)$$

where N_θ is defined by (6.14) and θ_m by (7.2); $\Phi_2(\phi)$ is given by (6.15)

and (6.16).

This simple representation cannot be expected to be able to give exact _quantitative_ agreement with all the hard-spheres results; the usefulness of the representation, with regard to the hard-spheres model, is that it adequately reproduces, and helps us to understand, all the _qualitative_ trends in these results. As an illustration of the inadequacy of our formulae to represent a general hard-spheres situation, it is clear that they must fail as the hard-spheres radius, R, becomes large; it follows from hard-spheres theory (and particularly from (A1) and (A2) in the Appendix) that, as $R \rightarrow \infty$, (7.2), (7.4) and (7.6) are replaced, respectively, by

$$R \rightarrow \infty \quad : \quad \sin\theta_m = \sin\theta_1/[1-4\mu\cos^2\theta_1/(1+\mu)^2]^{\frac{1}{2}} \quad , \qquad (7.8)$$

$$R \rightarrow \infty \quad : \quad \Delta = 0 \quad , \qquad (7.9)$$

$$R \rightarrow \infty \quad : \quad \gamma_v = 1 - \sin\theta_1\cosec\theta_m \quad ; \qquad (7.10)$$

also, C is then infinite,

$$R \rightarrow \infty \quad : \quad C = \infty . \qquad (7.11)$$

The association of a gas-surface pair in the hard-spheres model with a parametric set $(\mu, \theta_{m\pi}, \kappa, C)$, referred to above, may be accomplished in many ways; the one chosen here is based on fitting the curves of acs vs. θ_1. The acs are calculated from the empirical representation by substituting for t_2/t_1 from (6.6) in their definitions (see Section 2 and

and Refs.7-9):

$$\alpha_e = 2\,\overline{\gamma_v} - \overline{\gamma_v^2} \,, \qquad\qquad (7.12)$$

$$\alpha_n = 1 + (\overline{\cos\theta_2} - \overline{\gamma_v \cos\theta_2})\sec\theta_1 \,, \qquad (7.13)$$

$$\alpha_t = 1 - (\overline{\cos(\phi_2-\phi_1)\sin\theta_2} - \overline{\gamma_v \cos(\phi_2-\phi_1)\sin\theta_2})\,\mathrm{cosec}\,\theta_1 \,, \qquad (7.14)$$

where γ_v is given by (7.6) and where bars denote averages over the scattered distribution g_2: averages of function of θ_2 are calculated using (7.7) and those of functions of ϕ_2 using (6.15). The calculations are done on an IBM 7040, all the results being obtainable in closed forms[43]; the author's program evaluates about four cases per second, each case involving one set $(\alpha_e, \alpha_n, \alpha_t)$.

For each gas-surface pair, curves of $\alpha_e(\theta_1)$, $\alpha_n(\theta_1)$ and $\alpha_t(\theta_1)$ are calculated from (7.12)-(7.14) for the given value of μ and for many combinations of $\theta_{m\pi}$, κ and C; from these curves values of these parameters are chosen which give a reasonable fit with the corresponding curves from the hard-spheres model[8].

Some a priori statements are possible concerning the qualitative dependence of the empirical parameters on μ, R and S. It is expected[7-9] that none of the parameters depends strongly on S, the surface structure. Result (B) of Section 5 implies that C increases as μ increases, and results (B) and (F) that $\theta_{m\pi}$ increases as μ increases. At first sight it may seem that, as μ increases, κ should decrease, as do the widths of the scattered distributions[8];   however, even if κ were independent of μ,

these widths would decrease automatically as μ increases on account of

relations (7.2) and (7.4) combined with the increase of $\theta_{m\pi}$ with μ.

Finally, it is clear that C increases (the scattered beam narrows) as R

increases.

It turns out, not unexpectedly, that our representation is at its

best for situations concerned with small mass ratios, μ, and fairly rough

surfaces (that is, fairly small R);   it is at its worst for large μ and

large R, particularly at near tangential incidence, $\theta_1 \to \pi/2$.   The hard-

spheres results in Fig.2 are for μ = 0.1 and R = 0.9, and this is a

situation which is represented very well indeed by our formulae;   this is

clear from Fig.7, in which the empirical results for μ = 0.1, $\theta_{m\pi}$ = 18°,

C = 1.5 and κ = 1.0 are presented.   One other example is given, for which

the representation is only qualitative rather than quantitative.   Such an

example is provided by the case μ = 0.9 and R = 1.3, the hard-spheres

results for which are shown in Fig.8;   the corresponding empirical results

for μ = 0.9, $\theta_{m\pi}$ = 65°, C = 4.0 and κ = 1.0 are shown in Fig.9.

8.   DISCUSSION

It is clear from comparing Fig.2 with Fig.7 and Fig.8 with Fig.9 that

the empirical representation of the scattered distribution, $g_2(t,\theta,\phi)$,

presented here gives adequate representation of the hard-spheres model in

these cases, particularly for smaller μ and smaller R.   The discrepancy

in the shapes of the $\alpha_t(\theta_1)$ curves in Figs.8 and 9 is not serious, as
it is sometimes very difficult indeed to calculate $\alpha_t(\theta_1)$ for near-
normal incidence ($\theta_1 \simeq \pi$) in the hard-spheres model[8,9].   Curves of
$\alpha(\theta_1)$ such as those in Fig.8 are difficult to fit empirically because
of the large difference in magnitude between $\alpha_t(\pi)$ on the one hand and
$\alpha_e(\pi)$ and $\alpha_n(\pi)$ on the other.   The inadequacy of the empirical fit in
Fig.9 may be due mainly to inadequacy of (7.6) for large μ (in any case,
the hard-spheres results are certainly invalid for $\mu > \sim 0.8^1$).

The empirical $\alpha_t(\theta_1)$ curve in Fig.9 is interesting as it displays a
maximum for $\theta_1 < \pi$, similar to that of curve A in Fig.2 and discussed in
Result (H) of Section 5.   As it may not be clear how such a maximum is
possible in the empirical representation, we indicate how it arises.
Although the result holds for all μ, it is simplest to restrict the
argument to the case $\mu = 0$, because then $\gamma_v = 0$ and (7.14) reduces to

$$\mu = 0 : \quad \alpha_t = 1 - \overline{\cos(\phi_2-\phi_1)} \, \sin\theta_2 \csc\theta_1 \, . \tag{8.1}$$

The angle ρ_1 is defined by

$$\rho_1 = \pi - \theta_1 \, , \tag{8.2}$$

this being a convenient definition because $\rho_1 \to 0$ gives normal incidence;
it may be shown from (6.15), (6.16) and (8.2) that

$$\rho_1 \to 0 : \quad \overline{\cos(\phi_2-\phi_1)} = C\rho_1/2 - (3C^2-8) \, C\rho_1^3/48 + \ldots \tag{8.3}$$

For simplicity, it is assumed that

$$\overline{\sin\theta_2} \simeq \sin\theta_m \quad, \tag{8.4}$$

and from (7.2), (8.2) and (8.4) we obtain

$$\rho_1 \to 0 \quad : \quad \overline{\sin\theta_2} \simeq s_{m\pi} + (1-s_{m\pi}) \rho_1^2/2 + \ldots \tag{8.5}$$

Of course,

$$\rho_1 \to 0 \quad : \quad \mathrm{cosec}\theta_1 = 1/\rho_1 + \rho_1/6 + \ldots \quad, \tag{8.6}$$

and from (8.1), (8.3), (8.5) and (8.6) we obtain

$$\rho_1 \to 0 \quad : \quad \alpha_t \simeq (1 - Cs_{m\pi}/2) + C(C^2 s_{m\pi} - 4) \rho_1^2/16 + \ldots \tag{8.7}$$

Hence, if $C^2 s_{m\pi} > 4$, α_t has a minimum at $\theta_1 = \pi$, and, therefore, a maximum for $\theta_1 < \pi$. Although this simple argument applies only if $\mu = 0$ and if (8.4) is valid, it should by now be clear that sufficiently large values of C will always give such a maximum; in fact, by making C sufficiently large it is easy to make $\alpha_t(\pi) < 0$, although this is not possible physically.

The result, $\kappa = 1$, in both the empirical fits leads one to hope that κ may be set equal to unity for all the hard-spheres results; then the set of three hard-spheres parameters (μ,R,S) would correspond to the empirical set $(\mu,\theta_{m\pi},C)$. Although short, the Table summarises the two sets of results obtained so far for the hard-spheres model.

The empirical representation may also be compared with the detailed scattering patterns which are available[8,9] for the hard-spheres model. This comparison alone is not a useful test of the empirical formulae because these patterns have been made a basis for these formulae. However, the

values of the empirical parameters which give good representation of the

ac curves should also give good representations of the detailed scattering

patterns. Comparison of the empirical results in the Table with the

relevant scattering patterns reveals that this is, in fact, the case.

As expected, the a priori statements made in Section 7 about the

behaviour of the empirical parameters are verified wherever the results

allow.

A few words may be in order concerning the form chosen in (6.15) for

$\Phi_2(\phi)$. Because $\Phi_2(\phi)$ must be an even function of $(\phi-\phi_1)$ and must have

zero gradient at $\phi = \phi_1 + \pi/2 \pm \pi/2$, it is reasonable to try the form

$$\Phi_2(\phi) = N_\phi \exp(X(\theta_1) \cos(\phi-\phi_1)) , \qquad (8.8)$$

where $X(\theta_1)$ is a function of θ_1 only. Once this form is agreed on, then

one shows that

$$X(\theta_1) = - C \tan\theta_1 \qquad (8.9)$$

is the most convenient choice as follows: it is known that $0 < \alpha_t(\pi) < 1$,

and it may be shown that this simple fact implies that

$$\theta_1 \to \pi: \qquad X(\theta_1) \to C(\pi-\theta_1) , \qquad (8.10a)$$

where C is a constant; furthermore, zero beam-width at tangential incidence

requires that

$$\theta_1 \to \pi/2 : \qquad X(\theta_1) \to \infty . \qquad (8.10b)$$

It must be emphasised here, with reference to Section 4 and Fig.1, that in cases where a potential well, W, is present then representation of the scattered distribution, $g_3(t,\theta,\phi)$, in terms of the incident conditions (t_o,θ_o,ϕ_o) is considerably more involved than the representation discussed in detail above.   As explained in Section 4, our empirical representation of g_2 in terms of (t_1,θ_1,ϕ_1) refers only to conditions at the bottom of the well; (t_1,θ_1,ϕ_1) must first be written in terms of (t_o,θ_o,ϕ_o):

$$t_1 = t_o + t_w , \qquad (8.11)$$

$$\cos^2\theta_1 = (t_o\cos^2\theta_o + t_w)/t_1 , \qquad (8.12)$$

$$\phi_1 = \phi_o . \qquad (8.13)$$

The empirical representation is then used to relate g_2 to (t_1,θ_1,ϕ_1); finally, g_3 is related to g_2 by means of the following analogues of (8.11)-(8.13):

$$t_3 = t_2 - t_w , \qquad (8.14)$$

$$\cos^2\theta_3 = (t_2\cos^2\theta_2 - t_w)/t_3 , \qquad (8.15)$$

$$\phi_3 = \phi_2 . \qquad (8.16)$$

As discussed in Section 4, if $t_2\cos^2\theta_2 < t_w$ and $t_2 > t_w$, the gas atom is reflected back to the surface and the phenomenon we call hopping has taken place;   it is only when $t_2 < t_w$ that trapping occurs[44]:

hopping : $t_2\cos^2\theta_2 < t_w$ and $t_2 > t_w$,   (0.17)

trapping : $t_2 < t_w$.   (8.18)

The types of problems which may be undertaken using the empirical representation seem ideally suited to Monte Carlo techniques; the advantage of such a representation over ordinary trajectory calculations lies mainly in the relatively very short time required to calculate the result (t_3,θ_3,ϕ_3) from a given initial set (t_o,θ_o,ϕ_o). To undertake any problem, however, it is necessary to specify the five parameters $(\gamma_{v\pi\infty},t_i,\theta_m,\Delta,C)$, and it is not yet known how this may be done for a general case; the Table and the above discussions provide useful guides, however.

For certain problems, it may be possible to use the simplified hard-spheres representation as it stands in Section 7, when it would be hoped that three parameters $(\mu,\theta_{m\pi},C)$ would suffice for a complete description of events at the bottom of the well (t_w may or may not be set equal to zero); an example is hyperthermal free-molecule flow. In fact, this simplified representation has been used by Bird[45] in recent Monte Carlo studies of the hypersonic flow of gases over simple solids in the transition regime.

It is expected that application of the work presented here, for example to the two problems (A) and (B) raised in Section 1, will be the subjects of future publications.

ACKNOWLEDGMENTS

The calculations for this work were done on

(i)   a National-Elliott 803 in the Department of Electronic Computing, University of Aberdeen, Scotland.

(ii) an IBM 7094 in the Computation Centre of the Massachusetts Institute of Technology, under problem number M4443, and

(iii) an IBM 7040 in the Basser Computing Department of the School of Physics, University of Sydney, New South Wales.

The author is indebted to the Government of the Commonwealth of Australia for the award of a Queen Elizabeth II Research Fellowship, during the tenure of which this work was done.

REFERENCES

1. F. O. Goodman, J.Phys.Chem.Solids $\underline{23}$ (1962) 1269.

2. F. O. Goodman, J.Phys.Chem.Solids $\underline{24}$ (1963) 1451.

3. F. O. Goodman, in Rarefied Gas Dynamics (Academic Press Inc.,
    New York, 1966), Vol.II, p.366.

4. F. O. Goodman,

5. C. M. Chambers and E. T. Kinzer, Surface Science $\underline{4}$ (1966) 33.

6. R. A. Oman, Grumman Research Department Report RE-306 (1967).
    Unpublished work of Jackson and of Madix and Korus is
    cited here.

7. F. O. Goodman, in Rarefied Gas Dynamics (Academic Pres Inc.,
    New York, 1967), Vol.I, p.35.

8. F. O. Goodman, NASA Report CR-933 (1967).

9. F. O. Goodman, Surface Science $\underline{7}$ (1967) 391.

10. On the basis of results of Refs.1-3, it is reasonable to suppose
    that several thousands of trajectories would be necessary for
    each case with the surface at $0°K$; with a hot surface,
    however, this may well rise to several tens of thousands.

11. R. A. Oman et al., AIAA J. $\underline{2}$ (1964) 1722.

12. R. A. Oman, A. Bogan and C. H. Li, Ref.3, p.396.   See also
    Grumman Research Department Report RE-181J (1964).

13. R. A. Oman, Grumman Research Department Report RE-222 (1965).

14. R. A. Oman, Ref.7, p.83.

15. R. A. Oman, V. S. Calia and C. H. Weiser, Grumman Research
    Department Report RE-272 (1966).

16. R. M. Logan and R. E. Stickney, J.Chem.Phys. 44 (1966) 195.

17. R. M. Logan, J. C. Keck and R. E. Stickney, Ref.7, p.49.

18. R. M. Logan and J. C. Keck, MIT Fluid Mechanics Laboratory Report
    No.67-8 (1967).

19. J. E. Lennard-Jones, Physica 4 (1937) 941.

20. F. O. Goodman, Phys.Rev. 164 (1967) 1113.

21. J. C. Maxwell, in The Scientific Papers of James Clerk Maxwell
    (CUP, 1890), Vol.II, p.708.

22. J. J. Hinchen and W. M. Foley, United Aircraft Research Laboratories
    Report D910245-7 (1965).

23. B. Baule, Ann.Phys. 44 (1914) 145.

24. A. I. Yerofeyev, Inzh.Zh. 4 (1964) 36.   Translated into English
    by J. W. Brook, Grumman Research Department Translation
    TR-38 (1966).

25. For example, the hard-cubes model (Ref.16) is based on the present
    author's exact one-dimensional "rigid-box-model" calculation
    (Ref.26).

26. F. O. Goodman, J.Phys.Chem.Solids 26 (1965) 85.

27. S. Nocilla, in Rarefied Gas Dynamics (Academic Press Inc., New York
    1963), Vol.I, p.327.

28. F. C. Hurlbut, Ref.7, p.1.

28a. This and further work using Nocilla's model has been reported recently
    by F. C. Hurlbut, Entropie No.18 (1967) p.99.

29. M. Epstein, AIAA J. $\underline{5}$ (1967) 1797.

30. F. O. Goodman and H. Y. Wachman, J.Chem.Phys. $\underline{46}$ (1967) 2376.

31. L. M. Raff, J. Lorenzen and B. C. McCoy, J.Chem.Phys. $\underline{46}$
     (1967) 4265.

32. The definitions (13), (15) and (16) of Ref.31 differ fundamentally
     from the conventional definitions of energy $\underline{acs}$.

33. T. J. Healy, in Fundamentals of Gas-Surface Interactions (Academic
     Press, New York, 1967), p.435.

34. For example, see Ref.30 and the discussion on pp.530 and 531 of
     Ref.33.

35. R. Schamberg, The Rand Corporation Report RM-2313 (1959); ASTIA
     Document Number AD 215301.

36. S. A. Schaaf and P. L. Chambre, in Fundamentals of Gas Dynamics
     (Princeton Univ.Press, 1958), p.687.

37. In the notation of the present paper, a gas atom is scattered
     below the specular direction if $\theta_3 > \pi-\theta_o$ and above if
     $\theta_3 < \pi-\theta_o$.

38. See Equation (A16) in the Appendix and a result in Ref.24.

39. J. N. Smith, Jr., J.Chem.Phys. $\underline{40}$ (1964) 2520.

40. F. M. Devienne, J. C. Roustan and R. Clapier, Ref.7, p.269.

41. This formula is different from the empirical formula proposed in
     Ref.30; the present author's opinion is that the present
     formula is an improvement over the earlier one for both
     large and small t_1.

42. See, for example, E. Jahnke, F. Emde and F. Losch, Tables of higher

functions (McGraw Hill, London, 1960), Section IXB.

43. Modified Bessel functions may be called closed-form expressions

as far as modern computers are concerned.

44. This definition of trapping does not agree with that of previous

workers, for example Refs.6 and 18; these workers do not

seem to allow hopping to take place, defining all gas atoms

for which $t_2 \cos^2 \theta_2 < t_w$ as being trapped.   This may be one

of the reasons for the difficulty discussed in problem (B)

of Section 1.

45. G. A. Bird, to be published; this is an extension of the earlier

work in Refs.46 and 47.

46. G. A. Bird, AIAA J. <u>4</u> (1966) 55.

47. G. A. Bird et al., presented at the AIAA 6th Aerospace Sciences

Meeting, Jan.1968.   (AIAA Paper No.68-6).

APPENDIX

A hard-cylinders model for calculations at near-tangential incidence

In this appendix, a simplified version of the hard-spheres model is considered, the object being to confirm results (5.2), (5.3), (5.5) and (7.1) in the case of near-tangential incidence $(\theta_1 \to \pi/2)$. We consider a "hard-cylinders model" of the gas-surface interaction; the gas atoms are still modelled by point particles, but the surface spheres are replaced by cylinders, whose axes are parallel to each other and perpendicular to the plane of incidence (the plane containing the incident velocity vector and the surface normal). It is assumed that this hard-cylinders model does not differ qualitatively from the hard-spheres model at near-tangential incidence.

We consider first some results for a general hard-spheres (or hard-cylinders) collision, with reference to Fig.10. λ_1 and λ_2 are defined, respectively, as the acute angles which the incident and scattered velocity vectors, V_1 and V_2, of the point gas particle make with the tangent plane to the initially stationary surface sphere at the point of impact, all measurements being relative to a stationary frame of reference; these definitions are convenient as we are eventually interested in the limit $\lambda_1 \to 0$ (implying $\lambda_2 \to 0$ also). In this stationary frame, the relations among V_2, V_1, λ_2 and λ_1 are as follows:

$$V_2^2 = V_1^2 \left[1 - \frac{4\mu}{(1+\mu)^2} \sin^2\lambda_1 \right] , \tag{A1}$$

$$\tan\lambda_2 = \left(\frac{1-\mu}{1+\mu}\right) \tan\lambda_1 . \tag{A2}$$

We consider now an actual collision, and it is convenient to define the angles σ_1 and σ_2 as follows:

$$\sigma_1 = \theta_1 - \pi/2 , \tag{A3a}$$

$$\sigma_2 = \pi/2 - \theta_2 . \tag{A3b}$$

The collision geometry is illustrated by Fig.11; ψ is the angle between the radial plane, containing the axis of the cylinder and the point of impact, and the surface normal plane containing the axis of the cylinder. The σ_1 and σ_2 of (A3) and Fig.11 are related to the λ_1 and λ_2 of (A1), (A2) and Fig.10 as follows:

$$\lambda_1 = \sigma_1 + \psi , \tag{A4a}$$

$$\lambda_2 = \sigma_2 - \psi . \tag{A4b}$$

When the shielding effect on a surface cylinder by the adjacent cylinder is considered, it is clear from Fig.11 that the "allowed range" of ψ may be written, with the obvious sign notation,

$$\sigma_1 \to 0 : \qquad -\sigma_1 < \psi < \nu \tag{A5}$$

where it may be shown from simple geometric considerations that

$$\nu = (2\sigma_1/R)^{\frac{1}{2}} . \tag{A6}$$

Therefore, as $\sigma_1 \rightarrow 0$, $\nu \gg \sigma_1$ and nearly all incident gas atoms have ψ of order ν; in fact, those with ψ of order $\pm \sigma_1$ need not be considered in the lowest-order calculation here. For example, (A5) may be replaced by

$$\sigma_1 \rightarrow 0 : \qquad 0 < \psi < \nu \quad . \tag{A7}$$

Substituting for λ_1 and λ_2 in (A1) and (A2) from (A4) and using the relation $\psi \gg \sigma$, we may obtain

$$\sigma_1 \rightarrow 0 : \qquad \sigma_2 = \left(\frac{2}{1+\mu}\right) \psi \quad , \tag{A8}$$

$$\sigma_1 \rightarrow 0 : \qquad V_2 = V_1 \left[1 - \frac{2\mu}{(1+\mu)^2} \psi^2\right] \tag{A9}$$

The normalised distribution density function, $\Psi(\psi)$, of ψ for incident gas atoms is required:

$$\Psi(\psi) = R\psi/\sigma_1 \quad . \tag{A10}$$

Where bars now imply averages over the distribution (A10), the acs, α_e, α_n and α_t, may be calculated from (A8) and (A9) as usual:

$$\alpha_e = 1 - \overline{V_2^2}/V_1^2 \rightarrow \frac{4\mu}{(1+\mu)^2} \overline{\psi^2} \quad , \tag{A11}$$

$$\alpha_n = 1 - \overline{V_2 \sin\sigma_2}/V_1 \sin\sigma_1 \rightarrow 1 - \left(\frac{2}{1+\mu}\right) \overline{\psi}/\sigma_1 \quad , \tag{A12}$$

$$\alpha_t = 1 - \overline{V_2 \cos\sigma_2}/V_1 \cos\sigma_1 \rightarrow \left(\frac{2}{1+\mu}\right) \overline{\psi^2} \quad . \tag{A13}$$

The averages, $\overline{\psi}$ and $\overline{\psi^2}$, follow from (A6), (A7) and (A10):

$$\overline{\psi} = (8\sigma_1/9R)^{\frac{1}{2}},\qquad\text{(A14a)}$$

$$\overline{\psi^2} = \sigma_1/R.\qquad\text{(A14b)}$$

Substitution of (A14) into (A11)-(A13) yields

$$\sigma_1 \to 0 \;:\qquad \alpha_e = \frac{4\mu}{(1+\mu)^2}\,\frac{\sigma_1}{R},\qquad\text{(A15)}$$

$$\sigma_1 \to 0 \;:\qquad \alpha_n = 1 - \left(\frac{2}{1+\mu}\right)\left(\frac{8}{9R\sigma_1}\right)^{\frac{1}{2}},\qquad\text{(A16)}$$

$$\sigma_1 \to 0 \;:\qquad \alpha_t = \left(\frac{2}{1+\mu}\right)\frac{\sigma_1}{R}.\qquad\text{(A17)}$$

It is noted that (A15) with $R \simeq 1.1$ confirms, for near-tangential incidence, the result (5.2);   similarly, for near-tangential incidence, (A16) and (A17) confirm, respectively, (5.5) and (5.3).

We may also derive a simple expression for the behaviour of θ_m as $\theta_1 \to \pi/2$ ($\sigma_1 \to 0$);   this expression is used in Section 7.   It follows from (A8) and (A10) that the distribution of σ_2 is proportional to σ_2, and hence that the maximum in this distribution, at σ_{2m} say, occurs at the largest value, σ_{2max}, of σ_2;   using (A6)-(A8) we may obtain

$$\sigma_1 \to 0 \;:\qquad \sigma_{2m} = \sigma_{2max} = \left(\frac{2}{1+\mu}\right)\psi_{max} = \left(\frac{2}{1+\mu}\right)\left(\frac{2\sigma_1}{R}\right)^{\frac{1}{2}}\qquad\text{(A18)}$$

This is, effectively, the result (7.1), and it confirms (7.2) for near-tangential incidence. A similar result has been obtained by Yerofeyev[24].

The author has been unable to perform satisfactory calculations,
analogous to those in this appendix, for the case of near-normal
incidence; these would help in confirming or rejecting many of our
formulae.

Table 1. The empirical parameters for the two hard-spheres cases considered, resulting from a comparison of the empirical curves of acs versus θ_1 with the corresponding hard-spheres curves.

R	S	μ	$\theta_{m\pi}$	κ	C
0.9	square	0.1	18°	1.0	1.5
1.3	square	0.9	65°	1.0	4.0

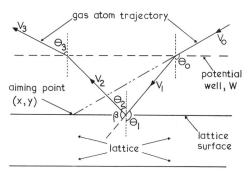

Fig. 1. A typical interaction in the theoretical lattice model. The gas atom
velocities V_0 and V_1 are in the plane of the paper, but V_2 and V_3 are
not in general; that is, $\phi_0 = \phi_1$ and $\phi_2 = \phi_3$ but $\phi_1 \neq \phi_2$. The lattice
parameters are S, T, l, m and ω; the gas atom parameters are $M, V_0, \theta_0,$
ϕ_0, x and y; the gas-lattice interaction parameters are λ, ϵ and
$W, W/\epsilon$ being a function of S and λ/l. β denotes the scattering angle
at the bottom of the well, and is the angle between V_2 and V_1.

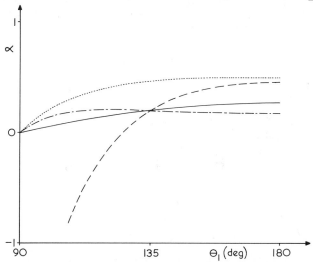

Fig. 2. Examples of accommodation coefficients obtained from the hard-spheres
theory: a_e (————), a_n (–––––––––) and a_t (..........) are shown
as functions of θ_1 for the case $\mu = 0.1$, R = 0.9, S = square, $\phi_1 = 0$;
curve A (–·–·–·–·–) is a_t (θ_1) for $\mu = 0.1$, R = 1.3, S = square,
$\phi_1 = 0$, and is an example of the maximum which occurs in a_t (θ_1) for
$\theta_1 < \pi$ in some cases.

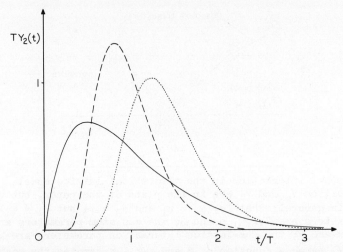

Fig. 3. The empirical distribution function, $TY_2(t)$, of effective temperatures of the scattered gas atoms, calculated from (6.7)-(6.9) for the three cases $t_1 = 2T$, $\gamma_v = 0.5$ (..........); $t_1 = T$, $\gamma_v = 0.5$ (------------); $t_1 = T$, $\gamma_v = 1$ (————).

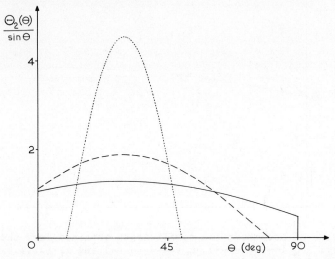

Fig. 4. The empirical distribution function, $\theta_2(\theta)\operatorname{cosec}\theta$, of polar angles of the scattered gas atoms, calculated from (6.11)-(6.14) for the three cases $\Delta = 20°$ (..........); $\Delta = 50°$ (------------); $\Delta = 80°$ (————); $\theta_m = 30°$ in all cases.

40–44

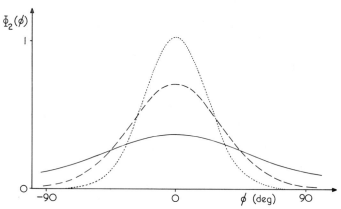

Fig. 5. The empirical distribution function, $\Phi_2(\phi)$, of azimuthal angles of the scattered gas atoms, calculated from (6.16) and (6.17) for the three cases $C = 2$, $\theta_1 = 150°$ (——————); $C = 2$, $\theta_1 = 120°$ (----------); $C = 4$, $\theta_1 = 120°$ (..........); $\phi_1 = 0$ in all cases.

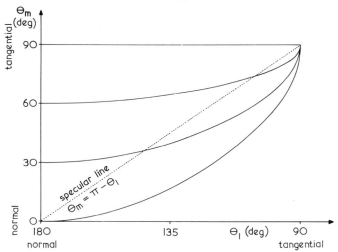

Fig. 6. θ_m as a function of θ_1 according to the empirical formulae (7.2) and (7.3) for the special case of the hard-spheres model and for the four cases $\theta_{m\pi} = 0$, $\pi/6$, $\pi/3$, $\pi/2$. The 'specular line' is the locus of points for which θ_m is the specular angle, $\theta_m = \pi - \theta_1$.

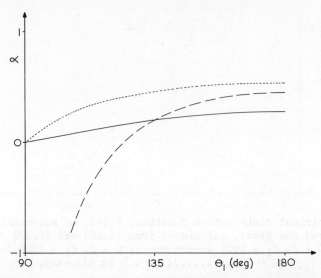

Fig. 7. Accommodation coefficients obtained from the empirical representation (7.12)-(7.14): a_e (————), a_n(—————) and a_t (..........) are shown as functions of θ_1 for the case $\mu = 0.1$, $\theta_{m\pi} = 18°$, C = 1.5, $\kappa = 1$; the corresponding results from the hard-spheres theory are shown in Fig. 2.

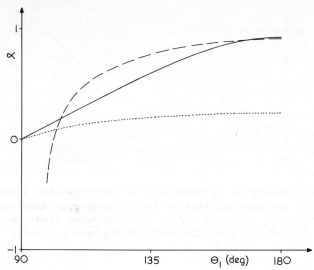

Fig. 8. Accommodation coefficients obtained from the hard-spheres theory: a_e (————), a_n(—————) and a_t (..........) are shown as functions of θ_1 for the case $\mu = 0.9$, R = 1.3, S = square, $\phi_1 = 0$.

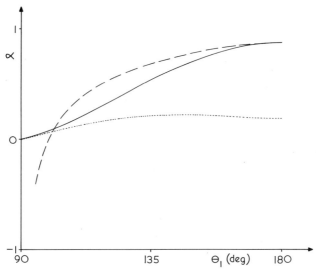

Fig. 9. Accommodation coefficients obtained from the empirical representation (7.12)-(7.14): a_e (——————), a_n (----------) and a_t (..........) are shown as functions of θ_1 for the case $\mu = 0.9$, $\theta_{m\pi} = 65°$, $C = 4$, $\kappa = 1$; the corresponding results from the hard-spheres theory are shown in Fig. 8.

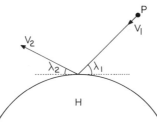

Fig. 10. A general hard-spheres collision. H is the initially stationary surface sphere and P is the point gas particle; λ_1 and λ_2 are, respectively, the angles at which the incident and scattered velocity vectors, V_1 and V_2, of P make with the tangent plane to the surface sphere at the point of impact, all measurements being relative to a stationary frame of reference.

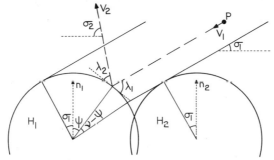

Fig. 11. A hard-cylinders gas-surface collision. H_1 and H_2 are surface cylinders, P is the point gas particle and n_1 and n_2 are surface normal planes containing the axes of the cylinders. λ_1 and λ_2 are explained in Fig. 10; σ_1 and σ_2 are, respectively, the angles which the incident and scattered velocity vectors, V_1 and V_2, of P make with a surface tangential plane. ψ is the angle between n_1 and the radial plane, containing the axis of H_1 and the point of impact of P with H_1; the shielding effect on H_1 by H_2 results, for small σ_1, in the restriction $-\sigma_1 < \psi < \nu$.

LOW-ENERGY MOLECULAR SCATTERING AS A TOOL FOR STUDYING GAS-SOLID INTERACTION POTENTIALS: COMPARISON OF THEORY AND EXPERIMENT

Robert E. Stickney

Department of Mechanical Engineering and
Research Laboratory of Electronics
Massachusetts Institute of Technology
Cambridge, Massachusetts

1. Introduction

The principal objective of this paper is to consider the information on gas-solid interaction potentials that may be obtained from measurements of the scattering of low-energy molecules (and atoms) from solid surfaces. We shall use the term scattering in the most general sense so that it includes all of the following interaction processes: diffraction, reflection, incomplete accommodation, and complete accommodation followed by desorption. The present definition of the low energy is that E_i, the energy of the incident molecule, is in the range $0 < E_i < 20$ eV; at higher energies a significant fraction of molecules may penetrate into the bulk of the solid, thereby introducing a number of additional processes which we do not wish to consider here. In order to emphasize the analogy between molecular scattering and low-energy electron diffraction (LEED), we shall refer to the present technique as low-energy molecular scattering (LEMS).

As illustrated by the papers presented at this symposium, studies of solid surfaces frequently are based upon measurements of the interaction of a beam of particles (photons, electrons, neutrons, ions, atoms, or molecules) with the sample. In most cases the beam is well collimated and nearly monoenergetic, since this leads to more clearly defined data. Measurements of the spatial and/or energy distributions of the scattered particles may be used to determine various properties of the surface if there is a satisfactory theory of the interaction to relate the data to the properties. Therefore, the success of a surface study based on a specific particle-solid interaction depends both on the availability of the experimental techniques for generating and detecting the beam, and on the existence of a theoretical interpretation of the data.

Only within the last few years have the experimental techniques associated with LEMS studies been advanced to the degree that it is now possible to generate intense beams that are well collimated and nearly monoenergetic, and to measure the spatial and energy distributions of molecules scattered from solid surfaces. A brief survey of these techniques is given in Section 2, and the existing experimental data are summarized in Section 3. The question of interpretation of LEMS data is considered in Section 4, and evidence is given in support of a simple classical model. Section 5 contains a comparison of experimental and theoretical information on the nature of gas-solid interaction potentials.

Clearly, the selection of the proper "probe" particle depends on the surface property to be investigated. For example, photons, electrons, or neutrons are more appropriate than molecules for diffraction studies of surface structure, since the deBroglie wave lengths of molecules (aside from hydrogen and helium) are not comparable to the lattice spacing except at inconveniently low temperatures (Fig. 1). However, molecules certainly are the natural choice for studies of gas-solid interaction potentials. Also, the penetration of molecules into a solid is far less than that for photons, electrons, or neutrons of the same energy. An attractive possibility is to include the Auger (1), LEED (2), and LEMS techniques into a single apparatus so that the composition, structure, and interaction potential, respectively, may be determined under identical conditions.

2. Recent Improvements of LEMS Techniques

The first requirement of an "ideal" LEMS apparatus is that the molecular beam be well collimated, nearly monoenergetic, and of high intensity. The recently developed nozzle beam (3) satisfies this requirement, and energies above 1 eV may be obtained either by arc heating (4) or by seeding (3,5,6). High purity gases are commercially available, and a variety of techniques may be used to attain further purification (7).

A second requirement is that the target be a single crystal of known crystallographic orientation. Fortunately, it is now possible to purchase a wide variety of materials in the form of oriented single crystals. There are a number of techniques for obtaining an atomically clean surface (8), and the clean state may be maintained either by continuous deposition of the target material (9), by operating at elevated temperatures, or by establishing ultra-high vacuum conditions.

"Ideal" detectors capable of determining both the spatial and energy (velocity) distributions of scattered molecules have been developed recently in several laboratories (10-14). Since velocity distribution measurements are difficult and time consuming, some investigators have settled for approximate measurements of the mean speed of the scattered molecules either by lock-in amplifiers with phase-sensitive detection (15-17) or through the relation between the flux and density of the scattered molecules (18-20). A mass spectrometer is the appropriate detector for gas-solid interactions involving chemical reactions (20-24).

3. Survey of LEMS Experiments Pertaining to Gas-Solid Interaction Potentials

Since the subject of this paper is the use of LEMS as a tool for studying gas-solid interaction potentials, we shall consider only those experiments that are relevant to this objective. More general surveys may be found elsewhere (24,25).

First consider the case in which the beam impinging upon the target consists of atoms (or of molecules which do not dissociate in the interaction).

As illustrated in Fig. 2, the gas-solid interaction potential may be represented by a map of the equipotential contours or by potential curves corresponding to each point on the surface (e.g., curves A and B). Clearly, the nature of the gas-solid potential is an important factor in such processes as energy and momentum transfer, adsorption and desorption, surface migration, and heterogeneous chemical reactions. Since theoretical predictions of gas-solid potentials are not yet satisfactory (Section 5), experimental studies are essential.

LEMS techniques have been employed by a number of investigators to determine the atomic binding energy, X, for a variety of gas-solid systems (26-30). The majority of these studies has been concerned with alkali-metal atoms (26) because they are easily detected by surface ionization. Since the bulk of the existing data is limited to polycrystalline metals under intermediate vacuum conditions, we recently have initiated an investigation of alkali metals on single-crystal tungsten under ultrahigh vacuum conditions (27). We are aware of only two studies[*] of adsorbates other than alkali metals: Ga on W(100) by Arthur (28), and Cd on polycrystalline W by Hudson and Sandejas (29). In both cases the detector was a mass spectrometer and ultrahigh conditions were attained. A most interesting conclusion from these careful studies of Ga/W and Cd/W is that X is essentially independent of coverage until nearly a complete monolayer is formed, whereas it is known that X depends strongly on coverage for alkali metals on W (31). No doubt this difference results from the ionic character of the alkali-metal adsorption bond which produces long-range interactions between the adatoms.

If the energy of the impinging atoms is sufficiently large relative to X that the majority of the atoms is scattered without being trapped in the gas-

---

[*] A third study of a more qualitative nature has been performed by Shelton and Cho (30).

solid potential, then the nature of the interaction depends on the shape of the potential as well as on the well depth, X. This condition is most likely to occur for the rare gases since they have the lowest values of X. Therefore, we would expect that LEMS studies of rare gases on various solid surfaces may yield information on the shape of the gas-solid potential. Experimental data of this sort have been obtained for a number of systems, and the general characteristics (25) may be summarized as follows, where the nomenclature is defined in Fig. 3:

Characteristic I: $\partial\theta_r'/\partial T_g \geq 0$ if the molecular beam has a Maxwellian velocity distribution. (If the beam is monoenergetic, then the corresponding characteristic is $\partial\theta_r'/\partial u_i \geq 0$ or $\partial\theta_r'/\partial E_i \geq 0$.)

Characteristic II: $\partial\theta_r'/\partial T_s \leq 0$ if the composition and structure of the target surface do not vary with T_s.

Characteristic III: $\partial\theta_r'/\partial\theta_i \geq 0$.

Characteristic IV: $\partial\theta_r'/\partial m_g < 0$ when $\theta_r' < \theta_i$ (i.e., subspecular scattering), whereas $\partial\theta_r'/\partial m_g > 0$ when $\theta_r' > \theta_i$ (i.e., supraspecular scattering).[*]

Characteristic V: The dispersion or half-width of a scattering pattern increases with the energy of adsorption (e.g., X in Fig. 2) and/or with m_g, the molecular mass.[*]

Characteristic VI: Supraspecular patterns (i.e., $\theta_r' > \theta_i$) are most likely to occur when both of the following conditions are met: (a) $T_g > T_s$ and (b) $\theta_i \ll 90°$.

---

[*]Although it appears that the scattering pattern depends both on m_g and on the adsorption energy [e.g., see Saltsburg and Smith (32)], it is difficult to determine the effects of each individually because both change when we select a different beam gas. Hence, characteristics IV and V are not as definite or as general as I, II, and III.

Characteristic VII: Nondiffuse or lobular scattering is most likely to occur
from target surfaces that are both smooth and free of gross contamination.

Characteristic VIII: The dispersion or half-width for nondiffuse scattering
is greater for the in-plane pattern than for the out-of-plane pattern.
(That is, the probability is low that molecules will be scattered out of
the plane defined by the incident beam and the target normal.)

It is remarkable that these characteristics are observed for an extremely
wide variety of both gases (e.g., rare gases, atmospheric gases, certain metal
vapors, H_2 and D_2, etc.) and solids (e.g., ionic and metallic, monocrystalline
and polycrystalline, with different atomic masses and surface conditions) (25).
On the other hand, for gas-solid systems having large adsorption energies and
large adsorption (sticking) probabilities, the scattering patterns are diffuse
(cosine distribution), as expected. Although these characteristics are based
on data for beam energies in the range $0 < E_i \lesssim 1$ eV, there is some recent
evidence (33) that they also apply at higher energies.

Now consider the more complex case in which the beam consists of molecules
rather than atoms. The gas-solid potentials illustrated in Fig. 4 are representative of systems for which the molecules may dissociate on the surface to
form atom-solid bonds that are stronger than the molecular bond (i.e., $2X > D$,
where D = dissociation energy). As indicated in the figure, the atomic and
molecular adsorption states may be separated by an activation energy barrier
of height E_a. In this case of activated adsorption, the probability that an
incident molecule will have sufficient energy to pass over the barrier is
proportional to $\exp(-E_a/kT_g)$, where it is assumed that the gas has a Maxwellian
distribution. Krakowski and Olander (23) have recently demonstrated for the
system H_2/Ta that a LEMS apparatus with variable T_g may be used to determine
E_a. LEMS techniques also may be employed to determine the atomic adsorption

energy, X, for cases in which the incident beams consist of molecules. (For example, see Steele's study (22) of the system O/W.) Since the potential curves shown in Fig. 4 depend on the point of impact of the incident molecule (Fig. 2), experimental data on E_a and X represent average* values for the system. Also, the experimental value of E_a corresponds to an average over the possible orientations of the incident molecule.

As in the case of atomic scattering, molecules that are scattered without being trapped in the gas-solid potential provide a possible source of information on the shape of the potential curves shown in Fig. 4. Nondiffuse scattering patterns have been observed for a variety of gas molecules, including H_2, D_2, N_2, O_2, NH_3, and H_2O. (See refs. 16, 24, and 25).

4. Hard-Cube Model of Low-Energy Molecular Scattering from Solid Surfaces

As stated in the preceding section, it is expected that the nondiffuse (lobular) scattering patterns often observed in LEMS studies constitute a potential source of information of the nature of gas-solid interaction potentials. What is needed is a collision theory which may be used to compute the gas-solid potential from experimental scattering data. Although the theory of binary collisions of gas molecules has been developed to a very high level (34,35), the treatment of gas-solid collisions appears to be much more complex since it is a many-body problem.** Because of the lack of a satisfactory theory of gas-solid collisions, we shall resort to the primitive "hard-cube" model developed by Logan, Keck, and Stickney (38 40). This model assumes an oversimplified form for the gas-solid potential but the semiquantitative agreement of the hard-cube predictions with experimental data is an indication that

---

*Rather than being simple area-weighted averages, we would expect that, by the nature of the experiment, the low energy points are weighted most heavily in the case of E_a, whereas the reverse is true for X.

**The existing theoretical treatments of gas-solid collisions have been reviewed recently by Trilling (36) and by Beder (37).

the assumptions may be valid approximations. (We shall return to this point in Section 5.) The model has been refined and extended by introducing a more realistic representation of the gas-solid interaction potential in terms of its unknown properties, such as the curvature of the equipotential surfaces (39), the depth of the potential well (39,41), and the slope or range of the repulsive potential (41). Although, in principle, it should be possible to evaluate these properties from LEMS data, we shall not consider this aspect of the problem at present because of its complexity and the limited amount of suitable data. Instead, we shall concentrate on the most elementary form of the hard-cube model with the primary purpose of obtaining qualitative information on the gas-solid potential.

The hard-cube model was devised as an attempt to obtain a qualitative explanation of the general characteristics of the scattering data described in Section 3. Since closed-form expressions for the scattering characteristics were desired, it was necessary to make a number of drastic assumptions:

Assumption I: The gas-solid potential is such that the repulsive force is impulsive and the attractive force is zero. That is, both the gas molecules and solid atoms are considered to be perfectly rigid and elastic.

Assumption II: The gas-solid potential is uniform in the plane of the surface (i.e., the equipotential curves (Fig. 2) are perfectly planar). Assumptions I and II may be represented by the single assumption that the gas molecule behave as hard (i.e, perfectly rigid and elastic) spheres whereas the solid atoms behave as hard cubes oriented with one face parallel to the surface. As indicated in Fig. 5, a particular characteristic of this model is that collisions do not affect the tangential component of velocity.

Assumption III: Each gas molecule experiences a single collision with one solid atom which behaves as a free particle (i.e. uncoupled from the other solid atoms) during the interaction.

<u>Assumption IV:</u>   The solid atoms have a Maxwellian velocity distribution.

With these assumptions it is possible to derive the following closed-form expression for the scattering pattern (flux distribution) for an incident molecular beam that is monoenergetic (40):

$$\frac{1}{u_{ni}} \frac{dR}{d\theta_r} = \pi^{-1/2} B_2 U (1 + B_1 \sec \theta_i) \exp(- B_1^2 U^2). \qquad (1)$$

(A list of nomenclature is given at the rear of this paper.) The corresponding expression for a Maxwellian beam is (39):

$$\frac{1}{u_{ni}} \frac{dR}{d\theta_r} = \frac{3}{4} B_2 (1 + B_1 \sec \theta_i) \left(\frac{m_s T_g}{m_g T_s} \right)^{1/2} \left(1 + \frac{m_s T_g}{m_g T_s} B_1^2 \right)^{-5/2}$$

$$(2)$$

By setting the derivative of Eq. (1) equal to zero, we obtain an expression involving θ_r', the angular position of the maximum of the scattering pattern (Fig. 3), which may be rearranged to give

$$U^2 = \frac{\cot \theta_r'}{B_1 B_2} + \frac{1}{2B_1(B_1 + \cos \theta_i)}. \qquad (3)$$

The corresponding expression for a Maxwellian beam (Eq. (2)) is

$$\frac{T_g}{T_s} = \mu \left(\frac{5}{2} U^{-2} - B_1^2 \right)^{-1} \qquad (4)$$

where U is defined by Eq. (3).

In previous papers (38 40) we have shown that the predictions of the hard-cube model agree qualitatively with the general characteristics of the experimental data listed in Section 3. In fact, it is quite remarkable that the agreement is semiquantitative in a number of cases (25, 38-40). Additional comparisons are presented below.

According to Eq. (4), θ_r' depends only on the <u>ratio</u> of the gas and solid temperatures, and not on either temperature by itself. In Fig. 6 we compare this prediction with the experimental data of Hinchen et al. (16) for a Maxwellian beam of Ar scattered from polycrystalline and single-crystal Pt targets. (The ordinate, $\Delta\theta$, is a measure of the degree to which the angular position of the maximum, θ_r', deviates from the specular angle, θ_i). The data represent a variety of values of T_g and T_s, and Hinchen et al. (16) report that a systematic dependence of $\Delta\theta$ on either T_s or T_g alone was not observed. In addition to predicting the correct temperature dependence, the hard-cube model also appears to be quite successful in predicting the magnitude of $\Delta\theta$ on an absolute basis for three values of θ_i (Fig. 6). According to Hinchen et al. (16), qualitative agreement also is observed for Ne, Kr, O_2, and N_2 scattered from Pt. Another conclusion that may be drawn from Fig. 6 is that the crystallographic orientation of the Pt surface has no noticeable effect on the scattering process.

As shown elsewhere (40), the hard-cube model predicts that θ_r' will asymptotically approach a high speed limit as $U \to 0$. This is illustrated in Fig. 7 for both monoenergetic and Maxwellian beams. (Notice that the parameter on the abscissa is equal to twice the reciprocal of that used in Fig. 6; the factor of two arises from equating E_i to $2kT_g$, the average impact energy of the molecules of a Maxwellian beam.) The data reported very recently by O'Keefe (33) for Ar scattered from a (100)W crystal are included in the figure, and the agreement with the hard-cube model is remarkably close. (The agreement would be even closer if the hard-cube curves in Fig. 7 had been computed for the μ of Ar/W rather than for Ar/Pt.) We have suggested previously (40) that the hard-cube model may not be valid at these higher energies because the assumption that the gas-solid potential is planar becomes increasingly weak as E_i increases.

Also included in Fig. 7 are data for other systems having approximately the same values for μ and θ_i. Although the agreement is quite good for $E_i/kT_s < 2$, at higher energies the data for Ar/Pt and Ar/Au deviate considerably from the hard-cube prediction. We should mention, however, that the scattering data for Au are somewhat anomalous because supraspecular patterns $(\theta_r' > \theta_i)$ have not been observed for any of the rare gases under a variety of test conditions (9). Using a nearly monoenergetic beam, Moran (13) recently has verified the Ar/Pt data in the range $E_i/kT_s < 2$ obtained by Hinchen et al. (16) with a Maxwellian beam (Fig. 7). Unfortunately, Moran did not obtain data in the range $E_i/kT_s > 2$ where the discrepancy between the hard-cube model and the data of Hinchen et al. is significant.

A more stringent test of the hard-cube model is presented in **Fig. 8.** The flux distribution was computed from Eq. (1), whereas the velocity ratio is given by (40):

$$u_r/r_i = \left(1 - \frac{4\mu}{(1+\mu)^2} \cos^2\theta_i\right)^{1/2} \tag{5}$$

where it is assumed that the incident beam is monoenergetic with speed u_i. The agreement with the experimental data of Moran et al. (12) for Ar/Pt is good except in the range $\theta_r < 20°$. If we were to assume that the incident beam were Maxwellian rather than monoenergetic, then the agreement for $\theta_r < 20°$ would be slightly better both for the flux distribution (see Fig. 2 in ref. 40) and velocity ratio (see **Fig.** 4 in ref. 39). Since Moran's beam is not completely monoenergetic, the actual beam is somewhere between the Maxwellian and mono-energetic cases.

More recently, Moran (13) has conducted experiments to test the assumption of the hard-cube model that the tangential velocity component of the incident molecules is unchanged in gas-solid collisions. On the basis of careful mea-surements of both the flux and mean speed distributions for Ar/Pt, he concludes that the mean tangential velocity of the scattered molecules is close to the incident beam value throughout the range of scattering angles that contains most of the molecules.

5. Discussion of Gas-Solid Interaction Potentials

The fact that the predictions of the hard-cube model agree, semiqualitatively, with existing LEMS data suggests that the assumptions underlying the model are valid as zeroth-order approximations of the gas-solid interaction potential. Since it is difficult to reconcile these assumptions with existing theoretical treatments and with other sources of experimental data on the gas-solid potential, we shall discuss each assumption in detail.

Assumption I: Impulsive repulsion and zero attraction.

The assumed shape (e.g., slope) of the repulsive potential will not influence the scattering process in the case of planar equipotentials (Assumption II) until it becomes sufficiently long-ranged to cause τ_c, the collision time, to be comparable with τ_v, the characteristic time of the lattice vibration. Notice that since the assumption that the solid atom is "free" (Assumption III) is equivalent to taking τ_v to be infinite, it is useless to introduce a more realistic repulsive potential unless accompanied by a modification of assumption III (e.g., by accounting for the coupling between lattice atoms). Conversely, since an impulsive repulsive potential corresponds to $\tau_c = 0$, it is useless to improve upon III without a simultaneous improvement of I.

It is a rather simple matter (39) to add a square-well potential to the hard-cube model as an approximation of the attractive portion of the gas-solid potential. Although this non-zero potential causes the scattering patterns to differ somewhat from those based on the present assumption of zero attraction, the general scattering characteristics are the same (39). Clearly, the attractive potential should be accounted for in detailed analyses, especially for gas-solid systems having large adsorption energies. A principal effect of attraction is to accelerate the incident molecules, thereby reducing the collision time and resulting in a more impulsive collision. However, this effect does not occur with the hard-cube model because the repulsive interaction is assumed to be impulsive regardless of attraction. Also, the smoothness of the equipotential surface (Assumption II) precludes the possibility of attractive forces focussing the incident molecules on to particular impact points.

Assumption II: Planar equipotential surfaces

The existing LEMS data strongly support the view that the gas-solid equipotential surfaces appear to be extremely planar (smooth) to a significant fraction of the incident molecules. For example, Moran's observation (13) that the mean tangential velocity of the molecules is essentially unchanged is rather convincing evidence that the mean tangential component of the gas-solid interaction forces is small. Also, the fact that the half-width of the out-of-plane scattering pattern is less than that of the in-plane pattern (42,43) is a direct indication of the planarity of the surface (39).

At present we are unable to find a satisfactory theoretical justification for assuming that the planar potential is the correct zeroth-order approximation. In fact, most current theoretical treatments of gas-solid collisions (36) are based on the pairwise model in which the interaction potential is represented by the summation of the individual interactions of the gas molecule with each solid atom. For example, Goodman (44) has studied the scattering of hard-sphere gas atoms from hard-sphere solid atoms having collision diameters that are consistent with values predicted by the pairwise model (45). The equipotential surfaces associated with this model are far from being planar, and the result is that the agreement of the hard-sphere scattering patterns with experimental data is poor in comparison with the agreement obtained with the hard-cube model. This apparent inconsistency between the pairwise model and the planar assumption of the hard-cube model has been examined by Goodman (42) in detail, but it has not yet been satisfactorily resolved.

There are several other indications that the pairwise model does not provide an accurate description of the interaction potential (45-48). Ehrlich and his co-workers have measured the diffusion (46) and binding energies (47) of tungsten adatoms on several crystal planes of tungsten, and they conclude that the dependence of these energies on the crystal plane of the substrate is not consistent with the pairwise model. A similar conclusion has been reported by Plummer et al.(48) for the binding energies of the 5d transition elements on tungsten. Both groups suggest that the failure of the pairwise model results from the redistribution or "smoothing" of the metal electrons at a free surface. (It is tempting to propose that electron smoothing causes the gas-solid equipotential surfaces to be smoother than predicted by the pairwise model, thereby providing some justification for the planar-potential assumption of the hard-cube model; however, this proposal would not be consistent with the observations that (a) the scattering patterns for nonmetallic surfaces (25) are quite similar to those for metallic surface, (b) the binding energy of an adatom on a solid is not constant over the surface, as indicated by the fact that the diffusion energy is a significant fraction [~0.3 for Ar on W (49)] of the binding energy, and (c) the results of Ehrlich et al. (50) indicate that the pairwise model is satisfactory for the rare gases on tungsten.)

Assumption III: Single* collsion with "free" solid atoms.

Although it is known that for high-energy collsions the solid atoms may be treated as "free" particles (see discussion in ref. 40), it seems reasonable to suspect that the coupling between lattice atoms may have a significant effect on low-energy $(E_i < 20$ eV$)$ scattering. As mentioned above in the discussion of Assumption I, Assumptions I and III are interrelated. Logan and Keck (41) have refined the hard-cube model to include both lattice coupling and

---

*As the mass ratio $(\mu = m_g/m_S)$ increases, there is a decrease in the probability that a single collision will be sufficient to reverse the direction of gas molecule. Although the original analysis (38) of the hard-cube model accounted for multiple collisions, the treatment was rather unrealistic and a closed-form expression for the scattering pattern could not be obtained. Hence, only single collisions were considered in the second analysis (39), thereby simplifying the problem to the degree that closed form solutions were obtained. Since the original analysis indicates that single collisions predominate when $\mu \overset{>}{\sim} 1/3$ the application of the second analysis should be restricted to this range.

a more realistic interaction potential. These refinements are important because the hard-cube model is not applicable to cases of high mass ratio, and it
tends to overestimate the influence of m_s on the scattering process.

Assumption IV: Maxwellian distribution of solid atoms.

Since the predictions of the hard-cube model are rather insensitive to the
assumed form of the distribution function of the solid atoms (51), this assumption is not a critical one.

6. Concluding Remarks

The techniques associated with molecular scattering experiments have been
developed to the degree that an impressive quantity of new data has been obtained
on the interaction potentials of a variety of gaseous species (34,35). Although
similar techniques are available for studying gas-solid interaction potentials,
they have been utilized in only a few investigations, primarily because of the
problems of obtaining both an atomically clean surface and an adequate signal-
to-noise ratio. Solutions to these problems have been found recently (Section
2), and it is clear that an increasing number of laboratories now are turning
to LEMS techniques as a tool for studying various gas-solid interaction processes,
such as nondiffuse scattering, adsorption and desorption, oxidation, and
catalysis. We may expect considerable advances to be made in these areas in
the near future, especially when the LEMS techniques are combined with existing techniques for determining the composition (1) and structure (2) of the
target surface.

As a single illustration of the possibilities of using LEMS data to
determine the nature of gas-solid interaction potentials we have adopted
a most elementary model and tested it against existing scattering data. The
agreement was sufficiently close to suggest that the underlying assumptions
of this hard-cube model may be valid approximations. In particular, the

agreement leads to the suggestion that the effective or average gas-solid
equipotential surfaces may be much smoother than predicted by the pairwise-
potential model which is the basis for many of the current theoretical treatments of gas-solid interactions. Since existing data on the tangential velocity component (13) and the out-of-plane scattering pattern (42,43) are nearly
direct proofs of the smoothness of the equipotential surface, the above-mentioned
suggestion would stand even if the remaining assumptions of the hard-cube model
were erroneous. An unresolved question is how can we reconcile the present
suggestion with the results of surface migration experiments which clearly
indicate that the gas-solid potential is not smooth (49), and binding energy
measurements (50) which show that the pairwise model is consistent with data
for rare gases on tungsten? The most probable answer is that the potential
only "appears" to be smooth to a significant fraction of the incident molecules.

There is an increasing amount of evidence that the hard-cube model is an
appropriate basis for developing a more exact model-such as that proposed
recently by Logan and Keck (41) - which will make it possible to determine
gas-solid interaction potentials from LEMS data. Clearly, this knowledge
will have a great impact on the field of heterogeneous catalysis.

Acknowledgments

This work was supported by the Godfrey L. Cabot fund and by the Joint
Services Electronics program (Contract DA28-043-AMC-02536 (E)). The author
gratefully acknowledges the many contributions to this work by J. C. Keck,
R. M. Logan, S. Yamamoto, and T. J. Lee.

Nomenclature

$$B_1 = \frac{1 + \mu}{2} \sin\theta_i \cot\theta_r - \frac{1 - \mu}{2} \cos\theta_i$$

$$B_2 = \frac{dB_1}{d\theta_r} = \frac{1 + \mu}{2} \sin\theta_i \csc^2\theta_r$$

E_i = Energy of incident gas molecule

E_a = Activation energy for dissociative adsorption

m_g = Mass of gas molecule

m_s = Mass of surface atom

T_g = Temperature of gas

T_s = Temperature of solid

u_i = Incident gas velocity (normal component = u_{ni})

u_r = Scattered gas velocity

U = $u_i/(2kT_s/m_s)^{\frac{1}{2}}$

θ_i = Angle of incident molecules

θ_r = Angle of scattered molecules

θ_r' = Angular position of the maximum of the scattering pattern

λ = deBroglie wave length

μ = Mass ratio, m_g/m_s

χ = Atomic adsorption (binding) energy

References

1. J. J. Lander, Phys. Rev. 91, 1382 (1953);   R. E. Weber and W. T. Peria, J. Appl. Phys. 38, 4355 (1967);   L. A. Harris, J. Appl. Phys. 39, 1419 and 1428 (1968);   P. W. Palmberg and T. N. Rhodin, J. Appl. Phys. 39, 2425 (1968);   P. W. Palmberg, in this volume.

2. J. J. Lander, in Progress in Solid State Chemistry, Vol. 2 (H. Reiss, ed.; Pergamon Press, 1965), pp. 26–116.

3. J. B. Anderson, R. P. Andres, and J. B. Fenn, in Molecular Beams (Advances in Chemical Physics, Vol. 10, J. Ross, ed.; Interscience Publ., N. Y., 1966), pp. 275–317.

4. E. L. Knuth, N. M. Kuluva, and J. P. Callinan, Entropie 18, 38 (1967).

5. J. B. French and D. R. O'Keefe, in Rarefied Gas Dynamics, Vol. 2 (J. H. deLeeuw, ed.; Academic Press, N.Y., 1966), pp. 299–310.

6. N. Abuaf, J. B. Anderson, R. P. Andres, J. B. Fenn, and D. G. H. Marsden, Science 155, 997 (1967);   J. B. Anderson, Entropie 18, 33 (1967).

7. G. Ehrlich, in Advances in Catalysis, Vol. 14 (Academic Press, N.Y., 1963), pp. 255–427;   R. M. Mobley, in Methods of Experimental Physics, Vol. 4, Part B (V. W. Hughes and H. L. Schultz, eds.; Academic Press, N.Y., 1967), pp. 318–328.

8. R. W. Roberts, Brit. J. Appl. Phys. 14, 537 (1963).

9. J. N. Smith, Jr. and H. Saltsburg, in Rarefied Gas Dynamics, Vol. 2 (J. H. deLeeuw, ed.; Academic Press, N.Y., 1966), pp. 491–505.

10. O. F. Hagena and A. K. Varma, Rev. Sci. Instr. 39, 47 (1968);   also see the paper by M. N. Bishara et al. in this volume.

11. P. B. Scott, M.I.T. Fluid Dynamics Research Laboratory Report No. 65–1, Feb. 1965.

12. J. P. Moran, H. Y. Wachman, and L. Trilling, in Fundamentals of Gas-Surface Interactions (H. Saltsburg et al., eds.; Academic Press, N.Y., 1967), pp. 461–479.

13. J. P. Moran, M.I.T. Fluid Dynamics Research Laboratory Report No. 68-1, Feb. 1968.

14. F. C. Hurlbut and K. Jakus, in this volume.

15. G. E. Moore, S. Datz, and E. H. Taylor, J. Catalysis $\underline{5}$, 218 (1966).

16. J. J. Hinchen, E. S. Malloy, and J. B. Carroll, United Aircraft Research Laboratory Report F910439-7, June 1967. (A portion of this work appears in Fundamentals of Gas-Surface Interactions (H. Saltsburg et al., ed.; Academic Press, N.Y., 1967), pp. 448–460.

17. S. Yamamoto and R. E. Stickney, J. Chem. Phys. $\underline{47}$, 1091 (1967).

18. R. T. Brackmann and W. L. Fite, J. Chem. Phys. $\underline{34}$, 1572 (1961).

19. J. D. McKinley, J. Phys. Chem. $\underline{66}$, 554 (1962).

20. J. N. Smith, Jr. and W. L. Fite, J. Chem. Phys. $\underline{37}$, 898 (1962); and in Rarefied Gas Dynamics, Vol. 1 (J. A. Laurmann, ed.; Academic Press, N.Y., 1963), pp. 430–453.

21. J. D. McKinley, J. Chem. Phys. $\underline{40}$, 120 (1964).

22. W. C. Steele, Tech. Rept. AFML-TR-65-343, Part II, Avco Missiles, Space and Electronics Group, Jan. 1967.

23. R. A. Krakowski and D. R. Olander, J. Chem. Phys. (in press).

24. G. Ehrlich, in Annual Reviews of Physical Chemistry, Vol. 17 (Annual Reviews, Inc., Palo Alto, Cal., 1966), pp. 295–322; J. N. Smith, Jr. and H. Saltsburg, in Fundamentals of Gas-Surface Interactions (H. Saltsburg et al., eds.; Academic Press, N.Y., 1967), pp. 370–391; F. C. Hurlbut, in Rarefied Gas-Dynamics, Vol. 1 (C. L. Brundin, ed.; Academic Press, N.Y., 1967), pp. 1–34.

25. R. E. Stickney, in <u>Advances in Atomic and Molecular Physics</u>, Vol. III

(D. R. Bates and I. Estermann, eds.; Academic Press, N.Y., 1967),

pp. 143–204.

26. F. L. Hughes and H. Levinstein, Phys. Rev. <u>113</u>, 1029, 1036 (1959);

M. D. Scheer and J. Fine, J. Chem. Phys. <u>39</u>, 1752 (1963), and <u>37</u>, 107

(1962); M. Kaminsky, Ann. Physik <u>18</u>, 53 (1966). (Additional references

are given in Kaminsky's paper.)

27. T. J. Lee and R. E. Stickney (work in progress).

28. J. R. Arthur, in <u>Report on 27th Annual Conference on Physical Electronics</u>

(M.I.T., 1967), pp. 188–192.

29. J. B. Hudson and J. S. Sandejas, J. Vac. Sci. Technol. <u>4</u>, 230 (1967).

30. H. Shelton and A. Y. H. Cho, J. Appl. Phys. <u>37</u>, 3544 (1966).

31. J. B. Taylor and I. Langmuir, Phys. Rev. <u>44</u>, 423 (1933).

32. H. Saltsburg and J. N. Smith, Jr., J. Chem. Phys. <u>45</u>, 2175 (1966).

33. D. R. O'Keefe, University of Toronto Institute for Aerospace Studies, Report

No. 132, July 1968.

34. J. O. Hirschfelder, <u>Intermolecular Forces</u> (Advances in Chemical Physics,

Vol. 12, Interscience Publ., N.Y., 1967).

35. J. Ross, <u>Molecular Beams</u> (Advances in Chemical Physics, Vol. 10, Inter-

science Publ., N.Y., 1966).

36. L. Trilling, in <u>Fundamentals of Gas–Surface Interactions</u> (H. Saltsburg et

al., eds.; Academic Press, N.Y., 1967), pp. 392–421.

37. E. C. Beder, in <u>Advances in Atomic and Molecular Physics</u>, Vol. III

(D. R. Bates and I. Estermann, eds.; Academic Press, N.Y., 1967),

pp. 206–290.

38. R. M. Logan and R. E. Stickney, J. Chem. Phys. <u>44</u>, 195 (1966).

39. R. M. Logan, J. C. Keck, and R. E. Stickney, in <u>Rarefied Gas Dynamics</u>,

Vol. 1 (C. L. Brundin, ed.; Academic Press, N. Y., 1967), pp. 49-66.

40. R. E. Stickney, R. M. Logan, S. Yamamoto, and J. C. Keck, in Fundamentals of Gas-Surface Interactions (H. Saltsburg et al., eds.; Academic Press, N. Y. 1967), pp. 422-434.

41. R. M. Logan and J. C. Keck, M.I.T. Fluid Mechanics Laboratory Publication No. 67-8, Oct. 1967.

42. H. Saltsburg, J. N. Smith, Jr., and R. L. Palmer, in Rarefied Gas Dynamics, Vol. 1 (C. L. Brundin, ed.; Academic Press, N. Y., 1967), pp. 223-238.

43. J. J. Hinchen and E. F. Shepard, in Rarefied Gas Dynamics, Vol. 1 (C. L. Brundin, ed.; Academic Press, N. Y., 1967), pp. 239-252.

44. F. O. Goodman, Surface Sci. 7, 391 (1967).

45. F. O. Goodman, Phys. Rev. 164, 1113 (1967).

46. G. Ehrlich and F. G. Hudda, J. Chem. Phys. 44, 1039 (1966).

47. G. Ehrlich and C. F. Kirk, J. Chem. Phys. 48, 1465 (1968).

48. E. W. Plummer, Ph.D. Thesis, Cornell University, Jan. 1968. Also see the paper by R. N. Rhodin, P. W. Palmberg, and E. W. Plummer in this volume.

49. R. Gomer, J. Phys. Chem. 63, 468 (1959).

50. G. Ehrlich, H. Heyne, and C. F. Kirk, in this volume.

51. R. M. Logan and R. W. Stickney (unpublished work).

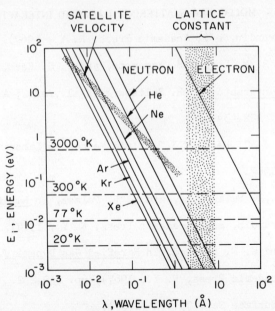

Fig. 1. Dependence of the deBroglie wave length, λ, on the energy, E_i, of various particles impinging upon a solid surface. The dashed lines of constant temperature are based on the relation $E_i = 2kT_g$.

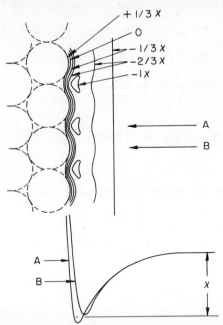

Fig. 2. Illustration of the gas-solid interaction potential for the case of a gas consisting either of atoms or of molecules that do not dissociate in the interaction. (In the latter case, the potential will be a function of the orientation of the incident molecules.)

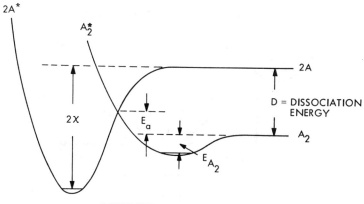

ACTIVATED ADSORPTION

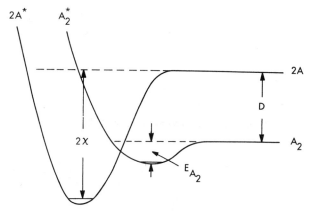

NONACTIVATED ADSORPTION

Fig. 3. Illustration of the gas-solid interaction potential for gas molecules that may dissociate in the interaction.

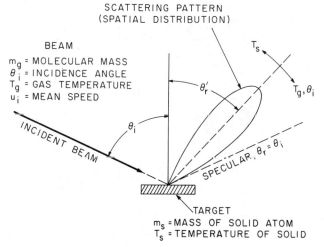

Fig. 4. Schematic of the scattering of a molecular beam from a solid surface. (The arrows in the upper right-hand corner indicate the
direction of change of θ'_r with increasing T_s, T_g, and θ_i.

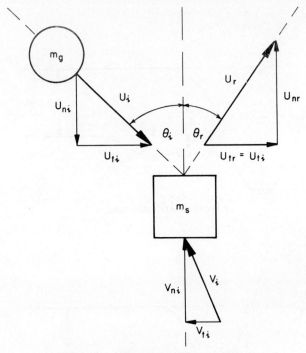

Fig. 5. The hard-cube model.

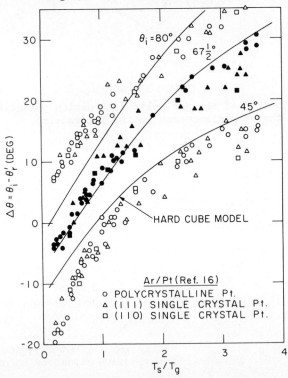

Fig. 6. Comparison of the hard-cube model with experimental data: Dependence
of $\Delta\theta$, the deviation of the angular position of the maximum of the
scattering pattern from the specular angle, on T_s/T_g and θ_i for Ar
scattered from polycrystalline and single-crystal Pt.

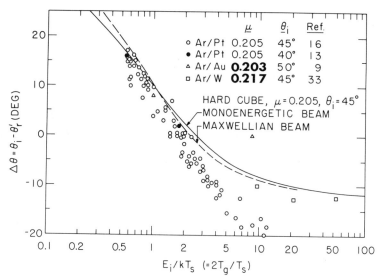

Fig. 7. Comparison of the hard-cube model with experimental data: Dependence of $\Delta\theta$ on E_i/kT_s for Ar scattered from several solids for conditions of similar values of μ and θ_i. (Note: $E_i/kT_s = 2T_g/T_s$ if we assume that E_i is equivalent to $2kT_g$, the mean energy of a Maxwellian molecular beam.)

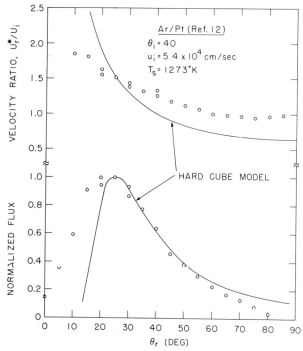

Fig. 8. Comparison of the hard-cube model with experimental data: Dependence of the velocity and flux on θ_r for Ar scattered from polycrystalline Pt. (Note: u_r^* is the mean speed of the scattered molecules.)

THE INTERACTION OF MOLECULAR BEAMS WITH SOLID SURFACES: DIATOMIC AND POLYATOMIC MOLECULES

Howard Saltsburg, Joe N. Smith, Jr., and Robert L. Palmer

Gulf General Atomics, Inc.
San Diego, California

I.   INTRODUCTION

The study of the scattering of atomic and molecular beams from solid surfaces has been recognized for some time as a powerful technique for the study of the primary process in the collision of a gas atom or molecule with a solid surface. In recent years, improvements in vacuum techniques, surface preparation techniques, and the development of the modulated molecular beam have combined to make possible a more systematic study of these collision processes than had been possible heretofore.

The principal experimental measurement has been that of the spatial distribution either of the scattered beam flux or number density and, in general, has been confined to the distribution in the plane of incidence, defined by the incident beam and the surface normal. A broad classification of experimentally observed scattering distributions into diffuse and directed scattering is possible. For diffuse scattering, the scattered flux is proportional to $\cos \theta_r$ where θ_r is the scattered angle. This angular distribution is independent of all other variables which characterize the system. The other broad class is directed scattering which includes specular scattering as a special case. The specular scattering distribution is derivable from the incident beam distribution by simply reversing the normal component of the incident particle momentum after the collision, leaving the tangential component unchanged. Thus, the scattered beam maximum occurs at the specular angle and the divergence of the scattered beam is that of the incident beam. Under appropriate conditions, diffracted beams also may be observed. All other directed scattering in which there is a well-defined maximum intensity at some point in space has been found to depend upon many specific parameters of the gas-solid system.[1]

One can characterize all these scattering distributions in terms of the degree of energy transfer between the gas and the solid (taking due account of residence time), and the net result is that any deviation from specular scattering which can be shown not to result from some artifact of surface structure implies that energy transfer has occurred. Thus, even if the beam is scattered at the specular angle but has a half width much greater than that of the incident beam, energy transfer is presumed to have occurred. Similar considerations apply to displacements of the scattered beam from the specular position.

Work Supported by AFOSR under Contracts
AF49(638)-1435 and AF49(638)-1599

In the past, in order to minimize the complexities of the system, primary emphasis was placed on the study of the scattering of the rare gases. The results of studies of the angular distributions of the scattered beam indicated that a limited number of parameters was sufficient to characterize the trends of the scattering distributions in spite of the complexity of the overall gas-solid system. In particular, knowledge of the relative velocities of the individual solid and gas particles (specified in terms of the solid and gas beam temperatures, T_S and T_B), the masses of the solid and gas particles M_s and M_g (usually as the ratio $\mu = M_g/M_s$), and the incident angle, θ_i, permits one to formulate simple models of the scattering process involving only the normal components of the relative motion.[2] Experimentally, it was found that as one progresses through the rare gas series from He to Xe the scattered beam is broadened and has a maximum intensity farther away from the specular angle. When $T_B < T_S$ the maximum lies closer to the normal (subspecular scattering), but when $T_B > T_S$ the maximum may lie at the specular angle (quasispecular) or closer to the tangent (supraspecular scattering). Further, as the incident angle approaches glancing incidence, the maximum intensity tends to deviate from the specular angle as a monotonic function of θ_i when $T_S > T_B$, but when $T_B > T_S$ the supraspecular maximum is a non-monotonic function of θ_i becoming subspecular at glancing incidence.[1,3-5] The simple models are surprisingly successful in describing the trends of these data.[2]

Although the initial studies of rare gas scattering did show that the gas masses were apparently relevant variables, more recent experimental studies have shown that the mass ratio predictions of the simple models are partially incorrect. A comparison of rare gas scattering from Ag, Au, and Ni (111) single crystal planes shows that although the trend of the variation of the deviation of the scattered beam maximum from the specular angle with increasing M_g (at constant M_s) is correctly predicted, the deviation with increasing M_s at fixed M_g is not correctly described.[6] Further, it is not possible to distinguish between effects of increasing M_g and the simultaneous increase of the van der Waals type of interaction between the gas and the solid surface. In an attempt to assess the role of a variation in this interaction energy, two scattering experiments have been reported in which the energy was varied while keeping μ constant or nearly so (with M_s constant). In both cases, for comparative purposes, it is assumed that the magnitude of the gas-solid interaction energy is related directly to the heat of physical adsorption, ΔH_p. Using this relative scale, a comparison of the scattering of a rare gas (Ne) with polyatomic gases (CH_4 and NH_3) keeping all other parameters fixed indicated that variations in ΔH_p had a significant effect on the scattering. The larger values of ΔH_p were associated with broader distributions which were also further displaced from the specular angle.[3] In a second study, an investigation of He and D_2 scattering from Pt was carried out. The differences which were observed were consistent with a stronger interaction between D_2 and a Pt surface, as compared to He and a Pt surface.[7]

In both these cases ΔH_p seemed to be a relevant parameter. However, it was also found that the magnitude of the incident kinetic energy relative to ΔH_p was important since increases in T_B were observed to mitigate the effects of ΔH_p in that more nearly specular and narrower distributions were observed at the higher beam temperatures.[3,5]

When epitaxially grown clean single crystals began to be utilized as scattering surfaces, significant effects of crystal topography and surface contamination were found, particularly for He and H_2. For He, a clean polycrystalline Au film was shown to scatter the gas in a diffuse fashion while contamination of a single crystal (which had displayed specularly directed scattering) broadened the distribution significantly.[4] For H_2, the contamination of the single crystal Au film not only broadened the distribution but also caused the peak to shift from the specular angle toward the normal.[5] It seemed appropriate therefore to re-examine the scattering of He and D_2 without the possible complication of polycrystallinity and contamination.

This report discusses the results of these studies using Ag (111) surfaces and the results of more recent studies using Ni (111) surfaces. As will be shown, significant complications are introduced by the internal energy states of the diatomic molecules when their moments of inertia are sufficiently small. Studies of the scattering of heavier diatomic and polyatomic species are also described and discussed in relation to the phenomenology of rare gas scattering.

II.   EXPERIMENTAL

The modulated molecular beam apparatus and the associated detectors and target surface preparation technique have been described in detail elsewhere.[1,3,4,6]

The measurements which are reported here are of the spatial distribution of the molecular beam (flux or number density) when scattered from an epitaxially grown single crystal film, either formed continuously during the scattering study to keep the surface clean as required for Ni (111) surfaces, or after growth of Ag (111) surfaces from which it has been found that the angular distributions of the rare gases are the same after growth as during growth.[3]

All the data refer to measurements in the plane of incidence and both scattered and incident beam angles are measured with respect to the target normal. The silver surfaces are grown on mica substrates held at 560°K and the nickel surfaces on mica substrates held at 700°K, and these temperatures are maintained throughout the scattering study.

III. SCATTERING OF LIGHT GASES ($\mu \ll 1$)

A. Mass Effects: The Scattering of He^3 and He^4 from (111) Ag [8]

In the limit of $\mu \ll 1$, the simple collision theories predict a minimal amount of energy transfer.[2,9,10] Therefore, one would expect that the He isotopes would exhibit prominent specular or quasispecular scattering. The angular distributions are shown in Fig. 1. Although the fractional change in μ is large, the absolute magnitude of the energy transfer is sufficiently small so that even large fractional changes in energy transfer do not appear to affect scattering behavior. The He^3 distribution exhibits a slightly more intense specular peak and a slightly narrower quasispecular peak, but it can be seen that the mass change of 33% has little effect on the angular distribution.

B. Rotational State Effects: H_2, HD, and D_2 Scattering from (111) Ag [8,11]

In contrast to the similarity of the scattering of the He isotopes, scattering of the hydrogenic molecules shows a significant divergence in behavior. While H_2 exhibits a sharply specular peak on top of a broad quasispecular peak as well as a third subspecular component, both HD and D_2 show a dominant sub-specular peak which is much broader than the incident beam divergence, and only under some conditions can a small contribution at the specular angle be resolved. Typical distributions are shown in Figs. 2, 3, and 4.

A previous comparison of He^4 and D_2 scattering from Pt^7 also showed differences in angular distributions which were interpreted in terms of an inelastic collision strongly affected by the interaction energy. Inclusion of the H_2 scattering distributions in the comparison shows that other factors are relevant.

Since the mass effects for $\mu \ll 1$ seem to be small, the similarity of HD and D_2 scattering (particularly the absence of appreciable specular scattering) together with the presence of the very prominent specular component in H_2 scattering led to the conclusion that these differences resulted from differences in the internal energy state structure of these molecules. Although the transitions between vibrational states of the hydrogenic molecules involve energies much greater than the relative collision energy of the incident beam and solid, transitions between some of the rotational states of these molecules involve energies comparable to the collision energy. Moreover, these rotational state transition energies are also comparable with the bulk Debye phonon energy of the Ag lattice. It was concluded that the most likely cause of the observed differences was the relative ease with which interaction between these rotational levels and the phonons of the Ag lattice could occur. This inelastic interaction is possible since the lowest allowed rotational transition in D_2 (the $J = 0$ to $J = 2$) requires an energy of 512 cals which is quite comparable to the energy of the bulk Debye phonon of Ag (~ 450 cals). The corresponding transition in H_2 requires twice the energy of the D_2 transition. Consideration of these magnitudes together with the populations of the appropriate rotational energy levels makes such an interaction quite probable. HD with the $J = 1$ to $J = 2$ transition requiring the same energy as the $J = 0$ to $J = 2$ transition in D_2 would therefore be expected to show similar inelastic interactions which would affect the scattering distribution in the same way.

It is unlikely that the incident kinetic energy of the beam is responsible for the transition since the scattering of $80^\circ K$ beams of H_2 and D_2 still show the same differences as do $300^\circ K$ beams and yet the $80^\circ K$ beams have insufficient kinetic energy to induce the transition in D_2 (Fig. 3).

Thus, for the hydrogenic molecules, it is not the small differences in interaction energy, ΔH_p, which dominate the _apparent_ differences in scattering, but rather the quantum effects. Two such effects seem to be operative: the de Broglie wave length of the particles, λ, is such as to enhance the specular scattering of the lighter gas since $\lambda > h \cos \theta_1$ (where h is the height of the surface roughness) is more easily satisfied and the rotational states of the heavier molecules (D_2 and HD) are more closely spaced, making inelastic interactions with the lattice phonons more likely. Both effects tend to widen the gap in the observed scattering behavior but the rotational transitions are the heavily predominant factor as evidenced by the small differences between He^3 and He^4 scattering.

It should be noted, however, that the specular peak in the H_2 distribution represents only $\sim 5\%$ of the <u>total</u> scattered beam flux and that in D_2 it is only $\sim 0.5\%$.[11] The major portion of the interaction of all the hydrogenic molecules with the surface is quite similar: the rotational state interactions appear to affect the existence of the small specular component in a dramatic and obvious manner, but their effect on the rest of the distribution is not resolved. The similarity is readily seen by exhibiting the data with comparable quasispecular or subspecular peaks rather than normalized to the specular peaks, as shown in Fig. 3. The similarity of H_2 and D_2 is apparent and, further, the distinction between the He isotopes and the hydrogenic molecules is also apparent in the region of the specular peak and is particularly obvious in the plane transverse to the plane of incidence (Fig. 5). The broader quasispecular peak of the hydrogenic species indicates that the role of the interaction energy ΔH_p may not be negligible and, in addition, that the scattering of the diatomic molecules may differ from that of the rare gases due to involvement of internal energy states even though this involvement is only easily resolved in the specular component.

C.   Rotational State Effects: H_2, D_2, and HD Scattering from (111) Ni

The differences between H_2 and HD or D_2 scattering which are observed using (111) Ag surfaces are attributed to phonon interaction with the solid since a single bulk Debye phonon of Ag can induce one of the rotational transitions of D_2 or of HD but cannot effect such a transition in H_2. Ni with a bulk Debye temperature, θ_D, of $456°K$,[12] which is approximately twice that of Ag ($\theta_D = 225°K$), should therefore show different scattering behavior since the Debye phonon energy is ~ 912 cal, much above the lowest HD and D_2 rotational transitions and still below the lowest allowed H_2 transition. The scattering distributions are shown in Figs. 6, 7, and 8. Comparison with He[4] scattering again shows that HD and D_2 are similar and He[4] and H_2 are similar. At elevated temperatures the D_2 distribution is quite similar to that of H_2, as previously observed on Ag. Compared with Ag (111), the specular component from Ni (111) is much more intense in D_2 and HD, implying a lesser inelastic collision probability based upon the general behavior outlined above.

The reason for the more intense specular component from a Ni (111) surface relative to a Ag (111) surface is not understood. It has been found that the scattering of all the rare gases from (111) Ni shows narrower and more nearly specular scattering than from either a Ag (111) or a Au (111) surface.[6] It is unlikely that this "sharpening" effect is crystallographic since the relative changes in He scattering are small. The overall narrowing observed with all the rare gases could be due to the increased "stiffness" of the Ni lattice relative to that of Ag. Such a parameter is included in the so-called "soft cube" model[13] to account for differences in scattering from Ag and Au and the more recent comparison of scattering from Au, Ag, and Ni only serves to emphasize the importance of the lattice properties of the solid in describing the collision process.

There are other possible differences between Ni and Ag. For example, the root mean square displacement of the solid atoms from their equilibrium position varies $\sim (T_S/M_s\theta_D^2)^{\frac{1}{2}}$ [14] and the ratio of such displacements for Ag/Ni is 1.33, implying a lesser degree of surface roughening in Ni. Such decreases in thermally induced roughening could be particularly important when the conditions for specular scattering of the light gases are considered.

It should be noted that although explicit calculations of scattering distributions of diatomic molecules have not yet been carried through, the results of energy transfer calculations[15] indicate that the proposed rotational state-phonon interaction is plausible. The results of thermal accommodation coefficient measurements for the He isotopes[16] and the hydrogenic molecular species[17] are also consistent with these energy transfer calculations.

The results of this study seem to imply, therefore, that there is an interaction between the rotational levels of the light diatomic molecules and the phonon levels of the solid, and the condition for this interaction is more easily satisfied with Ag than with Ni. Detailed discussions of phonon interaction are complicated by the inclusion of surface modes whose characteristic Debye temperatures are only about one half those of the bulk.[18]   Qualitatively, the trend toward more phonons for an H_2 transition than for a D_2 or HD transition is maintained, but a resonant type of interaction involving nearly exact matching of transition energies may not be an appropriate description.

IV.   SCATTERING OF HEAVY DIATOMIC AND POLYATOMIC MOLECULES: O_2, N_2, CO, CO_2 SCATTERED FROM (111) Ag

The basis of the argument to explain the differences between the degree of specular scattering of the various hydrogenic molecules lies in the magnitude of the energies involved in transitions between the rotational levels.   For these light gases, the energy levels have characteristic temperatures of 40 to 80°K. For the heavier gases, however, the rotational level spacings are very much smaller, having characteristic temperatures of only a few degrees Kelvin.[19] One would expect therefore that transitions between translation and rotation, induced by collision with the lattice, would occur rather freely as they do in gaseous collisions.[20]

For the gases O_2, N_2, and CO, all of comparable mass and heats of physical adsorption (see Table I), one would expect to find very similar scattering behavior.   Although significant effects of the rotational states might be expected to occur, any effects of the individual rotational states should not be able to be resolved in terms of scattering behavior.   Therefore, although different from rare gas scattering, the trends should be the same.   Similarly, if the scattering distributions follow the general patterns observed with rare gas scattering, CO_2 should, due to a larger mass and larger ΔH_p, exhibit a scattering distribution which is broader and further displaced from the specular angle than the other three.   The data for the angular distributions of these gases scattered from (111) Ag are shown in Fig. 9.   The distributions may be grouped into pairs, O_2 and N_2 being quite similar and CO and CO_2 also being quite similar.   These results are surprising in view of the observations of rare gas scattering.

The CO, CO_2 similarity is particularly surprising since both a mass and ΔH_p increase would move a 300°K rare gas scattering peak toward the normal and broaden it.   If anything, the opposite is observed.   CO does not appear to fit into the expected pattern.

A comparison of the scattering of CO with that of Ar and of O_2 with that of Ne [from Ag (111)] is shown in Fig. 10. Although the mass ratios for CO_2 and Ar are nearly the same, the Ar with a smaller ΔH_p is displaced more toward the normal than the CO_2. The O_2 and Ne distributions are also quite similar with the slightly broader O_2 peak displaced slightly more toward the normal, which is consistent with the rare gas behavior due to differences in ΔH_p and M_g. Quantitatively, however, the differences in scattering are smaller than one might have expected.

It appears, therefore, that the simple collision models probably do not apply to the scattering of diatomic and polyatomic molecules, most likely due to the participation of internal degrees of freedom in the interaction. The orientation of the incident molecule with respect to the surface should play some role even if the collision is languid[23] and although the rotational transitions involve energies too small to be resolved in such a scattering experiment, they might play a role due to the ease with which rotational-translational interactions may occur in such systems.

The scattering of increasingly energetic molecules does follow the trend observed with rare gases: as T_B increases, the beam dispersion decreases and the scattered beam moves toward the specular angle. When $T_B > T_S$, supraspecular scattering is observed. This is shown in Fig. 11 for N_2 scattered from Ag (111). It was not possible to use energetic O_2 incident beams. At O_2 beam temperatures above $800°C$, the silver surface was altered sufficiently to seriously broaden the He scattering. A $1500°K$ N_2 beam did not affect the surface sufficiently to alter the scattering of He.

It is interesting to note the highly-directed scattering of the oxygen-containing gases and it does imply that for low beam energies the reaction between the gas and stabilized solid surface (the surface after Ag deposition is completed) is improbable, but the reaction probability goes up sharply with increasing beam temperature. Experiments are in progress to study O_2 scattering during deposition to determine if the low reaction probability results from a stable film on the Ag (111) surface (which does not, however, affect rare gas scattering distributions). Absolute values of scattered flux were not determined and hence quantitative interpretation of temperature dependent O_2 scattering data cannot be made. The discrimination of the lock-in detection system against particles with long surface residence times might also complicate this type of analysis.

V.   SUMMARY AND CONCLUSIONS

Although the scattering of rare gas atoms by solid surfaces has been described in relatively simple terms utilizing a limited number of parameters, it is clear that a relatively complete theoretical description is not yet available. Extension to diatomic and polyatomic species is even more difficult due to effects of the internal structure of the molecular species.

For gases with small moments of inertia like H_2, D_2, and HD, it appears that the large rotational level spacings cause appreciable perturbation of the specular component of scattering due to a complex interaction of the phonons of the solid with the rotational and translational levels of the gas. The similarity of He^3 and He^4 scattering shows that mass effects for $\mu \ll 1$ are very small and the overall similarity of the hydrogenic molecules shows that both ΔH_p and the internal structure may be important factors leading to differences when compared with He scattering.

For heavier diatomic and polyatomic molecules, the trends observed with variations in M_g and ΔH_p for the rare gases do not seem to be observed. The effects of the internal structure (collisional orientation and rotational levels) probably play a role in the scattering, with significant rotational-translational interaction occurring. The previously quoted comparison of Ne, CH_4, and NH_3 scattering must be reviewed in the light of this evidence as it is no longer an obvious conclusion that the observed differences in scattering are due solely to effects of ΔH_p. The anomolous behavior of CO, if due to an incorrect assessment of the interaction energy, would however re-establish the validity of the comparison. That is, even though the scattering of diatomic and polyatomic molecules does not compare quantitatively with the rare gases for similar ΔH_p and M_g, these gases would as a group show the same trend with variations in M_g and ΔH_p as the rare gases, making the comparison between CH_4 and NH_3, at least, a reasonable assessment of the effect of ΔH_p.

REFERENCES

1. J. N. Smith, Jr., and H. Saltsburg in Fundamentals of Gas-Surface Interactions (Academic Press, 1967), p. 370.

2. R. E. Stickney in Advances in Atomic and Molecular Physics (Academic Press, 1967), Vol. 3, p. 143.

3. H. Saltsburg and J. N. Smith, Jr., J. Chem. Phys. 45, 2175 (1966).

4. J. N. Smith, Jr., and H. Saltsburg, J. Chem. Phys. 40, 3585 (1964).

5. J. N. Smith, Jr., and H. Saltsburg in Rarefied Gas Dynamics, 4th Symposium (Academic Press, 1966), Vol. II, p. 49.

6. J. N. Smith, Jr., H. Saltsburg, and R. L. Palmer in Rarefied Gas Dynamics, 6th Symposium (Academic Press), to be published.

7. S. Datz, G. E. Moore, and E. H. Taylor in Rarefied Gas Dynamics, 3rd Symposium (Academic Press, 1963), Vol. I, p. 347.

8. R. L. Palmer, H. Saltsburg, and J. N. Smith, Jr., to be published.

9. L. Trilling in Fundamentals of Gas-Surface Interactions (Academic Press, 1967), p. 392.

10. E. C. Beder in Advances in Atomic and Molecular Physics, Vol. 3 (Academic Press, 1967), p. 205.

11. H. Saltsburg, J. N. Smith, Jr., and R. L. Palmer in Rarefied Gas Dynamics, 5th Symposium (Academic Press, 1967), Vol. I, p. 223.

12. C. Kittel, Introduction to Solid State Physics, 3rd Edition, (John Wiley and Sons, 1966), p. 180.

13. R. M. Logan and J. C. Keck, M.I.T. Fluid Mechanics Laboratory Report 67-8, October 1967.

14. J. M. Ziman, Principles of the Theory of Solids (Cambridge University Press, 1964), p. 62.

15. P. Feuer, J. Chem. Phys. 39, 1311 (1963).

16. L. B. Thomas, private communication (1968).

17. K. Schäfer and D. Schuller, Ber. Bunsengesell. Physik. Chem. 70, 27 (1966).

18. E. R. Jones, J. T. McKinney, and M. B. Webb, Phys. Rev. 151, 476 (1966).

19. R. H. Fowler and E. A. Guggenheim, Statistical Thermodynamics (Cambridge University Press, 1949), p. 90.

20. B. Stevens, Collisional Activation in Gases (Pergamon Press, 1967), p. 41.

21. D. O. Hayward and B. M. W. Trapnell, Chemisorption (Butterworths, 1964), Table 27.

22. D. M. Young and A. D. Crowell, Physical Adsorption of Gases (Butterworths, 1962), Table 2.3.

23. F. O. Goodman, Rarefied Gas Dynamics, 4th Symposium (Academic Press, 1966), Vol. 2, p. 366.

Table I

Estimated Relative Values of ΔH_p Using
Data of References 21 and 22; Values
of ΔH_p Normalized to Ne

Gas	M_g	ΔH_p (kcal/mole)
O_2	32	3.3
N_2	28	3.3
CO	28	3.9
CO_2	44	5.9
Ar	40	2.4
Ne	20	0.9

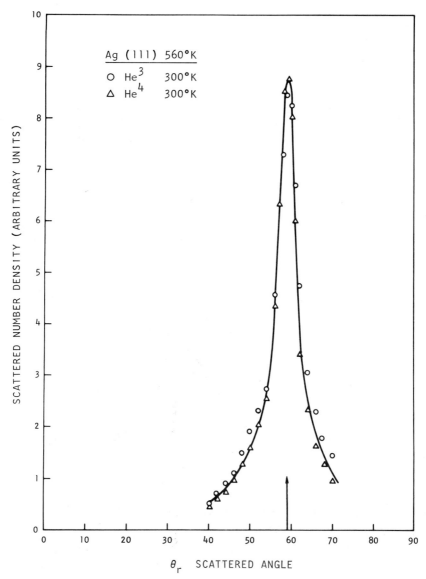

Fig. 1. He[3] and He[4] scattering from (111) Ag; $T_B = 300°K$, $T_S = 560°K$, $\theta_i = 59°$ (Ref. 9).

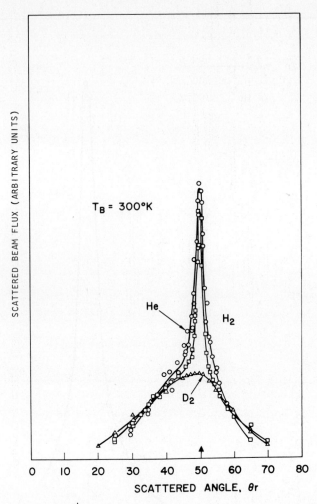

Fig. 2. H_2, D_2, and He^4 scattering from (111) Ag; $T_B = 300^{\circ}K$, $T_S = 560^{\circ}K$, $\theta_i = 50^{\circ}$ (Ref. 11).

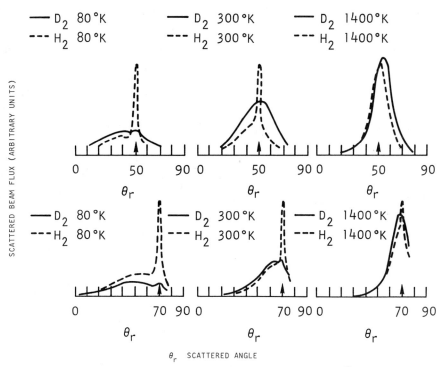

Fig. 3. H_2 and D_2 scattering from the (111) Ag; $T_S = 560°K$; T_B as indicated; $\theta_r = \theta_i$ denoted by arrow (data from Ref. 11).

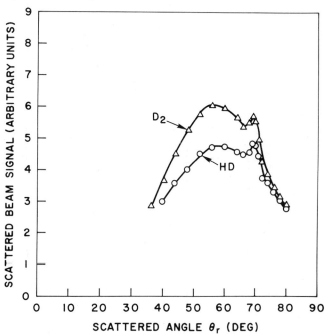

Fig. 4. HD and D_2 scattering from (111) Ag; $T_S = 560°K$; $T_B = 300°K$, $\theta_i = 70°$.

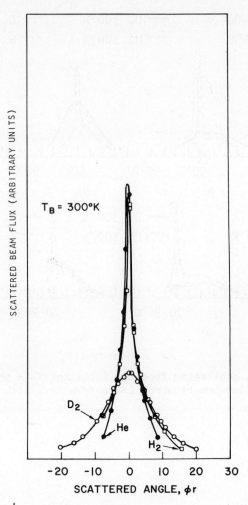

Fig. 5. H_2, D_2, and He^4 scattering from (111) Ag in the plane transverse to the incident plane at the specular angle. $T_B = 300°K$, $T_S = 560°K$, $\theta_i = 50°$ (Ref. 11).

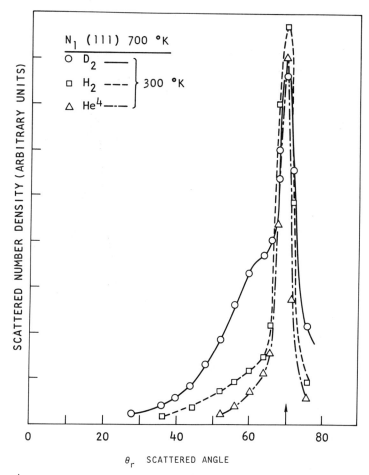

Fig. 6. He^4, H$_2$, and D$_2$ scattering from (111) Ni; $T_B \leqslant 300^{\circ}$K, $T_S = 700^{\circ}$K, $\theta_i = 70^{\circ}$.

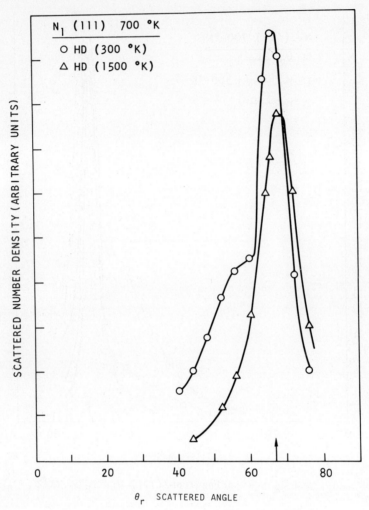

Fig. 7. HD scattering from (111) Ni; $T_S = 700^{\circ}K$, $T_B = 300^{\circ}K$, $1500^{\circ}K$, $\theta_i = 67^{\circ}$.

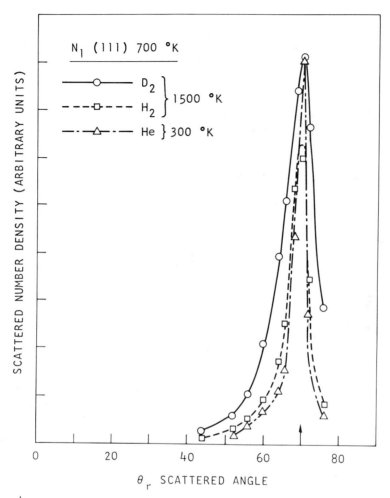

Fig. 8. He^4, D$_2$, and H$_2$ scattering from (111) Ni; $T_S = 700^\circ$K, T_B (H$_2$, D$_2$) = 1500°K, T_B (He^4) = 300°K, $\theta_i = 70^\circ$.

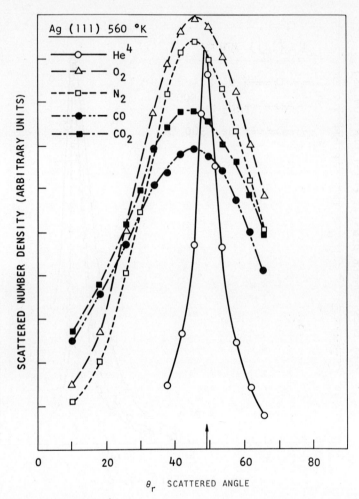

Fig. 9. O_2, N_2, CO, CO_2, He[4] scattering from (111) Ag; $T_B = 300°K$, $T_S = 560°K$, $\theta_i = 50°$.

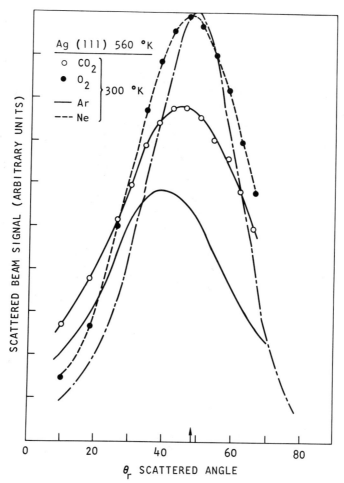

Fig. 10.   Ar, Ne, O_2, CO_2 scattering from (111) Ag; $T_S = 560°K$, $T_B = 300°K$, $\theta_i = 49°$.   (Data for Ar and Ne from Ref. 5.)

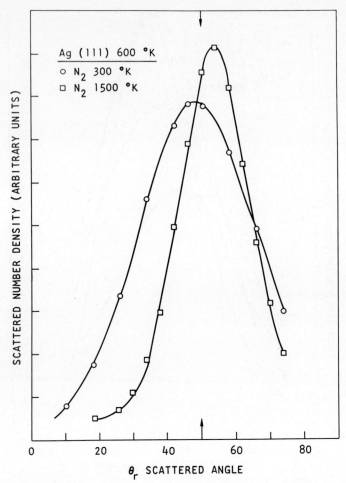

Fig. 11. N$_2$ scattering from (111) Ag; T$_S$ = 600°K, T$_B$ = 300°K, 1500°K, θ_i = 50°.

THEORY OF METALLIC ADSORPTION ON REAL METAL SURFACES[*]

J. W. Gadzuk[+]

Department of Physics, Solid State and Molecular Theory Group, and
Research Laboratory of Electronics
Massachusetts Institute of Technology
Cambridge, Massachusetts

I.   INTRODUCTION

During the past two years there have been significant advances in the theory of alkali adsorption on metals.[1-8] This is particularly gratifying, since the output of good quality experimental data is also increasing. It appears that we are at a sufficiently advanced stage to attempt a quantitative comparison between experimental data and "first principles" calculations. The aim of this paper is to present such a calculation, which is an amalgamation of previously published work[1,2,7] and some new ideas and then to compare the calculated values with measured data.

The physical situation considered here is that of an ion near the surface of a metal or electron gas. The metal will have a tendency to screen the field of the ion so that no net electric field exists within the conductor. The screening mechanism is a combination of two distinctly different processes.

In the first mechanism considered, the charged particle is envisioned as a massive, static point impurity which is screened within a finite distance by the nonperfect conductor. The details of this situation have been discussed previously by the writer.[6,7] The problem is modeled in the spirit of the "volume-impurity in an electron gas" theory of Langer and Vosko.[9] This is the type of screening mechanism that gives rise to the long-range Friedel oscillations of surface impurities.[4,6,7,11] In the region in which the ion-metal separation is much greater than the Fermi wavelength, this screening charge should produce effective fields equivalent to those of the classical image. For ions a few angstroms from a surface, however, the separation is of the same order as the Fermi wavelength and the image notion is of limited

---

[*] The work at the Research Laboratory of Electronics, M. I. T., was supported by the Joint Services Electronics Program (Contract DA28-043-AMC-02536(E)).

[+] Present Address: National Bureau of Standards, Washington, D. C.

usefulness. Details of the polarization screening will be discussed in
Section III.

The second kind of screening results from the formation of virtual bound
impurity states lying within the occupied portion of the conduction band.[1-3]
These states are analogous to the Koster-Slater impurity states of the type that
is responsible for localized magnetic moments of impurities in metals and on
surfaces.[12,13] Effectively, these states concentrate charge around the impurity
center, thereby reducing the net charge of the ion and producing no dipole
moment in lowest order.[2] The role of virtual-state screening in adsorption will
be discussed in Section IV.

The ideas presented here originate from the theory that metallic or ionic
adsorption is basically a surface analog of volume impurity effects. Hence we
draw heavily on ideas from impurity theory.

The structure of the paper is outlined as follows. In Section II, the
fundamental physical ideas of ionic adsorption are briefly reviewed. Section III
presents the theory of the dielectric screening of the surface impurity. In
Section IV, the problem of bound-state screening is treated. Section V contains
the prescription for combining the two types of screening in a self-consistent
manner. Self-consistent numerical results are presented and then compared with
experimental data. Section VI concludes the paper.

II. GENERALITIES

The salient features of an alkali impurity (adsorbate) interaction will
now be briefly reviewed. These have been discussed at great length by several
authors.[1-4,7,8]

As a hydrogenic atom or ion with a valence level near the Fermi energy of
a metal is brought near ($< \sim 10$ Å) a metal surface, the valence electron may
make a transition from the atomic to the metallic state. Subsequently, a
charged ion (impurity) results which now interacts with the metal in various
ways. The interaction will cause the original valence level to be shifted
and broadened so that $V_i \rightarrow (V_i + \triangle E) + i\Gamma$, where V_i is the original ionization
potential, $\triangle E$ is the shift of the real part of the energy, and Γ is an
imaginary part of the energy acquired from the finite lifetime of the atomiclike
state. The situations considered here are illustrated in Fig. 1. The change
in the valence (ns, n = 3 for Na, n = 4 for K, n = 6 for Cs) level as a
function of atom-metal separation is schematically displayed. At the
equilibrium separation, a portion of the perturbed impurity level lies below

the Fermi energy; hence, there will be partial occupation of the level by electrons shared by the atom and metal. This partial occupation of the virtual bound state gives rise to screening of the impurity, as discussed by Koster and Slater,[11] Anderson,[12] and Bennett and Falicov,[2] in a very localized manner. Electron charge is piled up at the impurity center, thereby decreasing the effective charge of the ion. Bennett and Falicov[2] estimate the resulting effective charge of an alkali on a reasonable metal surface to be approximately 85% of the pure ionic charge. As a first approximation, it is imagined that the partial neutralization by this mechanism results from piling up charge at the impurity, which reduces the monopole component of the impurity effective charge but has no effect on higher order multipoles. Once this sort of screening occurs, the result is an impurity with some effective charge, Z_{eff}, sitting on an electron gas surface. As with impurities in the volume,[9] the Coulomb field from this source will polarize the electron gas (electrons in the conduction band), thereby causing a redistribution in real space of these electrons which will screen out the effective charge field of the impurity within the metal.[6,7] Numerical results from the case of a structureless impurity at s' = 0 have been presented elsewhere.[6,7] These results were presented in the form of screening-charge density as a function of position, where the nature of the resulting dipole configuration formed from the positive-charged impurity plus negative-charged screening charge was displayed. These results were considerably different from those inferred from image charges. This is a consequence of the fact that the metal does not look like a perfect conductor, a necessary condition for the validity of the image method, when the separation between impurity and metal is of the order of the Fermi wavelength or some relevant screening length in the metal.

In Section III we shall calculate the screening charge as a function of atom-metal separation and effective ion charge. From this we can obtain dipole moments for various ion-metal combinations for different crystallographic planes. We adopt the point of view that the essential change in dipole from one plane to another is caused by different atom-metal separations, and hence different polarization charges, more than by different band-structure effects from one direction to another. Certainly, in the ultimate theory we shall have to consider the asymmetry in k space but, at present, we believe that these effects are not the dominant ones.

Once the screening charge is known, it is then possible to infer what the interaction between a charged particle and the metal is. Problems of

self-consistency arise as follows. First, in order to calculate the effective charge Z_{eff}, one must know both the nature of the particle-metal interaction and the values of $\triangle E$ and Γ shown in Fig. 1. But to calculate $\triangle E$ and Γ, one must know Z_{eff} and the type of interaction. Bennett and Falicov have carried out a self-consistent calculation for Z_{eff} in terms of an assumed image inter-action.[2] In the present paper, the self-consistent procedure outlined by Bennett and Falicov[2] is further extended. Section III is concerned with calculating the screening charge as a function of s' and Z_{eff}. Schematically we get $\delta n = \delta n(s', Z_{eff})$ and then the dipole moment $\mathscr{P} = \mathscr{P}(\delta n)$. From this calcu-lation we can obtain an effective interaction between a charged particle and the metal. Then using the procedure previously adopted[1] and the effective interaction calculation here, we can obtain expressions for $\triangle E$ and Γ which will be functions of s and Z_{eff} and functionals of δn, in the manner:
$\triangle E = \triangle E([\delta n(s', Z_{eff})], s, Z_{eff})$ and $\Gamma = \Gamma([\delta n(s', Z_{eff})], s, Z_{eff})$.
Knowing $\triangle E$ and Γ, we then use the procedure of Bennett and Falicov[2] to satisfy the self-consistency requirement on the effective charge $Z_{eff} = Z_{eff}(\triangle E, \Gamma, s, Z_{eff})$. The effective charge on the impurity is thus obtained self-consistently. This result is used in the original expression for δn to calculate self-consistent dipole moments that can be compared with experimental values. We should note that the effective impurity-metal separation s, is determined by δn and by the billiard-ball geometry of a particle on a particular crystallographic plane. We have been referring to two different separations, s and s'. The distance s is the true impurity-metal separation, whereas the distance s' is the distance between the impurity and the Z = 0 plane of the electron gas hich will be discussed. The relation between s and s' will be established.

III. DIELECTRIC SCREENING OF A SURFACE IMPURITY

The notion of screening is most familiar in classical plasma theory from which the Debye length derives. In the case of a quantum plasma, that is, the conduction-band electrons of a metal plus the massive positive cores, the equivalent parameter is the Fermi-Thomas screening length. Fortunately, the particular approximations, whether Fermi-Thomas, Hartree, Hartree-Fock, Random Phase Approximation (RPA) or Hubbard modified RPA,[14-16] are of no consequence to us until the final stage of numerical calculations. Thus the theory presented here does not have the usual RPA restrictions. We use the image technique presented elsewhere[7] to obtain expressions for the surface-impurity screening charge and dipole moments in terms of a generalized wave number-dependent dielectric function of a uniform, interacting electron gas without a surface.

First, we sketch some of the basic ideas pertaining to dielectric impurity screening and show how the simple results obtained here are related to the more elaborate Green's function theory previously derived.[7]

Consider a linear response theory of screening. A driving force, the electric field of the bare impurity, is applied to the electron gas. The electron gas responds in the sense that polarization charge is attracted to the bare impurity so that, far away, the net electric field of the combined system, impurity, and induced charge is zero. This may be shown for the case of an impurity deep within the metal. Total charge and electric potential energy are related through Poisson's equation

$$\nabla^2 V(\vec{r}) = 4 \pi e^2 n(\vec{r}). \tag{1}$$

The total charge number is the sum of the impurity charge plus screening charge. Hence

$$n(\vec{r}) = Z_{eff} \delta(\vec{r}) - \delta n(\vec{r}), \tag{2}$$

with the impurity at the origin, and $\delta n(\vec{r})$ the screening charge. Taking Fourier transforms of Eqs. (1) and (2), we obtain

$$q^2 V(q) = 4 \pi e^2 n(q) \tag{3}$$

$$n(\vec{q}) = Z_{eff} - \delta n(\vec{q}), \tag{4}$$

Within a linear response theory, the Fourier transform of the net potential is related to the transform of the bare potential and to the dielectric function as

$$V(q) = \frac{v_{im}(q)}{\epsilon(q,0)} \tag{5}$$

with $v_{im}(q)$ the Fourier transform of the bare impurity potential, and $\epsilon(q,0)$ the dielectric function in whatever approximation is adopted. Solving Eq. (3) and (4) for $\delta n(q)$ and using Eq. (5) gives

$$\delta n(q) = Z_{eff} - \frac{v_{im}(q)}{4\pi e^2/q^2} \frac{1}{\epsilon(q,0)}.$$

Consequently, the screening charge density in real space is

$$\delta n(\vec{r}) = \int \frac{d^3q}{(2\pi)^3} \left(Z_{eff} - \frac{v_{im}(q)}{4\pi e^2/q^2} \frac{1}{\epsilon(q,0)} \right) e^{i\vec{q}\cdot\vec{r}}. \tag{6}$$

For the case in which $Z_{eff} = 1$ and $v_{im}(q) = \dfrac{4\pi e^2}{q^2}$, a bare Coulomb potential, Eq. (6) reduces to

$$\delta n_{LV}(\vec{r}) = \int \frac{d^3q}{(2\pi)^3} \left(1 - \frac{1}{\epsilon(q, 0)} \right) e^{i\vec{q}\cdot\vec{r}}, \tag{7}$$

the result obtained by Langer and Vosko.[9]

The object of interest here is the manner in which this type of thinking can be made useful in surface events. An image technique is useful here. It has been shown[7] that in an electron gas with a specularly reflecting surface, one in which electron wave functions go as $\sin k_z z\, e^{i\vec{k}_T \cdot \rho_T}$ for $z \geqslant 0$ only with z normal to the surface and ρ_T transverse, the dielectric response of a semi-infinite gas with a surface could be put in correspondence with the dielectric response of a ful infinite gas with sine-wave electrons for the full region $-\infty \leqslant z \leqslant +\infty$. This realization makes calculations tractable. The correspondence comes about by considering Fig. 2. Here we have drawn the total charge density as a function of z for the full infinite gas. It has been shown that the resulting polarization charge coming from the potential of a bare impurity at a distance s' from the z = 0 plane and from the mirror image of this potential is equivalent to the polarization charge around an impurity at a distance s' into the metal; in other words, we have the physical situation that would result if the left-hand side of Fig. 2 were deleted. The physical origin of this effect is shown in Fig. 3. Here we consider an incident electron scattering from the bare impurity and propagating onward in the infinite gas. But for this event there is an equivalent event from the left hand side. It can be seen that the net electron history on either side of the z = 0 plane is that of an electron scattering from the impurity, and then from the specular surface.

Furthermore, it has been shown[7] that the dielectric function evaluated for electrons on a sine-wave basis was such that the quantity $(1 - \frac{1}{\epsilon}) \rightarrow 2(1 - \frac{1}{\epsilon})$ for the sine-wave case. The screening charge was then evaluated for the sample case of $s' = 0$ in which the two potentials shown in Fig. 2 merge into one spherically symmetric Coulomb potential. A feasibility argument will now be given to show the results of previous work.[7] If we make the substitution of the extra factor of 2, consider the z part of the Fourier transform in Eq. (7) as a sine expansion, and add a few necessary constants, then the surface impurity screening charge is given by

$$\delta n_s(\vec{r}) = 4i \int \frac{d^3q}{(2\pi)^3} \left(1 - \frac{1}{\epsilon(q, 0)}\right) e^{i\vec{q}_T \cdot \vec{\rho}_T} \sin k_z z$$

which resembles the expression previously published.[6] Writing the sine as a sum of exponentials,

$$\delta n_s(\vec{r}) = 2 \int \frac{d^3q}{(2\pi)^3} \left(1 - \frac{1}{\epsilon(q, 0)}\right) e^{i\vec{q} \cdot \vec{r}} \left(1 - e^{-2ik_z z}\right).$$

The first term of this integral is just the Langer and Vosko result multiplied by 2. A form of the mean-value theorem is then used as an approximate method for evaluating the second term. Retaining only the symmetric part and using a trigonometric identity yields

$$\delta n_s(\vec{r}) = 2 \sin^2\left(\frac{P_F z}{\sqrt{2}}\right) \int \frac{d^3q}{(2\pi)^3} \left(1 - \frac{1}{\epsilon(q, 0)}\right) e^{i\vec{q} \cdot \vec{r}}. \tag{8}$$

The integral is the spherically symmetric Langer and Vosko result given in Eq. (7), so we may write

$$\delta n_s(\vec{r}) = 2 \sin^2(\xi z)\, \delta n_{LV}(r) \tag{9}$$

with $\xi = P_F/\sqrt{2}$.

Now we are interested in treating the case for the impurity outside the surface of the electron gas. Again the image technique is useful. The external impurity problem is illustrated in Fig. 4. The dashed lines show the full bare potential, whereas the solid lines show the part of the bare potential which enters the electron gas. Only this part causes the polarization charge. In Fig. 5 the coordinates for the following calculations are shown. The Fourier transform of the imaged impurity potential of charge Z_{eff} shown in Fig. 4 is given by

$$V_{im}(\vec{r}-\vec{s}') = \int \frac{d^3q}{(2\pi)^3} \; v(q) \; e^{-i\vec{q}\cdot(\vec{r}+|\,\vec{s}'\,|\,\vec{1}_z)} = \int \frac{d^3q}{(2\pi)^3} \; v_{im}(\vec{q}) \; e^{-i\vec{q}\cdot\vec{r}} \qquad (10)$$

for $0 \leqslant \theta \leqslant \frac{\pi}{2}$ and,

$$V_{im}(\vec{r}-\vec{s}') = \int \frac{d^3q}{(2\pi)^3} \; v(q) \; e^{-i\vec{q}\cdot(\vec{r}-|\,\vec{s}'\,|\,\vec{1}_z)} = \int \frac{d^3q}{(2\pi)^3} \; v_{im}(\vec{q}) \; e^{-i\vec{q}\cdot\vec{r}} \qquad (11)$$

for $\frac{\pi}{2} \leqslant \theta \leqslant \pi$, where θ is the angle between $\vec{r}$ and the z axis, $v(q) = 4\pi Z_{eff}\, e^2/q^2$, and $v_{im}(\vec{q})$ is the Fourier transform of the bare potential with respect to a coordinate system whose origin is on the z = 0 plane rather than at the impurity center. Comparing the two integrals in Eqs. (10) and (11), we see that the Fourier transform of the bare impurity potentials are given by

$$v_{im}(\vec{q}) = \frac{4\pi Z_{eff}\, e^2}{q^2} \; e^{-i\vec{q}\cdot\vec{s}'} \qquad \text{for } 0 \leqslant \theta \leqslant \frac{\pi}{2} \,,$$

and

$$v_{im}(\vec{q}) = \frac{4\pi Z_{eff}\, e^2}{q^2} \; e^{+i\vec{q}\cdot\vec{s}'} \qquad \text{for } \frac{\pi}{2} \leqslant \theta \leqslant \pi.$$

Inserting those results into Eq. (6), making appropriate changes in the Fourier transform of the impurity charge density, and including the surface effects leading to Eq. (8) gives

$$\delta n_s(\vec{r}) = 2Z_{eff} \sin^2(\xi z) \int \frac{d^3q}{(2\pi)^3} \left(1 - \frac{1}{\epsilon(q,0)} \right) e^{i\vec{q}\cdot(\vec{r}-\vec{s}')}. \qquad (12)$$

for the screening charge when $0 \leq \Theta \leq \pi/2$. Comparing the integral of Eq. (12) with Eq. (7), we see that

$$\delta n_s(\vec{r}) = 2Z_{eff} \sin^2(\xi z) \; \delta n_{LV}(|\vec{r}-\vec{s}'|). \tag{13}$$

To go further we must make a small s' approximation. That is, we limit ourselves to the case in which s' is of the order of the Fermi wavelength. This is not so drastic, however, since (as pointed out previously[7] and in recent work on electron gas surface potentials[17]) an impurity at the z = 0 plane of an infinite barrier electron gas is equivalent to an impurity that is a distance $z_1 \approx 0.5 - 1$ Å from the surface of an electron gas with a finite barrier of the height found in real metals. Consequently, in the present model calculations, an impurity a distance s' from the infinite barrier is equivalent to one that is a distance $s' + z_i = s$ from a real barrier. In real life, the separation is approximately 1.5 Å. Thus the relevant s' in these calculations is less than 1 Å and the small s' approximation is reasonable. This will be discussed in more detail later.

The small s' approximation comes from $|\vec{r}-\vec{s}'| = (r^2 - 2s'r \cos \Theta + s'^2)^{1/2} \approx r\left(1 - \frac{s'}{r} \cos \Theta\right)$. Thus

$$\delta n_{LV}(|\vec{r}-\vec{s}'|) \approx \delta n_{LV}\left(r\left(1 - \frac{s'}{r} \cos \Theta\right)\right)$$

which when expanded becomes

$$\delta n_{LV}(|\vec{r}-\vec{s}'|) = \delta n_{LV}(r) - s' \cos \Theta \frac{\partial}{\partial r} \delta n_{LV}(r) + \frac{(s' \cos \Theta)^2}{2!} \frac{\partial^2}{\partial r^2} \delta n_{LV}(r) - \ldots \tag{14}$$

It has been noted that except in the Friedel oscillation region, the Langer and Vosko screening charge behaves much as an exponential, that is, $\delta n \sim e^{-\gamma r}$ with $\gamma = 1.3$ Å^{-1}. Consequently, the nth-order derivative is given by $(-1)^n \gamma^n n(r)$. With this insertion, the exact expression for Eq. (14) within the small s' approximation is

$$\delta n_{LV}(|\vec{r}-\vec{s}'|) = \delta n_{LV}(r) e^{-\gamma s' \cos \Theta};$$

hence, we now return to the exact expression for $\delta n_{LV}(r)$ in calculations requiring $\delta n_{LV}(|\vec{r}-\vec{s}'|)$. We can write the expression for the surface impurity screening charge, Eq. (13), as

$$\delta n_s(\vec{r}) = 2Z_{eff} \sin^2 (\xi r \cos \Theta) \, e^{-\gamma s' \cos \Theta} \delta n_{LV}(r). \tag{15}$$

This is the main physical result. Now we are in a position to calculate less abstract quantities. The first quantity of interest is the total amount of charge displaced on the right-hand side within linear response screening theory:

$$N_{Tot}(s') = \int_{RHS} d^3r \, \delta n_s(r)$$

$$= Z_{eff} 4\pi \int_0^\infty r^2 \, \delta n_{LV}(r) \left\{ \int_0^{\pi/2} \sin^2 (\xi r \cos \Theta) \, e^{-\gamma s' \cos \Theta} \sin \Theta \, d\Theta \right\} dr.$$

After considerable manipulation, this reduces to the single integral

$$N_{Tot}(s') = Z_{eff} 2\pi \int r^2 \, \delta n_{LV}(r) \left\{ \frac{1 - e^{-\gamma s'}}{\gamma s'} \right.$$

$$+ \frac{\gamma s'}{(2\xi r)^2 + (\gamma s')^2} (e^{-\gamma s'} \cos(2\xi r) - 1) - \frac{(2\xi r)e^{-\gamma s'} \sin(2\xi r)}{(2\xi r)^2 + (\gamma s')^2} \left. \right\} dr. \tag{16}$$

Equation (16) has been numerically evaluated, by using the tabulated values of $\delta n_{LV}(r)$ within the RPA, and the results plotted in Fig. 6 as a function of s' for two different electron gas densities corresponding to $r_s = 1.5$ and 4.5. Note that the dimensionless interelectron separation r_s is defined through the electron density, $1/\frac{4}{3} \pi r_s^3 a_o^3$, with a_o the Bohr radius. We shall return to these results later.

The next quantity of interest is the effective position of the screening charge, defined as

$$\langle z(s') \rangle = \frac{\langle M(s') \rangle}{N_{Tot}(s')}, \tag{17}$$

with $\langle M(s') \rangle$ given by

$$\langle M(s') \rangle = \int d^3r \, \delta n(\vec{r}) \, z$$

$$= Z_{eff} 4\pi \int_0^\infty r^3 \, \delta n_{LV}(r) \left\{ \int_0^{\pi/2} \sin^2 (\xi r \cos \Theta) \right.$$

$$e^{-\gamma s' \cos\theta} \cos\theta \sin\theta \, d\theta \Bigg\} dr.$$

Eventually, the single integral is obtained:

$$\langle M(s')\rangle = Z_{eff} 2\pi \int r^3 \, \delta n_{LV}(r) \Bigg\{ \frac{1}{(\gamma s')^2} - \frac{e^{-\gamma s'}(\gamma s'+1)}{(\gamma s')^2}$$

$$- \frac{1}{(\gamma s')^2 + (2\xi r)^2} \Bigg\{ e^{-\gamma s'} \Bigg(2\xi r \sin(2\xi r) - \gamma s' \cos(2\xi r)$$

$$- \frac{((\gamma s')^2 - (2\xi r)^2)}{((\gamma s')^2 + (2\xi r)^2)} \cos(2\xi r) + \frac{2(2\xi r)(\gamma s')}{((\gamma s')^2 + (2\xi r)^2)} \sin(2\xi r) \Bigg)$$

$$+ \frac{(\gamma s')^2 - (2\xi r)^2}{((\gamma s')^2 + (2\xi r)^2)} \Bigg\} \Bigg\} dr. \qquad (18)$$

Equation (18) has also been evaluated numerically using the Langer and Vosko results for $\delta n(r)$. Curves of $\langle M(s')\rangle$ as a function of s' for three values of r_s are drawn in Fig. 7. Equation (17), which yields the expectation value of position per unit charge, is drawn in Fig. 8 as a function of s'.

We shall make only a few comments here on the nature of these results, and return to a fuller discussion of their meaning and significance later. In all cases, the results have been drawn on a dimensionless scale, with the Fermi wave number used as the basic unit of measure. Since $p_F = 3.64/r_s$ ($\mathring{A}^{-1}$), the physical implications of these results are not quite as transparent when viewed in a casual manner. This will be illustrated in Section V.

IV. SCREENING BY VIRTUAL BOUND STATES

The second mechanism for impurity screening or neutralization comes about by the formation of virtual impurity states which derive from the unperturbed atomic state of the impurity.[1-4,13] These states are quite analogous to the impurity states first discussed by Koster and Slater.[11] The role of the virtual impurity state in adsorption theory has been outlined in great detail by the writer[1] and by Bennett and Falicov.[2] These states have also been experimentally observed by Hagstrum.[18] (See note added in proof.)

In the present study, our point of view is the following. As the atom or
ion is brought to the surface, the ns level shifts upward and broadens as
shown in Fig. 1. The upward shift, calculated elsewhere,[1] is approximately
0.7 eV for all alkalis, and is an extremely weak function of atom-metal
separation.[19,20] In the following calculations we shall use the value $\Delta E = 0.7$ eV
independent of all other parameters of the system. The width Γ is more
complicated. Previously, Γ was calculated[1] for K and Cs and found to be between
1 and 2 eV for equilibrium atom-metal separations. From Eq. 21 of the previous
paper[1] the expression for Γ is

$$\Gamma = \frac{2Z_{eff}a^4 q^2}{a_o \sqrt{2m}} \left(\frac{(\phi_e + E_F - V_i + \Delta E)^{1/2}}{(\phi_e + E_F)} \right) s^3 e^{-2as} (1 - 4/a^2 s^2 + 8/a^4 s^4), \qquad (19)$$

where ϕ_e is work function, E_F is Fermi energy, and a is a constant characterizing the particular adsorbate. For Na , a = 1.53 Å^{-1}; for K, a = 1.16 Å^{-1};
and for Cs, a= 0.99 Å^{-1}. A special note should be made about the meaning of
s in Eq. (19). This variable is the effective distance between the ion core
and the polarization charge that it induces, as suggested in Eq. 23 1/2 of the
previous paper,[1] and not just the atom-metal separation distance. The values
of V_i and a depend upon the particular alkali considered. The Fermi energy is
determined by the substrate in question. The work function is determined by the
material and the particular crystal face in question. Schematically we now
know that $\Gamma = \Gamma(Z_{eff})$ from which Z_{eff} must be determined self-consistently.
This is done by using the prescription of Bennett and Falicov.[2] Because of
the broadened level, there is some energetic overlap with the occupied states
of the metal conduction band. This amounts to sharing of electrons between
the impurity and the host, metallic bonding, and partial neutralization of the
impurity. Following Bennett and Falicov, the effective charge number on the
impurity of initial unit charge is

$$Z_{eff} = 1 - \int_{-\infty}^{\infty} \rho(E) F(E) dE \underset{T \to 0}{\to} 1 - \int_{-\infty}^{E_F} \rho(E) dE \qquad (20)$$

where F(E) is the Fermi function, $\rho(E)$ the effective charge number per unit
energy in the broadened impurity virtual level, and the T = 0 limit is taken
for simplicity. The customary procedure is to choose $\rho(E)$ to be a Lorentzian

$$\rho(E) = \frac{1}{\pi} \frac{\Gamma}{\left(E-E_F-\phi_{eo}+V_i-\Delta E\right)^2 + \Gamma^2} \tag{21}$$

although Plummer has shown that under certain circumstances involving transition metals, a Gaussian line shape may be more appropriate.[19] In the zero temperature limit that is appropriate for field-emission experiments, Eq. (21) substituted in Eq. (20) yields for the effective impurity charge number

$$Z_{eff} = 1 - \frac{1}{\pi} \left[\tan^{-1}\left(\frac{V_i-\Delta E-\phi_{eo}}{\Gamma(Z_{eff})}\right) + \frac{\pi}{2} \right]. \tag{22}$$

Equation (22) must be solved graphically to find the effective charge on a surface impurity when absorbed on different crystal faces of different materials. This will be done in Section VI.

V. SELF-CONSISTENT DIPOLE MOMENTS

We have now arrived at the point where, with one more physical assumption, we can calculate self-consistent dipole moments for the desired combination of alkalis, host metal, and crystallographic plane.

One of the requirements of the total charge displaced in the polarization process is that it amounts to exactly Z_{eff} so that the field of the bare impurity plus electrons partially occupying the virtual level is completely screened at large distances from the impurity. Since the value of $N_{Tot}(s')/Z_{eff}$ given by Eq. (16) does not equal a constant for all s', something is still missing from the present formalism. One of the problems is that we have not been able to satisfactorily account for the screening charge coming from the field lines emanating from the back side of the bare impurity. Another problem is that, because of some mathematical assumptions that are implicit in a linear response theory, we have lost some of the self-consistent screening effects included in the volume impurity screening theory. A plausibility argument is now offered for an attempted solution to this problem.

Consider Fig. 9 in which an impurity and some electric field lines have been drawn. For the bare impurity field that hits the surface within a circle of some critical radius r_c, the linear response theory adequately describes the screening. For the field lines outside this radius, screening is not well represented by the linear response theory. Gauss's law tells us, however, that the field arriving at the surface outside of r_c is relatively weak. Consequently, this part of the field does not penetrate far into the metal and to a good approximation is similar to surface polarization and thus to image fields. Consequently, we shall assume that the induced dipole is made up of two parts, one of which comes from the polarization process described in Section III, and the other which accounts for the remaining screening charge by an effective surface polarization contribution. This is written

$$\mathcal{D}(s') = \mathcal{D}_{vol}(s') + \mathcal{D}_{image}(s')$$

and

$$\frac{\mathcal{D}(s')}{Z_{eff}e} = \eta(s')\langle z(s')\rangle + s', \tag{23}$$

where $\eta(s') = \dfrac{N_{Tot}(s')}{Z_{eff}}$ is simply related to Eq. (16), $\langle Z(s')\rangle$ is given by Eqs. (17) and (18), and $Z_{eff}(s')$ by Eq. (22). In Fig. 10, Eq. (23) is drawn for several values of r_s as a function of s'. This expression and Eq. (22) for $Z_{eff}(s')$ will allow us to calculate the final dipole moments for real materials. Frequently when image dipoles are used, one takes the dipole length to be the distance between the real charge and its image. Then when computing work-function changes with the use of the Helmholtz equation, the results must be multiplied by 1/2 to take account of the fact that the relevant dipole length is the distance between the ion and the surface polarization charge at z = 0, and not the distance between the ion and its image. Equation (23) gives the true dipole length, and thus does not require the factor 1/2.

It is interesting to note the structure of this result. As the impurity is at the s' = 0 point, the full contribution to the dipole moment comes from volume polarization effects, in agreement with the results shown in Fig. 8. As the impurity is removed from the z = 0 plane, the polarization dipole increases slowly, and eventually merges into the dipole moment obtained from the classical image picture, as it must when the impurity metal separation becomes significantly larger than the Fermi wavelength. Since it may be physically misleading to give graphs in nondimensional form, in Fig. 11 we have drawn a graph of dipole length in angstroms versus separation in angstroms. Here the physical interpretation is more apparent. For the highest density gas, the one that seems most like a perfect conductor, the volume polarization effects die out most quickly and the polarization looks like surface charge and hence image dipoles. As the density of electrons decreases, the impurity field penetrates farther into the metal and hence the dipole formed by the impurity and its screening charge appears to be longer than the length inferred from pure image ideas.

VI.   CALCULATION COOKBOOK: NUMERICAL RESULTS, COMPARISON WITH EXPERIMENTAL RESULTS

So many things have been calculated, thus far, that widespread confusion probably prevails. The question remains: By using input parameters determined by the isolated atom and metal and the theory presented here, how do we get to "first principles" end results that can be compared with experimental results? The theoretical flow diagram shown in Fig. 12 provides the answer. The cookbook recipe for performing the calculation may be outlined as follows.

1. Choose an adsorbate. This specifies the values of V_i, the ionization potential; a, the effective Bohr radius; and r_i, the ionic radius.

2. Choose a metal. This tells us the electron density parameter r_s and the Fermi energy E_F.

3. Choose a crystallographic face. This specifies the value of the bare work function, ϕ_e.

4. The value of r_i and the crystallographic face allow us to calculate, with billiard-ball geometry, the distance between the impurity center and the last layer of metal determined by the position of the ion cores. This is the distance s, which we take to be the true atom-metal separation.

5. The sum of the work function and Fermi energy tells us that the depth of the well that a free electron metal would have to be. Using the surface-potential curves,[17] pick out the curve that gives the correct average internal potential. Then get the potential at z = 0 for this curve. Setting this potential equal to $\frac{-e^2}{4z'}$ determines the value of z' as outlined previously.[17] This distance is a correction to the infinite square-well electron gas used in the present calculations. This comes from the fact that an impurity at the z = 0 plane of the infinite barrier electron gas is equivalent to an impurity a distance z' outside an electron gas with a finite barrier of height $\phi_e + E_F$.

6. Knowing s and z', get s' = s - z'. s' is the impurity distance outside the infinite barrier electron gas needed to produce essentially equivalent effects to an impurity a distance s outside a finite barrier gas.

7. By using the specified value of r_s and the value s' given in 6, Eq. (23) and Fig. 11 give the value of dipole moment as a function of effective impurity charge, $\mathcal{D} = \mathcal{D}(Z_{eff})$.

8. Take the value $\mathcal{D}/Z_{eff}e + z' \equiv s_\Gamma$. Use the atomic parameters a and V_1, the metal parameters E_F and ϕ_e, $\Delta E = 0.7$ eV, and the just mentioned value of s_Γ in Eq. (22) to calculate $\Gamma(Z_{eff})$ which still requires specification of Z_{eff}.

9. Using the value of $\Gamma(Z_{eff})$ calculated above, the atom and metal parameters, and the self-consistency requirement given by Eq. (22), calculate the self-consistent impurity charge number Z_{eff}. We might note parenthetically that if it were desired to include metal temperature dependences, this is where we would do it. Rather than use the zero temperature Fermi function in Eq. (20), we could use the finite temperature generalization. This would have the effect of populating some of the electron levels above the Fermi energy, thereby reducing the effective charge Z_{eff}.

10. The calculated dipole moment, together with the calculated effective charge gives the final dipole moment M, which can be compared with the experimental result.

This procedure has been carried out for several atom-metal and crystallographic face combinations for which data are now available. The input data for the substrates is shown in Table 1. Table 2 lists the input data for the alkali atoms. Table 3 contains billiard-ball geometry formulas for the distance between the last lattice plane and the adsorbate center, where r_a = the lattice atomic radii calculated in the billiard-ball picture.

The grand reckoning appears in Table 4. Here we display the value of Γ from Eq. (19), the calculated Z_{eff} from Eq. (22), and the final dipole moment calculated theoretically for various combinations of alkalis, substrates, and crystallographic faces. The experimental values, from different sources,[3,21,22,27] appear in the next column in Table 4. The experimental dipole moments are calculated under the assumption that the Helmholtz equation is valid. This implies $M = \frac{\Delta\phi}{4\pi n} = 4.88\ Z_{eff}$ $\mathcal{D}$ (Debye units) with n, the number density of adsorbed particles; Z_{eff}, the charge number of the impurity charge; and $\mathcal{D}$, the dipole length in angstroms.

In order that the results may be displayed in a more graphical manner, the theoretical and experimental dipole moments are drawn as a function of ionization potential minus work function for several of the mentioned combinations of alkalis and metals in Figs. 13 and 14. (The writer is indebted to Dr. E. W.

Plummer for this suggestive display of the results.) Here the results are
seen in a dramatic fashion to be in uncannily good agreement. All trends of
dipole moment variations are similar both experimentally and theoretically.
The alkalis on Ta data of Fehrs is in complete agreement with the calculations.
The alkalis on Ni(110) of Gerlach is reasonably good as is Schmidt's K on W
data.

The new Na-Ni data presented at this conference by Gerlach and Rhodin[21]
actually provides the most stringent test of the theory. Because of the small-
ness of Na^+, the ion really nuzzles in close to the surface and the volume
polarization processes dominate. The positions of the shifted and broadened
virtual impurity states however, are such that one cannot naively predict the
variations in going from the (100) to the (111) face. Both experiment and
the detailed theory presented here agree however.

The agreement between experiment and theory is felt to be sufficiently
compelling to lead us to the conclusion that the physics of an alkali adsorbed
on a metal is fairly accurately envisioned as some combination of an impurity
with a partially occupied virtual bound state polarizing the conduction-band
electrons and producing a dipole configuration.

One of the most likely sources of error in the present theory has to do
with the fact that the screening seems to be much too efficient. Most traces
of the finite screening of the impurity seem to go as s' becomes greater than
1Å. This seems to be too short a distance. A possible cause of this may be
the adoption of the exponential decay of the short-range part of the Langer-
Vosko charge density in Section III. This may cut off the effects of finite
atom-metal separation too quickly.

Other sources of error may be found in some of the physical assumptions.
Probably it is not too good to assume that the charge that partially neu-
tralizes the impurity ion exists at the ion core center, as Bennett and
Falicov have also assumed.[2] In fact, the neutralizing charge lies somewhere
between the ion core and the surface, thereby providing an additional dipole
contributionn

Still another source of error is contributed by the Helmholtz equation.
It is not at all obvious that the dipole layer contribution to the work function
from an array of localized dipoles resembles the contribution that would
result from a smeared-out sheet whose charge density is the same as that of an
array of localized dipoles.

In the final analysis, it is unrealistic to suppose that we can get "back-of-the-envelope" first principles calculations of electronic surface phenomena that will give quantitative agreement with experiment to better than a factor of two, as obtained herein. The writer, who is a product of the Slater-Koster energy-band environment, has seen too many graduate students spend vast amounts of time on programming directed toward calculations on unperturbed metal electron properties to suppose that accurate results of metal electronic properties perturbed by both a surface and an impurity can be achieved without at least equal effort. Perhaps the tight-binding approach will prove fruitful. Since the present theory has the essential physics, although watered down to make calculations tractable, until someone is ready to invest at least as much time in doing surface electronic calculations as is done, at present, in quantitative volume electronic calculations, we must be satisfied with semi-quantitative agreement between experiment and theory.

Acknowledgment

The author appreciates the furnishing of experimental data by R. L. Gerlach and Professor T. N. Rhodin, of Cornell University, before its presentation at this conference.

Great thanks are also due Dr. Will Rudge, formerly with the Solid State Theory Group, for many favors.

The utmost thanks are extended to Keith Hartman at Cornell who redid the numerical calculations minus my numerical errors.

NOTE ADDED IN PROOF: Using the ideas of resonance tunneling through atoms in field emission experiments, the virtual impurity states discussed in this paper have unequivocally been observed and interpreted. Level widths of s states are of the order of 1 eV as predicted here. See the following papers:

1. C. B. Duke and M. E. Alferieff, J. Chem. Phys. $\underline{46}$, 923 (1967).
2. E. W. Plummer, J. W. Gadzuk, and R. D. Young, Solid State Comm. (in press).
3. J. W. Gadzuk, E. W. Plummer, and R. D. Young, Bull. Am. Phys. Soc. $\underline{14}$, 399 (1969).
4. J. W. Gadzuk, E. W. Plummer and R. D. Young, "Resonance Tunneling Spectroscopy of Adsorbed Atoms on Metal Surfaces: Theory, Experiment," (in preparation).

References

1. J. W. Gadzuk, Surface Sci. $\underline{6}$, 133 (1967); $\underline{6}$, 159 (1967).

2. A. J. Bennett and L. M. Falicov, Phys. Rev. $\underline{151}$, 512 (1966).

3. L. B. Schmidt and R. Gomer, J. Chem. Phys. $\underline{45}$, 1605 (1966).

4. T. B. Grimley, Proc. Phys. Soc. (London) $\underline{90}$, 751 (1967).

5. J. W. Gadzuk, Phys. Rev. $\underline{154}$, 662 (1967).

6. J. W. Gadzuk, Solid State Commun. $\underline{5}$, 743 (1967).

7. J. W. Gadzuk, Solid State and Molecular Theory Group, MIT, QPR $\underline{68}$, 66 (1968); $\underline{69}$, 81 (1968).

8. A. J. Bennett, J. Chem. Phys. $\underline{49}$, 1340 (1968).

9. J. Langer and S. J. Vosko, J. Phys. Chem. Solids $\underline{12}$, 196 (1960).

10. F. Stern, Phys. Rev. Letters $\underline{18}$, 546 (1967).

11. G. F. Koster and J. C. Slater, Phys. Rev. $\underline{95}$, 1167 (1954); $\underline{96}$, 1208 (1954).

12. P. W. Anderson, Phys. Rev. $\underline{124}$, 41 (1961).

13. D. M. Edwards and D. M. Newns, Phys. Letters $\underline{24A}$, 236 (1967).

14. D. Pines, Elementary Excitations in Solids (W. A. Benjamin, Inc., New York, 1964).

15. J. Hubbard, Proc. Roy. Soc. (London) $\underline{A243}$, 336 (1957).

16. J. Hubbard, Phys. Letters (Amsterdam) $\underline{25A}$, 709 (1967).

17. J. W. Gadzuk, Surf. Sci. $\underline{11}$, 465 (1968).

18. J. D. Hagstrum and G. E. Becker, "Band Structure of an Ordered Surface Monolayer," Report on Twenty-seventh Physical Electronics Conference, Massachusetts Institute of Technology, Cambridge, Mass., March 20-22, 1967, pp. 122-124.

19. E. W. Plummer, Ph.D. Thesis, Cornell University, January 1968 (unpublished). In this work it was suggested that a computational oversight in the previous calculation[1] of ΔD resulted in a value that was too low.

20. T. N. Rhodin (personal communication, 1968). Rhodin and Plummer have suggested that ΔE ~ 0.7 eV, rather than the published value of ~ 0.3 eV.

21. R. L. Gerlach and T. N. Rhodin, a paper to appear in the Proceedings of this conference.

22. D. L. Fehrs, Ph.D. Thesis, Massachusetts Institute of Technology, January 1968 (unpublished).

23. T. L. Loucks, Phys. Rev. $\underline{139}$, A1181 (1965).

24. L. Hodges, H. Ehrenreich, and N. D. Lang, Phys. Rev. 152, 505 (1966).

25. Handbook of Chemistry and Physics (Chemical Rubber Publishing Co.,
     Cleveland, Ohio, 42nd edition, 1961).

26. F. Herman and S. Skillman, Atomic Structure Calculations (Prentice-Hall,
     Inc., Englewood Cliffs, N. J. , 1963).

27. T. J. Lee, Ph.D. Thesis, University of Southampton (England), January
     1967 (unpublished).

Table 1. Input data characterizing the substrate material.

Property / Material	Lattice Constant	ϕ_e (eV)	E_F (eV)	r_s
W (bcc)	3.16 Å		7.5 [23]	~2.5
(100)		4.82 [3]		
(110)		5.85		
(111)		4.41		
Ni (fcc)	3.52 Å		8.15 [24]	~2.5
(100)		4.75 [21]		
(110)		4.2		
(111)		4.68		
Ta (bcc)	3.38 A			
(110)		4.73 [22]	~8.0	~2.5

Table 2. Input data characterizing the alkali adsorbate.

Property \ Material	V_i (eV) [25]	r_i (Å) [25]	a (Å)$^{-1}$ [26]
Na	5.12	.95	1.53
K	4.32	1.33	1.16
Cs	3.89	1.69	.99

Table 3. Results of billiard-ball calculations for various combinations of substrate material, alkali adsorbate, and crystallographic face.

Face \ Alkali S(Å)			Na	K	Cs
[W] bcc $\left(r_i = \frac{\sqrt{3}}{4}d\right)$	100	$\left((r_i+r_a)^2 - 8/3r_i^2\right)^{1/2}$		2.19	2.62
	110	$\left((r_i+r_a)^2 - 3/2r_i^2\right)^{1/2}$		2.26	2.60
	111	$\left((r_i+r_a)^2 - 32/9r_i^2\right)^{1/2}$		2.63	3.0
[Ni] $\left(r_i = \frac{\sqrt{2}}{4}d\right)$	100	$\left((r_i+r_a)^2 - 2r_i^2\right)^{1/2}$	1.31	1.88	2.34
	110	$\left((r_i+r_a)^2 - 3r_i^2\right)^{1/2}$	.431	1.4	1.99
	111	$\left((r_i+r_a)^2 - 1.33r_i^2\right)^{1/2}$	1.66	2.14	2.56
[Ta] bcc	110		2.15	2.54	2.94

Table 4. Calculated values of Γ, Z_{eff}, and M; experimental values of M; for various combinations of substrate material, alkali adsorbate, and crystallographic face.

Alkali - Substrate	Γ (eV)	Z_{eff}	M Theoretical (Debye units)	M Experimental (Debye units)
Cs - W_{100}	1.37	.8	6.29	~ 8.5 [27]
Cs - W_{110}	1.13	.88	9.12	~ 8.7
Cs - W_{111}	1.63	.76	4.52	
K - W_{100}	1.65	.76	4.19	5.75 [3]
K - W_{110}	1.33	.84	6.77	7.85
K - W_{111}	2.47	.70	2.51	4.75
Na - Ni_{100}	1.52	.69	3.44	3.6 [21]
Na - Ni_{110}	1.90	.65	2.07	1.85
Na - Ni_{111}	1.20	.69	4.26	3.7
Na - Ni_{110}	1.90	.65	2.07	1.85 [21]
K - Ni_{110}	1.61	.71	3.61	2.92
Cs - Ni_{110}	1.34	.76	5.51	4.05
Na - Ta_{110}	1.23	.70	4.26	4.43 [22]
K - Ta_{110}	1.21	.77	6.31	6.40
Cs- Ta_{110}	1.03	.82	8.72	8.58

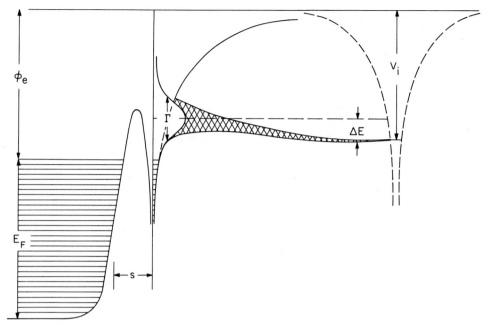

Fig. 1. Energy-level diagram revelant to a metal surface impurity problem.

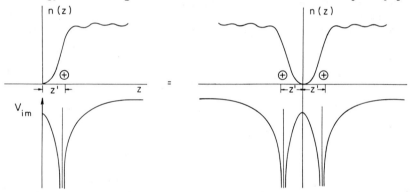

Fig. 2. Model for describing the dielectric response of an electron gas with
a surface in terms of the response of an infinite gas with an impurity slightly inside the gas.

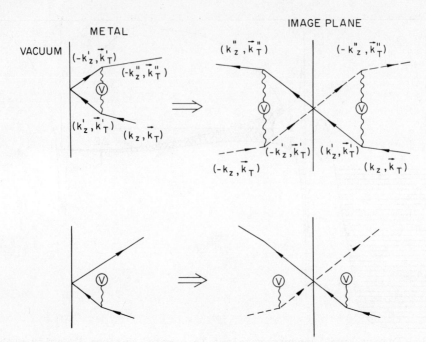

Fig. 3. Schematic diagrams showing the analogy between electrons scattering
from an impurity in the surface region of specularly reflecting surface and electrons scattering from imaged impurity potentials in the
infinite electron gas.

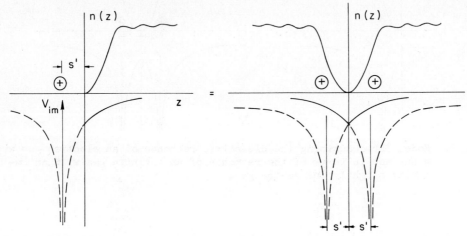

Fig. 4. Model for describing dielectric response of an electron gas with a
surface in terms of the response of an infinite gas with an impurity
slightly outside the gas.

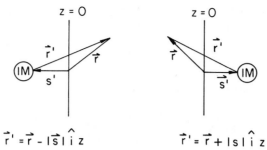

$$\vec{r}' = \vec{r} - |\vec{s}| \hat{i} z \qquad\qquad \vec{r}' = \vec{r} + |s| \hat{i} z$$

Fig. 5. Coordinates for performing off-center integrations.

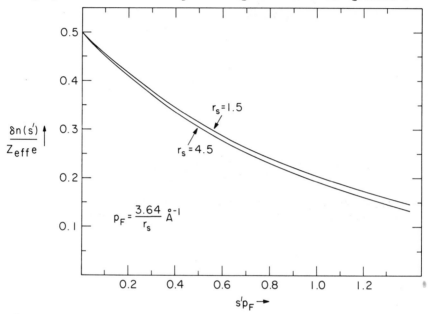

Fig. 6. Total volume screening charge as a function of nondimensional impurity-surface separation with metal electron density treated parametrically.

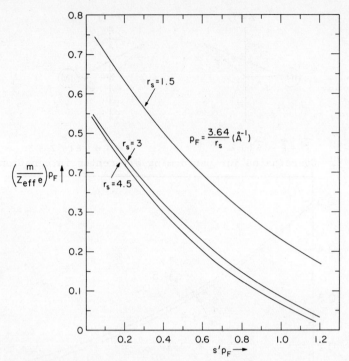

Fig. 7. Dipole moment of the impurity plus volume polarization screening charge only as a function of nondimensional separation.

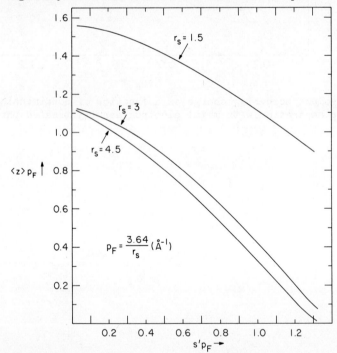

Fig. 8. Average position of volume screening charge as a function of nondimensional separation.

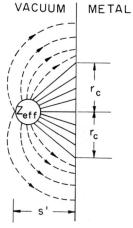

VACUUM | METAL

Fig. 9. Schematic impurity electric field lines showing the division between
volume and surface polarization.

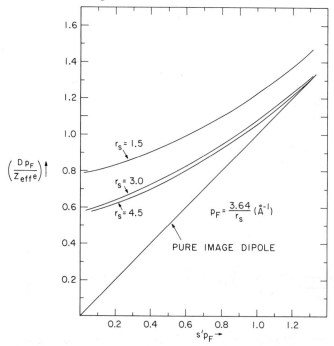

$$\left(\frac{D\,p_F}{Z_{eff}e}\right)$$

$r_s = 1.5$

$r_s = 3.0$

$r_s = 4.5$

$p_F = \dfrac{3.64}{r_s}\ (\text{Å}^{-1})$

PURE IMAGE DIPOLE

$s'p_F \rightarrow$

Fig. 10. Total nondimensional dipole length as a function of nondimensional
separation.

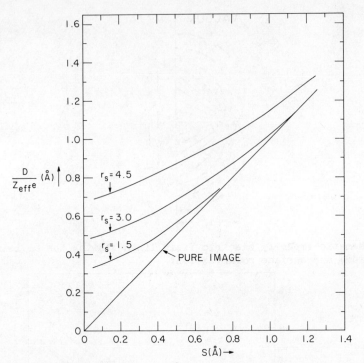

Fig. 11. Total dipole length ($\overset{\circ}{A}$) as a function of separation ($\overset{\circ}{A}$).

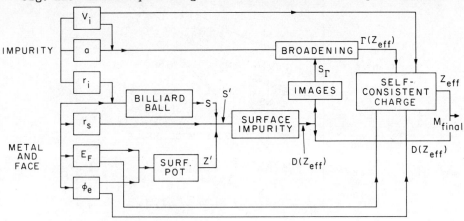

Fig. 12. Flow diagram describing calculational procedure.

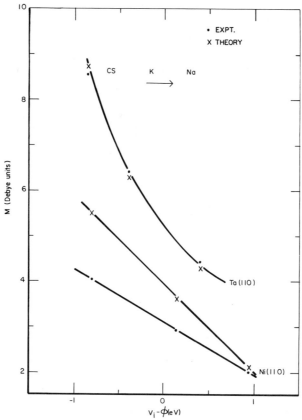

Fig. 13. Dipole moment versus $V_i-\phi_e$ for alkalis on a single face.

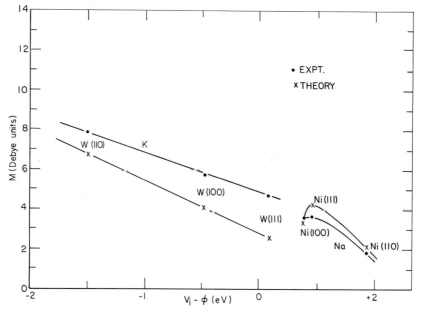

Fig. 14. Dipole moment versus $V_i-\phi$ for an alkali on different faces.

STUDIES OF GAS-SURFACE SCATTERING USING
TIME-OF-FLIGHT TECHNIQUES

K. Jakus and F. C. Hurlbut

University of California
Berkeley, California

INTRODUCTION

This paper constitutes a preliminary report on gas-surface interaction studies conducted by nozzle beam techniques in the newly constructed rarefied gas wind tunnel of the Division of Aeronautical Sciences at Berkeley. Nearly monoenergetic molecular beams were gated by a disk modulator and directed to the specimen surface. Gated time-of-arrival detection and signal integration were used to obtain time-of-arrival distributions of the scattered molecules as a function of the angular parameters, incident beam energies and other factors of the experiment. Nozzle beams of argon, nitrogen and helium and surfaces of single crystal and polycrystalline aluminum, cleaved crystals of lithium fluoride and single crystal silver were employed in this work. However, only that portion dealing with the reflection of argon from silver will be considered here, since the remainder of the data at the time of this writing has not been reduced.[1,2]

Our long range objectives have made desirable the study of gas-surface interaction for neutral incident particles having energies in excess of 0.1 eV and extending to such energies as may conveniently be reached. The advantages of energy and intensity afforded by nozzle beams in this circumstance are well known.

Although Kantrowitz and Grey[3] suggested the nozzle beam technique and established an elementary theory of operation in 1951 it has only been within the past ten years that we have seen intensive development of nozzle beam systems. Parker's[4] more precise theory has shown the general correctness of the original suggestion and the work of many experimenters[5,6] has served to verify, at least in a qualitative sense, that intense, well collimated and nearly monoenergetic beams may indeed be produced by nozzle beam techniques. However, certain complexities have recently appeared, chiefly having to do with the gas dynamic expansion preceeding the nozzle beam, which have as their practical effect the limitations of intensity and mono-chromatisity of the beam. These limitations relate to the effects of viscosity, of skimmer interaction and to the formation of polymers of the expanding gas and are dealt with in part in a recent study by Bossel.[7] Cognizance of these matters has been taken in the present study, as will be discussed in the next section.

EXPERIMENTAL APPARATUS

The gas particle beams were produced by skimming from free-jet expansions at positions on the axis giving low thermal energy spread and high particle flux at the collimator. As indicated in Figure 1, the entire apparatus was housed within the test section of the rarefied gas wind tunnel, the pump of that facility being used to pump the external flow. The tunnel and its capability in relation to the production of large scale free jet expansions is described at more length by Bossel. Stagnation heating to 1400°C was achieved through the resistance heating of a thin walled tantalum tube of 1 cm diameter. A 1 mm diameter orifice in the wall of the tube served as a nozzle for the free-jet. The source tube was supported on a traverse mechanism permitting its positioning at various distances from 38 to 65 nozzle diameters upstream from the skimmer, depending on temperature, molecular species and operating condition. The skimmer cone was constructed of stainless steel with 25° total interior included angle and 30° exterior included angle and having a 0.028 cm diameter hole at its tip. The skimmer cone was an extension of a conical support body extending approximately 25 cm from the cylindrical surface of the collimator chamber. Modulation was accomplished within the region of this chamber by a mechanical chopper of standard configuration located 35 cm upstream of the target surface.

Beam macroscopic velocities and parallel temperature distributions were determined through direct observation of the time-of-arrival signal at the detector, the target being withdrawn during this calibration period.

Collimation was arranged so that the target was illuminated by a beam of ∿ 1 cm × 0.3 cm cross section. The target surfaces, measuring 1 cm × 1 cm by 0.2 cm thick, or 1 cm × 1.5 cm oval by 0.2 cm thick, as in the case of silver, were mounted on a holder which permitted both radiative heating and heating by electron bombardment. The holder could be rotated through ± 90° as measured from the direction of the incident beam. The reflected beam passed into the ionizing volume of a flow through detector at a distance of 5 cm. The ionizer window was 1 cm × 0.4 cm by ∿ 0.3 cm deep and subtended a solid angle of 1.6×10^{-2} steradian as viewed from the surface. The detector could be rotated in a plane into any position about the target except for directions within ± 30° of the incident beam axis.

Great care was taken with the design of the signal train to insure adequate bandwidth and maximum freedom from systematic noise sources. The overall bandwidth was from D.C. to 90 kilo hertz at the -3 db points, sufficient to transmit all signals in this experiment without significant distortion. This matter has been discussed at more length by Amend.[8] The ion current was amplified by electron multiplier, high speed electrometer and decade voltage amplifier. The resultant raw signal was next introduced into a PAR 100 channel signal averager. Close attention was also given to the signal timing to minimize systematic errors from this area. The estimated uncertainty in timing was ± 2 microseconds. The processed signal was recorded on an X-Y recorder and observed on an oscilloscope screen. For more detail relative to apparatus and experimental procedures the reader is referred to Jakus, Reference 1.

The Incident Beam

Pressures in the stagnation region of up to 1 atmosphere at source wall temperatures of 1400°C were employed in obtaining distributions of beam intensity vs. nozzle skimmer separation distance. When the nozzle Reynolds numbers are sufficiently high, families of performance curves may be obtained resembling those of Figure 2. These show regions of high skimmer interference, regions of terminal Mach disk interference and also regions of satisfactory skimming.

Figure 2 was obtained in one such calibration plot for a source wall temperature of 1173°K. The Mach number has been shown as a function of the dimensionless distance x/D from the nozzle. Detailed studies of beam performance have been made by Bossel in which the relevant beam production criteria are discussed in detail and Amend has shown that near the peaks of these curves the Mach number is essentially as predicted. On the basis of these calibration studies, beam operating conditions and nozzle skimmer conditions were selected to give an essentially monoenergetic beam of good intensity. Beam fluxes ranged from 10^{18} molecules/steradian sec to 3×10^{18} molecules/steradian sec, corresponding to 1.83×10^{14} molecules/cm^2 sec and 5.5×10^{14} molecules/cm^2 sec at the target surface. Energies of the argon beams were found to lie in the range 0.07 to 0.21 electron volts. The mechanical chopper wheel contained 10 slots of 9.1° each. The wheel was rotated at $\sim$ 100 rps giving a trapezoidal shutter function of 350 microseconds width at its base. The time of arrival signal at the surface was found to reproduce the shutter function except for a modest rounding and slight broadening due to the thermal spread. Typically the beam thermal energy spread was of the order of 10^{-3} electron volts.

Surface Preparation

There have been a relatively large number of gas-surface interaction studies using epitaxially grown silver crystals in the recent past[9] and it appeared worthwhile to study the silver single crystal as cut from a bole for comparison. There are certain other advantages of the silver crystal; its ready availability,* the large body of information relative to the cutting, cleaning and polishing of the crystal and to the condition of the resultant surface, and finally, the relative ease of maintaining clean surfaces at slightly elevated temperatures. In the case of our present equipment the beam-off test section pressure was of the order of 10^{-8} torr or slightly lower, while the beam-on pressure was about 5×10^{-8} torr. The beam flux itself was about 2 orders of magnitude larger than flux due to background gas. Thus the principal source of contaminant was the beam gas itself. No special measures were taken in respect to the purity of the source gas which was led through gas service regulators from tanks and then directly to the stagnation region.

The single crystal was spark cut from the bole, mechanically polished, and then given a chemical polish in situ before closing the vacuum system. The target was cleaned by argon ion bombardment and then annealed and degassed at 773° Kelvin. However, some studies were performed with the target at room temperature prior to the initial degassing. Clearly the degree of resident contamination remains a matter of conjecture.

RESULTS AND DISCUSSION

Observations were made under steady conditions of beam energy and surface temperature and at a fixed angle of incidence. For each combination of these parameters a series of time-of-flight traces was obtained, there being

---

* We are indebted to Prof. G. A. Somorjai of the Department of Chemistry for our specimen.

a time-of-flight curve for each 15° of detector angle, in many instances on both the right and left sides of the incident beam axis. Some hundreds of time of flight curves were obtained in the course of the data taking. A representative family of data curves is shown in Figure 3 for an incident angle of 60°, surface temperature of 773° Kelvin and beam energy of 0.21 eV. The direct beam as seen at the surface is superimposed in the correct time relationship.

Results of these observations are presented in the next series of figures. In each instance the ratio of the amplitude of the reflected signal to the incident beam signal is plotted as a function of angle as measured from the surface normal. The amplitude of reflected signal can be seen to follow closely the cosine distribution in Figure 4 (beam 296° Kelvin, surface 296° Kelvin, degassed at 473° Kelvin), while in Figures 5, 6, and 7 an increasing degree of specularity may be seen. In Figure 5 with the surface temperature increased to 473° Kelvin and with increased beam energy we see a marked departure from the cosine distribution. In Figures 6 and 7, both for surface temperatures of 773° Kelvin, we also see marked lobality in the reflection pattern, with the maximum of the lobe moving toward the surface tangent, from above the specular direction to a position below the specular direction as the energy of the beam is increased. In Figure 8 we see continuing specularity as the beam energy has been diminished and the surface cooled to 473° Kelvin.

At this point a difficulty, already present in the reduction of the data for amplitude, becomes more apparent when we look at the arrival times in relation to the parent velocity distribution. The detector is of the flow-through configuration and careful attention was paid to the elimination of

interreflection and gas-kinetic relaxation of those particles passing directly
through the ionizer volume. However, a compromise had to be reached regarding
the interreflection of particles from the bulk of the flux emitted from the
target surface. In the present case a small fraction of particles which strike
other areas of the detector structure can make a first-bounce passage through
the ionizer volume at a delayed time. While the fraction is small, the cumulative effect is large enough so that a comprehensive analysis and correction
must be made. Fortunately, it has been possible to make a complete reconcili-
ation of these effects and to predict the correct surface temperature from
observations taken along the surface normal in the presumably diffusely scattered
case, Figure 9. The same technique was used to predict the arrival signal for
a lobular reflection case with results as in Figure 10. In order to apply
corrections to large numbers of traces a fiducial point has been selected, the
half amplitude point on the arrival time distributions, and the correction
applied to this point in all cases. The rising portion of the arrival pattern
is much easier to correct with confidence since it is composed primarily of
the fast and single bounce components while the falling portion is composed of
the slow particles and suffers from some admixture of second bounce particles.

The most probable speeds of emitted particles are roughly represented
by a proportionality to the inverse of the corrected arrival times of the half-
amplitude points and may be represented by the quantity L/t. In Figure 4 we
see that $1/t$ is essentially constant for all angular directions and has a
value corresponding with emission at the surface temperature. In Figures 5
the inverse arrival times may be seen to be constant over a range of reflection
angles near the surface normal but rise to a broad peak (higher velocities) and
then fall again toward the surface temperature value. Thus at the higher beam

temperature the appearance of the peak indicates a reduction in the effectiveness of the energy transfer in some regions of the reflection pattern.

At the higher surface temperature and lower beam energy, Figure 6,
the peak in L/t is significantly shifted away from the surface tangent
toward the specular angle. Figure 7 shows a sharp rise in L/t corresponding
more or less in location with that of the amplitude peak and then a drop as
the surface tangent is approached. Values at the final observation angle,
$\sim 75°$, corresponding with an effective emission temperature close to that of
the surface but somewhat above it. As the surface temperature and the beam
energy are reduced, Figure 8, the angular variation in L/t decreases
considerably.

In many regards these data exhibit systematic behavior common to
other observations of the reflections of argon from metal surfaces.[10,11]
These may be summarized in part for the distributions in reflected particle
density, as follows:

1. Lobal scattering is observed when the crystal surface
   has been degassed.

2. As the incident beam energy is increased the scattering
   patterns become narrowed and their maxima move toward
   the surface tangent.

Work remains to be done in the interpretation of the time of arrival
distribution since, for example, they have not been examined for a possible
systematic parameterization in terms of the Nocilla re-emission model.[12]
However, certain trends are observed. The re-emitted velocities are found to
be representative of the surface temperatures near the surface normal but
assume higher values in the neighborhood of the lobe maxima except for incident

angles near the surface tangent. At the higher surface temperature the velocity maxima lie distinctly farther from the surface tangent than at the lower temperature. Since these data are alone in the field at this time, other and independent observations will be required before the maxima in most probable velocity can be accepted as characteristic of a class of inter-actions. It should be noted that these observations tend to run counter to recent numerical studies by Goodman[13] using a three dimensional hard sphere model. However, that model would be expected to correspond more closely with reality at higher interaction energies.

ACKNOWLEDGMENT

        The authors wish to express their appreciation to the National Science Foundation for the support of this work.

REFERENCES

1. Jakus, K., "Experimental Determination of Velocity Distribution of Gas Molecules Scattered from Surfaces," Ph.D. Thesis, Univ. of Calif., Berkeley, in preparation.

2. Jakus, K. and F. C. Hurlbut, "Gas Surface Scattering Studies using Nozzle Beams and Time-of-Flight Techniques," to be published in Rarefied Gas Dynamics, Proceedings of the 6th International Symposium.

3. Kantrowitz, A. and J. Grey, Rev. Sci. Inst. 22, 328 (1951).

4. Parker, H. M., A. R. Kuhlthau, R. N. Zapata, and J. E. Scott, Jr., in Rarefied Gas Dynamics, F. M. Devienne, Ed., Pergamon Press, New York (1960).

5. Anderson, J. B., R. P. Andres and J. B. Fenn, in <u>Advanced Atomic and Molecular Physics</u>, D. R. Bates, Ed., Academic Press, New York, Vol. 1, p. 345 (1965).

6. French, J. P., AIAA J. <u>3</u>, 994 (1965).

7. Bossel, U., "Investigations of Skimmer Interaction Influences on the Production of Aerodynamically Intensified Molecular Beams," Ph.D. Thesis, Univ. of Calif., Berkeley, in preparation; also to be published as Univ. of Calif., Berkeley Aero. Sci. Lab. Report AS-68-6 (1968).

8. Amend, W., "Application of Statistical Moments in the Reduction of Super-sonic-Molecular-Beam Data," Ph.D. Thesis, Univ. of Calif., Berkeley, (1968); also published as Univ. of Calif., Berkeley Aero Sci. Lab. Report AS-68-4 (1968).

9. Smith, J. N., Jr. and H. Saltsburg, in <u>The Fundamentals of Gas-Surface Interactions</u>, H. Saltsburg, T. N. Smith, Jr. and M. Rogers, Eds., Academic Press, New York (1967), pp 370-391.

10. Hinchin, J. J. and E. S. Malloy, in <u>The Fundamentals of Gas-Surface Interactions</u>, H. Saltsburg, T. N. Smith, Jr. and M. Rogers, Eds., Academic Press, New York (1967), pp 448-460.

11. Hinchin, J. J., E. S. Malloy and J. B. Carroll, "Scattering of Thermal Energy Gas Beams by Metallic Surfaces, II," United Aircraft Report No. F910439-7 (1967).

12. Hurlbut, F. C. and F. S. Sherman, Phys. Fluids <u>11</u>, 486 (1968).

13. Goodman, F. O., Surface Science <u>7</u>, 391 (1967).

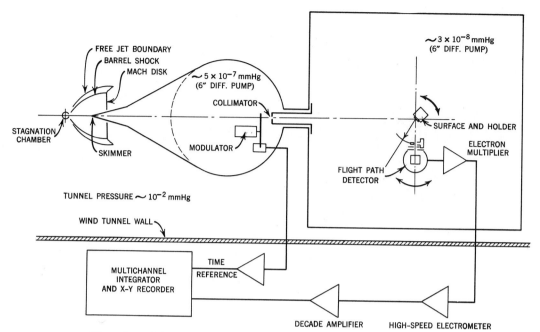

Fig. 1. Schematic diagram of surface scattering system.

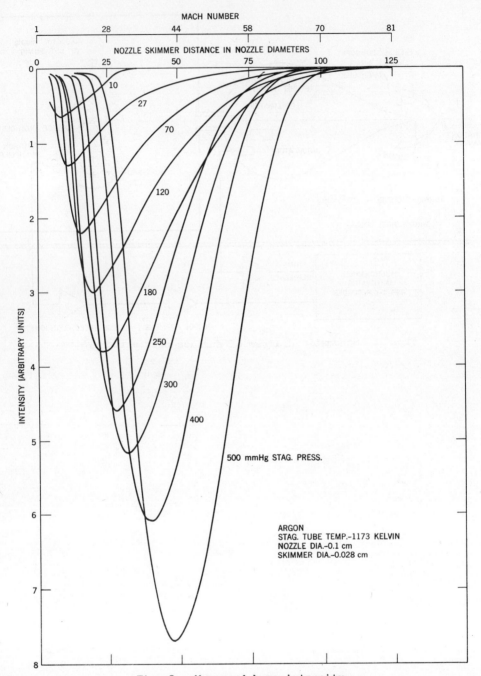

Fig. 2. Measured beam intensity.

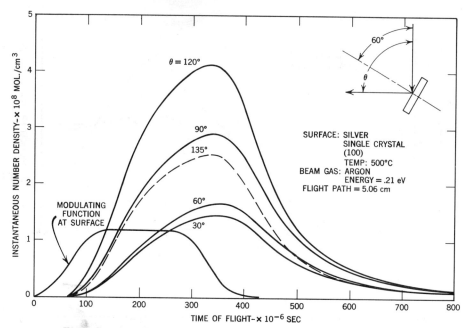

Fig. 3. Time-of-flight of reflected molecular beam.

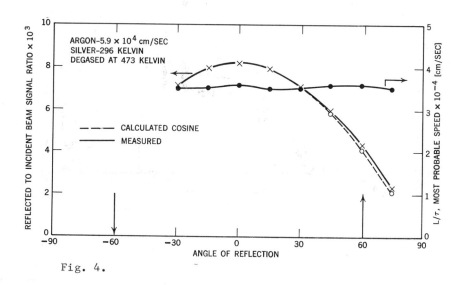

Fig. 4.

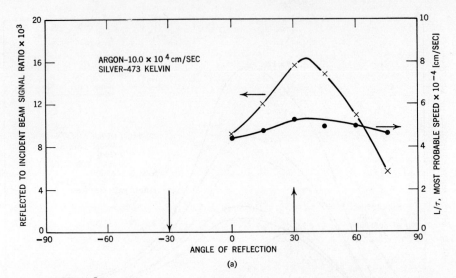

Fig. 5.a

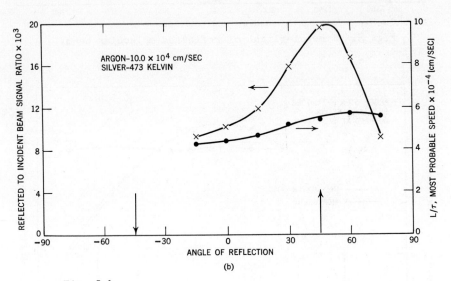

Fig. 5.b

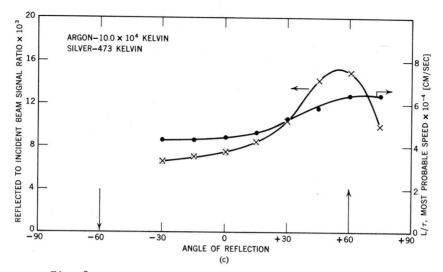

Fig. 5.c

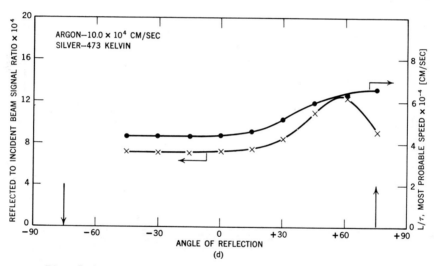

Fig. 5.d

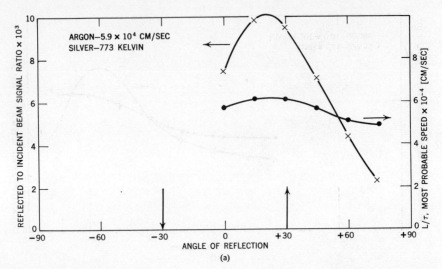

Fig. 6.a

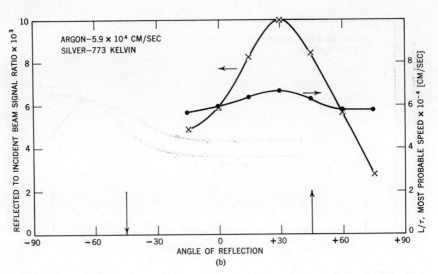

Fig. 6.b

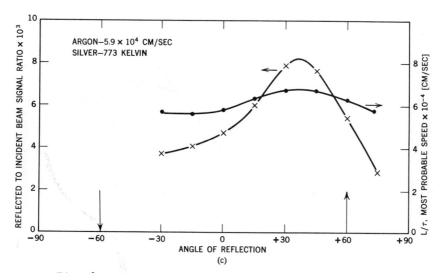

Fig. 6.c

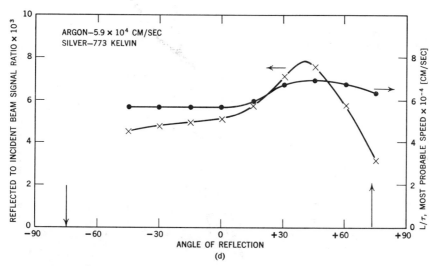

Fig. 6.d

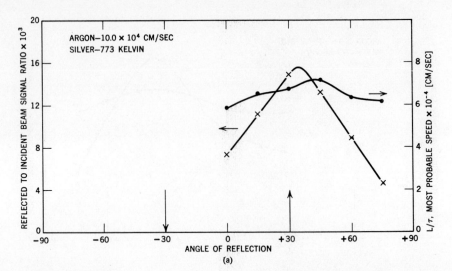

Fig. 7.a

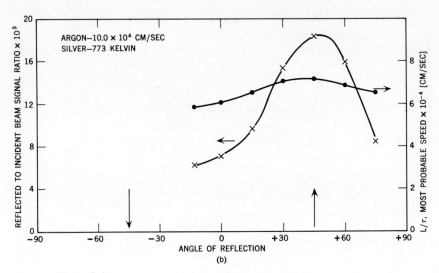

Fig. 7.b

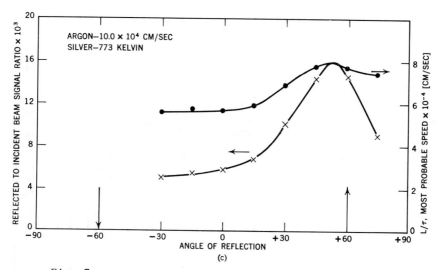

Fig. 7.c

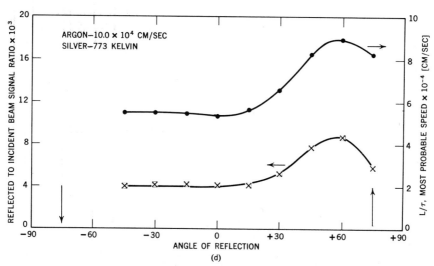

Fig. 7.d

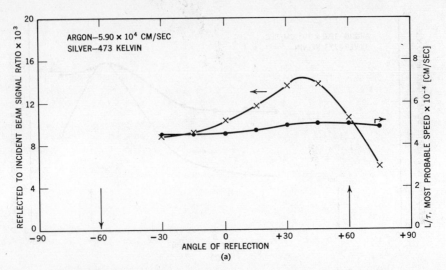

Fig. 8.a

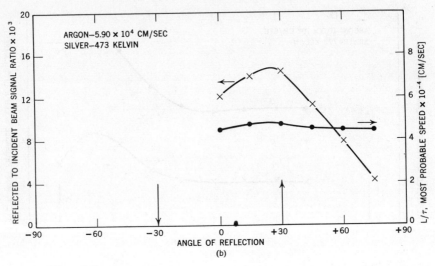

Fig. 8.b

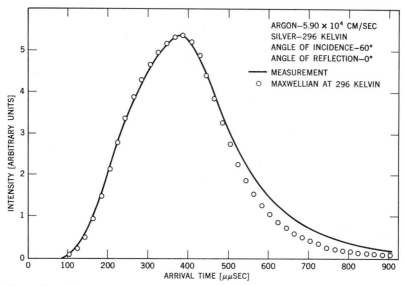

Fig. 9. Time-of-arrival signal of diffusely scattered atoms.

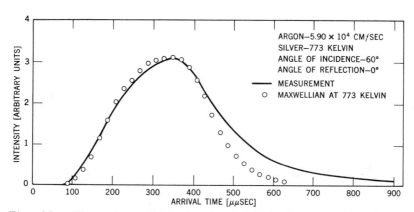

Fig. 10. Time-of-arrival signal of lobularly scattered atoms.

ANALYSIS OF GAS-SOLID SURFACE KINETIC MODELS USING LOCK-IN DETECTION OF MODULATED MOLECULAR BEAMS

D. R. Olander

Inorganic Materials Research Division, Lawrence Radiation Laboratory
Department of Nuclear Engineering
University of California, Berkeley, California

INTRODUCTION

Heterogeneous chemical kinetic studies by modulated molecular beam mass spectrometry have been made possible by the development within the past decade of three important experimental tools: high sensitivity, compact mass spectrometers; ultra-high vacuum techniques and clean, high speed vacuum pumps; and sources for the generation of intense molecular beams. The reactant gas interacts with the solid surface as a collimated beam of particles rather than as a randomly directed flux. The products of the interaction are monitored by a mass spectrometer located in the vacuum system. The molecular beam is chopped or modulated to permit detection by phase-sensitive amplification, which greatly improves the signal-to-noise ratio.

A major advantage of the molecular beam-mass spectrometric method of investigating heterogeneous chemical kinetics is that very reactive free radical intermediates of the reaction can be analyzed as readily as stable products. Knowledge of the nature and concentration of these intermediates is of considerable assistance in elucidating reaction mechanisms. Sampling by molecular beams in a high vacuum insures that the observed species represent primary products of the surface reaction, and not stable species resulting from wall or gas phase collisions prior to analysis.

Experiments of this type determine the identity of the primary products of the reaction, the fraction of the incident reactant molecules which ultimately return to the gas as a particular reaction product, and the average residence time of the product species on the solid surface. The controllable experimental parameters are: the

intensity of the molecular beam (or the equivalent pressure of the
gaseous reactant); the temperature of the molecular beam; the temperature of the solid surface; the composition of the beam, and the modulation frequency. The dissociation of hydrogen on tungsten and the
reaction of chlorine with nickel have been studied in this manner.[1]

The apparatus shown in Fig. 1 has been used to investigate the
hydrogen-tantalum system[2] and is typical of systems which have been
employed or are being constructed for studies of this type. The system
consists of two vacuum chambers. The chamber on the left of Fig. 1
contains the molecular beam source, the efflux from which is mechanically
modulated by a small rotating toothed disk. The small fraction of the
gas effusing from the source which passes through the small orifice
connecting the two chambers constitutes the modulated molecular beam.
The solid target and the mass spectrometer are shown in the larger
vacuum chamber on the right of Fig. 1.

The heart of the detection system is the phase sensitive or lock-in
amplifier. The output signal from the mass spectrometer first goes
through a preamplifier, then an intermediate amplifier where the entire
spectrum of frequencies is amplified, and then through a signal-tuned
amplifier which limits the signal to a narrow bandwidth centered on
the frequency of the chopped primary molecular beam. The signal is
then mixed with the first harmonic of the reference signal in a synchronous demodulator to produce a rectified ac signal which passes
through an integrating circuit to produce a dc output. When employed
in this manner, the lock-in amplifier produces two independent pieces
of information: the amplitude of the output signal and its phase
difference compared to some selected signal.

The absolute magnitude of the amplitude of the signal is generally not of interest. To relate the amplitude directly to the absolute rate of emission from the surface requires knowledge of the fraction of the emission from the surface intercepted by the electron beam of the mass spectrometer, ionization cross sections and extraction and transmission efficiencies of the mass spectrometer, and the gain of the electron multiplier. At best, such estimates are accurate to no more than a factor of 2-3.[3] However, the percentage change in the output signal amplitude at a particular mass number as experimental conditions are varied can provide quantitative kinetic information. This relative amplitude contains two factors; the "competition factor" represents an actual change in the average rate of emission of a particular species from the surface due to effects such as competitive reactions or diffusional processes. The second factor is determined by the response of the lock-in amplifier to the shape and amplitude of the ac portion of the modulated signal waveform. This "demodulation factor" is directly related to the ratio of the average residence time on the surface to the chopping time, and hence to the phase lag of the signal. The amplitude of the output signal is most conveniently measured relative to the signal when the surface interaction is sufficiently simple to be considered completely interpretable. In the case of simple dissociation of diatomic gases on metals, for example, the reference situation might be the output signal when the target is at temperatures high enough to instantaneously re-evaporate as atoms all of the molecular gas which chemisorbs.

The phase shift is most conveniently measured with respect to experimental conditions under which the residence time of the molecules on the surface is small compared to the modulation frequency (e.g., residence times on the order of 10^{-4} sec or less). For noncondensible gases, this could be a room temperature beam scattered from a room temperature target. Provided that the transit time between the target and the mass spectrometer is small, the shift in phase of the output signal of a product species is indicative of reaction times which are of the same order of magnitude as the modulation time. Generally, phase shift data provide information on the absolute magnitude of the rate constants of the elementary surface processes, whereas relative amplitude data (after correction for the demodulation factor) provide information on the ratio of rate constants due to competing surface reactions. In non-modulated (dc) experiments such as those of McKinley,[4] only the latter type of information is obtainable.

Yamamoto and Stickney[5] analyzed the response of the lock-in detector to molecular beams which undergo simple scattering at the solid target. Although the process of slow first order desorption was considered briefly, the emphasis was on the velocity distribution of the scattered molecules. Schwarz and Madix[6] have investigated a simple first order surface reaction formally similar to slow first order desorption. The effect of reactant and product transit times were considered in detail. In this study, we consider surface processes more complex than scattering or simple first order desorption, but assume molecular flight times from the chopper to the target and from the target to the mass spectrometer to be small compared to the modulation time. All species are assumed to be emitted from the surface at the temperature of the surface and in a diffuse manner. Since the ionization efficiency of the mass spectrometer is proportional to the inverse of the mean speed of the molecules, the product of the square root of the surface temperature, T_T, and the instantaneous output signal from the mass spectrometer, $S_i(t)$, is assumed proportional to the instantaneous rate of emission of a species from the surface $R_i^E(t)$:

$$\sqrt{T_T} \; S_i(T) = \alpha_i R_i^E(t) \tag{1}$$

The constant of proportionality α_i depends upon the fraction of the surface emission intercepted by the electron beam in the ionizer, the ionization cross section, extraction and transmission efficiencies of the mass analyzer, and electron multiplier sensitivity for species i. Intercomparison of different species requires calibration of the mass spectrometer, as indicated by Smith and Fite.[1]

I. Signal Processing by the Lock-in Amplifier

The narrow band first amplification stage of the lock-in detector passes only the fundamental component of $S(t)$. Since the process is periodic with an angular frequency ω, the choice of zero time is of no consequence, and the fundamental model can be expressed in terms of the sine function

$$\mathcal{F}\{S(t)\} \sim \beta \sin(\omega t + \delta) = A \cos(\omega t) + B \sin(\omega t) \tag{2}$$

where

$$\tan \delta = A/B$$
$$\beta = \sqrt{A^2 + B^2} \tag{3}$$

The action of the synchronous demodulator of the lock-in detector is equivalent to multiplication of $\mathcal{F}\{S(t)\}$ by the fundamental mode of a reference signal, $\sin(\omega t + \phi')$, where ϕ' denotes the adjustable phase

angle. The RC filter which constitutes the final stage of the lock-in
detector averages the signal from the demodulator. The dc output of the
detector as a function of phase setting can be written as

$$\mathcal{Q}\,(\phi') \sim \frac{\omega}{2\pi} \int_0^{2\pi/\omega} \beta\,\sin(\omega t + \delta)\,\sin(\omega t + \phi')\,dt \sim \beta\,\cos(\delta - \phi')$$

When $\mathcal{Q}\,(\phi')$ is maximized by setting $d\mathcal{Q}/d\phi' = 0$, the optimum phase angle
is denoted by ϕ, which is seen to be equal to δ. The amplitude at the
optimum phase angle is denoted by $\mathcal{Q}$ (the argument ϕ has been omitted)
and is proportional to β.

In the following analysis, it will be convenient to use the parameters
A and B instead of β and δ. The phase shift and dc amplitude of the
lock-in detector are related to A and B by

$$\tan \phi = A/B \tag{4}$$

$$\mathcal{Q} \sim \sqrt{A^2 + B^2} \tag{5}$$

The parameters A and B are analogous to the K and I functions used by
Yamamoto and Stickney,[5] Schwarz and Madix[6] and Harrison, Hummer and
Fite.[7] In general, these parameters are functions of the modulation
frequency, the rate constants describing the surface processes, and
parameters describing transit time effects if these are important. Since
transit time effects have been neglected in this study, Eq. (1) indicates
that it suffices to compute $\mathcal{F}\{R_1^E(t)\}$ or the fundamental mode of any
surface concentration on which the emission rate depends rather than
$\mathcal{F}\{S(t)\}$. In the last part of this study methods for generalizing the
results to situations in which molecular flight times are not negligible
will be considered briefly.

General Formulation

The periodic nature of the surface emission rate is due to the regular interruption of the primary molecular beam of reactant gas by a mechanical chopper. If the molecular transit time between the molecular beam source and the detector is small compared to the chopping time and if the beam diameter at the chopper location is small compared to the length of the arc that the beam axis traces on a chopper blade, the primary modulated beam can be represented by a square wave. Figure 2 shows an oscilloscope trace of a 145 Hz primary beam of molecular oxygen measured by a mass spectrometer 4.5 cm from the effusion source. The amplitude of the primary beam striking the solid, I_T° molecules/cm^2-sec, is proportional to the height of the square wave in Fig. 2. The modulated beam intensity is analogous to the reactant gas pressure in conventional rate experiments.

For molecular beams of non-condensible species such as H_2, O_2, or CO, the background gas in the vacuum chamber containing the target consists primarily of the same species as the beam itself. In addition to the modulated molecular beam, the target is also subject to a steady bombardment of reactant from the background gas. This contribution to the total reactant gas impingement rate is denoted by I_{bkg} molecules/cm^2-sec., and can be calculated from pressure measurements and simple kinetic theory formulae. The total rate at which each cm^2 of target is struck by reactant molecules is

$$I_T(t) = I_T^\circ [f^\circ(t) + 1/\rho] \qquad (6)$$

where $f^\circ(t)$ is a square wave of unit amplitude at the modulation frequency and ρ is the beam-to-background impingement rate ratio I_T° / I_{bkg}.

For first order surface processes, the background contribution in Eq. (6) does not affect the interpretation of the lock-in detector data, since it represents a dc signal to which the lock-in system does not respond. For non-linear surface processes, however, the background impingement increases the supply of adsorbed atoms on the surface with which the molecular beam can react. In this situation, the background contributes in an intimate manner to the product emission rate.

The flux of a species from the surface is related in a simple way to the concentration of that species or its precursor on the surface. The response of the lock-in system can be computed in terms of the time variation of an adsorbed species on the surface. A sketch of the change of the surface concentration of an adsorbed species during a complete modulation cycle is shown in Fig. 3. Here, t_c denotes the chopping time, which is the reciprocal of twice the modulation frequency or π/ω. During the "on" portion of the cycle $(-t_c < t < 0)$ the surface concentration grows due to the supply of reactants from the beam. During the "off" portion of the cycle $(0 < t < t_c)$ the surface population diminishes due to removal of the adsorbed layer by direct evaporation, reaction, or other loss processes. These mechanisms are of course operative during the "on" part of the cycle as well, but here the supply rate from the beam is greater than the removal rates. During the "on" period, the driving function $f^\circ(t)$ is constant at unity, and during the "off" portion of the cycle, the driving function is zero. The time variation of an adsorbed species can be calculated by solving the appropriate differential equations representing surface mass balances for all of the

adsorbed species. The resulting function is then Fourier analyzed and
the coefficients of the fundamental mode identified. In order to relate
the phase shift ϕ_i and the amplitude α_i to the periodic behavior of
the surface concentration, a specific model of the surface processes
must be chosen.

I. Slow, Competitive First and Second Order Surface Reactions

This process represents the atomization of a diatomic gas on
(typically) a metal surface. The elementary reactions can be written
as

$$X_2(g) \xrightarrow{} 2X(a)$$
$$X(a) \xrightarrow{k_1^E} X(g)$$
$$2X(a) \xrightarrow{k_2^E} X_2(g)$$

The first reaction represents dissociative chemisorption, which proceeds at a rate equal to the product of the sticking probability η_2
and the impingement rate given by Eq. (6). The second reaction represents direct evaporation of the adsorbed atoms, and the last
reaction represents bimolecular surface reconstitution and evaporation
of molecules. If the surface concentration of X is denoted by n, a
mass balance on the surface yields:

$$\frac{dn}{dt} = 2\eta_2 I_T^\circ[f^\circ(t) + 1/\rho] - k_1^E n - 2k_2^E n^2 \tag{7}$$

where k_1^E and k_2^E are the rate constants for the atom and molecule evaporation steps, respectively. Dimensionless parameters are defined by:

$$N = \frac{k_1^E n}{\eta_2 I_T^\circ} \tag{8}$$

$$\theta = 2k_1^E t \tag{9}$$

$$b = 16 \frac{\eta_2 I_T^o}{(k_1^E)^2/k_2^E} \tag{10}$$

and Eq. (7) written as

$$\frac{dN}{d\theta} = f^o(\theta) + 1/\rho - \frac{1}{2}N - \frac{1}{16} bN^2 \tag{11}$$

Equation (11) can be solved by direct integration. During the "on" part of the cycle, $f^o(\theta) = 1$, and the solution is:

$$N_{on}(\theta) = \frac{4}{b}\left\{[1+b(1+1/\rho)]^{1/2} \tanh\{\frac{1}{4}[1+b(1+1/\rho)]^{1/2}(\theta+C_{on})\} - 1\right\} \tag{12}$$

Similarly, for the "off" portion of the cycle, $f^o(\theta) = 0$, and the solution is:

$$N_{off}(\theta) = \frac{4}{b}\left\{[1+b(1/\rho)]^{1/2} \operatorname{ctnh}\{\frac{1}{4}[1+b(1/\rho)]^{1/2}(\theta+C_{off})\} - 1\right\} \tag{13}$$

The constants of integration C_{on} and C_{off} are determined by applying matching conditions representing the cyclic steady state nature of the process:

$$N_{on}(0) = N_{off}(0) \tag{14}$$

$$N_{on}(-\theta_c) = N_{off}(\theta_c)$$

which yield

$$[1+b(1+1/\rho)]^{1/2} \tanh\left\{\frac{1}{4}[1+b(1+1/\rho]^{1/2} C_{on}\right\}$$

$$= [1+b(1/\rho)]^{1/2} \operatorname{ctnh}\left\{\frac{1}{4}[1+b(1/\rho)]^{1/2} C_{off}\right\} \tag{15}$$

$$[1+b(1+1/\rho)]^{1/2} \quad \tanh \left\{ \frac{1}{4} [1+b(1+1/\rho)]^{1/2} (-\theta_c + C_{on}) \right\}$$

$$= [1+b(1/\rho)]^{1/2} \quad \operatorname{ctnh} \left\{ \frac{1}{4} [1+b(1/\rho)]^{1/2} (\theta_c + C_{off}) \right\} \tag{16}$$

These two equations must be solved numerically for C_{on} and C_{off} as functions of the parameters, b, ρ, and θ_c.

The rate of evaporation of X(a) (species X denoted by subscript 1) is:

$$R_1^E(t) = k_1^E n(t) \tag{17}$$

Combining Eqs. (8) and (17), shows that $\mathcal{F}\{R_1^E(t)\} \sim \mathcal{F}\{N(\theta)\}$.

The coefficients of the fundamental terms of the Fourier expansion of $N(\theta)$ are given by:

$$A = \frac{1}{\theta_c} \int_{-\theta_c}^{0} N_{on}(\theta) \cos\left(\frac{\pi\theta}{\theta_c}\right) d\theta + \frac{1}{\theta_c} \int_{0}^{\theta_c} N_{off}(\theta) \cos\left(\frac{\pi\theta}{\theta_c}\right) d\theta \tag{18}$$

$$B = \frac{1}{\theta_c} \int_{-\theta_c}^{0} N_{on}(\theta) \sin\left(\frac{\pi\theta}{\theta_c}\right) d\theta + \frac{1}{\theta_c} \int_{0}^{\theta_c} N_{off}(\theta) \sin\left(\frac{\pi\theta}{\theta_c}\right) d\theta \tag{19}$$

Inserting Eqs. (12) and (13) into (18) and (19) yields:

$$A = \frac{4}{\pi} \frac{1}{b} \left\{ \epsilon_{on} {}^J A_{on}\left(\frac{\epsilon_{on}\theta_c}{4\pi}, \frac{\epsilon_{on}C_{on}}{4}\right) + \epsilon_{off} {}^J A_{off}\left(\frac{\epsilon_{off}\theta_c}{4\pi}, \frac{\epsilon_{off}C_{off}}{4}\right) \right\} \tag{20}$$

$$B = \frac{4}{\pi} \frac{1}{b} \left\{ \epsilon_{on} {}^J B_{on}\left(\frac{\epsilon_{on}\theta_c}{4\pi}, \frac{\epsilon_{on}C_{on}}{4}\right) + \epsilon_{off} {}^J B_{off}\left(\frac{\epsilon_{off}\theta_c}{4\pi}, \frac{\epsilon_{off}C_{off}}{4}\right) \right\} \tag{21}$$

where

$$\epsilon_{on} = [1 + b(1 + 1/\rho)]^{1/2}$$

$$\epsilon_{off} = [1 + b(1/\rho)]^{1/2}$$

(22)

and the J's are the integrals:

$$J_{Aon}(x,y) = \int_{-\pi}^{0} \tanh(xu+y) \cos u \, du$$

(23)

$$J_{Aoff}(x,y) = \int_{0}^{\pi} \operatorname{ctnh}(xu+y) \cos u \, du$$

(24)

$$J_{Bon}(x,y) = \int_{-\pi}^{0} \tanh(xu+y) \sin u \, du$$

(25)

$$J_{Boff}(x,y) = \int_{0}^{\pi} \operatorname{ctnh}(xu+y) \sin u \, du$$

(26)

these integrals cannot be evaluated in closed form.

For this reaction sequence, it is convenient to relate the amplitude of the output signal to the value when $k_1^E \gg 1/t_c$, $k_1^E \gg k_2^E$, which generally occurs at high target temperatures. In dimensionless terms, these conditions imply $\theta_c \to \infty$ and $b \to 0$. The coefficients A and B reduce to 0 and $-4/\pi$, respectively, and:

$$[Q_1]_{\substack{b=0 \\ \theta_c=\infty}} = \frac{4}{\pi}$$

(27)

$$[\phi_1]_{\substack{b=0 \\ \theta_c=\infty}} = 0$$

(28)

These results could have been obtained directly from Eq. (11) by dropping the first and last terms.

Equations (15), (16), (20), and (21) have been utilized to calculate A and B as functions of b and θ_c for $1/\rho = 0$. The results are plotted in Figs. 4 and 5 as the phase shift and the amplitude attenuation, defined as

$$\frac{Q_1}{\left[Q_1\right]_{\substack{b=0 \\ \theta_c=\infty}}} = \frac{\pi}{4} (A^2 + B^2)^{1/2} \qquad (29)$$

These two plots show the effect of competition between the first and second order processes through the parameter b and the speed of the evaporation reaction compared to the modulation time through the parameter θ_c. If both the amplitude attenuation and the phase shift are experimentally accessible, both k_1^E and k_2^E can be determined individually.

I(a) Rapid Competitive First and Second Order Reactions

This system is a special case of the slow competitive first and second order reactions considered above. It is typical of the adsorption of hydrogen on refractory metals.[1,2] The residence time of hydrogen on the surface $(1/k_1^E)$ is small compared to typical modulation times, so that $\theta_c \to \infty$. However, over much of the target temperature range, atom evaporation and molecular reconstitution are of comparable magnitude, so that the parameter b is not zero. Since the phase shift is always zero, all information on the absolute magnitudes of k_1^E and k_2^E is lost. Only amplitude attenuation data remain to provide information on the ratio of these rate constants.

Since $\theta_c \to \infty$, θ also approaches ∞ and the hyperbolic tangent and cotangent terms in Eqs. (12) and (13) reduce to unity. The functions N_{on} and N_{off} become constants over their respective intervals and the coefficients A and B are easily evaluated from Eqs. (18) and (19).

There results:

$$A = 0$$

$$B = \frac{8}{\pi b} \left\{ [1 + b(1 + 1/\rho)]^{1/2} - [1 + b(1/\rho)]^{1/2} \right\}$$

The amplitude attenuation is

$$\frac{\left[a_i \right]_{\theta_c = \infty}}{\left[a_1 \right]_{\substack{b = 0 \\ \theta_c = \infty}}} = \frac{2}{b} \left\{ [1 + b(1 + 1/\rho)]^{1/2} - [1 + b(1/\rho)]^{1/2} \right\} \quad (30)$$

This equation (with $1/\rho = 0$) is well represented by the $\theta_c = 500$ curve of Fig. 4. Provided the model is correct, the data permit the parameter b to be evaluated from Eq. (30), and from b, the rate constant ratio $(k_1^E)^2/k_2^E$. In general, the amplitude attenuation does not vary either as the first power or the square root of the equivalent beam pressure, which is contained in the parameter b. These limits are approached asympotically at large and small values of b. Equation (30) demonstrates the importance of quantitative knowledge of the beam-to-background ratio in interpreting lock-in detector data for processes involving a second order surface step.

I(b)   Slow First Order Desorption

This limiting case is applicable to the desorption of cesium from tungsten, which has been studied by Perel, Vernon, and Daley.[8] Since no second order surface processes are present, $b = 0$, but since the residence time on the surface is comparable to the chopping time, θ_c

is finite. Since N_{on} and N_{off} must remain finite, the hyperbolic

tangents and cotangents in Eqs. (12) and (13) must approach unity.

They do so because C_{on} and C_{off} become large. Approximating $\tanh(x)$

by $1 - 2e^{-2x}$ and $(1 + x)^{1/2}$ by $1 + x/2$, the matching conditions Eqs.

(15) and (16) yield:

$$e^{-1/2C_{on}} = \frac{b}{4} \frac{e^{-\theta_c/2}}{1 + e^{-\theta_c/2}}$$

$$e^{-1/2C_{off}} = \frac{b}{4} \frac{1}{1 + e^{-\theta_c/2}}$$

The integrals involved in Eqs. (20) and (21) can be evaluated analytic-

ally to yield:

$$A = \frac{4}{\pi} \frac{\pi(\frac{1}{2}\theta_c)}{\pi^2 + (\frac{1}{2}\theta_c)^2} \tag{31}$$

$$B = -\frac{4}{\pi} \frac{(\frac{1}{2}\theta_c)^2}{\pi^2 + (\frac{1}{2}\theta_c)^2} \tag{32}$$

the phase lag is:

$$\phi_1 = \tan^{-1}\left(-\frac{\pi}{\frac{1}{2}\theta_c}\right) \tag{33}$$

and the amplitude attenuation is:

$$\frac{\left[\mathcal{Q}_1 \right]_{b=0}}{\left[\mathcal{Q}_1 \right]_{\substack{b=0 \\ \theta_c = \infty}}} = \frac{\frac{1}{2} \theta_c}{[\pi^2 + (\frac{1}{2} \theta_c)^2]^{1/2}} \tag{34}$$

Equations (33) and (34) are identical to the expressions obtained by Perel et al.[8] Their method, however, is not easily adapted to the general case in which second order processes compete with first order desorption.

II.   The Nickel-Chlorine Reaction

This relatively simple gas-solid reaction is the only heterogeneous reaction involving volatilization of the solid substrate which has been studied by modulated molecular beam techniques. Smith and Fite[1] report preliminary data on the amplitude and phase shift of the NiCl product signal. McKinley[4] has performed a dc beam experiment on the same system and reported amplitude data for both NiCl and $NiCl_2$. Both of these studies present signal amplitude data in arbitrary units, so that kinetic information can only be obtained from the amplitude attenuation factors discussed previously.

McKinley has shown that the amplitude of the NiCl and $NiCl_2$ signals are proportional to the beam strength, so that all reactions are first order with respect to chlorine concentration. His data indicate that the sum of the NiCl signal and twice the $NiCl_2$ signal is constant. This suggests that the instrumental constant α in Eq. (1) is the same for NiCl and $NiCl_2$, and that adsorbed chlorine leaves the surface only as one of the two nickel-bearing species. Based upon these observations, McKinley's model for the surface process is:

$$2Ni + Cl_2(g) \longrightarrow (NiCl)_2 \begin{cases} \xrightarrow{k_c p_c} 2NiCl(a) \xrightarrow{k_1^E} NiCl(g) \\ \xrightarrow{k_c(1-p_c)} NiCl_2(a) \xrightarrow{k_2^E} NiCl_2(g) \\ \qquad\qquad + Ni \end{cases}$$

where $1/k_c$ is the mean lifetime of the complex $(NiCl)_2$ on the surface and p_c is the probability of complex decomposition to adsorbed NiCl. The parameters k_c and p_c are analogous to the half-life and branching ratio used to describe radioactive decay. The rate constants k_1^E and k_2^E are the reciprocals of the mean lifetimes of the adsorbed species $NiCl(a)$ and $NiCl_2(a)$.

If the surface concentrations of $(NiCl)_2$, $NiCl(a)$ and $NiCl_2(a)$ are denoted by n_c, n_1, and n_2 respectively, balances on these three species are:

$$\frac{dn_c}{dt} = \eta_2 I_T^o [f^o(t) + 1/\rho] - k_c n_c \tag{35}$$

$$\frac{dn_1}{dt} = 2k_c p_c n_c - k_1^E n_1 \tag{36}$$

$$\frac{dn_2}{dt} = k_c(1 - p_c)n_c - k_2^E n_2 \tag{37}$$

The rates of emission of the two volatile species are:

$$R_i^E = k_i^E n_i, \quad i = 1, 2 \tag{38}$$

Using the dimensionless variables

$$\theta = k_c t \tag{39}$$

$$N_i = \frac{k_c n_i}{\eta_2 I_T^o}, \quad i = c, 1, 2 \tag{40}$$

The surface balances become:

$$\frac{dN_c}{d\theta} = f^o(\theta) + 1/\rho - N_c \tag{41}$$

$$\frac{dN_1}{d\theta} = 2p_c N_c - (k_1^E/k_c)N_1 \tag{42}$$

$$\frac{dN_2}{d\theta} = (1 - p_c)N_c - (k_2^E/k_c)N_2 \tag{43}$$

For the dc beam experiment,

$$R_i^E = \eta_2 I_T^o \left(\frac{k_i^E}{k_c} \right) N_i, \quad i = 1, 2 \tag{44}$$

and $f^o(\theta) = 1$, $dN_i/d\theta = 0$. The NiCl amplitude is

$$\bar{Q}_1 \sim 2(1 + 1/\rho) \, p_c \tag{45}$$

At high target temperatures, $\overline{\mathcal{A}}_1$ was observed to saturate, and $\overline{\mathcal{A}}_2$ vanished so that $p_c \to 1$ as $T \to \infty$. The amplitude attenuation factors are

$$\frac{\overline{\mathcal{A}}_1}{\left[\overline{\mathcal{A}}_1\right]_{T=\infty}} = p_c \tag{46}$$

and

$$\frac{\overline{\mathcal{A}}_2}{\left[\overline{\mathcal{A}}_2\right]_{T=\infty}} = \frac{1}{2}(1-p_c) \tag{47}$$

Thus, only one item of information is obtainable from the dc beam experiments: the variation of p_c with target temperature.

In the modulated beam experiment,

$$\mathcal{F}\{R_i^E\} = \eta_2 I_T^\circ \left(\frac{k_i^E}{k_c}\right) \mathcal{F}\{N_i\}, \quad i = 1,2 \tag{48}$$

Since Eq. (41) is identical to the differential equation describing a simple first order desorption process, the solutions $N_{c,on}(\theta)$ and $N_{c,off}(\theta)$ can be obtained from the considerations of Sec. I(b). Equation (42) can be integrated during the "on" and "off" periods using the appropriate expressions for $N_c(\theta)$ during each portion of the cycle. The constants of integration are determined by the usual matching technique. The Fourier coefficients of $N_1(\theta)$ are obtained from Eqs. (18) and (19), and $\mathcal{A}_1$ is

$$\mathcal{A}_1 \sim \frac{4}{\pi} p_c \frac{\theta_c}{(\pi^2 + \theta_c^2)^{1/2}} \frac{\theta_1}{(\pi^2 + \theta_1^2)^{1/2}} \tag{49}$$

where $\theta_1 = k_1^E t_c$ and $\theta_c = k_c t_c$.

At high target temperatures, $p_c \to 1$ and θ_c and $\theta_1 \to \infty$. The amplitude attenuation is :

$$\frac{a_1}{[a_1]_{T \to \infty}} = p_c \frac{\theta_c}{(\pi^2 + \theta_c^2)^{1/2}} \frac{\theta_1}{(\pi^2 + \theta_1^2)^{1/2}} \qquad (50)$$

The phase shift with respect to the high temperature NiCl signal is

$$\tan \phi_1 = -\pi \left(\frac{\theta_c + \theta_1}{\theta_1 \theta_c - \pi^2} \right) \qquad (51)$$

For $NiCl_2$, the analogous results are:

$$\frac{a_2}{[a_1]_{T = \infty}} = \frac{1}{2}(1 - p_c) \frac{\theta_c}{(\pi^2 + \theta_c^2)^{1/2}} \frac{\theta_2}{(\pi^2 + \theta_2^2)^{1/2}} \qquad (52)$$

$$\tan \phi_2 = -\pi \left(\frac{\theta_c + \theta_2}{\theta_2 \theta_c - \pi^2} \right) \qquad (53)$$

where $\theta_2 = k_2^E t_c$.

The amplitude attenuation ratios for the two volatile products are reduced by the product of two demodulation factors, which are absent from the dc equivalents, Eqs. (46) and (47). The phase shift formulae reflect the series resistances in each branch of the complex decomposition. If the dimensionless desorption rate constants θ_1 and θ_2 are much larger than θ_c, the phase lag is controlled by the rate at which the complex decomposes, and $\tan \phi_i = -(\pi/\theta_c)$, $i = 1,2$. On the other hand, if desorption is slow compared to complex decomposition, $\tan\phi_i = -(\pi/\theta_i)$, $i = 1,2$.

The significant advantage of a modulated beam experiment over a dc experiment is in the fact that the left hand sides of Eqs. (50) – (53) represent four independent and experimentally obtainable pieces of information. The data permit determination of p_c, k_c, k_1^E, and k_2^E at each temperature. The dc experiment, however, yields only p_c.

One disadvantage of the modulated beam technique can be inferred from Eq. (52). The two demodulation factors for $NiCl_2$ act in opposite directions from the competition factor $(1 - p_c)/2$. At high temperatures, the demodulation factors approach unity, but so does p_c. At low temperatures, where nearly all of the incident chlorine returns to the gas as $NiCl_2$, the demodulation factors are small. In either extreme, the modulated output amplitude goes to zero. $NiCl_2$ can be expected to be found only at intermediate temperatures, and probably at amplitudes considerably smaller than $NiCl$. At a modulation frequency of 100 Hz, Smith and Fite[1] never observed $NiCl_2$. The loss of amplitude and phase information on $NiCl_2$ renders interpretation of the $NiCl$ amplitude and phase shift data impossible, since Eqs. (50) and (51) contain three unknowns. The inability to detect modulated $NiCl_2$ at 100 cps points up the necessity of varying the modulation frequency in hunting for potential reaction products. Had lower modulation frequencies been tried, the product $NiCl_2$ might have been observed. However, the response of the lock-in amplifier deteriorates at low frequencies, and a digital signal-to-noise averager may be required.

III. Bulk Solution and Diffusion

In the preceding sections, the only removal mechanisms in the surface mass balances were evaporation or conversion to another adsorbed species. However, gases such as H_2, O_2, N_2, and CO show appreciable solubility in many metals, and at the elevated temperatures characteristic of many modulated beam experiments, diffusion of the dissolved gas in the target material may be rapid. Consequently, bulk solution and diffusion must be considered as a potential loss mechanism in the surface mass balance. Moore and Unterwald[3] have shown that solution and diffusion of hydrogen in tungsten and molybdenum dominated their flash filament experiments at temperatures above 1000°C.

Here we consider the case in which bulk solution-diffusion and rapid first order desorption are competitive processes. This restriction is valid for beams which adsorb without dissociation and do not react chemically with the substrate. This situation is also characteristic of the interaction of diatomic gases with metals at temperatures high enough such that bimolecular recombination is negligible compared to direct atom evaporation. At these temperatures, the mean life of an adsorbed atom is much smaller than the modulation time, so that the dn/dt term in the surface balance can be neglected. This simplification is analagous to the stationary intermediate assumption in homogeneous chemical kinetics.

Generally, modulated molecular beam experiments involve bombarding one side of a thin ribbon of target material with the beam. The back side of the target is subject only to bombardment by the

background gas. The higher impingement rate on the front face creates
a concentration gradient of the dissolved species through the target.
Such a net drain on the surface population results in a "competition
factor" reducing the amplitude of the output signal from the lock-in
amplifier (this component of the amplitude attenuation would be present
in dc beam experiments as well). However, even if the competition
factor were unity (by making the rear face of the target impervious
to the dissolved gas), a reduction in the amplitude of the lock-in
amplifier signal could result by demodulation due to the inability
of the relatively slow bulk diffusional process to keep up with the
periodic alteration of the supply of adsorbed atoms to the front
surface. Since most molecular beam targets are of the thin ribbon
variety, only the former case will be considered.

Let $c(x,t)$ denote the bulk concentration of dissolved gas (assumed
to be the same form as the adsorbed species) at position x within the
target and at a time t during one modulation cycle. Transport in
the target is governed by the diffusion equation:

$$\frac{\partial c}{\partial t} = D_v \frac{\partial^2 c}{\partial x^2} \tag{54}$$

where D_v is the bulk diffusivity of the dissolved species in the
target material and x is measured from the front face. We are interested in the process only after the initial transients due to
starting up the beam or heating the target to temperature have died
out. Consequently, the initial condition on Eq. (54) is replaced
by the cyclic steady state condition:

$$c(x,t) = c(x,t + 2t_c) \tag{55}$$

In order to join the surface kinetic problem to the bulk diffusion problem, we assume that the concentration of the dissolved species just beneath the surface is proportional to the instantaneous surface concentration. For the front face of the target, this yields:

$$c(0,t) = H\, n(t) \tag{56}$$

where H is the solubility measured from the adsorbed state. In principle, H can be determined by combining the adsorption isotherm with the usual solubility based upon gas pressure.

The surface mass balance for the front face of the target constitutes one boundary condition on Eq. (54). Neglecting the dn/dt term and using Eq. (56), this is

$$\eta\, I_T^\circ\, f^\circ(t) = \frac{k_1^E}{H}\, c(0,t) - D_v \left(\frac{\partial c}{\partial x} \right)_0 \tag{57}$$

Since the diffusion and evaporation processes are first order, the term reflecting the steady background $(1/\rho)$ has been neglected; this contribution is eliminated by the lock-in amplifier.

On the back face $(x = \ell)$ the supply term due to the beam is absent and the second boundary condition is:

$$0 = \frac{k_1^E}{H}\, c(\ell,t) + D_v \left(\frac{\partial c}{\partial x} \right)_\ell \tag{58}$$

While the diffusional term constitutes a loss to the front face, it represents a gain to the back face.

In order to solve Eq. (54) subject to (55), (57), and (58), the following dimensionless terms are introduced:

$$\theta = D_v t / \ell^2 \tag{59}$$

$$\xi = x/\ell \tag{60}$$

$$\mathcal{C}(\xi, \theta) = \frac{D_v}{\eta \, I_T^0 \, \ell} \, c \tag{61}$$

$$h = \frac{k_1^E \ell}{D_v H} \tag{62}$$

The diffusion equation and its associated conditions become:

$$\frac{\partial \mathcal{C}}{\partial \theta} = \frac{\partial^2 \mathcal{C}}{\partial \xi^2}, \quad 0 \leq \xi \leq 1, \quad -\theta_c \leq \theta \leq \theta_c \tag{63}$$

$$\mathcal{C}(\xi, \theta) = \mathcal{C}(\xi, \, \theta + 2\theta_c) \tag{64}$$

$$f^q(\theta) = h \, \mathcal{C}(0, \theta) - \left(\frac{\partial \mathcal{C}}{\partial \xi}\right)_0 \tag{65}$$

$$0 = h \, \mathcal{C}(1, \theta) + \left(\frac{\partial \mathcal{C}}{\partial \xi}\right)_1 \tag{66}$$

where $\theta_c = D_v t_c / \ell^2$ is the ratio of the chopping time to the characteristic diffusion time of the system.[*]

---

[*] In this situation, there are three characteristic times: the chopping time t_c, the diffusion time $t_D = \ell^2/D_v$, and the evaporation time, $t_E = 1/k_1^E$. The analysis of this section is restricted to situations in which t_D and t_c are of comparable magnitude but both are much larger than t_E. Had the dn/dt term in the surface mass balance been retained, the first term on the right of Eq. (65) would have been replaced by $h[\mathcal{C}(0,\theta)+(t_E/t_D)(\partial c/\partial \theta)_0]$. However, the last term of this expression is assumed to be small, since $t_E/t_D \ll 1$.

The lock-in amplifier processes a signal due to evaporation from the surface, or:

$$\mathscr{F}\{k_1^E \, n(t)\} = \eta \, I_T^\circ \, \mathscr{F}\{h \, C(0,\theta)\} \tag{67}$$

The surface mass balances have been written for a non-dissociative adsorption. For dissociative adsorption of a diatomic gas, the term $\eta \, I_T^\circ$ in Eqs. (57), (61), and (67) must be multiplied by two. The computed phase shift and amplitude attenuation, however, are unaltered.

The solution to Eqs. (63) - (66) is presented in Appendix A, where the coefficients A and B of the fundamental modes of the Fourier expansion of $hC(0,\boldsymbol{\theta})$ are shown to be:

$$A = \frac{2}{\pi} \sum_{n=1}^{\infty} hb_n \frac{(\alpha_n^2 \theta_c)\pi}{\pi^2 + (\alpha_n^2 \theta_c)^2} \tag{68}$$

$$B = -\frac{2}{\pi} \left\{ \frac{1+h}{2+h} - \sum_{n=1}^{\infty} hb_n \frac{\pi^2}{\pi^2 + (\alpha_n^2 \theta_c)^2} \right\} \tag{69}$$

where the b_n are given by:

$$b_n = \frac{2\alpha_n^2}{\alpha_n^2 + h^2 + 2h} \left\{ \frac{1}{h} \frac{1+h}{2+h} \left[\sin\alpha_n - \frac{h}{\alpha_n}(\cos\alpha_n - 1) \right] \right. \tag{70}$$

$$\left. \frac{1}{\alpha_n} \frac{1}{2+h} \left[(\cos\alpha_n + \alpha_n \sin\alpha_n - 1) + \frac{h}{\alpha_n}(\sin\alpha_n - \alpha_n \cos\alpha_n) \right] \right\}$$

and the eigenvalues α_n are the roots of the transcendental equation:

$$\left[\left(\frac{\alpha_n}{2}\right)\tan\left(\frac{\alpha_n}{2}\right) - \frac{h}{2}\right]\left[\left(\frac{\alpha_n}{2}\right)\cot\left(\frac{\alpha_n}{2}\right) + \frac{h}{2}\right] = 0 \qquad (71)$$

As long as the combined activation energy of the permeability $D_v H$ is less than that of the rate constant for evaporation k_1^E (which is typically 50 kcal/mole), bulk solution-diffusion effects become less important as the temperature is increased (i.e., h becomes large at high target temperatures). Experimentally, this situation can be observed as a levelling off of the emission rate from the surface at high target temperatures, which signals an end to the competition between the two removal mechanisms in favor of evaporation. Comparing the output signal to the signal in the absence of bulk solution-diffusion ($h \to \infty$, $A \to 0$, $B \to -2/\pi$), the amplitude attenuation and phase shift are:

$$\frac{Q_1}{[Q_1]_{h \to \infty}} = \frac{\pi}{2}(A^2 + B^2)^{1/2} \qquad (72)$$

$$\phi_1 = \tan^{-1}(A/B) \qquad (73)$$

The results of the computations are plotted in Figs. 6 and 7 for a range of the parameters h and θ_c.

The response of the system is predominantly controlled by the parameter h. If either the diffusion coefficient or the solubility is zero, $h \to \infty$, and bulk solution-diffusion does not affect the signal, irrespective of the value of the parameter θ_c.

As $h \to 0$ and $\theta_c \to \infty$ (infinite diffusivity, finite solubility), Eqs. (72) and (73) reduce to:

$$\frac{\left[Q_1\right]_{\substack{h \to 0 \\ \theta_c \to \infty}}}{\left[Q_1\right]_{h \to \infty}} = \frac{1}{2} \frac{2h\theta_c}{[\pi^2 + (2h\theta_c)^2]^{1/2}} \tag{74}$$

$$\phi_1 = \tan^{-1}\left(-\frac{\pi}{2h\theta_c}\right) \tag{75}$$

where

$$h\theta_c = \frac{k_1^E t_c}{H\ell} \tag{76}$$

Except for the factor of 1/2 in Eq. (74), these amplitude attenuation and phase shift expressions are identical to Eqs. (33) and (34) for slow first order desorption, with k_1^E in the latter replaced by $2k_1^E/H\ell$. The product $H\ell$ represents the equilibrium capacity of the target for dissolved gas; it is the number of atoms in a square centimeter of target area when the surface concentration is unity. The foil acts as a holding tank for a portion of the atoms in the beam.

If $h\theta_c \to \infty$ while $h \to 0$, the target is characterized by infinite diffusivity but zero capacity. In this instance, the amplitude attenuation reduces to 1/2 and the phase shift to zero. Physically this means that the incident beam is equally divided between the front and back faces of a transparent, zero capacity target. Evaporation from each face occurs at a rate one-half of the rate which would prevail if the target were impervious to the beam.

As the characteristic diffusion time becomes small compared to the chopping time $(\theta_c \to \infty)$, the phase shift approaches zero, but an amplitude attenuation of magnitude $(h+1)/(h+2)$ remains. This limit is well represented by the θ_c = 1000 curve on Fig. 6. Its minimum value is 1/2. This term represents the time-average fraction of the atoms supplied to the surface which leave by evaporation. The remainder is lost by diffusion through the foil and emission from the rear face. This amplitude attenuation can be computed directly from the boundary conditions of Eqs. (57) and (58) by setting $f^\circ(t)$ = 1/2 and the concentration gradient equal to $-[c(o)-c(\ell)]/\ell$. Thus, $(h+1)/(h+2)$, is the "competition factor" in the amplitude attenuation, and is analogous to Eq. (30) for competitive first and second order reactions and p_c in the nickel-chlorine reaction. There is no demodulation factor when the chopping time is large compared to the diffusion time, and the phase shift is zero when $\theta_c \to \infty$.

When the chopping time and the diffusion time are of the same order of magnitude, the amplitude attenuation is reduced below $(h+1)/(h+2)$ due to demodulation of the signal. During the "off" portion of the cycle, the sluggish response of the bulk diffusion process continues to supply atoms to the surface layer even though the beam is off. The evaporation of these diffusion-supplied atoms during the "off" cycle reduces the ac amplitude of the emission rate waveform to which the lock-in amplifier responds, and hence reduces the magnitude of the output signal. The phase shift is no longer zero.

As $\theta_c \to 0$, the signal is completely demodulated.

IV. Surface Diffusion

Localization of molecular beam impingement on a restricted area of the target implies that the amplitude and shape of the signal due to the reevaporated species can be affected by migration along the surface. If surface diffusion were very rapid compared to other loss mechanisms, the impinging beam would be instantaneously distributed over the entire heated area of the target instead of remaining in the region where the beam struck. Winterbottom and Hirth[9] have evaluated the effect of surface diffusion on the emission rate characteristics of hot molecular beam sources. Surface diffusion problems beset high temperature molecular beam experiments at both ends.

In this analysis, we assume that first order evaporation is the only other process with which surface diffusion is in competition. Consequently, the contribution of background gas impingement on the target can be neglected.

The total heated area of the target is assumed uniformly accessible to the mobile adatoms. Following Winterbottom,[9] edges and corners of the target (if any) are not considered to pose barriers to migration along the surface. The cool target supports, however, are assumed to present an impervious barrier to surface diffusion. The total heated area of the target (back as well as front faces if the target is a ribbon) is represented by a disk of radius r_0 at a uniform temperature. The radius r_0 is chosen such that πr_0^2 is equal to the total heated area of the target. The surface diffusion coefficient D_s and the evaporation rate constant k_1^E are assumed constant over the entire hot target area. The impinging molecular beam is considered to have its maximum intensity at $r = 0$, and to diminish with increasing radial distance according to a prescribed function $F(r/r_0)$. $F(0)$ is unity, and in most target-beam geometries, F decreases to zero at some $r < r_0$.

The surface mass balance can be written in dimensionless terms as:

$$\frac{\partial N}{\partial \theta} = \frac{1}{\xi}\frac{\partial}{\partial \xi}\left(\xi\frac{\partial N}{\partial \xi}\right) + f^\circ(\theta)\, F(\xi) - \lambda N, \qquad 0 \le \xi \le 1 \qquad (77)$$

The dimensionless terms in Eq. (77) are:

$$N = \left(\frac{D_s}{\eta I_T^\circ r_0^2}\right) n \tag{78}$$

$$\theta = D_s t / r_0^2 \tag{79}$$

$$\xi = r / r_0 \tag{80}$$

$$\lambda = k_1^E r_0^2 / D_s \tag{81}$$

The $\partial N/\partial \theta$ term has been retained in this analysis, since surface diffusion is generally significant at lower temperatures than bulk solution-diffusion; in the present case, the chopping, diffusion, and evaporation times may all be of the same order of magnitude.

The surface mass balance has been written for non-dissociative adsorption, but can be extended to diatomic gases by inserting a "2" in the denominator of Eq. (78).

The condition of cyclic steady state is:

$$N(\xi,\theta) = N(\xi, \ \theta + 2\theta_c) \tag{82}$$

and the boundary conditions are:

$$(\partial N/\partial \xi)_0 = 0 \tag{83}$$

$$(\partial N/\partial \xi)_1 = 0 \tag{84}$$

The output signal depends upon the instantaneous number density of the evaporating species in the ionizer of the mass spectrometer or other similar detector. The number density in the ionizer is the sum of contributions of emission from those portions of the target which offer a direct line of sight to the ionizer. Surface migration spreads the adatom population over a sizeable area of the target, some of which may be partially blocked off from direct view of the ionizer. Emission from such regions may not contribute as much to the total number density in the ionizer as the regions of the target at which the beam and detector are aimed. The instantaneous output signal from the detector is written as:

$$S(\theta) \ \sim \ \lambda \int_0^1 \xi \ G(\xi) \ N(\xi,\theta) \ d\xi \tag{85}$$

where $G(\xi)$ is a weighting function which accounts for the reduced contribution to the signal from off-center regions of the target (in particular, the back side of a ribbon target cannot contribute at all).

We assume that G is a function of radial position on the simplified disk-shaped target and is specified.

The amplitude of the output signal from the lock-in amplifier is proportional to $\sqrt{A^2+B^2}$, where A and B are the coefficients of the fundamental mode of the Fourier transform of $S(\theta)$. These coefficients are shown in Appendix B to be:

$$A = \lambda \sum_{n=0}^{\infty} \frac{k_n}{\alpha_n^2} \frac{(\alpha_n^2 \theta_c)\pi}{\pi^2 + (\alpha_n^2 \theta_c)^2} \tag{86}$$

$$B = -\lambda \sum_{n=0}^{\infty} \frac{k_n}{\alpha_n^2} \frac{(\alpha_n^2 \theta_c)^2}{\pi^2 + (\alpha_n^2 \theta_c)^2} \tag{87}$$

where:

$$k_n = \frac{1}{\pi} J_0^2 (\beta_n) f_n g_n \tag{88}$$

f_n and g_n are the coefficients of the Fourier-Bessel expansions of the functions $F(\xi)$ and $G(\xi)$. The β_n are the roots of $J_1(\beta_n) = 0$ (where $\beta_0 = 0$). The α_n are given by:

$$\alpha_n^2 = \beta_n^2 + \lambda \tag{89}$$

If $\lambda \to \infty$, surface diffusion is insignificant compared to atom evaporation; if $(\lambda\theta_c) \to \infty$ as well, the surface residence time is small compared to the chopping time. Under these conditions, $A \to 0$ and $B \to -\sum\limits_{n=1}^{\infty} k_n$. The amplitude attenuation is defined with respect to this limit:

$$\frac{Q}{[Q]_{\substack{\lambda \to \infty \\ (\lambda\theta_c) \to \infty}}} = \frac{(A^2 + B^2)^{1/2}}{\sum\limits_{n=0}^{\infty} k_n} \tag{90}$$

The phase angle is

$$\phi = \tan^{-1}(A/B) \tag{91}$$

For the purpose of illustrative numerical calculation, we have taken the beam shape factor $F(\xi)$ and the response function $G(\xi)$ to be:

$$F(\xi) = G(\xi) = \begin{cases} 1, & 0 \leq \xi \leq R \\ 0, & R \leq \xi \leq 1 \end{cases} \tag{92}$$

This represents the common situation in which the beam is collimated to impinge as a circular spot of uniform intensity on part of the total target area. The detector has been assumed to view only this spot as well. R is the ratio of the beam spot radius to the equivalent radius of the heated target.

The amplitude attenuation and phase shift are functions of the parameters λ, θ_c and R. Figures 8–11 show numerical results of the amplitude attenuation and phase lag for representative values of these parameters.

For either very large or very small values of the parameter λ, the amplitude attenuation and phase shift can be written as:

$$\frac{\mathcal{Q}}{[\mathcal{Q}]_{\substack{\lambda \to \infty \\ (\lambda\theta_c) \to \infty}}} = J(R) \; \frac{(\lambda\theta_c)}{[\pi^2 + (\lambda\theta_c)^2]^{1/2}} \qquad (93)$$

$$\phi = \tan^{-1}\left(\frac{\pi}{\lambda\theta_c}\right) \qquad (94)$$

As $\lambda \to 0$, $J(R) \to R^2$, which is just the ratio of the beam spot area to the total heated target area. This limit represents the situation in which surface diffusion serves only to instantaneously distribute the impinging beam over the target. At $\lambda \to \infty$, $J(R)$ approaches unity, and surface diffusion is slow compared to both the desorption and modulation times; the deposited atoms never migrate beyond the boundaries of the beam spot.

The upper curves of Figs. 8 and 10 are equivalent to the limit as $\lambda\theta_c \to \infty$. Here, the modulation time is slow compared to surface diffusion and desorption; the phase shift is zero and the amplitude attenuation is:

$$\frac{\mathcal{Q}_{(\lambda\theta_c) \to \infty}}{[\mathcal{Q}]_{\substack{\lambda \to \infty \\ (\lambda\theta) \to \infty}}} = \frac{\sum\limits_{n=0}^{\infty}\left[\frac{J_1(\beta_n R)}{\beta_n J_0(\beta_n)}\right]^2 \left(\frac{\lambda}{\lambda + \beta_n^2}\right)}{\sum\limits_{n=0}^{\infty}\left[\frac{J_1(\beta_n R)}{\beta_n J_0(\beta_n)}\right]^2} \qquad (95)$$

Equation (95) represents the "competition factor" for surface diffusion. While bulk solution-diffusion can reduce the surface population by no more than 50%, the reduction due to surface diffusion can be considerably greater since the limit of Eq. (95) as $\lambda \to 0$ is R^2.

The preceding results are based upon the assumption that the cooled periphery of the target is a perfect reflector for migrating adatoms. An equally acceptable boundary condition is to consider the outer edge of the target to act as a perfect sink for the diffusing species. In this case, Eq. (84) is replaced by $N(1,\theta) = 0$. The solutions for these two boundary conditions are very close for large values of λ (typically greater than 10); very few migrating adatoms reach the outer boundary before evaporating and the surface appears infinite in extent. However, there is a considerable difference in the curves at small λ. While the amplitude attenuation for $(\lambda\theta_c) \rightarrow \infty$ with a perfectly reflecting boundary approaches R^2 at small λ, it approaches zero in the same limit for a perfect sink at the outer boundary.

The implication throughout this discussion has been that surface diffusion represents a nuisance to molecular beam experiments because it reduces the magnitude of the measured signal. However, this need not be the case if one is interested in measuring surface diffusion coefficients at high temperatures. For a system in which only atom evaporation and surface migration can occur, measurement of the amplitude attenuation and phase lag in a well-defined beam-target-detector geometry can yield information on evaporation rate constants and surface diffusion coefficients.

In a similar manner, information on bulk solution and diffusion constants can be obtained by monitoring the emission from the back face of a ribbon target whose entire periphery has been rendered impervious to surface migration by suitable cooled barriers.

V. Effect of Molecular Transit Times

In the preceding analysis, the effects of molecular flight times were neglected. Relaxing this restriction requires two modifications.

First, the time variation of the beam intensity at the target cannot be taken as a square wave, as in Eq. (6). The correct function is[6]

$$I(t) = I_T^\circ \int_0^\infty G(t - \ell_1/z\, v_{MPr})\, P_3(z)dz + I_{bkg} \qquad (96)$$

where $G(t)$ is the "shutter function," or the fraction of the dc beam intensity I_T° which is transmitted by the chopper at time t. This function depends upon the details of the chopper construction and the beam size and shape. z is the dimensionless molecular speed, v/v_{MPr}, where v_{MPr} is the most probable speed of the reactant gas which constitutes the primary beam. ℓ_1 is the distance from the chopper to the target, and $P_3(z)$ is the normalized Maxwellian distribution function for the flux, $2z^3 \exp(-z^2)$. If ℓ_1/v_{MPr} is small compared to the chopping time t_c, the flight time in the argument of G in Eq. (96) can be neglected compared to t and $G(t)$ extracted from the integral. If in addition the beam is square modulated, $G(t)$ can be replaced by the square wave $f^\circ(t)$, and Eq. (6) is recovered.

Determination of the emission rate $R_1^E(t)$ by solution of the appropriate surface balances proceeds as before, except that $f^\circ(t)$ is everywhere replaced by the integral in Eq. (96).

Second, if the flight time of the product species between the target and the detector is not small, the signal fed to the lock-in amplifier is not correctly given by Eq. (1). Rather, it is,

$$\sqrt{T_T}\ S_i(t) = \alpha_1 \int_0^\infty R_i^E\ (t-\ell_2/z\ v_{MPi})\ P_2(z)\ dz \qquad (97)$$

Here $P_2(z)$ is the number density distribution function of a Maxwellian gas, equal to $(4/\sqrt{\pi})z^2 \exp(-z^2)$ and v_{MPi} is the most probable speed of the product species i emitted from the surface at the target temperature T_T. ℓ_2 is the distance separating the target and the detector. The detector has been assumed to be number density sensitive, which is characteristic of mass spectrometers with single transit ionizers of the electron bombardment type. If the second term in the argument of R_i^E is small compared to the chopping time, Eq. (97) reduces to Eq. (1).

The coefficients of the fundamental mode of $S(t)$ are determined as indicated by Eq. (2) and the amplitude and phase shift related to the coefficients A and B by Eqs. (4) and (5). In addition to the various parameters characterizing the surface processes, A and B will now be functions of the two time of flight parameters

$$X_r = \omega \ell_1/v_{MPr}$$
$$X_i = \omega \ell_2/v_{MPi} \qquad (98)$$

The calculations presented in this study are valid only if X_r and X_i are much smaller than unity. The effect of these parameters on the amplitude and phase shift for a simple first order surface reaction is considered in detail by Schwarz and Madix.[6]

APPENDIX A

Solution of Eqs. (63)-(66)

During a single "on" cycle ($-\theta_c < \theta < 0$), $f^\circ(\theta) = 1$ and Eqs. (63) to (66) can be written as:

$$\frac{\partial c_{on}}{\partial \theta} = \frac{\partial^2 c_{on}}{\partial \xi^2} , \quad -\theta_c \leq \theta \leq 0 \tag{A-1}$$

$$C_{on}(\xi, -\theta_c) = C_{off}(\xi, \theta_c) \tag{A-2}$$

$$-\left(\frac{\partial c_{on}}{\partial \xi}\right)_0 + h[C_{on}(0,\theta) - 1/h] = 0 \tag{A-3}$$

$$\left(\frac{\partial c_{on}}{\partial \xi}\right)_1 + h\, C_{on}(1,\theta) = 0 \tag{A-4}$$

where the cyclic condition of Eq. (64) has been replaced by a condition requiring that the concentration profiles at the beginning of the "on" cycle and at the end of the "off" cycle be identical. The solution to this set is:[10]

$$C_{on}(\xi, \theta) = \mu(\xi) + w(\xi, \theta) \tag{A-5}$$

$$u(\xi) = \frac{1}{h}\, \frac{1+h}{2+h} - \frac{1}{2+h}\, \xi \tag{A-6}$$

$$w(\xi, \theta) = \sum_{n=1}^{\infty} (a_{on,n} - b_n)\, X_n(\xi)\, e^{-\alpha_n^2(\theta + \theta_c)} \tag{A-7}$$

The eigenvalues α_n are given by Eq. (71) in the text, and the eigen-functions are:

$$X_n(\xi) \;=\; \cos(\alpha_n\xi) + \frac{h}{\alpha_n}\sin(\alpha_n\xi) \qquad\qquad (A\text{-}8)$$

The $a_{on,n}$ are the coefficients in the eigenfunction expansion of $C_{on}(\xi,-\theta_c)$:

$$C_{on}(\xi,-\theta_c) \;=\; \sum_{n=1}^{\infty} a_{on,n}\, X_n(\xi)$$

$$a_{on,n} \;=\; \frac{2\alpha_n^{2}}{\alpha_n^{2}+h^{2}+2h}\int_0^1 C_{on}(\xi,-\theta_c)\, X_n(\xi)\, d\xi \qquad (A\text{-}9)$$

The b_n are the coefficients of the eigenfunction expansion of $u(\xi)$:

$$u(\xi) \;=\; \sum_{n=1}^{\infty} b_n X_n(\xi)$$

$$b_n \;=\; \frac{2\alpha_n^{2}}{\alpha_n^{2}+h^{2}+2h}\int_0^1 u(\xi)\, X_n(\xi)\, d\xi \qquad\qquad (A\text{-}10)$$

The complete expression for b_n is given as Eq.(70) in the text.

During the "off" cycle $(0 < \theta < \theta_c)$, $f^\circ(\theta) = 0$ and Eqs. (63) to (66) become:

$$\frac{\partial c_{off}}{\partial \theta} \;=\; \frac{\partial^2 c_{off}}{\partial \xi^{2}}\;, \qquad 0 \le \theta \le \theta_c \qquad\qquad (A\text{-}11)$$

$$C_{off}(\xi,0) = C_{on}(\xi,0) \tag{A-12}$$

$$-\left(\frac{\partial C_{off}}{\partial \xi}\right)_0 + h\, C_{off}(0,\theta) = 0 \tag{A-13}$$

$$\left(\frac{\partial C_{off}}{\partial \xi}\right)_1 + h\, C_{off}(1,\theta) = 0 \tag{A-14}$$

The solution is:

$$C_{off}(\xi,\theta) = \sum_{n=1}^{\infty} a_{off,n}\, X_n(\xi)\, e^{-\alpha_n^2 \theta} \tag{A-15}$$

where X_n and α_n are the same as before, and $a_{off,n}$ are the coefficients of the eigenfunction expansion $C_{off}(\xi,0)$:

$$C_{off}(\xi,0) = \sum_{n=1}^{\infty} a_{off,n}\, X_n(\xi)$$

$$a_{off,n} = \frac{2\alpha_n^2}{\alpha_n^2 + h^2 + 2h} \int_0^1 C_{off}(\xi,0)\, X_n(\xi)\, d\xi \tag{A-16}$$

Utilizing the matching conditions, Eqs. (A-2) and (A-12), yields:

$$a_{on,n} - b_n = -\frac{b_n}{1 + e^{-\alpha_n^2 \theta_c}} \tag{A-17}$$

$$a_{off,n} = \frac{b_n}{1 + e^{-\alpha_n^2 \theta_c}} \tag{A-18}$$

Substituting Eq. (A-17) into (A-7) and (A-18) into (A-15) yields the solutions for $C_{on}(\xi,\theta)$ and $C_{off}(\xi,\theta)$. The desired forms of the solutions are the coefficients of the fundamental mode of the Fourier expansion of $hC(0,\theta)$ [Eqs. (18) and (19) with $N(\theta)$ replaced by $hC(0,\theta)$]. These are given by Eqs. (68) and (69) in the text.

APPENDIX B

Solution of Eqs. (77) and (82-84)

During the "on" cycle, Eq. (77) is:

$$\frac{\partial N_{on}}{\partial \theta} = \frac{1}{\xi} \frac{\partial}{\partial \xi} \left(\xi \frac{\partial N_{on}}{\partial \xi} \right) + F(\xi) - \lambda N_{on}, \quad -\theta_c \leq \theta \leq 0 \qquad \text{(B-1)}$$

and the boundary conditions are:

$$(\partial N_{on}/\partial \xi)_0 = 0 \qquad \text{(B-2)}$$

$$(\partial N_{on}/\partial \xi)_1 = 0 \qquad \text{(B-3)}$$

The solution to Eq. (B-1) which satisfies Eqs. (B-2) and (B-3) is:

$$N_{on} = \sum_{n=0}^{\infty} \left(b_n + c_n e^{-\alpha_n^2 \theta} \right) J_0(\beta_n \xi) \qquad \text{(B-4)}$$

where the b_n are given by:

$$b_n = f_n/\alpha_n^2 \qquad \text{(B-5)}$$

and α_n^2 is given by Eq. (89) in the text. The β_n are the zeros of the Bessel function of the first kind of order one. The coefficients f_n are:

$$f_n = \frac{2 \int_0^1 \xi F(\xi) J_0(\beta_n \xi) d\xi}{J_0^2(\beta_n)} \qquad \text{(B-6)}$$

During the "off" portion of the cycle, Eq. (77) is:

$$\frac{\partial N_{off}}{\partial \theta} = \frac{1}{\xi} \frac{\partial}{\partial \xi} \left(\xi \frac{\partial N_{off}}{\partial \xi} \right) - \lambda N_{off}, \quad 0 \leq \theta \leq \theta_c \qquad (B-7)$$

with boundary conditions analagous to Eqs. (B-2) and (B-3). The solution can be written as:

$$N_{off} = \sum_{n=0}^{\infty} a_n e^{-\alpha_n^2 \theta} J_0(\beta_n \xi) \qquad (B-8)$$

The unknown coefficients c_n and a_n are obtained from the matching conditions:

$$N_{on}(\xi, -\theta_c) = N_{off}(\xi, \theta_c) \qquad (B-9)$$

$$N_{on}(\xi, 0) = N_{off}(\xi, 0) \qquad (B-10)$$

These yield:

$$a_n = \frac{f_n}{\alpha_n^2} \frac{e^{\alpha_n^2 \theta_c}}{1 + e^{\alpha_n^2 \theta_c}} \qquad (B-11)$$

$$c_n = - \frac{f_n}{\alpha_n^2} \frac{1}{1 + e^{\alpha_n^2 \theta_c}} \qquad (B-12)$$

Inserting the solutions for N_{on} and N_{off} into Eq. (85) of the text and determining the coefficients A and B by replacing $N(\theta)$ in Eqs. (18) and (19) by $S(\theta)$ yields Eqs. (86) and (87) of the text. The coefficients g_n are given by Eq. (B-6) with $F(\xi)$ replaced by $G(\xi)$.

ACKNOWLEDGMENTS

The author would like to thank P. Concus and L. Wilson for their help in the computations.

This work was performed under the auspices of the United States Atomic Energy Commission.

REFERENCES

1. J. N. Smith and W. L. Fite, in Advances in Rarefied Gas Dynamics, J. A. Laurmann, Ed., Vol. 1, Suppl. 2, (Academic Press, New York, 1963), p. 430.

2. R. A. Krakowski and D. R. Olander, USAEC Report, UCRL-17336-Rev., University of California, Berkeley, to be published in J. Chem. Phys.

3. G. E. Moore and F. C. Unterwald, J. Chem. Phys. 40, 2626, 2639 (1964).

4. J. D. McKinley, J. Chem. Phys. 40, 120 (1964).

5. S. Yamamoto and R. E. Stickney, J. Chem. Phys. 47, 1091,(1967).

6. J. A. Schwarz and R. J. Madix, J. Catalysis, to be published.

7. H. Harrison, D. G. Hummer and W. L. Fite, J. Chem. Phys. 41, 2567 (1967).

8. J. Perel, R. H. Vernon, and H. L. Daley, J. Appl. Phys. 36, 2157 (1965).

9. W. L. Winterbottom, J. Chem. Phys. 47, 3546 (1967); W. L. Winterbottom and J. P. Hirth, J. Chem. Phys. 37, 784 (1962).

10. H. S. Carslaw and J. C. Jaeger, Conduction of Heat in Solids, 2nd. Ed. (Oxford Press, New York, 1959), p. 114, 118.

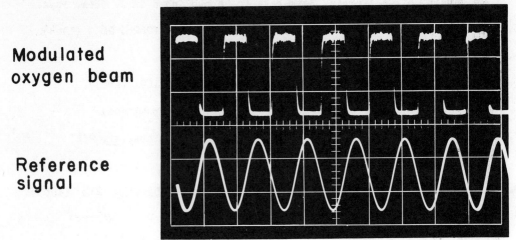

Fig. 1. Schematic diagram of molecular beam apparatus.

Modulated
oxygen beam

Reference
signal

Fig. 2. Oscilloscope trace of primary molecular beam modulated at 145 Hz.

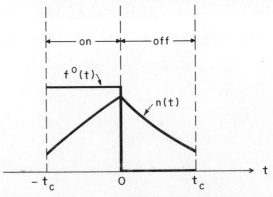

Fig. 3. Variation of surface concentration during a modulation cycle.

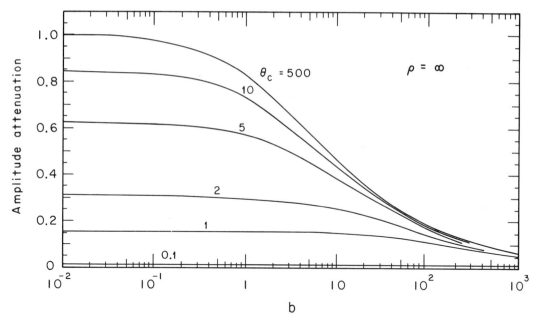

Fig. 4. Amplitude attenuation of lock-in detector signal for slow competitive
first and second order reactions with no background contribution.

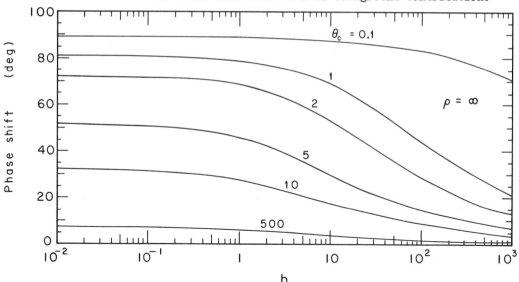

Fig. 5. Phase shift of lock-in detector signal for slow competitive first and
second order reactions with no background contribution.

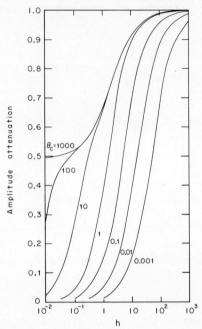

Fig. 6. Amplitude attenuation of lock-in detector signal for competitive rapid first order desorption and bulk solution-diffusion.

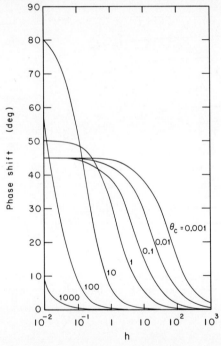

Fig. 7. Phase shift of lock-in detector signal for competitive rapid first order desorption and bulk solution-diffusion.

45–48

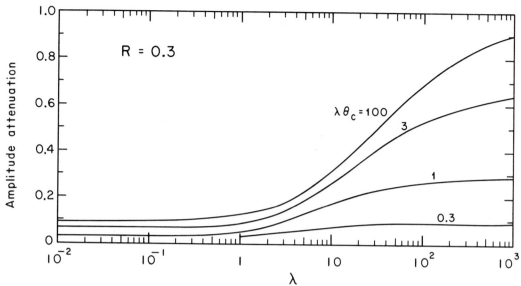

Fig. 8. Amplitude attenuation of lock-in detector signal for competitive slow
first order desorption and surface diffusion. R = 0.3.

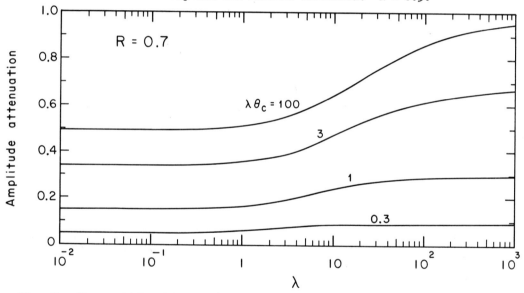

Fig. 9. Phase shift of lock-in detector signal competitive slow first order
description and surface diffusion. R = 0.3.

45-49

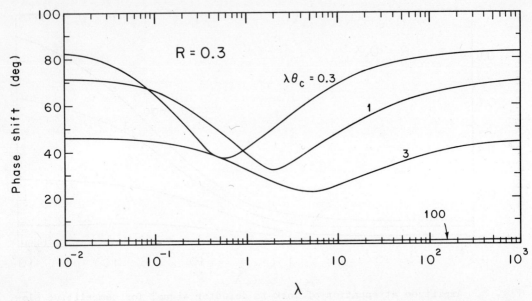

Fig. 10. Amplitude attenuation of lock-in detector signal for competitive slow
first order desorption and surface diffusion. R = 0.7.

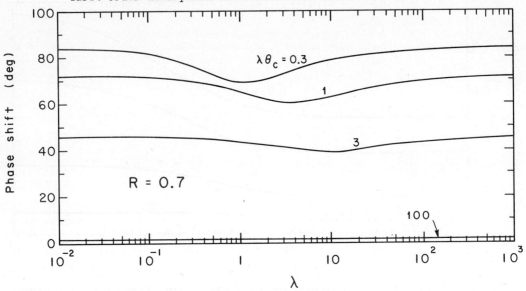

Fig. 11. Phase shift of lock-in detector for competitive slow first order desorption and surface diffusion. R = 0.7.

INTERACTION OF As$_2$, P$_2$ AND Bi MOLECULAR BEAMS WITH GaAs AND GaP ($\bar{1}\bar{1}\bar{1}$) SURFACES

J. R. Arthur

Bell Telephone Laboratories, Inc.
Murray Hill, New Jersey

While the interaction of Group III and Group V elemental vapors with the surfaces of III-V compounds is of fundamental interest in understanding crystal growth and evaporation of these materials, few adsorption studies of these systems have been reported.[1,2,3] It was recently observed[3] that Ga is adsorbed on GaAs (111) and ($\bar{1}\bar{1}\bar{1}$) surfaces with unit sticking probability below 750°K while the activation energy for desorption of Ga is 60 kcal/mole, approximately equal to the heat of vaporization of pure Ga$_{(\ell)}$. It was also found that the presence of adsorbed Ga markedly increased the sticking probability for As$_2$ vapor, and that this effect makes possible the growth of stoichiometric epitaxial films of GaAs by sublimation of the compound or by the "three temperature" technique.[3,4,5]

The interaction of Bi, Bi$_2$, P$_2$ and As$_2$ with GaAs ($\bar{1}\bar{1}\bar{1}$) and GaP ($\bar{1}\bar{1}\bar{1}$) surfaces has been studied by observing with a mass spectrometer the flux of molecules leaving a surface upon which a molecular beam of a particular vapor species was incident. Specifically, sticking coefficients were measured by comparing impinging and departing beam fluxes, surface lifetime was determined from the time dependence of the

desorbing beam when the impinging beam was pulsed, and
desorption rates were determined from the spectrometer signal
when a surface containing a known concentration of ad-molecules
was heated. The results show that while atomic species (Bi)
are adsorbed with near unit sticking probability, molecular
species (As_2,P_2,Bi_2) are predominantly reflected from the
crystal even at room temperature. The molecular adsorption
rate can, however, be greatly increased by the presence of
Ga atoms on the surface.

EXPERIMENTAL

The samples were rectangular plates 1.5×0.7×0.2 cm
cut from undoped, boat grown crystals (GaAs) or from a vapor
grown epitaxial crystal (GaP). The (111) and ($\bar{1}\bar{1}\bar{1}$) faces were
determined by suitable etchants[6], then the ($\bar{1}\bar{1}\bar{1}$) faces were
first mechanically polished and then chemically polished using
hot 1 H_2O:1 H_2O_2:H_2SO_4 for GaAs[7] and Cl_2-Methanol for GaP.[8]
After several microns were removed, each sample was washed in
distilled water and methanol, mounted in the Mo clamps of
the sample holder and immediately placed in the vacuum system.

Figure 1 shows the experimental system. Molecular
beams effused from two parallel Knudsen cells through a
mechanical chopper (magnetically coupled to an external motor)
and through collimating apertures to impinge on the sample.
The apertures were cooled to 77°K by contact with a liquid

nitrogen filled reservoir, and one of the Knudsen cells
(used for As$_2$ or P$_2$) was also surrounded by a cooled shield
and aperture. GaAs and GaP powders were used to supply
As$_2$ and P$_2$ beams since above 1000°K As$_2$ and P$_2$ are the
predominant vapor species over GaAs and GaP.[9,10] Vapor
leaving the sample passed through another cooled aperture
and into the ionization chamber of the mass spectrometer.[11]
The fraction of the beam which was not ionized then condensed
out on a cold surface surrounding the ionization chamber.
Thus the mass spectrometer signal was proportional to the
flux of vapor leaving the sample, at least for the condensible
species of interest here.

The vacuum system was ion pumped and background
pressures of 10^{-9} torr or less were maintained, the principal
impurities being H$_2$ and CO, neither of which was observed to
desorb from heated GaAs or GaP. The crystals were cleaned by
heating resistively to 900-925°K; the temperature was measured
both with an optical pyrometer and with a chromel-alumel
thermocouple tightly inserted into a 10 mil hole drilled in
the edge of each sample. It is recognized that the term
"clean surface" applied to compound semiconductors is fairly
ambiguous since a surface may be quite free from foreign atoms
and yet may not be stoichiometric due to preferential evaporation of one component. This is, in fact, observed on GaAs
heated above 925°K.[1,10,12] While MacRae was able to obtain

LEED patterns from the thermally cleaned GaAs ($\bar{1}\bar{1}\bar{1}$) which were
of good quality, we have observed that only faceted patterns are
obtained from thermally cleaned GaP.[13] Very likely the surfaces
used in this work were partially faceted; nevertheless the data
were quite reproducible for six different GaAs samples, and no
difference in results was observed after prolonged Ar ion
sputtering of the single GaP sample which was first studied after
thermal cleaning.

Two different experimental conditions were used:

1) A molecular beam was allowed to impinge on the
sample at a known flux for a measured time; the sample was
then heated at a programmed rate and the desorption flux of
various species was recorded as a function of temperature.

2) One or more beams were mechanically chopped and the
magnitude and time dependence of the reflected pulse were
recorded. If a clean surface is exposed to a flux of molecules
F at t = 0 and if a fraction (1-α) of the flux is immediately
reflected while α of the molecules are adsorbed with a mean
lifetime τ prior to desorption (assumed to be first order),
then the flux of molecules leaving the surface is

$$\dot{N} = (1-\alpha)F + \alpha F(1-e^{-t/\tau}) \text{ for } t \geq 0$$
$$\dot{N} = 0 \text{ for } t < 0$$

(1)

Clearly,

$$\alpha = 1 - \frac{\dot{N}_{(t=0)}}{\dot{N}_{(t \to \infty)}}$$

(2)

Also, the lifetime τ may be determined from the time dependence of $[\dot{N}_{(t)} - \dot{N}_{(o)}]$. The lifetime is related to the activation energy for desorption E_d by[14,15]

$$\tau = \tau_o \exp(E_d/RT) \tag{3}$$

At very low coverage E_d is constant and is determined from the slope of log τ vs. $1/T$. The mass spectrometer sensitivity and geometrical configuration were such that a flux of 10^{10} atoms/ cm^2/sec leaving the sample was detectable. Beam fluxes at the sample were usually 10-100 times greater, producing maximum surface concentrations of 10^{13} atoms/cm^2 or less. At these low coverages the ion current obeyed Eq. 1.

The relationship between ion current and evaporation flux was determined for Bi by depositing ~10 monolayers, 10^{16} atoms/cm^2, on GaAs, measuring the Bi$^+$ current at some known temperature, and then using the known vapor pressure of pure Bi[16] to calculate the evaporation flux. The impinging flux could then be determined by heating the substrate to a temperature at which Bi atoms striking the surface were immediately desorbed into the mass spectrometer; we assume that since the surface lifetime of the Bi atoms was several milliseconds, specular reflection of the beam was minimal.

The calibration for As$_2$ and P$_2$ was less accurate. It was assumed that the As$_2$ and P$_2$ evaporation rates from the GaAs and GaP samples were equal to the equilibrium rates.[9,10]

The spectrometer sensitivity factor calculated in this way agreed within a factor of two with that obtained by multiplying the Bi sensitivity by factors to account for the different ionization cross-sections and mass dependent transmission of the quadrupole analyzer. The results to be described do not depend critically on the As_2 or P_2 calibration.

RESULTS

Figure 2 and 3 show flash desorption spectra for Bi adsorbed at 300°K on GaAs at various initial coverages. The indicated coverage n was obtained from the time and incident flux assuming a sticking coefficient of unity. The area under each desorption curve is also equal to n. The two calculations of n are in agreement until n $\sim 4\times10^{14}$ atoms/cm^2; for larger n the integrated dosage is greater than the integrated desorption flux by a factor of two, indicating that the sticking coefficient was reduced to ~0.5.

As the coverage increased, three distinct peaks in the desorption rate developed which we associate with different adsorption states. When the Bi_2^+ current was monitored, only one desorption peak was observed which closely matched the low temperature Bi peak at 660°K and did not appear until n > 5×10^{14}. While the maximum desorption rate of the low temperature peak increased with initial coverage, the desorption rate between 550° and 640°K was essentially independent of coverage for $10^{15} < n < 10^{17}$ atoms/cm^2. We have therefore assumed that at these coverages the substrate was completely covered with crystalline Bi and that the desorption rate in this temperature range was equal to the evaporation rate for pure Bi; the assumption is supported by the comparison between activation energy for desorption, 43 kcal/mole, and enthalpy of vaporization of Bi, 45.5 kcal/mole.[16] We feel some confidence, therefore, in using the known Bi evaporation flux[16] to calibrate the spectrometer sensitivity.

The kinetic parameters for desorption of Bi from nearly clean GaAs were determined from the slope and ordinate intercept of an Arrhenius plot of desorption rate with initial coverage less than 10^{13} atoms/cm^2. By using only the low temperature tail of the curve the change in n was small compared to the change in the exponential term, assuming

$$\dot{N} = n\nu \, \exp(-E_d/RT). \tag{4}$$

For Bi on GaAs

$$\dot{N} = 1.3 \times 10^{14} n \, \exp\!\left(\frac{-51.6 \; \text{kcal/mole}}{RT}\right) \frac{\text{atoms}}{\text{sec}} \tag{5}$$

and for GaP

$$\dot{N} = 8 \times 10^{12} n \, \exp\!\left(\frac{-47.5 \; \text{kcal/mole}}{RT}\right) \frac{\text{atoms}}{\text{sec}} \; . \tag{6}$$

For initial coverage greater than 10^{14}, E_d decreased by 3 kcal/mole or more.

Figure 4 shows the mean surface lifetime of Bi atoms on GaAs and GaP. The straight lines follow for GaAs

$$\tau = 9\times10^{-15}\exp(\frac{50.3}{RT} \text{ kcal/mole}) \text{ sec} \qquad (7)$$

and for GaP

$$\tau = 1.5\times10^{-14}\exp(\frac{44}{RT} \text{ kcal/mole}) \text{ sec}, \qquad (8)$$

in reasonable agreement with the desorption data if

$$\dot{N} = n/\tau.$$

From Eq. 2 the sticking coefficient α for Bi atoms is given by the ratio of the initial discontinuity to the final value of the signal when the surface is abruptly exposed to the Bi beam.[*] Figure 5 shows the mass spectrometer signal when a GaAs crystal at 840°K was exposed to a chopped flux of 8×10^{11} atoms/cm^2/sec peak intensity. When the rising edge of the pulse is examined on an expanded time scale, it is apparent that when the shutter opened a small step appeared in the signal due to a reflected component. The signal increased according to Eq. 1 with $\alpha = 0.8$ and $\tau = 0.15$ sec. α varied only slightly with temperature from $\sim$0.85 at 300°K to 0.8 at 900°K.

---

[*]More properly, α is a differential condensation coefficient, since the sticking coefficient is usually defined as the ratio of adsorbed to impinging atoms measured at the end of some definite period. Under conditions where desorption occurs the sticking coefficient defined in this way will be smaller than our α.

Bi$_2$ was also present in the vapor beam,[16] however for Bi$_2$ on clean GaAs or GaP α was effectively zero at all temperatures. This was demonstrated by the reflected Bi$_2$ pulses which were identical with the incident pulses, and by the absence of a Bi$_2$ flash desorption peak. It seems likely that the reflected Bi component was actually produced by Bi$_2$ molecules which were dissociated by electron bombardment in the ionization chamber. Thus α_{Bi} is probably close to unity.

α for As$_2$ and P$_2$ molecules was also very small so that the reflected component of As$_2$ or P$_2$ pulses completely masked any time-dependent component; however when GaAs or GaP was exposed to As$_2$ or P$_2$ beams for >10^{15} atoms/cm^2 integrated flux, As$_2$ and As$_4$ or P$_2$ and P$_4$ were evolved upon subsequent heating. Figure 6 shows flash-desorption spectra for As$_2$ and P$_2$ deposited on GaAs. Desorption occurred over a broad temperature range, and a significant amount of association to As$_4$ or P$_4$ took place. In Fig. 6, the As$_2$ desorption rate increased above 750°K due to decomposition of the GaAs substrate; however the desorption rate in this region was somewhat greater for the surface originally dosed with As$_2$ than for clean GaAs. We attribute the difference to adsorbed As, and believe that the low temperature desorption peak was produced by decomposition of crystalline arsenic on the surface.

While the sticking coefficients of As_2 and P_2 were quite low, $\sim 10^{-2}$, they could be increased by certain surface treatments. If adsorbed Ga was present, α_{As_2} and α_{P_2} become unity. Figure 7 shows the As_2^+ signal from a GaAs surface at 300°K exposed to As_2 pulses. Between the second and third pulses ~ 0.1 monolayer of Ga was deposited which greatly reduced the amount of As_2 reflected in the subsequent pulse. In fact, it can be seen that no As_2 was initially reflected, but the reflected fraction increased with time as the adsorbed Ga was consumed, presumably by forming GaAs. The Ga was not desorbed under these conditions. α_{As_2} and α_{P_2} were also found to increase to ~ 0.2 when the GaAs or GaP substrate was heated above 800°K. We have previously presented evidence for a mobile Ga surface population in this temperature range[3] with which the impinging As_2 or P_2 molecules can react. The sticking coefficient of Bi_2 was <u>not</u> increased above 800°K, nor was it affected by adsorbed Ga, which is explained by the lack of Ga-Bi compounds.[17]

DISCUSSION

GaAs and GaP ($\bar{1}\bar{1}\bar{1}$) surfaces were quite similar in their reactions with Bi, Bi_2, As_2 and P_2 molecular beams. The atomic species, Bi, was adsorbed on both surfaces with high sticking probability and was desorbed at about the same rate from either material. The molecular species, on the other hand, were adsorbed very slowly or not at all; however the presence of adsorbed Ga produced a high adsorption rate for

As$_2$ and P$_2$ but not for Bi$_2$. A surface on which Ga and As$_2$ were deposited, when subsequently heated did not evolve As$_2$ until the temperature was high enough to decompose GaAs, indicating that the Ga did not simply provide active sites for As$_2$ adsorption but actually reacted chemically to form GaAs, i.e,

$$As_{2(g)} + Ga_{(ads)} \rightarrow GaAs + As_{(ads)}.$$

This reaction was probably followed by either

$$As_{(ads)} + Ga_{(ads)} \rightarrow GaAs$$

or

$$2As_{(ads)} \rightarrow As_{2(g)}$$

The difference between the reactivity of Bi and Bi$_2$ makes the species As$_{2(ads)}$ or P$_{2(ads)}$ seem unlikely.

The Bi flash desorption experiments showed that desorption occurred in three steps, with the most tightly bound atoms adsorbed up to a coverage of $\sim 10^{14}$ atoms/cm^2. Since Bi adsorption is evidently not an activated process the heat of adsorption is approximately equal to the desorption activation energy. Therefore dissociative adsorption of Bi$_2$ should be favored on energetic grounds since [16]

$$Bi_{2(g)} \rightarrow 2Bi_{(g)} \qquad \Delta H^o_{298} = 48.1 \text{ kcal/mole}$$

$$2Bi_{(g)} \rightarrow 2Bi_{(ads)} \qquad \Delta H \sim -100 \text{ kcal/mole}$$

(16)

The fact that Bi_2 did <u>not</u> adsorb probably indicates that the
surface sites are sufficiently separated   that both atoms
of the molecule cannot simultaneously interact with the
surface and hence dissociation requires a prohibitively
large activation energy.[18] A similar argument has been
presented to explain the low sticking probability of O_2 on
GaAs.[2]

The same conclusion can be drawn from the very
slow adsorption of As_2 and P_2; however for these two elements
the adsorption rate could be greatly enhanced by the presence
of adsorbed Ga. This phenomenon offers a kinetic mechanism
for the growth of stoichiometric GaAs and GaP layers from the
elemental vapors provided that the arrival rate of As_2 or P_2
is considerably greater than that of Ga. Epitaxial layers
of both materials have been grown on both GaAs and GaP substrates
from 875°-1000°K by sublimation of the compound, relying on
the higher vapor pressure of the Group V element to provide
an arrival rate 10-50 times the rate for Ga.[3] The low
growth temperature and clean surface conditions offer the
possibility of reducing the intrinsic defect concentration in
these films. Thus adsorption studies of these systems have had
the practical result of suggesting new crystal growth techniques.

REFERENCES

1.   A. U. MacRae, Surf. Sci. 4, 247 (1966).

2.   J. R. Arthur, J. Appl. Phys. 37, 3057 (1966).

3.   J. R. Arthur, J. Appl. Phys., to be published.

4.   K. G. Gunther, Naturwissenschaften 45, 415 (1958).

5.   J. E. Davey and T. Pankey, J. Appl. Phys. 39, 1941 (1968).

6.   H. C. Gatos, in "Progress in Semiconductors", Vol. 9, editors, Gibson, Kröger and Burgess, John Wiley and Sons, New York, (1965).

7.   F. A. Cunnell, J. T. Edmond and W. R. Harding, Solid State Electronics 1, 97 (1960).

8.   C. S. Fuller and H. W. Allison, J. Electrochem. Soc. 109, 880 (1962).

9.   C. D. Thurmond, J. Phys. Chem. Solids 26, 798 (1965).

10.   J. R. Arthur, J. Phys. Chem. Solids 28, 2257 (1967).

11.   Quad 250 manufactured by EAI, Inc., Palo Alto, California.

12.   M. F. Millea and D. F. Kyzer, J. Appl. Phys. 36, 308 (1965).

13.   A. Cho, J. R. Arthur and A. U. MacRae, to be published.

14.   I. Langmuir, J. Amer. Chem. Soc. 54, 2798 (1932).

15.   F. L. Hughes and H. Levinstein, Phys. Rev. 113, 1029 (1959).

16.   F. J. Kohl, O. M. Uy and K. D. Carlson, J. Chem. Phys. 47, 2667 (1967).

17.   R. P. Elliott, "Constitution of Binary Alloys, First Supplement", McGraw-Hill, New York (1965).

18.   G. Ehrlich, J. Chem. Phys. 31, 1111 (1959).

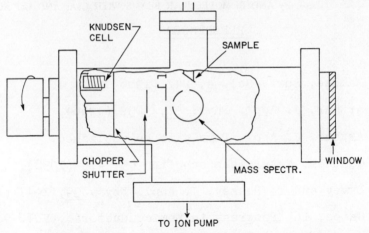

Fig. 1. Apparatus for studying adsorption on GaAs and GaP.

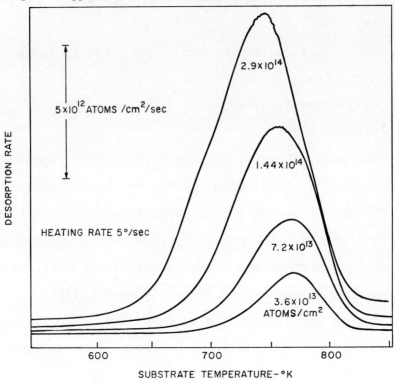

Fig. 2. Flash desorption spectra of Bi adsorbed on GaAs ($\overline{111}$) surface at various initial coverages.

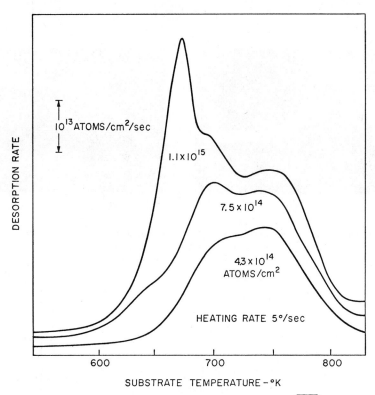

Fig. 3. Flash desorption spectra of Bi adsorbed on GaAs ($\overline{1}\overline{1}\overline{1}$) at higher coverages.

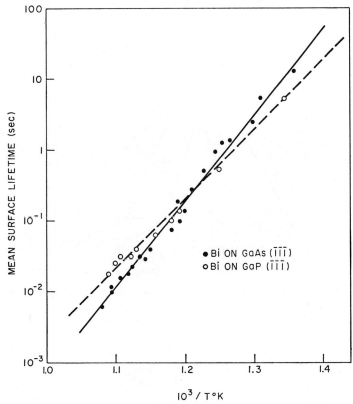

Fig. 4. Mean surface lifetime of Bi on GaAs and GaP ($\overline{1}\overline{1}\overline{1}$) surfaces.

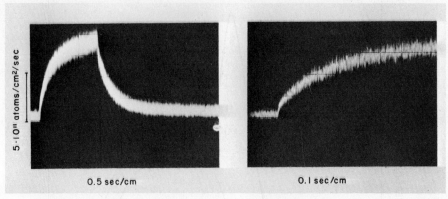

0.5 sec/cm   0.1 sec/cm

Fig. 5. Initial transient produced by Bi pulse desorbing from GaAs ($\overline{111}$) surface, T = 840°K, Bi flux = 8×10^{11} atoms/cm^2/sec.

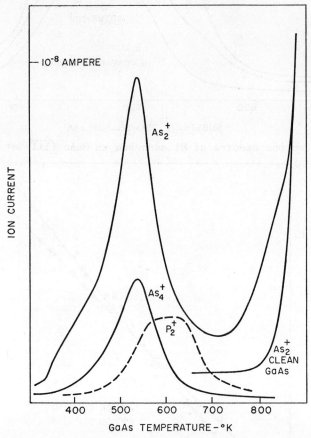

Fig. 6. Flash desorption spectra of As$_2$, As$_4$ and P$_2$ on GaAs ($\overline{111}$).

46-16

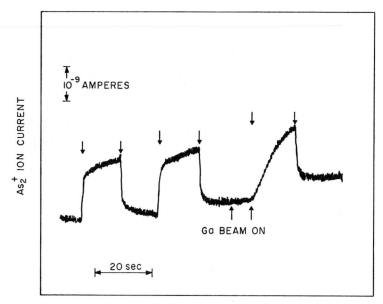

Fig. 7. Effect of adsorbed Ga on the reflection of As₂ from GaAs ($\overline{1}\overline{1}\overline{1}$). Arrows denote time when As₂ beam turned on and off.

MEASURED FLUX AND SPEED DISTRIBUTIONS OF ATMOSPHERIC GASES SCATTERED FROM POLYCRYSTALLINE COPPER AND ALUMINUM SURFACES*

M. N. Bishara, S. S. Fisher, A. R. Kuhlthau and J. E. Scott, Jr.

Department of Aerospace Engineering
University of Virginia
Charlottesville, Virginia

Introduction

A fundamental key to the experimental study of gas-surface interactions has been the development of the molecular beam. From its beginnings early in this century, improvements in molecular beam technology have consistently been followed by improvements in the understanding of the gas-surface interface. The emergence in the past decade of the nozzle type molecular beam [1] as an intense, nearly monoenergetic molecular source has opened new horizons to the study of surface scattering. These beams, in conjunction with the avail ability of increasingly sophisticated vacuum and electronic equipment, are enabling the experimenter to obtain scattering information which is of significantly greater detail than that previously available. Consequently, one can anticipate that the results of such experiments will be very effective in providing a description of the dynamics of the gas-surface interaction. Alternately, such beam-detector system may be useful as probes for the study of the properties of surfaces themselves.

Findings to be presented in this paper derive from the second in a series of experiments [2] in which nozzle type molecular beams are scattered from solid surfaces and measurements of the angular distribution of the flux and speed of the scattered molecules obtained. The scope of the present study is limited to thermal energy (0.06-0.28 eV) beams of argon and nitrogen and to polycrystalline copper and aluminum surfaces. The independent variables

*This work supported by the National Aeronautics and Space Administration (Grant NGR-47-005-046).

which have been studied are beam incidence angle, beam energy, beam gas, target
material, surface texture, target temperature, and target heat treatment.
Target surfaces are cleaned by heating in vacuo and/or heating in hydrogen.
Observations of the scattered beams are confined to the plane of incidence
(that plane containing the incident beam and the surface normal). Distributions
of scattered particle speeds are measured at selected angles in this plane.
Moments (weighted averages) of these distributions have been determined and
are presented in lieu of the distributions themselves. These moments are
a) the flux or (intensity) of scattered particles, b) their mean speed, and
c) the mean spread in particle kinetic energies (a measure of the half-width
of the speed distribution).

The ultimate pressure in the apparatus is not sufficiently low to avoid
considerable contamination of the target surfaces through adsorption of background gases, especially if the target is not heated. No in situ observations
of surface structure or contamination have been attempted. However, estimates of structure can be made from micrographs obtained before and/or after
tests. Furthermore, estimates of contamination can be made in terms of target
temperature and exposure history. Finally, surface nature may also be inferred
from properties of the scattered beams themselves.

Previous investigations with these gases and these surfaces have been
made using oven-type molecular beams and have been limited to measurement
of scattered flux distributions [3,4], measurement of momentum accommodation
coefficients [5-8], and some approximate time-of-flight ("phase shift") measurements [9]. These are the first measurements to be reported on the scattering
of atoms and molecules at thermal energies from copper and aluminum targets
where the velocity distributions of the scattered beams have been obtained.

Experimental Method

The apparatus used in these experiments is the same as that already described in Reference 2 (and other references cited therein). The experiments are carried out in a cylindrical vacuum chamber which is divided into three subchambers, two for forming and collimating the molecular beam and a third where the beam is scattered from the target and detected. The scattering chamber is oil-diffusion pumped and liquid nitrogen trapped. Chamber pressure is generally about 5×10^{-8} torr, rising to about 1×10^{-7} torr when the beam is admitted. The principal background constituent is H_2O.

A diagram of the experimental geometry is shown in Fig. 1. The mean intensity of the argon beam emanating from the nozzle at room-temperature is about 3×10^{16} atoms/cm^2-sec where it strikes the target, with 90% of these atoms having speeds within 7% of the mean speed, trajectories within 3° of the beam axis, and impact points within a 4 mm radius of the beam axis. For nitrogen, the beam intensity and speed spread are about 50% higher. For either gas at higher nozzle temperatures, the speed spread is again slightly higher. Nevertheless, under all operating conditions, the incident beams are essentially monoenergetic.

The scattered beam is detected with a universal-type ionization detector which is incorporated with a slotted-disc chopper to form a gated time-of-flight detection system [10]. This system can be rotated to selected scattering angles within the plane of incidence. The target and detector geometry are such that 90% of the particles entering the detector have trajectories within 3° of the nominal scattering direction.

The signal from the detector is essentially a direct transform of the speed distribution of the scattered particles measured in terms of times-of-flight from the chopper to the detector. After amplifying and signal averaging to reject noise, an approximate curve-fitting technique is applied to the recorded time-of-flight distributions to extract the desired moments. A discussion of this method may be found in Reference 2. This technique yields values of the moments which are within ± 3% of their true values.

Targets used in these experiments were cut from rolled sheets of 99.9% pure copper and Type 11-S aluminum (94% Al, 5% Cu). Four different copper and two different aluminum target surfaces were employed. Each surface was given a standard metallurgical polish and finished with 0.5 micron alumina. Following this, two of the copper targets were etched for 30 seconds in a 50% nitric acid solution. The two aluminum targets were etched for 30 seconds in a 50%-5% nitric-hydrofluoric acid solution. Prior to insertion into the vacuum chamber, all surfaces were rinsed in standard organic solvents and in distilled water.

Flux and Speed Measurements

The following notation will be used in presenting the results:

θ_i = angle of incidence, measured from the target normal;

θ_r = scattering angle, measured from the target normal;

T_o = nozzle temperature;

T_s = surface temperature;

I_r = scattered flux (particles/steradian-second);

$\bar{v}_r$ = mean speed of scattered particles;

$\bar{v}_s$ = mean speed for particles scattered with a Maxwellian speed distribution at temperature T_s, $\bar{v}_s = (9\pi k \, T_s/8m_g)^{1/2}$;

v_i = mean speed of incident beam,

$v_i \doteq (5kT_o/mg)^{1/2}$ for argon

$\doteq (7kT_o/m_g)^{1/2}$ for nitrogen;

m_g = particle mass;

S = normalized thermal spread in the energies of the scattered

particles defined as $[(\overline{v_r^2} - \overline{v}_r^2)/\overline{v}_r^2]/[(\overline{v}_s^2 - \overline{v}_s^2{}_0/\overline{v}_s^2]$;

I_{rl} = maximum scattered flux for purely diffuse scattering; I_{rl} is

estimated from the nearly diffuse-scattering surfaces.

wherever I_{rl} can be estimated, the quantity I_r/I_{rl} is plotted. The mean

speed is always plotted in the form $\overline{v}_r/\overline{v}_s$. For comparison, values of the

ratio $\overline{v}_i/\overline{v}_s$ are included in the margin of each figure. The quantity S is

a shape parameter of the speed distribution which for simplicity, will be

referred to simply as the "spread". If the scattered particle distribution

is Maxwellian, i.e. that distribtuion corresponding to effusion from an oven

(at any temperature) in which the particles have a Maxwellian velocity distri-

bution, then S = 1. For the incident beam, S_i (obtained by replacing $\overline{v}_r^2$ by

$\overline{v_i^2}$ and $\overline{v}_r^2$ by $\overline{v}_i^2$) is always small, ranging from 0.01 - 0.05 depending upon

the beam gas and nozzle temperature. Since S_i is always much smaller than

S, it has not been indicated in any of the figures.

All six targets at room temperature prior to any heating in vacuo, whether

the beam gas was argon or nitrogen, exhibited scattering patterns which were

quite diffuse with scattered particle speed distributions near those correspond-

ing to full thermal accommodation to the surface. The properties of argon

beams scattered from two such targets are shown in Fig. 2. Here $\theta_i = 60°$,

$T_o = T_s = 298°K$. The dashed lines correspond to diffuse, fully accommo-

dated scattering. Clearly, the measured flux distributions fall very near the

diffuse $(\cos \theta_r)$ curve and $\overline{v}_r$ values are quite near $\overline{v}_s$ for almost all the

scattered flux. The values of S are slightly less than unity at most scatter-

ing angles. Slight increases in $\overline{v}_r$ (toward $\overline{v}_i$) decreases in S (toward S_i) are

characteristic trends at large θ_r which will be discussed later. For nitro-

gen beam under the same conditions (data not shown) the agreement of the flux and mean speed with the dashed curves is better than that shown in Fig. 2

(even though the speed of the nitrogen beam is about 15% greater). Values

of $S \doteq 0.9$ are also observed with nitrogen for all scattering angles θ_r.

Properties of room temperature argon beams scattered from an unheated

etched copper surface with incidence angle θ_i as the varied parameter are

shown in Fig. 3. The normalizing factor I_{rl} has been chosen as that correspond-

ing to the maximum observed flux. Within the experimental uncertainty, the

measured properties of the scattered beams are independent of θ_i and agree

with those shown in Fig. 2. A slight back-scattering of flux is noted for

$\theta_i = 60°$. This back-scattering was consistently noted with the etched copper

targets and, after heating in vacuo, was found to occur with the etched aluminum

target. Its prominence generally increased with increasing θ_i but appeared

insensitive to beam gas or nozzle temperature changes. Its prominence for

aluminum (after initial heating) was reasonably independent of T_s, but for

copper, this tendency was masked by a moderate forward-scattered lobe for T_s

above 600°K. Unless the target was heated for prolonged periods at much

higher temperatures, however, the back-scattering reappeared upon cooling

to room temperature. Similar back-scattering was observed by Hurlbut [3]

with nitrogen beams at angles of incidence near 70° scattering from aluminum

and polished steel targets. While he attributed this to spurious scattering

from the target edge, such an explanation is highly improbable for the pre-

sent experimental arrangement. Moreover, it is not difficult to visualize

a rough surface model with steps, ramps or waves which would result in back-

scattering. Those surface sites which face more toward the incident beam are

preferentially illuminated by the incident beam and this preference increases

with increasing θ_i. The reduced accommodation of particles scattered near

the surface tangent (indicated by the shift of $\bar{v}_r$ toward $\bar{v}_i$ and S toward S_i

as $\theta_r \to 90°$) might also be explained on the basis of surface roughness;

the probability for scattered particles to have undergone multiple collisions with a roughened surface will be smaller for $\theta_r = 90°$.

Further evidence of the reduced accommodation of particles scattered near the surface tangent is provided by the data in Fig. 4. In this figure $\bar{v}_r/\bar{v}_s$ has been plotted versus θ_r for nitrogen beams at three different nozzle temperatures scattered from an unheated etched copper surface at room-temperature (for these experiments, the observed variations of I_r and S with θ_r were, within the data uncertainty, independent of T_o and identical to those shown in Fig. 3). At the two lower beam energies (lower T_o), $\bar{v}_r$ is seen to approach $\bar{v}_s$ at negative values of θ_r and to approach $\bar{v}_i$ monotonically with increasing θ_r. At the highest beam energy, even those molecules which are scattered at negative θ_r exhibit reduced thermal accommodation. Mean speed data for argon under similar experimental conditions (data not shown) display much the same behavior.

One may define a partial thermal accommodation coefficient for particles scattered at any given θ_r as follows:

$$\alpha_r(\theta_r) = (\overline{v_i^2} - \overline{v_r^2})/(\overline{v_i^2} - \overline{v_s^2})$$

where $\overline{v_s^2} = 4kT_s/m_g$. Extrapolating the data in Fig. 4 to $\theta_r = 90°$ and employing the measured values of S, one computes: $\alpha_r(90°) = 0.7$, 0.7 and 0.6 for $T_o = 300$, 600 and 930°K, respectively. The corresponding values obtained for $\alpha_r(-25°)$ are 1.02, 0.99 and 0.95, respectively. The decrease of α_r with increasing beam energy is interesting, but caution should be exercised in assigning this as a characteristic of the interaction since the higher energy molecules may well be more efficient in removing adsorbed gases from the surface.

Figure 5 shows the properties of argon beams of different energies scattered from an unheated aluminum target. Contrary to the flux distributions for nitrogen scattered from etched copper (associated with the speed data presented in Fig. 4), the flux distributions in this case are increasingly lobular with increasing beam energy. The lobe maximum occurs at a scattering direction between the surface normal and the specular direction (this is commonly referred to as "subspecular" scattering). At fixed θ_i the systematic increase in $\bar{v}_r$ with either increasing T_o or increasing θ_r is similar to that seen in earlier figures. The variation of S with θ_r is also similar to previous data (the strange behavior of S at large θ_r for $\theta_i = 30°$ and $T_o = 835°K$ is at the moment unexplained). The speed data in Figs. 4 and 5 suggest that tangential momentum is best conserved for those atoms scattered along the surface tangent, with such conservation better at large θ_i and higher beam energy.

A comparison of the effects of surface texture on the scattered beam properties is shown in Fig. 6 for room-temperature argon scattering from copper targets. For these tests, a single sheet of copper was formed into a channel-shaped section. Both outer walls of the channel were polished, but only one was etched. The target was mounted so that alternate surfaces could be exposed to the molecular beam merely by rotating the target holder through 180°. In this figure I_{rl} has been chosen as the maximum flux for the room temperature etched surface. At $T_s = 298°K$ a difference in flux distributions between polished and etched surfaces is clearly discernible. Heating either surface brings about an increasing redistribution of the scattered flux toward sub-specular lobes with the greater redistribution occurring for the polished surface. With these still semi-diffuse flux distributions, it is not surprising to find that $\bar{v}_r$ and S are relatively independent of both θ_r and surface texture. The decrease in $\bar{v}_r/\bar{v}_s$ with increasing T_s is indicative of thermal accommodation coefficient values less than unity. It is unwarranted, however, to attribute the fine structure of the variations of $\bar{v}_r/\bar{v}_s$ or S with θ_r to more than irregularities in the target and to experimental error. The most

accurate determinations of α_r can be made for the highest T_s in this figure.
Let $\bar{\alpha}_r$ be the average value of α_r over the range of θ_r observed. At 815°K,
$\bar{\alpha}_r$ = 0.55 for the etched surface and 0.60 for the polished surface.

In an attempt to clean these surfaces, the diffusion pumps were switched
off and the target was maintained at 640°K in hydrogen at 2 torr for 45 minutes
and at 810°K in hydrogen at 10^{-5} torr for 8 hours. Two minutes after termination
of the hydrogen flow, scattering data were recorded for the polished surface
at θ_r = 30° (other parameters as in Fig. 6). These showed an increase in
I_r of 10% over its previous value. However, this increase disappeared in less
than an hour.

Next to promote grain growth, the targets were heated to 1160°K at 10^{-7}
torr for 2.5 hours. With the target temperature reduced to 900°K, room-temperature argon was again scattered from both surfaces. Beams scattered from either
surface resembled one another but differed noticeably from those obtained
prior to heat treatment. The etched surface results are shown as one set of
curves (for T_s = 900°K) in Fig. 7. The height of the scattered flux lobe is
now considerably increased. While $\bar{v}_r$ and S are near their previous values
at θ_r = 0°, $\bar{v}_r$ now approaches $\bar{v}_i$ and S decreases markedly as $\theta_r \to$ 90°. The
slight decrease of $\bar{v}_r$ and S at lowest θ_r are also characteristics of lobular
scattering [2]. After four hours at 900°K, the target was heated to 1090°K
and maintained for 22 hours. The scattering patterns for the etched surface
after this heat treatment are also shown (T_s = 1090°K) in Fig. 7. A large
subspecular lobe now appears in the flux pattern and is accompanied by even
more pronounced excursions in $\bar{v}_r/\bar{v}_s$ and S. Upon cooling the target to room
temperature (T_s = 298°K in Fig. 7), the greatest portion of this lobe disappears;
yet the scattered beam is still easily distinguishable from those scattered from
this target at room teperature prior to the above heat treatment. Upon reheating the target to 1100°K, the previously observed lobe was restored, suggesting

that background gases physisorbed on the surface were dominating the scattering process at room temperature.

Other data have been obtained for aluminum, also. Properties of scattered argon beams for approximately thirty combinations of T_s, T_o, and θ_i in the range $300 \leq T_s$, $T_o \leq 850°K$ and $30 \leq \theta_i \leq 65°$ were measured using the two aluminum surfaces. On the whole, the data in Fig. 5 are typical of the patterns observed. Deviations from diffuse scattering were never large. The energy spread S was always near unity for $\theta_r < 60°$ and decreased slightly for $\theta_r > 60°$. The mean speed $\bar{v}_r$ was consistently near $\bar{v}_s$ at small θ_r and near $\bar{v}_i$ at $\theta_r = 90°$. The average thermal accommodation coefficient $\bar{\alpha}_r$ (defined earlier) decreased linearly from 1.0 to 0.4 as T_s was increased from $300°K$ to $850°K$ but showed no dependence on θ_i or T_o within the experimental uncertainty.

Discussion

The diffuse, almost fully accommodated scattering from copper and aluminum targets prior to heating in the vacuum chamber is similar in all respects to that previously observed with nickel and stainless steel targets [2]. Apparently, the scattering mechanism depends primarily upon the gases which are adsorbed on the target surface under such conditions and only secondarily on the target material. The diffuse flux distributions observed with argon scattering from aluminum agree with those found by Hurlbut [3], using nitrogen beams from a room-temperature oven source, and by Kuhlthau and Bishara [4], using nitrogen beams from a room temperature nozzle source. The normal momentum transfer data of Stickney and Hurlbut [6] indicate that room temperature argon is scattered diffusely from aluminum with full accommodation to the surface temperature in the range $400 < T_s < 800°K$. This indication is in contrast to the decreases in $\bar{\alpha}_r$ with T_s noted in the present results; this difference is possible due to the higher background pressures (ca. 10^{-5} torr) in their experiments. Abuaf and Marsden [7], employing nozzle type molecular beams and room-temperature aluminum surfaces, have determined normal momentum accommodation coefficients for argon beams in the energy range 0.064.4 11 eV. While concluding that the scattering must be diffuse at all beam energies, they observe a steady decrease in their coefficients from approximately 1.1 at 0.064 eV to 0.6 at 4.11 eV. Over the range of argon beam energies in the present tests (0.06-0. 18 eV for the argon beams), momentum accommodation coefficients computed from the data exhibit a similar decrease and agree with theirs within about 10%.

Leonas [9] has made measurements of scattered particle speeds using copper surfaces and thermal energy gases. He observed mean speeds for Ar and CO_2 scattering from copper which were indicative of adsorption of the incident beam on the target surface followed by evaporation. To our knowledge, no other comparable speed measurements for either copper or aluminum have been reported. However, several characteristics of the particle speed variations observed here for copper and aluminum have been observed in the scattering of similar gases from other surfaces. In previous investigations [2,11-15], the following two trends (to varying degrees of clarity) have been observed:

a) the mean speeds of particles scattered in directions near the surface surface normal are generally most typical of the surface temperature.

b) for targets where the scattering is highly directed, those particles scattered at angles between the flux peak and the surface tangent exhibit small accommodation of tangential momentum (this is particularly obvious from the data in Ref. 14).

Reduced thermal accommodation in conjunction with nearly diffused flux patterns such as those observed with the present aluminum and copper target (prior to an extensive anneal) has also been found with nickel and stainless steel targets [2].

Little in the way of a comparison of the present results to theory can be assembled since most of the scattering analyses are not applicable under the present circumstances. There is some basis for comparing the results in Fig.7 to predictions based on the so-called "soft cube" scattering model [16], but the appropriate computations have not yet been made.

Recrystallization and subsequent grain growth has been linked with all our observations of directed scattering from polycrystalline metals. While the original grain size for both the copper and the aluminum targets was about 0.01 mm, the copper which exhibited the patterns in Fig. 7 had grains near 1.0 mm in diameter. For aluminum the grain size increased to about 0.1 mm. Since grains of 0.1 mm exhibited highly directed scattering with nickel [2], oxide on the surface is probably responsible for the diffuse reflections from aluminum. Experiments in progress with single-crystal counterparts to these polycrystalline targets should clarify the relation of grain growth to the appearance of directed scattering.

The results reported in this paper pertain to atmospheric gases and engineering surfaces and, presumably, will be of more interest to the rarefied gas-dynamicist than to the surface scientist. The latter will undoubtedly be restricted in his interpretation by the limited specification of surface condition. Nevertheless, since the scattered beam properties tend to suggest the nature of the scattering surface, this technique may yet prove as valuable for probing surfaces as for studying gas-surface interactions.

Acknowledgements

Many individuals have contributed to this paper. We wish to acknowledge the assistance in the experiments given by students G. A. Pjura, Jr., W. F. Bailey, Jr., J. G. Gittings, and W. C. Hechtkopf. Sincere thanks are due Dr. J. J. Laidler of the Department of Materials Science for his collaboration in the metallurgical preparation of the copper specimens.

Above all, we take this opportunity to thank Dr. Otto Hagena whose efforts and abilities made these experiments possible.

References

1. For review papers which trace the development of nozzle-type molecular beams, the reader is referred to E. L. Knuth, Appl. Mech. Rev. 17, 751 (1964) or J. B. French, AIAA Journal 3, 993 (1965).

2. S. S. Fisher, O. F. Hagena and R. G. Wilmoth, J. Chem. Phys. (to be published).

3. F. C. Hurlbut, J. Appl. Phys 28, 844 (1957).

4. A. R. Kuhlthau and M. N. Bishara in Rarefied Gas Dynamics, Suppl. 3, Vol. 11 (Academic Press, 1966), p. 518.

5. A. R. Kuhlthau, Proc. 3rd Midw. Conf. on Fluid Mechn. (1953), p. 495.

6. R. E. Stickney and F. C. Hurlbut in Rarefied Gas Dynamics, Suppl. 2 Vol. 1 (Academic Press, 1963), p. 454.

7. N. Abuaf and D. G. H. Marsden in Rarefied Gas Dynamics, Suppl. 4 Vol. 1 (Academic Press, 1967) p. 199.

8. G. T. Skinner and B. H. Fetz in Rarefied Gas Dynamics, Suppl. 3, Vol 11 (Academic Press, 1966), p. 536.

9. V. B. Leonas, Zh. Priklad. Mekh. Tekh. Fiz. 2, 84 (1965).

10. O. F. Hagena and A. K. Varma, Rev. Sci. Instr. 39, 47 (1968).

11. S. Datz, G. E. Moore, and E. H. Taylor in Rarefied Gas Dynamics, Suppl. 2, Vol. 1 (Academic Press, 1963), p. 347.

12. J. N. Smith, Jr., and W. L. Fite in Rarefied Gas Dynamics, Suppl. 2, Vol. 1 (Academic Press, 1963), p. 430.

13. J. J. Hinchen and W. M. Foley in Rarefied Gas Dynamics, Suppl. 3, Vol. 11 (Academic Press, 1966) p. 505.

14. J. P. Moran, MIT Fluid Dynamics Res. Lab. Report T68-1 (1968).

15. J. P. Moran, H. Y. Wachman, and L. Trilling in Gas Surface Interactions (Academic Press, 1967), p. 461.

16. R. M. Logan and J. C. Keck, MIT Fluid Mech. Lab. Publ. 67-8 (1967).

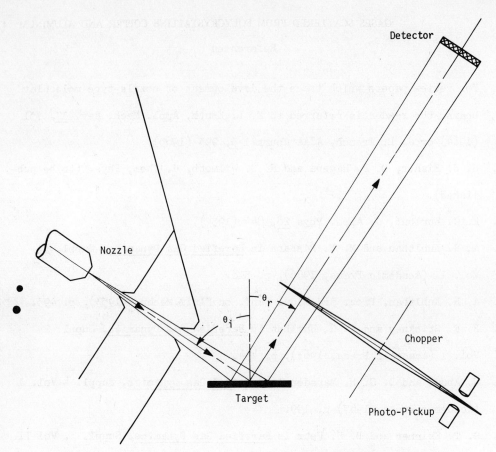

Fig. 1. Diagram of incident and Reflected Beam Geometries.

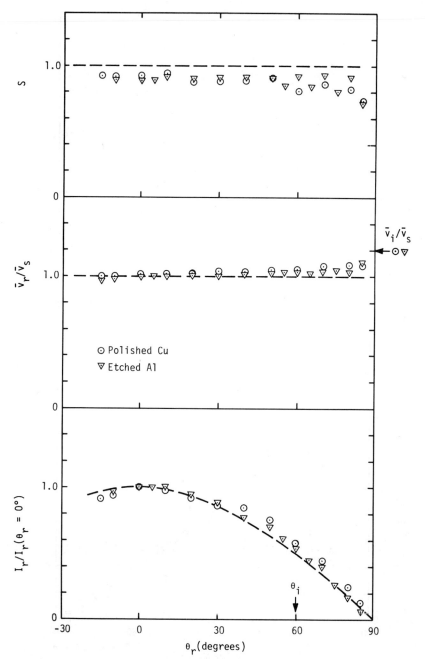

Fig. 2. Typical Flux, Mean Speed, and Thermal Spread Variations for Scattering from Unheated Targets. $\theta_i = 60°$, $T_o = T_s = 298°K$, Argon Gas.

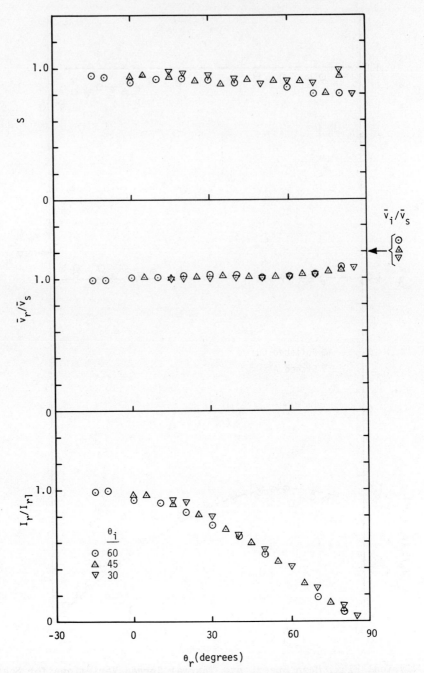

Fig. 3. Illustration of the Lack of Dependence of Beams Scattered from Un-heated Targets upon Angle of Incidence, θ_1. Etched Copper, $T_o = T_s = 298^\circ K$, Argon Gas.

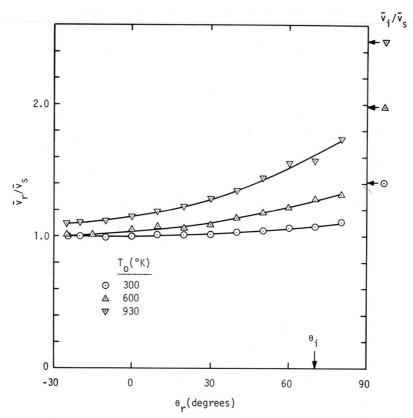

Fig. 4. Effect of Changing Incident Beam Energy on the Mean Speeds of Scattered Nitrogen Molecules (Flux Patterns Essentially Diffuse, Speed Distribution Essentially Maxwellian). Etched Copper, θ_i = 70°, T_s = 298°K.

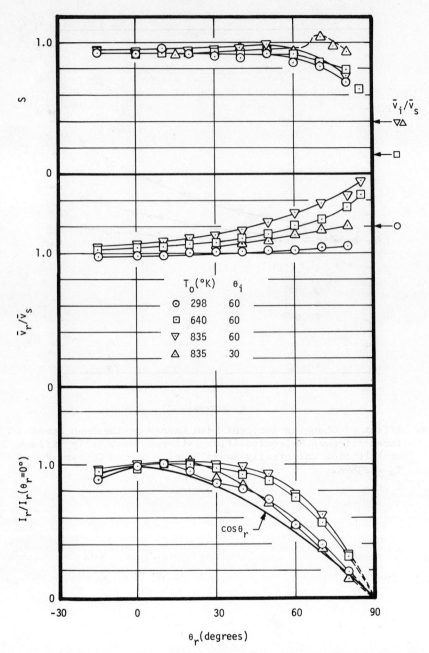

Fig. 5. Argon Beam Properties after Scattering from an Unheated Etched Aluminum Surface as a Function of Incident Beam Energy. $T_S = 298^\circ K$.

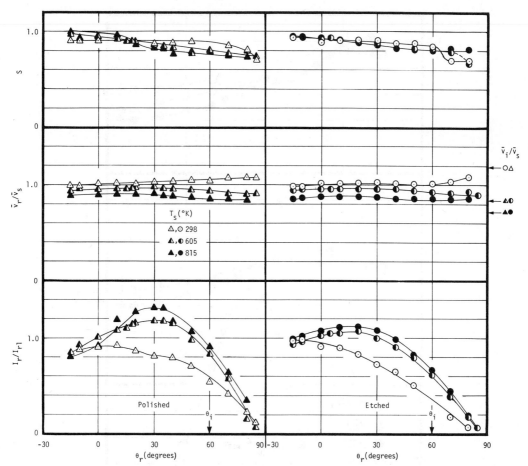

Fig. 6. Comparison of Argon Beams Scattered from Polished and from Etched
Copper Surfaces as a Function of Surface Temperature. $T_o = 298°$,
$\theta_i = 60°$ (no special heat treatment).

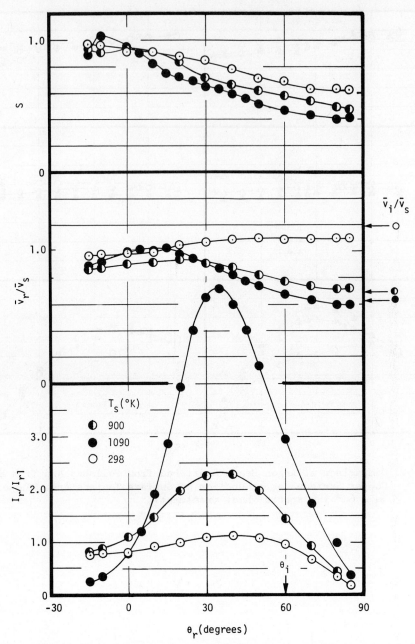

Fig. 7. Scattering of Argon From an Etched Copper Surface after Prolonged Target Heating at Elevated Temperatures. $T_o = 298°K$, $\theta_i = 60°$.

THE INTERACTION OF LOW-ENERGY ATMOSPHERIC IONS
WITH A WELL-CONTROLLED (100) TUNGSTEN SURFACE

J. B. French and R. H. Prince

Institute for Aerospace Studies
University of Toronto, Canada

1. INTRODUCTION

The study of electron ejection from a metal surface by impingement of ions has received considerable attention in the literature and is summarized by Massey and Burhop (1) and Kaminsky (2). Unfortunately, most of the work performed prior to 1950 is inconsistent due to the unspecified nature of the surfaces involved.

There are two reasonably distinct mechanisms by which electrons may be ejected from a surface as a consequence of ion bombardment. At energies above an ill-defined threshold of several hundred eV, ions eject electrons by a kinetic mechanism in which appreciable momentum is transferred from the ion to lattice atoms and subsequently to bound electrons in a manner that is still poorly understood. Nevertheless, this mechanism exhibits a strong dependence on incident energy in contrast to the mechanism at low incident ion energies which is the so-called "potential" ejection or Auger (radiationless) state transition. We are concerned here with the radiationless (Auger) transitions since the probability of radiation accompanying ion neutralization is extremely low ($\sim 5 \times 10^{-7}$) even for ions of thermal energies. This is because the characteristic radiative lifetimes ($\sim 10^{-8}$ sec) by far exceed the time spent by the ion in the vicinity of the surface ($\sim 10^{-14}$ sec). The ions must therefore be neutralized by processes involving tunnelling of metal electrons through the surface barrier. Figure 1. shows a sketch commonly used to describe the situation that exists when an ion approaches a metal surface. One electron (2) falls into the vacant atomic level by either process A (Auger Neutralization) or possibly by the two-stage process B (Resonance Neutralization and Auger de-excitation).

The energy released in this transition is then absorbed by a second electron (1). As long as the relative velocity between incident ion and metal surface is much smaller than the orbital velocities of atomic electrons, the neutralization process can be treated as quasi-stationery (adiabatic) by time-independent perturbation theory.

A number of theoretical models for this process exist, principally evolving form the experimental and theoretical work of Hagstrum (3-9) which in turn is based on the earlier work of Shekhter (10) and Cobas and Lamb (11). These models are first-order perturbation theories in which the Coulomb interaction between the two participating electrons is the perturbation which causes the transition. A slightly different approach is taken by Propst (12) in that the radiation field established by the neutralizing electron is the perturbation that excites the second electron. The role of the incoming ion is to provide a low-lying vacant electronic level for the Auger process. A new approach using the quantum mechanical theory of scattering has been derived by Wenaas and Howsmon (13) which removes some objectionable elements inherent in the former theories and appears to correctly describe the experimental data. The essential criterion is that the ionization potential, E_i, of the incoming ion must exceed twice the metal work function, i.e. $a = (E_i - 2\phi) > 0$ in order that an external (secondary) electron be produced.

Energy "a" is the maximum kinetic energy attainable by an ejected electron if the mechanism is independent of the ion kinetic energy, and is well substantiated by experimental work.

Aside from the work of Hagstrum, probably only that of Propst and Lüscher (14), Vance (15-17)*, and Mahadevan et al (18,19) can be considered sufficiently well specified with respect to gas adsorption. It is desirable for comparison with theory that the ion kinetic energy be as low as possible, and only Hagstrum has used incident ion energies below 50 eV. The only work pertinent to the present tungsten experiments is that of Propst and Lüscher who have published data for 50eV He^+, N_2^+, H_2^+ and O_2^+ ions incident on a clean polycrystalline tungsten surface. Vance and Mahadevan have published data for these ion species at a polycrystalline molybdenum surface at energies higher than the present work. There is excellent agreement with the extrapolation of these higher energy results to the pertinent low-energy regime presently considered.

The yields for diatomic ions are uniformly lower than a corresponding inert gas of similar ionization potential, and the theory of Auger emission while it can account reasonably well for the inert gas yields does not

*   The authors are indebted to D. W. Vance for communication of these results prior to publication.

include a mechanism describing the true behaviour of diatomic ions. It is this behaviour in particular which is of interest in this work, since the atmospheric species N_2^+, N^+ are considered. Propst and Lüscher (14) made the reasonable suggestion that since diatomic ions have the ability to vibrate and rotate, perhaps the neutralized ion is excited in these modes and that this energy sink is absorbing energy otherwise used in electron ejection. They suggested a procedure by which the inert-gas theory of Hagstrum could be modified to include the possibility of molecular vibration in the final state. A quantitative calculation of the N_2^+ - tungsten interaction is given in Appendix A which predicts reasonably well the experimental results of both Propst and Lüscher and the present work.

The experiment has illustrated and emphasized the effects of surface contamination; it has been the aim of this work to specify as comple- tely as possible the physical and chemical properties of the target material. Even with this ability it is necessary to perform the desired experiment on a sufficiently short time scale because of gas adsorption. A large amount of effort was spent in perfecting the apparatus to satisfy this requirement with the result that we are able to observe the transient effects of adsorption on the energy distribution functions which to our knowledge is the first time this has been done.

In summary, the intention of the experiment is to obtain data concerning the yields and energy distribution functions of secondary electrons produced at metal surfaces by the impact of atmospheric ions in the satellite velocity regime. In particular, we wish to infer from the observations the likelihood of molecular excitation as a result of neutralization, and to observe the effects of ion kinetic energy and surface contamination on the process.

2. EXPERIMENTAL APPARATUS

Figure 2 is a schematic diagram of the apparatus showing the location of internal components. Briefly the system operates as follows. Gas is admitted from storage bottles through a metering valve into the ion source, where ionization by electron impact is performed. Ions are extracted from the source and accelerated to an energy suitable for magnetic mass analysis by the 60 degree sector-field analyzer. The desired ion species is subsequently decelerated and focussed in two stages on the target T. Secondary electrons produced by this interaction are energy-analyzed by applying varying potentials to the collector sphere S, and observing changes in the collector sphere current I_s. The target assembly contains devices for both removing adsorbed gases and detecting their presence by means of a surface work function monitor based on the retarding-field diode technique. The divider network used to supply lens potentials has been referenced to the ion source, so that the ion kinetic energy at locations other than the target is independent of the anode potential. This convenient arrangement permits the ion beam energy (anode potential) to be continuously varied while maintaining lens focus. The behaviour of the intensity-energy curve suggests that at low energies there are space-charge repulsion effects in the final lens since the ion energies are fixed elsewhere in the system. At higher energies, the intensity is limited by the number of ions generated in the source, and the fraction of these transmitted by the mass spectrometer. In fact, above 20 eV kinetic energy, the beam intensity is almost constant at about 6×10^{-9} amperes (approx. 3×10^{10} ions/second) which is the "emission-limited" value.

The intensities obtained for the N^+ species are a few percent of those for N_2^+ due to the low ratio of cross-sections for 30 eV electrons. The kinetic energy spread of the beam was estimated by retarding the ions at the target. The distribution exhibits a decay towards lower energies character-istic of the source potential distribution such that the width at half-maximum is 0.4 eV, with 50% of all ions within a 0.5 eV range, and 90% within a 2 eV range of energies.

There are limitations to the apparatus that should be emphasized. Although the majority of ions are well-focussed at the target to perhaps a 3 mm. dia, spot, the variation of beam focus with ion energy is such that at about 10 eV the fraction of incident ions missing the target and striking the secondary electron collector has reached a few percent. The apparatus is then unsuitable for electron-emission yield studies since the typical values of γ_i are of this order. Furthermore, the beam focus will deteriorate at higher energies if strong negative potentials are applied to the electron collector sphere in the electron-retarding field analysis. These considerations have limited the study of electron energy distributions to those for ions of 15 eV kinetic energy or greater, for which negative potentials of up to 8 volts may be applied to the spherical collector without serious lens failure. The total yields may be obtained to slightly lower energies, however, since the sphere is electron-attracting in this mode. The lens failure is more rapid for the lighter ion species which suggests that the mechanism is at least partly magnetic in nature, most likely due to the stray getter-ion pump fields. The above constraints do not apply if the beam is to be used in a environment free of electrostatic fields for aerodynamic simulation, for example. For this purpose, flux simulation of up to 3×10^{11} ions/cm^2-sec is possible.

3. <u>EXPERIMENTAL PROCEDURE</u>

   <u>Instrumentation:</u>

   The philosophy followed in developing the instrumentation is that it is extremely desirable to perform all data-reduction operations simultaneously with the data-gathering. Furthermore, in order to directly compare the changes in distribution function resulting from a change in a physical parameter (ion energy, surface work function, etc.), the output data must be normalized to the ion beam current I_i. There are two approaches to this problem. One is to maintain constant beam current under all conditions by a feedback stabilizing system as used by Hagstrum (20), the other is to accept variations in I_i, but perform operations on the normalized current ratio $\rho = I_s/I_i$. There are disadvantages to either system, but it is felt that the technique to be described is the best to date.

   In the constant ion current system, a signal proportional to the true beam current I_i is used to control the ionizing electron current in the source and hence vary I_i. There is a limitation on the degree to which the ion current fluctuations may be reduced without oscillation occuring due to excessive closed loop gain. Hagstrum (20) reports that the variations in I_i during a sweep of collector potential are no greater than 0.2% by this method. A limitation however, is that if one wishes to compare distributions for conditions resulting in a wide ion current spread, a current level attainable by the most detrimental case must be used throughout, with the result that the quality of the data is not maximized in every case.

   In the other method of operation (normalizing with respect to a variable ion current) there is a difficulty in devising a fast, accurate analog divider to obtain the non-dimensional electron current ρ. Propst and Lüscher (21) describe a system of self-balancing servo loops which is neither flexible nor fast, requiring sweep times of 100 seconds and servo-motors of high performance. Analog division is performed in the present experiment by means

of a digital voltmeter with high precision and sampling rates, that was
constructed in this laboratory because the unique combination of desired
specifications was not commercially available. By this means, we obtain
a numerical readout of the current ratio $\rho($ and hence γ_i) as well as its
voltage analog within 0.1%, in conjunction with a sampling rate that permits
the entire distribution function to be obtained during a 10 second interval.
This time interval permits the observation of clean-surface experiments in
the presence of adsorbable gases without resorting to exotic vacuum techniques.

Figure 3 illustrates the data reduction system employed in the
present experiments. In addition to the real space currents within the
energy analyzer, there are additional virtual (displacement) currents which
exist by virtue of parasitic resistive and capacitive elements in the device.
The electron currents to be measured have a range of magnitudes of the order
10^{-12} to 10^{-10} amperes, while we wish to vary the sphere potential by $\pm$ 10
volts within a time compatible with clean-surface experiments, i.e. within
a ten second period. The use of guarding techniques can eliminate all but a
remnant sphere-to-ground capacitance of approximately 10 pf, which at the
above sweep rate contributes a displacement current of 10^{-11} amperes. Our
problem now is essentially how to apply a time-dependent waveform to a capacitor
(which of necessity requires a current waveform) without passing the charging
and discharging current through a DC ammeter attached to one plate of this
capacitor. The solution (Fig.3) is that we may apply this waveform by
capacitive coupling rather than direct DC coupling and furthermore if the
electrometer is placed across the resulting capacitance bridge, a nulling of
the bridge is possible when the ratios of capacitance and resistance are correct.

This method of bridge balancing (S - NULL) permits collector potential
variations of 1 volt/second with resultant spurious current as low as 10^{-14}
amperes, and is achieved by adjustment of the variable capacitance at three
levels until the value $(C_s + C_g)$ is reached, (the resistive arms are equal).
A similar solution allows nulling of the spurious target current due to C_s.
For this case we again form a bridge network, but since the capacitance is
already low, we prefer to adjust the resistive ratio. In addition, the waveform
applied to the bridge must be inverted, and this operation is ideally performed
by an operational amplifier inverter (6) of variable gain, (T - NULL).

Adsorption Experiments:

   A number of auxiliary experiments were performed to determine the
surface state during the experimental interval using the retarding-field
diode technique. By this method, the degree of surface coverage (and species)
may be determined by a correlation with the change in surface potential (or
work function). The direction of potential change is determined by the relative
importance of the various binding forces (2), but in most cases an adsorbed
monolayer results in a surface work function increase of between 0.2 and 1
volts. There have been a number of highly-pertinent LEED experiments recently
performed which enable the present surface diode experiments to indicate
quantitative results.

   Estrup and Anderson (22) have investigated the adsorption of hydrogen
molecules at a (100) tungsten surface, and correlate work function and surface
coverage, while Madey and Yates (23) describe the other gas of interest,
nitrogen. The primary motive is not to investigate the adsorption process in
detail, but rather to be certain that the gas species and degree of coverage
is established for every electron energy distribution function obtained. A
series of experiments were performed both with and without the beam in operation,
in which the target was flashed to a high temperature (approx. $2300^{\circ}K$) and
the time variation of work function observed.

   The change in surface potential is almost linear with surface coverage,
σ, so that the adsorption traces also represent the variation of σ with time
after flashing. (Fig. 4)

   The initial sticking probability is high, and from the quasi-linear
portion of the trace can be estimated to be about 0.67, in comparison with
that estimated by Estrup and Anderson (22) of 0.65.

Influence of Ion Beam:

   It may be argued that the adsorption of beam molecules on impact is
unlikely since their kinetic energy is large. There is evidence, moreover,
that the beam may actually reduce the net number of adatoms (most of which
come from the background gas) by virtue of this energy. In Fig.4 we show three
curves for operation with a nitrogen source, two representing conditions under
which the beam strikes the target at 30 and 15 eV kinetic energy respectively,
and a third for which the mass analyzer is used to prevent N_2^+ impact while the
source and background gas loads remain. The time constants characteristic of

hydrogen adsorption are seen to decrease in the above sequence, indicating not only that the ion beam is capable of "scrubbing" the surface, but that the process is energy-dependent as expected. It is possible to make inferences from these curves, by considering the change in slope, of the overall effective sticking probability resulting from ionic impact (24).

4.   RESULTS

EXPERIMENTAL STUDY OF AUGER NEUTRALIZATION OF N_2^+, N^+ AT A (100) TUNGSTEN SURFACE:

The energy distributions of electrons ejected from a clean (100) tungsten surface by N_2^+ ions of 15 to 30 eV incident kinetic energy are shown in Fig.5 and are typical of the data obtained. The curves are plotted immediately after flashing the target to approximately $2300^{\circ}K$ for several seconds and represent data for a surface clean to less than 5% of full coverage. It is interesting to note that the upper energy limit is in excellent agreement with the value a = $(E_i - 2\phi)$ predicted by energy conservation in the simple model described in Appendix A . The distribution is considerably sharper than that obtained by Propst and Lüscher for 50 eV N_2^+ ions at a polycrystalline tungsten surface (see Fig.11) and suggests that either kinetic broadening is significant at 50 eV or that there is a effect due to the single-crystal surface. The total yield is also considerably greater than in the previous work, but is nevertheless about one third that of an inert gas of similar ionization potential (Ar^+). It is this divergence that we attempt to explain by the introduction of molecular vibration energy sinks. We would expect that the $N_o(E_k)$ curves form a smooth sequence as ion energy is varied, without sudden departures. It is seen that the distribution narrows considerably as ion energy is decreased, and that violation of the simple limit $(E_i - 2\phi)$ is observed at higher energies. In addition, the peak location shifts to lower values of E_k, and the peak height passes through a minimum at about 20 eV incident ion energy.

These trends must be taken as evidence of kinetic broadening effects ignored in the simplified treatment of Appendix A . It has been impossible to obtain distributions and yields at ion energies much below 15 eV (nominal) due to defocussing of the ion beam at the target. The data presented here, however, are considered reliable and free from the effects of lens defects. The onset of lens failure is easily seen in the $N_o(E_k)$ distributions as a violation of the asymptotic approach to the energy axis at large E_k. The data

presented is well-behaved in this regard. The reproducibility of the data
is illustrated in Fig.6 which represents the influence of adsorbed hydrogen
for constant ion energy (30 eV). Curve (1) represents the clean condition
and two such curves are shown superimposed. The subsequent transient effects
of adsorption are represented by curves (2) to (7) for coverage up to an
equilibrium level. Correlation between time and hydrogen adsorption level
(or work function change $\Delta\emptyset$) is provided by the retarding-field diode curve
from Fig. 4 for the case of a 30 eV beam. The work function change $\Delta\emptyset$ is
tabulated for each curve in Fig.6 since the curves should be displaced to
the left by this amount in order to compensate for the varying contact
potential between target and sphere. Although the zero-field point established
by thermionic measurements is correct for the case of clean (100) tungsten,
it does not apply to the case of a gas-covered surface since the target work-
function is increased by an amount as high as $\Delta\emptyset = 0.8$ volts. It was felt
that the replotting of the distributions would be detrimental to the quality
of the presented data. In a sense, the shift in potential of the origin of
the Auger electron distribution is in itself a measurement of $\Delta\emptyset$ and hence
surface coverage. With this point in mind, the peak of the distribution not
only is drastically reduced by the adsorption of hydrogen but also is shifted
to lower energies. Similar considerations apply to the family of curves for
15 eV N_2^+ at a (100) tungsten surface (Fig.7). The effect of ion kinetic energy
on the distributions for hydrogen-covered (100) tungsten is shown in Fig.8.
The shift in retarding potential due to target adsorption is plainly visible
and may be seen to be very close to the surface diode measurement of 0.8 volts.
It is seen that the distributions for the clean and hydrogen-covered surface
change with ion kinetic energy in a similar manner with respect to maximum
energy (distribution broadness) and peak location. The peak height, however,
appears to pass through a maximum at about 25 eV, an effect consistently
observed. The variation of the total yield γ_i with ion kinetic energy is also
similar for the clean and hydrogen-covered surface (Fig.9), together with an
indicated yield obtained by Propst and Lüscher (14). The latter curves were
obtained by plotting the ratiometer output against source anode potential for
the spherical collector slightly electron-attracting. Alternately, the total
yield may be plotted as a function of time after flashing so that a family of
curves representing increasing hydrogen coverage may be obtained. The behaviour
of γ_i with time closely resembles the "reflection" of the target work function

variation (Fig.4) and suggests that for a given ion-surface combination
the yield is quasi-linear with surface work function. There is a decrease
in yield with decreasing energy that is reasonably gentle on this scale.

The total electron yield γ_i and its dependence on ion kinetic energy
for N^+ ions incident on a clean and hydrogen-covered (100) tungsten surface
are shown in Fig.10. It was not possible to obtain meaningful energy distributions due to the low signal-to-noise ratio which exists as a consequence
of the small N^+ beam intensitites ($\sim 10^{-11}$ amperes).

The total yields exhibit a behaviour and magnitude not unlike the
diatomic N_2^+ ion, though the ionization potential of N^+ is approximately 1 volt
less. We would expect the N^+ ion to behave in a manner similar to an inert gas
of comparable ionization energy and incident velocity. The electron yield for
Kr^+, for example, is about .04 (4) at a clean tungsten surface at an incident
energy of 40 eV. The comparison is crude since the nature of the interaction
potentials at close range will differ due to the high bonding energy of the
nitrogen atom, but the yields are nevertheless of comparable magnitude.

5.   DISCUSSION OF RESULTS

The comparison of the total yields with the work of Propst and Lüscher
deserves comment. The present yields are considerably larger and in addition
the distributions are narrower, even though the theoretical energy limit
($E_i-2\phi$) appears valid in both cases. This is surprising, since in the present
work, we find that at 30 eV the distributions have broadened considerably and
further broadening would be expected at the 50 eV level used by Propst and Lüscher.
The difficulty arises primarily because the surface used by the latter is poorly
specified, at least in the open literature. It is suggested that the polycrystalline surface employed is either predominantly (110) in strucutre ($\phi =$
5.2 eV), or if consisting of lower work function planes, is carbon contaminated
such that the effective work function is approximately 0.75 volts above the
present value for (100) tungsten ($\phi = 4.5$ volts). This implies that the value
of ($E_i-2\phi$) for their work is perhaps 1.5 volts lower than in the present case
(reducing γ_i) and that broadening of the distributions is also of this
magnitude.

It is true that prior to obtaining the nitrogen ion data, Propst and
Lüscher had obtained results for He^+ to compare with the earlier data of
Hagstrum (6). However, the $N_o(E_k)$ distribution for the (He^+-W) system lies
entirely above E_o (the lower limit of E_k is finite) so that changes in surface

work function of the above magnitude result in small changes in γ_i. Furthermore, the shift of the distribution function on the energy scale would go un-noticed since the zero-field point was located by the apparent origin of the function itself. In the measurement of the present data, inadequate surface cleaning or non-idealities in electron-collector geometry tend to reduce the total yields so that these factors are not relevant. It is natural to propose extra energy sources such as metastable ions, although recent work by Vance (17) indicates that extremely few survive the time-of-flight from source to target. In addition, the energy maximum of the $N_o(E_k)$ distributions strongly supports the value $(E_i - 2\phi)$ expected for a ground-state ion within the amount reasonably explained by kinetic broadening effects.

In the case of N_2^+, the absence of variation in γ_i with changes in electron impact energy suggests the beam is entirely in the ground state, although a small population in the low-lying A state $(^2\Pi_u)$ may go undetected. Significant populations in the higher states would result in larger increases in γ_i which are not observed (17). The lifetimes of these excited stated for N_2^+ have all been measured and are all at least one order shorter than the minimum possible transit time through the apparatus ($\sim 40\ \mu s$), obtained from a knowledge of lens potentials and assuming that the source residence time is zero. The latter is probably untrue, since it is determined by the complex, potential structure in the source. The time spent in the ion source may thus be an order of magnitude greater than the transit time through the instrument. We thus conclude that the beam is in the ground electronic state upon arrival at the target. There remains the possibility of vibrational excitation in the incoming ion beam and Moran and Friedman (25) have considered this problem. Since for homonuclear diatomic ions vibrational transitions are not permitted, the electronic transitions from the short-lived A state will retain their former vibrational quanta. The calculation of the vibrational populations formed in the electron impact process indicated that 90% of all incident ions are in the ground state electronically and vibrationally, and of the remaining 10% a large majority possess only a single vibrational quanta $(v = 1)$.

The excitation level of the N^+ ion beam is in some doubt. Vance reports that the value of γ_i is independent of the electron impact energy, although the published results (17) do not extend to ionization threshold. The metastable states $^1D_2$ and ^1S_o have lifetimes of 4.13 min. and 0.9 sec, respectively so that the apparent absence of these states from the beam suggests that the selection rules are obeyed even in low-energy electron impact ionization. This is surprising, and represents a discrepancy with recent measurements of charge transfer cross-sections. Both in the present case, and in the work of Vance there is no possibility of appreciable N_2^{++} concentration which would result in very large electron yields because of the large value of E_i ($\sim$ 43 eV).

It is important to realize that small concentrations of N^+ in the low-lying $^1D_2$ state would be undetected by measurement of γ_i alone. The measurement of the energy <u>distribution</u> of secondary electrons can, however, indicate the presence of excited ion states, by observation of the energy maxima. We have observed that the distribution for N^+-Mo extends perhaps 1 eV <u>beyond</u> that for $(N_2^+$ - Mo) a situation that is unexpected since the ionization potential for N_2^+ exceeds that of N^+ by about 1 eV. Regardless of the state of carbon contamination in the surface the value of ϕ is equal for both distributions, so that the apparent recombination energy for N^+ exceeds the ionization potential by about 2 volts. This is precisely the height of the metastable level $^1D_2$ above the $^3P_0$ ground state, so that there may be a small number of excited N^+ ions that are insufficient to alter the total yield appreciably, but sufficient to shift the energy maximum. This proposal may resolve the earlier conflict in Vance's work.

The above discussion has thus eliminated initial ion excitation as a possible alternative to a difference in work function in order to account for the higher yields in the present work compared to that of Propst and Lúscher.

Another explanation that is not easily refuted is that a significant fraction of incident ions undergo resonance neutralization (transition B,Fig.1) at low incident ion energies. This process requires that an unoccupied excited level is adjacent to the filled portion of the conduction band, i.e. $\phi < (E_i - E_x)$ $< E_o$. For nitrogen on (100) tungsten this limits the excited level to 4.6 eV$<$ $E_x < 11.0$ eV if energy level shifts are neglected. A most likely level is $^3\Sigma_u^+$ which gives rise to the forbidden Vegard-Kaplan Bands ($\sim$ 6 eV). The maximum energy of external electrons resulting from the Auger de-excitation of this level is $(E_x - \phi)$ $\sim$1.5 eV so that we could not support this hypothesis by observation of the high energy tail as in the case of Ne^+. The narrowness of the observed distribution does, however, add credence to this process, since its effect will be to populate the low-energy portion of the distribution beyond that expected on the grounds of pure Auger neutralization (see Fig. 11).

Finite instrumentation time response will broaden the experimental distributions , so that the narrowness of the data is considered significant. The theory could be reworked to include this possibility, assuming vibrational excitation both at the molecular ground state and at $^{3}\Sigma_{u}^{+}$, and superimposing the resultant distributions in a manner reflecting the relative probabilities of the two mechanisms.(i.e. a fitting parameter). A procedure similar to this was adopted by Hagstrum (5) to explain the behaviour of Ne^{+} at a tungsten surface, though in this case there was a considerable violation of the simple energy maximum $(E_{i}-2\phi)$.

The success of the pure Auger neutralization theory in matching Propst and Lüscher's distribution at 50 eV using Hagstrum's value of the escape parameter (Fig. 11) is most encouraging. It is even better than shown, since the energy scales used are not identical. Propst and Lüscher used the distribution itself to locate the zero-field point, whereas in reality the zero point lies to the left, so that the distribution appears to extend to negative energies. This anomaly is attributed to stray magnetic fields in the vicinity of the energy analyzer. In the present simple theory, our only fit to experiment is through the escape function f, formerly used by Hagstrum (5). It must be increased in order to match the results at 15 eV (by equal areas or yields). By increasing this parameter we are in essence postulating an increased focussing of excited metal electrons towards the surface normal, before they surmount the surface potential barrier. It could be argued that this is reasonable in a material possessing less disorder and preferred directions for electron motion, one of which is the normal itself. Hagstrum (8) finds that a value f=4.8 is necessary to fit data for He^{+} on (111) Ge, assuming the required large escape probability is entirely due to the anisotropy of P_{Ω}. The escape probability is also increased by the fact that the surface barrier is not planar, as well as the anisotropy effect. The bulge in the equi-potential surfaces outside the solid leads to a greater critical angle Θ_{c} than that predicted for a planar boundary. It is possible that a better fit could be obtained by fixing f=2.2 and including the effect of Auger de-excitation mentioned earlier, since the high-energy tail of both experimental distributions (Fig.11) is well-described by the simple theory of Appendix A.

6. SUMMARY

Experiments have been performed under stringent conditions of high vacuum to obtain new data concerning the Auger neutralization of atmospheric ion species at a well-characterized (single crystal) surface. The effects of surface adsorption on the electron energy distributions, which represent the means by which we have gained information about the neutralization process, have been observed transiently by the incorporation of modern instrumentation techniques. The data-reduction system employed is believed a new approach to the problem, and represents an alternative to existing methods, generally more complex.

A theory revision to account for the failure of the inert-gas formulation to adequately describe the neutralization of diatomic ions has provided encouraging quantitative agreement with experimental data. The hypothesis of vibrational excitation during the Auger neutralization process has been shown to be a plausible explanation of the experimental results using reasonable assumptions concerning the physical parameters pertinent to the mechanism. It has been suggested that the two-stage process is more probable than formerly believed for the low incident ion energies considered. Either this mechanism or the simple formulation of the revised theory prohibits agreement with the present data in both overall yield and energy distribution, although the salient features are represented in a fashion not at all possible using the inert-gas theory.

It is felt that the ultimate solution to the theoretical problem lies not in detailed modifications of an existing theory, but in an entirely fresh approach, based firmly on first principles, such as that presently considered for the simpler He^+ and Ar^+ cases by Howsmon and Wenaas (13). The difficulties to be experienced in applying a rigorous quantum-mechanical treatment to complex diatomic ion interactions cannot be over-stressed, however.

ACKNOWLEDGEMENTS

This project was supported by the Office of Naval Research, under Contract Nonr-4073(00) during ion beam development, the Air Force Flight Dynamics Laboratory, under Contract AF-33(615-3855) for the ion neutralization studies, and continuously by the Defence Research Board of Canada. This assistance is gratefully acknowledged.

APPENDIX A: <u>THEORETICAL DESCRIPTION OF THE AUGER NEUTRALIZATION OF</u> N_2^+ <u>AT A CLEAN</u>
<u>TUNGSTEN SURFACE</u>.

It was hypothesized earlier that the failure of the relatively successful
inert-gas theory to predict the observed electron yields for diatomic gas ions
was due to vibrational excitation of the neutralized molecule. Vibrational
excitation only is considered because the energy quanta are large compared to
those for rotation, but in addition the vibrational periods ($\sim 10^{-14}$ s) are
compatible with the time scale of the interaction. The rotational periods are
typically two orders greater in magnitude. The effect of the vibrational states
will be to produce a large increase in the density of the final states lying
close to the initial state. Transitions to these states will result in excited
electrons of lower average energy and hence lower probability of escape. The
electron yield should therefore decrease in accordance with experiment.

We will attempt to quantitatively calculate the new $N_o(E_k)$ distribution
without knowledge of the close-range interaction (i.e. large s) by modifying
the procedure used by Hagstrum (5,8) and proposed by Propst and Lüscher (14).
We assume that all state transitions are equally probable, and further stipul-
ate that the molecule is not dissociated by the interaction, ensuring ion
kinetic energy conservation. The energy available for excitation is at most
$(E_i' - \phi)$ so that only at large s is dissociation energetically possible; the
transition probabilities are low in this case and it is unlikely that the <u>entire</u>
energy excess is absorbed by the molecule alone. Small anharmonicity of the
vibrational oscillator may be included by writing the vibrational energy levels
as

$$E_v = hc\omega_e(v+\tfrac{1}{2}) - hc\omega_e x_e(v+\tfrac{1}{2})^2 + \ldots.$$

where v is the vibrational quantum number, ω_e is the vibrational wave number
(2360 cm^{-1}), $\omega_e x_e$ is the anharmonic constant (14.46 cm^{-1}), c is the velocity of
light and h is the Planck constant. Vibrational quantum numbers up to v=25 are
possible at large s_o, although this value may be exceeded if the transition occurs
at close range (in general, at high incident energies). With the consideration
of vibrational excitation, the vertical displacement from initial to final state
becomes the sum of E_k and E_v. For a given value of E_v, there is a probability
density function $N_v(E_v)$ which is proportional to the number of permissible
states of electron energy E_k subject to the inequality

$$(E_i - 2E_o) \leq (E_k + E_v) \leq (E_i - 2\phi)$$

This is an expression for the range of energy levels available for the final
state; a negative value of E_k is permissible, but implies that the electron

cannot escape the surface barrier. The density of final states available to excited electrons is $N(E)$ so that

$$N_v(E_v) = C\int_{E_F}^{(E_i-2\emptyset+E_o-E_v)} N(E)\ dE \qquad\qquad 0 \leq E_v \leq E_d$$

where the upper bound is obtained from the above inequality and is a function of E_v itself. The result of this calculation for a $\frac{1}{2}$-power $N(E)$ is a quasi-linear distribution favouring low vibrational excitation as expected which is truncated at the dissociation limit at a value of 20% of that at v=0. For E_v' above $(E_i-2\emptyset)$ the production of an external electron is not possible. The energy conservation equation now becomes (Fig.1)

$$E_1+E_2 = 2E = (E_k+E_v) + 2E_o - E_i'$$

so that in calculating the number of initial states resulting in an excited electron of energy E_k, the appropriate Auger transform is now $T[\frac{1}{2}(E_k+E_v+2E_o-E_i')]$. The inclusion of vibrational excitation results in a shift of the Auger transform to lower energies, providing the mathematical means by which the total electron yield is reduced. Since numerous values of E_v are possible for a given E_k, we must weight the Auger transform by $N_v(E_v)$ and integrate over all permissible values of E_v. Similarly, in calculating the number of final states available for a given E_k, we must consider that for each value there are a number of values of E_v such that the level has a degeneracy (v_k+1) where v_k is the vibrational quantum number corresponding to an energy of $(E_i-2\emptyset-E_k)$ or E_d, whichever is less. The formulation for the distribution of excited electrons prior to escape thus becomes

$$N_i(E_k+E_o) = G.(v_k+1).N(E_k+E_o)\int_{E_v} N_v(E_v)T[\frac{1}{2}(E_k+E_v+2E_o-E_i')]\ dE_k$$

where G is a normalizing constant so that we obtain one excited electron (which may or may not escape) for every incident ion. Using Hagstrum's parametric escape function $P_e(E_k)$ we obtain the observable free electron energy distribution function

$$N_o(E_k) = N_i(E_k+E_o).P_e(E_k)$$

The integration may be performed using E, the mean value of participating electrons as the variable of integration, while fixing (E_o+E_k).

The function is thus evaluated over the conduction band $0 \leq E \leq E_F$ in two half-range integrations corresponding to the split transform regions $0 \leq E \leq E_F/2$ and $E_F/2 \leq E \leq E_F$. The integration limits may not coincide with the latter limits since there are energy constraints on the variable of integration. Firstly, E_v is always positive which from the energy conservation relation implies that $E \geq (E_k+E_o-E_i+E_o)$. Since (E_k+E_o) is the excited electron energy, it must exceed the Fermi level E_F since all levels below this energy are filled. Thus the <u>lower</u> limit of integration is given by the latter inequality and not by E=0. Furthermore, we find numerically that for $E_k+E_o > 11.0$ eV, this lower limit exceeds $E_F/2$ so that only the second range of T(E) is required. Secondly, it has been assumed that $E_v < E_d$, placing an <u>upper</u> bound on E when $(E_k+E_o) < (E_i-2\phi+E_o-E_d)$, i.e. low enough to permit dissociation. In other words, for $(E_k+E_o) < 10.03$ eV, the upper bound on E is no longer E_F but $\frac{1}{2}(E_o+E_k+E_d-E_i+E_o)$. Thus there are three distinct ranges of integration depending on the excited electron energy, two requiring integration over the entire conduction band, and one over only the upper half. The vibrational distribution function $N_v(E_v)$ may be rewritten in terms of the variables E,E_k so that the entire integration over initial states is performed using E as variable with E_k fixed. The expression for the kinetic energy distribution function of Auger-excited metal electrons is thus

$$N_i(E_k+E_o) \propto (v_k+1).N(E_k+E_o). \left\{ \int_b^{E_F/2} N_v(E,E_k)\ (E)\ dE\ +\ \int_{E_F/2}^{b+E_d} N(E,E_k)\ (E_F-E)\ dE \right\}$$

$$E_F < (E_o+E_k) < (E_i-2\phi+E_o-E_d)$$

$$\propto (v_k+1).N(E_k+E_o). \left\{ \int_b^{E_F/2} N_v(E,E_k)\ (E)\ dE\ +\ \int_{E_F/2}^{E_F} N_v(E,E_k)\ (E_F-E)\ dE \right\}$$

$$(E_i-2\phi+E_o-E_d) < (E_o+E_k) < (E_F/2+E_i-E_o)$$

$$\propto (v_k+1).N(E_k+E_o). \left\{ \int_{E_F/2}^{E_F} N_v(E,E_k)\ (E_F-E)\ dE \right\}$$

$$(E_F/2+E_i-E_o) < (E_o+E_k) < (E_i-2\phi+E_o)$$

where b is the bound given by $b = \overline{E_o+E_k-E_i+E_o}$. By assuming the density of final states as $N(E_k+E_o) \propto (E_k+E_o)^{\frac{1}{2}}$, the resulting energy distribution function of excited metal electrons prior to escape is shown in Fig.12 normalized to one excited electron per incident ion. The distribution neglecting vibrational excitation is shown for comparison. Since only those electrons of energy greater than E_o have any probability of escape, we can see that the yield of free electrons will be drastically reduced from the inert-gas value.

The energy distribution of free electrons $N_o(E_k)$ is obtained by means of the parametric escape function shown in Fig. 12 for three values of the parameter f. The curve for f=1 is that for an isotropic angular distribution of excited electrons over all angles Θ_1. The curve for f=2.2 is that used by Hagstrum to fit his data for He^+ ions incident on a polycrystalline tungsten surface. The parameter f=3.0 provides the best fit of the theory to the present results. The corresponding $N_o(E_k)$ functions are shown in Fig.11 and a comparison is made with both the present results and those of Propst and Lüscher (14).

REFERENCES

1. H. S. W. Massey and E. H. S. Burhop, <u>Electronic and Ionic Impact Phenomena</u>, Oxford University Press, London (1956).

2. M. Kaminsky, <u>Atomic and Ionic Impact Phenomena on Metal Surfaces</u>, Academic Press, New York, (1965).

3. H. D. Hagstrum, Instrumentation and Experimental Procedure for Studies of Electron Ejection by Ions and Ionization by Electron Impact, Rev. Sci. Instr. <u>24</u>, 1122 (1953).

4. H. D. Hagstrum, Auger Ejection of Electrons from Tungsten by Noble Gas Ions, Phys. Rev. <u>96</u>, 325 (1954).

5. H. D. Hagstrum, Theory of Auger Ejection of Electrons from Metals by Ions, Phys. Rev. <u>96</u>, 336 (1954).

6. H. D. Hagstrum, Auger Ejection of Electrons from Tungsten by Noble Gas Ions, Phys. Rev. <u>104</u>, 317 (1956).

7. H. D. Hagstrum, Auger Ejection of Electrons from Molybdenum by Noble Gas Ions, Phys. Rev. <u>104</u>, 672 (1956).

8. H. D. Hagstrum, Theory of Auger Neutralization of Ions at the Surface of a Diamond-Type Semiconductor, Phys. Rev. <u>122</u>, 83 (1961).

9. H. D. Hagstrum and Y. Takeishi, Effect of Electron-Electron Interaction on the Kinetic-Energy Distribution of Electrons Ejected from Solids by Slow Ions, Phys. Rev. 137, A304 (1965).

10. S. S. Shekhter, Neutralization of Positive Ions and Ejection of Secondary Electrons, Zh. Eksper. Teoret. Fiziki 7, 750 (1937).

11. A. Cobas and W. E. Lamb, Jr., On the Extraction of Electrons from a Metal Surface by Ions and Metastable Atoms, Phys. Rev. 65, 327 (1944).

12. F. M. Propst, Energy Distribution of Electrons Ejected from Tungsten by He$^+$, Phys. Rev. 129, 7 (1963).

13. A. J. Howsmon and E. P. Wenaas, Theory of Auger Ejection of Electrons from Metals by Ions, Proceedings of the Fourth International Materials Symposium (1968), Berkeley, California.

14. F. M. Propst and E. Lüscher, Auger Electron Ejection from Tungsten Surfaces by Low-Energy Ions, Phys. Rev. 132, 1037 (1963).

15. D. W. Vance, Auger Electron Emission from Clean and Carbon-Contaminated Mo, Phys. Rev. 164, 372 (1967).

16. D. W. Vance, Auger Electron Emission from Clean Mo Bombarded by Positive Ions. II. Effect of Angle of Incidence, Phys. Rev. 169, No. 2, 252 (1968).

17. D. W. Vance, Auger Electron Emission from Clean Mo Bombarded by Positive Ions. III. Effect of Electronically Excited Ions, Phys. Rev. 169, No. 2, 263 (1968).

18. P. Mahadevan, J. K. Layton, D. B. Medved, Potential and Kinetic Electron Ejection from Molybdenum by Argon Ions and Neutral Atoms, Phys. Rev. 129, 2086 (1963).

19. P. Mahadevan, G. D. Magnuson, and J. K. Layton, Secondary-Electron Emission from Molybdenum Due to Positive and Negative Ions of Atmospheric Gases, Phys. Rev. 140, A1407 (1965).

20. H. D. Hagstrum, D. D. Pretzer, and Y. Takeishi, Focussed Slow Ion Beam for Study of Electron Ejection from Solids, Rev. Sci. Instr. $\underline{36}$, 1183 (1965).

21. F. M. Propst and E. Lüscher, Apparatus for the Study of Ejection of Auger Electrons from Solid Surfaces, Rev. Sci. Instr. $\underline{34}$, 574 (1963).

22. P. J. Estrup and J. Anderson, Chemisorption of Hydrogen on Tungsten (100), J. Chem. Phys. $\underline{45}$, 2254 (1966).

23. T. E. Madey and J. T. Yates, Jr., Kinetics of Desorption of the β-Nitrogen States Chemisorbed on Tungsten, J. Chem. Phys. $\underline{44}$, 1675 (1966).

24. R. H. Prince, Interaction of Low Energy Atmospheric Ions with Controlled Surfaces, UTIAS Report No. 133 (June 1968).

25. T. F. Moran and L. Friedman, Cross Sections and Intramolecular Isotope Effects in Ab-HD Ion-Molecule Reactions, J. Chem. Phys. $\underline{42}$, 2391 (1965).

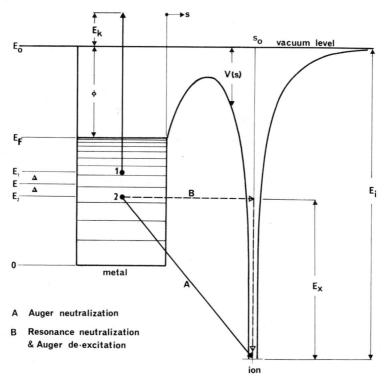

Fig. 1. Schematic diagram illustrating the Auger processes resulting in the neutralization of low-energy ions at a metal surface. (Approximately to scale for the N_2^+-tungsten system). Energy released by electron (2) as a result of either transition A or B is totally absorbed by electron (1) which may be excited beyond the vacuum level. (Secondary electron emission).

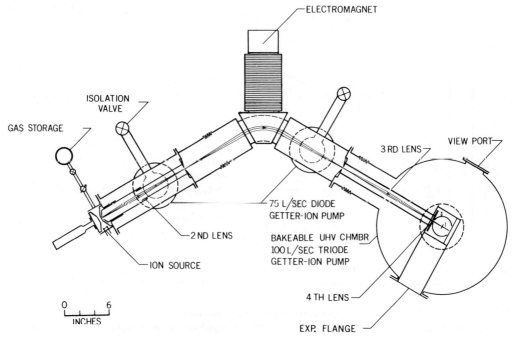

Fig. 2. A schematic diagram of the apparatus.

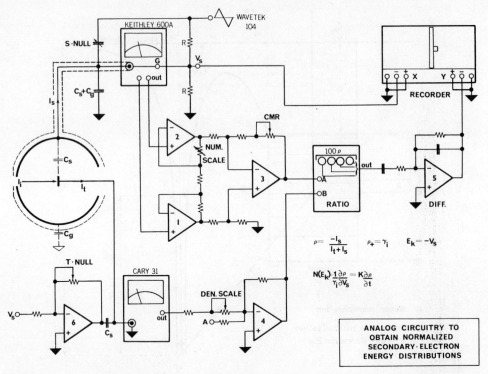

Fig. 3. Analog circuitry to obtain normalized secondary-electron energy distributions.

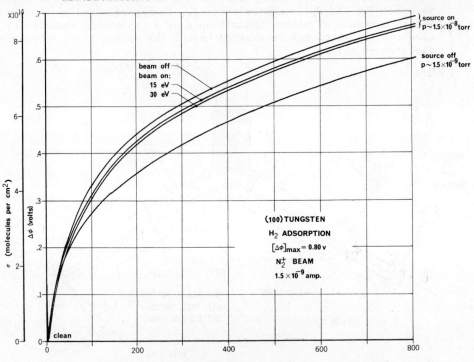

Fig. 4. Transient behavior of the (100) tungsten target work function due to hydrogen adsorption. (Diatomic nitrogen ion beam.)

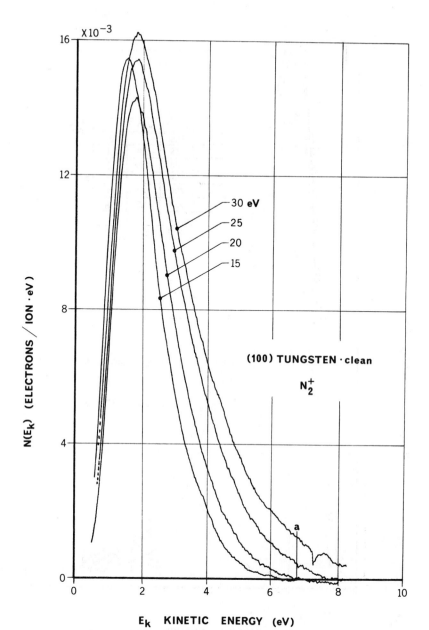

Fig. 5. Energy distribution of secondary electrons produced by Auger neutral-ization of 15, 20, 25, 30 eV N_2^+ ions at a clean (100) tungsten surface.

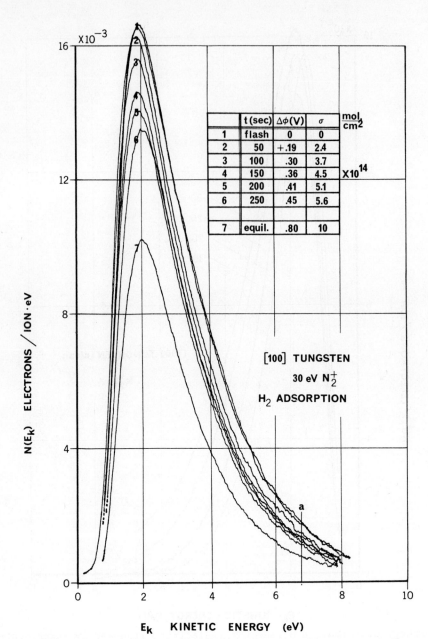

The table within the figure:

	t (sec)	Δφ(V)	σ	mol/cm²
1	flash	0	0	
2	50	+.19	2.4	
3	100	.30	3.7	X10¹⁴
4	150	.36	4.5	
5	200	.41	5.1	
6	250	.45	5.6	
7	equil.	.80	10	

[100] TUNGSTEN

30 eV N_2^+

H_2 ADSORPTION

Fig. 6. Energy distribution of secondary electrons produced by Auger neutralization of 30 eV N_2^+ ions at a (100) tungsten surface showing effect of
hydrogen adsorption.

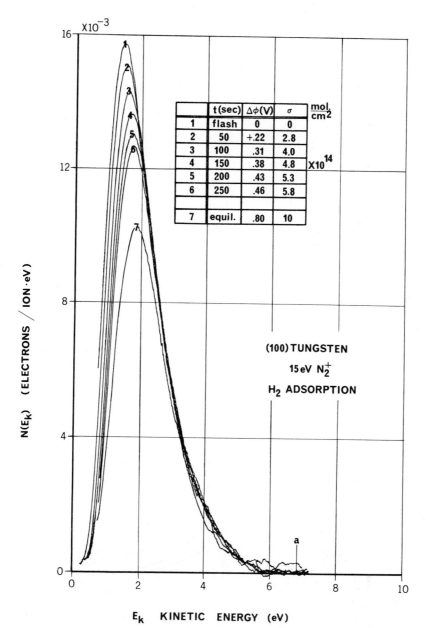

Fig. 7. Energy distribution of secondary electrons produced by Auger neutralization of 15 eV N_2^+ ions at a (100) tungsten surface showing effect
of hydrogen adsorption.

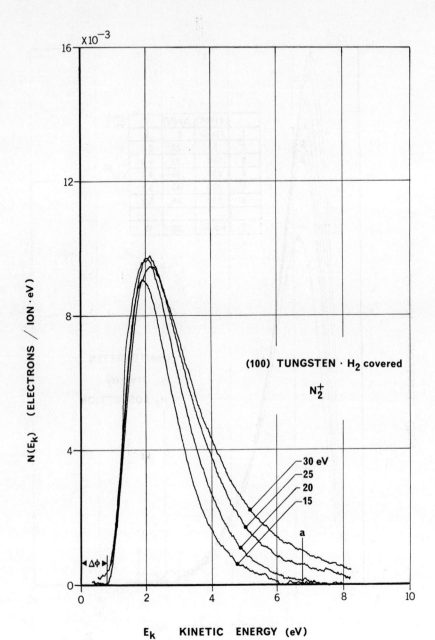

Fig. 8. Energy distribution of secondary electrons produced by Auger neutralization of 15, 20, 25, 30 eV N_2^+ ions at a hydrogen-covered (100)
tungsten surface.

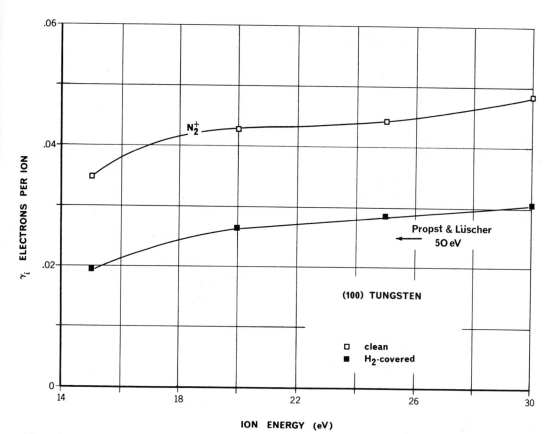

Fig. 9.   Dependence of electron yield on incident ion kinetic energy for
diatomic ions at a clean and hydrogen-covered (100) tungsten surface.

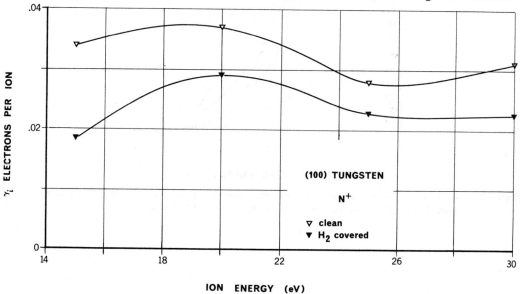

Fig. 10.   Dependence of electron yield on incident ion kinetic energy for N^+
ions at a clean and hydrogen-covered (100) tungsten surface.

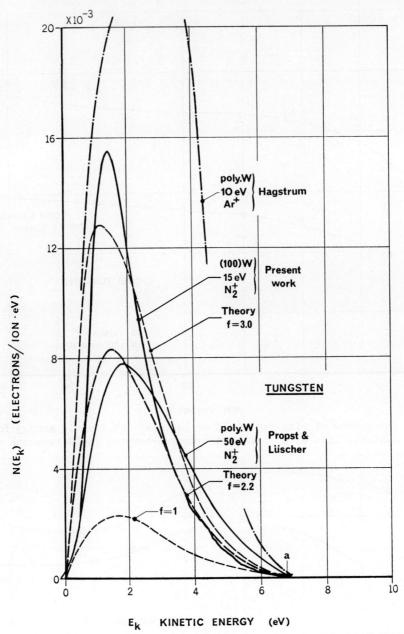

Fig. 11. A comparison of experimental and theoretical energy distribution
functions for the present work and that of Propst and Luscher (ref.14).

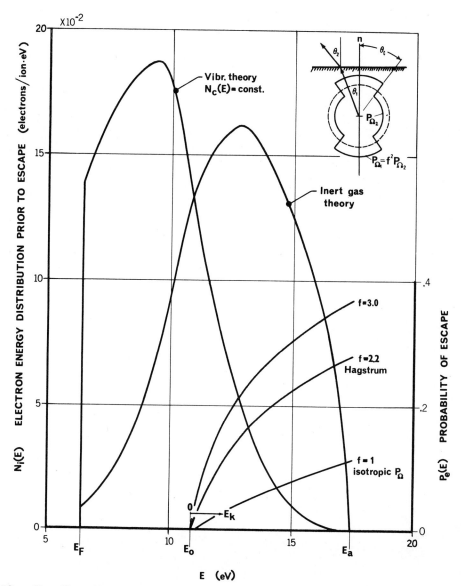

Fig. 12. The distribution in kinetic energy of electrons excited by the Auger
neutralization of N_2^+ ions at a tungsten surface prior to their escape.
Only those electrons at energies beyond E_o have finite escape probability. The form of the distribution function without including
the possibility of molecular vibrational excitation is shown for
comparison.

PHYSICAL ADSORPTION OF ATOMICALLY SMOOTH
SINGLE CRYSTAL PLANES*

Gert Ehrlich, Hansjurgen Heyne, and C. F. Kirk
General Electric Research and Development Center
Schenectady, New York

Much of the current interest in surface studies is focused on the behavior of strongly bound atoms and molecules at metal surfaces; such con-
centation of activity has led to remarkable progress. This is especially
striking for metal atoms, which represent perhaps the simplest type of
chemical adsorption.

Through the use of the field ion microscope[1], with its ability to reveal
the atomic configuration of surfaces on the scale of a few Angstroms, it has
even been possible to measure the binding energy of individual atoms on perfect crystal planes of different orientations. From the data for tungsten
atoms on tungsten[2], in Table I, surprising discrepancies appear between the
measured binding energies and expectations based on the traditional model
of pairwise atomic interactions. On atomically rough surfaces, such as the
(411), binding is actually lower than on a relatively smooth surface like
(211). This suggests that rearrangement of the electron distribution at
protruding surface atoms may play a dominant role in this very simple type
of chemisorption bond.

In contrast with the scope and intensity of the effort devoted to chemical
phenomena at surfaces, only little has been done to establish the atomic behavior of weakly bound systems on clean metals[3], for which dispersion forces
should play an important role. Studies of rare gases on tungsten have been

---

*Supported in part by the Air Force Office of Scientific Research,
Office of Aerospace Research.

carried out in the field emission microscope[4,5,6]; these revealed a strong dependence of the interaction upon the nature of the crystal surface exposed. Subsequent work on other body-centered cubic metals such as molybdenum and tantalum, as well as on hexagonal close-packed rhenium, indicated a similar sensitivity to surface structure.[7] On tungsten and molybdenum the behavior of the rare gases conforms semiquantitatively to what would be expected for adatoms interacting with the lattice through a Lennard-Jones potential; on tantalum and rhenium there appeared significant deviations from this simple rule. Only for tungsten, however, was the structure of the surface established on an atomic scale.

For face-centered cubic lattices, it is evident from the model in Fig. 1 that the atomic environment at the surface varies widely with orientation. That structure may be important in the physical interactions of rare gases with such systems has so far only been a matter of surmise. The aim of the present work has therefore been to establish such effects directly for the chemically more inert gases on face-centered cubic metals.

With weakly bonded gases the unexcelled resolution of the field ion microscope cannot be utilized as it has in the study of metal atoms; xenon on a tungsten surface, for example, is immediately removed during imaging. The electron field emission microscope[8] has sufficient resolution to discern processes on individual crystal planes and operates under less stringent conditions, which do not severely disturb the adsorbed layer. It is, however, limited in its ability to establish the atomic arrangement of the planes on which adsorption occurs. This is extremely important, as surfaces prepared by high temperature annealing have been shown to be marred by considerable disorder.[9,10]

To circumvent these difficulties, we have combined two techniques: surfaces of extremely high perfection are formed by field evaporation at low temperatures, and characterized by operating the projection microscope in the field ion mode; adsorption on these atomically smooth planes is then followed by studying the field emission of electrons. Here we present a preview of observations for two chemically inert systems interacting with atomically smooth rhodium: xenon, as an example of a noble gas with completely filled valence shell; nitrogen, as a molecular entity bonding only weakly on noble metals.

FIELD EMISSION FROM RHODIUM SURFACES

The electron emission image of a rhodium surface, obtained by field evaporation at $20°K$ and shown in Fig. 2a, differs from that of a thermally annealed surface. However, atomic arrangement and cleanliness of this surface are established by the ion image of the same tip in Fig. 2b obtained immediately thereafter.

By superposing the field ion micrograph on the electron image, the identity of any of the planes contributing to electron emission is easily established. This, as well as the great perfection of the surface, now allows us to discern the course of adsorption in considerable detail.

Xenon Layers

An atomically smooth surface, prepared by field evaporation and maintained at $77°K$, has been exposed to xenon at $p \sim 10^{-9}$ mm in Fig. 3. Immediately noticeable is the marked increase in emission along the [211] zone lines leading from the (111) to the (210) planes. As the amount of xenon on the surface increases so does electron emission. The work function drops and the region of the (210) and (321) planes assumes increasing importance. Gradually the area up to the[110] zone lines becomes emitting.   On further adsorption, the contribution of the (210)s to the electron emission diminishes, leaving the higher index plane in the area dominant. With increasing coverage the emission changes once more until the patterns resemble that of the clean surface.

Warming a saturated surface reverses this sequence of changes, until the clean metal is restored at T ~ 115°K. Evidently these emission patterns are indicative of a xenon distribution dictated by the desire of the adatoms to find the thermodynamically most favorable sites.

The behavior of xenon on a rhodium surface at 20°K is quite different. In the initial stages shown in Fig. 4, the emitting regions of the clean surface continue to dominate but assume a mottled appearance. Only at high coverages does the distribution of emission change to approach that at 77°K. The areas along the [211] zones, which are prominent for low xenon coverages at 77°K, are not distinctive at low temperatures. This contrast between emission patterns for uniform adsorption (at 20°K) and adsorption under reversible conditions demonstrates that the latter are indicative of preferential binding of xenon at specific crystal planes.

The areas on which the emission, and therefore the xenon, is concentrated during the early stages of adsorption, in which adatom-lattice interactions predominate over adatom-adatom effects, can now be identified. They belong to the group of planes like (432), (321) and (531). From the temperature of desorption we estimate a maximum binding energy of ~ 7 kcal/mole. Binding on the (210)s is weaker. Weaker still is the interaction between xenon and the (311) and (511) planes.

Nitrogen Layers

For a molecular gas more unusual behavior might be expected. Indeed, a different order of activity is apparent in the adsorption of nitrogen on an atomically smooth rhodium surface in Fig. 5. In a sequence at 135°K, exposure to nitrogen enhances emission from planes in the region of the (511) and (311). The (320) planes, as well as the planes around the (210) along the

[211] recede into darkness during the early stages. The (210)s themselves, however, remain emitting up to higher coverages. On completion of adsorption at this temperature emission is concentrated on planes such as (511) and (311) as well as around the (110). Upon raising the temperature, the gas is completely removed at T ~ 180°K.

There are some differences between adsorption and desorption. However the gross sequence of planes is reversed on warming; it follows that the main features of the emission at this temperature are indicative of an equilibrium distribution of nitrogen over the different planes of the rhodium surface.

At lower temperatures entirely different effects are encountered. Adsorption at 112°K or lower does not result in a highly structured pattern. Instead, with increasing exposure to nitrogen the emission is more uniformly distributed over the surface, until toward the end the entire tip appears bright. The detailed patterns obtained at T ~ 135°K therefore suggest significant differences in the concentration of nitrogen from one plane to the next.

In the interpretation of these changes it is vital to note the unusual effects of nitrogen upon the field emission characteristics. Adsorption on rhodium at 135°K hardly changes the work function. However, just as with tungsten[11,12] and nickel[11] the pre-exponential factor in the tunnelling equation is markedly lowered, causing a drop in overall emission. Bright areas in Fig. 5 therefore signify regions of little or no adsorption; planes which were emitting in the clean surface but disappear on adsorption have a strong affinity for nitrogen. From our observations the latter include the (321), (531) and (320) planes. The desorption energy from these planes is on the order of ~ 11 kcal/mole. Nitrogen bonding is less strong on (210); it is weakest on planes that dominate the emission after adsorption, that is on (511) and (311).

Although adsorption of nitrogen is more complex than that of rare gases, there are surprising similarities. Despite a sizeable difference in the binding energy, the variation in the strength of bonding from one plane to the next is similar for N_2 and Xe. There is at least one exception: for nitrogen the (320) appears more favorable than the (210); this trend is not apparent for xenon.

ADSORPTION AND SURFACE STRUCTURE

How does the marked dependence of binding upon the atomic details of the face-centered cubic lattice fit into our present understanding of surface forces? Adsorption of xenon is easiest to analyze, as dispersion forces should play an important role. Unfortunately, even here the theory is still in a vestigial state of development.

Following Bardeen[13], and Margenau and Pollard[14], we recognize that because of the finite relaxation time of the conduction electrons, adsorption on metals should have some resemblance to that on insulators. For the latter, pairwise summation of the interaction between adatom and lattice particles provides a reasonable account of the energetics. We have followed much the same course; the interplay between adsorbed atom and the lattice particles is approximated by a Lennard-Jones potential

$$V_r = \epsilon \left\{ \left(\frac{r^*}{r} \right)^{12} - 2 \left(\frac{r^*}{r} \right)^6 \right\},$$

where ϵ denotes the depth of the well. Provided we know r^*, the inter-atomic distance at the minimum of the pair potential, then the relative values of the total interaction $\mu = \sum_r V_r/\epsilon$ at different sites can be determined by summing over all lattice atoms.

Such calculations have till now been confined almost entirely to alkali halides and body-centered cubic metals[15]. We have therefore carried through estimates of the total interaction on a variety of planes in the face-centered cubic system, emphasizing those area preferred by xenon and nitrogen. The results of summations up to eight lattice distances from the adsorbed entity are listed in Table II.

Xenon

The relative binding energies of xenon, based on the assumption of Lennard-Jones interactions between gas and lattice agree reasonably well with our observations.   On this picture the low index planes should interact least with the rare gas, the strength of binding going inversely as the packing density. From the atomic model in Fig. 1, it is apparent that planes like (321) are characterized by kink sites which can accommodate Xe atoms. On such surfaces the calculated binding energies are high. These are precisely the areas on which xenon is found to aggregate in our experiments. Our measurements are not sensitive enough to discern differences between all of the high index planes. However, in accord with a Lennard-Jones model they are preferred over the (210).

It should be noted that despite this there are some apparent deviations from the 6-12 estimates. These seem to involve the (311), (210) and (320) planes. For xenon, the binding energy calculated for the first two is the same, but increases on going to the (320). Actually in our observations the (210) remains bright, whereas the (311) disappears. This suggests somewhat stronger bonding on the former, although this must still be put on a quantitative basis. Most interesting, the (321) does not become more prominent than the (210), as predicted by the calculations. However, the most important feature of the interactions between xenon and rhodium, the marked preference for rough planes, like (321), at the expense of smoother planes is in line with the estimates based on the assumption of Lennard-Jones behavior between xenon and lattice atoms.

Nitrogen
‾‾‾‾‾‾‾‾

Surprisingly enough the behavior of nitrogen fits into much the same framework. From its van der Waals radius ($r^* = 4.16$ Å), nitrogen should be analogous to krypton, with $r^* = 4.04$ Å. For krypton in turn it is evident from Table II that the predicted order of binding energies resembles that for xenon. From our observations we deduce relatively weak interactions on the (311) and its immediate surroundings - - nitrogen prefers the (320) and (321)s; the (210)s are somewhere between these planes. All this is in agreement with the calculated behavior.

The binding energy, however, is clearly outside the realm of dispersion forces. Nitrogen is comparable to krypton not only in size but also in polarizability. Dispersion forces should therefore be of smaller magnitude than in the interactions of xenon with rhodium; actually the desorption energy of nitrogen exceeds that of xenon by more than 50%. Weak chemical interactions with the substrate must therefore be invoked, and this conclusion is buttressed by the anomalous effects of nitrogen on electron tunneling.

In chemical bonding short range forces predominate. That despite this a Lennard-Jones potential provides a reasonable guide to the variation of bind-in energy with structure is intriguing. It will be important to establish quantitatively how well pair potentials predict actual binding energies. The success of a simple Lennard-Jones model in describing the behavior of xenon is less surprising. In such interactions long range forces are operative; these can overwhelm small contributions arising from a rearrangement of the electron cloud at the surface.

Although the present results are still qualitative, they clearly indicate the utility of combining the field-ion microscope, for preparation and characterization of surfaces, with field emission studies for probing adsorption. With the means for obtaining detailed information on the role of surface structure

in weakly bound systems now at our disposal, it will be of interest to explore the behavior of rare gases on atomically perfect surfaces of tantalum
and rhenium, on which notable deviations from Lennard-Jones behavior have
been found.

ACKNOWLEDGMENT

We are indebted to F. G. Hudda, formerly of this laboratory, for important contributions at the start of this work.

REFERENCES

1. E. W. Muller, Science $\underline{149}$, 591 (1963).

2. G. Ehrlich and C. F. Kirk, J. Chem. Phys. $\underline{48}$, 1465 (1968). These measure-
    ments have recently been confirmed by E. W. Plummer, Thesis, Cornell
    University (1968).

3. For references to much of this work, see D. Brennan and M. J. Graham,
    Trans. Roy. Soc. (London) $\underline{258}$, 325 (1965).

4. G. Ehrlich and F. G. Hudda, J. Chem. Phys. $\underline{30}$, 493 (1959).

5. R. Gomer, J. Phys. Chem. $\underline{63}$, 468 (1959).

6. W. J. M. Rootsaert, L. L. van Reijen, and W. M. H. Sachtler, J. Catalysis
    $\underline{1}$, 416 (1962).

7. G. Ehrlich, Brit. J. Appl. Phys. $\underline{15}$, 349 (1964).

8. R. Gomer, Field Emission and Field Ionization, (Harvard University
    Press, Cambridge, Mass. 1961).

9. T. H. George and P. M. Stier, J. Chem. Phys. $\underline{37}$, 1935 (1962).

10. D. W. Bassett, Proc. Roy. Soc.(London) $\underline{286A}$, 191 (1965).

11. G. Ehrlich and F. G. Hudda, J. Chem. Phys. $\underline{35}$, 1421 (1961).

12. A. van Oostrom, J. Chem. Phys. $\underline{47}$, 761 (1967); Philips Res. Rept. Suppl.
    $\underline{1}$, (1966).

13. J. Bardeen, Phys. Rev. $\underline{58}$, 727 (1940).

14. H. Margenau and W. G. Pollard, Phys. Rev. $\underline{60}$, 128 (1941).

15. For references in this area, see F. O. Goodman, Phys. Rev. $\underline{164}$, 1113 (1967).

TABLE I

Binding Energy of Tungsten Adatoms
on Tungsten Planes

Desorption Plane	Field $(V/\text{Å})$	(eV)	Binding Energy (eV)	
			Extp'l	Morse Potential Estimate
(110)	3.6_8	5.5	5.2_6	5.77
(211)	4.8_8	4.88	7.0_1	7.68
(310)	5.2_1	4.34	6.6_7	7.81
(111)	4.8_8	4.40	6.0_4	7.93
(321)	5.0_2	4.54	6.6_5	8.11
(411)	4.9_6	4.40	6.2_3	8.56

TABLE II

Lennard-Jones Interaction $-\sum\limits_{r}^{8} V_r/\epsilon$ on Rh

Plane	$r^* = 3.575\ \text{Å}$ (Xe)	$r^* = 3.365\ \text{Å}$ (Kr)
(100)	8.62	7.7
(110)	9.24	8.53
(311)	9.62	8.89
(210)	9.62	9.04
(320)	10.23	9.47
(321)	11.01	10.11

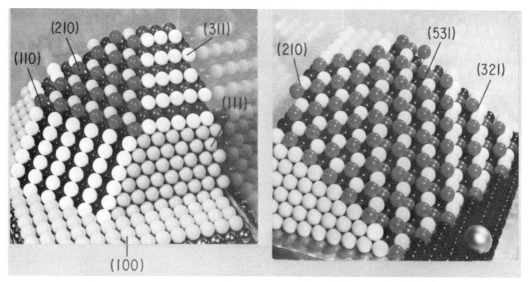

Fig. 1. Atomic configuration of planes in the fcc lattice.

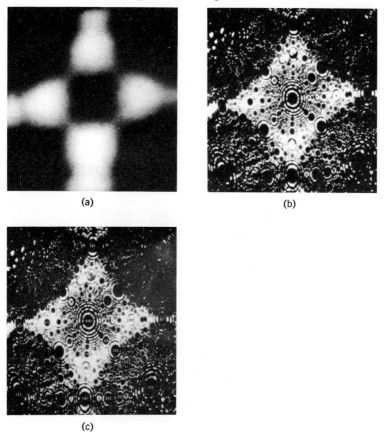

(a)

(b)

(c)

Fig. 2. Atomically smooth rhodium surfaces, formed by field evaporation at 20°K and 20.3 kV.

Fig. 2a. Field emission image at ~ 1.7 kV.

Fig. 2b. He ion picture of identical surface at 18 kV.

Fig. 2c. Planes of primary interest in ion micrograph.

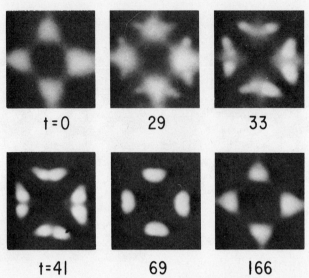

Fig. 3. Reversible adsorption of xenon on rhodium at T = 77°K. Total time in
xenon stream at p ~ 10⁻⁹mm indicated in minutes. Total work function
change φ = -1.2 eV.

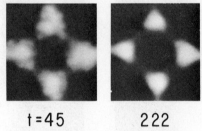

Fig. 4. Irreversible adsorption of Xe on Rh, at T ~ 20°K. Emission characteristics of patterns correspond to those at t = 33 and 166 obtained
at 77°K.

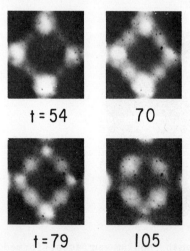

Fig. 5. Nitrogen on field evaporated rhodium at 135°K. At end of adsorption
$\Delta\varphi$ ~ 0 eV, yet tunnelling probability has been cut a factor of ten.

LEED AND ELLIPSOMETRY STUDIES OF PHYSICAL ADSORPTION
ON A (110) SILVER SURFACE AT LOW TEMPERATURES

J. M. Morabito, Jr.,* R. Steiger,** R. Muller, and G. A. Somorjai

Inorganic Materials Research Division, Lawrence Radiation Laboratory
Department of Chemistry and Chemical Engineering
University of California, Berkeley, California

I. INTRODUCTION

A clearer understanding of the atomic nature of physical adsorption on metals, i.e. its dependence on surface structure, adsorption energetics etc. is now possible due to the development and subsequent application of techniques such as field ion and field emission microscopy,[1,2,3,4] low energy electron diffraction (LEED),[5] ellipsometry[6] and others. LEED has revealed the formation of ordered surface structures during the chemisorption of many gases[7,8,9] at temperatures $\geq 25°C$, but the nature of ordering in physical adsorption has not yet been investigated to any appreciable extent. Ellipsometry can detect the presence of an adsorbed phase in coverages below the monolayer and as we shall show in this paper, it is particularly useful in obtaining quantitative information on the amount of gas adsorbed as a function of pressure (isotherms), temperature and time. Once isotherms as a function of temperature are measured, heats of adsorption can be calculated. Furthermore, the surface area occupied by the adsorbed molecules at monolayer coverage can be calculated, and hence the ratio of adsorbed molecules to surface atoms obtained. Therefore, the combination of structure sensitive LEED and coverage sensitive ellipsometry[10] allows one to study physical adsorption in terms of the molecular properties of the adsorbed phase on clean surfaces under carefully controlled conditions, i.e. ultrahigh vacuum (ca. 10^{-10} torr) and/ or in an ambient of pure gases.

The experimental techniques used in the past to measure adsorption isotherms on single crystal surfaces were restricted by experimental difficulties. Vacuum microbalance techniques[11,12] have been used in the pressure and temperature range where contamination effects from the residual gases present in the vacuum chamber are appreciable, making the results obtained questionable. The application of field emission or

---

*Visiting Scientist to Philips Laboratories, Eindhoven the Netherlands.

**Present Address: CIBA Photochemical Ltd., Fribourg, Switzerland.

field ion microscopy to the study of physical adsorption is useful in the
measurements of the work function, surface diffusion or of the effect of
surface structure on physical adsorption, etc., but accurate isotherms or
heats of adsorption on clean single crystal surfaces cannot be otained by
this technique. The combination of LEED and ellipsometry can then
provide information on physical adsorption which cannot be obtained by
any other combination of techniques.

It is for this reason that we have applied these two techniques and
mass spectroscopy to a study of the physical adsorption of several gases
(krypton, xenon, oxygen, methane, acetylene, ethylene, butane) on the
(110) crystal face of silver in the temperature range $-72°C$ to $-10°C$
and the pressure range of 10^{-10} torr to 10^{-6} torr. Mass spectroscopy
was useful in the analysis of the residual gases in the vacuum chamber
and to determine the composition of the gases used in the experiment. A
comparison of the results obtained by these techniques has provided
quantitative information on the nature (degree of order, energetics, etc.)
of physical adsorption on the clean silver (110) surface and on the
sensitivity of both techniques in the monitoring of gas adsorption.

II. EXPERIMENTAL

A. Low Energy Electron Diffraction

The system used was a modified Varian LEED apparatus which utilizes
the post acceleration technique and displays the LEED pattern on a
fluorescent screen. A liquid nitrogen cold trap was connected to the
diffraction chamber to reduce the partial pressure of the residual gases
to a minimum (see Fig. 1).

Pressure measurements of the gases (introduced into the LEED chamber
by means of a Granville-Phillips variable leak valve) were made with a

hot filament ionization gauge,[*] which was calibrated for nitrogen. No corrections have been applied for the other gases used in the adsorption studies, or for the difference (due to the geometry of our pressure measurements) between the recorded pressure and the actual gas flux impinging on the crystal surface. Although the measured pressures may be erroneous, their influence on the shape of the adsorption isotherms which were derived from relative measurements is negligible because of the linear dependence[13] between pressure and ion current between 10^{-6} and 10^{-9} torr.

The diffraction pattern can be directly photographed to obtain a permanent record of a given diffraction feature. The camera was placed on a rigid support fixed to the removable table containing the ellipsometer (see Fig. 1) in order to keep the geometry of the camera optics constant.

Intensity measurements of the individual diffraction spots were made with a spot photometer[**], having a fiber optics which allowed for variable apertures.

A crystal manipulator has been designed for the study of metal surfaces at low temperatures[14] (Fig. 2). The outstanding features of this particular design are: 360° rotation, translation both horizontally and vertically, good visibility of the fluorescent screen (ca. 85% as shown in Fig. 3) and easily accessible connections to a cryostat or a liquid nitrogen container. The manipulator is constructed of stainless steel, the silver single crystal being held in mechanical contact on a silver coated copper block by means of stainless steel washers. The crystal can be cooled to a constant temperature ($\pm$ 1°C)

---

[*]   Type UHV - 14 Nude Ion Gauge, Varian, Inc.

[**] Model 2000 Telephotometer, Gamma Scientific Inc.

in the range of -195°C to 0°C by cooling the copper block by means of an ultracryostat[‡] or by a controlled flow of liquid nitrogen. The crystal can also be heated by heating the block with a small alumina enclosed tungsten resistance heater (maximum temperature ca. 600°C). The crystal temperature is measured by a calibrated chromel-alumel thermocouple attached to the copper block.

B. Ellipsometry

The optical properties of an isotropic and absorbing medium (metal) are defined by two characteristic constants: refractive index (n) and absorption index (κ). These constants can be determined experimentally by measuring the change in polarization of reflected light.
This is done by resolving the electric vector of the incident polarized light into two components, E_p and E_s (parallel and perpendicular to the plane of incidence, see Fig. 4); the state of polarization can then be conveniently defined by the phase difference (Δ) between the parallel and perpendicular components upon reflection

$$\Delta = \delta_p - \delta_s \tag{1}$$

and the amplitude ratio

$$\tan \psi = \frac{E_s''/E_p''}{E_s/E_p} \quad (\text{E'' for reflected light}) \tag{2}$$

The electrogmagnetic theory of light predicts that the two scattered components are retarded in phase and reduced in amplitude to different extents upon reflection from a metal. As Δ and ψ can be measured[15] very accurately (in our case with an error of not more than 0.02°) by means of an ellipsometer, the constants n and κ of the pure metal surface can be determined.

[‡]Type LAUDA, liquid circulation (methanol), temperature range: -72° to +50°C.

In the presence of a surface layer, in which the optical constants change from those of the surrounding medium to those of the bulk metal, the reflection formulae are modified and can be derived from Maxwell's equations. In the case of a thin, uniform, non-absorbing and isotropic film on an absorbing substrate, the linear relationships

$$\Delta = \bar{\Delta} - \alpha\, d \tag{3}$$

$$\psi = \bar{\psi} + \beta\, d \tag{4}$$

(where $\bar{\Delta}$ and $\bar{\psi}$ are the phase shift and amplitude for the film-free metal) between Δ and ψ for the film-covered metal surface and the film thickness d can be derived.[16] The constants, α and β, in Eqs. (3) and (4) are functions of the complex refractive index $n_{cs} = n(1 + i\kappa)$ of the substrate (here the pure silver surface), the refractive index of the film, the angle of incidence, the wavelength of the incident light, and the refractive index of the incident medium.

A computer program[17] determines the values of Δ and ψ for different film thicknesses d by using the complex refractive index of the substrate, the refractive index of the film, the refractive index of the incident medium, the angle of incidence, and the wavelength of the incident light. These values of Δ and ψ are then compared with the measured values of Δ and ψ and the corresponding film thickness d calculated.

The equations evaluated by the program have been discussed in a previous communication.[17] These equations have been developed for a homogeneous single film with density and index of refraction independent of thickness. Although these conditions do not necessarily exist for films less than one monolayer in coverage, it has been shown experimentally, by using large molecules[18] (like lauric acid, stearic acid,

etc.) and, independently,[6] by comparison of ellipsometric measurements
with BET-adsorption measurements, that a linear relationship between Δ
and coverage in the sub-monolayer coverage region does exist. An extrapolated, macroscopic DRUDE[19] relationship can thus be used, within
experimental error, to yield absolute values of the thickness and
thus the cross sectional areas of adsorbed molecules. For the calculation of the optical thickness of the adsorbed phase, we have computed the
complex refractive index of the pure silver surface from the values
$\bar{\Delta}$ and $\bar{\psi}$ corresponding to the lowest pressure obtainable in the vacuum
chamber. This value has then been introduced into the above mentioned
computer program. The index of refraction of the film was that of the
liquid at the approximate temperature of measurement[†] (calculated from
density and molecular refraction if not available in the literature).

Finally, the requirement of an accurate alignment, combined with the
need to remove the optical instrumentation from the LEED system each time
the bakeout shroud is placed around the ultrahigh vacuum system, is an
important consideration in combining the two techniques - LEED and
ellipsometry. In order to avoid this alignment for each series of
measurements, the optical components[*] have been mounted on a movable
table[20] which can be rapidly placed over the frame of the LEED apparatus
(Fig. 1). The following features are important in this construction:

(a). Lateral and vertical translation, as well as rotation, are
possible for the telescope containing the analyzer.

(b). Selected glass (flat to 5 wavelengths and parallel to 1 minute
of arc over a diameter of 3/4" in the center) has been chosen for the
construction of the LEED windows.

[†] See appendix.

[*] Optical components taken from ellipsometer model L 119, Gaertner
Scientific Instruments, Chicago, Ill.

(c). The surface of the windows are normal to the light beams within 1° (which affects the measured polarizer and analyzer angles by not more than ± 0.01°.)

(d). The spread in the angle of incidence (± 0.15° in our case) is controlled by a pinhole in the focal plane of the telescope.

(e). The mechanical disposition of the optical system is such that other light sources (e.g. He-Ne-Laser, Na-lamp, etc.) can be used. All measurements discussed here were obtained at a wavelength of 5461 Å using a fixed quarter wave compensator[*] which produced elliptically polarized incident light.

(f). Due to the angle of incidence of 45° which was dictated by the geometry of the LEED chamber, the amplitude ratio varied less than the experimental error of 0.02°. Computations show that by changing the angle of incidence from 45° to 75°, the variation of the phase shift would be ca. 3-5 times larger for a silver surface covered with a 5 Å thick layer of n-butane, which would thus lead to a definitive improvement in the experimental conditions and extend the temperature range of the measurements into that of chemisorption. Heats of chemisorption could then be measured.

C. Gas Analysis

The residual gases in the diffraction chamber $(p_{total} \leq 10^{-10}$ torr) and those gases which were used for the adsorption studies have been analyzed by means of a quadrupole mass spectrometer[**] connected to the LEED diffraction chamber and placed below the sample (see Fig. 1). Mass spectra were recorded before and after each series of adsorption

---

[*] Type Senarmont, retardation = $\lambda/4$ for $\lambda = 5461$ Å.

[**] Type EAI, mass range 1-500 (divided into 3 ranges), sensitivity, ca. 10A/torr, minimum partial pressure of residual gases measured: ca. 10^{-13} torr.

measurements in order to establish that no gaseous impurities were present in concentrations greater than 1% of the gas flux.[*]

D.   Crystal Properties and Preparation

A single crystal rod (3/8") of pure silver[**] has been oriented by x-ray diffraction, sectioned by spark cutting and ground parallel to the (110) plane (deviation less than 1°). The crystal has then been mechanically polished. In order to obtain a good electron diffraction pattern as well as a maximum ellipsometric sensitivity, the crystal surface has been chemically polished by wrapping the oriented surface of the silver crystal on a soft polishing cloth impregnated with a solution of 100cc 0.2M KCN and 2 cc 30% H_2O_2 until an optically flat and highly reflecting surface resulted. The final cleaning of the crystal surface has been obtained by argon or xenon ion bombardment (at $p = 10^{-5}$ torr, 130-340 eV) and subsequent annealing of the resulting crystal damage at 150°C.

The physical quality of the investigated silver surface has been studied by electron microscopy and interference microphotography after a series of adsorption measurements. The electron micrographs (Figs. 5a and 5b) and the interference microphotograph (Fig. 6) both show that the microstructure of the crystal surface resulting from the chemical polishing has an average depth of approximately 600-900 Å while scratches which occur occasionally are ca 2000 Å deep. Figure 6 also shows that the investigated surface is slightly curved.

---

[*] All gases used for adsorption measurements were Matheson high purity gases ($\geq 99.8\%$).

[**] Ag 99.999% (Mat. Research Corp.), total impurity content: 11.16 ppm. Impurities as determined by mass spectroscopy: Cu 0.3 ppm. Cd < 1.0, Sn < 0.3, Au < 2.0, Ta < 1.0, Fe 0.8, Nd < 0.4, Pd < 0.4, C 1.24, O_2 0.7, N_2 0.2 ppm. Other impurities are present in amounts < 0.1 ppm.

The (110) surface of silver is well known[21] to be highly susceptible to thermal faceting, especially in the presence of oxygen.* Preliminary LEED experiments with the (110) surface showed signs of faceting at temperatures as low as 300°C in ultra high vacuum. Extra diffraction features due to the formation of additional crystal planes became clearly visible. Therefore, caution has to be exercised in heating the (110) surface in order to preserve the crystal orientation and maintain optimum surface order. The procedure adopted in our work was to heat the crystal to no higher than 85°C and then to cool it in the presence of the gas to be used in the adsorption study. Flushing the cooled surface with gas, followed by heating to 85°C in vacuo, in an effort to displace gaseous impurities trapped on the crystal surface and to condition the surface for the measurements to follow was adopted as standard procedure. This was done before each adsorption measurement.

The LEED diffraction pattern was also checked prior to a series of measurements as a precaution against any gross contamination. However, the diffraction pattern is fairly insensitive to the presence of adsorbed amorphous contaminants which may be present in amounts less than a monolayer (see Sec. V).

III. SURFACE CONTAMINATION PROBLEMS

Contamination from the residual gases present in the ultra-high vacuum can be a serious problem at low temperatures as reported by Lander.[5] The ellipsometry measurements indicated a rapid contamination on the (110) surface at -195°C even in an ambient of 8×10^{-11} torr- 8×10^{-10} torr. Figure 7 shows the build up of an adsorbed gas layer in

---

* In the presence of oxygen, at high temperatures $(T > 25°C)$ an ordered surface structure was formed on silver (110). The other gases used in the physical adsorption study did not induce the formation of ordered structures. The fact that silver is quite inert to most gases was one of the reasons we chose it for our study of physical adsorption.

terms of optical thickness, d, as a function time at -195°C and -40°C.
This contamination is most probably due to water vapor, carbon dioxide
and hydrocarbon fragments which were ever present in the recorded mass
spectra.   Indeed, the height of mass peak 18 did decrease to almost zero
after some time (approximately 16 min) when the crystal was cooled to
-195°C.   Lander[5] reports similar contamination effects at low temperature.
This unwanted contamination was less pronounced at higher temperatures
(as shown in Fig. 7 for -40°C).   At -40°C no detectable* contamination
occurs within the times of the adsorption measurements (10-20 min).   The
same was true for temperatures as low as -72°C, and therefore our adsorption measurements were restricted to temperatures $\geq$ -72°C.

Furthermore, the adsorption studies were carried out using a steady
flux of gas instead of a static system.   In this way, we could maintain
the purity of the gas even at low pressures ($\geq 10^{-10}$ torr).   Typical
results of the phase shift Δ for krypton at various pressures are shown
in Fig. 8.

IV.   ELLIPSOMETRY RESULTS

In order to measure the small amounts of gas (less than 10^{-8} g/cm^2)
adsorbed at the pressure range necessary to maintain a clean surface at
low temperatures, an extremely sensitive technique is required.   A sensitivity of .2 monolayer has been reported by Archer and Gobeli[22] in
their ellipsometric study of oxygen on silicon surfaces which is sufficient
to measure accurate adsorption isotherms (optical thickness vs pressure)
on clean single crystal surfaces.   Therefore, the phase change, Δ, between
the two component waves was measured as a function of pressure at several
temperatures as shown for krypton in Fig. 9a.   Optical film thicknesses

---

*The limit of our optical thickness measurements at 45° incidence
 is .3Å - .5Å.

were calculated from these measured values of Δ by using the previously discussed computer program. The adsorption isotherms for krypton are shown in Fig. 9b. That these measurements are relative to the clean surface is shown by the form of Eq. (3). $\bar{\Delta}$ is the measured phase change for the clean surface at temperature T and pressure p_o (in the order of 10^{-10} torr) before admission of the gas. Δ is the measured phase shift at temperature T and pressure p_1 of the gas. The flux and low pressure of the gas minimized the possibility of back streaming (mostly CO, Ar, H_2) from the vac-ion* pump which was due to sputtering on the titanium cathode. The appearance of CO, H_2 and Ar in the mass spectrometer was common at pressures $\geq 10^{-6}$ torr.

Similar measurements of Δ as a function of pressure were also made for xenon, oxygen, methane, acetylene, ethylene and butane. The isotherms plotted for these gases were similar in shape (Type I, Langmuir) to that of krypton which is shown in Fig. 9b.

Once the relationship between the optical thickness d and pressure p at various temperatures is established, $\log_{10} p$ vs $1/T$ can be plotted (as shown for krypton in Fig. 10a) and the value of ΔH_{st} determined at each coverage from the slope. The isosteric heats of adsorption are shown for Xe and Kr in Fig. 10b and for O_2, C_2H_2, C_2H_4, and CH_4 in Fig. 11.

A. Discussion of Ellipsometric Results

1. Heats of Adsorption

The initial heat of adsorption extrapolated to zero coverage $(\theta = 0)$ can be looked upon as a measure of the gas-surface interaction energy and is closely related to other physical properties of the metal and the gas such as polarizability, diamagnetic susceptibility, etc.

---

*Ultek, Inc.

For example, the initial heats of adsorption for xenon-tungsten[23,4] (9-10 kcal/mole), krypton-tungsten[23,4] (4.5-5.9 kcal/mole), argon-tungsten,[24,25] (1.9 kcal/mole) and for neon-tungsten[25,4] (.8 kcal/mole) are in the order of decreasing polarizabilities for the gas as reported by both Gomer and Ehrlich using field ion and field emission microscopy.

The initial heats of adsorption for the gases studied in this work are summarized in Table I, along with the polarizabilities and the heats of condensation for each gas (the polarizabilities for ethylene and acetylene are those parallel to the sigma bond). We have assumed linear extrapolation to $\theta = 0$.* From Table I we see, with the exception of xenon, that the initial heats are considerably lower or of the same magnitude (oxygen) as the heats of condensation. The heat of condensation is the upper limit to the lateral interaction energy among the adsorbed molecules. Therefore, the energies of the gas-surface interaction and those of the gas-gas interactions are of the same magnitude on the (110) crystal face of silver. Note also that the initial heat of adsorption for xenon on the (110) face of silver (Table I) is considerably greater than that of krypton as was the case on tungsten.[4,23] The polarizability of xenon is much larger than that of krypton and it is a larger molecule than krypton which may explain the large difference in the initial heats of adsorption for these gases. The polarizabilities of methane and acetylene are very similar, and the initial heats of adsorption for these gases are also similar. Furthermore, the magnitude of the initial heats for all the gases (Table I) indicates that the physical adsorption of these non-polar gases on silver (110) is primarily due to van der Waals (dispersion) forces.

---

*The initial heat may start at higher values, decrease in the very low film thickness ($\leq 1\text{Å}$) range and then begin to increase. We could not calculate any accurate heats below 1Å and therefore used linear extrapolation.

The isosteric heat of adsorption, ΔH_{st}, at any coverage is the sum of the potential and kinetic energy changes of the gas due to the adsorption process, and its variation with coverage (θ) depends on the cooperative* effects on the individual potential and kinetic energy terms. The potential energy change is due to gas-metal and gas-gas (lateral) interactions and is considerably greater than the change in kinetic energy. When the change in the lateral interactions with coverage is greater in absolute value than the change in the gas-metal interaction, an increase in the isosteric heat of adsorption with coverage occurs. This seems to be the case on (110) silver for all the gases studied in this work as shown in Fig. 10b for xenon and krypton and in Fig. 11 for oxygen, methane, acetylene, and ethylene. The heat of adsorption increases faster for xenon (Fig. 10b) than for krypton due to the higher polarizability of xenon. The increase in the heat of adsorption for methane is rather small, while for acetylene** the increase is rather large. For all the gases this increase in the heats of adsorption occurs at low coverage. For acetylene (Fig. 11) the gas-metal interaction could decrease to a low value as the monolayer approaches completion, but the gas-gas (lateral) interaction is increasing most sharply at this point; this is the probable reason for the maximum*** in the heat curve which is frequently observed in other work also.[26,27] Rhodin[11,12] calculated the isosteric heats of adsorption for nitrogen on the three low index crystal planes of copper and of argon on a zinc single crystal at low temperatures. An increase in the isosteric heats of adsorption with gas coverage was also observed.[11,12]

---

* The gas-metal interaction will decrease with surface coverage, while the gas-gas interaction will increase with coverage.

** The calculation of the variation in the heat of adsorption for acetylene is based on only two temperatures.

*** The maximum in ΔH_{st} occurs close to monolayer (3.8Å) coverage (Fig. 11).

The same interpretation can be applied to the ethylene curve and to the oxygen as well, but measurements at higher coverage were not possible.[*] Nevertheless, the heat of adsorption should approach the heat of condensation (which is indicated on the graphs, Figs. 10b and 11, for each gas) as the partial pressure of the gas approaches the equilibrium vapor pressure or as the thickness increases. It is to be expected then that physically adsorbed layers many molecular diameters thick behave like a two-dimensional liquid.[25]

2.   Optical Thickness

The adsorbed phase has been considered as a three-dimensional liquid characterized by a constant index of refraction and the validity of this assumption has been discussed in Section II. It would be very interesting then to compare the diameter of the molecule in the liquid state (d_o) to the thickness (d) of the adsorbed phase at the monolayer as measured by ellipsometry. The diameters of any molecule can be calculated from the density of the liquid by the equation

$$d_o = \left(\frac{M}{\rho\, N_A} \right)^{1/3} \tag{5}$$

where M is the molecular weight of the molecule, ρ the density, and N_A is Avogadro's number. The values for the density used in our calculations are the densities of the liquid phase at the experimental conditions of our measurements listed in the Appendix. The results are listed in column 3 of Table II. The measured thicknesses at monolayer coverage

---

[*] Higher thicknesses would require pressures approaching the vapor pressure which for all the gases studied was beyond 1 atm at the temperatures of our experiments. The pressure range of our experiments was limited to 10^{-6} torr. Therefore the anticipated maximum in the heat curve followed by a sharp decrease was not observed for the other gases.

should be close to this value. We have taken monolayer coverage to be the flat constant portion of the adsorption isotherms (which is listed in column 4) for each gas. We find that the agreement between the measured (d) and calculated (d_o) values is quite good.* Archer[28] reports that ellipsometric thicknesses calculated on silicon at monolayer coverage also agree with the dimensions of the adsorbed molecules (water, carbon tetrachloride, acetone) for surfaces etched at a maximum rate, but for other etching procedures the ellipsometric thicknesses at monolayer coverage were smaller than the diameter of the adsorbed molecule by factors as small as 1/2 to 1/3.

The measured thickness d can now be converted to the surface concentration (molecules adsorbed per cm^2) by the formula

$$\text{molecules/cm}^2 = \frac{d \, \rho N_A}{M} \quad \text{(cgs units)} \quad (6)$$

and compared to the number of atoms per cm^2 on the ideal (110) surface which is reported to be 8.5×10^{14}.[29] The ratio of the two is a coverage ratio or the number of silver atoms covered by each adsorbed molecule. Krypton, xenon and oxygen cover roughly two silver atoms. The larger n-butane molecule covers 4 silver atoms when adsorbed. For the hydrocarbons, the number of silver atoms covered seems to be related to the number of carbon atoms in the molecule. In chemisorption studies acetylene[30] is believed to associatively adsorbed by opening the π-bond and forming two carbon-metal bonds, e.g.,

$$\begin{array}{ccc} H & & H \\ | & & | \\ C & = & C \\ | & & | \\ M & & M \end{array}$$

---

* Using gas densities in calculating d_o there would be no agreement between the two.

This corresponds to 50% coverage also. The same is apparently true for ethylene on (110) copper as reported by Ertl[31] who found a diffraction pattern indicative of a 2:1 coverage. Therefore, a diffraction pattern of the physically adsorbed phase would be very helpful in distinguishing between possible surface ratios as calculated from the optical measurements. However, the adsorbed layer does not seem to be ordered which will be discussed in a later section.

3. Cross Sectional Areas

Cross sectional areas are parameters of great importance for surface area calculations. In order to calculate the effective area of coverage for any adsorbed molecule, one can assume the same packing as the plane of closest packing for the solid, i.e., hexagonal close packing for face centered cubic crystals, and a density which corresponds to the bulk phase at the temperature of the experiment. The adsorbed phase is usually taken to be solid or liquid like. Experimentally, cross-sectional areas are usually determined by adjusting surface areas as determined by other gases until they yield the same value as nitrogen which is assumed to give the most accurate surface areas. However, McClellan and Harnsberger[32] have recently reviewed the available data on cross-sectional areas and conclude that the size of the adsorbed molecule varies with adsorbent, temperature of adsorption, and choice of reference substance.

For the case of krypton, xenon, oxygen, acetylene and n-butane on (110)-silver, the adsorption isotherms were Type I (Langmuir), i.e., the optical thickness reached a constant value which can be taken at monolayer coverage. An example of this type of isotherm is shown in Fig. 12. It is possible then to determine when a monolayer forms on the clean single crystal silver surface with ellipsometry. Armbruster[33] found the same type of isotherm for nitrogen, carbon monoxide, argon and oxygen on a silver foil with a fairly high degree of preferred orientation (foil plane at angle of 15° to the (110) plane). She also found the amount of adsorbed gas to be substantially independent of the pressure over a large range. This apparent saturation of the surface as indicated by the flattening of the isotherm has been reported by both Langmuir[34] and Wilkins[35] for platinum foil also. Rhodin,[11,12] however, did not observe this flattening of the isotherm in his work on copper or zinc.

The constant optical thicknesses have been converted to a number of molecules adsobed per cm^2 by Eq. (6) as discussed in the previous section. The molecular area is the reciprocal of this number. The results obtained are shown in column 7 of Table II, and compared to the values reported in the literature (column 8) for metal substrates.[*] For krypton, xenon and oxygen, the literature values are reported at considerably lower temperatures than at our experimental conditions (column 2). Nevertheless, agreement is relatively good for these gases. For the case of n-butane and acetylene, however, the literature values are reported at approximately the same temperature as our experiments.

Therefore, an independent measurement of cross-sectional areas on clean single crystal surfaces is then possible with the technique of ellipsometry, especially for the more condensable, larger molecules. Application to noble gases is only limited by contamination effects at the lower temperatures (-195°C) necessary for such work.

---

[*] When values were available on several metals, an average was taken.

V.   LEED ANALYSIS OF PHYSICAL ADSORPTION

A.   Results

The low energy electron diffraction analysis of all the physical adsorption results did not reveal any ordering in the adsorbed phase, that is, there were no extra features in the diffraction pattern. Temperatures as low as -195°C and pressures as high as 10^{-3} torr were used in some of our studies. Before cooling to -195°C the gas was introduced at a pressure large enough to displace any contaminant present in the ambient gas which could interfere with the adsorbate. Lander[5] reports that the monolayer of xenon physically adsorbed on graphite at -183°C and 10^{-3} torr is ordered. The same conditions failed[*] to produce an ordered phase on (110) silver. It may be, however, that the electron beam is perturbing or in fact desorbing the weakly bound array. This would be very difficult to detect with the mass spectrometer because the gas pressures coming off the surface would be extremely small. A way to confirm this possibility would be to take optical measurements in the presence of the electron beam. Unfortunately, we could not do this because of the geometry of the diffraction chamber.

Although ellipsometry indicated that appreciable[**] adsorption had taken place at pressures below 10^{-8} torr and temperatures below 0°C, only at higher pressure ($\geq 10^{-7}$ torr) and low beam voltages did the intensity in the background of the diffraction pattern increase,

---

[*] There was, however, a very large increase in the background of the diffraction pattern, and a gradual blurring of the normal diffraction features due to the disordered nature of the adsorbed xenon. The other ordered structures reported by Lander are for substances which should be classified as weakly chemisorbed rather than physically adsorbed.

[**] Close to the monolayer at temperatures $\leq$ -40°C, see Fig. 12.

The intensity of the (00) spot was also constant until about 10^{-7} torr where it showed a decrease. Both the intensity in the background and of the (00) spot returned to their original values when the gas was removed which is indicative of the weak physical adsorption forces (heats of adsorption ≤ 6 kcal/mole) for all the gases studied in this work, i.e., the adsorbed phase could be removed by simply reducing the pressure or by gentle heating (reversibility). This return to the original intensity was also true for oxygen at the pressures used in this experiment.

B. Discussion

Ellipsometry is a more sensitive measure than is LEED of this disordered adsorption at coverages below and up to the monolayer. These measurements were performed with the post-acceleration type LEED apparatus and intensities measured with a telephotometer. Similar results may not be obtained with a Faraday cage which is an absolute measure of the intensity and perhaps a more sensitive indication of the intensity change due to adsorption at low coverage (below the monolayer). In addition these measurements were taken at beam voltage ≥ 50 eV. Measurements below 50 eV may indicate that LEED is sensitive to this disordered adsorption, but accurate measurements below 50 eV were not possible with our apparatus. It should also be possible to calibrate or compare the sensitivities of the Auger technique[36] with ellipsometry. Such a calibration would make the results obtained by both the Auger technique and ellipsometry more reliable on any given gas-metal system. The sensitivity of the Auger technique to gaseous impurities on the surface is still questionable.

The magnitude of the calculated heats of adsorption suggests that the adsorbed phase is nonspecific, i.e., the molecules in the adsorbed phase are not bound to any particular surface metal atoms, but move freely over the surface. This occurs when the energy required for such motion

is appreciably lower than the energy of desorption (approximately equal
to the initial heat of adsorption), or when the thermal energy of the
molecule (RT) is greater than the energy barrier (activation energy) for
such motion. The atomic arrangement in such an adsorbed phase is random
or disordered. However, at lower temperatures ($<<$ -195°C) and higher
pressures, the adsorbed phase is often localized to definite positions
(potential minima) on the surface. There is some restriction to free
movement, and the adsorbed phase more ordered. It may be then, by going
down to much lower ($<<$ -195°C) temperatures, that ordering will occur on
(110) Ag. However, Kruger[37] has studied the condensed layers of several
gases on copper and gold single crystals at liquid helium temperatures
and found optical constants which were more indicative of a liquid than
a solid. Nevertheless, it would be very interesting from just a physical
adsorption point of view to extend our measurements to much lower temperatures and higher pressures. In this way we could study the formation
of multilayers, two-dimensional condensation, phase changes in the adsorbed layer, and obtain information on how ordering on the surface is
related to the magnitude of the binding forces between the surface and
the gas, to the size and shape of the adsorbed molecules, and to pressure
(concentration) and temperature. The orientation of the surface may also
be an important consideration.

Finally, the fact that the LEED pattern remained remarkably unchanged
while the ellipsometer had registered the pressure of almost a monolayer
of adsorbed gas on the silver single crystal surface seems to indicate
that the gas atoms are adsorbed in "patches" on the metal substrate.
In this way a fraction of the metal surface is still free of adsorbed
gas and thus, the diffraction pattern is characterisitic of a clean (110)
surface. In addition, the fact that the isosteric heat increases at

fairly low coverages suggests that adsorbed molecules readily associate
(clustering) which substantiates the possibility of a "patch-like"
adsorption model.

VI. CONCLUSIONS

The combination of LEED and ellipsometry provides unique and
valuable information on the atomistics of physical adsorption. The
availability of well defined, clean single crystals, and the combination
of structure sensitive LEED with coverage sensitive ellipsometry has revealed that the adsorbed phase is completely disordered[*] on (110) silver.
The fact that a LEED analysis of all adsorption results did not reveal
any ordering in the adsorbed phase is not surprising due to the weak
binding observed between the adsorbed molecules and the surface. In
those cases of physical adsorption where the binding energy between the
adsorbed gas and the metal is so weak, ordering can perhaps only occur at
temperatures and pressures considerably below the triple point of the gas.
Under these conditions, the adsorbed gas is localized on the surface and
may form an ordered structure.

Ellipsometry is a more sensitive technique than is LEED for the
detection of disordered adsorption at low pressures (10^{-10} to 10^{-8} torr).

Ellipsometry is sensitive to coverages below the monolayer and, in
fact, can be used to predict when the monolayer forms on the single crystal
surface. This can be very useful in the calibration and interpretation of
LEED diffraction patterns and of the Auguer technique. In addition, the
measurement of adsorption isotherms on clean single crystal surface by
ellipsometry represents a new and independent experimental technique

---

[*] The electron beam may have perturbed any ordering which occurred in
the adsorbed layer, but this is difficult to affirm with certainty.

for calculating cross-sectional areas of adsorbed molecules, as well as coverage ratios on clean single crystal surfaces. These calculated cross sectional areas are independent of any standard such as nitrogen.

The magnitude of the initial heats of adsorption confirms that the physical adsorption of non-polar gases on silver is due to van der Waals forces. The increase of the heats of adsorption is due to lateral interaction energies which are increasing more rapidly with coverage than the decrease in the gas-metal interaction energy with coverage. The fact that this increase occurs at low coverages for all the gases studied suggests that a very small portion of the surface contains high energy sites. The energy of lateral interaction between these sites and the gas is greater than the lateral interaction between the gas molecules, but as the coverage increases, the remaining sites are considerably less energetic. The energy of lateral interaction between the gas molecules is now as great or greater than the energy of interaction between the gas and these remaining sites, resulting in the possibility of clustering or "patch-like" adsorption on the (110) silver surface.

Finally, the thickness of the adsorbed layer on the (110)-surface of silver is approximately equal to the diameter of the molecule. This suggests that the surface of the metal used in our experi ents is homogeneous (absence of surface imperfections such as cracks, flaws, etc.). However, a microscopic investigation revealed that the surface studied is relatively rough on an atomic scale.

This is the first attempt to study the physical adsorption of gases on a clean single crystal metal surface with the combination of LEED and ellipsometry. Our results definitely show that this combination will be very useful in the study of a large variety of surface phenomena.

ACKNOWLEDGMENT

This work was performed under the auspices of the United States Atomic Energy Commission.

REFERENCES

1. G. Ehrlich, Brit. J. Appl. Phys. 15, 349 (1964).

2. G. Ehrlich, Metal Surfaces (ASM, Metals Park, Ohio, 1962).

3. G. Ehrlich and F. G. Hudda, J. Chem. Phys. 33, 4 (1960).

4. R. Gomer, Discussions Faraday Soc. 28, 23 (1959).

5. J. J. Lander and J. Morrison, Surface Sci. 6, 1 (1967).

6. G. A. Bootsma and F. Meyer, Surface Sci., to be published.

7. J. J. Lander, Prog. Solid State Chem. 2, 26 (1965).

8. J. W. May, Ind. Eng. Chem. 57, (7), 19 (1965).

9. J. M. Morabito, Jr. and G. A Somorjai, J. Metals 20, 5 (1968).

10. A. J. Melmed, H. P. Layer, and J. Kruger, Surface Sci. 9, (3),
    476 (1968).

11. T. N. Rhodin, Jr., J. Am. Chem. Soc. 72, 4343, 5691 (1950).

12. T. N. Rhodin, Jr., J. Phys. Chem. 57, 143 (1953).

13. Saul Dushman, Scientific Foundations of Vacuum Technique (John Wiley
    and Sons, Inc., New York, 1962).

14. E. I. Kozak and J. M. Morabito, Jr., UCRL Report 18038, University
    of California, Berkeley, (1968), to be published.

15. A. B. Winterbottom, Trans. Faraday Soc. 42, 487 (1946).

16. E. Passaglia, R. R. Stromberg, and J. Kruger, Nat. Bur. Stand. Misc
    Publ. 256 (1964); F. L. McCrackin and J. P. Colson, J. Research Nat.
    Bur. Stand. 67A, 363 (1963); L. Tronstad, Trans. Faraday Soc. 31,
    1151 (1935).

17. R. Mowat and R. Muller, University of California, Berkeley, UCRL
    Report No. 17128, (1967).

18.   T. Smith, J. Opt. Soc. Am. (1968) to be published.

19.   R. Drude, Annalen der Physik and Chemie 36, 532 (1889).

20.   R. H. Muller, University of California, Berkeley, UCRL Report 18208 (1968); to be published in J. Sci. Instr.

21.   A. J. Moore, Metal Surfaces (ASM, Metals Park, Ohio, 1962).

22.   R. J. Archer and G. W. Gobeli, J. Phys. Chem. Soc. 76, 343 (1965).

23.   G. Ehrlich and F. G. Hudda, J. Chem. Phys. 30, 493 (1959).

24.   R. Gomer, J. Chem. Phys. 29, 2 (1958).

25.   R. Gomer, J. Phys. Chem. 63, 468 (1959).

26.   R. A. Beebe and D. M. Young, J. Phys. Chem. 58, 468 (1954).

27.   J. G. Aston, and H. Chon, J. Phys. Chem. 63, 1015 (1961).

28.   R. J. Archer, Ellipsometry in the Measurement of Surfaces and Thin Films, Sysmposium Proceedings, Washington, 1963.

29.   J. Benard , J. Oudar, and F. Cabane-Brouty, Surface Sci. 3, 359 (1965).

30.   G. C. Bond, Catalysis by Metals (Academic Press, New York, 1962) p. 282.

31.   G. Ertl, Surface Sci. 6, 208 (1967); 7, 309 (1967).

32.   M. L. McClellan and H. F. Harnsberger, J. Coll. Interf. Sci. 23, 577 (1967).

33.   M. H. Armbruster, J. Am. Chem. Soc. 64, 2545 (1942).

34.   I. Langmuir, J. Am. Chem. Soc. 40, 1361 (1918).

35.   F. J. Wilkins, Proc. Roy. Soc. (london) A164, 510 (1938).

36.   L. A. Harris, J. Appl. Phys. 39, 1419 1428 (1968).

37.   J. Kruger and W. J. Ambs, J. Opt. Soc. Am. 49, (12) 1195 (1959).

38.   J. Hildebrand and R. L. Scott, The Solubility of Nonelectrolytes, Dover Publications, New York, 1964.

TABLE I.

gas adsorbed	ΔH_{st} ($\theta=0$) (kcal/mole)	$\alpha \times 10^{24}$ (cm^3)	$\Delta H_{cond.}$ (kcal/mole)
Kr	1.4	2.46	2.31
Xe	6.1	4.00	3.02
O_2	1.8	3.88 (O^{2-})	1.63
CH_4	1.0	2.60	1.96
C_2H_2	1.0	2.43	4.27
C_2H_4	1.6	3.59	3.24

TABLE II.

1	2	3	4	5	6	7	8
gas adsorbed	surface temperature $<°C>$	d_o calc. $<Å>$	d_o measured $<Å>$	Molec./cm^2 x 10^{14}	Coverage ratio Ag : gas	Cross Section meas. $<Å^2/molec.>$	Cross Section Litt $<Å^2/molec.>$
Kr	-68	5.4	6.0	3.9	2.2:1	25.6	21.5 at -183°C, -196°C
Xe	-72	4.3	4.0	5.1	1.7:1	19.6	22.5 at -183°C, -196°C
O_2	-72	5.0	4.7	3.8	2.2:1	26*	17.4 at -183°C, -196°C
C_2H_2	-72	4.1	3.8	5.4	1.6:1	18.3	22.0 at 78°C
n-C_4H_{10}	-42	5.3	3.6	2.4	3.5:1	42	43.4 at -78°C

*Calculated from density at T_{crit} = -118.8°C.

Appendix: The following table indicates the densities and refractive indices used for the calculation of cross section areas and optical thicknesses as derived from the measured adsorption isotherms of different gases on the (110) face of silver single crystals

Gas	Temperature T°C	density d_4^T	refractive index, n_D	Literature
Kr	-63.8	0.908	1.26	For n_D see (a), for d_4^T see (b)
Xe	-70	2.8	1.46	For n_D see (a), for d_4^T see (b)
O_2	-118.8	0.43	1.22	For n_D see (d), for d_4^T see (c)
CH_4	-72	0.22	1.13	n_D calculated from d_4^T and molecular refraction
C_2H_4	-66	0.51	1.32	n_D calculated from d_4^T and molecular refraction
C_2H_2	-84	0.62	1.36	n_D calculated from d_4^T and molecular refraction
$n-C_4H_{10}$	-42	0.64	1.37	for $n_{D,T}$ see (e), d_4^T calculated from n_D and molecular refraction

a. H. Rudorf, Phil. Mag. 17 (6), 795 (1909).
b. Matheson Gas Data Book, 4th Ed. (Matheson Gas Corp., New Jersey), 1966.
c. International Critical Tables, Vol. III, p. 204, (1928).
d. Handbook of Chemistry and Physics
e. Landolt-Boernstein, Phys. Chem. Tabellen

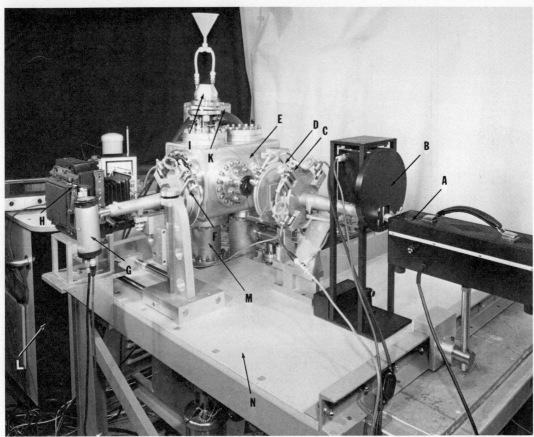

Fig. 1. Experimental apparatus: A. He-Ne laser; B. Light-chopper; C. Polarizer
circle; D. Compensator; E. LEED chamber; F. Analyzer circle; G. Photo-
multiplier; H. Camera; I. Low temperature manipulator; K. Liquid nitrogen cold trap; L. Cryostat; M. Quadrupole mass spectrometer; N. Mobile
table.

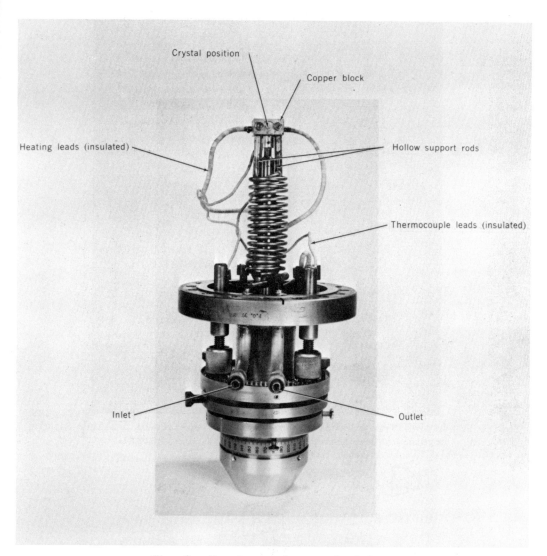

Fig. 2. Low temperature manipulator.

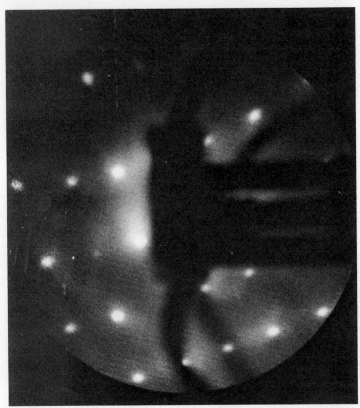

Fig. 3. LEED diffraction pattern of the (110)-surface of a silver single crystal. (Temperature - 195°C, 90 eV, 10^{-10} torr).

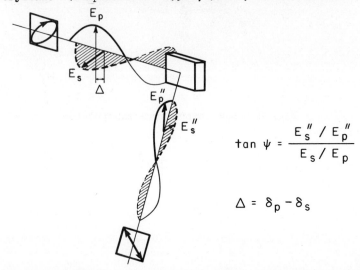

$$\tan \psi = \frac{E_s'' / E_p''}{E_s / E_p}$$

$$\Delta = \delta_p - \delta_s$$

Fig. 4. Schematic of incident elliptically polarized light and plane polarized reflected light.

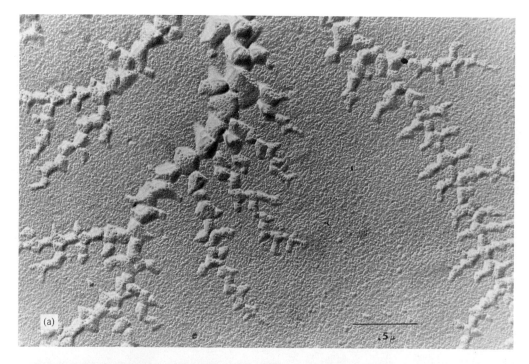

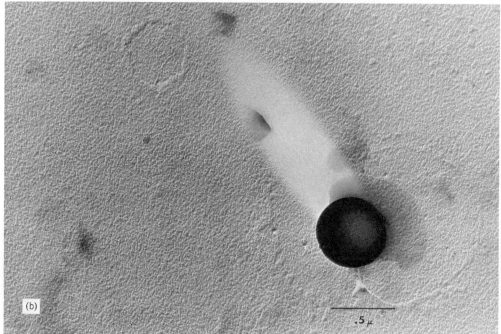

Fig. 5a and b. Electron micrographs of a carbon replica (linear enlargement 10,000) of the chemically polished (110) surface of a silver single crystal. (V = 60 keV, shadow depth under reference latex ball (ϕ .55μ) is .21μ.

Fig. 6. Interference micrograph of the etched silver cyrstal surface. Fringe spacing = 0.27 μ.

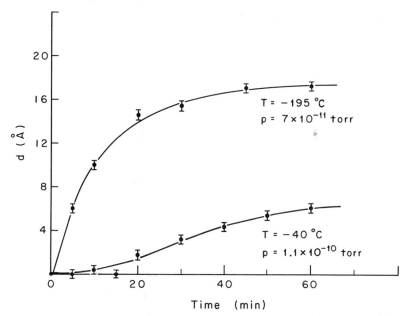

Fig. 7. Contamination curves in ultrahigh vacuum at -195°C and -40°C (assuming a refractive index of 1.25 for the calculation of the optical thickness d).

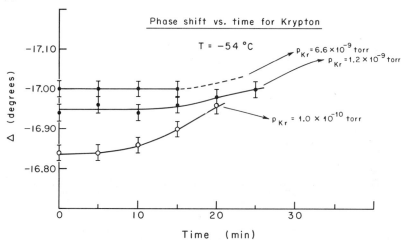

Fig. 8. Measured phase shift Δ (zone 90-135° for polarizer circle) versus time for the adsorption of Krypton on Ag (110) at different partial pressures of Krypton (T = -54°C).

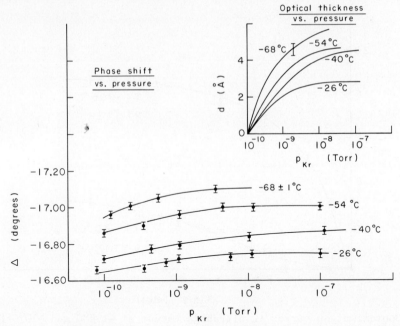

Fig. 9a and b. Phase shift Δ and optical thickness d versus pressure for Krypton at various temperatures.

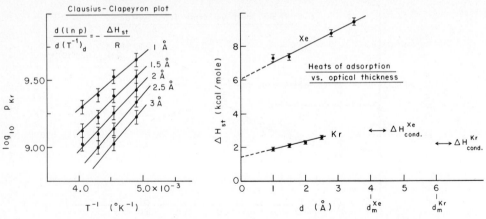

Fig. 10a. Clausius-Clapeyron plot for Krypton. Fig. 10b, isosteric heats of adsorption versus optical thickness for Krypton and Xenon. d_{Xe} and d_{Kr} correspond to the measured thickness of the monolayer, ΔH_{cond} = heats of condensation as found in the literature.

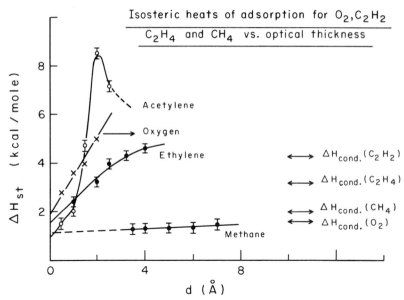

Fig. 11. Isoteric heats of adsorption for oxygen, acetylene, ethylene and
methane versus optical thickness as calculated from the respective
adsorption isotherms.

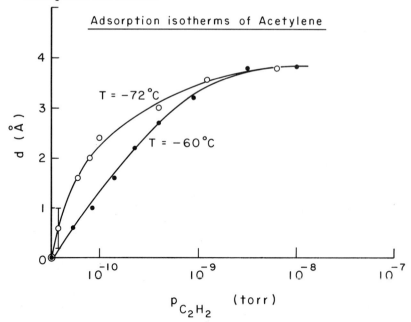

Fig. 12. Optical thickness d versus partial pressure of acetylene at
T = -60°C and -72°C on Ag (110).

HYDROGEN AND OXYGEN ON A (110) NICKEL SURFACE*

J. W. May and L. H. Germer

Department of Applied Physics
Cornell University
Ithaca, New York

Introduction

The reaction of oxygen with hydrogen is catalyzed by nickel. We have used low energy electron diffraction (LEED) to study a (110) nickel surface while this reaction was going on. Because adsorption of hydrogen and of oxygen individually lead to simple diffraction patterns, and because there are only two kinds of gas atoms participating, the hydrogen-oxygen reaction is especially suitable for investigation. The reaction of this (110) nickel surface with oxygen has already been extensively studied, but its reaction with hydrogen less extensively.

The thin nickel crystal used in this work was cut from a massive crystal and its ends were welded to nickel supports. It could be heated by current flowing through it. Its surface area was about 1 cm^2. Temperature was measured by a thermocouple welded to the back of the crystal near its center. A mass spectrometer was used to monitor the gases.

It is natural to divide the experiments into three groups: A: adsorption of hydrogen; B: reaction of oxygen with a surface covered by hydrogen; and, C: reaction of hydrogen with a surface containing oxygen. We consider first the adsorption of hydrogen upon the clean crystal surface, and later the reactions involving the two gases.

---

*Research supported by National Aeronautics and Space Administration Contract NGR 33-010-029 and American Iron and Steel Institute Contract No. 148. We gratefully acknowledge the Advanced Research Projects Agency for financial support of this project through the use of the Central Facilities of the Materials Science Center.

A. ADSORPTION OF HYDROGEN

When a clean (110) nickel surface is exposed to hydrogen at room temperature new diffraction spots develop at the h, k + 1/2 positions indicating
double spacing in the [01] surface direction which is a [100] direction of
the three dimensional crystal.[1,2] The new features come to maximum intensity
at hydrogen exposure of about 45 L (i.e. 45×10^{-6} torr sec) and do not change
with further exposure. This represents incidence of 640×10^{14} hydrogen molecules per cm^2 on the surface. Diffraction patterns are shown in Fig. 1.

In order to estimate the amount of adsorbed hydrogen, measurements were
made of the pressure-time integral on flash-off after long exposures to hydrogen, much beyond the 45 L exposure necessary to bring the h, k + 1/2 diffraction
features to their maximum intensities. In a typical experiment the hydrogen
valve was shut, and about 1 minute later the crystal temperature was raised
to about 500°K at the rate of about 50° per second. A single hydrogen pressure burst was observed, at about 330°K, and the pressure-time integral had
a value corresponding to 1.6 to 2.2 monolayers of hydrogen atoms. The precision of this estimate of coverage is not high.

The half orders of Fig. 1 can be caused by nickel atoms moved to new
lattice sites, by small displacements of nickel atoms, or by phase shifts
from hydrogen adsorbed in alternate troughs. If the first of these interpretations is correct, the double spacing in the [01] surface direction can be
attributed to reconstruction of the surface of the nickel crystal to form ridges
running parallel to the [10] direction, as shown in the upper marble model
of Fig. 2 (see also ref. 1, Fig. 2c and ref. 2, Fig. 3). It has previously
been pointed out that this reconstructed surface has the same number of missing first nearest neighbors per cm^2 as the smooth surface, and it should
therefore have only slightly more surface energy than the smooth surface.
The energy of reconstruction is doubtless very much smaller than the binding

energy of hydrogen atoms to the surface[3], which is nearly 3 eV. This strong binding energy can also be inferred[4] from the heat of adsorption which has been measured[2] to be 1.2 eV per molecule adsorbed as atoms.

The three photographs of Fig. 1 show that the h, k + 1/2 diffraction features are sometimes considerably extended parallel to the [01] surface direction, and the degree of this diffuseness has been found to be sometimes even greater than in Fig. 1a. We suggest that occasional breaks in the regularity of double spacing between ridges may be the result of surface impurities present in amounts too small to be detected from the LEED patterns prior to hydrogen admission, but yet adequate to block occasionally the ideal reconstruction of the surface. In support of this speculation we have the observation that the diffraction patterns become progressively sharper, i.e., less streaked, with many repetitions of argon ion bombardment, oxygen exposure, and hydrogen reduction[*]. The photographs of Fig. 1 are arranged in chronological order, showing this increase of sharpness with successive cleanings. Similar streaking of h + 1/2, k diffraction spots, attributable to impurities, has not been seen from nickel (110) surfaces that have been reconstructed by oxygen adsorption, probably because the much higher adsorption energy of oxygen is adequate to overcome the disturbing effects of minute impurities.

The streaking of Fig. 1 we attribute to irregularities in the double spacing, with perhaps occasionally only a single spacing between adjacent ridges (i.e., spacing of a_0 = 3.52 Å, rather than $2a_0$) so that the surface consists of double spacings with random admixture of a few single spacings, or perhaps even occasional triple spacings.

---

[*] When this crystal was first mounted, contamination on its surface could not be removed by severe heating in oxygen and hydrogen, but it was removed by argon bombardment. It is likely that small traces of this impurity were present after its detection was no longer possible from the LEED pattern. Its eventual removal accounted for the gradually improved sharpness of hydrogen patterns.

We have carried out a computer calculation of the pattern to be expected from occasional random interpolation of single spacings, and the results roughly suggest that this can account for the streaked character of some patterns.[*] The streaks are not symmetrical about their center positions and the calculation predicts that the asymmetry will change character with electron wave-length. This effect has been observed.

The occurrence of reconstruction of metal surfaces accompanying gas adsorption has been repeatedly questioned[5,6,7]. At the present state of the art the interpretation of the patterns of Fig. 1 is unclear. The reconstruction model of Fig. 2 (upper) can explain all the observed facts in simple fashion, but models are not based on reconstruction require special pleading to interpret many of the observations listed below:

1. The small size of each hydrogen atom makes distribution of these atoms over the inclined facets of the reconstructed surface, Fig. 2 upper model, seem quite reasonable. For the unreconstructed surface of Fig. 2 lower model, hydrogen atoms must be located in alternate columns only, and in these columns they must be fixed in positions with separations $(a_o/n2^{1/2})$ $\overset{\circ}{A}$ with n an integer. This seems highly artificial because the empty troughs of this unreconstructed model are _never_ filled even after extremely long exposure to hydrogen.

2. A (1×2) diffraction pattern will be produced by the upper reconstructed surface of Fig. 2 independently of H atoms adsorbed upon the inclined faces, and thus H atoms are not required to be frozen in position on these faces. Since hydrogen is evolved at $330°K$, leaving the crystal clean, it seems highly probable that hydrogen atoms behave like a two-dimensional gas at the temperature of $300°K$ at which observations were made. The unreconstructed model, on

---

[*] We are indebted to Mr. Harold Potter for programming the computer.

the other hand, requires that the H atoms are frozen, or at least spend most
of their time, in fixed sites and accurately lined up with each other in alternate parallel columns. Previous experimenters[8,9,10] have reported hydrogen
mobility on nickel surfaces. For example, we quote from ref. 9, page 1106,
"Diffusion observed in the temperature range from 250° to 280°K is undoubtedly
taking place within the chemisorbed layer".

3. Whenever the h, k + 1/2 diffraction spots are considerably streaked
the detailed features of the individual spots move "erratically" back and forth
with changing electron wavelength, which implies that ridge heights are not all
exactly double the normal interplanar nickel spacing, i.e., some admixture of
other facet sizes - - single heights occasionally or even triple heights
occasionally. It seems inconceivably difficult to imagine a reason for frozen
H atoms being accurately placed in every alternate [10] row, yet with some
faulting to give occasionally an intermediate complete row or even occasionally
another missing row. Irregular reconstruction explains the facts simply.

4. The reconstruction of Fig. 2A leaves unchanged the number of missing
nearest neighbors and is thus a "possible" rearrangement in the sense defined
by J. F. Nicholas,[11] wherease other rearrangements of the (110) surface are
not. (The (100) and (111) planes possess no "possible" rearrangement in the
Nicholas sense, and fractional orders do not result from hydrogen adsorption
on these surfaces[12,13]).

5. The high intensities of the fractional orders naturally suggest that
their origin is nickel atoms rather than hydrogen atoms. When hydrogen is
adsorbed upon (111) or (100) faces of nickel not only are there no new diffraction features as pointed out above, but one does not even notice any increase
in background brightness. On these more dense surfaces there cannot be reconstruction without considerable energy change.

6. Excellent (1×2) diffraction patterns have been obtained at 280 volts and at 355 volts (Fig. 3). The quality of these patterns is comparable with that from clean nickel at the same voltages, i.e., the high background scattering is beginning to obscure the pattern but is no worse from hydrogen covered surfaces than from clean surfaces. All the spots of Fig. 3 are comparably weakened by thermal vibrations (resulting from the shorter wavelength). This is a natural consequence of the reconstruction model.

7. In Fig. 4 is plotted the measured voltage dependence of the 0-1/2 diffraction beam from a hydrogen covered surface. The small marks at the bottom edge of the figure are calculated voltages of intensity maxima to be expected from the reconstruction model of Fig. 2 (top). A somewhat better fit with the experimental curve is obtained by assuming the ridge heights are smaller by 2.5 per cent, and calculated voltages of the maxima for this contracted model are marked by large arrows. One notes that the simple kinematical calculation is in fair, but not perfect, agreement with the observations. The agreement is too good for chance, but not completely satisfying. This situation is quite usual in all LEED observations of this sort. Similar measurements of surface spacing were reported by Germer and MacRae (ref. 2 page 1383), which they interpreted as confirmation of the reconstruction model. Note that in both sets of observations the agreement is obtained without correction for inner potenial. This supports the model.

8. Finally we wish to cite some field emission microscope observations of Wortman, Gomer and Lundy[9] in which they study hydrogen adsorption on nickel. The dominant change which hydrogen produces in the FEM pattern is in the (110) areas. Before adsorption these areas are not prominently marked, but after adsorption they have grown very large and very dark, about as prominent and as dark as the (111) area. For comparison we list here the changes produced by hydorgen adsorption on different crystal planes as observed by LEED

and by FEM (Table 1). Although we do not try to interpret the FEM patterns, it seems to us striking and significant that adsorption of hydrogen upon a (110) plane results in reduction of field emission to about the same low level that is found from (111) planes, but adsorption on other planes does not greatly change emission from them.

TABLE I

Changes Produced by Hydrogen
Adsorption on Nickel

Planes		LEED	FEM
	Fractional Orders	Reference	Ref. 9, Fig. 1,5,8 Dark Area
(111)	None	12	No change
(100)	None	13	No change
(110)	h,k+1/2	1,2, this paper	Much larger

B. REACTION OF OXYGEN WITH A SURFACE COVERED BY HYDROGEN

In these experiments, a (110) surface has first been given a hydrogen exposure of 45 L or more at room temperature, and is then exposed to oxygen, also at room temperature. New diffraction patterns develop as a result of the oxygen exposure. These are caused by a succession of 2-dimensional structures containing more and more oxygen. At the beginning of this sequence the patterns differ from those that are produced by oxygen on a clean surface, but the patterns in the latter part of the sequence seem identical.

Starting with such a diffraction pattern as those shown in Fig. 1, the first observed change due to oxygen exposure is the appearance of weak diffuse streaks at the h + 1/2, k positions, the sketch of Fig. 5a. We interpret this as due to domians of (2×1) -O[1/2] structure with the narrowness of each domain in the [01] direction giving the poor resolution in this direction.[14]

Alternatively the streaking could perhaps result from a special disorder in

the [01] direction. The presence of hydrogen has had two effects; it has

suppressed the (3×1) -O[1/3] structure that is the stable low temperature

arrangement of the first adsorbed oxygen on the clean surface,[15] and it has

given rise to the 1/2 0 horizontal streaks which never develop in the absence

of hydrogen. One notes (using a hard sphere model) that the transition from

reconstruction (1×2) to long islands of reconstructed (2×1) does not necessarily

involve motion by nickel atoms of more than one atom spacing.

The pattern of Fig. 5A is followed by weakening and diffuseness of the h,

k + 1/2 "hydrogen" spots and growing sharpness and strength of the new

h + 1/2, k "oxygen" spots, (diffraction pattern and sketch b of Fig. 5).

Later the spots due to hydrogen vanish and the "oxygen" spots become stronger.

The pattern is then like that of Fig. 6, or the sketch of Fig. 5c. It is

indistinguishable from the pattern that would be produced by an initially

clean surface that had received the same oxygen exposure, of the order of

0.8 L (see also ref. 2). We suggest that the streaks in the photograph of

Fig. 5, which are centered at h + 1/2, k + 1/2 positions are due to multiple

scattering at boundaries between (1×2) and (2×1) domains.

With further oxygen exposure the h + 1/2 k spots become streaked in the

[10] direction, due to random mixtures of (2×1) and (3×1)-O[2/3] sequences,

as with oxygen on a clean surface. Whether or not we can produce thermal

ordering[15] of these structures has not been tested. At an oxygen exposure

of the order of 5L these streaks are resolved into a sharp (3×1)-O[2/3]

pattern and with still further exposure a (9×4) pattern is developed, just

as if the surface had been initially clean.[16] The sequence of changes seems

to be determined by the total oxygen exposure, over the pressure range tested

from 0.4 to 9 × 10^{-8} torr. Furthermore, in experiments in which the ambient

hydrogen was varied after the initial coverage of the surface, it was found
that this hydrogen makes no difference either in the patterns themselves or
in their rates of development, from no hydrogen to hydrogen at a pressure of
10^{-5} torr. The fate of hydrogen that was initially on the surface has not
yet been determined.

C. REMOVAL OF OXYGEN BY HYDROGEN

Experimental Procedure

We have shown earlier[16] that, for a crystal containing dissolved oxygen
with an atomic oxygen/nickel ratio greater than about 5×10^{-4}, the equilibrium
surface contains half a monolayer of oxygen atoms giving the (2×1)-0[1/2]
diffraction pattern. An example of such a pattern in shown in Fig. 6. This
equilibrium surface can be produced at will by heating such a crystal for an
instant to 1100°K, and we have used this method to give us repeatedly an oxygen
covered surface for a number of sccessive experiments.* The crystal thickness
was about 0.28 mm and, for an oxygen/nickel ratio of 5×10^{-4}, the reservoir of
dissolved oxygen is sufficient to supply more than 10^{3} half layers of oxygen
to each of the two crystal surfaces.

Most of the experiment described in this section were carried out in a
very simple manner. The half monolayer covered surface is produced by flashing to 1100°K. The heating current is then decreased to allow the crystal to
come to the desired experimental temperature. When this equilibrium temperature

---

* This experimental procedure has, of course, the limitation that we can
   study oxygen removal from half monolayer coverage only. We have carried
   out less extensive work on surfaces having higher coverages. The conventional
   method of obtaining oxygen covered nickel requires starting with a clean
   surface at room temperature and then opening the valve connected to the
   oxygen supply allowing oxygen to enter the experimental chamber until coverage reaches the desired value, e.g. 0.8 L for half monolayer (ref. 16,
   Table I). By this procedure one cannot start each experiment with quite
   the same uniform coverage, because this depends upon always stoppin the
   oxygen exposure at exactly the right time. Developing the half monolayer
   coverage by diffusion has the great advantages of reproducibility, uniformity and time saving.

is reached, hydrogen is admitted by opening a valve with continual pumping by the Varian ion pump. It is observed that alternate diffraction spots along the vertical rows, marked by arrows in Fig. 6 (Miller indicates h + 1/2, k), becomes progressively weaker. At low temperatures the spots are changed in position indicating a phase transformation before they finally disappear. The oxygen in the surface layer has then been removed. The experimenter simply notes when the fractional order spots disappear by observing the diffraction pattern on the fluoresecent screen. Measurement of the time is quite precise because the diffraction spots being to weaken very slowly, with decrease of intensity becoming continually more rapid until they are gone. Observations were made at constant temperature as a function of hydrogen pressure, and then more extensive measurements at constant pressure as the temperature was varied.

Although the initial half monolayer coverage was always produced by a brief flash to $1100°K$, a temperature as high as this was not necessary. Heat-in the crystal to considerably lower temperatures produced also (2×1) surface structures; and the clean-off times were the same. This is illustrated by the data of Table II in which are listed observations of clean-off times, t_c, after different preheat temperatures. In each test the hydrogen pressure was 6×10^{-6} torr and the clean-off temperature $525°K$. Uniformity of the clean-off times is proof that we were dealing with a reproducible half monolayer.

TABLE II

Measurements of Clean-off Time, t_c, After
Different Preheat Temperatures, T_p

$T_p(°K)$	715	750	790	950	1100	1200
t_c(sec)	26	28	31	37	32	32

Results

In Fig. 7 is plotted the clean-off time t_c against hydrogen pressure at the fixed temperature 550°K. The observation that t_c varies inversely with the square of the pressure is supported by observations at other temperatures reported below. Note that the square law is in marked contrast to the results of the exposure of a hydrogen covered surface to oxygen, reported in the last section, where the changes produced are completely determined by the pressure-time product.

The dependence of clean-off time on temperature is shown in Fig. 8. Many of the tests were made at the pressure 6×10^{-6} torr, and these measurements are plotted on the figure as solid horizontal bars. The dotted bars represent tests made at other pressures reduced to the values to be expected for 6×10^{-6} torr by means of the relationship to $t_c p^2$ = constant. The agreement of data plotted in this way with the more extensive measurements made at $p = 6 \times 10^{-6}$ torr is further support of the square law pressure relation deduced from Fig. 7. The lengths of the horizontal bars of Fig. 8 represent estimated uncertainty in the temperature, due in part to downward dirft during each measurement and in part to variation of temperature along the crystal.

We wish to obtain from Fig. 8 the efficiency of removal of oxygen at different temperatures, defining efficiency, ϵ, as the ratio of the number of oxygen atoms removed by the time the surface is cleaned to the number of hydrogen molecules that have struck the surface during the time of cleaning. At higher temperatures there is diffusion of oxygen to the surface from the body of the crystal and the removal of these extra atoms accounts for the great increase in t_c from 600 to 800°K. At 800°K the crystal is kept covered by just half a monolayer and diffusion will never cause this coverage to be exceeded.[16] We have estimates of this diffusion from earlier experiments. If t_d is the time required for half a monolayer of oxygen atoms to diffuse to the surface from the interior, we can write $\epsilon = 5.70 \times 10^{14} (1 + t_c/t_d)/1.433 \times 10^{21}$ pt_c, where p is the hydrogen pressure in torr and the hydrogen temperature is 300°K. This is conveniently written,

$$\epsilon = 0.40 \ (1/t_c + 1/t_d)/10^6 p \qquad (1)$$

Values of ϵ calculated from this relation are plotted against $1/T$ in Fig. 9, with $1/t_c$ taken from the smooth curve of Fig. 8 and $1/t_d$ calculated from the diffusion coefficient $D_1 = 2.3 \times 10^{-5} \exp (-11,500/T)$ cm^2 sec^{-1}, with the relation $1/t_d = 1.1 \times 10^{10} D_1$ (see ref. 16). This calculated efficiency increases with increasing temperature from about 10^{-5} at 400°K to a maximum of 0.002 at 525°K, then decreases to almost 0.001, but finally rises again to about 0.02 at 800°K, the highest temperature tested. The dashed curve represents the result that is found if one neglects the oxygen atoms that diffuse to the surface from the interior, i.e., neglects the term $1/t_d$ in Eq. (1).

The observation of a range in which efficiency of removal of oxygen by hydrogen decreases with increasing temperature is sufficiently unusual to merit reconsideration of the earlier diffusion data. Because the measurements leading to the diffusion coefficient in ref. 16 are extremely scattered, we have calculated how much change in this equation would be required to alter the plot of Fig. 9 to eliminate this decrease of ϵ in the temperature range immediately above 525°K might perhaps be spurious, but that there must be at least a striking leveling off of efficiency in the range from 525 to 650°K, before it increases steeply at higher temperatures.

In watching progressive weakening and final disappearance of the fractional order diffraction spots, two effects have been noted which suggest a simple interpretation of the dip in the efficiency plot with increasing temperature.

The first of these effects is the relatively slow initial decrease of intensity with time and continual acceleration of the rate until the spots disappear, mentioned above. Since the diffraction intensity presumably is proportional to the square of the oxygen coverage, the removal of oxygen must start even more slowly than does the intensity decrease. We suggest that oxygen is attacked by hydrogen very inefficiently on the surface of an intact (2×1) oxygen nickel surface layer, but that after such attack has finally produced a small clean area the attack is greatly accelerated at the edges of the area.[*] On this hypothesis the removal of oxygen after the very beginning is almost entirely from the edges of clean areas and the rate can perhaps be assumed to be proportional to the sum of all the perimeters of clean area on the surface. As clean areas begin to coalesce the surface changes from a (2×1) oxygen-nickel surface with islands of clean metal to a clean surface with islands of (2 1) oxygen-nickel. This happens only above 450°K when surface diffusion is sufficiently easy for islands to form.

The second effect occurs below 450°K. There is then a qualitative change of the diffraction pattern with coverage as oxygen is removed. The pattern is, of course, initially (2×1). As the sharp h + 1/2, k diffraction spots begin to weaken they become streaked along the [10] surface direction and then the streaks are resolved into spots at one-third order positions giving a (3×1) pattern indicative of one-third of a monolayer of oxygen atoms. These spots then weaken until they finally disappear. But above 450°K this qualitative change in the pattern does not take place. The h + 1/2, k diffraction spots become weaker until they disappear without ever showing the (3×1) pattern, i.e., when removal of oxygen has brought the average coverage down to one-third of a monolayer the pattern remains, as before, that of the half monolayer. This

---

[*] Such widely different reaction efficiencies are exhibited in other chemical reactions, notably in the very slow reaction of oxygen with an intact basal plane of graphite and the rapid reaction with carbon atoms at the edges of a basal plane.

behavior agrees with our previous observation[15] that one-third of a monolayer

coverage can exist on the surface in two structures that are stable at different

temperatures, a uniform (3×1) structure over the entire surface stable up to

roughly 200°C according to our previous work, and bare surface containing (2×1)

islands above this temperature. The present observation that the initial (2×1)

can change to (3×1) before the oxygen is completely removed implies that short

range surface diffusion takes place even at the lowest temperatures of our

tests. The observation of (2×1) structure until the surface is clean, when

the temperature is above 450°K, gives convincing proof of islands surrounded

by clean surfaces.

We have a speculative explanation of the curious shape of the efficiency

plot of Fig. 9, which is based on the two deductions that have just been made.

In fact, we have two possible explanations either of which, or both operating

together, could load to the observed relation between efficiency and tempera-

ture. These explanations are based on the onset of easy surface diffusion

at 450°K, well below the temperature at which there is much diffusion of oxy-

gen in bulk of the crystal. We have not succeeded in making these theories

quantitative and any attempt to do so must take into account the anisotropic

character of the surface; surface diffusion is much easier along the [10] surface

direction than along the [01]. In the qualitative discussion given here

this anisotropy is neglected which is naturally not realistic.

Theory 1

The observations prove that during the removal of oxygen by hydrogen at

temperatures above 450°K there always comes a stage at which the remaining

oxygen is in the form of (2×1) islands. We assume that the attack upon these

islands by hydrogen is effective almost entirely along their perimeters (or

perhaps upon oxygen that has diffused out from islands upon the clean surface).

Low values of efficiency actually measured are compatible with high efficiency of removal of edge atoms and much lower efficiency for atoms within islands. If there is a perimeter energy (the two dimensional analog of surface tension) then one expects large islands to grow at the expense of small above 450°K. This will decrease the total perimeters and consequently the rate of removal of oxygen.

Theory II.

We have been led to the view that the beginning of oxygen removal, and only the beginning, results from the inefficient attack of hydrogen upon an intact (2×1) oxygen-nickel surface layer. The essential first breaks in the (2×1) layer might be healed and thus the removal of oxygen impeded at its very beginning. The source of new oxygen for this healing must be diffusion from the bulk of the crystal which is very slight at 450°K but possible sufficient. Oxygen atoms on the surface must be sufficiently mobile to allow new atoms to repair the gaps even when the new atoms arrive on the surface some distance away. By extrapolation of the diffusion plot of ref. 16 we get 1.5×10^5 sec as the time for a half monolayer of oxygen atoms to reach the surface at 450°K (4×10^9 atoms cm^{-2} sec^{-1} at 450°K). For comparison, the time required to remove half a monolayer of oxygen atoms at 450°K and hydrogen pressure of 6×10^{-6} torr is 100 sec. If the initial removal of oxygen to make clean surface were less efficient than the mean efficiency for the removal of the entire half monolayer by the factor $1.5 \times 10^5/100 = 1500$, and if surface diffusion were perfectly successful in bringing every atom arriving at the surface from the bulk to a freshly formed clean area, the surface half monolayer would remain entire at 450°K for an indefinite time.

All of the observations so far reported on the removal of oxygen by hydrogen were made upon a surface with an initial coverage of only half a monolayer. There are, therefore, not strictly comparable with any other published data

upon the reduction of an oxide of nickel. Published efficiencies for oxide reduction[19,20] have been lower than those we have given by many orders of magnitude. This difference is consistent with some very limited measurements that
we have made upon the reduction of coverages greater than half a monolayer.

For half monolayer coverage the surface was cleaned at $550°K$ and a hydrogen pressure of 6×10^{-6} torr by an exposure of about 200 L, ($\epsilon = 0.002$).
But we have found that the (3×1)-0[2/3] structure, containing only 33 per cent
more oxygen atoms, required exposure several times greater (at the same temperature and pressure) just to reduce the coverage to half a monolayer. And the
(9×4)-0[10/9] structure required cleaning exposures of several thousand L.
Evidently cleanoff is less efficient when the initial coverage is higher. This
is consistent with our observation, concerning cleanoff of just half a monolayer, that there was an acceleration of the rate of cleanoff as the oxygen
coverage fell. A similar acceleration effect has been noted during the
reduction of MoO_2 powder[20] and copper oxide.[21]

The dependence of the cleanoff time t_c on the square of the hydrogen pressure is strong evidence that cleanoff requires two separate steps, the first of
which must entail sticking of hydrogen to the oxygen covered surface, possibly
to form a surface hydroxide. In the second step, which also requires participation of hydrogen from the gas, water molecules evaporate from the surface:

$$H_2 + 2\,O_{ads} \rightleftharpoons 2OH_{ads} \qquad\qquad \text{fast}$$

$$H_2 + 2\,OH_{ads} \longrightarrow 2H_2O \qquad\qquad \text{slow}$$

This plausible sequence gives the rate of formation of water proportional to
the square of the hydrogen pressure, as observed. It is known that, when
oxygen coverage is high, hydrogen is not easily adsorbed[8] and the equilibrium
is shifted far to the left. When the coverage is low, however, equilibrium

might shift to the right and the rate be accelerated as observe.

Another factor tending to stabilize higher oxygen coverage is that at temperatures permitting easy surface diffusion, areas of clean surface must develop rarely if ever during the early stages of reduction. Thus attack by hydrogen is carried on almost entirely under the unfavorable condition of a continuous layer of nickel-oxygen structure. Longer times of reduction for higher oxygen coverage are readily understandable if reduction is truly easier at places where patches of oxygen structure about on clean areas. In reduction of massive oxides the autocatalytic effect of reduced metal is probably very common.[20,21] The general importance of this was recognized by Langmuir almost half a century ago.[22]

Acknowledgment

We wish to thank Dr. R. L. Gerlach for helpful comments.

REFERENCES

1. L. H. Germer and A. U. MacRae, Proc. Nat. Acad. Sci. 48, (1962) p. 997.

2. L. H. Germer and A. U. MacRae, J. Chem. Phys. 37, (1962) p. 1382.

3. T. L. Cottrell - The Strengths of Chemical Bonds, Butterworths, London (1958) Second Ed. Page 189. The binding energy of H on Ni is comparable to the bond energy of NiH.

4. D. O. Hayward and B. M. W. Trapnell - Chemisorption, Butterworths, London (1964) pg. 202 ff.

5. E. Bauer, Adsorption et Croissance Crystalline (C.N.R.S. Paris 1965).

6. E. Bauer, Surface Science 5, (1966) p. 155.

7. G. Ehrlich, Disc. Faraday Soc. 41, (1966) "--- diffraction measurements have been wildly interpreted as indicating a surface rearrangement in chemisorption---" p. 12. ----" This assumption, however, appears to be only tenuously related to experimentalfact.---" p. 68.

8. O. Beeck, Adv. In Catalysis 2, (1950) p. 151.

9. R. Wortman, R. Gomer and R. Lundy, J. Chem. Phys. 27, (1957) p. 1099. Photographs of Fig. 1, 5, and 8.

10. E. K. Rideal and F. Sweett, Proc. Roy. Soc. A257, (1960) p. 291.

11. J. F. Nicholas, J. Phys. Chem. Solids 24, (1963) p. 1279.

12. L. H. Germer, E. J. Scheibner and C. D. Hartman, Phil. Mag. 5, (1960) p. 233.

13. H. E. Farnsworth, R. E. Schlier and J. Tuul, J. Phys. and Chem. Solids 9, (1959) p. 57.

14. Because all domains are tied to the substrate lattice it is difficult to see why amplitudes should not be added over many domains to give good resolution. This point has been raised previously without satisfactory explanation. L. H. Germer and C. D. Hartman, J. Appl. Phys. 31 (1960) 2091.

15. L. H. Germer, J. W. May and R. J. Szostak, Surf. Sci. 7, (1967) p. 430.

16. J. W. May and L. H. Germer, Surf. Sci. 11, (1968) p. 443.

17. L. H. Germer and A. U. MacRae, J. Appl. Phys. 33, (1962) p. 2923.

18. A. U. MacRae, Surf. Sci. 1, (1964) p. 319.

19. M. D. Low and E. S. Argano, J. Phys. Chem. 70, (1966) p. 3115.

20. M. R. Hillis, C. Kemball and M. W. Roberts, Trans. Faraday Soc. 62, (1966) p. 3570.

21. R. N. Pease and H. S. Taylor, J. Amer. Chem. Soc. 43, (1921) p. 2179.

22. I. Langmuir, Trans. Faraday Soc. 17, (1922) pp. 619-620.

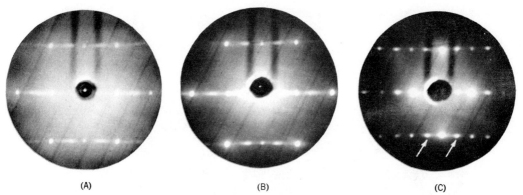

Fig. 1. Diffraction patterns from a (110) nickel surface after saturation exposure to hydrogen at room temperature, showing h, k + ½ diffraction spots that are not produced by a clean surface. Two of these are marked by arrows in Fig. 1c. (a) Streaks indicate presence of traces of an unidentified adsorbed impurity (see text). 139 eV. (b) Decreased impurity coverage. 125 eV. (c) Impurity concentration negligible. Note intensity of the half order spots. 150 eV.

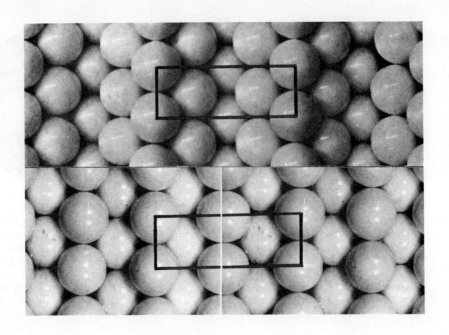

Fig. 2. Upper model--reconstructed nickel surface after hydrogen adsorption.
A (1x2) unit mesh is outlined. We assume that the hydrogen atoms are
adsorbed on the inclined rudimentary (111) faces and are mobile at
room temperature. Lower model--the normal nickel (110) surface. A
(1x2) unit mesh is outlined here also, and if hydrogen adsorption were
non-reconstructive, hydrogen atoms must be placed in the outlined area
in a way to give the (1x2) pattern. We show for comparison H and Ni
atoms drawn to scale, diameters 0.72 Å and 2.49 Å. The small hydrogen
atoms must be present, at least predominantly, in every second column
to give the h, k + ½ spots at all. Furthermore, the presence of strong
h, k + ½ spots (of which 1$\overline{2}$ and 1½ are marked by arrows in Fig. 1c)
requires fixed locations in these columns, and alignment between columns. The small size of hydrogen atoms is perhaps reason for rejecting
this model.

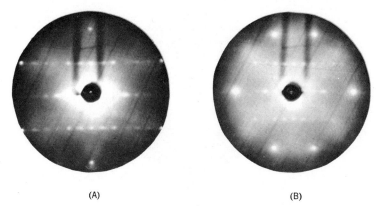

(A)   (B)

Fig. 3. Patterns showing strong h, k + ½ spots from hydrogen covered surfaces
at higher voltages. (A) 280 eV. (B) 355 eV. The high background,
including weak Kikuchi bands, is comparable with that from clean
nickel at the same voltages. Note that the half orders are relatively
as strong as in **Fig. 1.**

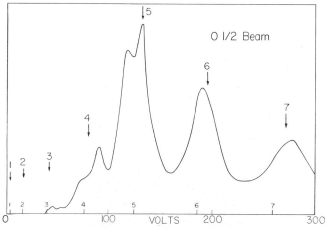

Fig. 4. Variation with beam voltage of the intensity of the 0 ½ diffraction
beam from a hydrogen covered surface at normal incidence. For the upper model of Fig. 2 the calculated voltages of the maxima of the
various orders are marked at the bottom of the figure. A better fit is
obtained (large arrows) for a contraction normal to the surface of 2.5
per cent. (Both are for refractive index unity.)

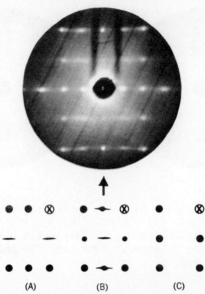

Fig. 5. A hydrogen covered surface (giving a LEED pattern like those of Fig. 1c) has been exposed to oxygen at room temperature. (A) Sketch showing the first appearance of h + ½, k streaks interpreted as due to (2x1)–O[½] domains that are narrow in the [O1] direction. (B) Sketch, and LEED pattern at 85 eV, after about 0.5 L oxygen exposure. Oxygen spots (h + ½, k) are stronger and sharper; hydrogen spots (h, k + ½) are weaker and more diffuse. (C) Sketch after oxygen exposure of about 0.8 L. Oxygen spots are strong and sharp, hydrogen spots are gone. Like diffraction pattern of Fig. 6. Cross in circle is 00 beam.

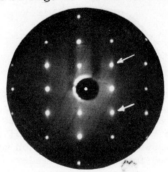

Fig. 6. LEED (2x1) diffraction pattern from a (110) nickel surface covered by half a monolayer of oxygen atoms, 135 eV. Half orders, including those marked by arrows, are not present when the crystal is clean.

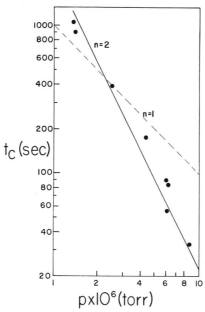

Fig. 7. Variation with hydrogen pressure p of the time, t_c, required to remove half a monolayer of oxygen atoms, at $550^\circ K$. The solid line represents $p^2 t_c = $ constant, and the dashed line $p t_c = $ constant.

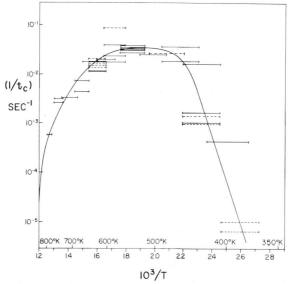

Fig. 8. Variation with temperature of the time, t_c, required to clean a (110) nickel surface originally covered by half a monolayer of oxygen atoms, at the constant hydrogen pressure of 6×10^{-6} torr. (Dashed lines represent tests at other pressures corrected to 6×10^{-6} torr by the square law relation of Fig. 7.)

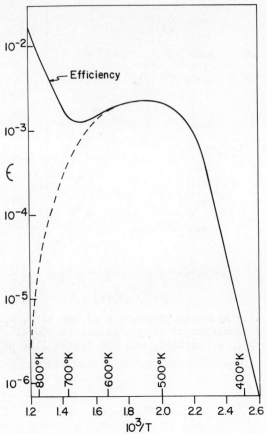

Fig. 9. Efficiency of removal, ϵ, of half a monolayer of oxygen by hydrogen at the hydrogen pressure 6×10^{-6} torr. It was necessary[16] to work with a crystal containing dissolved oxygen ($\sim 5 \times 10^{-4}$ atomic per cent). The dashed line represents the incorrect result that would be found if the oxygen diffusing to the surface from the bulk of the crystal were neglected.

THE INTERACTION OF SLOW ELECTRONS
WITH CO ADSORBED ON Ni(100)

R. A. Armstrong

Radio and Electrical Engineering Division
National Research Council
Ottawa, Canada

Introduction

The (100) surface of nickel has been studied many times. However, previous work has raised some questions about the chemisorption of CO. Mitchell and Sewell[1], using RHEED, found a different pattern of extra diffraction beams than the $c(2 \times 2)$ patterns observed by Park and Farnsworth[2] using a scanning LEED system. It was important to know whether the two techniques gave different results for the same structure. While this question was being investigated, Lichtman, Kirst and McQuistan[3] reported that CO did not adsorb on a Ni(100) surface which had been cleaned by heating extensively in uhv, but which had not been cleaned by oxidation and reduction or by ion bombardment. Lichtman et al based their conclusion on the absence of electron induced desorption. This negative conclusion suggested electron bombardment experiments with CO adsorbed on a Ni(100) surface cleaned by ion bombardment. These experiments showed that the weakly bound CO absorbed during large CO exposures was desorbed by slow electron impact. In addition, the electrons induced ordering of the more tightly bound CO and high intensity patterns of extra diffraction beams were observed, consistent with the RHEED observations. The ordered form of adsorbed CO was not desorbed by the electron beam.

Recently, Onchi and Farnsworth[4] have reported experiments which demonstrate that extensive heating is frequently inadequate to clean the Ni (100) surface because carbon diffuses to the surface during the heating and prevents CO adsorption. In the course of this work, they observed some of the extra beams of the pattern observed by Mitchell and Sewell.

Experimental

The exterior of the mostly glass LEED system used for these experiments is shown in Fig. 1. The diffraction spots appear on the fluorescent screen coated on the inner wall of the spherical part of the diffraction tube. Intensities were measured by placing a light pipe directly above the diffraction spots of interest. The light pipe was connected in turn to a photo-multiplier. Thus relative intensities could be recorded easily as a function of the incident electron energy as the diffraction spot moved under the light pipe, and the widths of the recorded peaks, ΔE, served as a useful measure of the sharpness of the diffraction beams. Intensities were also recorded as a function of time during adsorption.

The interior of the system is shown in Fig. 2. The nickel crystal was cylindrical, 7 mm in diameter and 7 mm long. It was mounted at the end of a shaft which could be rotated to change the angle of incidence. Diffraction patterns have been observed with angles of incidence as large as 85°. At these large angles of incidence only a part of the diffraction pattern could be observed because diffraction beams making angles greater than 135° with the incident beam could not be observed.

The crystal was heated by bombardment with 5 keV electrons from the gun mounted behind the crystal and temperatures were measured with a calibrated infra-red pyrometer. To study induced desorption and rearrangement the crystal was rotated 180° to face this gun and the gun was operated at low voltage. Ion bombardment cleaning was also carried out with the crystal facing the gun using a shielded auxiliary filament to ionize the gas.

Results

Adsorption of CO on the clean, annealed, room temperature Ni(100) surface was monitored by following the work function shift, the elastic reflection of electrons, and the appearance of extra diffraction beams.

The increase in work function, $\Delta\phi$, caused by CO exposure is shown in Fig. 3. The work function shift was measured from the shift of the retarding potential characteristic[5]. $\Delta\phi$ increased by ~0.15 V with an exposure of 10^{-6} Torr min and thereafter $\Delta\phi$ increased more slowly to a maximum of 0.25 V after an exposure of 1.6×10^{-4} Torr min.

The fraction of incident electrons reflected elastically, R, was frequently observed. Typical results are shown in Fig. 4. A characteristic maximum appeared at 16 eV and grew with CO exposure. In a manner similar to the increase in $\Delta\phi$, this maximum in R grew rapidly at first and then after an exposure of ~10^{-6} Torr min its rate of growth decreased. The curve shown for intermediate coverage corresponds to $\Delta\phi = 0.07$ V.

Reflection measurements were also useful in checking the state of the surface. An additional characteristic maximum appeared when either carbon diffused to the surface or when oxygen was present on the surface during oxidation experiments.

Quite broad diffraction spots appeared in addition to the Ni(100) diffraction pattern during CO adsorption and reached maximum intensity after an exposure of ~10^{-6} Torr min. The only pattern of extra diffraction beams observed was a 2×2 pattern with (h/2o) and (0k/2) missing for h, k odd as indicated in Fig. 5. This pattern corresponds to that observed by Mitchell and Sewell using RHEED[1].

The extra beams were brightest at high angles of incidence (60° to 70°) where high diffraction orders could be observed at relatively low energy. At high angles of incidence and at certain energies, weak forbidden beams appeared on the axis of the pattern parallel to the axis of rotation as indicated in Fig. 5 for $\theta > 0$.

The intensity of the (1/2-1/2) spot is shown as a function of energy in Fig. 6 and compared with the intensity of the same diffraction beam observed by Park and Farnsworth[2]. The positions in energy of the maxima agree quite well. Apparently the two adsorption structures are the same.

The diffraction pattern could be intensified by electron bombardment. The curves of the normalized intensity and energy width for the (1/2-1/2) beam shown in Fig. 7 were obtained by rotating the crystal to face the electron bombardment gun and bombarding for a minute or more at a time. After the bombardment gun was turned off, the crystal was returned to its original position. The intensity was measured and then the whole procedure was repeated until the data for a particular curve had been collected. The intensity of the (1/2-1/2) beam increased by more than ten times, while the intensity of the (11) beam increased by about 3 times. After intensification, the half-order and integral-order diffraction beams were of similar intensity.

In Fig. 7 it is seen that the intensification was progressively more rapid as lower bombardment energy was used. The rate of intensification at 75 eV (not shown) was somewhat slower than at 100 eV. These results indicated that there was a maximum rate of intensification at about 100 eV. The distribution of electrons over the crystal and the variation of this distribution with energy was unknown. Consequently, this maximum may be an experimental arti- fact. In any case, the intensification was certainly not a thermal effect since the pattern was also intensified by the probe electron beam operating for 16 hours at 1μA, 100 eV.

The decrease in energy width of the (1/2-1/2) beam, caused by electron bombardment, is shown in Fig. 7. The extra diffraction beams decreased in energy width and angular width as the intensity increased. At the same time, the integral beams decreased in width by 10% or less.

Electron bombardment resulted in the change in the reflection coefficient, R, plotted versus energy in Fig. 8. Both maxima moved to lower energy and the higher energy maximum became smaller. Since this latter maximum grew steadily during CO exposure with no change in position, this shift in position of the maximum indicated a change in state of the remaining adsorbed CO.

Electron induced intensification of the extra diffraction beams was studied after different exposures. For small CO exposures in the range where $\Delta\phi$ and R changed rapidly, the final intensity was roughly proportional to exposure. The maximum intensity was observed for an exposure corresponding to the knee in the $\Delta\phi$ curve. Further exposure resulted in further adsorption, as indicated by changes in both $\Delta\phi$ and R but produced no further increase in the final intensity of the half-order beams.

A number of experiments were carried out after the clean crystal had been given a CO exposure of 10^{-6} Torr min and the extra diffraction beams intensified by electron bombardment:

a) A further CO exposure of $\sim 10^{-6}$ Torr min resulted in additional adsorption is indicated by an increase in the maximum in R characteristic of CO adsorption. This increase in R was removed when the crystal was bombarded by electrons. This cycle of adsorption followed by electron induced desorption was repeated several times with essentially the same results.

The most revealing experiments involved heating the crystal.

b) Heating the crystal to temperatures below $\sim 300°C$ desorbed weakly bound CO and produced some further intensification which was always less than 20%.

c) Heating to $\sim 350°C$ caused the extra diffraction beams to disappear but changed R very little. Subsequent electron bombardment caused the extra diffraction beams to reappear and their intensity grew with further bombardment to nearly the intensity observed before heating.

The thermal desorption spectrum was observed before and after intensification. The low temperature maximum ($\sim 100°C$) was about 20% smaller when intensification had been carried out. Thermal desorption spectra are of little value for cylindrical crystals since the area of the surface of interest is a small fraction of the total surface. However, this result is consistent with the electronic desorption of weakly bound CO observed by other means. All the CO was desorbed for temperatures above ~600°C.

Discussion and Conclusions

This study has provided some answers to the questions of previous work and demonstrated a new interaction of electrons with CO adsorbed on Ni(100).

Each of the three methods used to monitor the state of the surface during CO exposure indicated that CO adsorbed on the (100) surface of nickel when it had been cleaned by ion bombardment.

The ordered adsorption structure agreed with the structure observed by Mitchell and Sewell using RHEED. The pattern of extra diffraction beams was nearly 2 × 2; there were, however, no extra beams in the [110] azimuths.

The curves of elastic reflection coefficient, R, versus electron energy were found to be a convenient method of monitoring the average state of the surface. Changes in R showed that slow electrons caused the desorption of weakly bound CO absorbed during large exposures and changed the form of the remaining CO.

The appearance of extra diffraction beams and the decrease in the intensity of the nickel beams during adsorption indicated that the CO also adsorbed in two, more tightly bound forms, one ordered and the other disordered. The simultaneous increase in the intensity and the sharpness of the extra diffraction beams with the duration of the electron bombardment showed that electrons interacted with the disordered CO to convert it to the ordered form. The extended bombardment experiments show that ordered CO was not desorbed by the electron beam.

These results indicate that a conversion interaction occurs on Ni(100) in which CO molecules are excited by electron impact and make transitions to the ordered state instead of leaving the surface. The conversion is not directly related to electron induced desorption since heating the CO covered Ni(100) surface to ~350°C desorbed the weakly bound CO and at the same time disordered the more tightly bound CO, and this remaining, tightly bound CO could again be converted to the ordered form of adsorbed CO.

This conversion interaction appears to be similar to the one observed by Menzel and Gomer[6], with a tungsten field emitter on which virgin CO was converted to β-CO by slow electron impact. Redhead[7] also found, for CO on tungsten, an electron induced conversion from a desorbable state to a state which was not desorbed. Anderson and Estrup[8] have studied the more complex system of NH_3 adsorbed on W(100) where the NH_3 is apparently dissociated with high probability by the action of the electron beam. For CO adsorbed on Pt(100), Tucker[9] found that electron induced desorption lowered the CO density on the surface and allowed a higher density, ordered from of CO to change to a lower density, ordered form. This latter case appears to be different than the others since the conversion from one form of adsorption to another did not depend directly on the excitation of the molecule by an electron but on a change of adsorbate density.

There is a growing body of results which show that slow electrons can induce profound changes in surfaces being studied and consequently care must be taken in the interpretation of LEED results. The high energy electron beams used in RHEED are less likely to distort results in this way since, in general, the cross-sections for excitation decrease with electron energy for energies greater than a few hundred volts and because lower electron currents are required.

Acknowledgments
    The author wishes to acknowledge fruitful discussions with P.B. Sewell, D. F. Mitchell and E. V. Kornelsen. The Ni(100) crystal was oriented and the surface was prepared by P.B. Sewell.

References

1. D. F. Mitchell and P. B. Sewell (unpublished).

2. R. L. Park and H. E. Farnsworth, J. Chem. Phys. 43, 2351 (1965).

3. D. Lichtman, T. R. Kirst and R. B. McQuistan, Phys. Letters 20A, 129 (1966).

4. M. Onchi and H. E. Farnsworth, Phys. Letters 26A, 364 (1968).

5. R. A. Armstrong, Can. J. Phys. 44, 1753 (1966).

6. D. Menzel and R. Gomer, J. Chem. Phys. 41, 3311, 3329 (1964).

7. P. A. Redhead, Supplemento al Nuovo Cimento, Serie I, 5, 586 (1967).

8. J. Anderson and P. J. Estrup, Surface Sci. 9, 463 (1968).

9. C. W. Tucker Jr., Surface Sci. 2, 516 (1964).

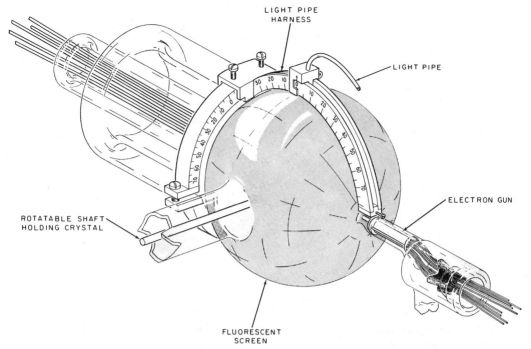

LIGHT PIPE
HARNESS

LIGHT PIPE

ELECTRON GUN

ROTATABLE SHAFT
HOLDING CRYSTAL

FLUORESCENT
SCREEN

Fig. 1. The exterior of the LEED system.

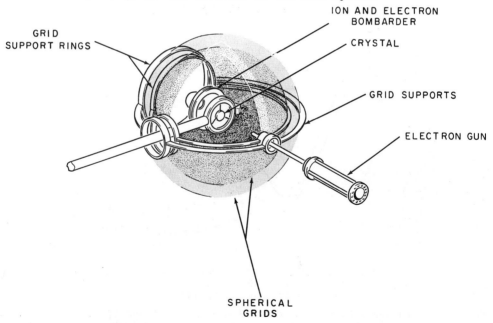

GRID
SUPPORT RINGS

ION AND ELECTRON
BOMBARDER

CRYSTAL

GRID SUPPORTS

ELECTRON GUN

SPHERICAL
GRIDS

Fig. 2. The interior of the LEED system.

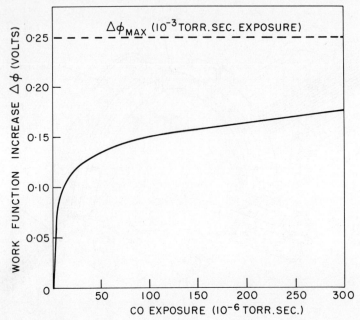

Fig. 3. The increase in the work function of the Ni(100) surface caused by CO exposure.

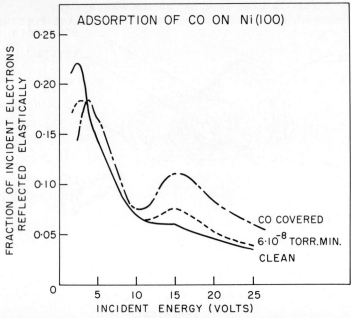

Fig. 4. The fraction, R, of the incident electrons reflected elastically as a function of the incident energy and the change in R caused by CO adsorption.

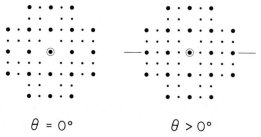

$\theta = 0°$ $\theta > 0°$

Fig. 5. The Ni(100)-CO diffraction pattern, 2 x 2 with (h/2 o) and (o k/2) missing for h, k odd.

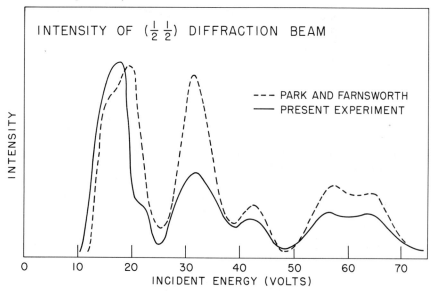

Fig. 6. The intensity of the $(\frac{11}{22})$ diffraction beam plotted as a function of energy compared with the results of Park and Farnsworth.

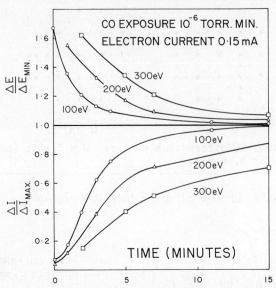

Fig. 7.   The increase in the intensity and sharpness of the $\left(\frac{11}{22}\right)$ diffraction beam plotted as a function of the duration of the 0.15 mA electron bombardment at a number of electron energies.

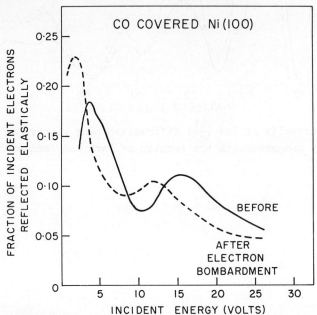

Fig. 8.   The change in the fraction, R, of electrons reflected elastically caused by electron induced desorption and ordering.

INFRARED AND MAGNETIC STUDY OF NITROGEN CHEMISORBED ON NICKEL

F. P. Mertens and R. P. Eischens

Texaco Research Center
Beacon, New York

The infrared spectrum of nitrogen adsorbed on nickel provides rigorous proof that the observed nitrogen is in the molecular form (1). The spectra also provide reasonable evidence that the adsorption is chemical in nature and that the chemisorbed molecule is in the linear structure, Ni-N≡N. However, magnetic data suggested that the adsorption of nitrogen on nickel resembles the physical adsorption of argon (2). Correlation between infrared and magnetic work is complicated by the difficulty of preparing silica-supported nickel samples which are capable of adsorbing nitrogen. The reproducibility problem is encountered even in work within a single laboratory. This makes it essential to compare magnetic and infrared data for cases where the infrared spectra show that nitrogen is actually adsorbed.

Apparatus was constructed which can be used to observe the infrared spectrum of chemisorbed molecules while simultaneously measuring the effect of the chemisorption on the magnetization of the adsorbent. The sample is suspended from a Cahn electrobalance by a quartz fibre which places it in the path of the infrared beam and between the faces of a Faraday type magnet.

This report describes the apparatus and the present status of the combined infrared-magnetic studies of the nitrogen-on-nickel system. Also described are experiments related to sample reproducibility, sample contamination, and the determination of the heating effect of the infrared beam.

EXPERIMENTAL

Infrared Spectrophotometer

A Cary-White Model 90 (Applied Physics Corporation) double beam, prism-grating spectrophotometer was used for all of the reported work. This instrument has a sample space which is 25 cm long and is large enough to conveniently provide space for the electromagnet. A duplicate sample is placed in the reference beam and the balance potentiometers of the spectrophotometer are adjusted to compensate for small differences between the samples. This gives a relatively straight background over the 3400-1300 cm^{-1} region.

A filter which cuts off radiation above 4000 cm^{-1} is inserted in front of the sample to minimize heating of the sample by the infrared beam. Even though a filter is used, determination of the extent of heating of the small nickel particles remains a problem because conventional thermocouple measurements are of limited utility. The change of magnetization with temperature provides a convenient method of measuring the heating effect and determining the effectiveness of the filter.

Sample Preparation

All of the silica-supported nickel samples were prepared by impregnating Cabosil with a solution of nickel nitrate, drying, and reducing with hydrogen at temperatures near 350°C.

Details of the reduction step will be discussed in a latter
section devoted to the effect of reduction conditions on the
properties of the supported nickel.

Prior to reduction the powdered sample, which has
been dried at room temperature for 24 hours, is pressed into
a self supporting disk. Pressures of 5,000 psi are used.
The die is 45 mm in diameter. A rectangular portion of the
disk (14 mm x 42 mm) is cut out and placed in a simple quartz
frame which is suspended from the weighing arm of the Cahn
balance.   The dimensions of the sample were selected to match
the infrared beam in order to eliminate an uneven heating of
the sample by the beam. The 14x42 mm samples weigh in the
range 70-130 mg.

Samples used only for infrared studies commonly are
prepared to contain about 9 wt % nickel after reduction. Samples
to be used for combined infrared and magnetic work are prepared
to have about 3 wt % nickel. The lower metallic contents allow
higher field strengths to be used.

Cahn RG Electrobalance

As obtained from the manufacturer, the Cahn balance
was enclosed in a glass envelope with a volume of about 5 liters.
The glass envelope was replaced with an envelope machined from
a solid piece of 304 stainless steel. This modification reduced the difficulties due to vibrations and decreased the
volume to about 500 cc.

The modified electrobalance is shown in Fig. 1. The dimensions of the machined cavity are approximately 7x7x20 cm. A 12 mm thick pyrex glass window is sealed to the steel envelope with an "O" ring. The assembly at the upper left is used as a port for inserting the samples. The top of the port is fitted with a glass window to allow observation of the sample when the port is closed. The quartz fibre, which holds the sample, is suspended through the cylinder at the lower left. The tare weight is suspended in the cylinder at the lower right. A 25 mm outlet (hidden by the torque motor) is used to connect to the vacuum and gas handling systems. The assembly at the top right contains a vacuum tube, resistors, and a Zener diode. These components, which were originally inside of the glass envelope, were placed outside of the steel envelope to minimize degassing problems. All insulation from wires inside the vacuum chamber, except that in the torque motor windings, was also removed.

When the vacuum is better than 10^{-4} torr the balance has a usable precision of 0.001 mg. The precision is almost as good at high pressure if the pressure is constant. However, there is an uncertainty of about 0.020 mg when a weight at 10^{-4} torr is compared to a weight at 100 torr. This difficulty is encountered in studies of nitrogen adsorption at room temperature because the adsorption is pressure sensitive up to 100 torr (1).

A typical sample will chemisorb about 0.400 mg of carbon monoxide. A monolayer of nitrogen would weigh the same as a
monolayer of carbon monoxide. However, at room temperature
the nitrogen uptake is only about 0.050 mg. Since nitrogen
chemisorption at room temperature is measured at high pressure,
an improvement in weighing would be desirable for nitrogen
studies. However, the present situation is tolerable. The
good precision at low pressures is useful in studies of uptake
of contaminants as the sample is exposed to vacua of the order
of 10^{-7} to 10^{-6}torr.

An important feature of the Cahn balance is that it
operates on a null principle so the sample position is constant relative to the infrared beam and the magnet.

Sample Cell

The sample cell is shown in Figure 2. The horizontal
portion of the cell, through which the beam passes, was also
machined from a 304 stainless steel cylinder. The cylinder was
24 cm long and 4.1 cm in diameter. The CaF_2 end windows are
sealed to the cell with Viton-A "O" rings. The ends of the cell
are cooled by circulating water through channels which were cut
out and then covered by soldering stainless steel rings over the
grooves. The cell is heated by a nichrome ribbon furnace im-
bedded in Marinite insulation. The rectangular shape of the cell
body allows the cell to easily fit between the 5.1 cm spacing of
the magnet pole pieces. A thermowell extends through the vertical cylinder into the body of the cell. The thermocouple

goes near to the top of the sample but is not in the path of
the infrared beam. The furnace is capable of heating the interior of the cell to 500°C. The narrow vertical tube at the
left is fitted with a Hoke bellows valve and is used as the
hydrogen outlet during reduction.

Magnet

The water cooled, electromagnet, which is shown in
Figure 3, was custom made by Alpha Scientific Labs, Inc. The
Faraday pole pieces are 72 mm in diameter. The power supply
can produce fields up to 5,000 gauss with a precision of 0.01
per cent. However, the maximum usable field strength is limited
by the necessity of keeping the sample hanging freely away from
the cell walls. With a 9 wt % nickel sample the maximum field
is 900 gauss. Field strengths of about 1800 gauss can be
used with samples containing 3 wt % nickel. The magnet is
keyed to steel tracks and an adjustment screw provides fine
control of the magnet position.

The field strengths were determined using Cabosil
disks containing ferrous ammonium sulfate and polycrystalline nickel powder. However, no attempt was made to express
the magnetization of the nickel in absolute units. The sample shape, imposed by the infrared requirements, is poorly
suited to quantitative Faraday magnetization studies. Instead, the magnetization isotherm method, introduced by Selwood (2), was used. The percentage decrease in magnetization

produced by nitrogen is compared to the decrease produced by hydrogen. In using hydrogen as a standard it is assumed that the hydrogen dissociates into atoms and each atom forms one bond to the surface. Since nitrogen does not dissociate, a molecule of chemisorbed nitrogen would be expected to produce a magnetization decrease which is one half that produced by adsorbing a molecule of hydrogen.

RESULTS AND DISCUSSION

Spectrum of Nitrogen on Nickel

The spectrum of nitrogen chemisorbed on a 2.8 wt % nickel-on-silica sample is shown in Fig. 4. This spectrum was observed at 28°C with a nitrogen pressure of 100 torr.

The sample disk weighed 124 mg. The area of one side of the disk was 5.2 cm^2. The nitrogen uptake was 0.05 $\pm$ 0.02 mg. The integrated intensity* per molecule of chemisorbed nitrogen is found to be 18$\pm$ 7 x 10^{-17} cm/molecule.

The integrated intensity for chemisorbed nitrogen is about the same as we observe for carbon monoxide on metals.

---

*The specific integrated intensity is 2.3/C $\int$ log T_o/I d$\bar{\nu}$ where C is the concentration in terms of molecules of ad-sorbed nitrogen in a volume of sample which is transfixed by one cm^2 of the infrared beam.

The value for chemisorbed carbon monoxide is about 200 times larger than that for gaseous carbon monoxide. It is significant that the specific intensity for chemisorbed nitrogen approaches the highest values in the range of infrared intensities.

The ratio of the number of molecules of chemisorbed nitrogen to the total number of nickel atoms is 0.03 ± 0.01. The maximum chemisorption of carbon monoxide gives a CO/Ni ratio of 0.25. Considering the precision of the measurements, it is reasonable to conclude that the chemisorbed nitrogen, shown in Figure 4, represents a surface coverage of about 10 per cent. The relatively low surface coverage obtainable for nitrogen at 28°C is pertinent to the interpretation of the magnetic measurements.

The CO/Ni ratio of 0.25 implies that at least one fourth of the nickel atoms are on the surface of the nickel particles. This is consistent with an average particle size of $36A^\circ$.

Magnetic Measurements

Table I compares the decrease in magnetization produced by adding N_2, Ar, and H_2 to the sample whose nitrogen spectrum is shown in Figure 4. The nitrogen and argon data were both obtained at pressures of 100 torr. The decrease in magnetization for nitrogen is the same as that observed for argon. This is in accord with results reported by Selwood (2).

The magnetization isotherm approach is designed to compare the magnetization changes on a molecule per molecule basis. Thus, it is necessary to take into consideration the low surface coverage for nitrogen when comparing nitrogen and hydrogen.

TABLE I

Effect of Adsorbed Gases on the Magnetization of Nickel

Adsorbed Gas	$\Delta M/M$ %	Coverage θ	$-\Delta M/M\theta$
N_2	-3.0	0.10	30
Ar	-3.0	----	---
H_2	-28.0	1.0	28

The $-\Delta M/M\theta$ column of Table 1 shows that the decrease in magnetization produced by a molecule of nitrogen is the same as that produced by an atom of hydrogen. This might be taken to indicate that nitrogen is chemically bonded to the nickel. However, we prefer a more limited conclusion to the effect that the small $-\Delta M/M$ value for nitrogen should not be taken as evidence that the nitrogen is physically adsorbed.

The results observed with argon provide one reason for favoring the limited conclusion. The weight uptake for argon is about the same as that for nitrogen. Thus the line of reasoning based on the $-\Delta M/M\theta$ values could also be used to prove that argon is chemisorbed. This is unrealistic and raises doubt as to the validity of the $-\Delta M/M\theta$ approach.

There is no detectable uptake of nitrogen or argon on silica samples which do not contain nickel. Adsorption on the silica would cause the $-\Delta M/M\theta$ values to appear smaller than the true values.

The decreases in magnetization produced by nitrogen and argon were easily reversed by evacuation at 25°C. Reversibility of the effect was used as a test to eliminate the possibility that the observed changes in magnetization were due to impurities such as oxygen, hydrogen, hydrocarbons, or water.

Reduction of Silica-Supported Nickel

Nickel samples which have a high adsorptive capacity for carbon monoxide may be easily prepared. Minor variations between samples are detected but these usually do not interfere with the proposed experiments. However, for nitrogen adsorption, differences between satisfactory and unsatisfactory samples can be caused by subtle factors which are not considered to be worthy of notice or are not detected. The greater sensitivity of nitrogen adsorption is probably due to the fact that the adsorption forces are weaker for nitrogen than for carbon monoxide.

An understanding of the factors involved in the chemisorption of nitrogen on nickel is a long range goal of the present work. At present, even an empirical knowledge of the critical preparative factors is important to the efficient study of the nitrogen-on-nickel system. On the basis of an extensive electron microscopic study VanHardeveld and VanMontfoort have concluded that nickel particle diameters in the range 15-70A° are critical to the ability to adsorb nitrogen (3). Nakata and Matsushita have shown that it is not necessary to start with nickel nitrate by preparing satisfactory samples from nickel

acetate (4). In the present work we have found that it is helpful to have a high rate of hydrogen flow during the initial
stages of the reduction.

In subsequent discussion the term "slow" refers to
samples reduced by hydrogen which was flowing at a rate of 40
ml/min. The hydrogen was introduced at 125°C, the temperature
was brought up to 360°C, and reduction continued at this temperature for 16 or more hours. The 40 ml/min rate is limited by
the throughput capacity of the palladium diffuser. The term
"fast" refers to a reduction procedure in which unpurified
cylinder hydrogen is used during the two hour period during
which the temperature is raised from 125°C to 360°C. During
this period the flow rate is 1000 ml/min. When 360°C is reached
the reduction is continued with palladium diffused hydrogen at
the slow rate. Weight change data indicate that in both cases
the reduction,is 90 per cent completed by the time the temperature
of 360°C is obtained.

The adsorbed nitrogen shown in Figure 4 was observed
on a sample prepared by the fast flow procedure. Satisfactory
samples have been consistently prepared by this method. Magnetic and infrared experiments on this sample were compared with
an unsatisfactory sample prepared by the slow flow method. The
latter was able to chemisorb less than one tenth as much nitrogen
as the Figure 4 sample. Despite the low capacity for nitrogen
adsorption the capacity for carbon monoxide was about three
fourths that of the fast flow sample.

Figure 5 compares the decrease in magnetization as a function CO/Ni ratio for the fast flow and slow flow samples. It is apparent that there are significant differences in the magnetic properties of the two types of sample and it is reasonable to assume that these differences are related to the differences in capacity for nitrogen adsorption.

Since the effect of chemisorption on magnetization is due to the decrease in the magnetic contribution of the surface atoms it would be expected that smaller particles would show the greatest relative decrease in magnetization. However, in Figure 5 the sample having the smaller capacity for carbon monoxide has the larger $-\Delta M/M$. Thus, the chemisorption capacity evidence and the magnetic evidence for particle size appear to be contradictory.

Extensive studies of the effect of particle size on the magnetization of silica supported nickel have been described by Selwood (2) and by Geus and co-workers (5,6). Their work indicates that it is not realistic to expect that all of the particles would be magnetically saturated under the experimental conditions used in the present work (28° - 1,800 gauss). This has been confirmed for the fast flow samples in a separate magnetic apparatus which could be used at fields up to 5,000 gauss. It was found that at any fixed CO/Ni ratio the value of $-\Delta M/M$ increased with increasing field strength. This shows that at higher fields more of the smaller particles are magnetizable.

The data of Figure 5 may be interpreted by assuming that the slow flow sample is composed of particles which are uniformly close to the average of 52Å while the fast ilow sample has two ranges of particle sizes, one larger than 52Å and the other smaller. The particles which are 52Å or larger are magnetized under the conditions used. However, the small particles in the fast flow sample are not magnetized. Thus, the magnet indicates a larger particle size for the fast flow sample even though the average particle size ($36\text{A}°$) as determined by carbon monoxide chemisorption is smaller.

If the nitrogen capacity of the fast flow sample involved the small particles, no decrease in magnetization would be produced by chemisorption of nitrogen. Thus, the large value of $-\Delta M/M\theta$ for nitrogen indicates that the sites for nitrogen adsorption are on the larger particles of this sample. If particle size is the only factor the required size must be larger than 52Å since the slow flow sample does not have as many particles of the required size.

Spectrum of Chemisorbed Carbon Monoxide

Figure 6 compares the spectrum of CO chemisorbed on the fast flow and slow flow samples. Spectrum A was obtained at a CO/Ni ratio of 0.15. This is almost full coverage for this sample. Spectrum B was obtained at a CO/Ni ratio of 0.17 for the fast flow sample and Spectrum C was obtained at full coverage (CO/Ni = 0.28).

All spectra have approximately the same distribution of the linear (band at 2075 cm^{-1}) and bridged forms (bands below 2000 cm^{-1}). The most obvious difference in the spectra is in the bridged band region where the maximum absorption of Spectrum A is near 1940 cm^{-1} and the maximum absorption in B is at 1910 cm^{-1}. Since the chemisorption of nitrogen involves only 10 per cent of the surface a completely different carbon monoxide spectrum would not necessarily be expected for the two types of sample. Thus, the difference in the bridged band frequencies might be related to the factors which are involved in nitrogen chemisorption sites. It is reasonable to assume that the bridged structure is affected by things such as the specific crystal face on which the carbon monoxide is adsorbed(7). However, at the present time the infrared interpretation of these bands is not sufficiently advanced to warrant definite conclusions on this point.

Extent of Sample Contamination

In the previous discussion it has been tacitly assumed that the difference between the fast flow and slow flow samples originate within the samples rather than through contamination from an outside source. This is reasonable since all of the experiments were done in the same apparatus. Even though outside source contamination was not considered to offer a reasonable explanation of the differences between the fast flow and slow flow samples some work was carried out to get an idea of the rate at which the nickel adsorbed contaminants while exposed to vacua in the range of 10^{-7}-10^{-6}torr. These studies were made possible by the high sensitivity of the gravimetric system. The approach was based on the concept that in an atmosphere of hydrogen at 350°C the nickel would be free of all adsorbed species except hydrogen.

A typical nickel-on-silica sample used in infrared studies has a nickel surface area of 0.5-1.0 square meters and a silica area of about 20 square meters. The nickel, which is prepared by the reduction procedure is extremely active and will avidly adsorb oxygen, hydrocarbons, and water. The silica will compete strongly for water and if only limited quantities of water are available most of it will be taken up by the silica.

The best vacuum obtained in the infrared magnetic apparatus was 1×10^{-7} torr and evacuations were generally to the range of 10^{-7}-10^{-6} torr. A General Electric Monopole residual gas analysis showed that the residual gas is mainly water and hydrocarbon vapor. Simple calculations show that at 10^{-6} torr an active residual gas would cover only a few hundredths of a per cent of the nickel surface. However, an infinite source such as back diffusion from an oil pump could produce serious contamination after long exposure.

After a 48 hour reduction at 350°C the temperature was lowered to 200°C and the hydrogen was removed. The weight was recorded with hydrogen present and during the evacuation. The sample was then allowed to stand exposed to the 10^{-7}-10^{-6} torr vacuum and the weight uptake was followed.

A monolayer of oxygen atoms or carbon atoms would weigh about 0.20 mg. This is 200 times the weighing precision under these experimental conditions. During an 8 hour period a 2.8 wt per cent nickel-on-silica sample showed a gradual weight uptake which totaled 0.025 mg. A blank silica sample, without nickel, had an uptake of 0.015 mg during a similar period. The difference of 0.01 mg is attributed to the adsorption of oxygen or hydrocarbon on the nickel. This corresponds to a surface coverage of 5 per cent. This neglects a possible loss in weight due to desorption of hydrogen which could amount to the equivalent of about 8 per cent of the weight of an oxygen monolayer.

A sample contamination of about 10 per cent could correspond to coverage of all sites capable of chemisorbing nitrogen if the contaminant were selectively adsorbed on these sites. However, this could not explain the low nitrogen capacity of the slow flow samples because in all cases the elapsed time for the nitrogen experiments was less than one half hour.

Nickel Particle Temperature

A long standing problem in the infrared study of adsorbed molecules has involved the question of the degree of heating of the sample by the infrared beam. This problem is especially acute with samples containing metal particles which are strong absorbers of radiation. This problem was recognized in early work with a Perkin-Elmer 12-C single beam spectrophotometer and calculations indicated that the metal particles could be 10-15°C warmer than the ambient temperature. Thus, it was common practice to insert a filter in front of the sample which eliminated radiation above 4000 cm-1 and was expected to decrease the heating effect by 70-80 per cent.

It is difficult to obtain a valid measurement of the temperature of the nickel particles, therefore, when the combined magnetic-infrared apparatus became available it was used for this purpose. This use is based on the effect of temperature on the magnetization of the nickel. The results are shown in Figure 7.

With the infrared beam blocked off a calibration plot of magnetization versus thermocouple temperature was made as the sample was alternately heated and cooled over the range of 20-80°C. The infrared beam was then unblocked and allowed to impinge on the sample. The magnetization was again recorded as a function of thermocouple temperature. The displacement between the two lines is a measure of the heating effect of the infrared beam.

In Figure 7, which was obtained in an experiment in which a filter was not used and the system was evacuated, the heating effect of the beam is about 16°C. Use of a filter decreases the effect to 4-5°C. In the presence of hydrogen at 760 torr the heating effect is only 3-4° when the filter is not used. These temperature determinations apply only to the Cary-White Model 90 operated at a glower current of

1.02 amps. This instrument probably represents the minimum beam heating problem because the beam does not have a sharp focus and is interrupted by the chopper which is between the sample and the source.

SUMMARY

At 25° C the adsorption of nitrogen on silica-supported nickel involves 10 per cent of the nickel surface. This adsorption produces a small but easily measured decrease in the magnetization of the nickel. The interpretation of this decrease is complicated by the fact that the magnitude of the effect is equivalent to that produced by argon but is also equivalent to the decrease expected on the basis of comparison with chemisorbed hydrogen. Because of this ambiguity it is concluded that the magnetic measurements are not inconsistent with infrared spectra which show that the nitrogen is chemisorbed.

Simultaneous infrared and magnetic studies are necessary because of the lack of reproducibility in preparing nickel samples which are capable of chemisorbing nitrogen. The difference in nitrogen adsorptive capacity is reflected in differences in magnetic properties and in carbon monoxide chemisorption. The reproducibility problem can be overcome by empirical experimental methods which involve a fast hydrogen flow during the initial stages of the reduction.

REFERENCES

(1)   R. P. Eischens and J. Jacknow, Third International Congress on Catalysis (1964) p. 627, North-Holland Publishing Company, Amsterdam.

(2)   P. W. Selwood, "Adsorption and Collective Paramagnetism" Academic Press (1962).

(3)   R. VanHardeveld and A. VanMontfoort, Sur. Sci., 4, 396 (1966).

(4)   T. Nakata and S. Matsushita, J. Catalysis 4, 631 (1965).

(5)   J. W. Geus, A. P. P. Nobel and P. Zwietering, J. Catalysis 1, 8 (1962).

(6)   J. N. Geus and A. P. P. Nobel, J. Catalysis 6, 108 (1966).

(7)   D. A. King, Sur. Sci., 9, 375 (1968).

Fig. 1. Stainless steel vacuum microbalance enclosure.

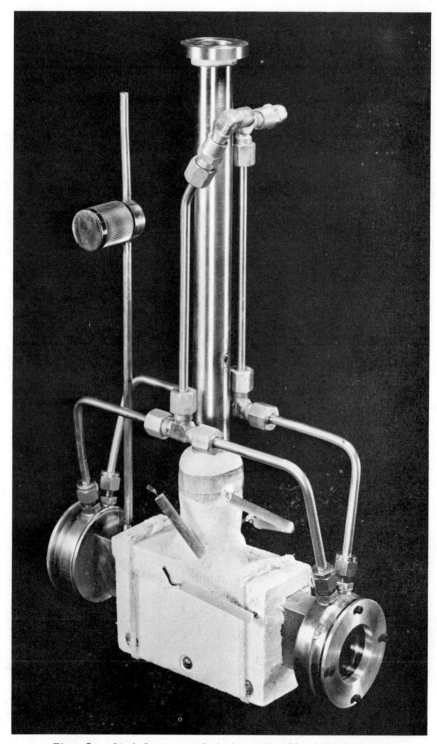

Fig. 2. Stainless steel infrared cell with furnace.

Fig. 3. Infrared cell connected to vacuum microbalance enclosure with magnet
in place.

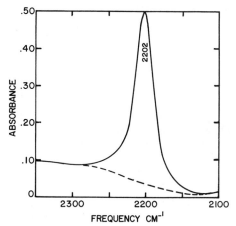

Fig. 4. Spectrum of nitrogen chemisorbed on nickel; pressure 100 torr at 28°C.

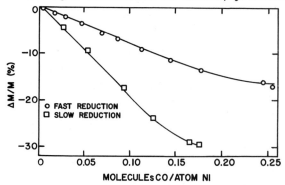

Fig. 5. Effect of chemisorbed carbon monoxide on the magnetization of nickel: comparison of samples prepared by fast and slow flow reduction procedures.

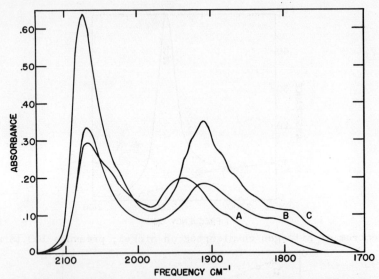

(A) SLOW FLOW SAMPLE CO/Ni = 0.17; (B) FAST FLOW SAMPLE CO/Ni = 0.15;
(C) FAST FLOW SAMPLE CO/Ni = 0.25.

Fig. 6.   Spectra of carbon monoxide chemisorbed on nickel: (A) slow flow
sample at CO/Ni ratio of 0.15; (B) Fast flow sample at CO/Ni ratio of
0.17; (C) fast flow sample at CO/Ni ratio of 0.28.

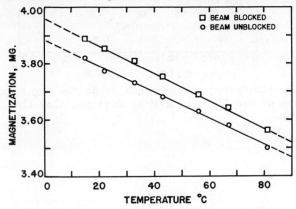

Fig. 7.   Effect of infrared beam on the temperature of silica-supported nickel.

MAGNETIZATION VOLUME ISOTHERMS OF SMALL NICKEL PARTICLES

W. J. Wosten, Th. J. Osinga, and B. G. Linsen

Unilever Research Laboratory
Vlaardingen, The Netherlands

1. Introduction

The interaction of gases with a Ni-surface at room temperature can be studied by measuring the decrease of the magnetic susceptibility of small Ni-particles, as found in Ni on SiO_2-catalysts. This method developed by Selwood et al.[1] starts from the assumption that for a superparamagnetic system of Ni-particles, the contribution of a Ni-surface atom to the total magnetic moment is lost when a chemical bond is formed. However, interpretation of these susceptibility measurements as a function of surface coverage of various gases becomes difficult as a result of the Ni-particle size distribution in these catalysts. For our studies we have therefore selected super-paramagnetic Ni-catalysts with only small Ni-particles and a fairly narrow distribution in size. Our purpose was to investigate whether the interpretation of the magnetization volume isotherms obtained is simpler and whether these small Ni-particles have special properties.

2. Apparatus and Materials

2.1 Reduction and gas adsorption apparatus

A gas-handling apparatus was used for the reduction of Ni-catalysts in purified H_2 at $450°C$, followed by desorption of the adsorbed H_2-gas at $400°C$ until a vacuum of 10^{-6} mm Hg is reached.

The gas-adsorption part consists of a burette, sample cell and a Texas Instruments quartz precision pressure gage. Reduction, evacuation and the gas-adsorption were all performed in the same sample cell. The purity of the gases for gas-adsorption was: H_2 99.999%, N_2 99.998%, O_2 99.6%, CO 99.5%.

2.2 Apparatus for the susceptibility measurement

The magnetic susceptibility was measured in a two-coil system analogous to the apparatus described by Selwood[1] and Geus et al.[2], each coil consisting of a primary (3,000) turns, height 29 cm, diameter 74 mm) and a secondary coil (2,000 turns, height 21 cm, diameter 55 mm). One of the coils contained the sample cell with the Ni-catalyst and the other one was used to compensate for the induction voltage of the first coil. The field strength H was 66 Oe and a frequency of 280 Hz was used. Considerable attention was paid to the stability of the compensation, by keeping the temperature of both coils constant within $0.02°C$. The sample cell could be removed from the coils during a gas-adsorption run to check the zero-point drift. The stability obtained was sufficient to detect changes in relative magnetization of the samples used well below 1%. For the detection of the signals, a precision lock-in amplifier was used. The absolute value of the susceptibility was measured using a sample of ferrous ammonium sulfate as a standard.

All measurements were performed at room temperature.

2.3 I.R. measurement

The I.R. measurements were carried out on a disk of the catalyst, pressed at a pressure of 20 tons and a diameter of 15 mm. The spectra was recorded with a Grubb-Parsons GS-4 "double beam" spectrometer. For the measurement, a closed glass cell was used connected to a gas-adsorption system.

2.4 Catalysts

The catalysts used were Ni on SiO_2 support and contained ca. 20% Ni-metal by weight. The support was Aerosil with a surface area of 170 m^2/g. After drying at $120°C$, the samples were reduced and evacuated. The reproducibility of the value of the magnetic susceptibility of several separately reduced samples was constant within 3%.

Two Ni-catalysts were used which differed in the maximum decrease in magnetization as found by H_2-adsorption.

The degree of reduction was calculated[3,4] from the volume of H_2 evolved after solution in 4 N H_2 SO_4. A correction was made for the H_2 adsorbed on the Ni-surface if the sample was not evacuated after reduction.

3. Ni-particle size determination

While Ni-catalysts always contain metal particles with a range of diameters, only a mean diameter can normally be obtained by a number of methods. The results of these methods will differ depending on the definition of the mean particle size (surface diameter, volume diameter). The narrower the particle size distribution, the smaller the difference between the values obtained by these methods will be. In this way, one can get an indication about the particle size distribution.

3.1 H_2-adsorption

The total surface are of the Ni-particles can be measured by H_2-adsorption at room temperature and one atmosphere pressure[3,4] if the assumption is made that each Ni-surface atom occupies an area of 6.33 $\overset{\circ}{A}{}^2$ and absorbs one H-atom. To obtain an average volume diameter from the specific Ni-surface area measured in this way, the particles were assumed to be hemispherical with the equatorial plane attached to the support.[4] The mean diameter, related to the diameter found by X-ray analysis[4], is then given by:

$$D_{H_2} = \frac{4310}{S_{ni}} \times \overset{\circ}{A}$$

in which S_{Ni} = the Ni-surface area calculated from the adsorbed quantity of H_2 in m^2/g Ni. The results are given in Table 1.

3.2 X-Ray Analysis

From X-ray analysis line broadening, an average particle size can be calculated. The crystallite size determined in this way is defined as $D_{X\text{-ray}} = \sqrt[3]{V}$, in which V is the mean volume of the Ni-particles. The results are given in Table 1.

3.3 Magnetic susceptibility

The absolute value of the magnetic susceptibility of the reduced and degassed Ni-catalyst was obtained by comparison with the known susceptibility of a paramagnetic salt. From this susceptibility, the mean diameter of the particles was found by means of:

$$D_{\chi} = \left(\frac{3 \ kT \cdot \chi}{I_{sp}^{2}} \right)^{1/3}$$

in which χ = the magnetic susceptibility per cm^3 Ni-metal in the sample and I_{sp} = the spontaneous magnetization for bulk Ni-atoms per cm^3 Ni-metal.

The results are given in Table 1.

4. Superparamagnetic Behavior

The superparamagnetic behavior of the catalyst used was checked by measuring the magnetic susceptibility as a function of temperature at constant field strength.[5] For superparamagnetic materials, the temperature dependence of the susceptibility is given by:

$$\chi = \frac{I_{sp}^{2}}{3 \ N \cdot kT}$$

which N = the number of Ni-particles per cm^3 Ni-metal.

From this formula, a linear relationship between χ and 1/T must be expected. The catalysts used in this investigation satisfy this requirement. An example is given in Fig. 1.

χ was found to be independent of the magnetic field strength up to 300 Oe which was the maximum possible variation in our equipment.

5. Magnetization-Volume Isotherms

5.1 H_2-adsorption

Hydrogen adsorption on Ni has been the subject of a great many papers[1,2;6,7].
This chemisorption is dissociative. For particles with the same radius and
magnetic moment, the relative decrease in magnetization caused by H_2-adsorption
is calculated by Selwood[8]:

$$\frac{\Delta M}{M} = - \frac{2 \epsilon \beta n_H}{\mu n} + \left(\frac{\epsilon \beta n_H}{\mu n} \right)^2$$

in which ϵ = the change of moment expressed as the number of electron spins
lost per atom adsorbed

β = the Bohr-magneton

n_H = the number of adsorbed H-atoms

μ = the magnetic moment of a bulk Ni-atom and

n = the number of Ni-atoms in the sample

For a constant field strength $\frac{\Delta M}{M}$ is identical with $\frac{\Delta X}{X}$. Normally,
the quadratic term of the above equation is neglected and a linear relation-
ship between $\frac{\Delta M}{M}$ and n_H is obtained. However, for Ni samples with a large
surface area, the second term causes deviations from this linear dependence
at high surface coverages. The results of our measurements for two catalysts
are given in Fig. 2 (line A) and Fig. 3 (line B).

From the slope of these lines, we can calculate the ratio $\epsilon\beta/\mu n$. From
the metal content of the samples, n is known, hence also $\epsilon\beta/\mu$. Using the
assumption that the magnetic moment of a surface atom is equal to that of a
bulk atom, it follows that $\mu = 0.6 \beta$, in other words ϵ can be calculated.
The results are given in Table 1.

5.2 O_2-adsorption

The high heat of adsorption characteristic of chemisorption may cause a high mobility in the crystal lattice, owing to which the reaction is not restricted to a unimolecular layer. A good example is the adsorption of oxygen on nickel.

The influence of oxygen adsorption on the magnetic susceptibility behavior has already been measured by several investigators[2,9,10]. A more or less linear decrease in magnetization was found, when the system was superparamagnetic. It was observed that the oxygen-adsorption was not restricted to the surface, but was extended to bulk nickel. However, complete oxidation does not occur.

In our case, a linear decrease up to a value of $\frac{\Delta M}{M}$ = 90% is observed if oxygen is admitted (Fig. 4). Extrapolation to $\frac{\Delta M}{M}$ = 100% gives a total amount of 200 ml (STP) of oxygen taken up per gram metallic nickel. This is exactly the amount of oxygen that is required for the formation of nickel oxide.

5.3 N_2-adsorption

The adsorption of nitrogen on nickel has already been studied by Eischens and Jacknow[11] by means of I.R.-adsorption measurements showing a strong absorption band at 2202 cm^{-1}. Recently Van Hardeveld and Van Montfoort[12] drew attention to the properties of very small Ni-particles with diameters in the 10–40 Å range with respect to nitrogen adsorption at room temperature. These authors also studied the adsorption of nitrogen by means of I.R.-absorption measurements. The adsorption of nitrogen on the small Ni-particles was explained by Van Hardeveld et al. by the existence of B_5-sites on the Ni-particles. The number of these sites is highest for 15–20 Å particles and decreases rapidly with increasing diameter. The strong polarizing field on these sites causes a shift of the Raman band of nitrogen from 2331 cm^{-1} to 2202 cm^{-1} due to the influence of the Stark effect. The type of nitrogen adsorption was classified by Van Hardeveld and Van Montfoort as physisorption.

Selwood[13,14] did not find an appreciable influence of nitrogen adsorption on the magnetic properties of nickel at room temperature, although there was a definite effect at -50°C.

From our experimental data, however, it appearaed that our samples adsorb N_2 in considerable quantities, namely 13 ml (STP) per gram metallic nickel. At the same time, a maximum decrease in magnetization of 5.4% was observed (Fig. 3). With argon we found an effect of less than 1%.

The recorded I.R. spectrum of N_2 adsorbed on our samples shows a band at 2202 cm^{-1} with a half width of $\Delta \nu_{1/2} = 28.5$ cm^{-1}, independent of the degree of coverage of the sample.

The log I/I_0 of the I.R. absorption band was measured at room temperature as a function of nitrogen pressure. The I.R. absorption increases until a N_2-pressure of 10cm Hg was reached. By pumping off N_2-gas the desorption of nitrogen could also be studied. The results are given in Fig. 5, which shows that equilibrium is almost reached within the time of measurement of 2 points (~ 5 min).

5.4 CO-adsorption

The magnetic susceptibility as a function of the amount of adsorbed CO has been studied by Selwood et al.[1] and Geus et al.[2]. They all confirm the conclusions of Eischens[15] that CO can be adsorbed in a linear form to one nickel atom or in a bridged form to two neighbouring Ni-atoms.

For our samples the decrease in magnetization (Fig. 2, line B) due to one molecule CO is equal to the decrease caused by the adsorption of one H atom. However, the maximum decrease which can be obtained is only one half of the maximum decrease obtained by H_2-adsorption. Further addition of CO only leads to the formation of $Ni(CO)_4$ in the gas phase, which gives an apparent adsorption (horizontal part of the curve B in Fig. 2).

For the experiments of line B, the CO was introduced in such a way that the pressure never exceeded 1 mm Hg until a coverage of 60 ml CO (STP) per gram metallic nickel was reached.

In the experiments of line D (Fig. 2) the CO-pressures were more than 10 mm Hg during introduction of CO. This causes the formation of $Ni(CO)_4$ in the gasphase before incomplete coverage of the surface took place, apparently causing a growth of the nickel particles by vapor-phase transport of nickel from smaller to larger particles. This particle growth is accelerated by adding H_2-gas. In a separate experiment, we found that addition of CO or CO + H_2 mixtures, causes a ten-fold increase in the susceptibility in 16 h due to an increase in particle size by a factor of about two.

This was confirmed by X-ray line-broadening experiments, which revealed an increase in particle size from 20 to 35 Å.

When the maximum decrease in magnetization during CO-adsorption was reached and the gas phase removed by pumping, addition of hydrogen caused a further decrease in magnetization up to the maximum value for pure hydrogen (Fig. 2, line E and C).

6. Discussion

A comparison of the mean particle sizes determined from H_2 adsorption X-ray and magnetic susceptibility measurement (Table 1) shows that the three values are very close together. It should be borne in mind that all these diameters are obtained in different ways. The mean particle size obtained by H_2-adsorption is a mean surface diameter so that the smaller particles have a large influence on the value of the mean diameter obtained. The mean diameter obtained from the magnetic susceptibility is more governed by the larger particles due to the fact that this mean diameter is related to the mean square volume. The mean particle diameter obtained by X-ray line broadening is related to the mean volume and must therefore be expected in between the other two values. From the fact that all these values are very close together, the conclusion may be drawn that the samples have a fairly narrow particle-size distribution and that large particles are absent.   From electronmicrographs the conclusion could be drawn that the largest particles had a diameter of 25 Å which is is agreement with the former conclusion.

The determination of ϵ is not very accurate owing to the influence of the particle size distribution even for a narrow one. If the magnetic moment for a surface atom is equal to that of a bulk atom, we would expect a value of about 0.6.

The values for ϵ obtained (Table 1) are, however, significantly lower and may therefore be interpreted as being an indication that the surface atoms in small nickel particles have a lower magnetic moment than the bulk nickel atoms, as already mentioned by Selwood[1].

In the literature, the oxidation of nickel at room temperature is mostly re-stricted to 2.3 layers of nickel atoms. The fact that we get an almost complete oxidation (Fig. 4) of the metallic nickel at room temperature is again an indi-cation of the presence of small crystallites. It also shows a high reactivity of the nickel metal available.

Combination of magnetic and I.R. measurements (Figs. 3 and 5) as a func-tion of surface coverage shows that a bond is formed between a N_2- molecule and a Ni-surface atom. Only part of the Ni-surface is active with respect to the adsorption of N_2 molecules, indicating a considerable surface hetero-geneity. The Ni-N_2 bond is weak because complete desorption of N_2 takes place after evacuation at room temperature (Fig. 5).

Another proof of chemical interaction between N_2 and a nickel surface at high temperature is obtained when N_2-H_2 mixtures are passed over the nickel catalysts. In this case, small amounts of NH_3 were detected at $450°C$.

In our opinion the adsorption of nitrogen must therefore be classified as chemisorption.

The decrease in magnetization due to adsorption of one molecule CO is equal to the decrease caused by adsorption of one H-atom. This indicates that

CO-molecules are adsorbed in a linear structure on the metal surface. However,

the maximum decrease in magnetization of a monolayer of CO is only half of that

at full H-coverage. The conclusion can therefore be drawn that in a monolayer

half of the Ni-surface atoms are bonded to CO, which may be explained by the

size of the CO-molecules. From liquid CO, assuming a closest packing of spheres,

a cross sectional area of $16.8 \overset{\circ}{A}$ can be calculated. In practice, however,

we may expect a cross sectional area smaller than this value because, as said

already, the CO-molecules are adsorbed linearly on the surface and a closest

packing of spheres is then no realistic picture. We may, however, conclude

that the cross sectional area is certainly larger than $6.33 \overset{\circ}{A}^{2}$, being the

average surface area occupied by one Ni-atom. Therefore, the amount of CO

molecules adsorbed will not be equal to the total amount of Ni-atoms exposed

in the surface. This conclusion is confirmed by additional chemi-sorption

of H_2 which doubles the decrease in magnetization (Fig. 2), because the small

H-atoms can be adsorbed on the other free Ni surface atoms. LEED studies of

Park and Farnsworth on a (110) Ni-surface[16] also show that one CO molecule is

present per two Ni atoms but as regards the (100) Ni surface[17], said authors

arrived at the conclusion that one CO-molecule is present per one Ni surface

atom.

The surface potential measurements of Siddiqi and Tompkins[18] on nickel

films covered with adsorbed CO and H_2, together with our experimental re-

sults lead to the following simplified one-dimensional model of a monolayer

of CO and H_2 on the Ni surface:

In reality, the configuration of adsorbed hydrogen and CO will be much more complicated.

Our general conclusion about the properties of small nickel particles is, that they show a high reactivity, especially with respect to oxygen and carbon monoxide and that a very pronounced surface heterogeneity occurs.

With modern instrumentation, a study of more clearly defined Ni surfaces is possible. These methods have the disadvantage that direct measurements of the adsorbed amount of gas are impossible owing to the small metal surface area available; moreover, they can only be carried out at very low gas pressures. Especially the results of our N_2 and CO adsorption studies show the usefulness of a combination of gas adsorption and magnetic measurements on Ni particles.

Acknowledgment

We thank Dr. J. Erkelens for the I.R. measurements and Dr. J. P. C. Holst for the X-ray investigations.

REFERENCES

1. Selwood, P. W., Adsorption and Collective Paramagnetism, Acad. Press, New York, 1962.

2. Geus, J. W., Nobel, A. P. P., and Zwietering, P., Catalysis 1, (1962) p. 8.

3. Linsen, B. G., Thesis Delft, 1964.

4. Coenen, J. W. E., Thesis Delft, 1958.

5. Bean, C. P., and Livingstone, J. D., J. Appl. Phys. 30, (1959) p. 126 S.

6. Broeder, J. J., Van Reijen, L. L. Sachtler, W. H. M., Schuit, G. C. A., Z. Electrochem. 60, (1956) p. 838.

7. Lee, E. L., Sabatka, J. A., and Selwood, P. W., J. Am. Chem. Soc. <u>79</u>, (1957) p. 5391.

8. Selwood, P. W., Actes Congr. Intern. Catalyse, 2e Paris <u>2</u>, (1960) p. 1795.

9. Leak, J., and Selwood, P. W., J. Phys. Chem. <u>64</u>, (1960) p. 1114.

10. Geus, J. W., and Nobel, A. P.P., J. Catalysis <u>6</u>, (1966) p. 108.

11. Eischens, R. P., and Jacknow, J., Proc. Intern. Congr. Catalysis, 3rd, Amsterdam <u>1</u>, (1964) p. 627.

12. Van Hardeveld, R., and Van Montfoort, A., Surface Science <u>4</u>, (1966) p. 396.

13. Selwood, P. W., J. Am. Chem. Soc. <u>80</u>, (1958) p. 4198.

14. Andrew, A. A., and Selwood, P. W., J. Catalysis <u>7</u>, (1967) p. 98.

15. Eischens, R. P., Pliskin, W. A., and Francis, S. A., J. Chem. Phys. <u>22</u>, (1954) p. 1786.

16. Park, R. L., and Farnsworth, H. E., J. Chem. Phys. <u>40</u>, (1964) p. 2354.

17. Park, R. L., and Farnsworth, H. E., J. Chem. Phys. <u>43</u>, (1965) p. 2351.

18. Siddiqi, M. M. and Tompkins, F. C., Proc. Roy. Soc. London, Ser. A, <u>268</u>, (1962) p. 452.

Table 1

Properties of the catalysts

Sample	%Ni	Degree of reduction	Ni-surface area m^2/g metallic Ni	D_{H_2} (Å)	χ (erg/Oe$^2 \cdot$ cm^3)	D_χ (Å)	$D_{RÖ}$ (X-ray) (Å)	
53	19.8	0.75	284	15	0.035	26	20	0.
10	21.0	0.96	266	17	0.022	25	25	0.

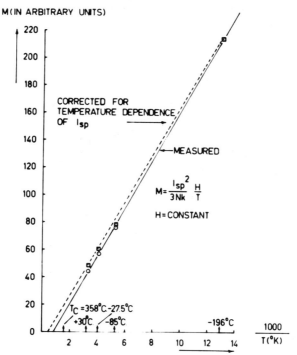

Fig. 1.   Temperature dependence of magnetization for catalyst 10.

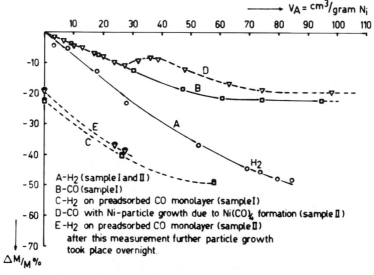

A-H_2 (sample I and II)
B-CO (sample I)
C-H_2 on preadsorbed CO monolayer (sample I)
D-CO with Ni-particle growth due to Ni(CO)$_4$ formation (sample II)
E-H_2 on preadsorbed CO monolayer (sample II)
   after this measurement further particle growth
   took place overnight.

Fig. 2.   Magnetization volume isotherms on Ni-catalyst 53.

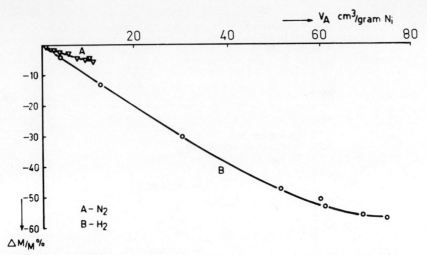

Fig. 3. Magnetization volume isotherms on Ni-catalyst 10.

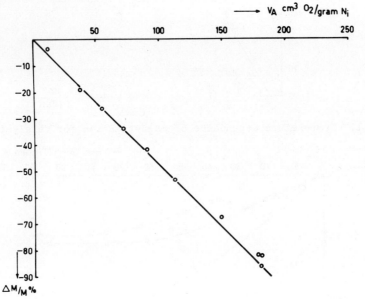

Fig. 4. Magnetization volume isotherms of O_2 on Ni-catalyst 53.

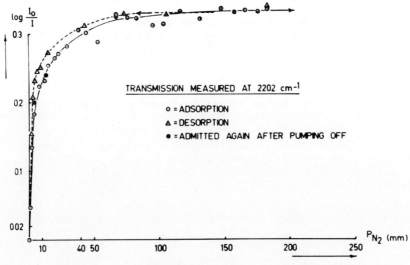

Fig. 5. I.R.-adsorption of N_2 on catalyst 10.

ALKALI ATOM ADSORPTION ON SINGLE CRYSTAL NICKEL SURFACES; SURFACE STRUCTURE AND WORK FUNCTION*

R. L. Gerlach** and T. N. Rhodin

Department of Applied Physics
Cornell University, Ithaca, New York

A. Introduction

A primary goal of surface adsorption studies is to predict the surface structures and electronic wave functions for a given adatom-surface system. With these structures and wave functions, important physical constants such as work function and desorption energy could, in principle, be calculated. Even for the relatively simple systems of alkali metal adsorption on clean metal surfaces this goal at present has not been attained.

B. Models of Atom-Surface Interactions

There have been very few measurements which have been directly related to surface structures formed by alkali adsorption on metal surfaces. From knowledge of metallic bonding it has generally been deduced that alkali adatoms form coherent surface structures on metal surfaces (at low coverages).[1-5] The alkali adatoms are believed to sit in contact with at least three substrate atoms. Two alkali adatoms are generally assumed never to sit on adjacent adsites which are less than twice the bulk radius of the alkali atom apart. Various distributions of alkali adatoms on these adsites have been assumed. In order to minimize the dipole repulsion energy, most models incorporate a square or hexagonal array of alkali adatoms.

Densities of alkali adatoms on metal surfaces have been observed which would more than fill the coherent adsites. This has led a few authors[3] to the conclusion that close-packed, incoherent surface structures are formed at high

coverages. Schmidt and Gomer[3] propose that when all the preferred adsites on a surface are filled, the addition
of more alkali atoms forces the alkali layer to assume an
incoherent structure to optimize the packing of alkali
adatoms on the metal surface.

Following the classical Cs on W studies of Taylor and
Langmuir,[1] the electronic structure of alkali metal adsorption on metal surfaces was treated in terms of an ion-metal
interaction based essentially on the classical image charge
concept. Various distributions of adatoms have been assumed
in order to calculate depolarization as a function of alkali
adatom coverage.[3,4,6-8] The classical theory has been successful in fitting work function vs. coverage data but does
not predict from first principles such basic parameters as
the degree of ionization of the adatom or the effective
dipole length.

Serious efforts have been made to formulate wave mechanical descriptions of alkali adatom-metal surface interactions.[9-14] It has been shown that the valence level of the
adatom is broadened and shifted by the interaction. Gomer
and Swanson[12] have pointed out that the adsorbate-substrate
bond will be ionic, polar-metallic or metallic if the broadened
and shifted atomic valence level is, respectively, above, at
the same level or below the Fermi level.

An important step toward formulation of a wave mechanical
description of alkali adatom-metal surface bonding is the
calculation of the dipole moment formed by the adatom-metal
surface complex. Recently, perturbation calculations of

alkali adatom-metal surface interactions have been made by Bennett and Falicov[13] and by Gadzuk.[14] In this approach, the metal is treated as free-electron-like with a smooth surface, and convenient atomic wave functions are assumed for the adatoms. The classical image potential is assumed to describe the interaction between the ion cores or atomic electrons and the metal. The Gadzuk[14] and Bennett-Falicov[13] calculations give the shift and the broadening of the atomic energy level as a function of the adatom-metal surface dis-tance. The average charge on an adatom may be calculated from this energy broadening and shift assuming a Lorentzian density distribution[13] for the adatom electronic energy level.

Before the dipole moments formed by the alkali atom-metal substrate system can be calculated, two questions must be answered. The first question is how far the alkali ion core is from the plane of the metal lattice sites. It is prob-ably valid to say that the effective radius of the alkali atom is between its ionic and bulk atomic radius. A second question is how the electrons are distributed in the surface region. It has been shown that the classical image charge concept is not accurate for a charge which is brought to within an atomic lattice dimension from the surface. In particular, calculations by Gadzuk[15] of electron screening in the surface region show that the electron image charge distribution is related to the density of free electrons in the bulk metal. A more definitive description of the charge distribution in the surface region is needed. Nevertheless,

the Bennett-Falicov or Gadzuk calculations of charge transfer in conjunction with an estimate of the effective dipole
length could lead to a reasonable value of the dipole moment.
However, perturbation theory applied to a free electron
metal with classical image potentials to describe the adatom-
surface interaction as used by Gadzuk and by Bennett and
Falicov may not be valid in many cases. In addition, the
model employing a smooth surface certainly will not apply
to alkali adsorption on high index planes. Also, it is
desirable to calculate other measureable parameters such as
binding energy as well as dipole moments in order to ade-
quarely test the various models.

A different approach is the application of molecular
orbital (MO) theory to surface bonding. A MO approach is
particularly desirable because surface geometry can be explicitly accounted for. One useful MO model is to consider that
an adatom and a few substrate atoms form a surface compound.
This model is certainly valid for many adatom-metal substrate
systems.[16,17].

A phenomenological molecular orbital (MO) approach has
been used by Gyftopoulos and Levine[2] and by Steiner and Gyftopoulos.[4] They equate the work function of a composite alkali'
atom-metal surface system with its electronegativity. Pauling's
rules and formulas for electronegativity, and dipole moments
of complex molecules are then applied to calculate the
work function and binding energy as a function of alkali coverage. This empirical approach is not satisfying from a physical

viewpoint but could be useful if some of the variables are indeed independent of crystallography and substrate material as predicted.

Serious efforts have been made at wave mechanical MO theories to describe adatom-metal surface interactions. In particular, binding energies of electronegative gases with metal surfaces have recently been estimated with various MO models.[16,17] It remains to be determined if similar quantum mechanical MO models will provide a good description of alkali adatom-metal surface systems.

Though much experimental work has been done on alkali adatom-metal substrate systems, more effort is required before an adequate wave mechanical description can be achieved. Although there have been many desorption energy and work function measurements,[1,3,18-25] few of them have been made on a series of clean, single crystal faces and for many, the absolute surface coverage was unknown. In addition, the method of measurement of desorption energy is questionable in many cases.[2] Practically no direct measurements have been made of the surface structures formed by alkali atom-metal systems. The net result is that with a few exceptions[3,18] there are insufficient data to test even the existing theories relating to work function and desorption energy.

C. Experimental Procedures

A study of the atomistics of alkali metal adsorption on the low index faces of Ni was undertaken to determine the

surface structures of a series of adatom-surface combinations and to obtain systematic work function and desorption energy data as a function of absolute surface coverage.
The studies were of Na on Ni(111), Ni(100) and Ni(110) and
of Na, K and Cs on Ni(110). Ni was chosen as a substrate
because the properties of its clean surface as well as its
behavior in the presence of O_2, CO and other gases are relatively well understood.

The observations were made in a direct display 3-grid
Varian LEED system including facilities for the direct
measurement of absolute surface coverage and a high flux
source of pure Na (K or Cs) vapor. The schematic arrangement
is illustrated in Fig. 1. The Na source consisted of a Pyrex
vial with a Ni heating coil wrapped around it. High purity
Na had been loaded into the vial in high vacuum. After bakeout the tip of the vial was broken off. A shield was placed
around the Ni crystal during deposition to insure that Na
was not deposited on the crystal leads. The vacuum remained
in the low 10^{-10} torr range during Na deposition and the
principal contaminant as measured by the mass spectrometer
was CO. After observing the LEED pattern, the Na was thermally desorbed from the Ni crystal with the crystal facing
the Ir surface ionization gauge. The hot Ir filaments ionized
a fraction of the desorbed Na and the collected ions gave a
measure of relative coverage. The gauge was calibrated for
absolute coverage through the use of a zeolite Na ion source.
The coverage measurements are accurate to about 3%. Relative
coverages had less than 1% error in those cases where special

care was taken in the measurements. A Kelvin probe was used to measure the contact potential and the work function was estimated in terms of a reference electrode of clean poly-crystalline tungsten.

Desorption energy measurements were made using the surface ionization gauge to measure the rate of thermal desorption. These results will be described elsewhere.

D. Results and Discussion

1. Structures of specific adsorbed alkali surfaces

Na on Ni(111) and Ni(100) and Na, K or Cs on Ni(110) form surface structures with some degree of order. These structures generally reflect the symmetry of the substrate surface, but there are certain basic similarities among them. Since the surface structures formed with Na on Ni(111) are simple and representative of most of the surface structures observed, they will be described here in some detail. Only a brief description of pertinate surface structures observed with Na on Ni(100) and with Na, K or Cs on Ni(110) will be presented here. A complete analysis will be given elsewhere.

(a) Ni(111)-Na. 110 V LEED photographs from Na on Ni(111) are shown in Fig. 2. The number at the bottom of each photograph is the atomic coverage*. Note from Fig. 2A-D that a diffuse ring is formed around the origin at low coverages which increases in diameter roughly proportional to $\theta^{+\frac{1}{2}}$. The general structure which is believed

---

*One atomic monolayer ($\theta = 1$) is defined as one adatom per atom in the outer layer of the substrate.

to produce this diffuse ring is a uniform distribution of
Na adatoms which are each coherently[26] positioned on three
Ni atoms. Such structures are shown in Fig. 3A and 4A
for $\theta = 0.17$ and 0.29, respectively. By uniformly distributed, it is meant that there are generally small deviations
from the average distance between nearest neighbor adatoms.
The LEED patterns from these structures were laser simulated
using reduced transparencies of the diffraction matrices
shown in Fig. 3B and 4B. The laser simulations shown in
the upper halves of Fig. 3C and 4C clearly agree with the 71V
LEED photographs in the lower halves of those figures and with
Fig. 2B and 2D, respectively.

It can be shown that a uniform distribution of Na adatoms which are incoherently positioned on the Ni(111) surface will also produce a diffuse ring around the 00 spot.
But strong evidence which will be discussed elsewhere favors
the coherent structure shown in Fig. 3A and 4A. The important thing, however, is that the Na adatoms must be
uniformly distributed to produce the diffuse ring.

In Fig. 2E-H for $\theta > 0.33$ it is seen that the diffuse
ring breaks up into a hexagonal array of spots, which
continue to move away from the origin with increasing
coverage. An inner set of hexagonal spots, which are
caused by multiple scattering, move in towards the origin
with increasing coverage. The proposed surface structure
leading to these diffraction photographs is shown in Fig. 5A.
A hexagonal array of Na adatoms is incoherently positioned on
the Ni(111) substrate with the same orientation as the outer

hexagonal layer of Ni atoms. The distance between Na
adatoms decreases continuously proportional to $\theta^{-\frac{1}{2}}$, for
0.33 < θ < 0.49. The predicted diffraction pattern for
θ = 0.40 is shown in Fig. 5B. Note the agreement with the
diffraction photograph in Fig. 3C.

To summarize, the Na adatoms on Ni(111) are uniformly
distributed at low coverages apparently due to strong
dipole repulsions. At high coverages, the Na adatoms form
an incoherent, hexagonal structure to maximize the packing.

(b) Ni(100)-Na. Na on Ni(100) produces a diffuse ring
at low coverages similar to that in Fig. 2A-D for Na on Ni(111).
Thus the Na adatoms are uniformly distributed at low coverages
on Ni(100) and are probably coherently positioned on four
Ni substrate atoms. At θ = 1/2, Na on Ni(100) forms a c(2x2)
structure.

(c) Ni(110)-Na, K or Cs. Some of the surface structures
formed with Na on Ni(110) have been reported by us elsewhere.[27]
We have shown that the structure depicted in Fig. 6 is
produced for 0.25 < θ < 0.32. Note that alternate troughs
on the Ni(110) surface contain incoherent rows of evenly
spaced Na adatoms. The rows of Na adatoms are randomly
shifted with respect to one another in the [110] direction.
The distance between Na adatoms in the [110] direction is
proportional to θ^{-1}. As the coverage is increased above
0.32 atomic monolayers, more troughs on the Ni(110) surface
fill with incoherent rows of evenly spaced adatoms. For
0.64 < θ < 0.71, each trough contains an incoherent row of

Na adatoms as shown in Fig. 7. Again the distance between Na adatoms in the $[110]$ direction is proportional to θ^{-1}.

K or Cs on Ni(110) produce structures which are very similar to those for Na on Ni(110). The principal difference is that the structures reflect the larger effective diameters of K and Cs.

2. Summary of structures

To summarize, the Na adatoms are uniformly distributed on the Ni(111) and Ni(100) surfaces at low coverages. These structures are probably coherent. There is a tendency to form incoherent structures at high coverages to maximize the packing of alkali adatoms. In particular, Na on Ni(111) forms a two-dimensionally incoherent, hexagonal structure at high coverages, since the Ni(111) surface is atomically smooth. Similarly, Na, K or Cs on Ni(110) forms one-dimensionally incoherent surface structures since the troughs in the $[110]$ direction are rather smooth.

3. Work functions of clean and adsorbed alkali surfaces

Work function vs. coverage curves for alkali adsorption on nickel are presented in Fig. 8. Important data from these work function curves are presented in Table 1.

The second column of Table 1 lists ϕ^o, the clean work function of the Ni surface. Note that

$$\phi^o_{Ni(100)} \gtrsim \phi^o_{Ni(111)} > \phi^o_{Ni(110)}.$$

The well-established charge redistribution model[28] predicts the sequence

$$\phi^{o}_{(111)} > \phi^{o}_{(100)} > \phi^{o}_{(110)}$$

for face-centered cubic metals. This qualitative prediction is correct except that it is questionable that

$$\phi^{o}_{Ni(111)} > \phi^{o}_{Ni(100)}.$$

Similarly the Steiner theory[4] predicts the work function values listed in column three. It is clear that these values are generally too high. In addition, the prediction that $\phi^{o}_{Ni(111)}$ is considerably greater than $\phi^{o}_{Ni(100)}$, is definitely not in agreement with these experimental results.

The work function $\phi_{m.d.}$ at the coverage corresponding to maximum density of adatoms on the Ni surface is listed in the fourth column of Table 1. It has been observed for some alkali adatom-metal surface systems that ϕ_m is rather close to the work function ϕ_f of the alkali metal being deposited. The approximation $\phi_{m.d.} = \phi_f$ has been incorporated in some empirical theories[2,4] for predicting work function vs. coverage curves. The values of ϕ_f for Na, K and Cs are listed in the fifth column of Table 1. Clearly, the approximation $\phi_{m.d.} = \phi_f$ is indeed very unreliable for some of these systems. For example, $(\phi_{m.d.} - \phi_f)$ is 0.6 eV for K on Ni(110).

4. Dipole moment calculations

Dipole moments calculated from the inital slopes of
the work function vs. coverage curves in Fig. 8 are listed
in the sixth column of Table 1. Note the general trends of
μ^o increasing with ϕ^o and with the sequence Na, K and Cs,
which is what is expected qualitatively from the Gurney
model.[9] The values of μ predicted by the Steiner theory[4] are
given in the seventh column of Table 1. The general trends
are correct, but the values of μ^o predicted for Ni(100) are
too large. Attempts to fit the work function curves with
the Steiner formula will be presented elsewhere. It is
possible that an empirical MO approach such as Steiner's
can be useful providing the empirical conditions are adequately defined.

Dipole moments calculated from the Gadzuk theory are
presented in the eighth column of Table 1. The shift and
broadening of the electronic level of the adsorbed alkali
atom were calculated from Eq. 10 and 22 of reference 15.
The effective charge q of the adatom was then calculated
assuming a Lorentzian energy distribution (see reference
13 for a discussion of this assumption). The dipole moment
μ^o was then estimated from[15]

$$\mu_G^o = (r_i + 1.12\, \rho^{-1/3})q$$

where r_i is the ionic radius of the alkali adatom and ρ is
the free electron density in $\overset{o}{A}{}^{-3}$. Two free electrons per
Ni atom have been assumed in this calculation.

The dipole moments estimated in this manner are of the right order of magnitude for Na on Ni(111) and Ni(100). The dipole moments estimated for Na, K and Cs on Ni(110) are clearly too large. But since the Gadzuk theory assumes a flat surface of a free electron metal, it should not be expected to apply to Ni(110). The Gadzuk theory does predict the correct trend for Na, K and Cs, though. The dipole moments estimated in this manner for Na on Ni(111) and Ni(100) appear reasonable considering that a simple model was developed by Gadzuk from first principles.

E. Summary and Conclusions

From the observed surface structures one must conclude that repulsive forces exist between adsorbed alkali atoms at all coverages (forming a single layer). In addition, from the filling of alternate troughs, on the Ni(110) surface with incoherent rows of alkali adatoms (see Fig. 6), it is clear that the forces between adatoms are anisotropic on the Ni(110) surface. In other words, the repulsion between alkali adatoms in the $[110]$ direction must be less than the longer range repulsion between alkali adatoms in the $[100]$ direction to explain why alternate troughs fill with rows of Na, K or Cs adatoms. Grimley[29] has computed an anisotropic interaction of this kind which may account for this behavior.

New approaches are required to effectively describe alkali adsorption on non close-packed surfaces, such as Na, K and Cs on Ni(110). Perhaps a sawtooth surface could be incorporated into the Gadzuk calculations to describe these surfaces. On the other hand, molecular orbital theory may provide a better description of these surface geometry effects.

This study is the beginning of an understanding of the surface structures formed by alkali adatom-metal surface systems. In its complete form, the work should be a significant contribution toward an understanding of the electronic structure and binding of these systems.

Acknowledgments

Principal support by the Air Force Office of Scientific Research and by the Advanced Research Projects Agency through the Cornell Materials Science Center is gratefully acknowledged. Assistance from K. Hartman with the computer programs and calculations is also acknowledged with appreciation.

*Supported by the Air Force Office of Scientific Research Grant AF-AFOSR-876-65 and the Advanced Research Projects Agency.
**Present address: Sandia Corporation, Albuquerque, N. Mex.

Bibliography

1. J. B. Taylor and I. Langmuir, Phys. Rev. 44 (1933) 423.

2. E. P. Gyftopoulos and J. D. Levine, J. Appl. Phys. 33 (1962) 67;
    Surface Science 1 (1954) 171, 225, 349.

3. L. Schmidt and R. Gomer, J. Chem. Phys. 42 (1965) 3573;
    45 (1966) 1605.

4. D. Steiner and E. Gyftopoulos, Report on the 27th Annual Conf. Phys.
    Elec. (Cambridge, 1967) 160.

5. V. N. Shrednik, Radiotekhnika i elektronika 5 (1960) 1203.

6. J. Topping, Proc. Roy. Soc. (London) A114 (1927) 67.

7. I. Langmuir, J. Am. Chem. Soc. 54 (1932) 2798.

8. N. S. Rasor and C. Warner, J. Appl. Phys. 35 (1964) 2589.

9. R. W. Gurney, Phys. Rev. 47 (1935) 479.

10. L. V. Dobretsov, Electron and Ion Emission (NASA Technical Transla-
    tion F-73, 1952).

11. T. Toya, J. Res. Inst. Catalysis, Hokkaido Univ. VI (1958) 308; VIII
    (1961) 209.

12. R. Gomer and L. W. Swanson, J. Chem. Phys. 38 (1963) 1613.

13. A. J. Bennett and L. M. Falicov, Phys. Rev. 151 (1966) 512.

14. J. W. Gadzuk, Surface Science 6 (1967) 133, 159.

15. J. W. Gadzuk, Ph. D. Thesis, Mass. Inst. Tech., Sept. 1967.

16. L. Jansen, Molecular Processes on Solid Surfaces, Edited by E. Draug-
    lis and R. Gretz (McGraw-Hill, N. Y., 1969).

17. T. B. Grimley, Molecular Processes on Solid Surfaces, Edited by E.
    Drauglis and R. Gretz, (McGraw-Hill, N. Y., 1969).

18. D. L. Fehrs and R. E. Stickney, Report on the 26th Annual Conf. Phys.
    Elec. (Cambridge, 1966) 287.

19. G. E. Moore and H. W. Allison, J. Chem. Phys. 23 (1955) 1609.

20. J. Anderson, W. E. Danforth and A. J. Williams, III, J. Appl. Phys. 34

55–16   R. L. GERLACH AND T. N. RHODIN

(1963) 2260.

21. W. H. Brattain and J. A. Becker, Phys. Rev. 43 (1933) 428.

22. V. M. Gavrilyuk and V. K. Medvedev, Soviet Phys. --Solid State 8 (1966) 1439.

23. R. G. Wilson, J. Appl. Phys. 37 (1966) 3161, 3170; R. G. Wilson and E. D. Wolf 37 (1966) 4458.

24. V. N. Shrednik and E. V. Snezhko, Soviet Phys. --Solid State 6 (1965) 2727.

25. V. M. Gavrilyuk, Y. S. Vedula, A. G. Naumovets, and A. G. Fedorus, Soviet Phys. --Solid State 9 (1967) 881.

26. E. Bauer, Proc. of the International Colloquium on Adsorption and Crystal Growth, Nancy, France, June 6-12, 1965.

27. R. L. Gerlach and T. N. Rhodin, Surface Science 10 (1968) 446.

28. R. Smoluchowski, Phys. Rev. 60 (1941) 661.

29. T. B. Grimley, Proc. Phys. Soc. 90 (1967) 751; 92 (1967) 776.

30. V. Fomenko, Handbook of Thermionic Properties (Plenum Press Data Div., N. Y., 1966).

Table 1

Comparison of Work Functions and Dipole Moments
for Alkali Adsorbed Single Crystal Nickel Surfaces

System	ϕ^o (eV)	$\phi^o_{Steiner}$ (eV)	$\phi_{m.d.}$ (eV)	ϕ_f (eV)	μ^o (Debye)	μ^o_s (Debye)	μ^o_G (Debye)
Ni(111)-Na	4.68 ± 0.1	5.56	2.32	2.28	7.4 ± 0.5	8.0	8.9
Ni(100)-Na	4.75	5.20	2.20	2.28	7.2 ± 0.5	8.0	8.9
Ni(110)-Na	4.20	4.84	2.28	2.28	3.2 ± 0.3	6.8	8.5
Ni(110)-K	4.20	4.84	1.6	2.2	5.3 ± 0.5	8.9	10.4
Ni(110)-Cs	4.20	4.84	1.5	1.8	7.0 ± 0.7	10.6	12.2

ϕ^o is the measured work function for the clean surface. $\phi_{m.d.}$ is the work function corresponding to the maximum surface density associated with one physical layer. ϕ_f is the best available value for the work function of a clean alkali metal surface from reference 30. μ^o is the computed effective dipole moment in the limit of zero coverage: $\mu^o = \frac{\Delta\phi}{2\pi n}\big|_{\theta=0}$. μ^o_s is the computed dipole moment from the Steiner[4] theory. μ^o_G is the computed dipole moment from the Gadzuk[15] theory.

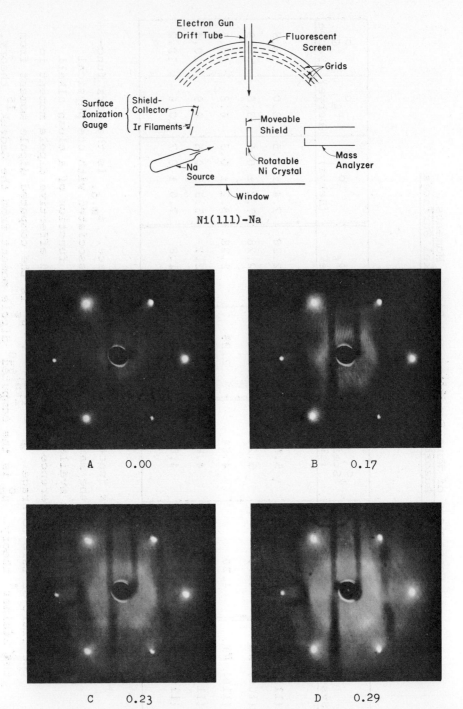

Fig. 1. Schematic drawing of the three-grid Varian LEED system with single Ni crystal, Na (K or Cs) source, Ir surface ionization gauge for determining coverage and mass spectrometer attached. The Ni crystal and shield are shown in position for vapor deposition.

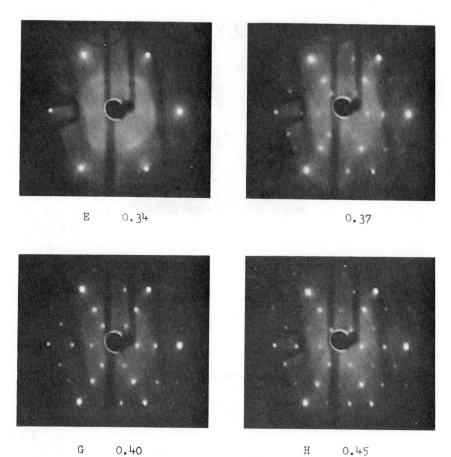

E   0.34   0.37

G   0.40   H   0.45

Fig. 2.   LEED photographs at 110V from Na on Ni(111). The number at the bottom of each photograph is the atomic coverage. The dark vertical
lines common to all photographs are caused by various pieces of
apparatus.

Ni(111)-Na

Proposed structure

0.17 atomic monolayers

Laser simulated diffraction

Laser diffraction matrix

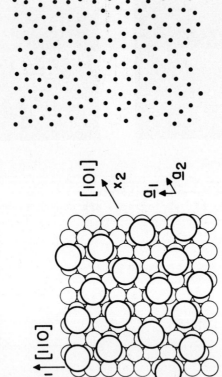

(a)

(b)

71 V LEED photograph

(c)

Fig. 3. Ni(111)-Na for $\theta = 0.17$. A. Proposed coherently positioned, uniformly distributed adatoms structure. An incoherently positioned, uniformly distributed structure would also produce the desired diffuse ring, but strong evidence favors this coherent structure. B. Laser diffraction matrix for simulating the Na overlayer in Fig. 3A. C. The upper half of this photograph is the laser simulated diffraction pattern and the bottom half is a 71V LEED photograph from Ni(111)-Na. The integer order spots in the laser simulation do not appear very bri   because no provision was made to simulate diffraction from the N   1) surface.

55—20

0.29 atomic monolayers

Proposed structure   Laser diffraction matrix   Laser simulated diffraction

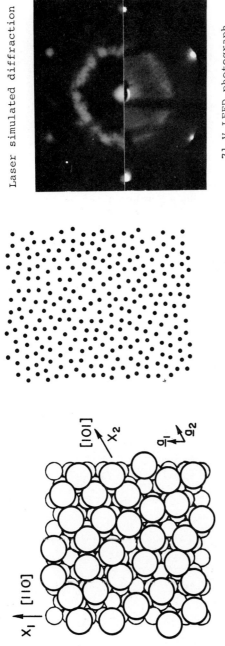

(a)   (b)   (c)

71 V LEED photograph

Fig. 4. Ni(111)–Na for θ = 0.29. A. Proposed coherently positioned, uniformly
distributed adatoms structure. An incoherently positioned, uniformly
distributed structure would also produce the desired diffuse ring, but
strong evidence favors this coherent structure. B. Laser diffraction
matrix for simulating the Na overlayer in Fig. 4A. C. The upper half
of this photograph is the laser simulated diffraction pattern and the
bottom half is a 71V LEED photograph from Ni(111)–Na. The integer
order spots in the laser simulation do not appear very bright because
no provision was made to simulate diffraction from the Ni(111) surface.

55–21

Ni(III) - Na

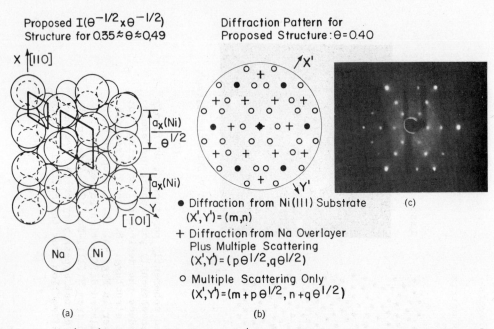

Proposed $I(\theta^{-1/2} \times \theta^{-1/2})$
Structure for $0.35 \approx \theta \approx 0.49$

$X \uparrow [110]$

$a_x(Ni)$
$\theta^{1/2}$

$a_x(Ni)$

Y
$[\bar{1}01]$

Na Ni

(a)

Diffraction Pattern for
Proposed Structure : $\theta = 0.40$

X'

Y'

• Diffraction from Ni(III) Substrate
  $(X',Y') = (m,n)$

+ Diffraction from Na Overlayer
  Plus Multiple Scattering
  $(X',Y') = (p\theta^{1/2}, q\theta^{1/2})$

○ Multiple Scattering Only
  $(X',Y') = (m + p\theta^{1/2}, n + q\theta^{1/2})$

(b)

(c)

Fig. 5. Ni(111)-Na for $0.35 < \theta < 0.49$. A. Proposed incoherent hexagonal
structure. B. Predicted diffraction pattern for $\theta = 0.40$. C. 110V
LEED photograph for $\theta = 0.40$.

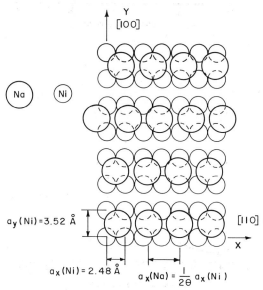

Proposed Structure for Ni(110)–Na
0.25 < θ < 0.32

Y
[100]

Na   Ni

a_y(Ni)=3.52 Å

[110]

X

a_x(Ni) = 2.48 Å a_x(Na) = $\frac{1}{2\theta}$ a_x(Ni)

Fig. 6. Proposed one-dimensionally incoherent structure for Ni(110)-Na with
0.25 < θ < 0.32. Alternate troughs on the Ni(110) surface contain incoherent rows of evenly spaced Na adatoms. The rows are randomly
shifted in the [110] direction.

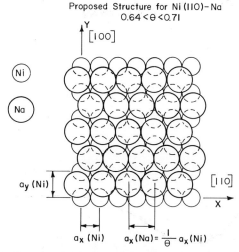

Proposed Structure for Ni(110)–Na
0.64 < θ < 0.71

Y
[100]

Ni

Na

a_y (Ni)

[110]

X

a_x (Ni) a_x(Na) = $\frac{1}{\theta}$ a_x(Ni)

Fig. 7. Proposed one-dimensionally incoherent structure for Ni(110)-Na with
0.64 < θ < 0.71. Each trough on the Ni(110) surface contains an incoherent row of evenly spaced Na adatoms. The rows of Na adatoms are
locked into an irregular hexagonal array.

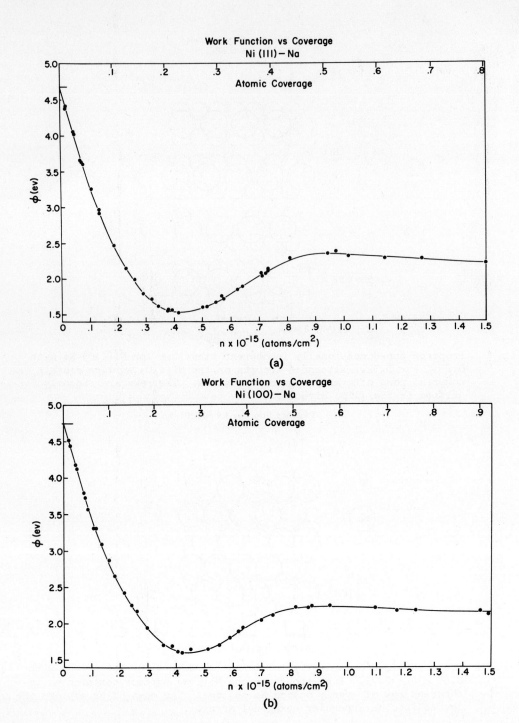

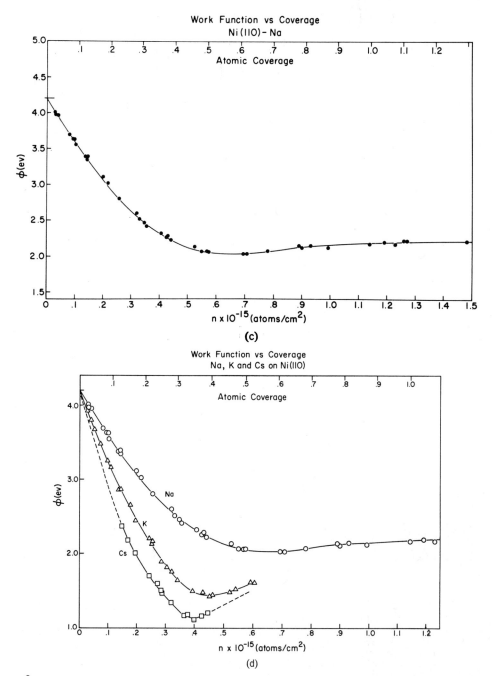

Fig. 8. Work function vs. coverage curves for Na on Ni(111), Ni(100) and for Na, K or Cs on Ni(110). The work function was measured with a Kelvin probe. The clean polycrystalline W reference electrode was assumed to have a work function of 4.55 eV.

THE INTERACTION OF NITROGEN ATOMS AND IONS WITH NICKEL SURFACES

J. C. Gregory* and D. O. Hayward

Chemistry Department
Imperial College of Science and Technology
London, England

INTRODUCTION

The interaction of molecular nitrogen with nickel surfaces in the form of evaporated films, FEM tips and single crystal faces has been studied by many workers (Beeck, Smith and Wheeler, 1940; Mignolet, 1950; Trapnell, 1953; Suhrmann and Schultz, 1955; Wagener, 1957; Ehrlich and Hudda, 1961; Madden and Farnsworth, 1961; King, 1968), using a variety of techniques, and all report that chemisorption does not occur in detectable amounts at room temperature and above, and at pressures less than 10^{-3} torr, although a weak adsorption was observed in many cases at lower temperatures or higher pressures. Infrared spectra (Eischens and Jacknow, 1965; Van Hardeveld and Van Montfoort, 1966) clearly show that the nitrogen molecule remains undissociated in this weak adsorption. The absence of strong, dissociative chemisorption of nitrogen on nickel surfaces indicates either that the process is endothermic or that the activation energy required is prohibitively high. It has been recently shown that nitrogen activated by electron bombardment is adsorbed on nickel surfaces at room temperature (Winters, Horne and Donaldson, 1964; Winters, 1966) and this was attributed to the presence of N and N_2^+ in the

---

* Presently: National Academy of Sciences Research Associate, Space Sciences Laboratory, NASA, Marshall Space Flight Center, Huntsville, Alabama, 35812, U.S.A.

gas phase. No activation energy is anticipated for adsorption of atoms and N_2^+ probably obtains the necessary energy for dissociation from the Auger neutralization of its charge at the nickel surface, (Propst and Luscher, 1963). If adsorption of ground-state nitrogen molecules is an endothermic process, as seems likely, the nitrogen adatoms will be in a metastable state and desorption will occur spontaneously and irreversibly as soon as the temperature is raised to a point at which the adatoms are able to migrate across the nickel surface and recombine.

This paper describes a study of the adsorption of activated nitrogen on nickel films and filaments, and the kinetics of the subsequent desorption at higher temperatures. This provides information on the various ways in which nitrogen atoms are bonded to the nickel surface.

EXPERIMENTAL

The apparatus was constructed of pyrex glass pumped by mercury diffusion pumps. The residual gas pressure was normally in the range $1 - 4 \times 10^{-10}$ torr after a 12-hour bakeout at 400°C. The partial pressure of active gases, however, was much lower than this. Total gas pressures were measured with a McLeod gauge, a Bayard-Alpert ionization gauge, and a high pressure ionization gauge (Schultz and Phelps, 1957). Both ionization gauges were calibrated against the McLeod gauge and also against a mercury manometer using a flow method. Partial pressures were analyzed with an omegatron mass spectrometer (Edwards High Vacuum Ltd.) incorporating a rhenium cathode coated with lanthanum boride.

Nickel wire was designated Grade I and supplied by Johnson-Matthey & Co., and nitrogen was prepared in the apparatus from previously outgassed sodium azide. No impurities were detected in this nitrogen with the mass spectro-meter.

Nitrogen atoms were produced in two ways; by dissociation of molecules on the surface of a hot tungsten filament (Nornes and Donaldson, 1966; Gregory and Hayward, to be published), and by a 300 W 20 Mc/s electrodeless discharge. No afterglow was observed with the discharge, presumably because of the rapid recombination of the nitrogen atoms at the baked-out glass walls of the system (Lewis, 1929), and adsorption had to be studied with the nickel in contact with the gas discharge. This meant that the adsorbing species included nitrogen ions as well as atoms, N_2^+ probably playing a dominant role as it is present in great excess over such species as N^+, N^{++}, N_3^+, N_4^+ (Knewstub and Tickner, 1962).

The one type of cell (figure 1), with minor modifications, was used throughout this work. It could be isolated from the ionization gauges, dosing system and pumps by means of magnetically operated glass valves (Dekker, 1954). Three types of experiment were performed with this cell and these are described in detail below.

Adsorption of discharge-activated nitrogen on films

In these experiments no potential leads were attached to the nickel filament, and the tungsten filament and associated pinch seal were absent. Films with an apparent area of 340 cm^2 were thrown from a 0.5 mm. nickel wire at pressures less than 1×10^{-9} torr. Nitrogen was dosed in to a pressure of ca. 1×10^{-2} torr. Such a dose contained ca. 10^{18} molecules. Mass spectrometric analysis, made at 10^{-6} torr, did not reveal any impurity, indicating the maximum concentration of such impurity to be 1 part in 10^4. A single dose, therefore, contained a maximum of 10^{14} molecules of impurity, which represents less than 0.1% of a monolayer on the film, and may be disregarded.

The electrodeless discharge was obtained by placing the work coil of the oscillator around the horizontal discharge tube shown in figure 1. It was operated for periods of 30 to 60 seconds, with the magnetic valves closed to isolate the discharge region from the gauges and the cold traps. Fresh doses of nitrogen were admitted when necessary to maintain the pressure approximately constant.

Partial desorption of the nitrogen was accomplished by heating the cell in an oil bath but the rate of heating was slow and the highest temperature attainable was 300°C.

The interaction of discharge-activated nitrogen with nickel filaments

The filaments were 0.5 mm. dia. 40 cms. long with an apparent surface area of 6.25 cm^2. They were thoroughly outgassed at as high a temperature as possible and during this process some film was unavoidably thrown onto the walls of the cell. The nickel filament surfaces so produced were almost certainly not clean but the operation of the discharge probably helped to remove much of the remaining surface contamination. Direct adsorption measurements could not be made due to competing adsorption of ions and atoms on the glass and nickel film surfaces in the cell. Adsorbed quantities were measured by flashing the filament to 1200°C. There was no pumping in the cell during desorption, (as occurs in most systems studied by the flash-filament technique), and the desorptions were therefore integral. It was assumed that very little gas remained in or on the filament after flashing to 1200°C; a reasonable assumption for this system.

Adsorption on a nickel filament of N atoms produced at a hot tungsten filament

Nornes and Donaldson (1966) have shown that nitrogen atoms are desorbed from a tungsten filament heated to temperatures higher than 2300° K in nitrogen at pressures of ca. 10^{-3} torr and below.

In our experiments the tungsten wire was 0.2 mm. in diameter and 25 cms. long, and was mounted in the same cell as the nickel filament with no obstruction between the two, (see figure 1). The temperature of the tungsten wire was measured with an optical pyrometer and was checked against resistance measurements. After black-body corrections the agreement between the methods was good, the maximum error in the range 200 - 2600° K being estimated at ±30°.

Measurement of the activation energy of desorption of nitrogen chemisorbed on nickel filaments

Activation energies were measured by first noting the rate of desorption, r_1, of nitrogen from the nickel filament at a fixed temperature, T_1°K, and then abruptly increasing the temperature to T_2°K and measuring the new rate of desorption, r_2. If it is assumed that the concentration of the adsorbed nitrogen does not change appreciably during the time required to establish the higher temperature and measure r_2, we obtain the following expression for the activation energy of desorption

$$E_{des} = 2.303 \ R \ \frac{T_1 T_2}{T_2 - T_1} \ \log_{10} \frac{r_2}{r_1} \quad \dots \dots \dots \dots \dots \dots \dots (1)$$

To obtain as abrupt a change in temperature as possible a field-emission tip controller based on a Kelvin bridge design described by Gomer and Zimmerman (1965) was modified and used to control the temperature of the nickel filament (diameter 0.5 mm; length 40 cms.).

The current through the filament and the voltage across a pair of potential taps (0.001" tungsten wire spot-welded to the filament) were displayed on two synchronous recorders. The temperature of the nickel wire was calculated from tabulated resistance data, (Gmelin, 1959). The pressure in the system was displayed on a third recorder. While gas was evolving from the nickel filament, the value of the control resistor on the filament controller was abruptly raised, causing the Kelvin bridge to unbalance. The out-of-balance current, suitably amplified, heated the nickel filament rapidly to the new selected resistance and temperature. By raising the filament temperature in a series of jumps, values of E_{des} at different coverages were obtained for both types of activated nitrogen.

To reduce the cooling effect of the tungsten rods supporting the nickel filament, the latter was thinned at both ends for about 0.5 cms. of its length.

RESULTS

Adsorption of discharge-activated nitrogen on nickel films

Adsorption was very rapid on a clean film at room temperature as shown in figure 2. The adsorption rate on baked-out glass under similar conditions is also shown for comparison. When the discharge was first switched on with a pressure of 4×10^{-2} torr in the cell, the discharge was extinguished in 5 seconds as the pressure dropped below 10^{-3} torr. The adsorption rate fell off after ca. 3×10^{18} molecules had been adsorbed. For a film of geometric area $340 \ cm^2$ and an estimated roughness factor of $3 - 6$, this corresponds to a coverage of between $4 - 2$ monolayers. The continued slow rate of adsorption after this may be due to deeper penetration of the substrate by a more energetic N species or to slower take-up in regions of the apparatus more distant from the discharge. The film was cooled to 78° K after adsorption of ca. 10^{16} molecules per cm^2. A further 1.2×10^{15} molecules per cm^2 were adsorbed, the uptake being measured after the film had returned to room temperature.

Desorption of nitrogen from the nickel film occurred in two stages. There was a fairly rapid evolution of gas between 100 and 200°C. The temperature was then held constant at 200°C and the rate of evolution quickly dropped to a very low value. Heating to higher temperatures produced a further evolution of nitrogen; and when the film was held at a constant temperature of 300°C, nitrogen was still being evolved after four hours.

It was estimated that 3×10^{17} molecules were desorbed during the first stage of the desorption and this corresponds to about 10% of the total uptake. The total amount desorbed was not measured.

Adsorption of activated nitrogen on nickel filaments and its subsequent

   desorption

The operation of the nitrogen discharge was found to raise the temperature of the nickel filament, due perhaps to recombination processes at the surface or eddy-current heating. For this reason the discharge was never operated for more than 60 seconds at one time, which was sufficient to keep the filament temperature below about 150°C. The temperature of the filament was raised by applying a constant voltage across its end and two desorption peaks were observed. A typical desorption spectrum is shown in figure 3. The first and smaller peak occurs between 140 and 200°C, and the second and larger peak between 250 and 400°C. Mass spectrometric analysis indicated both peaks to be composed of nitrogen.

For discharge activated nitrogen the average saturation or steady-state adsorption was achieved after 1.5 minutes discharge and was equal to about 10^{16} mols cm^{-2} assuming a roughness factor of 1.4 for the filament. This is equivalent to about 14 monolayers. The smaller lower temperature peak contained 1.7×10^{13} mols cm^{-2} which is equivalent to 2.4% of a monolayer and 0.17% of the total desorbed nitrogen. This is a smaller percentage than that obtained with the film and could be partly due to the heating of the filament during the discharge as described above.

Adsorption on a nickel filament of nitrogen activated by a hot tungsten filament

In this case the steady-state or saturation condition for adsorption on nickel depends on the temperature of the tungsten activator. Some adsorption curves at different activator temperatures are shown in Fig. 4.

With the tungsten activator at 2440° K saturation or steady-state was achieved after adsorption of about 0.9 monolayer equivalents. At 2620° K no equilibrium was reached but the rate of adsorption had dropped to about one-seventh of the initial value after adsorption of about 2.2 monolayers.

Desorption again produced two peaks. The estimated desorption from the low temperature peak was 8×10^{12} molecules cm^{-2}, which is equivalent to 1.2% of a monolayer.

Activation Energy for desorption from nickel filaments of (a) discharge-activated nitrogen and (b) nitrogen activated on a hot tungsten filament

The variation of activation energy for desorption E_{des} is shown in Fig. 5. E_{des} appeared to be roughly constant and equal to 28 ± 4 K cals $mole^{-1}$ for both kinds of activated nitrogen over the whole range of desorption of the second peak. The portion of the curve below 200°C represents in both cases less than 2% of the total adsorbed gas, and values of E_{des} in this region were generally lower but reproducibility was very poor and they are not considered reliable.

The high experimental error was mainly due to uncertainties in temperature measurement. As a typical temperature jump was of the order of 60 - 70°, an error of 10° in measurement of T_1 or T_2 could cause an error of about 15% in E_{des} (see equation 1). Also the change in temperature was not instantaneous, the heating period lasting about 3 to 5 seconds, so the assumption that the surface concentration does not change appreciably may not be valid. There is, in addition, the usual difficulty in drawing tangents to rate curves.

DISCUSSION

Results for the desorption of activated nitrogen from films and filaments are summarized in table 1. In each case there are two pressure peaks and the temperature ranges in which these occur are markedly similar. Also, there seems to be no difference between nitrogen activated in the discharge and nitrogen atomized at the tungsten filament so far as the activation energy for desorption of the second peak is concerned. It is therefore concluded that the same surface complexes are produced in all cases. However, there are marked differences in the populations of the first and second peaks (afterwards referred to as the α and β states respectively). For example, the maximum population of the α state on a nickel film is between 20 and a 100 times that on a filament, and such large differences cannot be explained simply in terms of the roughness of the film surface. Heating of the nickel filament during the discharge was mentioned earlier as one possible factor but this is unlikely to be important for results obtained using the tungsten filament. The surface of the nickel filament may not, of course, be clean and this would reduce the number of surface sites available for adsorption of N atoms. The surfaces of the nickel films, on the other hand, should be fairly clean initially as evaporation was carried out in a good vacuum. Thus, differences in population could easily be explained in this way if the α state consists of N atoms chemisorbed on the nickel surface. These may well be in a metastable state and the desorption temperature of 150 – 200°C would then represent the temperature range in which the adatoms first become mobile and able to recombine. Surface diffusion of nitrogen atoms chemisorbed on tungsten field emitters is first observed on different planes in the temperature range 130 – 230°C (Ehrlich and Hudda, 1961).

The β state is tentatively identified with nitrogen atoms which have penetrated below the first layer of nickel atoms and are occupying interstitial positions in the nickel lattice. Where the coverage rises to several monolayers (as in the case with the discharge activated gas, and with the tungsten filament experiments at higher temperatures), it is postulated that penetration occurs at several monolayers depth.

Examination of the properties of nickel nitride, Ni_3N, reveals several interesting parallels

(1)   The heat of formation from the elements at 20°C = 0.2±.1 K cal. mole^{-1}. (Hahn and Konrad, 1951), i.e. virtually zero.

(2)   Mathis (1951) estimates the temperature of onset of decomposition to be 290°C, and Trillat et al. (1957) describe an increase in magnetic susceptibility occurring at 250°C. In the present work the second

desorption peak (comprising 90 - 99% of the total gas) started to

evolve at 250°C and the rate rapidly increased in the range 250 - 300°C.

(3)   Trillat et al. (1957) and Terao et al. (1959) describe the expansion

of the nickel f.c.c. lattice at 150 - 175°C in the presence of ammonia.

At 175°C the hexagonal form of Ni_3N is obtained.   In the present work

the first desorption peak occurred at 150 - 180°C.

(4)   From Mathis' results it is possible to calculate the activation

energy for decomposition of Ni_3N obtained in his system.   This has

been done, and $E_{act.}$ = 25.5 K cals mole^{-1}.

The activation energy of the desorption in this work was found to be

28 ± 4 K cals mole^{-1}.

From this comparison it is maintained that a close analogy exists between

bulk Ni_3N and the nitrogen β state observed here.   It is postulated that the

takeup of nitrogen atoms into the nickel lattice at 20°C is not properly

ordered.   At around 150°C the nickel atoms possess enough energy to relax the

normal interatomic distances of the nickel f.c.c. lattice.   This enables most

of the nitrogen atoms to take up metastable equilibrium positions corresponding

in the limit to the octahedral coordination of Ni_3N.   A small proportion of N

atoms near the surface may be able to escape during this process and recombine

on the surface, thus contributing to the first pressure peak.

The long "tail" observed during desorption even at high temperatures

(>700°C) may be due to nitrogen atoms diffusing into the bulk, or to the

effusion of atoms which were deeply buried by a channelling mechanism,

(Zscheile, 1966).

ACKNOWLEDGMENTS

The authors wish to thank Prof. F. C. Tompkins, F.R.S., for many helpful

discussions.   One of us, (J.C.G.) gratefully acknowledges the support of the

J. Lyons Company Postgraduate Studentship which was held throughout the

execution of this work.

REFERENCES

Beeck, O., Smith, A. E., and Wheeler, A., Proc. Roy. Soc. A177, 62, (1941)

Dekker, R. W., J. Appl. Phys. 25, 1441, (1954)

Ehrlich, G., and Hudda, F. G., J. Chem. Phys. 35, 1421, (1961)

Eischens, R. P., and Jacknow, J., in "Proc. 3rd Intern. Congress on Catalysis," Amsterdam, 1964, Eds. W. M. H. Sachtler et al. (North Holland Publ. Co., Amsterdam, 1965), p. 627

Gmelin, Handuch der Anorg. Chem., Syst. 59 Nickel, Pt. D2, 253, (1959)

Gomer, R., and Zimmermann, D., Rev. Sci. Instr. 36, 1046, (1965)

Gregory, J.C., and Hayward, D. O., to be published

Hahn, H., and Konrad, A., Z. Anorg. Allgem. Chem. 264, 181, (1951)

King, D. A., Surface Sci. 9, 375 (1968)

Knewstub, P. F., and Tickner, A. W., J. Chem. Phys. 37, 2941, (1962)

Lewis, B., J. Amer. Chem. Soc. 51, 564, (1929)

Madden, H. H., and Farnsworth, H. E., J. Chem. Phys. 34, 1186, (1961)

Mathis, M., Bull. Soc. Chim. Fr., 443, (1951)

Mignolet, J. C. P., Disc. Faraday Soc. 8, 105, (1950)

Nornes, S. B., and Donaldson, E. E., J. Chem. Phys. 44, 2968, (1966)

Propst, F. M. and Lüscher, E., Phys. Rev. 132, 1037, (1963)

Schultz, G. J., and Phelps, A. V., Rev. Sci. Instr. 28, 1051, (1957)

Suhrmann, R., and Schultz, K., Z. Naturforsch. 10B, 517, (1955)

Terao, N., and Berghezan, A., J. Phys. Soc. Japan, 14, 139, (1959)

Trapnell, B. M. W., Proc. Roy. Soc. A218, 566, (1953)

Trillat, J. J., Tertian, L., Terao, N., and Lecomte, C., Bull. Soc. Chim. Fr., 804, (1957)

Van Hardeveld, R., and Van Montfoort, A., Surface Sci. 4, 396, (1966)

Wagener, S., J. Phys. Chem. 61, 267, (1957)

Winters, H. F., Horne, D. E., and Donaldson, E. E., J. Chem. Phys. 41, 2766, (1964)

Zscheile, H., Phys. Status Solidi, 14, K15, (1966)

TABLE I

Method of Activation	Nickel	Total uptake (molecules cm^{-2})	No. of molecules cm^{-2} in first peak	Temp. range $^\circ$C first peak	second peak
Discharge	film	10^{16}(for geometric area.)	10^{15} (for geometric area.)	100–200	>250
Discharge	filament	10^{16}	1.7 x 10^{13}	140–200	250–400
W filament at 2440°K	filament	6.4 x 10^{14}	8.0 x 10^{12}	140–200	250–400

CELL

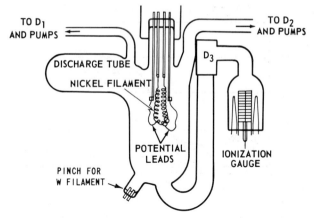

Fig. 1.

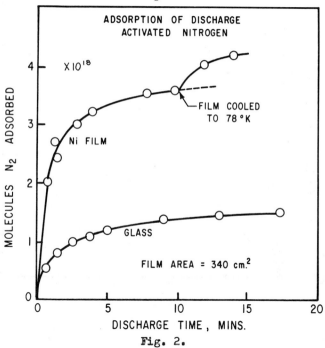

Fig. 2.

CONSTANT VOLTAGE DESORPTION OF DISCHARGE ACTIVATED NITROGEN FROM A Ni FILAMENT

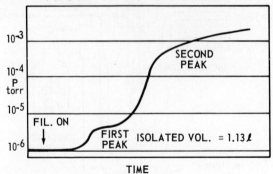

TIME

Fig. 3.

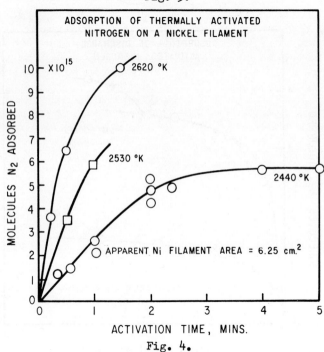

ADSORPTION OF THERMALLY ACTIVATED
NITROGEN ON A NICKEL FILAMENT

ACTIVATION TIME, MINS.

Fig. 4.

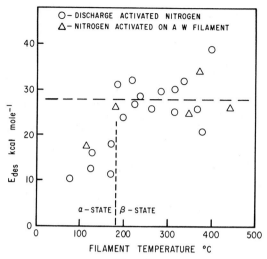

Fig. 5.

THE INTERACTION OF OXYGEN AND CARBON MONOXIDE WITH NICKEL FILMS: STICKING PROBABILITIES, SURFACE REARRANGEMENTS AND REPLACEMENT PROCESSES

A. M. Horgan and D. A. King

School of Chemical Sciences
University of East Anglia
Norwich, England

1. Introduction

Few studies have been made of the kinetics of the interaction of gas mixtures with clean metal surfaces during the initial stages of the process, when adsorption rates are very rapid. The interaction between carbon monoxide and oxygen on nickel films at very low pressures has been studied by Kawasaki et al.[1], but their apparatus was not designed to measure absolute rates of adsorption and reaction. We have designed an adsorption cell for the accurate evaluation of sticking probabilities of gases on metal films[2] (an adaptation of a previous design[3] utilizing the Wagener flow technique[4]) which is well-suited to studies of mixed adsorption.

The interaction of oxygen with clean metal surfaces has received considerable attention in recent years, and, in particular, LEED studies[5-8] and surface potential measurements[9-13] have yielded useful information on the adsorption, surface rearrangement and oxygen penetration processes which occur on nickel surfaces. However, no direct data have yet been published for the adsorption kinetics of oxygen on clean nickel surfaces and we have therefore made an extensive kinetic investigation of this system.

In this paper a survey is made of the results obtained from measurements of sticking probabilities, equilibrium isotherms, desorption spectra, replacement processes and reaction rates for oxygen and carbon monoxide interacting singly and consecutively with nickel films deposited under ultra-high vacuum conditions.

2.   Experimental

An apparatus (figure 1) has been designed for the accurate measurement of sticking probabilities by a reflexion detector technique.[2]   Using modern ultra-high vacuum techniques,[14] vacua of about 2×10^{-11} torr (nitrogen equivalent, measured by a Redhead modulated Bayard-Alpert ionization gauge with a thoria-coated filament) were achieved in the cell. Films were deposited from nickel filaments (0.5 mm diameter, Johnson Matthey Grade 1), at pressures of ca. 10^{-10} torr (nitrogen equivalent) onto a spherical 1 litre Pyrex glass bulb, and impurity-free oxygen or carbon monoxide (prepared by heating thoroughly out-gassed potassium permanganate or molybdenum hexacarbonyl, respectively) was allowed to flow into the cell at a determined rate through a specially construc-ted retractable diffuser of measured conductance, which is situated at the centre of the adsorption bulb to achieve a symmetrical distribution of the adsorbate over the film.

Pressures in the cell were measured using a modulated Bayard-Alpert ionization gauge and a 180° glass-head miniature mass spectrometer. A modulated ionization gauge is essential for the measurement of oxygen or carbon monoxide pressures since the residual pressure varies substantially with the exposure of the gauge to either gas, as shown by Redhead[15] and Singleton[16] and confirmed in our work. Sticking probabilities, s, for oxygen could, however, not be accurately determined with either type of detector in the region where $s < 10^{-3}$ due to the substantial pumping action of the hot cathode at high pressures, even at an emission current of 5μA. Independent experiments, using a Pirani gauge to monitor the pressure, showed that the pumping action due to a hot tungsten filament at ca. 10^{-4} torr of oxygen became appreciable at temperatures as low as 850°C. This effect could be overcome by operating the detectors intermittently, for a period (ca. 5 sec) sufficient to yield an indication of the pressure. However, to obtain accurate sticking probabilities for oxygen in the region where $s < 10^{-3}$, an alternative apparatus was used (see below).

The main feature of the reflexion detector technique is that the detector is placed so that gas molecules from the diffuser can only reach it after collision with the metal film. The sticking probability is evaluated from the steady state pressure P (torr) in the detector with gas flowing into the cell at a rate r_i (molecules sec^{-1} cm^{-2}) as follows:

$$s = \frac{\text{net rate of adsorption}}{\text{rate of bombardment}}$$

$$= \frac{r_i}{r_i + \text{reflexion rate}}$$

$$= 1/(1 + ZP/r_i) \tag{1}$$

where Z is the Knudsen collision factor. This method obviates the need for correction factors due to molecular beaming up the gauge side-arm[17], since gas entering the side-arm aperture in the present design is randomised. To prevent adsorption on the side-arm walls when the cell was maintained at low temperatures, a double-wall structure was used (figure 1) to prevent the inner tubing from coming into contact with the coolant.

To avoid the effects of the pumping action by a hot cathode at high oxygen pressures, an adsorption apparatus was used similar to the traditional design for following "slow" adsorption processes,[18] in which gas is added dose-wise to the adsorbent and subsequent pressure changes in the adsorption cell are followed using a Pirani gauge. Adsorption rates, and hence sticking probabilities, were evaluated from the slopes of the pressure-time curves obtained in this manner: the method is only accurate when $s < 10^{-4}$.

Results reported in this paper were all obtained on nickel films deposited without cooling the glass substrate, i.e. at about $320^{\circ}K$. Prior to runs at $373^{\circ}K$, films were sintered at this temperature for 1 hour.

3. Results and Discussion

3.1. Oxygen as Adsorbate

3.1.1. Initial Sticking Probabilities

Very high initial sticking probabilities, s_o, were observed. When the gas supply was opened ($r_i \approx 3 \times 10^{11}$ molecules cm^{-2}) to freshly prepared nickel films, the initial pressure rise registered by the ionization gauge was $<10^{-12}$ torr (the detection limit) with the cell maintained at 78°K, and 1.2×10^{-11} torr with the cell maintained at 290°K. Since the detection limit of the mass spectrometer is ca. 5×10^{-11} torr the latter reading could not be verified as a true oxygen pressure, and may be due to non-adsorbable impurities in the gas supply[19] (1 part in 10^8 would give a pressure rise of this order of magnitude). Thus from these pressures and equation (1), we can report values of s_o as > 0.999 at 78°K, and >0.99 at 290°K. These values are higher than those reported for many other gas-metal systems where the flash filament technique has been used[20], although we have reported values close to unity for the oxygen + tungsten system using the reflexion detector technique,[2] and Bell and Gomer[21] and King[22] have reported similarly high values for the carbon monoxide + tungsten system utilizing what were essentially also reflexion detector techniques. The advantages of the reflexion detector technique over the traditional flash filament technique for sticking probability measurements have been discussed in a recent paper.[2]

3.1.2. Variation of sticking probability with coverage

The variation of s with coverage at temperatures of 78, 195, 290 and 373°K is shown in figure 2.

At 78°K, s remains greater than 0.99 up to a coverage N of 25×10^{14} molecules cm^{-2} (geometric area) where a fall to 0.95 is noted, and finally at $N \approx 50 \times 10^{14}$ molecules cm^{-2} a rapid fall in s is observed as the film approaches saturation. This final drop in s can be attributed to the development of a finite equilibrium pressure, since on closing the gas supply to the cell at high coverages the oxygen pressure falls to a steady value. An equilibrium isotherm can be obtained in this manner, and a linear Temkin plot (i.e. log P vs. N) was obtained in the range 59 to 67 x 10^{14} molecules cm^{-2} and 10^{-9} to 10^{-5} torr. The adsorbate in equilibrium with the gas phase at 78°K will be referred to as the γ state. Even when the γ state is in equilibrium with the gas phase the probability that a molecule striking the surface be adsorbed remains close to unity, since when the gas supply is reopened after allowing the isolated system to attain a steady equilibrium pressure no 'instantaneous' pressure rise is observed. This result illustrates the existence of a mobile precursor state capable of migrating large distances over primary layer adsorbate at 78°K.

Further increments of oxygen can be taken up by the film at 78°K if, after adsorption to a steady equilibrium pressure, the cell (isolated from the pumps and gas supply) is allowed to warm up to room temperature and is then recooled to 78°K, as observed from surface potential measurements by Delchar and Tompkins.[12] The process can be repeated, each additional increment being smaller than the previous one. Adsorption at 78°K again takes place with s ≈ 1. This site regeneration process has been examined more closely by recording desorption spectra obtained on warming to room temperature after the addition of doses of various sizes at 78°K (figure 3). These curves reveal varying degrees of complexity, depending on the degree of coverage with oxygen and the degree of advancement of the oxygen penetration process: at low coverages they are monotonic, while at high coverages three maxima become apparent. A site-energy distribution curve for the γ state on a surface completely pre-saturated at 300°K was obtained by leaving a film in a dynamic oxygen pressure of 10^{-3} torr for 12 hours, then cooling to 78°K and saturating the film with oxygen at this temperature, and observing the desorption spectra obtained on removing the cold bath, with the cell open to the pumps. (Details of the method and the calculations and assumptions involved are given elsewhere.[23]) The curve is monotonic, with site energies below 16 kcal/mole and a peak at

13 kcal/mole, and is similar to that for the $\alpha\gamma$ states of nitrogen on Mo, Ti[24] and W,[23] except that on Ni the γ state is more sparsely populated.

The s against N plots at 195 to 373°K (figure 2) show completely different characteristics to the 78°K results, which can be attributed to the short life-time of the γ state at these temperatures. Three distinct stages can be iden-tified in the adsorption process:

1. Initially s decreases with increasing coverage from the clean film value of unity, reaching a minimum value at $N \approx 5 \times 10^{14}$ molecules cm^{-2}.

2. As the coverage is increased further, s increases to a broad maximum at ca. 15×10^{14} molecules cm^{-2}. A sharp drop is then noted at $N \approx 20 \times 10^{14}$ molecules cm^{-2}.

3. The third stage could only be observed with the standard apparatus equipped for dose-wise addition of gas. At $s \approx 10^{-5}$ (see figure 4) the sharp drop of s with increasing N is arrested and a "slow process" was observed involving the further adsorption of the equivalent of about half the amount already rapidly adsorbed. This slow process observed at 195 and 290°K is identical to the site regeneration process which occurs on warming up a 'sat-urated' film from 78°K: an approximate value of 10^{-5} can be estimated for s from the tail of the desorption spectra, e.g. from curves v and vi in figure 3.

3.1.3. Adsorption and Rearrangement Mechanisms

A phenomenological description of the processes which give rise to the above observations, and which also draws on calorimetric adsorption heat and surface potential results, is proposed here.

1. On a clean surface at 195 to 373°K, adsorption takes place by a Lang-muir-type adsorption process (s initially unity, falling continuously with coverage) into an immobile state (β'), with a heat in the region of 50 kcal/mole[25] and a negative surface potential, and which involves only minimal dis-tortion of the nickel lattice. Curvature in the plots of s against N at coverages less than 5×10^{14} molecules cm^{-2}, which could not be linearised by

introducing multiple collision corrections for the roughness of the outer film surface,[24,26] could be attributed to the existence of a precursor state (physically adsorbed) over filled β' sites. However, the curvature is not **very** marked at higher temperatures, indicating that the probability of hopping from a filled to an empty site is small, i.e. the precursor is rather ineffective.

2. At a coverage of 5×10^{14} molecules cm^{-2}, sections of the outer surface of the film (directly accessible to the gas phase) approach saturation in this β' state. It is suggested that adsorption in this state weakens the cohesion of the surface metal atoms to their neighbors, until a stage is reached when greater thermodynamic stability can be achieved (i.e. the surface free energy can be lowered) by displacement of the surface nickel atoms from their bulk lattice positions with the formation of a new surface structure, β_1. The overall adsorption heat in this state is _ca._ 110 kcal/mole[25], and the surface possesses a negative dipole outwards: $Ni \nearrow^{O} \searrow Ni$. In contrast to the β' surface, adsorption can occur into a weakly held, mobile state (γ) on a β_1 surface, and the sticking probability therefore rises once the β_1 structure has been nucleated, due to migration from β_1 patches on the outer surface of the film to unsaturated areas - largely the porous structure - of the film. The rate determining process in this region is the distance travelled in the mobile state before desorption, viz.

$$\bar{x} = a \exp[(E_d - E_m)/RT] \qquad (2)$$

where a is the hop distance and $(E_d - E_m)$ is the difference between the activation energies for desorption and migration. Since $(E_d - E_m) > 0$, $\bar{x}$, and hence s, increases with decreasing temperature (figure 2). As saturation is achieved in the β_1 state over the entire nickel surface at _ca._ 20×10^{14} molecules cm^{-2}, s falls rapidly with increasing coverage to below 10^{-3} before the onset of the third stage in the adsorption process. By comparison with other studies[27,28] it is estimated that the β_1 structure possesses approximately 1 oxygen atom per surface nickel atom.

3. Slow adsorption processes following rapid chemisorption have been observed in a number of systems[20,29] and generally attributed to the penetration of the adsorbate into the metal lattice. The slow process observed here (figure 4) can be compared with the slow change in surface potential for the oxygen-on-nickel system towards more positive values after the instantaneous change to a negative value which follows the addition of a dose of gas.[10-13] The β_2 structure formed in this process therefore involves the structure Ni $\overset{+}{\underset{O}{\diagdown}}$ Ni.

The sticking probabilities in figure 4 were obtained from the slopes of pressure against time curves for the slow decay following the addition of doses of gas to the adsorption cell at constant volume. The points shown in the figure were obtained over a range of pressures, from 1.5×10^{-3} to 2×10^{-4} torr: there is an expected scatter of points, but no regular variation of s with the pressure was obtained. Since s = adsorption rate/(ZP), the rate of the slow process is thus directly proportional to P.

No dependence of the rate on temperature was observed (figure 4) between 195 and 290°K. This is in fairly good agreement with the low activation energy of 1 to 1.5 kcal/mole reported for the process from surface potential results in the range 146 to 298°K.[10,13] However, from the desorption spectra in figure 3 and surface potential results at 78°K it must be concluded that the slow process does not occur at this temperature, and any proposed mechanism for the process must therefore provide an explanation for this temperature dependence at low temperatures only.

It is proposed that this surface rearrangement involving incorporation of oxygen atoms proceeds through an intermediate molecularly adsorbed state with an adsorption heat of ca. 8 kcal/mole (γ'); this state is distinct from the stable γ state (§3.1.2) which exists on a surface saturated at 300°K and 10^{-3} torr. Furthermore, the rate-limiting step is assumed to be the transformation of the γ' species held on an 'active' site to the β_2 structure, from which it follows that the γ' species is in equilibrium with the gas phase. If θ_i is the fractional coverage in the intermediate state γ' and θ_a the fraction of active sites, from absolute rate theory we obtain the following rate equation for this process:

$$r_p = \frac{kT}{h} N_s \, \theta_i \, \theta_a \, e^{\Delta S^{\neq}/R} \, e^{-E_p/RT} \qquad (3)$$

where N_s is the total number of surface sites cm^{-2}, $\Delta S^{\neq}$ is the activation entropy and E_p the activation energy for the penetration process. Assuming that the activation energy for desorption E_d (= heat of adsorption) for the γ' state is coverage independent, we can write, from the Langmuir isotherm,

$$\theta_i = P/(P + Ke^{-E_d/RT}) \qquad (4)$$

where $K = N_s \cdot (kT/h)(MT)^{1/2}/3.5 \times 10^{22}\sigma$, and P is in torr. ($\sigma$ is the probability that a molecule striking a vacant site is adsorbed, taken hereafter as unity.) Hence we have the general rate equation

$$r_p = \frac{kT}{h} N_s \, \frac{P}{(P + Ke^{-E_d/RT})} \, \theta_a e^{\Delta S^{\neq}/R} \, e^{-E_p/RT} \qquad (5)$$

When $\theta_i \ll 1$ we therefore have

$$r_p = \frac{3.5 \times 10^{22}}{(MT)^{1/2}} \, P\theta_a e^{\Delta S^{\neq}/R} e^{(E_d - E_p)/RT} \qquad (6)$$

which is valid at high temperatures when the γ' state is sparsely populated. Since, experimentally, the rate is independent of temperature in the range 195 to $300^{\circ}K$, E_d must have the same magnitude as E_p. Hence, we can write for the sticking probability in this temperature range

$$s = r_p(MT)^{1/2}/3.5 \times 10^{22}P$$

$$= \theta_a e^{\Delta S^{\neq}/R} \qquad (7)$$

From figure 4, the onset of the slow process occurs at $s \approx 10^{-5}$. Utilizing this value for $\theta_a e^{\Delta S^{\neq}/R}$ in equation (5) and a value of 8 kcal/mole for E_d and E_p, chosen within fairly close limits by the requirements for equal rates at 300 and $195^{\circ}K$, and a negligible rate at $78^{\circ}K$, the variation of the rate of the penetration process with temperature over the range 110 to $200^{\circ}K$ has been calculated (figure 5). It is seen that equation (5) is capable of providing a negligible rate at very low temperatures, and a temperature independent rate, proportional to pressure, at high temperatures. The intermediate γ' state is thus stable at 78°K, and is akin to the 'virgin' state of CO on W identified by Gomer[21]; equation (5) is therefore applicable to this system as well, and may indeed have general applicability to adsorption systems.

The value of the term $\theta_a e^{\Delta S^{\neq}/R}$, and its variation with coverage, remains a problem. The type of reconstruction process involved in the formation of the β_2 structure may proceed most easily at two-dimensional reaction boundaries, i.e. at the interface between reacted and unreacted surface, where the imbalance of surface tension forces provides a shearing force which could motivate the requisite movement of surface nickel atoms. The active sites are then situated at these boundaries, and, ignoring the entropy term, we have $\theta_a \approx 10^{-5}$ to 10^{-6} for the slow process, i.e. the number of active sites is ca. 10^9 to 10^{10} cm^{-2}. Alternatively, ignoring the θ_a term, the factor 10^{-5} to 10^{-6} could be attributed to a coverage dependent entropy of activation of -23 to -28 cal/deg/mole. No attempt is made here to choose between these possible alternatives.

The desorption spectra in figure 3 provide evidence for the existence of the γ' state and the clear distinction between this state and the γ state. Once the surface approaches saturation in the β_1 state, a small, sharp peak is noted in the desorption spectra (curves i to iii, figure 3): this is due to the desorption of the γ' state and its subsequent readsorption on uncovered portions of the film. At higher coverages some β_2 structure is formed on the surface so that adsorption at 78°K involves both the γ' and γ states and the desorption spectra (curves iv, v and vi) therefore show increased complexity: γ' (8 kcal/mole) desorbs first and begins to readsorb before the γ state (coverage dependent adsorption heat, peak at 13 kcal/mole) begins to desorb, providing the second maximum. The third peak could be attributed to a non-equilibrium distribution of the γ state on adsorption at 78°K, as described for anomalies in the desorption spectra for the $\alpha\gamma$ states of nitrogen on tungsten[23] and molybdenum.[24]

3.2.   Carbon monoxide as adsorbate

3.2.1.   Sticking probabilities.

As with oxygen the initial sticking probabilities were found to be very close to unity: s_o was > 0.998 at 78°K, and > 0.98 at 290°K. These are higher than values previously reported for this system, viz 0.6 (Oda[30]), 0.4 (Wagener[31]) and ca. 5 x 10^{-4} (Degras[32]). Two stages can be distinguished in the adsorption process from the variation of s with coverage at 78, 195 and 290°K (figure 6).

(1) At all temperatures s (≈ 1) is initially independent of coverage, indicating the existence of a precursor state capable of migrating large distances into the porous structure of the film over the primary layer adsorbate. The sticking probability then falls at a coverage dependent on the temperature.

(2) This fall is partially arrested at s = 10^{-2}, where a pronounced bend occurs irrespective of the adsorbent temperature. Details of this stage in the process are discussed in §3.2.3.

This behavior was also observed for CO on nickel films at 273°K by Oda, who has apparently also identified a third stage, a "slow process", using the traditional dose-wise gas addition apparatus[33], although Baker and Rideal report that it is insignificant, amounting to less than 2.5% of the total gas adsorbed[34]. It is certainly not comparable with the slow process observed in the oxygen-nickel system.

3.2.2.   Equilibrium pressures and desorption spectra.

The fall in s from unity at 7 x 10^{14} molecules cm^{-2} at 290°K is attributable to desorption from a species with an adsorption heat in the region of 20 kcal/mole, designated the α-species. When the gas supply to the film was closed at higher coverages slow pressure decays were noted, which are indicative of a process involving a species in pseudo-equilibrium with the gas phase coupled with slow adsorption into a more strongly bound state on the surface: in previous studies this type of slow process was identified with the redistribution of weakly bound adsorbate from the outer, more accessible surface of the film into sites of higher binding energy still available in the porous structure of the film[3,35]. In the present case however, the process appears only at relatively high coverages, when adsorbate has already been rapidly distributed into the pores by surface migration, so that it cannot be associated with a slow redistribution of adsorbate from one part of the film to another; an alternative explanation is proposed §3.2.3. At high coverages finite equilibrium pressures were observed after cessation of the gas supply, and linear plots of log P against N in the range 5×10^{-7} torr to 10^{-4} torr and 11.5 x 10^{14} to 14.5×10^{14} molecules cm^{-2} demonstrated the validity of the Temkin isotherm for αCO on Ni. At pressures below 10^{-5} torr the number of molecules in the gas phase is negligible compared with the number adsorbed, and isosteres adhering closely to the constant coverage condition were obtained by simply varying the temperature around the cell (up to 373°K) and recording equilibrium pressures. From a family of such isosteres, coverage-dependent adsorption heats were obtained for the α state from 24 kcal/mole to 17 kcal/mole; these values are in accord with the results of Beeck[36] and Baker and Rideal[36] who reported isosteric heats falling steadily from 35 kcal/mole to 4.5 kcal/mole with increasing coverage. Our results are, however, not in agreement with the recent results of Degras, who reports the desorption of a state with an adsorption heat of only 4 kcal/mole at 340°K[32,37].

A film saturated at room temperature was capable of adsorbing a further increment of CO at 78°K, amounting to about 10% of the total (coverages in figure 6 are not comparable as differences could be partly a result of variations in film thickness). This increment was shown to be distributed between

between two states, a α state with an adsorption heat of <u>ca.</u> 10 kcal/mole and the α state (which is incompletely filled at 290°K), by recording the desorption spectra shown in figure 7. These spectra show that the 78°K increment is not adsorbed in an equilibrium distribution, since some readsorption of the γ state occurs before substantial desorption of the α state is initiated (at about 200°K, figure 7). The γ state appears therefore to have similar properties to the metastable γ' intermediate found for O_2 on Ni.

3.2.3. <u>Second stage adsorption.</u>

It is supposed that in the second stage of the adsorption a strongly bound β state is occupied. During this process, and after its completion adsorption still proceeds into α sites as the ambient pressure increases, in accordance with Temkin isotherm behavior. This is summarized by the scheme:

$$CO(g) \underset{(ii)}{\overset{(i)}{\rightleftharpoons}} \begin{array}{c} CO(\alpha) \\ CO(\beta) \end{array}$$

The following equations apply when the gas supply to the cell is terminated, assuming that the adsorption rate is step (ii) is very much less than that in step (i).

Step (i) : $\log a\,P = k\,N_\alpha$ (Temkin isotherm)

i. e. $\dfrac{d\log P}{dt} = \dfrac{k\,dN_\alpha}{dt}$ (8)

Step (ii) : $\dfrac{dN_\beta}{dt} = -\dfrac{dN_\alpha}{dt} = ZPs$ (9)

Here s is the sticking probability for adsorption into the β state and N_α and N_β are the adsorbed amounts of α and β CO, respectively. Combining equations

(8) and (9) and integrating we have

$$1/P + kZst + 1/P_o$$

where the gas supply is closed at t = 0 and P = P_o. This scheme can then be

verified by evaluating s from the slope of the 1/P against t plot for the slow

decay using an experimental value for k, the slope of the isotherm, and checking

this value of s against the value obtained in the usual manner from equation (1).

For example, at 290°K and N = 10.1 x 10^{14} molecules cm^{-2} s was calculated from

a slow pressure decay as 3 x 10^{-2}, in fair agreement with the value obtained

from figure 6. This kinetic scheme also explains the further observation that

when the gas supply to the film was terminated frequently during a run the

second stage in the adsorption process was no longer apparent in the log s

against N plot.

In runs where the gas supply was not terminated s was found to fall

linearly with N in the range 10^{-2} to 10^{-3}. Since the existence of a mobile

precursor has been established and the efficiency of adsorption into this

precursor state is unity, the second stage must be reaction-controlled at the

β sites. As the rate is temperature independent, we can adapt equation (7)

to this reaction; the linear dependence of s on N indicates that the fraction

of active sites θ_a is in this case simply $(1-\theta_\beta)$, where θ_β is the fractional

coverage in the β state. Thus

$$s = (1-\theta'_\beta)e^{\Delta S^{\neq}/R}$$

and $\Delta S^{\neq}$ -9 cal deg^{-1} $mole^{-1}$.

3.2.4. Disproportionation.

No CO_2 was observed in the gas phase during the interaction of CO with

nickel films deposited onto smooth Pyrex surfaces at 78 to 290°K and at

pressures up to 10^{-5} torr. At 373°K a small amount of disproportionation was

observed at high coverages: thus at a CO coverage of 13.5 x 10^{14} molecules

cm^{-2} and a pressure of 10^{-6} torr with CO flowing into the cell, the CO_2

pressure developed in the cell was 10^{-8} torr. The rate of CO_2 production may
be calculated from the equation

$$\frac{dN_{CO_2}}{dt} = \frac{1}{A}\left(\frac{V}{kT}\frac{dP_{CO_2}}{dt} + FP_{CO_2}\right)$$

(10)

$$\approx F\, P_{CO_2}/A$$

where F (molecules torr^{-1} sec^{-1}) is the conductance of the diffuser, V the
volume of the cell and A the geometric area of the film. This gives a rate
under the above conditions of 10^7 molecules cm^{-2} sec^{-1}.

However, on nickel films deposited onto a Pyrex glass substrate which had
been thoroughly etched in hydroflouric acid, disproportionation was observed
even at 290°K. Thus, on a film sintered at 373°K, at high coverages and CO
pressures rates of 10^8 molecules cm^{-2} sec^{-1} were observed at 290°K.
Gregg and Leach similarly found that abraded nickel surfaces were
more active in disproportionating CO than electro polished surfaces[38] and
attributed this to either (1) a greater dislocation density or (2) a
relatively larger proportion of (111) planes, the only plane found by Leidheiser
and Gwathmey to form a deposit of carbon in CO[39], in the abraded surfaces.

3.3.   Interaction of carbon monoxide and oxygen

3.3.1.   Oxygen preadsorbed

The results in figure 8 illustrate the interaction of gaseous CO with
varying amounts of preadsorbed O_2 at 290°K. In no case was O_2 observed
in the gas phase. A nickel film on which O_2 adsorption had been allowed to
proceed to a coverage of 25×10^{14} molecules cm^{-2} and s = 10^{-3} took up a
further 2.5×10^{14} molecules cm^{-2} of CO, with negligible reaction to CO_2.
However, the formation of CO_2 was found to proceed more rapidly when smaller

amounts of O_2 were preadsorbed. Thus with 20×10^{14} molecules cm^{-2} of O_2
preadsorbed (O_2 adsorption was terminated with s = 0.2) a steady rate of
CO_2 production equal to 7×10^7 molecules cm^{-2} sec^{-1} (calculated from
equation (10)) was achieved, apparently independent of either the CO pressure
or the amount of CO adsorbed. With only 4×10^{14} molecules cm^{-2} of O_2
preadsorbed, adsorption being terminated prior to the sticking probability
minimum (figure 2), the rate of CO_2 production rose very substantially once
the surface approached saturation (figure 8 (iii)), reaching 2.4×10^{10}
molecules cm^{-2} sec^{-1} (cf. the CO inflow rate of 10^{11} molecules cm^{-2} sec^{-1})
when the run was stopped. Clearly the bulk of the preadsorbed oxygen is
capable of taking part in the reaction in this case.

The predominant form of adsorbed oxygen up to the sticking probability
minimum is the β' state, which does not involve incorporation or compound
formation with the surface nickel atoms and possesses a smaller binding
energy than the subsequent β_1 structure. It is therefore not unexpected
that $\beta'O_2$ is capable of reacting with CO, and the inactivity of the β_1
structure is similarly predictable. Thus, as the O_2 coverage is increased
and the β' to β_1 phase change approaches completion over the whole film, the
adsorbate becomes catalytically inert.

In our work, CO was always taken up by films after exposure to O_2,
whereas previous authors have reported no CO adsorption after saturation with
O_2[40,1]. This difference is a consequence of the greater degree of O_2 coverage
achieved in their work where higher O_2 pressures were used. Park and
Farnsworth, using the LEED technique, observed that a Ni (110) surface was
not capable of adsorbing CO after long exposures to O_2, but after short
exposures a new structure was observed with CO[41]; and Pitkethly has recently
demonstrated the reactivity of Ni (111) surfaces to CO after short exposures

to O_2.[8] The reaction to CO_2 after very short O_2 exposures has not been previously reported.

3.3.2. Carbon monoxide preadsorbed.

Results for the adsorption of O_2 and the displacement of CO, both as CO and as CO_2, from nickel films previously saturated with carbon monoxide at 195, 290 and 373°K are shown in figures 9, 10 and 11 respectively. The total amounts of CO and CO_2 displaced during the process can be calculated from the integration of equation (10):

$$N_{CO_2} = (F/A) \int_0^t P_{CO_2} \, dt \qquad (11)$$

The integration was performed graphically and results are tabulated in table 1. At 290 and 373°K the oxygen uptake is large and all the preadsorbed CO is, within experimental error, quantitatively displaced from the surface. At 195°K, however, only about 15% of the preadsorbed CO was displaced, and an equivalent small amount of O_2 was adsorbed: this does not, however, preclude the possibility of further reaction at this temperature with higher O_2 pressures. The reactivity of preadsorbed CO in the presence of gaseous O_2 has been established by Kawasaki et. al.,[1] although the appearance of CO in the gas phase was not noted. Park and Farnsworth observed that CO was quickly displaced from a Ni(110) surface by O_2, giving the usual sequence of NiO structures,[41] while Pitkethly found that on a Ni (111) surface interaction proceeds through a CO_2 pattern before the usual NiO pattern appears.[8] A comparison of figures 9 to 11 with figure 2 in the present work indicates that the O_2 adsorption mechanism on a Co - covered surface differs substantially from that on a clean surface.

At 290°K a plateau is reached in the rates of O_2 adsorption and CO and

CO_2 production over the region where 5 to 20 × 10^{14} molecules cm.$^{-2}$ of oxygen have been added (figure 10). In this region, the rates of CO and CO_2 formation were calculated from equation (11) as 1.6 × 10^{11} and 1.0 × 10^{11} molecules cm.$^{-2}$ sec.$^{-1}$, respectively. The rate of O_2 inflow was 4.3 × 10^{11} molecules cm.$^{-2}$ sec.$^{-1}$, and since O_2 is lost as CO_2 the net rate of O_2 adsorption was 3.8 × 10^{11} molecules cm.$^{-2}$ sec.$^{-1}$; thus in this coverage range, 1.4 molecules of O_2, on average, displace 1 molecule of CO from the surface, either as CO or as CO_2. The behavior at 373°K is similar, except that the replacement is more complex, showing a minimum in the plateau region, and the oxygen sticking probability is higher.

In all runs variations in the CO pressure were fairly closely followed by variations in the CO_2 pressure, which suggests that the mechanisms for the production of the two gases are closely allied. It is possible that both reactions proceed through a common activated complex of the four-centre type proposed to explain isotopic mixing in CO on W[42]:

$$O_2 \ (g) + CO \ (a) \rightarrow \begin{matrix} C\text{---}O \\ | \quad | \\ O\text{---}O \end{matrix} (a) \overset{\nearrow}{\underset{\searrow}{}} \begin{matrix} CO(g) + 2O \ (a) \\ CO_2(g) + O \ (a) \end{matrix}$$

However, it is not possible to rationalize all the results obtained in terms of a single, simple reaction mechanism.

4.   Summary

The initial sticking probability for O_2 on Ni films is very close to unity at 78 to 373°K, and at 78°K the sticking probability s remains close to unity up to high coverages. At 195 to 373°K three stages have been identified in the adsorption process: firstly, s decreases monotonically with increasing coverage; in the second stage, s increases with increasing coverage, reaching a broad maximum;   and the third stage involves a "slow process" with $s = 10^{-5}$ to 10^{-6} and resulting in a further substantial increase in the oxygen coverage.   Explanations are proposed for each of these stages in the adsorption process.

For CO on Ni films the initial value of s is again close to unity, and at 78 to 290°K this high value is maintained up to fairly high CO coverages. Two states are observed in the desorption spectra over the temperature range 78 to 290°K, designated γ and α, with adsorption heats in the region of 10 and 20 kcal/mole, respectively.   Low rates of disproportionation were observed at high coverages.

A film 'saturated' with O_2 at 290°K is capable of adsorbing a small increment of CO, but reaction to CO_2 is insignificant.   The reaction does, however, proceed rapidly if the O_2 preadsorption is terminated before the end of the first stage in the O_2 adsorption process.   When CO is preadsorbed it is quantitatively displaced from the surface by the subsequent addition of O_2, with the appearance of both CO and CO_2 in the gas phase.

Acknowledgments

Thanks are expressed to the Science Research Council for a grant to one of us (A.M.H.); the authors also gratefully acknowledge financial assistance from the University Grants Committee in the purchase of apparatus.

References

1. Kawasaki K., Sugita T. and Ebisawa S., J. Chem. Phys., $\underline{44}$, 2313 (1966).

2. Horgan A. M., and King D. A., Nature $\underline{217}$, 60 (1968).

3. Hayward D. O., King D. A., and Tompkins F. C., Proc. Roy. Soc., $\underline{A297}$, 305 (1967).

4. Wagener S., Brit. J. Appl. Phys., $\underline{1}$, 225 (1964).

5. MacRae A. U., Surface Science $\underline{1}$, 319 (1964).

6. Park R. L. and Farnsworth H. E., J. Appl. Phys., $\underline{35}$, 2220 (1964).

7. Farnsworth H. E. and Park R. L., Surface Science, $\underline{3}$, 287 (1965).

8. Pitkethly R. C., Private communication.

9. Anderson J. S. and Klemperer D. F., Proc. Roy. Soc., $\underline{A258}$, 350 (1960).

10. Quinn C. M. and Roberts M. W., Trans. Faraday Soc., $\underline{60}$, 899 (1964).

11. Quinn C. M. and Roberts M. W., Trans. Faraday Soc., $\underline{61}$, 1775 (1964).

12. Roberts M. W. and Wells B. R., Trans. Faraday Soc., $\underline{62}$, 1608 (1965).

13. Delchar T. A. and Tompkins F. C., Proc. Roy. Soc., $\underline{A300}$, 141 (1967).

14. Ehrlich G., Adv. Catalysis, $\underline{14}$, 255 (1963).

15. Redhead P. A., Vacuum $\underline{13}$, 253 (1964).

16. Singleton J. H., J. Chem. Phys., $\underline{45}$, 2819 (1966).

17. Hayward D. O. and Taylor N., J. Sci. Instr. $\underline{44}$, 327 (1967).

18. Rideal E. K. and Trapnell B. M. W., Proc. Roy. Soc., $\underline{A205}$, 409 (1951).

19. Hayward D. O., King D. A. and Tompkins F. C., Chem. Comm. 178 (1965).

20. Hayward D. O. and Trapnell B. M. W., 'Chemisorption', 2nd ed. (Butterworths, London 1964).

21. Bell A. A. and Gomer R., J. Chem. Phys. $\underline{44}$, 1065 (1966).

22. King D. A., Disc. Faraday Soc. $\underline{41}$, 63 (1966).

23. Hayward D. O., King D. A. and Tompkins F. C., Proc. Roy. Soc. $\underline{A297}$, 321 (1967).

24. King D. A. and Tompkins F. C., Trans. Faraday Soc., $\underline{64}$, 496 (1968).

25. Brennan D. and Graham M. J., Disc. Faraday Soc., $\underline{41}$, 95 (1966).

26. King D. A., Surface Science 9, 375 (1968).

27. Roberts M. W. and Wells B. R., Disc. Faraday Soc., 41, 162 (1966).

28. Müller J., Disc. Faraday Soc., 41, 186 (1966).

29. Bond G. C., 'Catalysis by Metals' (Academic Press, London & New York, 1962).

30. Oda Z., J. Chem. Phys. 25, 592 (1956).

31. Wagener S., J. Phys. Chem. 61, 267 (1957).

32. Degras D. A., Suppl. Nuovo Cim. 5, 408 (1967).

33. Oda Z., Bull. Chem. Soc. Japan 28, 281 (1955).

34. Baker M. Mc.D. and Rideal E. K., Trans. Faraday Soc., 51, 1597 (1955).

35. Hayward D. O., King D. A., Taylor N. and Tompkins F. C., Suppl. Nuovo Cim. 5, 374 (1967).

36. Beeck O., Adv. Catalysis 2, (1950).

37. Degras D. A., Trans. 3rd Intern. Vac. Congr., 2, 673 (1965).

38. Gregg S. J. and Leach H. F., J. Catalysis 6, 308 (1966).

39. Leidheiser H. and Gwathmey A. T., J. Am. Chem. Soc., 70, 1206 (1948).

40. Alexander E. G. and Russell W. W., J. Phys. Chem., 68, 1614 (1964).

41. Park R. L. and Farnsworth H. E., J. Chem. Phys., 40, 2354 (1964).

42. Madey T. E., Yates J. T. and Stern R. C., J. Chem. Phys., 42, 1372 (1965).

Table 1

Figure	Temp. (°K)	Preadsorbed $10^{-14}N_{CO}$*	Desorbed		
			$10^{-14}N_{CO}^*$	$10^{-14}N_{CO_2}^*$	$10^{-14}(N_{CO}+N_{CO_2})^*$
9	195	18	1.7	0.9	2.6
10	290	15	9.3	5.6	14.9
11	373	8.5	3.4	5.7	9.1

* molecules cm^{-2} (geometric area).

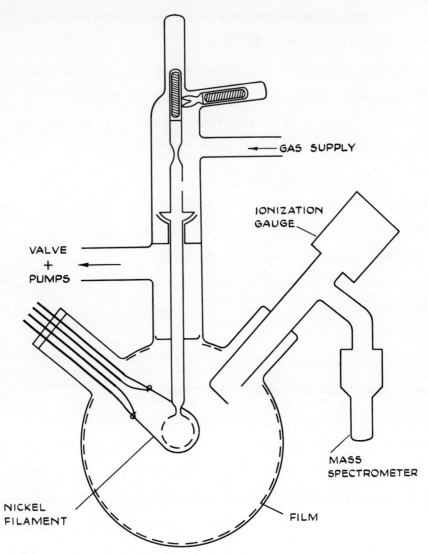

GAS SUPPLY

IONIZATION
GAUGE

VALVE
+
PUMPS

MASS
SPECTROMETER

NICKEL
FILAMENT

FILM

Fig. 1.   The adsorption cell.

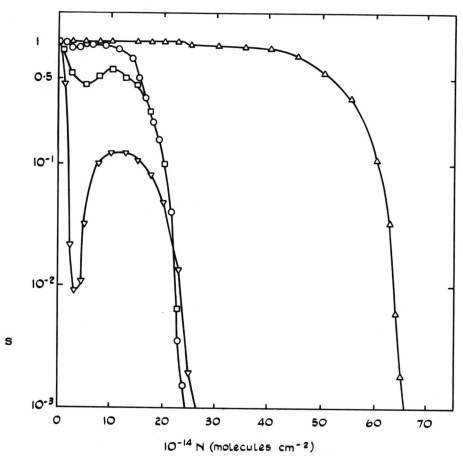

Fig. 2. Variation of sticking probability with coverage for oxygen on nickel films at: △ 78°K; ○ 195°K; □ 290°K; and ▽ 373°K.

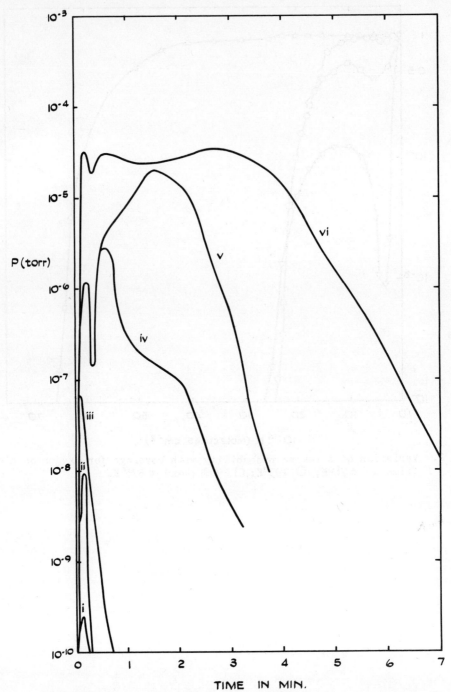

Fig. 3. Desorption spectra for oxygen on a nickel film. Doses added at 78°K.

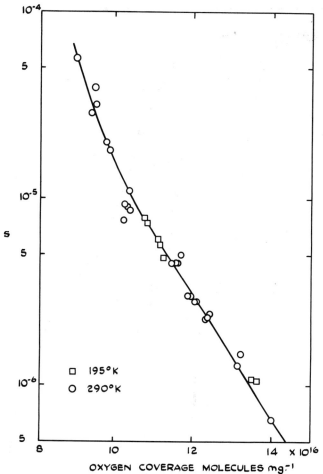

Fig. 4. Slow process for oxygen on nickel: results obtained with conventional dose-wise gas addition apparatus.

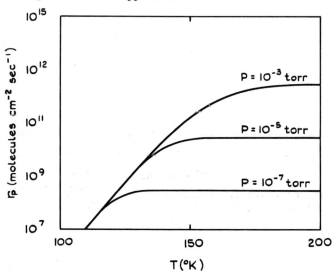

Fig. 5. Theoretical curves for the rate of the slow process from equation (5).

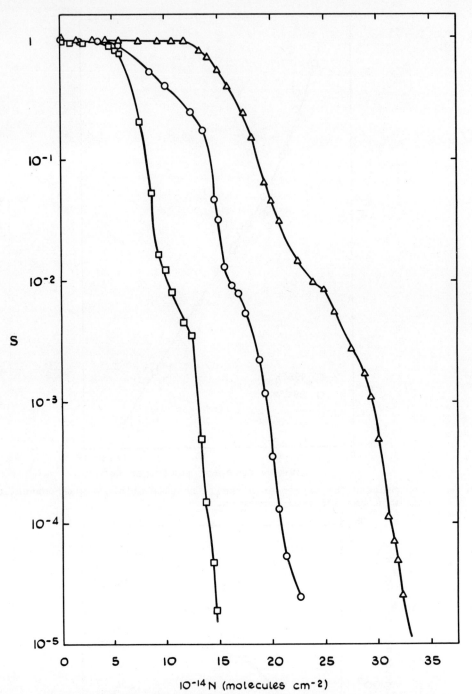

Fig. 6. Variation of sticking probability with coverage for carbon monoxide on nickel films at: △ 78°K; ○ 195°K; and □ 290°K.

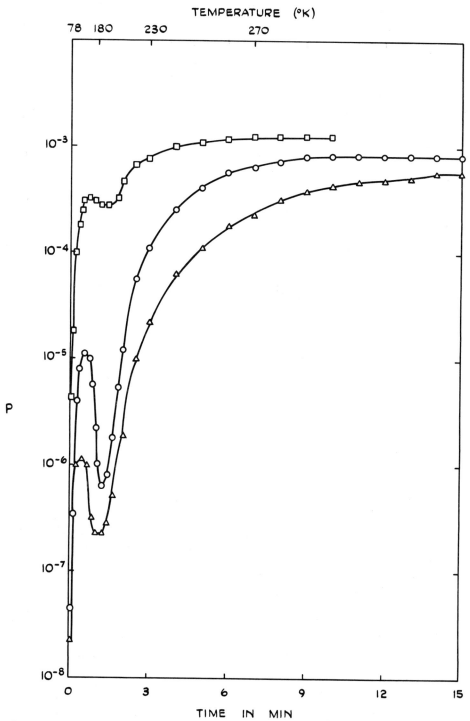

Fig. 7. Desorption spectra for carbon monoxide on a nickel film saturated at 290°K. Doses added successively at 78°K: $\triangle$ 0.9 x 10^{14}; $\bigcirc$ 1.8 x 10^{14}; and $\square$ 3.6 x 10^{14} molecules cm^{-2}.

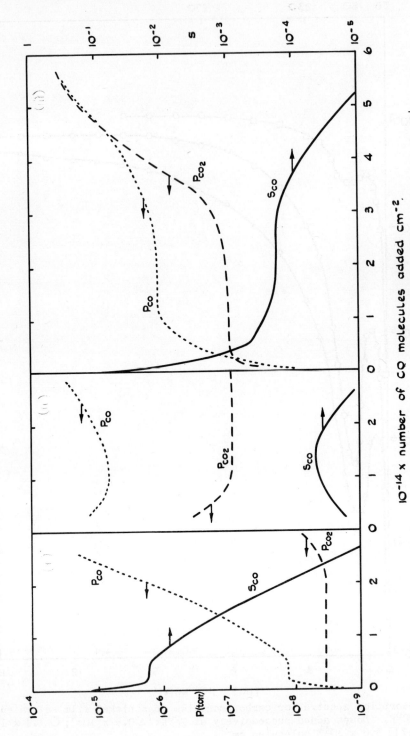

Fig. 8. Interaction of carbon monoxide with preadsorbed oxygen: (i) 25 x 10^14; (ii) 20 x 10^14; and (iii) 4 x 10^14 molecules cm^-2 oxygen preadsorbed.

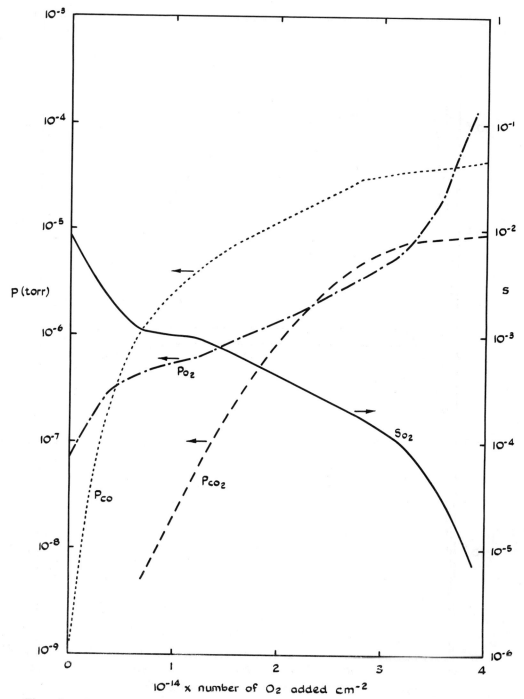

Fig. 9. Interaction of oxygen with preadsorbed carbon monoxide at 195°K.

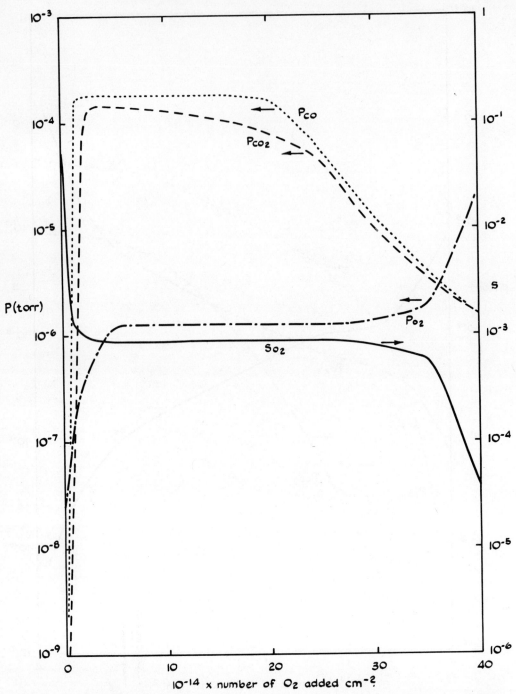

Fig. 10: Interaction of oxygen with preadsorbed carbon monoxide at 290°K.

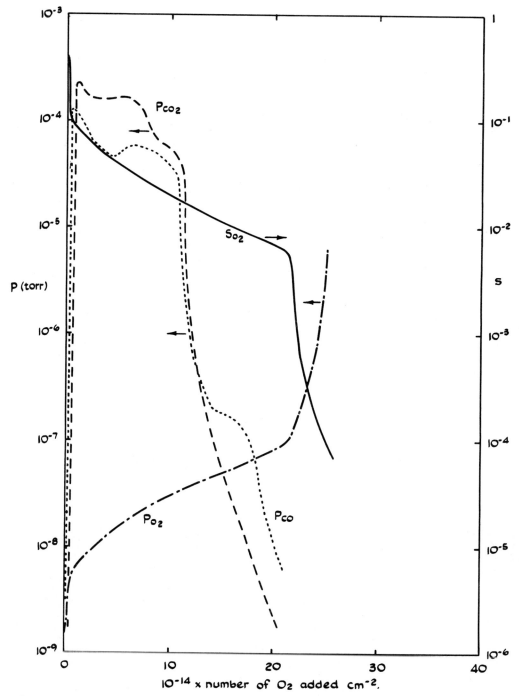

Fig. 11. Interaction of oxygen with preadsorbed carbon monoxide at 373°K.

57–31

Fig. 11. Distribution of oxygen with pressure and carbon monoxide and pressure.

IODINE FACETING OF THE NICKEL (210) SURFACE

Charles W. Tucker, Jr.

General Electric Research and Development Center
Schenectady, New York

Introduction

The formation of facets of different orientation than the original surface
due to the presence on the surface of a reactive gas is not only important to
the basic science of surfaces, but is also important technologically. Low-
energy electron diffraction (LEED) is particularly sensitive to faceting,
providing not only a clear-cut identification of the facets,[1] but often a good
idea of the crystallographic arrangement of the chemisorbed gas on the surface of the facet.[2] It is rather infrequent, however, that the growth or decay
of the facets can be observed in much detail particularly on the same sample.
Allpress and Sanders[3] followed the decay of oxygen-formed facets on silver
using the electron microscope. While much interesting information was
obtained, the method is destructive of the sample, hence the same area may
not be viewed continuously. The present work reports the decay of facets
formed on the nickel (210) surface by iodine and, aside from the scientific
interest of the work, illustrates the power of LEED in following the unfacet-
ing process continuously and in considerable crystallographic detail.

Experimental

The work being reported arose in the course of a general LEED study
of the interaction of the halogens, chlorine, bromine, and iodine with seven
different nickel single crystal orientations. The difficulties of working with
such reactive gases in an ultra-high vacuum system are well-known. However,
by using a flow system and maintaining fairly high pressures of the reactive

gas (ca. 5×10^{-6} torr), it is believed that such difficulties have been minimized, if not eliminated.

The LEED system used was a mercury pumped, all glass system,
except for bakeable valves, and has been described previously.[4] The (210)
crystal was cut with a diamond cut-off wheel from a single crystal rod produced by zone melting a nickel rod (Johnson, Matthey and Company, 99.999%
pure by spectrographic analysis). The (210) surface was prepared by
metallographic polishing and the crystal placed in the diffraction tube. After
various treatments including heating in vacuum, oxygen, and hydrogen and
argon ion bombardment a surface was obtained which gave diffraction patterns
free of extra reflections and maxima in the substrate reflection intensities
at the expected positions given by the fcc structure of nickel. Due to the
very open structure of the (210) surface these maxima are quite sharp[2] and
from experience provide a sensitive test of the cleanliness of the surface.
There was no evidence of facets on the clean surface and it was stable on
heating in good vacuum to the highest temperatures achieved, ca. 1300°C.

On reaction with iodine at ca. 5×10^{-6} torr under certain conditions a
number of ordered structures could be formed. The present paper is concerned, however, with a facet system which forms on slow cooling over a
thirty-minute period in the above iodine ambient from ca. 400°C to room
temperature. After this treatment the original (210) reflections are completely gone indicating that the entire surface has faceted. Thus the facet
system represents a new equilibrium configuration of the surface in the
presence of iodine and does not represent the ordering of steps already
present on the original (210) surface.

Low-Angle Facets

Upon cooling to room temperature after the above treatment and pump-ing out the excess iodine, the (210) surface is found to have broken up into a set of three facets. A pair of crystallographically equivalent facets which form a very low angle (ca. 3-1/2°) with the original (210) surface and a higher angle facet which turns out to be the (540) plane making a larger angle (ca. 12°) with the original (210) surface. The low-angle facets will be dis-cussed in this section and the (540) facets in the next.

When the as-formed facet system was heated for 15-minute intervals at successively higher temperatures, the low-angle facets rotated through ca. 20° about an axis normal to the original (210) surface before disappear-ing at a temperature of ca. 350°C. All observations were made at room temperature and the increments in heating temperature were ca. 50°C start-ing at 150°C. In order to facilitate analysis of the structures at each heating stage, diffraction patterns were taken every 5v from 50 to 200v and every 10v from 200 to 300v. The rotations of the low-angle facet system for each heating stage were roughly equal amounting to 4-5° for each step. It was noted that, aside from a gradual increase in one dimension of the unit cell edge by 10% and a gradual change in the interaxial angle by 20°, the diffrac-tion patterns from the low-angle facets were quite similar. That is, the intensities of corresponding diffraction spots changed rather little from one heating stage to the next, showing that the structure on the facet surface was not changing radically from one heating stage to the next, but rather was gradually distorting from the initial to the final configuration.

In the interest of brevity and because of the similarity of the diffraction patterns at the various heating stages, only one of the diffraction patterns will be discussed in detail. Figure 1a shows the diffraction pattern taken at 85v from the facet system after it had been heated 15 minutes at ca. 200°C.

Figure 1b shows a schematic of this pattern. While the pattern appears rather complicated, it can be understood in complete detail after a study of the patterns taken at the closely spaced voltage intervals mentioned above. In Fig. 1b the solid dots connected by the solid lines show the reflections from one of the low-angle facets while the open circles connected by the dashed lines are the reflections from its crystallographic mate. The facet specular or (00) reflections are designated by $(00)_F$. The other indices indicated in Fig. 1b at the intersections of the dashed and solid lines show the positions and indices of the reflections which would be present from the original (210) surface. The X's show the positions of the (30), (60), and (90) reflections from the (540) facet to be discussed in the next section.

The angle of inclination of the facets may be obtained in two ways. First, from the angular displacement of the $(00)_F$ beams of the facets from the position of the (00) reflection from the unfaceted (210) surface and second from the motion of the facet beams with changing electron wavelength.[1] Both methods were used for all facets observed and gave agreement within experimental error (ca. 1°). The periodicities and interaxial angles in the facet surfaces were obtained from measurements on the geometry of the diffraction patterns.

The unit cells which best fit the experimental observations for four heating temperatures in the range of 150 to 300°C are shown outlined in Fig. 2. In the figure the small open circles represent nickel atoms in the positions they would occupy in the top layer of an fcc crystal (210) plane. The larger open circles represent iodine atoms and only those iodine atoms are shown which mark the four corners of each unit cell. The circles are drawn to a scale such that the nickel atom size corresponds to its size in the metal (2.49 Å) and iodine to its usually accepted ionic size (4.32 Å). It will be noted that the iodine atoms at the corners of the top edge of each unit cell (i.e., at A_0, A_1, ... A_4) are positioned on <u>top</u> of the layer of nickel

atoms and equidistant from each of three nickel atoms in the surface. The

iodine atoms at the bottom edge of each unit cell, however, are lying <u>in</u> the

plane of the surface and are equidistant from three nickel atoms in the

second layer of the substrate. That is, the nickel atoms at the corners of

each unit cell are resting in saddle configurations with three-fold coordina-

tion with respect to the underlying nickel layer. It is interesting to note

that the iodine atom size (4.32 Å) just matches the spacing of the nickel atoms

along the diagonal of the nickel unit cell (4.31 Å). Thus this is a favorable

direction for forming rows of iodine atoms. On the other hand, the spacing

along the a_0-axis is only 3.52 Å and hence sterically unfavorable for rows

of iodine atoms as the closest approach of iodine atoms in the di-iodide[5] is

3.89 Å.

The angle of inclination with the original (210) surface of the unit cells

shown in Fig. 2 is 3.8° and this compares favorably with measured value

of 3.6° given by the two methods mentioned above. The unit cell dimension

perpendicular to the top edges of the four unit cells varies slightly from one

unit cell to the next but averages 13.0 Å which is almost exactly three times

the usual ionic size of iodine of 4.32 Å. It is, therefore, highly probable

that the arrangement of the iodine atoms on the surface of the low-angle

facets consist of rows of iodine atoms approximately parallel to the diagonal

of the nickel unit cell of the substrate. Similarly, the dimensions of the unit

cells along the long axes of the unit cells shown in Fig. 2 are strongly

related to the atomic size of iodine. A detailed study of the positioning of

the iodine atoms in the four unit cells shown in Fig. 2 makes it clear that

the arrangement of the iodine atoms gradually shifts from a slightly dis-

torted square configuration at the lowest heating temperature to a slightly

distorted triangular close-packed arrangement at the highest temperature

just before the facets disappear.

The above interpretation of the diffraction patterns requires that the iodine atoms which are not at the corners of the unit cells be arranged in an out-of-step or coincidence lattice arrangement[4,6] along the two unit cell axes. Such arrangements seem even more likely in the present case due to the fact that the iodine ion is so much larger than the metallic nickel atom (radius ratio 1.74). The coverages are high (~0.9) as one expects for coincidence lattice structures.[7] The behavior of the faceted structures in the vacuum system was in accord with the high coverages. While the clean (210) surface would gradually adsorb ambient gases from the vacuum system and desorb these with relatively mild heating, the iodine covered faceted surfaces showed no sign of adsorbed gases either in their diffraction patterns or upon heating after having remained in the vacuum system for many days. This inertness of the iodine covered facets toward ambient gases also suggests that the iodine is present in the ionic state as one would not expect an ionic surface to be as reactive as a metallic surface. Further support for the coincidence lattice interpretation is the observation that the various reflection intensities change so little from one structure to the next. If the iodine atoms were sitting in different crystallographic positions from one structure to the next rather strong intensity variations would be expected. For a gradually deforming coincidence lattice structure the expected intensity variations from structure to structure would be much weaker.

(540) Facets

In addition to the low-angle facets discussed in the preceding section another facet was produced simultaneously. There were rather few reflections produced by this facet. However, from the motion of the reflections from this facet with changing electron wavelength as they passed through the original position of the (00) reflection from the unfaceted (210) surface,[1] the angle of inclination of the facet was found to be 11.5°. The closest planes

to this angle of inclination are the (430) at 10.3° and the (540) at 12.1°.

However, the observed facet periodicity 7.36 Å seems closely related to

the value of 7.53 Å which is one-third the (540) periodicity of 22.6 Å, while

having no apparent relation to the (430) periodicity of 17.7 Å. The facet is

therefore identified as the (540) plane. It would be desirable to have more

detailed confirmation of this facet but the weakness and paucity of the reflec-

tions prevented more detailed study. It is worth noting that the 7.53 Å

spacing is fairly close to twice the distance of closest approach of iodine

ions in NiI_2 of 3.89 Å.

While the low-angle facets were undergoing the various changes in

orientation, size, and shape under the various heat treatments described in

the previous section, the (540) facets showed little if any change.

Discussion

The three facets formed by iodine on the nickel (210) surface, the two

crystallographically related low-angle facets plus the (540) facet, when put

together form either very shallow pits or pyramids with a highly obtuse

apex angle (ca. 170°). These are somewhat similar to the (731) facets

formed by oxygen under mild (low pressure) treatment on the rhodium (210)

surface.[2] The (731) facets only make an angle of 8.2° with the (210) surface.

While higher pressure oxygen treatments produced (100) and (110) facets on

the rhodium (210) surface and therefore a rather smaller apex angle (135°),

no facet system other than that described could be produced on the nickel

(210) surface. Since the highest iodine pressure used was that in the pre-

sent experiments (ca. 5×10^{-6} torr) and since McKinley[8] has reported that

bromine at 0.3 torr can cause faceting of even the major faces of nickel, it

seems likely that higher pressures of iodine would produce facet systems other than those observed in the present experiments.

Since the heating periods used in passing from one facet configuration
to the next were rather short (15 minutes), the question may be raised as
to whether the observed facet configurations are equilibrium arrangements
or not. Experiments using heating periods of 90 minutes showed that the
facets passed through essentially the same configurations before disappear-
ing at the same temperature. Thus the observed structures appear to be
equilibrium structures. The driving force for moving from one facet configuration to the next seems to arise from the loss of iodine from the surface
as the density of packing of the iodine decreases with increasing heating
temperature for the postulated structures.

Conclusions

From the preceding it is concluded that

1. Iodine at a pressure of ca. 5×10^{-6} torr causes the nickel (210)
surface to facet completely when the crystal is cooled slowly from
400°C in the iodine ambient.

2. The facet system consists of two crystallographically equivalent low-angle facets which make an angle of only 3.8° with the
original (210) surface and a (540) facet with an inclination of 12.1°.

3. On stepwise heating in the absence of an iodine ambient in the
range of 150 to 300°C, the low-angle facets rotate through ca. 20°
about the normal to the original (210) surface. All facets disap-
apear upon heating at 350°C.

4. The facet surfaces have a high coverage of iodine and the iodine
on the low-angle facets appears to be in a coincidence lattice configuration in two directions. The local arrangement of the iodine
atoms changes from a distorted square to a distorted equilateral
triangle during the heating sequence.

5. The facet configurations at each heating stage are equilibrium
structures.

References

A portion of this work was sponsored by the U. S. Air Force Materials Laboratory, Manufacturing Technology Division under Contract No. F33615-67-C1061

1.   C. W. Tucker, Jr., J. Appl. Phys. **38**, 1988 (1967).

2.   C. W. Tucker, Jr., Acta Met. **15**, 1465 (1967).

3.   J. G. Allpress and J. V. Sanders, Phil. Mag. **13**, 609 (1966).

4.   C. W. Tucker, Jr., J. Appl. Phys. **35**, 1897 (1964).

5.   R. W. G. Wyckoff, "Crystal Structures," Vol. 1 pp. 270-272, Interscience Publishers, New York, (1963).

6.   C. W. Tucker, Jr., J. Appl. Phys. **37**, 3013 (1966).

7.   C. W. Tucker, Jr., J. Appl. Phys. **37**, 528 (1966).

8.   J. D. McKinley, Surface Sci. **10**, 287 (1968).

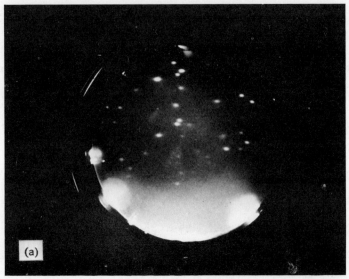

Fig. 1a. Diffraction pattern at 85v from iodine faceted (210) nickel surface after heating 15 minutes at 200°C.

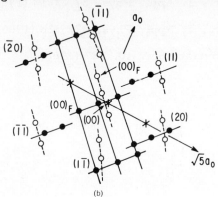

Fig. 1b. Schematic of diffraction pattern of Fig. 1a identifying various reflections. Solid dots connected by solid lines are reflections from
one of the low-angle facets and open circles connected by dashed lines
are reflections from its crystallographic mate. Facet specular reflections denoted by $(00)_F$. Indices at intersections of solid and
dashed lines denote positions of reflections which would be present
from original (210) surface. X's are reflections from (540) facet.

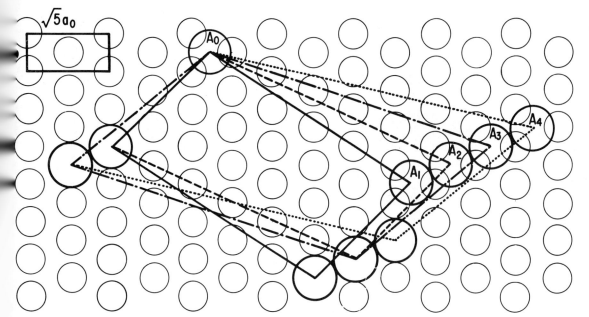

Fig. 2. Positioning of unit cells for low-angle facets on (210) surface. Small circles nickel atoms and large circles iodine atoms. Unit cells for various heating temperatures: —— $150^{\circ}C$, --- $200^{\circ}C$, — · — $250^{\circ}C$, ····· $300^{\circ}C$. Unit cell of (210) surface outlined in upper left corner.

ELECTRON IMPACT STUDY OF THE γ-N_2 STATE CHEMISORBED ON TUNGSTEN

John T. Yates, Jr., and Theodore E. Madey

Surface Chemistry Section
National Bureau of Standards
Washington, D. C.

I. INTRODUCTION

In this paper we are concerned with the effects of low energy electron impact ($<$ 100 eV) on nitrogen chemisorbed on a polycrystalline tungsten surface. Electron impact studies on surfaces containing chemisorbed species may be characterized by the rate and amount of electron impact per unit area, and most previous work in this field falls into either of the categories described below.

(1) Electron impact has been used to characterize various chemisorbed states by measurement of the threshold energy, cross section and energy distribution for positive ion production. This work has generally been carried out for short bombardment times at low current densities, ($\sim$ 0.1-50) x 10^{-6} A/cm^2, in order to minimize the influence of electron bombardment on the chemisorbed layer when viewed as a whole[1-6].

(2) Electron impact at higher current densities (10^{-4} - 10^{-3}) A/cm^2 has been used to cause major perturbations in the chemisorbed layer. This includes electronically induced state-conversion processes and extensive electronic desorption of ions and neutrals[7-9].

The work described here employs extensive electron bombardment in order to cause major changes in the chemisorbed layer. In order to judge the influence of electron impact on the chemisorbed layer, we have employed the flash desorption method which has been successful in distinguishing chemisorbed binding states on the basis of their characteristic binding energies. The use of a polycrystalline tungsten filament as substrate has temperature-homogeneity advantages in flash desorption. In addition, a polycrystalline surface permits the simultaneous exploration of electron impact phenomena on a variety of crystal planes. An extension of this exploratory work to single crystal

surfaces is anticipated.

Our interest in the effect of electron impact on the weakly-held

molecularly chemisorbed γ-N_2 state was prompted by recent papers by

Ermrich[9,10] and Ermrich and van Oostrom[11]. Ermrich found both by

field emission probe hole studies, and by retarding potential studies on

a (100) W macroscopic single crystal that γ-N_2 converted to a new state

(the χ-state) when electron bombarded. This conversion was accompanied

by a large increase in work function ($\Delta \phi^{max}_{(013)}$ = 4.5 eV; $\Delta \phi^{max}_{(100)}$ = 1.7 eV).

Furthermore, the field emission study showed that the transmission

probability for electron emission increased by ~ 8 orders of magnitude

when the χ-state was formed. The effect is specific for γ-N_2, and

clearly indicates that electronic excitation of chemisorbed species can

produce major changes in surface bonding.

Ermrich has shown that the χ-state begins to disappear on heating

to 400 K, and is completely gone at 800 K(as judged by the return of

the work function to a value nearly the same as that observed after ad-

sorption at room temperature). It was also shown that the work function

and transmission probability curves pass through maxima at identical

points during electron impact, and on this basis it was concluded that

extensive electron impact causes depletion of the χ-state, perhaps

by electronic desorption.

We have investigated the electron impact properties of γ-N_2 using

several different methods. Three kinds of information were obtained by

using a quadrupole mass spectrometer detector:

(1) Conventional flash desorption was used to measure

concentration changes in the various binding states as
a function of electron bombardment.

(2) Isotopic mixing effects due to electron impact were
examined in the chemisorbed layer.

(3) Positive ion production by electron impact on the surface layer was monitored by acceleration of the ions
into the mass spectrometer.

II. EXPERIMENTAL *

A. Apparatus

A bakeable Pyrex ultrahigh vacuum system pumped by a mercury
diffusion pump was employed in this work; this system had a limiting
pressure in the 10^{-10} torr range. Prior to adsorption, both $^{14}N_2$ and
$^{15}N_2$ (94.3% $^{15}N_2$; 5.5% $^{14}N^{15}N$) were stored at $\sim$ 0.2 torr in one liter
bulbs coated with Ni getter films to selectively remove impurities.
Spectroscopic grade O_2 was used without additional purification. Prior
to all adsorption experiments, the tungsten filament was cleaned by
flashing to 2440 K.

The experimental vessel is shown in Fig. 1. The sample was
in the form of a 0.28 cm diameter helical coil made of 0.0153 cm diameter
tungsten wire (Sylvania Electric Type NS-55), having a geometrical surface area of 1.18 cm^2. The center section of this coil (approximately
1/4 the total length) was subtended by two 0.0051 cm diameter tungsten
potential leads. Through the use of a Kelvin double bridge circuit and
an electronic temperature programmer, the temperature of the tungsten
helix could be raised almost linearly with time for slow flash desorption. The helix was mounted on stout 0.21 cm diameter tungsten
leads in order to promote rapid   cooling of the helix following a

---

*Certain commercial instruments and materials are identified to
adequately specify experimental procedure. This does not imply recom-
mendation or endorsement by NBS.

flash. The temperature of the helix could be monitored continuously
during cooling using a Keithley Model 503 milliohmmeter, and in control
experiments the helix temperature could be adjusted by ohmic heating to
a value identical to that achieved in electron bombardment.

The center section of the tungsten helix passed across the
line of sight of an Ultek Model 250 quadrupole mass spectrometer (QMS)
which was operated in the axial mode. Between the QMS and the helix were
two planar mesh grids woven of 0.0025 cm diameter tungsten wire; each
grid had an optical transparency of 98%. The helix could be electron
bombarded from an 0.0102 cm diameter thoria-coated tungsten emitter,
2.4 cm in length and located off the QMS axis, about 0.8 cm from the
helix.

Two modes of flash desorption were employed in this work.
Rapid flashing ($\sim 7 \times 10^3$ K per sec) from near 100 K could be carried
out, with the press immersed in liquid N_2, by application of $\sim$ 25 volts
(a.c.) to the helix[12]. Detection of the desorbed gas was carried
out by direct display of the output of the QMS multiplier on an oscilloscope. Rapid pumping, presumably by the cooled glass walls and the
metal leads, was observed for nitrogen under these conditions. A modulated Bayard-Alpert ionization gauge, separated from the experimental
vessel by about 15 cm of 3 cm diameter glass tubing, recorded pressure
pulses which were entirely different in form from the QMS signal. In
particular, the rapid pumping effect was not observed with the ionization
gauge. These differences were probably due to a long average transit
time from the experimental vessel to the ionization gauge, and may be
related to a long residence time of the nitrogen on the cooled parts of

the experimental vessel[13]. For this reason, all pressure pulse measurements were made with the QMS. The ionization gauge was useful in the
steady state mode for calibration of the QMS sensitivity which varied by
$\pm$ 10% throughout this investigation.

The wall effects at low temperature proved bothersome from
another point of view also. Following exposure of the cooled sample to
5×10^{-7} torr N_2 (the typical pressure at which adsorption was carried
out), subsequent evacuation of the system was a slow process. Typically,
it required 5 minutes of pumping to reach $\sim 3 \times 10^{-8}$ torr N_2, and 15 min.
to reach $\sim 4 \times 10^{-9}$ torr N_2. The "in vacuo" electron bombardment experiments discussed later were initiated when the N_2 pressure had dropped
below $\sim 2 \times 10^{-8}$ torr, as measured by the QMS.

In addition to the rapid flash measurements, it was desirable
to do slow flash desorption[14] in some portions of this study. The rate
of heating in these slow flash desorption experiments was 9.7 K per sec.
Because of wall effects at low temperatures, all slow flash desorption
experiments were made after bringing the walls of the vessel and the
tungsten helix to ~ 250 K by appropriate thermostatting of the glass press.

B. Electrical Parameters

For all electron impact experiments designed to produce
major changes in the chemisorbed layer the tungsten helix was maintained
at ground potential; the thoria emitter plus grid no. 1 and the upper
shield and grid no. 2 were at $-V_e$. Values of V_e in the range 45 - 75 V
were used for electron impact and typical electron emission currents, I_e,
were in the range $(175 - 400) \times 10^{-6}$ A. With the press heat-stationed
at 77 K, and with $I_e = 200 \times 10^{-6}$ A and $V_e = 75$ V, the steady state

temperature of the helix was 112 K; without electron bombardment or radiation from the emitter filament, the helix temperature at steady state was 96 K.

For observation of positive ion formation due to electron bombardment of the tungsten helix, the filament of the QMS was turned off, and all electrical conditions were maintained in the QMS except for the electron energy setting, which was reduced from 70 V to ~ 15 V, so as to maximize ion transmission. The tungsten helix was maintained at + 6 V, the electron emitter was at –75 V, grid no. 1 was at –85 V, and grid no. 2 was at ground potential.

III.   RESULTS

A.   Rapid Flash Desorption: Observations of the Effect of Electron Impact on γ-N_2.

In Fig. 2 are shown oscilloscope traces of the partial pressure of $^{15}N_2$ during rapid flash desorption following electron impact on a $^{15}N_2$ layer initially saturated at 96 K (exposure = 2.7 x 10^{-4} torr sec $^{15}N_2$)[15]. The electron bombardment energy was 45 eV. It can be seen from these traces that there is an electron-impact-induced depletion of the low temperature γ-N_2 state; from these and other rapid flash measurements, we have observed a related increase in the magnitude of the pressure pulse from the high temperature state(s). Because of changes of the sensitivity of the QMS between experiments, it is not possible to accurately compare the peak heights between photos. However, a plot of the ratio of the heights of the two desorption peaks versus the total electron impact charge transferred should be insensitive to QMS gain changes. In Fig. 3, such a plot is seen to be a smooth function of the extent of electron impact. Because of the radiant and electron impact heating of

the surface (from 96 K to 114 K in this experiment), it was necessary to

do a control experiment in which ohmic heating of the tungsten helix to

114 K was carried out. The dotted points, obtained after 20 minutes

heating are shown in Fig. 3; these points are plotted at the electron

charge equivalent to 20 minutes bombardment. Thus, thermal desorption

effects are either very small or zero under these conditions.

By comparison of several series of electron impact experiments

on γ-N$_2$, it was possible to make an estimate for the cross section, Q_γ,

for first order depletion of the γ-N$_2$ state; in this calculation we assume

that only one-half of the helix surface area is bombarded. For 45 eV

electron impact, $Q_\gamma = (9 \pm 2) \times 10^{-20} \text{cm}^2$. This is in fair agreement with

Ermrich's cross section measurements at the same electron energy for for-

mation of the χ -state on different planes ($Q_{(016)} = 2.7 \times 10^{-19} \text{cm}^2$;

$Q_{(013)} = 5.2 \times 10^{-19} \text{cm}^2$). We cannot exclude the possibility in our

measurements that Q_γ includes a contribution from electronic desorption

of neutral species as well as from the conversion to χ -nitrogen.

We have previously shown that γ-N$_2$ desorbs thermally without

isotopic mixing of ^{14}N$_2$ and ^{15}N$_2$ to form ^{14}N^{15}N, and this observation led

to an assignment of γ-N$_2$ as a chemisorbed molecular state[16]. It was of

interest to determine whether isotopic mixing occurred in the γ-N$_2$ state

upon electron impact. An equimolar mixture of ^{14}N$_2$ and ^{15}N$_2$ was adsorbed

at 96 K on the tungsten surface after cleaning by flashing in vacuo. In

Fig. 4a, the mass 29 desorption spectrum for rapid flash desorption is

shown; the small γ - ^{14}N ^{15}N peak is consistent with ~ 3% initial abundance

of ^{14}N^{15}N in the equimolar ^{14}N$_2$ + ^{15}N$_2$ gas mixture adsorbed. The relatively large high temperature $^{14}N^{15}N$ peak (β – nitrogen) is produced by complete

isotopic mixing in the β-state(s) as reported previously[16]. In Fig. 4b

the mass 29 desorption spectrum is shown for the same initial equimolar

adsorbed layer, except that 0.25 coulombs of 75 eV electron impact has

occurred prior to flash desorption. A marked increase in the surface

concentration of $^{14}N^{15}N$ in the high temperature peak is observed, while

the low temperature γ – $^{14}N^{15}N$ peak is decreased by $\sim$ 25%. This frac-

tional decrease in the concentration of γ – $^{14}N^{15}N$ is, within ex-

perimental error, equivalent to the depletion of the γ – N_2 state by

0.25 coulombs electron impact. Thus we see that within experimental

error, isotopic mixing does not occur in the γ – N_2 which remains behind

following electron bombardment. However, electron bombardment does

cause a state conversion process from γ – N_2 to a state which desorbs

at higher temperature, and extensive isotopic mixing is observed as

a consequence of this conversion process.

In a separate rapid flash experiment in which $^{14}N_2$ was chemi-

sorbed into the β-states at $\sim$ 350 K, and then $^{15}N_2$ was adsorbed into

the γ – N_2 state at 96 K, it was found that extensive amounts of $^{14}N^{15}N$

could be formed upon electron impact, and that this $^{14}N^{15}N$ was liberated

only in the high temperature region. A control experiment without

electron impact showed no isotopic mixing between the labeled γ – N_2

and β – nitrogen states upon rapid flash desorption.

Thus the isotopic mixing effect is a direct consequence of

electron impact conversion of γ – N_2 to a more strongly bound state. It

cannot be inferred from these experiments that electron impact itself

causes isotopic mixing, since the act of flash desorption may be the cause of the observed mixing.

B. Slow Flash Desorption: High Resolution Observations of the
Flash Desorption Spectra Above 300 K Following Electron
Impact on γ - N$_2$.

Rapid flash desorption spectra are of inherently low resolution compared to slow flash methods. However, slow flash desorption from $\sim$ 100 K is not feasible because of complications arising from liberation of nitrogen from cooled support leads. This complication is avoided to a large extent in the rapid flash method due to the relatively short time period for heat conduction from the filament to the cool support leads during the flash[12].

In order to gain the added resolution of the slow flash method, following electron impact near 100 K, it was necessary to first warm the walls, the support leads and the tungsten helix to $\sim$ 250 K by appropriate thermostatting of the press seal. During this warmup, γ - N$_2$ is liberated from the tungsten helix, and presumably also from the stout support leads. Some physically adsorbed N$_2$ is also liberated from the glass on warming. Upon reaching $\sim$ 250 K and when the background pressure has dropped below $\sim$ 1 x 10^{-8} torr, the desorption is continued by use of the temperature programmer.

In Fig. 5, a series of flash desorption spectra are shown for various amounts of electron bombardment on the surface initially saturated with γ - N$_2$ at 96 K. The cross-hatched desorption profile is characteristic of β - nitrogen, and shows the superposition of the β_2 and β_1 states in agreement with previous observations[14a,16,17]. When electron bombardment of the γ-N$_2$ state occurs prior to slow flash desorption, a slight enhancement of the β_2 and β_1 states is observed; in addition, new strongly chemisorbed states, designated λ -states are produced as shown. The λ -states begin to desorb near 800 K as compared to β_1-desorption which begins near 1050 K. The onset of desorption near 800 K indicates that these states are not the χ-state which Ermrich reports to disappear in the range 400 -800 K. The production of the λ -nitrogen states by electron impact on γ - N_2 may be enhanced (dotted spectrum) by carrying out the electron impact in the presence of gaseous N_2 at $\sim 5 \times 10^{-7}$ torr; presumably this is due to continual repopulation of the γ - N_2. This effect is shown in Fig. 6 where the ratio of the λ -state coverage to the initial $(\beta_1 + \beta_2)$ coverage is plotted versus extent of bombardment in vacuo and in background nitrogen. In both cases the surface concentrations of λ -nitrogen approach plateaus presumably due to depletion of γ - N_2 in the case of vacuum bombardment, and to occupancy of all available sites in the case of bombardment in the presence of background nitrogen.

At least two λ-substates are evident in the dotted desorption spectrum of Fig. 5 with the substate having the lower characteristic desorption temperature forming last. This apparent sequential filling of binding states is analogous to that found in $(\beta_2 + \beta_1)$ nitrogen[14a,16,17] and in other chemisorption systems[14]; the effect may be due in some cases to the influence of thermal redistribution processes as the temperature is raised prior to desorption.

C. Electron Bombardment of β - nitrogen at 300 K.

In Fig. 7, an experiment to detect the influence of extensive electron impact (75 eV, 2.42 coulombs), on the β - nitrogen states at

300 K shows that   within experimental error, no change in β - nitrogen
coverage occurs. However, lengthy exposure of the nitrogen-covered surface to the hot emitter filament, or bombardment for long periods of time
produces a small decrease in the area of the β_1-peak and the approximately
equivalent production of a nitrogen state desorbing at temperatures slightly
below the desorption temperature for β_1 - nitrogen. This effect is small
compared to the magnitude of the λ-state formation which may be achieved
if γ - N_2 is present, and is in fact identical to the effect observed by
Rigby upon deliberate CO poisoning of a nitrogen-saturated tungsten surface at 300 K[18]. In Rigby's work also, no change in total nitrogen
coverage was observed. Thus, we can conclude that 75 eV electron impact
is without effect on β_1 and β_2 nitrogen, at least   within the sensitivity
of this method, and that the small redistribution effect is caused by CO
impurities. The absence of an electron impact effect on β - nitrogen
agrees with Ermrich's field emission measurements on the β - nitrogen
state.

Since for adsorption we are using $^{15}N_2$ containing only $\sim$ 0.2%
$^{14}N_2$, it is possible by monitoring the desorption of mass 28 species to
approximately determine the extent of CO contamination responsible for
the above effect. Exposure of the nitrogen-saturated surface at 300 K
to the heated thoria emitter for $\sim$ 30 minutes in vacuo, followed by slow
flash desorption, yielded a CO desorption pulse equivalent to approximately 1% of the saturated nitrogen coverage. Contamination effects of
this order of magnitude are not unexpected in lengthy experiments in the
presence of a hot filament.

D.   Isotopic Dosing Experiments

A powerful method for studying interactional effects between binding states in the chemisorbed layer involves selective population of the individual states with different isotopic varieties of the adsorbate molecules[16]. Upon flash desorption from a surface isotopically dosed in this manner, interchange between the species in separate binding states may be followed with the mass spectrometer rapidly scanning all pertinent mass peaks during a single flash. Figures 8A and 8B illustrate this effect for the two β - nitrogen states; in Fig. 8A, an approximately equimolar mixture of $^{14}N_2$ and $^{15}N_2$ was chemisorbed into the β-states. On flash desorption, complete isotopic mixing occurred; this is indicated by the constant mole fraction of $^{14}N^{15}N$ liberated throughout the desorption interval ($^{29}X = 0.50$). In contrast to this, when the β_2-state is dosed with $^{14}N_2$, and the β_1-state with $^{15}N_2$, the profiles of the three desorption spectra for the liberated isotopic species of N_2 are dissimilar, and the value of ^{29}X changes over the desorption interval, as shown in Fig. 8B. The two β-substates display a memory for the particular isotopic species with which they were dosed. These experimental observations confirm earlier measurements of the same effect which were made using a slightly different procedure[16]. From considerations of migration energies for chemisorbed nitrogen, it is most likely that this effect arises from a spacial separation of patches of different crystallographic orientation on the polycrystalline tungsten surface, and that the presence of two β-substates is at least partially related to chemisorption on these different crystal planes[16].

Figure 8C illustrates the effect of dosing the β-states with

$^{15}N_2$, the γ - state with $^{14}N_2$, followed by electron bombardment of the isotopically dosed layer. Rapid flash desorption had previously indicated extensive production of $^{14}N^{15}N$ in the high temperature phase by this type of dosing experiment - the slow flash method provides a measure of the distribution of the $^{14}N^{15}N$ throughout the $\lambda + \beta_1 + \beta_2$ states. The nearly constant value of ^{29}X throughout the whole desorption interval strongly suggests that the γ - N_2 species are spread rather uniformly over the crystallographic regions responsible at least in part for the β_1 and β_2 states. This model is substantiated by the results shown in Fig. 5, where a slight overall enhancement of the β states is seen to occur initially upon electron impact of γ - N_2, and prior to extensive filling of the λ - states.

E.   Attempts to Observe the Intermediate χ-State.

Ermrich[9-10] has reported that the χ-phase produced by electron impact on γ - N_2 is not stable above about 400 K; heating in the range 400 - 800 K reduces the work function to values near those observed prior to electron impact. If this work function decrease is due to desorption of χ-nitrogen, as suggested by Ermrich and van Oostrom[11], the desorption should be observed in slow flash desorption measurements starting at 300 K. <u>Our measurements indicate that only α-N_2 is desorbed</u> <u>between 300 K and the onset of λ - nitrogen desorption at $\sim$ 800 K. Thus,</u> <u>the χ-state must undergo a thermal conversion to λ - nitrogen in the</u> <u>range 400 K - 800 K.</u> The peak maximum for the α-N_2 state occurs at 390 K, in good agreement with Rigby's observations[17]. Electron bombardment of γ - N_2 prior to slow flash desorption is accompanied by a decrease in the α-N_2 coverage from its maximum which was $\sim$ 3% of the β - nitrogen

coverage (saturated surface at 96 K). This effect on the sparsely populated α-state was not studied in detail.

Based on Ermrich's work function measurements, the χ-state
can also be depleted by extensive electron impact at ~ 80 K[9,10] and
Ermrich and van Oostrom[11] have suggested that prolonged bombardment
may result in desorption of at least part of the χ-state. In order to
test this hypothesis, we have produced the χ-state by electron impact
on γ - N_2 at ~ 100 K, and have then raised the tungsten temperature to
~ 300 K where additional electron impact was carried out in the absence
of γ -N_2. Upon slow flash desorption following the second bombardment,
no diminution in the concentration of λ - nitrogen was observed when
compared with experiments involving equivalent bombardment only of γ - N_2.
Thus, electron impact on χ-nitrogen at 300 K does not result in desorption.

The measurements reported in this section when combined with
Ermrich's observations suggest that the χ-state converts to the λ - state
by thermal means and also by an electron impact process. With the experimental methods available to us in this apparatus, we have not detected
an effect which can be attributed directly to the presence of the intermediate χ-state. This illustrates one of the deficiencies of the flash
desorption method, namely, that the method depends upon thermal destruction
of the chemisorbed layer, and thermal conversion processes may change the
distribution of the chemisorbed states prior to desorption.

F.   Search for Desorption of Positive Nitrogen Ions by Electron Impact.

A number of investigators[1-6] have shown that when adsorption
of diatomic molecules into certain molecular states occurs, electron
bombardment of these states leads to positive ion production. For this

reason, it was of interest to search for positive ion production upon electron bombardment of the molecular γ - N$_2$ state.

In order to search effectively for positive nitrogen ions, it was necessary to adjust the potentials of the QMS ion source for the maximum transmission of ions produced externally by electron bombardment of the tungsten coil sample. To accomplish this, a control experiment was performed in which the electron beam bombarded a layer of oxygen chemisorbed on the tungsten helix.

The sample was flashed clean in vacuum, and upon cooling to 300 K, was exposed to 1.8×10^{-4} torr sec of oxygen. With the QMS filament cold, and with the sample at + 6 V, V_e = 75 V, Grid 1 at $-V_e$ - 10 V, and Grid 2 at ground potential, the QMS ion source voltages were adjusted to maximize the mass 16 O$^+$ signal electronically desorbed from the sample; i_{16}^+ was 2.7×10^{-9} A for I_e = 120 μa. The maximum efficiency of the QMS system for measurement of oxygen ion production at 75 V was therefore $g_{16}^+ \text{(QMS)} = i_{16}^+/I_e = 2.3 \times 10^{-5}$ ions/electron. In our earlier study in which total ion currents were measured in a different apparatus, it was shown that an exposure of 1.8×10^{-4} torr sec of oxygen on a 300 K poly-crystalline ribbon led to an absolute ion efficiency $g^+ = 7.5 \times 10^{-7}$ ions/electron[19]. The "transmission factor" f of the mass spectrometer system is thus given by $f \approx g_{16}^+ \text{(QMS)}/g^+ \approx 31$.

Electron bombardment of the γ - nitrogen state produced by adsorption of ^{15}N$_2$ at 96 K yielded no detectable mass 15 N$^+$ (the minimum level of detectability was $\sim 2 \times 10^{-11}$ A, limited by current leakage in the QMS multiplier). Similarly, no ^{15}N$_2{}^+$ mass 30 was detected which could not be attributed to ionization of residual gas phase ^{15}N$_2$. In both cases, the maximum possible efficiency of ion production in the

electronic desorption of $^{15}N^+$ and $^{15}N_2^+$ is $g^+_{15,30}$ (QMS) $\lesssim 2 \times 10^{-11}$ A/1.2

$\times 10^{-4}$ A = 1.7×10^{-7} ions/electron. Taking into account the system

"transmission factor" f, the absolute efficiency for desorption of

$^{15}N^+$ and $^{15}N_2^+$ is $g^+_{15,30} \lesssim 5.5 \times 10^{-9}$ ions/electron. Using this figure,

an upper limit on the electronic desorption cross section Q^+ for ionic

desorption of γ - N_2 can be estimated from

$$Q^+ = \frac{g^+}{\sigma_\gamma}$$

where σ_γ is the coverage in the γ state. As Ehrlich as shown[12], the

maximum γ concentration is ~ 4×10^{14} molecules/cm^2, yielding $Q^+ \lesssim$

1.4×10^{-23} cm^2 for N^+ and N_2^+.

IV.   DISCUSSION

A.   Summary of Major Findings.

The discovery of the λ-state, when coupled to Ermrich's findings,

lead to the following schematic representation for the effect of electron

impact on the molecular γ - N_2 state chemisorbed on tungsten:

(1) γ - N_2 $\xrightarrow[\text{impact}]{\text{electron}}$ χ -nitrogen
(~ 100 K)

(2) χ - nitrogen $\xrightarrow[\text{impact}]{\text{electron}}$ $\left\{ \begin{array}{c} \text{enhanced } (\beta_1 + \beta_2) \text{ nitrogen} \\ + \\ \lambda \text{ - nitrogen} \end{array} \right\}$
(~ 100 K)

(3) χ - nitrogen $\xrightarrow{400 \text{ K} \lesssim T \lesssim 800 \text{ K}}$ $\left\{ \begin{array}{c} \text{enhanced } (\beta_1 + \beta_2) \text{ nitrogen} \\ + \\ \lambda \text{ - nitrogen} \end{array} \right\}$

Five other observations of importance are:

(1) γ - N_2 does not undergo isotopic mixing on electron impact.

(2)   Neither N^+ nor N_2^+ are detected on electron bombardment

of γ - N$_2$ ($Q^+ \lesssim 1.4 \times 10^{-23}$ cm^2 for N$^+$ and N$_2^+$ production).

(3) χ -nitrogen is not desorbed thermally.

(4) Electron impact on γ - N$_2$ produces a rather uniform distribution of strongly bound species throughout the λ and β states as judged by isotopic incorporation into these states.

(5) The cross section for depletion of γ - N$_2$ by 45 eV electrons is $(9 \pm 2) \times 10^{-20}$ cm^2.

B. Possible Assignments of the χ -State.

The central problem in interpretation of these and other results involves the identification of the χ - nitrogen state, and the understanding of the $\chi \longrightarrow \lambda$ conversion process. By analogy with gas phase processes, two possible kinds of molecular nitrogen species could be responsible either directly or indirectly for the χ-state, and each has been suggested as a possible species responsible for electron-impact stimulated adsorption:

(1) Metastable Nitrogen States, N$_2^*$

Ermrich and van Oostrom[11] have suggested that χ - nitrogen may be related to electronically excited molecular species similar to one or more triplet states of nitrogen which are known to be produced by electron impact in N$_2$ (g). They point out that the presence of such species possessing a bound level within the energy range of the conduction band would, according to the theory of Duke and Alferieff[20], produce enhancement of electron transmission as they have observed experimentally.

It is well known that most metastable species are

efficiently de-excited by interaction with metal surfaces, with the ejection of an Auger electron. This effect has been used[21] in fact, with a nickel surface to detect the particular triplet species which are invoked in the postulate of Ermrich and van Oostrom[11]. However, in particular cases it may be possible for an excited species to obtain some degree of stabilization by covalent bonding (chemisorption) with the metal; the net result would be that such a downward electronic transition could no longer be considered as a return to the ground electronic state of the adsorbate. By way of analogy, it has been reported[22] that the character of the excited a $^3\Pi$ state of CO (6.2 eV) is "strikingly similar to .... the ground state of the carbonyl group," ($>C = 0$), insofar as vibrational frequency and internuclear distance are concerned. While quantitative agreement of the bond dipole moment for a $^3\Pi$ CO and the carbonyl group is not good, both have a dipole moment much greater than ground state CO.

In this connection, it is of interest to note that vibrational excitation of the lowest triplet state of $N_2(g)$, $A^3\Sigma_u^+$, has been observed in electron impact by Winters[21]. The fundamental frequency is about 1290 cm^{-1} (0.16 eV) as compared to a ground state fundamental for N_2 of 2331 cm^{-1}. This corresponds to a 70% decrease in the force constant for this diatomic. Stabilization of such a species by chemisorption would be expected to produce a state (the χ-state ?) quite dissimilar from the parent γ - N_2 state.

(2) Nitrogen Molecule Ion, N_2^-

Ermrich has suggested that the large increase in work function during χ-state formation may be due to an adsorption-stabilized N_2^- species[10]. Winters[21] has suggested that a particular

double maximum in the excitation function in N_2(g) for the C $^3\Pi_u$ state

may be associated with the formation of a compound state (excited N_2^-),

with threshold near 11 eV. Ermrich's[9] threshold measurements for the

formation of χ-nitrogen indicate that the γ - N_2 conversion begins at

electron energies, V_T, near 5 eV. The <u>maximum</u> energy available for ex-

citation of a surface species is $(V_T + \phi)$ which in this case is $\sim$ 10 eV.

Energetically at least, it is possible that a surface compound C $^3\Pi_u$

state may be involved here; similar electron attachment processes for

other molecular states could also be responsible for the χ-state.

Electron attachment by N_2(g) in its ground state at electron energies

near 2.3 eV has, in fact, been reported by Schultz[23]; however, the

N_2^- is unstable, decaying to a free electron and vibrationally excited

N_2.

It should be emphasized that these analogies based on

the electronic properties of the gas phase molecule are only speculative.

The process of chemisorption can dramatically influence the electronic

properties of the adsorbate, as evidenced by the multiplicity of binding

states (and binding energies) observed in flash desorption, as well as

the shifts in infrared absorption bands of chemisorbed species[24].

C. Characteristics of the $\gamma \longrightarrow \chi \longrightarrow (\lambda + \beta)$ Process.

Ermrich's experiments and ours have shown that the $\gamma \longrightarrow \chi$

conversion is relatively inefficient compared to electron impact pro-

cesses in the gas phase. Thus, Ermrich[9] reports energy dependent

cross sections for χ - nitrogen formation in the 10^{-18}-10^{-19} cm^2 range;

electron bombardment of the order of one coulomb per cm^2 is necessary to produce major amounts of conversion in both our work and Ermrich's. This means either that the primary excitation process on the surface is relatively inefficient, or that efficient de-excitation processes are occurring. The fact that isotopic mixing between $^{14}N_2$ and $^{15}N_2$ does not occur in γ - N_2 upon electron impact on γ - N_2 is consistent with a Franck-Condon model for excitation and de-excitation, with transitions occurring in times short compared to molecular vibration times.

Referring to Fig. 8c, where isotopic dosing of the γ and β states, followed by electron impact, produced a rather uniform distribution of $^{14}N^{15}N$ throughout the $\lambda + \beta_1 + \beta_2$ states, it may be concluded that the γ - N_2 and β - nitrogen species are in close proximity to each other, in agreement with Ehrlich's view of the nature of γ - N_2[25]. It is also seen that the λ - states contain appreciable ^{15}N (as $^{15}N_2$ and $^{14}N^{15}N$) which must originate from the β states which were dosed with $^{15}N_2$. This suggests that during the flash, surface migration leads to mixing between the χ or the λ species and the presorbed β species. Such processes do not occur between the γ and β species because γ - N_2 desorbs prior to the onset of mobility of β - nitrogen.

It has been observed that desorption of positive ions occurs upon electron bombardment of a number of species chemisorbed molecularly on metal surfaces[1-6,19]. In this light, it is somewhat surprising not to see ions desorbed upon electron bombardment of the molecular γ - N_2 state. Two reasons may be offered. The γ - N_2 may be bound in an intimate "lying down" mode leading to rapid reneutralization of electron-bombardment-produced surface ions via transition to the ground state or an intermediate metastable level. Alternatively, there may be preferential excitation of the γ - N_2 to electronic states other than the repulsive part of an ionic potential curve.

References

1. P. A. Redhead, Can. J. Phys. 42, 886, (1964).

2. D. Lichtman and T. R. Kirst, Phys. Lett. 20, 7, (1966).

3. D. Lichtman, R. B. McQuistan, and T. R. Kirst, Surface Science, 5, 120 (1966).

4. J. T. Yates, Jr., T. E. Madey, and J. K. Payn, Nuovo Cimento (Supplemento) N.2, Serie I, Vol. 5, 558 (1967).

5. P. A. Redhead, ibid, 586 (1967).

6. P. A. Redhead, Appl. Phys. Lett. 4, 166 (1964).

7. D. Menzel and R. Gomer, J. Chem. Phys. 41, 3311, (1964).

8. D. Menzel and R. Gomer, J. Chem. Phys. 41, 3329, (1964).

9. W. Ermrich, Phillips Research Reports Suppl., No. 3 (1967).

10. W. Ermrich, Nuovo Cimento (Supplemento) N.2, Serie I, Vol. 5, 582 (1967).

11. W. Ermrich and A. van Oostrom, Solid State Communications, 5, 471 (1967).

12. For example see G. Ehrlich, J. Chem. Phys. 34, 29, (1961).

13. G. Ehrlich, Advan. Catalysis, 14, 255–427 (1963).

14. For example see P. A. Redhead, Trans. Far. Soc., 57, 641 (1961), or

14a. T. E. Madey, J. T. Yates, Jr., J. Chem. Phys. 44, 1675 (1966).

15. ^{15}N$_2$ was used in this work to avoid the possible confusion by mass 28-CO contamination. Experiments to be discussed later showed that only small amounts of contaminant CO are adsorbed after long exposure in the presence of hot filaments.

16. J. T. Yates, Jr. and T. E. Madey, J. Chem. Phys. 43, 1055, (1965).

17. L. J. Rigby, Can. J. Phys., 43, 532 (1965).

18. L. J. Rigby, Can.J.Phys., 42, 1256 (1964).

19. T. E. Madey and J. T. Yates, Jr., Surface Science, in press.

20. C. B. Duke and M. E. Alferieff, J. Chem. Phys. 46, 923 (1967).

21. H. F. Winters, J. Chem. Phys. 43, 926 (1965).

22. R. S. Freund and W. Klemperer, J. Chem. Phys. 43, 2422 (1965).

23. G. J. Schultz, Phys. Rev. 116, 1141 (1959).

     See also D. R. Bates, (ed.), "Atomic and Molecular Processes",
     Academic Press (1962), p. 455.

24. Few infrared studies have been done on surfaces that may be
     considered atomically clean. For one example of an ultrahigh
     vacuum study, see J. F. Harrod, R. W. Roberts and E. F. Rissmann,
     J.Phys.Chem. 71, 343 (1967).

25. G. Ehrlich, Proc. Third International Congress on Catalysis, Vol. I,
     p. 113-145 (1965).

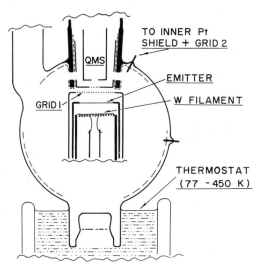

Fig. 1. Apparatus for Electron Bombardment of a Helical W Filament. QMS is a
quadrupole mass spectrometer used for detector in flash desorption and
for monitoring positive ion formation during electron impact on a chemisorbed layer.

RAPID FLASH DESORPTION OF $^{15}N_2$ FOLLOWING
ELECTRON IMPACT (45eV) AT 112 K

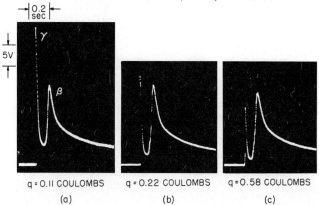

q = 0.11 COULOMBS	q = 0.22 COULOMBS	q = 0.58 COULOMBS
(a)	(b)	(c)

Fig. 2. Rapid Flash Desorption of $^{15}N_2$ Following Electron Impact at (45 eV) at
(45 eV) at 112 K.

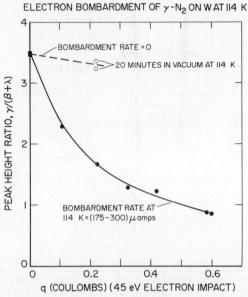

Fig. 3. Effect of Electron Bombardment (45 eV) on the γ - N_2 State.

RAPID FLASH DESORPTION FOLLOWING ELECTRON
IMPACT (75eV) ON EQUIMOLAR $^{14}N_2 + ^{15}N_2$ MIXTURE
(MASS SPECTROMETER ON $^{14}N\,^{15}N$ PEAK)

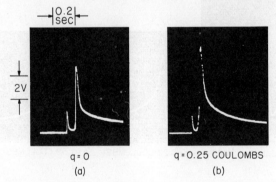

Fig. 4. Rapid Flash Desorption Following Electron Impact (75 eV) on Equimolar $^{14}N_2 + ^{15}N_2$ Mixture at 112 K. (Mass Spectrometer on $^{14}N^{15}N$ peak.)

FORMATION OF λ−NITROGEN STATE
BY ELECTRON IMPACT (75 eV) ON γ-N₂

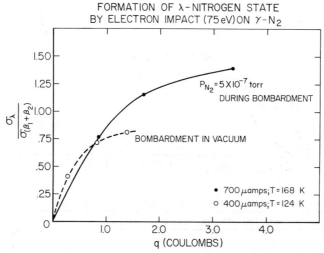

Fig. 5. Formation of λ - Nitrogen States by Electron Impact (75 eV) on γ - N₂. Cross hatched region is normal $\beta_1 + \beta_2$ chemisorption.

FORMATION OF λ−NITROGEN STATE
BY ELECTRON IMPACT (75 eV) ON γ-N₂

Fig. 6. Formation of λ - Nitrogen States by Electron Impact (75 eV) on γ - N₂. On the ordinate, σ represents the enhanced coverage of all species produced by electron impact, and desorbing above ∼ 800 K.

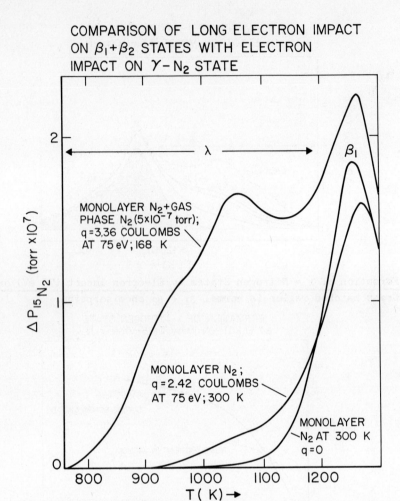

Fig. 7. Comparison of Extensive Electron Impact on $(\beta_1 + \beta_2)$ States at 300 K with Electron Impact on $\gamma - N_2$ State.

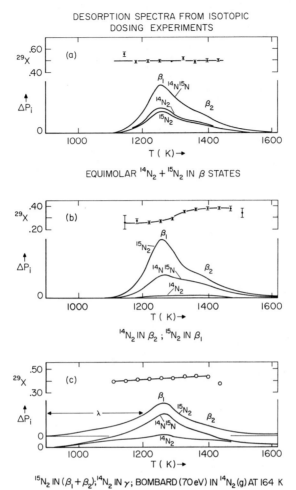

DESORPTION SPECTRA FROM ISOTOPIC
DOSING EXPERIMENTS

EQUIMOLAR $^{14}N_2 + {}^{15}N_2$ IN β STATES

$^{14}N_2$ IN β_2 ; $^{15}N_2$ IN β_1

$^{15}N_2$ IN $(\beta_1 + \beta_2)$; $^{14}N_2$ IN γ; BOMBARD (70 eV) IN $^{14}N_2$ (g) AT 164 K

Fig. 8.   Desorption Spectra From Various Isotopic Dosing Experiments. A. Equimolar $^{14}N_2 + {}^{15}N_2$ Mixture adsorbed in the β-states.
(^{29}X = mole fraction $^{14}N^{15}N$ desorbing) B. $^{14}N_2$ adsorbed in the β_2-state;
$^{15}N_2$ in the β_1-state. C. $^{15}N_2$ in ($\beta_1 + \beta_2$) - states; $^{14}N_2$ in γ - N_2
state. Electron bombard (70 eV) in $^{14}N_2$ (g) at 164 K.

THE INFLUENCE OF ADSORBATES ON THE TOTAL ENERGY DISTRIBUTION OF FIELD EMITTED ELECTRONS*

L. W. Swanson and L. C. Crouser

Field Emission Corp.
McMinnville, Oregon

INTRODUCTION

One of the motivations for field emission total energy distribution (TED) measurements of adsorbate-coated substrates stems from the possibility that the interaction potential of the adsorbate may alter the transmission probability and, hence the TED. This expectation has been given a theoretical basis by Duke and Alferieff[1] (D and A), who treated field emission through adsorbed atoms by an atomistic one-dimensional pseudo-potential model. In the case of metallic adsorption, their results predict an increase in transmission apart from the usual work function lowering. This approach is in contrast to other treatments which assume the adsorbed layer modified the Fowler-Nordheim equation either through the work function term[2,3] or the image potential term.[4] In each of these cases including the D and A theory a change in both the pre-exponential and exponential terms of the Fowler-Nordheim (FN) equation is predicted. The D and A theory is unique in the case of metallic adsorption since it also predicts additional structure in the TED curve when the valence level of the adsorbate falls below the substrate Fermi level.

Recent experimental results have shown the FN theory to be inadequate for describing the TED from the ⟨100⟩ directions of clean molybdenum[5] and tungsten[6] due to band splitting near the Fermi level. All other major crystallographic directions of both molybdenum and tungsten show no gross deviations from the Sommerfeld model, upon which the FN theory is based, in spite of the non-free-electron nature of the metals. The somewhat surprising success of the Sommer-

---
* This work supported by the National Aeronautics and Space Administration, Headquarters, Washington, D. C.

feld model in describing field emission from the transition metals is the result
of the narrow energy band (~0.2 eV) sampled by field emission. Accordingly
the specific effects of the emitter band structure are visible only when unusually
small Fermi surfaces are emitting[7,8] or when band gaps occur within ~0.2eV of
the Fermi level.

Historically the FN theory has been utilized with apparent success to describe the field emission characteristics of both clean and adsorbate coated surfaces. Recent use of probe techniques, which confine work function and energy
distribution measurements to single atomically smooth crytsal face, has eliminated the undesirable averaging and allows a more stringent test of the FN
theory. The evidence that work function changes due to adsorption are also reliably given by the FN equation is now substantial.

At present, two minor modifications to field emission calculated work functions of adsorbed layers have been considered. The first is due to Gomer[2], who
considered the detail shape of the contact potential within the adsorbed layer
and its effect on the FN equation. The graphical method employed by Gomer
showed that the apparent contact potentials obtained from FN analysis are lower
than actual due to the discrete nature of the dipole potentials of the adsorbed
atoms. This correction, which increases linearly with field becomes of significant consequence only at coverages less than 0.1 monolayer and field strength
in excess of 3×10^7 V/cm.

The second correction to field emission work function calculations stems
from the polarization of the adsorbate due to the applied electric field. This
correction has now been substantiated in a number of cases,[9,10,11] and has successfully explained the variation of emitting area upon adsorption. This effect, which
increases the work function ϕ, is linear with electric field F and thereby in
first order appears as an alteration to the preexponential term of the FN equation (3) such that the ratio of the adsorbate coated emitting area A to the clean

emitting area A_o is given by:

$$\log A/A_o = -4.20 \times 10^7 \phi^{1/2} g \pi \sigma \alpha/\epsilon \qquad (1)$$

where α is the adsorbate polarizability in cm^3, σ the adsorbate density in atoms/cm^2, g a factor ranging from 2 to 4 (depending upon the degree of localization of the induced dipole to the adsorbate), and ϵ is the ratio of the applied to actual field at the adsorbate F/F_o. One approximation of ϵ comes from the Topping[12] model in which the value of ϵ is given by $\epsilon = 1 + 9 \alpha \sigma^{3/2}$ and represents the reduction in field at a specified adsorption site due to the accumulated effect of the field induced dipoles in a square array of adsorbed particles. Experimental values of the preexponential changes in the FN equation for both electropositive[12] and electronegative[9,11] adsorbates have given reasonable values of α calculated in terms of equation (1). An additional modification of the FN equation has been given by van Oostron,[4] who considered the alteration of the image potential term by the adsorbate. In this case the adsorbate was treated as a structureless dielectric film of finite thickness which modified the image potential. This modification, which is in the same spirit as the first correction mentioned above, appears to be valid only for multilayer adsorption.

The above mentioned modifications to the FN theory due to adsorption do not predict changes in the TED which cannot be accounted for by corresponding changes in F and/or ϕ. Such is not the case for the D and A one-dimensional pseudopotential model of field emission. In this novel treatment of the effect of absorbed layers on field emission, it is shown that the adsorbed particle is able to affect resonance tunneling of the electrons when the valance level of the adsorbed particle is properly oriented with respect to the Fermi level. Under certain conditions of metallic adsorption additional structure in the TED is predicted along with large enhancements ($10 - 10^4$) in the preexponential term of the FN equation.

It thus should be possible to examine the reliability of the various modifications to the FN theory by simultaneous analysis of the FN equation and TED. In this study we have examined cesium and barium adsorption on a few crystal planes of molybdenum and tungsten with respect to their effect on the TED and current-voltage relationships throughout the monolayer coverage.

EXPERIMENTAL APPROACH

Before proceeding it will be helpful to list the relevant equations pertinent to the current voltage I(V) characteristics of field emitted electrons and the TED. For the Sommerfeld model, upon which the FN theory of field emission is based, the current density per unit total energy $J(\epsilon)$ (where ϵ is the energy E relative to the Fermi energy E_f) is given by[13]:

$$J(\epsilon) = J_o e^{\epsilon/d}/d(1 + e^{\epsilon/pd}), \qquad (2)$$

where $p = kT/d$ and J_o, the total current density (integrated over all ϵ) at $0°K$, is given by[14]:

$$J_o = \frac{1.54 \times 10^6 \, F^2}{\phi \, t^2 \, (y)} \, \exp \, [-6.83 \times 10^7 \phi^{3/2} \, v \, (y)/F] \qquad (3)$$

(The numerical form assumes J_o in A/cm^2, F in V/cm and ϕ in eV). The expression for the parameter d is

$$d = 9.76 \times 10^{-9} \, F/\phi^{1/2} \, t(y) \, (eV). \qquad (4)$$

The image correction terms $t(y)$ and $v(y)$ are slowly varying tabulated functions[14] of the auxiliary variable $y = (e^3 \, F)^{1/2}/\phi$. If the condition $\epsilon/pd < 0$ is met equation (2) can be expressed as

$$J(\epsilon) \cong J_o \, e^{\epsilon/d}/d \qquad (5)$$

In a retarding potential analyzer we note that ϵ can be related to the collector work function ϕ_c and the emitter-to-collector bias potential V_t in the form $\epsilon = \phi_c - V_t$, thereby allowing equation (5) to be written in the working form

$$\log J(\epsilon) = \log \frac{J_o}{d} + (\phi_c - V_t)/ 2.3d. \tag{6}$$

Thus, the value of d can be obtained from the slope $m = 1/2.3d$ of a plot of $\log J(\epsilon)$ versus V_t. Experimental values of $J(\epsilon)$ can be evaluated from the first derivative of the collector current I_c with respect to V_t

$$J(\epsilon) = -n\ dI_c/dV_t, \tag{7}$$

where n is an arbitrary constant of normalization.

It is further observed that a "Fowler-Nordheim plot" of $\log I_o/V^2$ versus $1/V$ yields a slope m_f which is related to the emitter work function ϕ_f and the geometric factor $\beta = F/V$ as follows:

$$m_f = -2.96 \times 10^7\ \phi_f^{3/2}\ s(y)/\beta\ (cm), \tag{8}$$

where $s(y) = 0.943$ over the range of y encountered in practice. The value ϕ_f relative to a reference value $\bar{\phi}$ can be readily established as shown by rewriting equation (8) as follows:

$$\phi_f = \bar{\phi}\ (m_f/m_o)^{2/3} \tag{9}$$

where m_o is the FN plot slope for the reference work function (usually the surfaces average) and the value β is assumed unchanged. Because of the difficulty in determining accurate values of ϕ and β, equation (8) cannot be employed for precise work function calculation; however, as pointed out by Young[15], the expressions for m_e and m_f can be combined to give

$$\phi_e = 3\ m_f\ t(y)/2\ m_e\ Vs(y), \tag{10}$$

where V is the anode voltage. Accordingly, within the framework of the Sommerfeld model a value of work function can be ascertained from combined energy distribution and FN plots from equation (10) which eliminated assumptions concerning $\bar{\phi}$ and β.

It will be our purpose to examine various experimental parameters obtained from TED and FN plots for self-consistency as a function of adsorbate coverage and attempt to interpet the results in terms of the theoretical modifications discussed in the preceeding section.

EXPERIMENTAL PROCEDURE

The experimental energy analyzer tube and procedure have been described previously.[6] Briefly the tube designed by van Oostrom[4], is a retarding potential analyzer equipped with magnetic deflection to align emission from a particular direction with a small aperture in the anode plate. Without magnetic deflection the analyzer tube gave a resoltuion of 15 to 20 mV. Both molybdenum and tungsten substrates were employed in this investigation. Attention was mainly given to those planes perpendicular to the wire axis owing to the slight alteration of the TED when magnetic deflection was employed. Zone melted single crystal wire was employed to fabricate emitters with the desired crystal direction along the wire axis.

The approximate geometric emitting area seen by the 1 mm diameter probe hole was approximately $3.4 \times 10^{-4} \, r^2 \, (cm^2)$, where r is the emitter radius. For emitter radii of 1 to 2×10^{-5} cm employed in this study the emitting area was the order of 3 to $13 \times 10^{-14} \, cm^2$; this corresponds to approximately 30 to 130 substrate atoms contributing to the probe emission from a (100) plane.

Deposition of cesium onto the substrate was accomplished by resistively heating to $1100°K$ a small bead of cesium zeolite fused onto a platinum wire. This method has proven to give a pure flux of cesium ions which can be focused onto the emitter. The barium source consisted of a short segment of iron clad barium wire mounted on Nichrome leads for resistive heating. A four lead tungsten loop supporting the emitter was utilized for heating the emitter to known temperatures. The results reported in this paper were obtained at $77°K$; this was accomplished by immersing the entire analyzer tube in a liquid nitrogen cryostat.

RESULTS

The results of basic significance to this study are (1) the agreement of the TED shapes with equation (2) throughout the coverage range, (2) the variation of A/A_o with adsorbate coverage and (3) the agreement between ϕ_f and ϕ_e as calculated by equations (9) and (10) respectively. The emitting areas are calculated both from the value of the intercepts of the FN plots and from FN slope data alone by a method described elsewhere.[16] Both methods yielded nearly identical A values in this investigation. Our method of analysis was therefore to treat the data according to the FN theory in a self-consistent manner and examine any discrepancies with respect to existing modifications of the FN theory for adsorption.

In obtaining values of ϕ_f via equation (9) for various planes or as a function of coverage it should be emphasized that a uniform field factor β is assumed. Thus, within the framework of the FN model disagreement between ϕ_f and ϕ_e can only be attributed to a difference between the local β_c and the average $\bar{\beta}$ field factors. The apparent change in β can be obtained from the relation $(\phi_e/\phi_f)^{3/2} = \beta_c/\bar{\beta}$.

Figures 1 to 7 show representative TED curves for the various systems investigated along with the theoretical curves derived from the value of d (and hence p) obtained from the best fit slope m_e of the low energy tail. The normalization constant n of equation (7) was adjusted to allow overlap in the latter region. The poor fit on the leading edge and near the peak is due to the limitations imposed by the analyzer resolution. A somewhat disturbing deviation from theory at low values of dI_c/dV_t on the low energy side of some of the experimental TED data is believed to be caused by an artifact of the analyzer and should be ignored. The values of dI_c/dV_t were obtained by monitoring the change in I_c for a 10 mV increment in V_t. Values of σ for the Cs/W,

Cs/Mo and the Ba/W systems were obtained from previously established relation-ships between ϕ and σ[17,18].

Barium on (111) Mo

The TED curves for barium on (111) Mo give a reasonably good fit to equa-tion (2) as observed in Figures 1 – 3. We observed no anomalous humps or shape changes throughout the coverage range that cannot be described by the equation (2) parameters. On the other hand, Fig. 8 shows that the apparent value of $\beta_c/\bar{\beta}$ is not constant, but is first larger and then smaller than unity as the coverage increases. The length of the vertical bars are the extreme values of $\beta_c/\bar{\beta}$ determined at different field strengths and probably represents the un-certainty in experimental values of m_e, since no systematic variation with field strength is observed in these results. Adherence to the FN model yields a local field enhancement $\beta_c/\bar{\beta}$ at $\sigma = 0$ of 14%, $\phi_e = 4.53$ eV and $\phi_f = 4.15$ eV. Several measurements in this laboratory have yielded the somewhat high value of $\phi_e = 4.5 \pm 0.1$ eV for the (111) plane of Mo.

In addition to the variation of $\beta_c/\bar{\beta}$ with σ, the Fig. 8 results also show a significant reduction in the apparent emitting area A/A_o which goes through a slight minimum with increasing σ. It is interesting to note that this mini-mum in A/A_o occurs near the minimum in the ϕ_f versus σ curve.

Cesium on (100) and (110) W

Representative plots of the TED data for cesium on (110) W shown in Figs. 4 and 5 also indicate a reasonably good fit to the theoretical expression given in equation (2) and show no anomalies in their shape with increasing cesium coverage. In contrast to the barium results, $\beta_c/\bar{\beta}$ increases significantly above unit in the mid-coverage range as observed in Fig. 9. As noted previously[6], the occurrence of a value $\beta_c/\bar{\beta} > 1$ for the clean (110) plane is physically unacceptable for this highly faceted region. We therefore suspect some inade-

quacy of the Sommerfeld model for emission from the clean $\langle 110 \rangle$ direction of W.
Also, in constrast to the Ba/Mo results, A/A_o values for both the (110) and
(100) planes are larger than unity at low coverages.

Cesium on (110) Mo

Results for cesium on (110) Mo have been limited to the low coverage region
and are shown in Fig. 10. The Mo results show a value of $\beta_c/\bar{\beta} \cong 1$ for $\sigma < 0.3 \times$
10^{14} atoms/cm^2, in agreement with the (110) W data. Also in agreement with
both the previously discussed systems, the TED data agrees reasonably well with
equation (2) in the limited range of σ investigated.

The most striking contrast with the (110) W results is the much larger increase in A/A_o for (110) Mo at low cesium coverage. It should be pointed out
that the value of $A_o = 1 \times 10^{14}$ atoms/cm^2 is considerably smaller than the (110)
W value of 46×10^{14} atoms/cm^2, in spite of the fact that both emitters possessed
almost identical radii. Within the framework of the FN theory the small value
of A_o can only be explained by a large local magnification; however, this explanation is undesirable in view of $\beta_c/\bar{\beta} < 1$. This raises suspicion that the FN
preexponential term may not fully describe clean (110) Mo emission as in the
case of (110) W.

DISCUSSION

From these preliminary results one may make several observations regarding
the adequacy of the FN model in its original and modified forms to describe
metallic adsorption. First, in each of the systems investigated the apparent
emitting area seen by the probe varied substantially with adsorbate coverage –
a result not explained by the unmodified FN theory. The polarization correction
described by equation (1) predicts a decrease in A/A_o with increasing value
$\phi^{1/2} \sigma$ and has been employed to evaluate an effective value of α for the adsorbed particles. Excluding the values of $A/A_o > 1$ at low cesium coverage
which clearly cannot be rationalized by the polarization model, we have evaluated

the apparent α throughout the monolayer coverage where $A/A_o < 1$ and give the results in Figs. 8 and 9. In spite of the classical and phenomenalogical nature of the polarization correction, it yields a reasonable and constant value of α throughout most of the monolayer coverage for barium and cesium. The values of A/A_o which occur for cesium on (110) Mo, (100) and (110) W at coverages < 0.1 monolayer may be attributed to a resonance tunnelling enhancement of the emission as described by D and A. Inasmuch as cesium at low coverage on the (100) and (110) planes of W and the (110) plane of Mo give a negative value for the difference between the adsorbate ionization potential V_i and work function $V_i - \phi$, one expects a small increase in coverage to align the broadened valence level of the adsorbate with the substrate Fermi level. It is under the latter conditions that maximum resonance tunnelling, manifested by an enhanced apparent emitting area and structure in the TED, is expected.[1] With increasing coverage $V_i - \phi$ becomes larger (in a positive sense) and resonance tunnelling diminishes due to a decrease in the interaction between the tunnelling electrons and the adsorbate valence band. The conspicuous absence of values $A/A_o > 1$ for barium on (111) Mo can be attributed to the larger positive value of $V_i - \phi \cong 1$ eV at zero coverage which positions the adsorbate valence level below the substrate Fermi level, thereby reducing adsorbate induced resonance tunnelling.

One may speculate further that because $V_i - \phi = -0.94$ for cesium on (110) Mo, as opposed to -1.95 for cesium on (110) W, the valence level of the former more favorably aligned with the substrate Fermi level at low coverage. This may account for the much larger values of A/A_o for (110) Mo. Further examination of the unusually small value of A_o for Mo should be performed before firm conclusions based on detailed comparison of the relative magnitudes of the A/A_o values should be made.

Although possible meanings of the variation in $\beta_c/\bar{\beta}$ with adsorbate coverage are manifold, it is basically a measure of the self-consistency between the energy

spectrum description of field emission as given in equation (2) and the I(V)
relationship given in equation (3). Specifically, within the framework of the
FN model β is a geometrical factor which can be influenced by atomic sized perturbations as demonstrated in the field ion microscope. In this sense, large
adsorbates such as cesium on a smooth (110) plane of W or Mo should cause an
increase in $\beta_c/\bar{\beta}$, particularly if clustering occurs; $\beta_c/\bar{\beta}$ should then decrease
as a smooth monolayer is formed. By the same token an atomically rough plane,
such as a (111), may not exhibit such an increase in $\beta_c/\bar{\beta}$, but rather a dee-
crease if the surface of the overlayer is geometrically smoother. Such is
apparently the case if one compares the gross features of the variation in
$\beta_c/\bar{\beta}$ with σ for barium on (111) Mo and cesium on (110) W. It is unlikely that
geometric effects alone can account for the exceptionally large values of
$\beta_c/\bar{\beta}$ in the midcoverage range of cesium on (110) W.

At this point it is important to emphasize that $\beta_c/\bar{\beta}$ is proportional to
$1/m_e^{3/2}$, thus an increase in m_e due to other effects should cause a reduction
in $\beta_c/\bar{\beta}$. According to the D and A theory,[1] resonance tunnelling effects for
adsorbed atoms whose valence level is aligned below the Fermi level are manifested in the TED curves by larger values of m_e. This effect would assert itself in our results as a diminution in $\beta_c/\bar{\beta}$ and may be a contributing cause to
the low and constant values for $\beta_c/\bar{\beta}$ at low cesium coverages on the (110) plane
of Mo and W where the previously mentioned increase in A/A_o occurs.

It is noteworthy that in spite of these results which give some measure of
support to resonance tunnelling at small or negative values of $V_i - \phi$ we find
no apparent field dependent structure in the TED curves in the corresponding
coverage interval as predicted by D and A. Rather, analysis of the Table I to
III tabulations of the data show $m_e \propto 1/F$ in general agreement with the FN model.

CONCLUSIONS

The results of this investigation which are preliminary in nature provide a measure of support for adsorbate induced resonance tunnelling. Primarily this support comes from the increase in the apparent emitting area, since the additional prediction of structure in the TED curve was not observed. The experimental evidence supporting resonance tunnelling occurs only when $V_1 - \phi$ is near zero, as expected. Over the major portion of the monolayer region the variation is apparent in the emitting area and can be described by the field polarization model. From this study it is apparent that field emission TED and $I(V)$ measurements of appropriate surfaces may be potentially useful in shedding light on the nature of adsorbate bonding modes. However, additional investigations of the TED of field emitted electrons for other systems are necessary before a clear picture of the relative importance of the various modifications to the FN theory due to adsorbed layers can be clearly established.

REFERENCES

1. C. Duke and M. Alferieff, J. Chem. Phys. 46, 923 (1967).

2. R. Gomer, J. Chem. Phys. 21, 1869 (1953).

3. R. Gomer, Field Emission and Field Ionization (Harvard University Press, Cambridge, Mass., 1961,) p. 50.

4. A. van Oostrom, Philips Res. Suppl. (Netherlands) 11, 102 (1966).

5. L. Swanson, and L. Crouser, Phys. Rev. Letters 19, 1179 (1967).

6. L. Swanson and L. Crouser, Phys. Rev. 163, (1967).

7. R. Stratton, Phys. Rev. 135A, 794 (1964).

8. F. Itskovich, Soviet Phys. JEPT 23, 945 (1966).

9. D. Menzel and R. Gomer, J. Chem. 41, 3311 (1964).

10. L. Schmidt and R. Gomer, J. Chem. Phys. 42, 3573 (1965).

11. A. Bell and L. Swanson, Surface Sci. 10, 255 (1968).

12. J. Topping, Proc. Rpyal Soc. (London) A114, 67 (1927).

13. R. Young, Phys. Rev. 113, 110 (1959).

14. R. Good and E. Muller, Handbuch der Physik, (S. Flugge, Springer-Verlag, Berlin 1956) Vol. 21, p. 188.

15. R. Young and H. Clark, Appl. Phys. Letters 9, 265 (1966).

16. F. Charbonnier and E. Martin, J. Appl. Phys. 33, 1897 (1962).

17. L. Schmidt, J. Chem. Phys. 46, 3830 (1967).

18. L. Swanson and R. Strayer, J. Chem. Phys. 48, 2421 (1968).

Table I. Results from Ba/Mo(111) where $\bar{\beta} = 1.334 \times 10^4$ cm^{-1}

σ ($\times 10^{14}$) atoms/cm^2	m_f volts	m_e volts^{-1}	V volts	A($\times 10^{-14}$) cm^2	ϕ_f eV	ϕ_f eV
5.5	7141	7.90	823	6.29	2.25	1.83
5.5	7141	5.95	1042	6.41	2.25	1.93
4.1	6201	8.20	755	4.61	2.04	1.68
4.1	6201	6.39	915	4.69	2.04	1.80
3.7	5973	6.58	779	3.36	1.99	1.95
3.7	5973	5.72	891	3.40	1.99	1.97
3.5	6217	6.79	731	3.06	2.10	2.11
3.5	6217	5.57	915	3.11	2.05	2.07
1.8	8476	5.36	1042	5.11	2.52	2.93
1.8	8476	4.48	1267	5.18	2.52	2.53
1.05	11270	4.46	1285	9.20	3.05	3.15
1.05	11270	3.42	1644	9.36	3.05	3.37
0.7	13740	2.65	1553	15.2	3.48	3.87
0.7	13740	2.65	1949	15.4	3.48	3.12
0.5	15830	3.42	1840	25.12	3.82	4.16
0.3	17270	2.92	1949	19.60	4.05	4.75
0.3	17270	3.08	2330	19.80	4.05	4.58
0	17910	3.15	2085	28.09	4.15	4.53

TABLE II. Results from Cs/W(110) where $\bar{\beta} = 2.022 \times 10^4$ cm^{-1}

σ ($\times 10^{14}$) atoms/cm^2	m_f volts	m_e volts^{-1}	V volts	A($\times 10^{-14}$)cm^2	ϕ_f eV	ϕ_e eV
0	19610	2.55	999	46.0	5.82	6.32
3.0	4620	5.17	555	3.7	2.22	2.67
2.6	4100	5.72	484	0.98	2.05	2.47
2.5	3850	5.64	484	1.3	1.96	2.35
2.4	3490	6.30	393	4.4	1.84	2.35
2.2	3010	6.21	368	4.9	1.67	2.20
1.5	2630	6.48	322	4.9	1.52	2.11
1.3	3190	5.80	393	6.6	1.73	2.35
0.3	10310	3.62	1158	42.0	3.79	3.75

Table III. Results from Cs/Mo (110) where $\bar{\beta} = 2.01 \times 10^4$ cm^{-1}

σ ($\times 10^{14}$) atoms/cm^2	m_f volts	m_e volts^{-1}	V volts	A($\times 10^{-14}$) cm^2	ϕ_f eV	ϕ_e eV
0.20	9174	3.731	1117	10.0	3.56	3.66
0.20	9433	3.484	1349	5.5	3.55	3.36
0.20	9433	4.065	1117	5.5	3.55	3.45
0.25	8771	4.386	1069	8.8	3.39	3.11
0.25	8771	4.367	997	8.8	3.39	3.34
0.25	8771	4.149	974	8.8	3.39	3.60
0.30	7462	3.953	997	5.7	3.04	3.18
0.30	7620	4.255	884	5.7	3.04	3.30
0	14960	2.857	1857	0.77	4.83	4.69

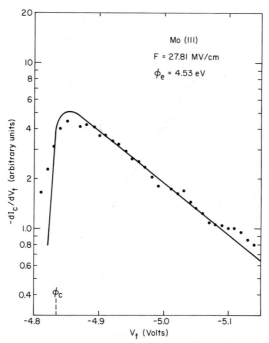

Fig. 1. Solid curve is derived from equation (2) for p = 0.048. Data points are experimental values of the TED for clean (111) Mo.

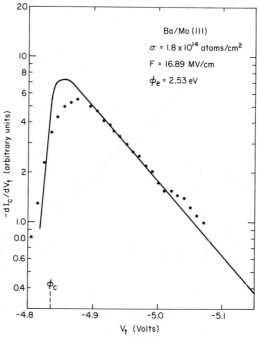

Fig. 2. Solid curve is derived from equation (2) for p = 0.068. Data points are experimental values of the TED for barium on (111) Mo.

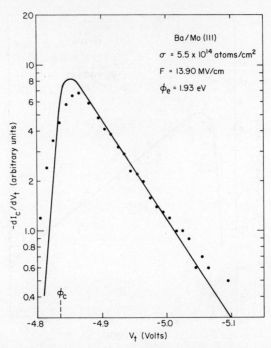

Fig. 3. Solid curve is derived from equation (2) for p = 0.091. Data points are experimental values of the TED for barium on (111) Mo.

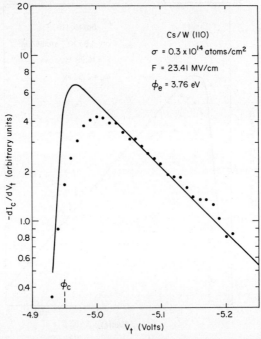

Fig. 4. Solid curve is derived from equation (2) for p = 0.060. Data points are experimental values of the TED for cesium on (110) W.

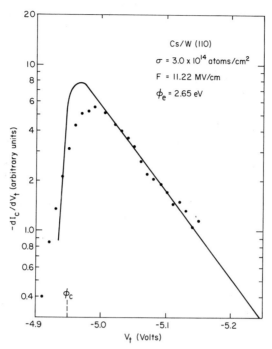

Fig. 5. Solid curve is derived from equation (2) for p = 0.079. Data points are experimental values of the TED for cesium on (110) W.

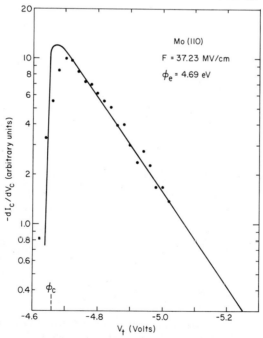

Fig. 6. Solid curve is derived from equation (2) for p = 0.044. Data points are experimental values of the TED for clean (110) Mo.

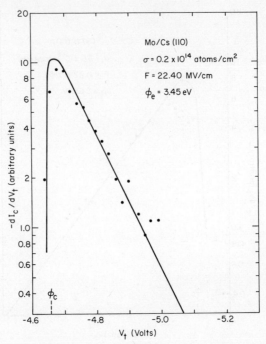

Fig. 7. Solid curve is derived from equation (2) for p = 0.062. Data points are experimental values of the TED for cesium on (110) Mo.

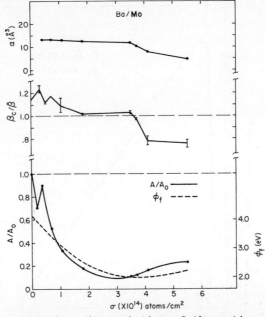

Fig. 8. The lower curves show the variation of the ratio of the adsorbate coated A to clean A_0 emitting area and work function ϕ_f with coverage σ. The middle curve gives the ratio $(\phi_e/\phi_f)^{3/2} = \beta_c/\bar{\beta}$ at various coverages. The upper curve shows the variation of polarizability a calculated according to equation (1). These data were obtained from barium (111) Mo where monolayer coverage occurs near $\sigma = 4.5 \times 10^{14}$.

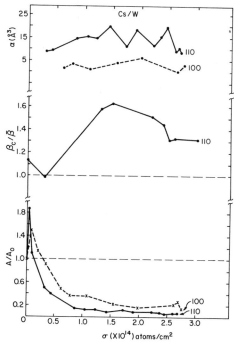

Fig. 9. The lower curves show the variation of the ratio of the adsorbate coated A to clean A_0 emitting area with coverage σ. The middle curve gives the ratio $(\phi_e/\phi_f)^{3/2} = \beta_c/\overline{\beta}$ at various coverages. The upper set of curves show the variation of α calculated according to equation (1). These data were obtained from cesium on (100) and (110) W where monolayer coverage occurs near $\sigma = 2.7 \times 10^{14}$.

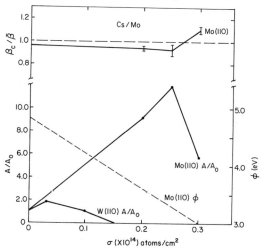

Fig. 10. The lower curves show the variation of the ratio of the adsorbate coated A to clean A_0 emitting area and work function with coverage σ. The upper curve gives the ratio $(\phi_e/\phi_f)^{3/2} = \beta_c/\overline{\beta}$ at various coverages. These data were obtained from cesium on (110) Mo. The value of A/A_0 for (110) W is given for comparison purposes.

SURFACE INTERACTIONS OF BROMINE AND OF HYDROGEN BROMIDE WITH TUNGSTEN

Bruce McCarroll

Research and Development Center
General Electric Co.
Schenectady, New York

I. INTRODUCTION

Unlike with tungsten iodides, a tungsten bromide (WBr_5) can be prepared by direct reaction of bromine with tungsten[1]. However, the surface reactivity of this halogen-metal system is not as yet clearcut. The significance of the results of field emission studies[2] for the bromine-tungsten system is obscured by possible contamination: The pattern obtained after bromine desorption looks very similar to the patterns for CO on tungsten[3]. Even ignoring the possible effects of such contamination, the interpretation of field emission data in terms of chemical reactions at the surface is difficult at best. For example, the granularity observed for the bromine patterns is not necessarily indicative of compound formation at the surface. Indeed, such granularity is observed for oxygen on tungsten at room temperature[4], but other data[5] indicate that the amount of oxide formation is small under these experimental conditions.

The surface interactions of such reactive vapors on a metal surface can be examined conveniently by first permitting the vapor to absorb and then by monitoring the desorbing reaction products with a time-of-flight mass spectrometer as the substrate temperature is increased. The two basic pieces of information obtained from this thermal desorption technique are the identities of the desorbing species and, very often, their desorption kinetics. These data then allow some deductions concerning the processes occurring during adsorption.

Previously reported examinations of the behavior of oxygen[5] and of iodine[6] on tungsten using this technique have provided some empirical observations on the interactions of these reactive vapors with a refractory metal. This report is a continuation of this series and thus details some of the results obtained for the bromine-tungsten and the hydrogen bromide-tungsten systems.

II. EXPERIMENTAL

A. Apparatus

The mass spectrometer and the ancillary gas handling systems are identical to that used in the iodine studies. Briefly, the ionization volume of the time-of-flight mass spectrometer is line of sight with the surface of a tungsten ribbon. This ribbon can be dosed by either of two gases, each issuing from quartz tubes with axial slots parallel to the front surface of the ribbon. Each tube is connected to a separate glass system, thus facilitating the control of two separate gases. One system is used for halogen materials, the other for oxygen. Prior to use, the systems are baked so that the background pressure is in the 10^{-10} mm range.

B. Materials

After repeated freezing and thawing in a stream of helium, a reservoir of degassed liquid bromine was prepared in a breakseal tube. Break seal ampules of liquid bromine were prepared by attaching this reservoir to an ultrahigh vacuum system, baking the system then purifying the bromine and filling ampules as described[6]. A mass spectroscopic analysis of the final product used in these experiments indicate ∼ 0.5 percent iodine as the only observable impurity.

Hydrogen bromide from commercially supplied 1.1 liter flasks was used after the gas had been through six freeze-pump-thaw cycles. Nonetheless, the HBr retained a 3 percent H_2O residual. As this is an exceptionally high impurity level, the results reported here for experiments with this material must be taken as tentative. However, as the qualitative behavior observed is believed to be correct, the results obtained with this gas are included in this report.

The oxygen gas was also supplied by the Matheson Company in 1.1 liter pyrex flasks with breakseals. The oxygen contained only ∼15 ppm impurities; krypton and nitrogen were the two largest.

The decarburized tungsten ribbon is the same one used in the iodine studies: The reaction surface consists mostly of (100) planes together with some (112)-type planes. Other experimental details are as previously described.

C. Data Reduction

Obtaining kinetic data from thermal desorption curves is still a formidable problem. The physical processes of desorption of gas atoms and molecules is often straightforward, but the observed pressure-time signals that experimentally reflect these processes are distorted by interactions of the desorbed species with the vacuum system. An analytic treatment of this problem is usually not feasible. Consequently, a computational approach based on the computer simulation of desorption curves is used.

The model used to analyze the data is described through the equations

$$\dot{n}(t) = -\nu_b \, n(t)^b \, e^{-E/kT(t)}, \quad b = 1,2 ; \tag{1}$$

$$E = E_0 - Xn(t)/n(0), \qquad X \geq 0 \quad ; \tag{2}$$

$$N(t) = \int_0^t e^{-S(t-p)} \, |\dot{n}(p)| \, dp, \, S \geq 0 \quad ; \tag{3}$$

$$I(E_0, X, \nu, S, \alpha; \, t) = |\dot{n}(t)| + \alpha N(t), \, \alpha \geq 0. \tag{4}$$

The first two equations described the desorption kinetics occurring on the sample ribbon as the temperature is raised according to a heating schedule $T(t)$. Equation (2) specifically allows for dependence of the desorption energy on the coverage. It is introduced in this form for convenience as there is presently no simple theoretical model to suggest a better form. Many of the complicating features of more sophisticated desorption models[7] are neglected.

Equation (3) describes the distortion of the desorption rate by the total finite pumping speed S of the vacuum system. It is just the solution of $\dot{N}(t) = F(t) - SN(t)$ for the gas scattered in a chamber with a source rate $F(t)$, assuming instantaneous mixing. The scattered gas density $N(t)$ is combined with the line of sight signal $|\dot{n}(t)|$ to yield the observed mass spectroscopic intensity I of Equation (4). As the scattering is presumed to depend upon the species scattered, the adjustable parameter $a \geq 0$ permits variation of the relative contributions of the two terms. Usually, a varies between zero and one.

Computer simulation with this model and visual matching of experimental data with a computer driven display facilitated deriving the kinetic and system parameters. Generally, much better fits between the simulated curves and the data are obtained when $\nu = kT/h$, rather than treating the frequency factor as a separate parameter.

The coverage dependent heat parameter X usually will not be quoted here; to do so may be misleading. During the calculations, all curves are normalized and the initial (normalized) coverage is set $n(0) = 1$. Consequently $n(t)/n(0)$ is a relative coverage monotonically descreasing from one. As X is derived on this basis, its magnitude is of no significance unless the absolute initial coverage is known. For the results reported here $X/E_o < 0.1$.

III. RESULTS AND CONCLUSIONS

A. Species Identification

Two types of measurements were made to determine the chemical species participating in the gas-surface interactions. The first involved analysis of the vapors desorbing from the surface as a function of tungsten ribbon temperature in a flow of room temperature bromine molecules. The second consisted of examinations of the species evolving upon heating the tungsten ribbon after room temperature adsorption of the halogen.

In both types of experiments, only bromine atoms and molecules are observed. No tungsten containing species (WBr_n , $0 \leq n \leq 6$) are found above the instrumental limitation of $\sim 10^{12}$ particles $\cdot$ cm^{-2}.

The flow experiments were conducted with ribbon temperatures between 300° and $\sim 2600°K$ and bromine impingement rates corresponding to pressures up to $8.7 \cdot 10^{-5}mm$. At the lower temperatures only bromine molecules are observed; at the higher temperatures there is some dissociation. Flash filament examinations of the surface species were made after room temperature exposure of the ribbon to bromine fluxes that could have deposited in excess of 10^4 monolayers in ten minutes. Flashing was done both in the presence of the dosing stream and in the absence of it. In either situation the results are identical: No tungsten halides.

The flow experiments with HBr utilized rates corresponding to pressures as high as $2.9 \cdot 10^{-3}mm$ and a ribbon temperature range of 300° to 2000°K. The only species observed from the ribbon are H_2, Br, HBr and Br_2 . The relative amounts of these materials varied with temperature, but it is not yet possible to specify the fraction of, say Br_2, that arises from reactions on the tungsten ribbon and that from the recombination of Br atoms on parts of the vacuum system.

Thermal desorption experiments on the HBr-W system yield the same four species. However, the origin of these species is more certain than for the flow experiments: The H_2 and Br-atoms originate from the ribbon, the HBr most probably originates there, and the Br_2 does not. This is discussed in detail below.

Examples of the desorption spectra seen for both the Br_2-W and the HBr-W systems are shown in Fig. 1 and Fig. 2, respectively. The Br_2 signals for both systems are artifacts caused by recombination on vacuum system components other than the ribbon.

B.   Thermal Desorption Experiments

    1.   Bromine

As with the iodine-tungsten system, the desorption spectra complicate as the initial coverage increases, as seen in Fig. 1. At low coverages only the higher energy state appears to be populated, and the kinetic
parameters are easily extracted from such desorption data. Using these
parameters, an attempt at resolution of the data taken at higher coverage
is made. Fig. 3 is an example of such an attempt. Note that the data are
plotted in the temperature domain. Only the solid curve is simulated,
the dotted one is obtained by subtracting the simulated desorption curve
from the data curve.

The resolution of the curve into these components is not unreasonable. An examination of other Br atom desorption curves taken
with high initial coverage also reveal the two knees seen in the ascending
part of the data curve in Fig. 3. The location of the maximum of the
difference (dotted) curve yield the same desorption energy on other spectra
taken during different experiments. The desorption energy for this β
state is estimated at 70 kcal.

Granting the existence of the α state with a maximum in the
region indicated by the arrow on the figure, the desorption energy associated with the α state is estimated at 55 kcal. Both is energy and the
β-state desorption energy are calculated under the assumption of a first
order process and through iteration of

$$E \; = \; RT_m \, \ell n (k \, RT_m^{\,3} / h \dot{T} E), \tag{5}$$

in which T_m is the temperature of maximum desorption and $\dot{T}$ is the heating
rate at T_m. The frequency factor normally seen in this recipe[6,8] has
been replaced by kT_m/h.

As mentioned, the Br_2 signal observed during thermal desorption is believed caused by recombination of atoms elsewhere in the vacuum system. This conclusion reached through elimination of the various surface process that could result in the desorption of Br_2 from the ribbon in the temperature range in which the signal is observed.

To eliminate the first process, an estimate of the ratio[9] of the probability for the first order bromine atom desorption p_a to the probability for a second order bromine molecule desorption p_m gives $p_a/p_m \sim 10^{23}/n_a$ at 1500°K. Even at high coverages of atoms, n_a, no second order desorption of bromine molecules should occur.

Desorption of bromine molecules at $\sim$ 1500°K with a first order kinetics implies molecular thermal stability on the surface with respect to dissociation at this temperature. Two methods of estimating such stability on the surface[6,9] give substantially the same result, Fig. 4. These ratios of first order desorption rates are calibrated against the corresponding curve for the oxygen-tungsten system: Use is made of the copious desorption of oxygen atoms from tungsten at T $\gtrsim$ 2100°K. Therefore, ratios below $\sim$ 2 · 10^{18} signify that first order molecular desorption may predominate; ratios above this value imply first order atomic desorption. For bromine, at temperatures above 900°K the diatomic species appears unstable on the surface with respect to dissociation and atomic desorption. Therefore, a first order process for bromine molecule desorption is rejected.

The third process considered for Br_2 desorption is the decomposition of a tungsten-bromine surface compound. The amounts of Br and Br_2 observed during a desorption was measured by comparison with the amount of oxygen desorbed from the same surface after saturation at room temperature. The techniques for such experiments and treatment of the data are as before[6].

Despite the poor precision of the data displayed in Fig. 5, it is apparent that the amount of Br_2 involved is less than a monolayer assuming it all originates from the surface. Even the combined bromine

atom-bromine molecule coverage is still on the order of a monolayer. The
slow increase in the calculated coverage is reproducible, but speculation
as to the cause is not warranted yet.

Invoking the vaporization behavior of the bulk tungsten bromides
certainly does not allow incisive predictions about the behavior of surface compounds. Nonetheless, it is worth mentioning that sublimation or
decomposition with liberation of Br_2 in vacuum and at temperatures below
$\sim 870°K$ characterize the behavior of the known tungsten bromides[1,10-12].
This behavior, taken with the results of the coverage experiments and the
high temperatures at which desorbing Br_2 is observed here appear to rule
out surface compound decomposition as the source of this bromine signal.

2.   Hydrogen Bromide

The results for the thermal desorption studies of the HBr-W
system are still of a preliminary nature. A reference to Fig. 2 indicate
the species of major concern are H_2, Br, Br_2 and HBr.

By use of the arguments given above, the Br_2 signal is taken
as spurious. That is, it does not originate from a surface process
occurring on the tungsten ribbon. On the other hand, all or at least
part of the HBr signal can originate from the ribbon. The results of
the rate ratio calculation for HBr are also displayed in Fig. 4, and
they indicate molecular stability on the surface at temperatures less
than $\sim 1650°K$. Below this temperature the first order desorption of
HBr as a molecular entity is preferred over the desorption of its immediate dissociation products, hydrogen and bromine atoms. Therefore, observation of an HBr desorption signal in the temperature range suggested by
Fig. 4 is not inconsistent with the thermal stability of this species on
the surface.

That the HBr does not result from a second order reaction of
of chemisorbed hydrogen and bromine atoms is indicated by the depletion
of such surface hydrogen by a second order process at a much lower temperature. Therefore, one of the reactants is missing at the time the HBr is
observed.

The adsorption of the hydrogenated atom into the same energy
states as the atom itself is not too unusual; it is also observed for the
adsorption of an oxygen-water mixture. The desorption spectra, Fig. 6
for the oxygen atom (m/e = 16) and water molecule (m/e = 18) show water
molecule occupancy of surface sites with almost the same energetics as
the oxygen atom, judging from the relative positions on the time scale.
The analysis indicated in Fig. 7 suggest two states 118 and 150 kcal for
oxygen atoms[13]. The corresponding high energy states for the water
molecule should be about the same.

To summarize, the chemistry appears uncomplicated by the appearance of tungsten containing species in the pressure regimes studied. At
these low pressures the surface chemistry of the bromine or tungsten
system consists of dissociative chemisorption. The iodine on tungsten
system behaves the same way, but with the retention of some molecular
iodine at the higher coverages. A corresponding retention of bromine
molecules is not observed.

The hydrogen bromide-tungsten system is somewhat peculiar. On
one hand the molecules undergo dissociative chemisorption, and the resulting adatoms behave independently during desorption. On the other, the
hydrogen bromide molecule seems also to adsorb as an entity, binding much
the same as a bromine atom. There is a similar parallel in the oxygen-
water-tungsten system.

REFERENCES

1. H. J. Emeléus and V. Gutmann, J.Chem.Soc. 2115 (1950).

2. M. J. Duell and R. L. Moss, Trans.Faraday Soc. 61, 2262 (1965).

3. G. Ehrlich and F. G.Hudda, J.Chem.Phys. 35, 1421 (1961).

4. See, for example,either R. Gomer and J. K. Hulm, J.Chem.Phys. 21,
    1177 (1953) or T. H. George and P. M. Steir, J.Chem.Phys. 37, 1935
    (1962).

5. B. McCarroll, J.Chem.Phys. 46, 863 (1967).

6. B. McCarroll, J.Chem.Phys. 47, 5077 (1969).

7. For example, K. Erents, W. A. Grant and G. Carter, Vacuum 15,
    529 (1965).

8. G. Ehrlich, Adv. Catalysis 14, 255 (1963).

9. G. Ehrlich, Trans.8th AVS Vac. Symp. (Pergamon Press, Inc.,
    New York (1962), Vol. 8, p. 126.

10. S. A. Shchukarev, G. I. Novikov and G. A. Kokovin, Russ.J.Inorg.
    Chem. 4, 995 (1959); [Zh.neorg.Khim. 4 2185 (1959)].

11. S. A. Shchukarev and G. A. Kokovin, Russ.J.Inorg.Chem. 9, 715
    (1964); [Zh.neorg.Khim. 9, 1309 (1964)].

12. R. E. McCarley and T. M. Brown, J.Am.Chem.Soc. 84, 3216 (1962);
    Inorg.Chem. 3, 1232 (1964).

13. These two values are comparable to the 114 and 144 kcal obtained
    by P. O. Schissel and O. C. Trulson, J.Chem.Phys., 43, 737 (1965)

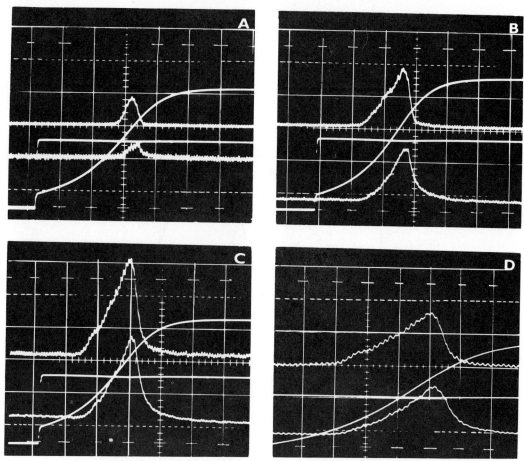

Fig. 1. Mass signals obtained on flashing a tungsten ribbon after its exposure to bromine vapor. The upper trace is the bromine (^{79}Br) signal, the lower is the bromine molecule (^{79}Br ^{81}Br) signal. Examples of low, intermediate and high initial coverage A, B, and C, respectively; 0.2 sec cm^{-1} sweep. D. High coverage, lower heating rate; 0.1 sec cm^{-1} sweep. 20 volt electrons. Also seen are curves proportional to the constant current through the ribbon and to the voltage drop across it.

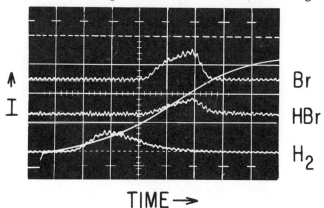

TIME →

Fig. 2. Mass signals (as labeled) obtained on flashing the ribbon at constant current after its exposure to hydrogen bromide vapor. Also shown is a trace proportional to the voltage drop across the ribbon during flashing. 0.1 sec cm^{-1} sweep; 20 eV electrons. A small Br$_2$ signal (not shown) is also seen in a location comparable to that seen in the bromine-tungsten experiments; see text.

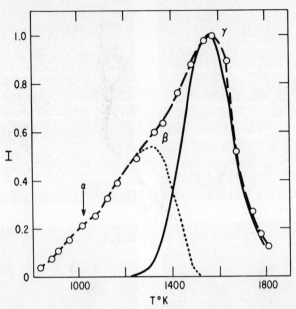

Fig. 3. Bromine atom desorption. Dashed curve, data taken from oscillographic recording; solid curve, computer simulated first order desorption, $\nu = kT/h$, $E_0 = 88$ kcal, $S = 20$ sec^{-1}, $a = 1$; dotted curve, difference between computer curve and data curve, $E_0 = 70$ kcal (Eq. 5). Arrow indicates probable location of a-state maximum, $E_0 = 55$ kcal (Eq. 5).

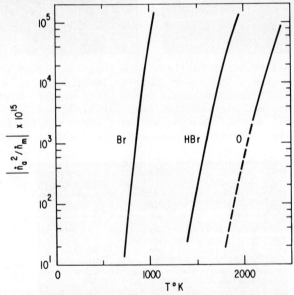

Fig. 4. Plots of the ratio between the evaporation rates $\dot{n}_a$ and $\dot{n}_m$ of the atomic constituents and the molecules. Br, bromine; HBr, hydrogen bromide; O, oxygen. The solid portion of the oxygen curve denotes the region of copious atom evolution.

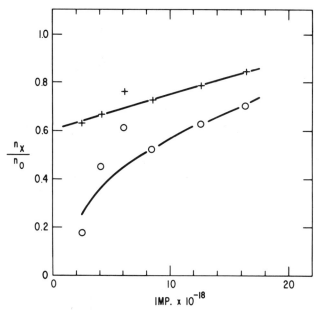

Fig. 5. Plot of the observed amounts of halogen species relative to maximum oxygen atom coverage on the same ribbon as a function of impingement of bromine molecules. Crosses, bromine atoms; circles, bromine molecules.

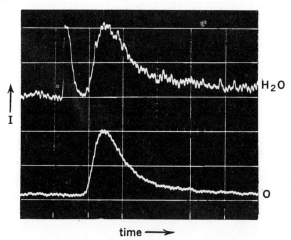

Fig. 6. Mass signals obtained upon flashing a tungsten ribbon exposed to oxygen and water vapor. Upper curve, water ($m/e = 18$); lower curve, oxygen ($m/e = 16$) 0.2 sec cm^{-1} sweep. Maximum temperature $\sim 2500^\circ$K.

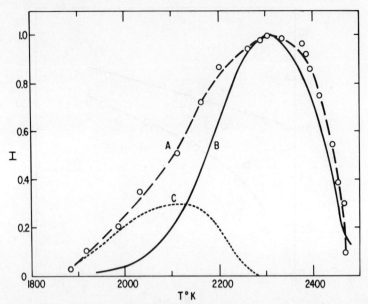

Fig. 7. Analysis of oxygen desorption data. Dashed curve, data taken from an
oscillographic recording of a desorption after maximum initial coverage. Solid curve, first order simulated desorption, E_0 = 150 kcal,
X = 18 kcal deg^{-1} assuming n_0 = 1 monolayer, a = 0. Dotted curve, difference between computer curve and data curve. First order desorption
energy E_0 = 118 kcal (Eq. 5).

STRUCTURE OF THE ANOMALOUSLY BRIGHT ATOMIC GROUPINGS SEEN IN FIELD ION MICROGRAPHS*

H. C. Tong and J. J. Gilman

Department of Metallurgy and Materials Research Laboratory
University of Illinois, Urbana, Illinois

I.   INTRODUCTION

A characteristic feature of field-ion micrographs is a set of unusually bright spots that lie along certain zonal lines of the images of metal crystals. It has been suggested by Müller[1] that the spots are associated with atoms that have moved slightly into metastable sites of low coordination number. The motion is induced by the applied field which polarizes the atoms, and the induced polarization energy stabilizes their temporary positions.

The purpose of this paper is to present micrographs of tungsten in which the fine structures of the groupings have been successfully resolved. This allows the interpretation of their nature to be placed on a more firm basis than before.

II.   EXPERIMENTS

Field-ion specimen tips were made from well-annealed tungsten wire that was electrolytically etched in 1-N sodium hydroxide solution with an alternating 4 V. applied at the electrodes. High purity helium was used for imaging. The vacuum system was baked and evacuated to a static pressure of 10^{-9} torr before making observations. Field evaporation was performed at both liquid nitrogen and liquid helium temperatures, and observations proceeded at the lower temperature. A well resolved micrograph of tungsten is shown in Fig. 1. A distinct band of bright dots appears near the center along the [100] zone.

These bright dots are distributed along the (011) plane edges toward two (010) and (001) poles. They fade out and reduce in size gradually toward the high index

*This research was supported by the U. S. Atomic Energy Commission under Contract AT(11-1)-1198.

plane region (mainly the region of (012) and (013) poles). Eventually, they become approximately the same size and brightness as the image dots of the normal surface atoms in that region. Closer examination of a few of the brightest spots in the first few edges of the (011) planes, shows that each large dot is composed of several smaller ones and most of them are triplets. For the best imaging conditions, these triplets consist of one bright spot associated with two small ones; and the brighter spot lies just outside the (110) plane edge, whereas the two small spots with equal intensity lie on the edge. No previously reported observations have revealed this structure. Therefore, a very careful study of the phenomenon has been made in order to get more exact information about the structures.

The detailed shapes of the above multiplets can be seen in Fig. 2. Their appearance is sensitive to the magnitude of the imaging field, and a series of pictures was taken and is shown in Figs. 2(a), 2(b), and 2(c). The relative contrast of the constituent spots of the triplets depends on the field. At the best imaging field, the outermost spots of the triplets is the brightest, but when the field is reduced slightly, the innermost pair tends to become brightest. Also, some bright singlets become triplets as shown in Fig. 2. On the other hand, when the field is increased, some triplets again become singlets corresponding to the outermost atom. When field is increased still further, all the bright spots become singlets, and the sizes of these bright spots continuously decrease as the field increases, until they become the size of the dots corresponding to normal surface atoms. The dots of the bright zone located near the higher index planes cannot be clearly resolved as multiplets. Their brightness and size gradually decrease to those of normal

atoms located in that region. If only the brightest spots (including the
multiplets) are considered, the reduction of brightness and size is
continuous; i.e., no sudden decrease occurs during a smooth change of the
field.

III. DISCUSSION

From Fig. 2 the bright dots near the (011) pole appear as multiplets
(mostly triplets) with certain definite shapes. Also, when the field is reduced
a few percent, single bright dots just below the triplets become triplets.
Therefore, it is concluded that the bright dots do not correspond to
individual atoms, but to clusters of atoms. The clusters are resolved
completely only at the proper imaging field. At higher fields than the best
one, the bright dots only represent the outermost atoms of the multiplets.
Consider the atomic arrangement of the triplet, A, B, and C shown in Fig. 3(a).
The electric field should be weak at positions B and C relative to their side
neighbors because atom A shields them somewhat from the positive charge
distribution. Therefore, the field-ion image of atoms B and C should not
appear if the other atoms along the ledge cannot be seen. Then, only the
field-ion image corresponding to atom A will be seen.

Müller has suggested that A-type atoms move into metastable sites D,
thereby accounting for the unusually bright dots along the zone lines.[1]
But the resolved multiplets of Figs. 1 and 2 indicate that what happens is
more complex. Therefore, we suggest the following mechanism. A multiplet
is assumed to be a single unit bonded molecularly at a metastable site in the
presence of the imaging field. Taking one of the triplets as an example, the

atom at D is the outermost and the atoms at D, E, and F become more positively

charged than do the atoms at A, B, and C under the same field. This accounts

for the appearance of the triplets. The same argument can be applied to other

multiplets. Using the quasi-chemical bond model,[2] and an additional

assumption, Müller gave the energy relationship for a single atom at a metastable

site[1] as:

$$\alpha_3 F^2 - \alpha_4 F^2 \geq \frac{\Lambda}{2}$$

where Λ is the evaporation energy and α_3 and α_4 are the polarizabilities of

the atoms at sites A and D respectively. F is the imaging field. The

multiplets may be classified into the four types shown in Fig. 4; and their

energy relationships are as follows:

triplet: $\alpha_3 F^2 - \alpha_4 F^2 \geq \frac{\Lambda}{6}$

quadruplet: $\alpha_3 F^2 - \alpha_4 F^2 \geq 0$

quintuplet: $\alpha_3 F^2 - \alpha_4 F^2 \geq \frac{\Lambda}{10}$.

From these relationships, multiplets are favored over singlets if α_3 is taken

to be greater than α_4 as was suggested by Müller.

The relative contrast among the constituent spots of the triplets shown

in Fig. 2 change dramatically when the electric field changes only a few percent.

Since the image intensity of an atom is determined not so much by its degree

of protrusion, as by its specific surface charge-density,[3] this fact suggests

that the local electron density-distribution is sensitive to the imaging

field.

It should be noted that observations of triatomic metastable molecules of niobium nitride have been reported.[4] Three atomic image spots combine to form a triad in which two atoms often have equal intensity. Also, the shapes of the triads change with the imaging voltage, and the triads field-evaporate as whole units. Thus they resemble the triplets that we have observed along the zone lines. This supports our suggestion that the present multiplets are multiatomic molecules at metastable sites.

REFERENCES

1. E. W. Müller, Zeitschrift fur Physikalische Chemie Neue Folge, 53, 1 (1967).

2. I. N. Stranski, Z. Physik. Chem. Abt. B. 11, 342 (1931).

3. E. W. Müller, Surface Sci. 7, 462 (1967).

4. T. T. Tsong and E. W. Müller, 13th Field Emission Symposium, Cornell University (1966).

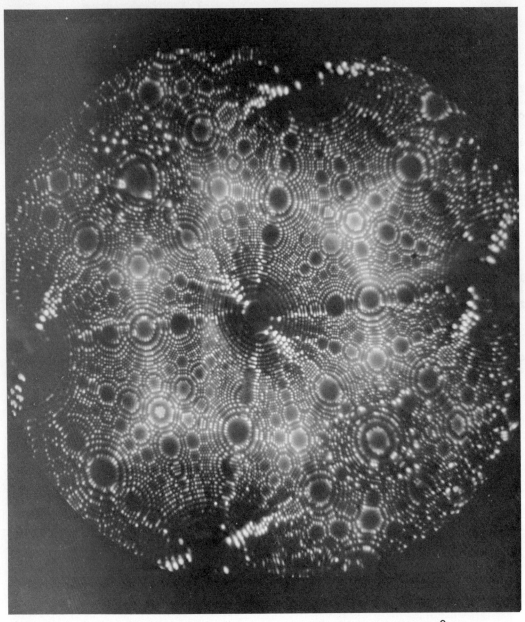

Fig. 1. Helium ion image of a tungsten hemisphere of radius 550Å. Five decorated (100) zone lines are shown around the five {100} poles.

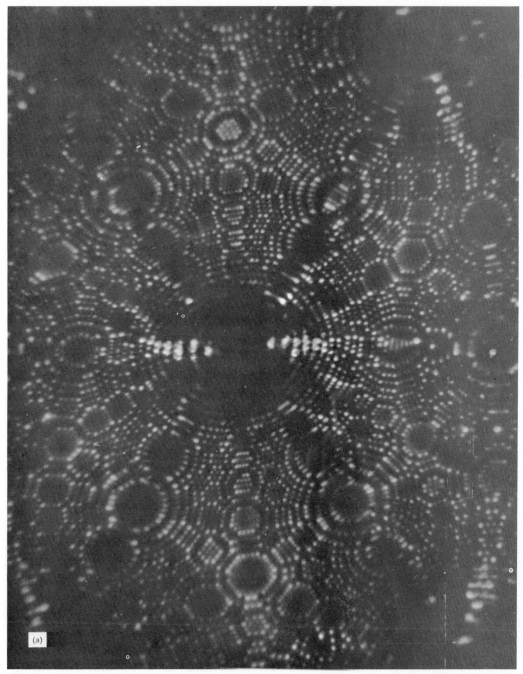

Fig. 2. The field ion image of the decorating dots along ⟨100⟩ zone line.
a. Tip at 16750 volts. b. Tip at 17250 volts. c. Tip at 17750 volts.

(b)

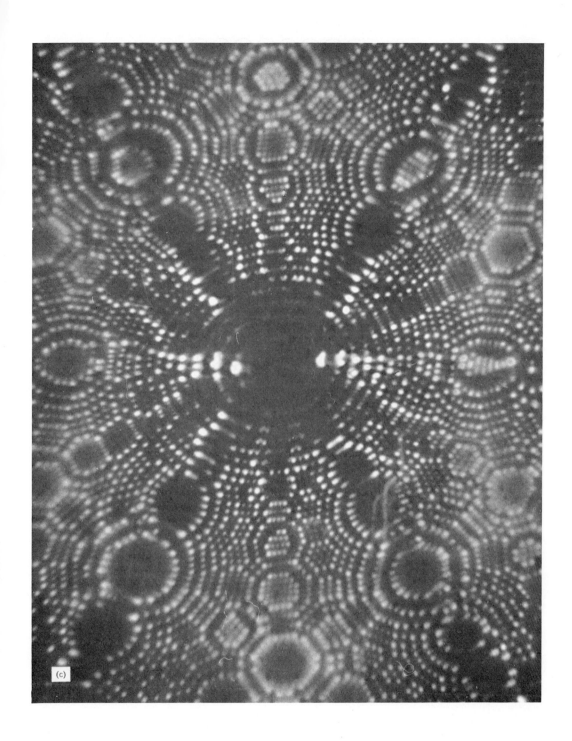

(c)

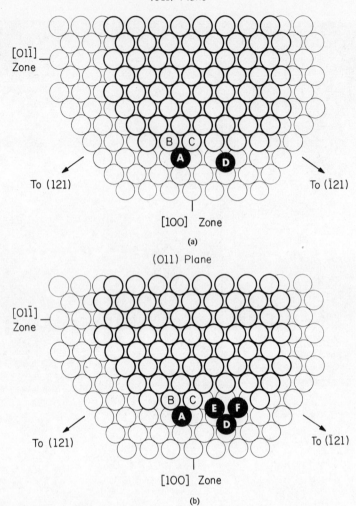

Fig. 3. a. Diagram of two successive (011) planes of the bcc tungsten structure. Atoms at A, B, and C are at normal atomic sites. Atom at D is at a metastable site, and forms a singlet. b. Diagram of two successive (011) planes of the bcc tungsten structure. Atoms at A, B, and C are normal atomic sites. Atoms at D, E, and F form a triplet, and lie in metastable sites.

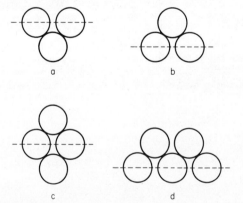

Fig. 4. Diagram of four type multiplets. a. Triplet. b. Triplet. c. Quadruplet. d. Quintuplet.

LEED INVESTIGATION OF THE CARBURIZATION OF TUNGSTEN SINGLE CRYSTALS ABOVE 2000°K

M. Boudart and D. F. Ollis

Department of Chemical Engineering
Stanford University, Stanford, California

INTRODUCTION

The decomposition of hydrocarbons on an incandescent tungsten filament has been studied by many workers since the early work of Andrews and Dushman (1-3). The present LEED work was undertaken to study the influence of crystal surface structure on the reactivity of tungsten and tungsten carbide toward the decomposition of light hydrocarbons above 2000°K.

A detailed investigation of the tungsten-hydrocarbon reaction using LEED is particularly interesting for two reasons. First, tungsten contamination by impurities usually attributed to carbon has been reported often (5-12); thus a related LEED study of the behavior of carbon on tungsten under reactive conditions appeared to be of general interest.

Second, we have recently examined the decomposition of methane, propane, cyclopropane, and tetramethylmethane on a tungsten wire under conditions where the sticking probability of the hydrocarbon could be directly measured (4). This work was done at wire temperatures of 2200-2650°K and reactant pressures of less than 10^{-3} torr. Since these conditions may be easily duplicated in the LEED system, a direct comparison of the reactivity of polycrystalline vs. monocrystalline surfaces is possible.

In this previous study, the sticking probability was defined to be the probability per gas-solid collision that the room temperature gas reactant striking the hot tungsten surface would undergo complete decomposition. The observed reaction product in the decomposition was the hexagonal (alpha) ditungsten carbide (W_2C) which is stable on the filament up to 2700°K (13). X-ray analysis of the carburized wires showed that this compound was the only carbide present. (No beta-ditungsten carbide or monotungsten carbide lines were observed).

The sticking probability could be calculated from the change in wire resistance which measured the amount of carbide formed and the total gas phase exposure. Typical results for the four gases used are summarized in Table I. These results show that the reactivity toward decomposition increases with reactant complexity, and that methane is the least reactive of all the gases investigated.

The expected behavior for light hydrocarbons striking a hot poly-crystalline filament is thus a reactive sticking probability of several parts in one hundred and the formation of the alpha-ditungsten carbide.

In the present work, we examined the results of experiments in which single crystals of tungsten exposing the (100), (110), or (111) faces were exposed to methane and cyclopropane under the same experimental conditions used in the polycrystalline study.

EXPERIMENTAL

High purity single crystals (99.998%) were cut to expose the desired face, then ground with diamond pastes and finally electro-polished to remove the strained surface. X-ray Laue back reflection photographs showed that all crystals were within approximately one degree of the desired orientation.

The methane (99.0% min) and cyclopropane (99.5%) hydrocarbons were from the identical cylinders used in the previous polycrystalline study.

RESULTS

W(100): The tungsten (100) surface unit cell is a square 3.16 $\overset{\circ}{A}$ on each side. The clean surface LEED pattern is shown in Figure 1a. The methane exposures required to generate the observed patterns are shown with the patterns in Figure 1. A variety of patterns were observed with increasingly severe temperature and exposure conditions. The most interesting and most stable pattern generated is shown in Figure 1e.

This (100) - (5x1) pattern resulted from heating the crystal at 2400°K for 180 secs in 2×10^{-6} torr methane. The corresponding surface structure could not be altered by heating in vacuo or in 10^{-4} torr of hydrogen for three hours. This pattern remained unchanged upon further exposure of 10 torr-sec.

(100 seconds at a pressure of approximately 10^{-2} torr methane).   A Laue
back reflection X-ray photograph showed only the original tungsten reflections.   X-ray scan measurements with Cu_{Ka} radiation showed several
very faint carbide lines, but examination under the microscope indicated that
these small amounts of carbide existed primarily at the boundaries of a few
large grains in the crystal.

W(110):   The centered rectangular unit cell for the clean W(110)
surface has dimensions a_o = 3. 16 Å and b_o = 4. 47 Å.   Its corresponding
diffraction pattern is shown in Figure 2a.   The minimum gas-time-temperature
exposures required to observe the various patterns are summarized in Figure
2.

Two high temperature patterns were observed, as reported by previous investigators (7, 12).   The most commonly observed LEED pattern
is shown in Figure 2c.   This pattern exhibits fifteen fold periodicity between
the (00) and ($\bar{2}$4) tungsten reflections and a three fold increase in periodicity
along the [22] direction.   For convenience we denote this pattern by rotated
(15x3) or R(15x3).   The corresponding structure was generated by heating at
2400°K in 10^{-5} torr methane.   More severe exposures up to 10 torr-sec
methane did not change the R(15x3) pattern.

Thus for the W(110) face as for the W(100) face, a high temperature
structure is observed which is very stable with respect to subsequent
carburization.

A second high temperature pattern found on the (110) face is shown in
Figure 2d.   This pattern has the same fifteen fold periodicity in the [24]
direction, but a fourfold larger periodicity in the [22] direction as is shown
clearly in Figure 2d. It was formed irregularly, usually after very high
temperature flashing of the crystal.   In one experiment, the melting point of
tungsten, 3680°K, was achieved.   When the crystal cooled, the unmelted
fragment exhibited this R(15x12) pattern almost up to the melted edges.

W(111): The (111) surface is formed by a plane of tungsten atoms arranged in hexagonal symmetry; the nearest neighbor distance in this plane is 4.47 $\overset{\circ}{A}$. The LEED pattern of the clean tungsten surface is shown in Figure 3a.

A thirty second exposure at 2400°K in 10^{-4} torr of methane or cyclopropane produced a W(111)-(6x6) pattern which we denote by (111)-6. This sixfold pattern is shown in Figure 3b. for cyclopropane carburization.

A severe exposure of this structure to 10^{-3} torr of cyclopropane for thirty or more seconds at 2400°K produced a new pattern which included three new specular beams shown in Figure 3c. These three new (00) beams lie about four degrees away from the original location of the (111) specular beam. Further, the original (111) specular beam does not appear for cyclopropane exposures which produced this triplet pattern. Thus the corresponding structure must consist of several layers of a new compound which has three equally preferred orientations on the (111) surface. Further exposure of 10^{-1} torr-sec in cyclopropane did not change the triplet beams, but did diminish the intensity of other features. Heating this triplet structure to 3000°K regenerated the (111)-6 pattern in Figure 3b.

Severe methane exposures on the (111)-6 structure did not generate the triplet specular beams in contrast to the behavior of cyclopropane.

The square pattern previously reported by Taylor (6) was not observed.

DISCUSSION

In summary, each crystal face exhibited at least one high temperature structure. All but one of these structures were meshes with the new unit cells parallel to the original tungsten surface. The unusual triplet pattern observed on the (111) face arises from a mesh with some major plane tilted about four degrees from the (111) surface. Further, the triplet pattern did not exhibit the (111) specular beam and is thus presumed to be due to a structure with a third dimension of at least several layers thickness.

These results are summarized in Table II.

STICKING PROBABILITIES

To compare these LEED results with the previous polycrystalline study, we discuss the reactivity of the high temperature structures listed in Table II. The sticking probabilities of methane and cyclopropane on these carburized single crystal surfaces was not measured by the resistance method used earlier, but with very simple assumptions, we placed useful bounds on their values.

Upper bounds may be calculated for the sticking probability where a long, severe exposure on a given carburized structure produced no further change in the LEED pattern. We assume that, prior to the exposure, the crystal contained only sufficient carbon to produce the surface structure. We further assume that, during the subsequent exposure, the crystal achieved full saturation of 0. 1 atomic % carbon at 2400°K (14).

This carbon content is clearly an upper bound on the amount of carbon which the crystal could have absorbed without a further change in the surface structure, e. g. , formation of bulk carbide. Knowing this amount of carbon, and the total incident reactant flux, we can calculate an upper bound for the reactive sticking probability.

Similarly, we can place a lower bound on the sticking probability for the one case where a three dimensional structure was formed, the (111)-triplet. We assume that at least five layers of a new compound (carbide) were formed in the minimum exposure required to generate the pattern, since the old (111) specular beam was no longer present at any voltage. We also assume the new compound to be ditungsten carbide in accord with the results of our poly-crystalline work (4). Knowing the amount of carbon required to form five layers of carbide plus the minimum cyclopropane exposure which generated the triplet pattern yields a lower bound for the sticking probability. This estimate is a lower bound because we have not counted the carbon dissolved in the tungsten crystal during the exposure.

Finally, the (111)-triplet structure was unchanged under a further cyclopropane exposure of 10^{-1} torr-sec at 2400°K. For a polycrystalline substrate, this exposure would have produced a carbide depth of five microns as calculated from the value for cyclopropane in Table I. The X-radiation half value layer has a thickness of about one micron, but a Laue back reflection photograph of the sample showed primarily tungsten reflections. Thus an upper bound for the cyclopropane sticking probability on the triplet structure appears to be somewhat less than the polycrystalline value.

The results of these sticking probability estimates are summarized in Table III. These bounds show that the reactivity of carburized (100) and (110) tungsten surfaces for the decomposition of methane is at least two orders of magnitude less than that found on a polycrystalline surface under comparable conditions. In absolute terms, the reactive sticking probability of methane on these two structures at 2400°K is something less than one in ten thousand.

The sticking coefficient of methane on the (111)-6 structure is also quite low compared to cyclopropane as evidenced by the fact that methane exposures were unsuccessful in generating the triplet pattern. The reactivity of cyclopropane on this (111) face is more nearly like the polycrystalline value according to the bounds calculated in Table III.

INTERPRETATIONS OF PATTERNS

We now consider the three most important structures observed in this study: W(100)-(5x1), W(110)-R(15x3), and W(111)-triplet. The first two of these structures are extremely stable under conditions where we had initially expected bulk carbide formation according to our polycrystalline study. It seems reasonable, however, to postulate that these patterns are caused by various ditungsten carbide monolayer structures. A carbide on tungsten orientation for each of these high temperature patterns is shown in Table IV. An orientation of this carbide on the (100) surface which, through multiple scattering, could generate a (5x1) pattern is found by distorting the carbide lattice 6°/₀ in one direction and 2°/₀ in a direction perpendicular to the first. Similarly, Bauer (7) has proposed an alpha-carbide orientation on the (110) surface which could give rise to the R(15x3) pattern, again invoking multiple scattering rather than the formation of very large unit cells with dimensions of the order 20-30 Å.

The triplet pattern is shown in Figure 3d at low voltage: from the motion of the bright reflections with wavelength, the pattern is decomposed into the superposition of three centered rectangles. The reciprocal lattice distances estimated from this photograph correspond to the unit cell dimensions for the (10.0) plane of alpha-ditungsten carbide. Thus we postulate that the triplet pattern is caused by a centered structure on the (10.0) plane of di-tungsten carbide which itself is inclined about four degrees off the (111) surface.

It seems worth noting that the first two cases involve monolayer lattices which exhibit coincidence sites for the tungsten atoms, that is, the carbide lattice and the tungsten lattice can be arranged to have tungsten atoms positioned at the same point in each lattice. Such a coincidence structure may also be postulated for the (111) triplet pattern: a 4.1 degree tilt for the postulated carbide orientation brings tungsten atoms into coincidence about every 26 Å. Although these considerations are somewhat speculative, it does not seem unreasonable to suggest that coincidence lattices may play an important role in determining the rate of carburization. In particular, carbide formation involves a diffusionless rearrangement of tungsten atoms with very little relative motion, which is also true for the recrystallization of a metal. In this latter rearrangement, the existence of these coincidence lattices at grain boundaries and their importance in the recrystallization rate of copper was first demonstrated by Kronberg and Wilson (15) in 1949. More recently, the existence of these same special boundaries in tungsten has been observed by Brandon et al. (16) in the field emmision microscope.

Finally, we note that the conclusions in this paper regarding the sticking coefficients are independent of these detailed interpretations of the LEED patterns.

SUMMARY

In summary, the sticking probability of two light hydrocarbons, methane and cyclopropane, on single crystal surfaces of tungsten has been studied with LEED. It was found that the reactivity of the carburized (100) and (110) surfaces toward the decompostion of methane is at least two orders of magnitude less than that found for polycrystalline carburized tungsten filaments. The diffraction patterns observed on these two faces has been interpreted as due to the formation of a monolayer or two dimensional structure on the original metal surface. By contrast, the (111) face when carburized by cyclopropane exhibited a pattern which is believed due to the formation of a true three dimensional structure as deduced by the disappearance of the original (111) specular beam.

It is suggested that each corresponding surface structure is due to the formation of carbide coincidence lattices, and further that the behavior of these coincidence structures may be important in determining the rate of carburization.

Further work to understand these differences in reactivity is in progress.

ACKNOWLEDGMENT

This work was supported by the Office of Scientific Research of the U.S. Air Force (AF 49 (638) 1423).

REFERENCES

1.   M. Andrews and S. Dushman, J. Franklin Institute, 192, 545 (1921).

2.   M. Andrews, J. Phys. Chem., 27, 270 (1923).

3.   M. Andrews and S. Dushman, J. Phys. Chem., 29, 462 (1925).

4.   M. Boudart, G. W. Harris, D. F. Ollis, Trans. Faraday Soc., to be published.

5.   R. M. Stern, Appl. Phys. Letters, 5, 218 (1964).

6.   N. Taylor, Surface Science, 2, 544 (1964).

7.   E. Bauer, Adsorption et Croissance Crystalline, Centre National de la Recherche Scientifique, Paris, 1965, p. 1; Surface Science, 7, 351 (1967).

8.   L. H. Germer, Fundamental Phenomena in the Materials Science, Vol. 2, Plenum Press, p. 3 (1966).

9.   J. W. May and L. H. Germer, J. Chem. Phys., 44, 2895 (1966).

10.   P. J. Estrup and J. A. Anderson, Surface Science, 8, 101 (1967).

11.   L. N. Tharp and E. J. Scheibner, J. Appl. Phys., 38, 3320 (1967).

12.   R. Baudoing and R. M. Stern, Surface Science, 10, 392 (1968).

13.   G. W. Orton, Phase Equilibria of the Tungsten-Carbon System, PhD Thesis, Ohio State University, 1961 (University Microfilm Mic 61-2839; Trans. A. I. M. E., 230, 600 (1964).

14.   H. J. Goldschmidt, J. A. Brand, J. Less Common Metals, 5, 181 (1963).

15.   M. L. Kronberg and F. H. Wilson, Trans. A. I. M. E., 185, 501 (1949).

16.   D. G. Brandon, B. Ralph, S. Raganathan and M. S. Wald, Acta Met., 12, (1964).

TABLE I

Polycrystalline Results (4)

Gas	Pressure (torr)	Temperature (°K)	Sticking Probability, γ
CH_4	$1.1 \cdot 10^{-3}$	2430	1.4×10^{-2}
C_3H_8	$1.2 \cdot 10^{-3}$	2395	2.6×10^{-2}
C_3H_6	$6.3 \cdot 10^{-4}$	2400	7.2×10^{-2}
$(CH_3)_4 C$	$2.4 \cdot 10^{-3}$	2415	*

*Diffusion controlled rate

TABLE II

Carburized Surface Structures

Mesh Parallel to Surface	Mesh Tilted
W(100)-(5x1)	W(111)-triplet
W(110)-R(15x3)	
W(110)-R(15x12)	
W(111)-6	

TABLE III

Relative Sticking Probability, β, on Carburized

Single Crystal Surfaces

Pattern	Gas	$\beta = \gamma_{(hkl)} / \gamma$ wire
W(100)-(5x1)	CH_4	$\beta < 10^{-2}$
W(110)-R(15x3)	CH_4	$\beta < 10^{-2}$
W(111)-triplet	C_3H_6	$10^{-2} < \beta < 1$

TABLE IV

Interpretation of LEED Patterns

Pattern	Carbide Orientation (α - W_2C) $\|$ W	
W(100)-(5x1)	(00.1) $\|$ (100)	[10.0] $\|$ [001]
W(110)-R(15x3)	(00.1) $\|$ (110)*	[10.0] $\|$ [$\bar{1}$15]*
W(111)-triplet	(10.0) $\overset{\sim}{\|}_{4°}$ (111)	[00.1] $\|$ [$\bar{1}$01]

* Bauer (7)

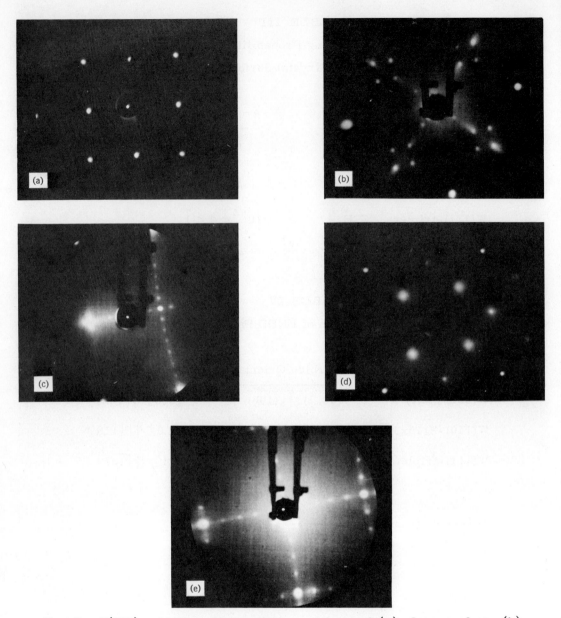

Fig. 1. W(100) patterns after methane exposures of (a) clean surface, (b) $5 \cdot 10^{-7}$ torr, 240 sec, 1560°K, (c) 10^{-5} torr, 90 sec, 1560 °K, (d) $5 \cdot 10^{-7}$ torr, 120 sec, 2000°K, (e) 2x10^{-6} torr, 180 sec, 2400°K.

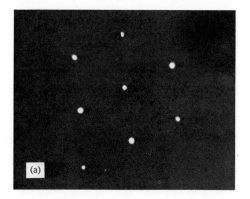
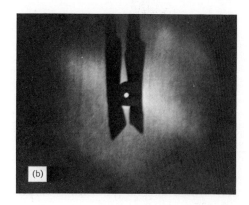
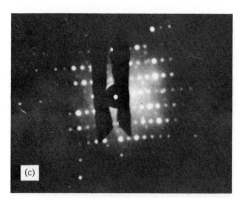
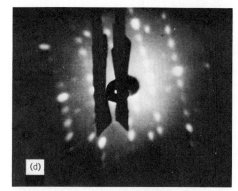

Fig. 2. W(110) patterns after methane exposures of (a) clean surface, (b) $4 \cdot 10^{-5}$ torr, 120 sec, 1000°K, (c) 10^{-5} torr, 90 sec, 2400°K, (d) flashing (c) to ~3000°K.

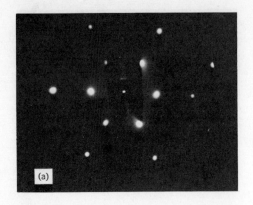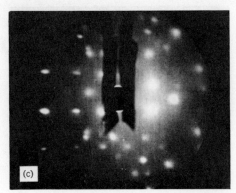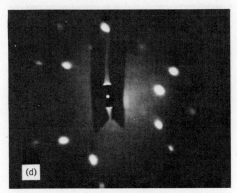

Fig. 3. W(111) patterns after cyclopropane exposures of (a) clean surface, (b) 10^{-4} torr, 30 sec, 2400°K, (c) & (d) 10^{-3} torr, 30 sec, 2400°K.

OBSERVATION OF SURFACE SUBSTRUCTURE BY HIGH ENERGY ELECTRON DIFFRACTION AT GLANCING INCIDENCE: TUNGSTEN (001) SURFACE*

James F. Menadue and Benjamin M. Siegel

Department of Applied Physics
Cornell University, Ithaca, New York

The ultra high vacuum electron diffraction apparatus described by Siegel and Menadue[1] has been used to study surface phenomena by the method of reflection high energy (40 kV) electron diffraction. The relationship between the electron beam and the specimen is shown in Fig. 1. The low glancing angle of incidence and high resolution obtained with well focused electron beams makes this method particularly suitable to monitor the topography of a surface. It is possible to follow variations in the topography brought about by treatments such as thermal etching or by any mechansim which would cause facets or asperities to grow or vanish.

The most prominent feature of glancing angle electron diffraction patterns are the streaks that usually run normal to the shadow edge cast on the fluorescent screen by a flat crystal. When the tilt and azimuth of a single crystal specimen are set so the Laue conditions are satisfied in the bulk, intense maxima are observed. In the neighborhood of these maxima and sometimes over a wide range of angles, high resolution may reveal a close array of sharp spots distributed along the vertical streaks. These spot groups are due, for an otherwise clean crystal, to a surface which is not perfectly flat or has a subgrain structure.

Some possible causes of vertical beam splitting are shown in Fig. 2.

A) Reflection from a perfectly flat single crystal surface produces single spots. One spot that is always visible is the specular reflection at a scattering angle just twice the angle of tilt of the crystal reflecting planes to the beam.

*This work has been supported by the Air Force Office of Scientific Research and the Advanced Research Projects Agency through the Materials Science Center at Cornell University.

B) Transmission through an asperity may produce a fine structure splitting due to double refraction.

C) and D) The beam may enter or exit through one face which is at a considerable angle to the surface plane, such as a {110} facet on a (001) surface. These cases result in large refractive shifts and for case C) with a sloping entrance face, a fine structure splitting.

E), F) and G) A subgrain structure in the crystal combined with low angle faceting will also produce splitting. Kato[2] has shown that facets must in general be larger than a few hundred Angstroms in order that well defined spots should appear. With facets of this size layers of foreign atoms in monolayer quantities will have little effect on the splitting, although they may greatly modify the intensities.

H) When there are two faces on a bulk crystal that are at a low angle α to each other with $\alpha < \theta$ the angle of incidence, as illustrated in Fig. 3, a splitting can occur. This condition would obtain if a crystal were cut at an angle α to a low index plane and tended to form facets parallel to that plane.

Experimental Observations on W(001) Specimen

A tungsten crystal was used to make observations on the types of splitting described above (Figs. 2 and 3). The specimen was prepared from a two pass zone refined rod 3 mm in diameter and was 1 mm thick. The surface was cut to within 0.2° of the (001) planes, alternating between chemical-mechanical and electro-polishing (in 0.1N NaOH at 10 volts) until the scratches had all but vanished, and a 5 sec. electropolish gave a brilliant mirror finish. An interferometer showed that the surface was flat within one fringe (3000Å) over 95% of its radius.

In our observations a diffraction pattern with complex groups of spots appeared the first time that the specimen was cleaned by flashing to 1500°C.

These spots became sharper after a cycling treatment in which the specimen was exposed to 10^{-6} Torr oxygen and heated to $800°C$ with a flash to $1800°C$ every three minutes for 30 minutes. This treatment is similar to that of Anderson and Danforth[3], who reported from LEED observations that the (001) surface will facet to {110} when heated in 10^{-7} Torr oxygen. Tracey and Blakely[4] have confirmed the faceting of the W(001) in the presence of oxygen. The basic appearance of the spot group did not change and a typical vertical scan and enlarged photograph of the spot group are shown in Fig. 4.

The splitting of the outermost intense spots of each group was found to be constant over a wide range of incident angle. As the beam was slowly raised or lowered off the diffracting face of the specimen, the spots in a group vanished in succession from one end or the other. The splitting was most extensive near a particular [10] azimuth, and decreased as the specimen was rotated to about $90°$ from the initial position, the spots vanishing in succession until there was just one spot left. Deflecting the beam sideways over the crystal face at this point could cause a second spot to appear. These observations suggest a subgrain structure of the tungsten crystal as discussed below.

At the azimuth for which the splitting was a minimum it was observed that the beam split into two with a separation depending on the angle of incidence characteristic of case H (Fig. 3). Figure 5 shows the vertical scan data recorded from this splitting. At other azimuths weak spots were seen to be moving among the strong "fixed" spots when the tilt angle or azimuth were varied, but they could not be followed far.

Methods of Calculating Beam Splitting

The deviation $\Delta\theta = \theta_2 - \theta_1$ of an electron beam passing through a crystal surface into or out of a region of mean inner potential Vo is plotted in Fig. 6. The curve is simply Snell's law, $\cos \theta_1 = n \cos \theta_2$, with the refractive index, n, given to a good approximation by $n^2 = 1 - \dfrac{Vo}{E}$, where E is the beam accelerating voltage. It is seen that the deflection $\Delta\theta$ becomes appreciable at low angles of incidence θ_1 and that there is a significant region of total internal reflection for electrons approaching the crystal surface from within.

The mean inner potential Vo was found by measuring the deviation of the Bragg maxima towards lower angles, i.e., towards the shadow edge. The crystal azimuth was chosen so the incident beam was not close to any low index sur‑ face direction which would result in multiple beam interference. Such multiple interference is minimal when the Kikuchi pattern in the neighborhood of the observation is simply the bright line corresponding to the Bragg reflection being observed. The effective values of Vo were observed for the 004, 006 and 008 reflections. They are given in Table I and compared with the value of 30.6 volts calculated by Anishchenko.[5] For the present work a value of 30 volts was assumed. The beam voltage E was 40 kVolt.

Double refraction, as in cases B and C, will produce a less obvious split‑ ting. If a beam is diffracted in a well defined wedge shaped crystal, under Laue diffraction conditions the existing beam will have been split into at least two. This splitting is the double refraction effect first recognized by Sturkey.[6] It is explicitly predicted by dynamic theory as was pointed out by Kato,[2] and it has been used by Cowley[7] and others to determine the struc‑ ture factors V_{hkl} for use in structure analysis.

The directions of exiting beams from any wedge shaped crystal may be determined by a construction on the dispersion surface, as illustrated in Fig. 7. Following Kato,[2] the exiting beam vectors $\bar{K}_{gi}$ which terminate on the reciprocal lattice point G in the usual dispersion diagram are translated parallel to their original direction so that they originate on the point E on the sphere S_E from which they were excited. As E is moved over the sphere the terminations of the $\bar{K}_{gi}$ trace out the shape of an apparent distribution of intensity in reciprocal space, which may be called a dynamic shape transform shown in Fig. 7 by the heavy lines near the dynamical reciprocal space point G_D. The construction has been extended in the diagram to two of the possible combinations of entrance and exit surfaces, and obviously as many cases as there are combinations of entrance and exit surfaces can be superimposed, provided that they are all formed on the one monolithic bulk crystal and thus have the same reciprocal lattice vector $\overline{OG}$. There should be no unexpected interferences effects if the different cases are separated on the crystal surface by more than the coherence width of the incident beam. In these observations the coherence width was the order of a thousand Angstroms and the area of specimen illuminated was very large by comparison (Fig. 1). For a subgrain at a tilt α to a reference grain, the whole dispersion surface construction would have to be rotated about the origin O of the reciprocal lattice by a similar angle α, and the new intensity distribution could then be superimposed as before. For a given incident beam at an angle θ to the Bragg reflecting planes, a sphere of radius $|\bar{K}_g|$ centered on E will pass through the combined intensity distribution curves and each intercept will define an emergent beam. The effects of adsorption of the electron beam have been ignored. Absorption will weaken, and may even further deviate some beams.

Characteristic Curves

There will be a significant amplification of the actual splitting in those cases where the beam is incident on a surface at a low glancing angle. The magnitude of these effects for the different cases, both calculated and observed, are shown in Fig. 8. The relative shape of these curves is characteristic of each case considered and they would seem to be sufficiently different in form to allow identification of an observed splitting. For the cases B and D (Fig. 2), the beam emerging from a sloping exit face will only be strong near a Bragg reflection. For those cases which involve one surface

parallel or nearly parallel with the Bragg reflecting planes, a mechanism involving total internal reflection of side beams as postulated by Kohra[8] will
result in "specularly" reflected spots always being present and the characteristic curves can be followed over a wide range of θ. Cases like G, localized on
particular grain boundaries, utilize very little of the beam and the spots will
generally be too weak to be seen.

The intensity distribution is sketched for case H in Fig. 3. From the
triangle formed by the exit surface normals $\bar{M}_2$ and $\bar{M}_3$, and the arc drawn from
the entrance point E, it can be shown that the splitting δ will be

$$\delta = \frac{2\alpha \; \Delta\theta_{tilt}}{(\theta_B + \theta_o)/2 - \Delta\theta_{tilt} + \alpha}$$

where θ_B is the Bragg angle, θ_o is the corresponding vacuum angle, and $\Delta\theta_{tilt}$
is the deviation of the incident angle θ from $(\theta_B + \theta_o)/2$. When the incident
angle $\theta \approx (\theta_B + \theta_o)/2$ the splitting will go to zero. The side beam reflection
mechanism mentioned above should keep both beam visible over a wide range of
tilt angles.

Figure 8 summarizes the splitting characteristic of the various cases and
gives the calculated curves and experimental data obtained.

Conclusion

Glancing angle high energy electron diffraction has been shown to be a
very sensitive means of observing substructure near the surface in essentially
flat single crystals. Observations on the fine splitting in the diffraction
streaks from a W(001) crystal indicated a subgrain structure in the crystal as
in case F (Figs. 2 and 8). The effect of rotating the crystal and of careful
deflection of the beam off the crystal lead to the conclusion that the sub-
grains were oriented with the tilt axes almost parallel, and that the grain
tilt increases across the crystal in a monotonic fashion. The average tilt
between each pair was 0.9' and an overall tilt was 9.3'. The resolution of the
high energy electron diffraction method in these observations was 30". A

Berg-Barrett X-ray photograph confirmed the presence of a lineage structure.

No large splittings of the character expected from {110} facets (cases C and D, Figs. 2 and 8) were observed, indicating that the oxygen and heat treatment had not produced any significant growth of facets as large as a few hundred Angstroms.

The angular deviation, α, of the substructure in this crystal of the type H (Figs. 3 and 8) was obtained by adjusting the parameter α in the equation for δ given above. A good fit was obtained for $\alpha = 7.4'$ using θ_B of the 004 reflection, indicating that this was the angle between the original polished surface and the (001) planes to which it was faceting, for the particular azimuth of observation.

REFERENCES

1. B. M. Siegel and J. F. Menadue, Surface Science 8, (1967) p. 206.

2. N. Kato, J. Phys. Soc. Japan 7, (1952), p. 397, 406.

3. J. Anderson and W. E. Danforth, J. Franklin Inst. 279, (1965) p. 160.

4. J. C. Tracey and J. M. Blakely, (to be published).

5. R. I. Anishchenko, Phys. Stat. Sol. 18, (1966) p. 923.

6. L. Sturkey, Phys. Rev. 73, (1948) p. 183.

7. J. M. Cowley, P. Goodman and A. L. G. Rees, Acta. Cryst. 10, (1957) p. 19.

8. K. Kohra, J. Phys. Soc. Japan 9, (1954) p. 690.

Table I.

REFLECTION	MEAN INNER POTENTIAL Vo	
	Effective	Theoretical
004	$28 \pm 2\frac{1}{2}$	
006	33 ± 4	30.6 Anishchenko (1966)
008	34 ± 6	

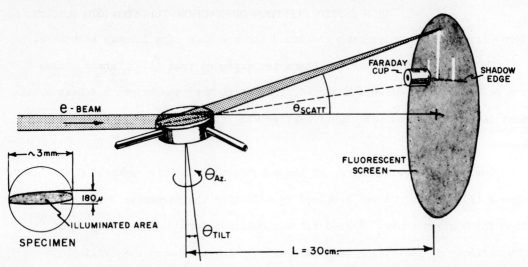

Fig. 1. The relationship between the specimen, the beam and the fluorescent screen and Faraday cup.

FACETS   SUB GRAINS

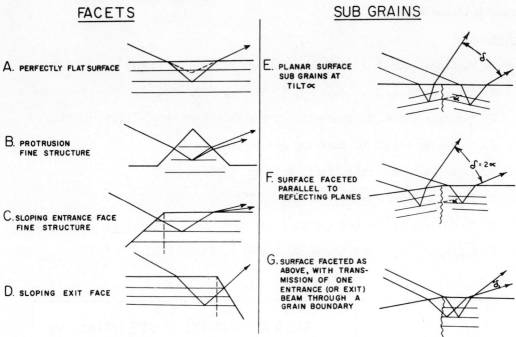

A. PERFECTLY FLAT SURFACE

B. PROTRUSION FINE STRUCTURE

C. SLOPING ENTRANCE FACE FINE STRUCTURE

D. SLOPING EXIT FACE

E. PLANAR SURFACE SUB GRAINS AT TILT ∝

F. SURFACE FACETED PARALLEL TO REFLECTING PLANES

G. SURFACE FACETED AS ABOVE, WITH TRANSMISSION OF ONE ENTRANCE (OR EXIT) BEAM THROUGH A GRAIN BOUNDARY

Fig. 2. Faceting and subgrain structure that can produce some of the possible beam splitting and spot groups.

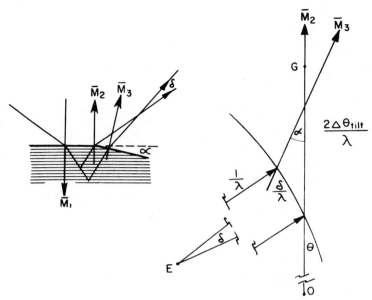

Fig. 3. Case H, involving a small angle a between surface planes very close to the Bragg reflecting lattice planes. (001). $\bar{M}_2$ and $\bar{M}_3$ are the exit surface normals.

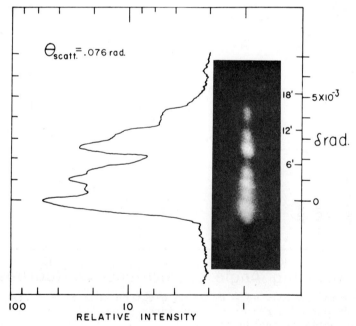

Fig. 4. Typical Faraday cup scan of a spot group, compared with a photograph of the scanned diffraction streak, enlarged 60 diameters. The Faraday cup entrance aperture was 50μ.

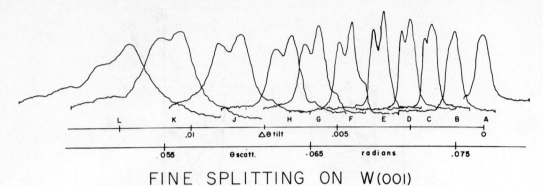

FINE SPLITTING ON W(OOI)

Fig. 5. Fine splitting observed for case H. Vertical scans taken at various angles of incidence.

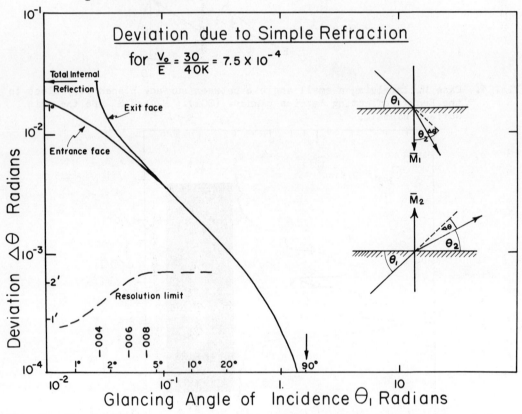

Fig. 6. The deviation $\Delta\theta = \theta_2 - \theta_1$ of an electron beam of E volts in passing into or out of a region of mean inner potential Vo volts as a function of the angle of incidence θ_1.

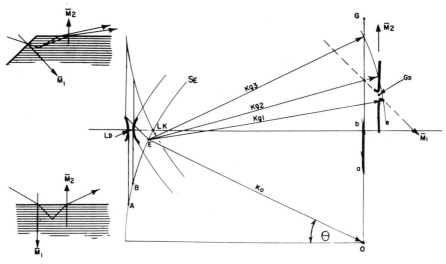

Construction on Dispersion Surfaces to Determine Splitting

Fig. 7. The dispersion surface construction for determining the directions of emerging beams (after Kato[2]).

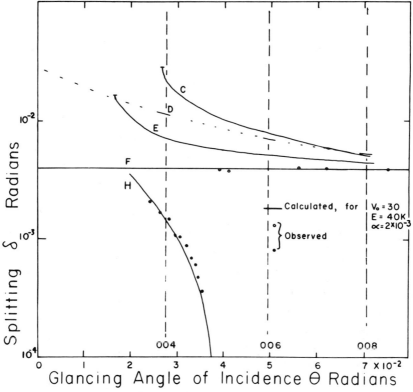

Fig. 8. Relative variation of beam splitting angle, δ, with variation of angle of incidence θ.

WORK FUNCTION AND SURFACE STRUCTURE CORRELATIONS IN THE ADSORPTION OF OXYGEN ON TUNGSTEN SINGLE CRYSTAL PLANES

J. C. Tracy[+] and J. M. Blakely

Department of Materials Science and Engineering
Cornell University, Ithaca, New York

INTRODUCTION

The study of gas-metal surface interactions is greatly enhanced by employing several experimental techniques simultaneously. For example, while low energy electron diffraction (LEED) determines the symmetry of the surface structure, it does not give the surface coverage nor define uniquely the exact atomic arrangement. Work function measurements as a function of coverage determine the dipole moment of the gas atom-surface configuration and in conjunction with LEED the mutual interaction between the adsorbed species in the layer. Coverage and sticking probability are obtained from pressure measurements during adsorption.

The tungsten-oxygen (W-O) system has been chosen for this study because the interaction is strong, single crystal surfaces of tungsten are easily prepared, and there is a wealth of experimental work[1] for comparison. Unfortunately it is theoretically very difficult to handle this system making quantitative comparisons of theory and experiment impossible. The need to study single crystal planes is well established.[2] The (110), (100), (112), and (111) surfaces of W have been investigated in the course of the present work. The interaction of oxygen with the (112) W surface is the best understood at this time and will be discussed here. The methods used to correlate work

[+]This paper is based on a thesis to be submitted to the Graduate School of Cornell University by J. C. Tracy in partial fulfillment of the requirements for the Ph. D. in Applied Physics.

function changes with surface structure for the (112) plane are typical of those being applied to the other surfaces. The adsorption of oxygen at room temperature on the (112) surface is described for coverages up to one monolayer. Consideration is made of the nature of the W-O bond and its interactions with other W-O bonds in the surface layer.

Experimental observations are made using the usual post acceleration LEED apparatus in conjunction with a Kelvin vibrating reed work function probe. This Kelvin probe measures the work function change continuously by utilizing the servomechanism of Palevsky.[3] Coverage was determined using the uptake method of Eisinger[4] with a quadrupole mass spectrometer. The crystals are in ribbon form measuring about 1 cm x 0.5 cm x 0.01 cm and are heated by passing current through them. The crystals were purged of carbon by the method of Becker et al.[5] The experimental details have been described[6] elsewhere.

RESULTS

Zero to 0.5 Monolayer Coverage

Figure 1 shows the change in work function ($\Delta\phi$) of a (112) W surface. $\Delta\phi$ is linearly dependent on the coverage up to 0.5 monolayers (0.5 θ_o). At higher coverages $\Delta\phi$ goes through a relative maximum at 0.75 θ_o followed by a relative minimum at full monolayer coverage. The saturation value of $\Delta\phi$ for exposure at room temperature is 1.1 eV but is not indicated because a reduction in sticking probability makes the coverage determination inaccurate beyond 1.3 θ_o.

The LEED results as a function of coverage are shown in Fig. 2. At coverages below one half monolayer, streaks form parallel to the $(11\bar{1})$ azimuths through the half order positions designated h ± 1/2, k where h and k are integers. There are however, no streaks through the integral order features, a very important result. As the coverage increases to 0.5 θ_o, the streaks become sharply peaked at the half order positions indicated above. This final p(2x1) pattern is shown in Fig. 2d. The intensity distribution along the streak is important in determining the way in which the p(2x1) oxygen structure grows from zero coverage. Figure 3a shows an example of such an intensity profile measured directly from the fluorescent screen at 0.35 θ_o. The dependence of the peak height and half width of streak profiles at different coverages is shown in Fig. 3b.

These streaks have been reported previously by Chang[7] and Germer.[8] They suggested that streaks result from scattering by islands of p(2x1) which are long in the $[11\bar{1}]$ direction and narrow in the $[1\bar{1}0]$ direction as shown in Fig. 4a. If this is the case, why are there no streaks through the integral order features? If the distance between the centers of the islands is such that not more than one island falls in the coherent area of an electron, then streaks result simply from particle size broadening and will exist through the integral orders. Such a model would also not account for the dependence of the streak halfwidth on coverage. To understand the pattern we must use the fact that the unit cell of the p(2x1) structure can be placed in two positions, both equivalent with respect to the substrate but out of step with each other by half the repeat distance in the $[11\bar{1}]$ direction. These positions are shown in Fig. 4b. At coverages less than one half a monolayer suppose there are a large number of these islands within the coherent area. There are then an equal number of islands in the two positions which are available to the p(2x1) unit cell on the substrate. The occurrence of islands in these two positions with equal probability is sufficient to explain the streaking of the half order beams while the integral orders remain sharp.

To illustrate this consider the diffraction from the following simplified model. Let $\underset{\sim}{a}_x$ and $\underset{\sim}{a}_y$ be the basic vectors of the substrate in the $[1\bar{1}0]$

and $[11\bar{1}]$ direction, respectively, (Figure 3). We will assume for simplicity

that all islands are rectangular and of the same dimensions $M|\underset{\sim}{a}_x|$ by $2N|\underset{\sim}{a}_y|$.

This model is sufficient to demonstrate most of the pertinent features of the

diffraction pattern. The islands then each contain MN scatterers arranged

in a p(2x1) structure. For isotopic unit scatterers, the scattering amplitude

f_i for an island is

$$f_i = e^{i\phi_{\underset{\sim}{k}}} \frac{\sin N \underset{\sim}{k} \cdot \underset{\sim}{a}_y}{\sin \underset{\sim}{k} \cdot \underset{\sim}{a}_y} \cdot \frac{\sin \frac{M}{2} \underset{\sim}{k} \cdot \underset{\sim}{a}_x}{\sin \frac{1}{2} \underset{\sim}{k} \cdot \underset{\sim}{a}_x} \qquad (1)$$

where $i^{i\phi_{\underset{\sim}{k}}}$ is a phase factor dependent on the scattering vector $\underset{\sim}{k}$. The inten-

sity at a particular $\underset{\sim}{k}$ scattered from the ensemble of islands is then given by

$$I = \sum_i \sum_j f_i f_j^* e^{i\underset{\sim}{k}(\underset{\sim}{r}_i - \underset{\sim}{r}_j)} \qquad (2)$$

where $\underset{\sim}{r}_i$ is the position vector to the characteristic point of the ith island.

Since all the f_i are the same, we obtain using (1)

$$I = \frac{\sin^2 N \underset{\sim}{k} \cdot \underset{\sim}{a}_y}{\sin^2 \underset{\sim}{k} \cdot \underset{\sim}{a}_y} \cdot \frac{\sin^2 \frac{M}{2} \underset{\sim}{k} \cdot \underset{\sim}{a}_x}{\sin^2 \frac{1}{2} \underset{\sim}{k} \cdot \underset{\sim}{a}_x} \sum_i \sum_j e^{i\underset{\sim}{k} \cdot (\underset{\sim}{r}_i - \underset{\sim}{r}_j)} \qquad (3)$$

The position vector r_i of an island may be written as

$$\underset{\sim}{r}_i = n_i \underset{\sim}{a}_x + 2m_i \underset{\sim}{a}_y + p_i \underset{\sim}{a}_y \qquad (4)$$

where p_i is unity for islands of type I and O for type II.

Since no observable broadening of the diffraction pattern is observed

in the y-direction even at low coverages, we will assume that N is large so

that the first interference function of (3) becomes N^2 when $\underset{\sim}{k} \cdot \underset{\sim}{a}_y = \pi \times$(integer),

the cases of interest in the present problem. To investigate the effect of the

island arrangement on the diffraction pattern we need to examine the sum

$$S = \sum_i \sum_j e^{i\underset{\sim}{k}(\underset{\sim}{r}_i - \underset{\sim}{r}_j)}.$$

We will suppose that the area of coherence of the incident electron beam is of width $A|\underset{\sim}{a}_x|$ and that within a particular coherence area there are a total of R islands, P of type I and Q of type II.

For $\underset{\sim}{k} \cdot \underset{\sim}{a}_y = 2\pi \times (\text{integer})$,
$$S = \left| \sum_i e^{-i\underset{\sim}{k} \cdot \underset{\sim}{a}_x n_i} \right|^2 \tag{5}$$

the summation being taken over all islands. Consider a random arrangement of islands...

For $\underset{\sim}{k} \cdot \underset{\sim}{a}_y = 2\pi \times (\text{integer})$ all islands are equivalent and by noting that the probability of an island occurring at any particular value of n_i is $\frac{R}{A}$ (5) becomes

$$S = \frac{R^2}{A^2} = \frac{\sin^2 \frac{1}{2} A \underset{\sim}{k} \cdot \underset{\sim}{a}_x}{\sin^2 \frac{1}{2} \underset{\sim}{k} \cdot \underset{\sim}{a}_x} \tag{6}$$

Thus the integral order beams have intensity $(RMN)^2$ and no streaking should be produced by the island arrangement.

For $\underset{\sim}{k} \cdot \underset{\sim}{a}_y = \pi \times (\text{odd integer})$, the half order positions,

$$S = \sum_i \sum_j \quad + \quad \sum_i \sum_j \quad - \quad 2 \sum_i \sum_j \tag{7}$$

i, j of type I   i, j of type II   i of type I
j of type II

For $\underset{\sim}{k} \cdot \underset{\sim}{a}_x = 2\pi \times (\text{integer})$

$$S = (P - Q)^2 \tag{8}$$

and for $\underset{\sim}{k} \cdot \underset{\sim}{a}_x \neq 2\pi \times (\text{integer})$

$$S = (P + Q) + \sum_{i \neq j} \sum \quad \sum_{i \neq j} \sum \quad - \quad 2 \sum_i \sum_j \tag{9}$$

type I   type II   i of type I
j of type II

For a random array the arguments $\underset{\sim}{k} \cdot \underset{\sim}{a}_x (n_i - n_j)$ will vary randomly over the sums and for $|\underset{\sim}{k} \cdot \underset{\sim}{a}_x - 2\pi \text{(integer)}| \geq \frac{2\pi}{A}$ the sums become zero.

Thus at the half order positions the intensity is $[NM(P - Q)]^2$, i.e., approximately zero and for $|\underset{\sim}{k} \cdot \underset{\sim}{a}_x - 2\pi \text{(integer)}| \geq \frac{2\pi}{A}$

$$I = N^2 (P + Q) \; \frac{\sin^2 \frac{M}{2} \underset{\sim}{k} \cdot \underset{\sim}{a}_x}{\sin^2 \frac{1}{2} \underset{\sim}{k} \cdot \underset{\sim}{a}_x} \tag{10}$$

For islands of small width (eq. M = 3) the observable half order features will be markedly streaked in the x direction.

The linear dependence of the work function change with coverage up to 0.5 θ_o is easily explained with the help of these LEED results. Since there is measurable intensity in the half order streaks at coverages as low as 0.05 θ_o, it is clear that the p(2x1) structure forms at very low coverages. The first oxygen atoms bound to the surface serve as nucleation centers for the growth of the islands. It is important to notice that during this growth there are only two surface phases, the clean phase and the p(2x1) oxygen covered phase. This further implies that the oxygen is mobile at some point, perhaps as a molecule before dissociation. There are no regions of intermediate oxygen density. These two phases have different work functions. Because the Kelvin method measures an area weighted average of the work function of the two phases[10] and the total area of the p(2x1) patches varies linearly with coverage up to 0.5 θ_o, $\Delta\phi$ is proportional to θ in this region.

The work function change $\Delta\phi$ is related to the dipole moment of the W-O bond ($= \int \underset{\sim}{r} \Delta\rho(\underset{\sim}{r}) \, dr^3$, where $\Delta\rho(r)$ is the change in charge density at $\underset{\sim}{r}$ after the formation of the bond) by the Helmholtz equation.

$$\Delta\phi = 4\pi n_s \theta \tag{11}$$

n_s is the dipole number density at one monolayer and θ is the dipole coverage in monolayers. The linearity of the measured $\Delta\phi$ means that the dipole moment of the bonds is constant in coverage. This is a direct consequence of the existence of only two surface phases. μ of an oxygen - tungsten surface bond in the p(2x1) structure is given by,

$$\mu = \frac{1}{4\pi n_s} \frac{d\Delta\phi}{d\theta} \tag{12}$$

and from the experimental data is equal to 0.40 Debye. This equation can be used to determine μ_o, the dipole moment of an isolated oxygen-surface bond from the slope at zero coverage only if the dipoles are not interacting in the limit of zero coverage. These experiments show that even in the limit of zero coverage the p(2x1) structure is formed and the adsorbed oxygen is therefore not widely dispersed on the surface. Thus, an attractive inter- action exists even at very low coverage.

The constant dipole moment means that if there is any depolarization of the dipoles due to mutual interaction, it too is constant. The 0.40 Debye value from (7) is only μ_o if there is no interaction in the p(2x1) structure. μ_o is most probably some greater value, since the very existence of the p(2x1) suggests the presence of an interaction. In order to form this struc- ture rather than some other one, the oxygen atoms joining onto the growing p(2x1) islands must experience a potential which is energetically more favorable than at a site in the middle of a clean region. Thus the islands attract the oxygen. Since the binding potential is different at these two sites, it is reasonable to assume that the resulting dipoles of the bonds is different. It is therefore not likely that 0.40 Debye is the isolated W-O dipole moment.

The p(2x1) symmetry and the half monolayer coverage led Chang[7] to suggest a model for this structure which is shown in Fig. 4c. While LEED evidence alone is not capable of specifying this structure, the work function results add some additional information. It has been shown[11] that partial or complete burial of oxygen in the (100) plane of tungsten causes a slightly

<u>negative</u> work function change.   Thus, the p(2x1) structure most probably

consists of oxygen sitting on top of the (112) surface and not <u>in it</u> as suggested

by the "reconstruction" model of the p(2x1) oxygen half monolayer structure

on W(110).[12]   The oxygen occupies those sites on the surface which would be

assumed by the next tungsten atoms in the bulk structure as this results in

the smoothest wave functions for the bound state and hence the lowest energy.

0.5 to 1.0 Monolayer Coverage

As indicated in Fig. 1a, the work function change in this region does

not vary linearly with coverage.   The dipole moment taken from (6) is plotted

against the coverage in Fig. 1b.   The nonlinear decrease in the dipole moment

beyond one half a monolayer indicates an increased interaction between the

dipoles.   The half order beams in the LEED pattern fade uniformly until at

one monolayer they have disappeared completely leaving a p(1x1) pattern.

This sequence is shown in Fig. 2.

In view of the proposed model for the p(2x1) structure, a model for the

p(1x1) structure with one monolayer of oxygen follows simply.   The oxygen

occupies those sites which would be filled by an additional (112) plane of

tungsten atoms.   Since this structure has the symmetry of the substrate,

there is only one allowed position.   The work function and LEED results

indicate that the p(1x1) structure is completed by randomly filling in those

sites left vacant in the p(2x1) structure.   Random filling produces the ob-

served nonlinear dependence of $\Delta\phi$ on θ.   As the oxygen is added beyond

0.5 θ_o it interacts strongly with its neighbors, causing the average dipole

moment to decrease.   At about 0.75 θ_o this total decrease of the neighboring

dipoles is greater than the dipole moment of the oxygen tungsten bond which

is created causing a decrease or "dip" in the $\Delta\phi$ vs. θ curves.   As the coverage approaches 1.0 θ_o, the dipoles around the vacant sites are already

so strongly depolarized by previously adsorbed oxygen that each new dipole

has a progressively smaller depolarizing influence. It must be remembered,

however, that the Helmholtz equation only gives the actual dipole moment

when every dipole is the same. This means that all dipoles must be in the

same environment which is clearly not the case for $0.5 < \theta < 1.0$. The quali-

tative argument is nevertheless still valid.

The LEED results further substantiate the suggested random filling

growth process. The half order beams remain fairly sharp as they gradually

weaken and background intensity increases. These features are character-

istic[13] of a random array of p(2x1) structure possessing long range order by

connection with the substrate. An island growth model on the other hand

would produce the same kind of streaks observed for the growth of the p(2x1).

Also that model requires that the work function change linearly with coverage.

In summary, for coverages less than 0.5 monolayers, oxygen covers

the W(112) surface in the form of islands of a p(2x1) structure on the order

of 10 atoms across. The work function changes linearly with coverage in

this region as it must for island growth, and gives a dipole moment of 0.40

Debye for the W-O bond in the p(2x1) structure. Beyond one half monolayer

the vacant sites in the p(2x1) structure are filled randomly to form a p(1x1)

structure at a coverage of one monolayer. The work function variation is

nonlinear in this region due to a strong depolarization of the layer by mutual

interaction.

DISCUSSION

The adsorbate-adsorbate (A-A) interaction has been considered re-

cently from first principles for the first time by Grimley[14] and Bennett.[15]

While it is not possible to apply their results quantitatively to the W-O system, the nature of the A-A interaction is clarified. The bond formed between a single adsorbed atom and the substrate perturbs the metal wave functions over a fairly long range as a result of the nonlocalized character of the states making up the metal band. Thus when a second atom is bound in the vicinity of the first, the metal wave functions it "sees" are different from those of the clean surface. The binding of the second atom will in turn perturb the wave functions in the vicinity of the first atom and so forth. A self consistent solution is clearly required. Grimley and Bennett have demonstrated that this interaction is oscillatory in the separation and falls off like R^{-3}. Grimley has further shown that there are special directions in the crystal along which the interaction is of much shorter range behaving like R^{-5}. While the oscillatory character and directional dependence of the interaction may cause the formation of particular structures, a comparison with experiment is not possible at this point. Bennett also considered a much shorter range A-A interaction, namely the direct overlap of the adsorbate orbitals. This interaction has been used qualitatively for some time in the consideration of hard sphere models of surface structures using atomic and ionic radii.

LEED has clearly demonstrated the existence of surface structures, even in cases of alkali adsorption on metals.[16] Theoretical treatments[17] of the dependence of the work function on coverage do not consider these structures but rather assume that mutual repulsions keep the dipoles uniformly distributed so the coverage is microscopically uniform. The adsorbed layer is considered to be a periodic array of continuously varying lattice constant. This is a good model for systems where the repulsive A-A interactions are able to overcome the barriers between adsorption sites. For W-O however, the binding energy[18] and well defined structures mean that such $\Delta\phi$ vs. θ treatments can not be applied.

Since LEED in conjunction with the coverage data gives a good idea of the structure, it is possible to relate the work function changes to the actual surface dipole moments at coverages when all dipoles are in the same environment. This is the case for oxygen on W(112) at 0.5 and 1.0 monolayers. The dipole moment μ, obtained from (6) is 0.40 Debye at 0.5 monolayer and 0.25 Debye at 1.0 monolayer. These are related on the basis of simple electric field depolarization to the isolated dipole moment μ_o and an effective polarizability α by

$$\mu = \mu_o - \alpha E_D \qquad (13)$$

where E_D is the electric field at one dipole due to all the other dipoles. A self consistent, analytic solution exists only when E_D is the same at every dipole. In this case E_D is proportional to μ and we have,

$$\mu = \frac{\mu_o}{1 + \alpha\beta} \qquad (14)$$

where β is given by $E_D = \beta\mu$. If each dipole is not in the same environment, this formula does not apply. The field factor β has been shown by Topping[19] to be equal to $9(n_s\theta)^{3/2}$ in c.g.s. units for both a square and triangular lattice, a fact which reflects the long range character of the electric field interaction between dipoles. By solving (9) for the two values of μ and the field factors $9(n_s\theta)^{3/2}$ for the p(2x1) and p(1x1) structures, we get $\mu_o = 0.55$ Debye and $\alpha = 5.5 \text{Å}^3$ by assuming the polarizability to be the same in both structures. These values serve mainly to describe the results in the framework of the model of electric field depolarization.

The interaction between dipoles is clearly more complex than a simple electric field depolarization. For alkali adsorption the electric field depolarization is much more plausible and indeed capable of yielding useful results.[17] The W-O bond is not so well understood. Some recent ion neutralization studies by Hagstrum[20] suggest that in the Ni-O system, oxygen is bound by electron exchange between the highest occupied level in the oxygen and nickel conduction band. There was no indication in the results that the electron affinity level was shifted and/or broadened sufficiently to be occupied by electrons from the nickel. Tungsten band structure calculations by Mattheiss[21] indicate that the bottom of the W conduction band is about 13 eV below the vacuum level for the (112) surface. This means the ionization level (13.6 eV) of the oxygen lies just at the bottom of the conduction band in the absence of any level shifting and broadening and it is reasonable to believe that a metallic typed bond will be formed between the oxygen and the tungsten surface. This type of bond could be polar as the result of asymmetric charge redistribution.

In conclusion it has been shown the work function changes as a function of oxygen coverage on a W(112) surface are closely correlated with the structural developments of the adlayer. Two structures form up to one monolayer coverage, the p(2x1) at one half monolayer which is completed by island growth, and the p̄(1x1) at one monolayer which is formed by random filling of sites vacant in the p(2x1) structure. The p(1x1) structure is characterized by depolarization strong enough to make the $\Delta\phi$ vs. θ curve have a relative minimum at one monolayer. The use of LEED in conjunction with work function and coverage determinations is clearly a powerful combination in the investigation of chemisorption phenomena.

ACKNOWLEDGMENT

This work was supported by the Advanced Research Projects Agency through the Materials Science Center, Cornell University, Ithaca, N. Y. In addition the support of the National Aeronautics and Space Administration under contract #NGR-33-010-029 and the American Iron and Steel Institute under contract #148 is acknowledged.

REFERENCES

1. See the complete bibliography recently presented by B. McCarroll, J. Chem. Phys. $\underline{46}$, 863 (1967).

2. J. W. May, Ind. Eng. Chem. $\underline{57}$, 19 (1965).

3. H. Pelevsky, R. K. Swank, and R. Grenchik, Rev. Sci. Inst. $\underline{18}$, 298 (1947); R. E. Simon, Phys. Rev. $\underline{116}$, 613 (1959); T. A. Delchar and G. Ehrlich, J. Chem. Phys. $\underline{42}$, 2686 (1965).

4. J. Eisinger, J. Chem. Phys. $\underline{29}$, 1154 (1958).

5. J. A. Becker, E. J. Becker, and R. G. Brandes, J. Appl. Phys. $\underline{32}$, 411 (1960).

6. J. C. Tracy and J. M. Blakely, submitted to Surf. Sci.

7. C. C. Chang, Ph. D. Thesis, Materials Science Center Report #720 Cornell University, Ithaca, New York (1967).

8. C. C. Chang and L. H. Germer, Surface Sci. $\underline{8}$, 176 (1967).

9. A. J. C. Wilson, X-Ray Optics, Wiley 1962.

10. C. Herring and M. H. Nichols, Rev. Mod. Phys. $\underline{21}$, 185 (1949).

11. Ya. P. Zingerman and V. A. Ishchuk, Soviet Physics Solid State (translation) $\underline{8}$, 728 (1966); $\underline{8}$, 2394 (1967); P. J. Estrup and J. Anderson, Proceedings of the Twenty Seventh Annual Physical Electronics Conference, MIT (1967).

12. L. H. Germer and J. W. May, Surface Sci. $\underline{4}$, 452 (1966).

13. A. Guinier, X-Ray Diffraction, W. H. Freeman (1963); P. J. Estrup and J. Anderson, Surface Sci. $\underline{8}$, 101 (1967).

14. T. B. Grimley, Proc. Roy. Soc. $\underline{90}$, 751 (1967).

15. A. J. Bennett, J. Chem. Phys. (in press).

16. R. L. Gerlach, this symposium.

17. J. R. MacDonald and C. A. Barlow, Jr., J. Chem. Phys. $\underline{39}$, 412 (1963); $\underline{44}$, 202 (1966); E. P. Gyftopolous and J. D. Levine, J. Appl.

Phys. $\underline{33}$, 67 (1962).

18.   D. Brennan, D. O. Hayward, and B. M. W. Trapnell, Proc. Roy. Soc. $\underline{A256}$, 81 (1960).

19.   J. Topping, Proc. Roy. Soc. $\underline{A114}$, 67 (1927).

20.   H. D. Hagstrum and G. Becker, Proceedings of the Twenty Seventh Annual Physical Electronics Conference, MIT (1967).

21.   L. F. Mattheiss, Phys. Rev. $\underline{139}$, 236 (1965).

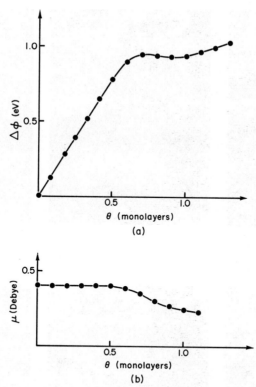

Fig. 1. a. Work function vs. coverage of oxygen adsorbed on W(112). The saturation value is 1.1 eV and is not indicated because the coverage determination beyond $\theta = 1.3\,\theta_o$ is very inaccurate. b. The dipole moment of W–O bonds as a function of coverage obtained from the Helmholtz equation (6) for W(112).

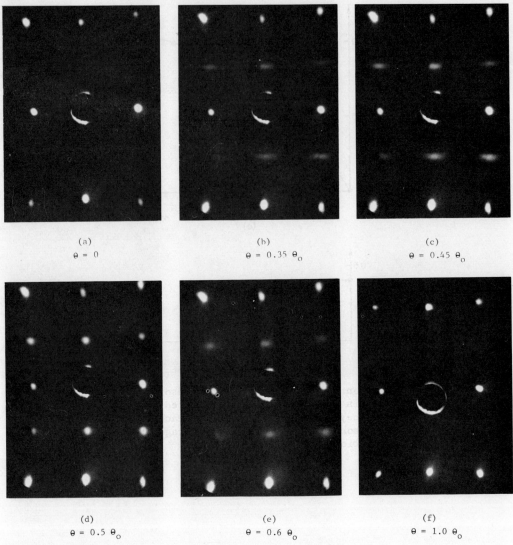

(a)
$\theta = 0$

(b)
$\theta = 0.35 \, \theta_0$

(c)
$\theta = 0.45 \, \theta_0$

(d)
$\theta = 0.5 \, \theta_0$

(e)
$\theta = 0.6 \, \theta_0$

(f)
$\theta = 1.0 \, \theta_0$

Fig. 2. LEED results for oxygen adsorbed on W(112) as a function of coverage. a. and f. taken at 59 eV beam energy; b., c., d., e., 53 eV.

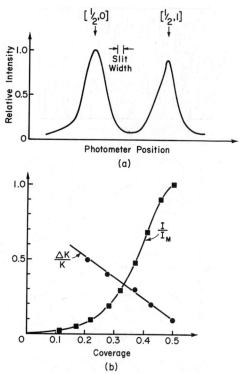

Fig. 3. a. Intensity profile of streaks of Fig. 2b measured directly from fluorescent screen with moveable spot photometer. The peaks correspond to the (1/2, 0) and (1/2, 1) positions. b. The breadth ($\frac{\Delta k}{k}$) and peak intensity ($\frac{I}{I_m}$) of the streaks at the half order positions as a function of coverage. The peak intensity is proportional to θ^2 up to about 0.4 θ_0 while the breadth is proportional to $1/\theta$ in the same region.

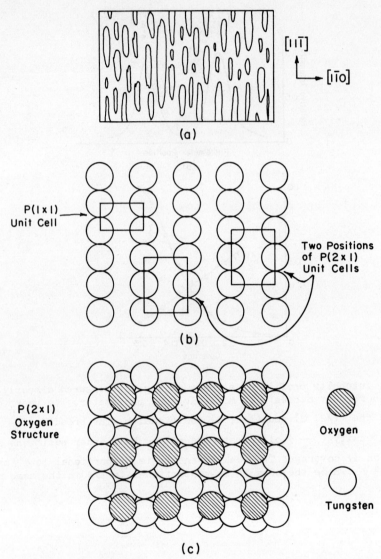

Fig. 4. a. A random arrangement of p(2x1) islands which might occur on the W(112) surface. b. The two possible out of step arrangements of the p(2x1) unit cell on the W(112) substrate. c. Arrangement of oxygen in the p(2x1) structure.

REFLECTION AND DISSOCIATION OF N$_2$O ON TUNGSTEN*

R. N. Coltharp, J. T. Scott, and E. E. Muschlitz, Jr.#

Department of Chemistry
University of Florida
Gainesville, Florida

Introduction

In recent years the modulated molecular beam method has been extensively applied in investigations of the scattering of atoms and molecules from solid surfaces[1], yet few attempts have been made in applying the method to studies of molecular dissociation on surfaces since the pioneering experiments of Smith and Fite[2] on the dissociation of hydrogen on tungsten. Nitrous oxide is particularly suitable for this type of investigation since the enthalpy change for the reaction, at 298°K,

$$N_2O(g) \quad \rightarrow \quad N_2(g) + O(g) \tag{1}$$

is only 40 kcal while the enthalpy change for the reaction,

$$N_2O(g) \quad \rightarrow \quad NO(g) + N(g) \tag{2}$$

is 103 kcal. The first reaction is therefore expected to be the predominant one, and this has been found to be the case for the gas phase thermal decomposition of nitrous oxide. Furthermore, the molecule is linear and the complete absence of spherical symmetry should provide an interesting contrast with hydrogen.

ᴧ

Supported by a National Aeronautics and Space Administration Institutional Grant to the University of Florida.

#

Visiting Fellow, Joint Institute for Laboratory Astrophysics, the National Bureau of Standards and the University of Colorado, 1968.

Apparatus

The modulated molecular beam apparatus used in these experiments is shown schematically in Figure 1. A beam of N_2O molecules is formed by a multi-channel source A consisting of a closely-packed array of glass capillaries 7 microns in diameter[3]. The beam is collimated by two slits each 0.040 inch wide and chopped mechanically at 90 herz. The tungsten surface is a polycrystalline ribbon T, 0.150 inch wide, located 12.5 cm from the beam source. The ribbon was heated electrically by passing a D.C. current through it. X-ray analysis of several aged samples showed the surface to consist predominantly of the (112) planes of small crystals, and that these surfaces were parallel to each other within one degree. The detector is an EAI quadrupole mass spectrometer mounted so that it could be rotated about the surface. The entrance slit of the spectrometer is located 1.2 cm from the surface and has an acceptance angle of 4 degrees. The ionizer of this instrument was modified in order to increase its sensitivity as a molecular beam detector as shown in Figure 2. In this arrangement electrons from the filament are accelerated by a positive potential of 60 volts on the grid. The outer can is operated at a small negative potential so that electrons are reflected back and forth through the molecular beam several times before being collected on the grid. An electron current to the grid of 10 ma was used in these experiments.

Although the incident angle of the molecular beam could be changed, all measurements were made at an incident angle of 50° from the normal to the surface. The detector could be rotated within 10° from the normal to the surface, and the data were taken at angles ranging from 10° to 80° from the normal. It was found necessary to avoid angles approaching 90° since the spectrometer begins to pick up some of the modulated beam that misses the target.

The signal from the quadrupole electron multiplier was first amplified by narrow band amplifier and the in-phase component measured by a PAR Model JB-4 lock-in amplifier. The chopper simultaneously interrupted a light beam which was detected by an NPN silicon light sensor thereby generating a reference signal.

The vacuum chamber was pumped by a 10-inch liquid nitrogen baffled oil diffusion pump. In the absence of a molecular beam the background pressure was 10^{-7} torr while in the presence of a beam background pressures rose as high as 10^{-6} torr. Since a pressure of the order of 10^{-9} torr is required to maintain a clean tungsten surface below 2500°K, the surface under investigation is covered with chemisorbed oxygen. Nevertheless, with beams of argon or nitrogen incident on the surface, the angular distribution of the reflected particles is diffuse (cosine law) at low temperatures, but a distinct lobular pattern begins to appear at temperatures of about 1800°K and as the temperature is increased further the lobe becomes sharper with its maximum in the neighborhood of the specular angle. These results are quite similar to those of Hinchen and Foley[4] who investigated the effect (on the angular distribution of argon atoms reflected from a platinum surface) of chemisorbed oxygen. They interpreted their results in terms of a gradual clean-up of the surface as the temperature is increased. However, the tungsten surface in this investigation cannot be considered completely clean below 2500°K.

Results and Discussion

Figures 3, 4, and 5 illustrate the transition from a diffuse to a lobular distribution of reflected N$_2$O molecules when a beam of N$_2$O is incident on the tungsten surface as the temperature is increased. The data in these figures are the directly observed mass spectrometer signals. In Figure 5, the data at all temperatures are shown on the same scale. In comparing these distributions with those for reflected argon or nitrogen from the same surface, several distinct differences are evident. Significant deviations from diffuse reflection for N_2O are not observed at temperatures below $2100^{\circ}K$, indicating a stronger interaction with the surface. Furthermore, the lobular distribution observed at high temperatures for N_2O is a markedly flattened one rather than having a distinct maximum in a specific direction, as is the case for Ar or N_2.

Dissociation of N_2O was found to be insignificant below $1800^{\circ}K$ and to increase rapidly between 1800 and $2500^{\circ}K$. The only dissociation product observed in the mass spectrometer was N_2. It is likely that the residence time of the oxygen atoms resulting from the dissociation is 10 milliseconds or longer in the temperature range of this investigation, which would prevent their observation by the technique employed. This conjecture is consistent with the observations of Steele[5] on the desorption of oxygen atoms from a tungsten surface.

The angular distribution of the dissociated nitrogen molecules was found to obey the cosine law throughout the entire range of temperatures investigated. Smith and Fite[2] found this to be the case also for the dissociation of hydrogen on tungsten. On the other hand, the reflected H_2 molecules exhibited a lobular distribution with a distinct maximum in contrast to the flattened lobe observed in this investigation for N_2O. Since the N_2 distribution was found to be diffuse, a reasonable assumption to make is that the velocity distribution of these molecules is a Maxwellian distribution characteristic of the surface temperature. The N_2 mass spectrometer signals can then be put on the same basis by first correcting for the N_2^+ ions formed by dissociative ionization of N_2O, then multiplying by the square root of the absolute temperature. The first correction could be avoided if the electron energy were kept below 17.0 ev, the appearance potential[6] for N_2^+ from N_2O, but above 15.6 ev, the ionization potential of N_2. However, the ionization efficiencies are much lower in this range of electron energies. It is assumed that the N_2^+/N_2O^+ ratio in the ionization of N_2O is independent of vibrational energy in the molecule, which may increase as the temperature of the surface is increased. However, part of this excitation will be lost by radiation before the molecule reaches the ionizer, since the vibrational transitions are optically allowed for N_2O. When these corrections are made, the results shown in Figure 6 are obtained, where the relative N_2 flux density at an angle of $40°$ from the normal as a function of the absolute temperature is shown. The N_2 intensity rises very rapidly at temperatures above $1800°K$, then levels off at temperatures above $2200°K$. Since the N_2O signal has practically disappeared at this point, the inference is that dissociation on the surface is nearly complete above $2500°K$.

References

1. R. E. Stickney, In "Advances in Atomic and Molecular Physics," Ed. by D. R. Bates and Emmanuel Estermann, Vol. 3, p.143, Academic Press, New York (1967).

2. J. N. Smith, Jr. and W. L. Fite, J. Chem. Phys. 37, 898 (1962).

3. J. C. Johnson, A. T. Stair, Jr., and J. L. Pritchard, J. App. Phys 37, 1551 (1966).

4. J. J. Hinchen and W. M. Foley, In "Rarified Gas Dynamics," Proc. 4th Intern. Symp., Toronto, 1964, Ed. by J. H. de Leeuw, Vol 2, p.505, Academic Press, New York (1966).

5. W. C. Steele, Avco Corporation Technical Report AFML-TR-65-343, Part II, Wilmington, Mass. (1967).

6. R. K. Curran and R. E. Fox, J. Chem. Phys. 34, 1590 (1961).

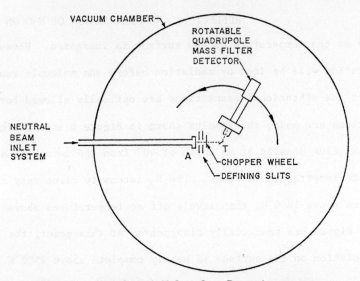

Fig. 1. Modulated Molecular Beam Apparatus.

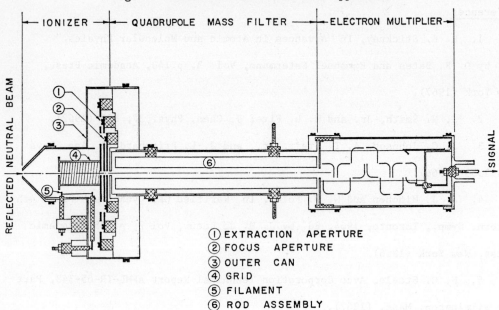

① EXTRACTION APERTURE
② FOCUS APERTURE
③ OUTER CAN
④ GRID
⑤ FILAMENT
⑥ ROD ASSEMBLY

Fig. 2. Quadrupole Mass Filter Molecular Beam Detector.

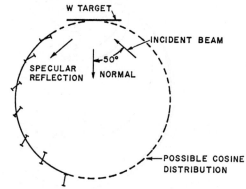

N$_2$O BEAM
W TARGET AT 300° K
LOOKING AT REFLECTED N$_2$O
ARBITRARY INTENSITY UNITS

Fig. 3. Angular Distribution of Reflected N$_2$O Beam at 300°K.

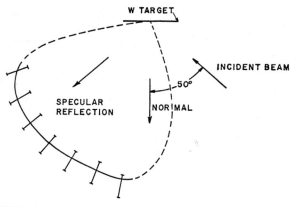

N$_2$O BEAM
W TARGET AT 2500°K
LOOKING AT REFLECTED N$_2$O
ARBITRARY INTENSITY UNITS

Fig. 4. Angular Distribution of Reflected N$_2$O Beam at 2500°K.

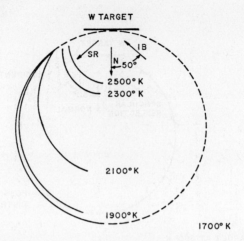

N₂O DISTRIBUTION AT VARIOUS TEMPERATURES

INCIDENT BEAM OF N_2O　　　　SR--SPECULAR REFLECTION

ARBITRARY INTENSITY UNITS　　　NN--NORMAL

　　　　　　　　　　　　　　　　IB--INCIDENT BEAM

Fig. 5. Comparison of the Angular Distributions of Reflected N2O at Various Temperatures. The data are normalized to the same incident beam intensity.

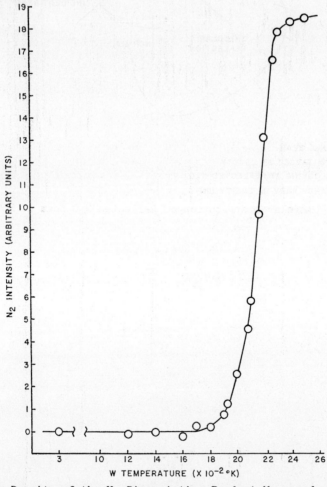

Fig. 6. Flux Density of the N_2 Dissociation Product Measured at 40° from the Normal to the Surface as a Function of Temperature.

THE ADSORPTION OF NITROGEN ON TUNGSTEN SURFACES

B. J. Hopkins and S. Usami

Surface Physics
The University
Southampton, England

1. Introduction

The experiments described in this paper are similar to those of Delchar
and Ehrlich[1] and embrace the same three crystal planes, (110), (100)
and (111), but also include measurements on the (112) and (116) planes.
The prime object of the work was to confirm and extend the surface
potential changes observed by Delchar and Ehrlich during the exposure
of these tungsten crystals to pure nitrogen gas.   Crude flash desorption
measurements are also reported.

2. Experimental

The crystals were discs sliced from a single crystal rod which had a
diameter of ~ 8 mm.   One surface of each disc was oriented to within
0.5° of the specified index and was both mechanically and electro-
polished prior to mounting in the experimental tube.   A tungsten/
tungsten-26% rhenium thermocouple was fitted to each crystal (with the
exception of the (110))and the assemblies mounted on tungsten rods.
Each crystal could be heated either by means of an electron beam focused
onto the reverse side or by straightforward radiation from the electron
source.   Using the latter only, a maximum temperature of about 1000°K
was possible: the electron bombardment could, of course, easily be made
to melt the crystals.   The arrangement in the experimental chamber
is shown in Fig. 1.   One of the crystals, the (110), and a sample of
polycrystalline foil were mounted onto stainless steel bellows so that
they could be positioned over either of two of the remaining four crystals
each.   Once positioned, an electromagnetic vibrator was applied to the
free end of the bellows and either the (110) crystal or the foil vibrated
in a plane perpendicular to its surface at a convenient natural frequency.   The
alternating signal due to the contact potential difference (c.p.d.) between the

two surfaces was amplified, integrated and fed to a three pens chart recorder.
Other pens on the same chart monitored the crystal temperature and the gas
pressure. The sensitivity of the c.p.d. detection system was about ± 10 mV.

The nitrogen was supplied from a 1 litre Pyrex bottle of spectroscopically pure gas via a bakeable leak valve and a liquid-nitrogen
trap. The gas purity, which was monitored by a mass spectrometer
fitted close to the experimental chamber, did not differ from the
residual gas composition which at about 2×10^{-10} torr comprised
H_2O (1×10^{-10} torr), CO_2 (5×10^{-11} torr) and CO (1×10^{-10} torr).
The effect on gas purity of an evaporated nickel getter was examined
and found to be negligible. For several days after the evaporation
of thoroughly outgassed nickel the partial pressures of H_2O and CO_2
were, respectively, x4 and x2 higher than the figures above, though
they ultimately fell back to that level. It is also of interest to
note that in recent measurements on evaporated nickel films made by
the authors, no adsorption of hydrogen was observed up to pressures
of 10^{-7} torr. The presence of hydrogen as an impurity in nitrogen is
therefore also unlikely to be appreciably affected by the presence of
a nickel getter.

No differences were observed for measurements made either with
or without the getter: nor for measurements with and without heated
filaments in the system. As a check on the purity of the gas actually
adsorbed onto the crystals, flash desorptions were made after exposure
of the (110), (112) and (111) crystals to 10^{19} molecules cm^{-2} and the
composition of the released gas analyzed in the mass spectrometer.
The results obtained are shown in Table 1.

The outgassing of the crystals involved about 150 h at $2500^{\circ}K$
together with intermittent flashing to $3000^{\circ}K$ for a few seconds at a

time.   At the end of this period the work functions of the clean crystals

were completely stable and the crystals could be flashed to $3000^{\circ}K$,

without the background pressure rising by more than 1 to 2 x 10^{-10} torr.

Identical heat treatment given to similar (110) and (100) crystals

gave, when examined by low energy electron diffraction, no evidence for

the existence of facets.   The appearance of the (110), (100) and (112)

crystal surfaces when viewed by optical and electron microscopy was very

similar to that before the heat treatment in vacuum and revealed large

flat featureless regions.   The (111) surface had, however, undergone

considerable changes.   The original single crystal surface had polygonised

and comprised about twenty crystallites roughly 2 mm in average diameter.

The Laué back reflection picture for this surface, Fig. 2, was now made

up of multiple spots though these were still accurately (111) oriented.

The new feature is, therefore, the presence of grain boundaries on an

otherwise well oriented (111) crystal.

The surface potential measurements were made in the following manner:

after the achievement of an ultra-high vacuum, with all of the crystals

in a clean state, nitrogen gas was admitted up to a pressure of about

0.5 torr and the c.p.d. between pairs of crystals plotted continuously.

The gas was then removed and an ultra-high vacuum obtained once again.

(A feature of nitrogen gas is the ease with which it can be removed

from the system.   The pump-down from 0.5 torr to 10^{-10} torr usually took

about 1 hour. This compares very favourably with gases such as oxygen,

hydrogen or carbon monoxide where exposure, even at 10^{-7} torr, necessitates

a much more extended pump-down.)   One of the surfaces was then flashed

clean in vacuum while surrounded with a radiation screen.   The clean

surface and the covered surface were re-exposed to nitrogen in exactly the

same way and the c.p.d. changes followed throughout.   The combination of

the two c.p.d. curves against exposure permits the determination of the
surface potential versus exposure curves for each of the two surfaces.
Clearly, in this method the reference is the gas saturated surface and,
provided that its surface potential at complete coverage is irreversible
with gas pressure, the results are reliable. At very high nitrogen
exposures ($\sim 10^{23}$ molecules/cm^2) reversible surface potential changes
of about 60 mV are observed on all surfaces. Beyond 10^{23} molecules/cm^2
the results are not unambiguous.

It is sometimes more convenient to express the results in terms of
work function changes on the various crystal planes rather than as surface
potentials. As a reference level we have followed our usual practice
of assuming the work function of clean polycrystalline tungsten foil
as 4.55 eV[2].

3. Results

The results of the surface potential measurements are shown in Fig. 3.
It will be seen that in all cases, with the exception of the (100) crystal
face, the final surface potential at high coverages is negative, i.e.
an increase in the work function. Further, appreciable potentials are
involved on all the crystal planes and the smallest change, that on the
(116) plane, amounts to 100 mV at a nitrogen expsoure of about 10^{23}
molecules/cm^2. The main results from the figure are summarized in Table 2.

Figure 4 and 5 show the results of flash desorption measurements
on the (100) and (111) crystals. The main limitation of these measurements
is, of course, that the edges of the crystal discs, the tungsten support
rod and the thermocouple will not have the orientation of the faces of
the discs. Indeed, the electron bombardment heating of the crystals
(at $<$ 1 kV electron energy) causes so much damage to the reverse sides

of the crystals that it is unlikely that much of this surface will contain the original orientation.   It is probably optimistic to suggest that more than 50% of the total flashed surface would be of the specified crystal orientation.   It would not have been surprising, therefore, if a multiplicity of states, rather similar to a polycrystalline wire sample, had been observed on all the crystals.   The simplicity of the results actually obtained is remarkable.

On the (110) plane, at all stages of exposure, a single desorption peak was observed.   Similarly on the (111), (116) and (112) single peaks were observed until exposures exceeding 10^{20}, 10^{20} and 10^{22} respectively were reached.   Beyond these exposures double peaks were observed.

The effects of heating on the adsorbed gas layers was examined for the (111) and (100) crystals.   Each crystal was completely covered with nitrogen and its temperature raised from $300^{\circ}K$ while the c.p.d. against a nitrogen saturated reference was followed.   The most accurate way of performing this experiment is to heat the crystals in steps, cooling to room temperature after each heating, in order to re-determine the c.p.d. The curves shown in Fig. 6 were obtained in this manner.   Without rather more sophisticated electronics and a double modulation of the vibrating electrode Kelvin measurements at temperatures exceeding about $800^{\circ}K$ are not very reliable, however they do confirm the general shape of the curves up to a temperature of about $900^{\circ}K$.

4.   Discussion of the Results

The main technical differences between the present measurements and those of Delchar and Ehrlich are as follows:

(a) the reference electrode used by Delchar and Ehrlich was a platinum foil which, it was claimed, does not adsorb nitrogen;

(b) very much milder outgassing of the crystals was used by Delchar
and Ehrlich which we doubt would be adequate, particularly for the removal
of carbon;

(c) their crystals appear to have been heated by electron bombardment
on their face side with electrons of 3.5 keV energy.   The reverse side
of our crystals, after electron bombardment by focused electrons of less
than 1 keV energy became very badly pitted;

(d) the gas purity was not monitored by Delchar and Ehrlich and there
is a possibility of contamination by water vapour, CO_2 or H_2 none of
which was pumped well by our nickel getter;

(e) the crystals of Delchar and Ehrlich were aligned less closely to their
stated orientations, within 4° compared with 0.5° in the present work.

In spite of these differences, the general agreement between the
surface potential measurements at $300^\circ K$ on the (100) and (111) planes
is good.   The depth of the minimum work function on the (100) plane
is slightly lower in the present work, 0.47 compared with 0.4 eV, but
the difference is barely significant.   Estrup and Anderson[3] observed
a work function decrease on the (100) plane of 0.65 eV at an exposure
which roughly corresponds to the minimum work function.   The subsequent
decrease in this value, which occurs at higher exposures, Estrup and
Anderson attribute to the gradual adsorption of carbon monoxide.   The
mass spectrometer soon confirmed that in the present measurements at
least, carbon monoxide is certainly not responsible for the surface
potential decrease, at high exposure.   The remainder of the (100)
curve appears to be in fair agreement up to the maximum exposure (about
10^{19} molecules/cm^2) reported by Delchar and Ehrlich.   The most important
difference on the (100) plane is in the flash desorption results: beyond
the minimum work function the present work reveals a double not a single
desorption peak.   The presence of two distinct nitrogen species is

confirmed by the stepwise desorption curve (Fig. 6). Ignoring, for the present, the changes between 300 and 700°K, it is clear that between 700 and 1000°K the less tightly bound state, responsible for the work function increase at high exposures, is removed. After heating to 1000°K the work function of the fully covered surface has returned to that of the minimum position. Between 1100 and 1400°K the strongly bound state giving rise to the work function decrease is also removed.

On the (111) surface and at an exposure of about 10^{17} the present surface potential is slightly more negative than that of Delchar and Ehrlich. Since, however, the surface potential is still increasing with exposure at that point the actual magnitude, between 0.17 and 0.22 volts has little significance. The presence of a double flash desorption peak is observed in both sets of measurements. The presence of these two states separated by about 200°K in desorption temperature is confirmed in Fig. 6, from which it also appears that they contributes roughly equally to the surface dipole.

We are, however concerned about the structural changes that took place on this particular (111) crystal during the extended high temperature outgassing period. The unusually high work function (Table 2), compared with our previous measurements on this plane[4] and which gave a value of 4.45 eV, may be a consequence of this structural change. We suspect that the severity of our outgassing has increased over the past few years and that this may be responsible for the changes on the (111) surface.

The greatest differences between the present measurement and those of Delchar and Ehrlich relate to the (110) orientation. It is quite clear from Fig. 4, that adsorption on the (110) surface took place at 300°K giving rise to a work function increase of 180 mV at an exposure of 10^{23} molecules/cm^2. This crystal was not fitted with a thermocouple, but the flash desorption experiment revealed a single peak at all exposures. At no stage in any of our measurements was a work function decrease observed due to nitrogen adsorption on the (110) plane.   This result is of some interest because it removes the basis for the Delchar and Ehrlich model for surface potential changes due to nitrogen on tungsten. Any adsorption on the (110) plane ought, by this model, to give rise to a work function decrease, because the (110), like the (100) plane, is atomically smooth.   This is unfortunate since the model has the virtue of simplicity and also the direction of the changes on the (112) and (116) planes fit the model well.

In conclusion we would point out that it has been possible to perform one of the experimental checks of the adsorption model of Robins, Warburton and Rhodin[5] for nitrogen on tungsten.   This involves heating the nitrogen covered crystal surfaces and observing the work function change.   The model predicts that above 650°K the work function will increase with temperature.   For neither of the crystals shown in Fig. 6 does this occur.

5.   Acknowledgments

Mr. Usami would like to thank the British Council for the provision of a maintenance grant.

6.   References

1.   T. A. Delchar and G. Ehrlich, J. Chem. Phys., 42, 2686, 1965

2.   B. J. Hopkins and J. C. Rivière, Proc. Phys. Soc. (London), 81, 590, 1963

3.   P. J. Estrup and J. Anderson, J. Chem. Phys., 46, 567, 1967

4.   B. J. Hopkins, K. R. Pender, S. Usami, Fundamentals of Gas–Surface Interactions, Academic Press, p.284, 1967

5.   J. L. Robins, W. K. Warburton and T. N. Rhodin, J. Chem. Phys., 46, 667, 1967

Table 1

The partial pressures of gases released from each of the crystal planes on flashing to $1500°K$ after exposure to 10^{19} molecules per cm^2. The difference between the nitrogen and the carbon monoxide pressures, mass 28, was estimated from the mass 12 peak using the measured cracking ratio: $C^+/CO^+ = 14\%$.

Plane	Partial Pressure (torr)				
	$\begin{matrix}12\\(C^+)\end{matrix}$	$\begin{matrix}18\\(H_2O^+)\end{matrix}$	$\begin{matrix}28\\(N_2^+)\end{matrix}$	$\begin{matrix}28\\(CO^+)\end{matrix}$	$\begin{matrix}44\\(CO_2^+)\end{matrix}$
(110)	2×10^{-10}	$< 1 \times 10^{-10}$	1.0×10^{-6}	$< 1.5 \times 10^{-9}$	2×10^{-10}
(112)	2×10^{-10}	$< 1 \times 10^{-10}$	1.6×10^{-6}	$< 1.5 \times 10^{-9}$	2×10^{-10}
(111)	1.5×10^{-10}	$< 1.5 \times 10^{-10}$	1.0×10^{-6}	$< 1.1 \times 10^{-9}$	1.5×10^{-10}

Table 2

Summary of the surface potential and work function changes on the various tungsten surfaces

Plane	Clean Work Function (eV)	Maximum Surface Potential * at Indicated Exposure (V)	Maximum and Minimum Work Function (eV)	Surface Potential at Exposure of 10^{23} (V)	Work Function at Exposure of 10^{23} (eV)
(110)	5.10 ± 0.01			−0.18 ± 0.03	5.28 ± 0.03
(112)	4.76 ± 0.01	−0.51 ± 0.02 (1 x 10^{22})	5.27 ± 0.02	−0.34 ± 0.03 **	5.10 ± 0.03 **
(100)	4.64 ± 0.01	+0.47 ± 0.02 (2 x 10^{15})	4.17 ± 0.02	+0.06 ± 0.03	4.58 ± 0.03
(111)	4.60 ± 0.01			−0.38 ± 0.03	4.98 ± 0.03
(116)	4.59 ± 0.01			−0.10 ± 0.03	4.69 ± 0.03
FOIL	4.55 ± 0.01 ***	+0.05 ± 0.02 (1 x 10^{15})	4.50 ± 0.02	−0.14 ± 0.03	4.69 ± 0.03

*   Negative sign indicates increase in work function

**   At exposure of 6 x 10^{23} molecules/cm^2

*** Assumed reference level [2]

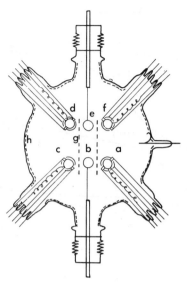

Fig. 1.   The experimental tube.   a, b, c, d and f are the (112), (110), (111),
(100) and (116) crystals, respectively; e is the polycrystalline
tungsten foil; g is a radiation screen that can be moved magnetically;
h is a conducting platinum screen "burnt" into the glass.

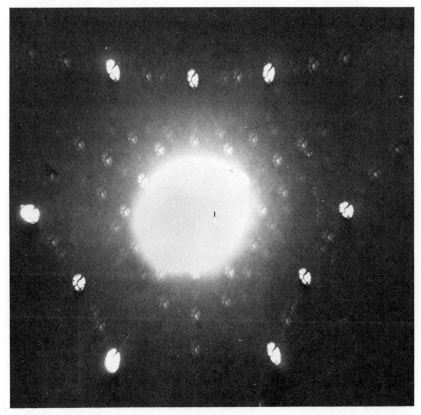

Fig. 2.   Laue back reflection photograph of the (111) surface after removal
from the experimental tube.

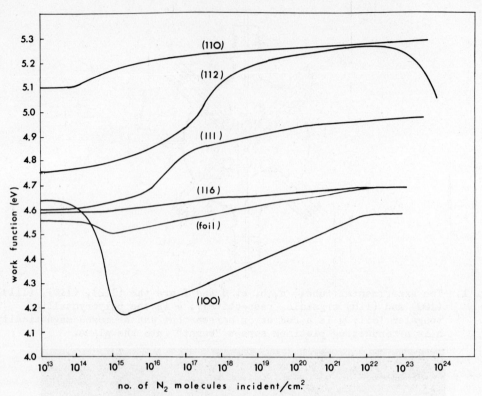

Fig. 3. Work function changes on the six surfaces during the continuous introduction of nitrogen.

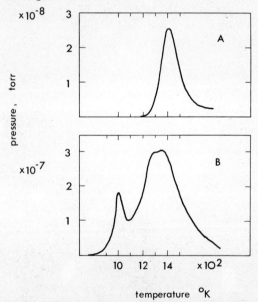

Fig. 4. Flash desorption results on the (100) surface. The crystal was heated linearly at a rate of 7°K/sec. Curve A is after an exposure of 4×10^{14} molecules/cm^2, B after an exposure of 1×10^{20} molecules/cm^2.

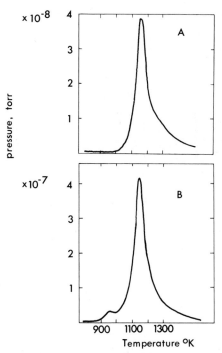

Fig. 5. Flash desorption results on the (111) surface. The crystal was heated linearly at a rate of 12°K/sec. Curve A is after an exposure of 3×10^{16} molecules/cm^2, B after an exposure of 1×10^{21} molecules/cm^2.

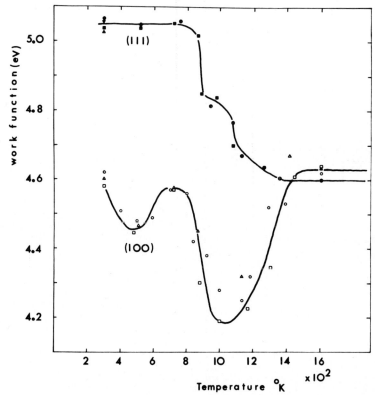

Fig. 6. The work function changes on the (100) and (111) surfaces due to heating the crystals after an exposure to 10^{23} nitrogen molecules/cm^2.

ADSORPTION PHENOMENA IN A POTASSIUM-HYDROGEN-TUNGSTEN SYSTEM AS OBSERVED BY THERMAL ACCOMMODATION OF INERT GASES AND HYDROGEN

Lloyd B. Thomas, William V. Best, and Howard L. Petersen

Physical Chemical Laboratory
University of Missouri
Columbia, Missouri

Introduction

The thermal accommodation coefficient (A.C.) of inert gases on metal filaments is known in a number of instances to be a sensitive indicator of the state of adsorption of the metal surface.[1,2,3] Description of the method and some early results commands space in most treatises or extensive review papers on adsorption. However, following the innovating work of J. K. Roberts and co-workers in the nineteen thirties, in which the strikingly low A.C. values of He and Ne on bare metals and the sensitivity of the A.C. of these gases to adsorption were discovered and first applied, use of the A.C. method of studying adsorption seems to have languished. In contrast there has been strong revival of interest in the theory of the A.C. of gases on clean surfaces and an accompanying interest in acquiring A.C. data for comparison and check. A next step, already beginning, is development of theory of the A.C. of indicator gases on adsorbed films. With adequate development of this, the magnitudes of the observed A.C. values should become significant in providing information concerning bonding and other characteristics of the films. The work presented here, from a sequence of investigations, uncovers some striking A.C. phenomena, leads to independent discovery of coadsorption of hydrogen and potassium and of upper film formation, and perhaps reveals aspects of such films not seen by other methods. Hopefully, this account may serve in a small way to help keep the method afloat pending revival, which it may well warrant.

Experimental Procedures and Results

The experiments reported below were done in glass systems capable of high vacuum, with necessary stopcocks confined to side manifolds separated by mercury cut-offs from the main system containing the McLeod gauge, liquid air traps, experimental tube, etc. Gases were admitted to the main system through baked, liquid air cooled charcoal traps. Filaments upon which the A.C. measurements were made were kept as simple as possible, with no potential leads or springs, and were about 20 cm long and between 1.5 and 2.5 mils in diameter. Extensive baking and flaming of the system was carried out, as necessary, until clean surface A.C. values of the inert gases, over extended periods after flashing, were achieved.

The first experiments reported were done in this laboratory by Petersen[4] in tubes set up to determine the A.C. of inert gases on potassium surfaces. Potassium capsules were broken into the system and the metal moved by flame to form a mirror over the tube walls. Filaments were heated to remove the original uneven coats, and relatively thin and uniform coats for bulk potassium measurements were applied by heating strips of the wall parallel to the filament with a small sharp-pointed flame. Flashing the filament about 2400°K renders the surface clean at cessation of the flash. In a thermostat bath at 25°C, with several hundredths of a millimeter of He or Ne present and a few milliamperes of current through the filament, a thermal steady state is reached in three to five minutes after the flash, which then allows A.C. determinations to be made. At 25°C potassium has a sublimination pressure 1.5×10^{-8} mm, and a filament requires about 1.4 minutes of impingement of K atoms at this pressure for complete coverage. Thus by the time A.C. measurements can be made, the filament is also at a steady state with regard to K atom adsorption-desorption. A.C. values at low (circa 2°) temperature excess of filament above wall, i.e. low ΔT, on filaments are quite steady with time, are reproducible, and have magnitudes

about 0.095 for He and 0.225 for Ne.   (These are quite close to the values

for bulk potassium. The corresponding values on bare tungsten are about

0.019 for He and 0.052 for Ne.) As ΔT is increased the A.C. decreases as

shown in Curves Helium (A) and Neon of Figure 1.   Identical curves were

found starting at high ΔT and coming down in ΔT or vice versa. The A.C. of

argon (not plotted) shows similar behavior in falling from 0.52 to 0.41 over

the ΔT range plotted. Furthermore, all three gases show minima in their A.C.

values at ΔT about 25 to 30 degrees, with prominent rise at higher ΔT. Since

the A.C. in general is a property which has been found to change quite slowly

with either filament temperature or, especially, ΔT, the 40% reduction of the

A.C. of both He and Ne for a 20° rise in ΔT was tentatively ascribed to loss

by the filament of an upper adsorbed layer of K atoms with binding energy

only slightly above the sublimation energy of the bulk potassium on the walls.

In the hope of gaining information concerning the cleanness of his potassium

surfaces, Petersen examined the effects of added H_2—a contaminant thought

possible by reaction of K with H_2O adsorbed in the system—on the A.C.

behavior of He on the surfaces. In one experiment he added 3×10^{-4} mm of

H_2, heated the filament to 700°K for 10 minutes, added 4.3×10^{-2} mm of He

and examined the behavior of the A.C. of He with changing ΔT. The points

(circles) along the He (B) curve of Figure 1 resulted, and again were closely

reversible. The He and H_2 were pumped out, fresh He was added, and the three

new points (triangles) fell precisely on the Helium (B) curve. Upon flashing

and re-examining, the behavior coincided with that shown in Helium (A). It

appears that hydrogen can be incorporated in the film in a stable enough way

to withstand evacuation, and this film shows quite different A.C. of He

characteristics.

Clarification of the phenomena described, which were incidental to
Petersen's main objective, were undertaken by Best.[5] One method of angling
for progress in these studies is to vary the procedures of preparation of the
tube for A.C. measurements, and to try to account for observed differences in
the A.C. results. Consider Curves A and B of Figure 2. Previous to both runs,
potassium was distilled about in the tube, which was open to the pumps. For
Curve A the filament was given a one hour flash at 2700°K, also open to the
pumps, He was admitted, and after the filament was held for 90 minutes at
$\Delta T = 30°$ the points 1 to 4 were taken. For Curve B the filament was given a
short flash, the current cut off for 80 minutes and the ten points taken in
the order indicated. Quite different behavior is indicated. For Curve C,
Figure 2, there were two liquid air traps between the tube and the vacuum
system. The potassium was baked out of the tube while open to the pumps into
the adjacent cold trap and after flashing the filament a clean surface A.C.
value was obtained (about 0.02). Some metal was then distilled back into the
tube (while pumping), He was admitted, the filament flashed, and after 8
minutes the first point of Curve C was taken at low ΔT. The A.C. was followed
for 300 minutes before finally coming to the value normally expected within
3 minutes. Considering the ready availability of K vapor, it is difficult to
attribute these variations in second layer formation as solely due to differences in the tungsten and potassium components (or helium) of the system.
(The variations were found later to be caused by varying amounts of traces of
H_2 in the system.)

In the experiments so far described the time required, after changing
the filament current, for a thermal steady state to be established (through
radial and axial conduction of heat from and by the filament) is longer than
that required for the adsorption-desorption steady state. By lowering the
bath temperature to 0°C the above situation is reversed. Times of the order

of 30 to 50 minutes are then required for monolayer coverage. Experiments at the lower sublimination pressures have been particularly fruitful in providing information on the adsorption processes under observation. In preparation of the runs of Figure 3 the filament was flashed and held at $\Delta T = 28.5°$ for 88 minutes with tube walls (bath) at 0.2°c. The first 7 points of Curve A were taken at 10 minute intervals at decreasing ΔT values. Then a $\Delta T = 25.09°$ was set, and after 15 minutes Point 8 was taken, again confirming strict reversibility of the process. For Curve B the first point, at zero time, is Point 8. At t = o the ΔT was dropped to 3°, and the surface started to acquire the upper layer, and continued to do so over a 41 minute period, until the A.C. returned to the value characteristic of 3° ΔT shown on Curve A. This appears to be reasonable indication that the A.C. behavior discussed is due to formation of a single, approximately complete layer of K atoms. It should be emphasized that the Curve C of Figure 2 behavior is due to inability of the lower film in that case to acquire an upper layer, which, if it had been like the lower film of Curve B, Figure 3, it would have accomplished in less than 2 minutes at the 32.6°C sublimation pressure.

All the preceding has been description of adsorption on a film already formed. By using lower bath temperature it becomes feasible to observe through the A.C. of He or Ne the whole adsorption process, starting with the bare tungsten. For the data of Figure 4 the bath was set at 0.2°C, the filament flashed for 3 minutes in the He to be used, and A.C. measurements were taken at intervals over a period of 170 minutes without disturbing the filament current. The A.C. is seen to rise to a peak value about 0.1, then fall for another 20 minutes to a minimum about 0.053 and then to rise and level at 0.091. If we assign the time to the minimum as that required to complete the first layer of potassium, the time for the second rise strongly suggests, taking into account the asymptotic manner in which one expects the approach to the steady state to occur as rate of desorption approaches rate of adsorption, a second complete layer of potassium atoms has been observed. At the time indicated by the arrow, X, the ΔT is increased to 23° and the A.C. drops to the value at the minimum. At Y the ΔT is lowered to 5.4° and the approach to the steady state is repeated. General observations from numerous such runs with various preparation procedures are: (1) The times to reach the minimum at a given bath temperature are closely reproducible and may be taken as a measure of the sublimination pressure of the potassium. (2) The peak heights vary for He from 0.7 to 0.13. (3) The final asymptotic value of the A.C. of He varies from that at the minimum up to 0.095. (4) If the ΔT is above about 20° there is no rise in the A.C. beyond the minimum. (5) The height of the peak and of the asymptotic A.C. value, the latter at low ΔT (much less than 20°), depends on the pressure of H_2 in the system. Preparation procedures which minimize the H_2 in the system lead to runs showing the lowest peak heights and the smallest rise past the minimum. Elimination of H_2 eliminates any rise in the asymptotic value above the minimum .

During the course of this work it became evident that whenever potassium was brought in contact with glass (except in the cold trap) the vacuum was not rigorously maintainable in the closed system. The ability of the McLeod gauge to maintain a "sticking" or "clicking" vacuum over long periods after cut-off from the pumps was lost when potassium was present. With potassium open in the system, even separated from the filament by cold trap, the A.C. of He observed upon attempting clean conditions was about 0.04, a value we had previously found for tungsten exposed to H_2.[3] In a separate investigation[6] we have studied the production of H_2 by alkali metals on glasses and find all the alkali metals liberate H_2 persistently from Pyrex glass when heated in contact. There is ample evidence that potassium does not compete with tungsten for H_2 adsorption, and pressures of H_2, of the order of 10^{-6} mm in the 1000 ml system, necessary for monolayer coverage of the filament, are to be expected. Beyond doubt, some H_2 was present in the system for all curves of Figures 1 to 4. Tubes were constructed utilizing mischmetal as a getter to eliminate or at least minimize the H_2 present and these showed an average A.C. value 0.072 at the peak and an average 4% rise from the minimum, at low ΔT, compared to the 70% rise shown in Figure 4. Other work[7] with stringent hydrogen elimination measures shows no rise beyond the minimum in the K-W-He system.

Brief mention is made below of experiments in which H_2 is used as the indicator gas and it is then also in plentiful supply as a potential constituent of the films. Potassium films allowed to accumulate on tungsten from the vapor, and bulk potassium metal deposited on tungsten both show an A.C. of H_2 value 0.064[*]. Accumulated films of potassium on tungsten, possibly containing hydrogen, show a reversible change in A.C. of H_2 with increasing ΔT, but in this case increasing ΔT gives increased A.C., opposite to that for He. In Figure 5, Points 1 to 12, taken at 5 minute intervals at increasing ΔT, appear to illustrate the above behavior, b ut in this set no attempt was made to reverse the ΔT steps. (Similar experiments show complete reversibility at least up to 16° ΔT at 32.6°C bath.) Previous experiments showed that irreversible changes in such systems begin to occur at a rapid rate with filament temperature a little above 100°C. In the case of Figure 5, following Point 12 at 30° ΔT the filament temperature was held at 112°C for 10 minutes, and then Point 13 was taken at 27° ΔT. The new film thus formed shows an A.C. of H_2 between .20 and .21 over the range of decreasing ΔT shown.

---

[*] This and all A.C. values given in this paper are uncorrected for the difference in end conduction of the filament in gas and in vacuum. Correction for this reduces the A.C. value about 14% in our experiments.

In an attempt to produce this film in complete coverage of the surface, by cycling the filament temperature from $35°C$ to $115°C$ and back several times, with bath at $32.6°C$, a maximum A.C. of H_2 value, 0.24, was found at a ΔT between $40°$ and $50°$, with the A.C. value diminishing above and below this ΔT range. Upon replacing the H_2 gas by He, taking care not to unduly disturb the film, the A.C. of He indicated identical behavior of this film to that shown in Figure 1, Curve He (B), and this again reverts, upon flashing, to the Curve He (A) behavior. Both of these films are thought to contain hydrogen, but their observed properties are distinct. Measurements of the A.C. of He, in a mixture containing 10% H_2, also indicate that the change in film nature from that indicated by Curve He (A) to that by Curve He (B), Figure 1, proceeds readily when the filament reaches $115°C$.

Roberts and co-workers preferred neon to helium as the indicator gas. We have found heretofore that the (A.C. of Ne)/(A.C. of He) ratio in situations investigated is 2.4 ± 0.3, and this compensates for the free-molecule thermal conductivity ratio to make both gases essentially equally effective in taking heat from filaments. Thus He, which has lower A.C., would have a slight advantage of less supression by the upper limit of unity, particularly when ratios of A.C. values are considered. In Figure 6 are plotted the A.C. of He and the A.C. of Ne, the latter divided by a factor of 2.1 to bring the two to the same ordinate on the right of the foot of the peak, against time of accumulation of potassium at $0.2°C$. These curves are for a nearly H_2 free system. It is noted that the A.C. of Ne is very high relative to that of He at the peak. Ne is thus more sensitive by a factor of 2.2 than He in this situation. This is presumably due to the larger polarizability of Ne and the resulting larger attractive force the Ne experiences in encountering the large dipole fields present when the surface is partially covered with potassium.

It has been pointed out that a ΔT about 25° eliminates the rise beyond the minimum of Figure 4. Further rise in ΔT causes K atoms to start to desorb from the first layer, thus accounting for the minimum in the A.C. when ΔT is increased above 25 to 30°, mentioned earlier. By heating the filament to a set temperature for a period, then cutting the current to a low value and starting a run of A.C. vs. time, one can retrace any portion of the curve ending at the minimum, and thus one gets the steady state degree of coverage at any temperature of filament and pressure of K. Roughly, with K sublimation pressure at 0°C, the filament retains three fourths of the K in the complete film at 380°K, half at 430°K, and one fourth at 700°K. One can obtain isosteric heats of adsorption over the full range of coverage by measuring the A.C. of He vs. temperature of filament at two sublimation pressures of K, selecting temperatures to give equal coverage (equal A.C.) and applying, with minor corrections, the Clapeyron type equation. For reliable results one needs a filament of constant temperature over the portion used (obtainable with fine potential leads on a long filament, for example). We have only confirmed feasibility of the method in our simple filament tubes. Similar limitations apply to the coverage-temperature figures given above.

Summary and Discussion

Two distinct types of film of K and H_2 coadsorbed on W are indicated by the experiments. One forms readily at 305°K and below and it picks up another weakly bound full layer of K from the vapor when the temperature of the film is near that of the K solid supplying the vapor. The other type requires a filament temperature about 385°K to form at a convenient rate. It also appears able to adsorb an upper layer of K at low ΔT, but this was not investigated in detail. The first type accounts for behavior of all curves in Figures 1 to 4 except the Helium (B) curve of Figure 1. The second type accounts for the behavior shown in the Helium (B) curve and by the points following Point 12, Figure 5. The A.C. of He and of Ne indicates the progress of formation of pure K on W films, but indicates no second layer of K forming on these films at ΔT as low as 2°.

At present little correlation of the A.C. value with film structure can be made and evidence for this must come from other sources. Bosworth and Rideal[8] have encountered coadsorption of K and H_2 on W in contact potential studies and postulate a film with electropositive K atoms and four times as many electronegative H atoms occupying neighboring sites on the W surface. This structure could conceivably attract an additional layer of K atoms. These authors and also Schmidt and Gomer[9], in connection with field emission studies, suggest existence of a W-H-K structure. The latter[9] suggest a configuration "in which the small H atoms are in their normal adsorption sites and K atoms are slightly displaced from theirs to establish some contact with both H and W atoms." Our results indicate, from the reproducibility of times to complete the films, that whatever the structure of the co-adsorbed film (first type) is, it contains the same number of K atoms per unit area as the pure K film on W.

The A.C. method as applied in these studies appears to be capable of indicating sensitively the occurrence and some rates of adsorption processes, provided that the A.C. values for the films encountered are well spaced. The linear relation between the change in A.C. and coverage often assumed for interpretation in past work is spectacularly refuted in formation of the coadsorbed and pure first films in the cases studied here -- compare the over-the-peak behaviors of Figures 4 and 6 with straight lines from the extrapolated beginnings (.04 and .02 respectively) of the curves to the minima. Although the A.C. value at present does not give reliable characterization of the films, information can be gained by altering conditions and finding the effect on the adsorption as viewed by the A.C. Many examples of this are illustrated in the paper. The A.C. method does have a distinct advantage over some other methods (field emission, for example) in that it can maintain surveillance over an adsorption process with a very minimal disturbance of the equilibrium condition and can tolerate pressures up to several tenths of a millimeter without foregoing the simplicity of the free-molecule range of heat conduction.

Acknowledgment

The authors wish to express thanks to AFOSR, NSF, **AROD** and DOD Themis Project for periods of support in this work.

References

1. A. R. Miller, The Adsorption of Gases on Solids (Cambridge University Press, Cambridge, 1949).

2. J. L. Morrison and W. E. Grummitt, J. Chem. Phys., 21, 654 (1953).

3. H. Y. Wachman, Thermal Accomodation Coefficient and Adsorption on Tungsten, Ph.D. Thesis, University of Missouri (1957).

4. H. L. Petersen, The Accomodation Coefficients of He, Ne, and Ar on Clean Potassium and Sodium Surfaces from 77° to 298°K, and a Study of Adsorption of Potassium on Tungsten, Ph.D. Thesis, University of Missouri (1958).

5. W. V. Best, Adsorption Studies in a Potassium, Hydrogen, Tungsten System, Ph.D. Thesis, University of Missouri (1962).

6. P. W. Blickensderfer, The Attack of Alkali Metals on Pyrex Glass as Related to High Vacuum Technique, Ph.D. Thesis, University of Missouri (1963).

7. D. V. Roach, A Study of the Thermal Accomodation of Helium, Neon, Argon, Krypton, and Xenon During Adsorption of Potassium and Cesium on Tungsten, Ph.D. Thesis, University of Missouri (1962).

8. R. C. L. Bosworth and E. K. Rideal, Proc. Roy. Soc. of London, A 162, 1 (1937).

9. L. D. Schmidt and R. Gomer, J. Chem. Phys., 43, 95, (1965).

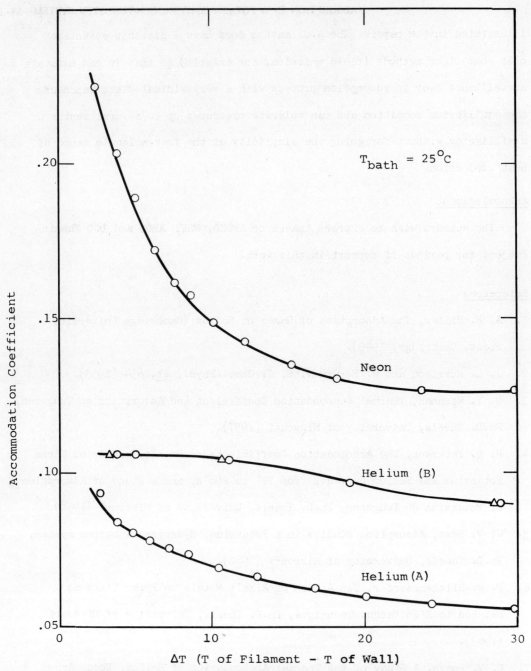

Fig. 1. Accommodation coefficients of neon and helium showing desorption of an upper film of potassium. The temperature of the bulk potassium in the tube is 25ºC.

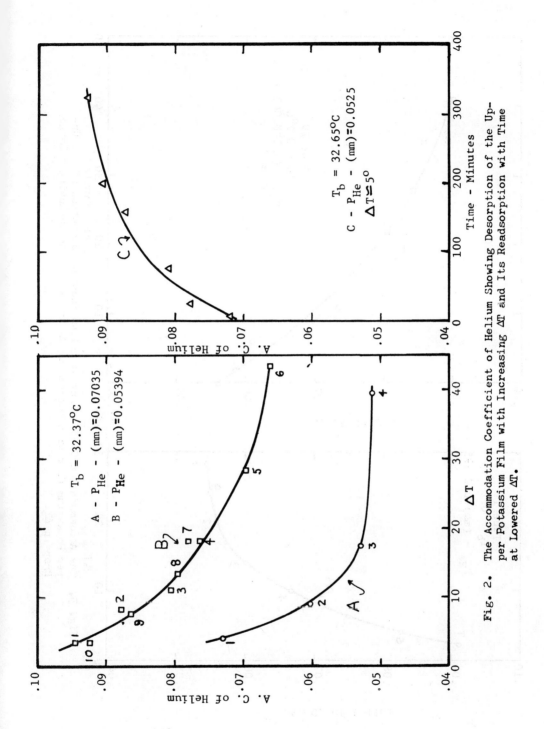

Fig. 2. The Accommodation Coefficient of Helium Showing Desorption of the Upper Potassium Film with Increasing ΔT and Its Readsorption with Time at Lowered ΔT.

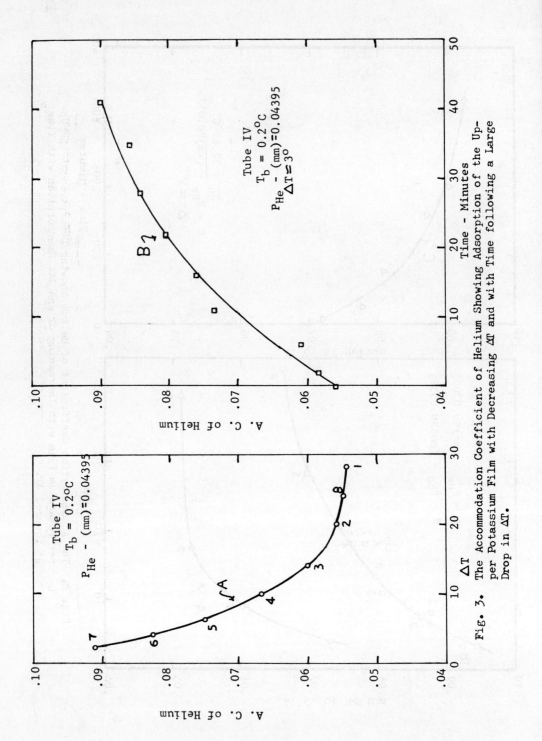

Fig. 3. The Accommodation Coefficient of Helium Showing Adsorption of the Upper Potassium Film with Decreasing ΔT and with Time following a Large Drop in ΔT.

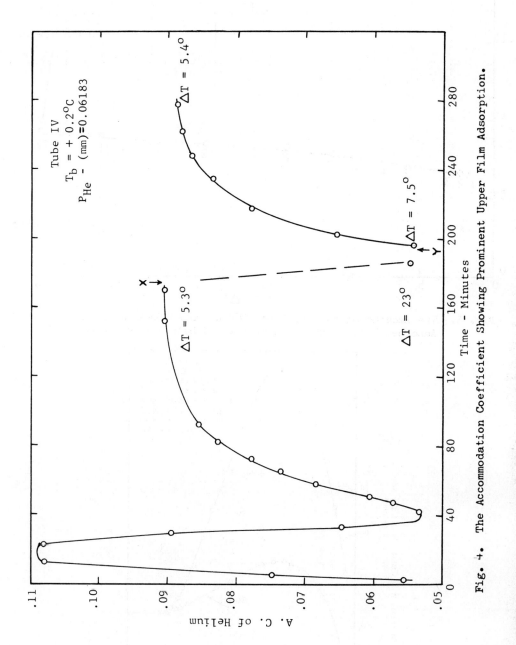

Fig. 4. The Accommodation Coefficient Showing Prominent Upper Film Adsorption.

68—15

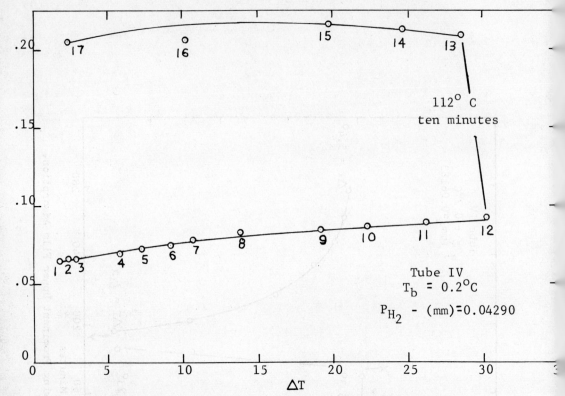

Fig. 5. The Accommodation Coefficient of Hydrogen Showing Alteration of the Adsorbed Film by Hydrogen at Lowered Bath Temperature.

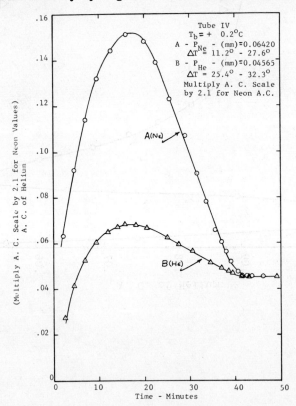

Fig. 6. The Accommodation Coefficients of Neon and Helium Plotted to Show Abnormal Ratio During Adsorption of Potassium.

LEED STUDY OF OXYGEN ADSORPTION AND FIRST STAGES OF OXIDE EPITAXY ON (110) SURFACES OF α- IRON

K. Molière and F. Portele

Fritz-Haber-Institut der Max-Planck-Gesellschaft
Berlin-Dahlem, Germany

Introduction

Surfaces of metallic single crystals are usually prepared as specimens for surface-chemical investigations by methods which may be called an "ill-treating" in some respects. In the most cases a crystal grown from the melt or by another appropriate method is cut mechanically or by spark erosion in a direction parallel to a certain (hkl)-plane as closely as possible. A following process of grinding and polishing by mechanical or chemical means produces a surface which is said to be an (hkl)-surface within an accuracy of ± some degrees, as confirmed by the X-ray back-scattering method. Obviously, such a surface will have an unknown real structure (facetting, steps, faults at places where dislocations and stacking-faults are coming out to the surface) also if the crystal is eventually treated afterwards by argon-bombardment and annealing in the UHV at high temperatures. Surely one has to expect that chemical properties of the surfaces (e.g. sticking probability with respect to adsorption of some gas molecules; rate of nucleation and epitaxy of chemical compounds deposited on the surface) are heavily influenced by the unknown real structure of the surface. To avoid some of these difficulties, the Berlin-Dahlem LEED-group, established end of the year 1964, started investigations using as specimens crystal surfaces grown by some natural processes guaranteeing an exact crystallographic orientation of the surface.

Among other methods, we tried to prepare our crystal surfaces by reduction of melted metal halides with hydrogen. This is a report on the properties of (110)-surfaces of needle crystals of α-iron, produced by this method, with respect to oxidation. A preliminary report on the results of these investigations has been given at the Clausthal Thin-Films Conference in 1965[1]). The investigations have been initiated by R. SZOSTAK in 1965 and continued by M. LAZNICKA[+]) in 1966/67, who prepared the ground for the present work[2]).

In the mean-time two investigations have been published by PIGNOCCO and PELLISSIER[3,4]) on the oxidation of iron-surfaces prepared by the conventional method. It may be interesting to compare the results of these authors with our recent results. We think we can show that the finer details of oxygen-adsorption and first stages of epitaxial oxide formation described in the present report cannot be detected if the crystal is prepared by the conventional method.

For the lack of dynamical methods the LEED patterns obtained in this work can only be interpreted on a pure geometrical basis, taking into account, of course, the possibility of multiple reflections. If structure models are given these should be accepted with the due criticism only as one possibility among others.

Preparation of the crystals

The crystals used in this investigation were produced by the well-known method for growing Fe-whiskers[5]): A melt of $FeCl_2$ is reduced at 700°C in a stream of a mixture of hydrogen with an inert gas. Under appropriate conditions[1]) the whiskers grow out to needle-crystals up to 0.5 mm in diameter. They have in most cases a hexagonal cross-section, the needle

---

[+]Working in the Fritz-Haber-Institute as a guest on leave of the Institute of Solid State Physics of the Czechoslovak Academy of Science.

axis being [111], the side faces of the type {1$\bar{1}$0}. (Sometimes one obtains also needles grown along [001] with side faces {100}.) These surfaces appear usually to be very smooth. Neither macroscopic steps nor indications of foreign crystallites (e.g. MnO) covering the surface can be detected on observation with an optical or scanning electron microscope, contrary to the finding of LAUKONIS[5]). The crystals kept for a very long time in dry air didn't show any indication of corrosion.

LEED experiments

We used a commercial LEED apparatus (VARIAN) equipped with a supplementary Ti-sublimation pump working once every day for a few minutes so that a basic pressure of 5-$7 \cdot 10^{-11}$ torr could be maintained. For adjusting the primary beam magnetically with respect to the very small crystal two pairs of coils were used (similar to Helmholtz-coils)[1]). In order to collect the part of the primary beam passing beside the needle crystal a cage was used constructed in such a way that the back-scattering from the cage could only get to the central part of the screen[1]).

The needle crystals, usually some mm's in length, were spotwelded carefully at the tip of a hairpin-shaped wire of very pure molybdenum, attached to the manipulator and serving as a heater. A thermocouple was also connected with the hairpin tip by spotwelding. The decay of temperature along the needle was measured with a micropyrometer in the higher temperature range and estimated by extrapolation in the lower range.

Because the crystals have a hexagonal cross section, as mentioned above, the electrons are not only reflected at the main surface which is usually adjusted nearly perpendicular to the incident beam. Superfluous spots which are due to reflections at the two faces inclined to the main surface by 120°

are to be seen on the diffraction patterns on horizontal lines connecting the stronger spots. They can, of course, be easily identified.

Preparation of the clean (110) surface

If the untreated crystal is heated in the LEED apparatus to moderate temperatures various complicated diffraction patterns are observed (not shown here) indicating perhaps multiple reflections due to coincidence lattices between the iron substrate and a surface film of unknown nature. If this film is removed by argon bombardment further heating produces simpler diffraction patterns showing that impurities are coming out of the inner of the crystal building up new surface structures. It could be shown later on that these diffraction patterns are similar to those which can be produced by oxidation of the clean surface. One can suppose, therefore, that oxygen is the main impurity solved in the inner of our crystals, which were kept for a very long time in dry air.

The α-iron cannot be cleaned, of course, by high temperature treatment because of the α-γ-transition point at 906°C. Heating to moderate temperatures in an atmosphere of clean hydrogen could perhaps remove the oxygen; but in this case we had then to solve the problem of getting rid of the hydrogen absorbed in the crystal. The method we employed for cleaning the crystal was, therefore, similar to the procedure of PIGNOCCO[4]), argon bombardment (250 eV) and subsequent annealing. If one repeats this cycle very patiently again and again raising the temperature of annealing to 600-700°C, the super-structures and the diffuse background becomes weaker and weaker. If the diffraction pattern after heating to 700° is showing only the iron spots and no diffuse background at all, we decide that the surface is the "clean" iron (110)-face (Fig. 1). In this state the diffraction pattern is very sensitive to smallest amounts of oxygen at room temperature.

Weak oxidation at room temperature

Oxygen at a pressure of 10^{-8} torr was led into the apparatus at room temperature in small portions, each portion defined by a time interval of 10-20 sec. (For the sake of brevity we make use of Germer's unit for oxygen exposure: 1 Langmuir = L = 10^{-6} torr.sec.) Some increase of the diffuse background was observed at first. With 0.5 L a c(2x2) superstructure began to be visible, showing maximum intensity with 1 L(Fig.2). The half-index spots were comparable then in intensity with the iron spots; they became weaker with further increase of oxygen exposure. With 2 L oxygen spots of a c(3x1) superstructure began to develop (Fig. 3), showing a streaky appearance in certain directions (e.g. [$3\bar{3}2$]). The c(2x2) spots were still visible in this stage; they disappeared with further oxygen exposure. The c(3x1) spots attained maximum intensity between 3 and 5 L (Fig. 4). (The splitting of the spots of the c(3x1) structure will be discussed later on.)

Patterns of a similar kind were also obtained by PIGNOCCO and PELLISSIER[4]). These authors observed maximum intensity of the c(2x2) spots after an oxygen exposure of 1.7 L; for the c(3x1) fractional order spots their value of oxygen exposure giving maximum visibility was remarkably high: 60 L. Our values as given above were 1 L and 3-5 respectively. It is surprising that Pignocco's figures are so much larger than ours. This may be perhaps, explained by the fact that Pignocco's patterns show apparently a high background of diffuse scattering due to surface faults which adsorb a high percentage of the impinging oxygen; the diffuse scattering prevents the visibility of the extra spots due to the superstructures at low oxygen exposures.

As concerns the structures giving rise to the c(2x2) and the c(3x1) diffraction patterns, we don't like to propose a definite model. There is no possibility as yet to decide the question: Surface reconstruction or not. But without respect to this alternative we may assume that the c(2x2) structure corresponds to an oxygen coverage of a quarter of a monolayer, in agreement with PIGNOCCO and PELLISSIER[4]). Concerning the c(3x1) structure, for which these authors have proposed a more complicated model with a 2/3 coverage, as to our opinion, a simple centered unit cell corresponding to one third of a mono-layer is much more probable because we could obtain the corresponding pattern already with a very low oxygen exposure.

In the case of the c(3x1) pattern we couldn't detect any indication of the facetting supposed by PIGNOCCO[4]). But, predominant at higher oxygen exposures, we observed a splitting of the fractional order spots in the c(3x1) patterns, each of these spots being decomposed into four components. Figure 4a and b show this splitting at oxygen exposure of 3.2 L and 5 L, respectively. Every quartet of spots consists of two doublets, obviously due to diffraction at sep-arate regions of the surface. The diffraction patterns of the two separate regions are symmetrical with respect to $[1\bar{1}0]$. One of the spots of either dou-blet is usually stronger than the other one. It could be observed now that the two spots of a doublet are moving along straight lines parallel to $[3\bar{3}2]$ or $[3\bar{3}2]$, respectively, in such a way, that the distance of the splitting is in-creasing continuously if the oxygen exposure is increased.

The situation may be more easily described if an oblique reciprocal unit cell is used, as shown in Fig. 5a. In this drawing the direction of movement of the stronger doublet components is indicated by arrows. The diffraction pat-tern corresponding to the c(3x1) structure (open circles in Fig. 5a) can be called a pattern of a (3x1)' structure[+]) with reference to the oblique unit mesh indicated in Fig. 5b. By extrapolation of the observed change of the doublet splitting with increasing oxygen dosis, one can imagine that the two stronger spots (big full circles) are further moving in opposite directions striving for a final state in which they meet each other half way (dotted circles). This final state would correspond to a (2x1)' structure as indicated in Fig. 5c.

[+] The dash shall indicate that the nomenclature is related to the oblique unit cell.

It was observed recently by GERMER et al.[6] that in the case of the
oxidation of a Ni(110) surface a transition from the (2x1) to the (3x1)
structure takes place in such a way that the splitting of the (2x1)-spots is
a continuous function of the oxygen doses; ERTL[7] found similar phenomena
in the case of the oxidation of Pd(110). Germer et al. explained their
observations making use of a theoretical investigation of FUJIWARA[8] concerning antiphase domains, by the assumption that sequences of certain units
(-O-Me- and -O-Me-O-) are chained together in a more or less uniform mixture
in the direction in which the splitting takes place; this direction is in
the case of the f.c.c. metals parallel to the dense packed lattice rows. --
In our case the interpretation is more difficult because in the direction
$[3\bar{3}2]$ there is no dense lattice row at all. If an explanation on similar
lines is feasible one can assume that in any intermediate state of oxidation
we have some uniform sequence of ribbons of the two structures of Fig. 5b and
c. The ribbons marked by streaks in the drawings are parallel to $[1\bar{1}3]$, i.e.
perpendicular to the direction of the splitting, $[3\bar{3}2]$. In the two structures
there are chains of units -Fe-O-Fe and -Fe-O-, respectively, in the direction
[001].--We must admit that the diffraction pattern of the hypothetical final
state of this transition, i.e., the (2x1)', could not be obtained. At higher
oxygen exposures the patterns show only the iron spots in a very high diffuse
background.

Weak oxidation at elevated temperatures

Figure 6 shows diffraction patterns obtained when the crystal is stepwise oxidized (again the O_2-pressure was 10^{-8} torr) at 190°C. Obviously the
behaviour is quite different now compared with the oxidation at room temperature. Firstly, the c(2x2) structure is developed again with maximum intensity
of the fractional order spots at an oxygen exposure of 1 L. At 1.5 L

satellites begin to be split off from the iron spots and the fractional
order spots in the [1$\bar{1}$0] direction (Fig. 6a). On further oxidation the c(2x2)
-reflections disappear and the spots of the c(3x1) begin to be visible; the
fractional order spots of this structure are also split up in the direction
[1$\bar{1}$0] (Fig. 6b: 5.5 L). At higher oxygen exposures (Fig. 6c: 15 L) only the
iron spots and their satellites are visible in a diffuse background. With
increasing oxygen exposure the satellites of the iron spots are migrating in
the directions indicated by arrows in the schematical drawing shown in Fig.
7. If the ratio $\frac{a}{b}$ (a and b as indicated in Fig. 7) is plotted against the
oxygen doses the experimental values of this ratio belonging to a given
temperature of oxidation may be connected by curves which seem to be quite
well reproducible. Obviously, also in this case the splitting is a continuous function of the amount of oxygen given to the surface at a constant
temperature.

If one would try to explain the satellites as being due to multiple
reflection on a coincidence lattice one would have to assume a surface film
with a lattice constant variable in the direction [1$\bar{1}$0] depending continuously
on the oxygen exposure. This would be difficult to understand in our
opinion. -- It will be preferable also in this case to try an explanation
similar to the lines given by the work of Germer et. al., as mentioned above.
But also here we don't have a dense-packed lattice row parallel to the
direction [1$\bar{1}$0] of splitting. At given oxygen doses and temperature one had
to assume then some uniformly mixed sequence of ribbons of different structures (say, c(3x1)); the direction of the ribbons would be [001] in this
case. --The two plateaus of the curve obtained at a crystal temperature of
220°C seem to show that there are two different kinds of sequences at lower
and higher oxygen exposure.

If the crystal is oxidized with more than 30 L one obtains a complicated superstructure of the type (12x5). In this state the migration of the satellites has come to an end. --Diffraction patterns corresponding to the c(1x5) structure reported by PIGNOCCO and PELLISSIER[4]) were not obtained in our work.

Strong oxidation; oriented deposition of oxide

As mentioned above, with oxygen exposures of more than 10 L at room temperature the diffuse scattering becomes very strong so that superstructure spots are no more visible. If the oxidation is further continued also the iron spots become weaker and with 80 L a hexagonal pattern of diffuse spots is developed (Fig. 8). In accordance with the statement of PIGNOCCO and PELLISSIER[4]) this diffraction pattern may be explained by an epitaxial deposition of FeO on the surface. Figure 9 shows schematically the grouping of the diffraction spots and the structural relations between the two lattices: (FeO(111)||Fe(110, FeO[10$\bar{1}$]||Fe[001]. The diffusity of the FeO spots may be due to some lack of exact orientation or due to a very small size of the crystallites. The lattice constant evaluated from the distances of the FeO-spots relative to those of the Fe-reflections seems to exceed the lattice constant of a compact FeO-crystal by 6%. For a thin (111)-layer of the rock-salt-lattice such a lateral expansion seems to be not unreasonable (cf. SZOSTAK and MOLIERE[9])).

If an oxide layer obtained with 100 L oxygen is heated for a short time to a temperature of 200°C the diffuse scattering is decreased and the FeO spots become stronger and sharper; at the same time they appear to be elongated in the [00$\bar{1}$]-direction of the iron lattice (Fig. 10a). By short heating to 270°C a splitting of the FeO spots in the [001] -direction can be observed (Fig. 10b); the grouping of the diffraction spots is shown schematically in Fig. 11a. The new spots neighbouring on the FeO reflections are lying on a straight line connecting the 11- and $1\bar{1}$-reflections of the iron lattice. It may be supposed that the lattice is now partly adapted to the Fe-substrate in the iron [001]-direction. Such an adaption seems to be possible because the difference of the two lattice constants in that direction is comparatively small, as shown schematically in Fig. 11b. --In the diffraction pattern some additional weak spots are visible which may be explained by multiple reflections. This effect can be usually expected if two (or more) lattices are superimposed one upon another, having small differences in their lattice parameters[10,11]). In Fig. 11a some of these multiple reflection spots are indicated by crosses. Some of the weak spots can only be explained by the assumption that all three lattices (Fe substrate, the adapted FeO-layer and the normal FeO-layer) give rise to triple (or higher order) reflections.--A pseudomorphism, i.e., an adaption of the lattice constants of a film oriented epitaxially on some substrate was assumed hypothetically by some authors (cf. [12]). This seems to be the first experimental evidence of this assumption.

If the crystal is further heated to temperatures above 300°C a remarkable change of the diffraction pattern occurs. The (111)-face of the FeO remains parallel to the iron surface, but the oxide film which was initially uniform at lower temperature is decomposed now into domains in which a rotation of the FeO lattice has occurred. The angle of rotation is about 5.3°. Separate regions with clockwise and counterwise rotation can be detected. The domains with uniform rotation become larger if the crystal is heated to 400-500°C. Figure 12 shows a pattern in which the rotation of the Fe layer is nearly uniform over the whole surface. In this case the crystal was heated to 450°C for a short time. In Fig. 13a the complicated pattern is reproduced schematically showing in addition to the reflections of the iron substrate (big

full circles) and the rotated hexagon of oxide spots (open circles) a lot of
weaker reflections. Apparently, multiple reflections (crosses) occur between
the regular (111)- film of FeO and some intermediate layer, the reciprocal
unit mesh of which is indicated in the figure by dashed lines connecting the
small full circles. One can suppose that in this intermediate layer the FeO
lattice is partly adapted to the Fe substrate by a shearing deformation such
that its lattice parameter coincides with that of the iron in the [001]
direction (Fig. 13b). The angle of shearing is equal to the angle of rotation
of the regular FeO lattice (5.3°). One of the axes of the deformed oxide
structure is parallel to the [001] direction of the iron, the other axis is
parallel to the [1$\bar{1}$0] direction of the regular FeO layer.

If the oxidation is carried through at a higher temperature ($\geq 200°$C) with
a very high amount of oxygen (>1000 L) the complicated details of the diffraction patterns of the rotated and adapted FeO films are concealed by
diffuse scattering. Figure 14 (200°C; 30 min at 10^{-6} torr) shows only the
reflections of a (111) FeO layer. The diffusity of the spots is probably due
to some lack of orientation of the crystallites. A very strong oxidation
produces additional spots which may be due to a spinel lattice, (γ-Fe_2O_3 or
Fe_3O_4)(Fig. 15 (a), (b): 300°C, 60 min at $5 \cdot 10^{-4}$ torr). If the crystal is
then annealed the spinel reflections begin to disappear at 360°C (Fig. 15c);
the spots are very diffuse in this stage. Annealing at 800°C gives a pattern
containing only the FeO reflections (Fig. 15d).

References

1. R. Szostak and K. Moliere, Proc. Intern. Symp. on Basic Problems in Thin Film Physics, Clausthal-Göttingen, 1965, p. 10.

2. To be published.

3. A. J. Pignocco and G. E. Pellissier, J. Electrochem. Soc. 112, (1964) p. 1188.

4. A. J. Pignocco and G. E. Pellissier, Surface Sci. 7, (1967) p. 261.

5. J. V. Laukonis, GMR-414, Oct. 1963.

6. L. H. Germer, J. W. May and R. J. Szostak, Surface Sci. 7, (1967) p. 430.

7. G. Ertl, private communication.

8. K. Fujiwara, J. Phys. Soc. Japan 12, (1957) p. 7.

9. R. Szostak and K. Molière, Z. Naturforschg. 22a, (1967) p. 1615.

10. E. Bauer, Surface Sci. 7, (1967) p. 351.

11. N. J. Taylor, Surface Sci. 4, (1966) p. 161.

12. D. W. Pashley, Advance Phys. 5, (1956) p. 173.

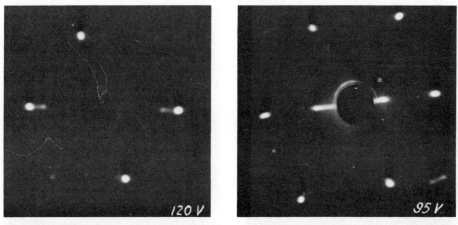

Fig. 1. LEED pattern of the iron (110) surface after cleaning by argon bombardment and annealing.

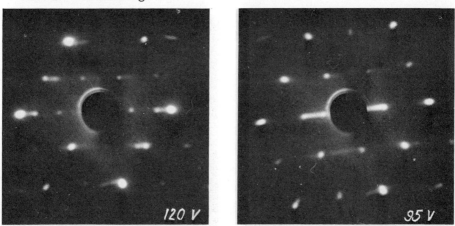

Fig. 2. Oxidation with 1 L at 20°C. c(2x2) pattern.

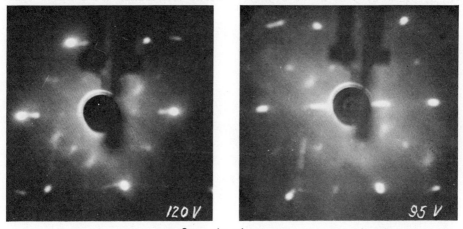

Fig. 3. 2 L at 20°C. c(3x1) in addition to c(2x2).

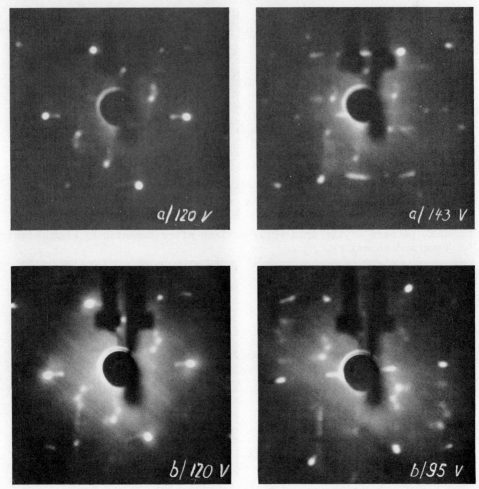

Fig. 4. a) 3,2 L b) 5 L; 20°C. c(3x1) pattern, fractional order spots split up.

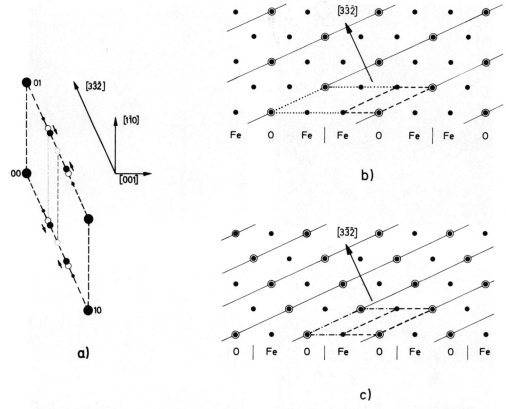

Fig. 5.   a) Oblique reciprocal unit mesh with spots split up in the $[3\bar{3}\bar{2}]$
direction. b) (3x1)' structure with 1/3 coverage. c) hypothetic (2x1)'
structure with 1/2 coverage.

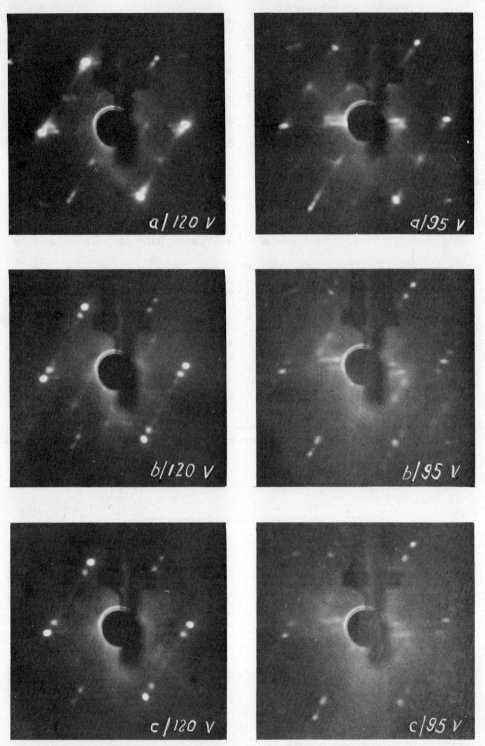

Fig. 6. Stepwise oxidation at 190°C. Splitting along [1$\bar{1}$0] . a) 1.5 L, b) 5.5 L, c) 15 L.

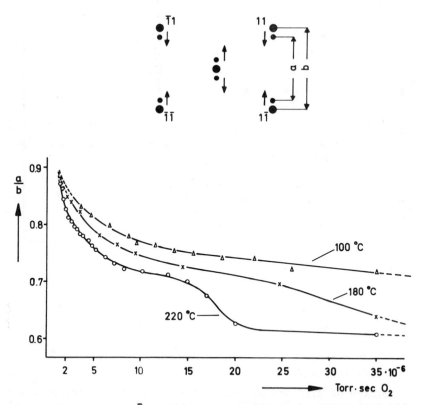

Fig. 7. Split in the [1$\bar{1}$0] direction as a function of oxygen exposure.

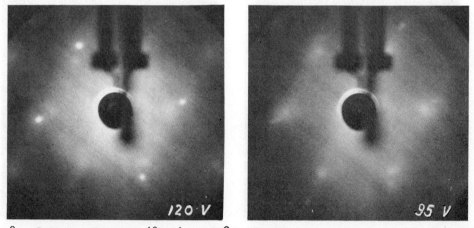

Fig. 8. Strong oxidation (80 L) at 20°C. Hexagon of diffuse spots indicating a (111) layer of FeO.

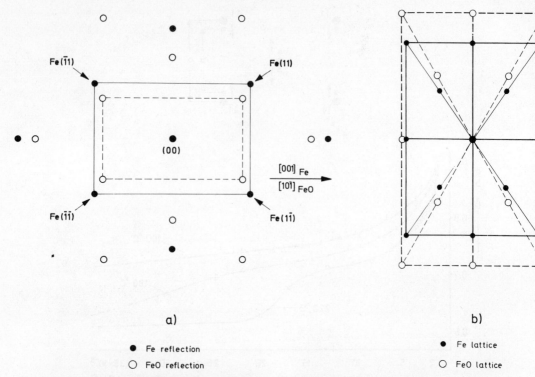

Fe(11)   Fe(11)

(00)

$\xrightarrow{\begin{array}{c}[001]\ \text{Fe}\\ \overline{[101]}\ \text{FeO}\end{array}}$

Fe(11)   Fe(11)

a)   b)

● Fe reflection   ● Fe lattice
○ FeO reflection   ○ FeO lattice

Fig. 9. (a) Scheme of diffraction spots of Fig. 8, (b) Geometrical relations of
the corresponding Fe and FeO lattices.

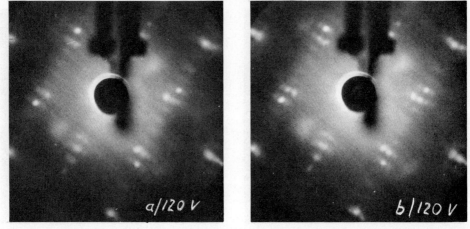

Fig. 10. Oxidation with 100 L at room temperature, crystal annealed at (a)
200°C, (b) 270°C.

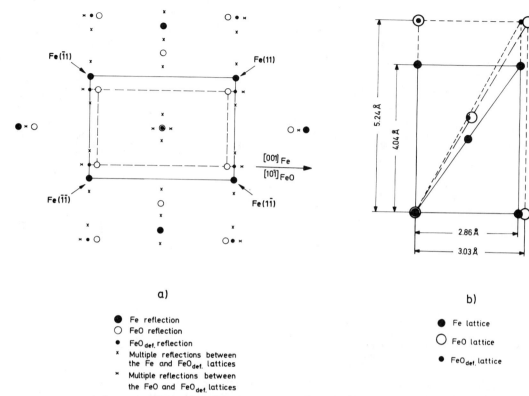

a)

b)

● Fe reflection
○ FeO reflection
• FeO$_{def.}$ reflection
× Multiple reflections between the Fe and FeO$_{def.}$ lattices
× Multiple reflections between the FeO and FeO$_{def.}$ lattices

● Fe lattice
○ FeO lattice
• FeO$_{def.}$ lattice

Fig. 11. (a) Grouping of the diffraction spots in Fig. 10b, (b) Geometrical relations of the corresponding Fe and FeO lattices.

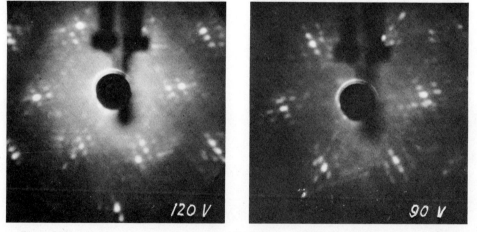

Fig. 12. Crystal annealed at 450°C after strong oxidation at 20°C.

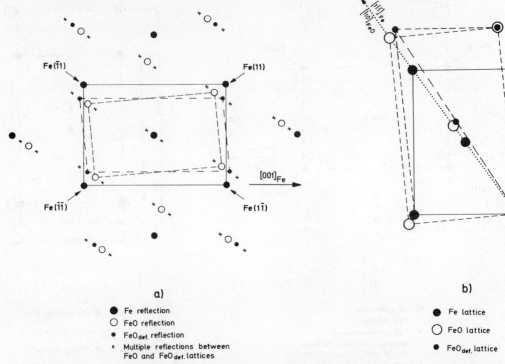

a)

- ● Fe reflection
- ○ FeO reflection
- • FeO$_{def.}$ reflection
- × Multiple reflections between FeO and FeO$_{def.}$ lattices

b)

- ● Fe lattice
- ○ FeO lattice
- ● FeO$_{def.}$ lattice

Fig. 13. (a) Grouping of the diffraction spots in Fig. 12. (b) Geometrical relations of the corresponding Fe and FeO lattices.

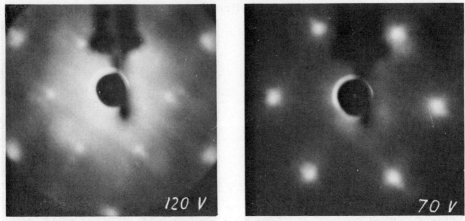

Fig. 14. Oxidation at 200°C, 30 min at 10^{-6} torr.

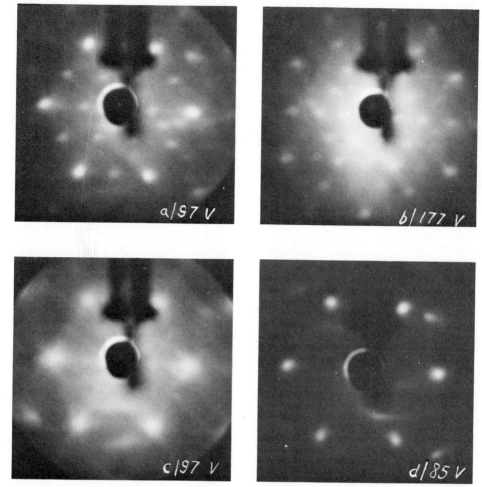

Fig. 15. (a,b) Oxidation at 300°C, annealing at 300°C: spinel structures, (c) annealing at 360°C, (d) annealing at 800°C.

LEED AND ADSORPTION STUDIES OF BERYLLIUM*

Richard O. Adams

The Dow Chemical Company
Rocky Flats Division
Golden, Colorado

INTRODUCTION

Beryllium with its many desirable properties has, with a few drawbacks, much to offer modern technology. Its use in the nuclear industry, especially as a moderator and reflector in reactors, is fairly well documented. Except for some oxidation studies and a few other scattered experiments, little work has been done to study beryllium surfaces. A program has been begun to investigate some beryllium surface phenomena and this is a report of the initial results of some of these studies. Two types of surfaces are being studied: The surface of thick films and the surface of single crystals. This discussion will be restricted primarily to the latter with only brief reference to some results obtained with films.

EXPERIMENTAL

The (0001) surface of beryllium was examined by low energy electron diffraction using apparatus of post acceleration type. Two crystals, 1 cm x 1 cm x 0.2 cm, were cut from a triple-pass zone refined beryllium single crystal. The (0001) plane was exposed parallel to the square face surface to within 2 degrees. Both crystals were ground and polished by the supplier, Nuclear Materials, Inc. In the course of the experiments the surfaces became badly etched. When this happened the crystal was removed from the apparatus and the surface lapped. Thus two different surface finishes were provided, one lapped, the other polished. No difference was observed in the results for these two different finishes, however.

Except for degreasing prior to placement in the LEED chamber, the crystals were used in the as received condition. It was felt that nothing would be

* Work performed under U. S. Atomic Energy Commission Contract AT(29-1)-1106.

accomplished by further lapping or polishing of the surface. Oxidation of a clean beryllium surface is extremely rapid until a few monolayers (4 or 5) of oxide are formed. After that the oxidation rate at room temperature is so slow as to be nonexistent. Even if the crystals had been lapped or polished by us to remove the accumulated oxide, by the time it was placed in the LEED chamber the oxide layer would be nearly as thick as it was before lapping.

The only chemical treatment of the surfaces was a standard degreasing in acetone, distilled water and alcohol prior to placing the crystals in the apparatus. No chemical etching or polish was done, nor were the surfaces electropolished. Following the usual system processing procedures, i.e., bakeout, etc., pressures in the low 10^{-10} torr range were achieved. The surfaces could be cleaned either by heating the crystal using electron bombardment of the crystal holder or by ion bombardment.

SURFACE STRUCTURE

Heating the crystal for 4 hours at 900°C produced the diffraction patterns shown in Figure 1. This is one of two patterns observed and will be referred to as the 12-spot pattern. The pattern is not what would be expected from a perfect beryllium surface although it does have the overall hexagonal pattern associated with the base plane of a hcp structure. The difference between these patterns and the expected ones is, of course, the presence of two concentric hexagonal rings instead of one.

The conventional $n\lambda = d \sin \theta$ relationship can be used to predict the positions of the diffraction pattern features. It is found that the positions of the features in the outer hexagonal ring correspond to positions predicted by this condition. The positions of these features move toward the center with increased electron energy as they should.

The features in the inner hexagonal ring are not predicted by the $n\lambda = d \sin \theta$ relationship if a perfect surface is assumed. For a perfect surface $a_o = 2.286\text{Å}$ and thus d = 1.98Å. Using $n\lambda = d \sin \theta$ on the inner ring d turns out to be 3.43Å, approximately one and one-half times the inner-atomic spacing of beryllium. Also the inner pattern is rotated 30° with respect to the bulk lattice. The features in this pattern move toward the center as the acceleration voltage on the electron is increased.

It is tempting to say that the additional spots are due to a simple superstructure on the beryllium surface. Figure 2 shows such a superstructure having hexagonal symmetry with a_o equal to 3.96Å. The black marbles in the figure represent the superstructure. This superstructure could be made up in at least three possible ways.

It is possible that the superstructure is actually made up of beryllium atoms forming a hexagonal lattice like that shown in Figure 2. Second, the surface could be covered by a tightly bound chemisorbed oxygen layer. The oxygen atoms could be placed in the positions shown for the superstructure of beryllium in Figure 2, or they could form a coincident lattice as described by Tucker.[1] The third possibility is that a layer of contamination left from the polishing exists on the surface. Such residual material could reside on preferred sites and form a superstructure. There are undoubtedly other explanations for a superstructure but these seem the most reasonable. Although a superstructure explanation is tempting there are some other alternatives which must be considered.

It is fairly well accepted that additional diffraction spots can be produced by multiple scattering of the electrons inside the bulk material. Unfortunately we can not say much about the possibility of multiple scattering in this case. To do so we need an analysis of the intensity variation with wavelength. Attempts to get such information were unsuccessful because of the non-reproducibility of the intensity data. It is known that beryllium etches thermally at high temperatures[2] and the etching is sufficient to cause drastic changes in the surface. Since the crystal was cleaned by heating to high temperature, the surface became badly etched. Following each cleaning the observed intensity distribution was different. Thus no meaningful intensity analysis could be made.

It is well known that beryllium oxidizes readily to form a very thin protective oxide layer. At room temperature in atmosphere the oxide layer grows to a few atomic layers instantaneously after which the rate of growth is negligible. The existence of a thin BeO layer on the surface must be considered.

Tucker[3] points out that if two lattices are rotated with respect to each other, each lattice can be treated independently. If a BeO lattice is rotated 30° with respect to a Be lattice and compressed slightly, it will form a coincident lattice with the Be. Rotation of the BeO lattice by 30° brings the rows of atoms in the BeO lattice with interatomic spacing of 2.698Å in

line with rows of atoms in the Be lattice with 3.96Å spacings. A compression
of 2% brings every fourth atom in the BeO lattice into coincidence with every
third atom in the Be lattice.

Elucidation of the 12-spot pattern can only be accomplished by consideration of
the treatment necessary to produce the second observed diffraction pattern.
Figure 3 shows this second diffraction pattern which will be referred to as the
6-spot pattern. The features in this pattern are in the position predicted by
the $n\lambda = d \sin \theta$ relationship for a perfect beryllium crystal. It is assumed
that this 6-spot pattern is the pattern attributable to a clean (0001) beryllium
surface.

The cleaning procedure used on the crystal prior to observing the 6-spot pattern
was rather elaborate and complete details can be found elsewhere.[4] It
consisted of heating the crystal in oxygen, vacuum and hydrogen at various
temperatures and pressures until a stable sharp pattern was observed.
Experimentation with cleaning showed that these sharp patterns could be produced simply by heating the crystal to 1000°C in good vacuum for a few minutes.
After each time the crystal was cleaned, the background intensity increased;
slowly at first but after many, 40 or 50, heatings the increase in intensity
became more rapid. After about 100 heatings to 1000°C the background intensity
became great enough to practically overwhelm the spots in the diffraction
pattern.

Figure 4(a) shows a photomicrograph of the beryllium surface after a short
period at 900°C. Figure 4(b) shows the same area on the same surface after
50 heatings to 1000°C. The change observable in the surface even at this
magnification is dramatic. There is little wonder that no meaningful intensity
data can be obtained.

We have been unable to produce the 12-spot pattern after the 6-spot pattern has
been produced by what could be called conventional means. Oxygen and other
gases have been allowed to adsorb on the crystal, and the crystal has been
heated at various temperatures in different gases (primarily oxygen) at
various pressures. The only change observed in the pattern was that commonly
attributed to adsorption of amorphous layers of gas.

The only way we have been able to produce the 12-spot pattern is to remove the
crystal, lap it and place it in the system. A mild heating will then produce
the 12-spot pattern. Heating to above 1000°C will then produce the 6-spot
pattern. From this we could conclude that the 12-spot pattern is due either

to oxide layer on the surface or to some contamination. The fact that the 12-spot pattern could not be produced by heating the crystal in oxygen make the latter possibility more probable.

ADSORPTION

Adsorption studies on the single crystal surfaces were done in the LEED apparatus using the 6-spot surface. The surface was prepared by heating to 1000°C for 5 minutes prior to admitting the gas to be studied. The common diatomic gases, nitrogen, hydrogen, oxygen and carbon monoxide were used. Of interest in these adsorption studies was whether or not these gases caused any change in the diffraction patterns.

When the clean surface was exposed to nitrogen, no change in the pattern was observed. There was no diminution in the intensity of the spots and there was no increase in the intensity of the background. Both these effects are commonly observed when a gas is adsorbed on a surface being examined by LEED. This would indicate that nitrogen is not adsorbed by the beryllium. Indications of this were also observed by the metering system admitting the gas. Experiments using films of beryllium as the adsorbing surface confirm that nitrogen is not adsorbed.[5] The same was observed with hydrogen: no adsorption was detected on either the crystal or the film.

Oxygen and carbon monoxide were observed to adsorb, oxygen with a high sticking probability and carbon monoxide with one somewhat lower. Adsorption of either gas caused no change in the surface structure as observed in both the 6-spot and 12-spot pattern. There was, however, the characteristic decrease in sharpness of the spots accompanied by an increase in background intensity. Thus indicating the adsorption of oxygen and carbon monoxide on beryllium to form an amorphous layer.

REFERENCES

(1) C. W. Tucker, Jr., J. Appl. Phys. $\underline{37}$ 528 (1966).

(2) D. W. White, Jr., and J. E. Burke, The Metal Beryllium. Cleveland: ASM, 1955.

(3) C. W. Tucker, Jr., J. Appl. Phys. $\underline{35}$ 1897 (1964).

(4) R. O. Adams, "Cleaning an Ideal Beryllium Surface," RFP-879, The Dow Chemical Company, Rocky Flats Division, Golden, Colorado, 14 Feb. 1967.

(5) J. T. Hurd and R. O. Adams, To Be Published.

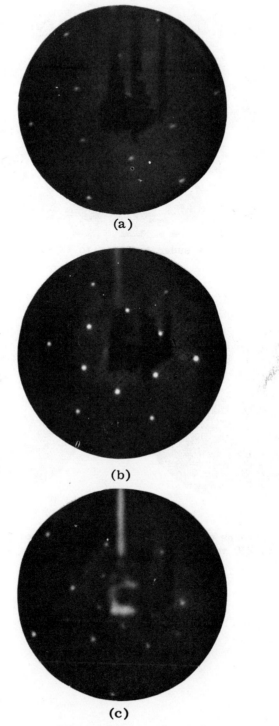

Fig. 1. LEED patterns of Be(0001) showing the 12-spot pattern at different electron energies: (a) 80 volts, (b) 135 volts, and (c) 230 volts.

Fig. 2.   Marble model of a superstructure to explain the observed 12-spot pattern.

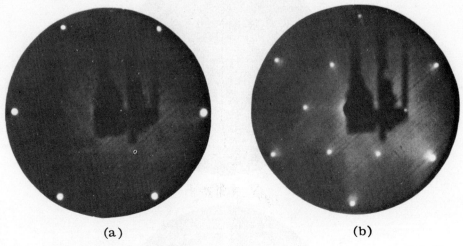

(a)   (b)

Fig. 3.   LEED patterns of Be(0001) showing the 6-spot pattern at (a) 80 volts, and (b) 230 volts.

(a)

(b)

Fig. 4. Optical photomicrographs of the Be(0001) surface (a) after four-hour
anneal at 900°C, and (b) after extensive heating at 1000°C. Magnification is 560.

A LEED STUDY OF THORIUM OVERLAYERS ON A TANTALUM (100) SUBSTRATE AND EVIDENCE FOR INITIAL PSEUDOMORPHISM FOLLOWED BY THORIUM CLUSTERING

J. H. Pollard and W. E. Danforth*

Bartol Research Foundation of The Franklin Institute
Swarthmore, Pennsylvania

1. Introduction

The adsorption of electropositive metal atoms on a metal substrate, which form a particularly important case of chemisorption, can give structures which can be rationalized from low energy electron diffraction studies. Two dimensional structures can, in many cases, be unambiguously assigned when LEED information is carefully correlated with coverage[1]. This allows a more precise characterization of the adsorbed layers and hence evaluation of the role played by the substrate in epitaxial growth, particularly under ultra high vacuum conditions. Correlation of LEED with other macroscopic properties such as work function, thermal desorption energies, can add information to improve the atomistic description of the nature of chemisorption.

The present paper discusses the results of some measurements of the adsorption of thorium on a Ta(100) surface at a temperature of 950°C. The properties of the first monolayer are described in detail and the nature of the epitaxially grown thorium discussed. The work is compared to earlier work of the Th-W(100) system[2,3] and some recent results we have obtained with this system.

2. Experimental Techniques

The Varian LEED equipment and thorium evaporator used are described in the earlier work with tungsten[2]. The tantalum specimen was 28mm × 3mm

* Work partially supported by the U.S. Atomic Energy Commission.

× 0.18mm cut from a single crystal rod and mechanically polished in the usual way. The electrolytic polishing, prior to final mounting in the crystal holder, was done in a sulphuric acid and hydrofluoric acid mixture with a platinum cathode[4] at a current density 0.1 amp/sq cm. A tungsten (100) ribbon prepared as earlier described was mounted across the Ta(100) specimen and 6 mm from it forming a cross allowing simultaneous retarding potential measurements from both Ta and W surfaces and hence giving a coverage calibration for thorium on tantalum.

The thorium evaporator was outgassed by heating to ∿ 1500°C for 100 hours before starting any thorium evaporations which were carried out at a pressure of 1.5×10^{-9} torr. Analysis with a Varian partial pressure gauge showed that 2/3 of the pressure was due to H_2, 1/3 to carbon monoxide. It was with this in mind that the tantalum was kept at 950°C during evaporation, a temperature high enough to prevent hydrogen accumulation on the surface. The evaporator conditions were chosen to give the best ratio for Th atom and gas atom arrival at the tantalum surface and a flux of approximately 10^{14} atoms/min sq cm of thorium resulted.

The coverage of thorium on the tantalum could also be found from the increase in thermionic emission of clean tungsten when the tantalum was flashed clean. This became necessary as the experiment progressed in order to eliminate as a possibility the evaporation of thorium from the tantalum surface at 950°C during deposition.

A correlation was made between thorium coverage and intensity of the $(\frac{1}{2}, \frac{1}{2})$ order beams by monitoring both intensity and retarding potential. LEED intensities of the (00) beam were made up to an equivalent coverage of 20 atomic layers of thorium. The presence of thorium on the tantalum ribbon was firmly established by an α-particle counting technique. An optical microscope study of the thorium covered surface was undertaken and revealed the presence of clusters which have been confirmed from an electron micrograph study.

3. Macroscopic Properties

The change in work function of the Ta(100) surface with increasing thorium coverage was measured by the retarding potential method and the results are shown in Fig. 1. It was compared to the thorium-tungsten system which has a minimum at 4.2×10^{14} atoms/sq cm[2] by simultaneous measurement and a coverage calibration thus obtained. The work function of Ta(100) was taken from the best value obtained by Protopopov et al.[5] and his value for Ta(110) agrees with a recent value obtained by Fehrs and Stickney[6] adding confidence to the assignment 4.15±0.02 eV. After allowance for the different thorium flux (from geometrical considerations) at the two ribbons one finds that a minimum in the work function occurs at a coverage of 3.4×10^{14} atoms/sq cm, somewhat lower than in the case of tungsten. The final value for monolayer coverage was 3.25 eV ± 0.04, a little lower than polycrystalline thorium but close to the value calculated by Steiner and Gyftopoulos[7] for a Th(100) surface. The change in work function is less for a thorium covered Ta(100) surface than that for W(100) and hence the dipole moment/atom of Th is less on a tantalum surface and hence one may expect less atom-atom interaction of the adsorbate than in the case of tungsten, a point we shall later consider. The lower-dipole moment 0.6 of its value on tungsten combined with the larger cell size of tantalum leads one, a priori, to believe that the thorium-thorium inter-action will be smaller and that tantalum may play a greater role in the forma-tion of the thorium overlayers than is found in the case of tungsten.

4. Observation of the LEED Patterns

The tantalum (100) surface was cleaned by resistive heating to approx-imately 2500°C for a few minutes. This procedure was found to give a satisfactory (1×1) pattern (Fig. 2) which was measured and gave a 2D cell size of

3.39±0.10Å in agreement, within experimental error, of the accepted value for

bulk tantalum of 3.30Å from X-ray data.

The evaporation of thorium onto the surface gave a c(2×2) pattern with

diffuse $(\frac{1}{2},\frac{1}{2})$ spots (Fig. 3) up to a coverage of 3×10^{14} atoms of thorium/sq cm.

The addition of more thorium changed the pattern gradually back to (1×1) which

was still good at a coverage of 18.4×10^{15} atoms of thorium/sq cm (Fig. 4) with

the same 2D cell size as that of the clean tantalum.

5.   LEED Intensity of $(\frac{1}{2},\frac{1}{2})$ Spots

A correlation between the intensity of the $(\frac{1}{2},\frac{1}{2})$ spots and coverage was

attempted by making retarding potential and intensity measurements simultane-

ously.   The results are presented in Fig. 5 showing a linear relationship

between the intensity and the coverage (proportional to the time of evapora-

tion) with a maximum intensity of 3.0×10^{14} atoms/sq cm.   The linearity suggests

the formation of completed islands[1] with the c(2×2) structure; the diffuse

spots resulted presumably from islands with one dimension small compared to

the coherence width of the electron beam ∿1000Å.   This also presents evidence

for a reasonable mobility of thorium over the tantalum substrate at 950°C.

The fall in intensity beyond the maximum suggests the formation of regions

of thorium with (1×1) registry with the substrate before the completion of the

c(2×2) structure which would occur normally at 4.6×10^{14} atoms/sq cm.

6.   LEED Intensity of (00) Beam

The variation of the intensity of the (00) beam with the electron energy

can yield information of the atomic interlayer distances.   In the case of

clean tantalum, the occurrence of multiple scattering prevents positive

identification of the major Bragg peaks, Fig. 6(a).   However, if the tantalum

is allowed to adsorb a fraction of a monolayer of hydrogen this destroys most of the peaks resulting from multiple scattering[8] allowing identification

of the integer order Bragg peaks, Fig. 6(b). The order numbers have been

assigned using an inner potential correction of 16 volts. Figure 7 shows a

plot of the (00) beam intensity versus electron energy as the thorium deposi-

tion proceeded. The initial evaporation of thorium did not greatly change the

appearance of the Bragg peaks although the multiple scattering peaks gradually

disappeared. The structure after 15 monolayers of thorium is very similar to

that shown for tantalum in Fig. 6(b) and would indicate that the thorium

interlayer distance is similar to that of tantalum.

7. Thorium Distribution

After the evaporation of thorium equivalent to 20 atomic layers, the

tantalum was removed from the LEED system and was placed on a nuclear track

plate (Kodak NTB2) for 20 hours with the thorium covered side facing the

emulsion. The plate was developed and the α-particle tracks were counted with

the aid of 250 lines/inch Ni mesh placed over the plate. The thorium decay

scheme[9] was used to determine the coverage which was found to be $15.3\pm1.5\times10^{15}$

thorium atoms/sq cm. The initial estimates of coverage from the tungsten

calibration was $18.4\pm1.8\times10^{15}$ atoms/sq cm and these two numbers agree fairly

well. The main purpose of the alpha track counting experiment was the estab-

lishment of a uniform coverage which was certainly obtained over each sq mm

of surface (Fig. 8(b)). This eliminates the possibility of significant

desorption or migration of thorium from the central portion of the ribbon.

8. Evidence for Thorium Clustering from Optical Microscopy

The ribbon after removal from the LEED system was examined in an optical

microscope up to magnification 2000. The existence of small clusters, which

were at first thought to be etch pits, were found to occur only on the thorium

covered side of the tantalum ribbon. The distribution of the clusters is

shown in Fig. 8(a). The density varied from 4×10^9 cm^{-2} near the end (Fig. 9)
to 8×10^8 cm^{-2} at the center of the ribbon (Fig. 10). The clusters only
occurred on the portion of the ribbon above 900°C during deposition and
varied in shape and size although they covered approximately 10% of the surface
area of the tantalum throughout. The largest clusters which occurred in the
center region had linear dimensions $\sim$10000Å, were predominantly rectangular,
and had at least one edge in the [100] tantalum substrate direction. No
structure on the edges of the rectangles could be detected indicating steep
sides or shallow hills. The clusters are visible because of differing reflection coefficients for thorium and tantalum.

9.   Electron Micrograph of the Clusters

An electron micrograph study using an organic replica 2-phase preshadowed
technique has revealed the clusters to be islands which vary in height between
100Å and 1000Å (Fig. 11). It would be anticipated from the known thorium
coverage, cluster density and size, that the thorium islands would have a
median height $\sim$500Å and this certainly falls in the range observed.

10.   Discussion

The evidence presented in this paper suggests that the growth of thorium
on a clean tantalum substrate above 900°C occurs in small clusters at nucleation centers after completion of a pseudomorphic first monolayer. The thorium
forms initially a c(2×2) structure which changes before completion to a (1×1)
structure which completes up to one monolayer as evidenced by the disappearance of center spots in the LEED pattern. It would appear possible for the
thorium to form a complete monolayer in registry with the tantalum from size
considerations because of the partially ionic character of the thorium-tantalum
bond with consequent reduction of the thorium radius. However, the addition
of further thorium cannot now go into 1×1 registry since in thorium-thorium

bonding the thorium has to assume its usual radius of 1.79Å.   The thorium

atoms were shown to have a high mobility from Section 5 during the completion

of the first monolayer and we may expect this to be true also for atoms being

adsorbed on the completed monolayer at 900°C.   The thorium thus migrates to

nucleation centers, where a three dimensional thorium crystal is formed.   The

apparent registry of the thorium islands with the substrate may be an indica-

tion that thorium migration takes place most readily along the low index

directions of the substrate surface, (i.e. [10] and [$1\bar{1}$]).

The lack of additional spots in the LEED pattern with a high thorium

coverage suggests the thorium clusters to have a random orientation to each

other.   The intensity of the (00) beam and its variation with thorium coverage

is now readily explained.   Very little change would be expected after comple-

tion of the first monolayer as indeed was seen to be the case.

The behavior of the Th-Ta(100) system is in contrast to that previously

observed for the Th-W(100) system[2,3].   Initially a c(2×2) structure forms;

thereafter a hexagonal structure similar to the (111) plane of bulk thorium

is observed.   We have repeated similar experiments with a W(100) ribbon and

when it is kept hot (∿900°C) during deposition we have obtained a 1×1 pattern

after 3 monolayers of thorium have been deposited indicating that the sub-

strate temperature during deposition is a critical factor in the structure of

the overlayers.   It seems likely that with the limited experiments we have

performed on tungsten that it behaves very similarly to that of tantalum.

This would, in fact, have important consequences in the theory of work function

changes in the Th-W system which have been derived from thermionic emission

data where presumably the tungsten was kept above 900°C in order to perform

the measurements.

In the evidence presented, the two basic theories of Frank and van der

Merwe, Finch and Quarel, and Rhodin and of Bassett, Menter and Pashley are somewhat reconciled[10]. It seems inescapable that the first monolayer is almost certainly of a two dimensional pseudomorphic form, this being possible because of the partial ionization of the adsorbate in direct contact with the substrate. However further deposition, under conditions that allow a reasonable mobility to the adsorbate atoms, forms a three dimensional absorbate crystal at a number of nucleation centers, possibly at emerging dislocation sites in the substrate.

It would be instructive to undertake experiments at lower temperature, however, the pressures at present obtainable during evaporation do not preclude the possibility of obtaining impurity stabilized structures. It seems likely that an accompanying Auger measurement[11,12] would give more information, however hydrogen would still present a problem in this connection.

ACKNOWLEDGMENTS

We are grateful to Mr. R. White at the Laboratory for Research on the Structure of Matter of the University of Pennsylvania for obtaining the electron micrographs.

REFERENCES

1. P. J. Estrup and J. Anderson, Surface Science 8 (1967) 101

2. P. J. Estrup, J. Anderson and W. E. Danforth, Surface Science 4 (1966)
    286

3. P. J. Estrup and J. Anderson, Surface Science 7 (1967) 255

4. G. W. Wensch, K. B. Bruckart and M. Conolly, Metal Progress 61 (1952) 81

5. O. D. Protopopov, E. V. Mikheeva, B. N. Sheinberg, G. N. Shuppe, Soviet
    Phys.-Solid State 8 (1966) 909

6. D. L. Fehrs and R. E. Stickney, Surface Science 8 (1967) 267

7. D. Steiner and E. P. Gyftopoulos, Report of 27th Annual Conference on
    Physical Electronics (1967) (Massachusetts Institute of Technology) p.160

8. R. L. Gerlach and T. N. Rhodin, Surface Science 8 (1967) 1

9. R. D. Evans, "The Atomic Nucleus" (McGraw-Hill Book Co., Inc., New
    York 1955) p.517

10. R. E. Thun, Physics of Thin Films 1 (1963) A discussion is given p.216

11. P. W. Palmberg and T. N. Rhodin, J. Appl. Phys. 39 (1968) 2425

12. L. A. Harris, J. Appl. Phys. 39 (1968) 1419

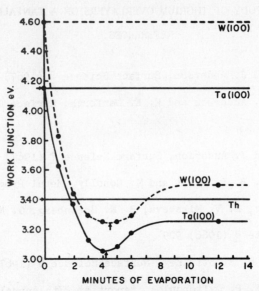

Fig. 1. A graph comparing the work function change of tantalum and tungsten as the thorium coverage is increased by the retarding potential method. The thorium flux at the tantalum is 0.87 of that at the tungsten from geometrical considerations.

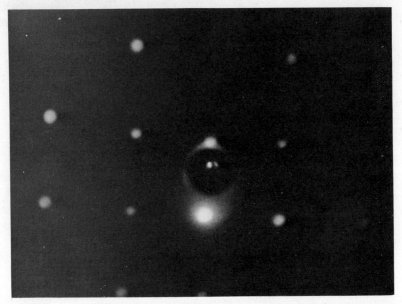

Fig. 2. LEED pattern of the clean Ta(100) surface, beam energy 110 volts.

Fig. 3. LEED pattern of maximum $(\frac{1}{2},\frac{1}{2})$ spot intensity at a thorium coverage of 3×10^{14} atoms/sq cm. Beam energy 58 volts.

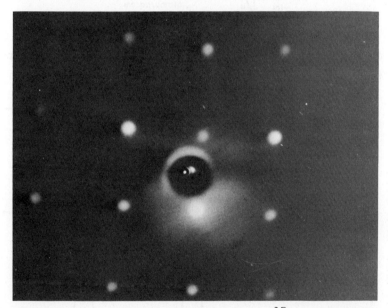

Fig. 4. LEED pattern for a thorium coverage 18.4×10^{15} atoms/sq cm. Beam energy 110 volts.

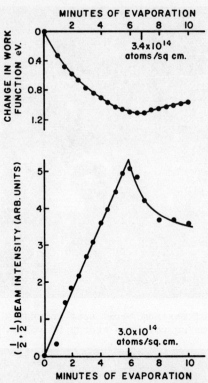

Fig. 5. A comparison between the work function change and the change in intensity of the $(\frac{1}{2},\frac{1}{2})$ spots as the thorium coverage is increased. The
LEED intensities were taken at a beam energy of 84 volts.

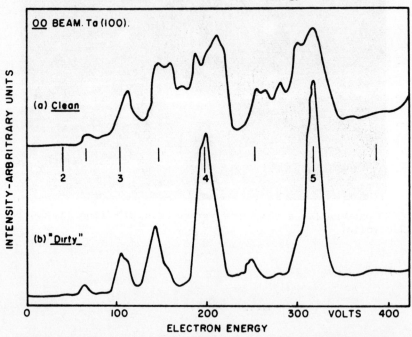

Fig. 6a. Intensity of the (00) beam versus electron energy for clean tantalum
(4° off normal incidence) showing many multiple scattering peaks.
b. Intensity of the (00) beam versus electron energy from a slightly
contaminated surface.

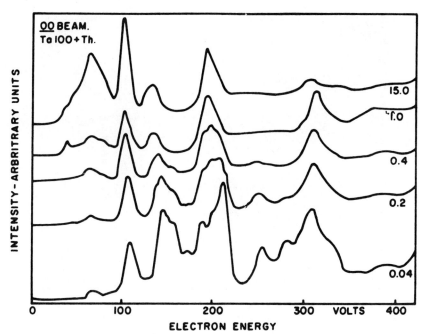

Fig. 7. The intensity of the (OO) beam versus electron energy for varying thorium coverage. The numbers indicate the monolayer equivalent coverage of thorium.

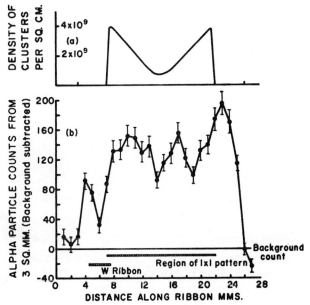

Fig. 8a. The density of thorium clusters along the ribbon. b. The a-particle count/3 sq mm along the ribbon. The region shielded by the tungsten monitor ribbon is shown.

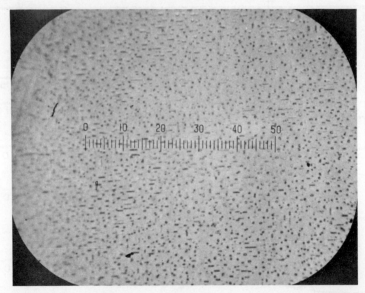

Fig. 9. A microphotograph of the surface 21mm along the ribbon.
1 scale division = 1 micron.

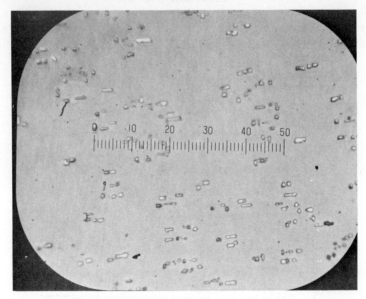

Fig. 10. A microphotograph of the surface 14mm along the ribbon.
1 scale division = 1 micron.

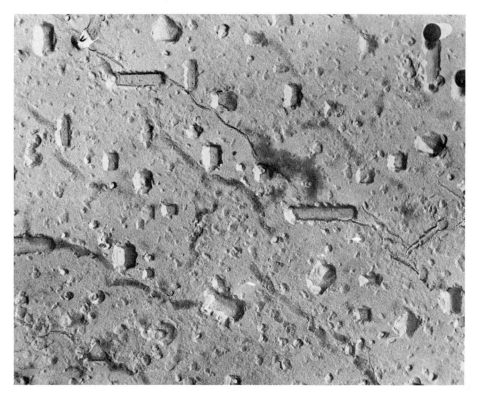

Fig. 11. An electron micrograph of the surface clearly depicting the structure of the thorium clusters taken ~16mm along the ribbon. An interpretation of the shadows in the plastic replica shows the clusters are hills on the tantalum surface. The small black circles are $2640\text{Å} \pm 60\text{Å}$ in diameter.

CLEAN Te SURFACES STUDIED BY LEED

S. Andersson, I. Marklund, and D. Andersson

Institute of Physics
Chalmers University of Technology
Gothenburg, Sweden

1. INTRODUCTION

A number of LEED investigations have been performed on surfaces of the single element semiconductors Si and Ge. Properties of the clean surfaces have been investigated and a great number of elements have been deposited on them. Surfaces of the other single element semiconductors Se and Te are also of great interest e.g. in studies of hetero-junctions. Recently we studied the Te(0001) surface obtained by epitaxic growth of Te and a Cu(111) surface[1]. The Te(0001) surface formed a rearranged Te(0001) - 2×1 structure. We found it of great interest to continue this investigation by a study of Te surfaces produced from Te single crystal samples. The Te($10\bar{1}0$) cleavage face and the Te(0001) hexagonal face are investigated from the point of view cleaning procedure, structure, effect of ion bombardment, and energy distribution of the back-scattered electrons. The clean surfaces obtained are suitable for studies of the inelastically backscattered electrons.

2. EXPERIMENTAL PROCEDURE

2.1 Apparatus

The low-energy electron diffraction system used in this investigation is a Varian LEED apparatus equipped with a three grid electron optics. This grid configuration provides a sharp cutoff when working in the diffraction mode and a reasonable energy resolution when performing differential energy analysis of the backscattered electrons. Our device has been described earlier[2] and a similar system has been used by Scheibner et al.[3] in their studies of the energy distribution of backscattered electrons from surfaces of tungsten graphite and copper. The base pressure in our system during operation was in the 10^{-10} Torr range.

2.2 Specimen preparation

The Te samples were obtained from Te single crystals of 99.999% purity. A single crystal cut in the form of a cube with two cube faces exposing the cleavage face Te(10$\bar{1}$0) could be repeatedly cleaved in ultra high vacuum by means of a wedge moved by a bellows manipulator.

Samples exposing the Te(000$\bar{1}$) surface were prepared by cutting the crystal in the required direction by spark erosion (orientation obtained within one degree). The mechanically polished samples were finally electrolytically polished in a solution of 6% KOH, 47% H_2O, and 47% C_2H_5OH, kept at -20°C. The low temperature is necessary to get a flat mirrorlike surface. At room temperature the surface becomes undulated and less well reflecting.

2.3 Cleaning of the Te(10$\bar{1}$0) surface

The Te(10$\bar{1}$0) surface obtained by cleavage in air showed no diffraction pattern after heating to 200°C for one hour. The only observation was a faint halo in the background indicating that Te atoms were present. The energy spectrum of the backscattered electrons revealed the existence of a surface contamination layer. Repeated argon ion bombardments (150 eV, $1\mu A/cm^2$ for 1 hour) and annealings (200°C, a few minutes) produced a clean surface as observed from the diffraction pattern and the energy spectrum, compared with our observations for the clean Te(10$\bar{1}$0) surfaces produced by cleavage in situ. The latter surface was somewhat more perfect giving narrower diffraction beams.

2.4 Cleaning of the Te(0001) surface

The Te(0001) surface obtained by electrolytical polishing produced no diffraction beams after heating to 200°C for one hour, but only the same faint halo in the background as observed for the Te(10$\bar{1}$0) surface. After a standard initial ion bombardment (150 eV argon ions, $1\mu A/cm^2$) and annealing (200°C, 30 min.) a bad diffraction pattern was observed containing fuzzy diffraction

beams and beams due to facets. Energy analysis of the backscattered electrons
showed (Fig. 1) that the bad pattern was not due to contamination but to bad
crystalline order. Successive ion bombardments with the above mentioned conditions for longer times finally produced an optimally clean surface. The required ion bombardment time was found to be sensitive to the initial surface
order. A less mirror-like and undulated surface needed 50 hours, while a mirror-
like, flat one only required 15 hours bombardment.

3. DIFFRACTION RESULTS

3.1 The Te(10$\bar{1}$0) surface

The diffraction patterns (Fig. 2a, b) from the clean Te(10$\bar{1}$0) surface
obtained by cleavage in situ or by ion bombardment annealing technique show
no fractional order diffraction beams and the measured surface lattice constants
agree with those from an ideal Te(10$\bar{1}$0) plane[4] (Fig. 3a). This is expected
because of the layered nature of the crystal containing chains of Te atoms
strongly held together by covalent forces. The interaction between different
chains is weak and no covalent bonds are broken during cleavage.

Surface disorder observed as streaks in the [10]* direction in the diffraction pattern (Fig. 2c) is due to a case of less good cleavage causing distribution of the lattice constant in the direction perpendicular to the chain. The
only distinct diffraction beams from facets came from analogous inclined Te
10$\bar{1}$0 surfaces. No other less inclined cleavage surfaces of type Te$\left\{ \text{h 1 } \overline{\text{h+1}} \text{ 0} \right\}$
develop areas large enough to produce any distinct diffraction beams, but
only narrow streaks giving evidence for a long range order in the direction of
the Te chains.

A strong asymmetry in the atomic scattering factor of Te was observed.
For primary electron energies of about 50 and 175 eV a deep minimum in the
background intensity of the diffraction pattern is observed (Fig. 2b) at
scattering angles of 145° and 150° respectively. Similar observations have

reported for other materials e.g. Sb^5, Bi^5 and I^6.

The intensity versus primary electron energy of the 00 and {10} diffraction beams agree well with the expected plane distribution in the Te [$10\bar{1}0$] direction. The specularly reflected 00 diffraction beam intensity is shown in Fig. 4 (angle of incidence: 5°). The arrows indicate the positions of the theoretically calculated Bragg peaks corrected with 6.5 V (involving an average inner potential of 10 V and a contact potential of 3.5V). The long arrows show the peaks due to scattering between planes of distance D=3.86Å and the small one at 136 eV scattering between subplanes of distance d=1.04Å (Fig. 3b). It is obvious that the scattering between equivalent atoms in different chains is dominating in the primary beam energy range used, as the intensity of the peak due to scattering between subplanes is very weak compared to the other peaks. The intensities of the {10} beams show a behavior similar to the 00 beam.

The effect of ion bombardment of the Te($10\bar{1}0$) surface was investigated. Surface damage was observed, but the diffraction pattern was never obliterated even when hundreds of atomic layers were sputtered away. Instead it was found that the damage reached a saturation level after a rather short bombardment time, e.g. 350 eV argon ions, $1\mu A/cm^2$ gave a saturation time of about 5 minutes. No further change was observed for a ten times longer bombardment. The damage was observed in the diffraction pattern as fuzzy diffraction beams alternating with sharper ones as the primary electron energy was changed. It was also observed that the fuzzy diffraction beams were resolved in double beams (Fig. 2d), when the bombardment was performed with 150 eV argon ions, $1\mu A/cm^2$ for a standard time of 60 minutes.

Splitting of the beams is due to faceting of the Te($10\bar{1}0$) surface in the direction perpendicular to the chains. The facet beams were only developed in the neighbourhood of the ordinary Te($10\bar{1}0$) diffraction beams. The indices of these facets were determined:

a. by observing the angle, α_{exp}, that the specularly reflected facet diffraction beam $(00)_F$ makes with the normal specular diffraction beam 00,

b. By observing the differential movement of the $(0k)_F$ diffraction beams when passing the 00 beam,

c. by observing successive orders $(0k)_F$ and $(0\ k+1)_F$, respectively, passing through the 00 beam.

In Table 1 the determined values of α_{exp} are given together with the angle α_h that the Te $\{h,\ 1,\ \bar{h}\mp\bar{1},\ 0\}$ planes make with the Te$(10\bar{1}0)$ plane (Fig. 3b); (Table 1 is given for $(00)_F$ observed around the 01 beam). The agreement is excellent and explains why the $(00)_F$ beams are observed only in the neighbourhood of the 0k beams. The facet planes are those giving the Bragg peaks in the 0k intensity distribution.

The methods b and c give the same result as a Tucker[7] has presented a suitable formalism for treating the cases b and c. As an example the use of method b will be given. It was observed that one $(0k)_F$ beam coincided with the 00 beam at 31 eV primary beam energy and moved $2.5°$ when the beam energy was changed to 27 eV.

Tucker's formula:

$$\frac{\frac{\partial\lambda}{\partial\phi}}{\frac{\lambda}{2}} = \cot\alpha;$$

where λ is the electron wavelength and $\partial\lambda/\partial\phi$ is the differential movement, gives an angle $\alpha_{exp} = 14.2°$ in good agreement with $\alpha_h - 14.0°$ for the facet Te$(41\bar{5}0)$.

The observed difference in damage produced by 150 and 350 eV argon ions is due to the fact that the 350 eV ions produce a less ordered damage by also causing disorder in the direction of the chains (the chains are broken) which is observed as a broadening of the diffraction beams in the $[01]^*$ direction.

The damage could easily be annealed; at $200°C$ the necessary annealing time is less than 2 minutes and at 100°C about 70 minutes. The resulting diffraction patterns and intensity distribution curves are well reestablished. The annealing mechanism is probably due to cooperating surface diffusion and evaporation.

3.2 The Te(001) surface

In contrast to the ideal diffraction pattern from the Te(10$\bar{1}$0) surface the diffraction pattern from the optimally cleaned Te(0001) surface contains fractional order diffraction beams (Fig. 5a). This is in agreement with our results from the Te(0001) surface obtained by epitaxic growth of Te on a Cu(111) surface[1] (Fig. 5d). However, in the latter case the observed 1/2 order beams are well defined spots while they here are elongated. The energy spectrum of the backscattered electrons (Fig. 1) is characteristic for a clean Te-surface, and we believe that this is a sensitive test (for reasons discussed in Section 4) that a contaminant is not responsible for the streaking of the beams. This streaking is instead thought to be due to disorder in the surface layer disturbing the long range order. A similar observation has been reported for the Ge(100) surface[8] and was explained by one-dimensional disorder in the surface.

Various annealing treatments and cooling of the sample to ⊤150°C by a liquid nitrogen cooling finger did not improve the surface order. Cooling, however, decreased the background, thus giving a better resolution of the diffraction pattern.

The observed 1/2 order diffraction beams are interpreted to be due to three equally probable 120° rotated domains of Te(0001) - 2×1 surface structures (Fig. 6). This model is in agreement with:

a. the observed diffraction pattern,

b. the saturation of all free bonds in the surface (one unsaturated bond for every Te atom in the surface), (Fig. 6),

c. the observed direction of streaks (Fig. 7). One direction is predominant for one type of the three possible reciprocal 2×1 unit cells. The sketched one

in Fig. 6 corresponds to disorder in the real lattice in the Te[1$\bar{1}$00]
direction. A possible solution is also drawn in Fig. 6. A displacement
of half the lattice constant in the [01] direction will be compatible with
the observed disorder.

The 1/2 order beams are not observed above a primary beam energy of 60 eV. Figure 5b shows the diffraction pattern at 96 eV with narrow integral order beams.

The intensity distributions of the 00 and {10} diffraction beams have been
studied. The result is shown in Fig. 8 for the 00 beam intensity versus primary beam energy. The arrows indicate the theoretically calculated values for
the integer order Bragg peaks and a d-value of b/3 = 1.98Å in the Te[0001]
direction (Fig. 3). The values are corrected with 6 V (an average inner potential
of 9V and a contact potential of 3 V) and correspond well to the dominant peaks
in the curve. The weaker peaks appear approximately at the energies at the
Bragg peaks in the {10} intensity curves and are thus due to multiple scattering.

The clean Te(0001) surface was insensitive to further standard ion bombardments. No damage could be observed from the diffraction patterns immediately
after bombardment; both fraction and integer order beams remained as clear as
before the bombardment. This is in contrast to the observation for the Te(10$\bar{1}$0)
surface (see above).

In the early stage of cleaning, weak diffraction beams originating from
facets could be observed but after optimal cleaning they never appeared again.
This means that they are due to the initial surface topography and not to ion
bombardment.

4.   ENERGY ANALYSIS OF THE BACKSCATTERED ELECTRONS

Energy spectra of the backscattered electrons for primary electron energies
of 50–250 eV have been obtained. Those from the vacuum-cleaved Te(10$\bar{1}$0) surface

have been used as characteristic for a clean Te surface. In Fig. 1 one can

see that the spectra obtained from the clean Te(0001) and Te(10$\bar{1}$0) surfaces

produced by ion bombardment agree well with that obtained from a cleaved Te(10$\bar{1}$0)

surface. This supports the diffraction observation that the surfaces are clean

(reproducible diffraction patterns and intensity distributions).

Besides the Te(0001) - 2×1 surface structure a temporarily appearing 4×1

structure was observed (Fig. 5c). The energy spectra obtained (Fig. 9) were,

however, changed in shape compared with those from clean Te-surfaces, but

no new peaks appeared. The intensities were reduced for both the 41.8 eV

ionization loss peak and the 34.4 eV loss peak due to double excitation of the

volume plasma. (These two peaks lie on a smooth background and are easy to

compare in magnified spectra.) It was found that the area under the 34.4 eV

loss peak was reduced by a factor 15 and that under the 41.8 eV loss peak by

a factor 5. The observed intensity reduction of the loss peaks is interpreted

as an effect of contamination, as the probability for N_{45} ionisation should not

change so drastically otherwise. No exact conditions for the appearance of the

4×1 structure could be obtained. Commonly it appeared after a long ion bombard-

ment followed by a long anneal at a low temperature of a short one at high tem-

perature, and it could thus be due to diffusion of a contaminant to the surface.

Because of the badly controlled conditions nothing can be said about the dis-

tribution of the contaminant on the surface and just below it. Further work is

going on to study these intensity reductions under more exact circumstances.

It has already been pointed out (Section 2.4) that the energy spectra can

give information about whether a bad diffraction pattern in the initial cleaning

stage is due to a contaminant or to bad crystalline order. This piece of infor-

mation is very valuable in order to understand how to produce a well-ordered

clean surface.

It should also be noticed that the energy spectra obtained immediately after ion bombardment of the Te surfaces were compatible with those after anneal. This is contrary to what we have observed for a Si(111) surface[9] where both the loss energy and intensity were strongly affected.

The energies of the observed loss peaks are in good agreement with Robin's results[10] (Table 2) from evaporated Te films. Furthermore we have observed a new loss peak at 7.6 eV which is readily observed for low primary electron beam energies (Fig. 10). Optical measurements[11] give a maximum in the reflexion coefficient at about 7.3 eV in agreement with our loss value. The calculated band structure for Te[12] gives a possible interpretation for the 7.6 eV loss; an interband transition $\Delta_3^1 - \Delta_1^4 = 7.4$ eV at $k_z = \pi/c$ where the corresponding bands are flat (high density of states).

Besides the loss peaks a peak due to Auger-ejected electrons (Fig. 11) is observed at 29.4 eV and corrected to 30.4 eV (the correction of 1.0 eV involves contact potential between the Te crystal and the analyzing grid system and non-ideal retardation of the grid system). This value is to be compared with 31.4 eV obtained for the Auger process 4d - 2(5s,5p) when using our experimental values of 41.8 and 5.2 eV for the single particle transitions, respectively (tabulated values[13] give 30.7 eV for the same transition assuming a work function of 4.5 eV).

ACKNOWLEDGMENTS

We are grateful to Fil. Mag. Jorgen Martinson for his valuable assistance in the experiments. Financial support from Malmfonden-Swedish Foundation for Scientific Research and Industrial Development and from the Swedish Natural Science Research Council is also gratefully acknowledged.

REFERENCES

1. S. Andersson, I. Marklund and J. Martinson, Surface Science to be published.

2. I. Marklund, S. Andersson and J. Martinson, Arkiv Fysik (1968) in print.

3. E. I. Scheibner and L. N. Tharp, Surface Sci. 8, (1967) 247.

4. R. W. G..Wykoff, Crystal Structures, Vol. 1 (Interscience, New York, 1960).

5. F. Jona, Surface Sci. 8, (1967) 57.

6. J. J. Lander and J. Morrison, J. Appl. Phys. 34 (1963) 3517.

7. C. W. Tucker, Jr., J. Appl. Phys. 38, (1967) 1988.

8. J. J. Lander and J. Morrison, J. Appl. Phys. 34, (1963) 1403.

9. S. Andersson and I. Marklund, to be published.

10. J. L. Robins, Proc. Phys. Soc. 79 (1962) 119.

11. H. Merdy, S. Robin-Candare and J. Robin, Le Journal de Physique 25 (1964) 223.

12. R. E. Beissner, Phys. Rev. 145 (1966) 479.

13. J. A. Beraden and A. F. Burr, Rev. Mod. Phys. 39, (1967) 125.

TABLE 1

Measured and calculated values of the angle between the Te $\{h, 1, \overline{h+1}, 0\}$ planes and the Te($10\overline{1}0$) plane.

α_{exp}	18°	14°	11°	9°
α_{h}	19°	14°	11°	9°
h	3	4	5	6

TABLE 2

Calculated and measured energy losses for Te.

Method	Energy losses (eV)						Reference
Theoretical	5.0	7.4					R. E. Beissner[12]
				15.6			J. L. Robins[10]
Optical reflectance measurement		7.3					H. Merdy et al.[11]
Electron scattering	5.2		11.8	17.0	33.8	41.5	J. L. Robins[10]
X-ray spectroscopy						39.8	J. A. Bearden A. F. Burr[13]
Electron scattering	5.2	7.6	11.8	17.1	34.4	41.8	Present results
Type of excitation	inter-band	inter-band	surface plasmon	single excitation volume	double excitation plasmon	N_{45} ionization	

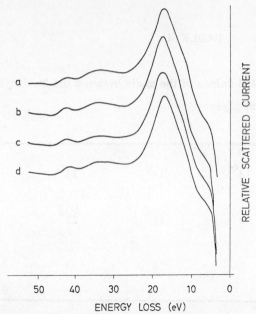

Fig. 1. Energy spectra of inelastically scattered electrons for a primary electron beam energy 125 eV (normal incidence) from: a) Te(0001) surface after the initial cleaning by ion bombardment and annealing. b) Te(0001) surface after the final cleaning by ion bombardment and annealing. c) Te(10$\bar{1}$0) surface cleaned by ion bombardment and annealing. d) Te(10$\bar{1}$0) surface obtained by cleavage. The curves are seen to be almost identical independent of the way of surface production. The curves are shifted by a fixed amount an an energy of 52 eV.

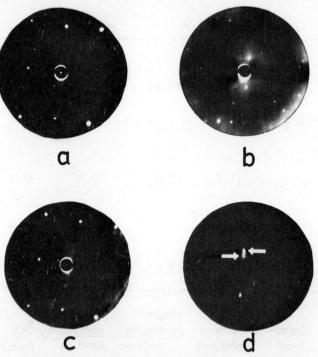

Fig. 2. Diffraction pattern from a Te(10$\bar{1}$0) surface. a) Clean surface. Primary beam energy 30 eV. b) Clean surface. Primary beam energy 175 eV. c) Clean surface but less good cleavage causing streaks in the pattern. Primary beam energy 30 eV. d) Split 00 beam due to ion bombardment damage. The sample is tilted 7°. Primary beam energy 38 eV.

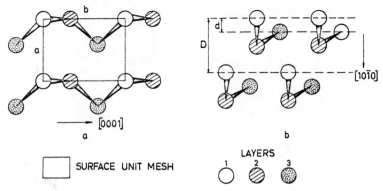

SURFACE UNIT MESH

LAYERS
1   2   3

Fig. 3. Schematic drawing of the Te structure. a) Te(10$\bar{1}$0) surface structure with a rectangular unit mesh, a = 4.46 Å, b = 5.93 Å. b) layer distribution in the [10$\bar{1}$0] direction.

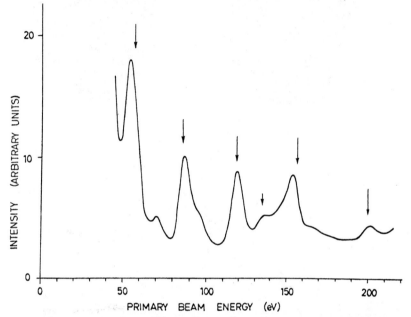

Fig. 4. Intensity of the 00 diffraction beam from a clean Te(10$\bar{1}$0) surface as a function of primary beam energy (angle of incidence 5°). The arrows indicate the calculated Bragg peak positions (corrected with 6.5 V) for inner potential and contact potential.

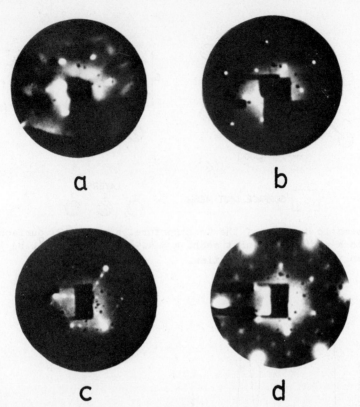

Fig. 5. Diffraction pattern from a Te(0001) surface. a) Clean surface. Primary beam energy 32 eV (retarding field mode, bias 100 V). b) Clean surface. Primary beam energy 96 eV. c) Contaminated surface. Primary beam energy 70 eV. d) Clean Te(0001) surface grown on Cu(111) surface. Primary beam energy 52 eV (retarding field mode bias 50 V).

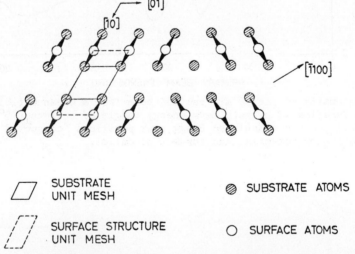

Fig. 6. Model of the proposed Te(0001) surface structure (one of the three possible rotation domains). To the right is drawn part of a possible domain which together with the domain to the left gives disorder in the [11] direction (the [$\bar{1}$100] direction in the substrate).

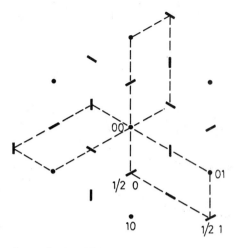

Fig. 7. Schematic drawing of the LEED pattern from the Te(0001) - 2 x 1 surface structure. The three 120° rotated unit cells are sketched.

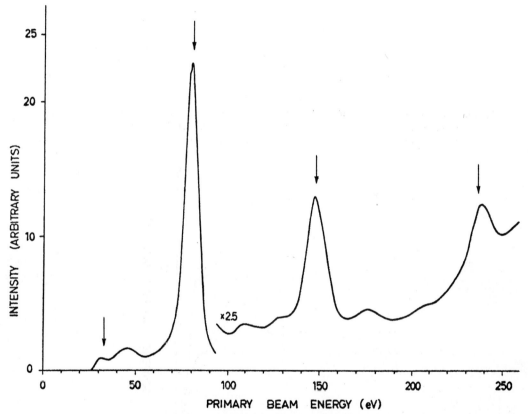

Fig. 8. Intensity of the 00 diffraction beam from a clean Te(0001) surface as
a function of primary beam energy (angle of incidence is 5°). The arrows indicate the calculated Bragg peak positions (corrected with 6 V
for inner potential and contact potential).

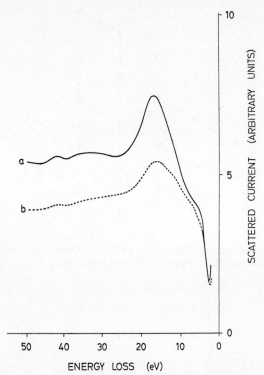

Fig. 9.   Energy spectra of inelastically scattered electrons for a primary
electron beam energy 125 eV from: a) clean Te(0001) surface with
2 x 1-structure. b) contaminated Te(0001) surface with 4 x 1-structure.

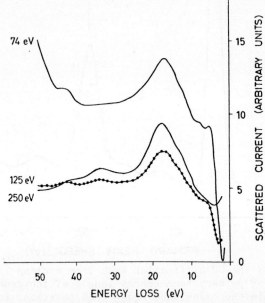

Fig. 10.   Energy spectra of inelastically scattered electrons for different
primary electron energies obtained from the Te(10$\bar{1}$0) surface produced
by cleavage. (The spectra are normalized to a primary beam current of
1 μA).

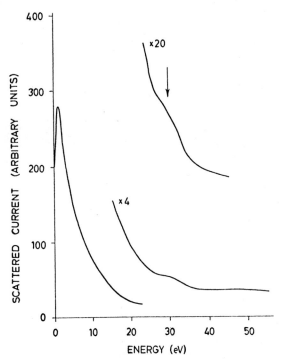

Fig. 11.   Low energy part of a spectrum of the back-scattered electrons from a clean Te surface. The Auger line is indicated.

LEED STUDY OF THE GROWTH OF TIN FILMS ON Nb(110)*

A. G. Jackson and M. P. Hooker

Systems Research Laboratories, Inc.
Dayton, Ohio

Introduction

The study of the tin-niobium system has almost entirely been prompted by its properties as a superconductor.[1] Bulk studies of the formation of various alloys of these two elements have shown that Nb_3Sn is the most commonly formed compound and is the easiest to form. Thin film studies of the formation of niobium stannide have been carried out, revealing the conditions for formation of Nb_3Sn and of non-stochiometric forms in which tin predominates.[2] Recently interest has been generated in determining the role of the surface and its structure in the super conducting phenomenon.[3,4] While this study is not concerned with superconductivity, LEED techniques are useful in supplying information about the surface structure and about some of the characteristics of the formation of Nb_3Sn.

From the point of view of LEED or thin film studies the Nb_3Sn system is of interest in itself as an example of a complex alloy system which can provide more information of the surface behavior of one metal on another.

In this paper we present the results of evaporating high purity tin onto the Nb (110) clean surface.

---

*Work supported by Aerospace Research Laboratories, Wright-Patterson Air Force Base, Ohio 45433

Experimental Apparatus and Sample

The LEED apparatus is of the display type with triple grid configuration. For high rates of evaporation, a boat suitably arranged with defining slits was used. For slow rates of evaporation a coil, onto which a drop of tin was melted, was used. The second evaporator was carefully calibrated by the standard crystal oscillator technique.

The Nb sample was cut and polished to bright finish and then X-ray oriented to ±1° of the (110) face as outlined by Haas. The **clean** Nb (110) surface was obtained by high temperature degassing of the sample and oxidation.[5] Figure 1 shows the clean pattern together with a marble model of the (110) surface.

Experimental Results

The experimental results are divided into two categories: rapid evaporation rate (> 10 lay/sec) and slow evaporation (2 layers/sec).

Rapid Evaporation. It has been[6] shown that high evaporation rates lead to amorphous films of tin. Rapid evaporation from the slit evaporator onto clean Nb (110) removes the clean pattern and leaves only a general diffuse "pattern" in which even the (00) beam was poorly resolved. This change of pattern always is produced for substrate temperatures as high as 600°C. Heating the sample to 650 ± 50°C., in any case, induces a new pattern (Fig. 2a) which has hexagonal symmetry. The shape of the spots, lack of definition, and high background intensity indicate that this structure is somewhat disordered. Heating to 700 ± 50°C. produces patterns which suggest that rotational disorder is prevalent (Fig. 2b). Heating to 950 ± 50°C., leads to a pattern which we interpret as Nb_3Sn (110), with evidence of other structures (Fig. 2c), and after heating at 1000° ± 50°C., niobium stannide is clearly in evidence.

Higher temperature heating to 1100 ± 50°C removes the Nb_3Sn pattern and restores the Nb (110) pattern. At 1800°C the clean Nb (110) is restored as evidenced by the clean pattern and the return of the usual specular beam intensity curve.

Evaporation onto the sample at 950°C results in the Nb_3Sn pattern, which can be then removed by heating to 1200 ± 50°C.

Slow Evaporation. (2 layers/sec) In this case the results differ from those for the rapid rate runs. At 20 layers the Nb (110) pattern becomes slightly diffuse, and at 40 layers, a new set of spots appears (Fig. 3) which is interpreted as due to Sn (110). In addition the usual Nb (110) spots are also present. Continued evaporation of up to 1000 layers does not significantly change this pattern. Heating to 900°C, however, induces rotational disorder similar to that mentioned in the high rate case, Fig. 2b, which changes with continued heating (Fig. 2c) to the rectangular pattern associated with Nb_3Sn (110). The clean Nb (110) is restored after heating to 1000°C±50.

Evaporations of 10 layers or more onto the sample at room temperature now produces a (3x1) pattern (Fig. 4) which is stable up to about 700°C. With the sample at 400°C, the same (3x1) pattern is produced, but changes to Nb_3Sn at 500°C, and the clean Nb (110) pattern is restored at 800°C. If the sample is held at 600°C during evaporation the resulting pattern is that of Nb_3Sn, which can be removed by heating to a temperature above 800°C.

These results indicate that Nb_3Sn can be easily formed under rather widely varying starting conditions. The significant point is the apparent difference in transition temperature. In low rate evaporations transition temperatures are lower than in high rate evaporations by as much as 400°C. This arises possibly because of the difference in type of film formed during the two kinds of evaporation. Rapid evaporation leads to large globules of tin which do not have crystallinity to any large extent. Heating this film

requires melting of the globules, diffusion into and across the Nb, and structure formation. The slow rate evaporation, on the other hand, forms many
small crystallites which are easily melted, requiring less heat or lower temperature. Hence there is an activation energy required in the first instance
which is not present is the second case.

Discussion

As mentioned above the type of tin film one obtains depends on the rate
at which the tin is evaporated. High rates produce large islands of tin,
while low rates produce small, well-formed crystallites. The crystal structure of tin above 13°C is tetragonal with the configuration of Fig. 5. Since
the unit cell has 4 atoms per cell, the basis is somewhat complex, but the
crystal structure does not have any space group extinctions. The non primitive motif may disallow certain beams, but the effect of the basis on the
LEED pattern is not simple as explained shortly.

The crystal structure of Nb_3Sn is β-tungsten and is illustrated in Fig.
6.   It is cubic but with a nonprimitive unit cell which contains 8 atoms. As
with tin the influence of the complex motif is not straight forward.

Reciprocal lattices constructed on the basis of the usual unit cells do
not account for all the patterns observed. In several patterns there appears
to be a combination of reciprocal lattices which can be derived from this
bulk cell and from a simple, geometrical, two-dimensional mesh derived by
simple considerations of atoms only in one plane and not immediately above
and below the plane. For example, on the basis of the bulk cell Nb_3Sn (110)
should have a simple rectangle as unit mesh with side ratio of $\sqrt{2}$. If, however, one naively slices Nb_3Sn along the (110) plane one obtains a body-
centered, two-dimensional mesh analogous to Nb (110). Hence one expects a

pattern similar to Nb (110), but of different size. But both patterns of niobium stannide are observed experimentally. The extinction pattern of Nb_3Sn (110) has intensity comparable to the simple, rectangular pattern at low voltage (<100cV), while at higher voltage the simple rectangular Nb_3Sn (110) pattern dominates.

This behavior suggests that the first few layers are not the same as the bulk, thereby giving rise to the extinction pattern, while deeper penetration of the beam samples the bulk structure, giving rise to the rectangular pattern. This can be interpreted in terms of an expansion of the plane separation between the first and second layers so that the influence of the atoms in the second layer is very small. In addition one can also conclude that this behavior is an example of kinematic and dynamic effects and their dependence on energy. Kinematically one expects the extinction pattern if only the first layer is considered. Bulk effects will only show up strongly at high voltages, where dynamical considerations dominate. In this view this case is an example of the transition between these two situations.

A valid question is what happens to the tin with high temperature heating? It is clear that diffusion of the tin into the Nb is occurring at lower temperatures since there is a definite temperature–time relationship. In the temperature range of 1100°C to about 1800°C diffusion into the sample is dominant. This is evidenced by a bending of the sample as it changes lattice parameter to that of Nb_3Sn, which is larger than Nb. By heating to temperatures above 1800°C, the sample returns to its original shape, indicating that the tin has diffused out to the surface and evaporated. This is reason-able if one considers that the vapor pressure of tin at temperatures above 1000°C is quite high ($>10^{-4}$ torr). Diffusion into the sample-holder-legs is possible, but the temperature of the legs never reaches above 800°C with the sample at 1800°C. Diffusion of significant amounts over short periods would be small. The complexity of this system warrants further investigation, especially in the difference in structure transformation temperatures. The results here are essentially the first steps in a long range study of alloy systems and ultra thin films.

As far as the behavior of the system goes, the sequence appears to be the following:

With high or low rate evaporation there exists a very low coverage state in which a hexagonal pattern results from an attempt by Sn to match substrate. High rate leads to amorphous layers which can be ordered by diffusing tin into Nb to produce Nb_3Sn (110). A low rate leads to Sn (3x1) which can be changed to Nb_3Sn (110) by heating the sample. Both procedures lead to Nb_3Sn after heating. The film is stable to about 1000°C; much above this, at about 1100°C, the tin diffuses into sample. Above 1800°C, tin diffuses out and is evaporated off the sample. In each case rotational disorder accompanies the changes from tin on clean niobium to the Nb_3Sn structures.

Since the optics we used is in the triple grid configuration, an apparatus was constructed for recording the energy spectrum using the technique of Scheibner et al.[7] It was hoped that Auger peaks due to tin could be seen as well as plasma losses which would identify trace amounts of tin present on the sample but not observable with LEED. Technical problems, however, prevented obtaining any Auger data.

In several runs the characteristic losses associated with tin were observed in addition to that of Nb when the pattern was aparently Nb (110). Such monitoring of the sample is very useful in studies of this sort because both the diffraction pattern and the (00) intensity curve indicated the sample was clean Nb (110).

Energy loss measurements were made for the case of tin on the sample in low evaporation rate form. Figure 7 shows the energy loss distribution for the Nb (110) surface with traces of gas present, probably CO and H_2. The losses are observable at 12eV (surface) and at 21eV. After a 30 sec evaporation at 2 layers/sec, the loss curve changed to that illustrated in Fig. 8. A single large peak is present, with a smaller one at higher energy. The losses are 14eV and 29eV respectively. These values agree closely with those listed by Powell[8] for the surface and the bulk loss for tin.

Figure 9 shows the spectrum after heating to 1900°C. The tin peaks are almost gone, the niobium peaks restored. Traces of tin are indicated by the small peak at 29eV, even though the pattern and the 00 intensity distribution would suggest a clean sample. More extensive data on the loss spectrum of each structure, as well as the Auger spectrum is being gathered.

This result points out again the need for sensitive measuring techniques which do not depend upon the structure of the material. Auger spectroscopy would establish the presence of amorphous constituents on the surface. Such information correlated with LEED data will serve to establish hard data on the influence of such amorphous regions on the formation mechanisms which produce the structures reported.

In conclusion we summarize by noting that there are many questions which remain to be answered concerning this system. Its behavior is quite complex in general and especially so with variations in temperature. This study has only provided data on the gross behavior of the system. Detailed results on the diffusion mechanism, gas interaction, presence and effects of impurities, stability of structures should be obtained to eliminate questions about this system. Studies are currently being pursued which will provide more data on some of these influences.

References

1.   Special issue of RCA Review, XXV (3).

2.   C. A. Neugebauer, J. Appl. Phys., 35(12), 3599 (1964).

3.   G. Bon Mardion, B. B. Goodman, A. Lacaze, Phys. Letts, 8(1), 15(1964).

4.   W. R. Bandy, C. R. Haden, T. D. Shockley, Phys. Letts, 24A (13), 711(1967).

5.   T. W. Haas, Surf. Sci., 5, 345 (1966).

6.   R. P. Riegert, Sloan Tech Note #8.

7.   L. N. Tharp, E. J. Scheibner, J. Appl. Phys., 38(8), 3320(1967).

8.   C. J. Powell, Proc. Phys. Soc. Lond., 76, 593 (1960).

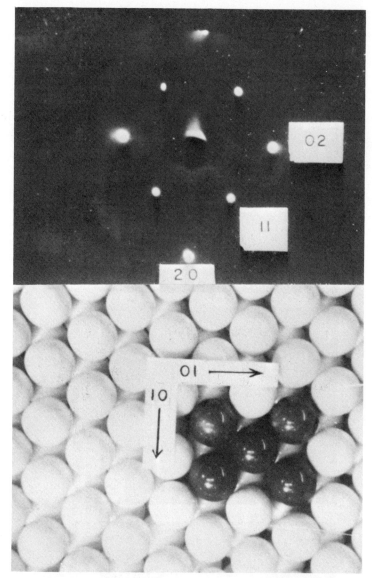

Fig. 1.   Photograph and marble model of the clean Nb(110) surface, showing the usual beam indices.   Voltage is 140 V.

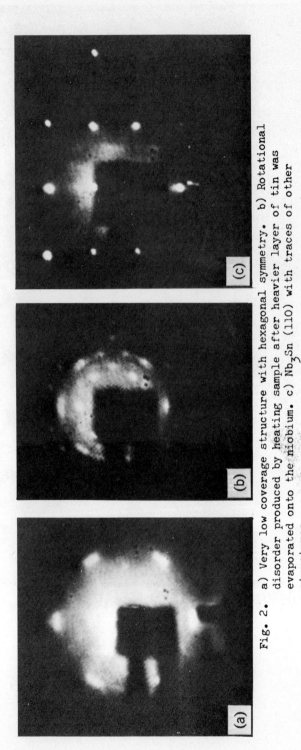

Fig. 2. a) Very low coverage structure with hexagonal symmetry. b) Rotational disorder produced by heating sample after heavier layer of tin was evaporated onto the niobium. c) Nb_3Sn (110) with traces of other structures.

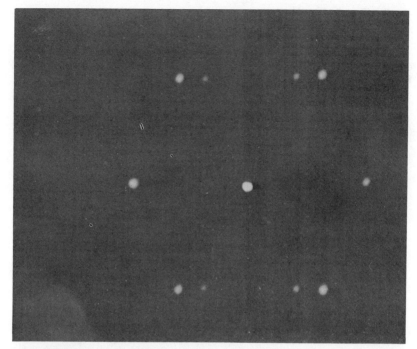

Fig. 3. Sn (110) pattern with that of Nb (110).

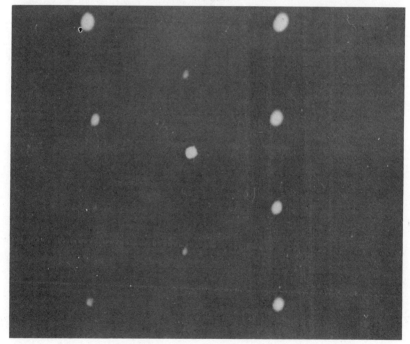

Fig. 4. A (3X1) pattern, probably due to tin.

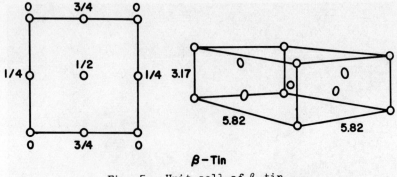

β−Tin

Fig. 5.   Unit cell of β-tin.

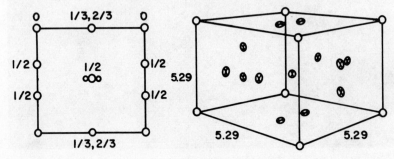

Nb$_3$Sn

Fig. 6.   Unit cell of Nb$_3$Sn.

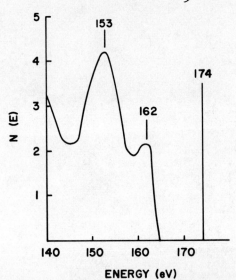

Fig. 7.   Energy loss spectrum for Nb(110) with primary beam of 174 volts.

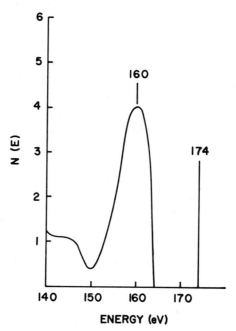

Fig. 8.   Energy loss spectrum due to tin evaporated onto Nb (110).

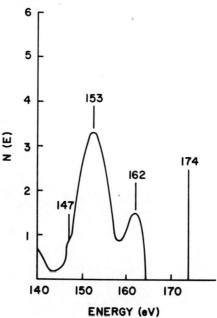

Fig. 9.   Energy loss spectrum after heating the sample to $1900^{\circ}C$, showing presence of tin and niobium peaks.

THE INTERACTION BETWEEN CHLORINE AND THE (100) SURFACE OF GOLD

D. G. Fedak, J. V. Florio, and W. D. Robertson

Hammond Laboratory
Yale University
New Haven, Connecticut

INTRODUCTION

When single crystals of gold are subjected to thermal and ion bombardment treatments which would normally ensure removal of surface contaminents, the (100) surface is observed to yield complex low energy electron diffraction (LEED) patterns. Typically, the patterns exhibit a periodicity of 5d × 20d, where d is the interatomic spacing, in lieu of the expected 1 × 1 or bulk-like configuration. A model for the arrangement of atoms responsible for the 5 × 20 periodicity has been described previously by Fedak and Gjostein[1]. The proposed structure consists of a hexagonal surface net superimposed on the cubic substrate resulting in a 5 × 20 coincident site lattice. Computation of the structure factor for this model using optical analogue techniques[2] demonstrates that the presence of all the fractional order beams is correctly accounted for. Apparently, these features derive from the modulation of the scattering factor (effective incident field) within the 5 × 20 coincident site cell.

Although the hexagonal surface layer was deduced to be an impurity stabilized phase[3, 4] all previous attempts to identify the surface contaminant(s) have proved to be unsuccessful. This difficulty prompted Mattera, Goodman and Somorjai[5] and Palmberg and Rhodin[6,7] to conclude that clean gold surfaces are reconstructed and, in the latter case, that the hexagonal surface layer is to be identified as pure gold.

In the present instance, results of continuing studies will be described which tend to support the original impurity surface postulate. As noted above, the Au (100) 5 × 20 surface structure forms irreversibly after either high temperature treatment or extended ion bombardment at room temperature. This periodicity transforms reversibly to a 1 × 1 or bulk-like structure at approximately 800°C. Provided a suitable source of impurities exists* these results are consistent with diffusion transport to the surface of relatively non-volatile species at elevated temperatures and a selective sputtering mechanism, respectively[4]. The validity of these mechanisms will be examined presently through the use of Auger electron emission spectroscopy.

As Harris[8] and Weber and Peria[9] have shown, if the second derivative of the inelastically scattered electron current is obtained as a function of energy, characteristic Auger transitions are detected which are a sensitive measure of small amounts of surface contaminants. Palmberg and Rhodin[7] have utilized this technique and found no evidence for impurity species for the case of epitaxially grown Au (100) thin film surfaces. This has led to the conclusion that clean (100) gold surfaces are reconstructed[10] while similar results and conclusions have recently been discussed for both Pt (100) and Au (100) surfaces[11].

In the present case the results of Auger analysis of single crystal Au (100) surfaces will be discussed and compared to the findings of Palmberg and Rhodin[7]. In this study an attempt has been made to make a positive identification of each characteristic transition which is observed, without perturbing the surface structure or composition. The possibility of undetected electron beam desorption, especially at higher electron energies, has been recognized and avoided through the use of a defocussed electron beam which samples a macroscopically large area of the surface.

To establish the origin of the various transitions observed in the Auger spectrum a number of methods have been used in an attempt to selectively affect individual transitions. One such approach has been to study the emission of positive alkali metal ions from the (100) surface at elevated temperature. Mainly, however, this study has concentrated on the effect of chlorine on the (100) surface structure. This study was instituted because with elements such as gold, if impurities are present, selective chemical interactions with halogens is a distinct possibility. Since gold halides are known to be the least stable of all the elemental halides the formation of impurity halides is favored. In the event these reaction products are less strongly bound to the gold surface than the impurity species alone there will be an increased probability for mass transport of reaction products away from the surface, at a given temperature. One such reaction, involving material transport of impurities away from the Au (100) surface through the formation of volatile impurity chlorides will be described presently.

EXPERIMENTAL CONSIDERATIONS

A modified 'Varian' three grid low energy electron diffraction apparatus was used in this study. The modifications necessary for Auger electron emission spectroscopy have been described by others; see, for example, the papers by Bauer and by Palmberg, this conference. In the present instance the same electron gun was utilized for both LEED and Auger studies. It was found that adequate sensitivity is obtained by using a 1500 eV, 5μA electron beam in the Auger work. The second and third grids of the electron optics were operated in tandem, i.e. both the DC retarding potential and the AC perturbing potential were applied simultaneously to both grids, in order to obtain improved energy resolution. The resultant energy-independent background signal which is induced on the collector due to capacitive coupling was eliminated electronically.

Initially, some spurious characteristic Auger transitions were detected. These features shifted in position (energy) depending upon the primary electron beam energy and current. These "ghost" peaks were found to be due to primary electrons incident upon the viewing window. The secondary electrons emitted at the window could not be eliminated by either manipulation of the electron gun focussing voltage or the sample position. If, however, the sample, first grid and drift tube were maintained at a potential of -90eV with respect to ground (i.e. the LEED chamber) the secondary electrons originating at the window were prevented from reaching the energy analyser and contributing to the true secondary electron energy distribution.

For a focussed 1500eV primary beam, desorption of adsorbed species may be appreciable while the affected area will remain unresolved at low (diffraction) energies due to the relatively larger beam diameter. For this reason the use of a defocussed beam with a diameter equal to the minimum sample dimension, 3mm., was explored. Use of a defocussed beam had no effect on the apparent sensitivity and resolution of characteristic Auger transitions. Meanwhile, in the defocussed mode an even distribution of current density was achieved. Even though the current density was still of the order of 10^{-4} A.cm^{-1} the effects of beam desorption on the structure, as determined from the LEED pattern, could be easily monitored.

For the detection of positive ions (alkali metal impurities leaving a hot surface are ionized upon penetration of the image force potential) the LEED screen-grid assembly was used as a collector. The identification of the components of the positive ion flux was accomplished by operating the mass spectrometer (E.A.I. "Quad 200") with the ionizing electron beam filament disconnected so that only ionized species were detected. The mass spectrometer intercepted only those ions which left the sample surface at an angle of 45° with respect to the surface normal and traversed a distance of approximately 10 cm., direct line of sight, to the mass spectrometer.

For adsorption studies, chlorine was obtained directly from a one liter Air Reduction Co. 'Research Grade' gas bottle which was connected to and dispensed by a bakeable, all-metal gas handling system. Repeated use of chlorine in the system resulted in a gradual rise in background pressure of HCl until this species became the predominant gas in the system at low pressures, as determined by the mass spectrometer. No deleterious effects upon system components due to the presence of chlorine have been noted during the course of these experiments.

RESULTS and DISCUSSION

The results of secondary electron energy analysis experiments will be presented below as plots of the second derivative of the total electron current (the first derivative of the true secondary electron energy distribution) as a function of electron energy. For the Au (100) 5 × 20 surface structure, shown in Fig. 1, the energy distribution shown in Fig. 2 is obtained. Characteristic transitions are found to occur at 41, 70, 146, 163, 186, 241 and 256 eV subsequent to ion bombardment cleaning of the surface.

Before ion bombardment, and immediately after electropolishing, a peak characteristic of carbon is observed at 276 eV. Heat treatment of the sample subsequent to ion bombardment results in the reappearance of neither a carbon peak nor of the formerly observed "ring" structures[4]. This may be due to the presence of background HCl gas.

The energy loss spectrum presented in Fig. 2 is in substantial agreement with that found by Palmberg and Rhodin[7] (hereafter referred to as P & R) for thin film crystals except for the fact that the relative magnitudes, measured peak to peak, or some of the characteristic transitions are significantly different. For example, the peaks at 41, 146, 163, 186 and 256 eV appear to be less pronounced, in the present case, when both sets of results are normalized for identical peak heights at 70 eV and 241 eV. Peaks at 59 and 100 eV are not observed in the present work. In spite of these differences the resolution and sensitivity appear to be similar in both instances.

Our results indicate that the transitions at 70 and 241 eV are due to gold since they increase and decrease simultaneously as the surface purity is changed. According to our findings, the transition at 146 eV is due to some contaminant, probably sulfur. The 163 eV peak may be characteristic of pure gold while the 186 eV transition coincides with the position at which chlorine is observed. The peak at 256 eV is due to potassium or gold or both potassium and gold. Evaporation of potassium from a xeolite source resulted in an increase in amplitude of this peak while the location of the minimum gradually shifted to 254 eV as monolayer coverage was attained. If this transition is due solely to potassium it is important to note that at no time during the course of our experiments was its amplitude reduced below that shown in Fig. 2. On the other hand, the distribution obtained by P & R contains a peak at this position which appears to be somewhat larger in amplitude.

Before proceeding to a discussion of the remaining transitions at 41 and 70 eV some results of studies of the emission of positive ions from gold (100) surfaces will be presented as they are related to the possible identification of the 256 eV transition. Fig. 3 shows the variation of the total positive ion current emitted from a (100) surface plotted as a function of $1/kT$. Notice that, at temperatures approaching 800°C., the evaporation flux is of the order of 10^{-3} monolayers per second, approximately 2×10^{-4} ampere seconds is equivalent to monolayers coverage defined in terms of gold surface atoms. This curve is virtually independent of time as identical results are obtained after prolonged high temperature annealing. In the temperature range explored the activation energy for the emission process is approximately 1.9 eV. This value is strongly suggestive of bulk diffusion as a rate limiting step and compares favorably with the value found by Ruedl and Bradley[12], 2.3 ± 0.1 eV, for the case of potassium ion emission from copper.

Fig. 4 illustrates the results of a mass spectrometer analysis of the positive ion current emitted at 750°C. It is apparent from this analysis that the major constituent of the positive ion flux is potassium, only trace amounts of sodium and rubidium or Na_2K are detected. No cesium was observed while lithium was not anticipated in this particular experiment. From these results it is apparent that significant amounts of potassium are likely to be present on (100) gold surfaces, especially at high temperatures. The latter possibility would have been tested using Auger emission spectroscopy had it not been impossible in the presence of the high currents, 60 to 70 A,

necessary to heat the single crystal samples. Finally, a number of extended

high temperature anneals were performed while the crystal was maintained at

+ or - 300 eV with respect to the surrounding system. In neither case was

there any effect on the 5 × 20 diffraction pattern or the Auger spectrum

obtained after cooling to room temperature. Brief high temperature anneals

in the presence of chlorine at a pressure of 1×10^{-7} Torr yielded negative

results also.

Now let us concentrate on a discussion of the transitions at 41 and 70 eV

shown in Fig. 2. In the present case the amplitude of the 41 eV peak is of

the order of 10% of the 70 eV gold peak. The P & R results, on the other hand,

exhibit a peak at this position which is approximately 30% of the 70 eV peak,

for the same resolution and sensitivity. In either case this characteristic

peak is the second largest. In the following section, the manner in which

the introduction of chlorine gas affects the characteristic transitions,

especially the 41 eV transition, will be outlined in detail.

THE ADSORPTION of CHLORINE

Attention will now be focussed on the interaction between chlorine and

the (100) gold surface. If chlorine is admitted into the LEED system and the

pattern observed it is found that at an exposure of 2 to 10×10^{-6} Torr. sec

(pressure uncorrected for gage constant, hot filament interactions) and at

room temperature, the 5 × 20 periodicity disappears irreversibly and one is

left with a diffraction pattern which possesses the spacing and symmetry

expected for a bulk-like surface. That is, the pattern shown in Fig. 1

transforms to the one shown in Fig. 5(a), 5(b) and (c) being typical of patterns observed at other electron wavelengths. This interaction between the Au (100) 5×20 surface phase and chlorine is not influenced by the incident electron beam at 20 eV since the disappearance of the 1/5,0 type reflections can be followed to completion and then the sample moved laterally to verify that the entire sample surface has undergone the same structural change. The 1/5,0 and 1,0 type reflections are observed to exhibit reciprocal relative changes in intensity. Increased exposure to chlorine beyond the point necessary to attain complete conversion to a 1×1 structure and maximum intensity of the 1,0 reflection results in a slow decrease in beam intensity and the development of very weak streaks in the positions expected for a c 2×2 diffraction pattern. When the chlorine pressure in the system is reduced these weak reflections quickly disappear due to beam desorption of chlorine from the surface, at a beam energy of 50 eV.

One can study the kinetics of these reactions by monitoring the intensity of the 1,0 beam as a function of time. Some of this data is shown in Fig. 6 wherein the intensity of the 1,0 beam is plotted as a function of exposure, for chlorine pressures in the range 1×10^{-9} to 1×10^{-7} Torr., the sample having been heated to 800°C. prior to each experiment to ensure the same initial conditions. As one can see from the curve, the conversion from 5×20 to 1×1 periodicity manifests itself in a variation of the intensity of a 1,0 type reflection which is proportional to the square of the exposure. This is taken to signify that the overall reaction is (1) independent of nucleation of saturation effects, (2) characterized by a constant (undetermined) sticking coefficient for chlorine and (3) such that every quantum of chlorine that interacts results in a corresponding identical change in intensity, independent of reaction extent. The last fact is, of course, consistent with a model of edgewise growth of 1×1 areas (or dissolution of 5 × 20 areas). The rate limiting step appears to be the rate of arrival of chlorine, diffusion to a receding edge being relatively fast. The square law dependence follows directly from the dependence of the diffraction intensity on the square of the scattering factor. It is, of course, entirely possible that the transformation to a 1 × 1 configuration is simply due to the adsorption of chlorine into a configuration characteristic of the symmetry and spacing of the underlying gold lattice.

This latter possibility is easily checked through examination of the Auger spectrum for the transformed surface, an example of which is shown in Fig. 7. This was obtained after an exposure to chlorine equivalent to 15 10^{-6} Torr. sec, which ensures complete conversion to a 1 × 1 configuration. An Auger transition due to adsorbed chlorine is evident at 187 eV. This particular peak was determined to correspond to a surface coverage $\leq$ 0.15 by following the adsorption reaction 2, see Fig. 6, to completion and by comparison with similar experiments with silicon[13]. All other characteristic peaks are observed to be unchanged except for the 70 eV gold peak which is slightly attenuated and the transition at 41 eV which has been caused to almost completely disappear.

If this surface is now subjected to electron bombardment (1500 eV, 10^{-4} A.cm^{-1}) the chlorine peak observed in Fig. 7 is reduced to a negligible value while the 1 × 1 structure is maintained and the 41 eV transition fails to reappear. On the other hand, if the crystal is heated the chlorine peak disappears and the 41 eV peak reappears. These two processes occur simultaneously as illustrated in Fig. 8. Heating the crystal to between 200 and 300°C. for 5 min. results in reversion to the diffraction pattern shown in Figure 1 and the characteristic Auger spectrum of Fig. 2. On the other hand, if reaction 2 is allowed to occur it is observed that adsorbed chlorine is desorbed, upon heating the crystal, as molecular Cl_2. No atomic chlorine or HCl is detected.

Clearly, adsorption of chlorine at room temperature does not proceed by growth of an adsorbed layer which possesses the symmetry and spacing of the underlying surface. Indeed, the evidence indicates that no chlorine need be retained at the completeion of the first stage, reaction 1, while the characteristic Auger transition at 41 eV is eliminated. This leads to the conclusion that adsorption of chlorine is followed by the reaction of the adsorbed gas with another adsorbed component (impurity), this species being the same one that is responsible for the stabilization of the hexagonal surface layer and the 5×20 surface periodicity. Apparently, a relatively volatile impurity chloride forms and is transported away from the vicinity of the surface, at room temperature.

In summary, under the conditions of the experiment, the point at which maximum intensity of an integral order diffraction beam is achieved, see Fig. 6, corresponds to a surface which is relatively free of chlorine and some as yet unidentified contaminant. This impurity species is responsible for the stabilization of the hexagonal surface layer. Both the impurity and the hexagonal surface layer can be made to reappear, in the absence of chlorine, by heating the gold crystal to approximately 250°C. Alternatively, both may be regenerated by ion bombardment of the 1×1 surface for extended periods of time. This observation supports the selective sputtering mechanism previously introduced[3,4].

Although positive identification of the impurity component(s) responsible for the hexagonal surface layer has not been achieved it is possible to speculate regarding its identity. Iron is a distinct possibility since it possesses an Auger transition near 41 eV. Upwards of 50 ppm of Fe are known to be present in gold of the purity used in preparation of the single crystals used in these experiments. In view of the results of the positive ion emission studies, potassium cannot be ruled out although a characteristic transition at 41 eV has not been verified. Another possibility is lithium, which has a characteristic electron energy loss close to 41 eV. More thorough positive ion emission experiments may help determine if any Li is present. Atomic packing considerations would indicate that, of all the above mentioned possibilities, Fe is a most likely candidate since it possesses an atomic diameter of either 2.48 or 2.57 Å depending upon its crystallographic form while the apparent atomic diameter within the hexagonal layer is 2.76 Å.

CONCLUSIONS

1) Experimental evidence indicates that clean gold surfaces are not rearranged, the "reconstructed" surface structure is impurity derived.

2) Mass transport of impurities away from gold surfaces through formation of weakly bound impurity chlorides occurs at room temperature.

ACKNOWLEDGMENTS

The authors would like to thank Professor Traugott E. Fischer for his invaluable contributions during the course of this investigation. One of us

(D.G.F.) is extremely indebted to Professor Gabor A. Somorjai, University of California at Berkeley, for making available the facilities of the Institute of Materials Research Division during the course of the preparation of this manuscript.

This research was sponsored by the U. S. Office of Naval Research.

Financial support for D.G.F. was provided through an International Nickel Company Research Assistantship.

Footnotes

*An analysis of Cominco '69' grade (99.9999+%) gold, courtesy of Palmberg and Rhodin has revealed the presence of the following impurities, in ppm. - Fe & Si = 50; Ca, Mg, Na, & K = 10; Ni & S = 5; Sn = 3; Al, O, N, & C = 2; Cu & P = 1; and Zn, Mn, Cr, V, Ti & Ag = 0.2.

REFERENCES

1) D. G. Fedak & N. A. Gjostein, Surface Sci., $\underline{8}$, 77 (1967).

2) D. G. Fedak, T. E. Fischer & W. D. Robertson, J. Appl. Phys. $\underline{39}$, 5658 (1968)

3) D. G. Fedak & N. A. Gjostein, Phys. Rev. Letters, $\underline{16}$, 171 (1966).

4) D. G. Fedak & N. A. Gjostein, Acta Met. $\underline{15}$, 827 (1967).

5) A. G. Mattera, R. M. Goodman and G. A. Somorjai, Surface Sci. $\underline{7}$, 26 (1967)

6) P. W. Palmberg & T. N. Rhodin, Phys, Rev, $\underline{161}$, 586 (1967).

7) P. W. Palmberg & T. N. Rhodin, J. Appl. Phys. $\underline{39}$, 2425 (1968).

8) L. A. Harris, J. Appl. Phys. $\underline{39}$ 1419 & 1428 (1968).

9)   R. E. Weber & W. T. Peria, J. Appl. Phys. 38, 4355 (1967).

10)   T. N. Rhodin, P. W. Palmberg & E. W. Plummer, this conference.

11)   P. W. Palmberg, this conference.

12)   E. Ruedl & R. C. Bradley, J. Phys. Chem. Solids, 23, 885 (1962).

13)   J. V. Florio, unpublished research.

Fig. 1.   Diffraction pattern typical of the Au (100) 5 X 20 structure, one domain, 49 eV.

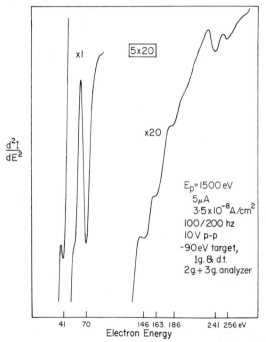

Fig. 2.   The second derivative of the total secondary electron distribution as a function of energy for the case of the Au (100) 5 X 20 structure shown in Fig. 1.

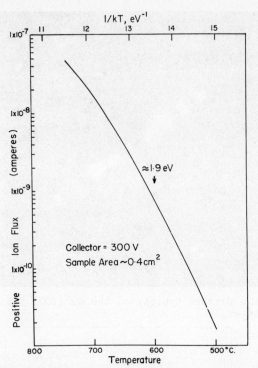

Fig. 3. The variation of the total positive ion current emitted by a (100) gold surface as a function of 1/kT.

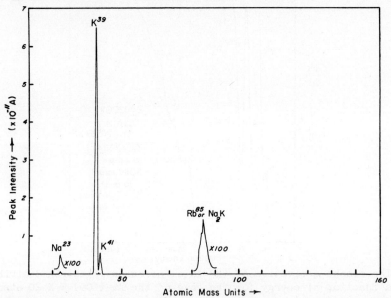

Fig. 4. The results of a mass spectrometer analysis of the positive ion current emitted from a (100) gold surface at 750°C.

Fig. 5. (a) LEED pattern characteristic of the transformation induced by an exposure to chlorine of 15 X 10⁻⁶ Torr.sec. This pattern should be compared to that shown in Fig. 1, 49 eV. (b) and (c) Patterns for the same conditions as in (a) except at 60 eV (b) and at 120 eV (c).

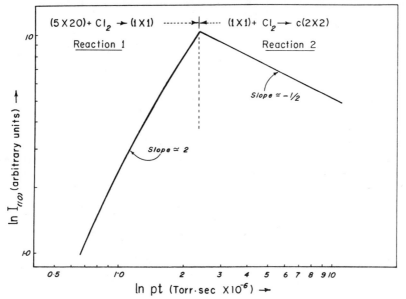

Fig. 6. The variation of the intensity of a 1,0 type beam as a function of exposure for various pressures, in the range 1×10^{-9} to 1×10^{-7} Torr, of chlorine.

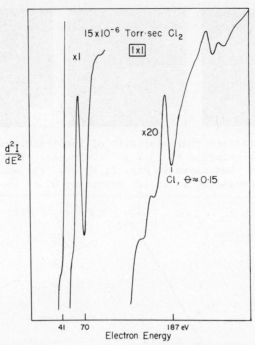

Fig. 7. The second derivative of the total secondary electron distribution as a function of energy for the case of the Au (100) 1 X 1 structure shown in Fig. 5.

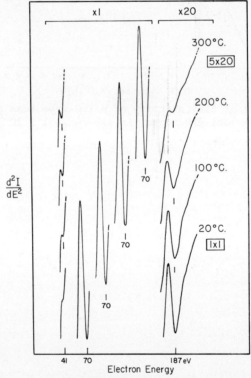

Fig. 8. The effect of increased temperature on the characteristic Auger spectrum shown in Fig. 7. Notice that at 300°C the distribution shown in Fig. 2 has been regenerated.

THE SURFACE STRESS OF GOLD AND ITS TEMPERATURE DEPENDENCE

J. S. Vermaak

Department of Physics, University of Port Elizabeth
Port Elizabeth, Republic of South Africa

D. Kuhlmann-Wilsdorf

Department of Materials Science and Department of Physics
University of Virginia, Charlottesville, Virginia

I. Introduction

Reliable data of either surface tension, or surface stress, of metals at
low or intermediate temperatures are virtually absent in the scientific literature. The only relevant data in this category concern surface tension at
temperatures close to the melting point. These were obtained either by the
technique of zero creep rate, introduced by Chapman and Porter[1] and further
developed by Udin, Shaler and Wulff[2], among others, or from observations of
the smoothing of surface scratches, as apparently employed first by Blakeley
and Mykura[3]. Moreover, considerable confusion regarding the difference
between surface tension and surface stress is evident from a number of
publications.

This state of affairs is most unsatisfactory, particularly in view of the
important role that surface tension and/or stress play in a variety of
phenomena of great interest such as sintering, void formation at grain-
boundaries during creep, and aggregation of supersaturated vacancies into
voids, may they be due to quenching, or radiation damage, or any other cause.
In the following an attempt at clarifying the issues shall be made, and
measurements of surface stress of solid gold over a wide temperature range are
reported.

II. Theoretical Considerations

1. General

As Shuttleworth[4] pointed out, the concept of surface stress, i.e. the
force exerted per unit length of exposed surface, was first introduced by
Cabeo[5] and was later stated more explicitly by Segner[6]. The concept of
surface energy, meaning the free energy per unit area of surface, is more
recent and was introduced by Gauss[7]. Subsequently, a better understanding of
these concepts has been sought in many publications. Preceding work devoted
to this problem can be classified into three major categories. Firstly, the
classical macroscopic approach. The most important contributions in this class
are mainly due to Gibbs[8], Tolman[9,10], Shuttleworth[4], Herring[11,12,13],
Frank[14], Taylor[15] and Mullins[16]. Secondly, atomistic theories using the

microscopic approach. The theoretical research of this kind prior to 1951 has been reviewed by Herring[17]. Later contributions were made by, among others, Ewald and Juretschke[18], Skapski[19], Girifalco and Weizer[20], Gazis and Wallis[21], Benson and Yun[22,23,24], Richman[25], Oriani and Sundquist[26], and Walton[27]. Thirdly, experimental observations. Reviews covering the experimental field have been written by Udin[28], Shaler[29], Shewmon and Robertson[30], Kuznetsov[31], and Bikerman[32].

Ever since Gibbs[8] pointed out that surface free energy, σ, and surface stress, f of a solid are not identical, meaning that, in general, a different amount of reversible work is required to form unit surface, than to increase a large surface by unit area through reversibly stretching it, there have been many discussions and some controversies regarding these concepts[12,29,32]. Actually, the reason why surface tension and surface stress are in general different is obvious when considering a large surface whose area is being increased from A to A + dA by reversibly stretching it. Assuming isotropy, we find that the work done, f dA, is partly converted into the surface energy of the extra area of surface generated, of magnitude σ dA, and partly into the change of surface tension accompanying the extension of the surface area, given by A dσ. Thus, in the isotropic case[4],

$$f = \sigma + A \frac{d\sigma}{dA} \qquad (1)$$

while, in general[12,16],

$$f_{ij} = \delta_{ij}\sigma + \frac{d\sigma}{de_{ij}} , \qquad (2)$$

where δ_{ij} is the Kronecker delta, i is the direction in the plane of the surface normal to the surface cut to which the force is applied, j is the direction in which the force acts, and e_{ij} is the strain.

It is intuitively obvious that $d\sigma/de_{ij}$, or $d\sigma/dA$, must vanish whenever the atomic density and the configuration of the average atom in the surface are independent of strain. This is the case for slowly deforming, planar surfaces of liquids, as has been pointed out repeatedly in the past. However, no adequate attention seems to have ever been paid to the influence that experimental conditions, employed in measuring σ, have on $d\sigma/de_{ij}$. In particular, it appears to have escaped recognition that measured values of f, unlike σ, must necessarily depend on experimental conditions. Thus, the surface stress is not a materials constant, at least not when it is defined simply as the force per unit length of exposed surface required to stretch a surface reversibly.

Although equations 1 and 2 apply to both, planar as well as curved surfaces, it is convenient in the further discussion to consider these separately.

2. Planar Surfaces

The stretching of a planar surface may occur by the more or less statistical movement of atoms from the bulk into the surface, and the simultaneous re-arrangement of the atoms which already were in the surface, or it may occur by shear. The former is the mode of deformation of liquid surfaces and the surfaces of amorphous substances. In the latter type of deformation, namely shear, which occurs in crystalline substances at all but very slow deformations at high temperatures, the surface area is increased by the formation of surface steps. These are never parallel to the surface area, so that the surface becomes stepped or rumpled and the surface area becomes correspondingly ill-defined.

Another fundamental difference between the slow deformation of planar liquid or amorphous surfaces and the shear deformation of planar solid surfaces is the fact that a definite yield stress must be attained before shear deforma-tion takes place, while the permanent shear strength of liquid and amorphous surfaces is zero.

As was mentioned before, $d\sigma/dA$ and $d\sigma/de_{ij}$, respectively, vanish and, thus, measured surface stress and surface tension are identical, unless either the average density of surface atoms, or their average chemical potential, or both, change as a consequence of straining. Under conditions in which the strain, e_{ij}, is completely plastic, the said condition is fulfilled. This is common for liquids and for amorphous substances when strained not too fast, and is closely approached in solids at high temperatures and/or very slow strain rates. Correspondingly, measurements depending on the slow deformation of planar surfaces under completely plastic conditions must yield data on surface tension, i.e. surface free energy, rather than surface stress.

The described conditions of almost completely plastic strains obviously apply in the zero creep method[1,2] which thus measures surface tension. This, however, does not mean that $d\sigma/de_{ij}$ vanishes for _elastic_ strains. Indeed, as will be shown in the experimental part of this paper, the surface tension of gold, and by inference of other metals, does depend markedly on elastic strain, actually in the sense that the surface atoms are in a state of compression so that $d\sigma/de_{ij}$ is negative, as had been anticipated by Orowan[33]. Due to this compressive stress, the lattice constant in the atomic layers beneath the surface of the wires is doubtlessly mildly reduced, providing the reaction for the term $d\sigma/de_{ij}$, while the externally applied load balances only σ. Sensitive X-ray or electron diffraction experiments would reveal this effect, it is confidently expected.

By contrast to slow strain rates, very fast strain rates suppress plastic deformation and, in the limiting case, cause the strains to be completely elastic. In this limiting case the true value of the surface stress, say f_o, would be determined, which is a materials constant just as much as the surface tension. In order to achieve the ideal condition of completely elastic strains, very small strain amplitudes would have to be utilized, in order to prevent plastic strain. Presumably, high-frequency surface waves could be employed for the purpose, and it is anticipated that by means of the "ripple method", i.e. the measurement of surface "tension" via the speed of surface waves, the dependence of macroscopic measured surface stress on the amount of plastic strain could in fact be determined. Thus, ripples of relatively low frequencies would yield data on surface tension, extremely high frequency ripples, data on true surface stress, and intermediate frequencies would yield intermediate values for the macroscopic surface stress.

3. Spherical Surfaces

As mentioned above, very sensitive determinations of lattice constants or mechanical density beneath surfaces should, in principle at least, reveal elastic compressive strains which would allow evaluating the true surface stress. However, at planar or only mildly curved surfaces the said effect is absent or extremely small. Much more favorable are conditions underneath strongly curved surfaces and, in particular, within very small spherical particles. In these, no voluntary strain is applied to the surfaces, and all strains are necessarily elastic. Under this condition, the surface stress must be completely balanced by a hydrostatic pressure within the particle.

Assuming isotropy, and assuming that the pressure inside of the particles is uniform (i.e. is not a function of the distance from the center of the particle) we find

$$2\pi r f = \pi r^2 p = \pi r^2 \frac{3}{K} \frac{\Delta a}{a} , \qquad (3)$$

where K is the compressibility, $\underline{a}$ is the normal lattice constant (or average distance between atoms) of the substance considered, and $\Delta\underline{a}$ is the difference between the actual value of $\underline{a}$ within the particle and its normal value in bulk material. From eq. 3 we obtain

$$f = \frac{3}{2} \frac{\Delta a}{a} \frac{r}{K} . \qquad (4)$$

This equation, then, affords a means by which the surface stress may be evaluated from measured values of $\Delta \underline{a}$ and r, utilizing known values of the compressibility, K.

Eq. 4 is equally applicable to solids and liquids, concave and convex surfaces. Correspondingly, previous determinations of surface "tension" from the equilibrium pressure in bubbles did, in fact, measure surface <u>stress</u>. However, at least two possible complications need still be considered: It is a foregone conclusion that σ must depend on r as well as on e . Correspondingly, if $\Delta \underline{a}/\underline{a}$ could be measured with any desired precision, no perfectly straight line would result if $\Delta \underline{a}/\underline{a}$ were plotted as function of 1/r since f, which is found from the slope of $\Delta \underline{a}/\underline{a}(1/r)$, must slowly change with r.

The reason why f must depend on r directly is that, for the average surface atom, the number of missing neighbor atoms increases with increasing convex surface curvature and decreases with increasing concave curvature. The dependence of f on e, and thus indirectly on r, follows from the very fact that f ≠ σ, i.e. that surface atoms are in a state of compression or dilatation, depending on whether dσ/de is negative or positive. Still, neither of these two effects is believed to be significant. Firstly, even in the smallest particles which may be investigated with present techniques (r $\overset{>}{\sim}$ 18 Å in the experiments to be discussed below), the curvature is still quite small compared to the curvature of a single atom. Secondly, the elastic equilibrium strain encountered in the measurements is always small (perhaps up to 0.3%) so that no drastic change in dσ/de is anticipated. This latter idea is in agreement with a recent theoretical calculation by Drechsler and Nicholas[34] according to whose results a 0.2% decrease of the lattice constant, due to hydrostatic pressure, say, would be expected to increase the surface tension of silver by roughly 1%. The conclusion that no observable deviations from linearity between $\Delta \underline{a}/\underline{a}$ and 1/r should occur as a consequence of the dependence of surface stress on hydrostatic pressure and on particle curvature, thus drawn, is indeed borne out by the experimental results discussed below.

The preceding theoretical considerations are summarized semi-quantitatively in Fig. 1.

III. Experimental Determination of Surface Stress

1. Experimental Procedure

The experimental methods employed in the measurements to be discussed, based on eq. 4, will be described in greater detail elsewhere.[*] Only a brief outline shall be given here: Very small, roughly spherical gold particles

[*] See J. S. Vermaak and D. Kuhlmann-Wilsdorf, J. Phys. Chem. 72, 4150 (1968).

were obtained by evaporating 99.999% gold onto amorphous carbon carrier films in a vacuum of 1×10^{-9} to 1×10^{-10} torr, and at a substrate temperature of $350°C$. Depending on length of evaporating time, the particle radii varied between $r \cong 18$ Å and $r \cong 500$ Å , and up to continuous films. Directly after evaporation, the specimen was slowly heated to approximately $750°C$ and kept at this temperature for about one hour. The purpose of this heat treatment was to allow the nuclei to approach equilibrium sizes and forms, so as to prevent uncontrolled changes during the measurements at high temperatures. Some faceting, mainly according to {111} planes, it seemed, occurred and may be seen in Fig. 2 which shows typical examples of the specimens investigated.

Thallous chloride, to be used as an internal standard for the lattice constant measurements, was either evaporated onto separate carrier films, or onto the gold specimens. The specimens, side by side with the thallous chloride films unless the TlCl had been evaporated directly onto the gold, were transferred to the heating stage of a Siemens Elmiskop IA electron microscope. At different temperatures, ranging between $50°C \pm 20°C$ and $985°C \pm 20°C$, the radii of the nuclei on a number of specimens for each temperature was determined from micrographs taken at a high magnification. The lattice constants were found by means of electron diffraction, using the lattice constant of the thallous chloride as a standard[35] while the stage was not heated. In order to obtain the highest possible precision, the diameters of the diffraction rings of the electron diffraction patterns were measured by means of a projector microscope equipped with micrometer screws. Instead of visual determinations, a photoresistor, with a very narrow slit mounted on the projector screen and connected to a recorder, were employed. The precision of the data thus obtained was easily sufficient to measure the peak shifts due to stacking faults[36]. The stacking fault density to which these peak shifts corresponded was close to 1 faulted plane per 1000 planes. Hence, only a small fraction of the nuclei contained stacking faults, as was also confirmed by visual examination.

2. Results

Table I summarizes the results obtained. In this table, T^* is the temperature as read for the heating stage, while T is the corrected temperature which is obtained by comparing $a_o(T)$, the value of the lattice constant extrapolated to $1/r = 0$ (i.e. to the bulk value) with tabulated values[37]. In this, $a_o(25°C)$ had to be taken as 4.0725 Å. Namely, the difference between T^* and T is due partly to lack of accurate calibration of the heating stage, partly to localized heating of the specimen in the electron beam. The fact that the said localized heating is slightly different for gold and thallous chloride is responsible for the difference between the tabulated value of the lattice constant for gold, of 4.0784 $\overset{\circ}{A}$ at 25°C, and the value assumed for $\underline{a}_o(25°C)$ quoted above. In other words, for simplicities' sake the value of the lattice constant of TlCl was taken to be correct as tabulated, while the unknown differential between the lattice constant of TlCl tabulated for 25°C, and its actual value as influenced by local heating, together with the unknown heating effect of the electron beam on the gold without deliberate heating of the stage, i.e. at a nominal 25°C, was lumped together by employing a slightly modified standard value of $\underline{a}_o(25°C)$. The uncertainty in the temperature determination from all causes is estimated at $\pm$ 20°C throughout the whole temperature range.

The values of the compressibility for gold as a function of temperature do not appear to have ever been measured directly. The data for K in table I are calculated on the basis of the experimental values of Young's modulus given by Köster[38], as K = 3(1-2ν)/E, assuming Poisson's ratio ν = 0.42, independent of temperature. It is in fact not known whether, or how well, Poisson's ratio stays constant with temperature, but it is unlikely that it changes much. In any event, the calculated value of K is not sensitively dependent on ν. No estimate as to the probable error from this source can reasonably be made.

The slope, S, of the plot of $\Delta\underline{a}/\underline{a}$ versus 1/r, given as S = d$\bar{a}$/d(1/r) in the table, was found from experimental data points. Fig. 3 gives as an example a plot of the measured data points pertaining to T = 780°C. The probable errors of the S values listed in Table I were calculated with the help of a B 5500 computer, employing the least squares method. ⇠ In this connection it may be remarked that the evaluation of the numerical data is quite tedious so that the electronic computer was used throughout.

In accordance with eq. 4, and the definition of S as S = d$\bar{a}$/d(1/r), the surface stress is found as

$$f = -3S/2Ka_o \qquad (5)$$

The error limits given for f in table I reflect not only the probable errors in S, but in addition the uncertainty in the magnification of the micrographs from which the radii are determined. This is estimated at about 5% or less so that the total expected error in f is believed to range between ~ 6% at 50°C to about 9% at 985°C. This favorable assessment of the probable error in f is supported by the gratifyingly small apparent scatter of the values of the surface stress as a function of temperature (see Fig. 4), and also by the fact that an earlier, independent determination of f without a heating stage[39] (i.e. at a specimen temperature near T = 50°C), using the same method, yielded f = 1170 dyn/cm compared to f = 1145 ± 70 dyn/cm in the present investigation.

The values of the surface tension, σ, listed in the table, are not too reliable. They are obtained by using the value of σ obtained by Buttner, Udin and Wulff[40], of σ = 1400 ergs/cm^2 at 1030°C, in conjunction with Eötvös' rule[41], according to which

$$\frac{d\sigma}{dT} = - (\frac{M}{\rho})^{-2/3} K_\sigma = - 0.46 \text{ dyn/cm°C} \qquad (6)$$

for gold, where M is the molecular weight, ρ the mechanical density, and K_σ an empirical constant of value K_σ = 2.1 to 2.2 erg/°C. The values of σ thus derived have an unknown probable error, as it is not known to what extent Eötvös' rule may be applied to solids.

The data in the last column of table I, listing dσ/de, are found as the differences between the tabulated values of f and σ.

IV. Discussion

The results presented in Table I and Figs. 3 and 4 are decisive in more than one respect. Firstly, Fig. 3 and similar plots obtained at the other temperatures listed show that the pressure effect due to surface stress exists without question, and is measurable when employing the best techniques available. - The only previous attempt at measuring the surface "energy" of metal by the method used in this paper that is known to the writers[42] did not meet with obvious success, presumably, however, due to no fault of the authors: Even a few years ago, experimental techniques had not progressed to the point that the present experiment could have been carried out successfully.

Secondly, within the limits of accuracy, the slopes of Δa/a versus 1/r are constant over the entire range of radii studied- i.e. from r ≅ 18 Å to r ≅ 500 Å. Consequently, the theoretical expectation that the surface stress is very nearly independent of hydrostatic pressure and surface curvature, at least within the range accessible in the present experiment, discussed in section II 3, is confirmed by experimental evidence.

Thirdly, it is definitely the surface stress, not the surface tension, i.e. not surface free energy, which is being measured, in agreement with the theoretical considerations in Section II. Namely, even at room temperature the value obtained for f is clearly below the high-temperature value of the surface tension determined by Buttner, Udin and Wulff[40], and the gap between that value of the surface tension and the data obtained in the present experiment widens well beyond any conceivable error as the temperature

increases. This seems to be a most important result since it strongly rein-
forces the theoretical considerations of Section II, and confirms the conclusion
that many earlier determinations of surface "tension", in particular those of
liquids from equilibrium pressures in bubbles, in fact concerned surface stress.

Lastly, we have obtained, for the first time ever, it seems, numerical data
for the value of $d\sigma/de$. In agreement with Orowan's theoretical deduction[33],
this parameter is indeed negative, indicating that the surface atoms are in a
state of compression. That, with the lesser constraints of the surface atoms
compared to the atoms in the bulk, this state of compression (i.e. the
magnitude of $d\sigma/de$) should increase with temperature may perhaps be understandable. However, one would have hesitated to make the corresponding prediction
before the results were obtained.

In the numerical evaluation of the data, isotropy was assumed throughout.
In fact, the surface tension of gold is not entirely isotropical, causing
faceting of the nuclei that is visible in Fig. 2. It is believed that {111}
faces were developed primarily. While the error due to this cause is believed
to be small, the values of the surface stress which were obtained must represent
some weighted average of the surface stresses over many orientations, and are
probably close to the values appropriate to {111} faces.

Acknowledgment

The financial support of this research by the United States Atomic Energy
Commission under Contract No. AT-(40-1)3108 is gratefully acknowledged.

References

1) J. C. Chapman and H. L. Porter, Proc. Roy. Soc. (London) A83, 65 (1910)

2) H. Udin, A. J. Shaler and W. Wulff, Metals Trans. 185, 186 (1949)

3) J. M. Blakeley and H. Mykura, Acta Met. 10, 565 (1962)

4) R. Shuttleworth, Proc. Phys. Soc. (London) A63, 444 (1950)

5) N. Cabeo, Philosophia Magnetica, Ferrara, lib. ii (1629) cap. 20

6) J. A. Segner, Comment, Soc. Reg. Gött. 1, 301 (1751)

7) C. F. Gauss, Werke 5, 31 (1830)

8) J. W. Gibbs, Collected Works 1 (Longmans Green and Company, New York, 1928)

9) R. C. Tolman, J. Chem. Phys. 16, 758 (1948)

10) R. C. Tolman, J. Chem. Phys. 17, 118 (1949)

11) C. Herring, The Physics of Powder Metallurgy, ed. W. E. Kingston
     (McGraw-Hill Book Co., New York, 1951) p. 143

12) C. Herring, The Structure and Properties of Solid Surfaces, ed. R. Gomer
     and C. S. Smith (University of Chicago Press, Chicago, 1952) p. 5.

13) C. Herring, Phys. Rev. $\underline{82}$, 87 (1951).

14) F. C. Frank, Metal Surfaces: Structure, Energetics and Kinetics (American
     Society for Metals, Ohio, 1962) p. 17.

15) J. W. Taylor, Acta Met. $\underline{4}$, 460 (1956).

16) W. W. Mullins, Metal Surfaces:Structure, Energetics and Kinetics (American
     Society for Metals, Ohio, 1962) p. 17.

17) C. Herring, Metal Interfaces (American Society for Metals, Ohio, 1962) p.1.

18) P. O. Ewald and H. Jretschke, The Structure and Properties of Solid Surfaces,
     ed. R. Gomer and C. S. Smith (University of Chicago Press, Chicago, 1952)
     p. 82.

19) A. S. Skapski, Acta Met. $\underline{4}$, 576 (1956).

20) L. A. Girifalco and W. G. Weizer, Phys. Rev. $\underline{114}$, 687 (1959).

21) D. G. Gazis and R. F. Wallis, Surf. Sci. $\underline{3}$, 19 (1964).

22) G. C. Benson and K. S. Yun, J. Chem. Phys. $\underline{42}$, 3085 (1965).

23) K. S. Yun and G. C. Benson, J. Chem. Phys. $\underline{43}$, 3980 (1965).

24) K. S. Yun and G. C. Benson, J. Chem. Phys. $\underline{44}$, 2548 (1966).

25) M. H. Richman, U. S. Atomic Energy Comm. Contract No. AT(30-1)2394
     Technical Report No. 25 (1967).

26) R. A. Oriani and B. E. Sundquist, J. Chem. Phys. $\underline{38}$, 2082 (1963).

27) A. G. Walton, J. Chem. Phys. $\underline{39}$, 3162 (1963).

28) H. Udin, Metal Interfaces (American Society for Metals, Ohio, 1952) p. 11.

29) A. J. Shaler, The Structure and Properties of Solid Surfaces, ed. R. Gomer
     and C. S. Smith (University of Chicago Press, Chicago, 1952) p. 120.

30) P. G. Shewmon and W. M. Robertson, Metal Surfaces: Structure, Energetics
     and Kinetics (American Society for Metals, Ohio, 1962) p. 67.

31) V. C. Kuznetsov, Surface Energy of Solids (H. M. Stationary Office, London
     1957).

32) J. J. Bikerman, Phys. Stat. Sol. $\underline{10}$, 3 (1965).

33) E. Orowan, Zeitschrift fur Physik, $\underline{79}$, 573 (1932).

34) M. Drechsler and J. F. Nicholas, J. Chem. Phys. Solids, $\underline{28}$, 2609 (1967).

35) Fink Inorganic Index of Powder Diffraction File, ASTM (1965).

36) M. S. Paterson, J. Appl. Phys. $\underline{23}$, 805 (1952).

37) Goldsmith, Waterman and Hirschhorn, Handbook of Thermophysical Properties
     of Solid Materials, MacMillan, 1, 232 (1961).

38) W. Köster, Z. Metallkunde, 39, 1 (1948).

39) C. W. Mays, M.S. Thesis in Physics, University of Virginia, 1967.

40) F. H. Buttner, H. Udin, and J. Wulff, Trans. AIME, 191, 1209 (1952).

41) G. v. Eötvös, Wied. Ann. 27 (42.9), 452 (1886).

42) T. DePlanta, R. Ghez and F. Piuz, Helvetica Physica Acta, 37, 74 (1964).

Table I   Summary of experimental data, as explained in Section III, 2).

T^* [°C]	$a_o(T)$ [Å]	$\dfrac{a_o(T) - a_o(25°)}{a_o(25°)}$	T [°C]	K [cm²/dyn]	$S = \dfrac{d\bar{a}}{d(1/r)}$ [Å²]	f [dyn/cm]	$\mathfrak{G}$ [erg/cm²]	$\dfrac{d\mathfrak{G}}{de}$ [dyn/cm]
25	4.0739	3.44×10^{-4}	50	6.23×10^{-13}	-0.194 ± 0.007	1145 ± 70	1851	-706
325	4.0939	5.25×10^{-3}	370	6.88×10^{-13}	-0.162 ± 0.008	864 ± 50	1703	-839
555	4.1099	9.18×10^{-3}	605	7.86×10^{-13}	-0.151 ± 0.009	700 ± 55	1596	-896
725	4.1231	1.24×10^{-2}	780	8.90×10^{-13}	-0.149 ± 0.007	609 ± 45	1515	-906
940	4.1380	1.61×10^{-2}	985	10.55×10^{-13}	-0.119 ± 0.008	410 ± 40	1421	-1011

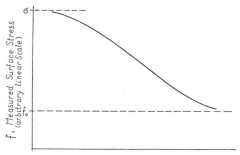

Fig. 1.   Semi-quantitative relationship between measured surface stress, f, and the strain rate, e, employed in the experiment by which f is being determined, for the case of planar surfaces, -liquid, amorphous or crystalline. In this figure, $d\sigma/de$ is assumed to be negative, corresponding to a state of compression in the surface. The value for f measured at zero strain rate is equal to σ, the surface tension, for the reason that all strains are plastic at vanishing strain rates, so that $\dfrac{d\sigma}{de}$ becomes zero. The asymptotic value at infinite strain rate, termed f_o, is a materials constant. It is the value which $f = \sigma + \dfrac{d\sigma}{de}$ assumes for completely elastic strains. In fact, the surface stress f_o acts at all times, but depending on experimental conditions a greater or lesser fraction of the term $\dfrac{d\sigma}{de}$ is supported by compressive (or dilatational) stresses in the atomic layers beneath the surface. The equilibrium pressure in small droplets or bubbles permits measuring f_o.

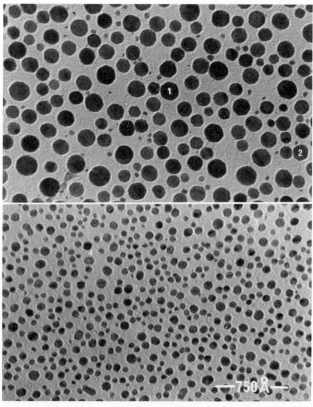

Fig. 2.   Transmission electron micrographs (Magnification 330,000 x) of typical gold nuclei grown in ultra-high vacuum on amorphous carbon substrates at 350°C, and thereafter slowly heated to approximately 750°C. Faceting is clearly visible in the nuclei at locations 1, 2, 3 and 4.

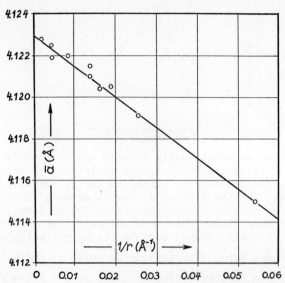

Fig. 3. Measured average lattice constants as a function of reciprocal radius of the nuclei, at a temperature of T = 780°C ± 20°C. The slope of the interpolation line, $S = d\bar{a}/d(1/r)$, equals -0.194 ± 0.007 Å, and its intercept with the ordinate is found as $\underline{a}_o(T)$ = 4.1231 Å ± 0.0002 Å.

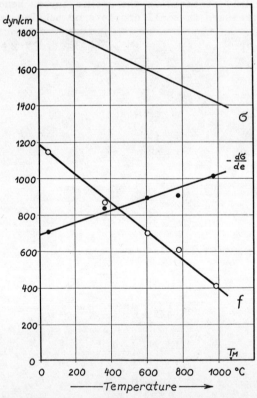

Fig. 4. Experimentally determined values of surface stress, f, surface tension, σ, and the term $\dfrac{d\sigma}{de}$ as a function of temperature as listed in Table I. The line indicating σ(T) is deduced from the value of $\sigma = 1400$ ergs/cm^2 at 1030°C, according to ref. 40, in conjunction with Eötvös' rule[41]. The values of $\dfrac{d\sigma}{de}$ are calculated as the difference between f and σ.

FIELD EMISSION AND FLASH DESORPTION STUDY OF THE SYSTEM MOLYBDENUM-CARBON MONOXIDE

M. Abon, B. Tardy, and S. J. Teichner

Institut de Recherches sur la Catalyse
Faculté des Sciences de l'Université de Lyon
Villeurbanne, France

INTRODUCTION

Extensive studies of carbon monoxide adsorption on tungsten have been performed by field emission microscopy (1 - 6), flash desorption (7 - 15), electron bombardment (4, 16, 17) and other techniques (18, 19). The corresponding results, discussed exhaustively by EHRLICH (20) show the complexity of the interaction of carbon monoxide with tungsten. Given the similarity of properties of molybdenum and tungsten, it is appropriate to sum up some results obtained with tungsten.

Carbon monoxide is adsorbed on tungsten without dissociation (2, 14, 20). Three states, at least, may be distinguished (20, 21). The first, a "virgin state", is formed at low temperature. At room temperature carbon monoxide is chemisorbed in states α and β. The β state, resulting from a stronger metal-gas bond, would account for an increase of the work function of the order of 0.7 eV. This global β state could be divided into several sub-states corresponding to different crystal planes on the surface of tungsten. The α state would be localised either between CO molecules adsorbed on sites β, or on less bonding crystal planes such as [110]. It would account for a small decrease of the work function. Carbon monoxide in α state would be bonded to a single atom $(M - C \equiv O)$ whereas in the β state two atoms $(\overset{M}{\underset{M}{>}}C = O)$ would be involved (22). All chemisorbed gas would be immobile below 600°K (21).

Adsorption of carbon monoxide on molybdenum outgassed in ultra high-vacuum has been studied to a much less extent (23 - 26). However the

interest of such investigations is easily understood if one reminds that molybdenum is widely employed in ultra high-vacuum devices and that carbon

monoxide is one of residual gases.

Electron bombardment (24) and thermal desorption of CO adsorbed on
polycrystalline molybdenum have shown, like for tungsten, several adsorbed
states. According to DEGRAS (25) state N° 1 consists of physically
adsorbed CO. Chemisorbed states N° 2 and 4 can be identified with the
states α and β for tungsten (25). The initial sticking probability on
molybdenum is of the order of 4×10^{-2} and the maximum coverage by CO is
smaller than by nitrogen (25, 27, 28). This is explained by i) long rearrangement time of chemisorbed states, ii) high probability of non bonding crystal
planes, iii) occupation of sites by carbon. This last hypothesis which
involves the dismutation of chemisorbed carbon monoxide would also explain
a progressive decrease of adsorptive capacity of polycrystalline molybdenum
filaments towards nitrogen, because carbon monoxide is an impurity in the
gas phase and cannot be eliminated by outgassing (23, 27, 28). However
the hypothesis of the carbon deposit on the surface is not confirmed by
experiments in which radioactive carbon has been used (26). The results
concerning chemisorption of CO on tungsten are also in favour of a non
dissociative behaviour on molybdenum.

EXPERIMENTAL

The field emission microscope (32, 33) used in this work has been
already described (29, 30). The ultra high-vacuum is achieved by oil
diffusion pump, a Biondi trap (31) and titanium getters. In the flash desorption device (34, 35) an omegatron allows the gas analysis. Metal valves
(Granville-Philips) are used, those connected with carbon monoxide supply
being of the variable leak type which enables the achievement of constant

pressure conditions. After baking and outgassing the whole system a residual pressure of 10^{-10} torr is registered. Very pure molybdenum (99, 95%) filament (diameter 0, 007 cm) and ultra pure grade carbon monoxide were employed. In the flash desorption experiments the variations of pressure during flashing were often recorded (Visicorder, Honeywell) simultaneously from omegatron (mass 28) and from an ionisation gage (grid current 1 mA only, to decrease the pumping action of the gage). The temperature of the molybdenum tip in the field emission microscope and of the filament in the flash desorption experiments was determined by optical micropyrometer, and also, in the second case, by resistivity measurements.

RESULTS

I. Field emission microscopy

Carbon monoxide at atmospheric pressure is introduced into the microscope after previous achievement of a pressure of 10^{-10} torr. The pressure is again reduced to 10^{-10} torr by pumping and baking. This procedure decreases the amount of residual gas impurities.

A constant pressure of CO (1×10^{-8} torr) is then established through the variable leak. The tip is first heated to 2000°K during one minute and the adsorption of CO is followed at room temperature as the function of time by measurements of the work function and recording of the emission pattern. The desorption of carbon monoxide is produced by heating the tip, during one minute, to successively increasing temperatures, the leak being closed and the pressure of the order of 10^{-9} torr.

The variations of the work function (Φ) are deduced from the variation of the slope (S) of the linear form of Fowler-Nordheim equation (33):

$$\log \frac{I}{V^2} = \log A - (B \, \Phi^{3/2}) \, \frac{1}{V} \qquad (1)$$

where V is the applied potential, I the emission current, A and B being approximately independent of V. The variations $(\Delta\Phi)$ are calculated from:

$$\Delta\Phi = \Phi_0 \left[(\frac{S}{S_0})^{2/3} -1 \right] \tag{2}$$

where Φ_0 and S_0 are values for decontaminated tip. The preexponential factor $(\log A)$ depends on the emitting area (33) and is calculated from:

$$\log A = \log \frac{I_0}{V_0^2} + \frac{S}{V_0}$$

where I_0 is the emission current for an arbitrarily selected value V_0. A mean value of $\log A$ is obtained from different values of V_0.

a) Emission patterns

The patterns A, B, C in figure 1 correspond to the adsorption of carbon monoxide, whereas patterns D, E, F correspond to the desorption. All patterns are recorded (Polaroid, 3000 ASA, exposure time 1 sec, $f = 8$) for $I = 10^{-7}$ A but for different values of V according to the data of table I. The pressure of CO is of 1×10^{-8} torr for A, B, C and 1×10^{-9} torr for D, E, F patterns.

Table I

Pattern in fig. 1		V (kV)	Conditions
A	Adsorp-	8.4	1 min. after outgassing at 1370°K during 1 min.
B	tion	9.3	15 min. after outgassing at 2000°K during 1 min.
C		10.7	4 hours after outgassing at 2000°K during 1 min.
D	Desorp-	9.4	after desorption during 1 min. at 950°K
E	tion	8.7	after desorption during 1 min. at 1100°K
F		8.5	after desorption during 1 min. at 1270°K

Outgassing at 1370°K (fig. 1A) or 2000°K gives the same pattern which therefore represents a decontaminated surface.

b) Adsorption

Figure 2 represents the results plotted according to equation (1).
Table II gives the time t of contact with CO after the initial decontamination
at 2000°K and the corresponding values of $\Delta\Phi$ and $\Delta \log A$ determined from
the straight lines of figure 2.

Table II

Lines (fig. 2)	A	B	C	D	E	F	G	H	I	J
t (min.)	2	5	10	15	20	30	45	240	1200	2160
Δ (eV)	0	0.1	0.2	0.35	0.45	0.6	0.7	0.7	0.7	0.7
$\Delta \log A$		0.1	0.3	0.4	0.5	0.5	0.1	-0.5	-0.6	-0.7

The pressure of CO is maintained constant (10^{18} torr) for all experiments up to I. For the last two experiments (I and J) the pressure during
the adsorption is of 10^{-7} torr but is decreased to 10^{-8} torr during the
measurement. The experiment J may therefore be considered as representing the state of saturation of the tip under 10^{-7} torr. Figure 3 represents the variations of the work function and of the logarithm of the
preexponential factor as a function of the time of adsorption of carbon
monoxide.

During the first 30 minutes the fast increase of Φ is accompanied by
an increase of log A and it affects essentially the plane [111] (pattern B of
figure 1) whose emission decreases markedly. After 45 minutes of adsorption $\Delta\Phi$ reaches a constant value (0.7 eV) but $\Delta \log A$ decreases now
notably (figure 3). The pattern C of figure 1 represents the emission after
240 minutes of adsorption. This pattern is identical to the pattern obtained
for the tip after 2160 minutes or after 60 minutes. Compared to pattern B
(after 15 minutes) the emission from region [111] is increased and the dark
zones which correspond to planes [110] and [211] are reduced. Compared

to pattern A (outgassed tip) the difference is not great, which shows that saturation by carbon monoxide increases the work function on all planes in a rather uniform way. This behaviour has been already observed for tungsten (2, 6, 21).

c) Thermal desorption

Figure 4 represents the Fowler-Nordheim plot obtained after heating during one minute to increasing temperatures the tip previously saturated by carbon monoxide. Table III gives the variation of Φ and log A calculated from the linear plot of figure 4.

Table III

Lines (fig. 4)	A	B	C	D	E	F
T°K	300	770	950	1100	1270	1370
$\Delta\Phi$ (eV)	0.7	0.7	0.65	0.45	0.25	0
Δ log A	-0.7	-0.6	-0.3	0.6	0.3	0

Figure 5 represents these variations as a function of the temperature. The emission pattern during desorption up to 800°K is still comparable to the pattern C of figure 1. Similarly the work function and log A practically do not change (fig. 5). The gas previously chemisorbed is therefore always present on the surface of the tip, except for a species which would be field emission inactive.

But after heating at 950°K the emission pattern D of figure 1 shows that some change now occurs whereas $\Delta\Phi$ begin to decrease and Δ log A is markedly increased. These observations can be interpreted by a rearrangement of chemisorbed carbon monoxide rather than by desorption.

After heating at 1100°K the emission pattern E of figure 1 changes very much and is now quite similar to pattern F of the outgassed tip. The values of $\Delta\Phi$ (0.45 eV) and of Δ log A (0.6) are close to the values observed previously after the first period of fast adsorption (fig. 3).

After heating at $1270°K$ the emission pattern F of figure 1 resembles very closely the pattern of a thoroughly outgassed tip. However figure 5 shows that the work function is still higher ($\Delta\Phi = 0.25$ eV) than for a decontaminated tip and also the preexponential factor does not decrease to the initial value. After $1370°K$ the gas is completely desorbed. The emission pattern is exactly the same as for pattern A of figure 1 and the work function as well as log A do not vary any more with further heating.

II. Flash desorption

Heating of the molybdenum filament at $2000°K$, after a vacuum of 10^{-10} torr has been obtained, releases an important amount of gases. Their analysis by omegatron shows then that the essential product is carbon monoxide with a small proportion of carbon dioxide. Once the pressure is again decreased to 2×10^{-10} torr, a constant pressure of carbon monoxide (2.1×10^{-8} torr) is established through the variable leak. Analysis by omegatron gives here evidence that carbon monoxide is the only constituent of the gas phase. After a certain time of contact the filament is flashed to $1520°K$, temperature which, as it has been shown previously, allows the desorption of all presorbed gas.

a) Flash desorption spectra

Figure 6 shows the flash desorption spectra. Spectrum A is obtained one minute after the first flashing, previously to the introduction of carbon monoxide. It shows only one desorption peak for the residual gas at 2×10^{-10} torr. The pressure recording sensitivity is 10 times higher than for spectrum B which corresponds to the previous adsorption of carbon monoxide at 2.1×10^{-8} torr in contact with the filament during one minute. Two desorption peaks are always recorded if the time of contact exceeds 30 seconds. This seems to account for the adsorption of CO under two principal forms of different bond strength. The comparison of the height of two peaks shows that the population of weakly bonded CO (low temperature peak) is smaller than the population of strongly bonded CO (high temperature peak).

Spectrum C has been recorded with omegatron (mass 28). Two peaks are also present but the second peak is much larger and less defined than the first.

b) Influence of the time of contact on the spectra

Only ionisation gage was employed in the experiments described here, where the CO pressure was in all cases equal to 2.1×10^{-8} torr and where only the time of contact with carbon monoxide varied. Figure 7 represents the heights of two peaks, calculated as the increase of the pressure Δp, with the time t of contact with carbon monoxide. The flashing temperature was in all cases equal to 1520°K. For both peaks a steady value of Δp is registered after 10 minutes of time of contact. The sticking probability, which is proportional to the slope of curves of figure 7, has an initial value of the order of 0.1, but decreases to some very low value after 10 minutes.

c) Presorption of carbon monoxide at higher pressures

Similar experiments to those previously described has been performed with the pressure of carbon monoxide being maintained at values of 2.3×10^{-7} torr, 2.3×10^{-6} torr and 2.3×10^{-5} torr. The temperature of flashing was in all cases equal to 1520°K. All flash desorption spectra for different times of contact have always the form of spectrum B in figure 6. The steady values of Δp, as in figure 7, are obtained after a time of contact which decreases when the pressure duringthe presorption of CO increases. These steady values of Δp (or the heights of both peaks) increase with the pressure of CO.

Table IV gives for these different pressures of presorption of CO the total number n_A of CO molecules adsorbed per cm^2 of the surface of molybdenum wire, once the steady value of Δp is reached, the approximate time of contact necessary for the steady value of Δp to be observed, the ratio n_A/n_M where n_M is the number of metal atoms per cm^2 (n_M is taken equal to 4.8×10^{14}) and finally $\Delta p_1/\Delta p_2$ which is the ratio of the steady values for both peaks.

Table IV

P_{CO} (torr)	2.1×10^{-8}	2.3×10^{-7}	2.3×10^{-6}	2.3×10^{-5}
n_A	1.10^{14}	3.6×10^{14}	5.5×10^{14}	6.75×10^{14}
n_A/n_M	0.21	0.75	1.15	1.4
t (min.)	10	5	2	1.5
$\Delta p_1/\Delta p_2$	0.46	0.27	0.27	0.3

The values of n_A (and hence of n_A/n_M) do not take into account the rate of pumping (less than 1 1/sec) and for this reason they are only indicative. The surface of the filament is considered as its geometric surface (the roughness factor is not taken into account) and n_M which corresponds to the plane [411] (34) reflects a mean value corresponding to the equipartition of the different planes on the surface. For this reason, the conclusion which may be obtained from values of table IV is that the coverage by carbon monoxide tends to be complete ($n_A/n_M \simeq 1$) when the pressure during its presorption increases to 2.3×10^{-5} torr. It may be also observed that if the number n_A is plotted against the pressure P_{CO} established during the presorption of carbon monoxide a Langmuir type isotherm is obtained. A second conclusion is that for pressures during the presorption of CO higher than 10^{-7} torr the ratio $\Delta p_1/\Delta p_2$ is almost constant and equal to 0.3 for the steady values of Δp. This seems to show that the population of two forms of adsorbed carbon monoxide, corresponding to two peaks, is maintained in a constant ratio even when this population increases with increasing pressure.

d) Flash-desorption spectrum for different temperatures of flashing

When the flashing temperature is of 1340°-1370°K no variation of pressure is recorded when the flashing is continued at some increased temperature. This previous temperature hence guarantees a total desorption of carbon monoxide by flashing. The flashing was therefore performed at lower temperatures and the corresponding spectra are shown on figure 8, where the carbon monoxide pressure during its presorption is of 4.8×10^{-8} torr and the time of contact is of one minute, in all cases.

Only one peak is recorded for flashing temperature below 700°K (spectrum A). The maximum of Δp is observed for 370°-400°K. The second peak is well observed for temperature of flashing above 800°K (spectrum B). The heights of two peaks are comparable whereas flashing at 1120°K shows an increase of the second peak. The variation Δp of this peak is no more monotonous but presents a transition point shown in figure 8C by an arrow. This may be interpreted by the existence in the temperature range of 750°-1370°K of not only one but of two forms of adsorbed carbon monoxide.

At 1320°K (spectrum D) all CO is practically desorbed because the form of this spectrum is almost identical to spectrum B of figure 6 corresponding to the temperature of flashing of 1520°K.

DISCUSSION

Chemisorption of carbon monoxide on metals is essentially of a covalent character, the ionic contribution (22, 36) being only of the order of 18% in the case of molybdenum (25). Physical adsorption cannot be taken into account for experiments performed at room temperature and above, at very low pressures. It is also very probable that the so-called virgin adsorption, characterised by fairly uniform dipole moment and surface concentration is not observed in the present experiments.

The increase of the work function registered during the initial adsorption of carbon monoxide in field emission experiments (fig. 3), which shows that dipole moments of the surface bonds have their negative end pointing outwards (37, 38), can only be connected with the formation of a rather tightly bound electronegative β layer. This is confirmed by the change in emission anisotropy from the clean emitter (fig. 1B). At the same time the logarithm of the preexponential factor increases. A comparable behaviour has been observed for tungsten (21) where adsorption of β-CO increases Φ and log A.

In the next step the work function tends to a stable value but log A now decreases and the final pattern (1C) is very similar to that shown by the clean emitter.

One may be tempted to correlate this behaviour with the formation of the α-CO layer which is known to decrease both Φ and log A (21). The stabilisation of the value of Φ in our experiments could then result from the compensation between a fast adsorption of α-CO and a much slower continuation of the adsorption of β-CO or also from the partial conversion of α-CO into β-CO (25).

However, desorption experiments in field emission microscope (table III) show that at 770°K, temperature which is twice the value necessary for desorption of the α-CO, at least for tungsten (4, 5), the log A still conserves its reduced value, incompatible with the departure of the α layer.

Variations of log A are connected with the variation of the emitting surface. A decrease of the preexponential factor is produced by a decrease of the emitting surface for crystal planes presenting small values of the work function (33).

Other explanations of the decrease of log A have also been proposed (4) but they hold only when a weakly bonded polarizable layer is present, which is not the case at 770°K. For this reason, the explanation proposed by HOLSCHER and SACHTLER (6) for tungsten, related to the rearrangement of the surface due to a corrosive chemisorption, seems to fit with our results. According to these authors, after the formation of a virgin layer (at low temperature) the intermetallic bonds are weakened and this allows a surface rearrangement which causes a tighter bonding of a part of adsorbed CO molecules by embedding them into a two-dimensional tungsten carbonyl structure. Tungsten atoms penetrate outward through the adsorbed CO layer which create new adsorption sites for an additional very weakly bonded layer on top of the chemisorbed layer.

The corroded layer, by disturbing the periodic potential at the surface, increases the scattering of electrons which has the same effect as a decrease in emitting area and of log A. According to this model the decrease of log A observed after 30 min. (fig. 3) may correspond to the start of the corrosion of the surface by CO adsorbed in a β layer.

Desorption at temperatures higher than 770°K produces the variation of log A and of the work function in the reverse sequence. This shows that the metal surface is reconstructed, the chemisorption complex being dedomposed rather than evaporated (6).

For a polycrystalline molybdenum Degras (24) has also observed an increase (0.2 eV) of the work function resulting from the adsorption of CO at room temperature. The difference with the value of $\Delta\Phi$ recorded in the present work may result from the fact that in the field emission the measurements of $\Delta\Phi$ concern mainly regions of low work function (33) which are probably the most reactive towards the adsorption of an electronegative species (37,38).

The weakly bonded layer which is adsorbed on the top of the "corroded" layer and which is well defined by the first peak (370°K-400°K) of the flash desorption experiments (figures 6 and 8) does not influence the work function,

the preexponential factor or the emission pattern. Indeed, all these parameters remain stable up to a desorption temperature of 770°K in the field
emission experiments. Above this temperature the reconstruction of the
metal surface due to the decomposition of adsorbed complex is responsible
for the second peak in the flash desorption experiments. Of course, pressure
peaks may not reflect the state of gas adsorbed on a surface in flash desorption experiments (39). In particular they may result from the conversion of
a weak binding state to a more stable form during warm up. In this respect,
comparison of desorption experiments performed in the field emission
microscope (no change below 770°K) and by flash desorption (a well defined
peak around 400°K) gives evidence that no spurious pressure peaks (not
related to separate states of binding of the gas) are registered in our
experiments.

The reversibility of adsorption-desorption cycle in the field emission
experiments gives besides evidence of the absence of any dismutation of
adsorbed carbon monoxide on the surface of molybdenum in the whole
temperature range.

The number of adsorbed molecules recorded, when the rate of adsorption decreases to a very low value, depends on the pressure (table IV). For
the highest pressure (2.3×10^{-5} torr) of carbon monoxide achieved in flash
desorption experiments the coverage exceeds the monolayer (1.4). Even
neglecting the approximation involved in this calculation, the previous value
is still reasonable if it is considered that two kinds of states contribute to
this adsorption i. e., the "corroded" state and the weakly chemisorbed state.
The almost constant ratio of the two peaks for any adsorption pressure
($\Delta p_1/\Delta p_2$, table IV) i.e., for any adsorbed amount tends to show that the two
peaks do not result from the conversion of a weak state into the more stable
state. Indeed, the conversion during warm up terminates usually, for

instance, through saturation of accessible sites, before the weaker state is completely evolved (39). But in this case a constant ratio of populations of two peaks would be difficult to understand.

REFERENCES

(1)   R. KLEIN - J. Chem. Phys. 1959, 31, 1306

(2)   G. EHRLICH, F. G. HUDDA - J. Chem. Phys. 1961, 35, 1421

(3)   L. W. SWANSON, R. GOMER - J. Chem. Phys. 1963, 39, 2813

(4)   D. MENZEL, R. GOMER - J. Chem. Phys. 1964, 41, 3329

(5)   A. A. BELL, R. GOMER - J. Chem. Phys. 1966, 44, 1065

(6)   A. A. HOLSCHER, W. M. H. SACHTLER - Discuss. Faraday Soc. 1966, 41, 29, 70

(7)   J. EISINGER - J. Chem Phys. 1957, 27, 1206

(8)   J. EISINGER - J. Chem. Phys. 1958, 29, 1154

(9)   R. E. SCHLIER - J. Appl. Phys. 1958, 29, 1162

(10) P. A. REDHEAD - Trans. Faraday Soc. 1961, 57, 641

(11) V. M. GAVRILYUK, V. K. MEDVEDEV - Soviet Physics Solid State 1963, 4, 1737

(12) G. EHRLICH - J. Chem. Phys. 1962, 36, 1171

(13) L. J. RIGBY - Can. J. Phys. 1964, 42, 1256

(14) T. E. MADEY, J. T. YATES, R. M. STERN - J. Chem. Phys. 1965, 42, 1372

(15) D. A. DEGRAS - Suppl. Al Nuovo Cimento 1967, 5, 408

(16) J. T. YATES, T. E. MADEY, J. K. PAYN - Suppl. Al Nuovo Cimento 1967, 5, 558

(17) P. A. REDHEAD - Suppl. Al Nuovo Cimento 1967, 5, 586

(18) T. E. MADEY, J, T. YATES - Suppl. Al Nuovo Cimento 1967, 5, 483

(19) B. J. HOPKINS, S. USAMI - Suppl. Al Nuovo Cimento 1967, 5, 535

(20) G. EHRLICH - Ann. Rev. of Phys. Chem. 1966, 17, 295

(21) R. GOMER - Discuss. Faraday Soc. 1966, 41, 14

(22) D. O. HAYWARD, B. M. W. TRAPNELL - Chemisorption 2nd Ed.
      Butterworths 1964

(23) A. A. PARRY, J. A. PRYDE - Brit. J. Appl. Phys. 1967, 18, 329

(24) D. A. DEGRAS, J. LECANTE - Suppl. Al Nuovo Cimento 1967, 5, 598

(25) D. A. DEGRAS - J. de Chim. Phys. 1967, 3, 405

(26) A. D. CROWELL, L. D. MATTHEWS - Surface Sci. 1967, 7, 79

(27) T. OGURI - J. Phys. Soc. Japan 1964, 19, 77

(28) R. A. PASTERNACK, H. U. D. WIESENDANGER - J. Chem. Phys.
      1961, 34, 2062

(29) M. ABON, S. J. TEICHNER - J. de Chim. Phys. 1966, 2, 272

(30) M. ABON, S. J. TEICHNER - Suppl. Al Nuovo Cimento 1967, 5, 521

(31) R. BIONDI - Rev. Sci. Instr. 1959, 39, 851

(32) E. W. MULLER - Zeits, Phys. 1936, 37, 838; 1937, 106, 541
      E. W. MULLER - Egreb. Exakt Naturwiss 1953, 27, 290

(33) R. GOMER - Field Emission and Field Ionization - Harvard 1961

(34) J. A. BECKER, C. D. HARTMAN - J. Phys. Chem. 1953, 57, 157
      J. A. BECKER - Adv. in Catalysis, 1955, 7, 135
                     Solid State Phys. 1958, 7, 379

(35) T. W. HICKMOTT, G. EHRLICH - J. Phys. Chem. Solids 1958, 5, 47
      G. EHRLICH - J. Appl. Phys. 1961, 32, 4

(36) F. C. TOMPKINS - Le Vide 1962, 17, 72

(37) C. P. CULVER, F. C. TOMPKINS - Adv. in Catalysis 1959, 11, 67

(38) A. EBERIIAGEN - Fortschritte der Physik 1960, 8, 245

(39) G. EHRLICH - Discuss. Faraday Soc. 1966, 41, 102

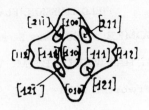

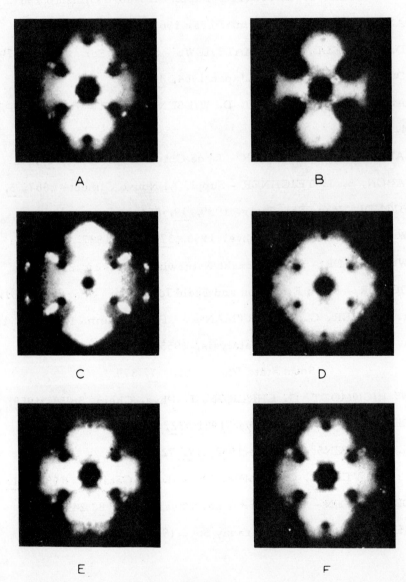

Fig. 1.

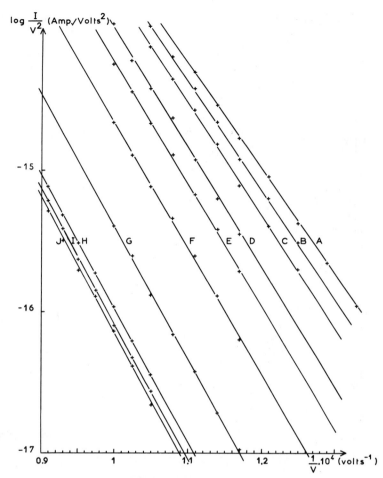

Fig. 2. Adsorption.

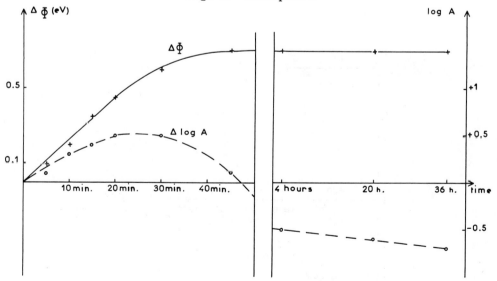

Fig. 3. Adsorption.

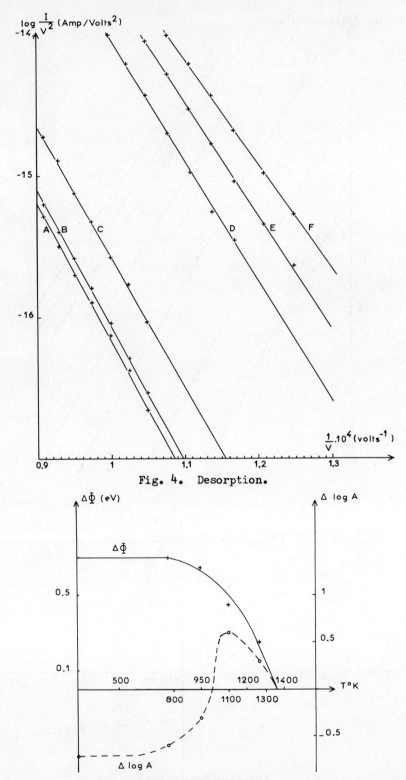

Fig. 4. Desorption.

Fig. 5. Desorption.

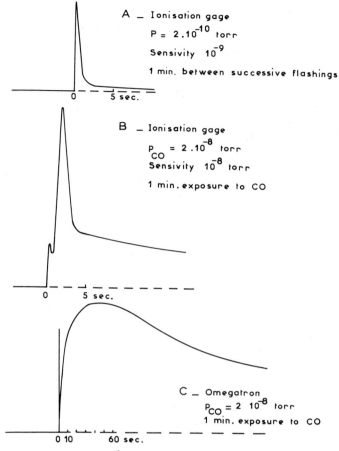

A — Ionisation gage

$P = 2.10^{-10}$ torr

Sensivity 10^{-9}

1 min. between successive flashings

B — Ionisation gage

$P_{CO} = 2.10^{-8}$ torr

Sensivity 10^{-8} torr

1 min. exposure to CO

C — Omegatron

$P_{CO} = 2 \cdot 10^{-8}$ torr

1 min. exposure to CO

Fig. 6. Desorption spectra.

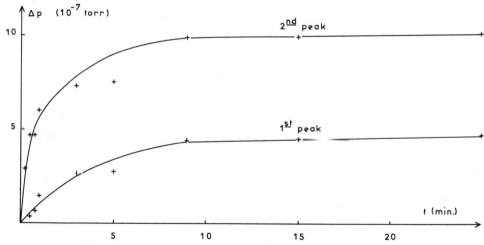

Fig. 7. Peaks height change with time of contact.

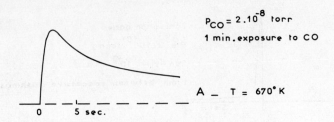

$P_{CO} = 2.10^{-8}$ torr

1 min. exposure to CO

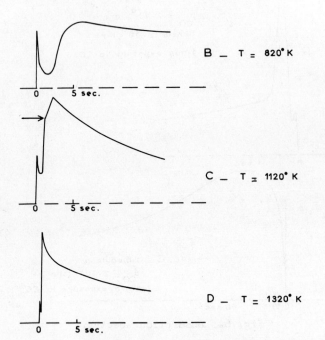

A _ T = 670°K

B _ T = 820°K

C _ T = 1120°K

D _ T = 1320°K

Fig. 8. Desorption spectra.

LEED STUDIES OF THIN FILM SILICON OVERGROWTHS ON α-ALUMINA

Chuan C. Chang

Bell Telephone Laboratories, Inc.
Murray Hill, New Jersey

INTRODUCTION

A number of studies of the epitaxial growth of silicon on α-alumina have been made recently[1-8] with the primary emphasis on the epitaxial relationships between overgrowth and substrate for different crystallographic substrate orientations. This paper reports on one type of investigation of this heteroepitaxial system that has not yet appeared in the literature: a LEED study of the α-alumina surfaces and their interactions with silicon during the early stages of formation of an epitaxial film.

In this program, the structural perfection of epitaxial silicon films as determined mainly by LEED analysis was studied for different substrate superstructures for the (0001), ($\bar{1}$012) and (11$\bar{2}$3) faces of α-Al_2O_3; the (hkil) notation refers to the hexagonal unit cell with a_o = 4.76 A and c_o = 12.99 A. These surfaces were selected partly on the basis of their frequency of use as substrates in previously reported work[1-8] and partly on the basis of important crystallographic features: These faces are, respectively, a slip plane, a cleavage plane, and the plane normal to the

growth direction of the crystal. An earlier, detailed study
of the (0001) surface has already been published.[9] Evidence
for a chemical reaction between silicon vapor and α-Al$_2$O$_3$ was
found and a preliminary analysis is presented.

EXPERIMENTAL

The experimental apparatus is a Varian 360 LEED
assembly (18" metal "bell-jar" type) with an EAI Quad 150
mass analyzer. 1×2×0.1 cm α-alumina substrates were cut
from Czochralski grown single crystal boules obtained from
the Linde Company. One face was polished optically flat and
within $\pm\frac{1}{2}°$ of the desired orientation, and 10,000 A of β-Ta
were then sputtered onto the unpolished face so that the crystal could be repeatedly and rapidly heated to over 1700°C by
resistive heating using about 100 watts power. Because of
the high thermal conductivity of α-alumina, the crystal temperature was assumed to be that of the tantalum backing as
measured with an optical pyrometer. Silicon was sublimed
from high purity single crystal silicon bars resistively
heated to a temperature near the melting point. Typical
deposition rates were about 80 A/min and the pressure was
maintained below 5×10^{-10} torr during deposition.

RESULTS AND DISCUSSIONS

1. <u>α-Al$_2$O$_3$ Surface Structures</u>

A freshly prepared crystal usually gives weak and
diffuse diffraction patterns before any heat treatment. Upon
heating to progressively higher temperatures (and then cooling

back to room temperature for LEED observations) a series of
different diffraction patterns is seen with each surface.
The diffraction patterns begin to sharpen near 600°C and at
some temperature between 900 to 1400°C depending on the particular sample, a final pattern characteristic of each crystallographic face is obtained. These final surface structures
are called equilibrium structures; they are reproducible from
sample to sample and have been observed with Verneuil as well
as Czochralski grown crystals and with specimens from different
suppliers. As shown later, there is good reason to believe
that the equilibrium structures represent clean surfaces.
Most of the numerous patterns observed before the appearance
of the equilibrium patterns never return again once they are
destroyed; they are quite possibly due to contamination.

Two dimensional lattices of the equilibrium structures for the three surfaces studied are described in the
third row of Table I, and the diffraction patterns are shown
in Fig. 1. These lattices are based on the primitive unit
cell of the aluminum lattice for each (hkil) plane; the symmetry of each lattice is shown in the second row of the table.

The lattices of the equilibrium structures for the
($\bar{1}012$) and (11$\bar{2}$3) are the relatively simple 2×1 and 4×5,
respectively; analyses   of the $\sqrt{31}$ equilibrium structure
on the (0001) appear elsewhere.[9,10] The symmetries of these
equilibrium structures are those of the parent planes.

Diffraction spots for the 1x1 lattice can be distinguished from extra spots due to superlattices by comparing diffraction patterns taken at high (e.g., 400 eV) and low (e.g., 100 eV) electron energies. In general the integral order spots are more prominent at higher energies, indicating that the superstructures are "thin" surface structures, of the order of 10 Å or less.

Surfaces displaying these equilibrium structures are "atomically smooth" over several thousand angstroms unlike the extensively faceted (and stable) surfaces that can be produced by etching with silicon under special conditions, as described below.

2.   Chemical Interaction of Silicon with α-Al_2O_3 Surfaces

Mass spectrometer analyses of material leaving α-alumina surfaces heated above 1000°C during silicon deposition show three major peaks at positions appropriate to Al, Si and SiO (with peak heights in the proportion 4:3:5, respectively, at 1100°C). No other peaks higher than 10% of the silicon peak were detected, and the data suggest the following overall reaction:

$$3Si(g) + Al_2O_3(s) \longleftrightarrow 3SiO(g) + 2Al(g).$$

This reaction differs from those proposed in the literature,[2,11,12] in that aluminum, and not its oxides, is produced. Experiments are presently in progress to obtain accurate values of the partial pressures in order to allow a thermodynamic analysis of the applicability of the reaction proposed above.

The surfaces studied etched uniformly (without faceting) when held above 1000°C and exposed to the silicon beam. However, if a surface with a deposit of silicon is heated to some temperature between 850 to 1450°C (higher temperatures with thicker films) silicon forms globules and desorbs, leaving behind a heavily pitted surface. Continued heating then results in a highly faceted surface. One consequence of this phenomenon is that a silicon deposit cannot be desorbed for purposes of reusing the substrate for a repeat experiment without changing the character of the substrate. A second consequence is that under conditions of epitaxial deposition the surface probably becomes severely etched near the silicon nuclei and may become very rough[2] before the film becomes continuous; the crystallography of the resulting interface can be quite different from the original surface.

Surfaces etched uniformly with silicon at temperatures above 1200°C always resulted in equilibrium structures. This is another indication that these structures represent clean surfaces. However, the possibility of diffusive impurities from the interior cannot be excluded,[9] and chemical analyses of equilibrium surfaces are now being conducted using Auger spectroscopy.

3. Film-Substrate Orientational Relationships

It was expected at first that the nucleation stage of film growth would also be studied with LEED. This turned out to be prohibitively difficult because the presence of isolated silicon nuclei caused the surface to charge up (when exposed to the LEED beam) to at least 1000 V so that no diffraction pattern could be obtained. To what extent this charging is caused by the large number of imperfections associated with the silicon nuclei and their silicon-alumina boundaries, or to geometrical factors (such as the shape of the silicon nuclei), or to other factors, is not well understood. Epitaxial films could be studied after coalescence which usually occurred at an average film thickness of about 1000 A. The observed epitaxial relationships are summarized in Table I, and are in general agreement with the literature[1,4,6,8]; while exceptions are noted below. All depositions were made on surfaces originally displaying equilibrium structures.

Silicon films grow with the (111) plane parallel to the (0001) plane even for surfaces cut several degrees off the (0001) orientation. The [$1\bar{1}2$] direction of silicon was found to be parallel to the [$10\bar{1}0$] direction of sapphire. Figure 2a displays a diffraction pattern showing spots from both the silicon film and the (0001) substrate. From the fact that the silicon pattern shows sixfold instead of its bi-threefold symmetry, it is concluded that there is extensive twinning in the epitaxial film.

Either the ideal 1x1 hexagonal structure or the $\sqrt{31}$ superstructure can be produced at will on the (0001) face.[9] Deposition onto the $\sqrt{31}$ structure is of special interest because of its double domains rotated by ±9°, which may be related to the double Si (111) domains (rotated by up to ±10.5°) observed by others.[1] Deposition onto either the 1x1 or the $\sqrt{31}$ structures usually produced films with no noticeable differences so that the surface superstructures were probably destroyed by the impinging silicon. However, in a narrow temperature range near 850°C, films with double domains rotated by about ±9° were produced, probably due to the rapid recovery[9] of the $\sqrt{31}$ structure at this higher temperature.

Aspects of the growth on the ($\bar{1}$012) face are in general agreement with the literature.[1,4,6,7,8] The exposed silicon surface consists of {100} planes and {113} facets; these are probably the facets visible in electron micrographs of carbon replicas and are sometimes assumed to be {111} facets.[11] A diffraction pattern from the faceted (100) film is shown in Fig. 2b.

On the (11$\bar{2}$3), the Si (111) is inclined about 7° to the surface so that it is parallel to the substrate (11$\bar{2}$4); this is shown in Fig. 2c. Unlike growth on the (0001), these films do not show twinning, as also reported by others.[1] It can be added that because the (11$\bar{2}$3) equilibrium structures

were always observed, even after extensive (uniform) etching
with silicon, the growth of silicon parallel to (11$\bar{2}$4) is not
a result of faceting to that plane.

CONCLUDING REMARKS

One objective of this study was to identify the
surface superstructures of each face of α-Al$_2$O$_3$ and to note
whether any properties of an epitaxial silicon film deposited
on that face could be related to that superstructure. Such a
correlation was in fact found when double rotated domains in
the silicon film were observed for the (0001) substrate under
growth conditions favoring the formation of the $\sqrt{31}$ structure; both the film and substrate domains were rotated by the
same amounts, approximately ±9 degrees about an axis normal
to the surface. Because of the rapid development of the $\sqrt{31}$
structure above 850°C,[9] the formation of the double silicon
domains could be expected (depending on the silicon deposition
rate) at all substrate temperatures above 850°C.

The advantages of the ability to study the surface
with LEED at each stage were quite apparent in these experiments. It was possible to determine whether the surface
damage from polishing was completely annealed out and whether
facets were developed after various treatments. Characterization of the surfaces by their equilibrium structures, which
(for the freshly prepared crystals) were sometimes not obtained
until the crystal was heated as high as 1400°C, assured that

silicon depositions were made on a known substrate surface.
These advantages greatly facilitated the efficient collection
of data, and with less uncertainties and in many instances
with the complete avoidance of time-consuming "trial and error"
methods.

Experiments are presently being conducted to investigate the electrical properties of silicon films deposited on
α-Al_2O_3 and to study the surface chemical composition and
structure using mass spectroscopy, electron microscopy and
Auger spectroscopy.

REFERENCES

1. (Review article) J. D. Filby and S. Nielsen, Brit. J. Appl. Phys., $\underline{18}$, 1357 (1967).

2. F. H. Reynolds and A. B. M. Elliot, Solid-State Electronics, $\underline{10}$, 1093 (1967).

3. F. P. Heiman and P. H. Robinson, Solid-State Electronics, $\underline{11}$, 411 (1968).

4. S. Namba, A. Kawazu and T. Maruyama, Sci. Papers I.P.C.R., $\underline{61}$ (3), 45 (1967).

5. D. J. Dumin, J. Appl. Phys., $\underline{38}$ (4), 1909 (1967).

6. H. M. Manasevit, R. L. Nolder and L. A. Moudy, Trans. Metall. Soc. A.I.M.E., $\underline{242}$, 465 (1968).

7. C. T. Naber and J. E. O'Neal, ibid., p. 470.

8. L. R. Weisberg and E. A. Miller, ibid., p. 479.

9. C. C. Chang, J. Appl. Phys., $\underline{39}$, 5570 (1968).

10. J. M. Charig, 4th International Materials Symposium, June, 1968, Univ. of California, Berkeley, California.

11. H. M. Manasevit, A. Miller, F. L. Morritz and R. Nolder, Trans. Metall. Soc. A.I.M.E., $\underline{233}$, 540 (1965).

12. R. W. Bicknell, B. A. Joyce, J. H. Neave and G. V. Smith, Phil. Mag., $\underline{14}$, 31 (1966).

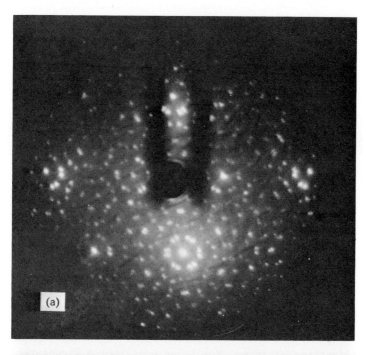

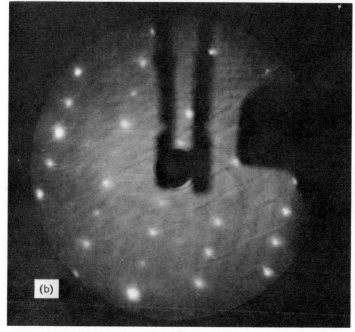

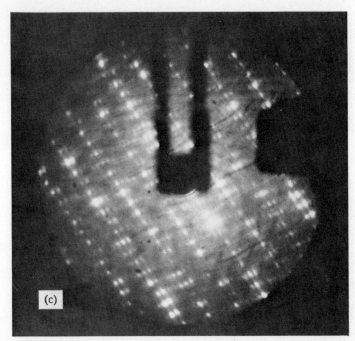

Fig. 1. Diffraction patterns from three equilibrium structures: (a) The $\sqrt{31}$ pattern from the (0001) face; the 0,0 beam is the bright spot below center surrounded by six bright doublets, 101 V. (b) The 2x1 from a ($\bar{1}$012), taken at normal incidence; half order spots are seen along the rows running from lower left to upper right, 79 V. (c) The 4x5 from a (11$\bar{2}$3) taken at normal incidence; ¼-th order spots are seen along the rows running from upper left to lower right, 112 V.

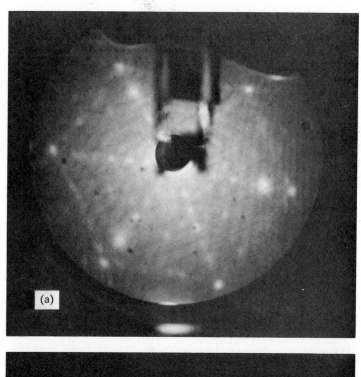

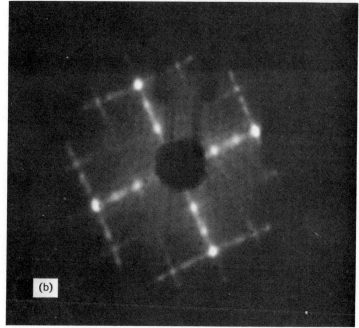

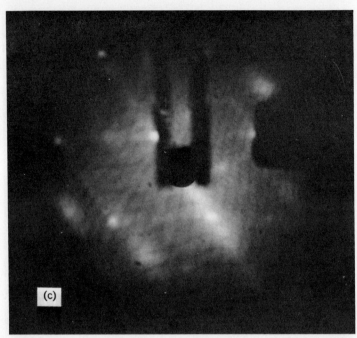

Fig. 2. Diffraction patterns from silicon films: (a) Si(111) on the
(0001); the six outermost spots are due to silicon; the three
large inner spots are from the substrate and show the extent of
lattice mismatch; normal incidence, 27 V. (b) Si(100) on the
($\bar{1}$012); a diffraction pattern from four {113} facets whose 0,0
spots lie beyond the ends of the bright cross, near the screen
edge (not visible), normal incidence, 42 V. (c) Si(111) on
the (11$\bar{2}$3); the 0,0 beam from the silicon (111) plane is slightly upper left from center; taken at normal incidence to the
(11$\bar{2}$3), 62 V. Spots from both silicon (small, round) and substrate (large, elliptical) are visible.

A LEED INVESTIGATION OF THE NITRIDATION
OF THE Si(111) SURFACE

Dr. R. Heckingbottom

Post Office Research Station, Dollis Hill
London, England

Introduction

The reaction between the Si(111) surface and both ammonia and nitrogen has been studied in a commercial LEED apparatus. The majority of the results refer to the Si(111)/ammonia system as it proved more suitable for a LEED study. Below 700°C the observations of Van Bommel and Meyer[1] were confirmed at room temperature exposure to ammonia caused only a slight deterioration of the Si(111)-7 pattern but on heating to 500°C all 1/7th order spots disappeared, leaving only 1st order spots. Above 700°C the previously reported LEED observations have been extended considerably and augmented with RHEED and both optical and electron microscopy studies.

Experimental

Silicon samples were cut to expose the (111) face and mechanically polished to 0.5 micron finish before being electropolished on one face; the sample dimensions were typically 1.5 cm × 0.7 cm × 0.1 cm. Before introduction into the LEED chamber, the samples were degreased by heating in a solution of a hydrogen peroxide in concentrated sulphuric acid. Any oxide formed at this stage was removed by immersing the sample in 40% hydrofluoric acid for a few minutes. This solution was then diluted at least tenfold with distilled water to eliminate any chance of "staining" or removing the sample. Finally the sample was rinsed in distilled water before mounting between tantalum supports. The sample surface appeared smooth at 800× magnification but, in general, no LEED patterns could be observed at this stage. Clear Si(111)-7 patterns were readily obtained by direct resistance heating of the sample at 900 - 1200°C in ultra-high vacuum.

Silicon from various sources and of several resistivity grades was examined. All samples behaved in a similar manner once the Si(111)-7 pattern was obtained. However samples heavily doped (milli-ohm cm resistivity) with antimony gave rise to the Si(111) - $\sqrt{19}$ pattern reported by Farnsworth[2] and Van Bommel[3] and attributed by the latter to nickel contamination. The appearance of this pattern was eliminated by immersing the samples in EDTA solution for several hours before introduction into the LEED machine behavior consistent with Van Bommel's explanation. Nickel contamination could arise in the present work from the electropolishing stage but, except in the above instance, the nickel was removed by the normal washing with aqua regia. The contaminated surface appeared to behave in the same manner as the clean silicon surface towards ammonia. This is analogous to the reported reaction of the two surfaces in phosphine.[3]

The composition of the gas phase was monitored by means of an MS 10 mass spectrometer mounted directly onto the LEED chamber. The ammonia was of 4 N purity and, apart from nitrogen and hydrogen, impurities in the chamber were kept below one part per thousand, with a typical operating pressure of 10^{-6} torr. The nitrogen, as supplied was nominally of 6N purity. After drying over sodium the only significant impurity was hydrogen and again, at 10^{-6} torr total pressure, other impurities were kept below 10^{-9} torr. Most work with nitrogen was carried out at a pressure of several torr however and no direct check on the gas composition was possible under these conditions.

The temperature of the sample was measured with an optical pyrometer, readings being corrected for the absorption by the glass window and the emissivity of silicon.[4] Below 700°C the temperature was estimated by extrapolation of a temperature heating current curve plotted for the region above 700°C. Most photographs were taken at room temperature and all below 700°C but direct viewing of the patterns (though considerably impaired) was possible up to 1150°C.

This extension of the normal temperature range of viewing was obtained by introducing a shield of tantalum foil arranged to cover all but a central strip 1 mm
wide on each face of the slice; the shield being about 1 mm from the sample at
all points. The amount of light in the chamber due to the hot sample was thus
significantly reduced. Though LEED observations were restricted to the 1 mm
strip, the field of view on the screen was only slightly less than normal. The
open strip on the reverse side allowed the sample temperature to be monitored
conveniently.

LEED Results

1. Ammonia

The initial work in ammonia confirmed Van Bommel's results.[1] A brief
summary is given in Fig. 1 which shows (a) the Si(111)-7 pattern from the clean
silicon surface, (b) the removal of all 1/7th order spots and a change in the
intensity of the 1st order spots after a brief ammonia treatment of 10^2 Langmuirs
at 500°C, (c) the Si(111)-8 N pattern formed after a similar exposure at 800°C
and (d) the doublet pattern which can be formed either by an exposure of 600 to
a 1000 Langmuirs at $\geqslant$ 1050°C or by heating the 1/8th order structure briefly in
a vacuum of 10^{-9} torr at a similar temperature. The pattern shown in Fig. 1(d)
was obtained by the latter method and the Si(111)-8-N pattern is also visible.
Further heating in vacuum at $\geqslant$ 1050°C restores the 1/7th order pattern. It
should be noted that in the present work the doublet pattern could not be obtained at temperatures as low as 900°C as reported earlier.[1]

At 700°C the Si(111)-8-N pattern formed relatively slowly and was only
poorly defined, thus after an exposure to ammonia of eighty Langmuirs only the
two hexagons typified by the spots at 8/8, 0 and 11/8, 0 were present against
a high background. At 800°C and above the reaction was much faster, the
stage described above being reached after only 5 Langmuirs and the pattern
was fully developed, as in Fig. 1(c), after only 25 Langmuirs, With prolonged

exposure at $800 - 1,000°C$ however the pattern deteriorated, for instance after 10^6 Langmuirs at $950°C$ it had completely dissappeared. On cooling the well-developed 1/8th order pattern to room temperature in ammonia a marked change in spot intensities occurred. At a typical pressure of 5.10^{-7} torr the change was complete within two minutes, the new pattern having only trigonal symmetry for a beam energy between 50 and 60 volts as illustrated for 57 volts in Fig. 2(a). Once formed, this trigonal pattern was quite stable over long periods e.g. 60 hours at 10^{-9} torr and it was also unchanged after an exposure to air at about 1 torr for one hour. The pattern was completely destroyed however by a similar exposure to air at one atmosphere. It was also sensitive to the electron beam even at 10^{-9} torr and after half an hour in the beam the pattern had deteriorated to the point illustrated in Fig. 2(b). The original Si(111)-8-N pattern, as shown in Fig. 1(c), was readily recovered from the condition shown in either Fig. 2(a) or 2(b) by flashing to $500°C$. Small rises in the pressure of hydrogen and nitrogen were observed on flashing the sample but no evolution of ammonia could be detected. The total pressure rise represented only a very low coverage which suggests that the sharp pattern was restored by ordering rather than evolution of the surface species.

Using the optical shield around the sample it became clear that the change between the 1/8th order and doublet patterns was not a simple temperature transition. Thus, starting with the sample at $1120°C$ and a surface giving the 1/7th order pattern, ammonia was introduced to about 10^{-6} torr. Initially the 1/8th order pattern formed but after about ten minutes the doublet pattern appeared as well. After nearly fifteen minutes the 1/8th order pattern had faded completely and extra inner doublets had appeared, as illustrated in Fig. 3. The unit mesh derived from this pattern is $\sqrt{3}\times$ larger and rotated by $30°$ from that suggested by the pattern in Fig. 1(d). The definition of the pattern varied but was always relatively poor. Prolonged treatment resulted in a marked deterioration of the pattern. Thus after on hour at $1120°C$ with

$P_{NH_3} = 10^{-6}$ torr the LEED pattern had deteriorated to the state shown in Fig. 4 where even the main doublets cannot be resolved.

2. Nitrogen

When ammonia was replaced by nitrogen the reaction with silicon was much less striking. At 10^{-6} torr no new patterns formed at any temperature and only slight deterioration of the Si(111)-7 pattern was detected. On increasing the pressure to 10^{-3} torr and heating to 900°C the 1/7th order pattern deteriorated considerably within a few minutes but no new pattern developed. However, after heating at 1130°C for half an hour with the nitrogen pressure at 40 torr, a well developed 1/8th order pattern was observed on pumping out. Under slightly different conditions - heating at 1150°C for fifteen minutes, again at several torr, a new doublet pattern, illustrated in Fig. 5 was formed. These reactions were not extensively investigated in view of the experimental conditions involved.

RHEED and Microscopy Results

It was apparent from the quality of the LEED patterns that at many stages in the reaction the growing layers were higly irregular. To provide greater insight into the nature of these layers the samples were also examined by reflection high energy electron diffraction (RHEED) and by microscopy. When viewed under both optical and electron microscopes the heterogeneous nature of much of the nitride was clearly visible. This is typified by the electron micrograph shown in Fig. 6. Hardly any of the nuclei appear to have extended surfaces parallel to the substrate, making it most unlikely that the LEED patterns came from them. It was also found that after the heat treatment required to obtain 1/7th order pattern the sample surface was often no longer smooth at 800× magnification - a more typical surface condition being illustrated in Fig. 7. Though such surfaces give good Si(111)-7 patterns the visible imperfections may well interfere with epitaxial growth.

The RHEED analysis showed that the growths were $\alpha - Si_3N_4{}^5$. As illustrated in Fig. 8 some of the nitride is randomly oriented but much of it is well ordered. From the spot pattern it was determined that the $10\bar{1}0$ face of $\alpha - Si_3N_4$ lay parallel to the substrate. As the unit mesh in this plane is rectangular one would except at least three and possibly six equivalent domains of the nitride. The 60° angle between two close packed rows in the silicon (111) surface was examined closely, photographs being taken at regular 6° intervals. This analysis indicated that there were in fact six domains present. The diagonal of the nitride unit mesh lay 9° from the close packed row in silicon and the nitride twinned about the close packed row. Allowing that spots may be visible from planes up to 3° away from the Bragg angle, due to the observed extension of the spots, the model described above predicts that sixteen nets should be visible on the set of photographs. Fifteen were observed, the only missing net corresponding to a plane of low atom density.

Discussion

At the outset it appeared that epitaxial growth of silicon nitride on the (111) face of silicon might well be possible. The unit mesh in the substrate surface plane is hexagonal and has a side of 3.84Å (or seven times this in the rearranged structure). In both α- and $\beta - Si_3N_4$ the 0001 face also has a hexagonal unit mesh; the mesh sides are 7.75Å and 7.61Å respectively - almost exactly twice the substrate value. Within the nitride unit mesh the silicon and nitrogen atoms do not show any particular matching with the substrate. It is frequently observed however[6] that the near coincidence of "keying" positions, corresponding in this case to the corners of the nitride unit mesh and every second atom in a Si(111) plane, is a sufficient condition for epitaxial growth. In practice it has been found that neither the LEED nor the RHEED patterns correspond to this matching of hexagonal meshes.

It is probable that the LEED and RHEED patterns are due to different types of nitride growth. Whereas the RHEED patterns correspond to planes in bulk $\alpha - Si_3N_4$ the LEED patterns do not. The RHEED signals is also likely to come predominantly from the nuclei observed in the microscope while, as remarked earlier, these nuclei are not a likely source of the LEED signal. The possibility that the hexagonal symmetry observed in the LEED doublet pattern is due to a supersition of patterns of two fold symmetry from the six domains of $10\bar{1}0$ orientation indicated by RHEED has been considered but no satisfactory matching of the patterns has been achieved. It should be mentioned however that occasionally, after ammonia treatment, more complex LEED patterns have been observed. This aspect of the work is as yet unsystematized but some of the patterns resemble that shown in Fig. 5, obtained by reaction in nitrogen. It is possible that these patterns will be more successfully interpreted in terms of a multi-domain model.

In summary, the use of the various techniques to examine the nitride suggests the following model. At about 900°C the ammonia reacts rapidly with the clean silicon surface and forms an ordered layer giving rise to the Si(111)-8-N LEED pattern. With further reaction under these conditions the LEED pattern deteriorates, nuclei clearly visible at 800× magnification appear and partially ordered growths of $\alpha - Si_3N_4$ can be identified by RHEED. At higher temperatures e.g. 1120°C, the sequence is similar except that the 1/8th order pattern first observed in LEED is succeeded by a doublet pattern which in turn deteriorates with further reaction. The evidence suggests that the essentially flat areas are the source of the LEED patterns and that these are overgrown as the nuclei multiply and grow. Probably, as discussed below, the two LEED patterns are due to relatively thin surface layers possibly in some cases only a monolayer or so thick. Certainly the patterns correspond to a nitride which is different in structure and possibly in composition from both the bulk nitrides.

1. Interpretation of the LEED patterns

The simplest interpretation of the Si(111)-8-N pattern is that it represents a surface layer with a large unit mesh, 8x the size of that in a bulk Si(111) plane. Even accepting this as a basis, one can say little about the structure of such a layer, for instance the coverage could be anything from 1/64 to 63/64 of a monolayer, assuming direct nitrogen involvement. One must also allow the possibility that a much smaller concentration of nitrogen could induce a rearrangement of the silicon atoms, analogous to the reported effect of nickel on silicon.[3] However the observations that the doublet pattern, with its longer repeat distance and hence closer atom spacing, can be formed while heating the 1/8th order structure at 10^{-9} torr, makes a high concentration of nitrogen in the surface layer more likely. On the other hand after an exposure of 25 Langmuirs one could conceivably have a similar number of nitride layers on the surface but the continuous development of the pattern during this period argues against the formation of several complete layers. Thus it is probable that during the development of the pattern, a single monolayer or at most a single structure a few layers thick is building up.

A slightly different model results if the 1/8th order pattern is interpreted in terms of multiple scattering. This eliminates the need for a large unit mesh in the surface layer. A smaller unit mesh implies a correspondingly higher minimum density of nitrogen atoms in the surface. If the proposed multiple scattering involves the substrate layer - as in the model examined in the next section - the coverage should again be of the order of a monolayer.

The interpretation of the doublet pattern is more straightforward. The doublets indicate the presence of two domains each with a hexagonal unit mesh of the side 5.04Å. The two unit meshes are rotated by 27° and 33° respectively relative to that of the substrate. The marked difference in maximum intensities of the weak spots at positions typified by 1, 0 and strong spots at those

typified by 1, 1 suggests the unit mesh is not primitive. Just this kind of
intensity variation would be given by a graphite type layer of hexagons. Such
a layer is illustrated in Fig. 9(b), with the unit mesh at the mean position-
rotated 30° with respect to the silicon (111) layer shown in Fig. 9(a). The
hexagon side is 2.91Å very close to the closest approach of nitrogen atoms
(2.82Å) in the bulk nitrides. This distance corresponds to two nitrogen atoms
joined to the same silicon atom. The way in which such a layer could overlay
the silicon to give two equivalent domains at approximately the correct orientation is shown in Figs. 9(c) and 9(d). Lines of near coincidence, along which the
misfit between keying positions is only 3.4%, occur as shown, at quite closely
spaced intervals. It will be appreciated that, in common with most cases where
only "keying" positions coincide, the model leaves many questions about the
bonding at the interface unanswered.

2. Multiple diffraction

In the doublet pattern shown in Fig. 1(d) several extra spots, ignored in
the analysis described in the previous section, are clearly visible. These features are readily explained if the simple kinematic theory is extended to include
double scattering.[7] In the Fig. 10 the extra spots due to double scattering
of the first order beams from both the Si(111) substrate and the overlayer giving the doublet pattern are shown. This limited amount of double scattering is
sufficient to explain all the extra features; where triplets rather than doublets
were observed, the third spot is in the position of a fairly prominent feature
in the 1/8th order pattern which is also still visible. It is not clear however
why there are not more spots from other allowed combinations of beams. One
further point to note is that although the doublet pattern does not form
immediately on exposure of the silicon to ammonia the double scattering model
implies the existence of the corresponding structure at only about monolayer
thickness. Equally, the absence of these satellite spots after further treat-

ment, e.g. Fig. 3 suggests the structure is not restricted to a monolayer.

The Si(111)-8-N pattern was examined in a similar manner. The two meshes, between which double and triple scattering were considered, were those which gave rise to the hexagonal arrays typified by the 8/8,0 and 11/8, 0 spots. These sets of spots were not only prominent but, more particularly they were the first features observed as the pattern developed. Most of the LEED observations were conveniently made at 50-60 volts and initially up to triple scattering was considered for all beams possible at 60 volts. This time the approach was unsuccessful. Though many of the bright spots e.g. 3/8, 0; 5/8,0 and 6/8, 0 could be correlated with doubly scattered beams many weaker spots e.g. 10/8 0 cannot be accounted for on this model. Even more striking was the persistence of the bright spots mentioned to very low voltages. Thus at 12 volts neither the 5/8, 0 nor the 6/8, 0 spots are predicted but both are in fact quite intense. The failure of this simple model suggests that there is either a genuine large superlattice in the surface layer or, if multiple scattering between simpler meshes is involved a model of the type proposed by Bauer[8] in which the two layers scatter simultaneously, as opposed to consecutively, is required.

Conclusion

One of the most important aspects of this work has been the study of heterogeneity in the growing layers. There has been direct LEED evidence for this - in Fig. 1(d) domains of different structure are viewed simultaneously, and on many occasions it has been visible under the microscope. At the same time many of the LEED patterns are of high quality indicating that the LEED technique is not particularly sensitive to some types of surface imperfection. The implications of this type of evidence for experiments in which only a LEED investigation are carried out are obvious. Even when several techniques are employed the correlation between any signal which is due to the whole surface and the LEED pattern, due only to the ordered areas, may be quite poor. In the present work the LEED evidence does not provide sufficient information about the physical perfection of the silicon surface to allow one to decide what causes the heterogeneity in the nitride. Thus it remains possible that the nucleation of α - Si_3N_4 occurs at defects in the substrate surface. However this note of caution should not be allowed to obscure the contribution that LEED makes to the understanding of the nitridation reaction. The fact that the first nitride layers formed on the substrate have different structures, as determined by LEED, from either bulk nitride suggests that even on an Si(111) face free from defects the subsequent nitride growth might be heterogeneous.

Acknowledgments

The author is most grateful to Mrs. M.A.G. Halliwell for carrying out the RHEED analysis. He would also like to thank Dr. M.M. Faktor for many stimulating discussions and Mr. T.H. Lemon for some experimental assistance. Acknowledgement is also made to the Senior Director (Development) of the British Post Office for permission to submit this paper.

References

1. A. J. Van Bommel and F. Meyer, Surface Science, 8, 381 (1967).

2. R. E. Schlier and H. E. Farnsworth, J. Chem. Phys., 30, 917, (1959).

3. A. J. Van Bommel and F. Meyer, Surface Science, 8, 467, (1967).

4. F. G. Allen, J. Appl. Phys., 28, 1510, (1957).

5. D. Hardie and K. H. Jack, Nature, 180, 332, (1957).

6. T. J. La Chapelle, A. Miller and F. L. Morritz, Progress in Solid State Chemistry, 3, 1, (1967).

7. N. J. Taylor, Surface Science, 4, 161, (1966).

8. E. Bauer, Surface Science, 7, 351, (1967).

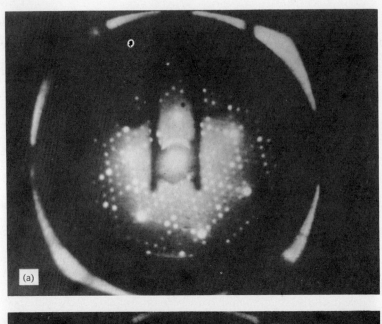

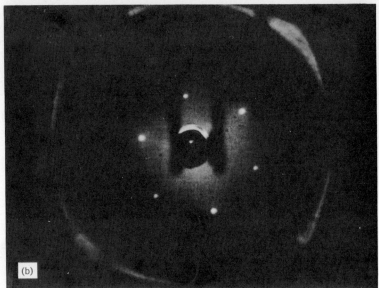

Fig. 1. Survey of preliminary results. a) The Si(111)-7 pattern from the
clean substrate surface, 50 volts. b) The same surface after an exposure to ammonia of 10^2 Langmuirs at 500°C, 50 volts. c) The
Si(111)-8-N pattern formed from the pattern in (a) after an exposure
to ammonia of 10^2 Langmuirs at 800°C, 56 volts. d) The doublet pattern formed in this case by heating the sample shown in (c) to 1050°C
in a vacuum of 10^{-9} torr for a few minutes, 57.5 volts. Note the
weaker 'satellite' spots near the main doublets and the central spot;
also the remnants of the Si(111)-8-N pattern.

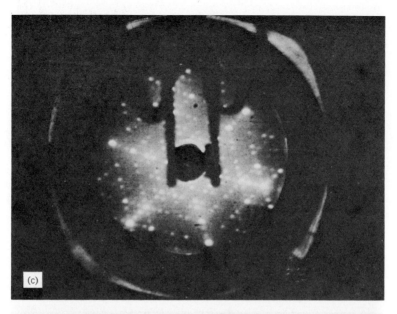

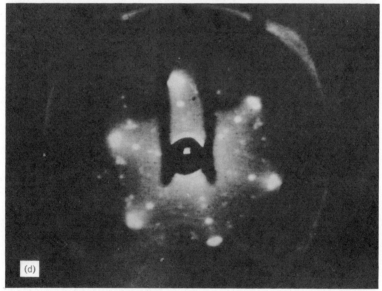

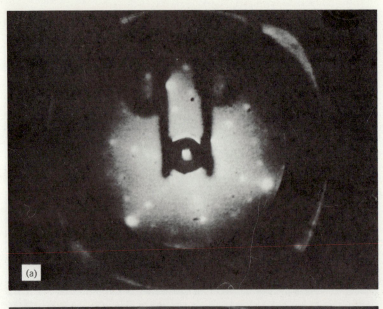

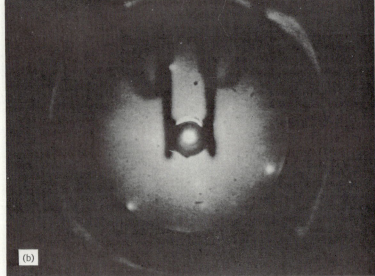

Fig. 2. a) The effect of ammonia at room temperature on the Si(111) 8-N pattern, 57 volts. b) The effect of the electron beam on the Si(111)
8-N pattern, 57 volts. b) The effect of the electron beam on the
Si(111)-8-N pattern, 53.5 volts.

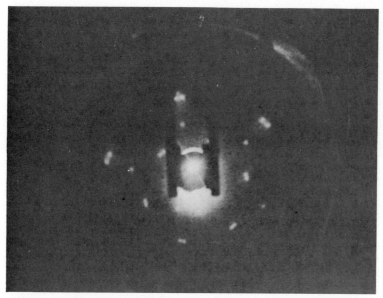

Fig. 3. The doublet pattern formed on heating to 1120°C in ammonia at 10^{-6} torr. The weaker inner doublets, closer to the centre than the silicon 1st order spots also present, are clearly visible, 37.5 volts.

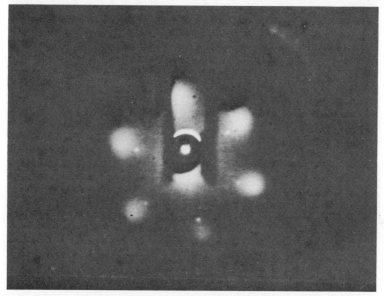

Fig. 4. The poor quality doublet pattern remaining after heating for 1 hour at 1120°C in ammonia at 10^{-6} torr, 60 volts.

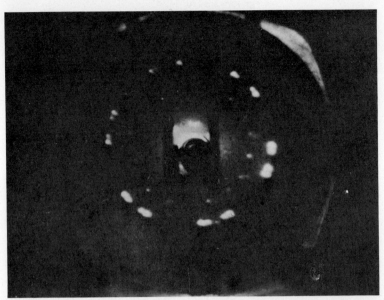

Fig. 5. The doublet pattern obtained by heating for fifteen minutes in nitrogen at several torr pressure and 1150°C, 51.5 volts.

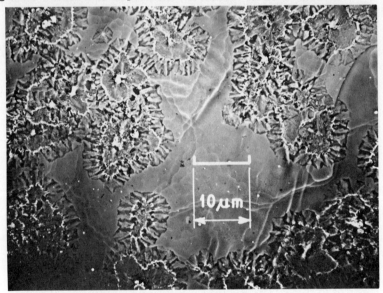

Fig. 6. Electron micrograph of the nitride nuclei formed after prolonged heating in ammonia, 1,000x in original.

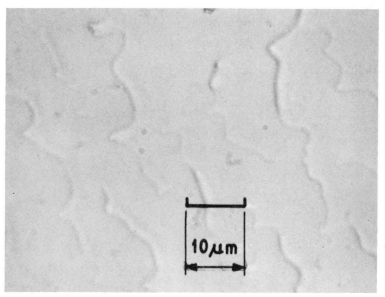

Fig. 7. Optical micrograph of a substrate surface which gave a good quality Si(111) -7 pattern, 800x in original.

Fig. 8. RHEED pattern from the nitride. The ring pattern corresponded to that for a - Si₃N₄. The spot pattern indicated that the 10$\overline{1}$0 plane of nitride lay parallel to the substrate surface.

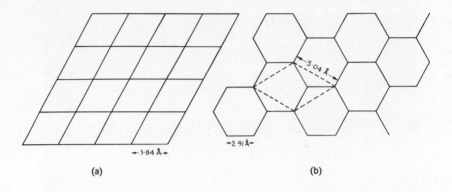

(a)　　　　　　　　　　　　　　(b)

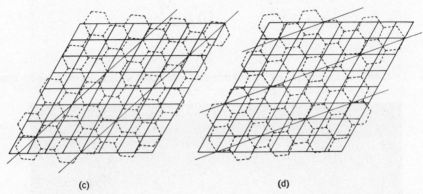

(c)　　　　　　　　　(d)

Fig. 9. Proposed surface structure responsible for the doublet pattern: a)
The unreconstructed Si(111) plane. b) The proposed hexagonal array
and its relation to the unit mesh. c) and d) Two equivalent ways of
matching the layers illustrated in (a) and (b). Lines joining the
'keying' positions of near coincidence in the two lattices are also
shown.

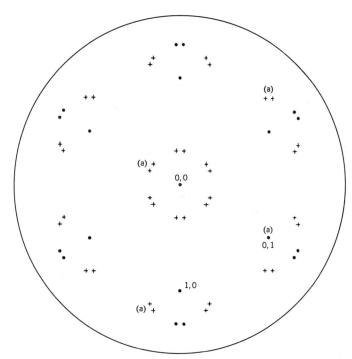

Fig. 10. Illustration of extra spots due to double diffraction of the first order beams from both the substrate surface and the doublet structure. Spots due to single diffraction are represented • ; those due to double diffraction x. Doubly diffracted spots designated (a) are derived from the singly scattered beam also designated (a).

THE REACTION OF OXYGEN WITH GERMANIUM AT ELEVATED TEMPERATURES

R. F. Lever
IBM Watson Research Center
Yorktown Heights, New York

INTRODUCTION

The fact that oxygen chemisorbs irreversibly on germanium,[1] desorbing at elevated temperatures as GeO vapor, greatly simplifies the study of the germanium-oxygen system. The overall reaction

$$2Ge(s) + O_2(g) = 2GeO(g)$$

can readily be observed either by weight loss[2,3] or by mass spectrometry.[4] This paper describes the results obtained by exposing an electrically heated single crystal slab of germanium to a low pressure oxygen ambient, and observing the steady state product, GeO vapor, by means of a mass spectrometer.

EXPERIMENTAL

Oriented slabs of single crystal germanium measuring $0.5 \times 0.25 \times 0.03$ inches were fused to .003" tantalum leads and mounted close to the ion source of a General Electric monopole mass spectrometer, as shown in Fig. 1. After lead attachment, and prior to mounting in the vacuum system, the crystal with leads attached was etched in a mixture of 1 part HF with 5 parts HNO_3. Following vacuum bake-out the crystals were heated near to the melting point for about 30 hours in an attempt to clean the germanium surface.[5] The crystal was heated by passing from 1 20 amps alternating current from a current-limited source. Such a heating arrangement can give rise to various temperature profiles, depending on the quality of electrical and thermal contact between the crystal and the tantalum leads. Initially, the ends were usually cooler

than the center, while after extended heating a relatively uniform ($\pm$ 5°C)
temperature was often observed. After very extended heating, hot spots sometimes developed at the contacts and, in any case, a tendency developed for the
ends of the crystal to be hotter than the center at higher temperatures. The
effects of temperature non-uniformity in the specimen were reduced somewhat
by mounting the crystal 0.125" behind a cold stainless steel annulus of 5/16"
internal diameter, so that the ion source did not receive any line-of-sight
molecules from the ends of the germanium specimen. The annulus was cooled by
blowing room temperature nitrogen through it, since water or liquid nitrogen
cooling was not found to be necessary.

The temperature of the crystal was observed by means of an infra red
spot photometer (Ircon 300) which employs a lead sulphide detector operating
in the wavelength range 2.0 to 2.6 microns. Intrinsic germanium at low
temperatures is somewhat transparent in this wavelength range, but for the
temperature region of interest in this paper (350-900°C) the calibration of
Jona and Wendt[6] is quite adequate. Allowing for both specimen non-uniformity
and calibration problems, the accuracy of temperature measurement is not
expected to be better than $\pm$ 20°C, although changes in temperature of the
order of 2°C are readily detected.

The ion source consists of a box of stainless steel or tantalum mesh
bombarded with electrons from thoria coated iridium filaments. It was usually
operated at 38 V and 1.5 ma electron current. This arrangement, though providing high sensitivity, has all the disadvantages of an ionization gauge in
that ions produced by electron bombardment desorption enter the mass filter
along with ions produced by the desired collision of electrons with gaseous
molecules. Towards the end of the experiments described in this paper, the
stainless steel box was replaced by a hemispherical tantalum mesh of 0.5"

diameter. Considerably higher sensitivity was obtained with this arrangement, and Ge^{++} and Ge^{+++} were observed in addition to Ge^{+} when the germanium was operated near to the melting point in ultra high vacuum conditions. Since the electron accelerating voltage was only 38 V, the multiply ionized species, which had intensities within a decade of the Ge^{+} intensity, must have arisen from the electron bombardment of freshly deposited germanium on the grid. In interpreting results from mass spectrometers having this type of ion source, the possibility of this type of interaction should be borne in mind.

Oxygen was admitted by means of a silver diffusion leak and a fine adjustment of the pressure acheived by means of an adjustable baffle which could reduce the pumping speed at the experimental chamber to about 2 liter sec^{-1}. The system was pumped by a 25 liter sec^{-1} ion pump supplemented by a titanium sublimation pump. Oxygen pressures up to 10^{-4} torr were maintained, the main impurities produced by the system being CO and CO_2. The amount of these depended on how often the sublimation pump was operated. No effect on the germanium-oxygen reaction could be observed on allowing the CO and CO_2 levels to vary between 10^{-4} and 3×10^{-2} of the oxygen level. The oxygen was monitored by means of a conventional Bayard-Alpert ionization gauge and by the 32 peak on the mass spectrometer. The GeO was monitored by observing the peak height of mass 90 corresponding to the most abundant isotope of germanium.

Two types of observation run were made. In the first, a constant oxygen pressure was maintained and the effect of temperature on the GeO peak was observed. In the second, a fixed temperature was established for each reading and the effect of varying the oxygen pressure was noted.

RESULTS

After the initial heat treatment, the magnitude of the germanium mass 74 peak was observed as a function of temperature and an activation energy for evaporation derived from the slope of the peak height versus T^{-1} plot. Values in the range 85-92 kcal were obtained in agreement with the published value for the equilibrium heat of vaporization of germanium.[7]

Results of constant pressure runs are typically as shown in Figs. 2 and 3. In the temperature range 350-500°C the GeO peak increased rapidly with temperature corresponding to an activation energy of 55-60 kcal mole^{-1}. No significant dependence on crystal orientation was noted, the difference in Fig. 2 between {110} and {100} being attributed to a change in the sensitivity of the electron multiplier between runs. In this temperature range the behavior of the system is dominated by the desorption of GeO, which is rate limiting, and is independent of oxygen pressure. At a temperature which depends on the oxygen pressure the GeO evolution becomes independent of temperature and directly proportional to the pressure. In the case of crystals of {100} orientation (two specimens) this behavior persists up to the melting point of germanium while for {110} crystals (five specimens) there was a steady drop from 30 to 40% in the temperature range 650-900°C. In the case of crystals of {111} orientation (six specimens) there is a sharp drop at about 750°C followed by a further steady decline as has already been reported.[4]

A series of constant temperature runs is shown in Fig. 4. The first order dependence of reaction rate on oxygen pressure is well displayed in the 550°C isotherm. The most noticeable characteristic of the other isotherms is their rapid approach to saturation. A complete family of isotherms as in Fig. 4 was taken only on two {111} samples and one {100} sample which showed similar behavior. All samples showed clear evidence of saturation, however, in that the steeply descending portion of the log (GeO peak) versus 1000/T plot did not vary with oxygen pressure. The more detailed isotherms illustrated in Fig. 4 were taken in an attempt to understand the behavior of partially covered surfaces (see discussion).

DISCUSSION

The range of pressure and temperature covered in these experiments may be divided into three regions. In the first, the high temperature-low pressure region, the GeO evolution is directly proportional to the oxygen pressure, and the reaction is limited by the rate at which oxygen sticks to a clean germanium surface. In the second, the low temperature-high pressure region, the rate of GeO evolution does not depend on the oxygen pressure, and the germanium surface may be regarded as fully covered with a monolayer of adsorbed oxygen or GeO. The third region lies between the first two and comprises the regime of partial coverage in which the rate of GeO evolution is limited by both the sticking of oxygen to clean germanium and the GeO desorption, operating in series.

The first region corresponds to that covered by the weight loss measurements of Anderson and Boudart[2] and Madix and Boudart.[3] Because of the greater detection sensitivity afforded by the mass spectrometric method, a wider range of oxygen pressures has been investigated, and the effect of variation of specimen temperature more readily observed. The mass spectrometric method, however, is far from being absolute and would be very difficult to make so, particularly using the ion source design illustrated in Fig. 1. A crude estimate of sticking coefficient can, however, be made if one assumes that the mass spectrometer is approximately equally sensitive to the mass 32 and 90 peaks. Allowance has to be made for the fact that, while the oxygen is incident on the ion source from all directions, the GeO is only incident from that solid angle subtended by the crystal at the ion source. This solid angle is approximately 0.1 steradians, giving an attenuation factor $\Omega/4\pi$ equal to 0.008. Both GeO and O_2 will have similar molecular velocities, the higher mass of the GeO being offset by its higher temperature, so that the only other major correction is due to the isotope effect, the 74 peak of germanium having an abundance of 36.6%. This results in a net attenuation factor of .003. That is, if one molecule of GeO left the germanium crystal for each oxygen molecule which impinged on it, the observed ratio of 90:32 peak heights would be approximately 3×10^{-3}. The actually observed ratios are in the range 5×10^{-5} to 10^{-4} corresponding to actual sticking coefficients between .016 and .033. Since the assumption of equal sensitivity is not likely to be accurate, we can only say that the observed peak heights give a sticking coefficient in essential agreement with the absolute determinations by weight loss measurement.[2,3] The fact that first order dependence on oxygen pressure can be observed to hold accurately over 3 decades strongly supports the supposition that the reaction is between gaseous oxygen and a clean germanium surface, although the possibility that some fixed part of the surface remains inert cannot be ruled out. The fact that 90:32 ratios greater than 10^{-4} were never observed strongly suggests, however, that at least the 10^{-4} readings correspond to a clean surface. The marked drop in sticking coefficient with temperature for the {111} orientation is of great interest, and has already been discussed.[4]

The second region corresponds to the LEED observations of Lander and Morrison,[12] who observed the pressure of oxygen required to extinguish the clean germanium LEED pattern as a function of crystal temperature. Their slope of 55 kcal mole^{-1} is in good agreement with that observed here as are their pressure values, showing that the LEED pattern disappears as the GeO evolution drops below its high temperature value. The activation energy may be compared with the thermodynamic heat of desorption as follows. Brennan, Hayward and Trapnell[8] measured the enthalpy of chemisorption of oxygen on evaporated germanium films in a Beeck type calorimeter and found a value of 132 kcal/mole which was relatively constant, at least up to a coverage of 0.4. Bues and vonWartenburg[9] and Jolly and Latimer[10] have examined the reaction

$$Ge(s) + GeO_2(s) = 2GeO(g)$$

and an enthalpy of reaction of 108 kcal mole^{-1} obtained.[10] This may be combined with the heat of formation of GeO_2 of 129 kcal mole^{-1},[8] to give an enthalpy of formation of GeO vapor of 10.5 kcal/mole. If one subtracts the reaction

$$Ge + 1/2 \ O_2 = GeO(g)$$

from the reaction observed by Brennan, Hayward, and Trapnell

$$Ge(s) + 1/2 \ O_2 = GeO \ (adsorbed)$$

one obtains an enthalpy of 55.5 kcal mole^{-1} for the reaction

$$GeO \ (adsorbed) = GeO(g)$$

This agrees with the activation energy for desorption of 55-60 kcal mole^{-1} observed in this paper. Had the activation energy been much greater than 55 kcal mole^{-1} this would have suggested a rather surprising extra activation step, while had the energy been much less it would have suggested that the desorption did not leave behind a clean surface.

The third region is illustrated in Fig. 4. The most noticeable characteristic is a fairly sharp approach to saturation as the pressure is raised.

It is instructive to compare this with the approach to saturation expected on a simple Langmuir model. If we assume a simple Langmuir adsorption where the adsorption sites are considered to be identical and non-interacting then the desorption rate of GeO may be written

$$R(GeO) = k_2 \; \theta \qquad\qquad (1)$$

where k_2 is a constant, strongly dependent on temperature, and θ the surface coverage. We define R_2 as the desorption rate characteristic of full coverage ($\theta = 1$) so that at high pressures

$$R(GeO) = R_2 = k_2 \qquad\qquad (2)$$

The rate at which oxygen sticks to the surface may be written

$$R(oxygen) = k_1 p \; f \; (\theta) \qquad\qquad (3)$$

where p is the oxygen pressure, k_1 is a constant which, in these experiments, seems to be essentially independent of temperature, and $f(\theta)$ some function of the coverage θ, being 1 when $\theta = 0$ and 0 when $\theta = 1$, it being assumed that oxygen does not stick to portions of the surface which are covered. We define R_1 as the sticking rate on a clean surface at a given temperature and pressure and may write, for $f(\theta) = 1$

$$R(oxygen) = R_1 = k_1 p \qquad\qquad (4)$$

In the steady state, the rate of sticking of oxygen in gram atoms $cm^{-2} sec^{-1}$ equals the rate of desorption of GeO in moles $cm^{-2} sec^{-1}$ and may be put equal to R.   Combining equations 1-4 we then have

$$R = R_1 \; f(\theta) = R_2 \theta \qquad\qquad (5)$$

In order to display the isotherm shape on a log-log plot as in Fig. 4 it is convenient to take the intersection of R_1 (i.e., the line $\log R_1 = \log k_1 + \log p$) and R_2 as the origin so that the reaction rate which we call J and the reduced pressure, which we call P may be written

$$J = R/R_2 \qquad P = R_1/R_2$$

Hence, substituting from (5) we have

$$J = R/R_2 = \theta$$

$$P = R_1/R_2 = k_1 p/k_2 = \theta/f\,(\theta)$$

The exact isotherm shape will then depend on $f\,(\theta)$. In Fig. 5 two cases are shown. In the first, it is assumed that the covered sites are contiguous so that, in general, an incoming molecule will be presented, as a whole, with either a clean or a covered surface the probability of being clean being simply $(1-\theta)$. The case $f\,(\theta) = (1-\theta)$ is shown as curve 1. In the second case, it is assumed that an oxygen molecule interrogates two sticking sites whose probability of being clean is unrelated, and will stick only if both are clean. In this case $f\,(\theta) = (1-\theta)^2$ and this is plotted as curve 2. Various other forms of $f\,(\theta)$ may be postulated[11] but would generally give J versus P curves similar to those shown.

Superimposed on Fig. 5 are the isotherms from Fig. 4. Curve 3 represents the 420°C isotherm, while curve 4 represents the 440°C and 465°C isotherms, which have similar shapes. It is clear that the experimental rates approach saturation more rapidly than would be expected from a simple Langmuir picture.

One may speculate on possible reasons for this. Three possible models are discussed below.

The first and simplest model is to preserve the Langmuir assumption of identical non-interacting desorption sites but to allow $f(\theta)$ to remain near unity until the surface is almost covered. This amounts to assuming that a molecule hitting a partly covered portion of the surface has a probability of chemisorption comparable to that of a molecule hitting a clean surface. It is possible to attribute the low sticking coefficient of oxygen molecules on clean germanium to a requirement that the oxygen molecule be favorably oriented at the moment of impact[2] and to suppose that molecules hitting oxygen covered portions of the surface can be held weakly enough to permit desorption as oxygen molecules, while having a diffusion length on the surface sufficient to encounter such clean sites as are present. Such a picture seems implausible but would require an examination of the energy distribution of the scattered unreacted oxygen molecules for proof. McKinley[13] has examined the temperature of unreacted chlorine in the nickel-chlorine surface reaction using the stagnation detector method described by Smith and Fite[14] and found that unreacted chlorine leaving the hot nickel surface was at the ambient temperature and not that of the surface. If the unreacted oxygen leaving the surface in this experiment were found to be at the surface temperature, then a model allowing surface migration before chemisorption would seem plausible.

The second model assumes that the low sticking coefficient on a clean surface may be due to a sparse distribution of sticking sites rather than a need for the molecule to be favorably oriented at the moment of impact. It is then further supposed that the chemisorbed species is capable of surface migration, which is quite plausible. If the assumption is then made that the heat of adsorption on the sticking sites is a few kilocalories smaller than on the majority of adsorption sites, then a simple explanation is provided for why the sticking sites could remain relatively clean even when the majority of adsorption sites is covered. This model is discussed in the first part of the appendix.

A third model invokes a parallel mechanism whereby oxygen can be chemi-sorbed onto a covered surface, thereby partially offsetting the decreasing availability of clean surface to stick to as the temperature is raised. Such a parallel mechanism could be visualized as the occurrence of a place inter-change, with some GeO bonded to the surface via the oxygen atom. An incident oxygen molecule could then react with the surface germanium atom desorbing GeO gas and leaving the original oxygen atom still chemisorbed to the surface. This model is discussed in the second part of the appendix.

Both of the models described in the appendix can be made to yield an isotherm having a sharper approach to saturation than a Langmuir isotherm, as shown in Fig. 6. The value of such speculations is dubious, however, in the absence of clear cut methods of disproving them. It is also worth remembering that the surfaces, though single crystal, are not microscopically homogeneous but faceted, as shown in Fig. 7.{111} facets can be observed to predominate on the crystal of {111} orientation while {110} facets appear on the {110} crystal. The faceting behavior does not appear, however, to be very reproducible. Really meaningful studies on steady state etching reactions require surfaces that remain essentially unchanged during the course of the reaction and display no obvious features under microscopic examination.

CONCLUSIONS

The mass spectrometric measurements described in this paper confirm previous observations[2,3] on the interaction of oxygen with clean germanium by weight loss measurement, and LEED observation[12] of the adsorption of GeO on the surface.   The sticking coefficient of oxygen on a clean surface does not show any significant temperature dependence below 720°C, suggesting that there is no activation step involved in the sticking process.   Above 720°C the sticking coefficient on {111} surfaces displays a marked drop which is not explained.   The reaction isotherms show a more rapid approach to saturation than might be expected from a simple Langmuir model.

ACKNOWLEDGEMENTS

I am greatly indebted to H. R. Wendt for technical assistance and to F. P. Jona and D. Jepsen for stimulating discussions.

APPENDIX

In both models, the reaction rate of oxygen with a set of sticking sites will be considered proportional to the number of such sites rather than the square or some other function of that number.   Similar results could be obtained using a more complex model for the oxygen-surface interaction, but this would hardly be justified at this point.

Small Energy Difference Model

We postulate two types of sites A and B, respectively.   These will have different sticking probabilities for oxygen when clean, and different desorption energies for GeO when covered.   We assume that the sticking probability for oxygen on clean B sites is negligible, that the adsorbent distribution between sites is in equilibrium and that the B sites are n times more numerous than the A sites.   Considering the equilibrium between sites, we have

$$\theta_B(1 - \theta_A) = k\theta_A(1 - \theta_B) \qquad\qquad A.1$$

where θ_A and θ_B are the fractional coverages and k the equilibrium constant. If the energy difference between sites is E kilocalories/mole and possible entropy differences are ignored, we may write

$$k = \exp\ (E/RT) \qquad\qquad A.2$$

Since adsorbent on A sites is less tightly bound than on B sites by E kcal mole^{-1}, the probability of desorption from an A site is higher by the factor k. However, since the B sites are n times more numerous, they will be more effective in desorption by a factor nk^{-1}. We may therefore write, using the notation already developed,

$$R\ =\ R_2(\theta_A + nk^{-1}\theta_B)/(1 + nk^{-1}) \qquad\qquad A.3$$

for desorption and

$$R\ =\ R_1(1-\theta_A) \qquad\qquad A.4$$

for adsorption

Hence, using the same notation as in the preceding discussion, we have

$$J\ =\ (\theta_A + nk^{-1}\theta_B)/(1 + nk^{-1}) \qquad\qquad A.5$$

$$P\ =\ J/(1-\theta_A) \qquad\qquad A.6$$

Isotherms may readily be generated by taking values of θ_B as a parameter generating values of θ_A from A.1 viz:

$$\theta_A^{-1}\ =\ 1 + k\ (\theta_B^{-1} - 1)$$

and hence values of J and P obtained. In order to obtain isotherms of the desired shape, one seeks, in effect, to maximize J at P = 1. Isotherms arbitrarily close to R_1 at P < 1 and R_2 at P > 1 can be generated if one is able to make n large enough and then optimize k. In general, one wants k to be large enough to favor coverage of B sites at the expense of A, but not so great that, despite the smaller coverage of A sites, their lower desorption energy causes A sites to dominate the desorption behavior. The experimentally observed sticking coefficient does not allow one to make n greater than 25 since, when the surface is clean, one oxygen molecule in 25 does in fact stick. For n = 25, the optimum value of k is 10. The isotherm corresponding to these values is shown in Fig. 6, Curve 3. From A.2, the energy difference required between A and B sites to give k = 10 at T = 750°K is 3.5 kcal $mole^{-1}$ which is quite small compared with the enthalpy of desorption, and therefore not implausible. Allowing a finite sticking probability to clean B sites, would, of course, cause the isotherm shape to approach more closely to the Langmuir case.

Place Exchange Model

In this model, the decrease in adsorption rate caused by the occupation of adsorption sites can be partially offset by postulating that the oxygen and germanium change places forming a surface-O-Ge-gas configuration and that this latter configuration can be attacked by gaseous oxygen. At high oxygen pressures, this form of adsorbent will disappear to give a saturation desorption rate characteristic of the surface-Ge-O-gas form. The model is best explained by writing down a series of reactions. [Ge] denotes bulk germanium, of constant activity, Ge(s) a surface germanium atom, GeO(a) an adsorbed GeO with oxygen facing the gas and GeO(b) an adsorbed GeO with a germanium atom facing the gas. We then write

$$2Ge(s) + O_2 \rightarrow 2GeO(a) \qquad\qquad\qquad B.1$$

$$J_1 = k_1 p \, (1-\theta_A - \theta_B)$$

$$[Ge] + GeO(a) \rightarrow Ge(s) + GeO(g) \qquad\qquad\qquad B.2$$

$$J_2 = k_2 \theta_A$$

$$GeO(a) \rightarrow GeO(b) \qquad\qquad\qquad B.3$$

$$J_3 = k_3 \theta_A$$

$$[Ge] + GeO(b) \rightarrow Ge(s) + GeO(g) \qquad\qquad\qquad B.4$$

$$J_4 = k_4 \theta_B$$

$$2[Ge] = 2GeO(b) + O_2 \rightarrow 2GeO(a) + 2GeO(g) \qquad\qquad\qquad B.5$$

$$J_5 = k_5 p \, \theta_B$$

The J's and k's represent the various reaction rates and rate constants, and p the pressure of oxygen, θ_A denotes the fraction of sites covered with GeO(a), θ_B the fraction covered with GeO(b), so that $1 - \theta_A - \theta_B$ represents the fraction of sites which are uncovered. As stated previously, the oxygen will be considered as reacting as a whole so that the probability of reacting with a surface species is directly proportional to its concentration, not the square of its concentration. The possible reaction of GeO(b) back to GeO(a) has been neglected. From the above, we may write the conservation conditions for GeO(b) and Ge(s) viz:

$$J_3 = J_4 + 2J_5 \qquad\qquad\qquad \text{B.6}$$

$$2J_1 = J_2 + J_4 \qquad\qquad\qquad \text{B.7}$$

Proceeding as before, we may write the total reaction rate

$$R = 2J_1 + 2J_5 = J_2 + J_3$$

$$\text{i.e.} \quad R = 2p\,(k_1(1-\theta_A -\theta_B) + k_5\theta_B) = (k_2 + k_3)\,\theta_A \qquad \text{B.8}$$

At low pressures, when θ_A and $\theta_B = 0$, $R = R_1$ and we have

$$R_1 = 2p\,k_1 \qquad\qquad\qquad \text{B.9}$$

and hence

$$R/R_1 = 1 - \theta_A + \theta_B(k_{51}-1) \qquad\qquad \text{B.10}$$

where $k_{51} = k_5/k_1$ and R_1 has the previously ascribed meaning. At high pressures, provided $k_5 \neq 0$, $\theta_A \to 1$ and hence

$$R_2 = k_2 + k_3 \qquad\qquad\qquad \text{B.11}$$

hence

$$R/R_2 = \theta_A = J \qquad\qquad\qquad \text{B.12}$$

using the same definition of J as previously. If $k_5 = 0$ then B.12 becomes

$R/R_2 = \theta_A/\theta_A^*$ where θ_A^* is the value of θ_A at saturation, but this case is of little interest, giving an isotherm shape of the same shape as the Langmuir hypothesis.   From B.9 and B.11 we have

$$P = R_1/R_2 = 2pk_1/(k_2 + k_3) \qquad\qquad\text{B.13}$$

which is a needed relation between P, the reduced pressure and p, the oxygen pressure appearing in B.1 and B.5.   Dividing B.10 by B.12, we also obtain

$$\frac{R_2}{R_1} = \frac{1}{P} = \theta_A^{-1} - 1 + \theta_B \theta_A^{-1} (k_{51} - 1) \qquad\qquad\text{B.14}$$

θ_B/θ_A may be obtained by substituting into B.3, B.4, and B.5 into B.6 and substituting for p in terms of P from B.13.   In addition we substitute J for θ_A from B.12.   We then obtain

$$\frac{1}{J} = \frac{1}{P} + 1 + \frac{1 - k_{51}}{k_{43} + k_{51}(k_{23} + 1)P} \qquad\qquad\text{B.15}$$

where $k_{ij} = k_i/k_j$.   The three coefficients in the above equation have a simple physical interpretation. k_{51} is the ratio of the sticking coefficient on a type B surface to that on a clean surface while k_{43} and k_{23} simply represent the desorption rate constants of the B and A species relative to the rate constant for interchange of A into B.

We see immediately that if $k_5 = k_1$ then B.15 reduces to $J^{-1} = P^{-1} + 1$ which is the simple Langmuir form shown as Curve 1 on Figs. 5 and 6.

In order to obtain a J value greater than that given by the Langmuir picture we require $k_{51} > 1$ since this makes J^{-1} less than $1 + p^{-1}$.

As can be imagined, three adjustable parameters give considerable scope for manipulation. However, a curve similar to those experimentally obtained can be derived by making $k_{51} = 2$, $k_{23} = 0$ and $k_{43} = 0.5$. Physically this would mean that the inverted O-Ge surface is twice as likely to react with oxygen as a clean surface, that it spontaneously desorbs with half the rate constant for place interchange and that the desorption rate constant for the non-inverted GeO is much smaller than the rate constant for inversion. This curve is shown in Fig. 6, Curve 2.

REFERENCES

1. A. J. Rosenberg, P. H. Robinson and H. C. Gatos, J. Appl. Phys. 29, 771 (1958).

2. J. B. Anderson and M. Boudart, J. Catalysis 3, 216 (1964).

3. R. J. Madix and M. Boudart ibid. 7, 240 (1967).

4. R. F. Lever, Surface Science 9, 370 (1968).

5. F. Jona, Appl. Phys. Letters 6, 205 (1965).

6. F. Jona and H. R. Wendt, J. Appl. Phys. 37, 3637 (1966).

7. D. R. Stull and G. C. Sinke, Thermodynamic Properties of the Elements, American Chemical Society, Washington, D. C. 1956.

8. D. Brennan, D. O. Hayward and B. M. W. Trapnell, J. Phys. Chem. Solids 14, 117 (1960).

9. W. Bues and H. von Wartenburg, Z. Anorg. Allgem. Chem. 266, 281 (1951).

10. W. L. Jolly and W. M. Latimer, J. Am. Chem. Soc. 74, 5757 (1952).

11. D. O. Hayward and B. M. W. Trapnell, Chemisorption Butterworths, London 1964.

12. J. J. Lander and J. Morrison, J. Appl. Phys. 34, 1411 (1963).

13. J. D. McKinley, Jr., J. Phys. Chem. 66, 554 (1962).

14. J. N. Smith and W. L. Fite in "Rarefied Gas Dynamics", Proceedings of the Third International Symposium on Rarefied Gas Dynamics, Paris, 1962. Academic Press Inc., New York, 1963.

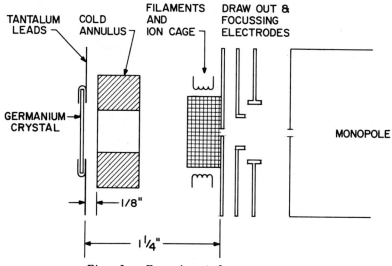

Fig. 1. Experimental arrangement.

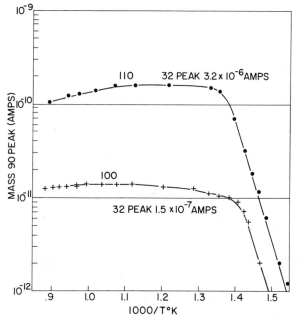

Fig. 2. Evolution of GeO versus 1000/T for two separate crystals: 1) a {110} crystal exposed to an oxygen ambient giving an ionization gauge reading of 8×10^{-6} torr; 2) a {100} crystal exposed to an oxygen ambient giving an ionization gauge reading of 1.1×10^{-6} torr.

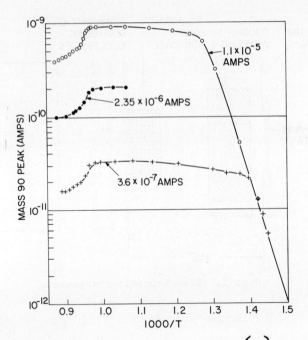

Fig. 3. Evolution of GeO versus 1000/T for a single {111} crystal illustrating the dependence of sticking coefficient on temperature. The curves are labelled with the height of the 32 peak, which was maintained constant. The 32 peak had a sensitivity of 0.4 amp torr^{-1} relative to an ionization gauge.

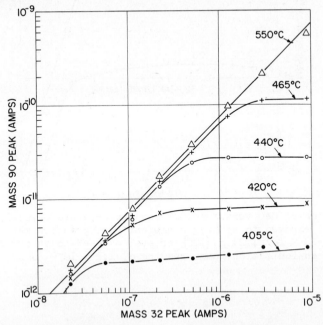

Fig. 4. Isotherms for a {111} crystal. Germanium oxide peak height as a function of oxygen peak height at constant temperature.

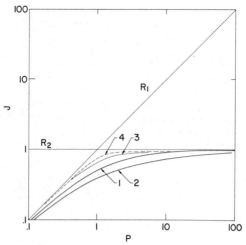

Fig. 5. Comparison of experimental isotherms with Langmuir model. Curves 1 and 2, possible Langmuir models. Curves 3 and 4, experimental curves showing the more rapid approach to saturation.

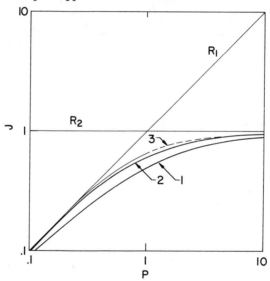

Fig. 6. Comparison of Langmuir curve (1) with two other theoretical curves using different models. (See Appendix).

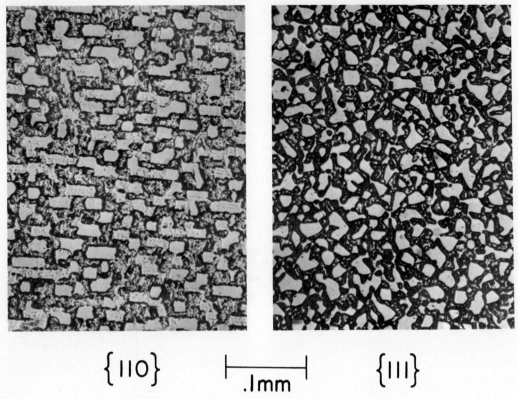

$\{110\}$ |——— .1mm ———| $\{111\}$

Fig. 7. Typical crystal surfaces after prolonged observation, showing the effect of oxygen etching.

CATALYTIC OXYGEN ATOM RECOMBINATION ON SOLID SURFACES

Joan Berkowitz

Arthur D. Little, Inc.
Cambridge, Massachusetts

I. Introduction

The heterogeneous recombination of simple atomic species at solid surfaces is in principle one of the most elementary of surface reactions. Like any heterogeneous catalytic process, atom recombination should be strongly dependent on the structure of the solid surface, although at present, there is no model that can be used to predict recombination rates quantitatively. The dependence of recombination kinetics on surface structure is in all probability the basic reason for the poor reproducibility of results within a given laboratory, and the even wider discrepancy of results between laboratories. Furthermore, when the surface under study is both reactive towards oxygen atoms, as well as catalytic towards atom recombination, the surface structure may change with time of exposure to a new surface with altered catalytic activity. The need for detailed surface characterization in parallel with measurements of recombination efficiency is pointed up in the preliminary results discussed below for quartz, Ti, Ta, and Mo, and in the comparison of results from a number of different laboratories on the "same" materials.

Oxygen atom recombination is highly exothermic and hence only occurs when a third body (gaseous molecule or solid surface) is present to absorb some of the energy released. During reentry of space vehicles at hypersonic velocities, high stagnation temperatures are generated which cause some of the oxygen in the air behind the shock wave to become dissociated. Depending on the conditions of temperature and pressure, recombination with release of thermal energy may occur in the gas phase and/or on the surface of the vehicle. If the time for an atom to diffuse across the boundary layer is long compared to the time for gas phase (homogeneous) recombination, equilibrium conditions will be established in the boundary layer. If, on the other hand, the diffusion time is short and there is essentially no opportunity for gas phase recombination, the boundary layer is said to be "frozen." In this case, a non-equilibrium concentration of atoms will prevail in the neighborhood of the vehicle surface, and heterogeneous recombination at the surface will contribute to the over-all heat transfer to the vehicle. Obviously, the maximum heat transfer from this source will occur if the surface is "fully cata-lytic"; i.e. if the rate of heterogeneous recombination is essentially equal to the rate of arrival of oxygen atoms. At the opposite extreme, heat transfer would be a minimum for a completely non-catalytic surface. The experiments described below are aimed at defining rates of recombi-nation from room temperature to 800°C for a number of metallic and non-metallic materials of practical interest. For those metallic materials in particular which strongly chemisorb oxygen atoms, the surface governing heterogeneous recombination will differ radically from that of a clean metal after a very short exposure to the oxidizing ambient. In later work, the nature of the surface will be defined more precisely.

II. Experimental Apparatus

A schematic diagram of the low pressure, fast flow apparatus used for the oxygen atom recombination studies is shown in Figure 1. Briefly, oxygen molecules in an argon stream are dissociated by means of a 2450 mc/sec generator with a microwave power output of 0-100 watts. Relative atom concentrations along the length of the flow tube are monitored photometrically by introducing a small amount of NO(g) into the flowing gas, upstream of the test section of the apparatus. The simultaneous presence of NO(g) and gaseous oxygen atoms results in a greenish after-glow whose intensity at any point is proportional to the oxygen atom concentration at that point.

As indicated in Figure 1, oxygen and argon are metered separately through sapphire ball-type flowmeters, and are introduced into the system via a common line. Argon flow rates can be varied between 200 and 12,000 cc (STP)/min, oxygen flow rates between 1-260, 10-1,900, and 200-12,000 cc (STP)/min by appropriate choice of flow line. Line pressure is measured with a Wallace-Tiernan gauge. For most of the experiments described below, the argon flow rate was set at 250 cc/min and the oxygen flow rate at 72 cc/min.

The oxygen-argon gas mixture is passed through a Raytheon 2450-Mc/sec powered Ophthos 125-w Evenson microwave discharge cavity where the oxygen is partially dissociated. A tubular sample of approximately the same diameter as the quartz housing (2 cm) and about 4 cm in length is inserted essentially as a liner about 50 cm downstream of the discharge. The static pressure up and downstream of the sample is measured with Meriam fluid ($\rho = 1.04$) differential manometers. The total pressure in most cases was close to 4 torr, and the pressure drop along the flow tube was undetectable. When relative atom concentration measurements are made, NO(g) is introduced through the upper jet, in minute well controlled quantities, by means of a series of reservoirs and a Nupro Model SS-2SA valve. The NO flow, which is proportional to the pressure on the upstream side of the valve, is monitored with a Pace P-7 transducer. In most experiments, light intensity is measured at two fixed photometer positions as shown, although in some cases the top photometer was moved along the flow tube for experimental determination of light intensity as a function of distance. The output from the two photometers is fed into a strip chart recorder.

To permit experiments at elevated temperatures, a portion of the flow tube is surrounded by a wire wound ceramic tube furnace. For ease in analysis of the data it is highly desirable to establish a long hot zone of constant temperature, with a sharp drop to room temperature at both ends. By adjustment of the furnace windings and installation of water cooled condensers at the ends of the flow tube external to the furnace, square temperature profiles were achieved up to 1200°C.

III. Analysis of Data

The greenish after-glow used to monitor oxygen atom concentrations in these experiments is due to the photochemical reaction:[1]

$$NO(g) + O(g) \rightarrow NO_2(g) + h\nu \qquad (1)$$

Any $NO_2(g)$ produced via reaction (1) immediately combines with residual oxygen atoms in an exceedingly rapid reaction that regenerates NO as follows:

$$NO_2(g) + O(g) \rightarrow NO(g) + O_2(g) \qquad\qquad (2)$$

The net result is that the $NO(g)$ concentration in the gas stream remains constant and that the light intensity diminishes along the flow tube in direct proportion to the oxygen atom concentration.

The decay in oxygen atom concentration along the flow tube is due not only to heterogeneous recombination at the walls, but also to homogeneous gas phase recombination in the bulk gas stream, and in some systems, to chemical reaction at the sample surface. The wall recombination is generally believed to be first order,[2] and at low oxygen atom concentrations, volume recombination is also pseudo-first order in oxygen atoms, proceeding through an ozone intermediate.[1] If chemical reaction is also first order, the fundamental rate equation governing the loss of oxygen atoms is:

$$-\bar{v}\,\frac{d(O)}{dt} = k_w\,(O) + k_R(O) + k_v(O_2)(C_{tot})(O) \qquad (3)$$

where $\bar{v}$ is linear flow velocity at position x along the flow tube, k_w is the first order rate constant for wall recombination, k_R the rate constant for chemical reaction, k_v the rate constant for volume recombination; (O) and (O_2) are respectively oxygen atom and molecule concentrations at position x, and C_{tot} is the total concentration of molecular oxygen and argon, both of which can serve as third bodies for homogeneous recombination. Equation (3) integrates to:

$$\bar{v}\,\ln\,\frac{I_{x_1}}{I_{x_2}} = (k_w + k_R + k_v O_2 C_{tot})(x_2 - x_1) \qquad\qquad (4)$$

where atom concentration has been replaced by light intensity, and the integration has been carried out between two positions x_1 and x_2 along the flow tube. Several measurements were made of light intensity as a function of distance for the unlined quartz flow tube at room temperature and served to verify the validity of the assumed first order kinetics.

The rate constant k_v is known from the literature[1] and the experiment is designed so that the correction for gas phase recombination is small. The extent of chemical reaction was assessed from the weight change of the sample during the time of the experiment. For all of the experiments reported below k_R was several orders of magnitude less than k_w, and no correction was necessary.

Absolute oxygen atom concentrations were measured, as the need arose, by NO_2 light titration. The flow of NO_2 is adjusted at constant oxygen atom flow until the light intensity measured by the photometer at a fixed position is at a maximum. At this point, by equations (3) and (4), the NO_2 flow rate is exactly half the oxygen atom flow rate. The NO_2 flow is then diverted for a known period of time through a weighed bottle of Indacarb, where it is quantitatively adsorbed. Thus, absolute atom concentrations can be determined.

IV. Results

From experimentally measured light intensities, flow rates, molecular oxygen concentration, and total pressure, combined with literature values of k_v, the rate constant k_w for surface recombination of oxygen atoms can be calculated via the sum of a number of equations of the form (4). It is more convenient, however, to report results in terms of a dimensionless parameter γ, the recombination coefficient, which is defined as the ratio of the number of atoms that leave a given surface as molecules to the total number of atoms that collide with the surface in the same period of time. The number of collisions per cm^2 per sec, N, is given by the kinetic theory expression:

$$N = 1/4 \ (0) \ \bar{c} \tag{5}$$

where:

$$\bar{c} = (8 \ RT/\Pi \ M)^{1/2} \tag{6}$$

In (6), R is the universal gas constant, T is the absolute temperature, and M is the atomic weight of oxygen. The loss of oxygen atoms from unit volume in one second, due to wall recombinations is $k_w(0)$. If diffusion is sufficiently rapid so that surface recombination uniformly lowers the volume concentration (0), the number of atoms that recombine per cm^2 per sec is $k_w(0) \ V/S$, where for a cylindrical section of flow tube length h, and radius r, $V = \Pi \ r^2 h$, $S = 2\Pi rh$, and $V/S = r/2$. Finally, the recombination coefficient γ can be calculated by means of the formula:

$$\gamma = \frac{k_w(0) \ V/S}{1/4(0) \ \bar{c}} = \frac{2 \ k_w r}{\bar{c}}$$

Quartz

When a tubular sample is inserted into the apparatus as indicated in Figure 1, recombination will take place on the quartz walls as well as on the sample surface. Hence the recombination coefficient for quartz must be known as a function of temperature in order to calculate γ values for the surface of interest from measured light intensity ratios. To obtain the necessary background information, a series of runs was made with the blank quartz tube. The tube was cleaned with an aqueous solution of nitric acid and HF (3 parts concentrated HNO_3 : 1 part HF : 3 parts water), rinsed with a copious quantity of distilled water and air dried. The system was then pumped down to 30 microns prior to starting the recombination measurements. Considerable time is often required for all parts of the system to reach a steady state—15 minutes for the argon and oxygen flows to stabilize, up to 45 minutes after the microwave power is turned on for the system to reach equilibrium, and 15 minutes on the average to stabilize the NO flow.

The quartz data are plotted as γ vs $1/T$ in Figure 2. The dark points represent the present work, and the results are compared with other data in the literature taken as a function of temperature. While all of our points are given, only averaged values are plotted from the other laboratories. Room temperature results summarized in Table I illustrate that typically γ values obtained in a single laboratory can vary by a factor of 2 to 5, while results between laboratories can differ by one or two orders of magnitude, in spite of the fact that all of the measurements were taken with some care. Our results show no systematic trend with water vapor content of the gas or with cycling from 800°C to room temperature. The one factor that has not been explored in any of the measurements to date, and which may be crucial, is the surface structure of the glass. For the purposes of this paper, the quartz data are only needed to correct the light intensity ratios obtained when samples of interest are inserted. Since the recombination coefficients measured for quartz are at least an order of magnitude lower than the recombination coefficients for the metallic samples discussed below, the correction factors are fairly small. A blank quartz run at temperature is made before and after each sample run, and the average value is used to calculate the corrections.

Titanium

Results for metallic titanium, washed successively with trichloro-ethylene, acetone, methanol, and distilled water, and for anodized titanium are plotted in Figure 3. The anodized sample, which is covered with a deep blue metallic looking interference film about 30Å in thickness, is seen to be a more efficient recombination catalyst than the untreated metal. The experimental activation energy over the temperature range investigated is 3.2 kcal/mole for both surfaces. The difference in the pre-exponential factor may be related to the concentration of a surface oxide intermediate.

Molybdenum

Results for metallic molybdenum over the temperature range 298-773°K are plotted in Figure 4, and correspond to an activation energy of 1.7 kcal/mole. At 1073°K, essentially all of the oxygen atoms were removed from the gas stream over the 4 cm sample length. The average weight loss at this temperature was 2.4×10^{-7} moles/sec. From NO_2 titration data, the oxygen atom flow rate is estimated as 2.1×10^{-6} moles/sec at the level of the top of the tubular molybdenum sample. If the molybdenum is removed from the surface linearly as $MoO_3(g)$, and it is recognized that oxygen atoms are much more reactive than oxygen molecules,[3] then the number of atoms removed by the chemical reaction would be 0.9×10^{-6} moles/sec, the remainder being lost by recombination. The corresponding

γ value would be at least 4×10^{-3}, which is consistent with the extrapolated results of Figure 4. At temperatures of 773°K and below, small net weight gains were observed and the major source of oxygen atom decay by far was surface recombination.

Tantalum

Results for tantalum are plotted in Figure 5, and although the number of data points are sparse, the temperature dependence of the recombination coefficients does seem to be much less at the high temperatures than at the lower.

V. Discussion

There are basically two modes of atom recombination on solid surfaces, or desorption of molecules from adsorbed atoms--(1) collisions between adsorbed atoms on the surface (Hinshelwood mechanism) and (2) collisions between adsorbed atoms and gaseous atoms (Rideal mechanism). For the metals under consideration--Ti, Mo, and Ta-- the binding energy for chemisorption of oxygen atoms estimated according to the method of Eley[4] is of the order of 160 kcal/mole for Ti and Ta, and about 140 kcal/mole for Mo. Although the estimates are very crude, the values when compared to the dissociation energy of O_2 of 118 kcal/mole, are indicative of strong dissociative chemisorption. Hence, the concentration of adsorbed atoms on the surface will be large. In fact, since no attempt was made to start with atomically clean surfaces, coverage probably exceeds a monolayer after a very short time of exposure to the oxidizing atmosphere, and the recombination rates observed should be characteristic not of clean metals, but of thin oxide layers. On oxide surfaces the equilibrium binding sites for atoms are thought to be widely separated compared to interatomic spacing, and although chemisorption may still be quite efficient in a partially dissociated gas stream, recombination is expected to be much lower than on clean metals.[5] Mass transport was ruled out as a limiting factor in our experiments both empirically by varying flow rate and sample size, and analytically by calculation of the diffusion limited k_w values for our system. Thus, while recombination coefficients of the order of unity would be expected for the clean metal surfaces at temperatures above $600°K$,[5] the observed values of 10^{-2} and below are not unreasonable for oxygen covered surfaces. More detailed interpretation of the results must await experimental characterization of the catalytic surfaces.

Acknowledgments

This work was supported by the U.S. Air Force Materials Laboratory Wright-Patterson Air Force Base, Ohio. Mr. Paul W. Dimiduk was Project Engineer. Invaluable assistance in the design and construction of the apparatus was provided by Mr. Peter Felsenthal. Mr. Charles Lukas conducted the laboratory experiments. It is a pleasure to acknowledge stimulating technical discussions of the work with Dr. D. E. Rosner of Aerochem and Dr. Frederick Kaufman of the University of Pittsburgh.

References

1. Frederick Kaufman, Ch. 1, in "Progress in Reaction Kinetics," Vol. 1, ed. by G. Porter, Pergamon Press, N.Y. (1961).

2. J. W. Linnett and D. G. H. Marsden, Proc. Roy. Soc. A234 489 (1956).

3. D. E. Rosner and H. D. Allendorf, J. Electrochem. Soc. 114, 305 (1967).

4. E. Eley, Disc. Faraday Soc. 8, 34 (1950).

5. G. Ehrlich, J. Chem. Phys. 31, 1111 (1959).

6. J. C. Greaves and J. W. Linnett, Trans. Faraday Soc. 55, 1346 (1959).

7. J. T. Herron and H. I. Schiff, Canad. J. Chem. 36, 1159 (1958).

8. L. Elias, E. A. Ogryzlo, and H. I. Schiff, Canad. J. Chem. 37, 1680 (1959).

9. L. Elias, J. E. Morgan, and H. I. Schiff, J. Chem. Phys. 33, 930 (1960).

10. C. Mavroyannis and C. A. Winkler, Canad. J. Chem. 39, 1601 (1961).

11. T. C. Marshall, Phys. of Fluids 5, 743 (1962).

12. (a) V. V. Voevodskii and G. K. Lavrovskaya, Doklady, Akad-Nauk, SSSR, 63, 151 (1948), C. A. 43, 1635d (1949); (b) Zhur Fiz. Khim, 25, 1050 (1951), C. A. 46, 2893c (1952).

13. F. Kaufman, Proc. Roy. Soc. A247, 123 (1958), J. Chem. Phys. 28, 352 (1958).

14. S. Krongelb and M. W. P. Strandberg, J. Chem. Phys. 31, 1196 (1959)

15. D. S. Hacker, S. A. Marshall, and M. Steinberg, J. Chem. Phys. 35, 1788 (1961).

Table I

SURFACE RECOMBINATION COEFFICIENT, γ, OF OXYGEN ATOMS AT ROOM TEMPERATURE

Surface	$\gamma \times 10^5$	Method	Reference
Pyrex	12	Side-arm (thermocouple) probes	Linnett and Marsden (2)
Pyrex	3.1-4.5	Side-arms (Wrede-Harteck guages)	Greaves and Linnett (11)
SiO_2	71	Side-arm (thermocouple probes)	(2)
SiO_2	16	Side-arm (Wrede-Harteck guages)	(11)
Pyrex	11	NO_2 titration	Herron and Schiff (16)
Pyrex	7.7	Isothermal calorimetry	Elias, Ogryzlo and Schiff (17)
Pyrex	1.7	NO_2 titration and NO afterglow	Elias, Morgan and Schiff (19)
Pyrex	7.5	NO_2 titration	Mavroyannis and Winkler (18)
Pyrex	50	Electron Paramagnetic Resonance	Marshall (8)
SiO_2	1.7	Capillary side arm	Voevodskii and Lavrovskaya (12)
Vycor	4.7	NO_2 titration; NO afterglow	Kaufman (1)
Pyrex	2	NO_2 titration; NO afterglow	Kaufman (15)
Quartz	32	Electron Paramagnetic Resonance	Krongelb and Strandberg (20)
Quartz	4.0	Electron Paramagnetic Resonance	Hacker, Marshall and Steinberg (21)
Quartz	2.7-7.8	NO afterglow	This work

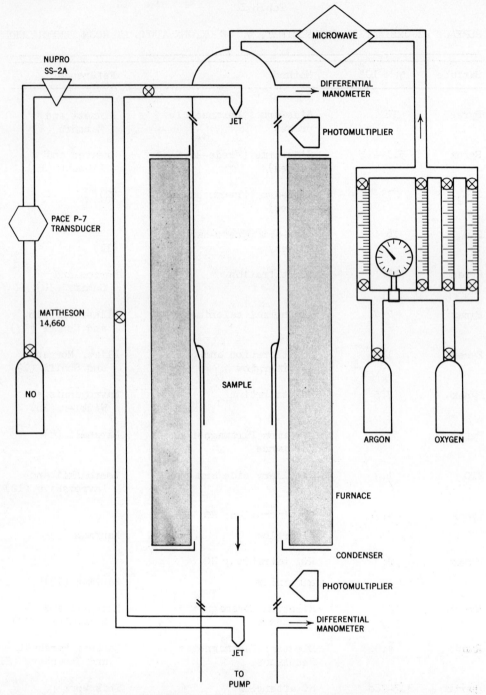

Fig. 1. Schematic Diagram of the Apparatus for Heterogeneous Oxygen Re-combination Studies.

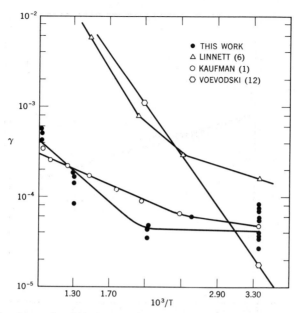

Fig. 2. Recombination Coefficients for Quartz as a Function of Temperature.

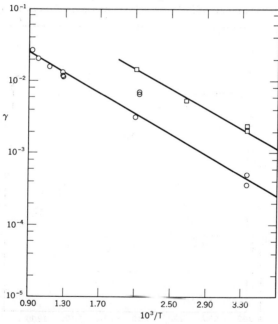

Fig. 3. Recombination Coefficients for Ti metal and Anodized Ti as a Function of Temperature.

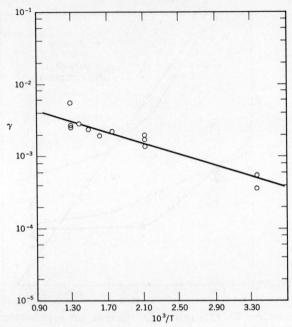

Fig. 4. Recombination Coefficients for Mo as a Function of Temperature.

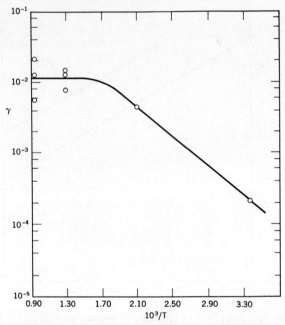

Fig. 5. Recombination Coefficients for Ta as a Function of Temperature.

STUDIES ON ULTRA-CLEAN CARBON SURFACES
I. CHARACTERIZATION OF SURFACE ACTIVITY OF
GRAPHON BY ROOM-TEMPERATURE OXYGEN CHEMISORPTION

P. L. Walker, Jr., R. C. Bansal, and F. J. Vastola

Materials Science Department
Pennsylvania State University
University Park, Pennsylvania

1. INTRODUCTION

Chemisorption of oxygen by carbons has been studied by a number of workers (1-21) using various carbon materials and a variety of experimental techniques. Several reviews (22-24) have also been written on the subject. However, there appears to be little agreement regarding the exact mechanism of the chemisorption process. Recently it has also been realized that a study of chemisorption kinetics may be helpful in elucidating the exact mechanism of the carbon-oxygen reaction.

Bonnetain and co-workers (8) studied the adsorption of oxygen on various forms of carbon in the pressure range 0.1-100 millitorr below 300°C. Bonnetain (25) studied the desorption of oxygen complexes under vacuum at 400-700°C. It was found in these studies that both adsorption and desorption rates followed the Elovich equation; i.e. plots of amount adsorbed or desorbed versus log of time were linear. Deitz and McFarlane (26), while studying the adsorption of oxygen on evaporated carbon films of high surface area at temperatures between 100 and 300°C and at pressures of the order of 100 millitorr, observed a rapid initial adsorption followed by a much slower adsorption. The slow step was molecular oxygen adsorption, the kinetics of which followed the logarithmic rate law and the first power dependence on pressure. Allardice (16), in his experiments with brown charcoal at temperatures between 25-300°C in the pressure range 160-760 torr, observed a similar two step adsorption. The rapid step was reversible; whereas, the slow step was irreversible chemisorption and followed the Elovich rate law.

Walker and co-workers (27) studied the kinetics of oxygen chemisorption on spectroscopic grade, highly crystalline natural graphite at an oxygen pressure of 760 torr and temperatures between 335 and 448°C by measuring the change in thermoelectric power of the graphite as chemisorption proceeded.

Unlike the situation which holds when the Elovich equation is applicable, they found the rate constant to be independent of surface coverage.

Carpenter and Giddings (17) and Carpenter and Sergeant (18), while studying the initial stages of oxidation of different varieties of coals at temperatures of 65, 85 and 105°C, found that chemisorption of oxygen for the first 5 minute period obeyed the Elovich equation. But as the time period of oxidation increased, the quantity of oxygen sorbed exceeded the amounts predicted by the Elovich equation. This was attributed to the creation of fresh adsorption sites by the desorption of oxidation products (CO_2, CO and H_2O).

Walker and co-workers (13,14), in their experiments on the chemisorption of oxygen on preoxidized Graphon samples in the temperature range 300-625°C at 0.5 torr oxygen pressure, observed a sharp increase in the amount of oxygen adsorbed at temperatures above 400°C, suggesting the presence of two types of active sites. These carbon samples were cleaned at 950°C in a vacuum of 10^{-5} torr. Recently (12) they studied the kinetics of chemisorption of oxygen on well cleaned Graphon samples (cleaned at 950°C in a vacuum of the order of 10^{-8} torr) in the pressure range 1-15 millitorr. The data once again suggested the presence of at least two types of active sites. The adsorption of oxygen was dissociative and the rate proportional to the square of the concentration of uncovered active sites.

It appears from the above brief resume that much of the work on the chemisorption of oxygen was carried out on poorly defined carbons and/or on carbon surfaces contaminated with significant amounts of previously adsorbed gases (such as oxygen and hydrogen). Some of the kinetic measurements were also carried out at comparatively higher temperatures, where effects due to gasification of the carbon are also significant.

The present paper describes an ultra-high vacuum system designed to
obtain a very clean carbon surface for low temperature chemisorption studies.
The paper also discusses results obtained on the room temperature chemisorption
of oxygen on an ultra-clean Graphon surface.

2. EXPERIMENTAL

2.1. <u>Materials</u>: The carbon used in this study was Graphon, the graphitized
carbon black Spheron 6 obtained from Cabot Corporation. The BET (N_2) surface
area of the original material was 76 m^2/g. Total impurity content is estimated,
by emission spectroscopy, to be below 15 ppm, with the major impurities being
Ti, Ca and Si. The sample was preoxidized to 16.6% weight loss at 625°C at
an O_2 pressure of 0.5 torr. This pretreatment increased the BET surface area
to 100 m^2/g. However, the area of the carbon sample which chemisorbed oxygen
at 300°C increased from 0.23 to 2.3 m^2/g as a result of this oxidation (15),
a sizeable increase of ten-fold.

The oxygen used was research grade obtained from Air Products and
Chemicals Company.

2.2 <u>Apparatus</u>: The apparatus (Figure 1) essentially consists of four different
units (i) a gas inlet unit, (ii) an evacuation assembly, (iii) the reactor
and (iv) a measuring unit.

The gas inlet system is a multiple aliquot expansion unit, using three
different aliquots of the gas for expansion at three different stages. The
first two expansions are carried out in the glass system. The pressures
after these expansions are measured using a Baratron differential manometer,
capable of reading accurately pressures down to 10^{-5} torr. The third
expansion of the gas is carried out by taking the aliquot between two stainless steel valves so that the organic matter from the grease may not affect

the high purity requirements of the system. The pressure of the gas after this expansion is measured by a mass spectrometer.

The degassing unit (evacuation assembly) is a Vacion pump backed by oil diffusion and mechanical pumps. The system is capable of giving a vacuum better than 10^{-9} torr.

The reactor is a double walled quartz tube with the sample suspended from a metal flange at the top by means of a quartz fiber. A Pt–Pt 10% Rh thermocouple is used to measure the temperature of the sample inside the reactor. The jacket of the reactor is evacuated continuously to 10^{-6} torr to reduce to a minimum the inward diffusion of the gas when the reactor is at elevated temperature. The temperature of the furnace is controlled by a Gardsman West temperature controller.

The adsorption-measuring unit has a residual gas analyzer, a Baratron differential manometer and a UHV micro-sorption balance. The residual gas analyzer is a mass spectrometer which has been placed directly on the reactor so that it can monitor gas pressures down to 10^{-10} torr without the necessity of a molecular leak. The highest gas pressure that the mass spectrometer can measure with this arrangement is 10^{-4} torr.

The Baratron differential manometer is used for adsorption measurements at pressures between 10^{-4} and 1 torr. The measurement of very small changes in pressure of the gas due to adsorption, at pressures above 1 torr, was not very accurate using the Baratron. The micro-sorption balance was, therefore, used for experiments above this pressure.

The micro-sorption balance (Cahn RG) has been enclosed in a stainless steel container connected to the ultra-high vacuum portion of the measuring unit through a stainless steel valve so that it conforms to the high purity requirements of the system. The sample is suspended from the beam of the balance in a double walled quartz hang-down tube, similar in construction to the reactor described above. The balance could measure accurately weight changes of the order of $2x10^{-6}$ g. However, the best vacuum obtained in the balance system was $5x10^{-8}$ torr.

2.3. Experimental Procedure: The carbon sample (ca 0.1g) is held in a quartz boat, suspended by means of a quartz fiber. It is degassed in vacuum at 1000°C for 10–12 hr until the residual gas pressure is 10^{-9} torr. The sample is then cooled in vacuum to room temperature (25°C). A known volume of oxygen is allowed to expand into the reactor, and adsorption is followed continuously by monitoring the gas pressure with the Baratron or the mass spectrometer. In order to minimize the change in oxygen pressure, larger volumes were introduced into the reactor system, specially for experiments at lower pressures. These volumes were so arranged that the drop in pressure of the gas due to adsorption is large enough to be accurately measured but sufficiently small so as to not cause a significant drop in adsorption rates.

Analysis of the gas with the mass spectrometer at different stages of adsorption at room temperature showed insignificant amounts of carbon monoxide or carbon dioxide, indicating little or no gasification of the carbon. Therefore, the decrease in pressure of oxygen in the system is due to its adsorption resulting in the formation of carbon-oxygen surface complex. However, in order to further verify the equivalence, the carbon-oxygen complex was decomposed by heating in vacuum at 1000°C and the amount of oxygen desorbed as CO and CO_2 measured. The amount of oxygen recovered agreed closely with the amount of oxygen previously chemisorbed at room temperature.

For experiments using the micro-sorption balance, the amount of oxygen adsorbed was measured directly as an increase in weight of the carbon sample. However, since the balance requires 2–3 min to stabilize after introduction of the gas, the first few adsorption measurements were ignored.

3. RESULTS AND DISCUSSION

Cleaning of Graphon by heating in ultra-high vacuum at 1000°C produces a highly reactive surface. The adsorbed oxygen is very stable at room temperature at pressures down to 10^{-9} torr and can be desorbed completely, only as oxides of carbon, at temperatures up to 950°C. The adsorbed oxygen is, therefore, irreversibly adsorbed. The amount of oxygen adsorbed is significantly greater than that reported in some of the earlier work on carbons evacuated at high temperatures (3,28). As the materials used in the earlier investigations were not well cleaned, some of the active sites remained contaminated with preadsorbed gases.

The rates of chemisorption of oxygen on Graphon at different starting pressures of the gas are shown in Figure 2. It is seen that chemisorption is very rapid in the beginning but that the rate slows down with time. The curves appear to level off, although a negligible decrease in pressure has occurred. Rate measurements were continued for up to 24 hr. It is also evident from Figure 2 that the rate of adsorption increases with increase in starting pressure of the gas.

Several kinetic expressions (12,16,20,26,29) have been applied, more or less satisfactorily, to the kinetics of the carbon-oxygen reaction. In recent years, many workers have found the rate equation

$$\frac{dq}{dt} = a \exp (-\alpha q) \tag{1}$$

where q is the amount adsorbed and a and α are constants, to be applicable to their adsorption data for a wide variety of systems. This equation, which is now commonly known as the Elovich equation, shall be used to interpret our results on room temperature chemisorption of oxygen. The integrated form of equation (1)

$$q = \frac{1}{\alpha} \ln (1 + a\alpha t) \tag{2}$$

indicates that a plot of q versus ln t should be a straight line if $a\alpha t \gg 1$.

The q-log t plots for chemisorption of oxygen on Graphon for some of the experiments at different starting pressures of oxygen are shown in Figure 3. It is seen that the plots have linear regions, but at higher pressures the straight lines are discontinuous and change slope at one, two or three points. It is interesting to note that the number of breaks in the q-log t plots and the time of appearance of a break depend on the oxygen pressure. For example, the plots show breaks at only one point for experiments at pressures lower than 50 millitorr, at two points for pressures between 50 and 164 millitorr. The plots at 164 millitorr pressure show three breaks. Plots at higher pressures again, show two breaks with the first break disappearing. A new break appears in the plots at 693 millitorr and higher pressures. The plot at 760 torr shows only one break, with the first three breaks missing.

Similar discontinuities were observed by Taylor and Thon (30) in the q-log t plots of the data of Sickman and Taylor (31) and of Strother and Taylor (32) for the adsorption of hydrogen on zinc oxide and of the data of Maxted and Moon (33) for the adsorption of hydrogen on clean platinum surfaces. It was suggested by Taylor and Thon (30) that if α is characteristic of the nature of sites involved in the adsorption, then the break in the linear plot indicates a change from one kind of site to another.

In order to see if these different linear regions are actually different kinetic stages of the same chemisorption process involving adsorption at different types of sites, instantaneous rates of adsorption, midway in each linear region, were calculated for each pressure for which data are available. These rates, normalized with respect to oxygen pressure, are presented in Table I. For each stage, the adsorption rate is essentially proportional to

the first power of oxygen pressure, which has been varied widely. Only at
760 torr is this relation not obeyed, with the rate being proportional to some
lower power of pressure. It is seen that the rate of oxygen chemisorption
decreases sharply in advancing from Stage I to Stage V; the rate for Stage V
being some 250-fold less than that for Stage I.

It is evident from the results presented above that the chemisorption
of oxygen on Graphon involves a number of types of sites. This receives
further support from the fact that each of these stages appears after a
definite amount of oxygen has been chemisorbed (Table II and Figure 3).
However, the time of appearance and the temporal range of existence of any
one particular stage is determined by the oxygen pressure. In general, any one
kinetic stage appears earlier and lasts for a shorter period as the pressure is
increased. For example, State III, which appears after the chemisorption of
ca 2.4×10^{18} atoms of O_2, is observed after 200 min at 50.6 millitorr pressure
and after only 3 min at 693 millitorr pressure. This is due to the higher
rates of chemisorption at higher pressures so that the sites are covered
much more rapidly. Similarly a lower kinetic stage disappears altogether
when the pressure is increased. For example, Stage I could not be observed
at pressures of 199 millitorr and above. Stages I, II, and III could not be
observed at 760 torr pressure. Chemisorption at these higher pressures is
so rapid that these stages have passed in the time period between the exposure
of the gas to the carbon and the first measurement. This is also evident from
the fact that the amounts of oxygen adsorbed in the very first measurements
at these pressures are more than the amounts chemisorbed in these stages.
This also explains the absence of higher kinetic stages in the low pressure
experiments. As the chemisorption is very slow at lower pressures, a higher
kinetic stage should appear only if the adsorption measurements are followed
for very long periods of time. In practice, however, this becomes difficult,
as the precision of measurement decreases due to the slow rate of adsorption.

It is evident, therefore, that the chemisorption of oxygen by Graphon involves definite kinetic stages and that all these kinetic stages are present in any one run. However, because of the limitations of the experimental technique, not all stages are detected in any single experiment.

Discontinuities in the q-log t plots for the carbon-oxygen system were not observed by previous workers in this field because of several reasons: (i) the carbon samples used were not sufficiently clean, (ii) studies were restricted to a limited range of temperatures and pressures, and (iii) the experimental techniques used were not very precise and accurate so that experiments could only be carried out for short intervals of time. In other words, these studies have been limited to adsorption on only a part of the complete surface.

These definite and different kinetic stages show the existence of five different types of sites, which are available for room temperature chemisorption of oxygen. The amounts of oxygen required to fill these five types of sites (surfaces) are given in Table II. It is interesting to note that the amount of oxygen required to fill any one of these surfaces is essentially the same at all pressures. Assuming that the carbon-oxygen complex consists of one oxygen atom per edge carbon atom and that the edge carbon atoms lie in $(10\bar{1}0)$ planes, i.e. each carbon atom occupies an area of 8.3 $\mathring{A}^2$, it can be shown that these different types of sites cover approximately 0.10, 0.12, 0.29, 0.66 and 0.96 m^2/g of the surface. The total surface area covered by these five groups of sites is 2.1 m^2/g. The active surface area of the sample, as determined by oxygen chemisorption at 300°C (15) is 2.3 m^2/g, which is quite close to the value obtained by room temperature chemisorption. This indicates completion of the adsorption process. This is further supported by the shape of the q-log t plots in the case of experiments at 9.9 and 760 torr. The plots eventually deviate from linearity and show decreasing slopes. This last stage with a decreasing slope represents completion of adsorption.

It is interesting to speculate briefly as to the different types of sites which would be expected at the edge of a graphite basal plane. Ideally the plane terminates with the carbon atoms in either a $(10\bar{1}0)$ or $(11\bar{2}0)$ configuration. Carbon-carbon distances of importance are 1.42, 2.46 and 2.84A. An O_2 molecule could also approach the surface with its bond essentially parallel to the c-axis of the graphite crystallites and form an activated complex with carbon atoms in adjacent layer planes. In this case, carbon-carbon distances of 3.35 and 3.62 Å are of importance. Also the basal plane need not be completely terminated in an ideal manner but can have carbon fragments protruding from it. This will present other carbon-carbon spacings to incoming O_2 molecules. The activated complex formed between O_2 and two surface carbon atoms would be expected to have different potential energy configurations, dependent upon the spacings between the carbon atoms. Thus, the activation energy for dissociative chemisorption of oxygen would be expected to vary. In fact, Sherman and Eyring (34) have shown, theoretically, that the activation energy for dissociative chemisorption of H_2 on carbon will vary with the carbon-carbon spacing.

This low temperature chemisorption technique shows promise of being a simple way of profiling the surface activity of graphites, thus making possible quantitative predictions of graphite behavior under different environmental conditions.

The influence of temperature on chemisorption on these different types of sites, the activation energies associated with each type, the variation of activation energy with surface coverage, the probable mechanism of chemisorption, and the results of desorption of oxygen from these different types of sites shall be discussed in subsequent papers.

This study was supported jointly by Small Industries Research, Commonwealth of Pennsylvania and the Atomic Energy Commission on Contract No. AT(30-1)-1710.

REFERENCES

1. Lobenstein, W. V. and Deitz, V. R., J. Phys. Chem., 59, 481 (1955).

2. Puri, B. R., Singh, D. D., Nath, J. and Sharma, L. R., Ind. Eng. Chem., 50, 1071 (1958).

3. Puri, B. R., Proc. Fifth Carbon Conf., Pergamon Press, New York, 1962, Vol. 1, p. 165.

4. Kiselev, A. V., Kovaleva, N. V., Polyakova, M. M. and Tesner, P. A., Kolloid Zhur, 24, 195 (1962).

5. Boehm, H. P., Hofmann, H., and Clauss, A., Proc. Third Carbon Conf., Pergamon Press, New York, 1959, p. 241.

6. Donnet, J. B. and Lahaye, J., Symposium on Carbon, Tokyo, Japan, VIII-16 (1964).

7. Lang, F. M., and Magnier, P., J. Chim. Phys., 60, 1251 (1963).

8. Bonnetain, L., Duval, X. and Letort, M., Proc. Fourth Carbon Conf., Pergamon Press, New York, 1960, p. 107.

9. Snow, C. W., Wallace, D. R., Lyon, L. L., and Crocker, G. R., Proc. Third Carbon Conf., Pergamon Press, New York, 1959, p. 279.

10. Vastola, F. J., Hart, P. J., and Walker, P. L. Jr., Carbon, 2, 65 (1964).

11. Walker, P. L. Jr., Vastola, F. J., and Hart, P. J., Proc. Symposium on Fundamentals of Gas Surface Interactions, San Diego, Calif., Academic Press, 1967, p. 307.

12. Hart, P. J., Vastola, F. J., and Walker, P. L. Jr., Carbon, 5, 363 (1967).

13. Lussow, R. O., Vastola, F. J., and Walker, P. L. Jr., Carbon, 5, 591 (1967).

14. Lussow, R. O., Ph.D. Thesis, The Pennsylvania State University (1966).

15. Laine, N. R., Vastola, F. J., and Walker, P. L. Jr., J. Phys. Chem., 67, 2030 (1963).

16. Allardice, D. J., Carbon, 4, 255 (1966).

17. Carpenter, D. L. and Giddings, D. G., Fuel, 43, 375 (1964).

18. Carpenter, D. L. and Sergeant, G. D., Fuel, 45, 311 (1966).

19. Zarifyanz, Yu. A., Kiselev, V. F., Lezhnev, N. N., and Nikitina, O. V., Carbon, 5, 127 (1967).

20. Barrer, R. M., J. Chem. Soc., 1261 (1936).

21. Evans, T. and Phaal, C., Proc. Fifth Carbon Conf., Pergamon Press, New York, 1962, Vol. 1, p. 147.

22. Walker, P. L. Jr., Rusinko, F. Jr., and Austin, L. G., Advances in Catalysis, Vol. 11, Academic Press, 1959, p. 133.

23. Smith, R. N., Quart. Rev. (London), 13, 287 (1959).

24. Culver, R. V. and Watts, H., Rev. Pure Appl. Chem., 10, 95 (1960).

25. Bonnetain, L., J. Chim. Phys., 56, 266 (1959).

26. Deitz, V. R., and McFarlane, E. F., Proc. Fifth Carbon Conf., Pergamon Press, New York, 1963, Vol. 2, p. 219.

27. Walker, P. L. Jr., Austin, L. G., and Tietjen, J. J., Chemistry and Physics of Carbon, Marcel Dekker, New York, 1966, Vol. 1, p. 327.

28. Dubinin, M. M., "Chemical Compounds on the Surface and Their Part in Adsorption", Moscow State University, p. 9, 1957.

29. Lendle, A., Z. Physik Chem., A172, 77 (1934).

30. Taylor, H. A. and Thon, N., J. Am. Chem. Soc., 74, 4169 (1952).

31. Sickman, D. V. and Taylor, H. S., J. Am. Chem. Soc., 54, 602 (1932).

32. Strother, C. O. and Taylor, H. S., J. Am. Chem. Soc., 56, 586 (1934).

33. Maxted, B. and Moon, C. H., J. Chem. Soc., 1542 (1936).

34. Sherman, A. and Eyring, H., J. Am. Chem. Soc., 54, 2661 (1932).

TABLE I

INSTANTANEOUS ADSORPTION RATES CALCULATED MIDWAY IN EACH LINEAR REGION

RATES NORMALIZED WITH RESPECT TO PRESSURE

Rate, atoms/g min^{-1} $\mathrm{millitorr}^{-1}$

Pressure Millitorr	Stage				
	I ($q = 0.6 \times 10^{18}$)	II ($q = 1.8 \times 10^{18}$)	III ($q = 4.5 \times 10^{18}$)	IV ($q = 9.6 \times 10^{18}$)	V ($q = 18.0 \times 10^{18}$)
0.77	2.1×10^{15}	--	--	--	--
5.76	2.5	--	--	--	--
11.6	2.7	2.6×10^{14}	--	--	--
22.9	2.0	2.9	--	--	--
50.6	2.0	2.9	--	--	--
99.2	--	3.0	--	--	--
164.	--	4.0	--	--	--
199.	--	3.9	2.2×10^{14}	--	--
302.	--	--	2.0	5.2×10^{13}	--
537.	--	--	2.2	6.5	--
693.	--	--	2.3	6.9	8.7×10^{12}
5438.	--	--	2.0	4.5	7.6
9930.	--	--	2.0	4.9	8.9
760×10^{3}	--	--	--	1.0	3.1

TABLE II

CHEMISORPTION OF OXYGEN IN DIFFERENT STAGES BY GRAPHON AT 25°C AND AT DIFFERENT PRESSURES

| Pressure Millitorr | Oxygen Chemisorbed atoms/g x 10^{-18} | | | | |
| | Stage | | | | |
	I	II	III	IV	V*
0.77	0.84	--	--	--	--
5.7	1.32	--	--	--	--
11.6	1.21	--	--	--	--
22.9	1.26	--	--	--	--
50.6	1.30	1.19	--	--	--
77.4	1.31	1.20	--	--	--
99.2	1.20	1.26	--	--	--
164.	--	1.38	3.36	--	--
199.	--	1.44	3.36	--	--
302.	--	1.38	3.42	--	--
537.	--	1.38	3.36	--	--
693.	--	1.44	3.48	8.10	11.1
5438.	--	--	3.48	8.10	11.1
9930.	--	--	3.42	7.92	11.2
760×10^3	--	--	--	7.50	11.7

* The amount of oxygen adsorbed on the fifth stage has been calculated by subtracting the total adsorbed in the first four stages from the saturation amount at 760 torr, that is, 25.2×10^{18} atoms/g.

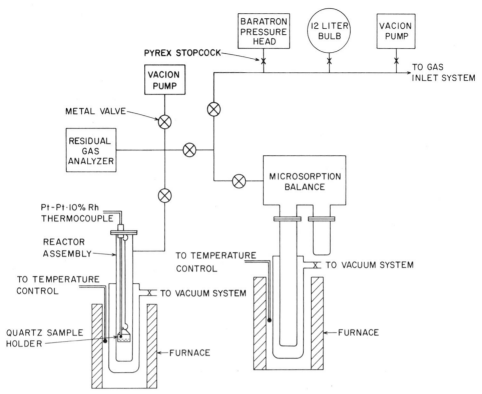

Fig. 1. Schematic drawing of apparatus used to study oxygen chemisorption on ultra-clean carbon surfaces.

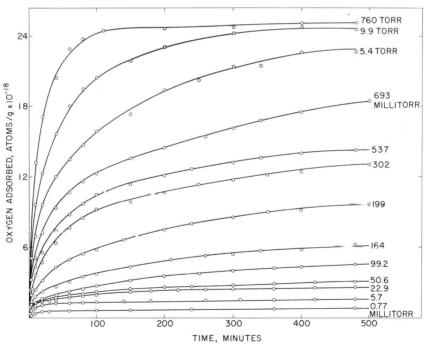

Fig. 2. Oxygen chemisorption on Graphon at 25°C and various oxygen pressures. Graphon was previously activated in oxygen to 16.6% weight loss.

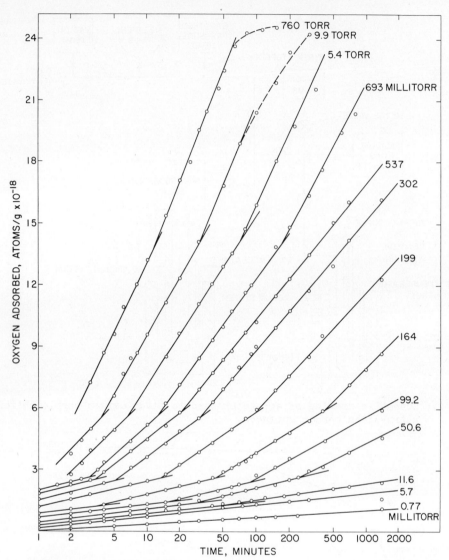

Fig. 3. Elovich plots of oxygen chemisorption on Graphon at 25°C and various oxygen pressures. Graphon was previously activated in oxygen to 16.6% weight loss.

INFLUENCE OF ANNEALING IN ULTRAHIGH VACUUM
AND OF OXYGEN ON FIELD EFFECT IN CdS SINGLE CRYSTALS*

R. Pinchaux, C. Sebenne, and M. Balkanski

Laboratoire de Physique des Solides de la Faculté des Sciences
Equipe de recherche associée au C.N.R.S.
Paris, France

INTRODUCTION

Surface properties of electrons in CdS single crystals have been investigated by different authors. First, BUBE[1] demonstrated the influence of oxygen on the photoconductivity spectrum. Some other qualitative studies followed. More recently, some quantitative investigations have been performed concerning the influence of different gases on the change of conductivity and its kinetics[2], the contact potential as a function of oxygen pressure[3] and the desorption energy of oxygen[4]. Moreover, field effect measurements have been reported[5] on etched samples without any heating under vacuum. Besides, it has been shown that CdS crystals, when submitted to an oxygen pressure, present a conductivity change, as a function of time, which depends on the initial surface treatment and on the thermal treatment under vacuum[6]. In order to explain these results, the presence of a gradient of energy levels near the surface has been invoked and a method for its computation has been proposed[9]. However, such an explanation had to be checked and the characteristics of the surface states due to such surface conditions had to be determined. It is the object of the present paper which reports the results obtained by combining d.c. and pulsed field effect measurements, annealing under vacuum and conductivity changes due to the action of oxygen.

EXPERIMENTAL CONDITIONS

The CdS samples used in these experiments are parallelepipeds about $1 \times .3 \times .05$ cm^3 obtained by cleaving in air different larger single crystals grown at the Aeronautical Research Laboratory (Wright Patterson Air Force Base). Their resistivity ranges from 1 to 5 Ω-cm. Low noise, low resistance electrical contacts are prepared using In-Ga alloy and are stabilised by successive discharges of a capacitor. The electrical bulk parameters are determined, using a five probe method, by Hall effect and conductivity measurements. The field effect capacitor

---

*Research supported by D.G.R.S.T. (France) under the contract n° 65FR174.

is constituted by one of the largest faces of the sample and a 25 μm
thick mica layer pressed on the sample and covered with evaporated
aluminium on the other side. The conductivity changes, either due to
oxygen chemisorption or to d.c. field effect, are recorded using a
Wheatstone type bridge allowing better than 10^{-4} relative accuracy.
Pulsed field effect measurements are performed using the classical
circuit described by RUPPRECHT[7] ; time constants lower than 10μ s
cannot be detected because of the unbalance of the system when a d.c.
current is going through the sample.

Every sample undergoes the following operations. First, it
is mounted in a cell which is connected to an ionic pump vacuum
system and evacuated. Then, a part of the vacuum system including the
sample is heated up to 400° C for 4 hours under a pressure of about
10^{-7} torr. Back at thermal and electrical equilibrium, at room
temperature, field effect measurements are performed under a pressure
of about 10^{-9} torr. Then, spectroscopically pure oxygen is abruptly
introduced at a pressure of about 10^2 torr. The resulting variation of
the sample conductivity is recorded as a function of time. After an
equilibrium has been reached in the presence of oxygen, field effect
measurements are carried out once more. This procedure is repeated at
least three times on every sample which is always kept in the dark.

In order to check whether the mica layer did perturb the
action of oxygen on the surface, an experiment has been carried out
using simply a sample and a parallel metallic plate as the electrodes,
without any dielectric in between. There is no observable difference
between the results of the two procedures.

RESULTS AND DISCUSSION

1 - Relation between surface charge and surface potential.

In order to find the characteristics of the surface states
from field effect measurements, it is necessary to know the relation
between the surface charge density and the surface potential as defined
from the band bending in the space charge region. When the semiconductor

contains a given constant concentration of donors or acceptors, that relation is fully determined for any value as soon as it is known for one[8]. But when this concentration varies within the space charge region, no general results are available and the computation of the surface potential has to be carried out for each donor distribution and each surface charge density. In the present case, the conductivity changes of CdS interacting with oxygen have been quantitively explained by a one dimensional donor gradient near the surface depending on the previously described treatments[6]. From the conductivity change, measured as a function of time, the donor gradient $N_D(z)$ can be computed[9]. If $n(z)$ is the free electron density at any point within the semiconductor, in the semi infinite n-type semiconductor approximation, and assuming that all the donors are ionized, the surface charge density is equal to the net excess of free electrons in the space charge region, that is :

$$\Delta N = \int_0^\infty \left[n(z) - N_D(z) \right] dz \tag{1}$$

The relation between surface charge and surface potential has then to be obtained by integrating Poisson's equation, which can be written, in the present case, using a reduced potential $v = \frac{eV}{kT}$ defined as the band displacement in energy due to the surface charge and positive for an accumulation layer :

$$\frac{d^2 v}{dz^2} = \frac{q^2}{kT \, \mathcal{K} \, \varepsilon_0} \left[N_D(z) - n(z) \right]$$

For a weak accumulation or a depletion layer, as in the present case, we can write, as an approximation :

$$n(z) = N_D(z) \exp v \tag{2}$$

That is, for Poisson's equation :

$$\frac{d^2 v}{dz^2} = \frac{q^2 N_D(z)}{kT \mathcal{K} \varepsilon_o} \left[1 - \exp v \right] \qquad (3)$$

Using a computer, that equation can be numerically integrated and the relation between the surface charge density ΔN and the reduced surface potential v_s obtained, after $N_D(z)$ has been determined.

Figure 1 shows the surface charge density change as a function of time of a typical sample after introduction of oxygen for three consecutive experiments as described in the preceeding part. The corresponding donor gradient, deduced from the three curves of Figure 1 are represented on Figure 2. The resulting relation between surface charge density and surface potential is represented on Figure 3, compared to the usual result for an homogeneous crystal containing the same constant donor density as the bulk of the actual sample.

The numerical values of the surface charge densities and surface potentials at equilibrium before and after oxygen introduction are given on table 1 for that sample and on table 2 for another sample, from a different crystal.

<u>Table 1.</u>

	n_{si} (cm^{-2}) negative charge	V_{si} (mV)	n_{sf} (cm^{-2}) positive charge	V_{sf} (mV)	$V_{si} - V_{sf}$ (m
2nd Experiment	.61 x 10^{12}	95	.92 x 10^{11}	− 75	170
3rd Experiment	.45 x 10^{12}	92	.72 x 10^{11}	− 65	157
4th Experiment	.41 x 10^{12}	98	.37 x 10^{11}	− 35	133

Table 2.

Experiment	n_{si} (cm^{-2})	V_{si} (mV)	n_{sf} (cm^{-2})	V_{sf} (mV)	$V_{si} - V_{sf}$ (mV)
1	$.73 \times 10^{12}$	89	$.15 \times 10^{12}$	-120	209
2	$.58 \times 10^{12}$	89	$.99 \times 10^{11}$	-80	169
3	$.63 \times 10^{12}$	102	$.64 \times 10^{11}$	-52	154

It is to be noted that the surface potential variation decreases from one experiment to the next and is approximately the same for corresponding experiments on different samples. The surface potential under vacuum V_{si} remains practically constant for all the samples and all the experiments and it is the surface potential in presence of oxygen which decreases from one cycle to the next.

2. D.C. Field effect.

As is well known, the total charge Q_s induced by the applied voltage in the semiconductor splits between Q_{scr} in the space charge region and Q_{ss} in the surface states :

$$Q_s = Q_{ss} + Q_{scr}$$

If C_g is the capitance of the sample electrode condenser, and V the applied voltage :

$$Q_s = C_g V$$

The conductivity change $\Delta \sigma_s$ of the sample is then related to V, if μ_{ns} is the electron surface mobility, by the relation :

$$\Delta \sigma_s = \mu_{ns} (C_g V - Q_{ss})$$

When the surface states are not involved, Q_{ss} is zero and the conductivity change $\Delta \sigma_s$ varies linearly with the applied voltage V if μ_{ns} can be considered constant. Figure 4 shows a typical example of the measured conductivity change as a function of the applied voltage for the considered sample (Table 1) at equilibrium under vacuum. Figure 5 represents the similar curve at equilibrium under oxygen. In both cases, a linear variation is observed for negative voltages, where no surface states are involved : its slope corresponds to the product $\mu_{ns} C_g$ and, since C_g can be measured independently, it shows that the corresponding electron surface mobility μ_{ns} is equal to the bulk mobility μ_B within an experimental error of about 10 per cent, essentially due to the experimental determination of the capacitance C_g. This result is in agreement with the difference of a few per cent between surface and bulk mobilities which is expected from Schrieffer's theory in the present cases of weak accumulation or depletion surface layers (Tables 1 and 2).

The difference between the linear variation and the actual variation for weakly negative and positive voltages, on Figure 4 and 5, corresponds to an occupation of surfaces states. Figure 6 represents this difference as a function of the reduced surface potential v_s deduced from the experimental curve of Figure 5 using Figure 3 to get the relation between the applied voltage and the surface potential.

Assuming that the surface states are characterised by a unique energy level distant of ΔE_t from the conduction band and by a density N_t, with different values of ΔE_t under vacuum and under oxygen and of N_t in each case after each experiment, we can use the Fermi statistics to determine the density n_t of occupied surface states, with a statistical weight equal to 1/2 :

$$n_t = N_t \left[1 + \frac{1}{2} \exp (E_t - E_F)/kT \right]^{-1}$$

That is, if W_B is the energy difference between the Fermi level and the conduction band in the bulk :

$$n_t = N_t \left[1 + \frac{1}{2} \exp (W_B - qV_s - \Delta E_t)/kT \right]^{-1} \qquad (4)$$

ΔE_t can therefore be determined from the curves of the type given on Figure 6 where the inflexion point corresponds to the cancellation of the second derivative relative to V_s of relation (4) ; that is :

$$W_B - qV_{s \text{ inflexion}} - \Delta E_t = kT \ln 2$$

Tables 3 and 4 give the results for ΔE_t and N_t, both after heating under vacuum and after chemisorption of oxygen for the samples and the experiments corresponding to tables 1 and 2.

Table 3.

Experiment	Vacuum		Oxygen	
	ΔE_t (eV)	N_t (cm^{-2})	ΔE_t (eV)	N_t (cm^{-2})
2	.055	.30 x 10^{11}	.21	2.5 x 10^{10}
3	.059	.25 x 10^{11}	.21	2.0 x 10^{10}
4	.053	.20 x 10^{11}	.20	1.0 x 10^{10}

Table 4.

Experiment	Vacuum		Oxygen	
	ΔE_t (eV)	N_t (cm^{-2})	ΔE_t (eV)	N_t (cm^{-2})
1	.055	1.3×10^{11}	.22	3.3×10^{10}
2	.056	$.9 \times 10^{11}$	.22	2.0×10^{10}
3	.083	$.8 \times 10^{11}$	.20	1.5×10^{10}

We can see that, if the densities of states both after heating under vacuum and after oxygen sorption decreases from one experiment to the next, around and below 10^{11} per cm^2, the energy level remains constant for any sample or experiment : the assumption of a unique discrete energy level in each case appears to be quite valid. These results can be summarized as follows :

After heating under vacuum, there is a donor level at .06 $\pm$ .01 eV below the conduction band.

After oxygen chemisorption, there is a deep acceptor level at .21 $\pm$ .01 eV below the conduction band.

3. Pulsed field effect.

The capture cross section A of surface states is related to the emission time constant τ_e by[7] :

$$\tau_e = (v_T \, A \, N_c)^{-1} \exp \frac{\Delta E_t}{kT} \tag{5}$$

and to the capture time constant, by :

$$\tau_c \simeq (v_T \, A \, n_{so+})^{-1} \tag{6}$$

where v_T is the average thermal velocity of the electrons, N_c the effective density of states in the conduction band and n_{so+} the density of free electrons at the surface right after the electric field has been applied.

The emission and capture time constants corresponding to the donor surface states at .06 eV cannot be measured because they are smaller than the limit of $10 \, \mu s$ of the experimental setup : it gives only an upper limit of 4×10^{-20} cm^2 for the capture cross section of these states which can therefore be neutral or positive centers.

Pulsed field effect under vacuum shows the existence of another donor surface level at most at .025 eV from the conduction band with an emission time constant τ_e equal to 15 ms and a capture time constant of 100 ms corresponding to a capture cross section of 10^{-24} cm^2 . Its density of states is lower than 10^{11} cm^{-2}. The emission time constant corresponding to the deep acceptor surface state at .21 eV is equal to 2 ms, and its capture cross section is 10^{-19} cm^2.

Pulsed field effect under oxygen shows the existence of another deep acceptor surface level, the energy of which being more than .30 eV below the conduction band and its density of states larger than 10^{12} cm^{-2}. The emission and capture time constants of these states, respectively equal to 25 s and 5 s, lead to a capture cross section of about 10^{-23} cm^2 .

4. Discussion.

The first important aspect of the results is the homogeneity obtained for the positions of the surface energy levels both under vacuum and under oxygen independently of the sample and the experimental cycle. Such a coherence constitutes the best proof of the validity of the assumptions and of the resulting computation methods. In particular, it shows, not only the existence of a donor gradient near the surface must be considered but also that its determination is obtained by a satisfactory method : after each experiment, a new donor profile is computed (Figure 2) and the resulting relations between surface charge and surface potential are quite different from one another and from the relation corresponding to a constant donor density (Figure 3). This hypothesis of a constant donor density would lead to a dispersion of a few tenths of eV in the determination of the surface energy level, much larger than the actual dispersion of .01 eV.

The slope of the curve (Figure 6) giving the occupation of surface states as a function of the reduced surface potential must be equal to $N_t/4$: in fact, the actual slope is about 20 % higher. Such an error is consistent with the error in the determination of N_t and does not impair the assumption of discrete energy levels at the surface : a surface energy band would lead to a smaller slope.

The change of the density of surface states from one experimen to another shows that the energy levels are associated with surface defects : the donor level at .06 eV is mainly associated with sulfur vacancies ; however association does not mean identification and previou studies[6] have demonstrated that the donor gradient near the surface cannot be explained by a simple process of generation and diffusion of sulfur vacancies during thermal treatment under vacuum. In the same way, the deep acceptor level is probably associated with oxygen : its density of states varies from one experiment to another and its capture cross section corresponds to a repulsive center. The same conclusion can be drawn about the other acceptor level, the density of which decreases sharply when the oxygen of the cell is evacuated : its small capture cross section and its very long time constant seem to indicate that this level is separated from the semiconductor by a potential barrier. It could be attributed to an equilibrium between physisorbed and chemisorbed oxygen.

BIBLIOGRAPHY

(1)   R.H. BUBE, J. Chem. Phys., 21, 1409, (1953).

(2)   C. SEBENNE and M. BALKANSKI, Surface Science, 1, 22, (1964).

(3)   K.J. HAAS, D.C. FOX and M.J. KATZ, J. Phys. Chem. Solids,
        26, 1779, (1965).

(4)   P. MARK, J. Phys. Chem. Solids, 26, 1767, (1965).

(5)   A. MANY and A. KATZIR, Surface Science, 6, 279, (1967).

(6)   C. SEBENNE and M. BALKANSKI, Surface Science, 5, 410, (1966).

(7)   G. RUPPRECHT, Annals New York Academy of Sciences, 101, 960, (1963).

(8)   A. MANY, Y. GOLSTEIN and N.B. GROVER, Semiconductors surfaces,
        North Holland Publiching Company, (1965).

(9)   C. SEBENNE and M. BALKANSKI, Surface Science, 5, 434, (1966).

(10) J.R. SCHRIEFFER, Phys. Rev., 97, 641, (1955).

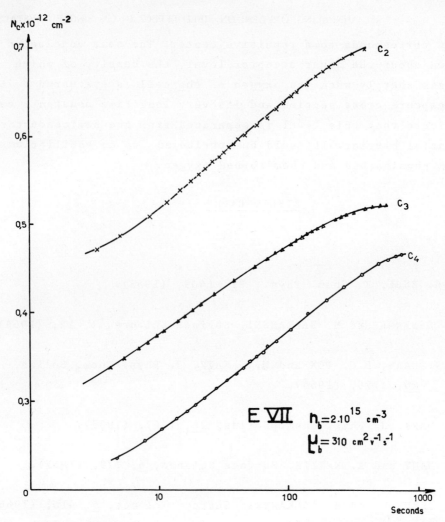

Fig. 1. Surface charge density change as a function of time after a constant
pressure of oxygen has been established at t = O for three successive
experiments on the same CdS crystal.

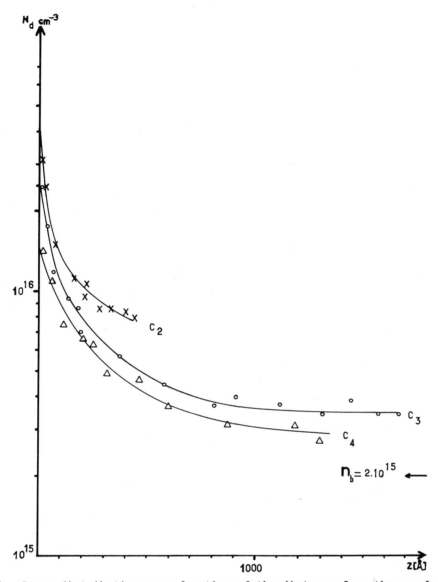

Fig. 2. Donor distribution as a function of the distance from the sample surface deduced from the kinetics of Fig. 1.

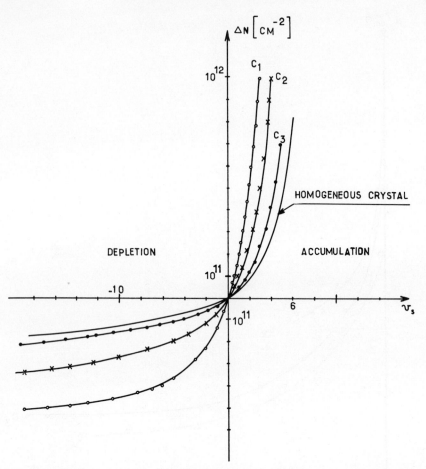

Fig. 3. Surface charge density at equilibrium as a function of reduced surface potential corresponding to the donor distributions of Fig. 2, and
to a constant donor density of 2×10^{15} cm^{-3}.

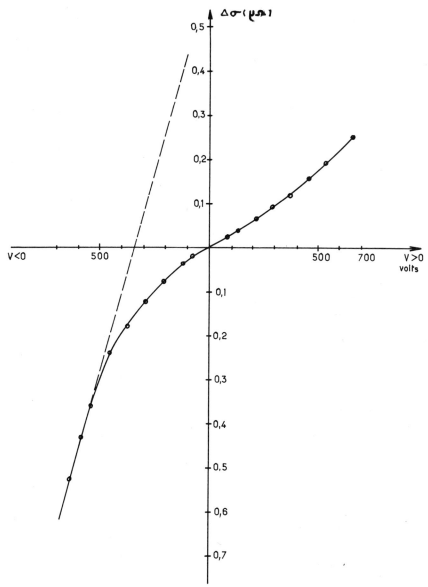

Fig. 4. Example of surface conductivity change as a function of applied voltage on a sample, after the second annealing under vacuum.

82–15

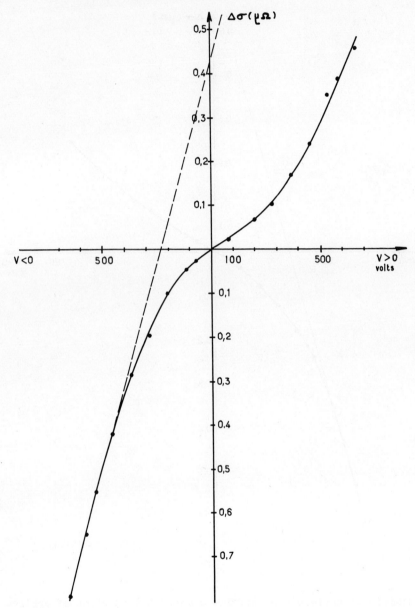

Fig. 5. Example of surface conductivity change as a function of applied voltage on the same sample after the second oxygen sorption.

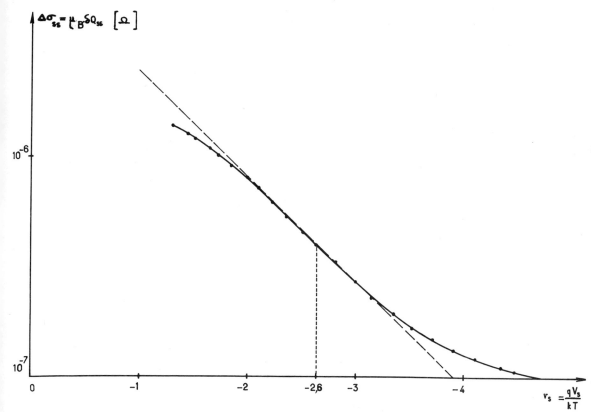

Fig. 6. Example of occupation of the surface states as a function of the reduced surface potential deduced from Fig. 3, (curve C₂) and Fig. 5.

IONIC DISORDER IN THE SURFACE OF EPITAXIAL
SILVER BROMIDE FILMS

F. Trautweiler, L. E. Brady, J. W. Castle, and J. F. Hamilton

Eastman Kodak Company
Rochester, New York

Continuous, oriented films of AgBr with either $\langle 200 \rangle$ or $\langle 111 \rangle$ surfaces have been made by vacuum deposition. The substrate was cleaved NaCl ($\langle 200 \rangle$ films) or cleaved mica ($\langle 111 \rangle$ films). Suitable substrate temperatures ranged from room temperature to 150°C. The $\langle 111 \rangle$ films show double-positioning, as is characteristic for metal films in that orientation. If the films are either annealed up to 250°C or held in a bromine atmosphere at room temperature for a few minutes; the double-positioning domains increase in size considerably and those of the same orientation coalesce. In a typical 0.1-μ-thick bromine-treated film, the surface area is about 30 times greater than the total area of the boundaries between domains.

The epitaxial AgBr films exhibit a large ionic conductivity at room temperature. An 0.5-μ-thick $\langle 200 \rangle$ film has 20 times the intrinsic conductivity of AgBr, a $\langle 111 \rangle$ film of the same thickness 100 times. Measurements of the conductivity as a function of film thickness show that in the $\langle 111 \rangle$ films the increased conductivity arises from two parts: a bulk conductivity, which is about 30 times higher than intrinsic conductivity, and a thickness independent contribution, which must be surface conductivity. No surface conductivity part was measurable in the $\langle 200 \rangle$-films, the bulk conductivity was about 20 times higher than intrinsic conductivity.

It is the purpose of this paper to report the temperature
dependence of the ionic conductivity of the $\langle 111 \rangle$-layers, to
interpret the surface contribution to this conductivity in terms
of a space charge region of interstitial silver ions in the subsurface, and to derive from this analysis the ionic disorder in
the $\langle 111 \rangle$ -surface and the formation energy of an interstitial
at the surface.

The subsurface of silver bromide contains a layer of space
charge, made up of interstitial silver ions. The electrical
field in this space charge region has been demonstrated and determined experimentally by analyzing the distribution of trapped
photoelectrons (which can be recognized as photographic latent
images) in the subsurface region of AgBr crystals.[1,2] The space
charge layer forms, because at the surface interstitial ions can
form without the simultaneous formation of vacancies, the compen-
sating charges being surface charges.[3,4] Vacancies also can form
independently, and if the free energy of formation of an interstitial is smaller than the free energy of formation of a vacancy
(as is the case in silver bromide), an excess concentration of
interstitials forms at the surface. These interstitials diffuse
into the crystal, building up a space charge layer and an
electrical field.   In thermal equilibrium the diffusion current
into the crystal is just compensated by the current due to the
field in the opposite direction. This situation is described
as a Gouy-Chapman[5,6] double layer, as long as the concentrations
involved do not require a consideration of the physical size of
the particles. The Gouy-Chapman theory is based on a Boltzmann
distribution of the charges. The result of the integration of

the Poisson equation is

$$V = \frac{2kT}{e} \ \log \ \tanh \left[\frac{1}{2}\left(Kx + b \right) \right] \tag{1}$$

in MKS-units where V is the potential as a function of distance x into the crystal, $K = \left(\frac{2e^2 C_0}{\epsilon \epsilon_0 kT} \right)^{1/2}$, the inverse Debye length, with C_0 the concentration of mobile charges in the bulk of the crystal, $b = \log \tanh \frac{eV_s}{4kT}$; the boundary conditions used were $V(x = o) = V_s$, and $\left(\frac{dV}{dx} \right)_{x=\infty} = 0$. The potential energies of the silver ions near the silver bromide surface are shown in Fig. 1. The energy of the silver ions in the $\langle 111 \rangle$ surface monolayer is indicated in the position corresponding to the results of the present analysis. This layer acts as a donor of interstitial silver ions, being responsible for the interstitial ion space charge. Since the silver ions in the crystal formally obey Fermi statistics, one can define an ionic Fermi level as the energy of a silver ion position, which has the probability 1/2 of being occupied by a silver ion. This level is also drawn in Fig. 1. Although the resemblance of Fig. 1 with an electronic band diagram of a solid is very convenient, it should be pointed out, that there are no bands for the ions, just levels. All the silver ions could assume the same energy at the same time.

The concentration of interstitial silver ions at the surface, $C_{i,s}$, is, because of the Boltzmann distribution

$$C_{i,s} = C_0 \ \exp\left(-\frac{eV_s}{kT} \right), \tag{2}$$

and the total number of space charges per unit area, n_q, is obtained from the Poisson equation

$$n_q = \int_{surface}^{bulk} \frac{\rho}{e}\, dx = \frac{\epsilon\epsilon_0}{e}\left(E_s - E_{bulk}\right) = \frac{\epsilon\epsilon_0}{e}\, E_s \tag{3}$$

The field at the surface, E_s, is the derivative of (1) at $x = 0$:

$$E = \frac{dV}{dx} = -\frac{2kT\mathcal{K}}{e}\,\frac{1}{\sinh(\mathcal{K}x + b)} \tag{4}$$

Eqs. (2), (3), and (4) may be combined to give an expression for n_q as a function of $C_{i,s}$

$$n_q = \sqrt{\frac{2kT\epsilon\epsilon_0 C_{i,s}}{e^2}} - \sqrt{\frac{2KT\epsilon\epsilon_0 C_0^2}{e^2 C_{i,s}}} \tag{5}$$

or inverted

$$C_{i,s} = \frac{n_q^2 e^2}{4kT\epsilon\epsilon_0} + C_0 + \sqrt{\left(\frac{n_q^2 e^2}{4kT\epsilon\epsilon_0}\right)^2 - C_0^2} \underset{(C_{i,s}\gg C_0)}{\approx} \frac{n_q^2 e^2}{2kT\epsilon\epsilon_0} \tag{6}$$

Since in the present case the concentration of interstitials at the surface is much larger than in the bulk, the approximate form of (6) will be used, which makes the knowledge of C_0 unnecessary.

The specific ionic conductivity of $\langle 111\rangle$-oriented epitaxial AgBr films is shown in Fig. 2 as a function of temperature for four films of different thicknesses. The conductivity can be divided into a surface contribution and a bulk contribution, as

shown in Fig. 3 for three selected temperatures. In Fig. 3 the
conductivity is plotted as a function of thickness of the film.
Generally no conductivity is observed until the sample is about
500 Å thick, probably because of the formation of noncoherent
islands. Then the conductivity shoots up and levels off to a
steady rise with increasing thickness. The slope of the straight
lines of Fig. 3 is taken as the bulk conductivity of the films,
and plotted in Fig. 2 as the extrapolated bulk conductivity, while
the intersection of the straight lines in Fig. 3 with the ordinate
is taken as the contribution due to the two surfaces. This surface conductivity is also plotted in Fig. 2. The separation of
surface and bulk conductivities appears to be quite successful,
judged from the appearance of curvature in the curve for the bulk
conductivity only. The bulk conductivity is still 30 times higher
than intrinsic conductivity, and we have no complete explanation
for that at present. The grain boundaries certainly contribute
part of the excess conductivity.

We interpret the surface conductivity as a consequence of
the space charge layer of interstitial silver ions in the subsurface. The surface conductance is related to the total number of
space charges by

$$G_\square = n_q e \mu_1 \qquad (7)$$

where μ_1, the mobility of interstitial silver ions, can be taken
from Müller's data:[7]

$$\mu_1 = \frac{48.7}{T} \exp\left(-\frac{0.15}{kT}\right) \qquad (8)$$

n_q is plotted in Fig. 4. It is of the order of $10^{13} \mathrm{cm}^{-2}$. In order to investigate the properties of the top layer of silver ions on the crystal, which is responsible for the excess interstitials, we will analyze the thermodynamical equilibrium between this top monolayer and the interstitials just underneath it. This interstitial concentration may be expressed as

$$C_{i,s} = N_i \frac{n_{Ag,s}}{n_{v,s}} \exp \frac{S_i}{k} \exp \left(- \frac{H_{i,s}}{kT} \right) \tag{9}$$

where N_i is the number of interstitial positions per unit volume in AgBr, $n_{Ag,s}$ the number of silver ions in the surface monolayer, $n_{v,s}$ the number of vacancies in this layer, S_i the vibrational part of the entropy of formation of an interstitial, taken to be the same at the surface as in the bulk $(3.40 \text{ k})^{7}$, $H_{i,s}$ the heat of formation of an interstitial from the surface silver ion layer.

$C_{i,s}$, as obtained from eq. (6) is also plotted in Fig. 4. In order to compare it with the expression (9) one has to make some assumptions about $n_{Ag,s}$ and $n_{v,s}$. The first approach is to equate $n_{v,s}$ with n_q, i.e. to assume that every interstitial in the space charge has left a vacancy in the surface layer. However, this model would require a number of silver ions in the top layer of $n_{Ag,s} = 3 \times 10^{16} \mathrm{cm}^{-2}$, which is 40 times larger than the total number of ions in a $\langle 111 \rangle$ -layer. On this basis, this first approach may be ruled out.

In a second approach we assume the ratio $\frac{n_{Ag,s}}{n_{v,s}}$ to be a constant, independent of temperature. Fig. 5 shows three possible configurations of an uncharged $\langle 111 \rangle$ layer where this ratio varies from ∞ to 1. The actual value found by equating (6) and (9) is 12.5, the heat of formation of an interstitial at the surface is $H_{i,s}$ = 0.35 eV.

Hence we conclude that the actual configuration of the $\langle 111 \rangle$ surface of an epitaxial AgBr film is very close to the first picture in Fig. 5: an almost complete layer of silver ions, with 8% vacancies in it, is covered by an approximately half filled bromide ion layer.

References

1. V. I. Saunders, R. W. Tyler, and W. West,
   Phot. Sci. Eng. 12, 90 (1968).
2. F. Trautweiler, Phot. Sci. Eng. 12, 98 (1968).
3. J. Frenkel, Kinetic Theory of Liquids
   Oxford University Press, New York 1946 p. 36.
4. K. Lehovec, J. Chem. Physics 21, 1123 (1953).
5. M. Gouy, J. de Physique (4) 9, 457 (1910).
6. D. L. Chapman, Phil. Mag. 25, 475 (1913).
7. P. Muller, Phys. Stat. Sol. 12, 775 (1965).

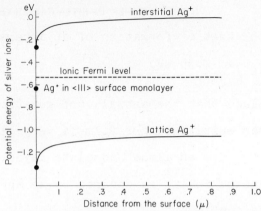

Fig. 1. Potential energy of the silver ions in the surface region of a
⟨111⟩-AgBr surface.

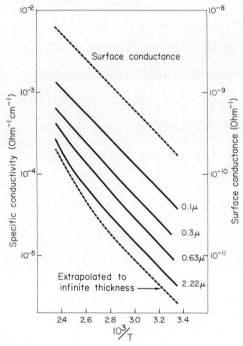

Fig. 2. Specific conductivity of ⟨111⟩-oriented epitaxial AgBr films as a
function of thickness and temperature, and surface conductivity of a
single ⟨111⟩ surface.

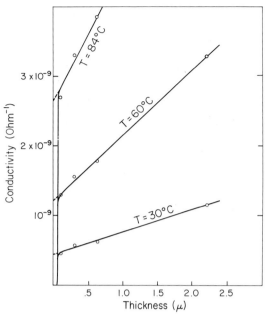

Fig. 3. Ionic conductivity of epitaxial $\langle 111 \rangle$ -AgBr films as a function of thickness.

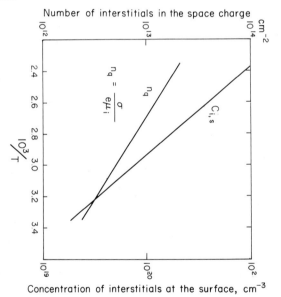

Fig. 4. Total number of interstitials in the space charge region, n_q, and concentration of interstitials right under the surface, $C_{i,s}$, as functions of temperature.

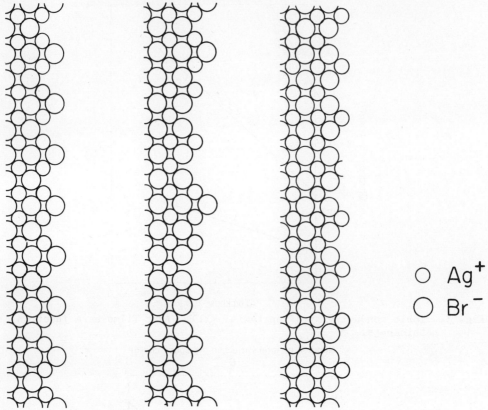

Fig. 5. Various possibilities for the arrangement of ions on an uncharged ⟨111⟩ AgBr surface. The ratio of occupied to unoccupied positions in the top silver ion layer is

(a) $\dfrac{n_{Ag,s}}{n_{v,s}} = \infty$;   (b) $\dfrac{n_{Ag,s}}{n_{v,s}} = 3$;   (c) $\dfrac{n_{Ag,s}}{n_{v,s}} = 1$.

LOW TEMPERATURE MOBILITY STUDIES OF ADSORBED HELIUM MONOLAYERS

J. G. Dash and G. A. Stewart

University of Washington, Seattle, Washington

I. INTRODUCTION.

Questions regarding the mobility of adsorbed monolayer films at low temperatures have engaged a number of researchers for more than three decades. Most of this work has been theoretical in nature, with relatively little direct and unambiguous experimental information either to confirm or deny characteristic features of mobile behavior. Considerable controversy has surrounded the issue. Lennard-Jones and Devonshire[1] have considered the two-dimensional band structure energy spectrum of non-interacting He^4 moving on the (100) surface of a LiF crystal. Their calculation indicated a group velocity at the lowest allowed energies consistent with an apparent mass increase of 8% over the free atom. Thus mobility was predicted for even the lowest states in their assumed potential, which was determined in conjunction with diffraction experiments Hill[2], in a calculation which assumed that the motion of an adsorbed atom over a surface is isomorphic to the hindered rotation of certain (single) hydrocarbon molecules, predicted a transition from mobile to localized behavior as the temperature drops below $V_0/10k$, where V_0 is the potential barrier height between minima. Ross and Steele[3] have used the He-Ar pair potential to calculate the He potential at the (100) surface of an FCC argon lattice and have approximated the potential at a single adsorption site by a V-bottomed well with appropriate slope near the bottom but with wall potentials extending to infinity. For this well they have solved the single particle Schroedinger equation and conclude that low coverage He adsorbed on such a substrate becomes highly localized, with virtually all the atoms in their quantum mechanical ground states below 20°K. These differing theoretical predictions demonstrate the need for definitive experimental evidence with regard to mobility.

II.   EXPERIMENTAL MOBILITY STUDIES: THEORETICAL MOTIVATION.

We have engaged in a series of experimental mobility studies[4] of adsorbed He monolayers and submonolayers utilizing the film heat capacity as a mobility probe in the range 0.1-4.2°K. The behavior of the heat capacity as a function of surface coverage and temperature in the limits of low coverage and low temperature is quite different for localized and mobile adsorption, making this technique a reasonably direct probe of the surface state of the adsorbate. For helium it is particularly useful since at these coverages and temperatures the vapor pressure above the film is sufficiently low that no desorption corrections are required in extracting the film heat capacity from the raw data.

In the low coverage limit, for completely localized adsorption, one expects the heat capacity to become independent of adsorbate areal density. In this limit each adsorbed atom is immobile on a particular site, as in the low lying levels of the substrate potential minima, and behaves essentially as a single particle without particle-particle interactions. Thus it may be compared to an independent oscillator (Einstein) model where each bound atom contributes individually to the total heat capacity. In addition, one expects for low enough temperatures to see an exponential decrease in heat capacity with temperature as kT becomes comparable with, and then falls below, the bound state level spacing. This is also characteristic of an Einstein specific heat.

For mobile behavior at sufficiently low coverages with non-interacting adsorbate atoms one expects the heat capacity to be independent of areal density and of magnitude k per atom. The condition for "sufficiently low coverage" is temperature and density dependent, being that $\rho\lambda^2<1$ where ρ is the areal density of the adsorbate and λ is the thermal de-Broglie wavelength, $\hbar/\sqrt{2\pi mkT}$. In this limit the system is like a two-dimensional ideal classical gas. At sufficiently low temperature, for any coverage such that $\rho\lambda^2\gtrsim 1$, complete mobility of non-interacting atoms on the surface produces for both Bose and Fermi systems a two-dimensional heat capacity which is linear in T.

For the degenerate two-dimensional ideal quantum gas, the heat capacity per atom is

$$\frac{C}{Nk} = \frac{\pi Mk}{6\rho\hbar^2} (2S + 1) T \qquad (1)$$

for both Fermi and Bose systems with appropriate spin S.

Expressed in terms of the Fermi temperature, T_F, for spin $\frac{1}{2}$ fermions, Eq. (1) becomes

$$\frac{C}{Nk} = \frac{\pi^2}{3}\left(\frac{T}{T_F}\right) \tag{2}$$

where $T_F = \pi\hbar^2\rho/mk$. A linear temperature dependence is characteristic of three dimensional non-interacting fermions but not of bosons due to the presence of the Bose-Einstein condensation. It is interesting to note in the low temperature mobile limit that the total heat capacity of two-dimensional fermion and boson systems is independent of N.

In addition to the gross differences in behavior of the heat capacity for the preceeding simple cases, recent calculations[5] for more involved models have shown that considerable modulation in the form of peaks and minima may also be present. In particular, consider the band structure model which arises from considering N_s equivalent adsorption sites. The proximity of these sites will cause the separate bound state levels to split into the N_s levels of a band. Dynamically this is equivalent to tunneling, with this mode of mobility dominantly influencing the heat capacity as kT drops below the band width. For purposes of showing the qualitative effect of the model on the heat capacity a two band model is sufficient, and is shown in Fig. 1. This model corresponds to a low-lying tunneling band and a higher thermal band. The spinless fermion heat capacity is shown in Fig. 2 for constant and equal densities of states in the two bands with Fermi temperature $T_F = 0.5\delta/k$. The case of $\Delta = 0$ is just the ideal quantum gas in two dimensions, and the linear dependence at low T is apparent. If an inter-band gap is present, modulation becomes apparent in the heat capacity. The lowest temperature peak for $\Delta = 2\delta$, 5δ, corresponds to maximum rate of excitation of the low lying tunneling band and occurs at temperatures of the order of δ/k. For finite Δ, there can be no excitation of the higher continuum band until the temperature is sufficiently large to excite states above the gap. As Δ becomes larger, this temperature increases and the second peak moves to higher temperatures. For the case of a continuum of states in the second band, full excitation corresponds to classical behavior and here the heat capacity saturates at k per atom, as expected. Space does not permit a full treatment[5] of the heat capacity for the band model, but clearly the relative magnitudes and positions of the peaks are dependent upon the ratio of densities of states in the two bands as well as on the ratio of energies Δ/δ.

It is of special interest to consider the entropy of a fully excited tunneling band, for this illustrates the limiting correspondence of the quantum mechanical band model with classical geometric theories of adsorption. The

heat capacity and entropy for a single tunneling band are
shown in Fig. 3 with the Fermi temperature at the tunneling
band mid-point again, corresponding to a half-filled band.
It will be noted that the entropy for full excitation of
the band saturates at $2Nk(\ln 2) = N_s k(\ln 2)$, the classical
value for the geometric configuration entropy, which is just
k times the number of ways of putting $N_s/2$ indistinguishable
atoms on N_s distinguishable sites with maximum occupancy
of 1 atom per site. This is to be expected, for in the high
temperature limit $(T > \delta/k)$ where the entropy becomes independent of temperature, there is a one-to-one correspondence
between the number of states accessible to spinless Fermions
and the number of geometric sites available, with the number
of ways of putting no more than one particle on a site being
just the Fermi-Dirac statistical count. This is the entropy
one would calculate for a simple Langmuir model of localized
adsorption, and would be thermodynamically stable to 0°K.
Here it arises in the high temperature limit of a fully
quantum mechanical model which is mobile via tunneling at
low temperature. The band entropy is not thermodynamically
stable at 0°K, but falls to zero as the number of occupied
states falls below the number of sites, i.e., when T falls
below δ/k. Thus, in the general band picture, the classical
configuration entropy is but an intermediate value through
which the entropy passes as the temperature drops toward
zero, and corresponds to full excitation of the lowest band.

This particular result is valid for spinless fermions, but the entropy also goes to zero at T = 0 for a Bose band model unless the lowest state is highly degenerate. Heat capacity measurements would probe this decrease in entropy below the classical configuration value.

Further modulation of film heat capacities could arise from the existence of separate low temperature phases co-existing on the surface. As is well known, two surface phases may coexist in equilibrium with the gas phase above the surface, and then one may have three phases along a line in the P-T plane. In the general case[6], neglecting desorption corrections as mentioned above, the heat capacity for such a system would include not only the separate contributions from the phases but two-dimensional compressional work terms and possible latent heat contributions resulting from the transformation of one phase into the other. These effects could also lead to coverage-dependent heat capacities. It is interesting to note that a two-dimensional Debye solid would have a low temperature heat capacity proportional to T^2.

III. EXPERIMENTAL CONFIGURATION.

The experimental configuration for heat capacity studies over the temperature range 0.1-4.2°K is similar to that shown in Fig. 4. Fig. 4 shows the contents of the sample chamber which is immersed in the helium bath and connected to a high

vacuum system. The calorimeter (A) is made of approximately

55 grams of 2-micron copper powder[7] which is pressed into

a copper container of wall thickness .030 inch. The con-

tainer is then sintered to provide thermal conductivity, and

a copper cap is silver-soldered to the top with approximately

eight inches of 1/16 inch O.D. copper capillary tube protruding

from the cap for filling and sealing. The B.E.T. surface

area of this calorimeter is approximately 1.4 m^2/gm of substrate,

providing adequate surface to volume ratio for separating

the film contribution from the total heat capacity of the system.

The heater (F) is made of 100 ohms of .005 inch diameter Evanohm

wire (Wilbur-Driver Company) which is fastened to the copper

exterior by means of General Electric 7031 insulating varnish

and epoxy. Typical heating power ranges from 1 microwatt at

0.1°K to 300 microwatts at 4.2°K, with heating times ranging

from 3 to 12 seconds. Heating times are controlled by a

decade counter which is synchronized to the 60 Hz. A.C. mains.

Two carbon resistors (E) are used as thermometers, a 56 ohm

Allen Bradley in the temperature range above 2°K and a 100

ohm Speer in the range 0.1-2.0°K. Resistance readout is on

a chart recorder following an A.C. Wheatstone bridge using

phase sensitive detection. The thermometers are calibrated

in the 1.2-4.2°K range against the He vapor pressure, using

exchange gas in the sample chamber, and against the magnetic

susceptibility of the chrome potassium alum salt pill (C)

below 1.2°K. Conventional adiabatic demagnetization from
11 Kg. is used with the salt pill to attain ultimate
temperatures less than 0.1°K, and the susceptibility is
determined using standard A.C. bucking coil techniques.
The susceptibility coils are mounted on the outside of
the sample chamber and are immersed directly in the helium
bath. Superconducting heat switch (B) is a lead filament
wound on a nylon rod. Control is by the fringing field of
an external 1.2 KG permanent magnet which straddles the
tail of the dewar.

The capsule is filled with He at liquid nitrogen temperatures after precoating with 1.3 monolayers argon. He film
contributions to the total heat capacity range from 5 to 25%,
with the lowest contribution occurring at 4.2°K where the
largest T^3 nuisance heat capacity from the bulk copper of
the empty calorimeter is present. The heat capacity of the
precoated calorimeter is measured separately and then subtracted from the total heat capacity with He to obtain the
film contribution.

IV.   HELIUM FILM HEAT CAPACITIES.

The heat capacity of adsorbed He films on argon-precoated
copper sponge has been obtained for several partial monolayer
coverages of He^3 and He^4. Fig. 5 is a full temperature
range plot from 0.2 to 4.2°K of 0.33 monolayer He^3. It is

apparent that there is a substantial contribution to the
heat capacity at all temperatures below 4.2°K. It is particularly interesting to note that C approaches the two-dimensional classical value of k/atom at 4°K. The decrease at
lower temperatures appears to be much more rapid than that
corresponding to the onset of statistical degeneracy. An
alternative model, currently being explored, suggests that
the general behavior is that of a two-dimensional Debye
continuum, with characteristic temperature in the neighborhood
of 7°K.

In the case of He^4, coverage-dependent broad maxima
have been observed for coverages of 0.62 monolayer and
below, as shown in Fig. 6. The heat capacity increases
with decreasing coverage and attains values in excess of k
per atom, larger than the classical value for full mobility
in two dimensions. At a coverage of 0.79 monolayer He^4,
the heat capacity was predominantly T^2, as in a two-dimensional
Debye solid. In this system such peaks are suggestive of
a two-dimensional phase change with an accompanying latent
heat. For the He system, a small bump is evident at
about 2°K (Fig. 5) at less than k per atom.

Sufficient data at various coverages has not been obtained yet for the He^3 system to ascertain whether this may also be due to a phase change.

REFERENCES.

1) J. E. Lennard-Jones and A. F. Devonshire, Proc. Roy. Soc. (London) A158, 242 (1937).

2) T. L. Hill, J. Chem. Phys. 14, 441 (1946).

3) M. Ross and W. A. Steele, J. Chem. Phys. 35, 3, 862 (1961).

4) J. G. Dash, D. L. Goodstein, W. D. McCormick, G. A. Stewart, Proc. 10th Int. Conf. Low Temperature Physics, Moscow, 1966, Viniti, 1968.

5) J. G. Dash and Michael Bretz, Phys. Rev. 174, 247 (1968).

6) D. M. Young and A. D. Crowell, Physical Adsorption of Gases, London, Butterworths (1962).

7) D. L. Goodstein, W. D. McCormick, J. G. Dash, Cryogenics 6, 3, 167 (1966).

8) D. L. Goodstein, J. G. Dash, W. D. McCormick, Phys. Rev. Letters 15, 10, 447 (1965).

Research supported by the National Science Foundation, Grant #GP 5693.

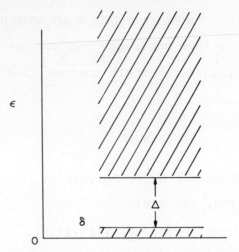

Fig. 1. Energy level diagram for a two-band model.

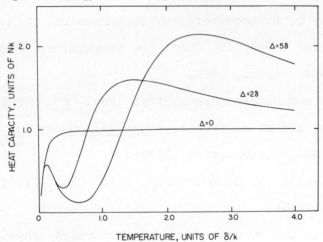

Fig. 2. Two-band spinless fermion heat capacity for coverage $\theta = 0.5$ and various interband gaps.

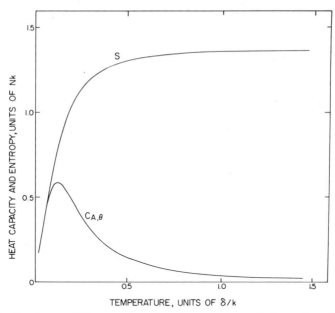

Fig. 3. Single band specific heat and entropy for spinless fermions at coverage $\theta = 0.5$.

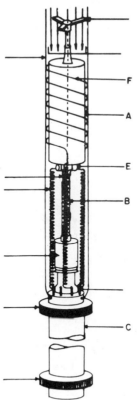

Fig. 4. Schematic of low temperature sample chamber.

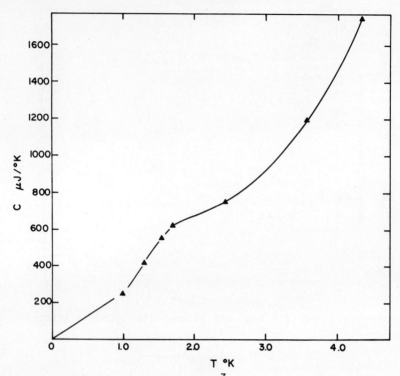

Fig. 5. Experimental heat capacity of He^3, $\theta = 0.33$, for temperatures below 4.3°K.

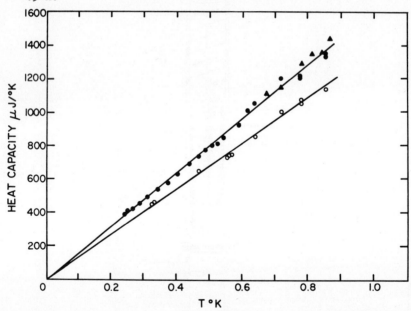

Fig. 6. Sub-monolayer heat capacity of He^4 for $\theta = 0.31$ (upper) and $\theta = 0.62$ (lower).

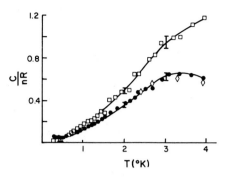

Fig. 7.

Index

The numerals in parentheses are reference numbers.